kine move
kym(o) wave
kypho bent
labi lip
lac milk
lacer torn
lachr tears
lamin(a) layer
lapar loins
lat(a) wide
later side
lemma sheath
leuco white
leva lift
levo left side
liga tied, bound
lipo fatty
lith stone
lumb loin
lut(e) yellowish
lys loose
macer soften
macr(o) large
macula spot
mal bad
mast breast
mastic chew
meat(us) passage
medi middle
medull marrow
mega(lo) large
megaly enlarged
melan black
mens month
ment mind, chin
mes(o) middle
met(a) change
micell little piece
micro small
mill one thousandth
mon one
morb disease
morph form
morul mulberry
mult many
myelo marrow, spinal cord
myo muscle
myop shortsighted
narc numbness
nas nose
nebul mist
necr dead
nephr kidney
neo new
neur nerve
neutr neither
nom name
noxi harmful
nuch nape of neck
nunci messenger

nutri feed, nourish
o(ov) egg
ob against
obtur stop up
oc against
occip back of head
occlu shut
ocul eye
odon tooth
olecran elbow
olfact smell
oligo few
ologist specialist
ology science of
oma tumor
oment fat skin
omni all
onco tumor
op reversed
opia vision
opt, opthalm eye
ora mouth
orth upright
os bone
ose full of
osis condition, disease
osity fullness
osmo pushing
ostium small opening
ostomy create an opening
oto ear
otomy surgical incision
ous full of
oxi sharp, acid
palp feel
pan all
papill nipple
pars part
partur give birth
patell little dish
path(o) disease
pauc few
ped(i) foot
pedia child
pelv basin
penia decrease, deficiency
pent five
pep digest
per through
peri around
pernic destructive
pes foot
pexy fixation, fastening
phag eat
phalan bone of finger or toe
phil attraction to
phleb vein
phob fear
phragm fence
pil(i) hair

pinn(a) wing
plant(a) sole of foot
plasia development, g
plasm something mol
plasty molding, rebui
pleur side, rib
plex network
pne breath, breathe
pneumo lungs
pod foot
poly many
pons bridge
popl back of knee
porphyr purple
post behind
pre before
presby old
pro in front, before
proct anus, rectum
propri one's own
proto first, original
proxim nearest
pseudo false
psor itch
pur(u) pure, pus
py pus
pyel(o) trough, pelvis
pyl gate
quadr, quatr four
quin five
radi ray, radius
raph suture, seam
resect cut off
ret(e) network
retr(o) behind
rhin nose
rrhea flow, discharge
sacchar sugar
salin salty
sangui blood
sarc flesh
scler hard
scoli curved
scopy observation
seb grease
sect cut
semi half
ser clear fluid
serr saw
sinus fold, hollow
sis act of
solu dissolve
soma body
sopor sleep
sphygm pulse
spic point, spike
spir breathe
spondyl vertebra
squam scale

stalsis constriction

supin lying on back
supra over
sutur seam
sym, syn with
synap union
syndesm ligament
synost bony articulation
sys with
systol contraction
tach rapid
tact touch
tard slow
tect roof
tel(o) end
teni band
tenu thin
ter three
tera monster
tetr four
thalam chamber
therap treatment
therm heat
tic belonging to
tom cut
ton tension
trabeculat has cross bars
trans across
trauma wound, shock
tri three
tripsy crushing
troch wheel
troph nourish
tuber swelling, knob
tympan drum
u(n) not
ultim last
unc hook
und wave
uni one
uria urine
utri bag
vac empty
vari change, swollen vein
vas vessel, duct
vert turn
vesic bladder
xiph sword
zyg yoke
zym ferment

HUMAN ANATOMY AND PHYSIOLOGY

SECOND EDITION

HUMAN ANATOMY
AND PHYSIOLOGY

JOAN G. CREAGER

Marymount University
Arlington, Virginia

WCB Wm. C. Brown Publishers

Book Team

Editor *Colin H. Wheatley*
Developmental Editor *Elizabeth M. Sievers*
Visuals/Design Consultant *Marilyn A. Phelps*
Designer *K. Wayne Harms*
Art Editor *Donna Slade*
Photo Editor *Michelle Oberhoffer/Mary Roussel*
Permissions Editor *Karen L. Storlie*
Visuals Processor *Kenneth E. Ley*

 Wm. C. Brown Publishers

President *G. Franklin Lewis*
Vice President, Publisher *George Wm. Bergquist*
Vice President, Publisher *Thomas E. Doran*
Vice President, Operations and Production *Beverly Kolz*
National Sales Manager *Virginia S. Moffat*
Advertising Manager *Ann M. Knepper*
Marketing Manager *Craig S. Marty*
Editor in Chief *Edward G. Jaffe*
Managing Editor, Production *Colleen A. Yonda*
Production Editorial Manager *Julie A. Kennedy*
Production Editorial Manager *Ann Fuerste*
Publishing Services Manager *Karen J. Slaght*
Manager of Visuals and Design *Faye M. Schilling*

Front & Back cover: Full Photographics

Front Inset images: Left: Vladimir Lange/The Image Bank; Right: Alexander Tsiaras/Photo Researchers, Inc.

Back Inset images: Left: Richard Anderson; Center: Sandy King/The Image Bank; Right: Superstock, Inc.

Cover and interior design by Sailer & Cook Creative Services
Production Coordinated by C. Jeanne Patterson

The credits section for this book begins on page 921, and is considered an extension of the copyright page.

Library of Congress Catalog Card Number: 90–82133

ISBN 0–697–12134–8

Printed in the United States of America by Wm. C. Brown Publishers, 2460 Kerper Boulevard, Dubuque, IA 52001

10 9 8 7 6 5 4 3 2 1

BRIEF CONTENTS

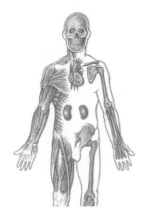
CONTENTS

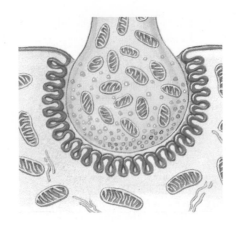

UNIT 3

CONTROL AND INTEGRATION

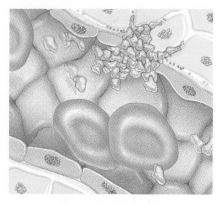

UNIT 4
HOMEOSTATIC SYSTEMS

UNIT 5

CONTINUITY OF LIFE

This book was written for students planning careers in the health sciences. It is also appropriate for students majoring in fields such as the social and behavioral sciences, education, health and physical education, and for those who have a personal interest in the structure and function of their own bodies. Many students and their professors lament the trend toward making each new edition of a text longer than the one before it. By using a conceptual approach and a concise writing style, and by limiting anatomical terms to those needed to explain physiology, I have tried to reverse that trend. In short, I have emphasized how things work and what happens when they don't.

CONCEPTUAL APPROACH

Three major concepts have guided the choice of material included in this book and how it is organized:

1. The concept that **structure and function are complementary** relates anatomy and physiology.
2. The concept of **increasingly complex levels of organization** emphasizes the importance of events at the molecular level in function at the higher cell, tissue, organ, and system levels.
3. The concept of **homeostasis—the maintenance of internal conditions within a narrow range of variation**—explains how functions are regulated and integrated in a living human being.

ORGANIZATION

This text has five units:

Unit 1 deals with levels of organization. It has chapters on the body as a whole and on function at the chemical, cell, and tissue levels.
Unit 2 concerns protection, support, and movement. It has chapters on skin and on the skeletal and muscular systems.

Unit 3 presents a unified view of control and integration. An introductory chapter on nerve tissue is followed by chapters on the central, somatic, and autonomic components of the nervous system, sense organs, and the endocrine system.
Unit 4 describes systems whose major role is to maintain homeostasis. These systems include the circulatory, respiratory, lymphatic and immune, digestive, and urinary systems. It also includes metabolism, nutrition, and fluid balance.
Unit 5 considers continuity of life. It has a chapter on reproduction and one on genetics and human life stages from embryonic development to aging and death.

Five interludes deal with topics of special interest—scientific methods, exercise and physical fitness, circadian rhythms, AIDS, and ethical decision making.

Reference plates displaying photographs of cadavers are found after Chapters 10 and 30.

End materials include **readings** appropriate for each chapter, a **glossary** with pronunciations and definitions of boldface terms in the text, and an **index** that is more extensive than those found in most textbooks. For terms having two or more words, each word may be defined separately in the glossary.

No space is wasted—the inside front cover contains definitions of prefixes, suffixes, and combining forms found in anatomical terms and the inside back cover contains tables of measurements, definitions of common medical abbreviations, and normal values for selected laboratory tests.

HOW TO USE THIS BOOK

Each chapter begins with a correlated layout of **outline** and **objectives.** Major heads and subheads appear adjacent to the objectives pertaining to those heads, giving students a quick overview of the chapter. Professors can easily tailor the text to their course by telling students which sections to study carefully and whether they may omit any sections.

PREFACE

In the body of the text, important terms are printed in **boldface type.** Pronunciations are given for all but the simplest terms. Vowels at the ends of syllables and those with macrons (¯) above them are long. Other vowels are short. Primary accents are marked ′ and secondary accents are marked ″. Students should pay particular attention to boldface terms.

Boxes, short passages separated from the main text by color bars, describe clinical applications. Screened boxes expand information being presented. The last box in each chapter lists relevant clinical terms usually not discussed in the text. Every chapter concludes with an **essay** that expands on an important clinical condition or considers a topic of special interest.

The career performance of health scientists, teachers, and many other users of this book will depend not solely on what one knows, but rather on how one thinks. Students can develop **critical thinking skills** as they read by looking for the following:

- inferences that can be drawn from what they read
- cause and effect relationships
- comparisons of similarities and differences
- new ways to solve problems.

I have used the adage that a picture is worth a thousand words. **Illustrations** in this text include 610 pieces of art and 197 photographs. Labels on many illustrations are aligned so a student can cover them and use the illustrations to review the names and locations of important structures.

Features at the end of each chapter reinforce learning. A **chapter summary** presents important points arranged by major subheadings from the outline. **Questions** are numbered by objective for easy references to appropriate sections of the text and are italicized when they call for critical thinking skills. **Problems** require synthesis of information from the current chapter, previous chapters, or other sources. Many pertain to clinical situations or familiar experiences.

ANCILLARY PACKAGE

This complete system of accessories makes it simpler for students to understand and for you to teach anatomy and physiology.

Slides, transparencies, and videotapes help students visualize. Study guides, study aids, and lab texts help them learn.

You'll guide your students through the most difficult concepts with ease and confidence.

- Instructor's manual
- Lab manual
- Customized lab manual
- Instructor's manual for lab manual
- *Anatomy & Physiology of the Heart* videodisc
- Terminology pronunciation audiotape
- 75 histology slides
- Classroom management software (TestPak 3.0)
- Student study guide
- 100 transparencies from Creager textual material
- 100 transparencies from additional anatomy & physiology sources
- Customized transparencies
- The WCB Anatomy & Physiology Video Series
- Study Cards for Human Anatomy & Physiology
- *Computer Review of Human Anatomy and Physiology* software
- *The Coloring Review Guide to Human Anatomy*
- *Atlas of the Skeletal Muscles*

ACKNOWLEDGMENTS

Many people have contributed significantly to this book, though I must take full responsibility for any errors it contains.

I thank Provost Alice Mandanis; Dean Robert Draghi; Computer Center Director Marty Coyne; the entire library staff; and many others at Marymount University for financial, intellectual, and moral support.

I am grateful to all the reviewers of this manuscript whose careful attention to details and helpful suggestions have made this book more accurate and more useful to students. Those reviewers follow:

Robert J. Boettcher
Lane Community College

Franklyn F. Bolander, Jr.
University of South Carolina

Leigh W. Callan
Floyd College

Brian Curry
Grand Valley State University

J. Mark Davis
University of South Carolina

Robert K. Flynn
The Stone Arabia School

Paul Holmgren
Northern Arizona University

Anne E. Lesak
Moraine Valley Community College

Willie L. Palmore
Tyler Junior College

Virginia Rivers
Truckee Meadows Community College

Kenneth S. Saladin
Georgia College

Robert J. Stone
Suffolk Community College

I especially appreciate the contributions of the book team and everyone at WCB who helped to turn a manuscript into a polished product. Working with the WCB staff has been the most rewarding of all my publishing experiences!

I also thank my colleague Thomas Adams for ideas about homeostasis, and my special friends, Barbara Day, Michelle Oberhoffer, and Carlyn Iverson, for their assistance and support. Finally, I am deeply grateful to my husband, John, who did everything else for the many months while I did nothing but this book.

Joan G. Creager

Joan G. Creager

HUMAN ANATOMY
AND PHYSIOLOGY

LEVELS OF ORGANIZATION

OBJECTIVES

1. Define anatomy and physiology,
   and explain why people study
   them.

2. Explain the concept of
   complementarity and its
   significance.

3. Describe the levels of complexity
   in the human body, and briefly
   describe how they are related.

4. Define the basic life processes, and
   list the organ systems that
   contribute to each process.

5. Describe the human anatomical
   position.
6. Define anatomical terms
   pertaining to position, planes, and
   sections.
7. Describe the divisions and cavities
   of the body, and list the major
   organs found in each cavity.
8. Distinguish between the body's
   external and internal
   environments.
9. Describe the location and function
   of the body's major fluid
   compartments.

10. Define homeostasis, and explain
    how control, communication, and
    self-regulation relate to it.
11. Explain how reflexes help to
    maintain homeostasis.
12. Describe how homeostasis is
    maintained at various functional
    levels.
13. Explain how positive feedback
    contributes to normal functions
    and to out-of-control situations.
14. Explain how homeostasis and
    health are related.

1

THE BODY
AS A WHOLE

The What and Why of Anatomy and Physiology

What You Will Study

Anatomy is the study of structure. **Physiology** is the study of function and how function is regulated. This book deals with human structure and function, and relationships between them. Anatomy and physiology are related to other branches of biology, especially embryology, or developmental biology, because structure–function relationships unfold during development. Anatomy and physiology are related to cellular and molecular biology, biochemistry, and biophysics because many body functions depend on cell functions, which follow chemical and physical principles. Anatomy and physiology are related to such diverse fields as genetics and ecology. Genetic information sets limits on function in conditions such as color blindness and Down's syndrome. Ecological factors such as environmental pollution also modify function.

Long before anatomy and physiology were studied formally, people were fascinated by the structure and function of their own bodies. Over a thousand years ago, Chinese acupuncturists tried to relate structure and function, but Aristotle in fourth century B.C. and Galen in second century A.D. were the first Westerners to leave records of such studies. Though Galen based his work on dogs and other animals, it stood for 1300 years until Andreas Vesalius, an anatomist working in Italy, made direct observations on human cadavers. Opposition from the state, the church, and even his own colleagues caused Vesalius to give up his studies and destroy many of his manuscripts.

If we call Vesalius the founder of modern anatomy, we might call seventeenth century English physician William Harvey the founder of modern physiology. Harvey is known for demonstrating blood circulation and developing experimental methods in physiology. Another giant in physiology, Johannes Müller, lived in nineteenth century Germany. He applied chemistry and physics to the study of physiology and fostered the development of many other outstanding physiologists. After Müller, physiology branched into two major areas—physical and chemical. Physiologists using physical methods devised ways to record and quantify physical changes pertaining to blood pressure, muscle contraction, and neural signals. Those using chemical methods studied changes in food as it was digested, and in blood as it passed between the lungs and the cells of other tissues.

The modern body of knowledge of human anatomy and physiology is vast and constantly growing. Fortunately, several concepts help to organize and integrate that knowledge, and make it easier to understand and remember. The **concept of complementarity** pertains to the relatedness of structure and function, and how they complement each other. This means that many structures have a certain shape that makes them particularly able to carry

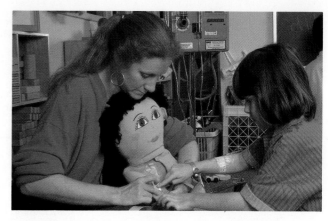

Figure 1.1
Knowledge of anatomy and physiology is used in many fields of science.

out certain functions. The **concept of organizational complexity** concerns the arrangement of body structures into functional levels of increasing complexity—chemical substances, cells, tissues, organs, and organ systems. Finally, the **concept of homeostasis** (ho″me-o-sta′sis) refers to the maintenance of body functions and internal conditions within a narrow, tolerable range. *This book uses a conceptual approach—wherever possible, it relates information to the concepts of complementarity, organizational complexity, and homeostasis.*

"Isn't it a shame that there aren't any great scientists like Harvey, Newton, and Einstein around today. All our diseases would be cured." Students often say such things to their professors, and the real shame is that professors don't always correct such misconceptions. Though cures have been found for many diseases, opportunities still exist for scientists to make important contributions. More scientists are making new discoveries today than ever before. Maybe because so many discoveries are being made, no individual's discoveries stand out like they might have a few centuries ago.

Why Study Anatomy and Physiology?

Many people study anatomy and physiology as a part of preparing for a career, but nearly all find it a fascinating way to learn more about themselves. Human anatomy and physiology is significant in many disciplines—health sciences, psychology, and education (Figure 1.1). Health science students use anatomy and physiology to understand medical treatments. Psychologists and educators use neural and sensory anatomy and physiology to understand behavior and learning. Students of physical education use physiological principles of exercise and nutrition. Students also learn about how their bodies work, why they need exercise, what constitutes a proper diet, and how they can contribute to maintaining their own health.

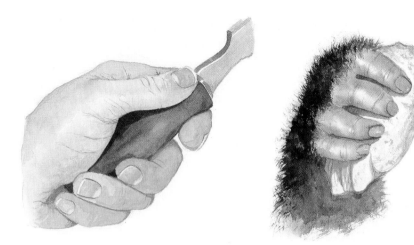

FIGURE 1.2
Certain higher apes, like humans, have apposable thumbs that help them to pick up small objects.

A thorough understanding of anatomy and physiology is fundamental to pursuing a wide assortment of health care careers. Public and private research centers offer opportunities at all levels—from research director to technician—and many universities provide opportunities for both research and teaching. For those who prefer more practical positions, health care offers many opportunities. Health care teams now include physicians, nurses, physician's assistants, nurse practitioners, and medical social workers. In addition, health care teams include a variety of technicians trained in medical laboratory and other diagnostic testing; emergency medicine; and physical, respiratory, radiologic, and other therapies.

Remarkable advances have been made in the diagnosis and treatment of human diseases. For example, computerized axial tomography (CAT) and magnetic resonance imaging (MRI) have greatly improved diagnostic precision for many disorders. New drugs are available to treat high blood pressure, high blood cholesterol, arthritis and other inflammatory disorders, and some mental disorders. Wastes can be removed from the blood of patients whose kidneys have failed, and clots can be dissolved in the blood vessels of heart attack victims. Surgical procedures and drugs that suppress immune reactions allow kidneys and some other organs to be transplanted. Lasers greatly increase surgical precision. Advances in genetic engineering may soon allow defective genes to be replaced with normal ones.

COMPLEMENTARITY OF STRUCTURE AND FUNCTION

Structure and function are complementary: structure influences function and function influences structure, as we can see by observing a hand. Watch the tendons in the back of your hand as you clench and relax your fist. These tendons are connected to muscles in your arm. If the bulky muscles were located in your hand and fingers, your finger dexterity would be greatly reduced. The sensation would be like constantly wearing a baseball mitt. Now, touch your thumb and index finger together. Only humans and a few other animals can perform such a movement. They are said to have **apposable** (ah-po'sa-bl) thumbs (Figure 1.2). Apposable thumbs allow us to pick up small objects, write, and perform many other highly specialized functions. Like the structures and functions of our hands, the structures and functions of most body parts are closely related and complement each other.

Human body function, like that of other living things, is the product of millions of years of evolution. During those years, previous generations survived because they could function and reproduce even when environmental changes occurred. For example, animals with apposable thumbs obtained food unavailable to those lacking such thumbs, survived, and produced offspring with equally useful thumbs. Eventually, the whole population came to have apposable thumbs. Other close relationships between structure and function—sweat glands, limbs adapted for fast running, or brains capable of directing more intelligent behavior that contribute to survival—also have evolved.

LEVELS OF COMPLEXITY

Our ultimate goal is to understand how the human body functions as a totally integrated organism. **Integration** is the coordination of the activities of various systems so that the body functions as a whole. To reach that goal, we will start with lower level functions—chemical substances in Chapter 2, cells in Chapter 3, tissues in Chapter 4—and proceed to higher level functions as we consider various body systems in later chapters. The body is formed from submicroscopic chemical structures—atoms, ions, and molecules. Molecules form organelles that aggregate to form cells. Groups of similar cells form tissues and tissues combine to form organs. Certain organs work together in systems, and systems working together make a whole organism (Figure 1.3). Here we are concerned with the human organism, but humans are part of more complex entities. Groups of humans form families, communities, and a global society. Humans also interact with other living things in ecosystems that make up the complex network of life on our planet.

Chemical Components and Cells

At the chemical level, the human body, like all other matter both living and nonliving, is composed of **atoms** and **ions,** which are charged atoms. Atoms combine to form **molecules.** Atoms, ions, and molecules participate in chemical reactions that contribute to all higher level body functions.

The **cell** is the basic unit of living material; whether bacterial or human, a cell can perform all the functions necessary to sustain life. Each cell is surrounded by a cell membrane that separates it from its environment, and the cell's organelles are suspended in a semiliquid **cytosol** (si′to-sol).

FIGURE 1.3
Levels of organization in the human body.

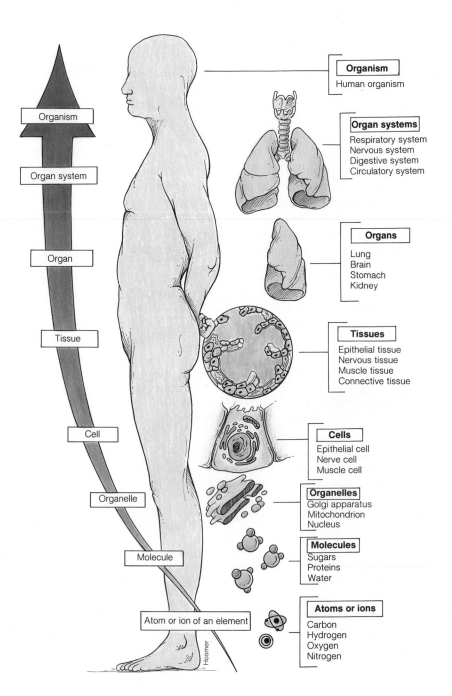

Though the human body develops from a single cell, it eventually contains about 75 trillion cells of 200 different kinds. During embryonic development, dividing cells undergo **differentiation,** during which they become structurally and functionally different. For example, muscle cells contract, gland cells secrete, and nerve cells conduct signals that convey information.

Tissues, Organs, and Systems

A **tissue** is a group of similar cells specialized to carry out particular functions. Tissues are grouped into four categories: epithelial, connective, muscle, and nerve tissues. **Epithelial tissue** is found in sheets that protect against abrasion and the entry of harmful substances. It forms the skin surface; lines passageways in the digestive, respiratory, and other systems; and forms parts of many glands. **Connective tissue** connects body parts. It forms bones, holds them together, and attaches muscles to bones and skin to muscles. Most connective tissue consists of cells surrounded by an organic fibrous matrix. **Muscle tissue** includes skeletal muscle, heart muscle, and smooth muscle in internal organs. It is specialized to exert force by contracting. **Nerve tissue,** which is specialized to carry signals, is found in the brain, spinal cord, and nerves.

Aggregation of cells to form tissues is only one step in the development of complexity in a multicellular organism such as a human. Different tissues combine to form an **organ,** a group of tissues organized to carry out a more general function than the tissues within it. For example, the stomach carries out part of the overall process of digestion. It contains components of each of the four tissue groups, and each component contributes to the organ's function. Epithelial tissue forms a protective lining and produces digestive juices and mucus. Connective tissue supports the stomach wall. Muscle tissue exerts forces that mix the food inside the stomach and propel it toward the intestine. Nerve tissue conducts signals that coordinate gland and muscle actions within the stomach and with other parts of the digestive system.

An **organ system** is a group of organs integrated by structure and functional to carry out one or more general processes. For example, the mouth, esophagus, stomach, intestine, and associated glands work together as the digestive system digests foods, absorbs nutrients, and excretes wastes. Activities of organ systems are integrated at the **organism** level of function.

Organ systems are a special kind of physical system. A **physical system** is any collection of objects that function together such as a bicycle, telephone, cell, organ, or body system. We often select body components—nerve cells, muscle tissue, or organs such as the kidneys—and study them as separate systems. Physical systems can be open or closed, depending on the flow of matter and energy. An **open system** exchanges both matter and energy with the environment. The whole human body, which takes food and oxygen, releases wastes, and gives off heat to the environment, is an open system. A **closed system** exchanges energy but not matter with its environment. A mixture of chemicals in a test tube giving off heat but not matter is a closed system.

ORGAN SYSTEMS AND BASIC LIFE PROCESSES

Human organ systems work together continuously to perform all the basic life processes necessary to maintain life. Basic life processes include *protection, movement, support, excitability* (ability to respond to stimuli), *transport, respiration* (taking in oxygen and getting rid of carbon dioxide), *ingestion* (eating), *digestion, excretion, reproduction,* and *integration* (coordination of all of the above activities).

The skeletal and muscular systems maintain posture and cause the body to move. The digestive and respiratory systems provide nutrients and oxygen, respectively, and the circulatory system delivers them to cells. These and other systems are regulated by various control processes performed by the nervous and endocrine systems. The body systems are summarized in Table 1.1 and illustrated in Figure 1.4.

TABLE 1.1 The Systems of the Body

System	Major Organs of the System	Processes Carried Out
Integumentary	Skin	Protection, regulation of body temperature
Muscular	Muscles	Excitability, movement
Skeletal	Bones	Movement, protection, support
Nervous and sensory	Sense organs, nerves, brain, and spinal cord	Excitability, control, integration
Endocrine	Glands such as pituitary, thyroid, and adrenal	Control, integration
Circulatory and lymphatic	Heart, blood and lymph vessels, blood and lymph	Transport, protection
Respiratory	Trachea, bronchi, and air passages of lungs	External respiration
Digestive	Mouth, stomach, intestines, liver, and pancreas	Ingestion, digestion, absorption, excretion
Urinary	Kidneys, passageways, and bladder	Excretion, regulation of internal environment
Reproductive	Sex glands and passageways	Reproduction

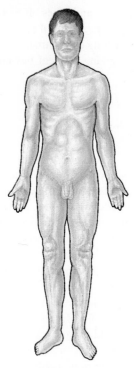

Integumentary system
Function: external support
and protection of body

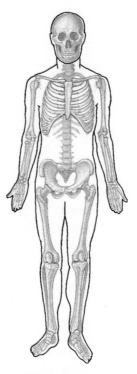

Skeletal system
Function: internal support and
flexible framework for body
movement; production of
blood cells

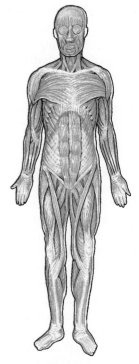

Muscular system
Function: body movement;
production of body heat

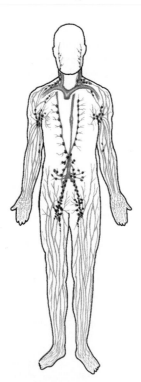

Lymphatic system
Function: body immunity;
absorption of fats; drainage
of tissue fluid

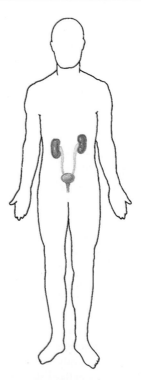

Urinary system
Function: filtration of blood;
maintenance of volume and
chemical composition
of the blood

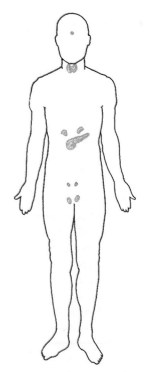

Endocrine system
Function: secretion of
hormones for
chemical regulation

FIGURE 1.4
Human body systems.

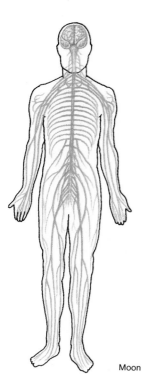

Moon

Nervous system
Function: regulation of
all body activities:
learning and memory

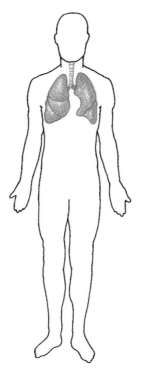

Respiratory system
Function: gaseous exchange
between external environment
and blood

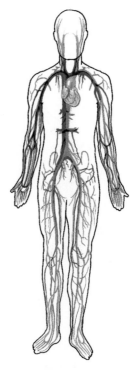

Circulatory system
Function: transport of life-
sustaining materials to body
cells; removal of metabolic
wastes from cells

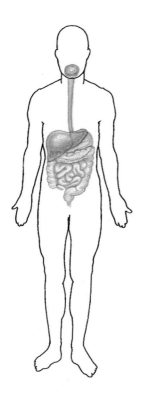

Digestive system
Function: breakdown and
absorption of food materials

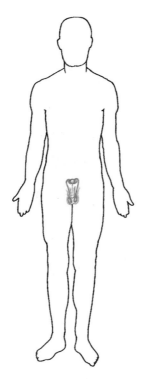

Male reproductive system
Function: production of male
sex cells (sperm); transfer of
sperm to reproductive system
of female

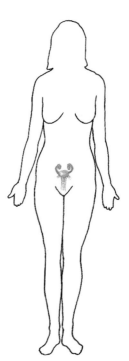

Female reproductive system
Function: production of
female sex cells (ova);
receptacle of sperm from
male; site for fertilization
of ovum, implantation, and
development of embryo
and fetus; delivery of fetus

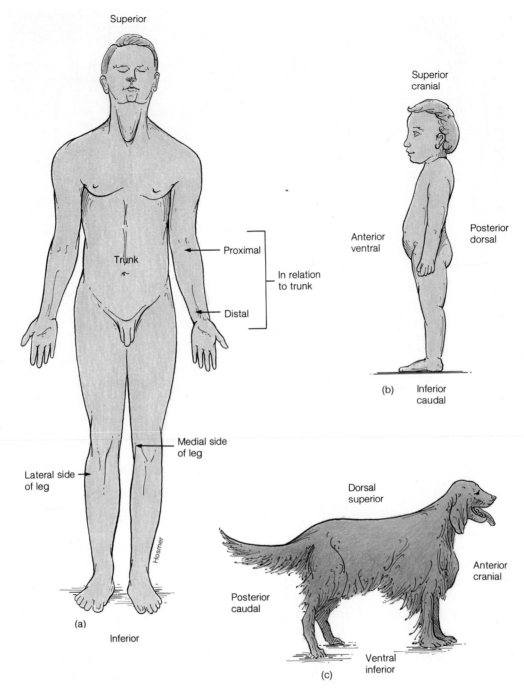

FIGURE 1.5
Terms of position illustrated on (*a*) the human body in anatomical
position, (*b*) a child in side view, and (*c*) an animal.

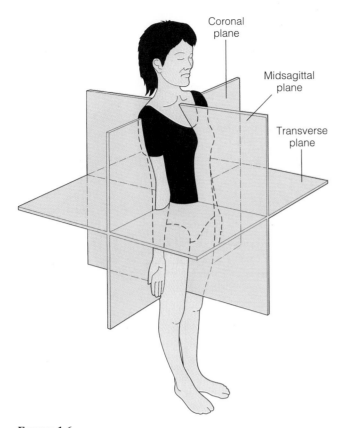

Coronal plane

Midsagittal plane

Transverse plane

FIGURE 1.6
Planes defining anatomical sections.

TABLE 1.2	Terms of Position	
Term	**Definition**	**Example**
Anterior (ante= before)	Toward the belly in humans; toward the head in other animals	The abdominal skin is anterior to the abdominal organs. A cat's head is anterior to its neck.
Posterior (post= hindmost)	Toward the back in humans; toward the tail in other animals	The skin on the back is posterior to the lungs. A dog's hind legs are posterior to its front legs.
Dorsal (dors=back)	Toward the back	The back is dorsal to the belly.
Ventral (vent= underside)	Toward the belly	The belly is ventral to the back.
Superior (super= above)	Toward the head in humans; toward the back in other animals	The head is superior to the chest. A cat's back is superior to its belly.
Inferior (infer= below)	Toward the feet in humans; toward the belly in other animals	The abdomen is inferior to the chest. A dog's belly is inferior to its back.
Medial (medi= middle)	Toward the midline	The mouth is medial to the cheeks.
Lateral (later=side)	Toward the side	The eyes are lateral to the nose.
Proximal (proxim= nearest)	Nearest the torso	The ankle is proximal to the foot.
Distal (dist= distant)	Farthest from the torso	The fingers are distal to the wrists.
Superficial	Near the surface	Skin is superficial to muscles.
Deep	Beneath the surface	The deep muscles lie beneath the superficial muscles.

THE BODY PLAN AND LANGUAGE OF ANATOMY

To describe the human body plan, precise, anatomical terminology is needed. Precise terminology allows people to communicate about complex subjects. In mathematics, we use terms such as sum, integer, set, infinity, and equality. In grammar, we use terms such as noun, phrase, comma, and paragraph. The need for specialized terms is even greater when we attempt to describe the human body. Without such terms, we might find ourselves talking about the "whatcha-ma-call-it" on the "far side" of the "thing-a-ma-jig." Think of the terms you learn as tools for understanding major concepts, and use the prefixes, suffixes, and combining forms on the inside covers of this text.

Terms of Position

The human body is considered to be in the **anatomical position** when standing erect facing the viewer with hands at the sides and palms forward (Figure 1.5). The anatomical position provides a frame of reference for using various terms of position to locate specific structures and describe spatial relationships between them. For example, in the human body, the head is superior to the trunk and the hands are distal to the arm. Terms of position are illustrated in Figure 1.6 and defined in Table 1.2.

Planes and Sections

In visualizing anatomical relationships, it is often helpful to imagine planes passing through the body (Figure 1.6). An imaginary slice through the body along a plane is called a section. A **sagittal** (saj'it-al) **section** divides the body into left and right portions. A **midsagittal section** divides the body into equal left and right halves. A **transverse section**

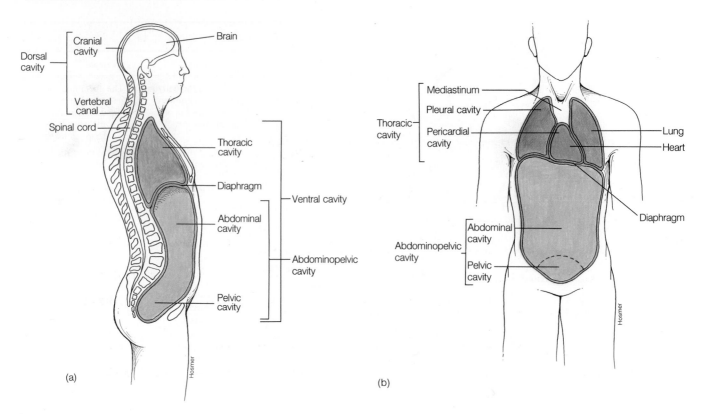

FIGURE 1.7
Body cavities: (*a*) lateral view and (*b*) frontal view.

(cross section) divides the body into superior and inferior portions. A **frontal section,** or **coronal** (ko-ro'nal) **section,** divides the body into anterior and posterior portions. These terms also can be used to describe sections through organs and other structures within the body.

Body Divisions and Body Cavities

The body can be divided into the **axial** (ak'se-al) **portion** (head, neck, and trunk) and the **appendicular** (ap''en-dik'u-lar) **portion** (arms, legs, shoulders, and hips). **Viscera** (vis'er-ah), or internal organs, lie within body cavities in the axial portion (Figure 1.7). The axial portion of the body has a **dorsal** (dor'sal) **cavity** and a **ventral** (ven'tral) **cavity.** The dorsal cavity includes the **cranial** (kra'ne-al) **cavity,** which contains the brain, and the **spinal** (spi'nal) **cavity,** which contains the spinal cord.

The **ventral cavity** includes the **thoracic** (tho-ras'ik) **cavity** and the **abdominopelvic** (ab-dom''in-o-pel'vik) **cavity,** which are separated by a broad muscle called the diaphragm. The left and right lateral portions of the thoracic cavity, the **pleural** (ploor'al) **cavities,** contain the lungs. A medial tissue mass called the **mediastinum** (me''de-as-ti'num) contains the heart, thymus gland, and part of the esophagus. The heart lies in the **pericardial** (per'i-kar'de-al) **cavity** within the mediastinum.

The abdominopelvic cavity includes the **abdominal cavity** and the **pelvic cavity.** These cavities are separated by an imaginary plane extending from the anterior pubic

bone to the posterior sacrum. Abdominal viscera include the stomach, small intestine, liver, spleen, most of the large intestine, and the kidneys. Pelvic viscera include the internal reproductive organs, the inferior portion of the large intestine, and the urinary bladder.

The abdominal cavity is divided into regions (Figure 1.8). The central **umbilical** (um-bil'ik-al) **region** surrounds the umbilicus (um-bil-i'kus), or navel, and has **lumbar** (lum'bar) **regions** to the left and right of it. Superior to the umbilical region is the **epigastric** (ep''i-gas'trik) **region,** with **hypochondriac** (hi''po-kon'dre-ak) **regions** to its left and right. Inferior to the umbilical region is the **hypogastric** (hi''po-gas'trik) **region,** with **iliac** (il'e-ak) **regions** to its left and right. Sometimes regions are designated simply as upper left, upper right, lower left, and lower right quadrants. Anatomists use these regions to describe internal organ locations and health care givers use them to describe the location of symptoms reported by people.

Embodied in the term *hypochondriac* is a bit of medical and human history. Literally, the word means under the cartilage (of the ribs). Its use to describe a person with imaginary ailments comes from the outdated notion that imaginary symptoms often arose from these regions; thus, someone who constantly complained of symptoms of illness became known as a hypochondriac.

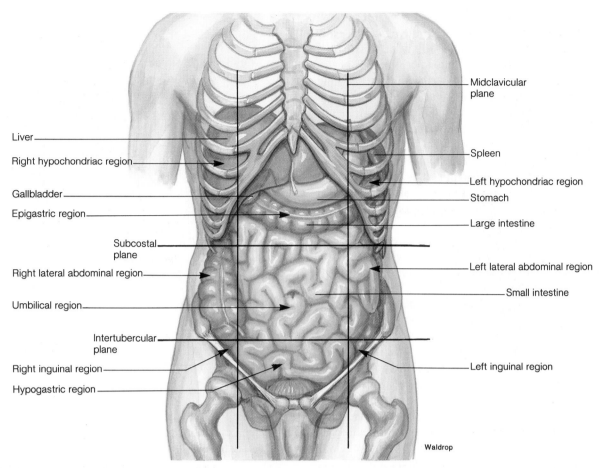

Midclavicular plane

Liver

Right hypochondriac region

Gallbladder

Epigastric region

Subcostal plane

Right lateral abdominal region

Umbilical region

Intertubercular plane

Right inguinal region

Hypogastric region

Spleen

Left hypochondriac region

Stomach

Large intestine

Left lateral abdominal region

Small intestine

Left inguinal region

Waldrop

FIGURE 1.8
Abdominal regions and the location of abdominal organs in relation to them.

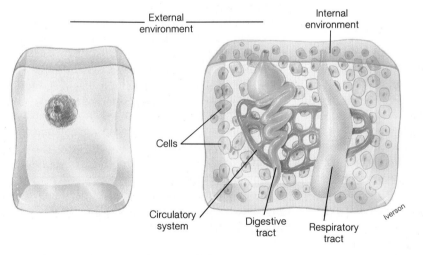

External environment

Internal environment

Cells

Circulatory system

Digestive tract

Respiratory tract

Iverson

(a) Unicellular organism

(b) Multicellular organism

FIGURE 1.9
(*a*) A single-celled organism exchanges substances between the cell itself and the external environment outside the cell. (*b*) In a multicellular organism, cells exchange substances between the cell and the extracellular fluid of the organism's internal environment. Various systems of the body transport substances between the organism's external and internal environments.

Internal Environment and Fluid Compartments

In addition to divisions, cavities, and regions containing particular organs, the body has internal fluids. Just as many protozoa and other single-celled organisms are surrounded by pond water, many human cells are surrounded by internal body fluids. For the protozoa, the pond water is the external environment; but for humans, the fluid surrounding cells is the **internal environment** (Figure 1.9) and the air normally surrounding the whole body is the **external environment.**

Human cells obtain nutrients and oxygen from and return wastes and carbon dioxide to their internal environment, which consists of **interstitial** (in"ter-stish'al) **fluid**—the fluid in tiny spaces (interstices) between cells. Substances also must be moved between interstitial fluid and the external environment, and several organ systems—digestive, respiratory, circulatory, and urinary—help with this task. The digestive system digests nutrients from the external environment. In fact, until food has been digested and absorbed, it remains outside the body, separated from the internal environment by epithelial membranes. Nutrients are absorbed into the blood and carried by the circulatory system to the interstitial fluid. Wastes released into interstitial fluid are similarly transported by the circulatory system to the kidneys and excreted to the external environment. The respiratory system similarly provides oxygen to the circulatory system and removes carbon dioxide from it.

Exchanges occur between **intracellular fluids** (fluid within cells) and interstitial fluid, and between interstitial fluid and **plasma,** the fluid portion of blood. Both interstitial fluid and plasma are **extracellular fluids.** The spaces occupied by intracellular fluid, interstitial fluid, and plasma are often referred to as **fluid compartments.** Regardless of which compartments they occupy, body fluids contain nutrients, wastes, and many other substances, and they are being exchanged between compartments constantly.

The total body fluid volume in all fluid compartments is about 40 liters in a young adult male weighing 70 kg (about 150 lbs). Normal values for body fluids and many physiological variables often are reported for young adult males because thousands of such measurements are available, having been made on military personnel since about 1940.

Body fluid consists mainly of water, though a variety of substances are dissolved or suspended in body fluids. Water makes up 55 to 60 percent of a young male's body weight and a slightly smaller proportion of a young female's body weight. The female body contains relatively more fat, which contains less water than other tissues. About two-thirds of body fluid is in the intracellular compartment and one-third is in the extracellular compartments. Of the extracellular fluids, about 80 percent is found in the interstitial compartment and 20 percent in the plasma compartment. Fluid volumes are summarized in Figure 1.10.

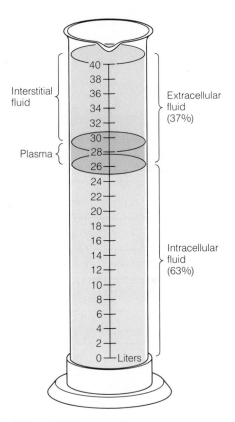

FIGURE 1.10
Fluid in each fluid compartment as a percent of total body fluid and as total volume of a compartment. (Total body fluid is assumed to be 55 percent of body weight and equal to about 40 liters.)

HOMEOSTASIS

We must marvel at the human body and how its many complex components perform carefully regulated functions. Our hearts beat throughout life—faster when we exert ourselves and slower when we are at rest—and always at a rate proportional to body needs. The heart rate is controlled by neural mechanisms, by chemical signals, and even by the amount of blood being returned to it. Regulation of the heart is only one of hundreds of homeostatic systems in the body.

Homeostatic (ho"me-o-stat'ik) **systems** maintain conditions within a narrow, tolerable range of variation or restabilize them after a disturbance. The concept of homeostasis (*homeo,* alike; *stasis,* to stand) was first applied to body fluids, but now refers to any physiologic condition that depends on a self-regulating control and communication process.

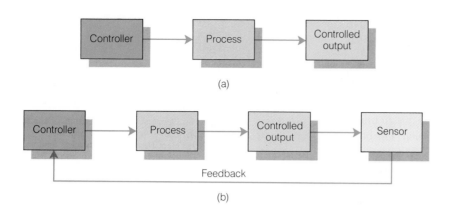

FIGURE 1.11
(*a*) An open cybernetic system with a controller, a process, and controlled output; and (*b*) a closed cybernetic, or feedback, system with a sensor and feedback in addition to the components of an open system.

Control, Communication, and Self-regulation

If you were to study how a racing bike works, you would find a mechanism to relay power from the pedals to the wheels, another for steering the bicycle, and a third one for stopping it. Each part contributes to the function of the whole bicycle. Likewise, the parts of the human body contribute to the function of the whole body, but the human body is far more complex than a bicycle. A bicycle lacks a means of communication between its parts and cannot regulate itself. A rider is required to control the speed and direction of movement—and even to maintain it in an upright position. In contrast, the human body relays many neural and chemical signals, and has many automatic control mechanisms—those that regulate heart function, breathing, body temperature, muscle contraction, and other processes. Control and communication mechanisms contribute to making the human body much more complex than a bicycle.

To *control* means to have power over, and to *communicate* means to transmit and receive information. Control requires some kind of communication. Plugging in an electric heater makes energy available to produce heat. Setting a temperature control regulates how much heat it produces, but control does not always imply regulation. To *regulate* means to fix or adjust the time, amount, degree, or rate of something. When we plug in an electric heater that lacks a temperature control, the heater will continue to release heat until we unplug it. We can control whether it produces heat, but we cannot regulate the rate of heat production. If the heater has a low–medium–high intensity control, we can regulate that rate. If it has a thermostat, it will regulate itself according to how we set the thermostat.

Systems having control and communication are **cybernetic** (si″ber-net′ik) **systems.** Plato used the term cybernetics, from the Greek word for governor, in his *Dialogs* to refer to the art of controlling society. **Biocybernetics** deals with control and communication in living systems. Cybernetic systems are classified as open or closed, according to the flow of information.

An **open cybernetic system** (Figure 1.11a) has a **controller,** which controls a process; a **process;** and **controlled output.** For example, a switch (controller) causes a light bulb to convert electrical energy to light energy (process) and, thereby, to give off light of certain intensity (controlled output). Information flows from the controller to control output, but the output provides no information to the controller. In other words, the light has no effect on the switch.

A **closed cybernetic system,** or **feedback system** (Figure 1.11b) has—in addition to a controller, process, and controlled output—a **sensor** that detects the level of output of the system, and **feedback,** or the flow of information from the sensor to the controller. We can convert the open light-and-switch cybernetic system to a closed cybernetic system by inserting a sensor that detects light intensity and a feedback signal that relays information to the controller. This closes the communication loop and provides feedback so the controller can regulate light intensity and keep it within a narrow range around a preset level. *Many physiological control systems, being closed cybernetic systems, have feedback and are self-regulating.* Such feedback systems will be illustrated throughout this text using the components shown in Figure 1.11b. In such systems, feedback, which minimizes variation, is called **negative feedback.** It aids in automatic self-regulation by reversing the direction of change and stabilizing the system.

Properties of Homeostatic Systems

In the human body, an important connection exists between physical and cybernetic systems. The body, a physical open system, is constantly subjected to disturbances from the external environment. In spite of these disturbances, the body's closed, negative feedback systems maintain stable internal conditions—that is, they maintain homeostasis.

Homeostatic systems in living organisms have five major components—a controller, a process, controlled output, a sensor, and negative feedback, as previously discussed for the closed cybernetic system in Figure 1.11b and as summarized in Table 1.3.

Homeostatic systems also have an operating set point and oscillation. The **operating set point** (OSP) is an imaginary average value for a physiologic condition. For body temperature in most humans, the OSP is 37° C and the homeostatic system that controls body temperature normally maintains it within a narrow range, usually between 36° C and 38° C. Conditions oscillate, or fluctuate, above and below the OSP. **Oscillations** are due to time lags—delays from the time a condition changes until the control system responds. Such a lag allows the blood sugar concentration to rise rapidly after a meal until the hormone insulin begins to decrease the blood glucose. A similar lag occurs after the blood glucose concentration drops and before hormones that increase it start to act. Lags also occur when hormones act for longer periods than necessary—persistent insulin activity causes the blood glucose level to drop below the OSP. Because of oscillation, negative feedback mechanisms maintain physiologic conditions in a narrow range around an OSP—not exactly at the OSP.

Incubators used to help premature infants maintain body temperature illustrate the principles of a mechanical negative feedback system. Suppose you are working in a hospital nursery and are responsible for making sure the air temperature in incubators is properly regulated (Figure 1.12a). The incubator OSP is set between 32° C and 35° C—higher for more premature infants and lower for near-term infants. The temperature in the nursery varies from 20° C to 22° C so the system must heat the air and maintain the air temperature in an acceptable range around the OSP even as the incubator continuously loses heat to the room environment. This situation is analogous to the physiological problem of maintaining human body temperature in a constantly changing environment where the body loses heat to a cooler environment or gains heat from a warmer environment.

The sensor continuously compares the actual incubator temperature with the OSP. When the air is cooler than the OSP, the sensor sends a feedback signal to the controller and the controller sends a signal to turn on the heater. While heat is being produced, the effect of the heat on the incubator temperature is being sensed by the sensor. As soon as the actual temperature becomes higher than the OSP, the feedback signal causes the controller to turn

TABLE 1.3	Components of Mechanical and Living Feedback Cybernetic Systems	
Component	**Examples of Components in Mechanical Systems**	**Examples of Components in Living Systems**
Sensors	Pressure gauges and electronic devices that detect change	Receptors of special senses, skin, muscles, tendons, and internal organs
Controllers	Switches and valves; electronic devices	Centers in the nervous system and pituitary gland; some systems have no apparent controller
Processes		
Structures	Heater, pump, or other mechanical device	Muscle, gland, or other anatomical structure
Functions	Heat production, pumping water, build up pressure, and so on	Muscle contraction to constrict a vessel or cause some other motion, secretion of sweat or another substance
Controlled output	Temperature, flow, pressure, or other variable	Body temperature, rate and depth of breathing, rate and strength of heartbeats, diameter of blood vessels, rate of secretion of sweat or other substances
Negative feedback	Signal from sensor to controller usually via wire	Signal from sensor to controller via neuron or chemical messenger

off the heater. After the heat has been turned off, the incubator begins to cool. The sensor continues to compare the actual temperature with the OSP. As soon as the actual temperature becomes lower than the OSP, the feedback signal causes the controller to turn on the heater. These processes are summarized in Figure 1.12b.

Suppose the temperature regulating system has been operating for several hours and you have been measuring the incubator air temperature at a point near the sensor every five minutes. A graph of your measurements should look like Figure 1.13. The initial increase in temperature

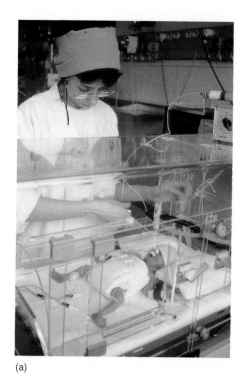

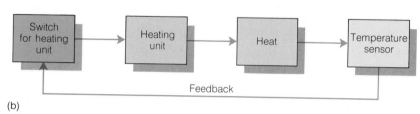

Feedback

(a)   (b)

FIGURE 1.12
(*a*) An incubator used to protect and maintain the body temperature
of premature infants. (*b*) A diagram of the incubator's control system.

is slow and becomes more rapid because heat is transferred first to the air nearest the heater unit and later to
air near the sensor. After the temperature reaches the OSP,
it oscillates around the OSP. Oscillations are large at first
because the heater has been on for some time and will continue to radiate heat after it is turned off. As the heater
stays on for shorter periods of time, it radiates heat for a
shorter time. Though oscillations decrease in size, they will
continue to occur because of time lags in the system.

Reflexes as Homeostatic Mechanisms

The incubator illustrates the properties of a mechanical
homeostatic system. Let us now consider how reflexes operate as homeostatic systems in humans. Neural reflexes
(Chapter 11) consist of signals from sensory receptors that
travel to control centers in the brain stem or spinal cord,
which elicit motor responses. The degree of motor response is monitored and negative feedback signals are relayed to the control center. Hormonal reflexes differ from
neural reflexes in that some signals are relayed by hormones traveling in the blood.

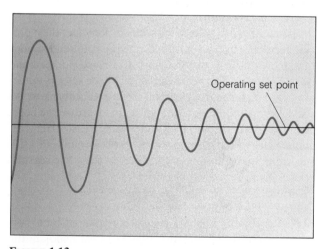

FIGURE 1.13
Changes in the incubator air temperature from beginning of operation
until oscillations in temperature around the operating set point have
reached a minimum.

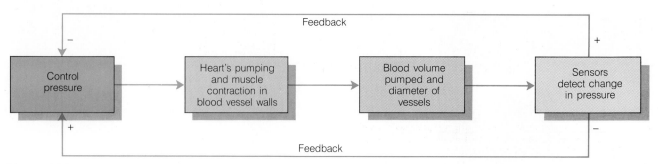

FIGURE 1.14

A simple neural control mechanism regulates arterial blood pressure. When baroreceptors sense a change in pressure, neural feedback signals are relayed to the vasomotor center in the brain. Depending on whether the sensors detected a rise or drop in pressure, the controller sends neural signals that adjust the volume of blood pumped by the heart and the diameter of blood vessels. The output of these changes is continuously monitored by the sensors, thereby maintaining arterial blood pressure within a narrow, tolerable range.

Regulation of Blood Pressure—A Neural Reflex

Arterial blood pressure is regulated by a neural reflex (Figure 1.14), which is controlled by the vasomotor center in the brain. Based on feedback from pressure sensors in certain arteries, this center sends signals that control the heart rate, the blood volume that the heart pumps, and arterial smooth muscle.

When sensors detect high blood pressure, they provide feedback to the controller, causing it to send signals that decrease the rate and quantity of blood pumped and increase the diameter of the arteries. When sensors detect low blood pressure, they provide different feedback to the controller, causing it to send signals that increase the heart rate and blood volume pumped, and decrease artery diameters. Arterial blood pressure varies around an OSP according to the level of pressure detected by the sensors.

Regulation of Hydrochloric Acid Secretion— A Hormonal Reflex

Secretion of acid in the stomach is regulated by a hormonal control mechanism (Figure 1.15). Food in the stomach causes the hormone gastrin to be secreted. Gastrin circulates in the blood and stimulates certain stomach lining cells to secrete acid. When enough acid is secreted, it acts by negative feedback to inhibit further gastrin release. In this system, gastrin-secreting cells act both as sensor and controller, and self-regulation of acid secretion is achieved by negative feedback.

Walter Cannon, a Harvard physiologist, developed the following propositions regarding homeostasis in 1930.

1. In an open system like the human body, continually subjected to disturbing conditions, constancy itself is evidence of agencies that maintain it.
2. If the body maintains near constancy, it does so because any tendency toward change is automatically met by increased effectiveness of factors that resist change.
3. A homeostatic regulating system may comprise cooperating factors acting at the same time or successively.
4. When a factor is found to shift a condition in one direction, it is reasonable to look for automatic control of that factor or for factors having an opposite effect.

These propositions still hold true today.

Homeostasis at Functional Levels

The body maintains homeostasis at several functional levels from the chemical to the whole body level. Molecular homeostasis involves chemical reactions and enzymes that keep the reactions going at a rate appropriate for a cell's needs. Cellular homeostasis maintains stability within cells. When nutrients are in short supply, some cells break down stored fats for energy. A product of these reactions inhibits fat synthesis.

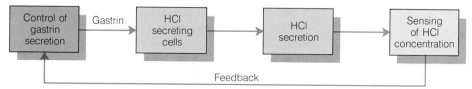

FIGURE 1.15

Control of hydrochloric acid secretion. Gastrin-secreting cells, which act as both sensors and controllers, detect stomach acidity and secrete the hormone gastrin (a chemical signal) according to the need for acid secretion. Hydrochloric acid secretion is proportional to the stimulation provided by gastrin and the acid output is continuously monitored by the sensors and adjusted by adjusting gastrin secretion.

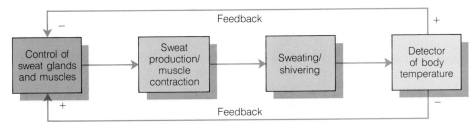

FIGURE 1.16

A simplified representation of the control system that regulates body temperature.

Certain local factors exert control at the tissue level, such as in muscle tissue. During strenuous exercise, muscles release heat, lactic acid, and carbon dioxide. These products act at the tissue level to dilate blood vessels, increasing the supply of nutrients and oxygen delivered to the active muscles.

How Body Temperature Is Regulated— A Whole Body System

Many, probably most, homeostatic systems operate at the organ–system or whole body level. In the system that regulates body temperature (Chapter 25), the sensors are temperature receptors in the skin and deep body tissues. The controller is a center in the brain. Depending on the feedback signals it receives, the controller sends signals to various structures whose functions change body temperature. These structures include sweat glands, blood vessel smooth muscle, skeletal muscles, and small muscles attached to hair follicles. Both skin and deep body temperature receptors continuously monitor heat production and the system's controlled output, and provide feedback (Figure 1.16).

Why do we feel hot and sweaty after exercising? During strenuous exercise, contracting muscles give off heat and the body temperature rises above its OSP, initiating processes that get rid of heat. Sweat glands secrete sweat. Smooth muscle in skin blood vessels relaxes and the vessels dilate, making the skin appear red. When the environment is cooler than the body, the body loses heat as sweat evaporates and heat is conducted to the skin surface. These processes are stopped by negative feedback when sensors find that the temperature is back to the OSP.

What causes us to shiver and get "gooseflesh" when we step out of the shower? As the body loses heat to the environment, the body temperature goes below its OSP. Smooth muscle in skin blood vessels contracts and vessels constrict so less heat is lost. Shivering involves involuntary muscle contractions during which muscle cells release heat. Gooseflesh is due to small muscles on hair follicles contracting and causing hairs to stand erect. (Erect hairs trap heat around the body in furry animals but not in humans.) A brisk rubdown adds to the heat produced from shivering. Soon, sensors find that the temperature is back to normal and feedback stops the processes that cause shivering and gooseflesh.

Hierarchical Structure of Control Systems

Just as the body has an anatomical hierarchy—cells, tissues, organs, and systems, it has a hierarchy of control systems at those levels working together to maintain homeostasis and to protect the body against adverse conditions. For example, in starvation, some tissues are broken down to provide energy for vital organs such as the heart and brain. Likewise, in extreme cold, the skin freezes as skin blood vessels constrict and conserve heat that keeps vital internal organs functioning.

Higher level control systems can regulate lower level ones. Lower level or local control systems regulate processes within cells, or between cells and fluids. The amount of waste that cells release into the blood regulates the rate at which oxygen moves from the blood to the cells. Intermediate level control systems can regulate an intermediate level condition, such as heart rate or blood pressure, or they can control a lower level function, such as delivering more blood to active muscles and less to inactive tissues according to feedback from the tissues. Higher level control systems also can regulate whole body conditions such as temperature, or control intermediate and lower level functions by feedback from them, as illustrated by the effect of stress on cortisol secretion.

Total body homeostasis requires homeostasis at all levels. It emphasizes interactions at various levels that keep basic life processes operating at optimum rates. Homeostasis, which typically depends on self-regulating control systems, must be maintained to maintain life itself.

Positive Feedback

Though most body control systems involve negative feedback, which reverses the direction of change in a process, a few involve **positive feedback,** which accelerates change. Using a light-and-switch analogy, positive feedback would cause the switch to further increase light intensity when it receives a signal that the intensity has increased. Positive feedback controls a system, but it does not produce self-regulation. Positive feedback is significant in two types of situations—those in which acceleration of a function is normal and those in which the acceleration creates an out-of-control disease or disorder.

Reflexes that empty body cavities such as coughing, sneezing, urination, defecation, ejaculation, and parturition (childbirth) are controlled by positive feedback. Stimuli that elicit these reflexes occur with increasing frequency until the body cavity is emptied, and then they abruptly cease. These processes are controlled, but they are not maintained within a narrow range. Thinking of regulating such processes about an OSP leads to absurd ideas such as stabilizing conditions midway through a sneeze or with a baby half born!

Some processes in illness or injury are controlled by positive feedback. Like other reflexes that empty body cavities, vomiting results from increasingly strong positive feedback signals. The first signs of nausea accelerate the salivation reflex. When saliva is swallowed, it adds to the nausea. The cycle of more nausea and more saliva and more nausea eventually leads to complete emptying of the stomach. Similarly, the chemical pathway in blood clotting involves positive feedback as the products of later reactions accelerate earlier reactions, thereby greatly increasing the availability of the end-product fibrin, the substance of a clot.

When positive feedback occurs in a situation normally regulated by negative feedback, the situation gets out of control and a self-accelerating, unstable disease condition such as a peptic ulcer results. An ulcer develops as stomach acidity fails to trigger negative feedback signals that ordinarily regulate acid secretion. Instead, positive feedback accelerates acid secretion. Acid damages the stomach lining and diffuses into the tissues where it stimulates nerve endings, which stimulate histamine release. Neural signals and signals from histamine further damage the stomach lining, cause more acid secretion, and damage blood vessels. Damaged blood vessels allow fluid to leak into the tissues and, eventually, severe blood loss occurs. As is typical of positive feedback in disease, one disorder leads to a sequence of other disorders and creates a vicious cycle.

> Some physiologists think alcoholism involves positive feedback, where each drink creates a desire for another drink. Alcoholics have another drink to feel better, yet another drink just leads to a bigger hangover. Drug addiction may involve a similar vicious cycle of positive feedback.
>
> We have all had experiences that upset us at first and made us more angry as we thought about them until anger filled our consciousness. We may have forced ourselves to think about something else and, in doing so, probably broke a positive feedback loop. From this experience, we can speculate that techniques to break positive feedback loops might be effective therapy for controlling certain behavioral disorders.

Homeostasis and Health

Homeostatic mechanisms are so important in maintaining normal conditions that one might think of health as a state of homeostasis, and illness or injury as a state of disturbed homeostasis. For example, a fractured bone or an injured joint disturbs the body's functions of movement and support. A loss of blood changes the total blood volume, and reduces the blood pressure and the heart's pumping efficiency. Invasion of the body by an infectious agent can cause fever and tissue damage.

When homeostasis is disturbed, the body activates mechanisms to restore it. A fracture or joint injury is repaired. New blood is made to restore the blood volume. White blood cells destroy invading organisms, and the body develops immunity to the disease organisms. Knowing how the body restores homeostasis helps us to devise medical treatments that assist this process. The effectiveness of treatments depends partly on how well the would-be healers understand normal physiological processes and their regulation.

CLINICAL TERMS

etiology (e''te-ol'o-je)   the study of the causes of disease

pathology (pah-thol'o-je)   the study of disorders of structure or function

pathophysiology (path''o-fiz''e-ol'o-je)   the study of the functional disorders produced by disease

sign   observable evidence of a disease

symptom (simp'tum)   evidence of a disease reported by the affected individual

syndrome (sin'drōm)   a set of signs and symptoms that occur together

ESSAY

HUMAN CHARACTERISTICS AND HUMAN DILEMMAS

We humans share many characteristics with other animals. Like most other animals, we can move; like other vertebrates, we have a spinal cord and vertebral column; and like other mammals, we have hair and mammary glands. With the anthropoid apes, we share apposable thumbs; and with the hominids, we share our land habitat, our bipedal (two-legged) gait, and our large cerebrum. In spite of these similarities, we are distinctly different from any other animals in important ways. We display the attributes of curiosity, imitation, attention, and memory to a higher degree and with greater flexibility than any other living animal. We use reasoning to adapt to and to modify our environment. We make and use tools, create mental abstractions, and are self-conscious, or aware of our own existence. We are unique in our abilities to create and use oral and written language, and to develop complex social traditions and codes of behavior.

Perhaps our most human characteristic is our **altruism,** or concern for the welfare of others beyond our own self-interests. Rene Dubos, in *Beast or Angel?,* claims that altruism has been recognized among humans for thousands of years. He cites as evidence the discovery of the remains of a 50-year-old man who died about 50,000 years ago. His bones showed that he had suffered from crippling arthritis for many years after he would have been unable to gather food or hunt game. Others of his clan must have cared for him—maybe risking their own survival.

This human desire to preserve human life and well-being is reflected in our use of tools and technology in medical science. We have created eyeglasses to correct vision, learned to use insulin and other therapies to control diabetes, and devised special diets to prevent mental retardation in individuals with hereditary metabolic diseases. With the aid of inventions such as heart pacemakers and artificial kidney machines, we can prolong the life of people who otherwise would have died.

Altruism in our biological past helped make group survival possible; however, Dubos says, "the really human aspect of altruism is not its biological origin or its evolutionary advantages, but rather the fact that humankind has now made it a virtue regardless of practical advantages or disadvantages." The advantages of technology have extracted a price by altering the composition of the human population. Like the arthritic man, thousands of people who might have died now survive as invalids or partially disabled individuals. More people with genetic defects survive and reproduce, and some pass their defects on to their children. Fewer infants die and the elderly live longer, so the total population grows faster. All these changes place great burdens on society's resources. In spite of the burdens, our altruism still makes us feel that preserving life is good.

When we apply our technology, we participate in our own evolution. For better or worse, *the human species is the first species capable of conscious participation in its own evolution.* Our big brains have given us this capacity, and at the same time, have saddled us with some big ethical questions. Are our brains "big enough" to know what is best for human evolution? Can we know long-term effects of a genetic variation? Is it a good idea to let parents choose the sex of their children? Does anyone have the right to decide whether another person should be allowed to reproduce? Our big brains and our altruism have led us into many human dilemmas. Will they lead us through them?

Questions

1. For any other three living organisms, name two characteristics they share with humans.
2. What is altruism? Give examples of altruistic behavior you have observed.
3. What human behaviors are lacking in altruism?
4. One might argue that the medical therapies sometimes allow people to survive, reproduce, and pass genetic defects along to their offspring. Should there be limits on such technologies?

CHAPTER SUMMARY

The What and Why of Anatomy and Physiology

- Anatomy is the study of body structure and physiology is the study of body function. A good understanding of anatomy and physiology is essential in many careers.

Complementarity of Structure and Function

- The concept of complementarity of structure and function means that structure and function are closely related and each influences the other. Relating structure and function is useful in organizing information about anatomy and physiology.

Levels of Complexity

- The human body functions as an entity with each of its parts contributing to the overall function. The cell, a fundamental unit of living organisms, contains chemical components; and cells are organized into tissues, tissues into organs, organs into organ-systems, and systems into the whole body.

Organ Systems and Basic Life Processes

- Basic life processes include movement, support, protection, excitability, ingestion, digestion, metabolism, excretion, transport, respiration, control, integration, growth, repair, and reproduction.
- Organ systems, their organs, and the processes they carry out are listed in Table 1.1.

The Body Plan and Language of Anatomy

- Anatomical position of the body is the position of standing erect with hands at sides and palms forward.
- Terms of position are illustrated in Figure 1.5 and defined in Table 1.2, and planes and sections are shown in Figure 1.6.
- Body portions and cavities are shown in Figure 1.7 and abdominal regions are shown in Figure 1.8.
- The internal environment of the body provides a fluid medium in which individual cells function. Fluid compartments include intracellular fluid inside cells, interstitial fluid around cells, and blood plasma, among which substances are constantly being exchanged.

Homeostasis

- Homeostasis is the maintenance of physiological conditions within a narrow, tolerable range. Homeostatic systems are control systems that allow small fluctuations of chemical or physical properties.
- A cybernetic system is a control system. Closed cybernetic systems have a controller, a process, controlled output, a sensor, and feedback from the sensor to the controller, which makes such systems self-regulating.
- In addition to the above components, homeostatic systems have an operating set point and display oscillation.

- Homeostatic mechanisms—which operate through neural and hormonal signals—regulate chemical reactions, intracellular processes, exchanges between cells and their environment, blood pressure and other intermediate level functions, temperature, and other whole body functions.
- In the hierarchy of physiological control systems, higher level systems control lower level functions and integrate functions.
- Positive feedback provides proper control of reflexes that empty body cavities and some mechanisms such as blood clotting that counteract illness or disease.
- When positive feedback replaces negative feedback, an out-of-control, vicious cycle results.
- Health might be equated with homeostasis.

QUESTIONS AND PROBLEMS

The questions at the end of each chapter are numbered to correspond with the objectives listed at the beginning of the chapter. Italics indicate that a question requires critical thinking skills beyond simple factual recall.

Questions

1. (a) What is the difference between anatomy and physiology?
   (b) What three concepts help to relate ideas in anatomy and physiology?
2. (a) What is the concept of complementarity?
   (b) Give three examples of complementarity.
3. *Arrange the levels of complexity of the body from the whole organism down to the atom.*
4. (a) What are the basic life processes that occur in the human body?
   (b) Which of the above processes are carried out by each of the human organ systems?
5. What is the human anatomical position?
6. *(a) Use all terms of position in a paragraph describing how human body parts are arranged.*
   (b) What kinds of sections are used in describing anatomy and what does each mean?
   (c) A patient sustains a stab wound in the back that collapses a lung. Through what tissues and membranes did the knife pass?
7. (a) Distinguish between the axial and appendicular portions of the body.
   (b) What are the components of the dorsal cavity and what organs are found within each cavity?
   (c) What are the components of the ventral cavity and what organs are found within each cavity?
   (d) Write sentences describing your imaginary symptoms that might cause pain in each abdominal region.
8. *How do the body's external and internal environments differ?*

9. What fluids are found in each fluid compartment?
10. (a) Define homeostasis.
    (b) *What is the function of each of the components of a typical homeostatic system?*
    (c) *Why is the self-regulation property of a homeostatic system so important?*
    (d) *How can you distinguish a closed cybernetic system from other control systems?*
    (e) *What are the properties of feedback systems, and why are they important in physiology?*
11. *How do neural and hormonal regulatory systems differ?*
12. (a) *How do mechanisms of homeostasis differ at various functional levels?*
    (b) *In what ways are higher level control systems related to lower level control systems?*
13. (a) *How does positive feedback control normal functions, and how does it contribute to out-of-control situations?*
    (b) *What kind of feedback is involved in each of the following: regulating the heart rate during exercise, accelerating labor, regulating breathing, and increased acid secretion in the stomach of an ulcer patient?*
14. *How is homeostasis related to health?*

Problems

1. For each component of a negative feedback system, predict what would happen if that component failed to work.
2. Design, and if possible, build a model of a negative feedback system.
3. Start a table of control systems categorized by levels as defined in this chapter. As you read this text, place each control system you read about in a proper category in your table.
4. In each of the following statements, identify the concept to which each statement relates and find the flaws, if any, in each statement:
    (a) The nervous system controls and integrates the functioning of other systems in the body.
    (b) The digestive system delivers food to cells of the body.
    (c) The reproductive system preserves the species from one generation to the next.
    (d) The human hand is constructed so that humans can do things that many other animals cannot do.
    (e) The stomach is well designed to serve its function as a storage bag.
    (f) Tissues are composed of several different organs.
    (g) Cells contain organelles, which are made of molecules, which are made of atoms.
    (h) Physiological conditions are maintained within a narrow tolerable range by positive feedback.

OBJECTIVES

1. Explain why an understanding of basic chemistry is needed to understand anatomy and physiology.

2. Describe how chemical building blocks and chemical bonds contribute to the structure and function of biological molecules.

3. Define the terms solvent, solute, and solution.
4. List the properties of water, and explain how water contributes to body functions and homeostasis.
5. Distinguish between a solution and a colloidal suspension.

6. Define acid, base, and pH, and relate the characteristics of acids and bases to their physiologic effects.

7. Briefly describe the properties of organic molecules, and name four kinds of complex biological molecules.

8. Describe the properties, categories, and physiologic importance of carbohydrates.

9. Describe the properties, categories, and physiologic importance of lipids.

10. Describe the properties, categories, and physiologic importance of proteins.

11. Describe the properties, categories, and physiologic importance of nucleotides and nucleic acids.

12. Briefly explain the laws governing chemical changes.
13. Describe and give examples of coupled reactions.

14. Relate the properties of enzymes to the regulation of chemical reactions.
15. Explain how temperature, pH, and the concentrations of substrate and enzyme affect the rate of enzyme reactions.

2

FUNCTION AT THE CHEMICAL LEVEL

WHY STUDY CHEMISTRY?

Today, when we visit a doctor, the doctor often orders laboratory tests on blood or urine to assess body functions. Nurses and other health care workers sometimes use test results to tailor patient care. Even as patients, we can understand our medical problems better if we know what our test results mean—that an abnormally high blood glucose level usually means diabetes or that abnormally high levels of blood cholesterol and other fats are associated with increased risk of heart attack. To understand laboratory test results, we need to understand how chemical processes contribute to normal body functions—and sometimes to medical problems.

Chemistry is concerned with matter—the properties and interactions of matter. **Matter** is anything that occupies space and has mass (substance), including air, water, rocks, and living things. Many physical and chemical properties of living organisms can be described in terms of physical and chemical properties of matter. Matter is composed of basic chemical building blocks. Just as letters of the alphabet can be combined in different ways to make thousands of words, chemical building blocks can be combined to make thousands of different substances. Although very few words in the English language contain more than 20 letters, complex chemical substances can contain more than 20,000 building blocks!

In the body, various chemical substances undergo changes in chemical reactions. The sum total of all the body's chemical reactions is called **metabolism** (met-ab′o-lizm). Metabolism includes the breakdown of nutrients for energy and the making of body substances. Other chemical changes occur as muscles contract, and nerves and hormones send signals.

As we saw in Chapter 1, homeostasis is an important organizing concept in anatomy and physiology. Homeostasis is possible only when each of a large variety of chemical processes is occurring at the right time, in the right place, and at the proper rate. Understanding the fundamentals of chemistry is important, not only because of these chemical processes, but because maintaining homeostasis often depends on events at the chemical level.

CHEMICAL BUILDING BLOCKS AND CHEMICAL BONDS

Chemical Building Blocks

Particles of matter are too small to be seen even with the best microscopes. Using various experiments, chemists have deduced certain characteristics of the particles. They have identified the **atom** as the smallest chemical unit of matter. Matter consisting of one kind of atom is called an **element.** Each element has specific properties (Figure 2.1). For example, the graphite in a pencil consists of a vast number of carbon atoms. The earth's atmosphere consists of gaseous elements such as oxygen and nitrogen. Sodium is a soft, metallic element.

Atoms of an element combine with other atoms of the same or different elements. The ability of carbon atoms to form long chains is important in the structure of living things. Chemists use letters to designate elements—C for carbon, N for nitrogen, Na for sodium (*Natrium* is the Latin word for sodium), and subscripts to indicate how many atoms of the element are present. Though oxygen and nitrogen can occur as paired atoms, O_2 or N_2, most atoms combine with atoms of other elements. One atom of carbon combines with two atoms of oxygen to form carbon dioxide (CO_2) and two atoms of hydrogen combine with one atom of oxygen to form water (H_2O). Two or more atoms combined chemically form a **molecule.** A few molecules consist of atoms of the same element like N_2, but most consist of atoms of different elements like CO_2. Molecules that contain atoms of two or more elements are called **compounds.** CO_2 and N_2 are molecules; CO_2 also is a compound.

More than 85 percent of the human body weight consists of four elements—carbon, hydrogen, oxygen, and nitrogen—mostly arranged in complex molecules. A molecule of the sugar glucose contains 24 atoms, $C_6H_{12}O_6$. Carbohydrates, proteins, fats, and nucleic acids can contain thousands of atoms.

Structure of Atoms

Atoms are the smallest particles that retain the properties of an element, but their subatomic components contribute to those properties. Among the many subatomic particles now known, we are concerned with only **protons, neutrons,** and **electrons** (Table 2.1). The mass of a proton and a neutron is arbitrarily designated as equal to one mass unit, and by comparison, electrons have a very small mass.

(a)

(b)

FIGURE 2.1
Different elements have different properties: (*a*) sulfur and (*b*) iron.

TABLE 2.1	Properties of Atomic Particles		
Particle	**Relative Mass**	**Charge**	**Location**
Proton	1	+	Nucleus
Electron	1/1836	−	Orbiting the nucleus
Neutron	1	None	Nucleus

Charges also are arbitrarily designated as negative for electrons and positive for protons. Neutrons are neutral—they have no charge. Protons and neutrons occupy the nucleus of the atom, whereas the electrons move in orbits around the nucleus. All atoms have an equal number of protons and electrons and are electrically neutral, and the atoms of a particular element have a specific number of protons that determines the element's **atomic number.** Atomic numbers range from one to over 100.

Electrons, being in constant motion, form an *electron cloud* around the nucleus. Some electrons display more energy than others, and their motion can be represented by concentric circles to suggest different energy levels within the cloud (Figure 2.2). Electrons with the least energy stay in orbits near the nucleus and those with more energy move to orbits farther from the nucleus. The negative charge of electrons holds them close to positively charged protons in the nucleus, and orbital motion moves them away from the nucleus. Orbital motion is analogous to the circular path made by swinging an object tied to a string.

Depending on the size of an atom, its electrons occupy one or more concentric circles or shells. An atom of hydrogen has one electron located in the innermost shell. An atom of helium has two electrons—the maximum number in the innermost shell. Atoms with more than two electrons always have two electrons in the inner shell and up to eight additional electrons in the second shell. The inner shell is filled before electrons occupy the second shell, the second shell is filled before electrons occupy the third shell, and so on. Very large atoms have several more electron shells, and some shells can contain more than eight electrons; but for the elements of physiologic importance, the outer shell usually is filled if it contains eight electrons.

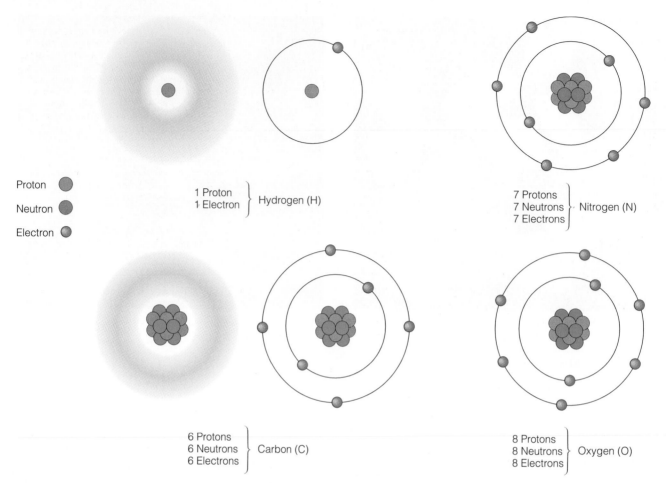

Proton ●
Neutron ●
Electron ●

1 Proton
1 Electron } Hydrogen (H)

7 Protons
7 Neutrons } Nitrogen (N)
7 Electrons

6 Protons
6 Neutrons } Carbon (C)
6 Electrons

8 Protons
8 Neutrons } Oxygen (O)
8 Electrons

FIGURE 2.2
The structures of some atoms commonly found in the human body.

Atoms with outer electron shells that are nearly full (have six or seven electrons) or nearly empty (have one or two electrons) have a tendency to form ions. An **ion** is a charged particle produced when an atom gains or loses electrons (Figure 2.3). When a sodium atom with one electron in its outer shell loses this electron, it has one more proton than electrons and becomes a positively charged ion called a **cation** (kat′i-on). (An easy way to remember that cations are positively charged is to notice that the top of the "t" in cation makes a "plus sign.") When a chlorine atom with seven electrons in its outer shell gains an electron, it becomes a negatively charged chloride ion called an **anion** (an′i-on). Sodium or chloride ions are chemically more stable than corresponding atoms because their outer electron shells are full.

In the body, many elements occur as ions (Table 2.2). Those with one or two electrons in their outer shell tend to lose electrons and form ions with charges of $+1$ or $+2$, respectively; those with six or seven electrons in their outer shell tend to gain electrons and form ions with charges of -1 or -2. Some ions, such as the hydroxyl ion (OH^-), contain more than one element.

All atoms of the same element have the same atomic number, but they may have different atomic weights. **Atomic weight** is the total number of protons and neutrons in an atom. For example, carbon atoms with six protons and six neutrons have an atomic weight of 12. Some naturally occurring carbon atoms have seven or eight neutrons and atomic weights of 13 or 14. Such atoms also can

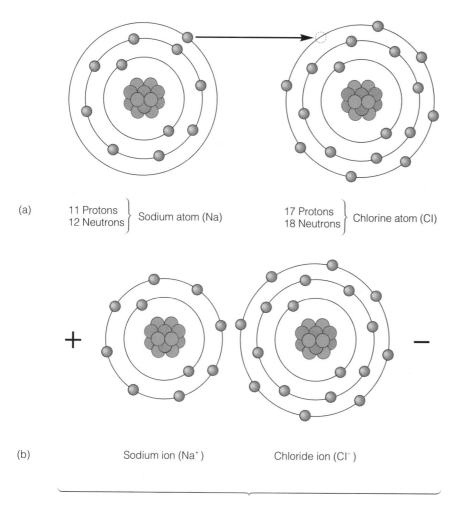

(a) 11 Protons } Sodium atom (Na)   17 Protons } Chlorine atom (Cl)
    12 Neutrons   18 Neutrons

(b)   Sodium ion (Na⁺)   Chloride ion (Cl⁻)

Sodium chloride

FIGURE 2.3
The formation of sodium and chloride ions:
(*a*) When sodium loses the single electron in
its outer shell, it becomes a positively charged
sodium ion. When chlorine, which has seven
electrons in its outer shell, gains an electron,
it becomes a negatively charged chloride ion.
(*b*) The oppositely charged sodium and
chloride ions attract each other electrically
and form a crystalline molecule of sodium
chloride.

be created in the laboratory. Atoms of a particular element that contain different numbers of neutrons are called **isotopes.** For an element that has naturally occurring isotopes, the atomic weight is the average atomic weight of the mixture of isotopes. Atomic weights can be decimal numbers, but a particular atom has a specific number of whole neutrons. Some isotopes are stable and others are not. Unstable isotopes tend to break apart, emitting radiation from their nuclei. Such emissions can be used to follow chemical processes in living things, but they also can harm them. (See Essay on Radioactivity at the end of this chapter.)

The properties of elements found in living things are summarized in Table 2.3.

TABLE 2.2   Ions Commonly Found in the Body

Anions	Cations
Sodium (Na^+)	Chloride (Cl^-)
Potassium (K^+)	Hydroxyl (OH^-)
Calcium (Ca^{2+})	Bicarbonate (HCO_3^-)
Hydrogen (H^+)	Phosphate (PO_4^{3-})
Magnesium (Mg^{2+})	

TABLE 2.3   Some Properties of Elements Found in Living Organisms

Element	Symbol	Atomic Number	Atomic Weight	Electrons in Outer Orbit	% of Body Weight	Biological Occurrence
Carbon	C	6	12.0	4	18	Forms "backbone" of all organic compounds
Hydrogen	H	1	1.0	1	10	Found in most biological molecules; H^+ important component of solutions
Oxygen	O	8	16.0	6	65	Found in most biological molecules; final electron acceptor in many energy-yielding reactions
Nitrogen	N	7	14.0	5	3	Found in proteins, nucleic acids, and many other biological molecules
Calcium	Ca	20	40.1	2	1.5	Essential component of bones and teeth; important in muscle contraction; controls many cellular processes
Phosphorus	P	15	31.0	5	1	Component of nucleic acids and energy-carrying molecules such as ATP; found in many lipids
Sulfur	S	16	32.0	6	<1	Component of many proteins and other important biological molecules
Iron	Fe	26	55.8	2	<1	Component of electron carriers and oxygen carriers
Potassium	K	19	39.1	1	<1	Important in conduction of nerve signals
Sodium	Na	11	23.0	1	<1	Ion in solutions; important in conduction of nerve signals and transport mechanisms
Chlorine	Cl	17	35.4	7	<1	Ion in solutions; synthesis of HCl
Magnesium	Mg	12	24.3	2	<1	Important in enzyme-catalyzed reactions in most cells; important in photosynthesis in plants
Copper	Cu	29	63.6	1	T*	Important in some energy yielding reactions; important in photosynthesis in plants
Iodine	I	53	126.9	7	T	Essential part of thyroid hormone molecules
Fluorine	Fl	9	19.0	7	T	Prevents microbial growth
Manganese	Mn	25	54.9	2	T	Found in enzymes or important in activating enzymes
Zinc	Zn	30	65.4	2	T	Important in activating some enzymes
Selenium	Se	34	79.0	6	T	Part of an antioxidant enzyme
Molybdenum	Mo	42	95.9	1	T	Part of several enzymes

*T = Trace amount found in human body

Chemical Bonds

Chemical bonds are forces between the outer shell electrons that hold the atoms of a molecule together. Three kinds of bonds commonly found in living organisms—ionic, covalent, and hydrogen bonds—vary in strength and help to determine how molecules behave in living organisms.

Ionic bonds form between oppositely charged ions as they are attracted to each other. For example, sodium ions, having a positive charge, combine with chloride ions, having a negative charge (Figure 2.3). Ionic bonds are relatively weak bonds. When molecules held together by them are put in water, the bonds easily break and the molecule **ionizes,** or forms ions. Sodium chloride in body fluids therefore exists as sodium and chloride ions.

Carbon, hydrogen, oxygen, and nitrogen atoms can be held together in molecules by **covalent bonds** in which electrons are shared instead of being gained or lost (Figure 2.4). A carbon atom with four electrons in its outer shell, can share an electron with each of four hydrogen atoms. At the same time, each of the four hydrogen atoms shares an electron with the carbon atom. Four pairs of electrons are shared, with each pair having an electron from carbon and an electron from hydrogen, and the outer shells of both atoms are filled. When forming covalent bonds, oxygen shares two electrons and nitrogen usually shares three electrons. Sometimes a carbon atom and another atom such as oxygen share two pairs of electrons to form a **double bond.** When writing structural formulas, chemists use a single line for a single pair of shared electrons and a double line for two pairs of shared electrons (Figure 2.4). Covalent bonds, which contain more energy than ionic bonds, exert more force and hold molecules together more tightly than ionic bonds. Molecules with covalent bonds are more stable in solutions because they tend not to ionize. Many molecules in living things are stable because they contain covalent bonds.

Weak covalent bonds called **hydrogen bonds** are particularly important in biological structures, where they typically bind hydrogen atoms to oxygen or nitrogen atoms. In such bonds, oxygen or nitrogen atoms attract electrons. Shared electrons are held closer to the atomic nucleus of oxygen or nitrogen, and are pulled away from hydrogen. The oxygen or nitrogen atom has a partial negative charge, and the hydrogen atom has a partial positive charge.

Hydrogen bonds contribute significantly to the structure and properties of large molecules such as proteins and nucleic acids, which consist of long chains of atoms. Hydrogen bonds help to maintain each molecule in its characteristic three-dimensional shape. They also help to account for polar and nonpolar regions within large molecules. **Polar regions** have charges because of ionization

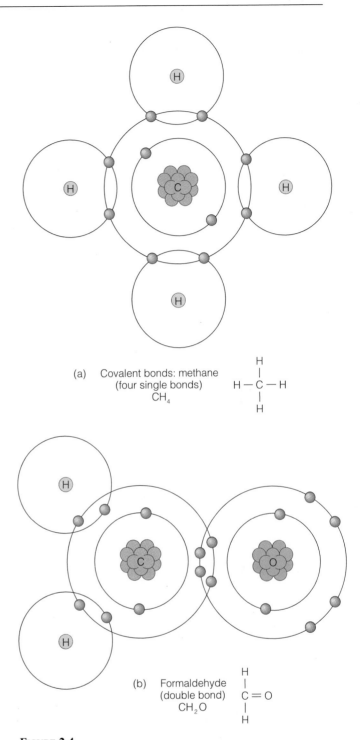

(a) Covalent bonds: methane (four single bonds) CH_4

(b) Formaldehyde (double bond) CH_2O

FIGURE 2.4
The sharing of electrons in covalent bonds: (*a*) In the gas methane, one carbon atom shares four electrons, one with each of four hydrogen atoms. The hydrogen atoms likewise share an electron with the carbon atom. (*b*) In formaldehyde, the carbon and hydrogen atoms share electrons as in methane, but carbon and oxygen each share two electrons. The sharing of two pairs of electrons forms a double-bond.

or because of uneven distribution of positive and negative components of a molecule. Such regions attract other charged molecules. **Nonpolar regions** are uncharged and usually are found below the surface of such molecules. Polar regions are **hydrophilic** (hi-dro-fil′ik), or water-loving, and nonpolar regions are **hydrophobic** (hi-dro-fo′bik), or water-fearing. Large biological molecules mix with water because of surface polar regions, but do not dissolve in it because of internal nonpolar regions.

Water mixes with biological molecules because of its hydrogen bonds and polarity. Water molecules have a positive (hydrogen) pole and a negative (oxygen) pole because shared electrons stay closer to the oxygen than to the hydrogen atoms and the hydrogen atoms lie to the side of the oxygen atom (Figure 2.5). The weak, partial charges allow the hydrogen side of one water molecule to form a hydrogen bond with the oxygen side of another water molecule.

MIXTURES, SOLUTIONS, AND COLLOIDAL DISPERSIONS

Mixtures and Solutions

Dining hall pranksters sometimes surreptitiously add salt to a sugar bowl to see what happens when others detect salt in their coffee. The sugar bowl, which originally contained pure sucrose, now contains a mixture of two kinds of molecules—sucrose and sodium chloride. Unlike a chemical compound in which the molecules contain atoms in specific proportions, a **mixture** consists of two or more compounds combined in any proportion but not chemically bound. A mixture retains the properties of the substances in it. For example, a sugar–salt mixture of any proportions will taste both sweet and salty, but the degree of each taste will depend on the relative amount of each substance in the mixture. Solutions and colloidal dispersions are examples of mixtures.

A **solution** is a homogeneous mixture of two or more substances in which molecules are evenly distributed and usually will not separate out upon standing. In a solution, the **solvent** is the medium in which one or more substances are dissolved and the **solute** is any dissolved substance—atoms, ions, or molecules. In the human body, water is the solvent in nearly all solutions. Typical solutes include the sugar glucose, small protein molecules, the gases carbon dioxide and oxygen, and ions.

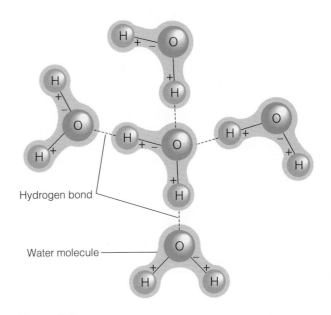

FIGURE 2.5
In hydrogen bonding between water molecules, the slightly negative oxygen region of one molecule is attracted to the slightly positive hydrogen region of another molecule.

TABLE 2.4 Properties of Water
1. Good solvent ability helps to dissolve substances for transport in body fluids.
2. Ability to form layers (because of high surface tension) helps to keep membranes moist.
3. Ability to store or release large quantities of heat (high specific heat) helps to regulate body temperature.
4. Distribution throughout the body provides a medium for chemical reactions.
5. Chemical reactivity allows components of water (H^+ and OH^-) to participate in many chemical reactions.

Water

Water, the solvent in body fluids, is so essential to life that humans live only a few days without it. As noted earlier, it comprises 55 to 60 percent of the total human body weight—more in infants and less in the elderly, and more in brain tissue and less in bone and fat. Several properties of water contribute to its importance in humans and other living things (Table 2.4). Because water is a polar compound, it acts as a good solvent and forms hydrogen bonds with other molecules of both water and other substances.

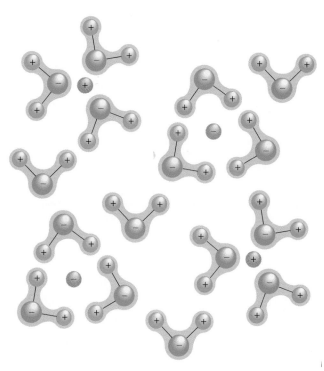

FIGURE 2.6
Water molecules surround positive and negative ions and help to hold these particles in solution.

FIGURE 2.7
The high surface tension of water allows some insects and spiders to literally walk on water.

Water is an especially good solvent for ions because the polar water molecules orient around ions, forming a *hydration shell*. The positive region of the water molecules surrounds negative ions and the negative region surrounds positive ions. Ions thereby become evenly distributed through water with water molecules interspersed between them (Figure 2.6). Ions and other substances that mix with water are easily transported through blood and other body fluids.

Water forms thin layers because its has high **surface tension;** that is, it forms a thin, invisible, elastic layer on membrane surfaces (Figure 2.7). Water molecules beneath the surface are attracted to each other in all directions, but no such attraction exists between water molecules and gas molecules in the air. Attractions between water molecules cause any water layer to contract until it occupies a minimum area. Surface water molecules crowd together with their hydrogen bonds projecting below the surface, creating sufficient tension to support waterstriders walking on a pond surface. On living membranes, high surface tension maintains the continuity of water layers, which is essential for gases such as oxygen and carbon dioxide to diffuse across membranes and for other important membrane functions.

Perhaps you have noticed that the climate usually is milder near the ocean or near another body of water than at other locations at the same latitude. This is partly due to water having a high **specific heat,** or the ability to absorb large quantities of heat energy with little change in temperature. This allows water to gain heat in the daytime and release it at night. Similarly, the high specific heat of water allows relatively large amounts of heat to be lost by the evaporation of small amounts of sweat, thereby regulating body temperature while conserving body water.

Finally, water serves as a medium for chemical reactions and it participates in many reactions. In synthetic reactions such as **condensation,** or **dehydration synthesis,** water is removed as two molecules form a larger molecule. Sugars combine to form complex carbohydrates and amino acids combine to form proteins by this kind of reaction. In degradative (breakdown) reactions such as **hydrolysis** (hi-drol'ĭ-sis), water is added as a large molecule is broken down into smaller molecules. For example, during digestion, large food molecules are broken down into simple sugars, amino acids, and other small molecules by hydrolysis.

TABLE 2.5   Ways of Expressing Concentrations in Solutions

Expression	Definition	Example
Percent by volume	Weight of solute in grams per deciliter (100 ml) of solution	Plasma* glucose: 100 mg% = 100 mg of glucose per deciliter of plasma
Molarity	Number of moles of solute per liter of solution	Plasma urea: 0.005 molar urea = 0.005 moles (5 millimoles) per liter of plasma
Osmolarity	Number of moles of solute per liter of solution times number of particles each molecule of solute produces	Plasma glucose, which does not ionize: 90 grams/liter = 0.5 molar solution = 0.5 osmolar solution; NaCl (which does ionize): 0.58 grams/liter = 0.1 molar solution = 0.2 osmolar (200 milliosmolar) solution
Equivalency or normality	Number of equivalents of solute per liter of solution	Plasma electrolytes such as Na^+ and Ca^{2+}: 327 mg Na^+/100 ml = 142 millimoles/liter of solution = 142 milliequivalents/liter of solution; 10 mg Ca^{2+}/100 ml = 1.5 millimoles/liter of solution = 3 milliequivalents/liter of solution

* Plasma is the liquid portion of blood.

Physiologists express concentrations of solutes in solutions in a variety of ways (Table 2.5). To understand these expressions, you may want to review the following terms.

A **gram molecular weight,** or **mole,** is a mass in grams equal to the atomic weights of the atoms in the substance. For example, the mass of a mole of NaCl is 58 grams, calcium hydroxide, $Ca(OH)_2$, is 74 grams, and glucose, $C_6H_{12}O_6$, is 180 grams. A mole of any substance—molecules, ions, or atoms—always contains 6.023×10^{23} particles. This is **Avogadro's number,** calculated by the Italian scientist Amedeo Avogadro in the early 1800s. We cannot count chemical particles like coins or pencils, but because of Avogadro's number, we know that if we combine a mole of OH^- ions with a mole of H^+ ions, we will get a mole of water.

A **gram equivalent weight,** or **equivalent,** is the mass of a substance that will neutralize all the charges in 1 mole of H^+ (or 1 mole of OH^-). Equivalents are often used to express ion concentrations. An equivalent of NaCl is the same as a mole of NaCl because the Na^+ will neutralize 1 mole of OH^- and the Cl^- will neutralize 1 mole of H^+. This applies to any ions with a **valence** of 1 (a charge of +1 or −1). An equivalent of calcium hydroxide is half a mole because each Ca^{2+} neutralizes two OH^- ions. Normality, essentially the same as equivalency, usually refers to acids and bases.

Colloidal Dispersions

Particles with diameters between 1 and 100 nanometers (a nanometer is one-billionth of a meter) are called **colloids** (kol'oidz). They are too large to form true solutions, but can form **colloidal** (kol-oid'al) **dispersions.** Such particles are suspended in a medium by opposing electrical charges, layers of water molecules around the particles, and other forces. Gelatin dessert is a colloidal dispersion with the protein gelatin dispersed in water. In the body,

colloidal dispersions consist of large protein molecules dispersed in water. Much of the **cytosol** (fluid or semifluid substance around organelles in cells) is a complex colloidal system. Proteins in plasma, which is the fluid portion of blood, also form a colloidal dispersion. Some colloidal systems have the ability to change from a semisolid **gel,** like gelatin that has set, to a more fluid **sol** state, like gelatin that has melted.

The properties of chemical substances, including water, solutions, and colloidal suspensions, set limits on physiologic possibilities. The behavior of molecules in living organisms is limited by the properties of the molecules themselves.

ACIDS, BASES, AND pH

Tart fruits and vinegar contain acids whereas antacids and drain cleaners contain bases. Except for the extreme acidity of the stomach, the body's external and internal environments are nearly chemically neutral—neither very acidic nor very basic (alkaline). Yet we need to know about acids and bases to understand how the body maintains nearly neutral conditions in spite of acids that accumulate as cells use nutrients.

Acids and bases readily form ions in water. An **acid** releases or donates hydrogen ions (H^+) to a solution and is a hydrogen donor, or proton donor. (A hydrogen ion is a proton.) A **base** usually accepts hydrogen ions from a solution, but some bases release hydroxyl ions (OH^-) into solutions. A base is a proton acceptor or a hydroxyl ion donor. The acidity of a solution increases with the H^+ concentration and its alkalinity increases with the OH^- concentration. In body fluids, H^+ is often released by organic acid, or carboxyl (−COOH) groups (COOH ionizes to COO^- and H^+). H^+ is accepted by OH^- to form water or by amino (−NH_2) groups to form ammonia (NH_3).

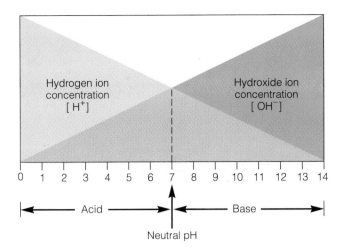

FIGURE 2.8
Solutions at pH 7 are neutral. Increasingly acid solutions have higher concentrations of H+ and lower pH values. Increasingly basic solutions have lower concentrations of H+ and higher pH values.

To express acidity or alkalinity, physiologists use the concept of **pH,** which is the negative log of the hydrogen ion concentration in moles per liter.

$$pH = -\log [H^+]$$

(Brackets [ ] denote concentration.) A solution at pH 7 is neutral, neither acidic nor basic, and contains equal numbers of H^+ and OH^- ions. Such is the case in pure, distilled water because one ten-millionth (10^{-7}) part of each mole of water is ionized, or exists as H^+ and OH^- ions. The H^+ concentration is 10^{-7} moles per liter and the solution's pH is 7.

The pH scale (Figure 2.8) relates proton concentrations to pH and is logarithmic, that is, the proton concentration changes by a factor of ten for each unit of the scale. The usual range of the pH scale is from 0 to 14, but most tissues and body fluids have a pH between 5 and 8 except in the stomach. Table 2.6 shows the pH of some body fluids, foods, and other common substances.

Changes in the pH of body fluids have profound physiologic effects. For example, pneumonia and other diseases that interfere with gas exchange in the lungs allow carbon dioxide, and H^+, to accumulate in the blood by the following reactions:

$$CO_2 \;+\; H_2O \rightarrow H_2CO_3 \rightarrow H^+ \;+\; HCO_3^-$$
carbon   water   carbonic   hydrogen bicarbonate
dioxide   acid   ion

An overdose of aspirin stimulates respiratory centers in the brain and increases gas exchange, causing too much carbon dioxide, and H^+, to be removed from the blood as the above reactions are reversed.

TABLE 2.6 The pH of Various Substances

Substance	pH*
Hydrochloric acid (1 molar)	0.0
Stomach hydrochloric acid (0.1 molar)	1.0
Gastric juice	1.0–3.0
Lemon juice	2.5
Vinegar, beer, wine	3.0
Orange juice	3.5
Tomatoes, grapes	4.0
Coffee	5.0
Urine	5.0–7.0
Milk	6.5
Saliva	6.3–7.3
Pure water (at 25° C)	7.0
Blood	7.35–7.45
Eggs	7.5
Ocean water	7.8–8.2
Household bleach	9.5
Milk of magnesia	10.5
Household ammonia	10.5–11.8
Oven cleaner	13.5
Sodium hydroxide (1 molar)	14.0

* pH of body fluids measured at 37° C.

Buffers, substances that resist pH change, help to prevent such changes. Carbonic acid, which can donate H^+, and bicarbonate, which can accept H^+, act as natural blood buffers, helping to keep the blood pH within a narrow, tolerable range.

Acids, bases, buffers, and disorders of acid–base balance will be discussed in more detail in Chapter 28.

Heat, such as in fever, causes water molecules to ionize. Increased ionization does not change the pH because H^+ and OH^- ions are still present in the same numbers, but the larger numbers of highly reactive ions present in cells may account for cell damage sometimes seen in fever.

Changing the pH of a solution can alter drug ionization, and facilitate or interfere with drug absorption. Being an acid itself, aspirin is only slightly ionized in the acidic stomach and is rapidly absorbed in its nonionized state. As aspirin is absorbed, it inhibits synthesis of mucus that protects the stomach lining. If aspirin is taken frequently, it sufficiently impairs mucus synthesis that stomach acid can damage the stomach wall and cause bleeding. Combining a buffer with aspirin neutralizes some of the stomach acid and reduces the risk of bleeding.

COMPLEX MOLECULES

Covering the fundamentals of general chemistry has paved the way for considering **organic chemistry,** the study of most carbon-containing compounds. Such compounds occur in living things, and their products and remains. The ability of carbon atoms to share electrons, and to form long chains and rings makes the number of possible organic compounds almost infinite.

The simplest carbon compounds are hydrocarbons, chains of carbon atoms with associated hydrogen atoms. The simplest hydrocarbon methane, has a single carbon atom, but gasoline and other petroleum products contain several carbons.

In addition to hydrogen, other atoms such as oxygen and nitrogen can bond to carbon chains, where they often form functional groups. A **functional group** is a molecular part that participates in chemical reactions and gives the molecule some of its properties. Functional groups demonstrate that structure and function are related even at the chemical level. Four categories of organic molecules with functional groups containing oxygen are alcohols, aldehydes, ketones, and organic acids (Figure 2.9). The functional groups of alcohols, called **hydroxyl** (hi-drox′il) **groups,** are found almost anywhere in a molecule. **Carbonyl** (kar′bon-el) **groups** form aldehydes at the ends of chains and ketones within chains. The functional groups of organic acids are called **carboxyl** (kar-box′il) **groups.** An **amino** (ah-me′no) **group** ($-NH_2$) contains nitrogen, but no oxygen. Amino groups, found mainly in amino acids of proteins, account for most of the body's nitrogen.

The amount of oxygen in functional groups is related to the energy they contain. Alcohols, which have lots of hydrogen atoms and few oxygen atoms, contain more energy than organic acids, which have less hydrogen and more oxygen. Molecules with large amounts of hydrogen are said to be reduced and energy can be extracted from them as they are oxidized. **Reduction** is the addition of hydrogen or the removal of oxygen. **Oxidation** is the addition of oxygen or the removal of hydrogen. Gasoline and other hydrocarbons, being extremely reduced, make good fuels. Glucose and other sugars, being relatively reduced, are important body fuel. Oxidized molecules like CO_2, a product of glucose oxidation, contain little energy.

Biological molecules, or organic molecules found in living organisms, fall into four major classes: carbohydrates, lipids, proteins, and nucleic acids. Enzymes, a group of proteins, are of sufficient importance to warrant special consideration. We will look at the properties, categories, and physiologic importance of each class of molecules. We will consider how they participate in chemical reactions in Chapter 24.

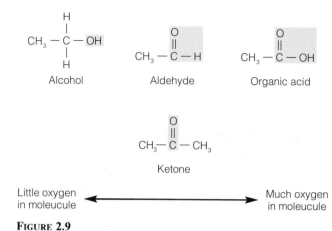

Little oxygen in moleucule ← → Much oxygen in moleucule

FIGURE 2.9
Oxygen-containing functional groups of organic molecules. The shaded portion of the molecule is the functional group. Molecules are arranged from most reduced (having the most hydrogen) to most oxidized (having the most oxygen).

CARBOHYDRATES

Foods such as bread, potatoes, table sugar, and honey contain large amounts of carbohydrate. **Carbohydrates** (kar-bo-hi′dratz) serve as the main source of energy for most living things, and humans are no exception. All carbohydrates contain carbon, hydrogen, and oxygen, generally in the proportion of two hydrogen atoms for each carbon and oxygen atom (CH_2O).

Monosaccharides (mon-o-sak′ar-īdz), the simplest carbohydrates, consist of aldehydes or ketones with several hydroxyl groups. Glucose, a six-carbon sugar found in human blood, can be represented as a chain, a ring, or a three-dimensional structure (Figure 2.10). Regardless of how they are drawn on flat paper, all molecules are three-dimensional. Other common monosaccharides include fructose and galactose, which are found in many carbohydrate foods, and ribose and deoxyribose, which occur in nucleic acids (Figure 2.11).

Glucose, fructose, and galactose are **isomers** (i′so-merz); that is, they contain the same numbers of each kind of atom ($C_6H_{12}O_6$), but the atoms are arranged differently. The aldehyde glucose and the ketone fructose are **structural isomers**—their structures differ significantly. Glucose and galactose, both of which are aldehydes, are **stereoisomers** (sta″re-o-i′so-merz). They differ only by the positions of the hydrogen atom and hydroxyl group on carbon 4. Certain isomers are mirror images like left and right hands and are designated L (for *levo* or left-handed) and D (for *dextro* or right-handed). Naturally occurring sugars are D-sugars, but mirror image L-sugars can be made in the laboratory. Because of their molecular shape, L-sugars cannot be broken down in the body. It has been suggested that L-sugars be used to make synthetic foods for people on weight loss diets, but such foods would be extremely expensive.

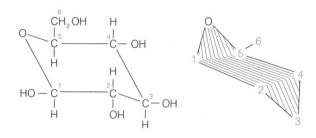

(a) Chain structure   (b) Folding   (c) Ring structure

(d) Three-dimensional structure

FIGURE 2.10

The structure of a glucose molecule can be represented in several ways: (*a*) a chain structure, (*b*) the folding of a chain to form a ring, (*c*) a ring structure, and (*d*) a three-dimensional structure.

Fructose   Galactose

Six-carbon sugars

Ribose   Deoxyribose

Five-carbon sugars

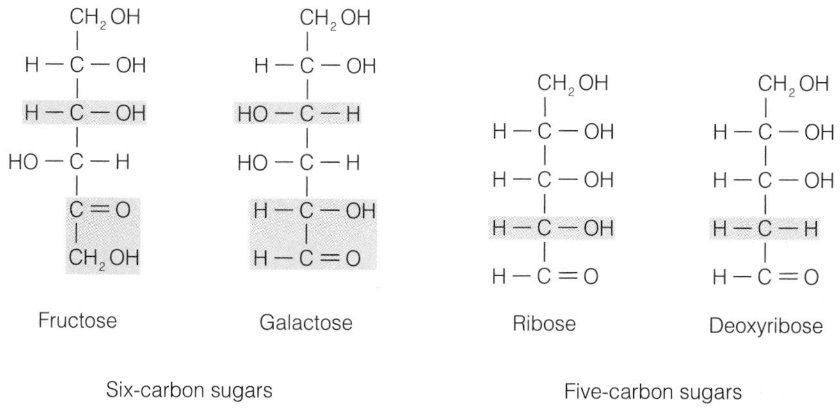

FIGURE 2.11

In addition to glucose, fructose and galactose also are six-carbon sugars. Ribose and deoxyribose are five-carbon sugars. The sites where these molecules differ are shown in color.

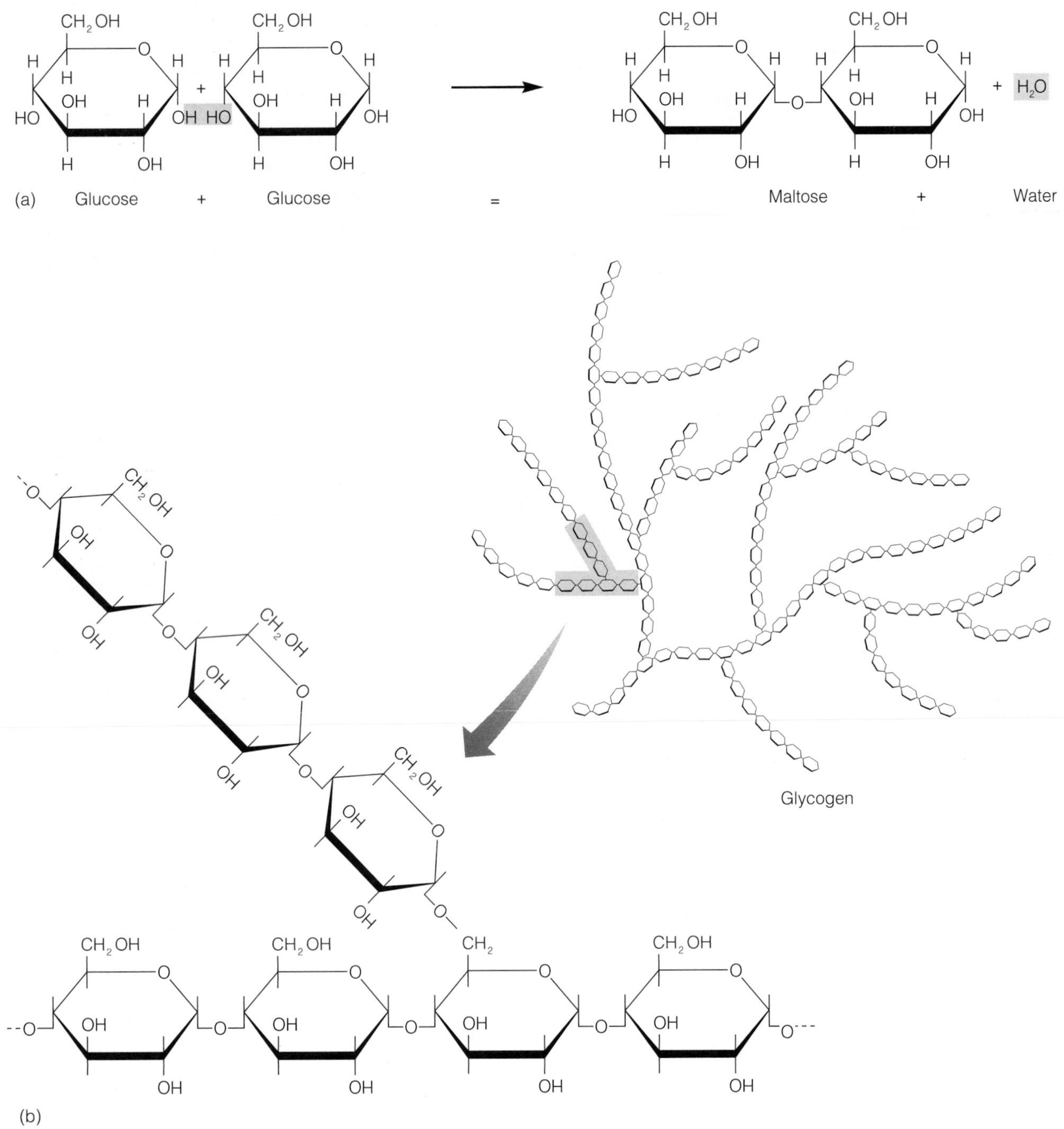

(a) Glucose + Glucose = Maltose + Water

Glycogen

(b)

FIGURE 2.12

(*a*) Two molecules of the monosaccharide glucose combine to form one molecule of the disaccharide maltose. The removal of water from hydroxyl groups of these molecules creates a glycosidic bond between the glucose units, and forms maltose. (*b*) The polysaccharide glycogen is composed of many glucose units linked together by glycosidic bonds.

Two monosaccharide molecules joined by dehydration synthesis form a **disaccharide** (di-sak′ar-īd), as in Figure 2.12a. Dietary disaccharides include sucrose, lactose, and maltose. Sucrose (table sugar) consists of glucose and fructose. Lactose (milk sugar) consists of glucose and galactose. Maltose contains two glucose molecules and is derived from the digestion of starch.

Polysaccharides (pol-e-sak′ar-īdz) are polymers of monosaccharides (Figure 2.12b). A **polymer** (pol′im-er) is a long chain of repeating molecular units. Common polysaccharides include starch, glycogen, and cellulose, all of which are glucose polymers. They can be distinguished by the arrangement of the chemical bonds that hold the units together and by which enzymes digest them. Human enzymes break down starch and glycogen, but cellulose passes through the digestive tract undigested as insoluble fiber or roughage.

Plants and animals use polysaccharides differently. Plants store the excess carbohydrate as starch, so potatoes, rice, grain products, and many other foods derived from plants contain starch. Animals use starch and sugars to make glycogen, which they store mainly in the liver and skeletal muscle. Plants use cellulose in cell walls and other structures. Many fruits and vegetables contain large amounts of cellulose. Celery, which is mainly cellulose and provides human cells with less energy than it takes to chew it, is more suitable for a weight loss diet than L-sugars—and considerably cheaper. Arthropods, such as lobsters, shrimp, and insects, make a complex polysaccharide called chitin, which serves as an external skeleton.

The properties of carbohydrates are summarized in Table 2.7.

LIPIDS

Butter and salad oil consist entirely of lipids; fatty meats and other fatty foods contain large quantities of lipids. **Lipids** are a chemically diverse group of substances—fats; steroids; and vitamins A, D, E, and K—that are relatively insoluble in water but soluble in solvents such as ether and benzene (Figure 2.13). Lipids easily mix with these solvents because both the lipids and the solvents are nonpolar. Lipids serve several functions in the body. Some form part of cell membranes and others store energy. The vitamins participate indirectly in various chemical reactions. Cholesterol is used to make bile salts that help digest lipids. Steroid hormones from the ovaries, testes, and adrenal glands regulate various body processes. Generally,

TABLE 2.7 Properties of Carbohydrates

Class of Carbohydrates	Examples	Description
Monosaccharides	Glucose	Six carbons, aldehyde functional group
	Fructose	Six carbons, ketone functional group
	Galactose	Six carbons, aldehyde functional group arranged differently than glucose
	Ribose	Five carbons, aldehyde functional group
	Deoxyribose	Five carbons, aldehyde functional group, one carbon has no oxygen
Disaccharides	Sucrose	Glucose and fructose
	Lactose	Glucose and galactose
	Maltose	Two glucose units
Polysaccharides	Starch	Polymer of glucose found in plants, digestible by humans
	Glycogen	Polymer of glucose stored in liver and skeletal muscles
	Cellulose	Polymer of glucose found in plants, not digestible by humans

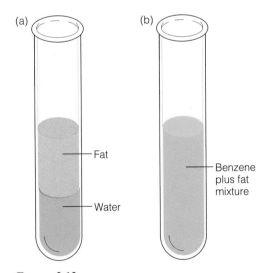

FIGURE 2.13
(*a*) When fat is added to a tube of water, the fat forms a layer on top of the water. (*b*) When fat is added to a tube of benzene or another fat solvent, the fat disperses through the solvent.

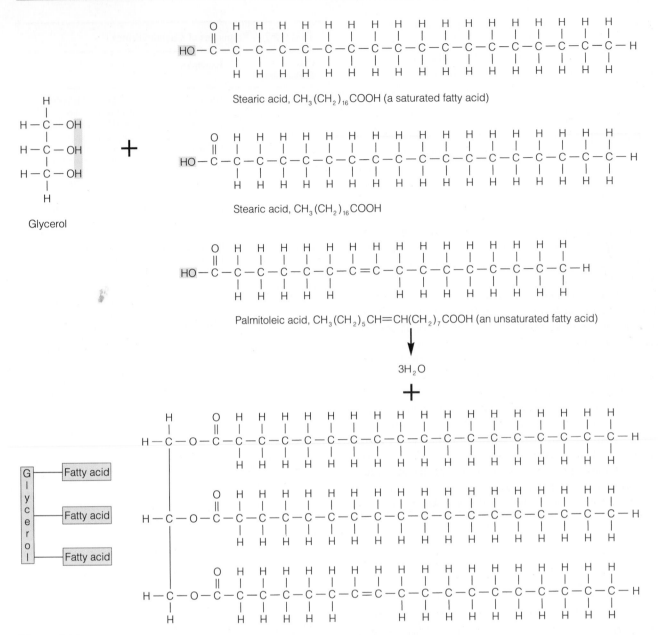

FIGURE 2.14
Glycerol and fatty acids, either saturated or unsaturated, are the
building blocks of a simple lipid. A molecule of water is removed as
each fatty acid is attached to glycerol by an ester bond.

lipids contain relatively more hydrogen and less oxygen,
and therefore, more energy than carbohydrates. Unless we
pay careful attention to the amount of lipids in our diet,
it is easy to eat more than we need. As we shall see later,
this can cause unwanted weight gain and increase the risk
of a "heart attack," or heart muscle damage because of
reduced blood flow.

Simple lipids, common in the diet, contain glycerol (a
three-carbon alcohol) and fatty acids (Figure 2.14). A
fatty acid is a long chain of carbon atoms with attached
hydrogen atoms and a carboxyl group at one end. Simple
lipids are made by condensation reactions in which one or
more fatty acids are joined to glycerol. Most are **triacylglycerols** (tri-as″il-glis′er-olz), formerly called triglycerides, having three fatty acids connected to glycerol. A

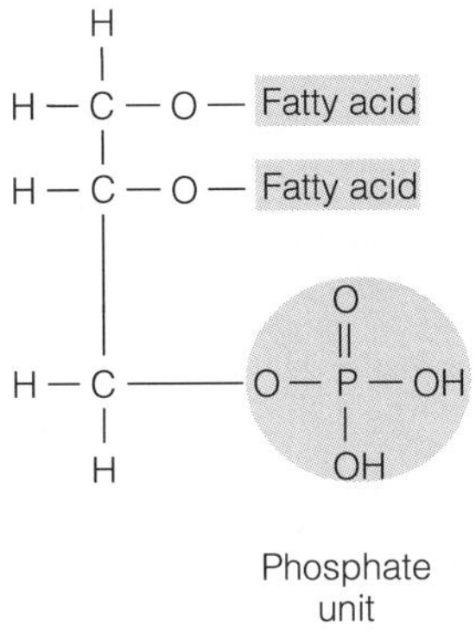

FIGURE 2.15
A phospholipid differs from a simple lipid by the substitution of a unit of phosphate for one of the fatty acids.

monoacylglycerol (monoglyceride) has one fatty acid and a **diacylglycerol** (diglyceride) has two fatty acids. Most mono- and diacylglycerols are formed by hydrolysis of triacylglycerols.

Fatty acids can be saturated or unsaturated. A **saturated fatty acid** is saturated with hydrogen—all its carbon atoms have the maximum number of hydrogen atoms bonded to them. An **unsaturated fatty acid** has lost one or more pairs of hydrogen atoms and a double bond is present between pairs of carbons that have lost hydrogen. Stearic acid is a saturated, and palmitoleic acid an unsaturated, fatty acid. Vegetable oils are high in unsaturated fatty acids, whereas animal fats contain more saturated fatty acids. Because diets high in fat, especially saturated fats, are associated with heart disease, limiting total dietary fat and consuming mainly vegetable fats may lower the risk of such disease.

Phospholipids (fos″fo-lip′idz), which are found in all cell membranes, are like simple lipids except that they have a phosphate group ($-HPO_4^-$) substituted for a fatty acid (Figure 2.15). The phosphate forms a polar, hydrophilic region, enabling the molecule to mix with water. On the surfaces of cell membranes, polar regions allow water to surround the membranes. On the surfaces of fat droplets being digested in the small intestine, they help water-soluble enzymes to mix with and digest fats.

Steroids have a characteristic four-ring structure (Figure 2.16). Cholesterol, the main body steroid, is found in all cell membranes and is used to make steroid hormones, bile salts, and vitamin D. A few of the more important steroid hormones are testosterone from the testes, estradiol and progesterone from the ovaries, and cortisol from the cortex of the adrenal glands (small glands located above the kidneys). Bile salts help to mix fats with watery digestive juices and vitamin D makes calcium available for building and maintaining strong bones. In spite of the body's many uses of cholesterol, this fatty substance often becomes deposited in heart blood vessels, increasing the risk of a heart attack.

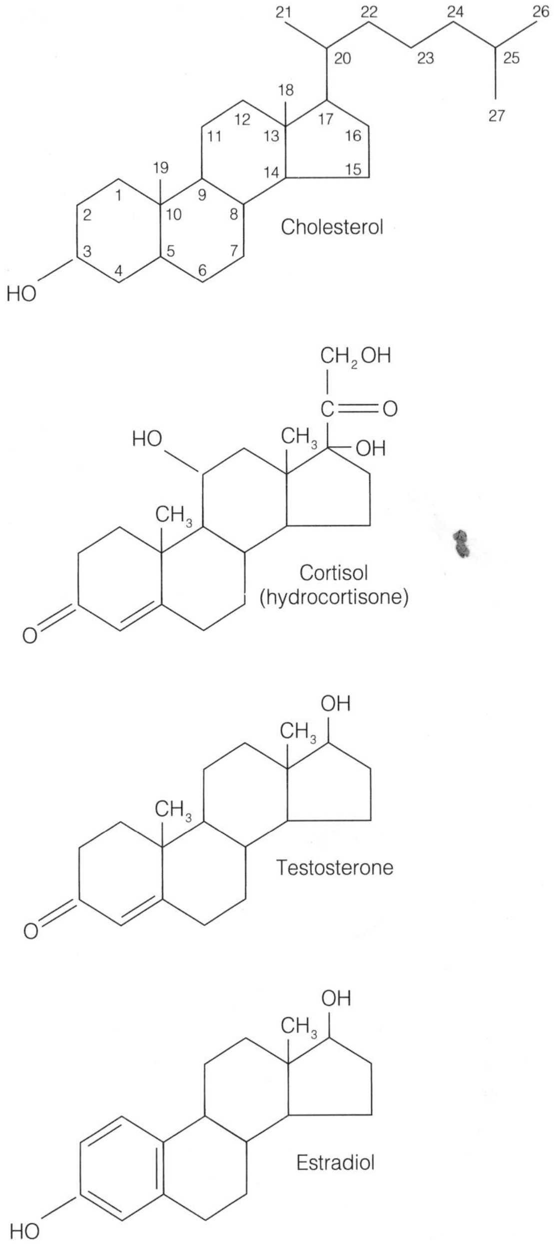

FIGURE 2.16
The steroid cholesterol is typical of all steroids, which have a particular arrangement of the carbon rings and the long nonpolar side chain. Cholesterol is the precursor molecule from which all other steroids, such as sex hormones, are made.

Limiting the lipids in your diet may reduce your risk of heart disease. Most vegetable oils are "good" fats because they contain no cholesterol and are high in unsaturated fatty acids, some of which are polyunsaturated (having two or more double bonds). Three-fourths of the fatty acids in safflower oil are polyunsaturated, whereas 55 percent of the fatty acids in corn and soybean oils are polyunsaturated. Though advertisers emphasize the value of oils high in polyunsaturated fatty acids, their low saturated fatty acid content (no more than 10 percent) may be more important in reducing the risk of heart disease.

The human body contains other lipids such as vitamins A, E, and K, and prostaglandins. Vitamin A is required for normal vision, vitamin E prevents saturation of unsaturated fatty acids in cell membranes, and vitamin K is essential for blood clotting. Prostaglandins, derived from a 20-carbon fatty acid, arachidonic acid, are important regulators of functions such as contraction and relaxation of muscle cells within organs (Chapter 16).

The properties of lipids are summarized in Table 2.8.

PROTEINS

Foods such as meats, eggs, dairy products, legumes (peas and beans), nuts, and seeds contain significant amounts of protein. Cereal grains such as wheat, oats, and rice also contain some protein. Protein is essential for maintenance of cell membranes and organelles, synthesis of enzymes, cell division, and other processes. Children need protein for growth and normal brain development. Young children deprived of protein grow more slowly than normal and can become mentally retarded.

Properties of Proteins

Proteins are composed of building blocks called **amino acids.** In addition to carbon, hydrogen, and oxygen, amino acids contain nitrogen and some contain sulfur. Each amino acid has at least one basic amino ($-NH_2$) group and one acidic carboxyl ($-COOH$) group. Naturally occurring amino acids are L-isomers. The body can synthesize about half of the 20 amino acids found in proteins, some of which are shown in Figure 2.17. Amino acids the body cannot synthesize, called the **essential amino acids,** must be supplied in the diet (Chapter 24).

The quality of dietary protein is determined mostly by whether it adequately supplies the essential amino acids. Most people in affluent countries obtain protein from meat, milk, eggs, and other animal sources, but adequate protein can be obtained from plant sources. Certain vegetable proteins complement each other by one of them supplying an essential amino acid that the other one lacks. For example, peas and beans lack tryptophan and sulfur-containing amino acids, but grains supply them. Grains lack isoleucine and lysine that legumes supply. Eating rice and beans, including soybean products, in the same meal supplies all the essential amino acids.

TABLE 2.8   Characteristics of Lipids

Class of Lipid	Characteristics
Simple lipids (fat)	Glycerol and three fatty acids form a triacylglycerol; mono- and diacylglycerols contain one and two fatty acids, respectively; fatty acids can be saturated (contain all the hydrogen they can) or unsaturated (have double bonds between carbons and lack some hydrogen)
Compound lipids	Contain other components in addition to glycerol and fatty acids; phospholipids contain a phosphate group instead of one of the fatty acids; found in cell membranes
Steroids	Contain a characteristic four-ring structure; include cholesterol, bile salts, vitamin D, and hormones from the adrenal and sex glands
Other	A variety of substances having some properties of lipids; include porphyrins (components of hemoglobin), vitamins A, E, and K, and prostaglandins (important regulatory molecules in cells)

A protein is a polymer of amino acids linked together by **peptide** (pep'tīd) **bonds,** bonds between the nitrogen of one acid's amino group and the carbon of another acid's carboxyl group (Figure 2.18). Two amino acids linked together make a **dipeptide,** three make a **tripeptide,** and many (4 to 100) make a **polypeptide.** Some amino acids have sulfhydryl ($-SH$) groups, functional groups that combine to form disulfide ($-S-S-$) linkages between amino acid chains, and maintain the three-dimensional shape of proteins.

Proteins have increasingly complex levels of structure (Figure 2.19). The *primary structure* is made up of a specific sequence of amino acids in a polypeptide chain held together by peptide bonds. The *secondary structure* is formed by coiling or folding a polypeptide chain. Coiled polypeptides form a helix like a spring, and folded polypeptides form a pleated sheet like a Japanese fan. Further folding and twisting into globular (irregular spherical) shapes or fibrous threadlike strands produces the *tertiary structure.* All proteins have these three structural levels, but certain very large proteins, such as hemoglobin, contain several tertiary units that aggregate to form a *quaternary structure.*

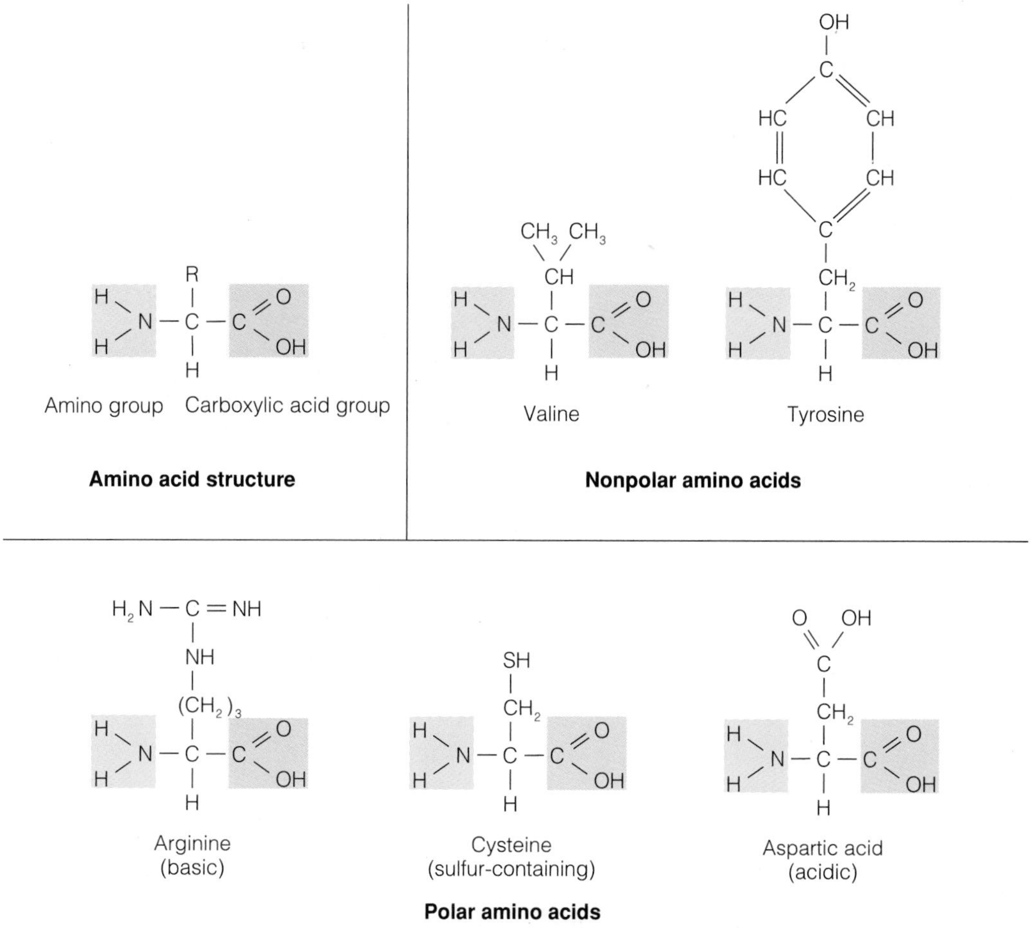

Amino acid structure

Amino group   Carboxylic acid group

Nonpolar amino acids

Valine   Tyrosine

Arginine
(basic)

Cysteine
(sulfur-containing)

Aspartic acid
(acidic)

Polar amino acids

FIGURE 2.17

All amino acids have at least one amino group and one carboxyl group
(shown in color). Nonpolar amino acids such as valine and tyrosine
contain one each of amino and carboxylic acid groups. Polar amino
acids can have an extra amino group and be basic as in arginine, a
sulfhydril group as in cysteine, or an extra acid group as in aspartic
acid.

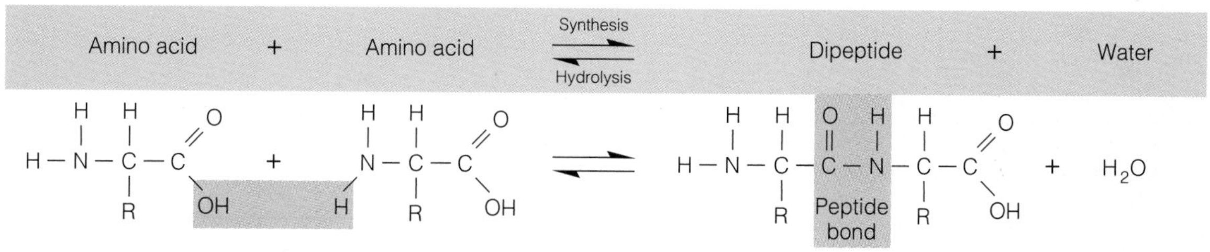

FIGURE 2.18

Amino acids are attached to each other by the removal of water—OH
from the carboxyl group of one amino acid and H from the amino
group of another amino acid—in dehydration synthesis. A peptide
bond forms between two amino acids from the carbon of a carboxyl
group to the nitrogen of an amino group.

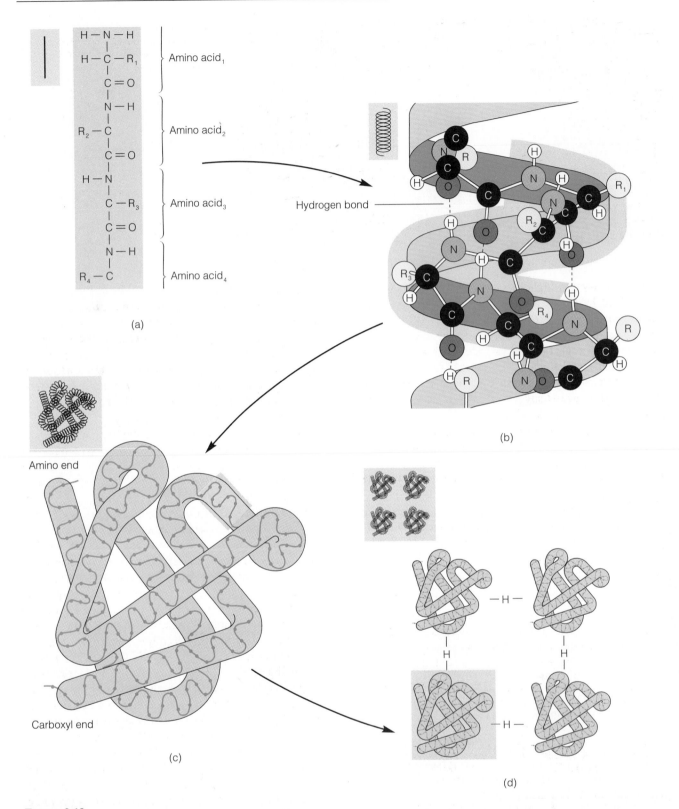

FIGURE 2.19

Proteins have at least three and sometimes four structural levels:
(a) The primary structure consists of a chain of amino acids connected
by peptide bonds. This is analogous to an uncoiled "Slinky®." (b) The
secondary structure often consists of a helix held together by hydrogen
bonds and disulfide linkages. This is analogous to a coiled "Slinky."
(c) The tertiary structure can consist of complex folding of the helical
structure. This is analogous to a tangled "Slinky." (d) Large proteins
sometimes have a quaternary structure, which consists of several
tertiary structures bound together. This is analogous to three or four
"Slinkies" tangled together. Tertiary and quaternary structures are
held together by the same kinds of forces that maintain the secondary
structure.

FIGURE 2.20
Proteins are denatured by heat. Frying an egg denatures the protein albumin, causing it to solidify and turn white.

Secondary and higher levels of protein structure are maintained by forces such as disulfide linkages and hydrogen bonds. Some proteins change shape as these weak forces are altered. That alterations in the shape of molecules can alter their function provides another example of how structure and function are related at the molecular level. That molecular alterations can be caused by changes in temperature or pH suggests that physical changes can control some physiologic processes.

Acidic or basic conditions and above normal temperatures disrupt the forces that maintain protein structure, causing **denaturation** (de-na″chur-a′shun) as shown in Figure 2.20. The severity and permanency of denaturation depends on the harshness of conditions. Slight acidity or a few degrees rise in temperature can temporarily separate the units of a quaternary protein or "unwind" the tertiary arrangement of a globular protein. The protein usually can function again when conditions return to normal and its structure is restored. Harsh conditions, such as boiling in acid, permanently disrupt all structural levels—even peptide bonds.

Denaturation has several practical applications. Heat, which sterilizes hospital equipment, and chemicals, which disinfect skin and inanimate objects, kill microorganisms by permanently denaturing their proteins. Cooking tenderizes meat and solidifies egg white by permanently denaturing proteins.

In so-called "permanent" hair waving, the waving lotion temporarily denatures hair proteins and winding hair on rollers flattens straight cylindrical hairs. Flattening causes hairs to curl and the neutralizer counteracts the waving lotion so the hair stays flattened—at least for a few months.

Fever and severe disturbances in the body's acid–base balance can denature proteins. If such conditions are treated promptly, denaturation can be temporary, but if treatment is too little or too late, denaturation can be permanent and deadly.

In addition to specific amino acid sequences and particular three-dimensional shapes, protein molecules have charged, polar hydrophilic (water-loving) regions and uncharged, nonpolar hydrophobic (water-fearing) regions. The presence of such regions causes molecular folding in which hydrophilic regions remain near the surface and hydrophobic ones get tucked inside the molecule. Water aversion of hydrophobic regions holds them close together and stabilizes the shape of the molecule.

Certain configurations of surface shape and charge on protein molecules form **receptors,** or **binding sites** (Figure 2.21), at which a particular molecule, atom, or ion can bind, or ligate, to the protein. A substance that binds to a receptor is called a **ligand** (li′gand). Some receptors are highly specific—they bind to one and only one ligand. Others are only relatively specific—they bind to any of a class of ligands with shapes and charges that complement those of the receptor. The affinity between receptors and ligands also depends on how tightly they fit together. Different ligands that bind to the same site can compete for receptors and receptors can become saturated. Which ligands bind to receptors depends on affinities, ligand concentrations, availability of receptors, and random collisions of ligands with receptors. These effects are demonstrated later with the discussion of enzymes and their inhibition.

Binding of ligands to protein receptors is very important in physiology: it allows communication at the molecular level and the information communicated, in turn, contributes to the control of a living system. Such control occurs in the actions of enzymes and hormones, reactions of the immune system, in neural functions, and in many other processes. Sometimes sensing of products of reactions provides negative feedback to the control, or ligand-binding, component of the system.

FIGURE 2.21
(*a*) The attachment of ligands to binding sites of a protein molecule. (*b*) A control diagram showing a possible way that binding ligands to protein receptors, or binding sites, controls cell function. Sensors and feedback are involved only in closed cybernetic control systems.

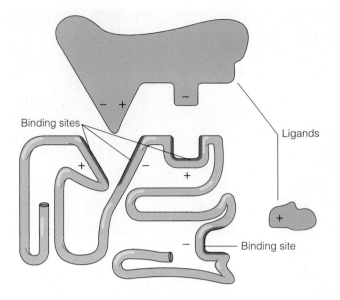

Binding sites

Ligands

Binding site

Protein molecule

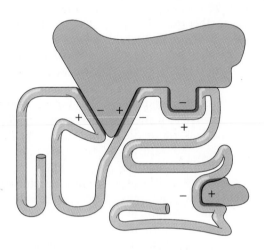

Protein molecule with ligands
attached to binding sites

(a)

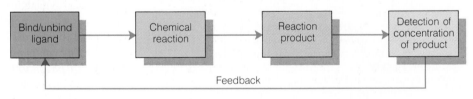

(b)

Classification of Proteins

Proteins can be classified by their functions in the body as structural proteins, motile proteins, enzymes, or regulatory proteins. **Structural proteins** form the structure of body parts. Various structural proteins form portions of membranes and organelles of all cells. The protein keratin contributes to the structure of the skin, hair, and nails. The protein collagen contributes to the structure of bone, cartilage, tendons, ligaments, and other connective tissues. Keratin and collagen, which form long, strong fibers, are classified as fibrous proteins.

Motile proteins contribute to movement from the cellular to the whole body level. They include actin and myosin, found in abundance in muscle cells. These proteins also appear in limited quantity in microtubules and other components of the internal skeleton of cells.

Enzymes (en′zīmz) control the rate of chemical reactions in the body. They are often globular and can change shape as they control a reaction. Some proteins in membranes function as enzymes and also contribute to the structure of cells.

In addition to enzymes, **regulatory proteins** include some hormones and proteins in cell membranes that exert control over cellular functions. Binding of a hormone ligand to a receptor on a membrane protein often serves to regulate cell function.

Other proteins are less easily classified. Antibodies protect against infections, fibrinogen and certain other proteins cause blood to clot, albumins help to maintain blood osmotic pressure, and some proteins transport lipids and other substances in the blood. Protein classification is summarized in Table 2.9.

TABLE 2.9	Classification of Proteins
Type of Protein	**Characteristics**
Structural	Form the structure of cell parts or cell products, including portions of cell membranes and organelles; keratin in skin, hair, and nails; collagen in bone, cartilage, and tendons
Motile	Fibrous proteins involved in muscle contraction and movement of microtubules within cells
Enzymes	Control the rate of chemical reactions in cells
Regulatory	Form receptors sites on cell membranes; act as hormones
Other	Antibodies that protect against infection, proteins that transport substances in the blood; albumins that help to maintain blood osmotic pressure

NUCLEOTIDES AND NUCLEIC ACIDS

Nucleotides are found in high energy compounds such as ATP, molecules called coenzymes that assist certain enzymes, and the nucleic acids DNA and RNA. A **nucleotide** (nu′kle-o-tīd) has three parts: (1) a nitrogenous base, so-called because it contains nitrogen and has alkaline properties; (2) a five-carbon sugar; and (3) one or more phosphate groups (Figure 2.22a).

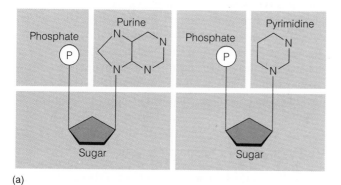

(a)

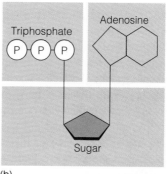

(b)

FIGURE 2.22
(*a*) A nucleotide consists of a nitrogenous base, a five-carbon sugar, and at least one phosphate. (*b*) The nucleotide ATP consists of adenine, ribose, and three phosphate units.

ATP and Other Simple Nucleotides

The nucleotide **adenosine** (ah-den'o-sēn) **triphosphate** (ATP) is a cell's main energy currency molecule because it can capture energy from food and release it in a form cells can use (Figure 2.22b). **Guanidine** (gwan'id-ēn) **triphosphate** (GTP) and **uridine** (u'rid-ēn) **triphosphate** (UTP) also can capture and release energy. Bonds between phosphates in these nucleotides are called **high energy bonds** because when hydrolzyed, they release much more energy than most covalent bonds, though they are otherwise no different from other covalent bonds. Enzymes control the hydrolysis of high energy bonds so as to make energy available at appropriate sites in cells as needed.

Nucleic Acids

Nucleic (nu-kle'ik) **acids,** long polymers of nucleotides, are information molecules that control heredity and protein synthesis. **Ribonucleic acid** (RNA) consists of one strand of nucleotides and **deoxyribonucleic acid** (DNA) consists of two strands arranged as a double helix. The helix is about 2 nm wide, makes a full turn every 3.4 nm, and its bases are separated by a distance of 0.34 nm. DNA and RNA have somewhat different building blocks (Table 2.10 and Figure 2.23). DNA contains four nitrogenous bases: adenine, cytosine, guanine, and thymine. RNA contains the same bases except that uracil is substituted for thymine. Of these bases adenine and guanine are purines, molecules with a double ring structure; thymine, cytosine, and uracil are pyrimidines, molecules with a single ring structure. RNA contains ribose; DNA contains deoxyribose, which has one less oxygen atom than ribose; and both contain phosphate.

In both DNA and RNA, sugar and phosphate molecules form the "backbone" of the strand from which nitrogenous bases protrude (Figure 2.23). The two strands of DNA are mainly held together by hydrogen bonds between pairs of bases on adjacent strands, in which adenine always pairs with thymine and cytosine always pairs with guanine. This linking of bases is called **complementary base pairing.** The same kind of complementary base pairing also occurs as information is transmitted from DNA to RNA in protein synthesis (Chapter 3). Complementary base pairing is a kind of communication—*it provides the sole means by which genetic information directs cellular activities.*

Any particular DNA or RNA strand contains several hundred nucleotides with bases in a particular sequence. This sequence of nucleotides, like the sequence of letters in words and sentences, conveys information that specifies what proteins an organism will have. Like changing a letter in a word, changing a nucleotide in a sequence can change the information it carries. The large numbers of nucleotides in nucleic acids makes possible a nearly infinite number of different base sequences—and a great many different pieces of information—in DNA and RNA.

The functions of DNA and RNA are to convey and express genetic information. DNA, located in cell nuclei and transmitted from one generation to the next, determines the heritable characteristics of the new individual by supplying the information for the synthesis of its proteins. Except for identical twins, each human being has a unique set of DNA molecules. RNA, which functions outside the nucleus, uses information from DNA to direct protein assembly. More will be said about these processes in Chapter 3.

BIOENERGETICS

To understand how the body works, we must first understand the fundamentals of **bioenergetics** (bi'o-en"er-jet'iks), the study of the flow of energy in living systems. Bioenergetics is a part of thermodynamics, the branch of science that deals with energy changes in all matter, living and nonliving. The aspects of bioenergetics considered here include laws governing chemical changes and coupled reactions.

Laws Governing Chemical Changes

Three laws—the first and second laws of thermodynamics and the law of mass action—are especially important in understanding energy flow in living systems. Because life cannot exist without energy, we need at least a simple introduction to these laws.

TABLE 2.10 Components of DNA and RNA

Component	Found in DNA	Found in RNA
Adenine	X	X
Guanine	X	X
Cytosine	X	X
Thymine	X	
Uracil		X
Ribose		X
Deoxyribose	X	
Phosphoric acid	X	X

Double strand DNA

Single strand RNA

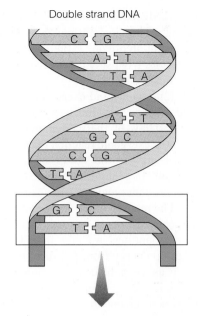

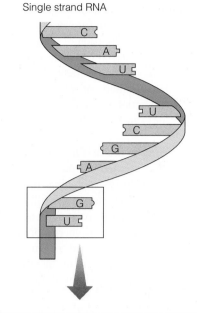

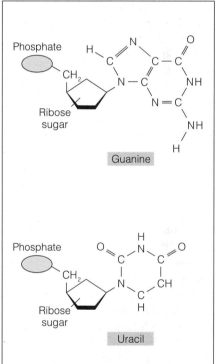

FIGURE 2.23
Nucleotide polymers are arranged in strands in DNA and RNA.
DNA contains two strands held together by complementary base
pairing—A with T and C with G. RNA contains a single strand of
nucleotides. Bases are connected to sugars in the formation of
nucleotides and to each other in the nucleic acid DNA. The bases
guanine and cytosine always pair with each other as do thymine and
adenine. Uracil replaces thymine in the nucleic acid RNA.

Building a perpetual motion machine, a machine that once started would run forever without an energy supply, has captured the attention of many inventors. No matter how ingeniously designed, such machines eventually run down because of friction between the mechanical parts. Careful measurements on such machines show that heat energy from friction exactly equals the energy used to start the machine—an example of the first law of thermodynamics.

According to the **first law of thermodynamics,** energy can be neither created nor destroyed, but it can be transferred in various ways, including rolling a rock down a hill (Figure 2.24). At the top of the hill, the rock has a quantity of **potential energy,** or stored energy. As it rolls down the hill, it releases **kinetic energy,** or energy in action, into the surroundings as heat and contains less potential energy when it reaches the bottom of the hill. Potential energy at the hilltop equals the potential energy remaining at the bottom plus the energy released as heat. Some energy was transformed into heat but none was lost, so the system obeys the first law of thermodynamics.

Like a rock rolling down a hill, some chemical reactions also release energy. The energy in the **reactants** (substances that enter a reaction) equals the energy in the **products** (substances remaining at the end of the reaction) plus energy released as heat. Such reactions are **exergonic** (ek″ser-gon′ik). They include reactions that occur during **catabolism** (kat-ab′o-lizm), the **degradation** (breakdown) of large molecules into smaller ones, in which energy is released from chemical bonds.

In contrast, some systems gain energy, as a rock would if someone carried it up a hill. Some chemical reactions also gain energy, and the energy in the reactants plus energy added from the surroundings equals the energy in the products. Such reactions are **endergonic** (end″er-gon′ik). They include reactions that occur during **anabolism** (an-ab′o-lizm″), the **synthesis** (building) of larger molecules from smaller ones with energy incorporated in chemical bonds.

Endergonic and exergonic reactions usually are coupled with one providing energy for the other, and both obey the first law of thermodynamics. Energy is transferred from the exergonic reaction to the endergonic one, and the total energy in the system of coupled reactions remains constant. In general, exergonic reactions are spontaneous whereas endergonic reactions are not.

The melting of ice (Figure 2.25) is a spontaneous endergonic reaction—an exception to the above statement. In surroundings warmer than $0°$ C, ice spontaneously takes up energy from the surroundings. Fluid water that forms as ice melts contains more energy than solid water in the ice. By the **second law of thermodynamics,** in any process involving spontaneous change from the initial to the final state, the randomness, or disorder, of the system and its surroundings increases. Such randomness is called **entropy** (en′tro-pe). Melting ice takes up energy and water molecules become more randomly distributed. (An irregularly shaped puddle is certainly less orderly than a block of ice.) Any spontaneous change is accompanied by an increase in entropy. Living organisms fight a constant battle against entropy—eventually losing the battle in death.

Many chemical reactions are reversible; that is, they can go either direction from A to B or from B back to A. If more A than B is present, the reaction will go toward B. If more B than A is present, the reaction will go toward A. The ability of reversible reactions to reach equilibrium and to be driven by the substance present in greatest concentration is called the **law of mass action.**

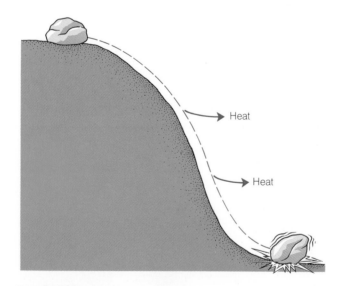

FIGURE 2.24
A rock at the top of a hill has high potential energy. It gives off heat energy as it rolls down the hill and has a lower potential energy when it comes to rest at the bottom.

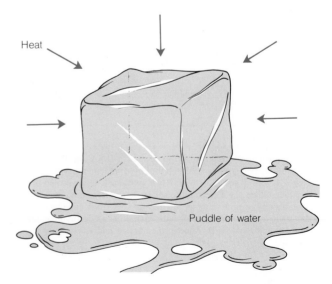

FIGURE 2.25
A block of ice gains heat energy from the environment. Melted ice, or liquid water, has less organization and, therefore, higher entropy than the block of ice.

Coupled Reactions

Endergonic reactions usually are coupled to, or occur simultaneously with, exergonic reactions. Frequent coupled reactions in living organisms include the capture and release of energy in ATP and oxidation-reduction reactions.

The formation of ATP when a cell captures energy from food,

$$ADP + P_i + energy \rightarrow ATP$$

(where P_i is inorganic phosphate), is coupled with the release of energy from glucose or other foods.

$$glucose + 6\ O_2 \rightarrow 6\ CO_2 + 6\ H_2O + energy$$

Similarly, the breaking of a high energy bond in ATP when a cell needs energy,

$$ATP \rightarrow ADP + P_i + energy,$$

is coupled with chemical reactions such as those that transport substances across cell membranes or cause muscles to contract.

In oxidation-reduction reactions, oxidation can occur by the addition of oxygen or the removal of hydrogen atoms or electrons, and reduction can occur by the removal of oxygen or the addition of hydrogen atoms or electrons. For example, during oxidation of a reduced organic molecule, hydrogen atoms are removed and used in reduction of a coenzyme, such as FAD.

$$succinate\text{-}H_2 + FAD \rightarrow fumarate + FAD\text{-}H_2$$

Similarly, certain coenzymes called cytochromes contain an iron ion. The iron is oxidized when it gains an electron.

$$Fe^{2+} + e^- \rightarrow Fe^{3+}$$

It is reduced when it loses an electron.

$$Fe^{3+} \rightarrow Fe^{2+} + e^-$$

In the final reaction in cellular oxidation-reduction, two hydrogen atoms combine with an oxygen atom to form water. One can think of such a reaction as the simultaneous oxidization of hydrogen and reduction of oxygen.

ENZYMES

The laws governing chemical changes apply to all chemical reactions, including those in living organisms. Most reactions are not feasible at physiologic temperatures, so **enzymes,** or **biological catalysts** (kat'al-istz), are needed to make them occur at rates fast enough to support life. Enzymes increase the rate of reactions. Most can catalyze a given reaction in either direction, and the relative concentrations of reactants and products determine the direction of reactions. To see how enzymes increase reaction rates, we need to know more about the properties of enzymes and the factors that affect enzyme reactions.

Properties of Enzymes

In general, spontaneous reactions occur without input of energy from the surroundings, but often at unmeasurably low rates. Though oxidation of glucose is thermodynamically a spontaneous reaction, it does not occur unless energy to start the reaction is available. The energy required to start such a reaction is called **activation energy.** Activation energy can be considered a hurdle over which molecules must be raised to get a reaction started (Figure 2.26). By analogy, a rock resting behind a hump at the

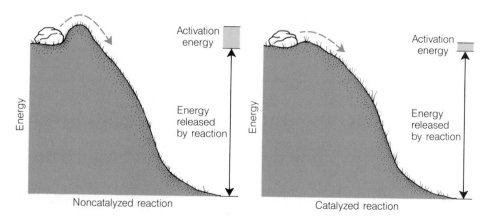

FIGURE 2.26

Enzymes catalyze reactions in living organisms by lowering the activation energy needed for a reaction to occur. This is analogous to a rock behind a small hump near the top of a hill. The rock will not move down the hill until enough "activation energy" has been added to raise it over the hump. Enzymes increase the rate of spontaneous reactions by reducing the activation energy required to start them. This allows biological reactions to proceed rapidly at temperatures that living organisms can tolerate.

top of a hill would easily roll down the hill if pushed over the hump. Activation energy is the energy required to lift the rock over the hump. Another way to activate a reaction is to heat the reactants, increasing their molecular movement. The quantity of heat needed to activate many reactions would denature proteins and evaporate liquids in living organisms. Enzymes lower the activation energy needed to start a reaction at body temperature and keep the reaction going at a sufficiently high rate to maintain life.

Enzymes, like catalysts in inorganic reactions, are not used up in the reactions they initiate. Though enzyme molecules eventually "wear out," probably by losing the molecular shape necessary for their catalytic properties, they can be used over and over again. Also, like inorganic catalysts, enzymes are not permanently affected by the reaction they catalyze.

Many enzymes are named by adding the suffix -ase to the name of the **substrate,** or substance upon which they act. For example, phosphatases act on phosphates, lipases act on lipids, sucrase digests the sugar sucrose, and peptidases break peptide bonds.

Enzymes provide a surface where reactions occur. Each enzyme has an **active site,** a surface binding site at which it forms a loose association with its substrate, or ligand (Figure 2.27a). After binding to the enzyme, the substrate undergoes chemical change and the product or products are formed (Figure 2.27b and c).

Sometimes a ligand similar to a substrate binds to the active site but fails to react. If the ligand forms a reversible attachment with the active site, **competitive inhibition** occurs (Figure 2.27d). Such inhibition is concentration dependent. Having the inhibitor occupy a few active sites part of the time, slightly slows the reaction rate. Having it occupy many sites most of the time, greatly slows the reaction rate.

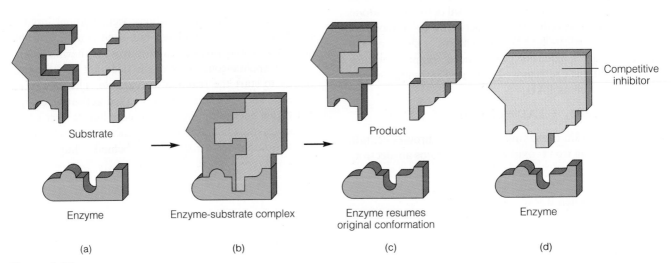

Substrate

Enzyme

Enzyme-substrate complex

Product

Enzyme resumes
original conformation

Competitive
inhibitor

Enzyme

(a)

(b)

(c)

(d)

FIGURE 2.27

(a) The surface of an enzyme has an active site on which its normal substrate fits. (b) The substrate and enzyme form a complex. (c) The substrate that fits the active site is converted to the product(s) of the specific reaction catalyzed by the enzyme.

(d) A nonsubstrate molecule that fits the active site and prevents the substrate molecules from occupying it can inhibit the action of the enzyme. Because it competes with the normal substrate, it is called a competitive inhibitor.

Sulfanilamide, a drug used to treat kidney infections, and penicillin, a widely used antibiotic, kill bacteria by competitively inhibiting their enzymes. Some bacteria use *para*-aminobenzoic acid (PABA) to make folic acid. When sulfanilamide, which is chemically similar to PABA (Figure 2.28), is more plentiful and occupies more active sites than PABA, it prevents the bacteria from making adequate amounts of folic acid. Penicillin inhibits the enzyme transpeptidase, preventing peptides from making cross linkages in bacterial cell walls. Because these enzymes are not present in human cells, the drugs do not inhibit human cell functions.

Enzymes also can be inhibited by **noncompetitive inhibitors** such as mercury, lead, and other heavy metals. Such inhibitors permanently attach to an enzyme at a site other than the active site and distort the enzyme so the shape of the active site is altered (Figure 2.29). Affected enzyme molecules are permanently inactivated and the reaction rate is greatly reduced regardless of the concen-

tration of the substrate. Mercurochrome is used to disinfect skin and mercuric chloride to disinfect inanimate objects, but substances containing mercury are highly toxic if ingested. Lead, which is found in some paints and in exhaust fumes from engines using leaded gasoline, replaces calcium in bones, interferes with hemoglobin synthesis, and causes brain edema (fluid accumulation). Lead also causes gout by allowing the waste product uric acid to accumulate in the body. The high incidence of gout among port wine drinkers in the nineteenth century has been attributed to lead solder in wine-making equipment.

Another extremely important characteristic of enzymes is their **specificity.** All enzymes catalyze a specific reaction and most act on a particular substrate, according to the shape and charges at their active sites. Enzymes that act on more than one substrate usually act on a group of substrates with a particular functional group or chemical bond. For example, a peptidase may break down many proteins, but it acts specifically on peptide bonds.

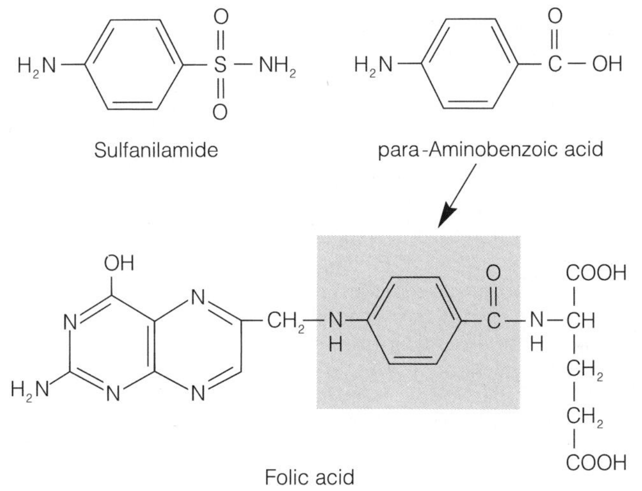

FIGURE 2.28
Some bacteria use *para*-aminobenzoic acid (PABA) to synthesize folic acid. Sulfanilamide, which is chemically similar to PABA, acts as a competitive inhibitor of enzymes that normally act on PABA.

Sulfanilamide

para-Aminobenzoic acid

Folic acid

FIGURE 2.29
In noncompetitive inhibition, an inhibitor binds to a site other than the active site. It distorts the molecular shape so the substrate cannot bind to the enzyme's active site.

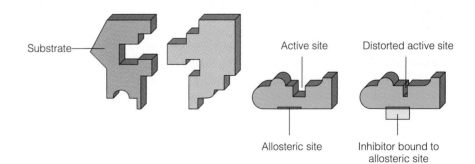

Substrate

Active site   Distorted active site

Allosteric site   Inhibitor bound to allosteric site

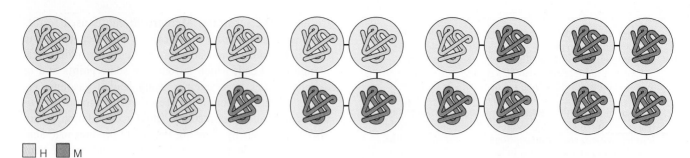

□ H ■ M

FIGURE 2.30
A schematic diagram showing subunits of the isozymes of lactic dehydrogenase.

Certain enzymes have two or more structurally different forms called **isozymes** (i'so-zīmz) that catalyze the same reaction. The enzyme lactic dehydrogenase has four subunits composed of heart (H) or skeletal muscle (M) types (Figure 2.30). Five isozymes, indicated by the number of H and M units they contain, are H_4, H_3M, H_2M_2, HM_3, and M_4. H_4 and H_3M isozymes make up over 90 percent of the lactic dehydrogenase in heart muscle, whereas HM_3 and M_4 make up about 75 percent of the lactic dehydrogenase in skeletal muscle and liver cells.

Finding cellular isozymes in blood can be used in identifying and monitoring of tissue damage. A rise in the blood level of any cellular enzyme indicates that some cells have been sufficiently damaged to pour their contents, including enzymes, into the blood. Finding H_4 and H_3M in blood indicates heart muscle damage, such as occurs in a heart attack. Periodically measuring the blood levels of these isozymes can be used to assess the amount of tissue damage and to determine whether such damage continues. Finding other lactic dehydrogenase isozymes indicates skeletal muscle damage, as might occur in overuse sports injuries, or liver damage, as might occur in infections.

Many enzymes require coenzymes or cofactors to catalyze a reaction. Coenzymes, such as NAD and FAD, are large molecules bound to or loosely associated with an enzyme. Many coenzymes are synthesized from vitamins—NAD from niacin and FAD from riboflavin—found in whole grains and some other foods. The vitamins are essential in the diet precisely because they are needed to make essential coenzymes. As we have seen, coenzymes often act as carriers of hydrogen atoms or electrons in oxidative reactions from which cells derive energy. **Cofactors** usually are inorganic ions such as copper, magnesium, and zinc, which are found in trace amounts in many foods. These ions often improve the fit of an enzyme's active site with its substrate, and their presence is essential to allow the reaction to proceed.

Factors Affecting Enzyme Reactions

Factors that affect enzyme activity, or the rate at which enzymes catalyze reactions, include temperature, pH, and concentrations of substrate, product, and enzyme. Like other proteins, enzymes are subject to alteration by heat and extremes of pH that can affect net charges on protein molecules, thereby altering enzyme activity.

Most human enzymes have an **optimum temperature,** the temperature at which they have the greatest activity, near normal body temperature (Figure 2.31a). Enzyme activity decreases gradually as the temperature drops below normal, but rapidly as the temperature rises above normal, especially above 42° C. Nearly all human enzymes are denatured at 50° C.

Similarly, many human enzymes have an **optimum pH** near neutral (Figure 2.31b). Their activity decreases gradually as the pH becomes either more acidic or more basic. Some human enzymes have an acidic or basic optimum pH. Pepsin, a peptidase that breaks down proteins in the stomach, has an optimum pH of 2. Two kinds of phosphatases, enzymes that remove phosphate groups from molecules, occur in human cells. Acid phosphatase has an optimum pH of 5.5 and alkaline phosphatase has an optimum pH of 9.0.

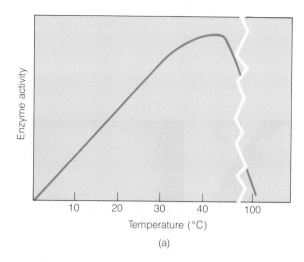

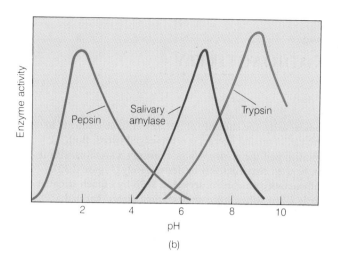

(a)

(b)

FIGURE 2.31

(a) Most enzymes found in human cells have an optimum temperature of 37° C. Lowering the temperature gradually reduces enzyme activity, but raising it above the optimum temperature rapidly reduces activity because it denatures the enzyme. (b) Though most enzymes have an optimum pH near neutral, some work best in acidic or alkaline conditions. Altering the pH from the optimum level also reduces enzyme activity, and sufficiently large changes also denature the enzyme.

The rate of enzyme catalyzed reactions also is affected by the concentrations of enzymes, substrates, and products. Though enzymes do not change the rate at which any particular molecule undergoes a reaction, the quantity of enzyme determines how fast the product accumulates if enough substrate is available. With 100 enzyme molecules acting on substrate, the product will accumulate ten times as fast as if only 10 enzyme molecules are acting. The reaction rate increases in proportion to the number of enzyme molecules present, reaching a maximum when all enzyme molecules catalyze a maximum number of reactions per minute. If the concentration of a substrate is too low to keep all the enzyme molecules working at maximum capacity, the substrate concentration will determine the reaction rate.

Substrates and products obey the law of mass action. Large quantities of substrate drive the reaction toward formation of product. Large quantities of product would drive the reaction toward reformation of substrate, but in most cellular reactions the product of one reaction quickly becomes the substrate of another reaction in a metabolic pathway.

A **metabolic pathway** is a sequence of enzyme controlled reactions that accomplish a particular task, usually within a cell. As we shall see in Chapter 24, different pathways exist to break down glucose, fatty acids, or proteins for energy and to synthesize glycogen, lipids, proteins, or the waste product urea. Cells have various ways to regulate enzyme activity, and regulating the rate of one enzyme reaction often controls the rate of an entire pathway. For example, suppose a pathway begins with substrate A and goes through **intermediates** B, C, and D, ending with product E. In **feedback inhibition,** the product of a reaction several steps along a metabolic pathway (E) acts to inhibit an early reaction in the pathway (A → B). Feedback inhibition is more common in microbes than in humans, but in one instance in humans, accumulation of the amino acid alanine determines whether pyruvic acid (a product of glucose metabolism) is used for energy or is used to make more alanine.

RADIOACTIVITY

Radioactivity, the capacity of isotopes of some elements to emit radiation and energy, is a two-edged sword. Both radiation and the great amount of energy associated with it can be used constructively or destructively (Figure 2.32). Radioactivity was first observed in 1896 by French scientist Marie Curie, when some photographic plates that had inadvertently lain next to a piece of uranium ore were developed and found to be clouded. Study showed that the uranium must have given off radiation similar to light, but invisible and capable of penetrating material light could not. This quality of uranium (and many other elements) is now called radioactivity.

Three kinds of radioactive emissions—alpha and beta particles, and gamma rays—are emitted from unstable nuclei of atoms of radioactive elements. *Alpha particles* consist of two protons and two neutrons. *Beta particles* are electrons from the disintegration of a neutron into a proton and an electron. *Gamma rays* are short wavelength electromagnetic radiation. Emissions always occur at a constant rate—the element's *half-life* ($t_{1/2}$), the time for half the atoms in a sample to emit radiation and disintegrate.

Radioactive emission causes several kinds of cellular damage. **Ionizing radiation** forms highly reactive ions often by splitting water molecules into H^+ and OH^- ions. The OH^- ions can combine to form hydrogen peroxide, H_2O_2, an unstable product that releases charged oxygen, O_2^-. Both charged oxygen and free H^+ damage enzymes and cellular molecules. **Cosmic radiation,** which passes through the earth's atmosphere from outer space, can cause mutations (changes in DNA) that kill cells; make cells cancerous; or, if in egg or sperm cells, can transmit defective information to the next generation.

Different emissions vary in their effects. Alpha particles, though large and potentially harmful, move slowly and do not penetrate the skin. Smaller, faster beta particles penetrate only a few millimeters into tissues and are relatively harmless. Fast moving gamma rays, such as X rays, easily enter and damage tissues. Elements with long half-lives have great potential for damage if incorporated into tissues. For example, strontium (^{90}Sr), which is sometimes found in milk, has a half-life of 28 years. (The superscript indicates the atomic weight of the isotope.) Being similar to calcium, this isotope concentrates in bones and emits radiation for years. Nuclear power plant wastes have half-lives longer than any available containers.

Radioactive substances are used in diagnosis and treatment. X rays are used to diagnose skeletal and chest disorders, and follow X-ray opaque substances through the intestine, kidneys, and blood vessels (Figure 2.33). Other methods trace small amounts of radioactive substances in tissues. Giving radioactive iodine (^{131}I) orally and measuring

(a) (b)

FIGURE 2.32
Radiation can be used (*a*) constructively to locate a fracture in a bone or (*b*) destructively in a nuclear explosion.

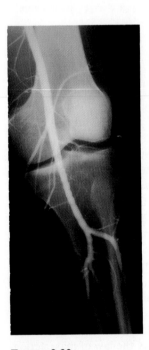

FIGURE 2.33
When dyes that block the passage of X rays are injected into the blood and X-ray photographs are taken, the blood vessels can be clearly seen.

subsequent thyroid gland radioactivity is used to assess thyroid function (Figure 2.34). Larger doses can be used to destroy cells in overactive glands. Because this isotope has a half-life of only 8.1 days, it poses no long-term hazard. Radioactive substances can be placed close to a tumor where their emissions destroy malignant cells with little risk to normal

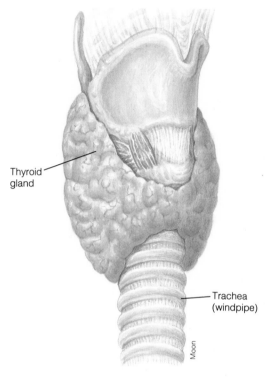

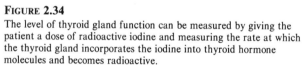

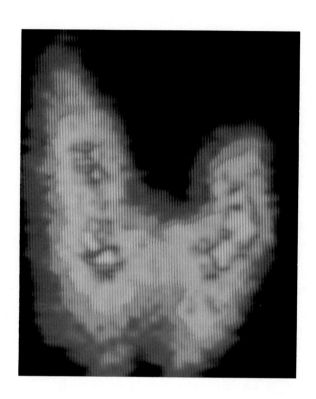

FIGURE 2.34
The level of thyroid gland function can be measured by giving the patient a dose of radioactive iodine and measuring the rate at which the thyroid gland incorporates the iodine into thyroid hormone molecules and becomes radioactive.

cells. Radium, which emits alpha particles, and cobalt, which emits both beta particles and gamma rays, are used in this way. X rays also destroy malignant cells.

The benefits of radiation in diagnosis and treatment must be weighed against its risks. Small doses of radiation over several years can cause as much damage as a single dose equal to all the small doses. Doses of radiation over time may cause a sequence of small changes in DNA, eventually leading to a mutation with observable and possibly deleterious effects. Radiation accelerates aging even though specific effects are not easily observed. Individuals subjected to radiation—researchers, medical personnel, or patients—must be protected against the cumulative effects of radiation.

Understanding the properties of radioactive materials is important in the study of physiology and its medical applications for three reasons: it explains some mutations and other kinds of cell damage, it can selectively destroy malignant cells, and it can help diagnose an ever-increasing variety of disorders. For each of radioactivity's benefits, there are potential risks. Radiation can destroy cancers or cause them. Over time, it can kill organisms or produce new species. It can generate useful power or destroy our world through a nuclear war. How responsibly we will use knowledge of radioactivity remains to be seen.

Questions
1. What is radiation and how was it discovered?
2. What properties distinguish each kind of emission from radioactive material?
3. In what ways can radiation benefit humans?
4. In what ways can radiation harm humans, and how can such radiation be avoided?

Chapter Summary

Why Study Chemistry?

- Basic chemistry is important in anatomy and physiology because body functions involve functions at the chemical level.

Chemical Building Blocks and Chemical Bonds

- An element is a fundamental kind of matter; the smallest unit of an element is an atom.
- A molecule consists of two or more atoms chemically combined; a compound consists of two or more different kinds of atoms chemically combined.
- The most common elements in the body are carbon (C), hydrogen (H), oxygen (O), and nitrogen (N).
- Atoms consist of positively charged protons and neutral neutrons in the nucleus, and negatively charged electrons, which have a much smaller mass than protons and neutrons and which orbit the nucleus.
- The number of protons in an atom is its atomic number, the total number of protons and neutrons is its atomic weight.
- Ions are atoms that have gained or lost one or more electrons; isotopes are atoms of the same element that contain different numbers of neutrons.
- Molecules are held together by chemical bonds: ionic bonds form by the transfer of electrons and the attraction between oppositely charged ions. Covalent bonds form by sharing pairs of electrons; hydrogen bonds are weak covalent bonds.

Mixtures, Solutions, and Colloidal Dispersions

- A solution is a mixture of one or more solutes evenly distributed through a solvent.
- The properties of water are summarized in Table 2.4.
- A colloidal dispersion has particles too large to form true solutions suspended in a medium by electrical charges, layers of water molecules, and other forces.

Acids, Bases, and pH

- Acids release H^+ ions and bases accept H^+ ions or release OH^- ions.
- The pH of a solution expresses its acidity or alkalinity; pH of 7 is neutral, below 7 is acidic, and above 7 is basic.

Complex Molecules

- Living organisms contain four major classes of complex organic (carbon-containing) compounds: carbohydrates, lipids, proteins, and nucleic acids.
- Molecules with important functional groups include alcohols, aldehydes, ketones, and organic acids, including amino acids.

Carbohydrates

- Carbohydrates consist of carbon chains with most carbon atoms having an associated alcohol group and one carbon having an aldehyde or a ketone group.
- The simplest carbohydrates are monosaccharides, which can combine to form disaccharides and polysaccharides.

Lipids

- Lipids include fats made of glycerol and fatty acids; phospholipids that also contain phosphoric acid; and steroids such as cholesterol, bile salts, and some hormones and vitamins.

Proteins

- Proteins, polymers of amino acids linked by peptide bonds, can be coiled and folded into large complex shapes held together by hydrogen bonds, disulfide linkages, and other forces.
- Proteins have receptors, which bind specific ligands according to their affinities and concentrations.
- Proteins form part of the structure of cells; act as enzymes; and contribute to other functions such as motility, transport, and regulation.

Nucleotides and Nucleic Acids

- A nucleotide consists of a nitrogenous base, a sugar, and one or more phosphates. Nucleotides serve as energy-capturing molecules, coenzymes, and building blocks of nucleic acids.
- Nucleic acid components are summarized in Table 2.10.

Bioenergetics

- The laws governing chemical change include the first and second laws of thermodynamics, and the law of mass action.
- Coupled reactions include pairs of reactions in which one molecule gains energy as another loses it, or in which one molecule is oxidized as another is reduced.

Enzymes

- Enzymes act as biological catalysts to increase the rate of chemical reactions in living things by lowering activation energy.
- Each enzyme has an active site to which a substrate attaches, and each catalyzes a specific reaction or reaction type.
- Rates of enzyme reactions are affected by temperature, pH, concentrations of substrate and product, and the concentration of the enzyme itself.
- Inhibitors of enzymes compete with normal substrates for the active site, or bind to another site and permanently inactivate an enzyme.

QUESTIONS AND PROBLEMS

The questions at the end of each chapter are numbered to correspond with the objectives listed at the beginning of the chapter. Italics indicate that a question requires critical thinking skills beyond simple factual recall.

Questions

1. Why is some knowledge of chemistry important in anatomy and physiology?
2. (a) Define atom, element, molecule, compound, proton, electron, neutron, atomic number, atomic weight, cation, anion, and isotope.
   (b) Use chemical formulas to illustrate the differences among ionic, covalent, and hydrogen bonds.
   (c) What are the important properties of polar compounds?
3. *How are solvents, solutes, and solutions related?*
4. *(a) In what ways is water important to living things?*
   (b) What disorders do you think are most likely to develop if humans are deprived of water?
5. *(a) How do solutions and colloidal dispersions differ?*
   b) What different methods do physiologists use to express concentrations?
6. (a) Define acid, base, and pH.
   (b) How might altering the pH of a body fluid affect body functions?
7. (a) Define organic chemistry and name four kinds of complex biological molecules.
   (b) What functional groups are commonly found in biological molecules?
8. (a) What are the distinguishing properties of carbohydrates?
   (b) How do mono-, di-, and polysaccharides differ?
   (c) In what ways are carbohydrates important in body function?
9. (a) What are the distinguishing properties of different kinds of lipids?
   (b) In what ways are lipids important in body function?
   (c) How do saturated and unsaturated fatty acids differ?
10. (a) What are the distinguishing properties of proteins?
    (b) Why are receptors and ligands important?
    (c) How are proteins classified?
    (d) Define essential amino acid, peptide bond, and denaturation.
    (e) What are the four levels of protein structure?
11. (a) Briefly describe the structure and function of ATP, nucleotide coenzymes, DNA, and RNA.
    (b) What properties of nucleic acids contribute to their ability to store and transmit information?
12. *(a) What laws govern chemical changes?*
    (b) How can a mixture of chemical substances suddenly begin to give off heat?
13. What two processes are coupled in energy transfer reactions? In oxidation-reduction reactions?
14. (a) Define catalyst, active site, specificity, and isozyme.
    (b) What properties of enzymes contribute to their ability to catalyze reactions?
    (c) In what ways are enzymes physiologically important?
    (d) How do competitive and noncompetitive inhibition of enzymes differ?
15. *(a) How do temperature and pH affect the rate of enzyme-controlled reactions?*
    (b) How do the concentrations of enzyme, substrate, and product affect reaction rates?

Problems

1. What is the pH of a solution with a hydrogen ion concentration of 0.001? Of 0.000000001?
2. Some antibiotics exert their effects on enzymes. Do some library research on the actions of antibiotics and their effects on microorganisms and humans.
3. New uses for enzymes in diagnosis and treatment are constantly being developed. Find out, and if possible report to your class, what enzymes are now in most widespread use.
4. Many genetically determined disorders of metabolism are due to the absence or low activity of a particular enzyme. Research this topic, paying particular attention to genetic disorders you have seen among family or friends.
5. Devise a way to determine whether solutions in each of three test tubes contain carbohydrate, lipid, protein, or more than one of these substances.

OBJECTIVES

1. Describe the organization and general function of cells, and explain how cell functions relate to higher level functions.

2. Use the fluid-mosaic model to relate the structure and function of cell membranes.

3. Distinguish between active and passive transport across cell membranes.

4. Describe the processes of simple diffusion, facilitated diffusion, and osmosis, and explain their significance.

5. Briefly describe filtration and bulk flow, and explain their significance.

6. Use the sodium-potassium pump to explain the process of active transport and its significance.

7. Distinguish between endocytosis and exocytosis.

8. Distinguish between cytoplasm and cytosol.

9. Briefly describe the structure of each organelle, and explain its function and significance.

10. Explain how cellular functions are integrated.

11. Summarize the steps in mitotic cell division, and explain its significance.

12. Summarize the steps in protein synthesis, and explain its significance.

3

FUNCTION AT THE CELLULAR LEVEL

ORGANIZATION AND GENERAL FUNCTIONS

Prior to the invention of the microscope, those who tried to study the human body were seriously hampered by their inability to see that the body was made of cells. As we know now, cells are the smallest units of living things and the body contains many kinds of interdependent cells with different shapes, sizes, and functions. In general, cells consist of internal structures called **organelles** (or''gah-nelz') suspended in a semifluid **cytoplasm** and surrounded by a **cell membrane** (Figure 3.1). Each organelle and each enzyme in the cytoplasm performs a specific function; the cell membrane defines the boundaries of a cell and regulates the movement of substances into and out of the cell.

To understand human structure and function, one must have some understanding of cellular structure and function because many functions occur at the cellular level. Some activities, such as digestion of food and transport of food molecules and gases, are largely extracellular, but the processes whereby food molecules and oxygen are used to capture energy in ATP take place in cells. All cells need ATP to maintain themselves and to carry out their roles in the overall functions of the body.

Many processes in addition to energy extraction occur at the cellular level. Muscles exert force because of movements of molecules within cells. The nervous system relays signals as ions move across cell membranes. The urinary system removes wastes and adjusts concentrations of substances in the blood by the movement of substances across

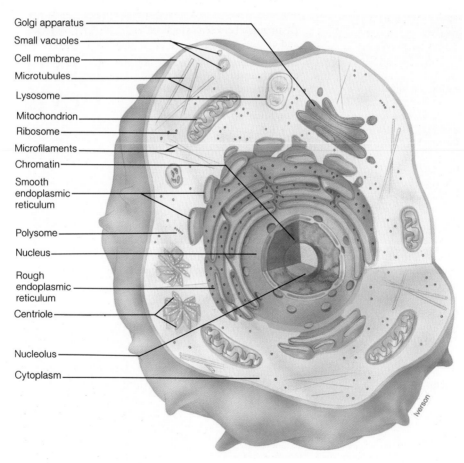

FIGURE 3.1
An animal cell and its major organelles.

the membranes of kidney cells. The endocrine system regulates body functions through gland cells synthesizing and secreting hormones. Nearly all chemical reactions that take place in the body are controlled by cellular enzymes. Finally, many cells of the body undergo cell division as the body grows, and new individuals are produced from the union of specialized cells called eggs and sperm; thus, all body functions depend on processes that occur at the cellular level.

Early microscopic studies of living things showed that they were divided into many smaller units. By the mid-nineteenth century, such studies led to the **cell theory:** all living things are composed of cells. The development of electron microscopy since the 1930s has made it possible to magnify cells up to 1 million times—1 thousand times that of light microscopes. The scanning electron microscope now provides three-dimensional views of the surfaces of cells and organs at magnifications of about 20 times that of the light microscope.

Biochemical and biophysical studies of cells have complemented the findings made with microscopes, and many new techniques have been developed to study cell function (Figure 3.2). Carefully ruptured cells can be centrifuged at different speeds in different media and their organelles separated and studied independently. The functions of enzymes in one organelle can be distinguished from those of other organelles by this technique. Molecules in living cells can be made radioactive or fluorescent, and can be followed through metabolic processes.

Electrophoresis, the movement of substances in an electric field, can separate molecules on the basis of their electrical charges. Spectroscopy, a technique in which radiation is passed through or reflected from particles, can be used to study bond distances in molecules, molecular shapes, and magnetic properties of atoms and molecules. One such technique, X-ray diffraction, was used to determine the structure of DNA. Magnetic resonance imaging (MRI), a newer technique, can be used to determine the precise location of processes in living cells.

As we shall see, the basic concepts defined in Chapter 1 apply to cells. Cellular functions derive from molecular and organelle functions, and contribute to higher level functions. Organelles demonstrate the concepts of complementarity and homeostasis. In this chapter, we will consider the functions of cell membranes, cytoplasm, and organelles, and the process of mitotic cell division.

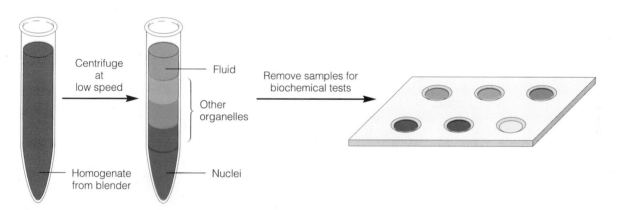

FIGURE 3.2

Many highly specialized techniques are now available for studying cells. Submitting cells to methods such as ultrasound or grinding them in a kitchen blender results in a homogenate, or "soup," of cell parts. The homogenate is centrifuged at low speed to separate large stuctures such as nuclei. The remaining homogenate is centrifuged at progressively higher speeds for longer periods of time to separate lighter particles. Each time a precipitate forms, the particles in it can be subjected to various biochemical and physical tests to learn more about the nature of the particles.

CELL MEMBRANES

The **plasma membrane,** or **cell membrane,** defines the boundaries of a cell, and membranes within cells define the boundaries of certain organelles. These membranes all have the same structural components (Figure 3.3). According to the **fluid-mosaic model,** a membrane consists of a semifluid lipid matrix in which proteins are embedded to form a mosaic. (A mosaic is a composite of many small pieces.) The lipid matrix is made of **phospholipids** arranged in two layers. The polar hydrophilic phosphate regions of the molecules are oriented toward the extracellular and intracellular surfaces, which are exposed to aqueous fluids. The nonpolar hydrocarbon ends point toward each other in the interior of the membrane. The lipid matrix also contains cholesterol, which contributes to the fluidity of the membrane at body temperature. Membrane proteins are interspersed among the lipids. **Integral** proteins extend partially or completely through the lipid layers, and **peripheral** proteins are loosely attached to membrane surfaces.

Sugars combine with some membrane proteins to form **glycoproteins** (gli″ko-pro′te-inz) and with some membrane lipids to form **glycolipids** (gli″ko-lip′idz). Glycoproteins and glycolipids cover about 7 percent of a cell's extracellular surface, or *glycocalyx*. These sugary molecules act as cell recognition sites, which are important in cell-to-cell interactions. For example, molecules called antigens, which are on the surface of red blood cells, identify the cells—and the person's blood type—as A, B, or AB. They also form binding sites that allow various chemical substances such as hormones to attach to the cell's surface and influence cell function.

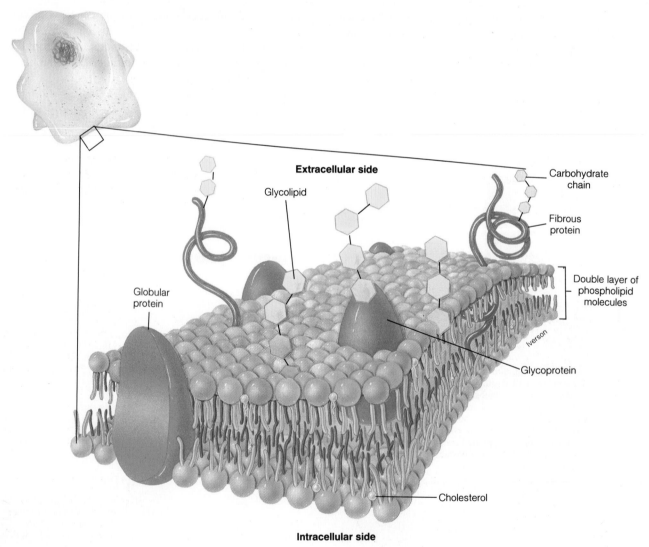

Extracellular side

Glycolipid

Carbohydrate chain

Fibrous protein

Double layer of phospholipid molecules

Globular protein

Iverson

Glycoprotein

Cholesterol

Intracellular side

FIGURE 3.3

Each cell is surrounded by a plasma membrane (cell membrane) that separates it from neighboring cells. The widely accepted Singer–Nicholson fluid-mosaic model shows the plasma membrane to consist of two layers of lipids with protein molecules interspersed through it in a mosaic pattern. Integral proteins are nearly surrounded by lipids, whereas peripheral proteins extend from the outer or inner surfaces of the membrane. The glycocalyx, or cell coat, contains glycoproteins and glycolipids.

The primary function of the plasma membrane is to control the passage of substances into or out of the cell, and both its lipids and proteins participate in this process. The plasma membrane regulates the environment within the cell and helps to maintain intracellular homeostasis.

The lipid portion of the membrane allows lipid-soluble substances to move across it, but passively prevents water-soluble substances from doing so. Many substances both inside cells and in the interstitial fluids around them are water soluble, so membrane lipids, which are insoluble in water, provide an ideal boundary between the cell and its environment. Phospholipids have polar and nonpolar ends as explained in Chapter 2. Their hydrophilic polar ends are exposed to aqueous solutions but their hydrophobic nonpolar portions prevent water and substances dissolved in it from crossing the membrane.

Some integral proteins form channels (already shown in Figure 3.3) just large enough to allow ions and water molecules to pass through the membrane. Living membranes are highly permeable to water. The proteins of channels are free to change shape, to open and close channels, and to move laterally in the semifluid lipid matrix. They regulate the movement of substance into and out of cells by several mechanisms to be described later in this chapter and in subsequent chapters.

In addition to regulating movements of substances, proteins of the plasma membrane perform other functions. They give the membrane structural stability. Some proteins are enzymes that regulate chemical reactions that take place in the membrane or on its intracellular surface.

Membranes associated with organelles also control movements of substances into and out of the organelles, but they perform other functions as will be explained for each organelle.

While many Americans are trying to lower cholesterol and other animal fats in their diets, researchers have discovered that too low blood cholesterol can be harmful. They observed that Japanese people on a strictly vegetarian diet who also had high blood pressure were much more likely to suffer from a rupture of brain blood vessels, a condition commonly called a "stroke." One explanation is that without sufficient cholesterol to maintain sturdy cell membranes, high blood pressure can damage cells that form blood vessel walls, especially in the brain.

PASSAGE OF MATERIALS ACROSS MEMBRANES

A living cell is a dynamic entity with substances constantly moving in and out of it across the plasma membrane. Understanding how these movements occur is essential to understanding how a cell functions. Polar substances, such as water and small ions, are believed to move across membranes by passing through channels composed of membrane proteins. Nonpolar substances, such as lipids and other uncharged particles, dissolve in the membrane lipids and diffuse through them (Figure 3.4a). Still other substances are moved through the membrane by carrier molecules.

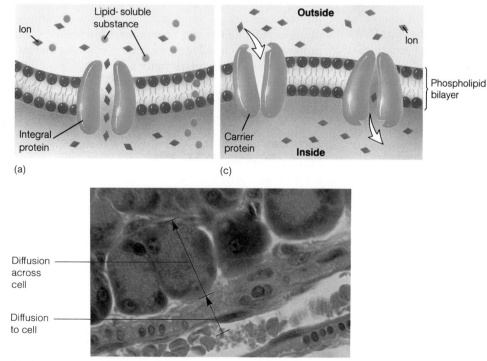

(a)

(c)

(b)

FIGURE 3.4

(*a*) Lipid-soluble substances diffuse through the phospholipid bilayer. Water, ions, and small water-soluble molecules diffuse through small pores formed by some of the integral proteins. (*b*) Diffusion of small diffusible molecules from extracellular fluid near a blood vessel to nearby cells can supply the cells with adequate nutrients if the distance is not too great and the surface-to-volume ratio of the cells is sufficiently large. (*c*) A possible model to explain facilitated diffusion. The carrier molecule, which is imbedded in the membrane, may have several binding sites. The carrier moves in the membrane so as to "push" diffusing substance across the membrane. The same mechanism could explain movement of substances in either direction across the membrane.

The manner of movement of an ion or molecule across a membrane is determined by a combination of the following factors: particle size, electrical charge, relative concentrations of the substance on the two sides of the membrane, lipid solubility, and the availability of carrier molecules in the membrane. The mechanisms by which substances move across membranes include passive processes that require no energy from the cell and active processes that do require energy from the cell. These processes are described in a general way in this chapter. Special movements of ions associated with the excitable membranes of muscle and nerve cells will be described in Chapters 9 and 11, respectively.

Passive Processes

Passive transport processes involve movements of substances down a concentration **gradient,** that is, from a region of higher concentration to a region of lower concentration, without cells supplying energy. Passive processes include simple diffusion, facilitated diffusion, osmosis, filtration, and bulk flow. In contrast, **active transport** processes require energy from ATP to move substances across the cell membrane. These processes include active transport, endocytosis, and exocytosis.

Simple Diffusion

Molecules are in constant motion and have kinetic energy proportional to the temperature. **Simple diffusion** is the net movement of molecules from their region of higher concentration to their region of lower concentration. For example, if one drops a lump of sugar into a cup of coffee, random movement of sugar molecules will eventually distribute the molecules evenly through the coffee without stirring. At first, the concentration of sugar molecules is greatest at the lump and least at the rim of the cup. Thus a concentration gradient exists with a range of concentrations from the high concentration at the sugar lump to the low concentration near the rim of the cup. Diffusion occurs because of the random motion of molecules in a liquid or gas. The net movement of molecules is down the concentration gradient. At equilibrium when a gradient no longer exists, random movement continues but no net movement occurs.

The time required to reach equilibrium by diffusion increases with molecule size and the distance between the regions of high and low concentration. For ions and small molecules, equilibrium is reached across a distance of 1 micrometer (μm) in 0.06 seconds, across 10 μm in 6 seconds, across 100 μm in 11 minutes, and across 1000 μm (1 mm) in 18 hours. Cells lying within 10 to 20 μm of a blood vessel receive small molecules in a matter of seconds, but cells located farther from blood vessels receive nutrients more slowly (Figure 3.4b).

Cells with diameters no greater than about 20 μm typically have a high **surface-to-volume ratio,** the amount of surface area relative to the volume. A sufficiently high surface-to-volume ratio allows small molecules and ions to enter and diffuse throughout the interior of a cell in a matter of seconds. The cells receive adequate nutrients and rid themselves of wastes. Large cells sometimes achieve an acceptable surface-to-volume ratio with unusual cell shapes. For example, muscle cells (fibers) can be very long (up to 0.5 m), but they have a relatively small diameter. Substances diffuse rapidly over the distance from the cell surface to its center.

When diffusion occurs across a living plasma membrane, the rate of diffusion is determined, not only by the concentrations of substances on the two sides of the membrane, but also by *membrane permeability.* The membrane itself severely limits diffusion. Another limiting factor is the **unstirred water layer,** a 100 to 400 μm thick layer of relatively stationary water molecules coating membrane surfaces. Because living membranes allow some substances to pass through and prevent others from doing so they are said to be **selectively permeable.**

Substances can move through plasma membranes either by diffusing through the lipid layer or by passing through the small channels provided by integral proteins. The rate of diffusion through the lipid layer is affected by the solubility of the diffusing substance in lipid, temperature, and the concentration gradient. Nonpolar substances such as fatty acids, steroid hormones, and gases cross the membrane rapidly by dissolving in the nonpolar fatty acids in the membrane.

The rate of diffusion through membrane channels is affected by the size and charge of the diffusing particles and by the size of the channels and the charges on their surfaces. Studies of small particle movement through channels indicate that the channels are less than 0.8 nm in diameter—too small to be detected with an electron microscope. Only water, small water-soluble molecules, and ions such as H^+, K^+, Na^+, Ca^{2+}, and Cl^- pass through these channels, and special channels with a certain size, configuration, and charge exist for each ion type.

Facilitated Diffusion

Facilitated diffusion is the diffusion of a substance across a membrane with the assistance of a carrier molecule. Glucose and some amino acids and vitamins enter cells by facilitated diffusion at rates faster than is possible by simple diffusion. Like simple diffusion, facilitated diffusion involves movement down a concentration gradient without expenditure of ATP. Unlike simple diffusion, facilitated diffusion requires a carrier molecule, a membrane-bound protein that binds to one or a few specific

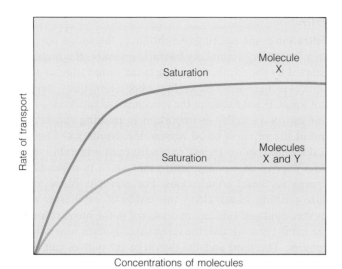

FIGURE 3.5

In facilitated diffusion, the maximum flow of a substance into a cell occurs when the carrier molecules become saturated, that is, when they are carrying the diffusing substance as fast as they can. When two substances transported by the same carrier are present in equal concentrations, at any one moment, half of the carrier molecules will be transporting one substance and half the other substance. Assuming equal affinities of the substances for the carrier, the total number of molecules transported will remain the same.

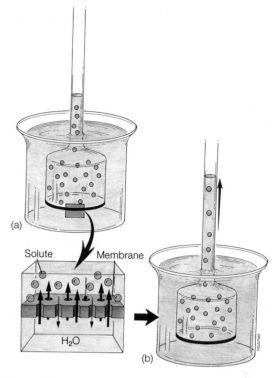

FIGURE 3.6

A laboratory demonstration of osmosis: (*a*) at the beginning of experiment and (*b*) after osmosis has occurred. Water molecules move by osmosis from their region of higher concentration to their region of lower concentration in much greater numbers than they move in the opposite direction. The membrane in the diagram, like a living membrane, is permeable to water but not to protein molecules.

molecules and assists in their movement across the membrane. A proposed mechanism of facilitated diffusion is that the carrier acts like a revolving door, providing a channel for the movement of substances across a membrane (Figure 3.4c).

Because a membrane contains only a limited number of carrier molecules, facilitated diffusion is subject to carrier saturation and competition between structurally similar molecules for a carrier (Figure 3.5). **Carrier saturation** occurs when all the available carrier molecules are transporting their specific substance as rapidly as possible. For example, if a cell had 10 glucose carrier molecules and each could transport 10 molecules per second, the carriers would be saturated when they move 100 glucose molecules per second. Because of the limited number of carrier molecules, no further rate increase is possible.

Competition occurs between the substance normally transported and another substance with similar shape, charge, and affinity for the carrier. Suppose that another sugar has equal affinity for the glucose carrier molecules on certain cells. When both sugars are present in the same concentration, the carrier molecules would transport about the same number of molecules of each sugar. Using the above example again, the membrane would transport 50 glucose molecules and 50 molecules of the other sugar per second, but never more than 100 molecules per second.

Even when different proportions of transported substances or substances with different affinities are used, the total number of molecules transported in a given time remains constant.

Osmosis

Water passes through cell membranes rapidly in both directions by **osmosis,** the diffusion of water across a selectively permeable membrane from a region of higher water concentration to one of lower water concentration. To explain osmosis, we will use a hypothetical situation in which a membrane permeable only to *water* separates two compartments. One compartment contains pure water and the other contains relatively less water and some protein molecules that cannot cross the membrane (Figure 3.6). Osmotic pressure exists because of differences between the two solutions. Water is free to move and it diffuses into the compartment containing the protein. The pressure necessary to prevent such movement is called **osmotic pressure.**

The property of solutions that leads to osmotic pressure is called **osmolarity** and is measured in osmols (Osm). The number of osmols per liter of a solution depends on the number of dissolved particles in the solution. In solutions where every molecule and ion contributes to osmolarity, molar and osmolar concentrations are directly related. A 1 molar solution of a nonionizing substance such as glucose has osmolarity of 1 Osm. A 1 molar solution of a salt such as NaCl, which ionizes nearly completely into 2 ions per molecule, has osmolarity near 2 Osm.

Osmotic pressures in living systems, which are much smaller than this, are expressed in **milliosmols** (mOsm). A milliosmol is one-thousandth of an osmol. In body fluids, some molecules ionize partly or not at all, and others bind to protein molecules, so the actual osmotic pressure in body fluids is less than would be predicted from the concentrations of the substances in them. Most body fluids have an osmolarity of about 300 mOsm/l.

Physiologists often use intracellular fluid as a standard and determine pressure differences between it and other fluids. Solutions with no pressure difference when compared with intracellular fluids are **isosmotic** (i″sos-mot′ik), those in which pressure is higher outside the cell are **hyperosmotic** (hi″per-os-mot′ik), and those in which pressure is lower outside the cell are **hyposmotic** (hi″pos-mot′ik).

Tonicity (to-nis′ĭ-te), which is related to, and sometimes confused with, osmolarity, is determined by observing the behavior of cells in a solution. A solution surrounding a cell is **isotonic** (i″so-ton′ik) when it causes no change in cell volume. The solution is **hypertonic** (hi″per-ton′ik) to the cell if the cell shrinks as water moves out of it into the solution; it is **hypotonic** (hi″po-ton′ik) to the cell if the cell swells or bursts as water moves from the solution into the cell (Figure 3.7). Where a selectively permeable membrane allows only water and *no* solutes to cross the membrane, osmolarity and tonicity are the same. In biological systems where some solutes *do* cross the membrane, osmolarity and tonicity are not the same.

If you live where winter snows snarl traffic, you probably know that road-maintenance crews often scatter salt on highways to melt the snow. Salt combines with water in snow to form a solution with a particular osmotic pressure. It melts snow—not by making it any warmer outside—but by decreasing the freezing point of the solution. Because the number of particles in a solution decreases its freezing point by a predictable amount, changes in the freezing point can be used to determine a solution's osmolarity. The freezing point of a 1 osmol solution is −1.86° C, that of a 500 mOsm solution is half that, or −0.93° C.

Filtration

Filtration is the pushing of substances through a barrier. In a laboratory funnel, **hydrostatic pressure,** the pressure exerted by the solution standing in the funnel, pushes substances through filter paper. In the body, filtration—though not a significant factor in the movement of substances into and out of all cells—is important in pushing substances out of the smallest blood vessels, the capillaries. Though it does not require energy expenditure at a membrane, it does depend on blood pressure that, in turn, depends on energy for heart contractions. For example, blood pressure pushing blood along the inside of capillaries also pushes fluid and substances dissolved in the blood through the capillary walls into the interstitial fluids or into kidney tubules. The fluid and the dissolved substances can pass through or between the cells of capillary walls.

Bulk Flow

In the body, both gases and fluids move by bulk flow. **Bulk flow** is a streaming of molecules caused by a pressure difference. As a person breathes, gas molecules in air stream into and out of the lungs by bulk flow. As the heart contracts, blood flows through blood vessels, and water and other substances in the blood move by bulk flow. Once moving in a given direction, molecules tend to flow in streams rather than by random movement. Among water molecules, bulk flow is probably enhanced by hydrogen bonds between water molecules. Bulk flow of water molecules also causes **solvent drag,** an increased rate of movement of substances dissolved in the water. As the water molecules flow along, they literally drag dissolved particles with them. Bulk flow and solvent drag are important in the movement of substances between blood and interstitial fluids.

Active Transport

In contrast to the passive processes described above, **active transport** moves substances against concentration gradients from regions of lower to higher concentration. This is analogous to rolling something uphill and it requires the cell to use energy from ATP. In fact, cells may spend as much as half of the energy they use in a resting state for active transport. Active transport requires membrane proteins that are both enzymes and carriers. Carriers have specificity in that each binds to and transports a single substance or a few closely related ones.

FIGURE 3.7
A cell placed in (*a*) isotonic solution will have no net movement of water because the same amount of water moves in each direction. A cell placed in (*b*) hypotonic solution will gain more water than it loses; the cell will swell and eventually burst. A cell placed in (*c*) hypertonic solution will lose more water than it gains; the cell will shrink and the membrane will shrivel. This is called crenation. The effects of these solutions on red blood cells are shown in the accompanying scanning electron micrographs. Arrows in the beakers indicate direction and relative amounts of water movement. (Cells in (*b*) are less highly magnified than the cells in the other micrographs.)

(a)
Isotonic

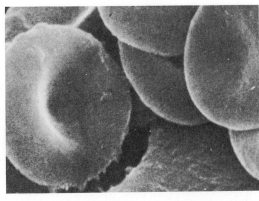

(b)
Hypotonic

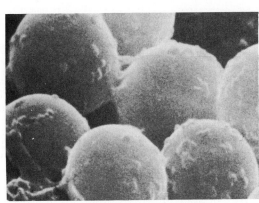

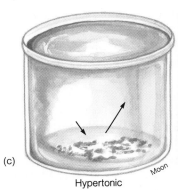
(c)
Hypertonic
Moon

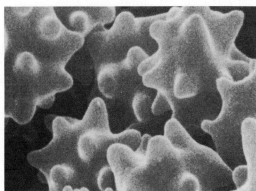

FIGURE 3.8

A model of the sodium-potassium active transport system. (*a*) A protein complex (a carrier and the enzyme ATPase) in the membrane binds Na$^+$ from inside the cell. (*b*) Na$^+$ binding activates ATPase in the presence of Mg^{2+}, ATP is hydrolyzed and the cleaved phosphate group binds to the protein complex. (*c*) Phosphate binding causes a change in the shape of the complex so the Na$^+$ binding site rotates to the outside surface of the membrane and releases Na$^+$. (*d*) The now exposed K$^+$ binding site binds K$^+$. (*e*) Having released Na$^+$ and bound K$^+$, the protein complex carries K$^+$ into the cell as it returns to its original shape. (*f*) The K$^+$ binding site then releases K$^+$ into the cell.

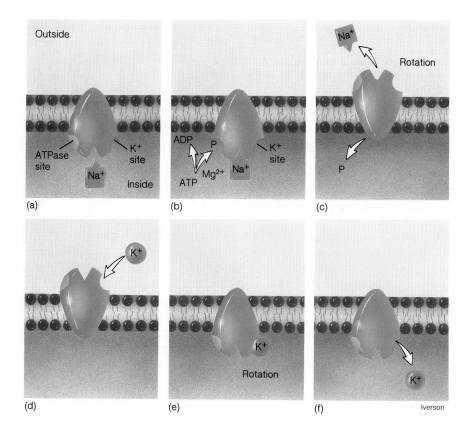

Active transport is essential for normal animal cell function because passive processes such as diffusion and osmosis cannot concentrate substances against a gradient. Active transport is the main means by which cells establish such gradients, and several other processes depend in one way or another on maintaining gradients. For example, active transport of sodium ions contributes to polarization of nerve and muscle cell membranes. Active transport of chloride ions in the kidneys creates a gradient that is essential for concentrating urine and conserving body fluids. Active transport of iodide ions in the thyroid gland concentrates those ions so they are available for the synthesis of thyroid hormones. Active transport of calcium ions is important in several processes—muscle contraction, transmission of signals from one neuron to the next, and in a variety of cellular control mechanisms. By concentrating substances against a gradient, active transport plays an important role in homeostatic mechanisms in many body systems.

We will use the **sodium-potassium pump** to illustrate active transport (Figure 3.8). Most animal cells maintain concentration gradients of Na$^+$ and K$^+$ in which the Na$^+$ concentration outside the cell is about 150 mmol/l and the concentration inside the cell is about 10 mmol/l, and in which the K$^+$ concentration inside the cell is about 140 mmol/l and the concentration outside the cell is about 4 mmol/l. Under these conditions, Na$^+$ constantly diffuses down its concentration gradient into cells and K$^+$ constantly diffuses down its concentration gradient out of them. Active transport counteracts diffusion by pumping Na$^+$ out of cells and K$^+$ back into them. As much as half of the body's resting energy use may be used to operate the Na$^+$–K$^+$ pump.

The sodium-potassium pump operates as follows: the membrane protein binds to a Na$^+$ ion at the inner membrane surface (Figure 3.8a). This binding activates the enzyme ATPase. In the presence of Mg^{2+} ions, ATP is hydrolyzed, releasing sufficient energy to power the pump and a phosphate group that binds to the protein (Figure 3.8b). Binding of phosphate causes the protein to change shape, placing the Na$^+$ binding site at the outer membrane surface and releasing Na$^+$ (Figure 3.8c). The K$^+$ binding site is exposed on the outer membrane surface and K$^+$ from the interstitial fluid binds to it (Figure 3.8d). The protein returns to its original shape carrying the K^+ ion to the inner membrane surface (Figure 3.8e), and K^+ is released into the cell (Figure 3.8f). One each of the Na^+ and K^+ ions are used in the example, but three Na^+ and two K^+ ions are transported each time the membrane protein goes through the sodium-potassium pump cycle.

Endocytosis and Exocytosis

In addition to the processes that move specific ions and small molecules directly across membranes, other processes move less specific and larger quantities of materials by the formation of vesicles, small bodies surrounded by membrane. In **endocytosis** (en"do-si-to'sis), vesicles form by invagination of a cell membrane to surround substances on the outside of the cell. Such vesicles pinch off from the plasma membrane and enter the cytosol. In **exocytosis** (eks"o-si-to'sis), vesicles inside the cell fuse with the plasma membrane and extrude their contents from the cell. Both endocytosis and exocytosis require the cell to expend energy. Much of this energy probably is used by contractile proteins of the cell's cytoskeleton that move vesicles to and from the plasma membrane.

In endocytosis, a substance outside the cell binds to the plasma membrane and causes the plasma membrane to invaginate and surround the substance. Once the substance is completely surrounded by plasma membrane to form a vesicle, the vesicle pinches off from the plasma membrane and becomes an intracellular vacuole. Endocytosis has the following features:

1. Substances are enclosed in membrane-bound vesicles and discharged into the cytosol in a matter of seconds to minutes.
2. Large amounts of plasma membrane enter the interior of the cell by endocytosis.
3. Plasma membrane taken into the cell usually is returned to the cell surface. Cells appear to have sorting mechanisms that return membrane segments to the particular sites from which they came.
4. Endocytosis can serve to accumulate substances in the cytosol, in larger vacuoles by the fusion of smaller vacuoles, in lysosomes, or in the Golgi apparatus.

Three different types of endocytosis have been identified: (1) phagocytosis, (2) fluid pinocytosis, and (3) adsorptive endocytosis. These processes differ in the contents of the vacuoles and the manner in which the vacuoles are formed.

In **phagocytosis** (fag"o-si-to'sis), cells, such as certain kinds of white blood cells, incorporate large vacuoles and with them large amounts of the plasma membrane (Figure 3.9). These vacuoles may contain microorganisms that have invaded the body or debris from tissue injury. The contents of these vacuoles are digested by enzymes from organelles called lysosomes. Lysosomes containing hydrolytic enzymes fuse with the membrane of a vacuole and release their enzymes. Sometimes materials from a vacuole are incorporated into lysosomes. Eventually, the substances are digested and released. Much of the membrane that surrounded the vacuole is returned to and fuses with the plasma membrane.

In **fluid pinocytosis** (pi"no-si-to'sis), cells take in small vacuoles of interstitial fluid by the same mechanism as in phagocytosis. Many kinds of cells obtain solutes from interstitial fluid by pinocytosis. The solutes enter the cell in concentrations directly proportional to their concentrations in interstitial fluid. Some cells take up a volume of fluid equivalent to one-fourth of their own volume per hour. Such cells accumulate a large quantity of fluid; they also internalize a massive quantity of plasma membrane. Studies of this process have shown that much of the fluid is regurgitated and that most of the membrane is returned to the cell surface.

In **adsorptive endocytosis,** or *receptor-mediated endocytosis,* specific substances bind selectively to plasma membrane receptors and are transported into the cell in vacuoles, probably by the action of some kind of carrier (Figure 3.9, steps 2, 3, 4). This process allows the cell to concentrate a substance against the concentration gradient. Segments of plasma membrane having such receptors are incorporated into vacuoles, but they are apparently recycled to the external surface with the receptors intact. These receptors bring the substance bound to them inside the cell and rapidly return to the cell membrane to pick up more of the substance. Cholesterol and iron enter cells in this manner.

A protein polymer called **clathrin** (klath'rin) has been shown to form vesicles on the cytoplasmic (inner) surface of membranes. This protein apparently controls traffic across the cell membrane—both inward and outward. It also is believed to play a role in sorting incoming substances and aiming them toward appropriate intracellular sites.

In exocytosis, the opposite of endocytosis, cellular substances within vesicles are released from the cell by the fusion of the vesicle with the plasma membrane (Figure 3.9, steps 6, 7, 8). Cells typically use exocytosis to rele͟͟v secretory products packaged in vesicles by the Golgi a͟ paratus or concentrated directly into vesicles in the cy-

FIGURE 3.9
Phagocytosis.

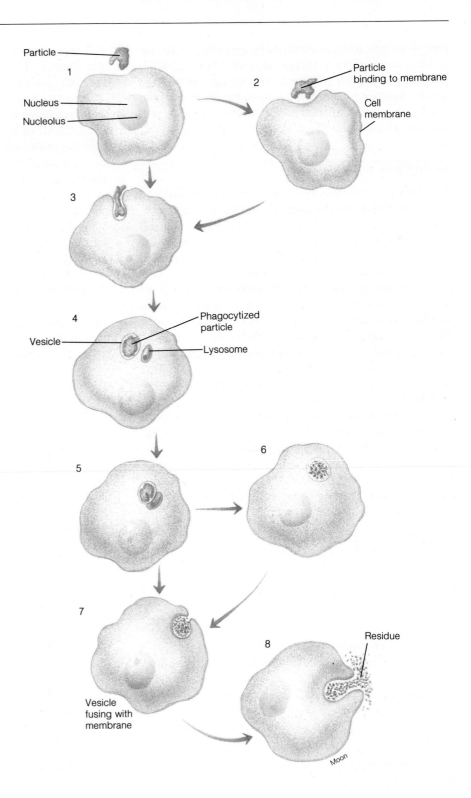

toplasm. Once secretory vesicles are formed, they move to and fuse with the plasma membrane and the contents of the vesicles are released into interstitial fluids. Nerve signals and some other stimuli seem to trigger the release of vesicles, probably by causing an increase in the concentration of free calcium ions in the cell. The calcium ions can cross the plasma membrane from extracellular fluids or they can be released from organelles. In sufficient concentration, the calcium ions stimulate fusion of vesicles with the plasma membrane. This fusion results in the release of the secretion and in an increase in the surface area of the plasma membrane.

When a cell undergoes endocytosis, the cell membrane surface area gets smaller; when it undergoes exocytosis, the surface area gets larger. Clearly, a predominance of one process over the other would change the surface area of the cell and also its volume. In most cells, the membrane surface area and the cell volume remain almost constant, so cells that undergo exocytosis and endocytosis must do so in a balanced manner. When plasma membrane is lost, as occurs when lysosomal enzymes digest some of the vesicular membranes, the cell synthesizes new segments of plasma membrane. These segments fuse with the existing plasma membrane and restore the normal amount of surface area. Conversely, when secretory cells gain plasma membrane, a burst of pinocytosis following secretion returns the membrane to its normal surface area.

Transport across cells can involve endocytosis on one cell surface and exocytosis on the opposite surface. For example, epithelial cells that line the small intestine take up some substances from the lumen of the intestine by endocytosis. The substances are soon released from the basal end of the cell by exocytosis. Though the details of this process are not clearly understood, it is known that the cells return plasma membrane from the lumen surface back across the cell to its original site. They also maintain the appropriate surface area on the basal surface. These observations suggest that cells must have some means of sorting segments of plasma membrane.

Movement of Substances and Homeostasis

All processes that move substances across membranes (Table 3.1) affect concentrations on both sides of the membrane. Sometimes the movements directly help to maintain conditions within the cell or its immediate environment in a narrow and tolerable range. Other movements, such as movements of ions that cause a neuron to send a signal or a muscle to contract, are necessary for cells to contribute to the normal function of a tissue, organ, system, or whole body.

TABLE 3.1 Mechanisms of Movement of Substances Across Cell Membranes

Process	Characteristics
Passive Processes	*Movement along a Gradient; Cell Expends No Energy*
Simple diffusion	Net movement down a gradient with or without a membrane, requires no energy
Facilitated diffusion	Net movement of molecules down a gradient, requires a carrier molecule but no energy
Osmosis	Movement of water along its own gradient; occurs across a selectively permeable membrane that prevents some molecules from diffusing
Filtration*	Movement of molecules along a hydrostatic pressure gradient
Bulk flow*	Rapid movement of water molecules along a gradient, probably because of hydrogen bonds between molecules; solvent drag occurs with bulk flow
Active Processes	*Movement Usually against a Gradient; Cell Expends Energy*
Active transport	Movement of substances across a membrane, usually against a gradient, requires energy expenditure and a carrier molecule
Phagocytosis	Formation of large vacuoles in which large particles enter a cell
Fluid pinocytosis	Formation of small vacuoles in which extracellular fluid and its solutes enter a cell
Adsorptive endocytosis	Concentration of specific substances in vacuoles
Exocytosis	Extrusion of secretions by the fusion of a vesicle with the plasma membrane

*These processes occur only in certain sites and are not general mechanisms by which substances enter and leave most cells.

CYTOPLASM

Scientists who first observed cells called the contents **protoplasm.** They identified the **nucleus** and **cytoplasm,** the substance surrounding the nucleus. We now know that cytoplasm contains many discrete structures called *organelles* (including the nucleus), which carry out specific intracellular functions. Cytoplasm also contains small particles suspended in a thick fluid called **cytosol** (si'to-sol). Cytosol consists of dissolved substances, enzymes, and several kinds of granules—most of which contain glycogen or fat. The cytosol and dissolved substances, such as ions, gases, and nutrient molecules, make up the intracellular fluid. Though many cellular enzymes are confined to organelles, those that control the first sequence of reactions in the metabolism of glucose are found in the cytosol as we shall see in Chapter 24.

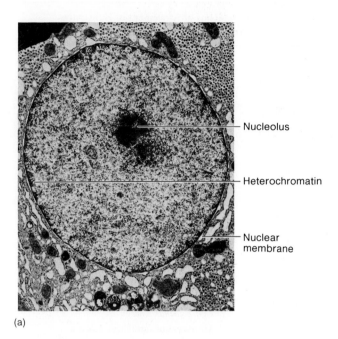

(a)

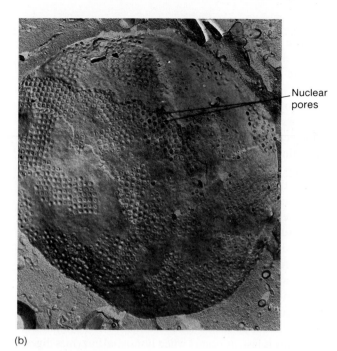

Nuclear pores

(b)

FIGURE 3.10
Electron micrographs of cell nuclei showing (*a*) the nucleolus, chromatin, and the double nuclear envelope, and (*b*) a freeze-fractured nuclear envelope showing nuclear pores.

ORGANELLES

Just as the stomach, lungs, and heart are functional units of body systems, organelles are functional units of a cell. The organelles of a typical cell are illustrated in Figure 3.2.

Nucleus

As the control center of the cell, the **nucleus** (Figure 3.10) is essential for a cell to survive and divide. The nucleus is surrounded by a folded, double layer of membrane called the **nuclear envelope.** Extending through this envelope are many large pores that can be closed by a protein associated with them. Opening and closing pores allows RNA to leave the nucleus and certain complex molecules to enter it. The **nucleoplasm** (nu′kle-o-plazm″) in the nucleus includes the nucleolus (or two or more nucleoli) and the chromosomes. The **nucleolus** (nu-kle′o-lus), which lacks a membrane, has an irregular shape that changes as the cell goes through stages of growth and division. It is the site of assembly of organelles called ribosomes.

The **chromosomes** (Figure 3.11), which condense into short rods during cell division, exist as more diffuse, elongated fibers called **chromatin** at other times. Chromatin exists in two forms: *euchromatin* consists of active DNA used by the cell to make proteins, and more condensed *heterochromatin* consists of inactive DNA. Human cells contain 46 chromosomes—22 pairs of autosomal chromosomes and 2 sex chromosomes (XX in females and XY in males). Chromosomes are known to consist of DNA wrapped around a protein core with surface proteins in certain locations.

DNA is important in two cellular processes. First, it dictates the nature of RNAs and proteins that a cell can synthesize and thereby controls the functions of the cell. Second, it transmits information for cellular control from one generation to the next—from parents to offspring in eggs and sperm.

If all of the DNA in all of a human cell's chromosomes were arranged end-to-end, the strand would be about 1.5 meters long. To pack this DNA into a nucleus requires a very compact arrangement. The DNA helix is coiled, and the coils are folded on themselves. Because of the supercoiled arrangement, determining the sequence of information stored in DNA still presents a challenge. Now that techniques exist to identify the sequence of bases in nucleic acids, a massive project for identifying all information in human DNA is under way. Having such information would be extremely useful in diagnosing and treating genetic diseases.

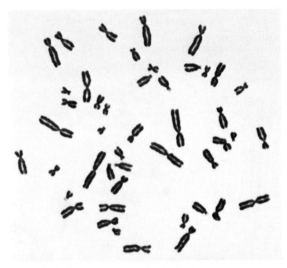

FIGURE 3.11
Chromosomes of a normal human cell. This is the historic photograph by which Drs. Joe Hin Tjio and Albert Levan demonstrated that human cells contain 46 chromosomes.

Endoplasmic Reticulum and Ribosomes

The **endoplasmic reticulum** (en''do-plas'mik rĕ-tik'u-lum), or ER, is an extensive network of interconnected flattened vesicles and tubules bounded by membranes of the same basic structure as the plasma membrane. In some photomicrographs, the ER appears to be continuous with the plasma membrane and the nuclear membrane. Tubules of ER serve as a transport system for proteins and possibly for other large molecules synthesized by a cell.

ER can be rough or smooth (Figure 3.12). Rough ER is coated with ribosomes, where protein synthesis occurs. Lipid and cholesterol synthesis, and some lipid breakdown occurs in smooth ER. Cells vary in the relative amounts of rough and smooth ER they contain. Pancreatic cells have large amounts of rough ER because they produce large quantities of protein enzymes for secretion, and liver and fat cells have large amounts of smooth ER because they are active in lipid metabolism.

Ribosomes (ri'-bo-sōmz), small bodies lacking membranes, are found on the surface of the rough ER or free in the cytosol where they serve as sites of protein synthesis. Chemically, ribosomes consist of ribonucleic acid (RNA) and protein. Ribosomes attached to the ER synthesize proteins to be secreted, such as some hormones and digestive enzymes. Researchers believe that as such proteins are synthesized, they move through the membrane to the lumen (inner cavity) of the ER to the Golgi apparatus and are eventually secreted. Ribosomes found free in the cytosol make proteins for the cell's own use. More details about ribosome structure and function are provided with the discussion of protein synthesis later in this chapter.

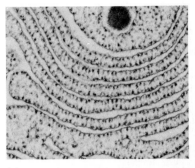

(a)

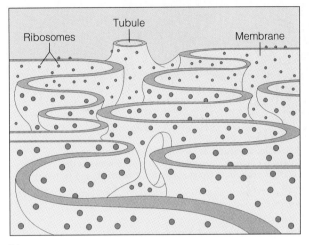

(b)

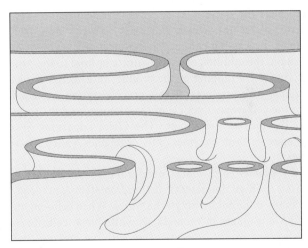

(c)

FIGURE 3.12
(*a*) Endoplasmic reticulum as seen in an electron micrograph, (*b*) diagram of rough ER, and (*c*) diagram of smooth ER.

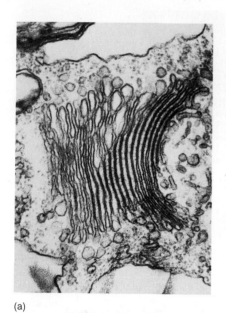

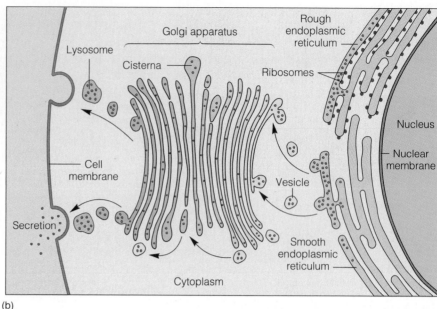

(a)

(b)

FIGURE 3.13

(*a*) The Golgi apparatus as seen in an electron micrograph and (*b*) enlarged diagram showing the relationship between the endoplasmic reticulum, Golgi apparatus, lysosomes, and secretory granules.

Golgi Apparatus and Lysosomes

The **Golgi** (gol′je) **apparatus** (Figure 3.13), usually located near the nucleus, is a stack of vesicles called **cisternae** (sis-ter′ne), which processes newly synthesized proteins. Processing can include removal of a segment of a protein, the addition of carbohydrate or lipid to the protein, and the wrapping of the product in a membranous envelope. The products of this processing are extruded from the cell when a vesicle of the Golgi apparatus fuses with the plasma membrane. The Golgi apparatus also packages certain enzymes into organelles called lysosomes.

Lysosomes (li′so-sōmz), membrane-bound organelles with an average diameter of 0.4 micrometers (μm), contain many different hydrolytic enzymes. When cells engulf foreign substances in vacuoles, as white blood cells do in infection and injury, lysosomes fuse with the vacuoles and pour their enzymes into the vacuoles. The enzymes digest the foreign matter. Even cells that do not engulf foreign substances contain lysosomes. As long as such cells remain healthy, the lysosomal membranes apparently prevent the release of enzymes. When a cell is damaged or dies, it accumulates acid, which causes lysosomes to rupture and their enzymes to digest the remains of the cell.

Mitochondria and Peroxisomes

Frequently called the powerhouses of a cell, **mitochondria** (mi″to-kon′dre-ah) (Figure 3.14) are the sites of most of the oxidative reactions that transform energy into a form usable by cells. Cells contain hundreds to thousands of these rod-shaped or ellipsoid organelles that range in size from 0.5 to 1.0 μm in diameter and from 3 to 4 μm in length. In general, the most metabolically active cells contain the largest numbers of mitochondria.

Structurally, a mitochondrion is composed of an outer smooth membrane and an inner folded membrane, each of which has a typical membrane structure. **Cristae** (kris′te), folds in the inner membrane, contain enzymes and other molecules that capture and store energy in ATP. The **matrix** (center) of a mitochondrion contains enzymes that break down fatty acids and pyruvic acid (from glucose) so enzymes in the cristae can capture energy from these important nutrient molecules.

The mitochondrial matrix contains DNA and ribosomes, the origins and functions of which have puzzled biologists. It now appears that millions of years ago, primitive free-living organisms that used oxygen to metabolize food were incorporated into larger single-celled organisms. The once free-living organisms with their own DNA and ribosomes have evolved into mitochondria. The original relationship provided protection for the "mitochondria" and energy for the larger organisms.

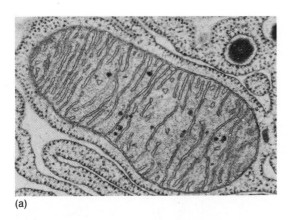

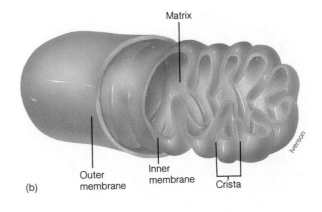

FIGURE 3.14

(*a*) A mitochondrion as seen in an electron micrograph and (*b*) a diagram showing the typical arrangement of inner membranes.

Recognized only since the early 1950s, **peroxisomes** (pĕ-roks′ĭ-sōmz) are membrane-covered organelles about 0.5 μm in diameter. Like mitochondria, peroxisomes contain oxidative enzymes that destroy certain toxins. Some reactions release hydrogen peroxide, which is broken down to oxygen and water by the peroxisome enzyme catalase. Certain human genetic defects involving absent or defective peroxisomes cause severe symptoms such as seizures, sight and hearing loss, uncoordinated movements, and liver and kidney disease; thus, defects at the cellular level can have profound effects on whole body function.

Cytoskeleton, Centrioles, Cilia, and Flagella

Cells have an internal **cytoskeleton** (Figure 3.15a) that gives them rigidity and that allows movement of whole cells and particles within cells. The cytoskeleton consists of a complex network of microtubules and microfilaments, both of which consist of protein devoid of any membrane covering. **Microtubules** are hollow fibers of the protein **tubulin,** with a diameter of 20 to 25 nanometers (nm). **Microfilaments** are much smaller fibers (5 nm in diameter) made mainly of the protein **actin.** This network appears to hold the nucleus in place in a cell and to provide binding sites to anchor protein molecules, organelles, and other cellular particles once thought to be randomly distributed in the cytosol. In addition to their roles as support structures, microtubules and microfilaments contribute to cellular movements. The behavior of these structures in living cells now can be observed with **video enhanced contrast microscopy** (VECM), which makes use of a computer, high contrast photography, and a light microscope. Before VECM, these structures could be observed only by electron microscopy using killed, fixed cells that no longer displayed movement. It now appears that within cells, microtubules function like bones and contractile microfilaments like muscles.

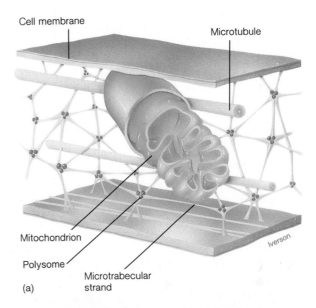

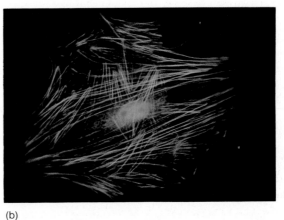

FIGURE 3.15

(*a*) A diagram of a proposed structure of the cytoskeleton.
(*b*) Microtubules visualized by reacting with immunofluorescent antibodies against the protein tubulin.

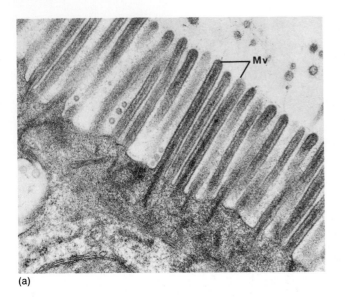

(a)

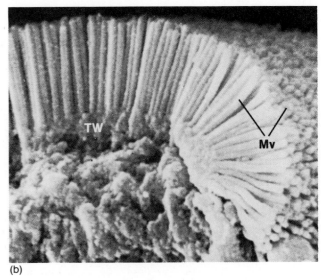

(b)

FIGURE 3.16

Microvilli in the small intestine as seen with the (*a*) transmission and (*b*) scanning electron microscopes.

From *Tissues and Organs: A Text Atlas of Scanning Electron Microscopy* by R. G. Kessel and R. Kardon. W. H. Freeman and Company © 1979.

FIGURE 3.17

A centriole (*a*) as seen in an electron micrograph and (*b*) the arrangement of bundles of three microtubules in a centriole. Paired centrioles oriented at right angles (*c*) as seen in an electron micrograph and (*d*) in a diagram.

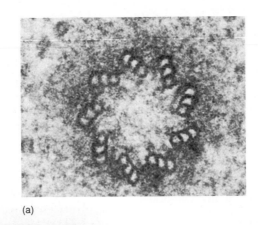

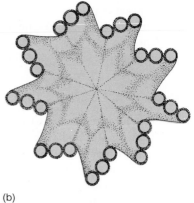

(a)

(b)

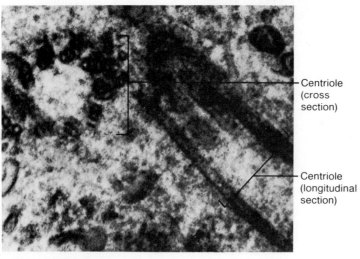

Centriole (cross section)

Centriole (longitudinal section)

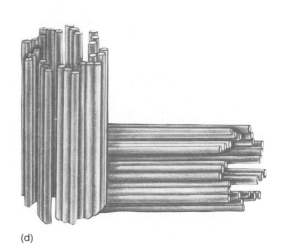

(c)

(d)

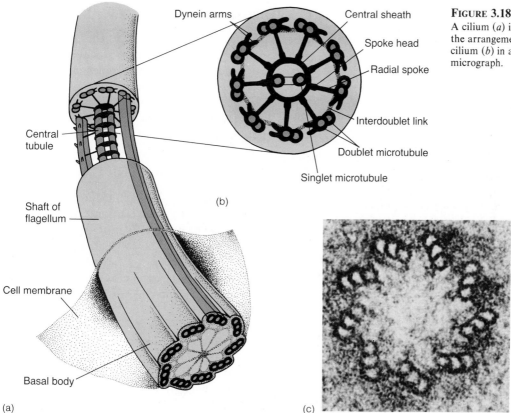

Dynein arms

Central sheath

Spoke head

Radial spoke

Central tubule

Interdoublet link

Doublet microtubule

Singlet microtubule

(b)

Shaft of flagellum

Cell membrane

Basal body

(a)

(c)

FIGURE 3.18

A cilium (*a*) in a longitudinal diagram, and the arrangement of microtubules within a cilium (*b*) in a diagram and (*c*) in an electron micrograph.

Microtubules form the spindle along which chromosomes move as they separate during cell division and the tracks along which various particles move within cells. For example, along microtubules in the axons (fibers) of nerve cells, particles have been observed to move in both directions (Figure 3.15b). The moving particles include mitochondria, large vesicles of surplus cell membrane from endocytosis, and small vesicles that contain molecules of neurotransmitter, a chemical substance that relays signals from one neuron to the next. How movement along microtubules occurs is not fully understood, but it appears that several molecules of a motor protein with enzyme activity are attached to the microtubules. These molecules seem to use energy from ATP to create motion of particles around a microtubule similar to that of a vehicle with "caterpillar" wheels.

Microfilaments are concentrated just beneath the membrane in many cells. They participate in amoeboid movement, particle engulfment during phagocytosis, the separation of daughter cells in cell division, and the movement of very small folds of membrane called microvilli (Figure 3.16). Microvilli projecting from the surface of cells lining the small intestine increase the rate of absorption of digested food.

Centrioles, cilia, and flagella consist primarily of microtubules. **Centrioles** (Figure 3.17) are paired structures always oriented at right angles to each other in nondividing cells. During cell division, they duplicate and the new pairs, which consist of one new and one old centriole, move to opposite sides of the cell. As the centrioles move, microtubule bundles associated with them elongate by the addition of tubulin to form the mitotic spindle along which chromosomes move.

Centrioles also give rise to cilia and flagella. **Cilia** are short, hairlike cytoplasmic projections that beat in waves and are found on the surface of some cells, such as those that move mucus along passageways in the respiratory tract. **Flagella,** also cytoplasmic projections, are longer than cilia but similar in structure. They usually cause movement of entire cells, such as a human sperm. Both cilia and flagella have nine pairs of microtubules around their periphery and a single pair in their center (Figure 3.18). The centriole persists as a basal body, which appears to anchor the cilium or flagellum to the cell. Because cilia and flagella can move after being severed from their basal bodies, the basal body appears not to be essential for movement.

TABLE 3.2   Characteristics of Organelles

Organelle	Characteristics
Nucleus	Large body surrounded by a nuclear envelope that consists of a double membrane with large pores; contains nucleoplasm, chromosomes, and one or more nucleoli; functions as the control center of the cell
Chromosomes	Long strandlike bodies within the nucleus consisting of DNA and protein; contain genetic information that controls the activities of cells
Nucleolus	Irregularly shaped body in the nucleus, site of RNA synthesis
Endoplasmic reticulum	Network of interconnected vesicles and tubules bound by membranes; serves as transport system for molecules synthesized by the cell; rough ER has ribosomes on the surface, smooth ER lacks ribosomes and is the site of lipid metabolism
Ribosomes	Small round bodies lacking membranes found on ER or in cytosol, consist of RNA and protein, function in protein synthesis
Golgi apparatus	Large membrane-bound structure consisting of a stack of vesicles; serves as a storage and processing organ for cell products, may remove a segment of a protein, add carbohydrate to a protein, wrap protein in a membrane; extrudes secretions from cell; forms lysosomes and secretory granules
Lysosomes	Membrane-bound bodies that contain hydrolytic enzymes, enzymes are released into vacuoles or into dead or injured cells
Mitochondria	Rod-shaped or spherical bodies that have a smooth outer membrane and a folded inner membrane, contain enzymes for reactions that capture energy in ATP
Microtubules	Hollow fibers composed of the protein tubulin, provide motility and internal support for cells; form centrioles, cilia, and flagella
Microfilaments	Very thin, solid structures composed of the protein actin, account for motility and contractility of cells
Centrioles	Paired structures that give rise to spindle fibers in cell division and give rise to flagella and cilia; consist of nine sets of three microtubules
Cilia	Short hairlike projections from the surface of certain cells; axis consists of a single pair of central microtubules and nine pairs of peripheral microtubules; propel movement of particles along cell surfaces
Flagella	Relatively long projections from the surface of certain cells, axis has same microtubule arrangement as cilia; cause whole cells to move

Cellular Inclusions and Extracellular Materials

In addition to the structured organelles, cells contain membrane-bound vacuoles of ingested substances or stored fats, and nonmembrane-bound granules of glycogen or other substances it has synthesized. Some cells also make and secrete extracellular materials such as mucus or collagen, and contribute to the basement membrane that is deposited beneath epithelial tissues.

Integration of Organelle Functions

The roles of organelles in cellular functions are summarized in Table 3.2. Many organelle functions are closely associated with the enzymes they contain so that organelles work together to carry out cell functions and maintain cellular homeostasis.

Many mechanisms for regulating cell function exist, and most involve protein binding sites, or receptors. These receptors usually are located on outer membrane surfaces, but some such as those concerned with the actions of steroid hormones are located inside the nucleus. Certain proteins have several specific receptors, each of which bind a different ligand. Binding of ligands to receptors is a reversible process, so binding and unbinding can function as a control process like turning on and off a switch. Receptors also provide a means by which cells can recognize, and thereby respond to, specific ligands that act as stimuli, so ligand–receptors binding also is a communication process. Receptors binding ligands is a fundamental control and communication process at the cellular level.

The properties of receptors—specificity, affinity, competition, and saturation—allow them to regulate cell function in a variety of ways. One way concerns the development of structural complexity. Such complexity is essential if molecules are to form living cells. If assorted cellular proteins were mixed in a solution, many would bind to each other spontaneously as random molecular movements cause collisions between the molecules. Because specific binding sites on proteins cause nonrandom combinations, molecular aggregations form. For example, protein subunits aggregate to form microfilaments and microtubules, and protein and RNA combine to form ribosomes.

The ability for spontaneous self-assembly of cell components is important in living systems. Bicycle parts would never assemble themselves into a bicycle no matter how the parts were laid out or how long one waited, but many complex structures within cells do, in fact, assemble themselves. In other words, they produce order from entropy. The significance of self-assembly of cell components is that once the sequence of amino acids in proteins has been established from information in DNA, no further information is required. The proteins take on particular shapes and acquire binding sites that lead to spontaneous self-assembly of many cell components.

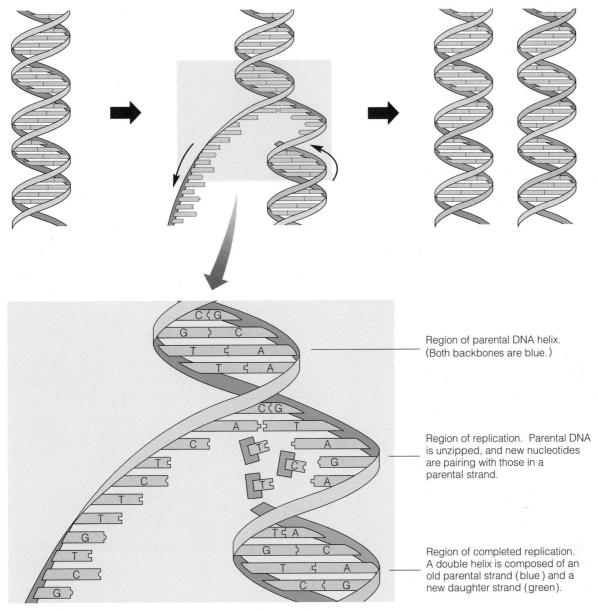

Region of parental DNA helix. (Both backbones are blue.)

Region of replication. Parental DNA is unzipped, and new nucleotides are pairing with those in a parental strand.

Region of completed replication. A double helix is composed of an old parental strand (blue) and a new daughter strand (green).

FIGURE 3.19
The replication of DNA. After separation of parent strands, replication of only one strand is shown. The other strand also is replicated but in the opposite direction. In each new molecule of DNA one strand is "old" DNA from the parent cell and one strand is recently synthesized "new" DNA. This is called semiconservative replication.

CELL DIVISION

The ability of cells to divide and produce new cells is an essential property of living things. Without this ability, even small cuts and scratches would not heal, and fertilized eggs would never develop into new individuals. Repair of injuries, growth, and development all involve cell division and all are essential to normal body function. Cell division involves three processes: (1) **replication** (duplication) of DNA, (2) **mitosis,** nuclear division, and (3) **cytokinesis,** division of the cytoplasm. These processes will be reviewed briefly here. Meiotic cell division that produces eggs and sperm will be reviewed in Chapter 29.

DNA Replication

In preparation for cell division, all the DNA in the parent cell is replicated. **DNA replication** (Figure 3.19) begins as segments of the nucleotide strands of each double helix of DNA separate and expose strands of unpaired bases. Each strand is replicated separately and a small segment of RNA primer starts, or primes, the process. Each nucleotide strand serves as a **template,** or pattern, for the synthesis of a partner strand. The enzyme **DNA polymerase** (pol-im′er-ās) binds to the primer and its paired base in the template. Phosphorylated nucleotides are plentiful in the surrounding nucleoplasm, and a nucleotide containing a complementary base combines with the exposed base adjacent to the enzyme-primer complex. Energy from a high energy bond in the nucleotide allows the enzyme to join the nucleotide to the primer. The enzyme moves along the template; another nucleotide binds to the complex and is added to the new strand of DNA. The process continues until a new molecule of DNA exactly like the original partner of the template strand is synthesized.

Each new chromosome consists of one old and one new strand of DNA, that is, it undergoes **semiconservative replication** (Figure 3.19). This was demonstrated by providing radioactively labeled nucleotides to cells while they were replicating DNA. In such experiments, half the radioactive material appears in each chromosome, so each chromosome must contain half the newly synthesized DNA. One strand of a DNA molecule is the **sense strand,** that is, information in it is used to direct protein synthesis. The other strand is the **missense strand,** which functions only during replication. Each sense molecule serves as a template for the missense molecule, and each missense molecule serves as a template for the sense molecule.

DNA replication is an amazingly error-free system of information transmission with only about one error in the placement of 10 billion bases. This is analogous to typing the text of about 12 books like this one without making a single typographical error. Cells appear to have two ways to reduce errors during DNA replication. First, the fact that complementary bases fit together most easily makes correct placement more likely than erroneous placement. Second, DNA polymerase or an enzyme very similar to it "proofreads" the new strand as it is being synthesized and can remove an erroneous base while it is still at the end of the strand.

Even after a double helix is completed, other enzymes sometimes can remove errors and replace them with appropriate bases. The most puzzling aspect of this process has been to discover how cells can tell which strand in a helix is the new, erroneous one. Researchers have found that bases in certain sequences become methylated (acquire a $-CH_3$ group) soon after synthesis. If DNA is studied immediately after synthesis, researchers can tell old methylated strands from new nonmethylated ones. Whether cells identify new, possibly erroneous strands in the same way is not known.

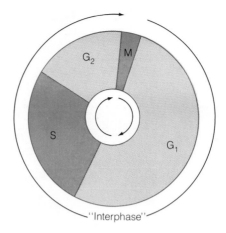

FIGURE 3.20
The cell cycle. Interphase includes G_1, S, and G_2 phases of the cycle. The time spent in phase G_1 is variable, but the other stages are of relatively constant length for a given type of cell. M stands for mitosis.

Mitosis and Cytokinesis

Cells capable of dividing go through a **cell cycle** (Figure 3.20). In stage G_1 (gap or growth stage 1), cells are carrying on normal metabolic activity and are neither dividing nor preparing to divide. G_1 can last from hours to years. Highly specialized cells such as nerve and muscle cells in adults have lost their ability to divide and remain in stage G_1 for years. In contrast, cells such as those in the deepest layers of the epidermis and in the lining of the digestive tract continue to divide throughout life. They regularly go through the cell cycle.

Though the trigger for cell division is not fully understood, it now appears that a protein called *cyclin* (si′klin) activates a kinase enzyme that adds phosphate groups to other proteins. Though it is not known exactly how the phosphorylations act, their effect is to initiate cell division. During division, cyclin concentration and kinase activity decrease sharply and gradually increase after division is complete.

Whatever the triggering mechanism, cells that can divide do so periodically. As a cell prepares to divide, it enters the S (synthesis) stage in which DNA is replicated, some proteins are synthesized, and the centromere (the structure that holds new strands together) is duplicated. After 6 to 8 hours in the S stage, the cell enters the G_2 stage, which lasts from 2 to 5 hours. Together, stages G_1, S, and G_2 are called **interphase.** The G stages are periods during which no events related to DNA replication or cell division are known to occur.

After the G_2 stage, the cell undergoes **mitosis,** the actual division of the nucleus. Mitosis is a continuous process, though biologists divide it for convenience into the following main stages: **prophase, metaphase, anaphase,** and **telophase.** Events in these stages are depicted in Figure 3.21, and events throughout the cell cycle are summarized in Table 3.3.

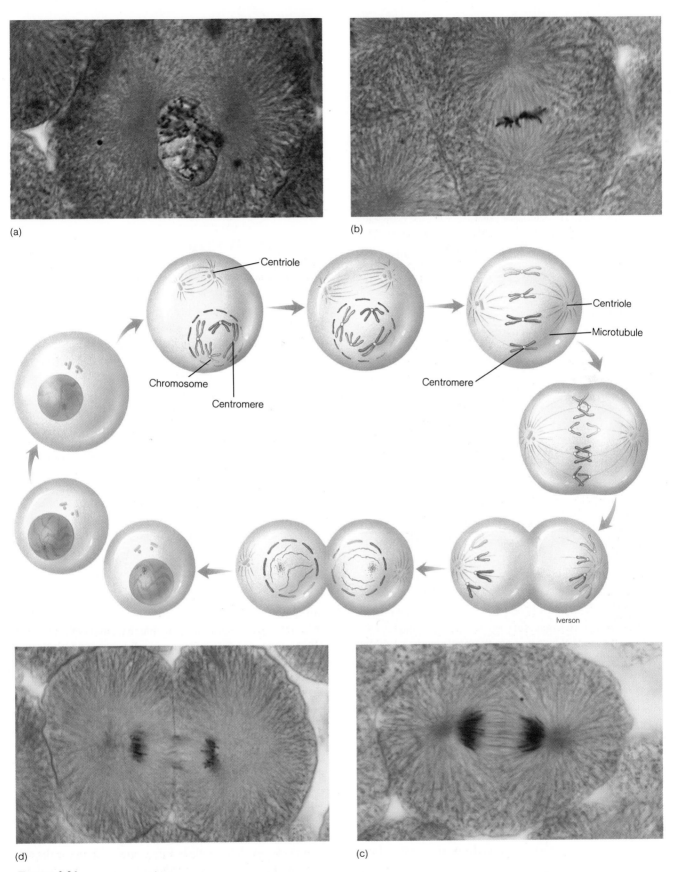

(a)

(b)

Centriole

Chromosome

Centromere

Centriole

Microtubule

Centromere

Iverson

(d)

(c)

FIGURE 3.21

Mitosis is a continuous process by which the nucleus of a cell divides.
The parts of that process include (*a*) prophase, (*b*) metaphase,
(*c*) anaphase, and (*d*) telophase.

TABLE 3.3   Events in the Cell Cycle

Stage	Events
Interphase	
G₁	Metabolism, no known events related to cell division
S	Synthesis of DNA to replicate chromosomes and replication of kinetochores
G₂	Metabolism, no known events related to cell division
Prophase	Chromosomes condense and become shorter and thicker
	Nucleolus and nuclear envelope disappear
	Centrioles divide and each pair moves to opposite sides of the cell
	Spindle fibers form between centrioles
Metaphase	Chromosomes move to center of spindle
	Kinetochores of each chromatid (chromosome of replicated double chromosome) attach to spindle fibers
Anaphase	A force developed by spindle fibers pulls on kinetochores; the chromatids separate and move toward opposite ends of the spindle
	Chromatids are now called chromosomes
Telophase	One of each kind of chromosome arrives at each of the poles of the cell
	Nucleolus and nuclear envelope reappear
	Spindle fibers disappear
	Chromosomes unfold and become longer and thinner
Cytokinesis	Occurs concurrently with anaphase and telophase
	Cytoplasm becomes furrowed between the two new nuclei
	Growth of plasma membrane completes separation of new cells

Cytokinesis, division of the cytoplasm, occurs concurrently with anaphase and telophase. The plasma membrane forms a **cleavage** furrow between the nuclei as a contractile ring of actin microfilaments and associated myosin molecules exert mechanical force. This force ultimately divides the cytoplasm into approximately equal parts, each part surrounding a nucleus.

Significance of Cell Division

During cell division, each new cell receives one of each chromosome found in the original cell. The process of mitosis allows the total number of cells in an organism to increase while the number and kind of chromosomes in the cells remain constant.

In embryonic development, cell division is accompanied by **differentiation,** the specialization of cells to form different kinds of tissues as they undergo division. Differentiation is guided by factors present in adjacent cells, by genetic information, and by regulatory molecules that control protein synthesis. Though cells retain all of their DNA, euchromatin becomes functional while heterochromatin has no effect on cell control. For example, DNA that directs synthesis of enzymes for energy metabolism functions in nearly all cells, but DNA that controls synthesis of a neurotransmitter functions only in certain nerve cells. Likewise, DNA that controls synthesis of actin and myosin is more active in muscle cells than in other cells.

The significance of mitotic cell division is twofold. It provides for each cell to have the same genetic information, yet during embryonic development, cell division is regulated to allow tissue differentiation.

PROTEIN SYNTHESIS

Proteins are important as part of the structure of cells, enzymes, carriers in transport systems, antigens and antibodies, and regulatory molecules. Protein synthesis is essential for growth and for replacement of cell components. How cells make proteins is a complex process that has been elucidated by painstaking research within the last few decades.

Cell Components in Protein Synthesis

As we have seen, DNA conveys genetic information to new cells and directs the synthesis of all kinds of proteins, including enzymes. Because enzymes control the synthesis of nonprotein cell components, DNA directly or indirectly determines the nature of all cell components. DNA acts in protein synthesis by relaying information to several kinds of RNA that work together to assemble proteins.

Process of Protein Synthesis

Protein synthesis is initiated by factors such as steroid hormones entering the nucleus and binding to receptors that activate a segment of DNA. The helix of activated DNA unwinds, exposing unpaired bases, and the sense strand serves as a template for RNA synthesis. Protein synthesis involves two processes—**transcription,** in which RNA is synthesized from DNA, and **translation,** in which protein is synthesized according to information in RNA (Figure 3.22).

The terms transcription and translation have the same meaning in describing protein synthesis in their less technical usage. For example, a stenographer transcribes shorthand notes into typewritten words; both are in the same language. Likewise the cell transcribes information in DNA to RNA; both are in "nucleotide" language. A linguist translates one language to another—English to French, Spanish to English, for example. Likewise the cell translates information in mRNA to an amino acid sequence, that is, from the language of nucleotides to the language of amino acids.

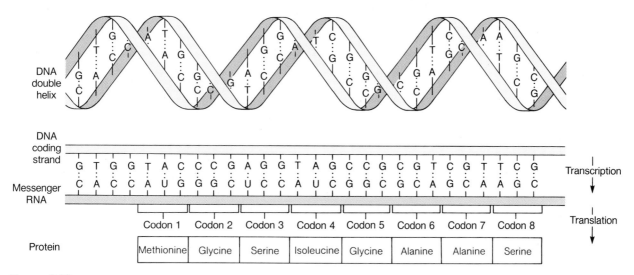

FIGURE 3.22

An overview of the processes of transcription and translation.

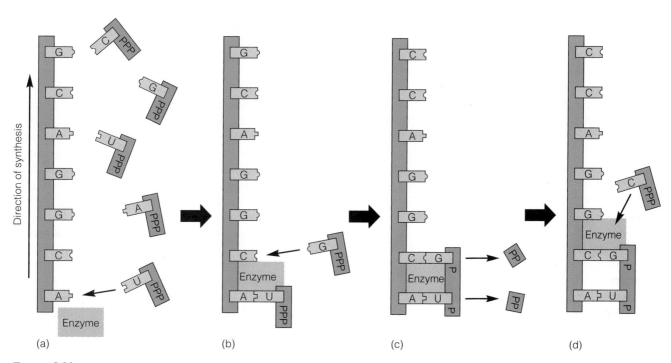

FIGURE 3.23

The transcription of RNA from template DNA: (*a*) An enzyme binds to the first exposed base and the appropriate phosphorylated nucleotide attaches to the DNA by base pairing. (*b*) The enzyme moves to the next base and a second nucleotide attaches. (*c*) The nucleotides bond together, releasing pyrophosphate, to form the first two units of RNA. (*d*) The process is repeated until a code in DNA indicates the end of a molecule of RNA.

Transcription (Figure 3.23) occurs in the nucleus where phosphorylated RNA nucleotides are plentiful. The enzyme RNA polymerase binds to the exposed template DNA, the appropriate phosphorylated nucleotide joins the DNA base-enzyme complex. For example, the nitrogenous base uridine attaches by base pairing to the adenine of DNA and pyrophosphate (P-P) is released. The enzyme moves to the next DNA base and the appropriate phosphorylated nucleotide joins the complex. The phosphate of the second nucleotide is linked to the ribose of the first nucleotide, forming the first link in the "backbone" of a new molecule of RNA. Energy to form this link comes from the hydrolysis of ATP. The enzyme continues to move along the DNA and this process is repeated until the end of a DNA template is reached and the RNA molecule is completed. Some form of punctuation exists in DNA to indicate the end of a sequence that will perform a particular function.

Second base of codon

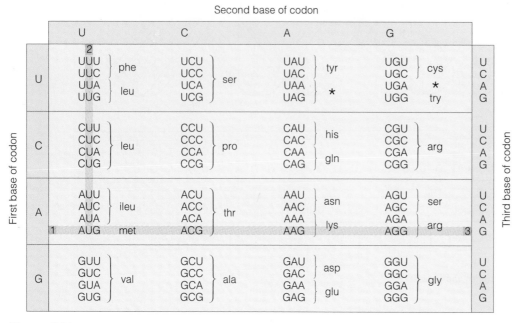

ala = alanine
arg = arginine
asn = asparagine
asp = aspartic acid
cys = cysteine
gln = glutamine
glu = glutamic acid
gly = glycine
his = histidine
ileu = isoleucine
leu = leucine
lys = lysine
met = methionine
phe = phenylalanine
pro = proline
ser = serine
thr = threonine
try = tryptophan
tyr = tyrosine
val = valine
★ = termination
codon

FIGURE 3.24

The genetic code consists of codons of messenger RNA. Most codons specify amino acids. The methionine codon always signifies the start of a protein, and one of three termination codons signifies the end of a protein. To find the codon for methionine, follow the shaded blocks numbered 1, 2, and 3.

The RNA base sequence is determined by base pairing with the DNA template. Pairing occurs as in DNA replication, except that in RNA, uracil, and not thymine, pairs with adenine. Base pairing is critical in protein synthesis because it transfers information from DNA to RNA and between RNAs. Three kinds of RNA are synthesized in the nucleus from information in DNA—messenger RNA, ribosomal RNA, and transfer RNA. Each RNA consists of a single strand of nucleotides and each is formed by transcription.

Messenger RNA (mRNA) is synthesized in specific length segments. Ultimately, an mRNA segment includes the exact number of bases to convey genetic information from DNA for the synthesis of a polypeptide. Many original mRNA molecules contain sequences of nucleotides not used to code amino acids in a protein. Such molecules are clipped into segments, unneeded segments discarded, and coding segments reunited. This is done in the nucleus by enzymes that apparently can recognize which segments to delete and which to retain. Once synthesized mRNA moves through a nuclear membrane pore and binds to a ribosome. Such an mRNA is ready to code the sequence of amino acids in a protein.

Ribosomal RNA (rRNA) combines with ribosomal structural proteins to form ribosomal subunits. To form a subunit, a previously made protein enters the nucleus and combines with nucleolar rRNA. Ribosomal subunits in animal cells are of two sizes: a large 60S subunit and a small 40S subunit. (The S refers to the sedimentation coefficient of the subunit, which is determined by its size, shape, and density.) The subunits pass through nuclear membrane pores and enter the cytosol. Any mRNA present in the cytosol binds to a 40S subunit and then binds to a 60S subunit to form a complete ribosome. The 40S subunit facilitates amino acid placement as a protein is made, and enzymes of the 60S subunit catalyze peptide bond formation between amino acids. Some ribosomes bind to endoplasmic reticulum and others cluster as *polyribosomes* in the cytosol.

In translation, information in mRNA is read in three-base sequences. These base triplets called **codons** specify particular amino acids or act as punctuation marks to indicate the beginning and the end of a polypeptide. At least one codon exists for each of the twenty amino acids found in proteins, and several codons exist for some amino acids. The full set of codons is called the **genetic code** (Figure 3.24). This code is the same, or nearly the same, in all organisms, and its universality allows us to apply findings from other organisms to our understanding of information transmission in human cells.

Bringing amino acids to the protein synthesis site requires **transfer RNA** (tRNA). Numerous tRNAs are synthesized in the nucleus and move to the cytosol. A tRNA typically consists of 75 to 80 nucleotides arranged in a cloverleaf shape with an amino acid binding site at one end and a three-base sequence called an **anticodon** at the other end (Figure 3.25). Each tRNA transports a specific amino acid, which must be activated by phosphorylation from ATP hydrolysis before it can bind to tRNA. At a ribosome, the tRNA anticodon attaches to an mRNA codon by base pairing. Any specific tRNA transports only the amino acid specified by the codon to which the tRNA's anticodon attaches.

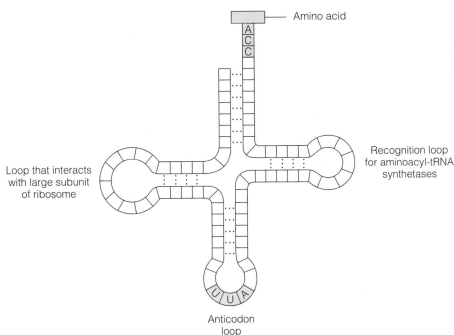

Amino acid

Loop that interacts
with large subunit
of ribosome

Recognition loop
for aminoacyl-tRNA
synthetases

Anticodon
loop

FIGURE 3.25
A molecule of transfer RNA has a binding
site for a specific amino acid and an
anticodon that base pairs with the codon for
the specific amino acid.

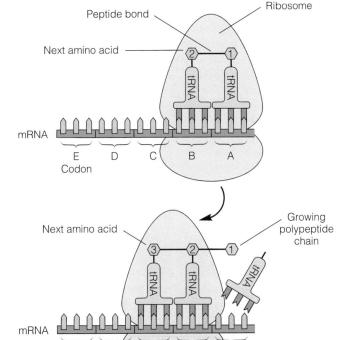

Peptide bond

Ribosome

Next amino acid

mRNA

E   D   C   B   A
Codon

Next amino acid

Growing
polypeptide
chain

mRNA

E   D   C   B   A
Codon

FIGURE 3.26
Translation occurs on ribosomes and makes
use of various kinds of RNAs previously
transcribed. The mRNA aligns on a
ribosome. As an appropriate tRNA arrives,
its anticodon base pairs with the first codon.
When a second tRNA arrives and base pairs
with the second codon, a peptide bond forms
between the amino acids. As this process is
repeated, first used tRNAs are released,
peptide bonds form, and the polypeptide
chain grows.

All RNAs are reusable, but a precise quantity of
mRNA is made according to the cell's needs for a particular enzyme or other protein. Control processes apparently assure that cells make appropriate quantities of
particular proteins so higher level functions occur in an
orderly fashion. Once an mRNA has joined a ribosome,
the ribosome initiates translation and provides a protein
assembly site. Translation (Figure 3.26) occurs in the following steps: When mRNA is aligned on the ribosome (a),
the first tRNA arrives (b). The small ribosomal unit ori-
ents tRNA so its anticodon can base pair with the mRNA
codon. Such base pairing allows mRNA codons to specify
the sequence of amino acids in the protein by specifying
the sequence of tRNA anticodons, which carry the amino
acids. When the second amino acid arrives (c), an enzyme
in the large ribosomal subunit forms a peptide bond between the two amino acids (d). ATP hydrolysis provides
energy to form the bond. As the ribosome moves along the

mRNA, a third amino acid arrives and the first tRNA is released into the cytosol (e). This tRNA can transport another amino acid molecule. A second peptide bond is formed (f) and the process continues until a terminator codon is recognized by the ribosome.

When the finished protein is released, the mRNA can synthesize another molecule of the same protein. In fact, several ribosomes can "read" different parts of a molecule of mRNA at the same time. Protein synthesis is summarized in Table 3.4.

Protein chemists now know enough about protein synthesis to make it happen in cell-free mixtures in a laboratory reaction chamber, provided the mixture contains the proper quantities of genes, amino acids, organelles, enzymes, and ATP. Protein production can be greatly increased by keeping mRNA in a chamber, continuously adding ATP and GTP to the chamber, and removing the protein product from it. This procedure is especially useful for making proteins that are toxic to cells, such as drugs to kill infectious agents or cancer cells. It also can be used to make proteins that are broken down by cells before they can be extracted.

Posttranslational Processing

Once a polypeptide chain has been synthesized according to the information in mRNA, it can be modified in a variety of ways. Methionine, the first amino acid in any polypeptide, can be removed or modified. Other amino acids in the chain can be modified by the addition of methyl, phosphate, hydroxyl, or other groups. Proteins destined to be secreted are often transported in the endoplasmic reticulum to the Golgi apparatus for processing. A portion of the polypeptide chain can be removed, as occurs in the processing of insulin. Carbohydrates or lipids can combine with the protein to form glycoproteins or lipoproteins, respectively. Proteins secreted in an inactive form are activated as needed. Pancreatic proteolytic enzymes are activated when they reach the small intestine. Blood clotting factors are activated by substances released from injured tissues.

Control of Protein Synthesis

Cells have genetic information to synthesize thousands of proteins, but they make only the proteins they need, and those only a few at any one time. For example, though pancreatic cells contain information to make any protein, different cells make insulin, glucagon, and digestive enzymes. Even cells that make digestive enzymes probably do not make large quantities of several different ones at the same time. For the proteins a cell normally makes, regulation of when and how much protein is made can be imposed at any point in the process of protein synthesis. Regulatory mechanisms work together to cause the right

| TABLE 3.4 | Summary of the Process of Protein Synthesis |

Transcription

1. Bonds between strands of DNA break and expose a portion of the DNA template.
2. An enzyme attaches to the first exposed base in DNA while phosphorylated nucleotides are available in the nucleoplasm.
3. The phosphorylated nucleotides are paired with appropriate bases in DNA and bound together using energy from attached phosphates.
4. The process continues until a complete molecule of mRNA is produced.

Translation

5. Messenger RNA moves from the nucleus to the cytosol and one end binds to the smaller unit of a ribosome; ribosomal units combine.
6. First and then second tRNAs carrying their respective amino acids combine with mRNA by base pairing of codons and anticodons.
7. The first two amino acids are linked by a peptide bond; energy from hydrolysis of ATP is used to form the bond.
8. The third tRNA combines with mRNA and the first tRNA is released.
9. Another peptide bond is formed.
10. Amino acids are carried to their proper positions and peptide bonds are formed until an entire polypeptide is synthesized.
11. When the ribosome recognizes a terminator codon, the polypeptide is released from the ribosome.

amounts of the right proteins to be made at the right times for optimal cell function, and, thus, for whole organism function.

The rate of synthesis of any particular protein has been thought to be influenced mainly by how fast the cell synthesized the proper mRNA. Recent evidence suggests that how fast the cell degrades the mRNA may be more important. It is already known that the mRNAs for a cell's most abundant proteins are stable long-lasting molecules, whereas the mRNAs for scarce proteins are unstable and short-lived.

CLINICAL TERMS

anaplasia (an″ah-pla′ze-ah)   reversion of cells to an undifferentiated form

dialysis (di-al′ĭ-sis)   separation of molecules of different sizes by causing the smaller ones to diffuse through a selectively permeable membrane as occurs in kidney dialysis

dysplasia (dis-pla′se-ah)   an alteration in the size, shape, and arrangement of cells

hyperplasia (hi″per-pla′ze-ah)   an abnormal increase in the number of cells in a tissue

hypertrophy (hi-per′tro-fe)   an abnormal increase in the size of cells in a tissue

metaplasia (met″ah-pla′ze-ah)   a change of one kind of cells into another kind of cells

neoplasm (ne′o-plazm)   abnormal growth of cells forming a tumor, which can be malignant

ESSAY
CANCER

Throughout the world, more than 2 million people die of cancer each year. In the United States, the death rate is about 1.8 deaths per thousand people per year, second only to the death rate of 4.2 per thousand from all circulatory diseases. Cancers of three organs account for nearly half of all cancer deaths—24 percent lung cancer, 13 percent large bowel cancer, and 9 percent breast cancer. Death rates do not tell the whole story; many deaths are preceded by long, painful, and expensive illnesses.

During growth or repair of normal tissues, cell division proceeds in an orderly fashion. Regulatory factors stop cell division when the organs reach normal size or when repair is complete. A factor called *contact inhibition* halts cell division when cells come in contact. In laboratory cultures, normal cells stop dividing when they have formed a single-layered sheet over a piece of glass, but cancer cells, which lack contact inhibition, continue to divide and pile on top of one another.

Cancer, or **malignancy** (mah-lig'nan-se), is characterized by invasive growth in addition to uncontrolled cell division (Figure 3.27). Excessive cell division can produce a nonmalignant, or benign, encapsulated tumor, or it can produce a malignant growth. Benign tumors cause mainly mechanical damage, but malignant growths **metastasize** (me-tas'tah-sīz), or invade, adjacent tissues. The word cancer means crab; like the claws of a crab, malignant cells spread out in all directions. They often leave their original site, travel in blood or lymph, and invade distant tissues until the body has many sites of malignant growth. A sequence of changes occur in cells as they go from normal to malignant. They induce other tissues to secrete various stimulating factors to which they respond, and they stimulate blood vessels to grow into the tumor. A diagnostic test for bladder cancer detects one such stimulating factor in urine. Malignant cells also acquire enzymes that digest proteins and other substances in adjacent tissues.

The causes of cancer are not fully understood, but many are caused by **carcinogens** (kar-sin'o-jenz), or cancer-causing substances, such as tars in cigarette smoke, asbestos, nickel, certain dyes, some food additives, air pollutants, alcohol, and even some substances naturally present in foods. Some studies suggest that tobacco causes 30 percent of cancer deaths and that food substances may cause as many as 35 percent of cancer deaths. Food carcinogens may generate oxidizing agents that make cells malignant, but antioxidants in other foods may counteract their effects. To minimize carcinogens, people should eat fewer calories, less fat, and less smoked, salted, or nitrate-preserved foods. To maximize anticancer substances, they should eat more foods rich in fiber or vitamin A and vegetables from the cabbage family, which appear to contain substances that counteract carcinogens.

Cancer can be caused by ultraviolet rays, X rays, oncogenic viruses (viruses that contribute to the development of cancer), and oncogenes (genetic information that contributes to the development of cancer). Caucasians who

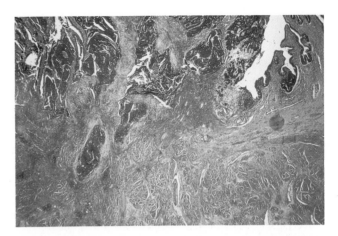

FIGURE 3.27
A photomicrograph showing the characteristics of cancer cells. These cells are from a carcinoma of the bladder.

FIGURE 3.28
Skin cancers are common among Caucasians who spend large amounts of time in the sun.

live in very sunny climates (where they are exposed to excessive ultraviolet radiation) have a high incidence of skin cancers (Figure 3.28). Leukemia (a cancer of the white blood cells) occurs ten times as often among radiologists (who are exposed to X rays) as among other physicians. Japanese who were exposed to nuclear bomb radiation in World War II show five times the incidence of leukemia as the rest of the population. Oncogenic viruses induce cancer in cells they infect by altering host cell DNA so the cells divide uncontrollably. Oncogenes, which are normally repressed, can be expressed and cause cancer if cellular conditions change.

Determining how cancer arises is difficult because of our limited understanding of how normal cell division is regulated, but two theories—mutation and selective gene action—are worthy of mention here. According to the mutation theory, cancer is caused by mutations, or changes in the DNA of a cell. Agents known to cause mutations—viruses, radiation, and certain chemical substances—also cause cancer. The increased incidence of cancer with age may be due to the cumulative effects of several mutations over many years. The mutation theory fails to explain occasional cases of spontaneous **remission,** disappearance of evidence of disease. The selective gene activation theory proposes that certain normally repressed genes become active and lead to uncontrolled cell division. By this theory, a remission might occur if the genes were again repressed.

Cancer cells probably arise far more often than they cause disease. Because cancer cells have receptors that the body's immune system recognizes as foreign, they may be destroyed before they become invasive. Only cells that divide too fast or evade detection by masking their receptors cause malignancy.

Many cancers might be prevented by avoiding carcinogens such as tobacco smoke and other pollutants. Deaths can be prevented by early diagnosis and treatment. For example, liver cancer can be diagnosed early by finding an increase in alpha-fetoprotein in a patient's blood. This substance is associated with rapid cell division and is normally present in the blood of a fetus.

Cancer can be treated with surgery, radiation, and chemotherapy. Surgery is used when malignant cells are localized, and radiation is used when it can be selectively applied to malignant cells. Chemotherapy, the use of drugs that interfere with DNA synthesis or cell division, works when cancer cells are widely dispersed. Chemotherapy damages all dividing cells, but has a greater effect on rapidly dividing cells.

A serious problem in treating cancer with drugs is that cancer cells, like infectious agents, develop drug resistance. In one kind of resistance, a drug kills most cancer cells and causes regression of a tumor. A few cells resist the drug and survive. The tumor grows as these drug-resistant cells divide. Such resistance can be counteracted by using drug combinations so that cells resistant to one drug will be killed by another.

Resistance to drug combinations occurs when tumor cells are genetically resistant to nearly all drugs. Such cells probably produce proteins that prevent drugs from entering cells or prevent them from damaging the cells if they do gain entrance.

Experiments are underway to see if chemotherapy can be combined with the administration of growth factors, which are proteins that normally regulate the rate of cell divison in some tissues. One such factor stimulates production of white blood cells, which fight infection. Because chemotherapeutic agents typically destroy white blood cells, being able to stimulate their replacement would make it possible to give higher doses of drugs intended to kill cancer cells without killing all of the white blood cell producers, too.

A novel treatment involving both chemotherapy and radiation begins with administering a synthetic molecule, similar to hemoglobin, containing several atoms of the element boron. This drug enters cancer cells to a far greater extent than normal cells. After receiving the boron drug, the patient is irradiated with slow neutrons, a kind of deep penetrating, low-energy radiation that splits boron atoms, causing malignant cells to explode without damaging normal cells.

Immunotherapy is beginning to be used to cause the immune system to destroy a particular patient's cancer cells. Vaccines against common kinds of cancer cells may someday be available.

Questions

1. How might loss of contact inhibition lead to cancer?
2. Distinguish between metastasis and remission.
3. What are some common environmental carcinogens and how might you avoid exposure to them?
4. How might the body prevent a malignant cell from dividing?
5. What are some advantages and disadvantages of different kinds of cancer therapy?

CHAPTER SUMMARY

Organization and General Functions

- Cells are the basic functional unit of living organisms; the human body contains many different kinds of cells, each having a particular function.
- Cells typically have organelles suspended in cytoplasm surrounded by a cell membrane.

Cell Membranes

- The plasma (cell) membrane consists of a lipid bilayer with protein molecules interspersed as integral or peripheral parts of the membrane.
- The cell membrane regulates the passage of substances into and out of the cell, and also receives and transmits information to the inside of the cell.

Passage of Materials Across Membranes

- Materials move in and out of cells passively by simple diffusion, facilitated diffusion, osmosis, filtration, and bulk flow.
- Passage of materials occurs actively by active transport, exocytosis, and endocytosis.
- Passive and active processes are summarized in Table 3.1.

Cytoplasm

- Cytoplasm includes organelles and cytosol, a liquid that contains many dissolved substances.

Organelles

- Organelles are structures within cells that carry out specific functions as summarized in Table 3.2.

Cell Division

- Mitotic cell division involves DNA replication, mitosis (nuclear division), and cytokinesis (division of the cytoplasm).
- These processes occur in a cell cycle.

Protein Synthesis

- DNA in the nucleus provides information for protein synthesis; mRNA relays information to ribosomes to which amino acids are transported by tRNA.
- Information is transferred from DNA to mRNA by transcription (Figure 3.22).
- Proteins are assembled according to information in mRNA by translation (Figure 3.26).
- Some proteins are modified after synthesis by posttranslational processing.
- Protein synthesis is controlled by histones, steroid hormones, and other factors.

QUESTIONS AND PROBLEMS

The questions at the end of each chapter are numbered to correspond with the objectives listed at the beginning of the chapter. Italics indicate that a question requires critical thinking skills beyond simple factual recall.

Questions

1. (a) What is the cell theory?
   (b) *What is meant by the statement that many body functions take place at the cellular level?*
2. (a) Describe the structure of a cell membrane as proposed by the fluid-mosaic model.
   (b) *What would be the effect on cell membrane function if any of the following components were absent or nonfunctioning: phospholipids, proteins, glycoproteins, and glycolipids?*
3. *What properties distinguish active and passive transport processes?*
4. (a) How do each of the following occur: simple diffusion, facilitated diffusion, and osmosis?
   (b) *What is the significance of each of these processes?*
   (c) Why are these processes said to be passive?
5. Compare filtration and bulk flow.
6. (a) Describe the mechanism of active transport.
   (b) Why is it said to be an active process?
   (c) *What is the significance of active transport?*
7. (a) How are exocytosis and endocytosis similar, and how are they different?
   (b) *What is the significance of each of these processes?*
   (c) Why are these processes said to be active?
8. How does cytoplasm differ from cytosol?
9. (a) List the organelles and other internal components of cells and give the function of each.
   (b) *What deficiencies in function would a cell lacking one of the following display: nucleus, mitochondria, ribosomes, Golgi apparatus, and cytoskeleton?*
10. *How are cellular functions integrated?*
11. (a) Summarize the steps in mitotic cell division.
    (b) *What is the significance of cell division?*
12. (a) What cell components are involved in protein synthesis?
    (b) *Distinguish between transcription and translation.*
    (c) Summarize the process of protein synthesis from DNA to the finished protein.
    (d) *Determine at least one possible sequence of bases in DNA, mRNA, and tRNA that would produce the following protein segment: alanine, glutamic acid, serine, valine, arginine, leucine, tryptophan.*
    (e) *For the DNA in question (d) above, (1) cite a point mutation that would not affect the protein structure, (2) cite a point mutation that would affect the protein, and (3) explain the effects of adding or deleting a base at the second base in the DNA nucleotide sequence.*
    (f) What happens in posttranslational processing?
    (g) *How is protein synthesis controlled?*

Problems

1. Calculate the osmolarity in osmols of (a) a solution of 1 gram molecular weight/liter of glucose and (b) a solution of 1 gram molecular weight/liter of sodium chloride.
2. Do some library research on cell membranes or organelles to find out what recent advances have been made.
3. Find out what new methods are being used to diagnose and treat cancer.

OBJECTIVES

1. Define the term *tissue,* and list the four categories of tissues and their major functions.

2. Describe how the three germ layers form from a fertilized egg, and name the tissues derived from each layer.

3. Identify the characteristics of epithelial tissues, and explain how these characteristics are related to the location and function of the tissues.

4. Distinguish between mucous and serous membranes by structure, function, and location.

5. Distinguish between exocrine and endocrine glands, and describe the structure, function, and location of the various exocrine glands.

6. List four classes of connective tissues, and describe their major characteristics.

7. Distinguish among the types of connective tissue proper by structure, function, and location.

8. Distinguish among the types of cartilage by structure, function, and location.

9. Distinguish between compact and cancellous bone by structure, function, and location.

10. Distinguish among the three types of muscle tissue by structure, function, and location.

11. Distinguish between neurons and neuroglia by structure, function, and location.

4

TISSUES

DEFINITION AND CLASSIFICATION OF TISSUES

As explained in Chapter 3, the cell is the basic unit of structure and function in living things. Cells of multicellular organisms such as humans are specialized in function and they do not work independently.

Imagine that the body is analogous to a large, modern hospital. The hospital provides many different services, but no single employee provides all of them. For example, as a member of a surgical team, you would be responsible for certain tasks. You would not be expected to provide meals for patients or sweep the halls. Similarly, each cell in an organism carries out particular functions, though the cell might be capable of other functions. To carry the above analogy further, you would be one of a group. Each group would have certain duties, as do nurses, laboratory technicians, surgeons, housekeepers, and kitchen workers. Likewise, organisms have groups of cells with similar structure and function.

Groups of cells that have the same structural characteristics and perform the same functions are called **tissues.** In this chapter, we will consider the structure and location of different kinds of tissues and see how their structures and functions are related. The many kinds of tissue are grouped into four general categories: (1) epithelial tissue, (2) connective tissue, (3) muscle tissue, and (4) nervous tissue.

Epithelial (ep''i-the'le-al) **tissues** cover surfaces, line cavities, and form the major portion of many glands. They protect the surfaces they cover against abrasion and harmful agents. Some epithelial tissues make and release secretions and others allow for the absorption of nutrients.

Connective tissues are a diverse group of tissues with cells imbedded in a matrix. The **matrix** is a network of fibers in which **ground substance** is deposited. The composition of ground substance varies by tissue, being hard and inflexible in bone; solid, but somewhat flexible in cartilage; and soft and jellylike in adipose (fat) tissue and some other connective tissues. Blood is a connective tissue lacking fibers and having a liquid ground substance. Most connective tissues support and protect the body.

Muscle tissues, which exert force as they contract, help to produce many kinds of movement. They move food through the digestive tract, pump blood, and move body parts or the whole body. Muscle tissue functions are explained in Chapter 9.

Nervous tissues respond to stimuli, relay signals, and coordinate conscious and unconscious activities. A **signal** is an event that conveys information. The functions of nervous tissues are explained in Chapter 11.

Structure and function are complementary in all tissues. Epithelial tissues usually exist in thin layers. They carry out functions such as absorption, protection, and secretion. The matrix of connective tissues is adapted for

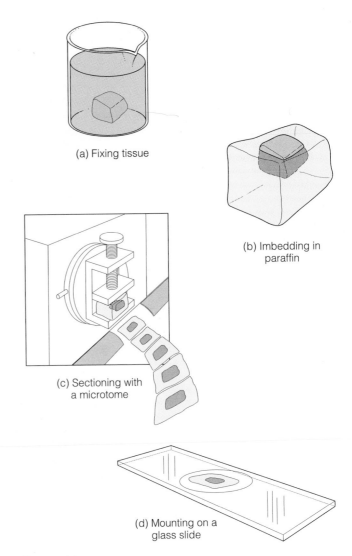

(a) Fixing tissue

(b) Imbedding in paraffin

(c) Sectioning with a microtome

(d) Mounting on a glass slide

FIGURE 4.1
Preparation of a tissue sample for microscopic study.

supporting other structures, holding tissues together, and storing fat. Muscle tissues are structured for exerting tension, and nerve tissues are arranged for responding to stimuli and relaying signals.

The study of tissues is called **histology** (his-tol'o-je). Most of the photomicrographs in this chapter represent tissues that were prepared according to established histological procedures designed by **histologists,** scientists who study tissues. After a tissue is dissected from an animal, it is fixed (preserved) in formaldehyde or some other fixative, imbedded in paraffin, sectioned with a microtome into slices 5 to 8 μm in thickness, mounted on a glass slide, and stained with one or more dyes for microscopic study (Figure 4.1). Tissue is prepared by a similar set of procedures for study with an electron microscope.

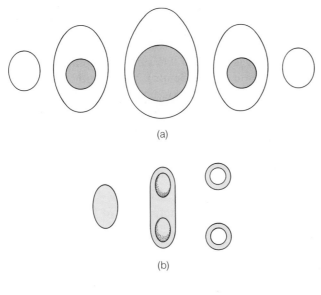

FIGURE 4.2
See if you can determine what everyday objects (large enough to be visible without magnification) these sections represent. After you think you know, refer to Figure 4.3.

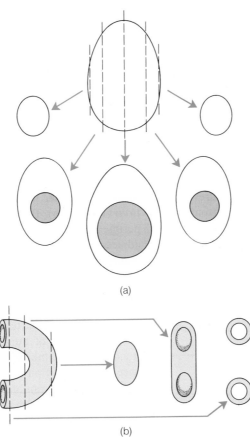

FIGURE 4.3
These objects are (*a*) a hard-boiled egg and (*b*) a piece of elbow macaroni.

Sometimes a pathologist must quickly examine a tissue for cancer cells, as when lymph nodes from a patient undergoing surgery for breast cancer arrive in the laboratory. Such tissues are frozen in liquid nitrogen and sectioned inside a **cryostat** (kri′o-stat), a refrigerated cabinet that maintains both the tissue and the microtome below freezing.

To interpret what we see in tissue sections, we must mentally reconstruct three-dimensional objects from sections through them. (Try your skill by studying Figure 4.2 before you look at Figure 4.3.)

DEVELOPMENT OF TISSUES

All cells, tissues, organs, and systems of the body develop from a single cell—the fertilized egg. Within the nucleus of that egg is all the information necessary to initiate growth and development of all body parts.

The fertilized egg (Figure 4.4) divides by mitosis, rapidly becoming 2, 4, 8, 16, and more cells in a solid ball called a **morula** (mor′u-lah). Within a week, some cells rearrange to form a hollow ball, the **blastocyst** (blas′to-sist), which has an outer cell layer (destined to become part of the placenta) and an inner cell mass. During the second week of development, the inner cell mass forms a plate of cells called the **embryonic disk,** which gives rise to the embryo itself. Cells above the embryonic disk form the **amniotic** (am″ne-ot′ik) **cavity,** and those below form the **gut cavity.**

Embryonic disk cells form two **primary germ layers.** The layer nearest the amniotic cavity is **ectoderm** (ek′to-derm) and the one nearest the gut cavity is **endoderm** (en′do-derm) (*ecto,* outer; *endo,* inner). During the third week of development, the embryonic disk differentiates into a head and a tail end. Ectodermal cells near the tail proliferate and grow forward between the two existing layers forming the middle **mesoderm** (mes′o-derm), the third primary germ layer.

So far, we have seen the early steps in cell **differentiation.** As embryonic cells divide, they become different from one another largely because different genes in their chromosomes exert specific effects. Each primary germ layer develops into tissues with specialized structures and functions (Table 4.1). Nervous tissue is derived from ectoderm, and connective and muscle tissues are derived from mesoderm. In contrast, some epithelial tissues arise from each germ layer. We will see how tissues combine to form organ-systems as we study those systems.

TABLE 4.1 Tissues and Organs Derived from Each of the Primary Germ Layers

Ectoderm	Mesoderm	Endoderm
Epithelium of skin and its derivatives	Bones, cartilage, and other connective tissues, including blood	Epithelium of digestive and respiratory systems, except mouth and anus
Nervous system	All kinds of muscle	
Pituitary gland		
Adrenal medulla	Organs of urinary and reproductive systems, except linings of cavities	Epithelium of thyroid, parathyroid, thymus, liver, and pancreas
Epithelial parts of most sense organs		
Lining of mouth and anus	Epithelial coverings of organs, and linings of body cavities and blood vessels	Epithelial lining of part of bladder and urethra
Pineal gland		
	Adrenal cortex	

EPITHELIAL TISSUES

Of all body tissues, epithelial tissue is easiest to observe. The skin surface is epithelial tissue, or **epithelium** (ep″ĭ-the′le-um). Your mouth, nose, ears, and other body openings are lined with epithelium. Less apparent are the internal epithelial tissues. Many organs of the digestive, respiratory, urinary, and reproductive systems are lined with epithelium and all blood vessels are lined with it. Finally, most tissue of glands contains epithelial cells that secrete the products of the glands.

Characteristics of Epithelial Tissues

Regardless of their origin, epithelial tissues have several common characteristics. First, their cells divide by mitosis, but in multilayered epithelial tissues like the outer part of the skin, only cells in the bottom layer divide. Older cells move toward the surface and change shape.

Second, epithelial tissues attach to underlying connective tissue by a noncellular secreted **basement membrane** consisting of glycoprotein from epithelial cells and a meshwork of collagen fibers from connective tissue.

Third, epithelial tissues lack blood vessels. They obtain nutrients by diffusion from underlying connective tissues, which are generally well supplied with blood vessels. In addition to cell membranes, the nutrients must diffuse through the basement membrane and, in multilayered tissue, through several cells, too.

Finally, the cells of epithelial tissues are tightly packed and held together by *tight junctions* and *adhering junctions* (Figure 4.5). Tight junctions extend around the entire perimeter of a cell making a very tight seal between it and adjacent cells. This seal is created by fusion of the cell membranes of adjacent cells as their membrane protein interlock like the teeth of a zipper. Tight junctions are found in the intestinal epithelial lining. Adhering junctions can be zonular (extend around the entire cell perimeter) or macular (scattered about the membrane). These junctions consist of bundles of immature keratin fibers (tonofilaments) imbedded in glycoprotein deposits between the cells. Epithelial cells in the epidermis (outer layer) of the skin can have all of the above kinds of junctions in a sequence with the tight junction nearest the surface, the zonular adhering junction next, and macular adhering junction nearest the base of the cell. Because of their arrangement zonular adhering junctions are called **intermediate junctions** and macular adhering junctions are called **desmosomes** (des′mo-sōmz) or **spot desmosomes.** As we shall see in Chapter 5, adhering junctions between adjacent cells account for the spiny appearance of epidermal cells in fixed tissue.

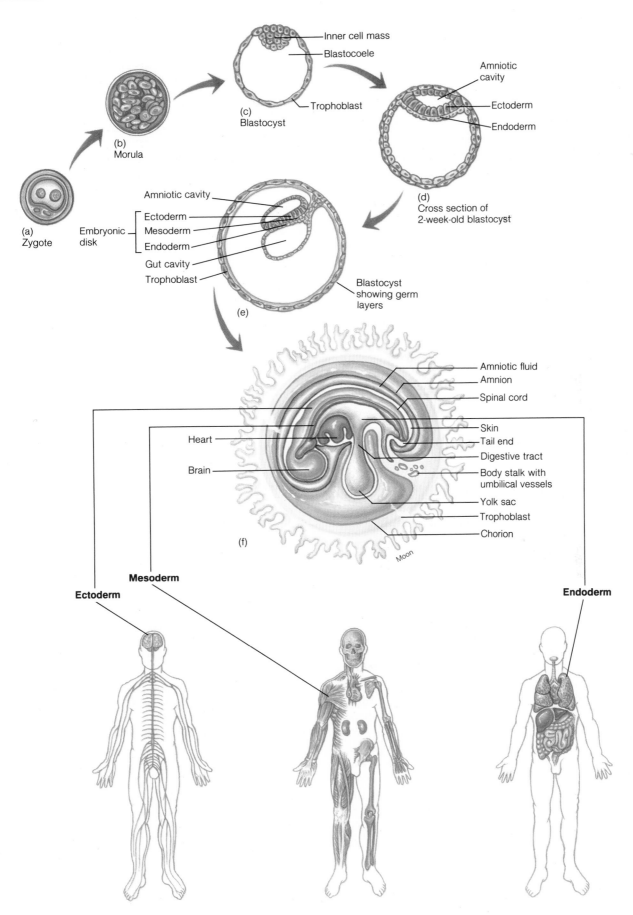

FIGURE 4.4

A diagram of the early stages of development of a human embryo.

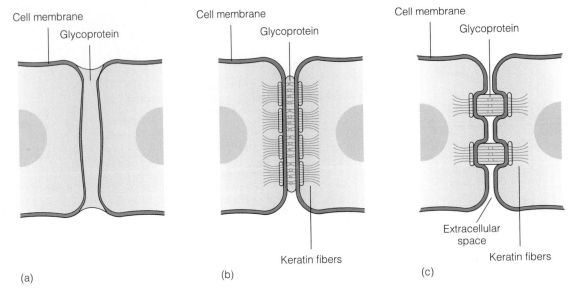

FIGURE 4.5
How epithelial cells are held together: (*a*) tight junction; (*b*) zonular adhering, or intermediate, junction; and (*c*) macular adhering junction, or desmosome.

Another connection between cells called a **gap junction** is found in smooth muscle and in many kinds of embryonic cells. It consists of tubular passageways or channels that allow direct communication between the cytoplasm of neighboring cells. Such gap junctions are large enough to allow passage of ions and small molecules such as amino acids, sugars, steroids, and nucleotides. Gap junctions can be opened by the messenger molecule cAMP and closed by calcium ions entering a cell.

Structure, Location, and Function

The cells of epithelial tissues come in three basic shapes— flat or **squamous** (skwa'mus), cube-shaped or **cuboidal** (ku-boi'dal), and cylindrical or **columnar**—usually tightly packed in sheets. Sheets with one layer of cells are **simple epithelium** and those with two or more layers are **stratified epithelium.** Stratified epithelium is classified by the shape of outer layer cells. **Pseudostratifed epithelium** (*pseudo*, false) appears to have more than one layer. On closer examination, it turns out to be a single sheet of cells of different heights with all cells touching the basement membrane. **Transitional epithelium** has cells that can change shape in response to mechanical stretching. The locations and functions of each class of epithelial tissues are given in Table 4.2, and structural characteristics are summarized in Figure 4.6.

TABLE 4.2 Structure, Locations, and Functions of Epithelial Tissues

Structure	Locations	Functions
Simple squamous	Lining of blood vessels, lining of body cavities, some parts of kidney tubules	Protection, absorption
Simple cuboidal	Secretory portion and ducts of some glands, part of kidney tubules	Secretion, protection
Simple columnar	Lining of gastrointestinal tract, excretory ducts of some glands	Absorption, protection, secretion
Pseudostratified columnar	Lining of trachea, upper respiratory tract, and parts of male reproductive system	Protection, secretion
Stratified squamous	Epidermis, lining of mouth, esophagus, and vagina	Protection, secretion, limited absorption
Stratified cuboidal	Ducts of sweat glands (a rare tissue)	Protection
Stratified columnar	Part of lining of male urethra (a rare tissue)	Protection
Transitional	Lining of ureter and urinary bladder	Protection

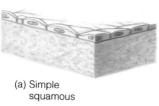

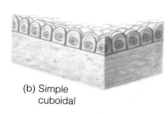

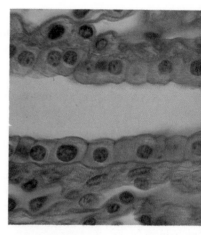

(a) Simple
squamous

(b) Simple
cuboidal

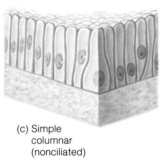

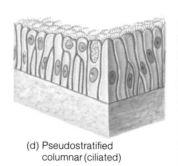

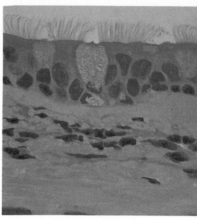

(c) Simple
columnar
(nonciliated)

(d) Pseudostratified
columnar (ciliated)

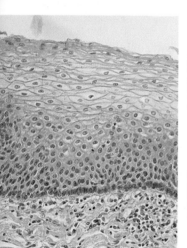

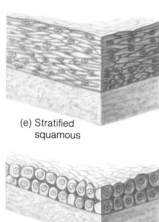

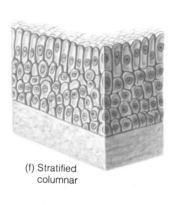

(e) Stratified
squamous

(f) Stratified
columnar

(g) Stratified
cuboidal

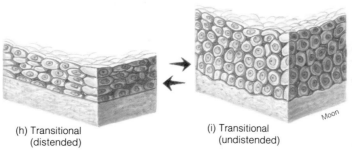

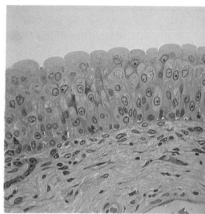

(h) Transitional
(distended)

(i) Transitional
(undistended)

Moon

FIGURE 4.6
Classification of epithelial tissues.

Simple Epithelia

Simple squamous epithelium is a single layer of thin, flat, many sided cells with central nuclei bulging from the surface. This widely distributed tissue protects underlying tissues. It lines parts of the urinary, respiratory, and male reproductive systems. As **mesothelium** (mes″o-the′le-um), it lines the abdominal, pleural, and pericardial cavities and covers organ surfaces within these cavities. As **endothelium** (en″do-the′le-um), it lines the heart, and all blood and lymph vessels. Many substances pass through endothelium in small blood and lymph vessels.

Simple cuboidal epithelium is a single layer of cube-shaped cells with central nuclei. It is found in parts of the kidney, and in the ducts and secretory portions of many glands. In sheets, it protects underlying tissues, and in glands, it makes and releases secretions.

Simple columnar epithelium is a single layer of tall, cylindrical cells with their nuclei near the cell base. This tissue, which lines the gastrointestinal tract from the stomach to the anus, protects and allows absorption of nutrients. Cells that contain **secretory granules** secrete a protective lubricating fluid called **mucus** (mu′kus).

Pseudostratified Columnar Epithelium

A tissue with cells of different heights all resting on the basement membrane and nuclei at different levels within a single layer is called **pseudostratified columnar epithelium.** This tissue contains **goblet cells,** specialized, unicellular glands that secrete mucus. Active cells have a large cup-shaped secretory vessel above a thin cell base, giving them the appearance of a goblet. The tissue lines and protects the upper respiratory tract and much of the reproductive tracts. Its mucous secretions lubricate these passageways. In the upper respiratory tract and uterine tubes, the epithelium is ciliated. In the respiratory tract, the cilia help to push debris coated with mucus out of the lungs or nasal cavity toward the pharynx, where it can be ejected or swallowed. In the uterine tubes, the cilia help to move an ovum (egg) toward the uterus.

Stratified Epithelia

Stratified squamous epithelium contains many layers of cells, but the outer layer always consists of squamous cells. The **epidermis** (ep″ĭ-der′mis), or outer portion of the skin, is stratified squamous epithelium (Chapter 5) as are the linings of the vagina and esophagus. The primary function of this tissue is protection, though the epidermis can absorb some oily substances and contains glands that produce various secretions.

Stratified epithelia other than the squamous type are rare in the human body. Such tissues have at least two layers, and they protect surfaces. **Stratified cuboidal epithelium** has cube-shaped surface cells and is found in the ducts of the larger sweat glands. **Stratified columnar epithelium** has cylindrical surface cells and is found in parts of the male urethra and on the covering of the epiglottis, which keeps food out of the respiratory tract.

Transitional Epithelium

Some epithelia do not maintain a constant shape. **Transitional epithelium** has several layers of cells that can change shape when the tissue is stretched. Found in the ureters and urinary bladder, the cells of this tissue become thin and flat as the bladder fills and tension increases, and thick and rounded when the bladder is emptied and tension decreases.

Membranes

At several sites in the body, a sheet of epithelium and its underlying connective tissue form serous or mucous membranes. **Serous** (se′rus) **membranes** line cavities and cover the surface of organs in the abdominopelvic, pleural, and pericardial cavities. Serous membranes (Figure 4.6) secrete a clear, watery fluid similar to interstitial fluid. This fluid lubricates membranes, allowing organs to move with little friction as the heart beats, the lungs draw air in and out, and the digestive tract propels food.

Mucous (mu′kus) **membranes** line the internal surfaces of passageways that lead to the outside of the body and are found in the digestive, respiratory, urinary, and reproductive systems. All mucous-secreting epithelial tissues along with their underlying connective tissue are mucous membranes. These membranes protect the organs they line; their mucous secretions moisten and lubricate the membrane.

Glands

In addition to forming sheets, epithelial tissues also contribute to the formation of glands, structures that produce secretions. Glands can be unicellular or multicellular. Unicellular glands are modified epithelial cells. Multicellular glands are more complex and develop from pouches of epithelium that grow inward from the surface. Glands that lose contact with surface epithelium become ductless, or **endocrine** (en′do-krin), glands, but those that retain contact become ducted, or **exocrine** (ex′o-krin), glands.

Endocrine glands (Chapter 16), such as the pituitary and thyroid, release secretions called hormones into interstitial fluids. These secretions diffuse into blood vessels, through which they are transported to other parts of the body.

Unicellular glands, though ductless, are exocrine because they release secretions into the lumen (inner cavity) of an organ. Goblet cells are unicellular glands.

Multicellular exocrine glands (Figure 4.7 and Table 4.3) are classified in four ways—by the number of ducts, the shape of secretory portions, the nature of secretions, and the manner in which secretions are released. Glands are **simple** if they have a single duct and **compound** if they have branched ducts. Glands are **tubular** if their secretory portions consist of a tube or several branched or coiled tubes. They are **alveolar** (al-ve′o-lar), or **acinar** (as′in-ar), if their secretory portions consist of one small alveolus (sac)

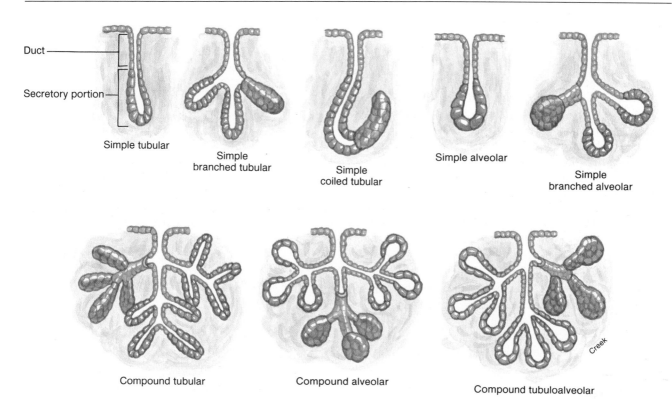

Duct

Secretory portion

Simple tubular

Simple
branched tubular

Simple
coiled tubular

Simple alveolar

Simple
branched alveolar

Compound tubular

Compound alveolar

Compound tubuloalveolar

FIGURE 4.7
Types of multicellular exocrine glands.

TABLE 4.3   Structure, Location, and Secretions of Exocrine Glands

Example of Gland	Structure	Location	Manner of Secretion	Nature of Secretion
Goblet cell	Unicellular	Mucous membranes	Merocrine	Mucus
Intestinal	Simple tubular	Small intestine	Merocrine	Enzymes
Sweat glands	Simple coiled tubular	Skin	Merocrine	Sweat
Sebaceous glands	Simple alveolar or simple branched alveolar	Skin	Holocrine	Oil and wax
Liver	Compound tubular, but very complex	Abdominal cavity	Merocrine	Bile
Pancreas	Compound alveolar	Abdominal cavity	Merocrine	Enzymes
Mammary glands	Compound alveolar	Chest area	Merocrine	Milk
Salivary glands	Compound alveolar and tubuloalveolar	Jaw area	Merocrine	Saliva

or clusters of alveoli. They are **tubuloalveolar** if they have both tubular and alveolar secretory components.

Secretions can be mucous, serous, or mixed. Mucous secretions contain the viscous glycoprotein mucus. Serous secretions are watery such as sweat and milk.

Finally, exocrine glands are classified according to how they produce secretions. The cells of **merocrine** (mer'o-krīn) glands synthesize their secretions and discharge them by exocytosis. The cells of **holocrine** (ho'lo-krīn) glands accumulate secretory products, and eventually, the whole cell and its products are released from the tissue.

When the only information about secretory cells came from studies with light microscopes, histologists (*histo,* tissue) identified a class of glands they called apocrine. They thought they saw the apical portions of cells breaking off and being released with the secretions. Evidence from studies with electron microscopes shows that what were thought to be apocrine glands are, in fact, merocrine. A few histologists still contend that mammary gland secretory cells lose tiny membrane fragments and cytoplasmic lipid droplets directly rather than by exocytosis.

Connective Tissues

Connective tissue is aptly named—in general, it connects body parts together. It forms bones and holds them together. It binds muscles to bones and skin to muscles. A portion of embryonic mesoderm becomes **mesenchyme** (mes'eng-kīm), which gives rise to all adult connective tissues. Connective tissues consist of cells surrounded by an extracellular fibrous matrix filled with ground substance of varying consistency.

Characteristics and Classes of Connective Tissue

The four general classes of connective tissue based on characteristics of the matrix and ground substance are connective tissue proper, cartilage, bone, and blood. **Connective tissue proper** contains various types of fibers and has a semifluid ground substance. **Cartilage** has a fibrous matrix with fairly solid ground substance, and **bone** has a fibrous matrix with very solid ground substance. Blood with its fluid ground substance will be discussed in Chapter 17.

Connective Tissue Proper

Connective tissue proper includes five kinds of tissue (Table 4.4), each with its own types and arrangements of fibers:

1. **Loose,** or **areolar, connective tissue** has loosely packed, irregularly arranged fibers.
2. **Dense connective tissue** has more abundant fibers, often arranged in bundles.
3. **Reticular connective tissue** has fibers in a reticulum (network) that supports other structures in an organ.
4. **Adipose** (ad'ip-ōs) **tissue** has few fibers and many cells filled with fat.
5. **Pigmented connective tissue** lacks fibers and contains pigment granules.

Loose connective tissue (Figure 4.8) is found between the skin and skeletal muscles, and beneath most epithelial linings of organs. This abundant tissue contains a variety of cell types, cell products, and fibers. The most numerous cells are **fibroblasts** (fi'bro-blastz), which synthesize and secrete the protein fibers of connective tissue. They also synthesize and secrete the glycoproteins found in ground substance. *Hyaluronic* (hi''al-ur-on'ik) *acid,* the predominant glycoprotein, gives ground substance a jellylike consistency. The enzyme hyaluronidase, produced by some

TABLE 4.4	Characteristics of Connective Tissue Proper	
Tissue Type	**Characteristics**	**Location**
Loose (areolar)	Fibroblasts and many other kinds of cells; all three kinds of fibers in a loose network	Between skin and muscles, and beneath most epithelial linings
Dense		
Irregular	Feltwork of mostly collagen fibers; fibroblasts are the predominant cell	Dermis of skin and membranes that cover bone and cartilage
Regular	Large bundles of mostly collagen fibers; few cells, mostly fibroblasts; yellow ligaments contain elastic fibers	Fascia, tendons, and ligaments
Reticular	Network of reticular fibers	Framework of liver, spleen, and lymph nodes
Adipose	Large numbers of fat cells with large fat vacuoles; only a few fibers of each type; an actively metabolic tissue	Scattered through loose connective tissue and as a distinct tissue under skin; around heart, kidneys, and joints; and in bone marrow
Pigmented	Irregularly shaped cells that contain pigment granules	A rare tissue found only in eyes (iris, ciliary body, and choroid)

microorganisms, digests hyaluronic acid and makes it easier for the microorganisms to invade tissues.

Other cell types in loose connective tissue include tissue macrophages, mast cells, lymphocytes, and other white blood cells. **Tissue macrophages** (mak'ro-fāj''ez), or **histiocytes** (his'te-o-sītz''), are phagocytic cells found mainly in connective tissues. **Mast cells** release heparin and histamine. Though heparin from most sources prevents blood clotting, heparin from mast cells has little effect on clotting and its significance is unknown. Histamine, released at the time of cell injury, is important in initiating the inflammatory response. (See the essay at the end of this chapter.) **Lymphocytes** (lim'fo-sītz), which migrate from blood into loose connective tissue following injury, are involved in immunity as explained in Chapter 21. Other white blood cells that enter connective tissue phagocytize foreign substances that have entered the tissue.

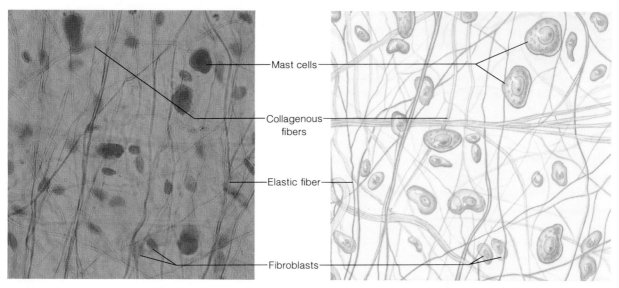

FIGURE 4.8
Cells and fibers of loose connective tissue.

Mast cells closely resemble white blood cells called basophils. Histologists once thought that mast cells were basophils that had migrated from the blood to loose connective tissue. It may be that mast cells and basophils are derived from the same precursor cell. Finding intermediate cells in a kind of chronic leukemia lends support to this idea.

Human mast cells lack serotonin, but those of rats and mice contain it. Though serotonin relays neural signals in the brain, its role in rat and mouse mast cells is not known.

Collagen can be made tougher or more tender. When the hair and epithelial cells are removed from animal hides, a thick sheet of connective tissue consisting mainly of collagen remains. Treating this collagen with tanning agents hardens it into leather. Conversely, the toughest meat can be tenderized by prolonged boiling, which adds water to tough collagen, turning it into tender gelatin. Gelatin, the protein in gelatin desserts, is an easily digestible dietary protein. Because it lacks the essential amino acid tryptophan and contains only a small amount of tyrosine, gelatin must be supplemented with other proteins for adequate nutrition.

Loose connective tissue and most other kinds of connective tissue contain collagen, elastic, and reticular fibers. **Collagen fibers,** the strongest fibers, consist of parallel, coiled molecules of the whitish protein **collagen** (kol'ah-jen). The arrangement of collagen molecules in overlapping strands gives collagen fibers their high tensile strength and flexibility. Collagen fibers do not branch and they do not stretch. **Elastic fibers** consist of thinner molecules of a fibrous, yellow protein **elastin** (e-las'tin). Elastic fibers branch to form a network capable of stretching and recoiling to its original shape. These fibers make connective tissues resilient—an important property in the walls of arteries and in the skin beneath the epidermis. **Reticular fibers** are thin, delicate collagen fibers coated with glycoprotein. They support capillaries, and nerve and muscle fibers.

Dense connective tissue contains closely packed fibers regularly or irregularly arranged. In **dense irregular connective tissue,** which contains many fibroblasts and all types of fibers, large numbers of collagen fibers are arranged irregularly in a tough feltlike mat. This tissue is found in the dermis of the skin and in membranes that cover bones and cartilage. **Dense regular connective tissue** contains only a few fibroblasts and collagen fibers are arranged in large, parallel bundles. It is a glistening, white tissue found in the following forms: **fascia** (fash'e-ah) consists of sheets covering muscles, **tendons** hold muscles to bones, **aponeuroses** (ap''o-nu-ro'sēz) are sheetlike tendons, and **ligaments** attach bones to each other. Structurally, ligaments are similar to tendons, except that yellow

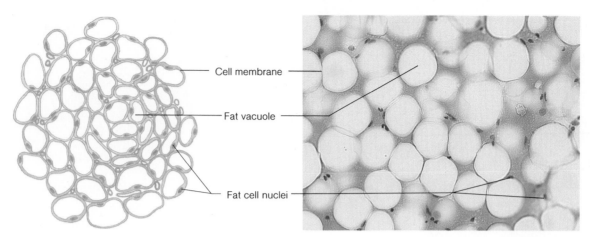

Cell membrane

Fat vacuole

Fat cell nuclei

FIGURE 4.9
Adipose tissue.

ligaments contain large numbers of yellow elastic fibers. Though dense regular connective tissue can have blood vessels, most of the above tissues lack blood vessels and receive nutrients by diffusion.

Reticular connective tissue consists of a reticulum (network) of fibers formed by **reticular cells,** but identical to reticular fibers made by fibroblasts in loose connective tissue. Reticular connective tissue forms the framework for soft tissue organs such as the liver, spleen, and lymph nodes.

Adipose, or fat, **tissue** contains only a few fibers of each type and many fat cells (Figure 4.9). Each cell has a large, central, fat-filled vacuole that pushes the nucleus and cytoplasm to the cell perimeter, and gives it the name "signet ring cell." Adipose tissue is scattered through loose connective tissue and exists separately in fat layers beneath the skin; around the kidneys, heart, and joints; and in the marrow cavities of long bones. It protects, insulates the body against heat loss, and serves as a storage depot for fat until needed for energy. Adipose tissue is metabolically very active because fat is continuously being stored and mobilized (Chapter 24).

Though found only in certain structures in the eyes (iris, choroid, and ciliary body), *pigmented connective tissue* consists of irregularly shaped cells that contain pigment, usually melanin.

Most researchers agree that fat cells can divide during infancy and recommend that overfeeding be avoided to minimize the number of fat cells an infant develops. They differ about whether fat cells can divide throughout life, only at puberty, or not at all after infancy. A cell may fill with fat when a person takes in too much food, become depleted of its fat when a person takes in too little food. A depleted fat cell may become less differentiated, divide, and start a new fat cell cycle when food is available. For health care professionals interested in understanding and controlling human **obesity,** resolving this issue is extremely important.

Cartilage

Like other connective tissues, cartilage consists of cells, fibers, and ground substance. The cells of developing cartilage, called **chondroblasts** (kon′dro-blastz) (*chondro,* cartilage), produce fibers and ground substance until they eventually become trapped in their own products. The trapped cells, now called **chondrocytes,** occupy cavities called **lacunae** (lah-ku′ne). Two or more cells in a lacuna are the progeny of a chondroblast that first occupied it. Cartilage lacks blood vessels and its thickness is limited by the quantity of nutrients diffusing from other tissues to the cartilage cells. The ground substance of cartilage contains proteoglycans, molecules of protein and specific carbohydrate units such as hyaluronic acid and condroitin sulfate, which give form to the tissue. Hyaluronic acid also helps hold cells of a tissue together and holds water in tissues. Chondrocytes contain enzymes that synthesize and degrade proteoglycans. Proteoglycan turnover (breakdown and resynthesis) usually takes months to years, but injury moderately accelerates turnover. Even with accelerated turnover, injured cartilage is exceedingly slow to heal, partly because nutrients are available only by diffusion. Three kinds of cartilage—hyaline, elastic, and fibrous—are recognized by the nature and arrangement of their fibers (Figure 4.10).

Hyaline (hi′ah-līn) **cartilage,** which is bluish-white in fresh preparations, has clearly visible chondrocytes in lacunae, but the ground substance and the thin, weak collagen fibers are visible only with special staining techniques. Hyaline cartilage, the weakest but most widely distributed kind, is found at the ends of the ribs, and in the nose, larynx, trachea, and bronchi. Articular cartilages on the ends of bones where they form joints with other bones are hyaline cartilages with unusually strong collagen fibers.

Elastic cartilage is slightly yellowish and resilient because of the presence of interlacing elastic fibers. It is found in the external ear, the Eustachian tube (between

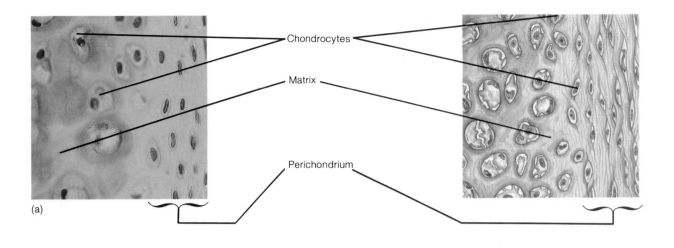

(a)

Chondrocytes

Matrix

Perichondrium

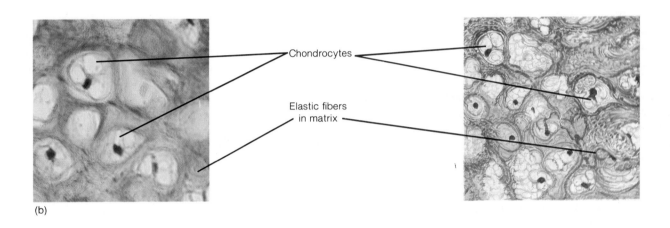

(b)

Chondrocytes

Elastic fibers
in matrix

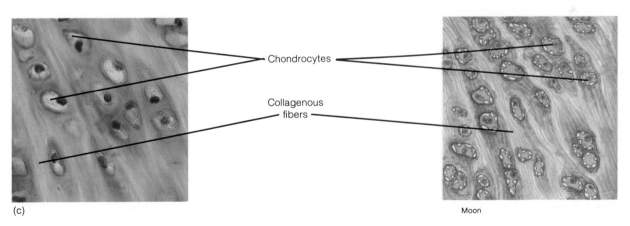

(c)

Chondrocytes

Collagenous
fibers

Moon

FIGURE 4.10
(a) Hyaline cartilage, (b) elastic cartilage, and (c) fibrocartilage.

the middle ear and the pharynx), and in the larynx (voice box) and epiglottis. Cartilage in the epiglottis helps keep food out of the respiratory tract. In the remainder of the respiratory tract, it helps keep the airway open.

Fibrous cartilage, the strongest kind of cartilage, contains many dense collagen fibers and a limited amount of ground substance. It is found in weight-bearing areas such as in disks between vertebrae and in the pubic symphysis, the anterior joint between the pubic bones. The cartilaginous intervertebral disks normally act as shock absorbers. If they are sufficiently compressed, the cartilage can be distorted and protrude toward a nerve or the spinal cord. This is called a slipped or herniated disk. It can cause neck, back, or leg pain, depending on its location.

Bone

Being the hardest of connective tissues, bone consists of cells, collagen fibers, and a dense mineralized ground substance. Bone-forming cells called **osteoblasts** (os'te-o-blastz'') secrete fibers in which minerals are deposited. As the cells become trapped in the mineral deposits they become mature bone cells called **osteocytes.** Bone also has **osteoclasts,** which dissolve bone. Together, osteoblasts and osteoclasts constantly alter bone density and reshape bones with osteoblasts being more active in the young and osteoclasts more active in the elderly. Bone is well supplied with blood vessels, the arrangement of which distinguishes **compact bone** and **cancellous** (kan-sel'us) **bone** (Figure 4.11). In most long bones, compact bone forms the outer layer and cancellous (spongy) bone lies between the compact bone and a central hollow marrow.

Compact bone contains **Haversian** (ha-ver'shan) **systems,** also called **osteons** (os'te-onz). A Haversian system consists of a **Haversian canal** that contains blood vessels and nerves; **lamellae** (lah-mel'e), or concentric layers, of matrix and ground substance; and osteocytes in lacunae. For the most part, Haversian systems are arranged parallel to the long axis of a bone, though there are some connections between adjacent systems. Blood is supplied to the Haversian systems by **nutrient arteries** that usually pass obliquely through the surface of a bone in a **nutrient canal.** The lacunae lie mostly between the lamellae, and the osteocytes within them have cytoplasmic processes that extend into **canaliculi** (kan''ah-lik'u-li), tiny crevices in the matrix. Nutrients diffuse from blood to the cytoplasm of osteocytes in the canaliculi, and wastes diffuse from the cytoplasm in canaliculi to the blood.

Cancellous bone contains osteocytes in lacunae, collagen fibers, and ground substance like compact bone, but it is weaker and lighter than compact bone. Instead of Haversian systems, it has an irregular meshlike arrangement of thin plates of bone called **trabeculae** (trah-bek'u-le), which often extend into the marrow cavity. The long bones of the arms and legs are strong but relatively light because they have an outer layer of compact bone and an inner irregular layer of cancellous bone.

The ground substance of bone is crystalline **hydroxyapatite** (hi-drok''se-ap'ah-tīt), $Ca_{10}(PO_4)_6(OH)_2$ or $Ca(OH)_2 \cdot 3Ca(PO_3)_2$. Certain factors can alter hydroxyapatite. Replacement of Ca^{2+} by Na^+ or K^+ in sufficient amounts decreases bone strength, but replacement of OH^- by Fl^- increases it. Fluoride treatments prevent tooth decay by strengthening tooth enamel, a material similar to, but denser than, bone. Hydroxyapatite, being a charged molecule, can bind ions such as lead and strontium. Binding lead, an environmental pollutant from leaded paints and fuels, keeps the lead from denaturing enzymes and other proteins. Binding strontium, a radioactive environmental pollutant from atomic reactions, irradiates bone marrow and can cause mutations in blood cells.

The composition of bone is constantly undergoing turnover as substances are deposited in bone and reabsorbed from it. For example, as much as 0.8 gram of calcium is exchanged between bone cells and the surrounding extracellular fluids daily. Though the amount of calcium in bone remains relatively constant under normal physiological conditions, some calcium ions are constantly being removed from bone and others are taken from the blood to replace them. Phosphorus is likewise removed and replaced. If the blood concentrations of calcium or phosphorus fall below normal, the minerals can be removed from bone to restore the concentrations to their normal range.

MUSCLE TISSUES

All **muscle tissue** has the ability to exert force when it contracts, and in most instances, contraction produces movement. The body contains skeletal, smooth, and cardiac muscle (Figure 4.12). More details about skeletal and smooth muscle appear in Chapters 9 and 10 and about cardiac muscle in Chapter 18.

Skeletal muscle tissue forms masses called muscles, which attach to bones and move the whole body or its parts. Its action is **voluntary**—it can be consciously controlled by a person and it requires neural stimulation to contract.

Each skeletal muscle consists of many unbranched muscle **fibers** arranged in bundles. Each fiber, a single cell that extends the entire length of a muscle, contains many nuclei, all peripherally located. The **sarcoplasm** (sar'ko-plazm), or cytoplasm of a skeletal muscle cell, contains an orderly arrangement of actin, myosin, and other proteins that play roles in contraction. The arrangement of these molecules accounts for the striations (stripes) seen in muscle with a light microscope.

Smooth muscle is found in the walls of the digestive tract, uterus, urinary bladder, blood vessels, and respiratory passages. Its action is generally **involuntary**—its cells do not require neural stimulation to contract and they are not generally under conscious control. Smooth muscle cells are spindle-shaped cells smaller than skeletal muscle cells.

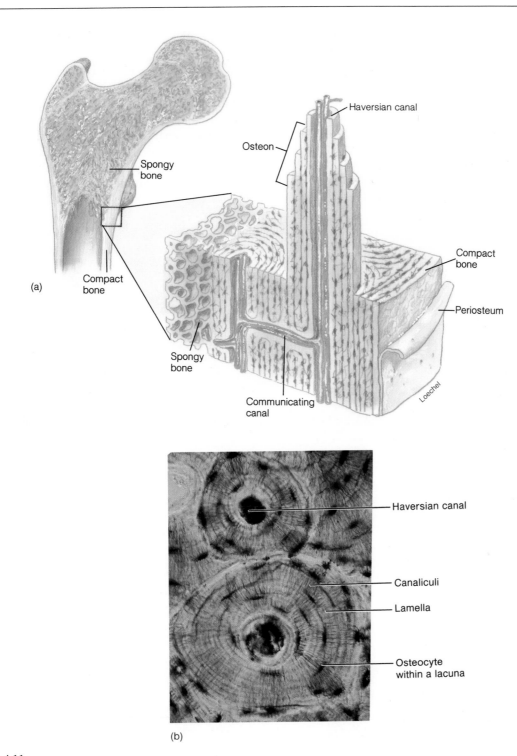

(a)

Spongy bone

Compact bone

Osteon

Haversian canal

Spongy bone

Communicating canal

Compact bone

Periosteum

Loechel

(b)

Haversian canal

Canaliculi

Lamella

Osteocyte within a lacuna

FIGURE 4.11

Compact bone: (*a*) diagram, and (*b*) photomicrograph.

They contain the same proteins as skeletal muscle cells, but the proteins are arranged with much less order and no striations are seen microscopically. Smooth muscle forms bands or sheets within the walls of organs. It contracts more slowly and sustains contraction longer than skeletal muscle.

Cardiac muscle is found only in the heart and in the walls of large blood vessels very near the heart. It is involuntary like smooth muscle and striated like skeletal muscle. Cardiac muscle has branching fibers and specialized cell junctions called **intercalated** (in-ter′kah-lāt-ed) **disks.**

NERVOUS TISSUE

Nervous, or nerve, **tissue** consists of two basic cell types—neurons that conduct signals and neuroglia that perform other functions (Figure 4.13). **Neurons** have a **cell body** (soma), an **axon,** and usually several **dendrites.** Typically, axons conduct signals away from and dendrites conduct them toward cell bodies. A nerve typically consists of a bundle of axons, although some nerves also contain dendrites. Neurons make up the signal conducting portion of

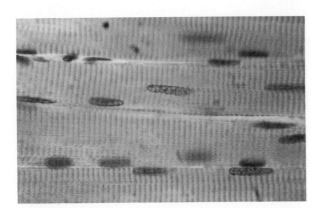

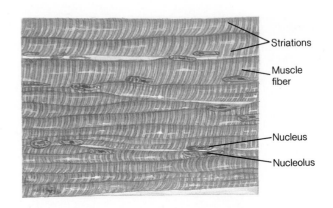

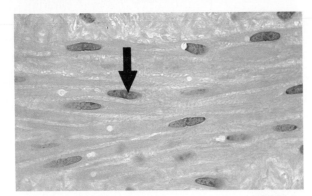

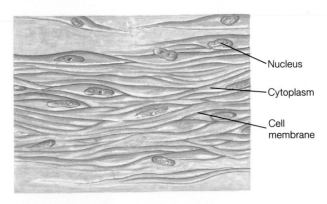

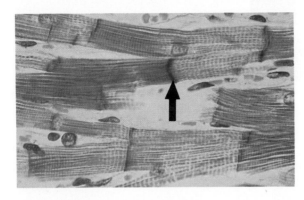

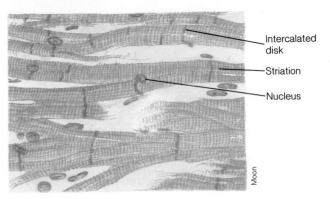

FIGURE 4.12
(a) Skeletal muscle, (b) smooth muscle, and (c) cardiac muscle. The arrow associated with smooth muscle shows an elongated nucleus. The arrow associated with cardiac muscle shows an intercalated disk.

the brain, spinal cord, and nerves (Chapter 11). **Neuro-glial** (nu-rog'le-al) **cells,** also called glial cells, are literally the "glue" of the nervous system (*glia,* glue). Some glial cells provide a supporting network for neurons. They also may help to nourish neurons. Most neuroglial cells are in the central nervous system (brain and spinal cord), but **Schwann cells** are associated with neurons in the peripheral nervous system. Schwann cells produce a fatty substance called myelin that insulates many axons.

CLINICAL TERMS

atrophy (at'ro-fe)   a decrease in size, or wasting, of a cell, tissue, or other part

biopsy (bi'op-se)   inspection of living tissue, usually by removal of a piece of the tissue for study

carcinoma (kar''sin-o'mah)   a malignant growth made up of epithelial cells

carcinoma-in-situ   a growth made up of epithelial cells lacking invasive properties

necrosis (nek-ro'sis)   death of a portion of a tissue

sarcoma (sar-ko'mah)   a malignant growth made up of cells resembling embryonic connective tissue

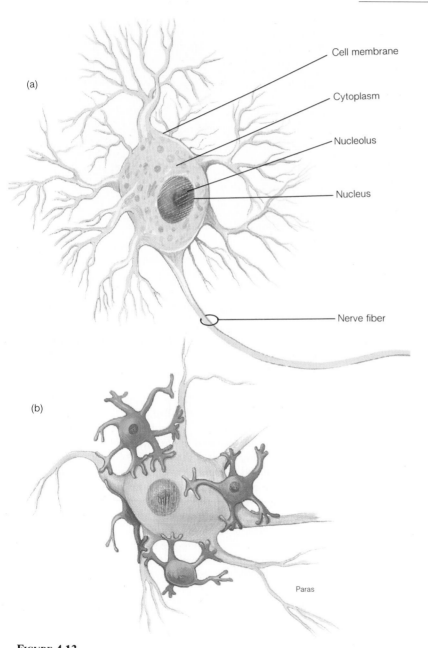

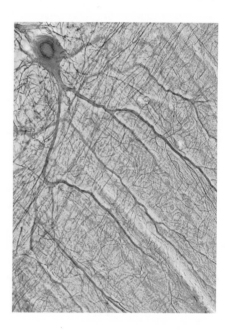

Cell membrane

Cytoplasm

Nucleolus

Nucleus

Nerve fiber

(a)

(b)

Paras

FIGURE 4.13

(*a*) Neurons and (*b*) neuron with neuroglia.

INFLAMMATION

One evening after a hard day, you are slicing fresh vegetables for your dinner. The knife slips and you cut your hand—but not very deeply. The bleeding soon subsides, and you put on a bandage. Soon the area around the cut gets red, swells, and grows hot. It has become inflamed.

Inflammation (Figure 4.14)—often identified by its visible effects of redness, swelling, heat, and pain or itching—is a natural response to tissue damage and the initial step in healing. Though we will study inflammation in more detail in Chapter 21, a basic knowledge of it will be helpful in understanding diseases discussed through Chapter 20.

Events at the cellular and tissue levels during inflammation (Figure 4.15) begin with cell damage and release of histamine (his'tah-mēn), which diffuses into nearby blood vessels. Histamine increases vessel permeability, allowing fluid to move from the blood to the region of the injury where it causes swelling. Histamine also dilates (increases the diameter of) blood vessels. Increased blood flow just beneath the epidermis causes the skin to become red and hot. The blood brings blood clotting factors, nutrients, and other important substances to the injured area, and removes wastes and some of the excess fluids.

Infections such as the common cold, insect bites, and allergies like hay fever cause tissue injury. Histamine release causes the itching of insect bites and the red, watery eyes and runny nose of hay fever. Drugs called antihistamines alleviate such symptoms by reducing the permeability and diameter of blood vessels. Unfortunately, some antihistamines cause unpleasant side effects such as sleepiness, dizziness, disturbed coordination, digestive disturbances, and thickening of mucous secretions.

Returning to the inflamed cut, fluid that enters injured tissue carries chemical factors that clot blood and adjacent interstitial fluids. This stops blood loss and prevents infectious organisms or toxins in injured cells from spreading.

Pain associated with injury may be due to the release of bradykinin (brad''e-ki'nin), but how it stimulates pain receptors in the skin is unknown. Pain, a valuable sensation, leads us to protect an injured area against further damage. When pain lasts longer and is more severe than is needed to warn us, drugs such as aspirin can be given to alleviate it. It is likely that such drugs interfere in some way with pain signals reaching the brain rather than with the bradykinin directly.

Inflamed tissues battle microbes. They release leukocytosis-promoting (loo-ko-si-to'sis) factor, which stimulates leukocytosis, the release of large numbers of leukocytes (white blood cells) from blood-cell-forming tissues.

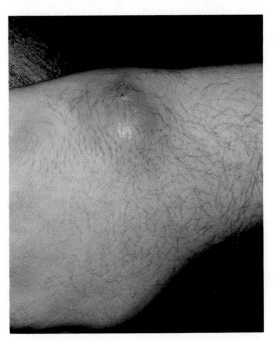

FIGURE 4.14
Redness and swelling are signs of inflammation.

Many leukocytes squeeze through blood vessel walls by a process called diapedesis (di''ah-pē-de'sis) and enter interstitial fluid around a wound. Being phagocytic, most leukocytes move by chemical attraction called chemotaxis (ke''mo-tak'sis) to microorganisms that have entered the wound and to debris from dead cells. In the ensuing battle, microbes are destroyed but many phagocytes also die. The accumulation of dead phagocytes and their ingested material forms a white or yellow fluid called pus. Pus continues to form until the infection has been brought under control. Should these natural defenses be overwhelmed, infectious organisms enter the blood and spread throughout the body. Antibiotics, drugs that inhibit microbial growth, often are used to minimize the chances of an infection spreading from the site of serious injuries.

In addition to some blood cells acting as phagocytes, other blood cells produce an immune response. This response helps to some degree to overcome the initial infection, but it is even more valuable in recognizing and destroying organisms to which it has previously responded. (Essay continues on page 114.)

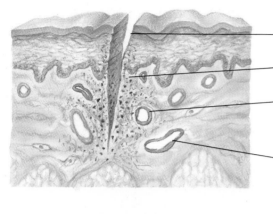

1. A cut allows bacteria to get beneath the surface.

2. Histamine and bradykinin are released by damaged cells.

3. Capillaries become dilated, bringing more blood to the tissue and making the skin red and hot.

4. Capillaries become more permeable, allowing fluids to accumulate and cause swelling.

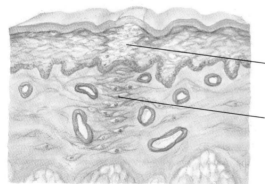

5. Blood clotting occurs, and a scab forms.

6. Bacteria multiply in cut.

7. Phagocytes enter tissue by moving through the wall of a blood vessel (diapedesis).

8. Phagocytic cells are attracted to bacteria and tissue debris (chemotaxis) and engulf them.

9. Larger blood vessels dilate, further increasing blood supply to tissue and adding to heat and redness.

10. As dead cells and debris are removed, epithelial cells proliferate and begin to grow under the scab.

11. Scar tissue (connective tissue) replaces cells that cannot replace themselves.

FIGURE 4.15
Steps in the processes of inflammation and healing.

Healing begins even before inflammation subsides. Epithelial cells in the basal epidermal layer proliferate replacing epidermis that was destroyed. Beneath the epidermis, fibroblasts proliferate and produce fibers and ground substance of a connective tissue called **scar tissue.** Scar tissue replaces damaged muscle and nerve cells, which themselves cannot divide. Though scar tissue does not contract like muscle cells or carry signals like nerve cells, it does provide a strong, durable "patch" that allows the remaining normal tissue to function. Deep cuts often are sutured (stitched together) to minimize the amount of scar tissue and prevent disfigurement.

Several factors affect the healing process. The tissues of young people heal more rapidly than those of older people. Cells of young people divide more quickly, their bodies are generally in a better nutritional state, and their blood circulation is more efficient. As you might guess from the many contributions of blood to healing, good circulation is extremely important in the healing process. Certain vitamins also are important in the healing process. Vitamin A is essential for the division of epithelial cells, vitamin C is essential for the production of collagen and other components of connective tissue, and vitamin K is required for blood clotting.

After several days, your cut is no longer painful, red, or swollen. All that's left is a thin scar to remind you to be more careful when using a knife.

Questions

1. How are the main signs of inflammation related to the body's effort to respond to tissue damage?
2. How do antihistamines and antibiotics affect inflammation?
3. Distinguish between diapedesis and chemotaxis.
4. What is scar tissue?
5. How are vitamins involved in wound healing?

CHAPTER SUMMARY

Definition and Classification of Tissues

- A tissue is a group of cells that work together to perform a particular function.
- The four categories of tissue are epithelial tissues, which protect, absorb, and secrete; connective tissues, which connect parts of the body and help to support and protect; muscle tissues, which exert force, contract, and make movement possible; and nervous tissues, which receive and carry signals or protect signal-conducting cells.
- Histologists fix, section, and stain tissues for study under the microscope.

Development of Tissues

- Development proceeds from the fertilized egg to the morula, blastocyst, and embryonic disk.
- By the end of the third week of human development, the three germ layers—ectoderm, mesoderm, and endoderm— have become established.

Epithelial Tissues

- Epithelial tissues form sheets of cells that cover or line organs, divide by mitosis, are attached to underlying connective tissue by a basement membrane, and receive nutrients by diffusion because they lack blood vessels.

- Cells of epithelial and certain other tissues are held together by tight junction, zonular and macular adhering junctions, and gap junctions.
- Epithelial tissues are classified by cell shape and the number of cell layers: Simple epithelia consist of one layer of cells and can be squamous, cuboidal, or columnar. Stratified epithelia consist of more than one layer of cells and are named according to the shape of the outermost layer of cells, which can be squamous, cuboidal, or columnar. Pseudostratified epithelia consist of one layer of cells of differing heights. Transitional epithelium consists of cells that can change shape.
- The structures, locations, and functions of epithelial tissues are summarized in Table 4.2.
- Membranes consist of a sheet of epithelium and the underlying connective tissue. Serous membranes are lubricated by a clear fluid that reduces friction between the surfaces of the internal organs they cover. Mucous membranes produce mucus and protect the internal surfaces of cavities in organs of the digestive, respiratory, reproductive, and urinary systems.
- Glands, structures that produce secretions, include endocrine glands, which lack ducts and release their secretions into the interstitial fluids, and exocrine glands, which have ducts and release their secretions through the ducts to the surface of an organ.
- The structure, location, and secretions of exocrine glands are summarized in Table 4.3.

Connective Tissues

- The four classes of connective tissues are blood, connective tissue proper, cartilage, and bone. All consist of cells surrounded by extracellular materials. The major characteristics of each class are as follows: Blood consists of various types of cells with fluid ground substance. Connective tissue consists of cells in a fibrous matrix with a jellylike ground substance. Cartilage consists of cells in a fibrous matrix with a semisolid ground substance. Bone consists of cells in a fibrous matrix with a solid, mineralized ground substance.
- The characteristics of the types of connective tissue proper are summarized in Table 4.4.
- Cartilage consists of chondrocytes in lacunae and a fibrous matrix with chondroitin sulfate ground substance.
- The structure, function, and location of various types of cartilage are as follows: Hyaline cartilage has thin collagen fibers and is found in low stress areas such as the ribs, nose, respiratory system, and on articular surfaces of bones. Elastic cartilage has elastic fibers and is found in the external ear and in parts of the respiratory system. Fibrous cartilage contains many dense collagen fibers and is found in high stress areas such as between vertebrae.
- Bone consists of osteocytes, fibers, and mineralized ground substance. Compact bone, which forms the outer layer of most bones, contains Haversian systems (which carry blood vessels and nerves), lamellae, and canaliculi. Cancellous (spongy) bone lacks Haversian systems and is arranged in an irregular meshwork with trabeculae extending into the marrow cavity. It forms the inner part of most bones and supports with a minimum of weight.

Muscle Tissues

- Skeletal muscle is voluntary, has multinucleated fibers, is striated, and is attached to bones, which it causes to move. Smooth muscle is involuntary; has uninucleate, spindle-shaped fibers; is nonstriated; and is found in internal organs. Cardiac muscle is involuntary, has multinucleated fibers, is striated, and is found in the heart, where it contracts throughout life.

Nervous Tissue

- Neurons consist of a cell body, an axon, and several dendrites. They carry signals through the nervous system. Neuroglial cells are the various types of cells that support, insulate, and maybe nourish neurons.

QUESTIONS AND PROBLEMS

The questions at the end of each chapter are numbered to correspond with the objectives listed at the beginning of the chapter. Italics indicate that a question requires critical thinking skills beyond simple factual recall.

Questions

1. (a) Define the term *tissue*.
   (b) List the distinguishing characteristics of each of the four major tissues.
   (c) *What is the significance of tissue differentiation? Hint: How would the body be different if all its cells were alike?*
2. Describe the development of the human embryo through the establishment of the three germ layers, and list the tissues derived from each layer.
3. (a) How are epithelial tissues classified?
   (b) Name an example of each type of epithelial tissue and give its function and location.
   (c) How are epithelial and other tissues held together?
   (d) *What would be the consequences of epithelial cells failing to stick together to form sheets?*
4. Describe the structure, function, and location of serous membranes and mucous membranes.
5. (a) *How do endocrine and exocrine glands differ?*
   (b) How are exocrine glands classified?
   (c) Give an example of each type of exocrine gland, and describe its structure, location, and function.
6. What are the distinguishing characteristics of each of the four classes of connective tissue?
7. (a) Give an example of each type of connective tissue proper, and describe its structure, function, and location.
   (b) *What might happen if loose connective tissue replaced dense connective tissue or vice versa?*
8. (a) Give an example of each type of cartilage, and describe its structure, function, and location.
   (b) *Can hyaline cartilage replace other kinds of cartilage? Why or why not?*
9. What are the distinguishing characteristics of compact and cancellous bone?
10. Describe the structure, function, and location of each of the three types of muscle tissue.
11. How are neurons and neuroglia distinguished?

Problems

1. The absence of any one of the three primary germ layers in an embryo would lead to death of the embryo. Determine which organs and tissues would be missing if an embryo lacked (a) ectoderm, (b) mesoderm, or (c) endoderm.
2. Design experiments to determine (a) whether a secretion is holocrine or merocrine, and (b) whether mammary glands produce apocrine secretions.
3. Use your knowledge of the inflammatory process to explain to a nonscientist the possible effects of taking antihistamines.

Interlude 1

Scientific Methods

Science and Its Methods

Broadly defined, **science** is that branch of human endeavor that attempts to understand the natural universe. By defining science in this way, we have assumed that the natural universe has order and that such order can be discovered and expressed in natural laws such as the law of gravity. Scientists do, in fact, make these assumptions, and they proceed from them to find ways to discover order in the universe and the laws that govern it.

The number of scientific methods in use may be almost as great as the number of scientists using them because each scientist develops his or her own ways of investigating problems (Figure I1.1). All approaches require careful **observations,** and most fall in one of these categories of scientific methods:

1. Descriptive studies that collect and order facts.
2. Studies that search for causes of observed effects.
3. Studies that proceed from recognizing the significance of unexpected findings.
4. Studies that test hypotheses or theories. In science, a **hypothesis** is a tentative explanation of certain events stated for the purpose of collecting evidence to support or refute it. A **theory** is a hypothesis (or group of related hypotheses) for which there is a large body of supporting evidence.
5. Efforts to incorporate theories into higher order laws.

Most early studies in anatomy and physiology were descriptive. Galen (Claudius Galenus of second century A.D. Rome) collected and ordered facts about human body structure and inferred possible functions for the structures. As physiology developed, emphasis shifted to looking for cause and effect relationships. William Harvey, a seventeenth century English physician, observed that blood flows because of the heart's pumping action and concluded that the body must have

Figure I1.1
Scientists investigate problems.

a closed system of vessels through which the same blood circulates over and over.

An outstanding example of recognizing significant unexpected findings is Scottish bacteriologist Alexander Fleming's discovery of penicillin in 1928. Instead of discarding cultures of bacteria in which a mold had inhibited bacterial growth, Fleming recognized that the mold might contain a substance capable of inhibiting bacterial growth in humans. The substance was eventually purified and named penicillin for the mold *Penicillium notatum* that made it.

Many physiological studies today test hypotheses about some aspect of an organism's function. In fact, hypothesis testing is so widely used that it is sometimes called *the* scientific method. This interlude concerns mainly hypothesis testing.

The Scientific Method of Hypothesis Testing

In the hypothesis testing method for solving a scientific problem, most investigators use the following steps:

1. Define and limit the problem.
2. Formulate a hypothesis and make a prediction.
3. Design and conduct an experiment, making appropriate observations.
4. Analyze and interpret the data collected from the experiment.
5. Draw conclusions.

Let's look at each of these steps in more detail.

Define and Limit the Problem

A problem for scientific study must deal with the natural world, be clearly and precisely defined, and be sufficiently limited in scope that the investigator can study it. Finally, it must be possible to formulate a testable hypothesis based on the problem.

Scientific methods deal only with natural conditions and events. They can be used in astronomy, the study of natural properties and movements of stars and planets, but not in astrology, the foretelling of future events in human society from the positions of stars and planets. Physiologists can study any aspect of the function of a living organism, but they would never attempt to study a mythical beast.

When scientists begin to define a problem, they often survey reports of what is already known. Then, they can plan their observations to increase knowledge rather than to repeat studies already done; but sometimes it is important not to accept the conclusions of others. For example, if explorers had accepted the notion that the world was flat, many voyages would not have occurred and progress in understanding our planet would have been curtailed greatly.

Also, it is important to limit a problem to a manageable size. Good physiologists would never attempt to study whole body function or the overall function of a body system, organ, or even a cell. They would select more manageable problems such as how a signal affects the membrane of a neuron, how glucose enters a cell, how hemoglobin carries oxygen, or how the liver makes glycogen. As you read the remaining chapters of this book, you will find that many chapters address the overall functions of a particular body system. You should keep in mind that the information in each chapter represents the integrated results of many studies of specific problems.

Formulate a Hypothesis and Make a Prediction

The primary purpose of scientific experiments is to test hypotheses, and progress in science is made by making and testing hypotheses. The hypothesis in any given study (1) must be an explanation for the defined problem and (2) must be testable. A good hypothesis is one that offers the most reasonable explanation or the simplest solution to a problem. A testable hypothesis is one for which evidence can be collected to support or refute that hypothesis. A **prediction** is the expected outcome or consequence if the hypothesis is true. Before beginning an experiment to solve a particular problem, the investigator should formulate a hypothesis and make a prediction. Scientific experiments test hypotheses by determining the correctness of predictions derived from the hypotheses.

A hypothesis and prediction can be stated in an "if . . . , then . . ." format:

Hypothesis: If an animal uses sight to find its way through a maze,

Prediction: then blindfolded animals will fail to find their way through the maze.

If an experiment shows that a prediction is true, that the blindfolded animals fail to find their way through the maze, then the evidence collected is said to support the hypothesis. Note that even when the evidence *supports* a hypothesis, it does not necessarily prove the hypothesis.

Evidence might support a hypothesis, but not prove it for two reasons. First, the prediction may be true for a reason other than that specified in the hypothesis. Blindfolded animals might fail to find their way through a maze, not because they could not see, but because they normally use smell to guide them and no smell was present. The prediction is still true, but not necessarily because the hypothesis is true. Second, though a prediction is true and supports a hypothesis, it doesn't prove the hypothesis because another experiment might disprove the hypothesis. Many supporting observations usually are required before a hypothesis is accepted. Even so, it is not proven.

Consider the following hypothesis and prediction: If living tissues are composed of cells, the tissues observed in any experiment will be composed of cells. When scientists first looked at tissues under a microscope, they observed small units they named cells.

All of the tissues they studied contained cells and no tissues lacked cells, so all of the evidence supported the hypothesis. If tissue not composed of cells had been found, such evidence would have refuted the hypothesis. Though the hypothesis would still be true for many tissues, it would not have been proven because exceptions had been found. Many investigators have tested this hypothesis without finding significant exceptions, so it has been elevated to a theory—the **cell theory**—that all living things are composed of cells. Biologists find the cell theory a useful generalization about living things.

Design and Conduct an Experiment

Given the hypothesis and prediction, an experiment must be designed to test the hypothesis and to collect evidence to show whether the prediction is true. To design an experiment, one must determine the specific conditions under which the experiment will be run, what observations will be made, and how and when they will be made.

We can use the actual work of the Dutch scientist Christiaan Eijkman who studied the disease beri-beri in 1893 to illustrate the design of an experiment. When Eijkman began to study beri-beri, a disease that causes degeneration of the nervous system and paralysis, many people believed it was caused by bacteria. While making other studies on chickens in his laboratory, Eijkman fed the chickens rice. Those fed polished rice developed a disease that closely resembled beri-beri in humans, but chickens fed unpolished rice with the husk remaining on the grains did not develop the disease. This observation led him to develop the following hypothesis and prediction:

Hypothesis: If beri-beri is a dietary deficiency caused by the lack of a factor in rice husks,

Prediction: then feeding chickens polished rice will produce the condition, whereas feeding them unpolished rice will prevent the condition.

The two groups of chickens used in Eijkman's original experiment were maintained in different pens under identical conditions except for their diets.

Eijkman performed a simple experiment to test his hypothesis (Figure I1.2). He obtained healthy chickens, divided them into two groups, and kept them under identical conditions for two weeks, except for their diets. He fed one group polished rice and the other group unpolished rice.

Eijkman made certain assumptions. An **assumption,** information or an idea taken for granted, can be **explicit** (clearly recognized) or **implicit** (not clearly recognized). We all make implicit assumptions—that restaurant food is safe to eat or that the cable won't break when we step into an elevator—but in scientific experiments we try to avoid them. Eijkman made the explicit assumption that there was no difference between the two groups of chickens. He purposely obtained a single group of healthy chickens and arbitrarily assigned them to two groups. He made the implicit assumption that no chicken had an unrecognized disease or other condition that made it different.

In designing a scientific experiment, it is essential that as many assumptions as possible be made explicit; otherwise, the experiment may not test the intended hypothesis. Suppose that one of Eijkman's chickens had been infected with undetected bacteria and that a disease spread through one group of chickens during the experiment. Eijkman could have made his implicit assumption that none of the chickens carried a disease more explicit by holding them for a few weeks and watching for signs of disease before beginning the experimental diets.

Another essential aspect of an experiment is to attempt to recognize all variables that might affect the outcome of the experiment. A **variable** is any factor that can change. For example, all of Eijkman's chickens should have been of the same breed, the same size, and the same age. The conditions in the two pens where the chickens were kept should have been identical—the same amount of sunlight, the same ventilation, and the same daily temperature variations. All these factors are **control variables,** factors that can change but that are prevented from changing during the experiment. An **experimental variable** is the one factor that purposely differs between groups in the experiment—the factor about which the hypothesis and prediction were formulated. Eijkman's experimental variable was the diet—either polished or unpolished rice.

Once an experiment has been designed, it must be carried out as designed, and all observations must be made and recorded accurately and precisely. If problems or unusual situations are encountered, these must be noted carefully, especially if they cause one to depart from the planned experiment.

Eijkman fed the chickens their assigned diets for two weeks exactly as planned. At the end of the period, he observed that many of the chickens fed polished rice had developed signs of beri-beri, but none of the chickens fed unpolished rice had developed such signs. His prediction was true, and it supported his hypothesis. Unpolished

rice somehow prevented the development of beri-beri in the chickens. Eijkman still did not know how this happened, but he could formulate more hypotheses. Polished rice might have lowered the chickens' resistance to infection. Rice husks might contain something that inhibited bacterial growth. He also could use a larger number of chickens in future experiments to reduce the chance of obtaining positive results by coincidence, but experiments with chickens might not apply to humans.

Studying the effects of polished and unpolished rice on a large number of humans ordinarily would have been difficult. When Eijkman was studying the problem, beri-beri was common among prison inmates in Java. Furthermore, the inmates' diets were quite similar, except that in some prisons, the inmates received polished rice and in others, they received unpolished rice.

Eijkman made use of data available from 100 prisons to test his hypothesis on humans. He classified diets in three categories: (1) at least 75 percent polished rice, (2) at least 75 percent unpolished rice, and (3) a mixture of polished and unpolished rice. He could not control variables, such as the age of prison buildings and ventilation, but he allowed for them in his analysis of the data.

Analyze and Interpret the Data Collected from an Experiment

Once an experiment is completed, the observations must be studied to see whether the prediction is true. Tables,

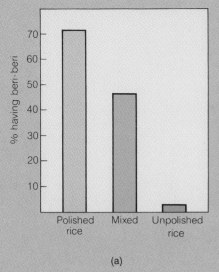

(a)

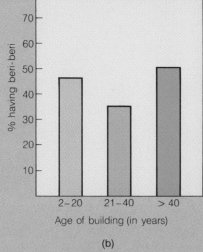

(b)

(c)

FIGURE I1.3
(*a*) The percentage of inmates suffering from beri-beri according to rice content of the diet, (*b*) the percentage of inmates suffering from beri-beri according to the age of buildings in which they were housed, and (*c*) the percentage of inmates suffering from beri-beri according to the adequacy of ventilation in buildings.

graphs, figures, and models are useful in making sense of observations. Scientists often use statistics to describe and summarize their findings and to determine how likely it is that their observations could have been due to chance.

Analyzing the prison data, Eijkman found that beri-beri occurred in only one of 27 prisons where the diet contained unpolished rice and that the percentage of inmates with the disease increased in direct proportion to the polished rice content of the diet (Figure I1.3a). Neither the age of the buildings where prisoners were housed nor the ventilation in the buildings were directly related to the incidence of beri-beri (Figures I1.3b and c). These observations on prisoners are much more reliable than those on chickens because many prisoners in 100 different prisons were used instead of two pens of chickens.

Draw Conclusions

The main conclusion of an experiment is a statement of whether the observations support the hypothesis. Sometimes the conclusions have to be qualified and often the conclusions of one experiment raise new problems and suggest new hypotheses.

Eijkman's observations of beri-beri in chickens supported his hypothesis, but they applied only to the chickens. The prison data provided evidence that polished and unpolished rice had the same effects on humans as it had on chickens. Though neither experiment explained how unpolished rice prevented beri-beri, both experiments suggested to other researchers that a factor that prevented beri-beri might be found in rice hulls. Subsequent experiments have shown that factor to be the vitamin thiamine, which is present in unpolished rice and other whole-grain foods.

PREPARE A REPORT

Scientists are obligated to report new information as the findings of one experiment are often used to develop another experiment. Had Eijkman not reported his results, another scientist at some future time might have made the same discoveries. Eijkman did report his results though—results that led to many new experiments. Because the results of these experiments were reported, scientists eventually learned that the disease could be treated with thiamine and prevented by providing a diet that contained adequate amounts of the vitamin.

The format of a report generally follows the same steps as the experiment itself. It should begin with a statement of the problem and a brief review of what is already known about it. The hypothesis and prediction should be clearly stated and the experiment should be described with sufficient detail that another scientist could repeat it exactly. The data should be reported precisely and completely, and the analysis and interpretation should be derived directly from the observations. Finally, the conclusion should be supported by the data. Suggestions for further research may be included.

The writing style should be clear and concise, and should be illustrated with tables, figures, and graphs. Such illustrations make it easier for another scientist to evaluate the data to determine whether the data support the conclusion of the author.

Problems

1. Design an experiment using the methods described here.
2. Exchange experimental designs with a classmate. Each should make suggestions for improving the other's experiment.
3. Read several reports in current research journals, and identify the components—problem, hypothesis, design, observation and data collection, analysis, and conclusions—in each report.

PROTECTION, SUPPORT, AND MOVEMENT

5

THE INTEGUMENTARY SYSTEM

ORGANIZATION AND GENERAL FUNCTIONS

Your skin separates you from the rest of the world. It covers the entire surface of your body and is the body's largest organ, accounting for 15 percent of the body weight. In an adult, the skin has a surface area of about 1.8 square meters and varies in thickness from 0.5 mm on the eyelids to 5 mm or more on the back between the shoulder blades, on the palms of the hands, and on the soles of the feet.

The **skin** and **skin derivatives**—hair, nails, and glands—make up the **integumentary** (in-teg-u-men′tar-e) **system.** The skin has an outer **epidermis** (ep″ĭ-der′mis) and an inner **dermis** (*derm* or *cut,* skin). The dermis is tightly fastened to the epidermis by a basement membrane and to underlying muscles by subcutaneous loose connective tissue. Except where heavy calluses have thickened the epidermis in response to pressure, the dermis is thicker than the epidermis.

Subcutaneous tissue is sometimes called the **hypodermis** (hi″po-der′mis) because it is literally under the skin. Hypodermic injections are introduced into this layer by passing a needle through both the epidermis and dermis before forcing the contents of a syringe into the tissue. Generally, the hypodermis is well supplied with blood vessels, so medications given hypodermically are absorbed quickly.

The skin protects underlying structures against abrasion and dehydration, and its sweat glands allow fluid loss. By regulating sweating and the amount of blood flowing through dermal blood vessels, the skin also helps to regulate body temperature and maintain homeostasis. The pigment melanin in the skin helps to protect against ultraviolet radiation. Millions of sensory receptors in the skin receive stimuli from the environment. The body makes responses to the stimuli that help to maintain homeostasis (Chapter 13). The skin also plays an important role in immunity and other body defense mechanisms (Chapter 21).

To an alert health care professional, the skin provides much information about the physiological condition of a patient. The rashes caused by diseases, such as measles and chicken pox, and by some allergic reactions are so distinctive that they can be used to make a definitive diagnosis. Hair loss and dry skin or cracked, scaly patches on the skin around the mouth can indicate vitamin deficiencies. In light-skinned people, blue or excessively red skin can indicate respiratory or circulatory disorders and jaundiced (yellow) skin suggests liver disease. Skin that is hard and dry, bronzed, or warm and moist may be related to endocrine gland disorders. Finally, noting the texture and wrinkling of skin helps to indicate whether the patient has given his or her true age.

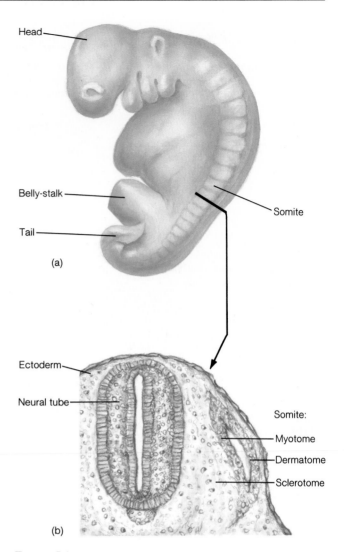

FIGURE 5.1
(*a*) Development of somites, shown in a whole mount of a four-week-old human embryo, and (*b*) the components of somites, shown in transverse section.

SKIN DEVELOPMENT

Of the embryonic germ layers (Chapter 4), ectoderm and mesoderm contribute to the development of the skin. Ectoderm gives rise to the epidermis. During the first 7 weeks of development, the epidermis consists of a single layer of cells, but at birth it has developed many layers. Sensory receptors arise from neural ectoderm. Skin derivatives—hair, nails, and glands—arise from epidermal ectoderm, even though many migrate into the dermis.

Mesoderm gives rise to the dermis. By the third week of embryonic development, blocks of mesoderm called **somites** (so′mĭtz) develop between the ectoderm and the endoderm. Somites differentiate into dermatomes, which give rise to the dermis of the skin; myotomes, which give rise to muscles; and sclerotomes, which give rise to bones (Figure 5.1). Isolated dermatomes grow together to form a complete layer beneath the epidermis.

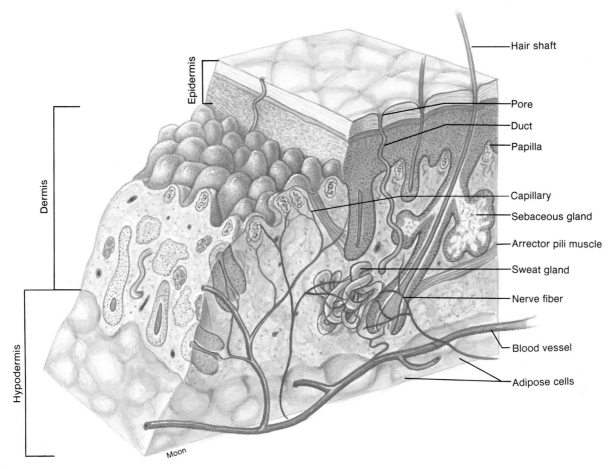

FIGURE 5.2
Structures found in the skin.

Nails and hair follicles form during the third month, and skin glands form during the fourth month of development when all of the basic components of skin are present. By the sixth month, hair follicles produce very delicate fetal hair called **lanugo** (lah-nu′go), which usually is shed prior to birth. Sebaceous (oil) glands secrete a cheesy substance called *vernix caseosa* (ver′niks ka-se-o′sah). This substance protects the skin from its watery environment and is still present at birth.

SKIN TISSUES

The tissues of the skin, epidermis and dermis, are clearly delineated (Figure 5.2). They fit together in a wavy configuration similar to the middle layer of a piece of corrugated cardboard. The dense connective tissue of the dermis gradually merges into the subcutaneous loose connective tissue and this border is less well delineated. Adipose cells are more numerous in subcutaneous tissue than in the dermis. As epidermal hair follicles and glands grow, they extend into the dermis.

Epidermis

The outer part of the skin, the epidermis, is composed of stratified squamous epithelium. It contains four, or in thick skin, five, different strata (Figure 5.3).

The *stratum basale* (stra′tum bas-a′le) lies on a basement membrane next to the dermis, and its cells divide continuously throughout life. About one-fourth of stratum basale cells are **melanocytes** (mel′ah-no-sītz), cells with cytoplasmic processes and that synthesize the pigment **melanin** (mel′ah-nin). Other cells in this layer become **keratinocytes** (ker-at′ĭ-no-sīt), the predominant epidermal cell. Keratinocytes acquire melanin by phagocytizing it from the tips of melanocyte processes. As basal cells divide and keratinocytes move upward, they take on characteristics identified with each successive stratum and are eventually sloughed from the surface at the same rate that new cells are produced. Complete renewal of the epidermis takes 30 to 45 days—it is more rapid in young people than in the elderly.

FIGURE 5.3
Layers of the epidermis.

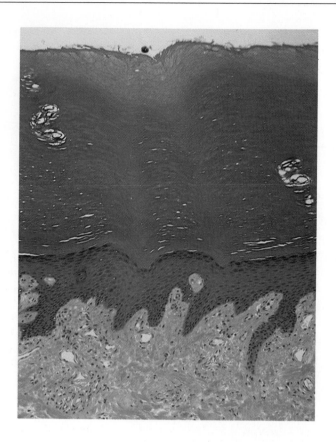

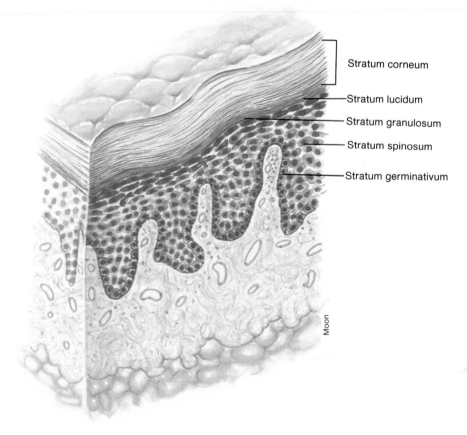

Stratum corneum

Stratum lucidum

Stratum granulosum

Stratum spinosum

Stratum germinativum

As keratinocytes reach the *stratum spinosum* (spin-o'sum), they have a spiny appearance in fixed tissues. The spines represent points of adherence between cells that remain after fixation and staining. Spinosum cells have oval nuclei and they synthesize protein filaments, which eventually become waterproof keratin. Though most cell division takes place in the stratum basale, some cell division continues in the stratum spinosum. Because cell division occurs in both, these layers are sometimes referred to collectively as the *stratum germinativum* (jer-min-a-ti'vum).

Cells within the *stratum granulosum* (gran-u-lo'sum) are somewhat flattened. They contain darkly staining granules of **keratohyalin** (ker"ah-to-hi'ah-lin), an intermediate substance in the synthesis of keratin. Langerhans cells, which play a role in specific immunity in the skin (Chapter 21), are found in this layer and in the stratum spinosum.

Only in thick skin, such as that found on the palms of the hands and the soles of the feet, do keratinocytes form a translucent layer of dead cells, called the *stratum lucidum* (loo'sid-um). Cells become flattened and lose their nuclei and keratohyalin forms a transparent substance called **eleidin** (el-e'id-in).

The *stratum corneum* (kor-ne'um) consists of about 25 layers of dead, dry squamous cells. Surface cells are constantly shed. Underlying cells have lost their nuclei and all cytoplasmic organelles, but retain contact with adjacent cells by desmosomes. Keratin in these cells waterproofs the skin surface, and the cells themselves form a mechanical barrier against light waves, microorganisms, and most other substances that come in contact with the skin.

The epidermis contains no blood vessels and few nerve endings (sensory receptors). Cells are nourished by nutrients that diffuse from blood vessels in the dermis. Because nutrients diffuse slowly, only cells within a short distance of the dermis stay alive. Factors that contribute to the death of outer cells include loss of the nucleus, shortage of nutrients, and probably the accumulation of keratin. Most sensory receptors lie in the dermis, so pressure, heat, and cold stimuli must pass through the epidermis to reach the receptors. Touch and pain receptors, which extend into the epidermis, are closer to the surface and more easily stimulated.

The terms *thick skin* and *thin skin* refer primarily to the number of strata in the epidermis and their relative thickness; however, there are other distinguishing characteristics:

1. Hair follicles are present only in thin skin.
2. The stratum lucidum is present only in thick skin, though some histologists believe that proper microscopic techniques may eventually show a thin layer of stratum lucidum in thin skin.
3. The stratum granulosum is a thin layer in thin skin, and the stratum corneum can be quite delicate.

Skin technically classified as thin can be thicker than skin classified as thick by its strata. For example, thin skin on the back can contain a thick layer of dermis, and any thin skin subjected to pressure or abrasion can thicken as the rate of mitosis in the basal layer increases and a **callus** (kal'us) forms.

Dermis

Directly beneath and held in close association with the epidermis by the basement membrane is the dermis (refer back to Figure 5.2). Being a connective tissue, it contains fibers and ground substance. **Dermal papillae** (pah-pil'e) project into folds in the epidermis in a tongue-and-groove arrangement that helps to anchor the epidermis to the dermis. Ridges made by dermal papillae account for fingerprints. The dermis is the durable part of the skin. In fact, leather is the toughened dermis of an animal skin after the epidermis has been removed.

The dermis has two layers, the **papillary layer,** which contains papilla and capillaries, and the **reticular layer.** The reticular layer contains sebaceous (oil) glands, sweat glands, fat cells, and larger blood vessels. Both contain sensory receptors and both consist of connective tissue in which collagen and elastic fibers run in many directions. The reticular layer is thicker and contains more fibers than the papillary layer. Some collagen fibers in the reticular layer lie in parallel bundles. Separations between these bundles form lines of cleavage. Surgical incisions made along these lines are less likely to gape and therefore heal more rapidly.

The loose connective tissue of the subcutaneous layer merges with the more dense reticular layer. Both vary in thickness according to the amount of fat they contain. Subcutaneous tissue, being loosely packed with a jellylike matrix, provides an ideal site for introducing medications by injection.

Both stretching and aging affect the dermis. During pregnancy or a period of excessive weight gain, the dermis stretches and can be ruptured. Such ruptures leave thin lines called stretch marks visible through the epidermis. During aging, the number of elastic fibers in the dermis decreases, thereby reducing the elasticity of the skin. Reduced elasticity leads to sagging of the skin and formation of wrinkles. The degree of elasticity can be observed by pinching the back of the hand and observing the time for the skin to return to its normal position.

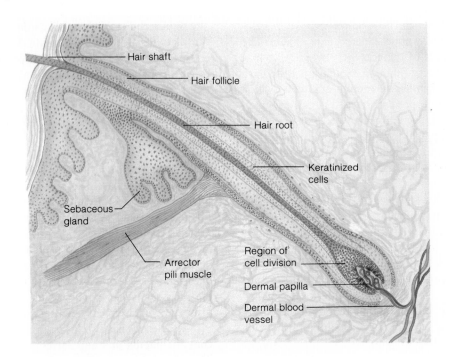

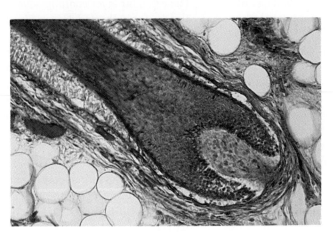

FIGURE 5.4
The structure of a hair.

SKIN DERIVATIVES

Skin, which is a single large organ, contains several specialized structures—hair, fingernails, toenails, sebaceous glands, and sweat glands. Even the mammary glands (Chapter 29) are highly specialized sweat glands. We call these structures skin derivatives because they develop from the epidermis.

Hair

Except for lips, eyelids, the palms of hands, and the soles of feet, the body is covered with hairs, or **pili** (pi'li). Scalp hair protects the scalp from overexposure to the sun's rays,

and eyelashes and tiny hairs inside the nose and ear canals help to prevent foreign particles from entering those organs.

Each hair (Figure 5.4) consists of a *hair shaft* that extends beyond the skin surface and a *hair root* that lies within the hair follicle. The *hair follicle* (fol'li-k'l) itself is epidermis that has grown down into the dermis to surround a connective tissue papilla. The epidermis is arranged in two layers, the internal and external sheaths. A sebaceous gland, a small muscle, and nerve endings are associated with each hair.

Cells of the stratum basale in the hair bulb at the follicle base produce new hair cells and cause hair growth by pushing old cells through the follicle. Each hair grows

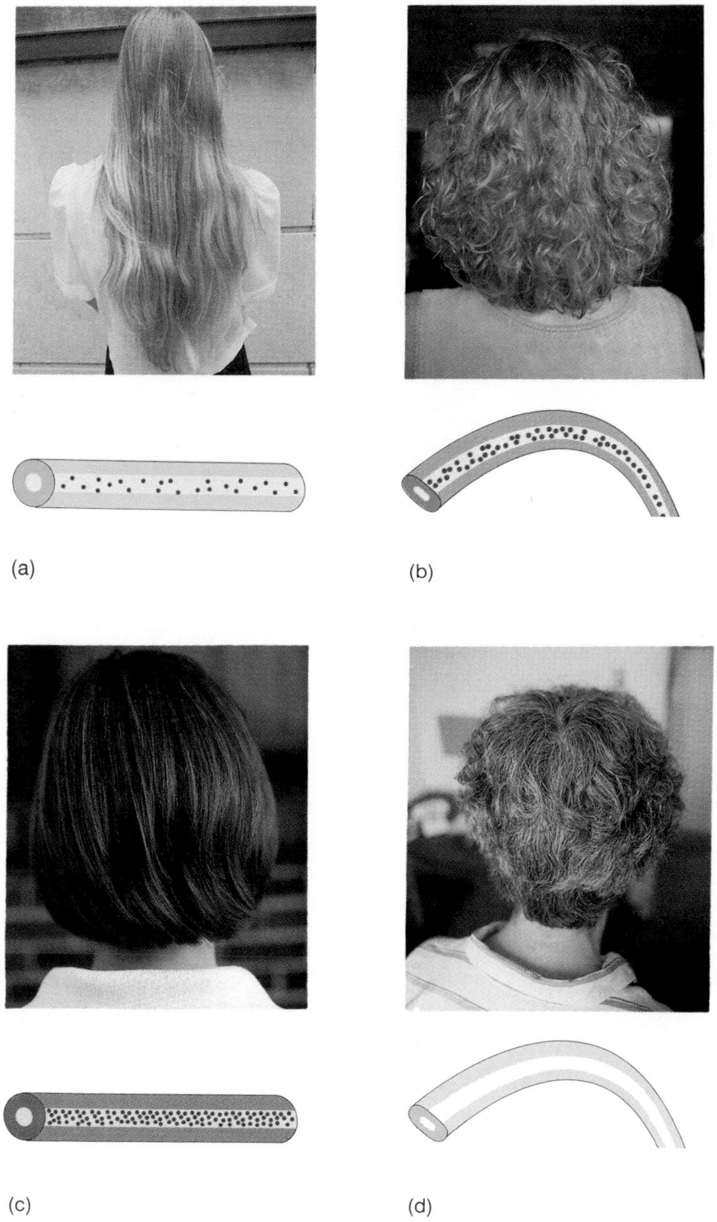

(a)

(b)

(c)

(d)

for two to three years at about 0.3 to 0.5 mm per day (about 10 to 18 cm per year). Then the hair rests for three to four months during which the old hair loses its attachment to the follicle and falls out of the scalp. Soon a new hair is growing in the follicle. Blood vessels at the base of the hair bulb nourish hair cells. Nutrients diffuse upward through the follicle but never reach the ends of a long hair. Most of the length of hair shafts consists of dead cells, the recipients of assorted rinses and conditioners we apply to our hair. Each day as many as a hundred hair shafts break, but they are continuously replaced by regrowth of hair in the follicles. Failure of hair to grow from scalp follicles results in baldness. Baldness can be caused by genetic factors, hormonal imbalances, scalp injuries and diseases, dietary deficiencies, and radiation and chemotherapy.

Both the amount of pigment in hair and whether it is curly or straight are genetically determined. In cross section, hair appears as a hollow tube with pigment deposited in the walls around a central air space. Dark hair has a large amount of melanin, blond hair a small amount, and red hair a medium amount. White hair has little or no pigment because the cells in the hair bulb have ceased to produce it. In the absence of pigment light of all wave lengths is reflected from the air spaces within the hair, so it appears white. Hair appears gray because of a mixture of pigmented and nonpigmented hairs. Straight hair is symmetrically round, whereas curly hair is somewhat flattened. These characteristics of hair are illustrated in Figure 5.5.

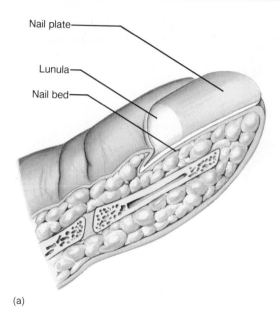

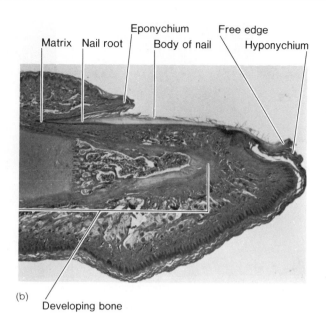

FIGURE 5.6
The structure of a nail.

Because hair contains ten times the concentrations of certain trace elements as blood or urine, hair is potentially useful for diagnostic tests. Heavy metals such as lead, arsenic, cadmium, and mercury concentrate in the hair of people exposed to these toxic substances. Some diseases result in abnormal concentrations of substances in the hair. In cystic fibrosis, hair contains an excess of sodium and a deficiency of calcium; in phenylketonuria, it is deficient in magnesium and calcium; and in malnourished individuals, it is deficient in zinc. Though hair analysis can be done by reputable laboratories, urine, blood, and other tissues are more often used.

Hair analysis has been exploited by pseudoscientists who make the assumption that nutritional deficiencies can be detected by hair analysis. These charlatans do an analysis at an inflated price and sell their victims an expensive diet of "health foods."

Nails

A **nail** (Figure 5.6) consists of a *nail plate* attached to a *nail bed*. It is partially covered at its proximal end by **cuticle,** or **eponychium** (ep''o-nik'e-um), and has a thick, white **lunula** (loo'nu-lah) or little moon near the cuticle. The **hyponychium** (hi''po-nik'e-um) is a thick layer of stratum corneum under the free edge of the nail. Dirt easily accumulates in the groove between the nail and hyponychium. Nails consist of modified, highly keratinized cells of stratum corneum. Nail growth occurs by division of cells beneath the cuticle at the base of the nail. As a nail grows, keratin is deposited and the cells become true nail cells.

Growth occurs at a rate of about 2 mm per month—a little faster in fingernails than in toenails. Nails help to protect fingers and toes, and humans sometimes use nails as tools, such as for picking up small objects.

Sebaceous Glands

Most **sebaceous** (sĕ-ba'shus) **glands** are associated with hair follicles (Figure 5.7), so they are found wherever there is hair. They are simple, branched alveolar glands that produce an oily, holocrine secretion called **sebum** (se'bum). Sebum, which consists mainly of lipids, helps keep hairs soft, pliable, and waterproof. Regular brushing of hair helps to spread sebum to the ends of the hairs and gives them a natural sheen.

Sweat Glands

Sweat glands, also called **sudoriferous** (su''dor-if'er-us) glands, are of two types, eccrine and apocrine. In spite of their names, both kinds of glands produce merocrine secretions. **Eccrine** (ek'rin) sweat glands, widely distributed over the body, produce a watery secretion called *sweat.* Sweat is important mainly in temperature regulation, but also serves to excrete small amounts of sodium chloride and urea from the body. **Apocrine** (ap'o-krīn) sweat glands, found mostly in the armpits and groin area, produce a white, cloudy secretion, which contains organic substances that give rise to odor. Sexual and other emotional stimuli cause contraction of cells around the secretory portion of these glands and releases sweat.

FIGURE 5.7
Sebaceous glands and sweat glands.

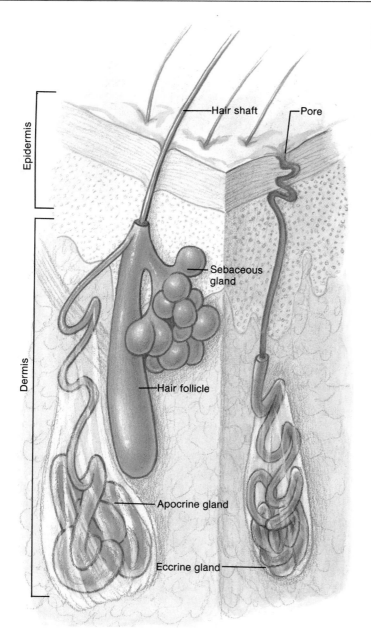

In certain parts of the skin, such as that in the external ear canal, certain sweat glands are modified to form **ceruminous** (sĕ-roo′min-us) **glands.** These glands secrete **cerumen** (sĕ-roo′men), or ear wax. Some release their secretions onto the skin surface and others release them into the ducts of adjacent sebaceous glands. Cerumen is a sticky substance that helps to trap foreign substances before they invade any deeper into the ear. Cerumen itself can build up and become impacted in the ear canal, interfering with hearing. Such wax accumulations should be removed by a professional and a wax solvent solution may be recommended for use periodically to prevent recurrence of wax buildup.

PHYSIOLOGY OF THE SKIN

Your skin is subjected to the influences of a host of factors in the external environment—microorganisms, toxic substances, objects that abrade and tear it, sunlight, heat, cold, wind, rain, and even radiation from outer space. The skin performs a variety of functions that help to maintain homeostasis. It acts as a barrier to most substances in the environment, but it can selectively absorb a few substances—it uses sunlight to synthesize vitamin D. It receives stimuli interpreted in the brain as touch, pressure, heat, cold, and pain. The body's responses to these stimuli help to regulate body temperature and to avoid injury. Should an injury occur, the skin undergoes an inflammatory reaction (Chapter 4) and repairs the damage.

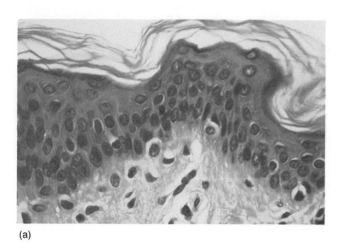

(a)

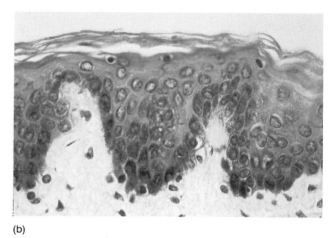

(b)

FIGURE 5.8
More melanin is synthesized in (*a*) black than in (*b*) white skin.

Protection

The skin has barriers to invasion at three levels. The surface of the skin forms the first barrier. Though the skin surface is teeming with microorganisms that feed upon skin secretions, the intact skin surface prevents these organisms and other foreign substances from entering the body. Also, some of the secretions are poisonous to microorganisms. The waterproof layer of keratin in the outer epidermis is a second barrier. It prevents all water-soluble substances from entering the body. The basement membrane (Chapter 4) forms a third barrier against substances entering the body.

Pigment Formation and Skin Color

The pigment melanin is synthesized by melanocytes (Figure 5.8) in the stratum basale, and some of it is released and phagocytized by keratinocytes as noted earlier. The amount and distribution of melanin is a significant factor in skin color. Though sunlight stimulates melanin production, too much ultraviolet radiation can damage cells, as you know if you have had a painful sunburn. Melanin absorbs some of this radiation, so increased melanin synthesis provides some protection against sunburn. Exposure to the sun can produce a deep tan in light-skinned individuals, but it also dries the skin and increases the risk of skin cancer. In addition, ultraviolet rays increase the rate at which elastic fibers are lost; hence, the rays accelerate aging of the skin. Each summer tan brings closer the day of dry skin and wrinkles.

Genetic factors determine the natural skin color. Though all races have about the same number of melanocytes, races differ genetically in the amount of melanin that those cells produce. Greater amounts of melanin result in darker skin, and the absence of a certain gene in albinos prevents melanin synthesis entirely. Another pigment, **carotene** (kar′o-tēn), found in the upper layers of the epidermis, is responsible for yellow skin tones.

Blood circulating in the dermis accounts for pinkness of skin in lighter skinned people. When blood vessels are dilated and more blood passes through the dermis, the skin reddens appreciably. Blood vessels in the skin dilate under strenuous exercise and emotional stress. During strenuous exercise, the body temperature increases and blood vessels dilate, thereby increasing heat loss from the blood through the skin. If a light-skinned person suffers embarrassment, the body's emotional response causes facial blood vessels to dilate, and the person blushes. Lack of oxygen in the blood causes the skin to have a bluish hue called cyanosis, and accumulation of breakdown products of worn-out red blood cells causes it to have a yellow hue called jaundice.

Temperature Regulation

Skin plays an important role in the homeostatic mechanism that regulates body temperature, as described briefly in Chapter 1 and treated in more detail in Chapter 25. The skin allows the body to lose heat by four processes: **evaporation, conduction, radiation,** and **convection.** Sweating, the single most important factor in heat loss, conducts water and heat to the skin surface; a large amount of heat is lost as the sweat evaporates. Dilated blood vessels carry a larger volume of blood through the dermis. Heat is conducted through the skin (as through pipes in a heating system) and radiates (as from a radiator) into the environment. Also, regardless of the body temperature, heat is constantly convected (transferred) from the skin to passing air currents.

Conversely, the skin conserves heat by greatly reducing sweat secretion and by constricting blood vessels. Shivering, involuntary contractions of skeletal muscles, releases heat that compensates for heat loss. Small involuntary, smooth muscles called the *arrector pili* (ah-rek′tor pi′li) attached to the base of hairs in the skin also contract. In mammals with heavy fur, these contractions cause hair to stand out from the skin and trap warm air around the body, but they have little effect on human body temperature. They merely produce "goose flesh" and remind us that our ancestors may have been furrier, and warmer, than we are.

Absorption

Protective barriers limit absorption through the skin, so for most substances, absorption is not a significant avenue of entry into the body. Small quantities of a few substances do enter the blood by cutaneous absorption. Low molecular weight gases such as oxygen and carbon dioxide diffuse through the normal barriers. Fat-soluble substances such as vitamins A, D, E, and K, and steroid hormones diffuse through fatty materials, particularly sebaceous gland secretions from which they are absorbed. Toxic substances such as some insecticides also can be absorbed through fatty materials in the skin. Wearing gloves and protective clothing while spraying insecticides is essential to avoid exposure to such toxic substances.

Synthesis

The skin synthesizes several substances that remain in the skin—keratin, melanin, and carotene. In addition, as dehydrocholesterol from dietary animal fats passes through skin blood vessels, it is converted to cholecalciferol, or vitamin D. A small amount of ultraviolet light—far less than that needed to cause a sunburn—is sufficient to cause this reaction. Cholecalciferol is converted to active vitamin D in the kidneys. Vitamin D helps to regulate calcium and phosphorus metabolism; thus, it is essential for the development of strong bones. Children exposed to ultraviolet rays in sunshine require smaller amounts of dietary vitamin D than those not exposed to sunshine. Children deficient in vitamin D develop the disease rickets, which interferes with normal calcium absorption and bone formation.

Sensory Reception

Large numbers of various sensory receptors are located in the skin where they play an important role in detecting changes in the environment. They are discussed with other sensory receptors in Chapter 13.

PHOTOAGING

Recent studies of **photoaging,** the deleterious effects of ultraviolet radiation on skin, have demonstrated that it differs from chronological aging. Photoaged skin is deeply wrinkled with a leathery, yellow surface and contains a variety of premalignant and malignant cells. The most striking feature of photoaged skin—a massive amount of thick, tangled elastin fibers—is seen in stained tissue sections (Figure 5.9). Wrinkling is directly related to damage

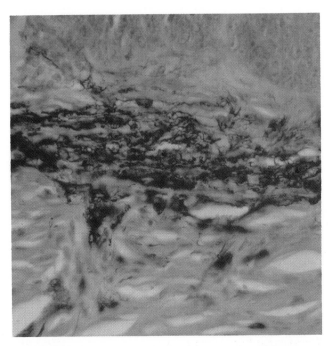

FIGURE 5.9
Photoaging.

to elastin fibers, which normally give the skin its resilience. Photoaged skin also has enlarged sebaceous glands and a thicker than normal dermis. Blood vessels dilate and become tortuous early in photoaging and later degenerate so that the upper dermis and epidermis are severely deprived of nutrients.

Photoaging can be prevented to some degree by applying broad spectrum sunscreens with a sun protection factor of 15 or more; however, such screens do not block all the damaging ultraviolet rays. Though the effects of photoaging had been believed to be irreversible, new evidence shows that some repair is possible once exposure to ultraviolet rays is stopped. New collagen is deposited in the upper dermis pushing the damaged elastin downward. Treatment with retinoic acid appears to accelerate the repair process, but the side effects of this agent are not fully known.

Within 24 hours of moderate exposure to ultraviolet radiation, some epidermal cells undergo unusual changes to become **sunburn cells.** In stained preparations, these cells are shrunken with dense, glossy cytoplasm and a deeply stained, condensed nucleus. How sunburn cells are formed is not clearly understood. One proposed mechanism is that ultraviolet radiation releases superoxide, which severely damages DNA and may activate lysosomal enzymes. Sunburn cells often appear in the basal layer of the epidermis and migrate quickly toward the surface from which they are sloughed off like other epidermal cells. The significance of sunburn cells is being debated by researchers. Some think they are simply dead or dying cells that have no special significance. Others think they represent a mechanism by which the epidermis rids itself of potentially cancerous cells.

Life styles have a lot to do with skin disorders. Sunburns may be responsible for the recent rise in the incidence of **melanoma** (mel''ah-no'mah), a serious form of skin cancer that can be fatal if it metastasizes. People who have had a blistering sunburn in their twenties are three times more likely to develop melanoma than those who have never had such a burn. An indicator of the risk of developing melanoma is the number of moles on the skin. Moles on the lower legs indicate a much higher risk of melanoma than moles on other parts of the body.

Many people who want to sunbathe or who must work in the sun use chemical sunscreens to block ultraviolet rays. A few who think they have a rash because they are allergic to sun are, in fact, allergic to a chemical in the sunscreen. Applying more of the sunscreen makes the rash worse.

Bruises, or black-and-blue spots, are due to rupture of small blood vessels under the skin surface. Such bruises are a plague of some sports enthusiasts as the following examples illustrate. Joggers get "black heel" from friction against running shoes, golfers get "black palm" from gripping their clubs, and tennis players get "tennis toe" from jamming their toes into the end of their shoes.

DISORDERS OF SKIN

Acne, an exceedingly common disorder of sebaceous glands and the skin around them, affects over 80 percent of teenagers and can affect adults as well. Acne arises when male sex hormones stimulate sebaceous glands to increase in size and secrete more sebum. These hormones are produced by the adrenal glands as well as the testes, so acne occurs in both males and females. When sebum accumulates, microorganisms feed on it, causing the duct of the gland and surrounding skin to become inflamed.

Acne occurs in varying degrees of severity (Figure 5.10). The least severe is characterized by comedos, or "blackheads," in which hair follicles and associated sebaceous glands become plugged with sebum and keratin. Blackheads are black because sebum turns black when it is exposed to oxygen in air. In more severe cases, the plugged duct can rupture and release secretions into surrounding tissues, break the skin, and initiate inflammation. Bacteria can infect the area and cause more inflammation, more tissue destruction, and scarring. In the

FIGURE 5.10
Acne begins with (*a*) a "blackhead," in which the duct of a sebaceous gland is plugged with excessive secretions. (*b*) Some blackheads become infected with microorganisms, inflamed, and filled with pus.

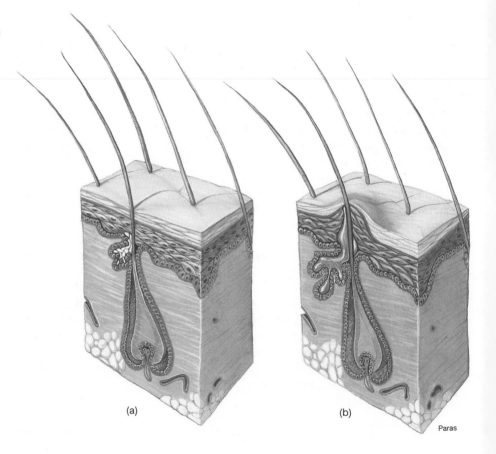

(a)

(b)

Paras

most severe cases, lesions (breaks in tissue) appear over many parts of the body and some become encysted in connective tissue.

Treatment includes frequent cleansing of the skin and the use of topical medication to reduce the risk of infections. Sometimes acne sufferers are advised to avoid fatty foods, but a connection between diet and sebum production is not at all well established. Dermatologists often prescribe oral antibiotics in low doses in an attempt to control bacterial infections in the lesions; however, continuous antibiotic use depletes natural intestinal microbes and can give rise to antibiotic-resistant bacteria. Drugs similar in structure to vitamin A are sometimes used to treat severe and persistent acne, but must not be given to pregnant women because they can damage the unborn child. They seem to work by inhibiting sebum production, and their effects last for several months after termination of treatment. The drugs also seem to prevent and possibly reverse wrinkling in aging skin, but several brand names now can have serious side effects such as intestinal bleeding. In most cases, acne disappears or decreases in severity as the body adjusts to hormonal changes of puberty, and homeostasis is reestablished in the functioning of the sebaceous glands.

Impetigo (im″pē-ti′go) is a bacterial infection of the skin, usually caused by streptococci or staphylococci. The body's inflammatory response produces pus-filled bumps, or pustules, on the skin surface. The bacteria are easily spread if pustules rupture. Such infections can be effectively treated with antibiotics.

Measles and **chickenpox** are viral infections involving the whole body that also cause skin lesions. Measles produces a dry rash and chickenpox causes pustule formation. No cure exists for these diseases, but a vaccine is available to prevent measles. Measles, a serious and sometimes fatal so-called childhood disease, could be eradicated if all infants worldwide were to receive the vaccine.

Eczema lesions are patches of reddened, raw, watery skin. Irritation and hypersensitivity to certain substances are possible causes. Treatment is usually aimed at removing the cause, but topical medications are used for temporary relief of symptoms.

Decubitus (de-ku′bit-us) **ulcers** (bed sores) are caused by a combination of irritation and inadequate blood circulation. Without sufficient blood flow to the dermis, some cells of the epidermis die. Bedridden individuals are particularly prone to these lesions, but their occurrence and severity can be reduced by changing the person's position regularly to reduce irritation and increasing blood flow by massaging near bony prominences and other areas likely to be affected.

Dermatitis, literally inflammation of the skin, is a sign of another disorder such as allergy, vitamin deficiency, and certain other diseases. Treatment is generally directed toward identifying and treating the problem that caused the dermatitis.

CLINICAL TERMS

abrasion (ah-bra′zhun) rubbing or scraping off of skin or mucous membrane

alopecia (al″o-pe′she-ah) hair loss

athlete's foot a superficial fungus infection of the skin of the foot

bulla (bul′ah) a large blister or fluid-filled vesicle

carbuncle (kar′bung-kl) an inflammation of subcutaneous tissue that eventually discharges pus

contusion (kon-tu′zhun) a large bruise, often affecting the brain or lungs

crust a layer of solid matter formed by dried secretions

cyst (sist) a sac filled with solid or semisolid material

dermatology (der″mah-tol′o-je) the study of diseases of the skin

ecchymosis (ek″ĭ-mo′sis) a small bruise

erythema (er″ĭ-the′mah) redness of the skin due to inflammation and engorgement of capillaries

excoriation (eks-ko″re-a′shun) loss of superficial skin, often by scratching

fissure (fish′ūr) a groove or cleft; a crack in the skin

keloid (ke′loid) a growth of dense tissue often appearing during the healing of an injury to the skin

keratolytic (ker″ah-to-lit′ik) pertaining to a substance that dissolves keratin

macule (mak′ūl) a discolored spot on the skin

nevus (ne′vus) a new growth of skin of congenital origin; a birthmark

nodule (nod′ūl) a small swelling or protruberance

papule (pap′ūl) a small, solid elevation on the skin

petechia (pe-te′ke-ah) small red spots formed by effusion of blood into the skin

pruritus (proo-ri′tis) intense itching

psoriasis (so-ri′ah-sis) a skin lesion consisting of scaly red patches

pustule (pus′tūl) a small, pus-filled elevation

vesicle (ves′i-k'l) a small, fluid-filled sac; a small blister

wart a nonmalignant epithelial growth caused by a virus

Essay

BURNS

Serious **burns** can be life-threatening, and burn treatment is a complex medical problem. Studying burns shows how the body's homeostatic mechanisms can be disturbed, and studying burn treatment shows how homeostasis can be restored. Each year in the United States, about 75,000 people are hospitalized—30,000 in intensive care units—for severe burns over 30 percent or more of their body surface. Even with intensive care, over 10,000 people (one-third of them children under 15) die of burns annually, making burns the third leading cause of accidental death.

Burns can be caused by heat, chemicals, electricity, or radiation, all of which destroy proteins and kill cells. Severe sunburn and other radiation burns are possible because they do not hurt while they are occurring—only later on. Infection is a common and serious complication of burns because the damaged skin fails to serve as a barrier to infectious agents.

The severity of burns is determined by depth and extent of cell destruction. In **partial-thickness burns,** some epithelial cells remain and can divide to replace lost epithelium. In **full-thickness burns,** all epithelial cells are destroyed and skin grafts are necessary to repair the damage of a burn. The extent of burns is estimated by the "rule of nines" (Figure 5.11), in which the body surface area is divided into areas that each account for 9 percent (or a multiple of 9 percent) of the total surface area. Such estimates of the damaged surface area are useful in determining how much fluid a burned person needs to replace the volume seeping from damaged tissues.

In addition to local effects, severe burns cause systemic (whole body) effects. Fluids seep out of tissues unprotected by skin, and proteins and fluids leak from damaged blood vessels. As blood osmotic pressure and blood volume decrease, shock, a severe reduction in blood pressure, can occur. The body conserves fluid by decreasing urine production, but this allows wastes to accumulate in the tissues. The bodies of burn victims have a higher than normal metabolic rate while the destroyed tissue is being replaced, and if sufficient protein and calories are not supplied, malnutrition results.

Removal of eschar, dead skin over a severely burned area that can harbor infectious organisms, has been a problem in burn treatment. Until recently, this tissue was allowed to separate from the burned area spontaneously over a period of several weeks. Now it can be removed a few days after injury with a laser scalpel that also reduces blood loss by cauterizing blood vessels. After eschar is removed, thin sheets of healthy epidermis from an uninjured body region are grafted to the burned area where they proliferate and protect underlying tissues from fluid loss and infection. In extensive burns where the victim has too little skin remaining, other materials are used. Skin from other humans or animals has been used, but it usually is rejected by the recipient's immune system. Synthetic, or artificial, skin has recently been made from carbohydrates and protein fibers from cattle. This skin is not rejected and it prevents fluid loss while fibroblasts grow into it and form scar tissue.

Questions

1. How do burns disturb homeostasis?
2. What are some common causes of burns, and what precautions can prevent burns?
3. What is eschar, and why is it removed?
4. What kinds of grafts are available to protect burned surfaces?

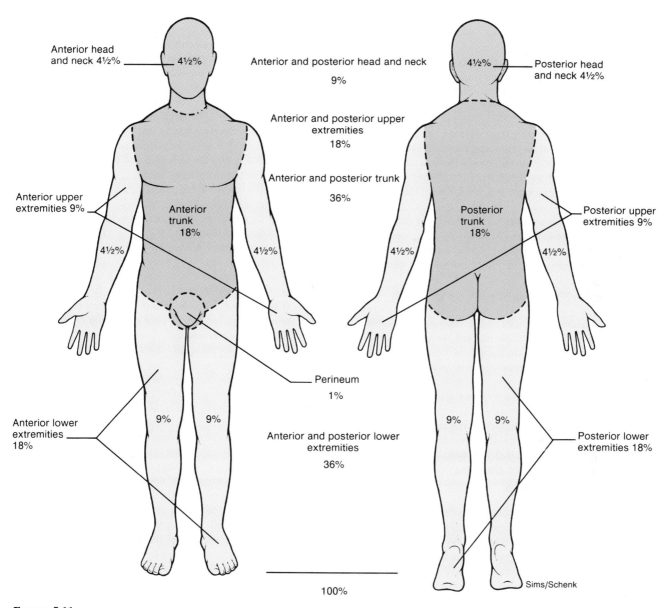

Anterior head and neck 4½%

4½%

Anterior and posterior head and neck
9%

Anterior and posterior upper extremities
18%

Anterior and posterior trunk
36%

Anterior upper extremities 9%

Anterior trunk 18%

4½%

4½%

Posterior head and neck 4½%

4½%

Posterior trunk 18%

Posterior upper extremities 9%

4½%

4½%

Perineum
1%

9%

9%

9%

9%

Anterior lower extremities 18%

Anterior and posterior lower extremities
36%

Posterior lower extremities 18%

100%

Sims/Schenk

FIGURE 5.11
The extent of burns is estimated by the rule of nines.

CHAPTER SUMMARY

Organization and General Functions
- The skin consists of epidermis and the dermis.
- Its functions include protection, prevention of excessive fluid loss, regulation of body temperature, and reception of stimuli from the environment.

Skin Development
- The epidermis of the skin; skin derivatives (hair, nails, and glands); and sensory receptors are derived from ectoderm.
- The dermis of the skin is derived from mesodermal dermatomes.

Skin Tissues
- The epidermis consists of an innermost stratum basale, the cells of which continuously divide so cells move outward and are transformed as they become part of the stratum spinosum; stratum granulosum; and in thick skin, the stratum lucidum.
- The outermost stratum corneum consists mostly of dead cells filled with waterproof keratin.
- Thin skin lacks a stratum lucidum and has a reduced stratum granulosum, but has hair follicles that are lacking in thick skin.

Skin Derivatives
- Hair grows in epidermal follicles and serves a minor function in protecting the scalp from radiation, and the eyes, nose, and ear canals from foreign objects.
- Nails consist of modified stratum corneum and are occasionally used as tools.
- Sebaceous glands are associated with hair follicles and produce sebum.
- Eccrine sweat glands secrete watery secretions through pores in the skin and assist in temperature regulation, whereas apocrine sweat glands secrete milky apocrine secretions into hair follicles in the armpits and groin area.

Physiology of the Skin
- The skin has mechanical and chemical barriers that protect the body from various environmental factors.
- Skin color is due to the presence of melanin—the synthesis of which is genetically determined, but can be increased by sunlight.
- Skin absorbs small gas molecules, fat-soluble vitamins, and steroid hormones; synthesizes keratin, melanin, and carotene; and participates in the synthesis of vitamin D.
- Various skin functions—especially those associated with temperature regulation—help to maintain homeostasis.

Disorders of Skin
- In acne, male sex hormones stimulate sebum secretion, and glands and surrounding tissues subsequently become infected.
- Other lesions result from infections, irritation, or reduced blood flow. They include impetigo, eczema, decubitus ulcers, and dermatitis.

QUESTIONS AND PROBLEMS

The questions at the end of each chapter are numbered to correspond with the objectives listed at the beginning of the chapter. Italics indicate that a question requires critical thinking skills beyond simple factual recall.

Questions
1. (a) What are the two main parts of the skin?
   (b) What are the main functions of the skin?
2. List the structures in the skin derived from ectoderm and those derived from mesoderm.
3. (a) Name and describe briefly the strata of the epidermis from the lower layer outward.
   (b) Name and describe briefly the layers of the dermis.
4. How does thin skin differ from thick skin?
5. (a) Describe briefly the structure and function of hair.
   (b) Describe the structure of a nail.
   (c) Describe the structure of a sebaceous gland and the nature of its secretions.
   (d) How do merocrine and apocrine sweat glands differ in structure and function?
6. *How do skin barriers protect the body from microbial invasions?*
7. *(a) How is skin color produced, and how does pigment protect the body?*
   (b) What is the role of the skin in regulating body temperature, and why is such regulation called a homeostatic mechanism?
   (c) What kinds of materials can be absorbed through the skin?
   (d) What substances are synthesized in the skin?
8. *How does light affect the skin?*
9. (a) Discuss the cause, effects, and treatment of acne.
   (b) List three types of skin lesions other than acne, and describe the symptoms they produce.

Problems
1. If the skin is broken and microbes penetrate the stratum corneum, what remaining skin barriers and other defense mechanisms might prevent a generalized infection? (Use information from all chapters you have studied.) What are some of the limitations of the use of various topical medications?
2. One severely burned person is placed in a plastic isolation tent maintained at 30° C with high humidity and receives a high protein diet. Eschar is removed by laser surgery and synthetic skin is grafted 2 days after the burn. Another severely burned person is placed in an ordinary hospital room and receives a standard diet. Eschar is allowed to separate spontaneously and pig skin is grafted 2 to 3 weeks after the burn. Explain the physiological effects of each treatment and how they might influence recovery.

OUTLINE

OBJECTIVES

1. Explain how the structure of compact bone and cancellous bone differs.

2. Explain how intramembranous and endochondral bone formation differ.

3. List four types of bones and describe the structure of a long bone.
4. Define the terms that describe bone markings.

5. Summarize the functions of bone.
6. Describe how bone is maintained, and how exercise, mechanical stress, vitamins, and hormones affect bones.

7. Describe the cause, effects, and treatment of rickets, osteomalacia, and osteoporosis.

8. Briefly describe different kinds of fractures, and explain how fractures heal and how healing can be facilitated.

6

BONE TISSUE AND BONES

BONE TISSUE

We sometimes say that a person is "just skin and bones," knowing, of course, that it is not literally true. These two body components—skin and bone—provide a good starting place for studying body systems. Skin, as we saw in Chapter 5, forms the outer covering of the body. Bones, as we shall see in this and in the next two chapters, are rigid supporting structures that form a framework for the body and allow movement at joints.

Bone is a connective tissue in which the matrix is hardened by deposited mineral called hydroxyapatite (Chapter 4). Large quantities of collagen fibers in the matrix give the bone tensile strength and the minerals give it rigidity; thus, the matrix contributes to the support and protection functions of the skeletal system.

The body contains two kinds of bone tissue as noted in Chapter 4. Compact bone is found on the surface of most bones. Cancellous, or spongy bone, lies beneath the compact bone and, in many bones, surrounds a hollow marrow cavity (Figure 6.1). Mineral deposits are more orderly and tightly packed in compact bone than in cancellous bone, but they both contain hydroxyapatite. The organic matrix of the body's strongest bones is about 90 percent collagen, a sturdy, durable protein, and the remainder consists of chondroitin sulfate, hyaluronic acid, and other polysaccharides.

Compact bone is organized structurally into osteons, or Haversian systems, consisting of concentric circles of bone called lamellae around a **Haversian canal.** The Haversian canal, itself, contains a blood vessel and a nerve fiber. The concentric circles are formed by osteoblasts, which secrete fibers and mineral until they become trapped in lacunae. Such trapped osteoblasts are mature cells called osteocytes. Each lacuna is connected to adjacent ones by narrow passageways called **canaliculi,** which contain fragile cytoplasmic processes of osteocytes and interstitial fluid. Nutrients diffuse from blood vessels in Haversian canals through the network of cytoplasmic processes to each osteocyte, and wastes move in the opposite direction.

Cancellous bone consists of bars of bone called **trabeculae** (trah-bek′u-le) strategically arranged to help the bone resist stress. These bars are only a few cells thick and many have sharp ends called **spicules** (spik′ūlz) extending into a marrow cavity. Osteocytes occupy lacunae and have canaliculi, but these structures are not organized into osteons. Cancellous bone adds greatly to the strength of a bone with a minimum of added weight.

The formation of hydroxyapatite, the main mineral in all bone depends on certain local conditions. Local blood and tissue fluid where bone is being formed, originally or during remodeling, must have a critical minimum amount of calcium and phosphate ions. This minimum is expressed as the Ca^{2+}. The ion product is the concentration of Ca^{2+} times the concentration of inorganic phosphate ions (P_i), including PO_4^{3-}, HPO_4^{2-}, and $H_2PO_4^-$. Once a

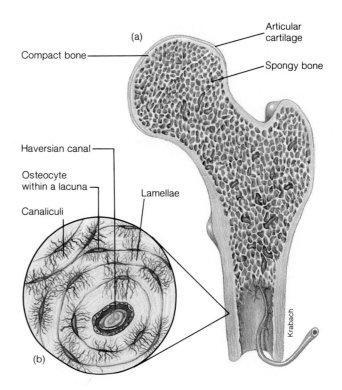

FIGURE 6.1
(a) Surface compact bone is covered by articular cartilage at joints. Spongy bone occurs beneath the compact bone and surrounds the marrow cavity. (b) A diagram of compact bone.

few tiny crystals are present, those crystals catalyze further crystallization even when the ion product is below the critical minimum for initiating crystallization.

BONE DEVELOPMENT

Certain bones of the trunk such as the vertebrae, ribs, and scapulae develop from the sclerotomes, portions of embryonic mesodermal somites (Chapter 5). Limb bones develop from other portions of somites. Though some tissue from the somites migrates to the head, most bones of the head develop from other (nonsomitic) mesoderm. **Ossification** (os″ĭ-fi-ka′shun) and **osteogenesis** (os″te-o-jen′ĕ-sis) are synonyms referring to the process of bone formation. **Calcification** (kal″sĭ-fi-ka′shun) refers specifically to the deposition of calcium as a part of the bone-forming process.

Ossification can occur in two different ways. Bones derived from nonsomitic mesoderm usually are laid down directly as bone within a connective tissue membrane in a process called **intramembranous** (in″trah-mem′brah-nus) ossification. Bones derived from somites usually begin development as cartilage models and the cartilage is later replaced by bone in a process called **endochondral** (en″do-kon′dral) ossification (bone formation within cartilage). Though the processes of endochondral and membranous ossification differ, the bone that results is the same. Most bones of the arms and legs are endochondral, whereas the flat bones of the skull are membranous. Bones of the face,

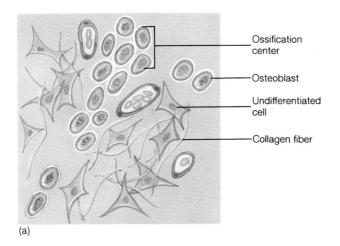

(a)

Ossification center

Osteoblast

Undifferentiated cell

Collagen fiber

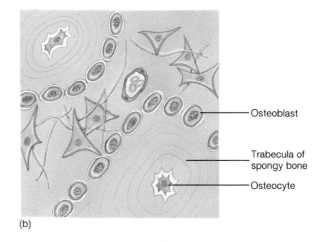

(b)

Osteoblast

Trabecula of spongy bone

Osteocyte

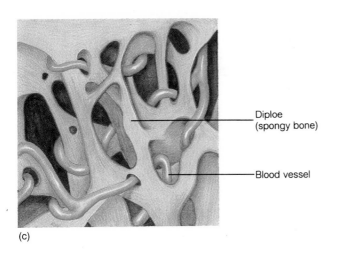

(c)

Diploe (spongy bone)

Blood vessel

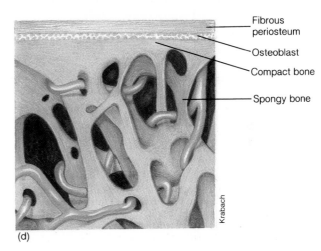

(d)

Fibrous periosteum

Osteoblast

Compact bone

Spongy bone

Krabach

FIGURE 6.2

Intramembranous bone formation: (*a*) osteoblasts beginning to form bone in a membrane, (*b*) trabeculae (bars of bone) form around trapped osteocytes, (*c*) blood vessels grow into diploe made of cancellous bone, and (*d*) compact bone forms plates, which are surrounded on both surfaces by periosteum.

base of the skull, and vertebrae are partly endochondral and partly membranous.

Though each bone has its own developmental schedule, a few have begun to ossify (become bone) by the eighth week of embryonic development and many more by the twelfth week. All bones continue to grow after birth and ossification is not complete in all bones until about age 25.

Intramembranous Bone Formation

Development of a connective tissue membrane in the region where a bone will form marks the beginning of intramembranous ossification (Figure 6.2). Numerous small blood vessels and undifferentiated cells grow into the membrane. Some cells differentiate into bone-forming cells called **osteoblasts** (os'te-o-blastz''), which secrete collagen. The collagen from many osteoblasts forms a network of fibers called **osteoid** (os'te-oid) within the membrane. As minerals are deposited in osteoid, it be-

comes more dense, trapping osteoblasts so they neither divide nor deposit bone. These cells, now called **osteocytes** (os'te-o-sītz), are the metabolically active cells of mature bone. The peripheral membrane on both the inner and outer surfaces becomes the **periosteum** (per''e-os'te-um), the membrane that covers the surface of a bone.

During their development, membranous skull bones grow and change shape according to changes in the size and shape of the brain. Osteoblasts near the outer periosteum continue to form bone even though those near the inner periosteum and growing brain have become trapped and cannot form new bone. **Osteoclasts** (os'te-o-klastz''), cells that digest bone, are present near the inner periosteum. These cells digest old bone and enlarge the cranial cavity at the same time osteoblasts deposit new bone on the outer surface. This allows the skull to grow with little change in the thickness of the bone. Openings called foramina appear in the skull because blood vessels develop first and bone develops around the vessels.

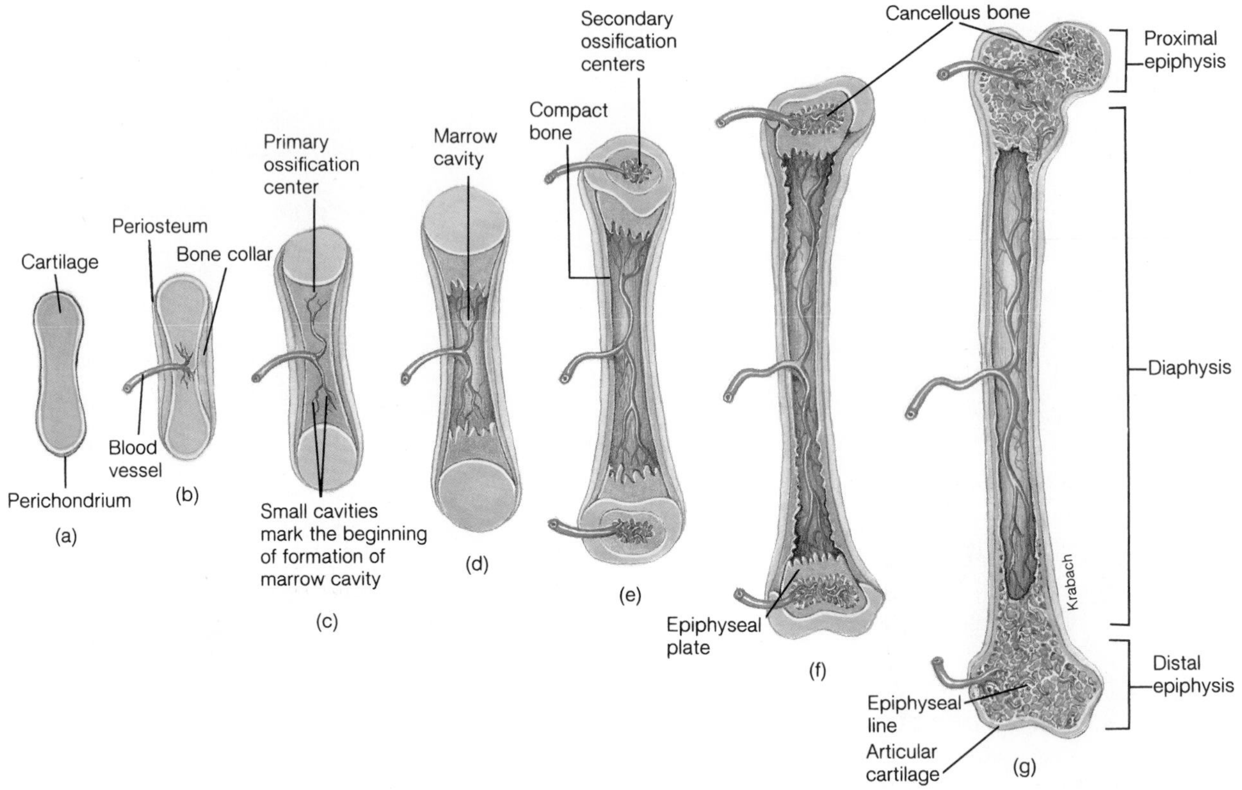

FIGURE 6.3

Endochondral bone formation: (*a*) a cartilage model, (*b*) formation of collar, (*c*) beginning of primary ossification, (*d*) formation of marrow cavity and entry of blood vessels into it, (*e*) formation of secondary ossification centers, (*f*) cartilage remaining at articular surfaces and epiphyseal plates, and (*g*) fully formed bone with epiphyseal lines.

Endochondral Bone Formation

As a bone begins to form by endochondral ossification, as shown in a long bone in Figure 6.3, hyaline cartilage, having the general shape of the future bone, is laid down. This cartilage model is covered by a membrane, the **perichondrium** (per″ĭ-kon′dre-um).

The first evidence of endochondral ossification appears in the **diaphysis** (di-af′ĭ-sis), or middle part of a long bone, at the *primary ossification center.* Chondrocytes proliferate, increasing the cartilage mass, and the cartilage calcifies. Because nutrients cannot diffuse into calcified cartilage, the chondrocytes die. As chondrocytes die, undifferentiated cells in the perichondrium become osteoblasts. The osteoblasts deposit a perforated collar of compact bone around the disintegrating calcified cartilage. The perichondrium of the bony collar is now periosteum because it covers bone. Blood vessels grow through the perforations in the collar and toward the calcified cartilage.

As the cartilage continues to disintegrate, a primitive **marrow** or **medullary** (med′u-lār″e) **cavity** forms. Undifferentiated cells infiltrate the cartilage, align themselves on cartilage remnants, and differentiate into osteoblasts. The osteoblasts enlarge the bony collar and form spicules of cancellous bone within the marrow cavity. Some cancellous bone is reabsorbed by osteoclasts, enlarging the marrow cavity, and many blood vessels grow into the marrow cavity.

The above processes occur in long bones as early as the eighth week of development and continue far into childhood. The ossification process begins at the middle of the diaphysis of each bone and spreads in both directions toward the epiphyses at a rate characteristic for the bone.

About the time of birth, *secondary ossification centers* are established in the epiphyses (ends) of long bones. Calcification and disintegration of cartilage, proliferation of osteoblasts and blood vessels, and deposition of collagen and minerals occur in the same way as in primary ossification centers. The secondary centers are separated from the primary center by **epiphyseal** (ep″ĭ-fiz′e-al) **plates,** which consist of the remains of the cartilage model. Each epiphysis also has a sheet of articular cartilage covering the end of the bone where it articulates with (comes in proximity to) another bone in a joint.

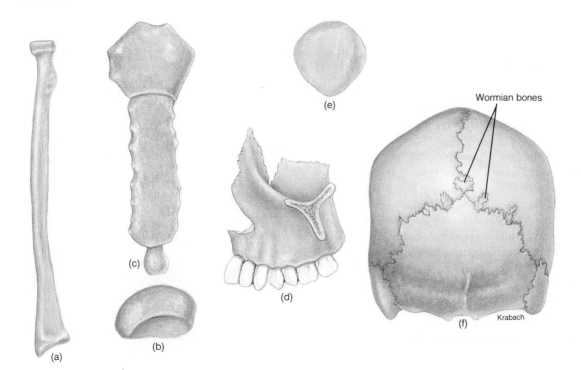

FIGURE 6.4
Types of bones: (*a*) long bone, (*b*) short bone, (*c*) flat bone,
(*d*) irregular bone, (*e*) sesamoid bone, and (*f*) sutural bone.

Bone Growth

Have you ever wondered how bones grow while they maintain sufficient strength to support a child's body? To answer that question, let's look at events that occur in secondary ossification centers and in epiphyseal plates. During growth, chondroblasts in the plates divide by mitosis and elongate the bone, increasing the distance between the epiphyseal plates and lengthening the bone. Ossification continues at both ends of the primary center and in the secondary centers. The latter makes the bone ends rigid and protects the growing cartilage.

Toward the end of puberty when growth is nearly complete, sex hormones suppress division of cartilage cells and the epiphyseal plates become thinner. When growth is complete at about age 17, the cartilage cells cease to divide, but ossification continues for about eight years until the diaphysis unites with the epiphyses, leaving only the **epiphyseal line** at the site where longitudinal growth once occurred. During growth, bones also increase in diameter by the action of osteoblasts beneath the periosteum. At the same time, osteoclasts on the inner surface of the bone enlarge the marrow cavity and keep the thickness of the bone nearly constant.

Children and adolescents are prone to skateboard and other accidents that can shear an epiphyseal plate in a leg bone. If this happens, bone growth in the leg can be slowed or accelerated, resulting in legs of different lengths.

CHARACTERISTICS OF BONES

Bones come in a variety of shapes and sizes, but they share many characteristics. All bones have a sturdy collagen matrix filled with mineral deposits. Their shapes are appropriate for the kinds of support and protection they supply. All bones have a blood supply that carries nutrients to their cells and removes wastes from them. At the same time, each bone has certain definitive characteristics. A person familiar with these characteristics could assemble a human skeleton from a box of bones, placing each bone in its proper anatomical position.

Types of Bones

Bones are classified by shape as long, short, flat, or irregular (Figure 6.4). In addition to the diaphysis, *long bones* typically have two **epiphyses** (ep-if′is-ēz), or end portions. Most are slightly curved; the curvature absorbs shocks and distributes stresses. Long bones have a large central marrow cavity almost as long as the diaphysis. Arm and leg bones, such as the humerus, radius, ulna, metacarpals, femur, tibia, fibula, and metatarsals (Chapter 7) are long bones. *Short bones* often are approximately cube shaped and are made of peripheral compact bone and central cancellous bone with very small marrow spaces. The carpals (wrist bones), tarsals (ankle bones), and phalanges (finger and toe bones) are short bones. *Flat bones* have a sandwichlike structure where the "bread" is compact bone and the "filling" is cancellous bone called **diploe** (dip'lo-e) with interspersed marrow spaces. The sternum, ribs, and bones of the cranium are flat bones. *Irregular bones* are bones that do not fit into any of the aforementioned categories. They include vertebrae, the hyoid bone, and some of the facial bones. Irregular bones have peripheral compact bone and central cancellous bone with varying numbers of marrow spaces.

Two other types of bone are defined by location. **Sesamoid** (ses'ah-moid) **bones** are imbedded in tendons where excessive pressure is exerted. The patella (kneecap) is the only sesamoid bone regularly present in the human body, but some individuals have additional sesamoid bones in the tendons of the shoulder. **Sutural** (su'tu-ral) **bones,** also called *wormian bones,* are found between joints called **sutures** (su'churz) of some, but not all, skulls. Both sutural and unusual sesamoid bones develop from extra isolated ossification centers that form during embryological development.

Structure of a Long Bone

A long bone (Figure 6.5) has a diaphysis and two epiphyses, as already described. The diaphysis is primarily compact bone, whereas the epiphyses have an outer layer of compact bone and an inner layer of cancellous bone. Marrow is found in the **medullary cavity** and within spaces in cancellous bone. The marrow cavity is lined with **endosteum** (en-dos'te-um), a membrane similar to the periosteum.

Bone marrow is of two types: red and yellow. **Red bone marrow** is a **hematopoietic** (hem"ah-to-poi-et'ik) **tissue;** that is, it can produce blood cells. In infants and young children, red bone marrow is found in many bones. In adults, it is found mainly in the ribs, sternum, vertebrae, and the coxal (pelvic) bones—areas of the body that have the highest temperature. **Yellow bone marrow** gradually replaces much red bone marrow as a child matures, coming to fill the medullary cavities in most long bones of adults. Normally, yellow marrow consists mainly of fat cells and makes no blood cells. When new blood cells are needed quickly, such as after sudden, severe blood loss, some yellow marrow begins to produce blood cells and again becomes red marrow.

Bones have an extensive blood supply. A **nutrient artery** enters a bone through a hole called a **nutrient foramen** (fo-ra'men) and passes to the marrow cavity. Branches of the vessel pass obliquely through the matrix of compact bone in channels called **Volkmann's canals.** Smaller branches enter the **Haversian canals** in the center of osteons. Numerous small epiphyseal vessels enter long bones near their ends, and branches of these vessels also supply osteons. Furthermore, the periosteum has a rich blood supply.

Bone Markings

Bones have surface markings that reflect their functions (Figure 6.6 and Table 6.1). The ends of most bones have joint surfaces such as **depressions** and rounded **processes.** The processes on one bone fit into the depressions in an adjacent bone. Bones have processes where muscles and tendons attach, and have **fissures** (grooves or openings) for the passage of blood vessels, lymph vessels, and nerves.

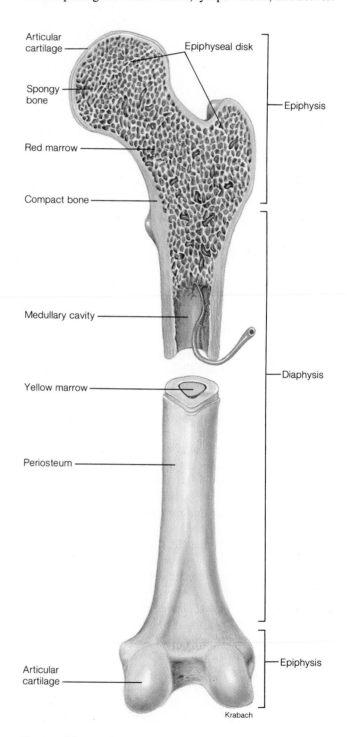

FIGURE 6.5
The structure of a long bone.

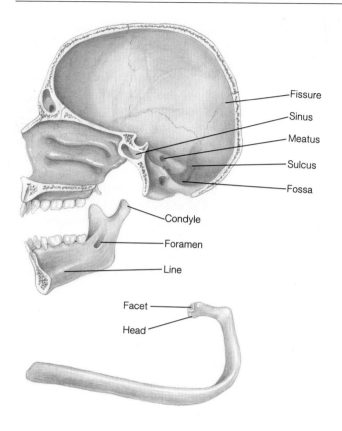

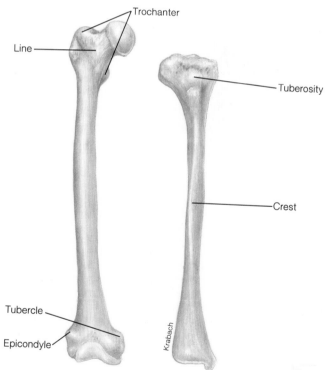

FIGURE 6.6
Bone markings.

TABLE 6.1	Bone Markings
Openings	
Fissure	A slit between two bones through which nerves or blood vessels pass, such as the superior orbital fissure of the sphenoid bone
Foramen	A hole within a bone through which nerves or blood vessels pass, such as the foramen magnum of the occipital bone
Meatus	A tubelike passageway within a bone, such as the auditory meatus of the temporal bone
Sinus	A cavity within a bone, such as the frontal sinus
Depressions	
Fossa	A simple depression or hollowing in or on a bone, such as the mandibular fossa of the temporal bone
Sulcus	A groove that may contain a blood vessel, nerve, or tendon, such as the malleolar sulcus of the tibia
Joint Processes	
Condyle	A large convex protrusion at the end of a bone, such as the medial or lateral condyles of the femur
Head	A round protrusion separated from the rest of a bone by a neck, such as the head of the humerus or femur
Facet	A flat smooth surface, such as a facet of a rib
Processes for the Attachment of Ligaments, Tendons, and Muscles	
Crest	A prominent ridge on a bone, such as the iliac crest of the coxal bone
Epicondyle	A second protrusion above a condyle, such as the medial or lateral epicondyles of the femur
Line	A less prominent ridge on a bone, such as the linea aspera of the femur
Tubercle	A small round protrusion, such as the greater tubercle of the humerus
Tuberosity	A large, round, and usually roughened protrusion, such as the ischial tuberosity of the coxal bone
Trochanter	A large protrusion found only on the femur

PHYSIOLOGY OF BONE

Instead of being inert structures, bones are metabolically active and well supplied with blood from which they obtain nutrients such as calcium, phosphate, and vitamins. They also are influenced by several hormones.

Bone Functions

Upon studying a skeleton, one might conclude that bone is merely an inert supporting substance. Bone does support the body well—it withstands tension and compression as well as reinforced concrete—but it performs other functions, too. It protects internal organs such as the brain, spinal cord, heart, and lungs. Bones contribute to movement by providing attachment sites for muscles that extend across and move joints. Bones also serve as reservoirs for minerals such as calcium, phosphorous, and sodium, which are used in various physiological processes. Finally, **marrow,** a tissue within hollow bone marrow cavities, produces blood cells.

Bone Maintenance

Bones are constantly being remodeled—a little bone tissue added along one surface and a little reabsorbed along another. By this means, certain areas of bones are thickened and reinforced in response to exercise and mechanical stress as discussed below. The processes of bone formation and reabsorption occur throughout life. In childhood and adolescence, more bone is formed than is reabsorbed, and these young bones contain large amounts of collagen. In adulthood up to about age 40, the processes are in equilibrium; but after age 40, more bone is reabsorbed than is formed. By old age, the marrow cavity has enlarged and minerals and collagen have been lost from the matrix, making the bones brittle and subject to fracture. As we shall see, hormones and nutritional factors also affect the maintenance of bone.

Exercise and Mechanical Stress Affect Bones

Keeping bone healthy depends, in part, on exercise and mechanical stress—both of which increase the volume of blood flowing through bone and the rate of exchange of substances between bone and blood. Exercise and moderate mechanical stress also prevent formation of excess fibrous connective tissue that can cause joint stiffness and reduced mobility. These factors illustrate the "use-it-or-lose-it" principle—placing stress on bones keeps them hard and resilient.

Mechanical stress contributes to bone remodeling by creating pressure on the crystalline hydroxyapatite. Such pressure causes convex (outwardly curving) surfaces to become negatively charged and attract positive ions, including H^+ in extracellular fluids. Removing the H^+ makes the extracellular fluid alkaline and increases the activity of the enzyme *alkaline phosphatase* (fos'fah-tās) in osteoblasts and osteocytes. This enzyme makes calcium phosphate and fosters bone formation. Lack of pressure causes concave (inwardly curving) surfaces to become positively charged. The extracellular fluid becomes acidic and increases the activity of the enzyme *acid phosphatase* in osteoclasts. This enzyme digests calcium phosphate and fosters bone degradation. Together, these processes thicken bone that is under pressure (where outward curving occurs) and reduce the thickness of bone that is not under pressure (where inward curving occurs).

Vitamins and Hormones Affect Bones

In addition to the effects of exercise and mechanical stress on bone maintenance, vitamins and hormones play important roles in both the growth and maintenance of bone. Three vitamins (A, C, and D) and many hormones are known to be involved in this complex regulatory activity. Though the actions of vitamins and hormones are discussed in more detail in Chapters 25 and 16, respectively, their effects on bone are summarized here.

Vitamins A and C are required in synthetic processes. Vitamin A facilitates synthesis of chondroitin sulfate, a substance intermingled with collagen fibers that gives plasticity to the organic bone matrix. Vitamin C (ascorbic acid) facilitates the addition of hydroxyl groups to the amino acid proline and the formation of cross-linkages between collagen molecules; cross-linkages add tensile strength to the bone matrix.

Both weak and excessively strong collagen contribute to human disease. Weak collagen, which has too few cross-linkages, can be due to genetic defects, vitamin C deficiency (scurvy), and other causes. Its effects include weak bone that leads to spinal and other deformities, weak connective tissue that leads to tearing of joint tissues and blood vessel walls, and skin lesions. Excessively strong collagen, which has so many cross-linkages that collagen becomes rigid, can be due to genetic defects and aging. Its effects include pulmonary fibrosis in which lungs lose elasticity, joint stiffness, and loss of skin elasticity.

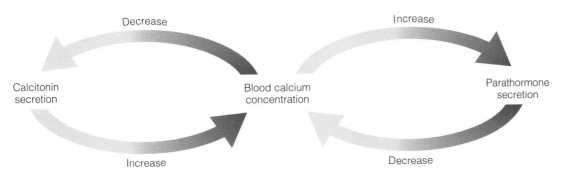

FIGURE 6.7
The blood calcium concentration is regulated by the blood concentrations of parathormone and calcitonin, which are regulated by negative feedback from the calcium ion concentration. Blood calcium concentration is proportional to blue color intensity of arrows.

Vitamin D, after activation by enzymes in the kidney or liver, travels in the blood to the intestinal mucosa where it facilitates Ca^{2+} absorption by inducing synthesis of a calcium carrier protein and increasing the active transport of Ca^{2+} into mucosal cells. Though vitamin D is necessary to maintain adequate blood Ca^{2+}, an excess can cause bone destruction by stimulating osteoclasts to remove Ca^{2+} from bone. Vitamin D synthesis is stimulated by parathormone, or parathyroid hormone (PTH), and inhibited by high blood Ca^{2+} concentration.

Parathormone increases blood Ca^{2+}. Low blood Ca^{2+} stimulates its release and high blood Ca^{2+} inhibits its release by negative feedback (Figure 6.7). Parathormone, in addition to activating vitamin D, causes the kidneys to return Ca^{2+} to the blood, maintaining sufficient blood Ca^{2+} to meet the needs of all cells. Though parathormone can remove Ca^{2+} from bone to maintain adequate blood Ca^{2+}, it normally facilitates bone deposition by making Ca^{2+} available to osteoblasts.

The hormone calcitonin, which opposes parathormone, reduces the blood Ca^{2+} concentration by decreasing osteoclast activity, by preventing formation of new osteoclasts, and by increasing osteoblast activity, at least for a few days. Calcitonin is more effective in children because their osteoclasts are far more active than those in adults. In children, osteoclasts release up to 5 grams of Ca^{2+} into extracellular fluids daily; but in adults, osteoclasts release only 0.8 grams. Calcitonin acts for one to two hours as a Ca^{2+} regulator until its effects are overridden by parathormone.

Growth hormone, which stimulates protein synthesis in general, has special effects on bone. It stimulates division of chondroblasts and osteoblasts, causes osteoblasts to secrete collagen, and promotes Ca^{2+} absorption from the intestine. Thyroid hormones and insulin also appear to work with growth hormone to promote collagen synthesis.

The steroid hormones cortisol and aldosterone from the cortex of the adrenal glands, testosterone from the testes, and estradiol from the ovaries also affect bone metabolism. Cortisol can cause collagen degradation with concurrent reabsorption of Ca^{2+} into the blood. Aldosterone stimulates the reabsorption of Na^+ and other minerals from the kidneys, and makes these minerals available for deposition in bone. (Though bone contains mainly hydroxyapatite, it also contains small quantities of other minerals.) Testosterone promotes protein synthesis during bone development, calcium retention, and the deposition of Ca^{2+} in bones. Estradiol promotes protein synthesis and increased bone density by Ca^{2+} accumulation. It can cause osteoclasts to convert to osteoblasts and, thereby, foster bone deposition.

In summary, several factors control bone physiology. Exercise and mechanical stress shape bones and maintain mobility. Parathormone maintains adequate blood Ca^{2+}, partly by activating vitamin D. Other hormones and vitamins facilitate development, growth, or maintenance.

BONE DISORDERS

Bones are subject to a number of structural problems that lead to inadequate support or impaired movement of the body. These conditions can be caused by injury or disease.

In **rickets,** which occurs in children, and **osteomalacia** (os"te-o-mah-la'she-ah), which occurs in adults, bones fail to calcify because of a vitamin D deficiency (Figure 6.8a). Too little calcium is absorbed from the food, even when the diet is adequate, and the blood calcium concentration drops. Homeostatic mechanisms, especially parathormone action, raises the blood calcium concentration by removing calcium from the bones. Fluctuations of blood calcium outside a narrow, tolerable range can be life threatening. In children, calcium sometimes never reaches

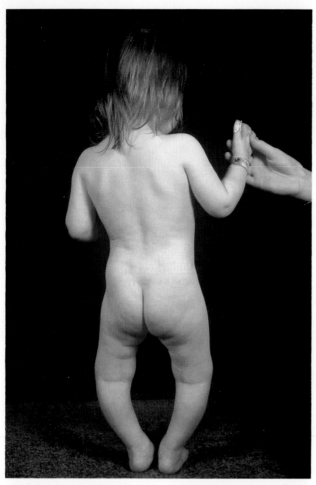

(a)

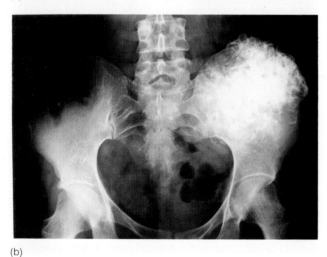

(b)

FIGURE 6.8
(a) A child with rickets, and (b) a patient with osteoporosis.

the bones. Calcification of a normal collagen matrix fails to occur because the concentrations of Ca^{2+} and P_i ions fail to reach the minimum levels for mineral deposition at many sites in bones. Osteomalacia and rickets are treated with dietary vitamin D.

Certain kidney diseases and some inherited defects also sometimes prevent calcification. When diseased kidneys allow excess excretion of calcium, homeostatic mechanisms maintain a nearly constant blood calcium concentration by removing calcium from the bones to replace that loss through the kidneys. In inherited vitamin D-resistant rickets, the kidneys excrete excessive amounts of phosphate, thereby preventing it from being incorporated into the bones. Vitamin D is not effective in treating this disorder. Though administering it increases calcium absorption, the calcium is of no use unless phosphorus is available to be deposited at the same time in the bones.

Osteoporosis (os''te-o-po-ro'sis) is a disease in which bone loses both minerals and protein fibers from its matrix (Figure 6.8b). It is especially common in postmenopausal women. For unknown reasons, white women are particularly susceptible to osteoporosis, both white and black men are somewhat susceptible to it, and black women are particularly resistant to it. Individuals taking hydrocortisone are especially prone to bone loss because hydrocortisone blocks the formation of new bone and decreases calcium absorption. Normal bone maintenance is impaired and some existing bone is lost. Diabetic individuals also are prone to bone loss. Those who take oral hypoglycemic medications are more severely affected than those who receive insulin injections. Even without any of these medications, many elderly people suffer from osteoporosis.

In osteoporosis, the vertebral column is most subject to degeneration and the affected individual becomes shorter in stature and hunched. Other bones also are weakened and fractures are common. Sometimes in elderly persons, a bone fractures and causes a fall—rather than a fall causing a fracture.

Several studies have been made of factors that might cause or prevent osteoporosis. Estrogens given to postmenopausal women significantly retard bone loss, though they may also foster breast cancer in susceptible women. Exercise that creates mechanical stress, especially weight bearing exercise also retards mineral loss, but extra dietary calcium may not do so. Smoking and consuming more than two alcoholic drinks per day each double the risk of developing osteoporosis.

Many physicians recommend estrogen replacement after menopause to reduce the risk of osteoporosis. Others recommend high calcium intake—at least 1000 mg per day for everyone and 1500 mg for postmenopausal women—to reduce the risk of osteoporosis. This recommendation assumes that extra dietary calcium might prevent bone loss and that it would not likely cause any

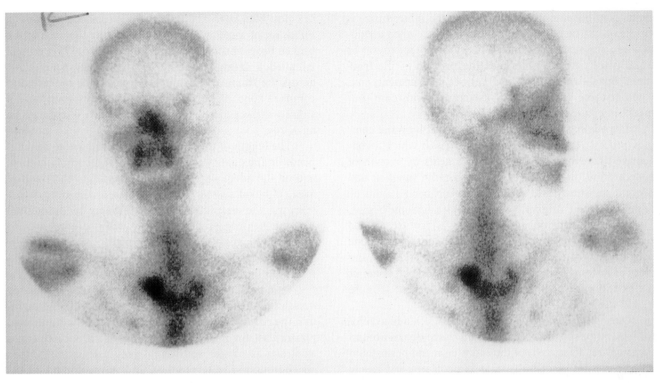

FIGURE 6.9
A bone scan showing the incorporation of a radioactive substance into
the bone marrow.

damage, though these assumptions are not necessarily
valid. Large amounts of dietary calcium may suppress vitamin D absorption (by suppressing parathormone secretion) and contribute to calcium-based kidney stones.

A bone scan (Figure 6.9) is often done on individuals who
have had bone cancer or who may have metastasis from
another site to the bones. In a bone scan, a dose of a
radioactive substance is given, and radiographs are taken over
the next few minutes while the substance is being incorporated
into the bone marrow. The radioactive substance, which shows
up as black dots on the developed film, is greatly concentrated
in malignant tissues because these tissues take up the
radioactive substance more rapidly than normal tissues.

FRACTURES AND HOW THEY HEAL

Fractures, or breaks in bones, are common bone injuries
that are more often due to twisting than to compression.
Though we may think of fractures as mainly affecting

bones, the tissues around the bones usually are damaged
as well. Skin and muscles can be bruised and lacerated.
Ligaments and tendons can be torn. And blood vessels and
nerves can be damaged.

Fractures are classified by whether the bone projects
through the skin and by the nature of the bone damage
(Figure 6.10). In a **closed,** or **simple, fracture,** the skin is
not broken; whereas in an **open,** or **compound, fracture,**
the skin over the fracture is broken and the end of a broken
bone may protrude through the skin and muscles. A person
with a compound fracture is at serious risk of infection.
A fracture is **complete** when the bone is broken into two
parts or **incomplete** (greenstick) when the bone is cracked.
Greenstick fractures are more common than complete
fractures in children because their bones, which contain
more collagen fibers and less mineral than adult bones,
are somewhat flexible. In a **comminuted** (kom'in-ūt''ed)
fracture, the bone is broken into several fragments.

Pain and swelling accompany fractures as they do
many other kinds of injuries. Consequently, it is difficult
to diagnose a fracture without using an X ray unless a

protruding bone or clear deformity of the injured part is visible or can be felt through the skin. Even when a fracture can be detected without an X ray, X-ray photographs usually are taken to determine the position of the fractured bone, whether other fractures are present, and whether there are fragments of bone or periosteum that might interfere with healing.

The healing of a fracture is a relatively long and complex process. Unlike many other injuries in which a connective tissue scar forms, a fracture heals by growth of bone around the broken ends. To maintain function and prevent deformity, the ends of a broken bone should be aligned in their original position before healing begins. This is called **reduction,** or setting, of a fracture. **Closed reduction** is done by manipulating the ends of the bone beneath the skin; **open reduction** is done surgically and often involves the use of rods or pins within the bone, or plates and screws along the surface of the bone (Figure 6.11).

During the healing process (Figure 6.12), a blood clot forms, and phagocytic cells clean up tissue debris around the fracture. Then, cells of the endosteum and periosteum differentiate into chondroblasts and osteoblasts, and invade the clot. New cartilage forms and is replaced by bone in the shape of a collar, which forms around the ends of the broken bone on the inner and outer surfaces. This growth is called a **callus** (kal'us). New cancellous bone grows across the fracture line, and eventually, it is converted to compact bone. Bone remodeling continues until only a slightly thickened area around the fracture is apparent in an X ray.

The length of time required for a fracture to heal depends on the quality of the reduction and immobilization, and on the adequacy of the blood supply to the fracture area. If blood vessels were damaged at the time of the fracture, the circulation to the area may be diminished and the healing process slowed. Because early weight bearing improves blood circulation and accelerates healing, an individual is encouraged to place some weight on a fractured limb as soon as the limb can withstand it.

Immobilization of fractured bones and associated muscles that exert force on the bones is essential to maintain the reduction. Fiberglass has replaced the traditional plaster cast for many fractures because it is three times as strong as plaster and weighs only half as much. It is also porous and waterproof, so air can reach the skin be-

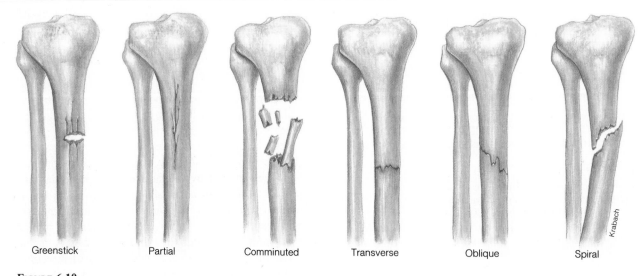

Greenstick   Partial   Comminuted   Transverse   Oblique   Spiral

FIGURE 6.10
Some different kinds of fractures. A *greenstick* fracture is incomplete, and the break occurs on the convex surface of the bend in the bone. A *partial (fissured)* fracture involves an incomplete longitudinal break. A *comminuted* fracture is complete and results in several bony fragments. A *transverse* fracture is complete, and the break occurs at a right angle to the axis of the bone. An *oblique* fracture occurs at an angle other than a right angle to the axis of the bone. A *spiral* fracture is caused by twisting a bone excessively.

neath it, which reduces itching, and the cast itself can be washed. Traction, a pulling force on a limb, can be used to maintain its proper position while healing. Electrical stimulation is being used to stimulate bone healing, especially in nonunion fractures (fractures in which bone ends fail to unite in the normal period of time).

CLINICAL TERMS

achondroplasia (ah-kon″dro-pla′ze-ah) a genetic defect leading to dwarfism

acromegaly (ak″ro-meg′ah-le) enlargement of the bones of the face, hands, and feet due to excess growth hormone in adult life

Colles' fracture a fracture of the lower end of the radius with posterior displacement, which often occurs when a person tries to "break a fall"

osteitis deformans (os″te-i′tis de-for′manz) an inflammation of bone in which some skeletal parts become thickened and others thinned; Paget's disease

osteoblastoma (os″te-o-blas-to′mah) a tumor produced by blood-forming cells

osteogenesis imperfecta (os″te-o-jen′ĕ-sis im-per-fek′tah) a genetic defect that leads to improper collagen synthesis and, thus, to defects in skin, cartilage, tendons, and bones

osteogenic sarcoma (os″te-o-jen′ik sar-ko′mah) a malignant growth of bone tissue

osteomyelitis (os″te-o-mi″ĕ-li′tis) an inflammation of the marrow cavity and usually the surrounding bone tissue

pathological fracture a spontaneous fracture in a bone weakened by disease

Pott's fracture a fracture of the lower fibula accompanied by injury to the articulation of the fibula with the talus.

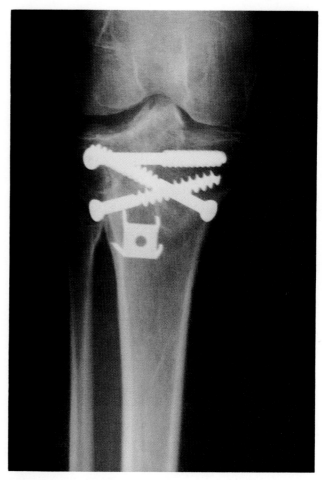

FIGURE 6.11
In open reduction of a fracture, screws, plates, and rods are often used to hold bone pieces together.

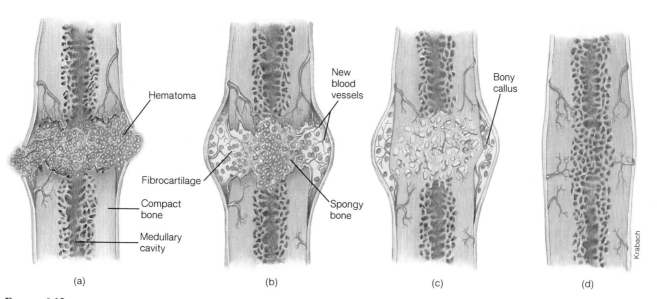

FIGURE 6.12
The healing of a fracture. (*a*) Blood escapes from ruptured blood vessels and forms a hematoma. (*b*) Spongy bone forms in regions close to developing blood vessels, and fibrocartilage forms in more distant regions. (*c*) Fibrocartilage is replaced by a bony callus. (*d*) Osteoclasts remove excess bony tissue, making new bone structure much like the original.

ESSAY
PAGET'S DISEASE

Described in 1877 by the English surgeon Sir James Paget, **Paget's disease** (osteitis deformans) is characterized by excessive and abnormal remodeling of bone in active phases interspersed with quiescent phases (Figure 6.13). During active phases, marked reabsorption is followed by disorderly bone replacement. Certain bones are especially likely to be affected—pelvic bones, skull, vertebrae, and femur. Other limb bones are less often affected. The rate of progression of Paget's disease in an affected bone has been estimated to be between 7 and 16 mm per year.

Though the disease is often asymptomatic, it affects 3 to 4 percent of humans over age 40—more in Great Britain and Australia, and fewer in Japan and China. The disease affects men more often than women and has been observed to occur in several members of the same family.

Paget's disease has been attributed to inflammation, genetic defects, immunologic factors, malignancy, hormones, and other factors, but its cause is still not entirely clear. The most convincing evidence today focuses on a slowly developing viral infection of osteoclasts, possibly the same virus that causes measles.

No treatment was available for Paget's disease until the 1950s when calcitonin, which inhibits bone reabsorption, was first used. This hormone reduces alkaline phosphatase activity and decreases excretion of hydroxyproline, a collagen component. Within a few weeks, pain is alleviated in patients with painful disease. Calcitonin apparently has no effect on bone formation. If treatment is stopped, a period of bone reabsorption occurs that is more rapid than it was in the untreated state. A second agent used to treat Paget's disease, etidronate disodium (EHDP), inhibits reabsorption of bone by binding to hydroxyapatite crystals preventing them from either growing or dissolving. Other effects of EHDP are the same as those of calcitonin, except that patients treated with EHDP may be especially susceptible to fractures. The antibiotic mithramycin successfully prevents bone reabsorption in Paget's disease, decreases bone pain, and fosters deposition of nearly normal bone. Because of the extreme toxicity of mithramycin, it is reserved for victims of Paget's disease who do not respond to other treatments.

In addition to bone damage itself, Paget's disease may contribute to a variety of other disorders, all of which can also be seen in the absence of Paget's disease. These disorders include stress fractures, several kinds of malignancies, osteomyelitis, joint degeneration, and neurological disorders often due to bone deposits that compress the brain or spinal cord.

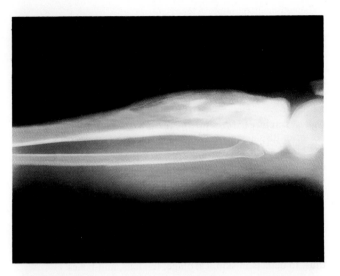

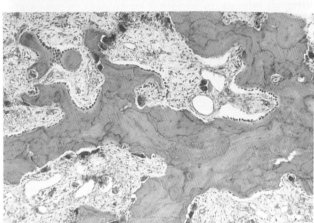

FIGURE 6.13
Disorderly bone in Paget's disease.

Questions
1. What are the main signs of Paget's disease?
2. What kinds of causes have been proposed for this disease?
3. How can Paget's disease be treated?
4. To what other disorders does Paget's disease contribute?

CHAPTER SUMMARY

Bone Tissue

- Compact bone is characterized by osteons, with cytoplasmic processes that deliver nutrients from blood in Haversian canals to osteocytes.
- Cancellous bone lacks osteons, but has sturdy bars of bone that provide great strength with little weight.
- Deposition of hydroxyapatite depends on certain minimum Ca^{2+} and P_i concentrations.

Bone Development

- In intramembranous bone formation, osteoblasts deposit bone in connective tissue membranes.
- In endochondral bone formation, a cartilage model is laid down, blood vessels grow into the perichondrium of the primary ossification center, and cells differentiate into osteoblasts. As cartilage disintegrates, it is replaced by bone in a collar at the primary ossification center and similar processes occur in secondary ossification centers until skeletal growth is complete.
- Both membranous and endochondral bones are constantly remodeled by osteoblasts laying down bone on one surface and osteoclasts digesting bone on another surface.

Characteristics of Bones

- Bones are classified by shape as long, short, flat, or irregular, and by location as sesamoid or sutural bones.
- A long bone consists of a diaphysis and two epiphyses, a medullary cavity lined with endosteum, and an outer membrane called the periosteum.
- Bone markings are summarized in Table 6.1.

Physiology of Bone

- Bone supports, protects, provides for attachment of muscles and movement, stores minerals, and makes blood cells.
- Exercise and stress increase blood circulation through bones, and cause strengthening of bones at high stress points and thinning of bones at low stress points.
- Vitamins A and C increase the strength of the bone matrix.
- Vitamin D, parathormone, and calcitonin work together to maintain the blood Ca^{2+} concentration in a normal range, thereby providing adequate Ca^{2+} for bone deposition.
- Certain other hormones promote bone growth and maintenance.

Bone Disorders

- Rickets in children and osteomalacia in adults are diseases in which bones fail to calcify because of a vitamin D deficiency. Genetic defects and kidney diseases also prevent calcification.

- Osteoporosis, a disease in which bone mass (both collagen and mineral) is lost, results from a variety of causes, especially cessation of estrogen secretion. It occurs most often in postmenopausal white women. It can be reduced in severity by estrogen replacement and exercise that creates mechanical stress.

Fractures and How They Heal

- Fractures are classified as simple (closed) or compound (open), and as complete, incomplete, or comminuted. They are set by open or closed reduction, and heal by callus formation around the fracture site.

QUESTIONS AND PROBLEMS

The questions at the end of each chapter are numbered to correspond with the objectives listed at the beginning of the chapter. Italics indicate that a question requires critical thinking skills beyond simple factual recall.

Questions

1. *How does compact bone differ from cancellous bone?*
2. (a) *How does intramembranous bone development differ from endochondral bone development?*
   (b) *What events occur during the remodeling of a bone?*
3. (a) How are bones classified by shape?
   (b) How do sesamoid and sutural bones differ from other bones?
4. (a) What terms are used to describe bone markings?
   (b) *Work with a partner to identify bones by each giving the other five descriptive terms about a bone.*
5. What are the functions of bone?
6. (a) *How is bone maintained?*
   (b) *How do exercise and mechanical stress affect bones?*
   (c) *Why do bones need vitamins and hormones?*
7. *What are the causes, effects, and treatments for (a) rickets, (b) osteomalacia, and (c) osteoporosis?*
8. (a) List five different kinds of fractures and describe their characteristics.
   (b) *How does a fracture heal?*
   (c) *What are the advantages of early weight bearing on a fractured bone?*

Problems

1. Two individuals have similar leg fractures. One received a walking cast, and the other received a regular cast and crutches. Which person's fracture would heal faster and why?
2. Devise a new method for immobilizing a fracture in an ambulatory person that would be equally effective, but less cumbersome than a plaster or fiberglass cast.

OBJECTIVES

1. Describe the basic plan of the skeletal system, and list its major functions.

2. Locate and identify the bones of the axial skeleton found in the head, and describe their structural characteristics.

3. Locate and identify the bones of the axial skeleton found in the trunk, and describe their structural characteristics.

4. Locate and identify the bones of the pectoral girdle and upper appendage, and describe their structural characteristics.

5. Locate and identify the bones of the pelvic girdle and lower appendage, and describe their structural characteristics.

6. Describe variations in the structure of bones due to age and sex.

7. Describe the causes and effects of scoliosis, lordosis, and kyphosis.

7

THE SKELETON

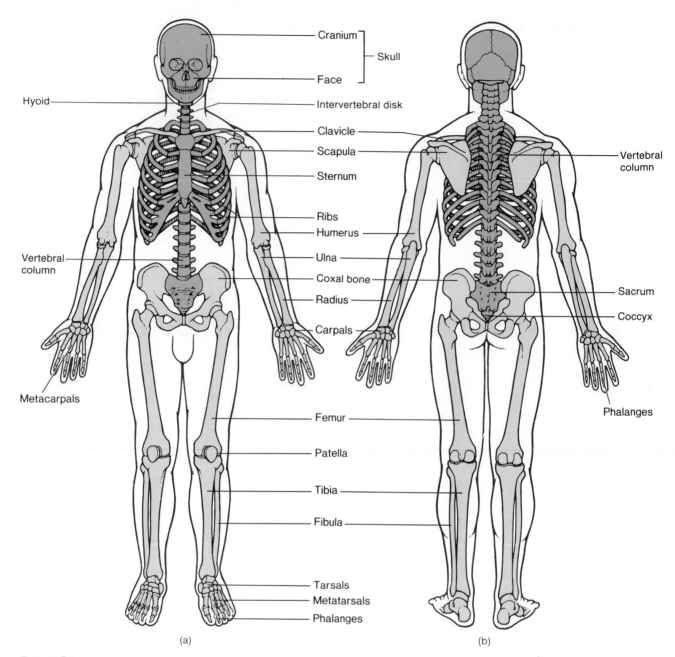

FIGURE 7.1

The human skeleton: (*a*) anterior view, and (*b*) posterior view. The bones of the axial skeleton are shown in orange; those in the appendicular skeleton are shown in yellow.

ORGANIZATION AND GENERAL FUNCTIONS

The **skeletal system** consists of the 206 bones of the body, the joints between them, the ligaments that hold bones together, and all cartilage associated with bones. Though the skeleton (Figure 7.1) functions as a whole; for study, it is divided into two parts, the **axial skeleton** and the **appendicular skeleton.** The axial skeleton includes the skull, vertebrae, ribs, sternum, and hyoid bone. The appendicular skeleton includes the pectoral girdle and bones of the upper appendages (arms), and the pelvic girdle and bones of the lower appendages (legs). The pectoral girdle, which holds the arms to the trunk, includes the paired scapulae and clavicles. The pelvic girdle, which attaches the legs to the trunk, consists of paired coxal bones. The bones of the upper appendage include the humerus, radius, ulna, carpals, metacarpals, and phalanges. The bones of the lower appendage include the femur, tibia, fibula, tarsals, metatarsals, and phalanges. All bones of the appendicular skeleton are paired—one each on the left and right sides of the body. They are summarized in Table 7.1. (In this chapter, the name of a bone appears in **boldface** where it is described. The names of parts of the bone appear in *italics*.)

The skeletal system serves several functions. Some bones protect internal organs. For example, your **skull** forms a hard cover for your brain, and your ribs and sternum (breastbone) form a protective cage around your heart and lungs. Other bones make a rigid framework that supports the rest of your body. In addition to supporting and protecting the body, bones provide surfaces for muscle attachment. When you move, muscles pull on bones and usually cause bending at joints. You can move your whole body, as when you run or swim; or you can move a body part, as when you write or talk. Bone tissue performs still other functions. It provides a reservoir of calcium and other minerals, and is actively involved in maintaining calcium balance in the body. Red bone marrow, a tissue inside certain bones, produces blood cells (Chapter 17).

> For anatomists, bones are useful as anatomical landmarks for locating other organs. This is very beneficial to students who master the terms that pertain to bones because many of the terms are used again to locate or describe other organs.

The axial skeleton (Table 7.2), as its name implies, forms an axis down the midline of the body. The bones of this axis are the skull and **vertebrae.** Lateral to the superior vertebrae are the bones of the thorax, the **ribs** and **sternum,** which form a sturdy cagelike structure that protects the heart and lungs. The skull includes eight bones of the **cranium** (kra′ne-um), which support and protect the brain, and 14 **facial bones,** which give structure to the face. These bones are shown from different perspectives in Figures 7.2, 7.3, and 7.4. Other bones considered part of the axial skeleton include the **hyoid** (hi′oid) and the **middle ear bones.**

TABLE 7.1 Bones of the Appendicular Skeleton				
	Upper Appendage		**Lower Appendage**	
Component	*Anatomical Name*	*Common Name*	*Anatomical Name*	*Common Name*
Girdle	Pectoral	Shoulder	Pelvic	Hip
Proximal segment	Humerus	Upper arm	Femur	Thigh
Middle segment	Radius, ulna	Forearm	Tibia, fibula	Leg
Distal segment	Carpals, metacarpals, phalanges	Wrist, hand, fingers	Tarsals, metatarsals, phalanges	Ankle, foot, toes

TABLE 7.2   Bones of the Axial Skeleton

Skull	22 bones		Middle ear bones	6 bones
Cranial bones	8 bones		Hyoid	1 bone
Frontal	1		Vertebral column	26 bones
Parietal	2		Cervical vertebrae	7
Temporal	2		Thoracic vertebrae	12
Occipital	1		Lumbar vertebrae	5
Sphenoid	1		Sacrum	1
Ethmoid	1		Coccyx	1
Facial bones	14 bones		Thorax	25 bones
Maxilla	2		Ribs	24
Palatine	2		Sternum	1
Nasal	2			
Lacrimal	2		**Total bones of axial skeleton**	**80 bones**
Inferior nasal conchae	2			
Zygomatic	2			
Vomer	1			
Mandible	1			

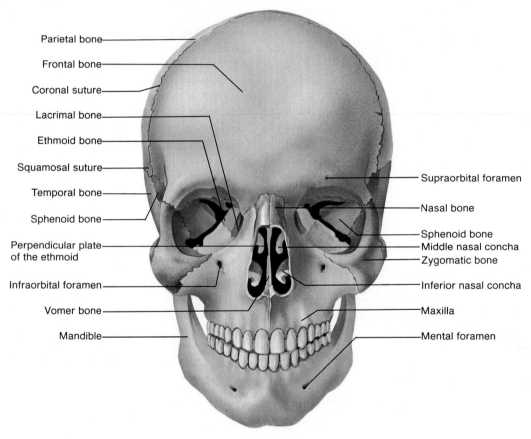

FIGURE 7.2
Anterior view of the skull.

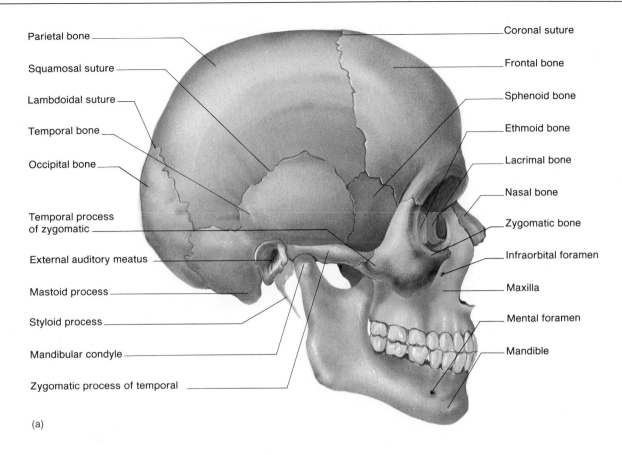

Parietal bone

Squamosal suture

Lambdoidal suture

Temporal bone

Occipital bone

Temporal process
of zygomatic

External auditory meatus

Mastoid process

Styloid process

Mandibular condyle

Zygomatic process of temporal

Coronal suture

Frontal bone

Sphenoid bone

Ethmoid bone

Lacrimal bone

Nasal bone

Zygomatic bone

Infraorbital foramen

Maxilla

Mental foramen

Mandible

(a)

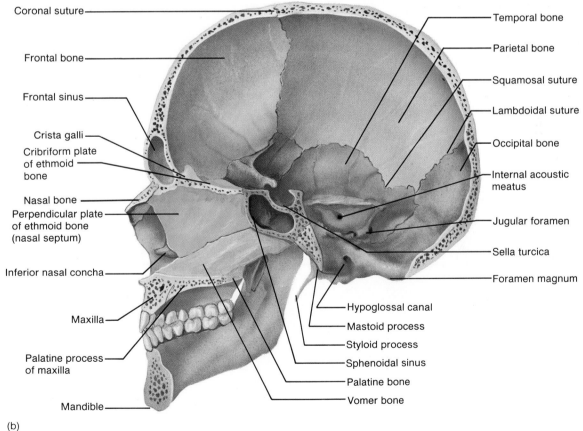

Coronal suture

Frontal bone

Frontal sinus

Crista galli

Cribriform plate
of ethmoid
bone

Nasal bone

Perpendicular plate
of ethmoid bone
(nasal septum)

Inferior nasal concha

Maxilla

Palatine process
of maxilla

Mandible

Temporal bone

Parietal bone

Squamosal suture

Lambdoidal suture

Occipital bone

Internal acoustic
meatus

Jugular foramen

Sella turcica

Foramen magnum

Hypoglossal canal

Mastoid process

Styloid process

Sphenoidal sinus

Palatine bone

Vomer bone

(b)

FIGURE 7.3
Lateral views of the skull: (*a*) external surface of the right side of the
skull, and (*b*) internal surface of the right side of the skull, after
making a sagittal section lateral to the midline.

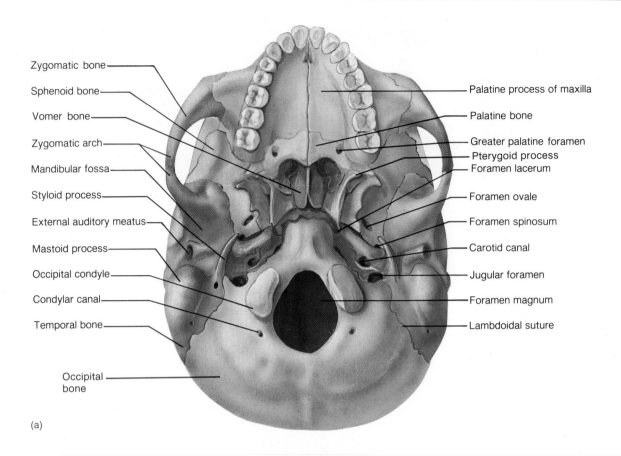

(a)

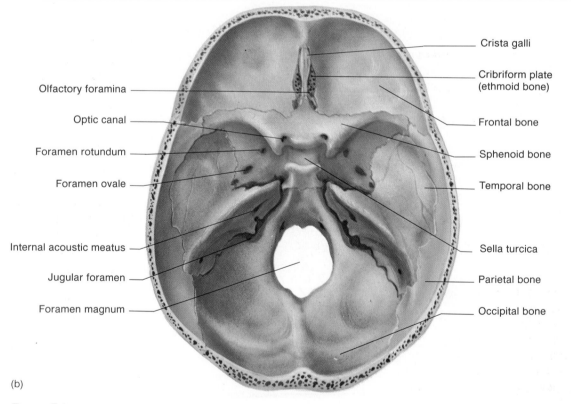

(b)

FIGURE 7.4

Views of the base of the skull: (*a*) inferior view of the external
surface, and (*b*) superior view of the internal surface, after removal of
the upper part of the cranium.

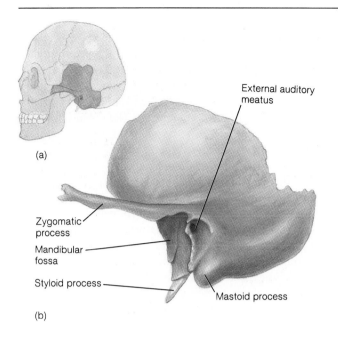

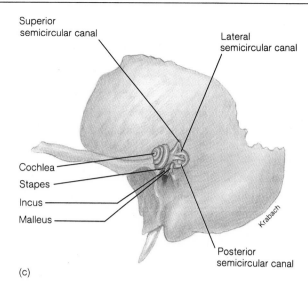

FIGURE 7.5
The temporal bone: (*a*) location of bone in the skull; (*b*) lateral view of the external surface of the right temporal bone; and (*c*) lateral view of the temporal bone, after removal of some of the petrous portion to expose the middle ear bones, cochlea, and semicircular canals.

BONES OF THE HEAD

Bones of the Cranium

The bones of the cranium enclose the brain—the frontal at the forehead, the parietals at the crown, the occipital at the back of the head, and the temporals on the sides. The ethmoid and sphenoid bones form the floor of the cranium.

The **frontal** (frun′tal) bone is a large, curved plate. Its *supraorbital* (soo″prah-or′bit-al) *margin* turns under sharply and forms the roof of each eye orbit (socket). Above each orbit is a *supraorbital foramen,* an opening through which blood vessels and nerves pass to the tissues of the forehead.

The **parietal** (par-i′et-al) bones on each side of the cranium lie posterior to the frontal bone. They join in the midline to form the roof of the cranium.

The **occipital** (ok-sip′it-al) bone joins the parietals and forms the base of the cranium. It has a large hole, the *foramen magnum.* The brainstem becomes the spinal cord at the point at which it passes through this foramen. Blood vessels also pass through this foramen. On either side of the foramen are the *occipital condyles* (kon′dīlz) that articulate with the first cervical vertebra. Superior to each

condyle are the *hypoglossal* (hi″po-glos′al) *canals,* through which the hypoglossal cranial nerves leave the cranium. Posterior to each condyle are the *condyloid* (kon′dil-oid) *canals,* through which veins carrying blood from the brain pass. Rough ridges on the occipital bone mark the attachment sites of neck muscles.

The **temporal** (tem′por-al) bones (Figure 7.5) join the parietal bones and extend downward to form the sides and part of the base of the cranium. The *zygomatic* (zi″go-mat′ik) *process,* which extends from the *squamosal* (skwa-mo′sal) *portion,* joins the zygomatic bone to form the zygomatic arch (cheek bone). The *mandibular fossa* (man-dib′u-lar fos′ah) is a depression, which forms part of the joint by which the mandible (lower jaw) is attached to the cranium. The *external auditory meatus,* the opening for the ear canal, leads to the middle ear where the **malleus, incus,** and **stapes** are located.

The *cochlea* (kok′le-ah) and *semicircular canals* are surrounded by the *petrous* (pet′rus) *portion* of the temporal bone. The *internal auditory meatus,* an opening in the petrous portion, provides a passageway for the cranial nerve from the cochlea and semicircular canals to the brain.

The *mastoid* (mas'toid) *process,* which extends from the *mastoid portion* of the temporal bone, serves as the point of attachment of several neck muscles. The mastoid portion itself contains air spaces that drain into the middle ear cavity; infection in the air spaces is called *mastoiditis* (mas''toi-di'tis). The *styloid* (sti'loid) *process* provides a point of attachment for muscles and ligaments of the hyoid bone, neck and tongue.

The irregularly shaped **ethmoid** (eth'moid) bone (Figures 7.2 and 7.6) forms part of the anterior portion of the cranial floor and part of the nose. The cranial portion contains the *cribriform* (krib'rif-orm) *plate* (*crib,* sieve), through which sensory nerve fibers from the nasal epithelium pass to the brain. It also contains the *crista galli* (kris'tah gal'e), or cock's comb, a triangular process to which membranes that cover the brain are attached.

The inferior portion of the ethmoid consists of the *perpendicular plate,* a large part of the nasal septum, and the *lateral masses* located between the nasal cavity and the orbits of the eyes. Each lateral mass has hollow areas called sinuses and contains two scroll-shaped, membrane-covered bones, the *superior nasal concha* (kong'kah) and the *middle nasal concha.* The conchae increase the surface area inside the nasal cavity; their surface membranes contain receptors for the sense of smell.

The **sphenoid** (sfe'noid) bone (Figures 7.4 and 7.7) is a large, relatively flat bone with winglike projections. Both the *greater* and the *lesser wings* form part of the cranial floor. The *pterygoid* (ter'ig-oid) *processes* extend downward to form part of the lateral wall of the nasal cavity. The sphenoid, which has a common border with most other cranial bones, acts as a keystone to hold the cranium together. The medial portion of the bone is hollow and contains the large *sphenoidal* (sfe-noi'dal) *sinus* (Figure 7.3).

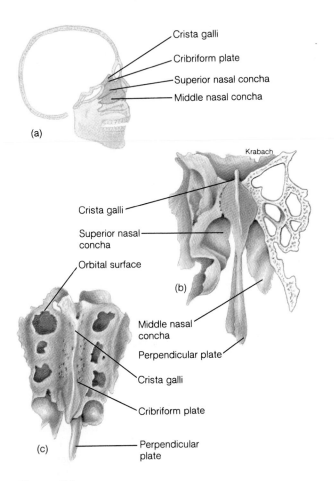

FIGURE 7.6
The ethmoid bone: (*a*) medial view of the skull, with shaded area showing the location of the ethmoid bone; (*b*) enlarged, anterior view; and (*c*) enlarged, superior view.

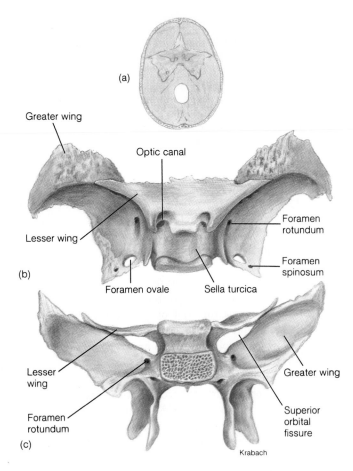

FIGURE 7.7
The sphenoid bone: (*a*) superior view of the internal floor of the skull, showing location; (*b*) enlarged, superior view; and (*c*) enlarged posterior view.

Its superior surface has a depression, the *sella turcica* (sel'ah tur'sik-ah), or Turkish saddle, in which the pituitary gland lies. Just anterior to the sella turcica is a shallower depression where some optic nerve fibers from each eye cross to the opposite side of the brain and pass through the *optic foramina* of the lesser wings. Also on each side of the sphenoid is a *foramen rotundum* (ro-tun'dum) and a *foramen ovale* (o-val'e), through which branches of the trigeminal nerve pass.

Sutures and Fontanels

The joints of the skull (Figure 7.3a), except for those that attach the mandibles, are **sutures,** or immovable joints (Figure 7.8). The *coronal* (ko-ro'nal) *suture* joins the frontal and parietal bones. The *squamosal sutures* join the parietal and temporal bones on each side of the head. The *lambdoidal* (lam-doid'al) suture joins the occipital and parietal bones. The *sagittal suture* (best seen in Figure 7.8) joins the parietal bones in the midline.

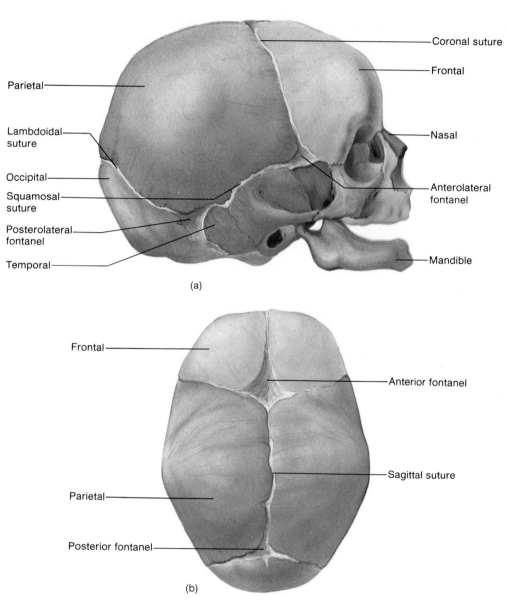

(a)

(b)

FIGURE 7.8
Fontanels of the fetal skull: (*a*) lateral view, and (*b*) superior view.

In the human infant, skull sutures are not completely immovable. At certain places where skull bones have not completely filled in the membrane surrounding them, there are **fontanels** (fon''tah-nelz'), or "soft spots." Fontanels close when bone is laid down in the membrane, and the edges meet and fuse in sutures. The *anterior,* or *frontal, fontanel,* the largest of the fontanels, closes at 18 to 24 months of age. The smaller *posterior,* or *occipital, fontanel* closes by about two months. Of the irregularly shaped fontanels, the *anterolateral,* or *sphenoidal, fontanels* usually close by three months and the *posterolateral,* or *mastoidal, fontanels* by 12 months.

Fontanels allow slight compression and sometimes distortion of the head during the birth process, thereby easing delivery. Distortions disappear shortly after birth and the plates of bone move back to their normal positions. Fontanels also allow the cranium to grow.

Facial Bones

The **facial bones** include six paired bones—the maxilla, palatine, nasal, lacrimal, inferior nasal conchae, and zygomatic—and two unpaired bones—the vomer and mandible (Table 7.2 and Figures 7.2, 7.3, and 7.4).

The **maxillae** (mak-sil'le), which are fused by a midline suture, articulate with every other facial bone except the mandible (Figure 7.9). Each maxilla forms a lateral surface of the nose and the inferior part of an eye orbit. The maxillae contain *maxillary sinuses* and *alveoli,* bony sockets for the upper teeth (Figure 7.3). The *palatine* (pal'ah-tīn) process contributes to the hard palate. The *infraorbital foramen* allows a nerve and an artery to pass through bone of the orbit.

Paired **nasal** bones, each about the size of a fingernail (Figure 7.9), forms the bridge of the nose. They are joined by sutures to the maxillae, the frontal bone, and each other.

Nearly as small as the nasal bones, the **lacrimal** (lak'rim-al) bones (Figure 7.9) contribute to the lateral wall of the nasal cavity and the medial surface of each orbit. Each has a foramen through which the lacrimal duct carries tears to the nasal cavity.

The **palatine** bones (Figure 7.10) articulate with the palatine processes of the maxillae, completing the hard palate.

The paired, scroll-shaped **inferior nasal conchae** (Figure 7.10) are similar to the ethmoid conchae. The inferior conchae articulate with the maxillae and have a large surface area. The highly vascular membrane that covers the conchae helps to warm and moisten air as it is inhaled.

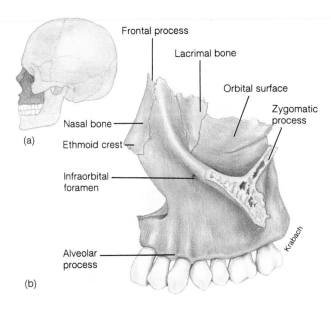

FIGURE 7.9
(*a*) Location of maxilla, lacrimal, and nasal bones; and (*b*) lateral view of these bones.

The **vomer** (vo'mer), a single shovel-shaped bone, forms the inferior portion of the nasal septum between the ethmoid and the maxillae (Figure 7.3b).

The paired **zygomatic** bones (Figures 7.2 and 7.3a) articulate with the zygomatic processes of both the maxillae and the temporal bones. The zygomatic bones and processes form the cheek bones.

The **mandible,** or jaw bone (Figure 7.11) consists of a body and two rami (projections). It contains alveoli for the lower teeth and paired foramina that serve as passageways for nerves and blood vessels. A *mental foramen* is located on each side of the mandible below and between the bicuspid alveoli, and a *mandibular* (man-dib'u-lar) *foramen* is located on the medial surface of each ramus.

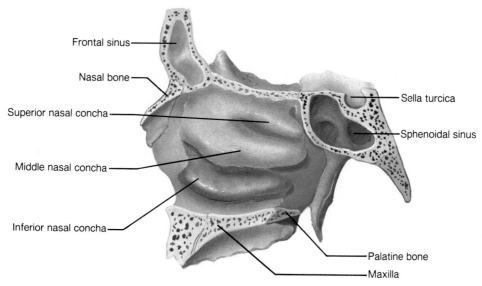

FIGURE 7.10
Sagittal section showing the internal, lateral wall of the nasal cavity.

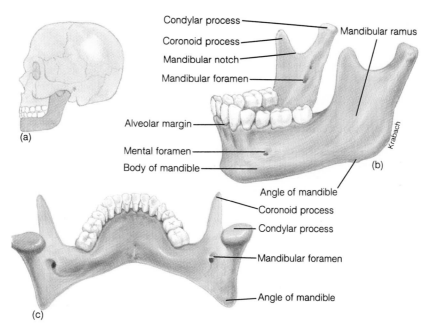

FIGURE 7.11
The mandible: (*a*) location, (*b*) lateral view, and (*c*) posterior view.

The mandible articulates with the squamosal portion of the temporal bone (Figure 7.3a) at a depression called the mandibular fossa. The *mandibular condyle* fits into this fossa, forming a sliding hinge joint—the only movable joint in the skull. The *coronoid* (kor'o-noid) *process* is the point of attachment of the temporal muscle that moves the mandible.

Features of the major bones of the head are summarized in Table 7.3.

Sinuses

Several hollow air sacs called **sinuses** (Figure 7.12) are found within facial and cranial bones. The frontal bone contains a pair of sinuses, the sphenoid bone and the maxillae each contain a single sinus, and the ethmoid bone contains many small sinuses. Each sinus is lined with mucous membrane and has a small opening through which mucus can drain into the nasal cavity. The sinuses give resonance to the voice and make the skull lighter than if it were composed of solid bone.

TABLE 7.3 Features of the Major Bones of the Head

Bone	Features		Bone	Features
Frontal (Figure 7.2)	Supraorbital margin—an arched ridge inferior to the eyebrow Supraorbital foramen—an opening through which blood vessels and nerves pass		Sphenoid (Figure 7.7)	Greater wing—lateral portion that forms part of the eye orbit Lesser wing—thin superior projection that also forms part of the eye orbit Pterygoid process—inferior projection that forms part of the lateral nasal wall Sphenoidal sinus—one of several mucous membrane-lined spaces within the body of the sphenoid Sella turcica—saddle-shaped depression in which the pituitary gland lies Optic foramen—opening in orbit through which the optic nerve passes Foramen rotundum—opening through which a portion of the trigeminal nerve passes Foramen ovale—opening through which another portion of the trigeminal nerve passes
Occipital (Figures 7.3, 7.4)	Foramen magnum—an opening at the base of the skull Occipital condyle—convex process that forms a joint with the first cervical vertebra Condyloid canal—opening through which a vein passes from the brain			
Temporal (Figure 7.5)	Squamosal portion — Zygomatic process—projection that joins the zygomatic bone / Mandibular fossa—depression that forms a joint with the mandible Petrous portion — Middle ear bones and cochlea imbedded in it Mastoid portion — Mastoid process—protuberance behind the ear to which muscles attach Styloid process—spikelike process in front of mastoid process to which muscles attach External auditory meatus—opening through which the auditory canal passes Internal auditory meatus—opening on the inside of the skull through which cranial nerves pass		Maxilla (Figures 7.9, 7.10)	Zygomatic process—lateral portion forms part of the orbit Palatine process—posterior portion forms part of the hard palate Alveolar process—contains the alveoli or tooth sockets Infraorbital foramen—opening just below orbit through which nerves and blood vessels pass
Ethmoid (Figure 7.6)	Cribriform plate—plate of bone containing many small holes through which branches of the olfactory nerve pass Crista galli—projection to which meninges attach Perpendicular plate—vertical portion that forms part of the nasal septum Lateral mass—lateral portion of bone filled with air spaces Superior nasal concha—part of the lateral wall of the nasal cavity Middle nasal concha—also part of the lateral wall of the nasal cavity Ethmoid sinus—air space lined with mucous membrane, found in lateral mass		Mandible (Figure 7.11)	Ramus—portion projecting upward from the body to form a joint with the temporal bone Mental foramen—opening on outer surface between bicuspids through which nerves and blood vessels pass Mandibular foramen—opening on inner surface through which nerves and blood vessels are distributed to teeth Mandibular condyle—end of ramus that forms joint with mandibular fossa of temporal bone Coronoid process—a more anterior projection of the ramus to which the temporal muscle attaches

Though mucus can drain out of sinuses, infectious organisms and irritants in the nasal cavity can move into the sinuses. This frequently occurs in colds and allergies, and results in an inflammation called **sinusitis.** If the membranes swell and prevent drainage, fluid pressure in the sinuses causes a sinus headache.

Special Bones

Strictly speaking, bones of the middle ear and the hyoid bone are neither cranial nor facial, but they are in the head area. The middle ear bones, or **auditory ossicles,** are located in a chamber within the petrous portion of each temporal bone. Each set of three bones transfers vibrations from the tympanic membrane (eardrum) to the oval window, from where they are transferred to fluids in the cochlea, the organ of hearing (Chapter 15). The ossicles (Figure 7.13) are the **malleus** (hammer), attached to the tympanic membrane; the **incus** (anvil), and the **stapes** (stirrup), adjacent to the oval window.

The **hyoid** bone (Figure 7.14) is a U-shaped bone located in the neck just below the angle of the mandible. Though it does not articulate with any other bone, it is attached by ligaments to the styloid process of the temporal bones. The hyoid has a *body* and two hornlike projections, the *greater cornua* and the *lesser cornua.* (Each projection is a cornu.) The body and all the cornua serve as points of attachment for neck and tongue muscles.

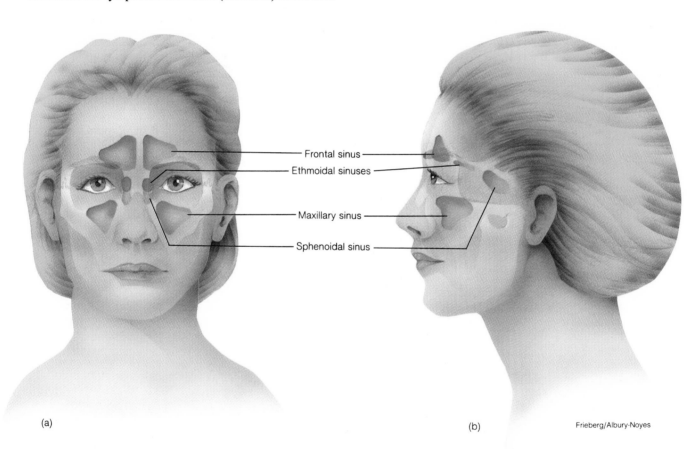

(a)

(b)

Frieberg/Albury-Noyes

FIGURE 7.12
The sinuses: (*a*) anterior view, and (*b*) lateral view of sagittal section.

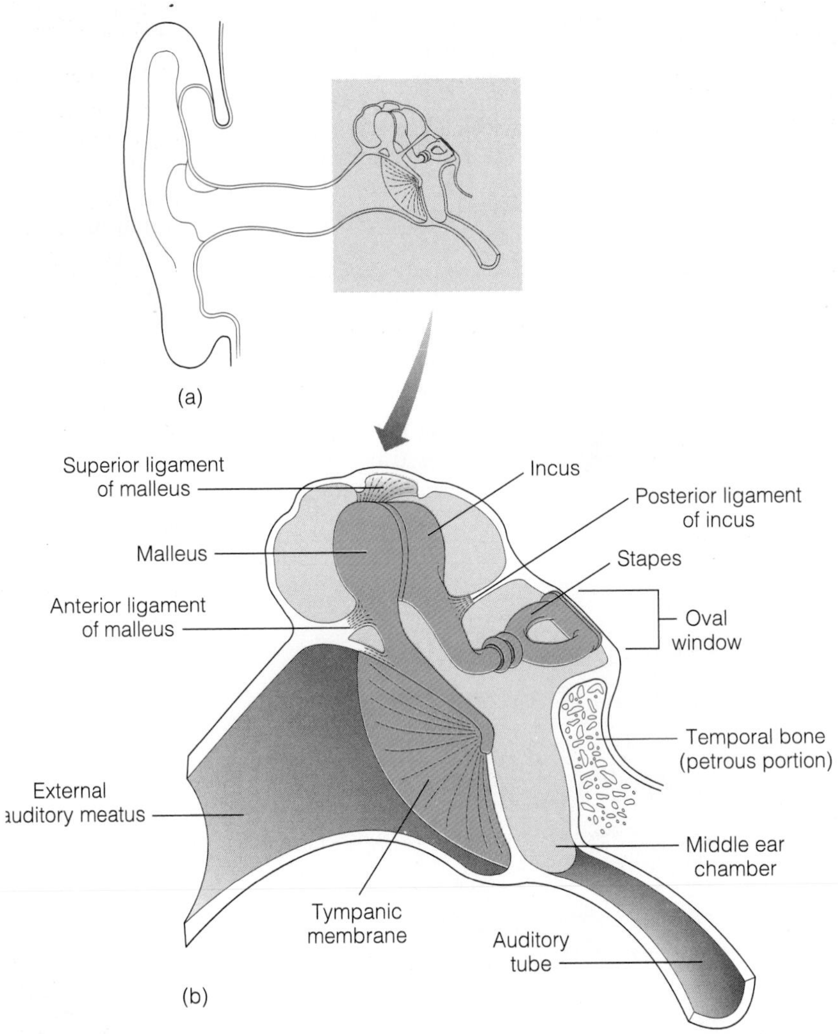

FIGURE 7.13
Auditory ossicles: (*a*) location, and
(*b*) detailed view of the ossicles.

(a)

Superior ligament
of malleus

Incus

Posterior ligament
of incus

Malleus

Stapes

Anterior ligament
of malleus

Oval
window

External
auditory meatus

Temporal bone
(petrous portion)

Middle ear
chamber

Tympanic
membrane

Auditory
tube

(b)

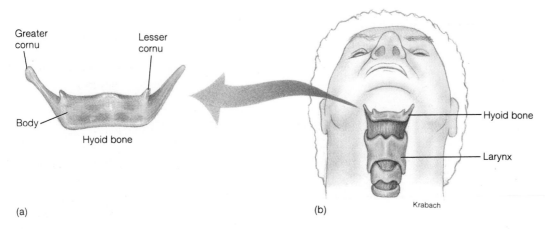

Greater
cornu

Lesser
cornu

Hyoid bone

Body

Larynx

Hyoid bone

Krabach

(a)

(b)

FIGURE 7.14
The hyoid bone: (*a*) anterior view, and (*b*) position in neck.

BONES OF THE TRUNK

Bones of the axial skeleton found in the trunk include the vertebral column and the bones of the thorax, the ribs and sternum.

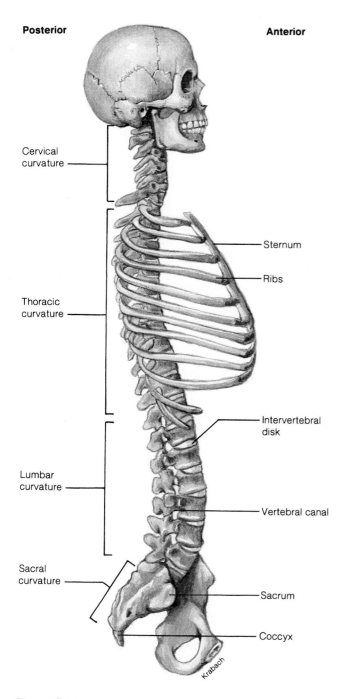

FIGURE 7.15
The ribs, sternum, and vertebral column in lateral view showing natural curvature of the vertebral column.

Vertebral Column

The **vertebral column** (Figure 7.15) consists of 24 vertebrae plus the sacrum and coccyx. Of the vertebrae, seven are in the cervical region, 12 are in the thoracic region, and five are in the lumbar region. The sacral and coccygeal regions each consist of a single bone, the sacrum and coccyx, respectively.

The vertebral regions have natural curvatures seen in lateral views. Cervical and lumbar curvatures are convex (bulging anteriorly); thoracic and sacral curvatures are concave (bulging posteriorly). These curvatures increase the weight-bearing and shock-absorbing capacities of the vertebral column. Because infants lack the convex curvatures, their vertebral columns have a continuous anterior curvature. As the convex curvatures develop, infants become able to hold their heads erect, sit, and stand.

A typical vertebra (Figure 7.16) has an anterior *body* (centrum) and several structures that create a bony posterior arch around the spinal cord. *Pedicles* (ped'ik-elz)

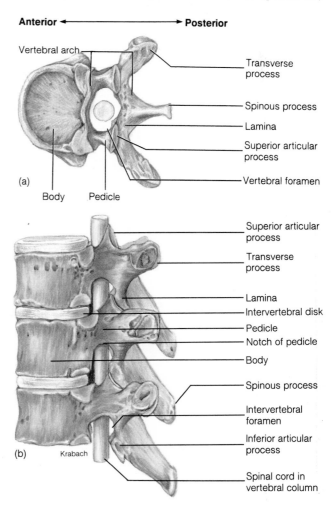

FIGURE 7.16
A typical vertebra: (*a*) superior view, and (*b*) left lateral view showing articulations.

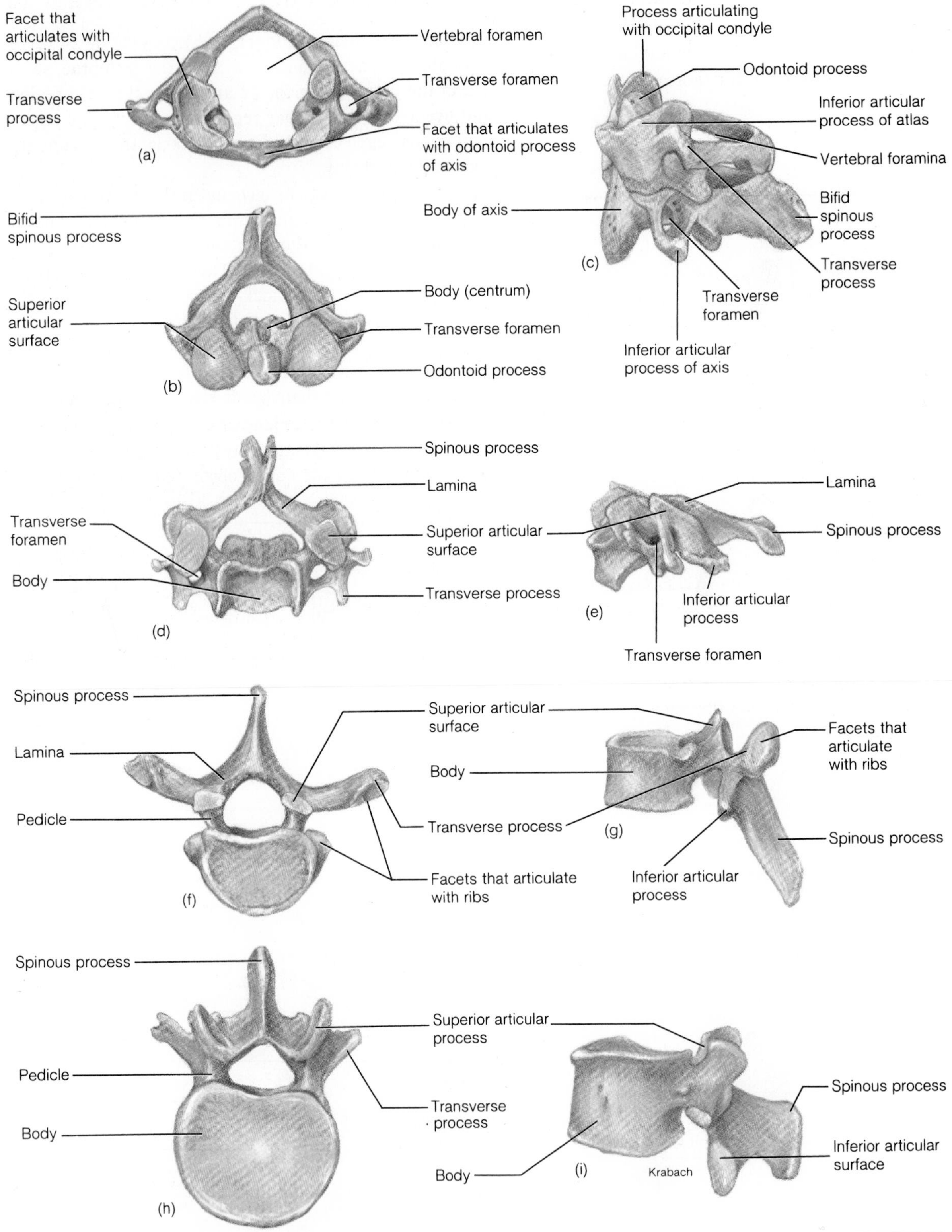

Facet that articulates with occipital condyle

Vertebral foramen

Transverse foramen

Transverse process

Facet that articulates with odontoid process of axis

(a)

Process articulating with occipital condyle

Odontoid process

Inferior articular process of atlas

Vertebral foramina

Bifid spinous process

Transverse process

Body of axis

Transverse foramen

Inferior articular process of axis

(c)

Bifid spinous process

Superior articular surface

Body (centrum)

Transverse foramen

Odontoid process

(b)

Spinous process

Lamina

Superior articular surface

Transverse process

Transverse foramen

Body

(d)

Lamina

Spinous process

Inferior articular process

Transverse foramen

(e)

Spinous process

Lamina

Pedicle

Superior articular surface

Body

Transverse process

Facets that articulate with ribs

(f)

Facets that articulate with ribs

Spinous process

Inferior articular process

(g)

Spinous process

Pedicle

Body

Superior articular process

Transverse process

Body

(h)

Spinous process

Inferior articular surface

Body

(i)   Krabach

FIGURE 7.17

Characteristics of certain vertebrae: (*a*) atlas, (*b*) axis, (*c*) articulation of atlas and axis, (*d*) superior view of a cervical vertebra, (*e*) lateral view of a cervical vertebra, (*f*) superior view of a thoracic vertebra, (*g*) lateral view of a thoracic vertebra, (*h*) superior view of a lumbar vertebra, and (*i*) lateral view of a lumbar vertebra.

form lateral walls of the arch and *laminae* (lam'in-e) fuse in the midline, making the posterior part of the arch. A *spinous process* slants inferiorly from the arch. Muscles attached to spinous processes contribute to back movements. Together, the vertebral foramina of the 24 vertebrae constitute the *vertebral canal* in which the spinal cord lies. Extending laterally from pedicles are paired *transverse processes*. Between adjacent vertebrae on both the left and right sides is an *intervertebral foramen* through which a spinal nerve passes from the spinal cord.

Joints are found in the vertebral column between the bodies of adjacent vertebrae and between the articulating processes. Each vertebra has a pair of *superior articulating processes* and a pair of *inferior articulating processes*. These processes of one vertebra touch those of adjacent vertebrae like overlapping shingles on a roof (Figure 7.16b). Other joint surfaces called *facets,* located on the bodies and transverse processes of thoracic vertebrae, articulate with ribs.

In addition to these general characteristics, the vertebrae of each region have specialized characteristics (Figure 7.17). The superiormost **atlas** has facets that articulate with the occipital condyles. The atlas lacks a centrum because during embryological development it became the *dens,* or *odontoid process* of the **axis.** It serves as a pin about which the atlas rotates.

Cervical vertebrae have short transverse processes, each with a *transverse foramen* through which blood vessels and nerves pass. Except for the atlas, each has a bifid (two-pronged) spinous process.

Each **thoracic vertebrae** has a long spinous process and thick, stubby transverse processes. Most transverse processes and bodies of these vertebrae have facets that articulate with ribs.

The large, chunky **lumbar vertebrae** have relatively short spinous processes, relatively long transverse processes, and large superior and inferior articulating processes with lateral articulating surfaces. Lateral articulations form strong, slightly movable joints that allow forward and backward bending.

Inferior to the lumbar vertebrae is the **sacrum** (sa'krum), which consists of five fused vertebrae, and the **coccyx** (kok'siks), which consists of four fused vertebrae (Figure 7.18). *Transverse lines* on their anterior surfaces show where fusion occurred.

About 5 percent of otherwise normal humans have more or less than the normal number of vertebrae, though humans and nearly all other mammals—including the giraffe—have seven cervical vertebrae. Variations in the number of human vertebrae occur mainly in the lumbar and sacral regions. In one common variation, the fifth lumbar vertebra is incorporated into the sacrum. In another, the first sacral vertebra exists separate from the rest of the sacrum. These conditions often cause no symptoms and are discovered when X rays are made to study a different problem.

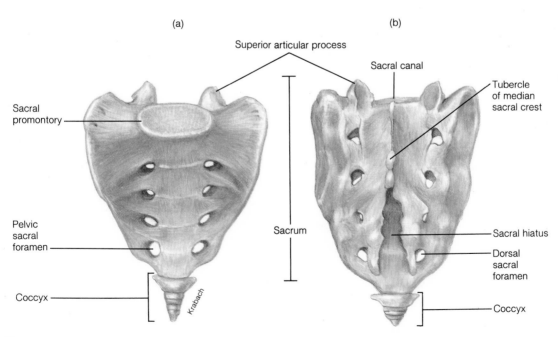

FIGURE 7.18
The sacrum and coccyx: (*a*) anterior view and (*b*) posterior view.

The sacrum also has four pairs of passageways for nerves and blood vessels called *pelvic foramina* on the anterior surface and *dorsal foramina* on the posterior (dorsal) surface. Also on the posterior surface are the *lateral sacral crests* and the *median sacral crest* to which various muscles attach. The *sacral canal* in the midline between the surfaces is a continuation of the vertebral canal. During embryonic development, the spinal cord extends into this canal, but in adults, it extends only to the first or second lumbar vertebra because the vertebral column grows to a greater length than the spinal cord. The lower vertebral canal and the sacral canal contain spinal nerve roots that have branched from the spinal cord. The *sacral hiatus,* an area in which the spinous processes generally do not fuse, is covered with fibrous connective tissue. The *sacral promontory* is an important landmark in obstetrics for pelvic measurements.

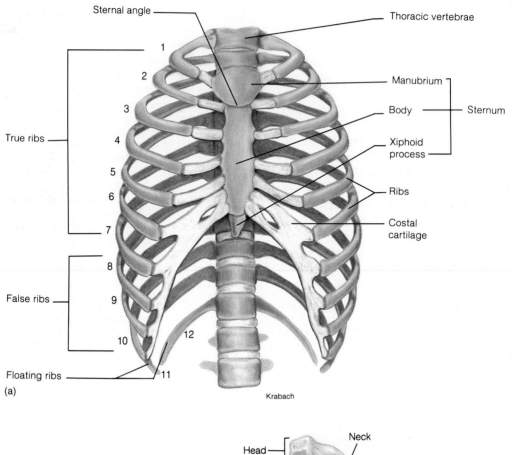

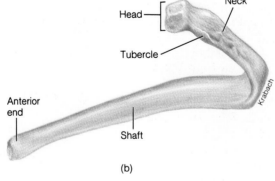

FIGURE 7.19
(*a*) The ribs and sternum, and (*b*) a detailed view of a rib.

The triangular coccyx, the vestige (remnant) of a tail, is attached by ligaments to the sacrum. It provides attachment points for pelvic muscles and acts as a shock absorber when a person sits down. Falls or blows to the coccyx often result in painful fractures or dislocations.

Anesthetics can be administered through the fibrous membrane of the sacral hiatus to block pain sensations originating in the pelvic area. Such anesthesia, called *caudal block* or *spinal block,* is safe to use during childbirth because little of the anesthetic enters the mother's bloodstream, and effects on the fetus are minimal.

Thorax

The chest, or **thorax** (tho'raks), consists of a bony cage that surrounds and protects the heart and lungs. Its bony components are the sternum and ribs (Figure 7.19a).

The **sternum** (ster'num), or breastbone, is located in the anterior midline of the thorax. It consists of three parts—the superior *manubrium* (man-u'bre-um); the *gladiolus* (glah-di'o-lus), also called the *body* or "little sword"; and the inferior *xiphoid* (zi'foid) *process.* The lateral edges of the sternum articulate with the ribs. The manubrium articulates with the clavicles and the first two pairs of ribs and the gladiolus articulate directly or indirectly with the next eight pairs of ribs. The xiphoid process does not articulate with ribs; rather, it serves as an attachment point for certain abdominal muscles.

Of the 12 pairs of **ribs,** one pair extends from each thoracic vertebra. The superiormost seven pairs are called *true ribs* because they articulate directly with the sternum by means of *costal cartilages,* small pieces of hyaline cartilage. The inferiormost five pairs of ribs are called *false ribs* because they fail to articulate directly with the sternum. Of these the upper three pairs articulate by cartilage with the costal cartilage of the seventh pair of ribs. The inferior two pairs are called *floating ribs* because their anterior ends do not articulate with either cartilage or the sternum. Of course, these ribs do not literally float—they are attached to vertebrae and body wall muscles.

Each individual rib (Figure 7.19b) has a *head,* a *neck,* and a *shaft.* The heads of ribs two through nine have two *facets,* which articulate with the body of the corresponding thoracic vertebrae. The heads of the first rib and of the last three ribs have only one facet. The *tubercle* (too'ber-k'l) of each of the first nine ribs has a facet that articulates with the transverse process of a thoracic vertebra; the last three ribs have no such facets. Except for the floating ribs, the anterior ends of the shafts of ribs are extended by pieces of costal cartilage.

Features of trunk bones are summarized in Table 7.4.

Bone	Features
TABLE 7.4 Features of Bones of the Trunk	
Vertebra (Figure 7.16)	Body (centrum)—anterior, flat, round mass that is the weight-bearing structure Vetebral foramen—hole in center of the vertebra formed by the body, pedicles, and laminae; passageway for the spinal cord Pedicles—projections extending posteriorly from the body Laminae—posterior parts of the vertebra from which spinous process projects Intervertebral foramen—opening between the vertebrae through which spinal nerves pass Superior articulating processes—upward (superior) protrusions from the laminae Inferior articulating processes—downward (inferior) protrusions from the laminae Transverse processes—lateral protrusions from the laminae
Atlas (Figure 7.17)	Facets—indentations that articulate with the occipital condyles
Axis (Figure 7.17)	Odontoid process—peglike protrusion around which the atlas rotates
Cervical vertebra (Figure 7.17)	Transverse foramina—holes in the transverse processes through which an artery, a vein, and some nerves pass
Thoracic vertebra (Figure 7.17)	Facets—smooth surfaces with which the ribs articulate
Sacrum (Figure 7.18)	Pelvic foramina—pairs of openings on the anterior surface through which blood vessels and nerves pass Dorsal foramina—pairs of openings on the posterior surface through which blood vessels and nerves pass Median sacral crest—a protrusion in the dorsal midline formed by the fusion of spinous processes Sacral canal—a continuation of the vertebral canal Sacral hiatus—an area covered by fibrous connective tissue where spinous processes failed to fuse
Sternum (Figure 7.19a)	Manubrium—superiormost portion of the sternum Gladiolus—large middle portion of the sternum Xiphoid process—inferiormost portion of the sternum
Ribs (Figure 7.19b)	Head—projection from the posterior portion that articulates with the body of the thoracic vertebra Neck—narrow portion just below the head Shaft—main part of the rib Facet—smooth surface on the head that articulates with a thoracic vertebra Tubercle—process just below the neck that articulates with the transverse process of a thoracic vertebra

BONES OF THE APPENDAGES

In contrast to the axial skeleton, which is more significant in supporting the body and protecting internal organs, the appendicular skeleton plays a more significant role in movement.

The appendicular skeleton (Table 7.5) includes the **pectoral** (pek′tor-al) **girdle,** with attached **upper appendages,** and the **pelvic girdle,** with attached **lower appendages.** All bones of the appendicular skeleton are paired—one of each pair on the left and the right side of the body.

Pectoral Girdle

Two paired bones, the scapula and the clavicle, make up the pectoral girdle (Figure 7.20). These bones and their associated muscles form the shoulders.

The **clavicle** (klav′ik-l), or collarbone, is a slightly curved, rod-shaped bone extending from the manubrium of the sternum to the acromion process of the scapula. Muscles attach to it at the distal *conoid tubercle* and the proximal *costal tuberosity.*

The **scapula** (skap′u-lah) is a large, flat bone with many processes and projections. On the anterior surface is the *subscapular fossa,* where a large muscle attaches. Medial to this fossa is the *vertebral border* extending from the *superior angle* to the *inferior angle.* These angles provide convenient anatomical landmarks on the scapula, as do its borders and other markings. On the lateral edge are the *acromion* (ak-ro′me-on) *process,* which articulates directly with the clavicle, and the *coracoid* (kor′ak-oid) *process,* which attaches to the clavicle by strong ligaments that stabilize the pectoral girdle. The *glenoid* (gle′noid) *fossa* is a depression into which the head of the humerus fits. On the posterior surface is a large *spine* that terminates laterally in the acromion process. Both the *supraspinous fossa* and the *infraspinous fossa* are attachment sites of large muscles that move the humerus at the shoulder. Other muscles extend from the spine and vertebral border to the vertebral column, attaching the pectoral girdle to the axial skeleton. They are, in fact, the *only* attachment between the scapula and the vertebrae because no joints exist between these bones.

Upper Appendage

The bones of the upper appendage include the humerus in the upper arm; the radius and ulna in the forearm; and the carpals, metacarpals, and phalanges in the wrist and hand (Figure 7.1).

TABLE 7.5 Bones of the Appendicular Skeleton	
Pectoral girdle	4 bones
Scapula	2
Clavicle	2
Upper appendage	60 bones
Humerus	2
Radius	2
Ulna	2
Carpals	16
Metacarpals	10
Phalanges	28
Pelvic girdle	2 bones
Coxal bones	2
Lower appendage	60 bones
Femur	2
Tibia	2
Fibula	2
Patella	2
Tarsals	14
Metatarsals	10
Phalanges	28
Total bones of the appendicular skeleton	**126 bones**

The **humerus** (hu′mer-us) is a long, relatively thick bone with a large smooth head at the proximal end and a number of projections at the distal end (Figure 7.21). The *head* articulates with the glenoid fossa; between the head and the *shaft* is a slight narrowing of the bone called the *anatomical neck* and a common fracture site called the *surgical neck.* Lateral to the head are the *greater* and *lesser tubercles,* which serve as attachment points for shoulder joint ligaments. The *intertubercular groove* is a depression in which the tendon of the long head of the biceps brachii muscle rests. (Ligaments join bones together and tendons join muscles to bones.) Other markings on the shaft are the *deltoid tuberosity* (del′toid too″ber-os′it-e) where the deltoid muscle attaches and the *radial groove* along which the radial nerve passes.

The distal (elbow) end of the humerus has *medial* and *lateral epicondyles* (ep″ik-on′dīlz) where muscles that move the forearm attach. The *capitulum* (kap-it′u-lum), or "little head," articulates with the radius, and the *trochlea* (trok′le-ah) articulates with the ulna. The *coronoid* and *olecranon* (o-lek′ran-on) *fossae* are depressions into which the coronoid and olecranon processes of the ulna fit.

The **radius** and the **ulna** (Figure 7.21) in the forearm have prominent *nutrient foramina,* where blood vessels enter the bone tissue. The proximal *head* of the radius articulates with the capitulum of the humerus. The *radial tuberosity* serves as an attachment for the biceps brachii muscle. At the distal end of the radius, the *styloid process* extends along the lateral surface of one of the carpals; it

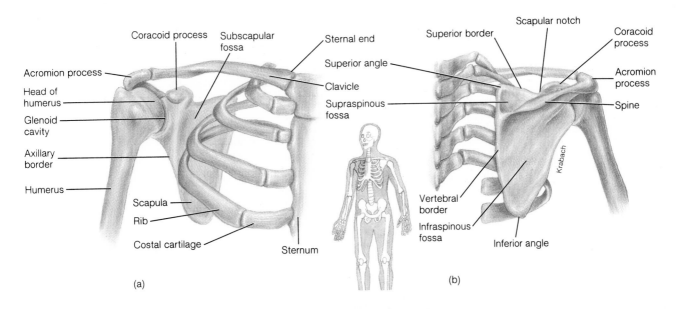

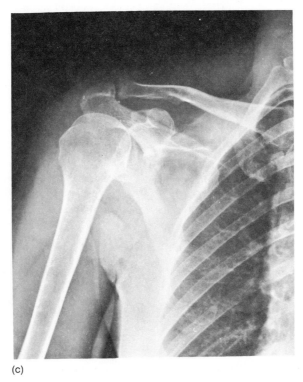

FIGURE 7.20
The pectoral girdle: (*a*) anterior view of right side; (*b*) posterior view of right side; and (*c*) X ray of pectoral girdle, scapula in anterior view.

can be felt through the skin as a bump on the thumb side of the wrist. The distal *head* of the ulna has a *styloid process,* which can be felt as a bump on the little finger side of the wrist. The proximal end of the ulna has a long projection, the *olecranon,* with a *trochlear notch* that articulates with the humerus. In addition to articulations with the humerus and carpals, the radius and ulna also articulate with each other. The head of the radius fits into the *radial notch* of the ulna allowing rotation of the radius about the ulna. No matter how the radius is rotated it always lies on the thumb side of the forearm.

The elbow is sometimes referred to as the "crazy bone," and we suffer sudden sharp pain when we bump it. The pain is due to stimulation of the ulnar nerve, which passes unprotected over the medial epicondyle of the humerus and into the forearm.

The bones of the wrist and hand include eight **carpals** (kar'palz), five **metacarpals** (met"ah-kar'palz), and 14 **phalanges** (fa-lan'jēz), shown in Figure 7.21. The carpals (shown in the figure) form the bony structure of the wrist.

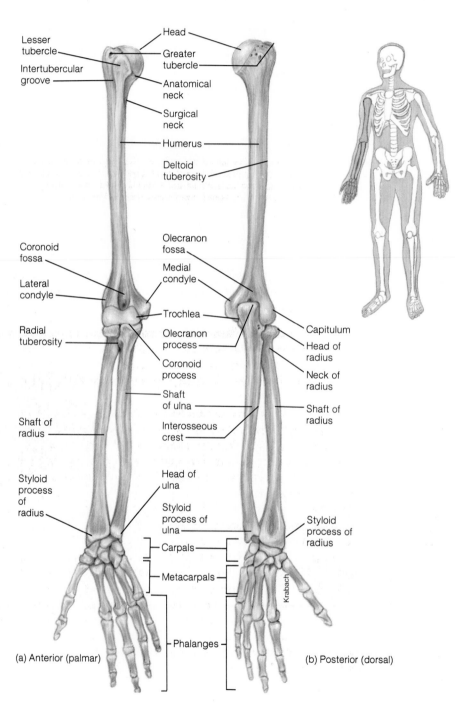

FIGURE 7.21

The right arm and hand: (*a*) anterior view, and (*b*) posterior view.

The most proximal bones articulate with the radius and ulna. Several of the carpals articulate with other carpals, and the more distal ones articulate with the metacarpals. The metacarpals give structure to the palm of the hand and the proximal thumb. The phalanges are named according to their position in the digits (fingers)—proximal, middle, and distal. The thumb has no middle phalanx.

Pelvic Girdle

The two **coxal** (kok'sal) **bones** make up the pelvic girdle (Figure 7.22). Each bone arises during embryonic development by the fusion of the ilium, ischium, and pubis. The **ilium** (il'e-um) forms the superior and lateral portion of each coxal bone. You can feel its superior edge when you

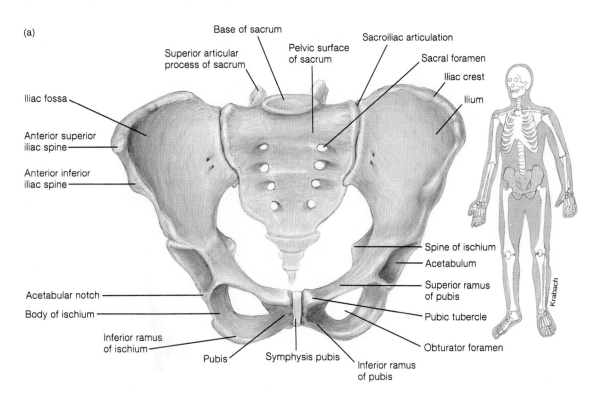

(a)

Iliac fossa
Anterior superior iliac spine
Anterior inferior iliac spine
Base of sacrum
Superior articular process of sacrum
Pelvic surface of sacrum
Sacroiliac articulation
Sacral foramen
Iliac crest
Ilium
Spine of ischium
Acetabulum
Superior ramus of pubis
Pubic tubercle
Obturator foramen
Acetabular notch
Body of ischium
Inferior ramus of ischium
Pubis
Symphysis pubis
Inferior ramus of pubis

Krabach

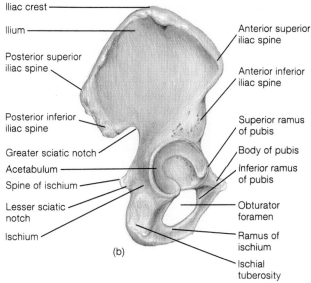

Iliac crest
Ilium
Posterior superior iliac spine
Posterior inferior iliac spine
Greater sciatic notch
Acetabulum
Spine of ischium
Lesser sciatic notch
Ischium
Anterior superior iliac spine
Anterior inferior iliac spine
Superior ramus of pubis
Body of pubis
Inferior ramus of pubis
Obturator foramen
Ramus of ischium
Ischial tuberosity

(b)

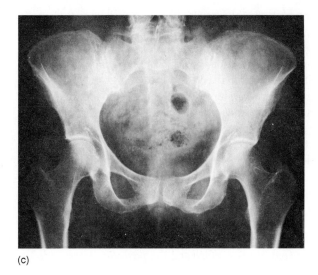

(c)

FIGURE 7.22
(a) Anterior view of the pelvic girdle, (b) lateral surface of the right coxal bone, and (c) anterior X-ray view showing head of femur in acetabulum.

place your hands on your hips. The **ischium** (is'ke-um) forms the inferior portion of the coxal bone—the part you sit on, and the **pubis** (pu'bis) forms the anterior portion. The *spines, crests, fossae,* and *lines* on the coxal bone are attachment points for muscles of the abdomen, hip, and thigh. A ligament across the *greater sciatic* (si-at'ik) **notch** makes a foramen through which branches of the sciatic nerve pass.

On the internal surface of the coxal bones are two articulating surfaces. The left pubis articulates with the right pubis at the *pubic symphysis* (sim'fis-is). This joint is immovable except during late pregnancy and delivery (Chapter 8). The ilium joins with the sacrum to form the *sacroiliac* (sa"kro-il'e-ak) *joint,* also a relatively immovable joint. So, the coxal bones are joined to each other and to the sacrum, thereby forming a sturdy weight-bearing structure that supports the trunk. Various rami of the ischium and pubis join to form the *obturator* (ob'too-ra"tor) *foramen,* a large membrane-covered opening through which blood vessels and the obturator nerve pass.

On the lateral surface of each coxal bone is the *acetabulum* (as"et-ab'u-lum), the socket of the hip joint. Parts of each of the three embryonic bones fuse in the acetabulum, and outgrowths of these bones along with their cartilage extensions form the rim of the acetabulum.

Lower Appendage

The bones of the lower appendage (Figure 7.1) are the femur, tibia, fibula, patella, tarsals, metatarsals, and phalanges.

The **femur** (fe'mur), or thigh bone (Figure 7.23), has a *head, neck,* and *shaft.* The head fits into the acetabulum, the neck is the attachment site of capsular ligaments that hold the head in the joint socket, and the shaft forms the bulk of the bone. The *fovea capitis* (fo've-ah kap'it-is) is the attachment site of an intraarticular ligament—a ligament found within the joint capsule (Chapter 8). At the proximal end are the *greater* and *lesser trochanters* (tro-kan'ters), attachment sites for thigh muscles. The *intertrochanteric* (in"ter-tro"kan-ter'ik) *line* on the anterior surface and the *intertrochanteric crest* on the posterior surface are attachment points for ligaments. In addition to the *nutrient foramen* on the posterior surface are the *linea aspera* (lin'e-ah as'per-ah) and *popliteal* (pop-lit'e-al) *surface,* where muscles attach. At the distal end are *medial* and *lateral condyles,* which form parts of the knee joint surface. *Epicondyles,* located proximal to each condyle, serve as muscle attachment sites. The *patellar* (pat-el'ar) *surface* and the *posterior intercondylar fossa* also are present.

The **patella** (pat-el'ah), a sesamoid bone, is imbedded in the tendon of the quadriceps femoris muscle. It is closely associated with the knee joint (Chapter 8).

The tibia and fibula (Figure 7.23) are bones of the leg. The heavier **tibia** (tib'e-ah) is the weight-bearing bone. At its proximal end, *lateral* and *medial tibial condyles* articulate with the femur. At its distal end, a *medial malleolus* (mal-e'o-lus) extends along part of the medial surface of the talus. The lighter **fibula** (fib'u-lah) lies lateral to the tibia and articulates with it at both its proximal and distal ends. The fibula has a *lateral malleolus* that extends over part of the lateral surface of the ankle. The malleoli show through the skin as bony projections on either side of the ankle. In many ankle fractures, a malleolus breaks off.

The ankle and foot contain seven **tarsals** (tar'salz), five **metatarsals** (met"ah-tar'salz), and 14 **phalanges.** Of the tarsals, the **talus** (ta'lus) is the main articulation with the leg and the **calcaneus** (kal-ka'ne-us) forms the bony structure of the heel. Proximal tarsals articulate with each other and distal tarsals articulate with metatarsals. In turn, metatarsals articulate with the proximal phalanges. Like the thumb, the first digit of the foot (the great toe), has only two phalanges.

The foot has two longitudinal arches and one transverse arch (Figure 7.24) maintained by tendons of certain leg muscles and strong foot ligaments. The *medial longitudinal arch* spans the calcaneus, talus, navicular, cuneiforms, and three medial metatarsals. The *lateral longitudinal arch* spans the calcaneus, cuboid, and two lateral metatarsals. The *transverse arch* spans the cuboid, cuneiforms, and the proximal ends of the metatarsals. Arches support the body's weight and act as shock absorbers during walking and running.

Features of the bones of the appendicular skeleton are summarized in Table 7.6.

Electricity, which has been used for nearly a century as a convenient energy source, until recently has been rejected as a mode of therapy. Now it appears that electricity can be used to stimulate bone growth. Though not yet a part of the routine treatment of bone disorders, a few orthopedic surgeons are using electrical stimulation in at least two situations. They apply it to fractures that have failed to heal after two or more months. And they apply it to joints, especially hip joints, in which bone is necrotic (dying). Electrical stimulation also is used for other purposes as described in the essay at the end of Chapter 11.

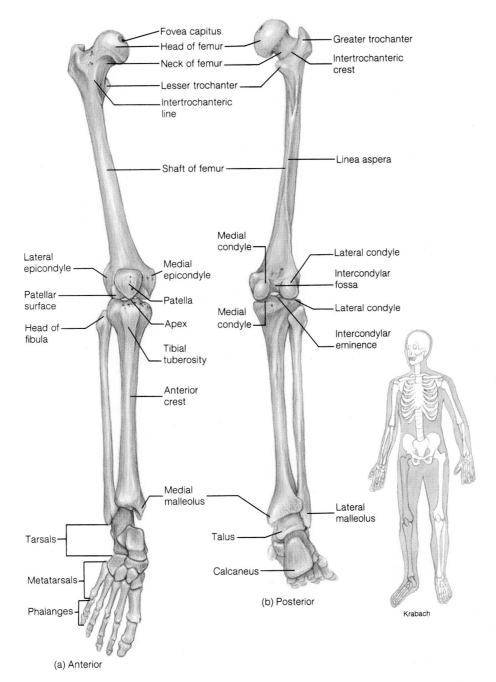

Fovea capitus
Head of femur
Neck of femur
Lesser trochanter
Intertrochanteric line

Greater trochanter
Intertrochanteric crest

Shaft of femur

Linea aspera

Lateral epicondyle
Patellar surface
Head of fibula

Medial epicondyle
Patella
Apex
Tibial tuberosity

Anterior crest

Medial condyle
Medial condyle

Lateral condyle
Intercondylar fossa
Lateral condyle
Intercondylar eminence

Medial malleolus

Lateral malleolus

Tarsals

Talus

Metatarsals

Calcaneus

Phalanges

(a) Anterior

(b) Posterior

Krabach

FIGURE 7.23
The right leg and foot: (*a*) anterior view, and (*b*) posterior view.

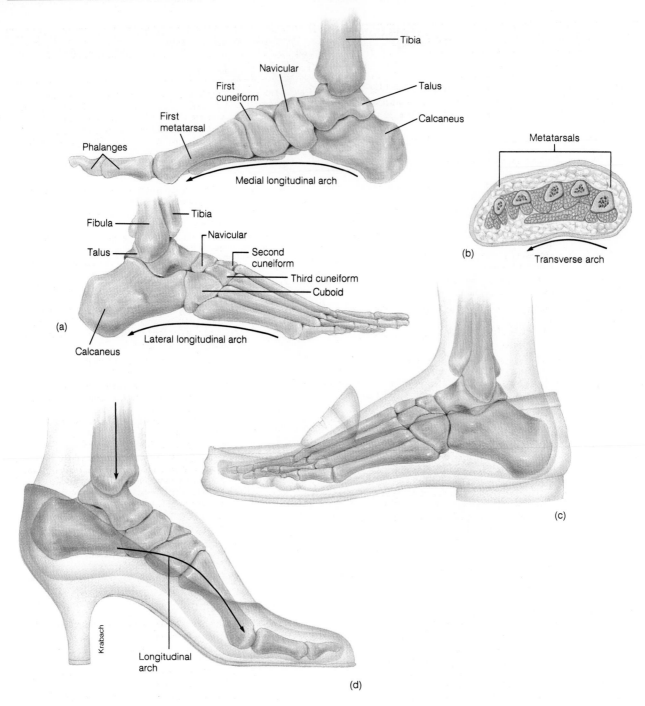

FIGURE 7.24
Arches of foot: (*a*) longitudinal arches of the right foot, (*b*) transverse arch of right foot (with toes removed), and weight distribution while wearing (*c*) low-heeled shoes and (*d*) high-heeled shoes.

TABLE 7.6   Features of Bones of the Appendicular Skeleton

Bone	Features	Bone	Features
Scapula (Figure 7.20)	Subscapular fossa—hollow anterior surface Vertebral border—edge nearest the vertebrae Superior angle—point at which the superior and vertebral borders meet Inferior angle—point at which the axillary and vertebral borders meet Glenoid fossa—depression that articulates with the humerus Acromion process—projection at lateral end of spine that articulates with the clavicle Coracoid process—anterior projection from the superior border Spine—sharp ridge running across the posterior surface Supraspinous process—depression superior to the spine Infraspinous process—depression inferior to the spine	Coxal bone (Figure 7.22)	Iliac crest—superior boundary of the ilium Iliac fossa—concave anterior surface of the ilium Anterior superior iliac spine—projection of the anterior end of the iliac crest Posterior superior iliac spine—projection at the posterior end of the iliac crest Anterior inferior iliac spine—projection below the anterior superior spine Posterior inferior iliac spine—projection below the posterior superior spine Ischial spine—projection superior to the ischial tuberosity Ischial tuberosity—large inferior process that bears the weight in sitting position Obturator foramen—opening surrounded by the pubis and ischium Greater sciatic notch—notch on posterior surface of the ilium Acetabulum—depression formed by fusion of the ilium, ischium, and pubis and development of the bony and cartilaginous rim; articulates with the head of the femur
Clavicle (Figure 7.20)	Conoid tubercle—distal projection to which muscles attach Costal tuberosity—proximal projection to which muscles attach		
Humerus (Figure 7.21)	Head—smooth, spherical projection on the proximal end Anatomical neck—groove just below the head Surgical neck—region distal to the anatomical neck, so named because it is frequently the site of a fracture Greater tubercle—anterolateral projection near the head Lesser tubercle—anterior projection just distal to the anatomical neck Intertubercular groove—long depression between the tubercles where a tendon of the biceps brachii muscle lies Deltoid tuberosity—rough area near the middle of the shaft where the deltoid muscle attaches Radial groove—depression in deltoid tuberosity through which the radial nerve passes Medial epicondyle—medial projection on the distal end Lateral epicondyle—lateral projection on the distal end Capitulum—rounded knob between the condyles that articulates with the radius Trochlea—projection on distal end that articulates with the ulna Coronoid fossa—depression above trochlea that articulates with the coronoid process of ulna Olecranon fossa—depression superior and posterior to the trochlea that articulates with the olecranon process of ulna	Femur (Figure 7.23)	Head—rounded proximal projection that articulates with the acetabulum Neck—constriction just distal to the head Fovea capitus—indentation on the head where a ligament attaches Greater trochanter—inferiolateral projection near the head Lesser trochanter—small projection inferiomedial to the greater trochanter Intertrochanteric line—anterior line between the trochanters to which a ligament attaches Intertrochanteric crest—posterior rough area between the trochanters to which a ligament attaches Linea aspera—lengthwise ridge on the posterior surface to which muscles attach Popliteal surface—rough area on the posterior surface to which muscles attach Medial condyle—medial projection at the distal end Lateral condyle—lateral projection at the distal end Medial epicondyle—small projection proximal to the medial condyle Lateral epicondyle—small projection proximal to the lateral condyle Patellar surface—depression between condyles beneath the patella Posterior intercondylar fossa—posterior depression between the condyles
Radius (Figure 7.21)	Head—process at the proximal end that articulates with the capitulum of the humerus and the radial notch of the ulna Styloid process—projection on the lateral surface of the distal end Radial tuberosity—roughened surface near the head on which the biceps brachii muscle inserts	Tibia (Figure 7.23)	Lateral tibial condyle—depression on the proximal end that articulates with the lateral condyle of the femur Medial tibial condyle—depression on the proximal end that articulates with the medial condyle of the femur Medial malleolus—medial projection at the distal end that articulates with the talus
Ulna (Figure 7.21)	Head—rounded process at the distal end Styloid process—projection at the distal end (toward little finger side of hand) Olecranon process—proximal projection that forms the elbow Radial notch—concavity into which the head of the radius fits	Fibula (Figure 7.23)	Lateral malleolus—lateral projection at the distal end that articulates with the talus

Skeletal Differences Due to Age and Sex

The closing of fontanels is but one of several skeletal changes associated with age. Such changes are not surprising because bone is a metabolically active tissue, and bones are subjected to various stresses throughout life.

During growth, striking changes in skeletal proportions occur. At birth, an infant's head is about one-fourth the body length, but an adult's head is only about one-eighth the total height. This difference results from rapid head growth before birth and rapid body growth in infancy and childhood. Curvatures of the vertebral column change from the anterior concavity of the infant to the alternating convex-concave curvatures of the adult. The contours of facial bones also change, giving adults sharper features than infants. As growth is completed, epiphyseal plates of growing bones disappear.

The skeleton continues to change even after growth is complete. Bone deposition occurs on the outer bone surfaces, particularly near joints, while absorption occurs on the inner surfaces. Both the external dimensions and the marrow cavities increase in size. Surface processes become more pronounced with aging, sometimes limiting joint mobility. Bone deposits near the articular cartilages severely limit joint movement. Such bone deposits, along with deterioration of articular cartilages, occurs in **osteoarthritis** (os''te-o-ar-thri'tis), a common joint disorder of the elderly (Chapter 8).

Marked differences between female and male skeletons are found in the pelvis (Figure 7.25). The female pelvis is wide and shallow, whereas the male pelvis is narrow and deep. The *inlet of the true pelvis* is wide and nearly round in the female but fairly narrow and heart-shaped in the male. Both the *false pelvis* and the *pubic arch* are much wider in females than in males. The basinlike female pelvis provides support for a developing baby and its wide inlet facilitates delivery. These and other sexual differences in the skeleton are summarized in Table 7.7.

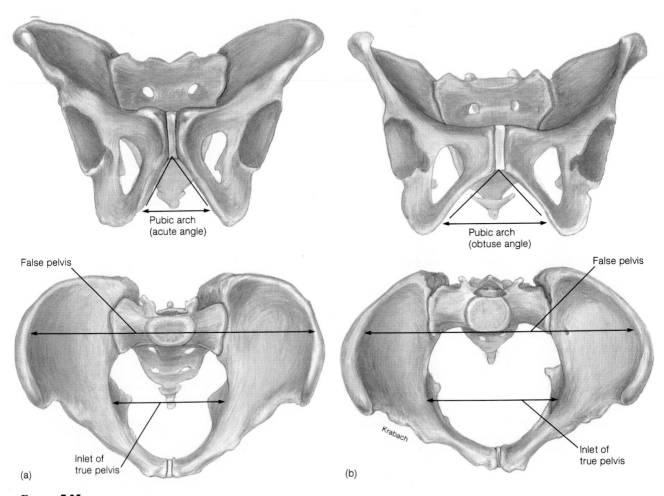

(a)

(b)

Figure 7.25
The differences in the structure of the pelvis related to sex: (*a*) anterior view of a male pubic arch and superior view of male pelvic girdle, and (*b*) anterior view of female pubic arch and superior view of female pelvic girdle.

TABLE 7.7   Sexual Differences of Skeletons

Part	Difference
Skull	The female skull is lighter in weight, and the muscle attachments are less pronounced than in the male skull. In the female, the facial area is more rounded, and the jaw usually is proportionately smaller than in the male.
Sacrum	The female sacrum is wider and the sacral curvature extends more sharply posterior than the male sacrum.
Coccyx	The female coccyx is more movable than the male coccyx.
Pelvis	The female coxal bones are thinner and lighter in weight, and muscle attachments are less pronounced than are male coxal bones. The female ischia are more flared, and the acetabula are further apart and smaller than those of the male.
Pelvic cavity	The female pelvic cavity is shorter and wider in diameter than the male pelvic cavity.

CLINICAL APPLICATIONS

Extreme curvatures of the vertebral column (Figure 7.26) include lateral (side-to-side) curvature called *scoliosis* (sko″le-o′sis), convex curvature of the thoracic region (swayback) called *lordosis* (lor-do′sis), and concave curvature of the thoracic region (hunchback) called *kyphosis* (ki-fo′sis).

Scoliosis, the most common curvature, can be due to asymmetrical, pronounced weakness of back muscles or to a hemivertebra (half a vertebra resulting from the failure of one side of the body and arch to develop). Such conditions usually are congenital (present at birth), but how and why they occur is largely unknown. Having one leg longer than the other can mimic scoliosis when the affected individual is standing, but the vertebral distortion disappears when the person is sitting or lying down. Scoliosis can be treated with braces or surgery.

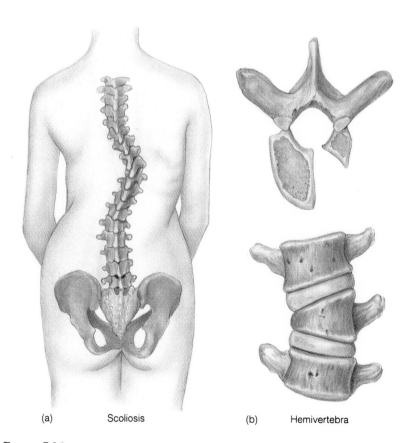

(a)   Scoliosis   (b)   Hemivertebra

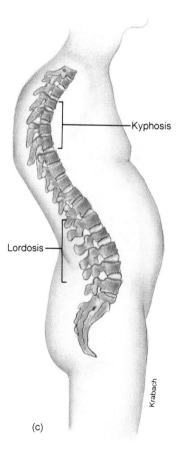

Kyphosis

Lordosis

Krabach

(c)

FIGURE 7.26
Extreme curvatures of the vertebral column: (*a*) scoliosis; (*b*) hemivertebra, which can cause scoliosis; and (*c*) lordosis and kyphosis.

Lordosis usually occurs in the lumbar region and is more often acquired than present at birth. Many obese patients develop lumbar lordosis and low back pain because increased weight of the abdominal contents shifts the body's line of gravity. Weight loss corrects this condition. Some pregnant women temporarily suffer from the same condition.

Kyphosis commonly occurs in the thoracic region, especially in elderly women, where it is referred to as a "dowager's hump." Osteoporosis (Chapter 6) may contribute to kyphosis and to other abnormal curvatures as well.

THE ICOSAHEDRON: A BIOLOGICAL SUPPORT SYSTEM

We sometimes tend to think of bones as forming a static support system, but the idea of the icosahedron as a biological support system, proposed by Levin (1981), offers a more dynamic view. An icosahedron is a solid geometrical form with 20 faces or surfaces each consisting of an equilateral triangle (Figure 7.27a). Close packing of spheres approximates the icosahedral structure and represents shapes seen in nature (Figure 7.27b). Certain viruses, colonies of unicellular organisms, and many other organisms clearly display icosahedral structures. Many joints, including nearly all freely movable joints, also have an icosahedral structure (Figure 7.27c).

Icosahedra have certain properties that make them suitable for the framework of living things:

1. Because of their triangular faces, they do not bend easily when subjected to compressive forces as rectangles do (Figure 7.28a and b).
2. Loads are distributed equally in all directions over the surface of icosahedra. For example, geodesic domes (half an icosahedron) need no central supports; if the load at the top of the dome were directed downward, supports would be needed.
3. Icosahedra have high strength for their weight. If an icosahedron and a hollow cube were each built of the same strength steel rods, greater force would be required to collapse the icosahedron than the cube.

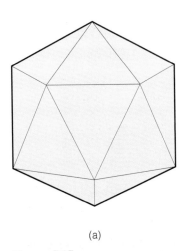

(a)

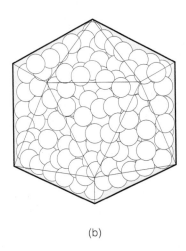

(b)

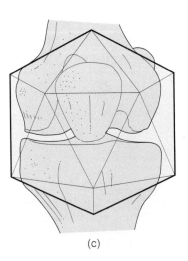

(c)

FIGURE 7.27
(a) An icosahedron, (b) close packing of spheres to form an icosahedron, and (c) a human knee joint showing an icosahedral shape.

CLINICAL TERMS

amputation (am″pu-ta′shun)   removal of part or all of a limb

craniotomy (kra″ne-ot′o-me)   any surgical procedure that involves cutting into the bones of the cranium

flat foot   the loss of curvature of one or more of the arches of the foot

laminectomy (lam″in-ek′to-me)   removal of the posterior arch of a vertebra

orthopedic (or″tho-pe′dik)   pertaining to the treatment of bone and joint disorders, and the correction of deformities

Joints having icosahedral structure display these same properties:

1. They are not easily distorted by compressive forces.
2. Loads are distributed over their surfaces.
3. They have high strength for their weight.

An especially important aspect of the icosahedron model of a joint is the absence of compressive loads across joints. Ligaments and other joint structures form an icosahedron over which compressive loads are distributed. Thus, the weight of the body on knees and ankles does not exert compressive force on the bones. If loads were distributed across joints like they are in masonry structures, the resulting forces would fracture bones (Figure 7.28c).

Experimental studies of elbow, knee, and ankle joints have been done on human subjects using X ray and an arthroscope, which allows observation of joint structures in living subjects. X rays show distinct spaces between the bones that form a joint even when loads are applied to them. Arthroscopic studies also show that the load is not transferred from one bone to another in a joint. No compressive loading across the joint could be demonstrated in any of these studies.

Thinking of joints as icosahedra is a revolutionary idea. Like many other new ideas, it has not yet become widely accepted; however, the icosahedral model of joint structure provides a much better basis for explaining joint function than the more static models of joints as support columns or simple levers.

Questions

1. What is an icosahedron?
2. Why is it a good supporting structure?
3. What is unusual about thinking of some joints as icosahedra?

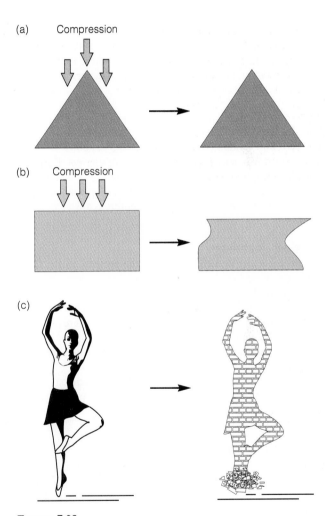

FIGURE 7.28
Compressive force applied to a triangle (*a*) is distributed through the structure, but force applied to a rectangle (*b*) causes bending of the support columns. (*c*) Comparison of compressive forces in a ballerina and a brick tower of comparable dimensions. If compressive forces were transferred across joints, the ballerina's foot would crumble just as the base of the tower did.

CHAPTER SUMMARY

Organization and General Functions

- The skeletal system—all bones and cartilage of the body—includes the axial skeleton, which supports the head and trunk, and the appendicular skeleton, which supports the appendages and attaches them to the trunk.
- The general functions of the skeletal system are support, protection, storage of calcium and phosphorus, provision of surfaces for the attachment of muscles, and formation of blood cells.
- The axial skeleton consists of bones of the skull, the vertebral column, thorax, hyoid, and middle ear bones (Table 7.2).

Bones of the Head

- Features of bones of the head are summarized in Table 7.3.
- Most bones of the cranium are shaped to protect the brain. Some bones have processes and depressions where muscles attach, and most have openings through which nerves and blood vessels pass.

Bones of the Trunk

- Features of bones of the trunk are summarized in Table 7.4.
- The vertebral column is a major supporting structure of the trunk. With it, the ribs and sternum provide a protective cage around the heart and lungs.
- Bones of the trunk have processes and depressions where muscles attach, and some have openings through which nerves and blood vessels pass.

Bones of the Appendages

- The appendicular skeleton consists of the pectoral girdle, the pelvic girdle, and the bones of the upper and lower appendages (Table 7.5).
- Features of these bones are summarized in Table 7.6.
- These bones have processes and/or depressions where muscles attach and articular surfaces where they form joints with other bones. Some have openings through which nerves and blood vessels pass.
- The girdles attach the appendages to the axial skeleton
- The appendicular skeleton serves many of the same functions as the axial skeleton but is more important in movement.

Skeletal Differences Due to Age and Sex

- After birth, the trunk and appendages grow at a more rapid rate than the skull; and after growth is complete, bones continue to be reshaped and remodeled, and some degeneration occurs.
- The female pelvic girdle is larger, shallower, and more bowl shaped than the male girdle. Other differences are summarized in Table 7.7.

Clinical Applications

- Abnormal curvatures of the spine can be lateral curvature (scoliosis), convex thoracic curvature (lordosis), and concave thoracic curvature (kyphosis).

QUESTIONS AND PROBLEMS

The questions at the end of each chapter are numbered to correspond with the objectives listed at the beginning of the chapter. Italics indicate that a question requires critical thinking skills beyond simple factual recall.

Questions

1. (a) What are the functions of the skeletal system?
   (b) What are the main divisions of the skeletal system, and how are they related to each other?
2. *(a) Cite five examples of complementarity of structure and function in the bones of the head.*
   (b) What is a suture?
   (c) What is a fontanel, and where are fontanels found?
   (d) What is a sinus, and where are sinuses found?
3. (a) How do vertebrae differ in structure?
   (b) Compare the structures of different types of ribs.
   (c) How do ribs connect the vertebra with the sternum?
4. (a) Which bones are part of the pectoral girdle? The upper appendage?
   (b) Classify the structural characteristics of the bones of the pectoral girdle and upper appendages as openings, depressions, articular surfaces, or surfaces for the attachment of other structures.
   (c) Give five examples of complementarity of structure and function among the bones of the pectoral girdle and upper appendage.
5. (a) Which bones are part of the pelvic girdle? The lower appendage?
   (b) Classify the structural characteristics of the bones of the pelvic girdle and lower appendages as openings, depressions, articular surfaces, or surfaces for muscle attachment.
   (c) Give five examples of complementarity of structure and function among the bones of the pelvic girdle and lower appendage.
   (d) From what bones are the arches of the foot formed, and what is the function of arches?
6. *(a) How is the skeleton of an adult different from the skeleton of an infant?*
   (b) How is the structure of the male skeleton different from the female skeleton?
   (c) What is the significance of the specialized structure of the female pelvis?
7. *How do various curvatures of the spine arise, and how do they affect body function?*

Problems

1. X rays of bones of the hand are sometimes used to study growth. Research this topic and determine how the technique could be used to determine whether a child's growth rate is abnormal.
2. Find out more about the treatment of abnormal curvatures of the spine.

OBJECTIVES

1. Relate articulations to the skeletal system, and describe their general functions.
2. Classify joints by function.
3. Classify joints by structure.
4. Describe the structure, function, and location of synovial joints; and describe the properties and function of synovial fluid.
5. List and define the kinds of movement that occur at synovial joints.
6. Describe the structure, function, and locations of bursae and tendon sheaths.
7. Describe the characteristics and movements of the joints of the axial skeleton.
8. Describe the characteristics and movements of the joints of the appendicular skeleton.
9. Discuss the alterations of the structure and function of joints associated with the following conditions: (a) osteoarthritis, (b) dislocations, (c) sprains, (d) strains, and (e) bursitis.

8

ARTICULATIONS

ORGANIZATION AND GENERAL FUNCTIONS

Have you an elderly relative who suffers from arthritis? We get the word *arthritis,* meaning inflammation of a joint, from the word *arthrosis.* **Arthroses** (ar-thro′ses), also called **articulations** (ar-tik″u-la′shunz) or **joints,** are the regions at which a bone joins another bone or a piece of cartilage.

CLASSIFICATION OF JOINTS

Functional Classification

We usually think of joints as sites of movement, but not all joints are movable. Joints are classified functionally according to how much movement they allow. An immovable joint is called a **synarthrosis** (sin″ar-thro′sis). A slightly movable joint is called an **amphiarthrosis** (am″fe-ar-thro′sis). A freely movable joint is called a **diarthrosis** (di″ar-thro′sis).

Joint function is associated with the relative need for support and movement at any particular site. Immovable joints, such as the sutures of the skull, fasten bones together and form rigid supporting structures. Slightly movable joints are found between vertebrae, where they provide both solid support and some movement. Freely movable joints bind and support bones but they allow one bone to move freely in relation to another. For example, a pivot joint between the first and second cervical vertebrae (atlas and axis) allow the head to turn. Ball-and-socket joints in the shoulders and hips allow for a wide range of movements of the limbs. These and other kinds of movable joints allow us to perform many kinds of activities from heavy construction work to the intricate manipulations made by eye surgeons.

Structural Classification

Joints also are classified structurally according to the manner in which bones are held together and whether the connective tissues around the bones form a fluid-filled cavity. They include fibrous, cartilaginous, and synovial joints. Most fibrous joints are immovable and most cartilaginous joints are slightly movable, but there are some exceptions as will be noted. Synovial joints are freely movable, and they have a fluid-filled joint cavity. Because they account for a large number of human movements, their structures and functions are emphasized in this chapter.

Fibrous Joints

In a **fibrous joint,** strong connective tissue fibers extend from one bone to the other and hold the two bones together tightly (Figure 8.1). Such joints lack a joint cavity. Little or no movement occurs at such a joint. Fibrous joints include sutures, gomphoses, and syndesmoses.

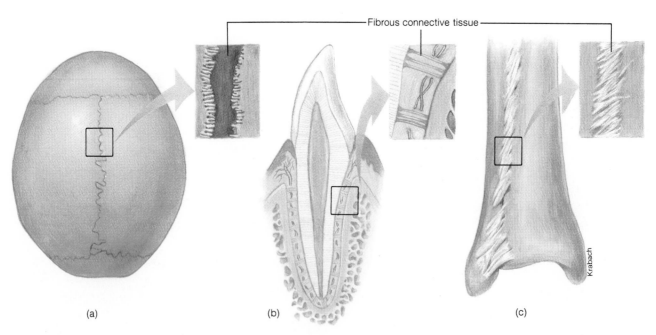

Fibrous connective tissue

(a)   (b)   (c)

Krabach

FIGURE 8.1
Types of fibrous joints: (*a*) suture, (*b*) gomphosis, and (*c*) syndesmosis.

Sutures, which are found in the skull, are immovable joints formed by fibers of dense connective tissue between bones. In a few instances, sutures present during development are replaced with bone. The frontal bones which fuse in the midline during infancy are an example. Such a joint is a **synostosis** (sin″os-to′sis).

Gomphoses (gom-fo′sez) are immovable joints between teeth and tooth sockets in the maxilla and mandible. The strong connective tissue fibers that hold teeth in their sockets are called **periodontal ligaments.**

Syndesmoses (sin″des-mo′sis) are flexible, very slightly movable joints found between the distal ends of the tibia and fibula near the ankle and between the distal ends of the radius and ulna near the wrist. Connective tissue in such joints forms an **interosseous ligament.** It binds the bones, but leaves a small space between them so a small amount of movement is possible. Trauma from athletic injuries or automobile accidents often tears these ligaments and forces excessive movement at syndesmoses. Until the separated bones are manipulated back to their normal positions, the victim experiences chronic pain.

Cartilaginous Joints

In a **cartilaginous joint,** bones are held together by a piece of cartilage interposed between two bones and fastened tightly to them by connective tissue fibers (Figure 8.2). Such joints include symphyses and synchondroses.

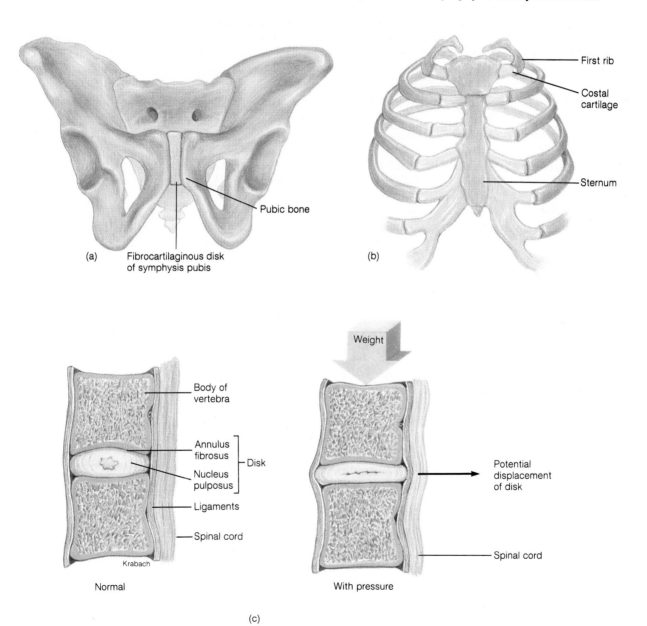

(a) Pubic bone
Fibrocartilaginous disk of symphysis pubis

(b) First rib
Costal cartilage
Sternum

Weight

Body of vertebra
Annulus fibrosus — Disk
Nucleus pulposus
Ligaments
Spinal cord
Krabach
Normal

Potential displacement of disk
Spinal cord
With pressure

(c)

FIGURE 8.2
Types of cartilaginous joints: (*a*) symphysis, (*b*) synchondrosis, and (*c*) an intervertebral symphysis showing compression.

A **symphysis** (sim'fis-is) is a slightly movable joint in which the articular surfaces of the bones are covered with hyaline cartilage and a piece of fibrous cartilage is interposed between them. Collagen fibers from the fibrous cartilage bind each bone to the cartilage, indirectly binding the bones together. Ligaments also connect the bones and reinforce the joint.

Symphyses are found between the vertebrae, where they contribute to the strength of the vertebral column and, at the same time, allow slight movement. Your ability to twist and bend your back demonstrates the degree of movement of these joints. The fibrous cartilage in an intervertebral joint has special weight-bearing and shock-absorbing properties (Figure 8.2c). This cartilage, also called a **disk,** consists of an inner gelatinous pulp, the **nucleus pulposus** (pul-po'sis), and an outer set of concentric rings of fibers, the **annulus fibrosus** (an'u-lus fi-bro'sis). When you stand erect, the disk is slightly compressed and flattened. As you walk, the pulpy portion acts as a shock absorber.

The **pubic symphysis,** which joins the coxal bones anteriorly, is movable only during late pregnancy and delivery when hormones have induced stretching of the joint fibers. Movement of this joint during childbirth eases the passage of the infant through the birth canal.

A **synchondrosis** (sin''kon-dro'sis) is a slightly movable joint with hyaline cartilage interposed between the bones. The attachment of the first costal cartilage to the sternum is a synchondrosis. The temporary joints between cartilage and bone in ossification centers of growing bones also are synchondroses. They become synostoses, represented by epiphyseal lines (Chapter 7), when growth is complete.

Synovial Joint

A **synovial joint,** or **diarthrosis** is bounded by ligaments and has a **joint cavity** filled with lubricating synovial fluid (Figure 8.3). The cavity is surrounded by the **articular capsule,** which consists of an outer layer of dense connective tissue attached to the periosteum of the bones of the joint and an inner **synovial membrane,** which consists of loose and elastic connective tissues. The outer layer helps to maintain bone alignment and ligaments outside the capsule further stabilize the joint. Muscles attached to the bones of a given joint extend across the joint, causing movement when they contract.

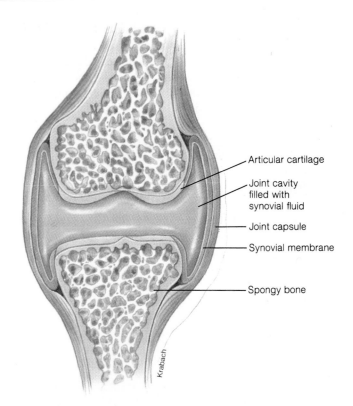

Articular cartilage
Joint cavity filled with synovial fluid
Joint capsule
Synovial membrane
Spongy bone

FIGURE 8.3
A synovial joint.

The ends of the bones within the joint cavity are covered with **articular cartilage,** a strong, smooth hyaline cartilage. The bones do not normally touch one another and the interposed synovial fluid allows the bones to move freely and smoothly. **Synovial fluid** is a viscous fluid secreted by the synovial membrane. It contains a high concentration of the lubricant hyaluronic acid. It also contains phagocytic cells that remove debris (from joint wear) and microorganisms (should they reach the cavity).

Acrobats (Figure 8.4) have more than the normal amount of movement at most of their joints, but the joints themselves are no different from those of less limber people. Acrobats increase the degree of motion by gradually stretching ligaments at various joints as a part of their regular practice routines.

People who claim their thumbs are "double jointed" only have unusually long ligaments at such joints. Anyone not born with such ligaments probably can acquire them—if they want them enough—by regularly exercising the joints and gently pushing the thumbs slightly farther than they normally move.

FIGURE 8.4
An acrobat doing stretching exercises to increase joint flexibility.

SYNOVIAL JOINTS AND THEIR MOVEMENTS

The majority of body joints are diarthroses, freely movable joints, located in the shoulders, hips, elbows, knees, wrists, ankles, hands, and feet.

Six types of synovial joints—gliding, hinge, pivot, ellipsoid, saddle, and ball-and-socket—are recognized by the shape of the bones within them and the movements they allow (Figure 8.5).

In a **gliding joint** (Figure 8.5a), the simplest synovial joint, the articular surfaces are flat or slightly curved and permit slight back-and-forth or side-to-side movements. No rotation occurs because the bones are bound together closely by ligaments. Gliding joints are found between the carpal bones, between the tarsal bones, and between the articulating processes of the vertebrae.

In a **hinge joint** (Figure 8.5b) the convex surface of one bone fits into the concave surface of another bone, allowing movement mainly in one plane. The elbow and knee joints are examples of hinge joints. To demonstrate the movement of a hinge joint, bend your arm at the elbow so your hand moves toward your shoulder. This movement, called **flexion,** decreases the angle between the bones of a joint. Now straighten your arm. This movement, called **extension,** increases the angle between the bones of a joint. Note that flexion and extension are opposite (antagonistic) movements and that both occur in a single plane.

In a **pivot joint** (Figure 8.5c), a rounded or pointed bone fits into a depression or opening in another bone, allowing motion in one plane. The radioulnar joint (between the radius and the ulna) and the atlantoaxial joint (between the atlas and the axis) are examples of pivot joints. The primary movement at a pivot joint is **rotation,** a circular movement around an axis. To demonstrate rotation, stand in anatomical position (Chapter 1) and move your arm so the palm of your hand is anterior. This movement is **supination** (soo″pin-a′shun). Then move your forearm so that the palm of your hand is posterior. This is **pronation** (pro-na′shun). In these movements, the radius rotates about the ulna. You also can demonstrate rotation by turning your head from side to side, causing the atlas to rotate about the dens of the axis.

In an **ellipsoid** (e-lip′soid) **joint,** or **condyloid** (kon′diloid) **joint** (Figure 8.5d), an oval (egg-shaped) condyle of one bone fits into an elliptical depression of another bone. This kind of joint allows movement in two planes, side-to-side and back-and-forth. Joints between metacarpals and phalanges are ellipsoid joints. To demonstrate side-to-side movement, spread your fingers apart and move them back together again. The spreading movement is **abduction,** or movement away from a midline (an imaginary line down the middle finger in this case). Moving your fingers back together is **adduction,** a movement toward a midline. Abduction and adduction are opposite movements within one plane. To demonstrate back-and-forth movement, bend (flex) a finger toward the palm of your hand and straighten (extend) it. To demonstrate **hyperextension,** bend your fingers backward toward the back of your hand.

In a **saddle joint** (Figure 8.5e), the articular surfaces and the kinds of movement are similar to those of an ellipsoid joint, but the joints are more freely movable. The carpometacarpal joint of the thumb is a saddle joint. Notice how much more freely your thumb with its saddle joint moves than do your fingers with their ellipsoid joints.

In a **ball-and-socket joint** (Figure 8.5f), the most freely movable joint, a "ball" on one bone fits into a "socket" on another bone. Three kinds of movement occur at a ball-and-socket joint: rotation, abduction-adduction, and flexion-extension (Figure 8.6). First, the ball can rotate in the socket. You can demonstrate this by holding your arm out in front of you with the elbow extended and rotating your whole arm back and forth around its own axis (an imaginary line down the midline of the arm). Second, the ball-and-socket joint allows abduction and adduction.

FIGURE 8.5
Types of synovial joints.

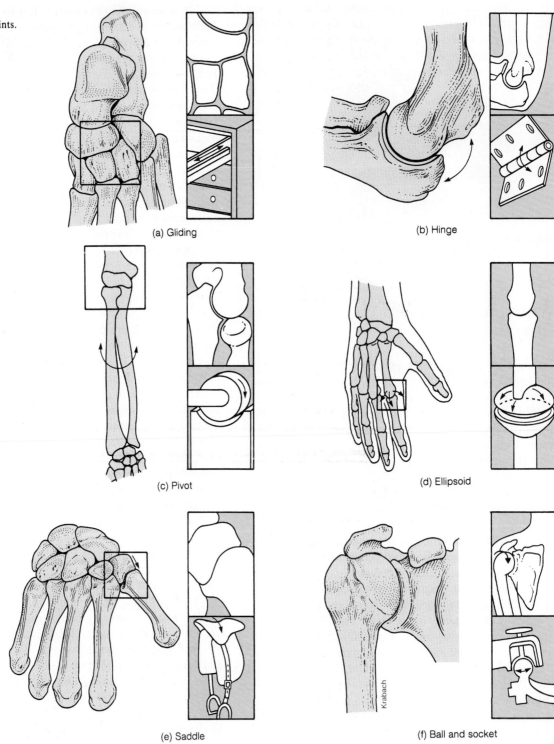

(a) Gliding

(b) Hinge

(c) Pivot

(d) Ellipsoid

(e) Saddle

(f) Ball and socket

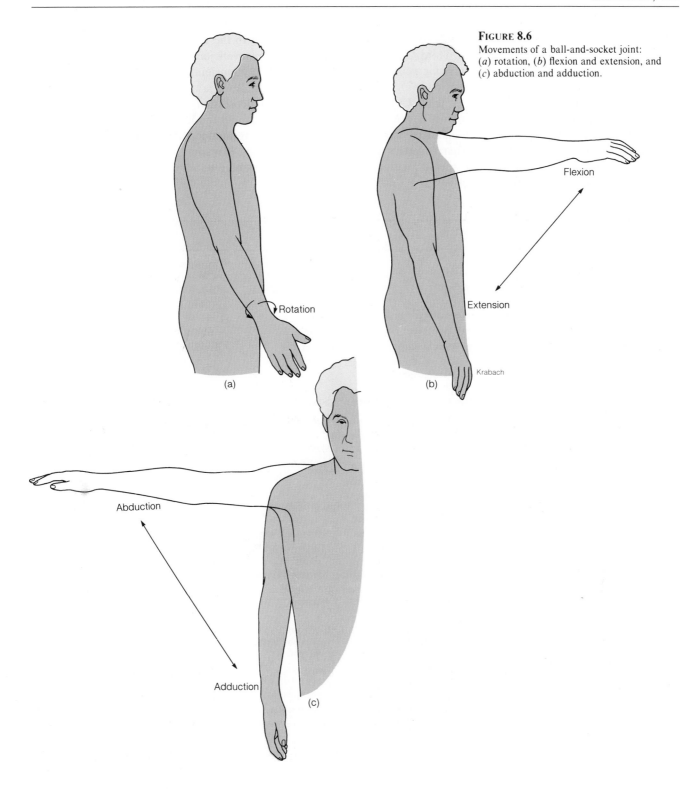

FIGURE 8.6
Movements of a ball-and-socket joint:
(*a*) rotation, (*b*) flexion and extension, and
(*c*) abduction and adduction.

Raise your arm up and away from your side to abduct it and bring it close to your side to adduct it. Third, this joint allows flexion and extension. When you swing your arm forward, you flex the shoulder joint. When you swing your arm backward, you extend the shoulder joint. **Circumduction** is a circular movement that combines flexion, extension, abduction, and adduction. You can demonstrate circumduction by moving your arm so your fingertips describe a circle and your arm describes the surface of a cone. Table 8.1 summarizes the types of joints.

In addition to the movements described for each of the types of synovial joints, special kinds of movement occur at only certain joints (Figure 8.7). Movements of the mandible, or lower jaw, include protraction–retraction and elevation–depression. **Protraction** is thrusting the mandible forward so the lower teeth extend beyond the upper teeth and **retraction** is returning the mandible to its normal position. **Elevation** is closing the mouth and **depression** is opening it. At the ankle joint, **dorsiflexion**

TABLE 8.1   Characteristics of Types of Joints

Type of Joint	Properties	Mobility	Examples
Fibrous joints (Figure 8.1)	Bones held together by fibers; no joint cavity	None or slight	
Suture	Fibers hold bones together tightly	None	Coronal and sagittal sutures
Gomphosis	Fibers hold teeth in sockets	None	Teeth in sockets
Syndesmosis	Fibers bind bones less tightly	Slight	Tibiofibular and radioulnar joints
Cartilaginous joints (Figure 8.2)	Cartilage interposed between bones; fibers hold bone and cartilage together; no joint cavity	Slight	
Symphysis	Fibrous cartilage between bones; fibers and ligaments stabilize joint	Slight	Intervertebral joints, and pubic symphysis
Synchondrosis	Hyaline cartilage interposed between bones	Slight	Joint between first costal cartilage and sternum; joints between bone and epiphyseal cartilage during growth
Synovial joint (Figure 8.3)	Joint has cavity, capsule, synovial membrane, and synovial fluid; bone ends in joint covered with articular cartilage; ligaments stabilize joint and muscles extend across joint, usually by tendons	Free	
Gliding	Flat or slightly curved articular surfaces	Side-to-side and back-and-forth movements	Joints between carpals and between tarsals
Hinge	Convex and concave articulating surfaces	Movements in only one plane	Elbow and knee joints
Pivot	Round end of one bone fits in depression in another bone	Rotation	Joint between radius and ulna and between atlas and axis
Ellipsoid (condyloid)	Oval condyle fits into an elliptical depression in another bone	Side-to-side and back and forth movements	Joints between metacarpals and phalanges of fingers
Saddle	Projection of one bone fits in saddle-shaped depression in another bone	Same as ellipsoid, but freer movements	Joint between metacarpal and phalanx of thumb
Ball-and-socket	Ball-like projection of one bone fits into socket-like depression in another bone	Most freely movable of all joints; allows movement in three planes and rotation of bone on its own axis	Shoulder and hip joints

(dor″si-flek′shun) is drawing the foot toward the shin and **plantarflexion** (plan-tar-flek′shun) is extending the foot away from the shin. Walking on one's heels requires dorsiflexion, whereas walking on tiptoes requires plantarflexion. Also at the ankle joint, **inversion** is rotating the sole of the foot inward and **eversion** is rotating it outward.

Movements and their definitions are summarized in Table 8.2.

Joint function can be severely impaired by immobility as tendons and ligaments lose elasticity and contractures develop. A **contracture** (kon-trak′tūr) is a reduction in the movement of a joint, often due to shortening of muscles acting on the joint. These effects are seen in people who are bedridden or otherwise fail to get enough exercise. Impairment can be minimized by range of motion exercises—moving each joint through its complete arc of movement in each plane in which it moves. Passive exercises can be done for a bedridden person by someone else moving each limb through its range of motion.

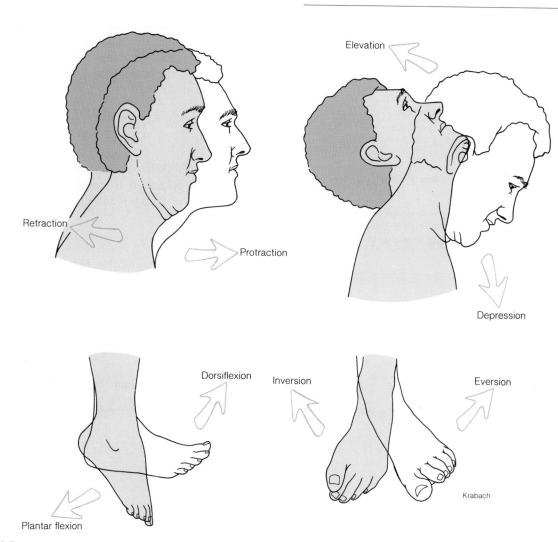

FIGURE 8.7
Special kinds of movements.

TABLE 8.2   Movements and Their Definitions

Pairs of Antagonistic Movements

Flexion—the decreasing of the angle between two bones	Extension—the increasing of the angle between two bones
Abduction—the movement of a body part away from a midline	Adduction—the movement of a body part toward a midline
Medial rotation—the turning of a bone on its own axis toward the midline of the body	Lateral rotation—the turning of a bone on its own axis away from the midline of the body
Supination—the placing of the palm of the hand in anatomical position	Pronation—the placing of the palm of the hand away from anatomical position
Elevation—the raising of a body part	Depression—the lowering of a body part
Protraction—the thrusting forward of a body part	Retraction—the withdrawal of a body part
Dorsiflexion—the bending of the foot toward the shin (tibia)	Plantarflexion—the bending of the foot away from the shin (tibia)
Inversion—the rotation of the sole of the foot inward	Eversion—the rotation of the sole of the foot outward

Unpaired Movements

Hyperextension—the excessive extension of a body part in the opposite direction of flexion	Circumduction—movement describing a circle

BURSAE AND TENDON SHEATHS

Bursae (ber'se) and **tendon sheaths** are saclike structures similar to joint capsules (Figure 8.8). They are made of connective tissue, lined with synovial membrane, and filled with synovial fluid. Bursae, found between the skin and an underlying bone or between a tendon and a bone, reduce friction and permit easy movement of one part on another. A few of the more prominent bursae include the prepatellar bursa between the skin and the patella, the olecranon bursa in the elbow, and the subscapular bursa beneath the scapula. Bursae also are found in the hips, heels, and other sites. Tendon sheaths are tubelike structures similar to bursae that surround tendons under pressure, such as those on the back of the hand and along the anterior surface of the foot.

JOINTS OF THE AXIAL SKELETON

Joints of the axial skeleton are primarily supporting structures, but some contribute to movement as well. They include the joints of the skull and the joints of the trunk.

Joints of the Skull

Most skull joints are immovable, fibrous sutures found between the bones of the cranium and the face. The only movable joints in the skull are the **temporomandibular**

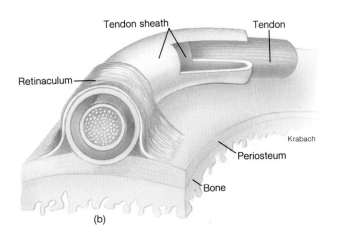

FIGURE 8.8
(*a*) Bursae, and (*b*) a tendon sheath.

(tem″po-ro-man-dib′u-lar) **joints,** synovial joints of the mandible. In these joints, a condylar process of the mandible articulates with the mandibular fossa of the temporal bone. Movements include elevation-depression, protraction-retraction, and a small amount of side-to-side movement as food is ground between the molars.

Joints of the Trunk

Several kinds of joints are found in the vertebral column itself—joints that articulate the ribs with the vertebral column and the sternum, and joints that articulate with the girdles. All are included among joints of the trunk.

Joints of the Vertebral Column

Joints of the vertebral column support the body and allow certain specialized movements. Joints between the bodies of all vertebrae are symphyses and are mainly support structures. The **atlanto-occipital joints** (articulations of the atlas with the occipital condyles of the skull) are ellipsoid joints. They allow a rocking motion such as occurs in nodding the head "yes" (Figure 8.9a). The **atlantoaxial joint** (articulation of the atlas and the axis) is a pivot joint. It allows side-to-side rotation such as in shaking the head "no" (Figure 8.9b). Other joints between articulating processes of vertebrae are gliding joints. In the cervical region these joints allow rotation, flexion, and hyperextension of the neck. In the thoracic and lumbar regions, they allow similar but less flexible movements.

Joints Associated with Ribs

The ribs articulate with the thoracic vertebrae at synovial joints between rib facets and the vertebral arches. Most ribs also articulate with the sternum, either directly or via cartilage (Chapter 7). Some of these joints are cartilaginous and some have small synovial sacs at their articulation with the sternum. Limited movement occurs at these joints during breathing.

Joints That Attach Girdles to the Axial Skeleton

The synovial **sternoclavicular joints,** which attach the pectoral girdles to the axial skeleton, are gliding joints between the proximal ends of the clavicles and the manubrium of the sternum. To detect movement at these joints, place the fingers of one hand over the manubrium and the proximal ends of the clavicles and shrug your shoulders.

The **sacroiliac joints** lie between the lateral edges of the sacrum and the posterior edges of each ilium, a portion of the coxal bone. These partly fibrous and partly synovial gliding joints have limited movement through early adulthood and during late pregnancy, but become almost entirely fused by midlife.

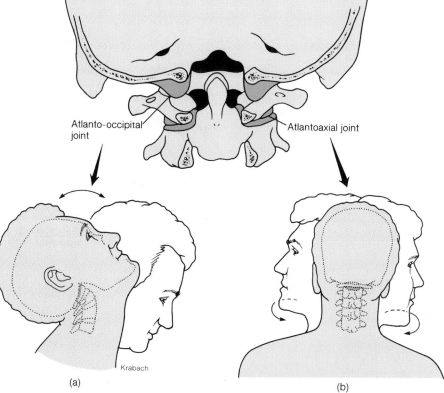

Figure 8.9
(*a*) Nodding the head, and (*b*) rotating the head.

Atlanto-occipital joint

Atlantoaxial joint

Krabach

(a)

(b)

JOINTS OF THE APPENDICULAR SKELETON

Joints of the Pectoral Girdle

In addition to the sternoclavicular joint just described, the pectoral girdle also has an **acromioclavicular** (ak-ro″me-o-klav-ik′u-lar) **joint** between the clavicle and the ac-

romion process of the scapula. This gliding joint allows elevation, depression, protraction, and retraction of the shoulder.

Joints of the Upper Appendage
The Shoulder Joint

The **shoulder joint** (Figure 8.10) is a ball-and-socket joint encased in a joint capsule partially surrounded by ligaments. The dorsal *coracohumeral* (kor″ah-ko-hu′mer-al)

FIGURE 8.10
The shoulder joint: (*a*) joint cavity, (*b*) ligaments, and (*c*) glenoid cavity and associated structures.

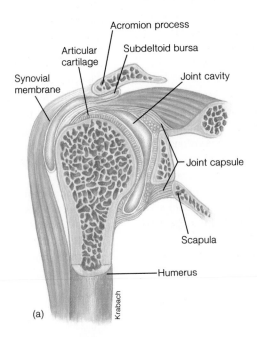

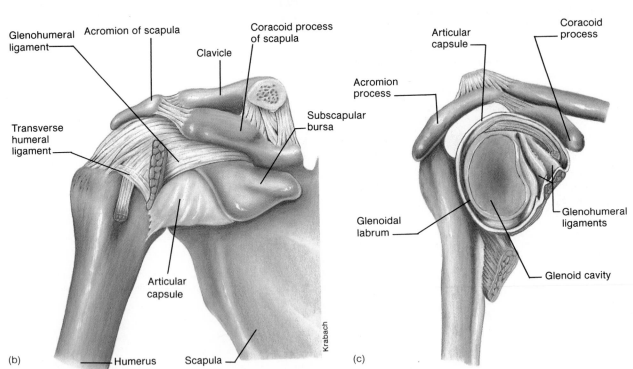

ligament connects the coracoid process of the scapula to the greater tubercle of the humerus. The ventral *glenohumeral* (glen″o-hu′mer-al) *ligaments* consist of three bands of fibers extending from the glenoid cavity to the lesser tubercle and anatomical neck of the humerus. The lateral *transverse humeral ligament* is a thin, narrow band of connective tissue connecting the greater and lesser tubercles of the humerus. (The long head of the biceps brachii, a major arm muscle, passes under this ligament.) The *glenoidal labrum* (gle′noid-al la′brum), which consists of fibrocartilage, forms a rim that deepens the glenoid cavity. Together, the fibrous joint capsule and the muscle tendons passing over the joint form a **rotator cuff,** which is often injured by baseball pitchers "winding up" to throw a ball.

The locations of several bursae associated with the shoulder joint are indicated by their names—subacromion and subcoracoid, beneath the acromion and coracoid processes of the scapula, respectively; the subdeltoid, beneath the deltoid muscle; and the subscapular, beneath the tendon of the subscapularis muscle.

Movements at the shoulder joint include flexion-extension, abduction-adduction, circumduction, and medial and lateral rotation. Though the ligaments and the more superficial muscles associated with this joint provide both protection and excellent mobility, the relatively shallow rim of the glenoid cavity allows the bones to be easily dislocated (moved out of alignment)—a very painful event.

The Elbow Joint

The complex **elbow joint** (Figure 8.11) contains three articulations within one joint capsule—the radius and ulna each articulate with the humerus, and the radius and ulna also articulate with each other at the proximal radioulnar joint. Two important ligaments—the medial *ulnar collateral ligament* and the lateral *radial collateral ligament* support and strengthen the joint capsule. The *annular ligament* encircles the head of the radius and attaches to both the radial collateral ligament and the trochlea of the ulna.

Articulations of the radius and ulna with the humerus form a hinge joint where movement is limited to flexion and extension. The articulation between the radius and the ulna is a pivot joint in which the radius rotates about the ulna within the bounds of the annular ligament. This latter movement results in supination and pronation of the hand.

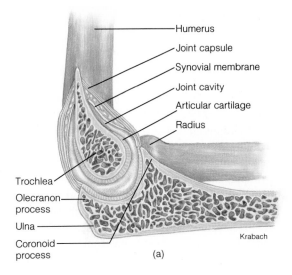

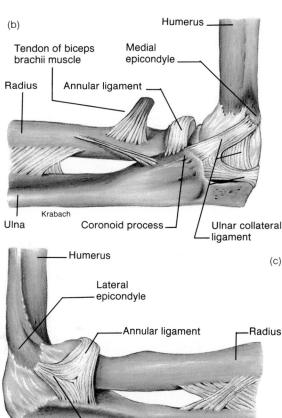

FIGURE 8.11
The elbow joint: (*a*) joint cavity, (*b*) ulnar collateral ligament and associated structures, and (*c*) radial collateral ligament and associated structures.

Joints of the Wrist and Hand

Joints of the wrist include another radioulnar articulation, and articulations of the radius and the ulna with adjacent carpals (Figure 8.12). Like its proximal counterpart, the distal radioulnar joint also contributes to rotation of the radius about the ulna. The radiocarpal and the ulnocarpal joints allow flexion, extension, hyperextension, abduction, adduction, and circumduction of the hand with respect to the forearm.

Joints of the hand include the carpometacarpal joints, the metacarpophalangeal joints, and the joints between the phalanges of the digits. Flexible movement at the carpometacarpal joint of the thumb makes the thumb apposable, but movement at the other carpometacarpal joints is less flexible. The metacarpophalangeal joints and the joints between phalanges in all digits provide great dexterity of finger movements.

Joints of the Pelvic Girdle

In addition to the sacroiliac joints described earlier, the pelvic girdle also has an anterior midline joint, the **pubic symphysis** which unites the two coxal bones. Together, the joints of the pelvic girdle stabilize the pelvis, making it capable of supporting the body in an upright position.

Joints of the Lower Appendage

The Hip Joint

The **hip joint** (Figure 8.13) is formed by the articulation of the head (ball) of the femur and the acetabulum (socket) of the coxal bone. The joint capsule is surrounded by strong ligaments that stabilize the joint. Within the joint cavity itself, the *ligamentum teres* (lig-am-en′tum te′res) runs from the fovea capitus on the head of the femur to the capsule and helps hold the femur in the socket. The rim of the acetabulum is made deeper by the *acetabular*

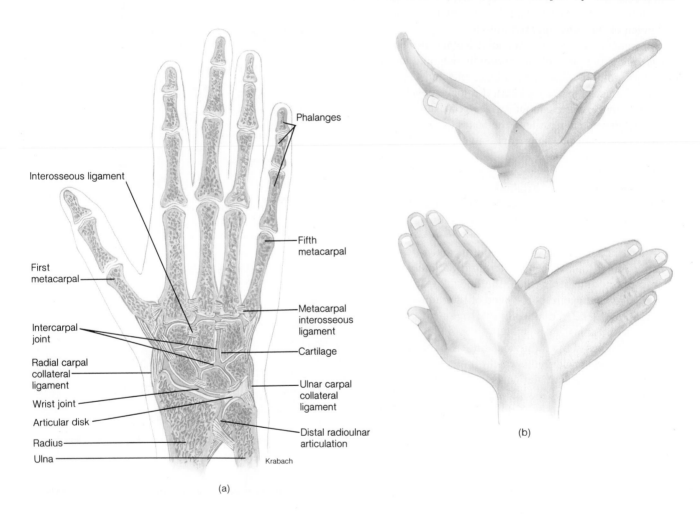

Phalanges

Interosseous ligament

Fifth metacarpal

First metacarpal

Metacarpal interosseous ligament

Intercarpal joint

Cartilage

Radial carpal collateral ligament

Wrist joint

Ulnar carpal collateral ligament

Articular disk

Radius

Distal radioulnar articulation

Ulna

Krabach

(a)

(b)

FIGURE 8.12
(*a*) Joints of the wrist and hand in cross section, and (*b*) movements of the wrist. What are the names of these movements?

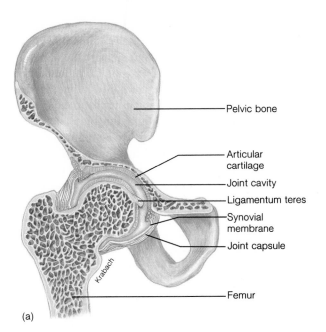

(a)

FIGURE 8.13
The hip joint: (*a*) joint cavity, (*b*) anterior view of ligaments, and (*c*) posterior view of ligaments.

Pelvic bone

Articular cartilage

Joint cavity

Ligamentum teres

Synovial membrane

Joint capsule

Femur

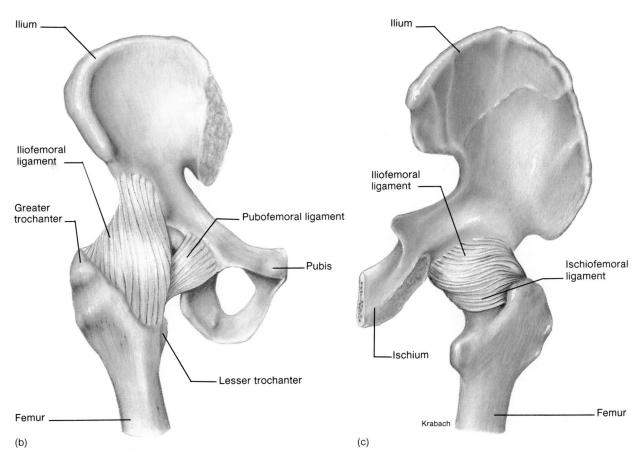

Ilium

Iliofemoral ligament

Greater trochanter

Femur

(b)

Pubofemoral ligament

Pubis

Lesser trochanter

Ilium

Iliofemoral ligament

Ischium

(c)

Ischiofemoral ligament

Femur

FIGURE 8.14

The knee joint: (*a*) joint cavity, (*b*) anterior view of ligaments, and (*c*) posterior view of ligaments.

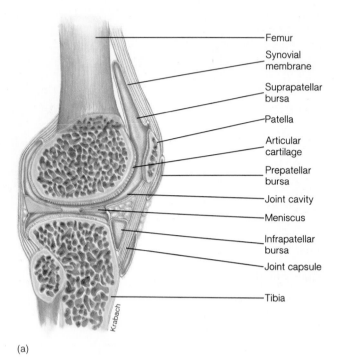

Femur
Synovial membrane
Suprapatellar bursa
Patella
Articular cartilage
Prepatellar bursa
Joint cavity
Meniscus
Infrapatellar bursa
Joint capsule
Tibia

Krabach

(a)

labrum, a horseshoe-shaped fibrocartilage. Three important ligaments—the *iliofemoral,* the *ischiofemoral,* and the *pubofemoral*—strengthen and support the joint capsule. The iliofemoral ligament extends from the anterior inferior iliac spine to the intertrochanteric line between the trochanters of the femur. The other ligaments extend from the coxal bone to the fibers of the joint capsule.

The same movements occur at this ball-and-socket joint as at the shoulder: flexion (drawing the knee toward the abdomen), extension (thrusting the whole leg behind the body), abduction (moving the whole leg away from the body), adduction (moving the whole leg toward the midline), circumduction (circular movement), and medial and lateral rotation of the thigh about its own axis (moving the knee medially or laterally with respect to the thigh).

The Knee Joint

A particularly complex and unstable joint, the **knee joint** (Figure 8.14), is the largest synovial joint in the body. The joint capsule is encased in ligaments between bones and tendons from muscles of the thigh and leg. Inside the joint capsule, the relatively flat surfaces of the condyles of the femur articulate with equally flat condyles of the tibia. Interposed between the articular surfaces are concave disks

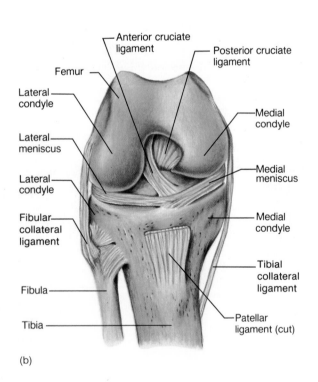

Anterior cruciate ligament
Posterior cruciate ligament
Femur
Lateral condyle
Medial condyle
Lateral meniscus
Lateral condyle
Medial meniscus
Fibular collateral ligament
Medial condyle
Fibula
Tibial collateral ligament
Tibia
Patellar ligament (cut)

(b)

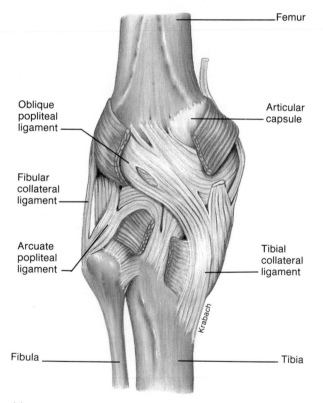

Femur
Oblique popliteal ligament
Articular capsule
Fibular collateral ligament
Arcuate popliteal ligament
Tibial collateral ligament
Fibula
Tibia

Krabach

(c)

of fibrous cartilage, the *medial meniscus* (men-is′kus) and the *lateral meniscus*. The thick outer edges of the menisci help to hold the articular surfaces in their proper positions. *Cruciate* (kroo′she-āt) *ligaments* (crossing ligaments) resist twisting and turning movements and help to prevent anterior displacement of the tibia.

Several more superficial ligaments surround the capsule and serve to stabilize the joint. The *tibial collateral ligament* connects the medial condyles of the femur and tibia, and the *fibular collateral ligament* connects the lateral condyle of the femur to the head of the fibula. The *oblique popliteal ligament* connects the lateral condyle of the femur to the head of the tibia, and the *arcuate popliteal ligament* connects the lateral condyle of the femur to the head of the fibula. The large *patellar ligament* and the patella (imbedded in the tendon of the quadriceps femoris muscle) are located anterior to the joint.

The synovial cavity of the knee joint is extensive and irregular in shape. It extends proximally from the patella beneath the quadriceps femoris muscle to form a large bursa called the suprapatellar bursa. Other bursae, near but not part of the joint cavity, include the infrapatellar bursa distal to the patella and the prepatellar bursa beneath the skin anterior to the patella.

Movements of the knee joint are mainly flexion (moving the leg toward the back of the thigh) and extension (straightening the leg). When the knee is flexed, the tension on the collateral ligaments is reduced, and a slight amount of rotation and circumduction is possible. Quick pivot turns, such as are required in football, basketball, and some other sports often result in knee injuries. These injuries include torn ligaments, dislocation of the bones of the joint, and damage to the menisci.

Nearly immovable joints between the tibia and the fibula are the **superior tibiofibular joint** just below the knee and the **inferior tibiofibular joint** near the ankle. The *interosseous membrane* holds the tibia and fibula together between these joints. These structures stabilize the fibula in relation to the tibia.

Joints of the Ankle and Foot

Joints of the ankle and foot (Figure 8.15) include articulations of the malleoli of the tibia and fibula with the talus, articulations of tarsals with each other, tarsometatarsal joints, metatarsophalangeal joints, and joints between phalanges. At the ankle, the tibiotarsal and the fibulotarsal joints allow dorsiflexion, plantarflexion, and limited circumduction. Articulations of the talus with other tarsals allows inversion and eversion of the foot. Other joints of the foot are similar to, but usually less flexible than those in the hand.

The major joints of the body and their movements are summarized in Table 8.3.

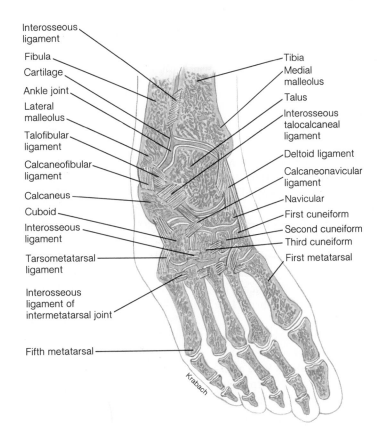

FIGURE 8.15

Synovial joints of the ankle and foot in cross section. What are the names of the movements already shown in Figure 8.7?

TABLE 8.3	Summary of Major Joints and Their Movements	
Joint	**Type**	**Movement**
Most skull joints	Sutures	Immovable
Temporomandibular	Hinge	Elevation, depression, protraction, retraction
Atlanto-occipital	Ellipsoid	Back-and-forth and side-to-side movements
Atlantoaxial	Pivot	Rotation
Intervertebral		
Bodies	Symphyses	Slight movement
Arches	Gliding	Rotation, flexion, hyperextension
Costovertebral	Gliding	Slight movement during breathing
Sternocostal	Gliding and synchondroses	Slight movement during breathing
Sternoclavicular	Gliding	Slight movement when shrugging shoulders
Acromioclavicular	Gliding	Elevation, depression, protraction, retraction
Sacroiliac	Synarthroses	Slight movement up to midlife
Shoulder	Ball-and-socket	Flexion, extension, abduction, adduction, circumduction, medial and lateral rotation
Elbow	Hinge	Flexion, extension
Radioulnar	Pivot	Rotation of radius about ulna
Wrist	Ellipsoid	Flexion, hyperextension, abduction, adduction, circumduction
Carpometacarpals		
Fingers	Ellipsoid	Back-and-forth and side-to-side movements
Thumb	Saddle	Free back-and-forth and side-to-side movements
Hip	Ball-and-socket	Flexion, extension, abduction, adduction, circumduction, medial and lateral rotation
Knee	Hinge	Flexion, extension; when fixed some circumduction and rotation
Superior tibiofibular	Gliding	Stabilizes fibula, almost no movement
Inferior tibiofibular	Syndesmosis	Stabilizes fibula, almost no movement
Ankle	Hinge	Dorsiflexion, plantarflexion, limited circumduction
Intertarsals	Gliding	Inversion, eversion
Tarsometatarsals	Gliding	Back-and-forth and side-to-side movements, but much more limited than similar joints in hands; form arches of foot

CLINICAL APPLICATIONS

Joint disorders include joint degeneration in the form of arthritis and joint injuries in the form of dislocations, sprains, and strains. Inflammation of bursae, called bursitis, is a joint-associated disorder.

Osteoarthritis is the most common form of arthritis. It is a progressive joint disease involving degeneration of articular cartilages (Figure 8.16) accompanied by new bone growths in the form of spurs in the joint cavity. Though these spurs interfere with joint movement, inflammation is not as severe as in rheumatoid arthritis (see essay).

Osteoarthritis appears to be associated with aging as it is rarely seen before age 40, but affects over 85 percent of people over 70. The disease probably starts with alteration of the collagen fiber network and the absorption of excessive amounts of water into articular cartilage. Normal cartilage is about 70 percent water, but what car-

tilage remains in arthritic joints contains an even higher percent of water. The matrix begins to deteriorate and loses both collagen molecules and associated polysaccharides. Mechanical stress further deforms the matrix, and lessened elasticity prevents it from returning to its normal shape. As the disease progresses, affected joints become enlarged, less mobile, and often painful. The disease usually is not crippling unless it affects weight-bearing joints.

Gouty arthritis, or **gout,** results from prolonged, excessive levels of uric acid, a waste product of nucleic acid metabolism. Uric acid accumulates in joint tissues, especially in the great toe, causing severe inflammation and pain. Gout can be treated with drugs that reduce uric acid formation or increase its excretion.

A **dislocation** is a displacement of the articular surfaces of a joint usually accompanied by damage to the surrounding ligaments. Dislocations result from falls, blows, or extreme exertion and are most often seen in the joints of the thumb, fingers, knee, or shoulder. Picking up

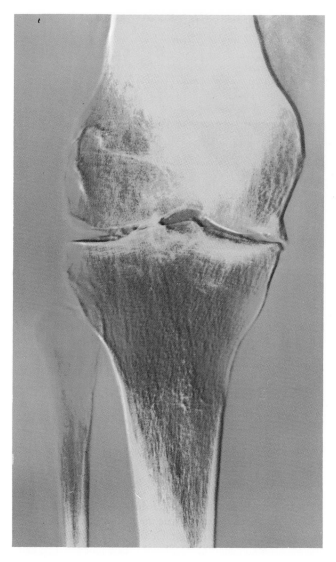

FIGURE 8.16
Joint damage from osteoarthritis.

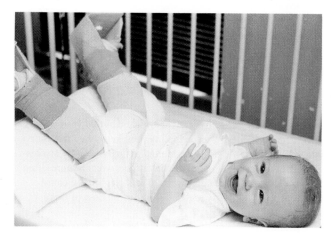

FIGURE 8.17
A child in traction for the treatment of congenital hip dislocation.

may be unable to use the joint. A less serious injury, a **strain** involves overstretching of muscles without tearing of tissues.

Bursitis (bur-si′tis), inflammation of a bursa, usually is caused by an injury or infection, or occurs as a complication of rheumatoid arthritis. It most commonly affects the subscapular bursa near the shoulder, the olecranon bursa at the elbow, or the prepatellar bursa of the knee. Once a bursa is inflamed, excessive use of the nearby joint often precipitates another painful episode. In chronic bursitis, calcium deposits build up in the bursa and cause extreme pain and restriction of movement. Treatment includes rest and analgesics following an acute attack, and later physical therapy to prevent loss of function. Hydrocortisone injections sometimes are used to reduce inflammation, and forceful manipulation under anesthesia can be used to restore movement in seriously immobilized joints.

a small child by one arm can dislocate the child's shoulder. Symptoms of dislocation include swelling, pain, and loss of motion. Treatment consists of returning the articular surfaces to their normal positions and immobilizing the joint while healing takes place.

Occasionally a child is born with **congenital hip dislocation,** in which the rim of the acetabulum are not deep enough to retain the head of the femur in its proper position. One or both hips can be affected. To treat this disorder, the child is placed in traction (Figure 8.17) while the acetabula develop strength before the joints are allowed to bear weight.

A **sprain** is a twisting of a joint in which the joint capsule, tendons, and muscles can be torn, but in which dislocation does not occur. Nerves and blood vessels also can be injured. The affected individual suffers severe pain and

CLINICAL TERMS

ankylosis (ang″kil-o′sis)   abnormal immobility of a joint

arthralgia (ar-thral′je-ah)   pain in a joint

arthrology (ar-throl′o-je)   a study of joints and the disorders affecting them

arthroplasty (ar′thro-plas″te)   surgery on a joint

bunion (bun′yun)   swelling of a bursa in the great toe

foot drop   the hanging of the foot in a plantar-flexed position, often due to prolonged immobility

luxation (luk-sa′shun)   dislocation of joint components

spondylitis (spon″dil-i′tis)   an inflammation of a vertebra

spondylosis (spon″dil-o′sis)   ankylosis or immobility of a vertebral joint

subluxation (sub″luk-sa′shun)   a partial dislocation of joint components

ESSAY
RHEUMATOID ARTHRITIS

Among the several forms of arthritis, **rheumatoid** (roo'mat-oid) **arthritis** is most often crippling. Unlike osteoarthritis, which afflicts the elderly, rheumatoid arthritis usually strikes between the ages of 30 and 40, and affects women more often than men. Early onset and severe effects combine to cause many years of disability and suffering for victims of this disease.

Manifestations of rheumatoid arthritis are numerous, but the joints of the hands and feet are nearly always affected. Early morning joint stiffness is seen in nearly all patients. Initially, the disease affects only one or a few joints, but additional joints are affected, usually bilaterally—the same joints on the left and right sides of the body—as the disease progresses. Joints become deformed and their functions impaired (Figure 8.18), sometimes fusing and becoming immovable. In addition to affecting joints, rheumatoid arthritis can cause inflammation of tendons, subcutaneous tissue, skin, and even blood vessels. In chronic disease, relatively symptom-free periods are interspersed with painful attacks.

Rheumatoid arthritis is a systemic autoimmune disease. The autoimmune response is complex—it involves "mistaken identity" of a person's own proteins as foreign substances, triggering the manufacture of autoantibodies, or antibodies that attack the person's own tissues as if they were infectious agents (Chapter 21). In rheumatoid arthritis, the autoantibodies attack various tissues, especially the synovial membranes of joints, and produce inflammation and destruction.

Autoantibodies called **rheumatoid factor** can be found in the blood plasma and synovial fluid of rheumatoid joints, and in subcutaneous nodules (lumps) often seen in individuals with the disease. Sometimes the factor is detected in the blood plasma of relatives of rheumatoid arthritis victims. This suggests that the propensity to develop the disease may be inherited, but that some people who have the rheumatoid factor (and, thus, a gene that leads to its formation) fail to develop the disease.

Though some antigenic substance causes the autoimmune response, why it does so is not yet understood. In some cases, the disease appears to be triggered by a virus; in others, by a bacterial agent; and in still others, by some unknown factor. Recent studies of rheumatoid synovial cells in tissue culture have demonstrated that some of these cells produce collagenase, an enzyme that digests collagen. The relationship of this enzyme and its role in collagen reabsorption to the autoimmune reaction is not yet known. Immunologic studies also have shown that a defect in the function of a kind of lymphocyte called a T-cell may contribute to the development of rheumatoid arthritis.

No cure exists for rheumatoid arthritis, so treatment is directed toward reducing inflammation and slowing the rate of joint damage. Hydrocortisone and other steroids are used to

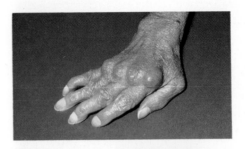

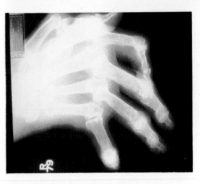

FIGURE 8.18
The effects of rheumatoid arthritis.

lessen inflammation and, thereby, reduce joint damage. Because long-term hydrocortisone therapy weakens bone structure and has other undesirable side effects, aspirin and other nonsteroid agents are used to decrease inflammation and reduce pain. They combat the most incapacitating effects of the disease with a minimum of undesirable side effects. Finally, physical therapy is used to keep joints movable and to maintain as near normal joint function as possible.

When rheumatoid arthritis, other forms of arthritis, or injuries lead to crippling degeneration of a joint, the natural joint sometimes can be replaced with a prosthetic device (a manufactured device to replace a body part). Joint replacement is most frequently performed on weight-bearing joints to enable a person to walk again, but it also can be done on finger joints to restore dexterity. Both hip and knee joints can be replaced now, and methods for replacing ankle joints are being developed. As the art of joint replacement advances, more and more victims of rheumatoid arthritis and other crippling conditions will benefit.

Questions

1. In what ways is rheumatoid arthritis more crippling than other forms of arthritis?
2. What symptoms are diagnostic for rheumatoid arthritis?
3. What is rheumatoid factor?
4. How is rheumatoid arthritis treated?

CHAPTER SUMMARY

Organization and General Function
- Articulations, also called joints or arthroses, are the points at which bones are joined or come in close proximity to each other, or at which a bone is joined to cartilage.
- Joints help to support the parts of the skeletal system, and many also make it possible for one bone to move in relation to another.

Classification of Joints
- Functionally, joints are immovable synarthroses, slightly movable amphiarthroses, and freely movable diarthroses.
- Structurally, joints are fibrous, cartilaginous, or synovial.
- Fibrous joints are held together by fibers of connective tissue that extend from one bone to another. Sutures, located in the skull, and gomphoses, which hold teeth in sockets, are immovable. Syndesmoses, with interosseous ligaments between forearm or leg bones, have very limited movement.
- Cartilaginous joints are held together by a piece of cartilage interposed between the bones and bound to them by connective tissue fibers; such joints are slightly movable. They include symphyses between vertebrae and between pubic bones, and synchondroses between some ribs and the sternum, and at the epiphyseal plates of growing bones.
- Synovial joints are surrounded by a joint capsule and ligament, have a joint cavity lined with synovial membrane and filled with synovial fluid, and have the cartilage-covered articular surfaces of the bones extending into the cavity. They include all freely movable joints.
- Synovial fluid is secreted by the synovial membrane and serves to lubricate the surfaces inside a synovial cavity.

Movements at Synovial Joints
- The types of synovial joints are gliding joints, which allow back-and-forth and side-to-side movements; hinge joints, which allow bending in a single plane; pivot joints, which allow rotation of one bone in relation to another; ellipsoid, or condyloid joints, which allow back-and-forth and side-to-side movements; saddle joints, which allow freer movements than at ellipsoid joints, and ball-and-socket joints, which allow very free movements, including abduction, adduction, flexion, extension, and circumduction.
- Specific kinds of movement are defined in Table 8.2.

Bursae and Tendon Sheaths
- Bursae and tendon sheaths are connective tissue sacs closely resembling synovial cavities found at points of excessive friction.

Joints of the Axial Skeleton
- Immovable or slightly movable joints of the axial skeleton include sutures of the cranium and facial bones, cartilaginous joints of the vertebral column and ribs, and several joints of the sacrum.
- Freely movable joints of the axial skeleton include the hinge joints of the mandibles, the pivot and condyloid joints that allow movements of the head, and some of the joints between the ribs and the vertebral column.

Joints of the Appendicular Skeleton
- Joints of the pectoral girdle include the sternoclavicular joint and the joint between the clavicle and the acromion process of the scapula; they allow limited movement.
- Joints of the upper appendage include the ball-and-socket joint of the shoulder, the hinge joint of the elbow, and a variety of joints in the wrist and hand.
- Joints of the pelvic girdle include the sacroiliac joint and the pubic symphysis, both of which are relatively immovable.
- Joints of the lower appendage include the ball-and-socket joint of the hip, the hinge joint of the knee, and a variety of joints in the ankle and foot.

Clinical Applications
- Osteoarthritis is a degenerative disease in which the articular cartilages of joints disintegrate, and bony spurs develop in joint cavities; these changes cause pain and limited movement in affected joints.
- A dislocation is the displacement of the articular surfaces of a joint with damage to the ligaments, whereas a sprain involves damage to joint structures and soft tissues without displacement of the articular surfaces and a strain involves stretching of muscles or tendons without major joint damage.
- Bursitis is an inflammation of a bursa from infection, from injury, or as a complication of rheumatoid arthritis.

QUESTIONS AND PROBLEMS

The questions at the end of each chapter are numbered to correspond with the objectives listed at the beginning of the chapter. Italics indicate that a question requires critical thinking skills beyond simple factual recall.

Questions
1. (a) What is a joint?
   (b) What are the major functions of joints?
2. What are the three functional types of joints found in the body?
3. (a) What are the structural characteristics of fibrous joints, and where are such joints found?
   (b) What are the structural characteristics of cartilaginous joints, and where are such joints found?
   (c) *How do cartilaginous joints differ in structure from fibrous joints?*
4. (a) What is a synovial joint?
   (b) What are its functional characteristics?
   (c) Where are synovial joints found?
   (d) What is synovial fluid, and what is its function?
5. List the six types of synovial joints, and describe the types of movement that are possible in each type.
6. How do bursae and tendon sheaths differ from a synovial joint and from each other?
7. (a) Classify the joints of the axial skeleton as synarthroses, amphiarthroses, or diarthroses.
   (b) What are the functions of the joints of the axial skeleton?
   (c) *Predict the effect of limiting movement at any three kinds of joints in the axial skeleton.*

8. *(a) How are the structures and functions of the shoulder and hip joints alike, and how are they different?*

   (b) How are the structures and functions of the elbow and knee joints alike, and how are they different?

   (c) Predict the effect of limiting movement at any three kinds of joints in the appendicular skeleton.

9. (a) What are the effects of osteoarthritis?

   (b) How can the degree of damage to a joint be used to distinguish between dislocations, sprains, and strains?

   (c) What is bursitis?

Problems

1. Devise exercises you can do to demonstrate the kinds of movement that can occur in the jaw, neck, vertebral column, shoulder, elbow, wrist, hip, knee, and ankle.

2. Several surgical techniques for replacing injured or degenerated joints have been developed. Do some library research on joint replacement and, if possible, report your findings to your class.

3. Find out why range of motion exercises are important, and distinguish between active and passive exercises.

OUTLINE	OBJECTIVES
General Properties of Muscle	1. Describe the general properties of muscle, and relate them to body functions.
Muscle Structure	2. Describe the gross and microscopic structure and specific properties of skeletal muscle, and relate them to body function.

Contraction at the Molecular Level
  Sliding Filament Theory
  Excitation-Contraction Coupling
	3. Explain the sliding filament theory of the contraction process, and describe how contraction is regulated at the molecular level.

The Motor Unit
  Normal Function of Motor Units
  Drugs at the Myoneural Junction
	4. Describe the components of a motor unit, and explain how motor units function in muscle contraction.

Mechanics of Muscle Contraction
  Kinds of Contractions
  Twitch
  Summation of Contractions
  Treppe
  Fatigue
  Length, Tension, and Velocity
    Relationships
  Graded Muscle Activity
	5. Explain how mechanical factors function in muscle contraction.

Muscle Metabolism
  Metabolic Processes
  Heat Production
	6. Explain how metabolic processes contribute to muscle function and heat production.

Fast and Slow Muscle Units Distribution of Muscle Units	7. Distinguish between fast and slow muscle units by function and distribution.

Smooth Muscle
  Structure of Smooth Muscle
  Types of Smooth Muscle
  Contraction of Smooth Muscle
  Smooth Muscle Tone
	8. Describe the gross and microscopic structure of smooth muscle, and explain how smooth muscle differs from skeletal muscle in structure and function.

Cardiac Muscle	9. Briefly describe the properties of cardiac muscle, and compare them with smooth and skeletal muscle.

Clinical Applications
  Involuntary Muscle Contractions
  Altered Muscle Size
  Myasthenia Gravis
	10. Describe the causes and effects of involuntary muscle contractions, altered muscle size, and myasthenia gravis.

Essay: Muscular Dystrophy Chapter Summary Questions and Problems	

PHYSIOLOGY OF MUSCLE TISSUE

GENERAL PROPERTIES OF MUSCLE

You hardly think about your muscles until pain signals from overuse remind you of them. Nonetheless, you depend on muscles for every movement you make, even for standing and sitting. You depend on muscle in internal organs to move food along your digestive tract and to control excretion of body wastes. Your very life depends on heart muscle to keep your blood circulating.

The body contains three types of muscle—skeletal, smooth, and cardiac. Most of the more than 600 skeletal muscles are attached to bones and extend across joints. The contraction of skeletal muscles exerts force on the bones and causes movement. Smooth muscle is found in blood vessels, where it brings about changes in vessel diameter, and in the walls of internal organs, where it pushes substances along passageways. Cardiac muscle is found in the heart, where it exerts force that propels blood.

Muscle tissue has four important functional properties: contractility, extensibility, elasticity, and excitability. **Contractility** is the ability to produce a force; production of a contractile force may or may not involve shortening of muscles, but it always creates tension. **Extensibility** is the ability of a muscle to undergo stretching. **Elasticity** is the ability of a muscle to return to its original shape after it has either contracted or has been stretched. Extensibility and elasticity are due, not only to muscle fibers themselves, but also to elastic fibers in connective tissue around muscles.

Excitability is the ability to respond to stimuli. The membranes of muscle cells, like those of nerve cells, are excitable, and they display many of the same membrane properties as nerve cells. All kinds of muscle respond to neural stimuli, but smooth and cardiac muscle also respond to certain kinds of chemical stimuli. In this chapter, we will emphasize events that occur in stimulated muscle cells, leaving the details of membrane excitability for Chapter 11.

When skeletal muscles contract, they exert tension, which maintains posture and moves the body or its parts. They also release heat. We will look first at properties of muscle related to movement, and later at muscle metabolism and heat production.

MUSCLE STRUCTURE

Whole muscles consist of muscle tissue and connective tissue arranged in a highly organized pattern (Figure 9.1). A sheet of connective tissue covers the whole muscle and each bundle of fibers within it. Connective tissue also attaches muscles directly to bones and forms structures such as **tendons,** cords that connect muscles to bones, or **aponeuroses** (ap''o-nu-ro'sēz), flat sheets that connect muscles to other tissues.

Each whole muscle contains many bundles of fibers and is covered with a connective tissue sheath called the **epimysium.** The epimysium usually is at least partially surrounded by another connective tissue called **fascia.** Each **fasciculus,** or bundle of fibers within a muscle, is surrounded by connective tissue called the **perimysium.** Each muscle fiber is a long, cylindrical, multinucleated cell 10 to 100 μm in diameter that can extend the full length of a muscle (up to 30 cm). It is covered with a connective tissue called the **endomysium.** Within each fiber are many small **myofibrils** (mi''o-fi'brilz), and each myofibril contains many still smaller **myofilaments** (mi''o-fil'a-mentz).

The functional unit of a myofibril is a **sarcomere** (sar'ko-mēr), a compartmental arrangement of thick and thin filaments (Figure 9.2a). Under high magnification, bands, zones, and lines can be seen in a sarcomere, which extends from one horizontal line called a Z line to the next—a distance of about 2 μm in a resting sarcomere. Thick filaments are located entirely within the A band and thin filaments are anchored to and imbedded at Z lines. Only thin filaments are present in the I band and only thick ones in the H zone, but both are present in the A band. These bands and zones provide reference points for studying muscle contraction and relaxation. They also account for the **striations** seen in microscopic views of skeletal and cardiac muscle.

Cytoplasmic structures in muscle fibers also have special names (Figure 9.2b). The plasma membrane of a muscle fiber is called the **sarcolemma** (sar''ko-lem'ah). Unlike many other plasma membranes, it has a thick, negatively charged glycocalyx (collection of extracellular materials). Glucose is transported across the sarcolemma into muscle cells, and lactate, a metabolic product, is transported out of them. Movements of sodium, potassium, and calcium ions across the sarcolemma are carefully controlled. The cytoplasm of a muscle fiber is called **sarcoplasm** (sar'ko-plazm). A highly organized system of sacs and tubules, the **sarcoplasmic reticulum** (sar''ko-plaz'mik re-tik'u-lum), corresponds to the endoplasmic reticulum in other cells. **Transverse tubules,** or **T tubules,** are folds of sarcolemma filled with extracellular fluid. They create channels through which the extracellular environment extends deep inside muscle fibers. Near the T tubules, the sarcoplasmic reticulum ends in saclike structures called the **terminal cisternae.** A T tubule and two adjacent terminal cisternae form a **triad.**

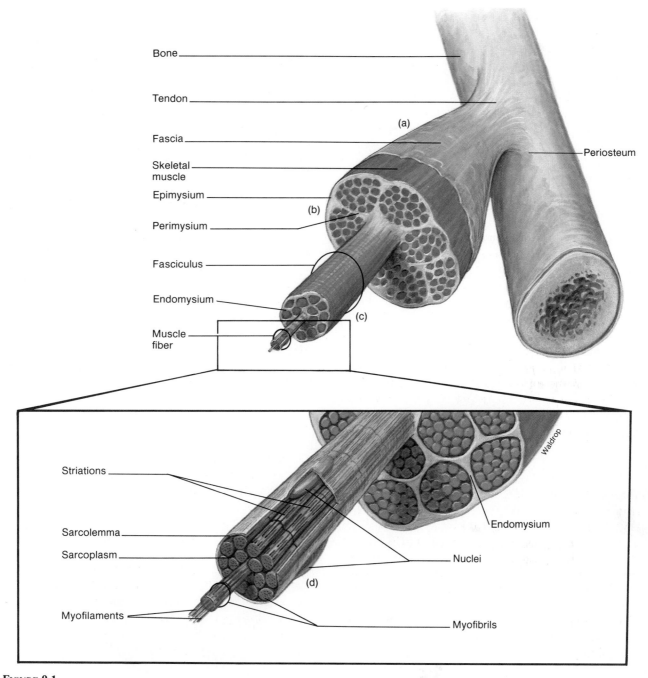

FIGURE 9.1

(*a*) A skeletal muscle is composed not only of muscle tissue but also connective tissue, and is typically connected to a bone by a tendon. (*b*) The muscle surface is covered by connective tissue called fascia, epimysium lies beneath the fascia, and perimysium extends between fasciculi (bundles) of muscle tissue. (*c*) Individual muscle fibers in fasciculi are separated by endomysium. (*d*) Each muscle fiber is bounded by the sarcolemma and contains many myofibrils, which are composed of myofilaments.

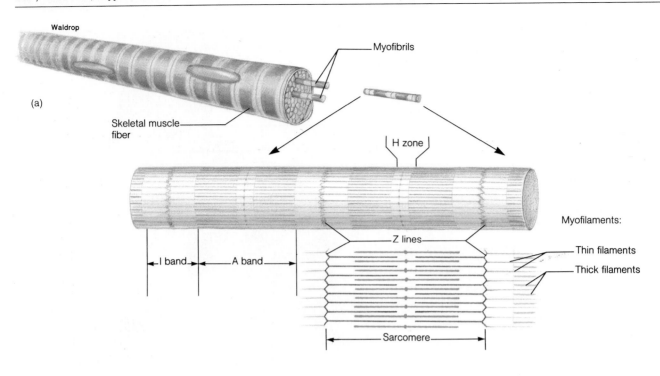

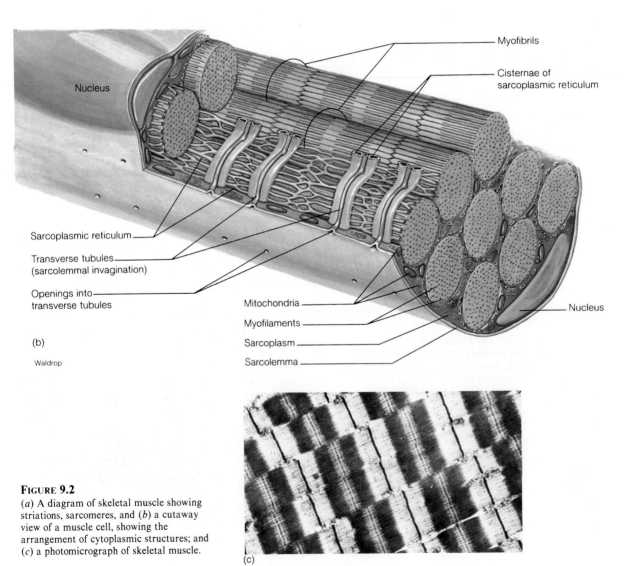

FIGURE 9.2
(a) A diagram of skeletal muscle showing
striations, sarcomeres, and (b) a cutaway
view of a muscle cell, showing the
arrangement of cytoplasmic structures; and
(c) a photomicrograph of skeletal muscle.

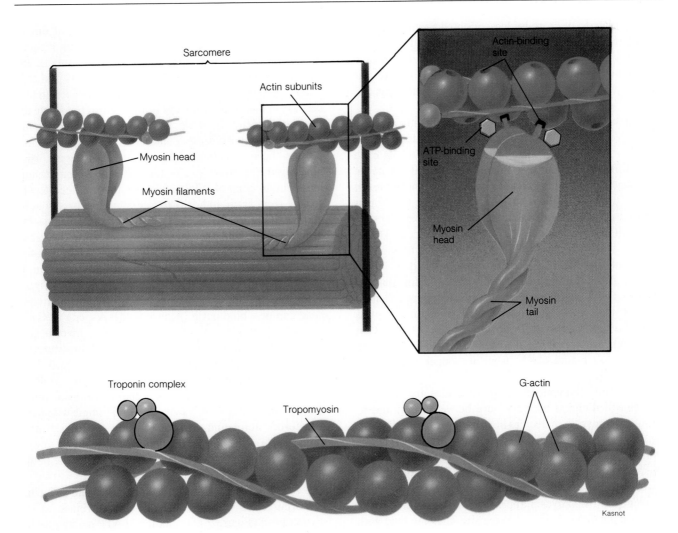

FIGURE 9.3
The structure of thick and thin filaments, including the locations and orientations of sites where ATP and actin bind to myosin heads.

CONTRACTION AT THE MOLECULAR LEVEL

Skeletal muscle is stimulated to contract (produce tension) when signals from nerves spread throughout the sarcoplasmic reticulum. Such signals cause the terminal cisternae to release calcium ions, which diffuse through the sarcoplasm and bind to a regulatory protein. Such calcium binding initiates contraction. Subsequent active transport of calcium ions back to the sarcoplasmic reticulum initiates relaxation.

Sliding Filament Theory

Though contraction of skeletal muscles can move the whole body, the contraction process itself occurs at the molecular level. The **sliding filament theory** of contraction, proposed in the 1950s by H. E. Huxley and based on electron micrograph studies, is generally accepted today though some details of the process are still not known. According to this theory, muscle contraction occurs when thin filaments slide over thick ones, thereby increasing the overlap of filaments and shortening the sarcomeres without changing the length of the filaments themselves.

Structure of Thick and Thin Filaments

To understand the contraction process, we must first consider the properties of myofilaments (Figure 9.3). Myofilaments contain four important proteins, **myosin** (mi'o-sin), **actin** (ak'tin), **troponin** (tro-po'nin), and **tropomyosin** (tro''po-mi'o-sin). Myosin is found in thick filaments; actin forms the basic structure of thin filaments, and troponin and tropomyosin are associated with it. Myosin and actin are **contractile proteins**; that is, they create tension in the contraction process. Troponin and tropomyosin are **regulatory proteins**; that is, they help to control the contraction process.

The thick filaments each consist of several hundred myosin molecules held together mainly by hydrogen bonds and hydrophobic forces. Each myosin molecule has two globular heads, or **cross-bridges,** oriented toward the ends of A bands. Cross-bridges have two functional sites. One site acts as a ligand (a part of a molecule that can bind to another molecule). It attaches to binding sites on the protein actin. The other site binds ATP. Finally, cross-bridges have ATPase activity.

Like the thick filaments, the thin filaments also are composed of protein molecules. The three proteins, actin, tropomyosin, and troponin, found in thin filaments each have different functions. Actin in a muscle cell is **F actin,** a polymer of globular units of **G actin.** Two molecules of F actin are wound about each other in a helix that forms the core of a thin filament. Actin has surface binding sites to which myosin binds in the contraction process. Tropomyosin, so named because it was once thought to be a precursor of myosin, is a fibrous, double-stranded helical protein. Overlapping tropomyosin molecules form a continuous flexible structure. Troponin, an aggregation of smaller molecules sometimes called **troponin complex,** binds Ca^{2+}. When the troponin complex lacks Ca^{2+}, it inhibits ATPase activity in the heads of myosin molecules. When it has bound Ca^{2+}, it can no longer inhibit that enzyme.

In a thin filament, the actin molecules are arranged as two coiled chains with a groove between them and fibrous tropomyosin molecules lie in a groove created by the coiling of a double strand of actin molecules. When a muscle fiber is relaxed, tropomyosin apparently covers binding sites on actin and prevents them from binding with myosin. When an action potential releases Ca^{2+} to bind to troponin, tropomyosin appears to slide into the groove between the actin chains, exposing sites that can bind with myosin.

Thick and thin filaments are arranged as hexagons within each myofibril—six thin filaments surround each thick filament and form a small hexagon, and six thick filaments surround each small hexagon and form a large hexagon. In such hexagons, every thin filament lies adjacent to a row of cross-bridges on each of three myosin filaments so that actin and myosin can easily interact during contraction.

Molecular Movements During Contraction

According to the sliding filament theory, molecular movements account for a **contraction cycle** with attach, pull, and release phases. Cross-bridges of myosin attach to actin and pull the thin filaments toward the H zone, as the cross-bridges swivel in an oarlike movement, and then release their hold on actin. This **attach-pull-release cycle** occurs asynchronously—first one and then another cross-bridge

attaches to actin. It is repeated over and over again during the contraction of a whole muscle. Many small, discrete movements at the molecular level produce a smooth and continuous sliding of the thick and thin filaments past each other. Though some of the details of the contraction process are conjectural, the steps in the contraction cycle are based on evidence from biochemical analysis, electron microscopy, X-ray diffraction, and other techniques.

As the contraction cycle (Figure 9.4) begins, ATP binds to the myosin head in the presence of magnesium ions. The binding of ATP causes the myosin head to dissociate from actin and drift to a 90° angle with the actin filament (Figure 9.4a). ATPase in the cross-bridge hydrolyzes ATP, and the energy released holds the cross-bridge like a stretched spring ready to swivel (Figure 9.4b). Ca^{2+} binds to troponin, which causes tropomyosin to expose the myosin binding sites on the actin (Figure 9.4c), and the activated actin binding site on the myosin head can now bind actin (Figure 9.4d). As long as Ca^{2+}-troponin deflects tropomyosin from actin, myosin continues to bind to it. Steps (a) through (d) constitute the attach phase of the contraction cycle.

Binding of the myosin head to actin releases ADP, P_i, and energy. The energy causes a change in the shape of the myosin head so that it swivels to a 45° angle with actin and forces actin and myosin to slide over each other (Figure 9.4e). The ends of the actin filaments move toward the center of the A band. This step constitutes the power stroke, or pull, part of the cycle.

When ADP and P_i have been released and swiveling has occurred, a second molecule of ATP binds to the myosin head (Figure 9.4f). Binding of ATP inactivates the actin binding site and frees the myosin head (Figure 9.4g). The myosin head returns to its original position and is ready to undergo another cross-bridge cycle (Figure 9.4h). Steps (f) through (h) constitute the release part of the cycle. No additional energy from ATP is required for the release cycle.

As a result of several repetitions of the contraction cycle, sarcomeres shorten as fixed-length filaments slide over one another. The H zones shorten and may disappear, but the width of the A band remains constant. The most significant event in this process is that filaments slide because of cyclic changes in the shape and position of myosin heads. When this process shortens sarcomeres, the contraction is said to be **isotonic.** When, instead, a muscle supports a heavy load or pushes against an immovable object, it exerts tension but the sarcomeres may not change length, and the contraction is said be **isometric,** as will be explained in more detail later.

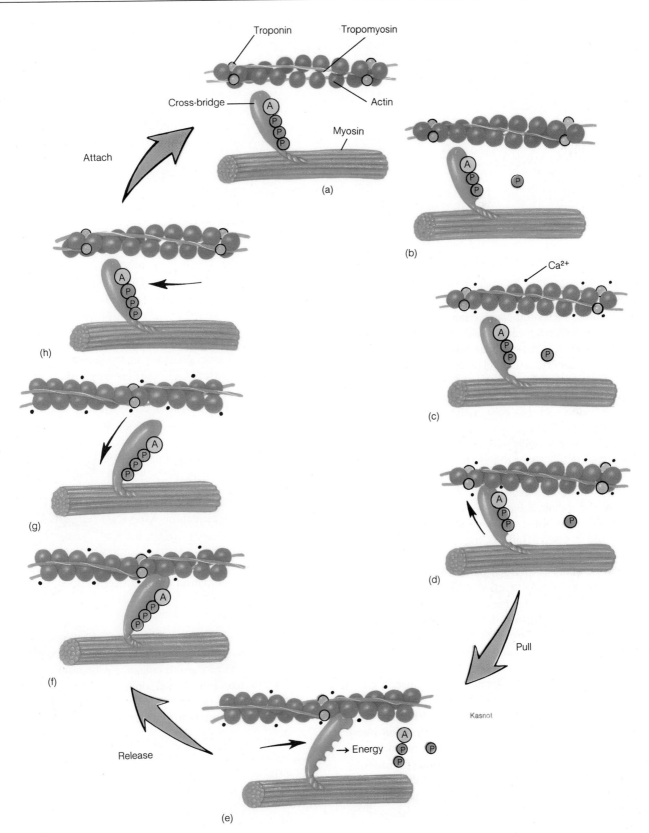

FIGURE 9.4
The steps in the contraction cycle.

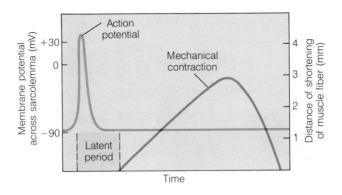

FIGURE 9.5
Relationship of sarcolemmal action potential to muscle contraction. The process of excitation—contraction coupling takes place during the latent period.

Excitation-Contraction Coupling

The direct connection between action potentials in the sarcolemma and the sliding of myofilaments in contraction is called **excitation-contraction coupling.** This coupling is accomplished through controlled release of calcium ions, which exert a powerful regulatory effect. A cytoplasmic Ca^{2+} concentration as low as 1 micromole per liter stimulates contraction, though the Ca^{2+} concentration in the sarcoplasmic reticulum is 2000 times that. That Ca^{2+} activates contraction has been confirmed by experiments using aequorin, a bioluminescent protein from jellyfish that emits light when it binds with Ca^{2+}. When muscle fibers are injected with aequorin and stimulated under proper conditions, they glow during the time between stimulation and the production of tension.

Like the membrane of a neuron, the sarcolemma is an excitable membrane. (Though membrane excitability is discussed in more detail in Chapter 11, it is considered briefly here.) The fluids inside and outside of cells have different numbers of charged particles. Such differences create a charge across the membrane called a **membrane potential,** which can be measured in millivolts (mV). Resting muscle cells have a membrane potential of about −90 mV. When they are stimulated, a signal called an **action potential** is propagated along the membrane in all directions at about 5 meters per second. The action potential usually lasts about 2 milliseconds (msec), but it can last as long as 4 msec. (A millisecond is one-thousandth of a second.) The signal has passed along the entire membrane and down into T tubules before contraction starts. The time between initiation of the action potential and initiation of contraction is the **latent period** (Figure 9.5).

Excitation-contraction coupling occurs during the latent period by the following mechanism. Once an action potential reaches the sarcolemma, it travels along T tubules to the region of the terminal cisternae of the sarcoplasmic reticulum where Ca^{2+} is stored. The action potential in the T tubule membrane increases the permeability of the terminal cisternae to Ca^{2+} and Ca^{2+} diffuses into the sarcoplasm near the sarcomeres. How this permeability increases is not fully understood, but it is possible that a chemical messenger molecule (probably inositol phosphate in skeletal muscle) initiates mechanical opening of calcium pores in the sarcoplasmic reticulum. Once released, Ca^{2+} diffuses only a short distance to reach calcium binding sites on troponin. When sufficient Ca^{2+} binds to troponin, tropomyosin moves, myosin binding sites on actin are exposed, and contraction can occur.

When stimulation of the muscle cell membrane stops, the terminal cisternae cease to release Ca^{2+}. An ATP-driven calcium pump in the sarcoplasmic reticulum continuously transports Ca^{2+} from the sarcoplasm back inside the sarcoplasmic reticulum. After Ca^{2+} release ceases, the calcium pump quickly lowers the Ca^{2+} concentration in the sarcoplasm. As the sarcoplasmic Ca^{2+} concentration decreases, Ca^{2+} dissociates from troponin, and tropomyosin again blocks myosin binding sites on actin. Contraction ceases. The events in excitation-contraction coupling can be summarized as follows:

1. The action potential travels along the T tubules to the sarcoplasmic reticulum.
2. Stimulation of the sarcoplasmic reticulum causes Ca^{2+} to leak into the sarcoplasm.
3. Ca^{2+} diffuses to and binds with troponin, thereby allowing contraction to occur.

Contraction, which is slow to start, is even slower to stop. It continues for 50 to 100 msec after the last action potential until the calcium pump sufficiently lowers the sarcoplasmic Ca^{2+} level so that tropomyosin can block myosin binding sites on actin.

It may seem surprising that muscle relaxation requires some energy—at least indirectly. The breaking of actin-myosin binding in a contracted muscle fiber is a passive process, but it can occur only when energy from ATP

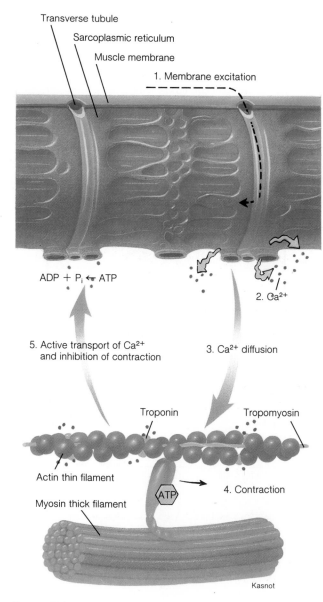

Transverse tubule

Sarcoplasmic reticulum

Muscle membrane

1. Membrane excitation

$ADP + P_i \leftarrow ATP$

2. Ca^{2+}

5. Active transport of Ca^{2+} and inhibition of contraction

3. Ca^{2+} diffusion

Troponin

Tropomyosin

Actin thin filament

ATP

Myosin thick filament

4. Contraction

Kasnot

FIGURE 9.6

The role of calcium ions in the contraction and relaxation processes: (1) Excitation of the membrane. (2) Calcium ions released from sarcoplasmic reticulum. (3) Calcium ions diffuse through sarcoplasm and some bind to troponin and cause tropomyosin to move away from the myosin binding sites of actin. (4) Contraction occurs. (5) Calcium ions are actively transported back to the sarcoplasmic reticulum and tropomyosin again covers the binding sites on actin; covering the binding sites inhibits contraction.

is available to operate the calcium pump. Though this pump operates continuously, a contracted muscle fiber cannot relax until sufficient Ca^{2+} has been removed from troponin. Also, the binding of ATP to myosin is required to separate myosin and actin, and allow relaxation. ATP so bound is subsequently hydrolyzed and its energy used in later contraction cycles. Excitation, contraction, and relaxation—and the role of calcium ions in these processes—are summarized in Figure 9.6.

Events in the contraction cycle and the pumping of calcium can be used to explain **rigor mortis,** the rigidity of muscles that occurs several hours after death. When an organism dies, cellular metabolism slowly comes to a halt; no new ATP is made, but existing ATP pumps Ca^{2+} and allows muscles to relax. Over the next several hours, depending on the environmental temperature and other conditions, ATP is depleted, the calcium pump ceases to operate, and sarcoplasmic Ca^{2+} binds to troponin. Because ATP must bind to myosin before cross-bridges from a previous contraction cycle can break, muscles lacking ATP become rigid. Eventually, enzymes from lysosomes or from bacteria that may have invaded the muscle degrade muscle proteins. Rigor mortis disappears 15 to 25 hours after death when sufficient proteolytic action has occurred to destroy the cross-bridges.

THE MOTOR UNIT

A **motor unit,** the functional unit of contraction, consists of a single motor neuron, including all of the branches of its axon, and the muscle fibers those axon branches innervate (Figure 9.7). Each muscle fiber is supplied with a small terminal branch of an axon. The axon ends are unmyelinated and lie in grooves in the sarcolemma. The sarcolemma beneath nerve endings is called the **motor end plate**; the membrane within a motor end plate is folded to form **subneural clefts.** The terminal ends of the axon and the motor end plate together constitute the **neuromuscular junction,** or **myoneural** (mi″o-nu′ral) **junction.**

Normal Function of Motor Units

An action potential in a motor neuron causes all its axon terminals to release a chemical substance called acetylcholine (ACh). ACh diffuses to receptor sites on the motor end plates of all muscle fibers of that motor unit, causing the motor end plate to generate a membrane potential called the **end plate potential.** In skeletal muscle, when enough acetylcholine is released, as it usually is, this potential, in turn, initiates an action potential in the sarcolemma. The action potential is propagated in all directions along the sarcolemma and down into the T tubules. Signals at the end plate are always excitatory, and a single action potential in a motor neuron generally causes skeletal muscle contraction.

In addition to receptors for acetylcholine, the membranes of the motor end plates also contain the enzyme **cholinesterase** (ko″lin-es′ter-ās), which destroys acetylcholine just as it does in certain postsynaptic neurons. When acetylcholine is broken down by cholinesterase, the action potential subsides and the membrane at the end plate returns to its resting potential until it receives another signal from the motor neuron. Processes involved in muscle contraction are summarized in Table 9.1, and their control is summarized in Figure 9.8.

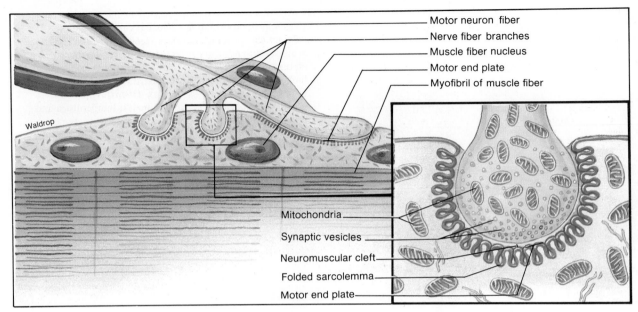

(a)

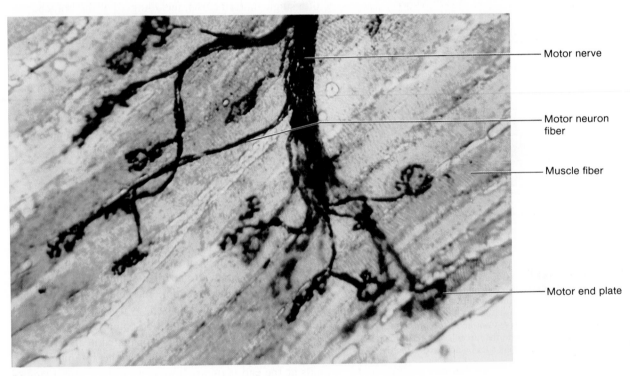

(b)

FIGURE 9.7

How an axon innervates a muscle fiber: (*a*) In a motor unit of an axon, each branch of a motor nerve goes to a muscle fiber where the axon terminal and the sarcolemma form a myoneural junction. The axon terminal lies in a groove in the sarcolemma; the portion of the sarcolemma beneath the axon is the motor end-plate. (*b*) Micrograph of a neuromuscular junction.

TABLE 9.1   Processes Involved in Muscle Contraction from Stimulation through Contraction to Relaxation

1. An action potential reaching axon terminals of a motor neuron initiates the release of acetylcholine from synaptic vesicles.

2. Acetylcholine diffusing to and binding to receptor sites on the motor end plate creates a local end plate potential.

3. The end plate potential initiates an action potential that is propagated in all directions along the sarcolemma and into the T tubules.

4. Depolarization of the T tubules causes calcium ions to leak out of the terminal cisternae of the sarcoplasmic reticulum that surrounds the myofibrils.

5. Calcium ions bind to receptor sites on the troponin complex.

6. The troponin complex causes tropomyosin to recede into a groove between chains of actin molecules, exposing binding sites on actin. Myosin heads with hydrolyzed ATP (ADP + P_i + energy) bind to actin. This is the attach phase of the contraction cycle.

7. Using energy from ATP, the myosin heads swivel and pull the actin filaments closer to the middle of the sarcomere. This is the pull phase of the contraction cycle.

8. The ATP binding sites on myosin are exposed and new molecules of ATP bind to myosin; the actin binding sites are inactivated and the myosin heads return to their original positions. This is the release phase of the contraction cycle.

9. The contraction cycle is repeated continuously as long as ATP is available and sufficient calcium is bound to troponin to allow myosin binding sites on actin to remain exposed.

10. The action potential lasts about 2 msec, but the contraction cycle continues for 50 to 100 msec as long as adequate calcium is bound to troponin.

11. When the action potential ceases, calcium release from the sarcoplasmic reticulum also ceases. The calcium pump removes calcium from the sarcoplasm and depletes calcium bound to troponin.

12. Tropomyosin again blocks binding sites on actin; myosin can no longer bind to actin so contraction ceases and the muscle fibers relax.

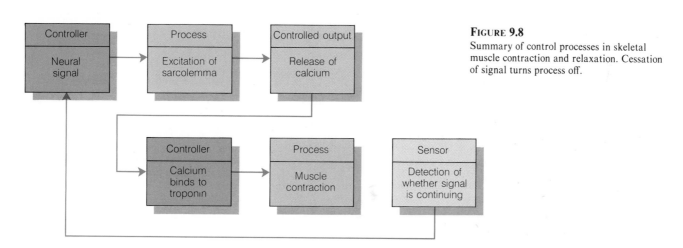

FIGURE 9.8
Summary of control processes in skeletal muscle contraction and relaxation. Cessation of signal turns process off.

Drugs at the Myoneural Junction

A variety of drugs and diseases interfere with events at the myoneural junction. Curare (D-turbocurarine), a poison put on arrowheads by some South American Indians, can be used to prevent muscle contraction during surgery. It paralyzes by competing with acetylcholine for motor end plate receptor sites. A 5 to 30 mg intravenous dose of curare completely paralyzes skeletal muscles, but has little or no effect on acetylcholine receptors in the autonomic nervous system (Chapter 14). Its effects last from 20 to 40 minutes until enough of the drug is metabolized or excreted by the kidneys to allow acetylcholine to compete successfully for receptor sites. Curare is not destroyed by cholinesterase, but its effective period can be shortened by cholinesterase inhibitors, which reduce the rate of acetylcholine breakdown and increase its ability to compete with curare. Large doses of curare can kill, usually by paralyzing the respiratory muscles. Other drugs have replaced curare in most surgical procedures.

Nerve gases, pesticides, and toxins also affect neuromuscular junctions. Some nerve gases and certain pesticides inactivate cholinesterase and cause spasmodic muscular contractions by prolonging acetylcholine activity. Botulism toxin from the bacterium *Clostridium botulinum,* sometimes found in inadequately processed canned goods, paralyzes muscles by inhibiting acetylcholine release from axon terminals.

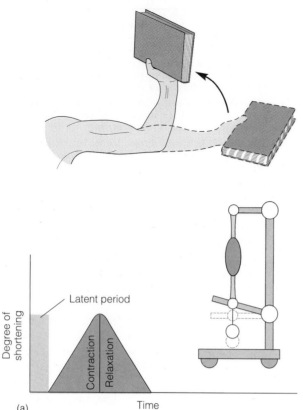

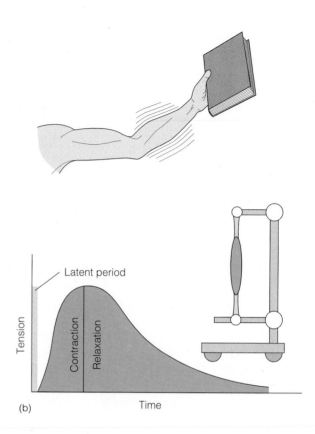

FIGURE 9.9

Differences between isotonic and isometric contractions: (*a*) An isotonic contraction. Movement of arm while lifting book (*top*), laboratory apparatus with muscle attached to a pivot that moves when muscle shortens (*middle*), and isotonic twitch shown graphically to illustrate the long latent period and short relaxation period (*bottom*).

(*b*) An isometric contraction. Arm holding a book in a fixed position (*top*), laboratory apparatus with muscle attached to a spring that measures muscle tension (*middle*), and an isometric twitch shown graphically to illustrate the short latent period and long relaxation period (*bottom*).

MECHANICS OF MUSCLE CONTRACTION

Much of what is known about the mechanics of muscle contraction has been learned through laboratory experiments on isolated muscle fibers. Such experiments involve two opposing forces, tension and load. Muscle **tension** is the force produced when a muscle contracts. **Load** is the force exerted on a muscle by a weight. For example, when you pick up a book, the force your arm muscles produce is tension and the book is the load.

Kinds of Contractions

Contractions can be isotonic or isometric (Figure 9.9). When the tension in a muscle is sufficient to lift a load, the contraction is said to be **isotonic** (same tension). The fibers exert a constant tension and they change length as they lift the load. Muscles that move your body or a part

of it generally undergo isotonic contraction. In contrast, when the tension in a muscle is used to support a load or push against an immovable object, the contraction is said to be **isometric** (same length). The length of the fibers remains the same, but the tension changes. Most complex body movements are a combination of isotonic and isometric contractions.

Twitch

A **simple twitch,** which has been studied in isolated muscle fibers in the laboratory, is the response of a skeletal muscle to a single stimulus. Twitches can be isotonic or isometric, depending on whether the load is lifted or the tension is increased in a fixed-length muscle. Three phases of a simple twitch are (1) the latent period, (2) the period of contraction, and (3) the period of relaxation. The **latent period** in an isotonic twitch, which includes isometric contraction during the development of tension, increases in

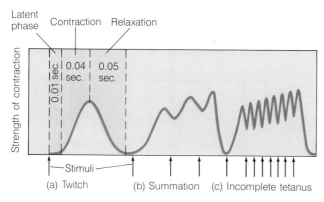

FIGURE 9.10

Effect of varying the frequency of stimuli; (*a*) twitch, (*b*) summation, and (*c*) incomplete tetanus.

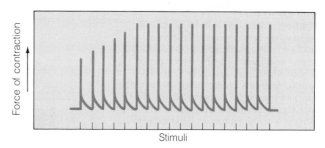

FIGURE 9.11

The phenomenon of treppe.

length as the load increases. Elastic tension builds as the muscle develops tension, and shortening of the muscle and lifting of the load begin only after the tension becomes greater than the load.

The **contraction period** is the period during which the contraction cycle occurs repeatedly. Its total length is approximately 35 msec for both isotonic and isometric contractions (the time is indicated by the rising lines on the graphs in Figure 9.9). The time varies, however, according to the particular muscles involved and, in isotonic contractions, according to the load. Isotonic contractions require more energy than isometric ones because more repetitions of the contraction cycle are required to lift a load than to maintain tension. Therefore, a person exercising to lose weight should do isotonic exercises.

The **relaxation period** occurs after the action potential ceases as troponin loses calcium. Relaxation after an isotonic twitch is more rapid than after an isometric twitch because the load pulls the muscle back to its original length. In an isometric twitch, the muscle returns to its original length as the calcium pump moves calcium to the sarcoplasmic reticulum.

Summation of Contractions

Using electrical stimulation, a muscle can receive several stimuli before the contraction period elicited by the first stimulus has ended. (Recall that the contraction phase lasts 25 to 50 times as long as the action potential in the muscle membrane.) Multiple stimuli applied at a frequency that allows complete relaxation to occur between successive stimulations produce a series of simple twitches (Figure 9.10a). Multiple stimuli applied close enough together to allow only incomplete relaxation between stimulations produce summation. In **summation** (Figure 9.10b), tension increases with each successive stimulus up to the point

of maximal contraction (the greatest tension the muscle can produce). Summation can occur in two ways: (1) by activating more motor units in a muscle (**motor unit summation**) or (2) by greater shortening of fibers of previously stimulated motor units (**wave summation**). When a muscle begins to contract, some force is used to "take up slack" in the elastic muscle fibers.

Increasing the frequency of stimulation increases tension by summation until maximum tension, or **tetanus** (tet'an-us), results. **Incomplete tetanus** occurs when the stimuli are frequent enough to allow only very brief relaxation between contractions (Figure 9.10c). **Complete tetanus** occurs when the stimuli are so frequent that no relaxation occurs between stimuli. The tension at complete tetanus is three to four times as great as that produced by a simple twitch. Most contractions associated with body movements and the lifting of loads are tetanic contractions.

Treppe

In contrast to rapidly delivered stimuli in tetanus, maximal stimuli (stimuli strong enough to produce maximal contractions), delivered at a frequency that allows complete relaxation between stimuli, produce **treppe** (trep'eh) as shown in Figure 9.11. Each of the first few contractions exerts more tension than the preceding one, creating a staircase effect. Two possible explanations have been offered for treppe: (1) The resistance to movement among muscle filaments is reduced with repeated contractions, at least for the first few contractions. (2) Calcium ions become more available to bind to troponin during the first few contractions. Accelerated release of calcium ions might be caused by a small decrease in potassium and a small increase in sodium inside the cell.

Treppe has been used to explain the effects of warming up exercises by athletes, but this has not been confirmed. A more likely explanation of increased strength of contractions following warming up exercises is that the enzyme ATPase works faster as the temperature of the muscle increases.

FIGURE 9.12

Positions of thick and thin filaments, and relationship between muscle length and maximal isometric tension: (*a*) Muscle at 70 degree equilibrium length exerts about 60 degree of maximal tension. In such slack muscle, filaments overlap maximally so that no more tension can be produced. (*b*) Muscle at 100 percent equilibrium length exerts 100 percent of maximal tension. Filaments are positioned so that the maximum number of crossbridges can function and maximum tension is possible. (*c*) Muscle at 130 percent equilibrium length exerts about 70 percent of maximal tension. These percentages vary in different muscles. In such stretched muscle, thick and thin filaments are pulled apart so that no myosin heads can bind to actin and no tension is possible.

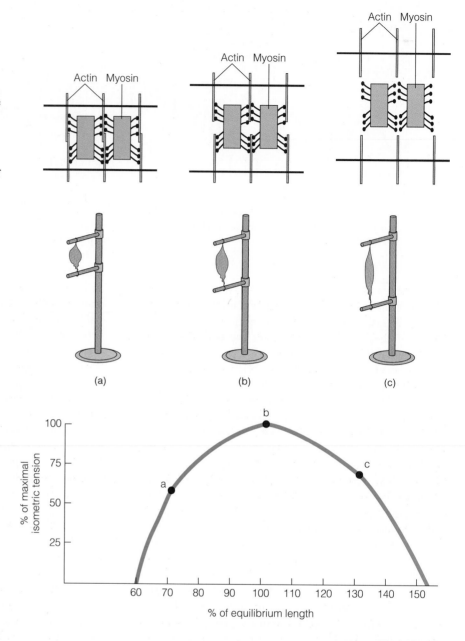

Fatigue

Muscle **fatigue,** a condition in which contractions become weaker, has been attributed to the inability of muscle fibers to make enough ATP. Because muscles must receive both oxygen and nutrients from the blood to make ATP, strenuous exercise that increases the demand for oxygen and nutrients can contribute to fatigue. How muscle fatigue is related to muscle metabolism and the physiology of different muscle units called red muscle and white muscle will be discussed in a later section.

Recent observations suggest that, in addition to metabolic factors, accumulation of Ca^{2+} in T tubules can contribute to muscle fatigue. Though the calcium pump returns most Ca^{2+} to the sarcoplasmic reticulum, a mechanism that pumps Ca^{2+} to the T tubules appears to exist. Calcium in the T tubules somehow interferes with excitation-contraction coupling.

Length, Tension, and Velocity Relationships

The relationship between muscle length and tension, the **length-tension curve,** can be demonstrated in the laboratory (Figure 9.12). Muscle (*b*) is suspended at its equilibrium length (its natural length after removing it from its bony attachments in an animal) and can produce maximal tension. Muscle (*a*) is suspended at 70 percent of its equilibrium length and produces about 60 percent tension, and muscle (*c*) is suspended and stretched to 130 percent of its equilibrium length and produces about 70 percent tension. (These lengths represent the limits of length change in living muscle.) A muscle at equilibrium length produces greatest tension and either slackness or stretching reduces the amount of tension it can produce. Most muscles below 70 percent or above 130 percent of equilibrium length produce no active tension at all, though elastic elements in them can produce passive tension.

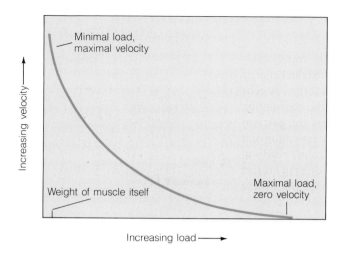

Length-tension variations depend on filament positions, shown at the bottom of Figure 9.12. In muscle (a), thin filaments must increase their overlap of thick filaments, or "take up the slack," within each sarcomere before they can develop active tension. Only a limited number of additional contraction cycles can occur before fibers have reached maximum overlap and sarcomeres have reached maximum tension. In muscle (b), no effort is wasted in taking up slack and filament overlap is sufficient to pull efficiently. In muscle (c), thin filaments fail to overlap thick ones so myosin cannot bind to actin and no force can be exerted by the muscle. Under normal physiological conditions, muscles are neither slack nor stretched, but the studies described here have shown that the natural length of a muscle is generally the length at which it produces greatest tension.

The velocity of muscle contraction (the amount of shortening in a given period of time) decreases as the load on it increases (Figure 9.13). Velocity is maximal with no load except the weight of the muscle itself. It decreases as the load increases until the load is so great that the velocity is zero; no shortening occurs and the contraction becomes isometric. Though the muscle produces tension, it is unable to lift the load. Time required to overcome the load before lifting begins is part of the latent period; it is not considered in determining contraction velocity.

Graded Muscle Activity

Graded muscle activity, or varying degrees of contraction of a muscle, depends, not only on the tension produced by each individual muscle fiber, but also on the number of fibers contracting at a given time. Motor units controlling precise movements have about 10 muscle fibers per motor unit; those controlling gross movements have 600 to 1000 (sometimes more) fibers per unit. Though an action potential in a motor neuron stimulates all of the muscle fibers innervated by that neuron, different neurons stimulate different numbers of fibers. And, the strength of a contraction varies in proportion to the number of fibers stimulated.

According to the **all-or-none law** of muscle contraction, any fiber that is stimulated contracts maximally. Though individual fibers contract maximally, bundles of fibers supplied by several neurons can display motor unit summation, and fibers that receive multiple stimuli over time can display wave summation. For a whole muscle, the number of fibers contracting is determined by the number of motor units activated and the number of fibers per unit. For example, during minimal activity like walking slowly, only a few motor units in a muscle being used are activated. As the exercise level increases (from walking to running), more motor units are activated, or recruited. Such **recruitment** is the main factor in increasing the strength of contractions in graded muscle activity.

The number of fibers per unit, though important in any graded activity, is especially important in controlling very precise movements. The situation is analogous to a large room lighted by many light bulbs. If a single switch controls all of the bulbs, they are either all on or all off. If there are many switches and each switch controls a few bulbs, a wide range of light intensity can be obtained by turning on different numbers of switches. Humans can make exceedingly finely graded, precise hand and finger movements—handwriting, typing, or playing a guitar or piano—because the relevant muscles contain many motor units, each with only a few fibers.

Graded muscle activity also maintains posture. Asynchronous motor unit firing, the contraction of fibers of first one and then another motor unit, creates a small degree of tension over a long period of time. (If firing were synchronous, fibers of all motor units would contract at the same time.) Fibers of different groups of motor units take turns maintaining posture so no group becomes fatigued. In skeletal muscles, a slight degree of contraction called **muscle tone** is produced by relatively infrequent asynchronous firing of a few motor units.

Muscle Metabolism

Muscle action, as we have seen, involves the transformation of chemical energy from ATP into mechanical energy for the swiveling of myosin heads in contraction. A continuous supply of ATP is needed to keep the contraction cycle going. Energy from ATP also powers the calcium pump to return calcium to the sarcoplasmic reticulum and the Na^+–K^+ pump to maintain the sarcolemma ready to receive an action potential. Metabolic processes in muscle must generate large quantities of ATP to keep all these processes operating.

Metabolic Processes

Oxygen and nutrients are carried in the blood to all cells, including muscle cells, where they can be used in cellular metabolism to store chemical energy in the high energy bonds of ATP (Chapter 2). Metabolism in muscle cells has some special attributes.

When an organism is at rest or exercising moderately, oxygen and glucose are plentiful. Large quantities of ATP are synthesized from ADP and P_i in the mitochondria in **aerobic metabolism** (Figure 9.14a). Some energy captured in ATP is transferred to another energy storage molecule called **phosphocreatine** (fos″fo-kre′at-in), or **creatine phosphate**, which consists of a molecule of the nitrogenous substance creatine combined with phosphate by a high energy bond:

ATP + creatine ↔ phosphocreatine + ADP

Phosphocreatine and creatine kinase, the enzyme that catalyzes the above reaction, are plentiful in both skeletal and heart muscle. When heart tissue is damaged in a heart attack, the enzyme leaks into the blood, where its concentration can be measured to estimate the degree of tissue damage.

In addition to regulating contraction as described above, Ca^{2+} also stimulates oxidation in muscle mitochondria. It acts by increasing the rate at which pyruvate (a metabolic product of glucose) enters mitochondria and by stimulating the activity of some of the oxidative enzymes in mitochondria.

During strenuous exercise, oxygen is used as rapidly as it is delivered to the muscles, and ATP levels increase to their maximum. Even so, too little energy is available to maintain maximally strenuous activity for more than a minute or so. Energy stored in phosphocreatine is quickly transferred back to ATP, but this adds only about three seconds to the duration of contraction. Then, the muscle must switch to anaerobic metabolism.

Glucose + $6H_2O$ + 2ATP + $6O_2$ ⟶ $6CO_2$ + $12H_2O$ + 40ATP

Net: 38 ATP

(a)

Glucose + 2ATP ⟶ 2 Lactic acid + $2CO_2$ + 4ATP

Net: 2 ATP

(b)

Figure 9.14

Summary reactions for (a) aerobic metabolism and (b) anaerobic metabolism.

In **anaerobic metabolism,** (Figure 9.14b) when muscles lack sufficient oxygen to metabolize glucose aerobically, they metabolize it anaerobically and obtain a net of 2 ATP per glucose molecule—only one-nineteenth that obtained in aerobic metabolism. Because so little energy is obtained from each glucose molecule large quantities of glucose are required.

If muscle fibers were dependent entirely on oxygen delivered moment by moment, the ATP supply would be depleted and contraction would cease in less than one second. Three mechanisms, myokinase activity, oxygen storage in myoglobin, and the oxygen-debt mechanism, prolong contraction when oxygen from the blood is insufficient to support aerobic metabolism.

The enzyme **myokinase** (mi″o-kin′ās) moves a phosphate and high energy bond from one ADP to another: 2ADP → ATP + AMP. It makes more energy available as ATP but causes AMP to accumulate. Cells must subsequently add two phosphates and two high energy bonds to AMP to make new ATP.

Myoglobin (mi″o-glo′bin), a pigment in muscle similar to hemoglobin in red blood cells, binds one molecule of oxygen (hemoglobin binds four). Myoglobin indirectly increases the rate at which oxygen diffuses into muscle fibers from the blood by binding oxygen and lowering the concentration of free oxygen in intracellular fluid. Muscle fibers vary in myoglobin content, but even in fibers with large quantities of myoglobin, it plays only a small role in prolonging contraction.

During exercise, the oxygen supply to muscle fibers increases as blood vessels in muscle dilate and the heart pumps a greater volume of blood per minute. This mechanism fails to meet the demands of muscle fibers for ATP when activity is strenuous. Oxygen from myoglobin supports contraction for a brief time, but then metabolism becomes anaerobic. Pyruvate from glycolysis is converted

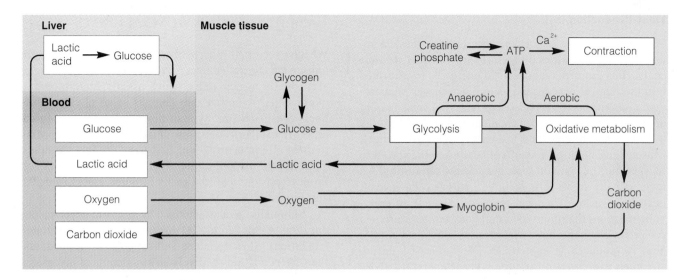

FIGURE 9.15
Main steps in muscle metabolism.

to lactic acid. Much of the lactic acid diffuses into the blood, but enough remains in muscle cells to lower the pH of muscle tissue. The lower pH releases additional oxygen from hemoglobin in blood, but it also inhibits enzymes in cells and makes anaerobic metabolism a self-limiting process. At best, anaerobic metabolism provides energy for only a few seconds of maximal exertion.

When we exercise strenuously, our bodies incur an **oxygen debt,** an amount of oxygen required to metabolize lactic acid and resaturate the myoglobin with oxygen. We "pay" the oxygen debt after a period of exercise when we breathe rapidly and deeply, and use up to six times the normal amount of oxygen. As this debt is paid, lactic acid in heart muscle cells is converted to pyruvate and metabolized there. Most of the lactic acid in skeletal muscle is released into the blood, transported to the liver, and converted back to glucose. Trained athletes incur a smaller oxygen debt than untrained individuals for a given amount of exertion because training increases the amount of blood that the heart can pump per minute and usually causes muscle hypertrophy. Improved heart performance allows the heart to pump more blood to active muscles if it is needed to supply them with oxygen and nutrients, and remove wastes. Hypertrophied cells contain somewhat more myoglobin and can store a bit more oxygen.

In addition to compensating for inadequate oxygen delivery, muscle tissue also compensates to some degree for inadequate nutrient delivery. During rest, when muscle tissues receive more glucose than they use directly for energy, they use some to store energy in phosphocreatine and some to synthesize glycogen. During exercise, the glycogen can be broken down to glucose. Along with initiating contraction, Ca^{2+} released from the sarcoplasmic

reticulum binds to **calmodulin** (kal-mod′u-lin), an intracellular calcium-binding protein. This binding activates enzymes, some of which lead to increasing the ATPase activity of myosin and some of which catalyze the breakdown of glycogen. The total quantity of glycogen stored in all muscle cells combined provides enough glucose to supply energy for a person to run about 10 miles. The glycogen stored in the liver supplies energy for another 3 miles.

Fatty acids stored in the body's fat deposits also supply energy for muscle contraction. When a person is exercising at 50 to 60 percent of maximum effort, about half of the energy used by muscles comes from fatty acids and about half comes from glycogen as long as glycogen lasts. When a person is exercising at 85 percent of maximum effort, almost all of the energy comes from glucose derived from glycogen. Only glucose can be metabolized during strenuous exercise because too little oxygen is available to use fatty acids, which are metabolized aerobically. Epinephrine, a hormone released from the adrenal gland during stress, causes muscles to start using fatty acids instead of glucose as their energy source. Epinephrine, like calmodulin-Ca^{2+}, makes glucose available at the same time that oxygen depletion has reduced fatty acid metabolism. When glycogen becomes depleted, however, muscles depend entirely on energy from fatty acids and can function only as fast as oxygen can be supplied. When strenuous exercise ceases and stress is reduced, oxygen becomes relatively more available and epinephrine is no longer secreted. Muscles then obtain more energy from fatty acids. These metabolic processes are summarized in Figure 9.15. The implications of muscle metabolism in sports are considered in Interlude 2.

Heat Production

We feel hot after strenuous exercise because muscles release specific quantities of heat as they contract. Muscle contraction requires that chemical energy be transformed into mechanical energy and, as in other energy transformations, some energy is released as heat. The total amount of chemical energy released equals the mechanical energy used by cross-bridges and the energy released as heat. Muscles give off large amounts of heat when they produce tension, but they give off some heat even when resting. All metabolically active cells, including muscle cells, give off heat from various chemical reactions. Heat from metabolic reactions maintains a normal body temperature in a cooler environment, but sometimes too much or too little heat is produced (Chapter 24). Heat released during strenuous exercise can raise the body temperature above normal, and heat released when muscle tension increases and muscle shortening occurs during shivering helps to maintain body temperature during extreme cold.

FAST AND SLOW MUSCLE UNITS

We have seen that muscle fibers are organized into motor units, each controlled by a particular neuron, but muscles also contain muscle units, which usually contain several motor units. A **muscle unit** consists of all of the muscle fibers of a particular type in any whole muscle. Muscle units are categorized as slow units or fast units according to their rate of contraction. Fast units are further subdivided according to the rate at which they fatigue as fast fatigue-resistant and fast easily fatigued units (Table 9.2 and Figure 9.16).

Slow units are characterized by a slow-to-develop, long duration twitch (up to 200 msec from excitation through relaxation). Slow units are well supplied with blood, and their fibers contain relatively large amounts of myoglobin and many mitochondria. Such units can carry out aerobic metabolism at a rapid rate. They store only small amounts

TABLE 9.2   Properties of Muscle Units

Property	Slow Unit	Fast Fatigue-Resistant Unit	Fast Easily Fatigued Unit
Twitch	Slow and of long duration	Fast and of short duration	Fast and of short duration
Axon diameter	Small	Intermediate to large	Large
Activity of myosin-ATPase	Low	High	High
Blood supply	Extensive	Extensive	Moderate
Quantity of myoglobin	Large	Large	Small
Number of mitochondria	Many	Many	Few
Primary type of metabolism	Aerobic	Aerobic	Anaerobic after a few seconds
Quantity of glycogen	Small	Medium	Large
Resistance to fatigue	Excellent	Moderate	Poor

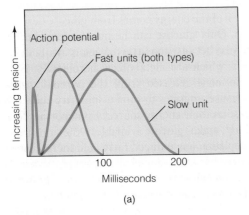

(a)

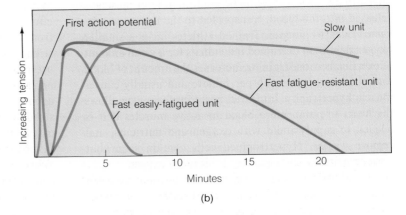

(b)

FIGURE 9.16

Comparison of muscle unit types: (a) nature of twitches and (b) nature of tetanic contractions.

of glycogen, but their ample blood supply delivers sufficient oxygen, fatty acids, and glucose to them so that glycogen stores usually are not needed. The rapid metabolic rate and the slow speed of contraction makes slow units resistant to fatigue; aerobic metabolism easily produces ATP as fast as it is used.

Fast fatigue-resistant units are characterized by a rapidly developing, short duration twitch (less than 100 msec from excitation through relaxation). Like slow units, fast fatigue-resistant units are well supplied with blood and, therefore, oxygen and nutrients. Their fibers contain a relatively large quantity of myoglobin, many mitochondria, and a moderate quantity of glycogen. Such units carry out aerobic metabolism at a rapid rate. Though contraction is rapid, the rapid metabolic rate and ample glycogen stores provide adequate ATP to sustain contraction over at least a moderate period of time.

Fast easily fatigued units are characterized by a rapidly developing, short duration twitch (less than 100 msec from excitation through relaxation). They have a modest blood supply and their fibers contain relatively few mitochondria and small amounts of myoglobin, but much glycogen. Such units metabolize anaerobically after the first few seconds of contraction. Their rapid speed of contraction and their dependence on anaerobic metabolism make these units easily fatigued.

Distribution of Muscle Units

Most muscles contain units of all three types, but many muscles have one predominant type of unit. Muscles that contain many slow units are called **slow muscles,** or **red muscles.** They appear redder than other muscles because of their high myoglobin content and their extensive blood-filled capillaries. (The drumstick and other dark meat of the Thanksgiving turkey are red muscles.) Slow muscles, such as calf muscles in the leg and many back muscles, sustain contractions that maintain posture. Slow muscles also respond to endurance exercises to a greater degree than fast muscles.

Muscles that contain large numbers of fast easily fatigued units are called **fast muscles,** or **white muscles.** Though not really white, they are not as red as other muscles because of their low myoglobin content and their less extensive capillary supply. (The white meat of a turkey consists of white muscle.) Fast muscles, such as those that move the eyeballs and fingers, are responsible for quick, precise movements that need not be sustained over a long period of time.

Many muscles, such as the larger limb muscles, are difficult to categorize as slow or fast. When such a muscle contracts, the slow units are recruited first; if more forceful movements are required, the fast units also are recruited. This sequential recruitment of units helps to explain why leisurely activity can be sustained for a long time, whereas strenuous activity can be sustained only for a short time. The fast easily fatigued muscle units become unable to produce sufficient ATP and they accumulate lactic acid. When fast fatigue-resistant muscle units also are present, they are able to maintain function during longer periods of strenuous activity.

SMOOTH MUSCLE

Unlike skeletal muscle, which usually is attached to bone, **smooth muscle** is found in the walls of internal organs and blood vessels. Smooth muscle contains actin, myosin, and tropomyosin like skeletal muscle, but lacks troponin. Its contractions are less precise and more sustained than those of skeletal muscle and usually are under involuntary control. It responds to internal physiological events that are not under conscious control.

Structure of Smooth Muscle

The fibers of smooth muscle are much smaller than those of skeletal muscle and their filaments are organized differently. Each fiber consists of a single, spindle-shaped cell 2 to 5 μm in diameter and 50 to 200 μm in length (Figure 4.12). A smooth muscle fiber is bounded by a sarcolemma, contains a single nucleus, has a poorly developed sarcoplasmic reticulum, and lacks T tubules. In contrast to the bundle arrangement of skeletal muscle fibers, smooth muscle fibers are arranged in sheets.

Smooth muscle contains thick myosin filaments, thin filaments containing actin and a small amount of tropomyosin, and intermediate filaments. Intermediate filaments, found only in smooth muscle, appear to form an elastic framework for the cell. Thick and thin filaments are oriented more or less parallel to the long axis of the cell, with thin filaments anchored to the sarcolemma by organelles called **dense bodies.** Sarcomeres are too poorly defined to create striations visible under the microscope. The contractile units of smooth muscle consist of ten or more thin filaments surrounding each thick filament, which has myosin heads projecting toward the thin filaments.

Types of Smooth Muscle

Smooth muscle is classified as either multiunit or visceral, but in many body sites, it displays gradations between the two types.

Multiunit smooth muscle is found in the iris and ciliary muscles of the eye and in the walls of the larger blood vessels. It consists mainly of discrete fibers that act independently of one another, although some fibers are connected by gap junctions, fused membranes across which action potentials easily pass (Chapter 4). Multiunit smooth muscle is controlled almost entirely by signals from the autonomic nervous system (Chapter 14). It rarely contracts spontaneously and rarely responds to chemical substances in the blood or tissues.

Visceral smooth muscle, also called **unitary smooth muscle** or **single-unit smooth muscle,** is found in the walls of hollow, internal organs, such as the stomach and intestines, and in the walls of small blood vessels. It is composed of sheets of cells interconnected by gap junctions. Because action potentials are easily transferred from cell to cell across gap junctions, the whole muscle sheet contracts as a unit, hence the name unitary. Although most visceral smooth muscle is innervated by autonomic neurons, some has no nerve supply. Visceral smooth muscle is capable of contracting spontaneously and of responding to mechanical; chemical; and, if innervated, neural stimulation.

Contraction of Smooth Muscle

Smooth muscle typically begins to contract 50 to 100 msec after it has been stimulated, reaches a peak contraction in about 500 msec, and gradually declines in contraction strength over the next 1 or 2 seconds. Smooth muscle contraction is slower than skeletal muscle contraction because smooth muscle ATPase has lower activity and the calcium pump operates at a slower rate.

Smooth muscle cells can be stimulated to contract in a variety of ways and to varying degrees. In some instances, they can be stimulated to contract by neural signals that elicit action potentials as in skeletal muscle. In other instances, smooth muscle contracts spontaneously or is stimulated by hormones, stretching of fibers, or local changes in the composition of extracellular fluids. In contrast to skeletal muscle fibers, which contract maximally when stimulated and which illustrate the all-or-none law, smooth muscle fibers contract proportionately to the degree of stimulation they receive. Stimulation need not be sufficient to elicit an action potential, and a small degree of stimulation can produce weak contractions.

Multiunit and visceral smooth muscle differ in their manner of contraction. Multiunit smooth muscle cells can contract spontaneously without a neural signal or can be induced to contract by neural signals. When numerous cells (typically 30 or more) are simultaneously stimulated, an action potential (signal that elicits contraction) can pass along the sarcolemma of all of the cells. Visceral smooth muscle can generate spike potentials and action potentials with plateaus. A **spike potential** can occur spontaneously or can be elicited by neurotransmitters, hormones, or other substances in the blood or tissues. An **action potential with a plateau,** which also can be elicited by various means, persists for a much longer time than a spike potential. The plateau, which may be due to slow action of the calcium pump, is typically 300 to 400 msec, but can be as long as 30 minutes. Long plateaus are seen in sheets of muscle where cells are connected by gap junctions and contract in long, slow waves.

Once an action potential is generated in visceral muscle, it spreads as an electrical current across the gap junctions between cells. This spreading of action potentials causes waves of contraction to pass through an entire visceral muscle mass. Such waves of contraction called **peristalsis** (per"is-tal'sis) are responsible for moving substances through the intestinal tract and ureters (tubes that carry urine from the kidneys to the bladder).

Stretching of visceral smooth muscle usually causes it to generate action potentials. Such potentials result from a combination of slow wave potentials and lowering of the resting potential (making it less negative) by stretching. When visceral smooth muscle is first stretched, it responds by increasing tension; if it is held in a stretched position, the tension gradually decreases. For example, when the contents of the intestine stretch the walls, peristalsis occurs and moves the contents away from the distended area. Similarly, filling the urinary bladder with urine stretches the bladder walls and initiates signals that are interpreted as a need to urinate. If the bladder is not voided, tension gradually decreases and allows further filling, at least up to a certain capacity. The ability of visceral smooth muscle to adjust tension to prolonged stretching is called **plasticity.** Because of plasticity, length-tension relationships are much more variable in visceral than in skeletal muscle. As we shall see in later chapters, various hormones and local tissue factors affect smooth muscle in blood vessel walls and in many hollow organs.

Variations in what elicits smooth muscle contraction and how much contraction occurs are due mainly to properties of smooth muscle cell membranes. As in skeletal muscle, smooth muscle also requires calcium ions for contraction, but the source and action of the calcium ions differs. Some calcium ions are released from the small

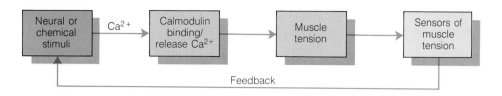

FIGURE 9.17
Control of smooth muscle contraction.

sarcoplasmic reticulum, but some enter the cytosol from extracellular fluids by passing through calcium channels in the sarcolemma. When calcium enters the cytosol, it binds to calmodulin, a protein similar to troponin. Calmodulin-Ca^{2+} activates a protein kinase enzyme, which catalyzes the binding of a high energy phosphate to myosin. Only after myosin has been phosphorylated can it bind with actin and produce tension. The mechanism of tension production appears to involve oarlike movements of myosin heads similar to those in skeletal muscle, though the filaments are organized differently.

Calcium is removed from calmodulin in the relaxation process by calcium pumps. Pumps in the sarcoplasmic reticulum, which are similar to, but much less numerous than, those in skeletal muscle, store Ca^{2+} in the sarcoplasmic reticulum. Pumps in the sarcolemma move Ca^{2+} out of the cell. Pumping Ca^{2+} out of cells effectively relaxes smooth muscle because cells have a high surface-to-volume ratio—much membrane relative to the number of contractile filaments. This mechanism would be ineffective in skeletal muscle because skeletal muscle cells have much less surface membrane relative to their numbers of contractile filaments. Calcium pumps in smooth muscle are relatively inefficient, so relaxation is slow and contraction lasts 10 to 30 times longer in smooth muscle than in skeletal muscle.

In smooth muscle, neurons branch out over the surface of the muscle sheet, but do not come close to the sarcolemma as they do in skeletal muscle. Axon terminals have beadlike swellings called **varicosities** (var"ik-os'it-ēz), which contain vesicles of neurotransmitter. As neurotransmitter is released, it diffuses through interstitial fluid to receptors on the sarcolemma of several adjacent cells. Smooth muscle cells can have both excitatory and inhibitory receptors for either or both of the neurotransmitters acetylcholine and norepinephrine. A cell's response, which depends on the interaction between the neurotransmitter and receptor, results in a junction potential. Binding of a neurotransmitter to an excitatory receptor creates an **excitatory junction potential** (EJP). Contraction is initiated by summation of EJPs. Binding of a neurotransmitter to an inhibitory receptor creates an **inhibitory junction potential** (IJP). Compared to end plate potentials in skeletal muscle, junction potentials in smooth muscle can last 100 times as long, can be excitatory or inhibitory, and are subject to summation. (End plate potentials are always excitatory and not subject to summation.)

Smooth Muscle Tone

Smooth muscle is rarely completely relaxed or completely contracted, but is usually maintained at some level of partial contraction that constitutes **smooth muscle tone,** or **tonus** (ton'us). Though skeletal muscle tone is maintained by neural signals (Chapter 11), smooth muscle tone is maintained by all of the factors that can elicit smooth muscle contraction. Smooth muscle tone constitutes a baseline for blood vessel diameters and tension in the walls of internal organs. Contraction is adjusted to increase or decrease blood flow, or to move substances through hollow organs (such as food through the digestive tract).

The functioning of smooth muscle and how it is controlled is summarized in Figure 9.17.

CARDIAC MUSCLE

The **myocardium** (mi"o-kar'de-um), or muscle layer of the heart, consists mainly of contractile cells called **cardiac muscle** (Figure 4.12c). Some of these cells have been modified to conduct signals that regulate contractions as is explained in Chapter 18. Like skeletal muscle fibers, cardiac muscle fibers contain thick myosin filaments and thin myofilaments of actin, tropomyosin, and troponin. They also have a well-developed sarcoplasmic reticulum and T tubules. In contrast to the long, multinucleated fibers of skeletal muscle, the fibers of cardiac muscle consist of uninucleated cells that undergo extensive branching and that are coupled by gap junctions (Chapter 4) at the **intercalated** (in-ter'kah-lāt-ed) disks. The electrical resistance across intercalated disks—only about one four-hundredth of that across the external membranes of cardiac muscle fibers—allows action potentials to propagate easily from one myocardial cell to the next. The ease of action potential propagation and the extensive branching of cardiac fibers allows fibers of the atria and fibers of the ventricles each to behave as a **syncytium** (sin-sish'e-um), a multinucleate cellular mass that acts as a functional unit. If a site in either syncytium is sufficiently stimulated to initiate an action potential, that action potential spreads and triggers a wave of contraction through the entire syncytium. Each syncytium obeys the all-or-none principle described for skeletal muscle.

Skeletal and smooth muscle are compared in Table 9.3, and cardiac and skeletal muscle are compared in Table 9.4.

TABLE 9.3   A Comparison of Skeletal and Smooth Muscle

Characteristic	Skeletal Muscle	Smooth Muscle	
		Multiunit	*Visceral*
Mechanism of excitation	Nerve signals	Nerve signals	Nerve signals, hormones, local tissue factors, stretching, and spontaneous contractions
Main calcium source	Sarcoplasmic reticulum	Extracellular fluids	Extracellular fluids
Calcium regulator	Troponin	Calmodulin	Calmodulin
Arrangement of filaments	Orderly in sarcomeres	Somewhat random, actin attached to dense bodies	Somewhat random, actin attached to dense bodies
Conduction of action potentials	Via sarcolemma and T tubules	Rare, if they occur only over part of sarcolemma	Via cell membranes with electrical conduction over gap junctions
Myoneural junctions	Axon terminals synapse with motor end plates	Varicosities of axons near smooth muscle cells	Varicosities of axons near smooth muscle cells
Contraction velocity	Rapid	Slow	Slow
Mode of control	Voluntary	Involuntary	Involuntary

TABLE 9.4   A Comparison of Cardiac and Skeletal Muscle

Characteristic	Cardiac Muscle	Skeletal Muscle
Nature of control	Autonomic nervous system	Somatic nervous system
Arrangement of fibers	Branched	Unbranched
Microscopic appearance	Striated	Striated
Nuclei per fiber	One	Many
Intercalated disks	Present	Absent
Arrangement of T tubules	One per sarcomere, located at Z lines	Two per sarcomere, located at A-I junctions
Duration of action potential	150 to 300 msec	1 to 2 msec
Contraction time	150 to 300 msec	About 40 msec, but varies in different muscles
Absolute refractory period	150 to 300 msec	1 to 2 msec

CLINICAL APPLICATIONS

Involuntary Muscle Contractions

Various kinds of involuntary contractions sometimes occur in skeletal muscles. They include natural responses such as shivering (to be discussed later), and abnormal responses such as spasms, cramps, fibrillations, and fasciculations.

In a muscle **spasm,** fibers of a muscle produce twitchlike contractions that can be forceful and painful. Muscle spasms are caused by substances such as metabolic wastes in body fluids and can be alleviated by massage. Massage increases blood circulation to the affected muscle and helps to remove wastes.

A muscle **cramp** involves tetanic contractions of muscle fibers. One explanation for cramps involves positive feedback: exercise increases signals from the muscle back to the spinal cord, and the spinal cord sends even more signals to the muscle until involuntary tetanic contraction is produced. Another possible cause of muscle cramps is depletion of ATP in the affected muscle. Probably the most widely accepted explanation for muscle cramps is an imbalance in body water and salt that results from profuse sweating during strenuous exercise. Athletes and people who work in hot environments often drink isosmotic fluids such as Gatorade® to replace salt as well as water, probably decreasing the risk of muscle cramps.

Fibrillations are spontaneous twitchlike rhythmic contractions of individual muscle fibers that are not visible through the skin. Fibrillations typically occur a few days after all the nerves to a muscle have been destroyed. Such muscle fibers, which have been released from nervous control, begin to generate action potentials spontaneously. Fibrillations cease after a few weeks as the muscle cells atrophy, or degenerate.

Fasciculations (fas-ik″u-la′shunz) are contractions of all fibers of a motor unit following an abnormal action potential in a motor neuron. Sometimes fasciculations produce sufficient contraction to be seen through the skin. Trauma to motor nerves and poliomyelitis often cause fasciculations.

Altered Muscle Size

Certain conditions cause muscle to increase or decrease in size. Muscle **hypertrophy** (hi-per′trof-e), an increase in muscle size, is due to an increase in the size of fibers (cells) and the number of mitochondria they contain, but not to an increase in the number of fibers. Isometric exercises such as weight lifting, which cause the muscle to produce at least 75 percent of its maximum tension, increase muscle size and strength. Though men typically have heavier,

FIGURE 9.18
Exercise causes muscles to increase in size.

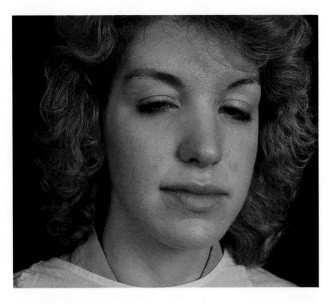

FIGURE 9.19
A myasthenia gravis patient.

stronger muscles than women, there seems to be no sexual difference in the response of muscles to strength-building exercise. Such exercises cause the muscles of men and women to increase in size in direct proportion to the original diameter of the muscle (Figure 9.18). Isotonic contractions like swimming, walking, or running have little effect on muscle strength, but they do increase muscle endurance by increasing the blood supply to muscle fibers and, thus, their capacity for metabolic activity.

Many young men and a few young women are so anxious to have big powerful muscles either to impress others or to excel in athletics that they take hormones called **anabolic steroids** (an-ah-bol'ik ste'roidz) to increase muscle mass. These drugs were originally developed in Nazi Germany allegedly to create an army of superstrong soldiers. Similar to testosterone, the hormone responsible for normal male muscular development, anabolic steroids can suppress normal testosterone secretion and cause shrinkage of male sex organs and enlargement of breasts. They also can lead to acne, teenage baldness, liver cancer and other damage to the liver and kidneys, electrolyte imbalances, mental disorders, and even death. In spite of their hazards, body builders and athletes now spend millions of dollars per year to obtain these illegal drugs that they must continue to take to maintain muscle development.

Muscle **atrophy** (at'ro-fe), a decrease in muscle mass, is due to a decrease in the size of fibers and the number of mitochondria. It can be caused by lack of use or denervation. **Disuse atrophy** from immobilization of a limb in a cast or bed rest is reversible by exercise. **Denervation**

atrophy is reversible only by regrowth of axons (Chapter 11). Electric stimulation of muscles can be used to prevent atrophy until reinnervation of muscles occurs.

Muscular dystrophy (dis'tro-fe) is the name of a group of related, inherited diseases that involve degeneration of muscle fibers. It is discussed in the essay at the end of this chapter.

Myasthenia Gravis

Myasthenia gravis (mi"as-the'ne-ah gra'vis), a disease characterized by muscle weakness (Figure 9.19), is caused by autoantibodies, antibodies made against the body's own proteins (Chapter 21). Evidence for the role of autoantibodies in the destruction of acetylcholine receptors came, in part, from studies with bungarotoxin, a protein of cobra venom that binds irreversibly with acetylcholine receptors. When surgically removed muscle tissue is exposed to bungarotoxin in the laboratory, the amount that binds to the tissue indicates the number of functional receptors. Muscle from myasthenia patients binds only 30 percent as much bungarotoxin as muscle from normal people; the nonfunctional receptors have antibody bound to them. What triggers autoantibody formation is still unknown and the disease cannot be cured yet. It can be treated with cholinesterase inhibitors such as neostigmine and Mestinon, which allow sufficient acetylcholine to accumulate at myoneural junctions to stimulate functional receptors. Newer therapies for this disorder include immunosuppressive drugs such as prednisone and Imuran and a procedure for removing antibodies from the blood called plasmapheresis.

CLINICAL TERMS

convulsion (kon-vul′shun)   involuntary muscle contraction
electromyography (e-lek″tro-mi-og′raf-e)   a procedure for recording electrical changes in muscle tissues
fibrosis (fi-bro′sis)   replacement of degenerated muscle cells with fibrous connective tissue
fibrositis (fi″bro-si′tis)   inflammation of fascia and other fibrous connective tissues

myalgia (mi-al′je-ah)   pain in one or more muscles
myology (mi-ol′o-je)   the study of muscle
myopathy (mi-op′ath-e)   any muscle disease
myositis (mi″o-si′tis)   inflammation of a skeletal muscle
myotonia (mi″o-to′ne-ah)   increased irritability and contractility of a muscle accompanied by decreased ability to relax

ESSAY
MUSCULAR DYSTROPHY

Muscular dystrophy is a general name for several inherited diseases in which skeletal muscles lose strength as muscle tissue degenerates. Dystrophy means degeneration, and as the name implies, muscle cells shrink and are eventually destroyed in the course of the disease. Duchenne muscular dystrophy (DMD), the most common form, is inherited as a sex-linked recessive characteristic and affects males almost exclusively. DMD occurs in about one in 3500 male births. It usually is diagnosed between the ages of two and 10. Muscle degeneration begins in the hips, spreads to the legs, abdomen, and spine, and progresses relentlessly until death occurs in early adulthood. Other less debilitating forms of muscular dystrophy are usually diagnosed in adolescence or early adulthood. These include limb-girdle dystrophy, which affects shoulder, hip, and upper limb muscles; facioscapulohumeral dystrophy, which affects facial and shoulder muscles; and myotonic, or distal, dystrophy, which affects muscles of the hands, forearm, and lower leg.

Though much remains to be learned about muscular dystrophy, recent findings have greatly increased our understanding of DMD. In mid-1987, the specific gene responsible for DMD was identified. By the end of the year, the protein made by the normal allele of the gene, but not by the alleles in DMD patients, was identified and named *dystrophin.* Because the protein is found in exceedingly small amounts (0.002 percent of total muscle protein) in normal muscles, researchers believe it is a control protein associated with the T tubules and the sarcoplasmic reticulum. One important function of dystrophin must be to prevent uncontrolled diffusion of calcium ions from the sarcoplasmic reticulum because such diffusion occurs in DMD patients. The excess of calcium ions in intracellular fluids is believed to activate the enzyme phospholipase A, which dissolves muscle fibers. As the body attempts to repair this damage, it replaces muscle tissue with fibrous connective tissue (Figure 9.20). As connective tissue accumulates in a muscle, it interferes with blood circulation through the tissue and more muscle cells die from lack of oxygen and nutrients. Similar changes in other forms of muscular dystrophy may be related to other genetic defects.

Knowing that muscle cells of DMD patients lack dystrophin has not led to a cure for the disease, but it may become possible to control connective tissue accumulation, to

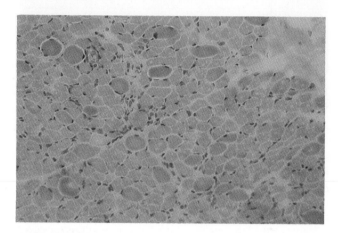

FIGURE 9.20
Muscle tissue from an individual with muscular dystrophy.

supply dystrophin to defective muscle cells, or even to replace the defective gene with a normal one. A new treatment for DMD is being tried. Transplants of healthy muscle cells containing the dystrophin gene are being surgically inserted into degenerating muscles and patients treated to prevent rejection of the transplants. Even if successful, such transplants will not cure the disease, but they will replace muscle cells that degenerate. Physical therapy and the use of braces that can overcome some disabilities caused by muscle wasting are the standard treatments. Because muscular dystrophy causes progressive crippling and premature death, genetic screening is important to allow carriers of defective genes to exercise reproductive options. Tests to detect carriers of some kinds of muscular dystrophy are presently available, and research is underway to develop more tests and to improve the reliability of those already available.

Questions

1. What are the main characteristics of Duchenne muscular dystrophy?
2. How is dystrophin related to muscular dystrophy?
3. How is muscular dystrophy treated?
4. How can muscular dystrophy be prevented?

CHAPTER SUMMARY

General Properties of Muscle

- Three types of muscle tissue are found in the human body: skeletal, smooth, and cardiac.
- All muscle tissue has four important properties: contractility, excitability, extensibility, and elasticity.

Muscle Structure

- Skeletal muscles are attached to bones or other structures by connective tissue fibers, tendons, or aponeuroses.
- A whole skeletal muscle contains several layers of connective tissue sheaths and many muscle fibers.
- Each fiber (cell) contains many myofibrils, and each myofibril contains many thick filaments of myosin and many thin filaments of actin, troponin, and tropomyosin.
- A fiber is bounded by a sarcolemma and contains sarcoplasmic reticulum, sarcoplasm, and transverse tubules.
- Myofilaments are organized into sarcomeres, and T tubules lie adjacent to terminal cisternae of the sarcoplasmic reticulum.

Contraction at the Molecular Level

- Molecular movements in the contraction cycle according to the sliding filament theory are summarized in Figure 9.4.
- Contraction is regulated by movements of calcium ions as summarized in Figure 9.6.

The Motor Unit

- Action potentials in motor neurons cause the release of acetylcholine, which binds to receptors at the motor end plate.
- Stimulation of the motor end plate receptors generates an action potential in the sarcolemma and T tubules, which causes Ca^{2+} to diffuse from the sarcoplasmic reticulum.
- Ca^{2+} binds to troponin and thereby couples excitation and contraction.
- The calcium pump causes relaxation by moving Ca^{2+} back to the sarcoplasmic reticulum.
- These processes are summarized in Table 9.1.
- Several drugs interfere with contraction by disrupting events at the myoneural junction.

Mechanics of Muscle Contraction

- Muscle contractions can be isotonic (same tension) or isometric (same length).
- A simple twitch, a response to a single stimulus, has a latent period, a contraction period, and a relaxation period.
- Summation of electrical stimulation can lead to incomplete or complete tetanus, depending on the frequency of the stimuli.
- Muscles exert the greatest tension at their equilibrium (resting) length, and lesser tensions when slack or stretched.
- Tension is greatest when the maximum number of myosin heads can exert force.
- Contraction velocity is inversely proportional to load in isotonic contractions.
- Graded muscle activity is possible mainly because of recruitment of motor units.

Muscle Metabolism

- Muscle metabolism can be aerobic or anaerobic, depending on the amount of oxygen available.
- Phosphocreatine, myoglobin, and the oxygen debt mechanism prolong the contraction of muscles during short bursts of strenuous activity when oxygen is in short supply.
- Oxygen debt is paid as lactic acid is metabolized aerobically, following strenuous exercise.
- Muscles release large quantities of heat when they contract.

Fast and Slow Muscle Units

- Skeletal muscle fibers are categorized as slow units, fast fatigue-resistant units, and fast easily fatigued units.
- The properties of muscle units are summarized in Table 9.2.

Smooth Muscle

- Smooth muscle, found in the walls of blood vessels and hollow organs, is categorized as multiunit or visceral.
- The contraction process is slower in smooth muscle than in skeletal muscle because of the low rates of activity of myosin-ATPase and the calcium pump.
- Multiunit smooth muscle rarely displays action potentials, but visceral smooth muscle displays two kinds of action potentials, spike potentials and action potentials with plateaus.
- Visceral muscle also spontaneously generates slow wave potentials, which can give rise to action potentials, and it displays plasticity.
- Calcium ions are essential for contraction of smooth muscle, but much calcium diffuses into the cells from extracellular fluids and only some comes from the sarcoplasmic reticulum.
- Calcium action is mediated by calmodulin instead of troponin and calcium controls phosphorylation of myosin rather than binding sites on actin filaments.
- Myoneural junctions in smooth muscle consist of axon terminals that release neurotransmitter into the extracellular fluids and receptor sites on adjacent muscle cell membranes.
- Neurotransmitters bind to excitatory or inhibitory receptors and generate excitatory or inhibitory junction potentials, respectively.
- Visceral smooth muscle maintains long-term steady contractions called tonus; tonus can be modified by the above stimuli and such modifications regulate blood supply and movement of substances in the digestive tract.

Cardiac Muscle

- Cardiac muscle has the same proteins as skeletal muscle, is striated and involuntary, and its cells contain many branches and gap junctions called intercalated disks.
- Cardiac muscle contraction occurs as a wave through a syncytium and is regulated by specialized signal conducting cells.

Clinical Applications

- Abnormalities in muscle activity include spasms, cramps, fibrillations, and fasciculations.
- Muscles can also undergo hypertrophy, atrophy, or dystrophy.
- Myasthenia gravis is an autoimmune disorder that causes muscle weakness.

QUESTIONS AND PROBLEMS

The questions at the end of each chapter are numbered to correspond with the objectives listed at the beginning of the chapter. Italics indicate that a question requires critical thinking skills beyond simple factual recall.

Questions

1. What general properties distinguish muscle tissue?
2. *(a) How is skeletal muscle organized from the contractile protein level to the whole muscle level?*
   (b) Which proteins are contractile proteins, and which are regulatory proteins?
3. *(a) How does the sliding filament theory explain muscle contraction?*
   (b) How is contraction regulated at the molecular level?
   (c) What is excitation-contraction coupling, and what is its significance?
4. (a) What are the functional properties of a motor unit?
   (b) How can events at the myoneural junction be altered by drugs and disease?
5. *(a) How do isotonic and isometric contractions differ?*
   (b) What happens in each of the phases of a simple twitch in an isotonic contraction? In an isometric contraction?
   (c) What is summation and how does it lead to incomplete and complete tetanus?
   (d) Why is treppe not an example of summation?
   (e) What impairment of muscle function accounts for muscle fatigue?
   (f) How is graded muscle activity accomplished in skeletal muscle?
6. *How does muscle metabolism differ during rest and during strenuous activity?*
7. *What are the functional differences between slow muscle units, fast fatigue-resistant muscle units, and fast easily fatigued muscle units?*
8. *(a) How does smooth muscle differ from skeletal muscle?*
   (b) How does multiunit smooth muscle differ in function from visceral smooth muscle?
   (c) Explain how excitation-contraction coupling and events at the myoneural junction in smooth muscle differ from those processes in skeletal muscle.
   (d) How do factors other than neural stimuli affect visceral smooth muscle?
   (e) What is smooth muscle tone, and how is it maintained?
9. *How does cardiac muscle differ from skeletal muscle and from smooth muscle?*
10. *(a) What are the causes of various involuntary contractions?*
    (b) How do muscle hypertrophy, atrophy, and dystrophy occur?
    (c) What causes myasthenia gravis, and how can it be treated?

Problems

1. Determine the percentage of maximal contraction in a muscle at 60, 80, 120, and 140 percent of its resting length. (Hint: See Figure 9.12.)
2. If a muscle can lift a maximum load of 100 grams, estimate the contraction velocity as a percentage of maximum velocity with a load of 50 grams and with a load of 0 grams.
3. Do some library research to further describe the different properties and actions of calmodulin and troponin.
4. Make a list of all naturally present chemical substances that can affect muscle physiology, and relate the properties of each substance to its function.
5. Design an experiment that you can safely do on yourself to determine which of your skeletal muscles fatigue most quickly and which fatigue most slowly.

OBJECTIVES

1. Name the major groups of skeletal muscles.

2. Briefly describe the development of skeletal muscle.

3. Relate the properties of levers to muscle action.
4. Explain muscle terminology, and list characteristics used to name muscles.

5. Locate and describe the actions of the major muscles of the head and neck.
6. Locate and describe the actions of the major muscles of the trunk, including the vertebral column, abdominal wall, and pelvic floor.

7. Locate and describe the actions of the major muscles of the pectoral girdle, shoulder, elbow, and wrist.
8. Locate and describe the actions of the major muscles of the pelvic girdle, hip, knee, and ankle.

10

SKELETAL MUSCLES AND THEIR ACTIONS

OVERVIEW OF MUSCLES

It is most fortunate for carrying on your daily activities that you need not know the names and actions of all your muscles to be able to use them. If you are planning a health related career, however, you will need some knowledge of the locations and actions of specific muscles. What you learned about the physiology of muscle tissue (Chapter 9) paved the way for just such a study.

The human muscular system consists of over 600 individual skeletal muscles, some of which are shown in Figures 10.1 and 10.2. Muscles usually are described in groups according to their location. Muscles of the axial skeleton include those of the scalp, face, neck, and trunk. Trunk muscles, in turn, include those of the vertebral column, abdominal wall, and pelvic floor. Muscles of the appendicular skeleton include muscles that help hold the pectoral and pelvic girdles in position relative to the axial skeleton muscles and those that move the joints of each of the limbs.

Only a small group of the many human muscles are considered in this chapter. They were selected to illustrate muscle actions in the movements of certain joints or because of other important functions. Of the muscles we will study, most are paired—one on each side of the body.

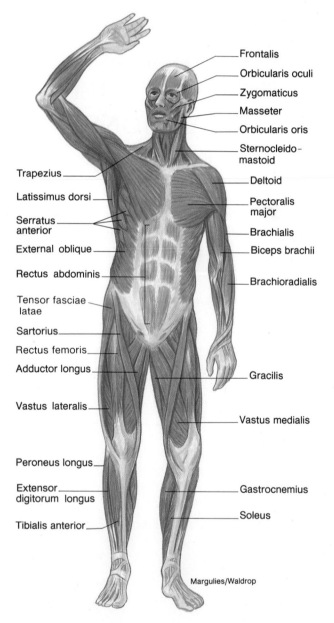

FIGURE 10.1
Some muscles of the human body (anterior view).

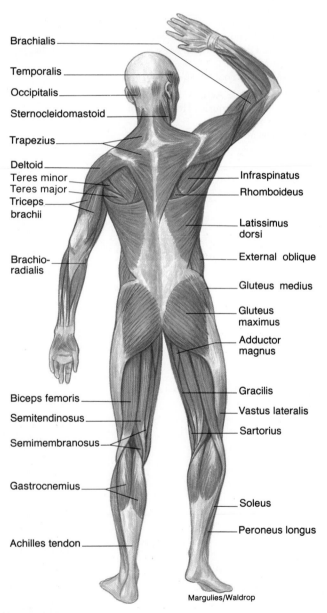

FIGURE 10.2
Some muscles of the human body (posterior view).

DEVELOPMENT OF SKELETAL MUSCLE

Most skeletal muscles develop from **myotomes** (mī'o-tōmz)—blocks of mesoderm in the embryonic trunk area—and from less highly organized mesoderm in the head area (Figure 10.3). Myotomes are part of somites as described in Chapter 5. Each myotome gives rise to many muscle fibers, or muscle cells. Each fiber is a single, multinucleated cell derived from the fusion of several embryonic muscle cells called myoblasts. Although myoblasts are capable of dividing, adult skeletal muscle cells have lost the ability to divide.

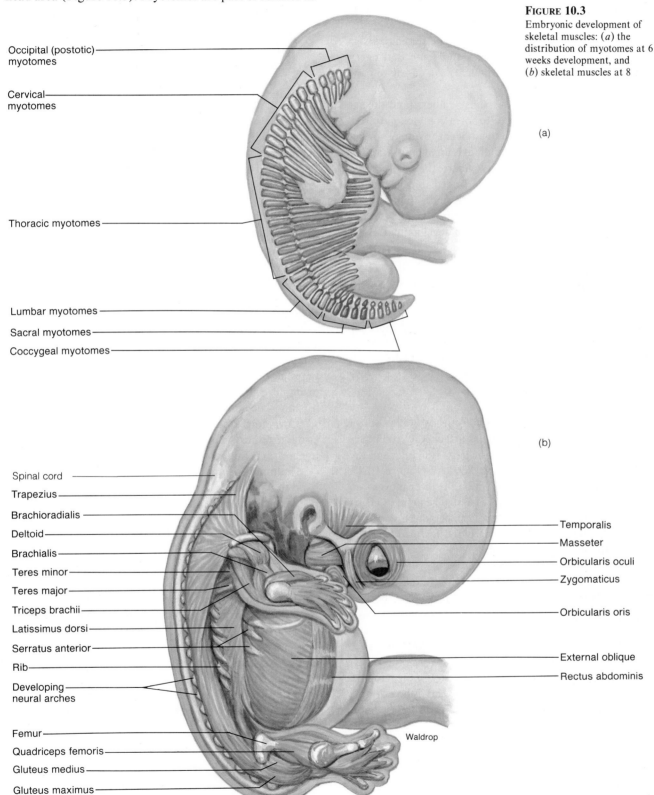

FIGURE 10.3
Embryonic development of skeletal muscles: (*a*) the distribution of myotomes at 6 weeks development, and (*b*) skeletal muscles at 8

Occipital (postotic) myotomes

Cervical myotomes

Thoracic myotomes

Lumbar myotomes

Sacral myotomes

Coccygeal myotomes

(a)

(b)

Spinal cord

Trapezius

Brachioradialis

Deltoid

Brachialis

Teres minor

Teres major

Triceps brachii

Latissimus dorsi

Serratus anterior

Rib

Developing neural arches

Femur

Quadriceps femoris

Gluteus medius

Gluteus maximus

Temporalis

Masseter

Orbicularis oculi

Zygomaticus

Orbicularis oris

External oblique

Rectus abdominis

Waldrop

As myotomes develop, they elongate and extend anteriorly toward the midline of the body or distally into the limb buds. The orientation of developing muscles is influenced by the presence of cartilage models of bones. Each myotome is associated with a sclerotome, the portion of a somite that gives rise to vertebrae. Between each of the vertebrae are spinal nerves, which branch from the spinal cord. One spinal nerve becomes associated with each developing myotome. Its fibers include some that carry signals from the muscle to the brain and some that carry signals from the brain to the muscle, causing it to contract.

As development continues, muscle fibers (and the associated nerve fibers) migrate to form complex patterns. The nerve fibers also branch to adjacent myotomes in an overlapping pattern. Even after extensive migration of myotomes and branching of nerve fibers, the embryologic origin of a muscle usually can be traced to a particular myotome by identifying the muscle's main spinal nerve. A muscle and the nerve fibers that carry signals to and from it form a functional unit early in development. Such units persist throughout life.

PROPERTIES OF MUSCLES

Learning about muscles is more interesting and more efficient if we first study the classes and principles of levers and the basic terminology of muscle actions.

Levers

The bones and joints of the body form a complex set of levers that move when the muscles attached to the bones exert force. A **lever** has three basic parts: the fulcrum; the point of effort; and the resistance, or load (Figure 10.4). The **fulcrum** is the point about which the lever moves—in the body, the fulcrum is usually a joint. The **point of effort** is the attachment of a muscle to a bone where the force is exerted. The **resistance** (load) is the force to be overcome or the weight to be lifted.

Levers are divided into three classes, depending on the placement of the lever components. In a **first-class lever,** the fulcrum is between the effort and resistance. For example, when the posterior neck muscles contract, they overcome the weight of the anterior skull and rock the head back (Figure 10.4a). In a **second-class lever,** the resistance is between the fulcrum and the effort. The use of leg muscles to stand on tiptoes is a possible example (Figure 10.4b), though some anatomists claim there is no example of a second-class lever in the body. In a **third-class lever,** the effort is between the fulcrum and the resistance (Figure 10.4c). This is the most common kind of lever in the body. Though lever action is important in many movements, some movements cannot be classified by lever type.

Two principles of lever action are important in movement:

1. Levers amplify force or velocity of movement.
2. The length of the lever arm from the point of effort to the resistance determines the amount of amplification of force and velocity.

A playground teeter-totter illustrates these principles. Placing a heavy child on one end of the teeter-totter will cause a lighter child on the other end to go up as the heavy child goes down. The force applied by the heavy child is amplified as it is transferred to the other end of the board. The velocity of movement is also amplified. The greater the difference between the weights of the two children, the faster the lighter child will ride up on the board. For each child, the total force applied is the product of the weight of the child (resistance) times the distance between the fulcrum and the child. We can apply these principles when we take two children of unequal size to a playground. We simply move a teeter-totter board with respect to its fulcrum until the two children balance before they begin to use the board.

Related to these principles is the observation that a short effort arm and a long resistance arm increase the range of motion of a joint. The large change in the angle between the bones at the elbow joint during flexion is an example. Similar wide ranges of motion occur at shoulder, hip, and knee joints.

Swinging a hammer shows how these principles operate together. Effort from muscles contracting in the upper arm is applied at some distance from the resistance (the hammer). Both force and velocity are amplified—the hand that holds the hammer moves with greater force and faster than the portion of the arm just distal to the elbow.

Muscle Terminology

Most muscles are attached to bone or connective tissue at two points—the least movable **origin** and the most movable **insertion.** The movement a muscle produces is called its **action.** Many muscles extend across joints and move the joint when they contract (Figure 10.5). For example, the biceps muscle on the anterior surface of the arm extends from its origin on the scapula across the elbow joint to its insertion on the radius. When it exerts force on the forearm, it causes flexion at the elbow. Flexion and other actions were summarized in relation to joints in Table 8.2 (Chapter 8).

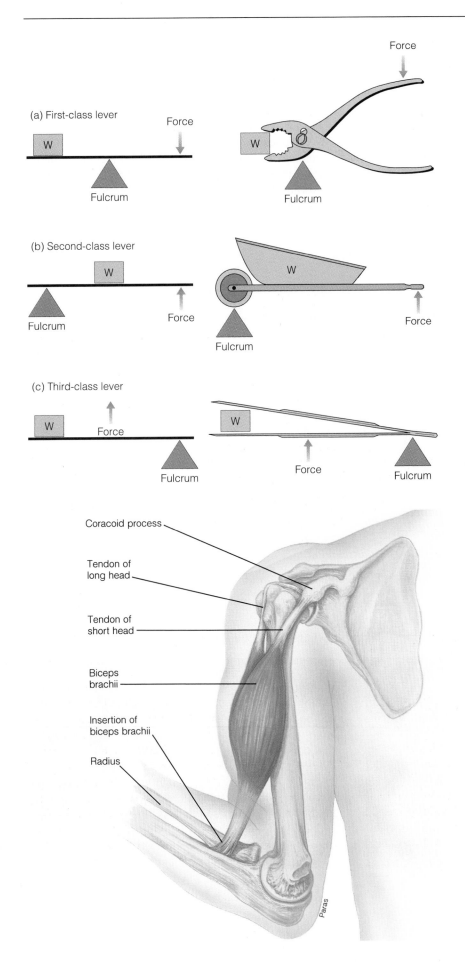

(a) First-class lever

Force

W

Fulcrum

Force

W

Fulcrum

(b) Second-class lever

W

Fulcrum

Force

W

Fulcrum

Force

(c) Third-class lever

W

Force

Fulcrum

W

Force

Fulcrum

FIGURE 10.4
Classes of levers. W = weight.

Coracoid process

Tendon of
long head

Tendon of
short head

Biceps
brachii

Insertion of
biceps brachii

Radius

Paras

FIGURE 10.5
Relationship of a muscle, the biceps brachii,
to the skeletal system.

In the preceding example, the biceps muscle causes a particular action and is, therefore, the **agonist** (ag'o-nist), or prime mover. Muscles that work with an agonist are called **synergists** (sin'er-jistz); muscles that exert an opposite force are called **antagonists** (an-tag'on-istz). The triceps muscle on the posterior surface of the arm is an antagonist to the biceps. It extends (straightens) the arm, and in doing so becomes the agonist. Because a single muscle can exert a force in only one direction, an antagonistic muscle is required to cause movement in the opposite direction. When an agonist and its synergists contract, the antagonists must relax for movement to occur. Most body movements are produced by actions across joints similar to the ones just described.

The **innervation** of a muscle—that is, what nerve carries signals to and from it—usually indicates which myotome gave rise to it. (Myotomes are derived from somites, as explained in Chapter 4.) Though not derived from myotomes, facial muscles develop from mesoderm that is largely innervated by the facial nerve (cranial nerve VII). Knowing the location of nerves in relation to muscles is important to prevent damage in surgical procedures.

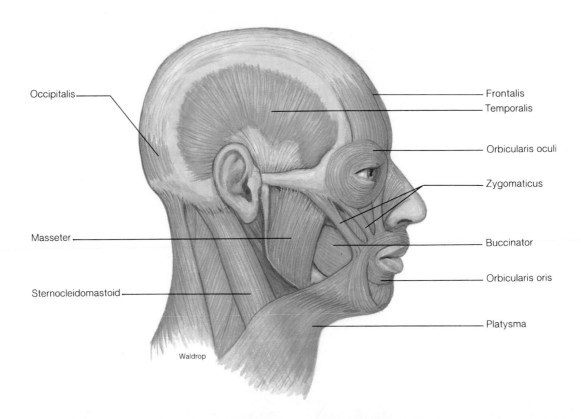

Occipitalis

Frontalis
Temporalis

Orbicularis oculi

Zygomaticus

Masseter

Buccinator

Orbicularis oris

Sternocleidomastoid

Platysma

Waldrop

FIGURE 10.6
The muscles of facial expression.

Naming of Muscles

Characteristics used in naming muscles include (1) the action they perform, (2) shape, (3) origin and/or insertion, (4) multiple points of origin, (5) location, (6) size, and (7) direction of their fibers. Muscles named for their action are flexors, extensors, abductors, adductors, or levators. A levator is a muscle that elevates a part of the body. Muscles named for their shape include the trapezius (shaped like a trapezoid) and the rhomboideus (shaped like a rhomboid). The origin and insertion appear in the name of the sternohyoid muscle; its origin is on the sternum and its insertion is on the hyoid. Muscles having more than one origin are said to have "ceps," or heads: a biceps has two; a triceps, three; and a quadriceps, four. The tibialis anterior and the tibialis posterior muscles are named for their locations anterior and posterior to the tibia, respectively. The gluteus maximus, the large muscle that forms a buttock, is named partly for its size. The external oblique and transverse abdominis are named partly for the direction of their fibers. Associating characteristics of muscles with their names helps to remember the names.

MUSCLES OF THE AXIAL SKELETON

Head and Neck Muscles

Grouped according to their functions, muscles of the head and neck are responsible for facial expression, mastication (chewing), moving the tongue and assisting in swallowing, and moving the whole head.

The muscles of facial expression (Figure 10.6) lie beneath the skin of the scalp, face, and throat. You can determine the actions of several of these muscles by studying the figure and then attempting to contract each muscle as you look at yourself in a mirror. You raise your eyebrows and wrinkle your forehead by contracting the frontal part of the *occipitofrontalis* (ok-sip''it-o-fron-ta'lis). You close your eyes or mouth by contracting the appropriate *orbicularis* (or''bik-u-la'ris) muscle. You use the major and minor *zygomatic* (zi''go-mat'ik) muscles to smile and the *depressor* muscles to create a sad or "down-in-the-mouth" expression. The *platysma* (plah-tiz'mah) lies just beneath the skin, where it gives the neck a smooth appearance and assists in pulling the angle of the mouth downward, as in pouting. The deeper *buccinator* (buk'sin-a''tor), or trumpeter's muscle, aids in puckering the lips.

The muscles of mastication (Figure 10.7) are used in chewing. The *masseter* (mas-se'ter) is the primary muscle in closing the mouth by elevating the mandible. The *lateral pterygoids* (ter'ig-oid) lower and protract the mandible, and cause the mandible to move from side to side. The *temporalis* (tem-po-ra'lis) and *medial pterygoids* are synergists of the masseter, and the medial pterygoids are antagonists of the lateral pterygoids. The buccinator, mentioned above, produces a sucking motion and helps keep food between the teeth. Third molar ("wisdom tooth") eruption can cause spasm of the masseter and temporalis muscles. This condition, called **trismus** (triz'mus), can be so severe that the person cannot open his or her mouth.

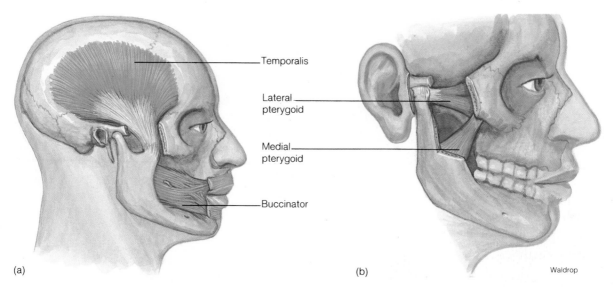

Temporalis

Lateral pterygoid

Medial pterygoid

Buccinator

(a)

(b)

Waldrop

FIGURE 10.7
Muscles of mastication: (*a*) superficial muscles, and (*b*) deep muscles.

The muscles of facial expression and mastication are innervated by the facial nerve (cranial nerve VII). The locations of the muscles and branches of the cranial nerve that supply them are illustrated by a "left hand rule" in Figure 10.8.

Certain muscles move the tongue (Figure 10.9) in mixing food, swallowing, and speaking. The *genioglossus* (je″ne-o-glos′us) extends from the chin (genio) to the tongue (glossus); it depresses the tongue and pulls it forward. The *stylohyoid* (sti″lo-hi′oid) extends from the styloid process of the temporal bone to the hyoid bone; it elevates the hyoid, to which the hyoglossus also is attached. The *hyoglossus* (hi-o-glos′us) is also attached to the tongue, so together these muscles elevate the tongue. The *styloglossus* (sti-lo-glos′us), which extends from the styloid process of the tongue, retracts the tongue. These and other muscles work together to move the tongue in mixing food and pushing it out of the mouth and into the pharynx. They also are important in moving the tongue in articulating words.

It is not essential that a muscle be attached to bone to be able to exert force. When a muscle contracts, it exerts a pulling force along its longitudinal axis and a pushing force perpendicular to its longitudinal axis (Figure 10.10). The tongue provides a good example in that it contains several muscles not attached to bone. Tongue muscle fibers are aligned in three mutually perpendicular planes. When muscle fibers in one plane contract, they create a bulging rigid framework. Muscles in other planes exert force about this framework.

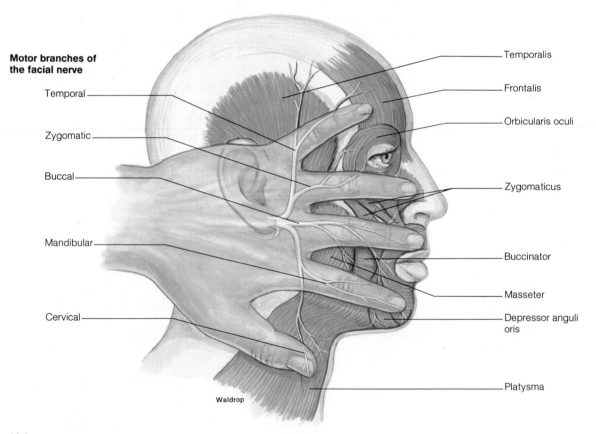

Motor branches of the facial nerve

Temporal

Zygomatic

Buccal

Mandibular

Cervical

Temporalis

Frontalis

Orbicularis oculi

Zygomaticus

Buccinator

Masseter

Depressor anguli oris

Platysma

Waldrop

Figure 10.8

Placing the left hand with fingers abducted along the right side of the face shows the location of certain facial muscles and the facial nerve branches that innervate them.

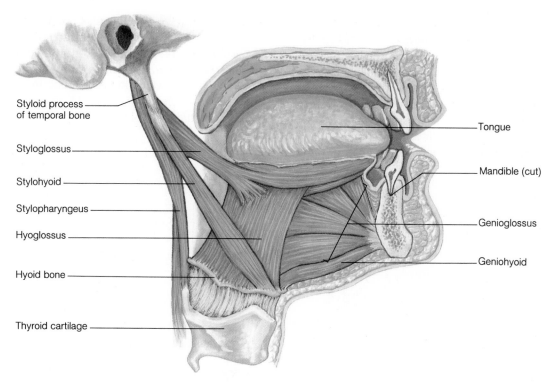

FIGURE 10.9
Muscles of the tongue.

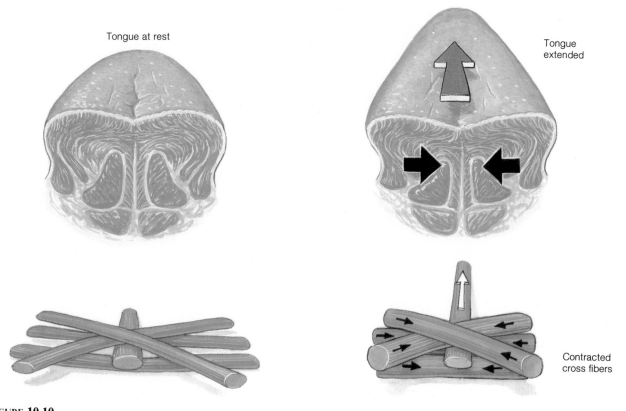

FIGURE 10.10
Muscles can exert force without being attached to bones. Contractions exert pull along a muscle's longitudinal axis and push perpendicular to that axis.

Several neck muscles are involved in moving the head (Figure 10.11). The *sternocleidomastoid* (ster″no-kli″do-mas′toid) extends from the sternum and clavicle (cleido) to the mastoid process of the temporal bone. When the pair of these muscles contract, they pull the chin toward the chest; when only one contracts, it rotates the face toward contracting muscle. A posterior muscle, the *splenius capitis* (sple′ne-us kap′it-us), is an antagonist of the sternocleidomastoid. Contraction of the pair of these muscles tilts the head backward. Contraction of only one rotates the head toward the contracting muscle. Two deeper muscles, the *semispinalis* (sem″e-spi-na′lis) *capitis* and the *longissimus* (lon-jis′im-us) *capitis* (Figures 10.12 and 10.13) extend the neck (bend the head back). When these muscles contract only on one side of the body, they extend the neck obliquely toward the contracting muscles.

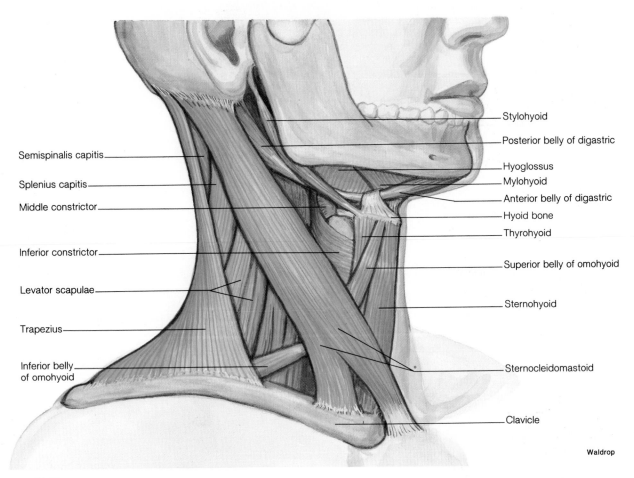

Stylohyoid

Posterior belly of digastric

Hyoglossus

Mylohyoid

Anterior belly of digastric

Hyoid bone

Thyrohyoid

Superior belly of omohyoid

Sternohyoid

Sternocleidomastoid

Clavicle

Semispinalis capitis

Splenius capitis

Middle constrictor

Inferior constrictor

Levator scapulae

Trapezius

Inferior belly of omohyoid

Waldrop

FIGURE 10.11
Superficial muscles of the neck.

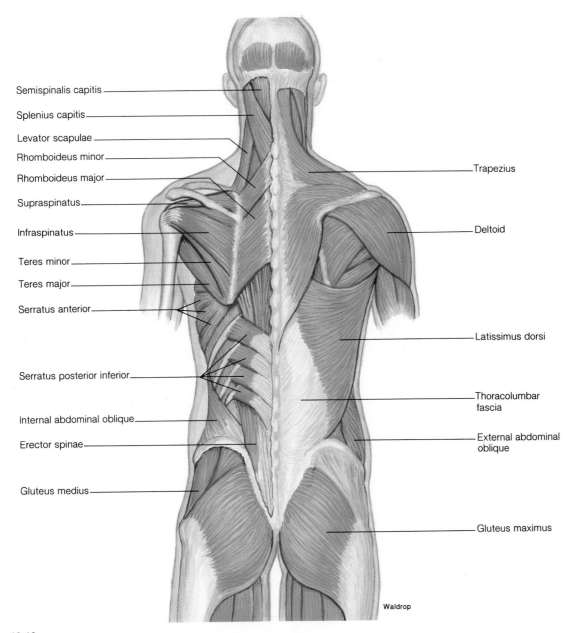

Semispinalis capitis

Splenius capitis

Levator scapulae

Rhomboideus minor

Rhomboideus major

Supraspinatus

Infraspinatus

Teres minor

Teres major

Serratus anterior

Serratus posterior inferior

Internal abdominal oblique

Erector spinae

Gluteus medius

Trapezius

Deltoid

Latissimus dorsi

Thoracolumbar fascia

External abdominal oblique

Gluteus maximus

Waldrop

FIGURE 10.12
Superficial muscles of the neck and back, with most superficial muscles shown on the right and the next deeper layer of muscles shown on the left.

Trunk Muscles

Muscles of the trunk include those of the back (Figures 10.12 and 10.13) and the rib cage. The pectoral muscles and several other superficial trunk muscles will be discussed with the muscles of the pectoral girdle because their actions are associated with the shoulder joint. The deeper muscles are associated with movements of the spine. The *sacrospinalis* (sa''kro-spi-na'lis), a major trunk muscle and an extensor of the spine, includes the lateral *iliocostalis* (il''e-o-kos-ta'lis), the intermediate *longissimus,* and the medial *spinalis*. The iliocostalis itself has three parts—the *lumborum* (lum-bo'rum), the *thoracis* (thor-a'sis), and

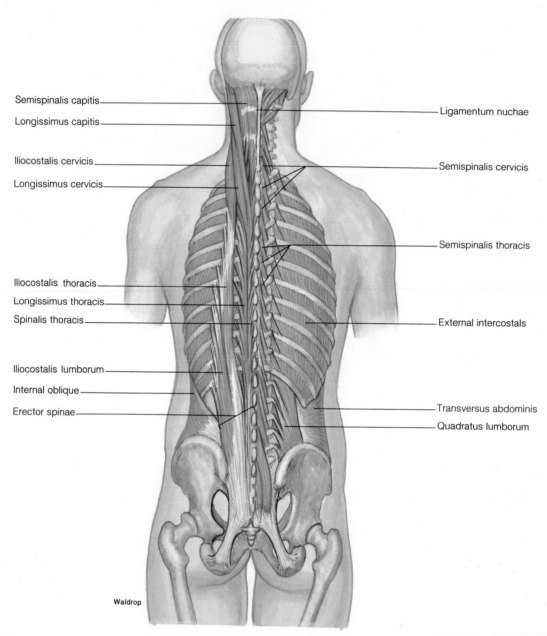

Semispinalis capitis

Longissimus capitis

Iliocostalis cervicis

Longissimus cervicis

Iliocostalis thoracis

Longissimus thoracis

Spinalis thoracis

Iliocostalis lumborum

Internal oblique

Erector spinae

Ligamentum nuchae

Semispinalis cervicis

Semispinalis thoracis

External intercostals

Transversus abdominis

Quadratus lumborum

Waldrop

FIGURE 10.13

Some intermediate and deep muscles of the back, after removal of most of the muscles in Figure 10.12.

the *cervicis* (ser've-sis)—each named for the area of the vertebral column in which it is found. The longissimus also has three parts—the *thoracis, cervicis,* and *capitis.* Portions of this muscle rotate the spine slightly on its own axis or bend it laterally from side to side. Such lateral movement is abduction—it moves the spine away from the midline of the body. Another trunk muscle, the *quadratus* (kwod-ra'tis) *lumborum,* also extends, rotates, and abducts the spine.

The muscles of the rib cage (Figure 10.14) help to move air in and out of the lungs. The *external intercostals* (in-ter-kos'talz) and *internal intercostals* are wide, short muscles, each having its origin on a superior rib and its insertion on the next inferior rib. The external muscles elevate the ribs and the internal ones depress them. The internal unpaired *diaphragm* separates the thoracic and abdominal cavities. When it contracts, it increases the volume of the thoracic cavity and draws air into the lungs.

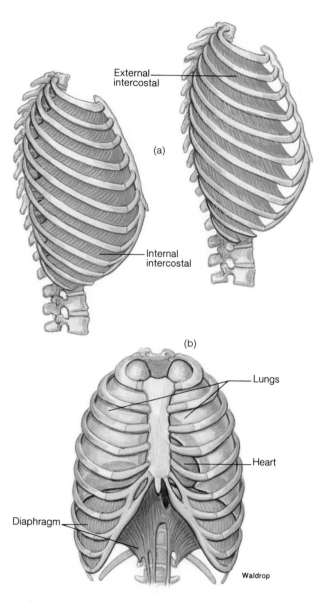

FIGURE 10.14
Muscles of the ribs are associated with breathing: (*a*) the external and internal intercostals, and (*b*) the diaphragm.

Abdominal Wall Muscles

The muscles of the abdominal wall (Figure 10.15) compress the abdomen. The *rectus abdominis* (rek'tus abdom'in-us), a long, flat midline muscle, is separated from its pair by midline connective tissue, the *linea alba* (lin'e-ah al'bah). Horizontal lines across this muscle, the *ten-* *dinous inscriptions,* are vestiges of embryonic divisions between myotomes. This muscle flexes the spine and acts as an antagonist to the sacrospinalis. The other abdominal muscles form the lateral abdominal walls. Their arrangement in layers (like a three-ply tire) increases the strength of the wall. The outermost *external oblique* runs anteriorly and inferiorly from the ribs to the pelvic area. The

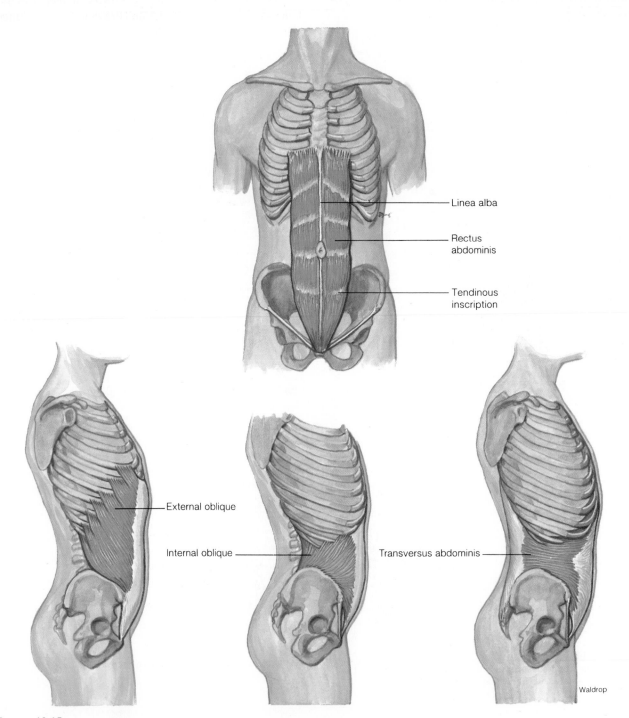

Linea alba

Rectus abdominis

Tendinous inscription

External oblique

Internal oblique

Transversus abdominis

Waldrop

FIGURE 10.15
Muscles of the abdominal wall.

middle *internal oblique* runs anteriorly and superiorly from the iliac crest to the ribs and xiphoid process. The inner *transverse abdominis* runs horizontally from the iliac crest, lumbar fascia, and rib cartilages to the linea alba. These muscles compress the abdomen and assist in defecation, coughing, and childbirth. They also help to maintain posture and can contribute to twisting motions and forced expirations.

Pelvic Floor Muscles

The muscles of the pelvic floor (Figure 10.16) form the pelvic diaphragm, which supports pelvic organs. The major muscles of this group are the *levator ani* (le-va'tor a'ni) and the *coccygeus* (kok-sij'e-us). If these muscles weaken, **prolapse,** or protrusion of pelvic organs through openings, can occur. Uterine prolapse often occurs in women who have had multiple pregnancies, especially if they have unrepaired injuries during childbirth.

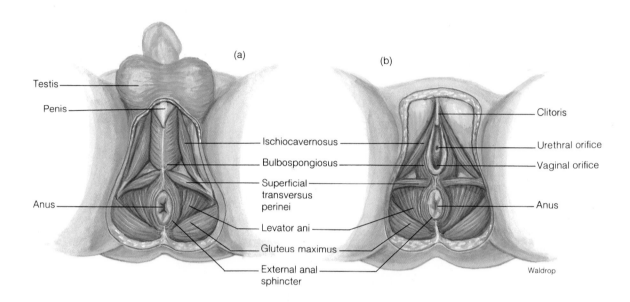

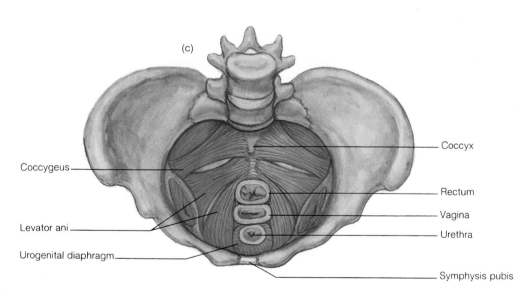

FIGURE 10.16
External muscles of the pelvic floor and perineum: (*a*) male, (*b*) female, and (*c*) internal muscles of the female pelvic floor.

MUSCLES OF THE APPENDICULAR SKELETON

Pectoral Girdle and Shoulder Muscles

Muscles that move the pectoral girdle and hold it to the trunk are closely associated with those that move the shoulder joint (Figure 10.17). Among the muscles that move the pectoral girdle is the *trapezius* (trap-e′ze-us), a large trapezoidal muscle with its origin on the thoracic vertebral column and its insertions on the spine and acromion process of the scapula and the distal end of the clavicle. Because of its large size and several insertions,

this muscle has several actions. The whole muscle rotates the scapula and abducts the arm. The superior portion elevates the glenoid fossa, the middle portion adducts the scapula, and the inferior portion depresses the scapula.

Beneath the trapezius are the *levator scapulae* (skap′-u-li), which elevates the scapula, and the *rhomboideus* (rom-boi′de-us) *major* and the *rhomboideus minor.* Both rhomboideus muscles adduct the scapula; the rhomboideus major also rotates it upward slightly. Still deeper is the *serratus* (ser-a′tus) *anterior,* which extends from the ribs to the scapula. When the ribs are fixed, it pulls the scapula forward and is used in reaching for objects. When the scapula is fixed, it elevates the ribs.

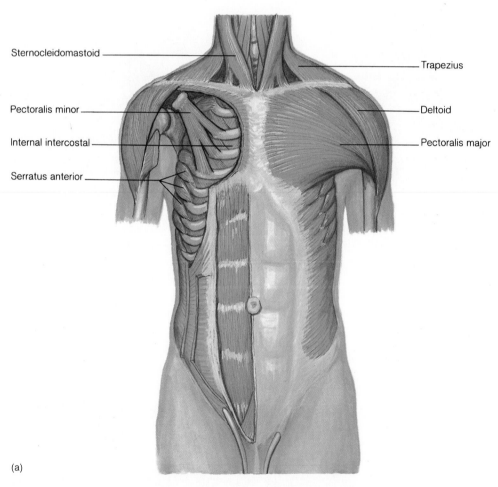

(a)

FIGURE 10.17
Muscles of the chest and shoulder.

Anterior to the pectoral girdle is the *pectoralis* (pek''to-ra'lis) *minor,* which extends from the ribs to the coracoid process of the scapula. It is smaller than, and lies mostly beneath, the pectoralis major (discussed below). When the ribs are stationary, the pectoralis minor depresses the scapula and rotates it anteriorly. When the scapula is stationary, it elevates the ribs.

Several large muscles act to cause movement at the shoulder joint. The large chest muscle, the *pectoralis major,* extends from the clavicle, sternum, and ribs to the humerus. It flexes, adducts, and medially rotates the arm.

Part or all of this muscle is sometimes congenitally absent. It is removed in a radical mastectomy, a surgical treatment for breast cancer. The *deltoid,* a triangular muscle that forms the fleshy part of the shoulder, extends from the clavicle and scapula to the humerus. It abducts the arm. The *latissimus dorsi,* a broad, flat, sheetlike muscle, extends from the lower thoracic and lumbar vertebrae, sacrum, and iliac crest to the humerus. It extends, adducts, and medially rotates the arm. Swinging the arm along side the body requires alternate contraction of the pectoralis major and the latissimus dorsi.

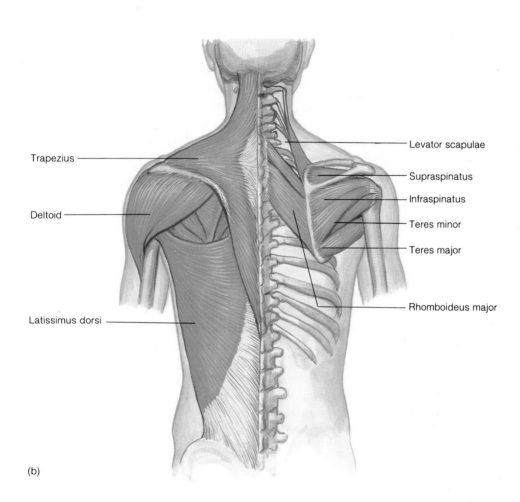

Trapezius

Deltoid

Latissimus dorsi

Levator scapulae

Supraspinatus

Infraspinatus

Teres minor

Teres major

Rhomboideus major

(b)

Beneath the deltoid, four muscles attached to various scapular surfaces maintain stability and help to move the shoulder (refer back to Figure 10.12). The *supraspinatus* (soo''prah-spi-na'tus), which lies superior to the spine of the scapula, assists the deltoid in abducting the arm. The *infraspinatus* (in''frah-spi-na'tus) lies inferior to the spine of the scapula; it rotates the arm laterally. The *subscapularis* (sub-skap''u-la'ris) occupies most of the undersurface of the scapula and helps to rotate the arm. The *teres* (te'rēz) *minor* lies inferior to the infraspinatus and also rotates the arm laterally. Together the supraspinatus, infraspinatus, subscapularis, and teres minor form the rotator cuff, which stabilizes and strengthens the shoulder joint. The *teres major* lies between the inferior border of the scapula and a point distal to the lesser tubercle of the humerus. It assists with extension, adduction, and medial rotation of the arm.

Arm Muscles

Muscles of the upper arm (Figure 10.18) generally have their origins on the scapula or humerus, and their insertions on the radius or ulna. The *biceps brachii* (bi'seps bra'ke-i), which lies between the scapula and the radius, flexes and supinates the forearm. The *brachioradialis* (bra''ke-o-ra-de-a'lis) is a synergist of the biceps. The *brachialis* (bra-ke-al'is) runs from the humerus to the ulna beneath the biceps and flexes the forearm. The *triceps brachii* originates on the scapula and humerus, and inserts by a large tendon on the olecranon process. It extends the forearm. Two small muscles, the *supinator* (soo-pin-a'tor) and the *pronator* (pro-na'tor) *teres,* run from the humerus and ulna to the radius. As their names imply, they supinate and pronate the forearm, respectively.

Wrist, Hand, and Finger Muscles

A sample of the many muscles of the wrist and hand are shown in Figure 10.19. The six discussed here have their origins on the humerus, though they have little action at the elbow. Because many wrist and hand muscles originate above the elbow, broken bones in the wrist often require a cast that extends above the elbow to immobilize wrist muscles.

Two flexors on the anterior forearm, the lateral *flexor carpi radialis* (flek'sor kar'pe ra''de-a'lis) and the more medial *flexor carpi ulnaris* (ul-na'ris), run parallel to the radius and ulna, respectively, and both flex the wrist (carpi). The radialis also abducts the hand and the ulnaris adducts it.

Two extensors (antagonists of the flexors) on the posterior forearm, the lateral *extensor carpi radialis longus* and the *extensor carpi ulnaris,* extend the wrist. They also act synergistically with the flexors, the lateral muscles abducting and the medial muscles adducting the hand.

Among the muscles acting on the hand and fingers, the *flexor digitorum profundus* (dig-it-or'um pro-fun'dus) flexes the wrist and phalanges, and the *extensor digitorum* extends the wrist and phalanges. Several muscles intrinsic to (entirely within) the hand control flexion and extension of joints between phalanges and allow spreading and closing of the fingers. Muscles of the thumb allow a wide range of rotational movement.

Pelvic Girdle and Hip Muscles

As noted earlier (Figure 10.13) some muscles extend from the pelvic girdle superiorly along the vertebral column. These muscles support the trunk and help to maintain posture. They also stabilize the pelvic girdle and help to keep it firmly attached to the sacrum.

Most muscles that act at the hip joint (Figure 10.20) originate on the ilium or pubic bone, and insert on the femur or tibia. The *iliopsoas* (il''e-o-so'as), sometimes considered to be two muscles, the *iliacus* (il-i'ak-us) and the *psoas major,* flexes the thigh. This muscle is often stretched or torn by the strenuous movements of ballet dancers. The *rectus femoris* (fem'or-is) is also a flexor of the thigh. Because its insertion is on the tibia by way of the patellar tendon, it is also an extensor of the leg.

The *gluteus maximus* (glu-te'us mak'sim-us), the heaviest muscle in the body, forms most of the buttock. It extends the thigh and rotates the femur laterally. Anterior to, and partly beneath, the gluteus maximus are the *gluteus medius* (me'de-us) and the *gluteus minimus* (min'im-us). Both abduct the thigh and help to stabilize the femur in the acetabulum. The *tensor fasciae latae* (ten'sor fash'e-e la'te), a muscle imbedded in the fascia on the lateral surface of the hip, also abducts the thigh.

A group of three muscles on the medial thigh, the superficial *adductor longus* (ah-duk'tor long'gus), and the deeper *adductor magnus* (mag'nus) and the *adductor brevis* (bre'vis), originate on the pubis and pull the femur toward the midline. As the names indicate, these muscles adduct the thigh. A small superficial, straplike muscle, the *gracilis* (gras'il-is), adducts the thigh and flexes the leg as well.

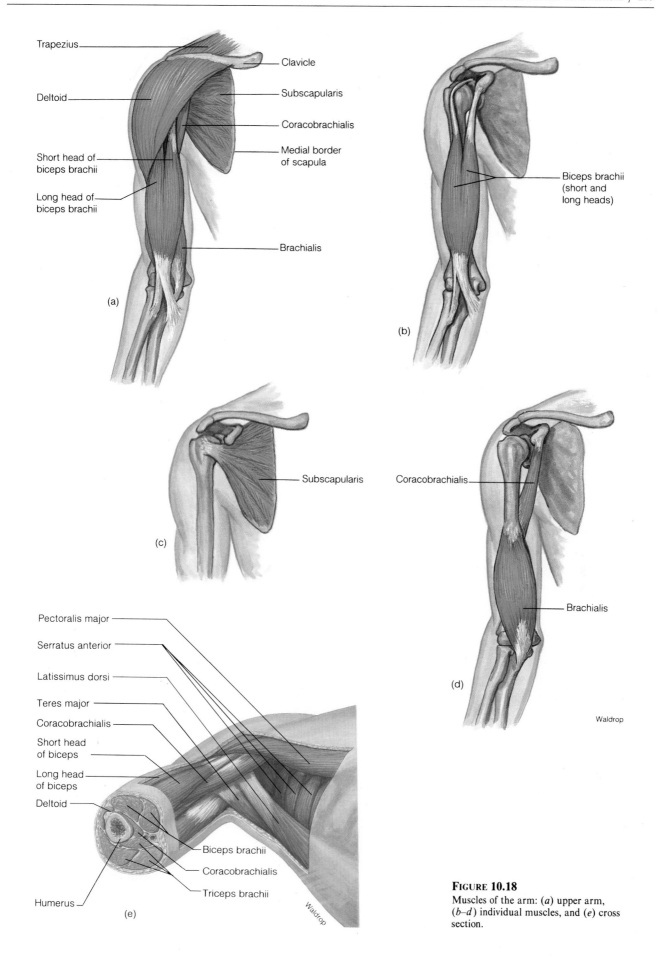

Trapezius

Clavicle

Deltoid

Subscapularis

Coracobrachialis

Short head of biceps brachii

Medial border of scapula

Long head of biceps brachii

Brachialis

(a)

Biceps brachii (short and long heads)

(b)

Subscapularis

(c)

Coracobrachialis

Brachialis

(d)

Waldrop

Pectoralis major

Serratus anterior

Latissimus dorsi

Teres major

Coracobrachialis

Short head of biceps

Long head of biceps

Deltoid

Biceps brachii

Coracobrachialis

Humerus

Triceps brachii

(e)

Waldrop

FIGURE 10.18

Muscles of the arm: (*a*) upper arm, (*b*–*d*) individual muscles, and (*e*) cross section.

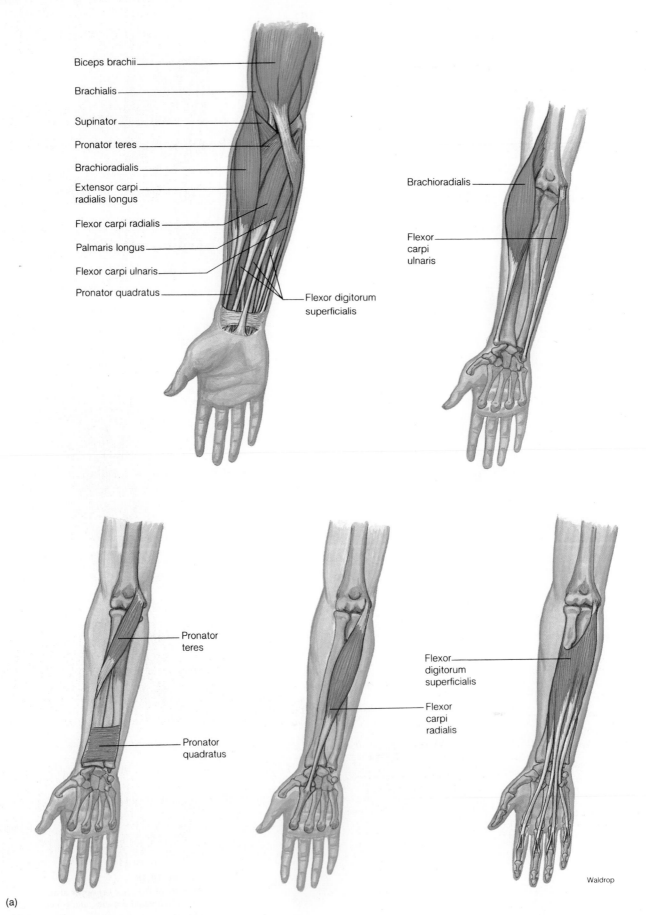

(a)

FIGURE 10.19

Muscles that move the wrist and hand: (*a*) anterior view, (*b*) posterior view, and (*c*) cross section.

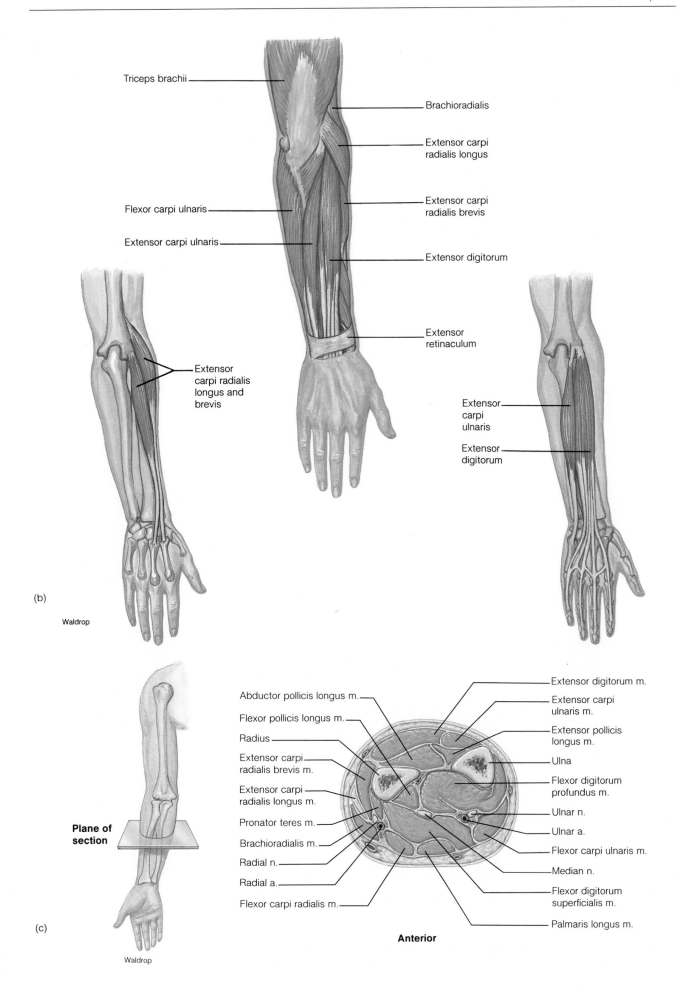

Triceps brachii

Brachioradialis

Extensor carpi radialis longus

Flexor carpi ulnaris

Extensor carpi radialis brevis

Extensor carpi ulnaris

Extensor digitorum

Extensor carpi radialis longus and brevis

Extensor retinaculum

Extensor carpi ulnaris

Extensor digitorum

(b)

Waldrop

Plane of section

(c)

Waldrop

Abductor pollicis longus m.

Flexor pollicis longus m.

Radius

Extensor carpi radialis brevis m.

Extensor carpi radialis longus m.

Pronator teres m.

Brachioradialis m.

Radial n.

Radial a.

Flexor carpi radialis m.

Extensor digitorum m.

Extensor carpi ulnaris m.

Extensor pollicis longus m.

Ulna

Flexor digitorum profundus m.

Ulnar n.

Ulnar a.

Flexor carpi ulnaris m.

Median n.

Flexor digitorum superficialis m.

Palmaris longus m.

Anterior

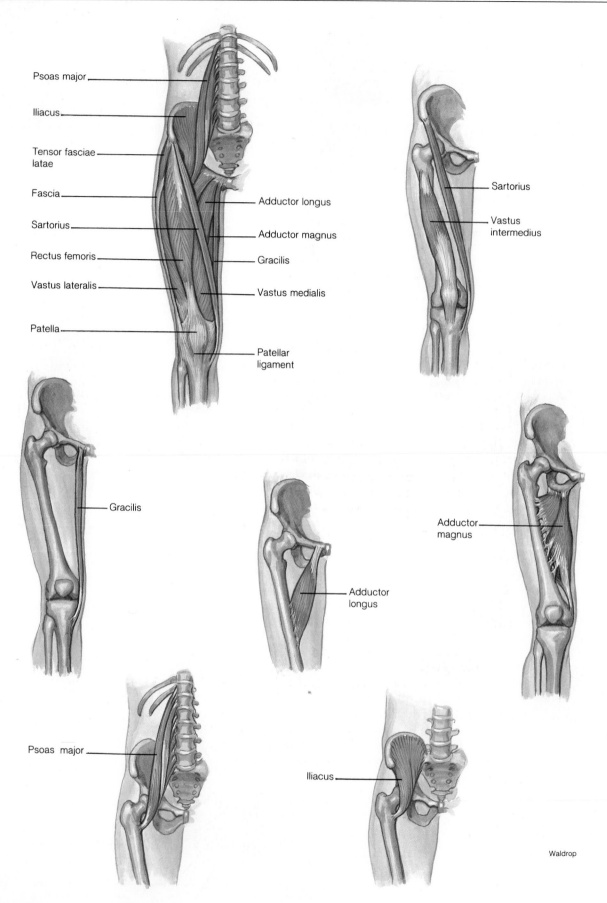

Psoas major

Iliacus

Tensor fasciae latae

Fascia

Sartorius

Rectus femoris

Vastus lateralis

Patella

Adductor longus

Adductor magnus

Gracilis

Vastus medialis

Patellar ligament

Sartorius

Vastus intermedius

Gracilis

Adductor longus

Adductor magnus

Psoas major

Iliacus

Waldrop

(a)

FIGURE 10.20
Muscles of the right pelvic girdle and thigh: (*a*) anterior view,
(*b*) posterior view, and (*c*) cross section.

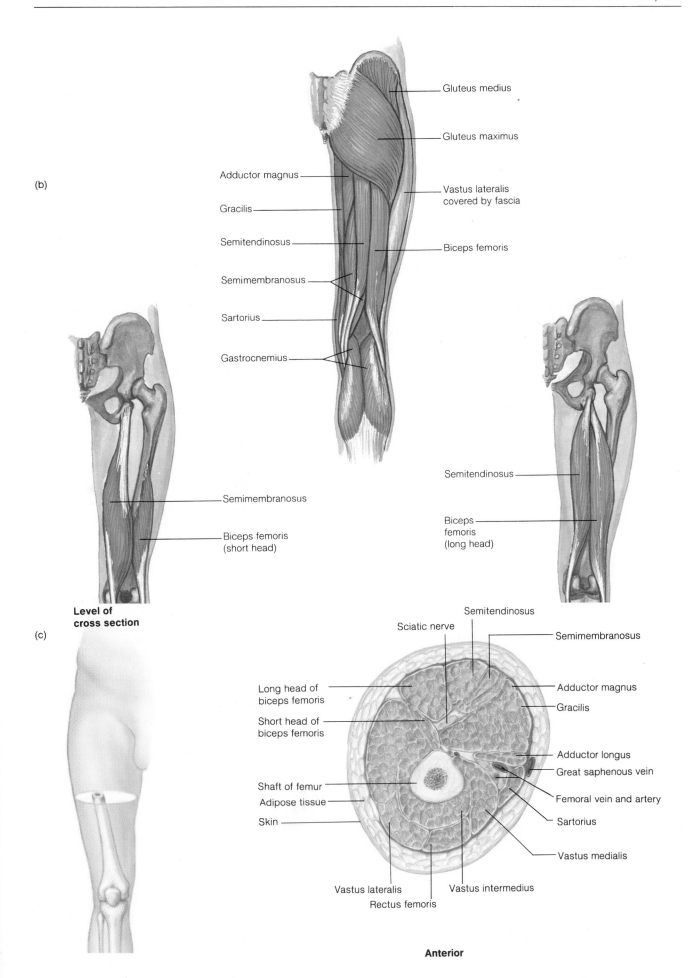

(b)

Gluteus medius

Gluteus maximus

Adductor magnus

Vastus lateralis
covered by fascia

Gracilis

Semitendinosus

Biceps femoris

Semimembranosus

Sartorius

Gastrocnemius

Semimembranosus

Semitendinosus

Biceps femoris
(short head)

Biceps
femoris
(long head)

**Level of
cross section**

(c)

Semitendinosus

Sciatic nerve

Semimembranosus

Long head of
biceps femoris

Adductor magnus

Gracilis

Short head of
biceps femoris

Adductor longus

Great saphenous vein

Shaft of femur

Femoral vein and artery

Adipose tissue

Sartorius

Skin

Vastus medialis

Vastus lateralis

Vastus intermedius

Rectus femoris

Anterior

The large gluteus maximus muscle of the buttock provides an extensive absorption area and is frequently used as the site of intramuscular drug injections. Because the sciatic nerve and its branches are distributed through the medial portion of the muscle, drugs should be injected only into superiolateral portion of the muscle (Figure 10.21).

Leg Muscles

The *quadriceps femoris* (kwod're-seps fem'or-is)—which includes the rectus femoris (see previous text), *vastus lateralis* (vas'tus lat"er-a'lis), *vastus intermedius* (in"ter-me'de-us), and *vastus medialis* (me"de-a'lis)—shown in Figure 10.20, flexes the thigh and extends the knee. The vastus intermedius is visible only in Figure 10.20c because it lies beneath the rectus femoris. The gracilis also flexes the leg and adducts the thigh, as noted previously. Another muscle of the anterior thigh is the *sartorius* (sar-to're-us), a straplike muscle that flexes both the thigh and the leg. It is used in crossing the legs and in assuming the cross-legged yoga or tailor's position.

The *hamstring* group of muscles occupy the posterior thigh. They are so-named because in a pig, their tendons are used to suspend a ham during curing. They include the *biceps femoris, semitendinosus* (sem"e-ten-din-o'sus),

and *semimembranosus* (sem"e-mem-bra-no'sus), shown in Figure 10.22. One head of the biceps originates on the ischium and the other on the femur. This muscle also has two insertions, one on the fibula and one on the tibia. Because of this pattern of origins and insertions, it acts across two joints; it extends the thigh and flexes the leg. The semitendinosus and the semimembranosus both extend the thigh.

Ankle and Foot Muscles

As with the wrist and hand, only a few of the many muscles of the ankle and foot are considered here (Figure 10.22). On the anterior surface of the leg, lateral to the tibia, is the *tibialis* (tib"e-a'lis) *anterior.* It dorsiflexes and inverts the foot. On the posterior surface in the calf of the leg is the large *gastrocnemius* (gas"trok-ne'me-us). It originates on the condyles of the femur and inserts on the calcaneus by way of the *Achilles tendon.* It extends the leg and plantarflexes the foot. Beneath the gastrocnemius is the *soleus* (so'le-us), also inserted by way of the Achilles tendon and also a plantarflexor of the foot. An even smaller muscle, the *tibialis posterior* (Figure 10.22c), extends, plantarflexes, and inverts the foot. Two muscles on the lateral surface of the leg, the *peroneus* (per"o-ne'us) *longus* and the *peroneus brevis,* plantarflex and evert the foot.

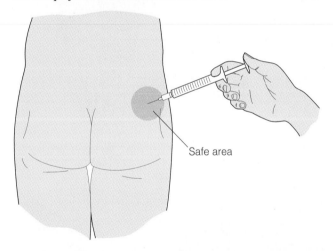

Safe area

FIGURE 10.21
The gluteus maximus muscle offers a large absorptive area for drug injection, but to avoid nerve injury, injections should be given only in the safe area indicated.

The Achilles tendon is named after the mythical Greek hero, Achilles. According to the legend, when Achilles was a baby, his mother dipped him in the river Styx, holding him by the heel of one foot. He became an invincible warrior until the Battle of Troy, when his heel, the one part of his body not protected by the river water, was pierced by an arrow. Tearing the Achilles tendon often occurs during sudden jumping and running in games such as tennis and squash. It is a serious athletic injury, not for the reasons offered in the myth, but because it renders the gastrocnemius muscle useless. Without the gastrocnemius, one is unable to stand on the injured leg. The action of this muscle through the lever of the foot (Figure 10.4b) keeps us from falling on our faces.

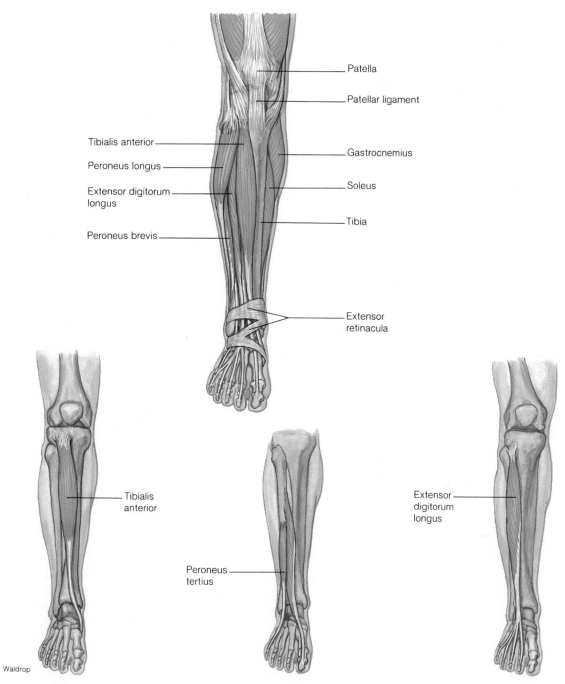

Patella

Patellar ligament

Tibialis anterior

Peroneus longus

Extensor digitorum
longus

Peroneus brevis

Gastrocnemius

Soleus

Tibia

Extensor
retinacula

Tibialis
anterior

Peroneus
tertius

Extensor
digitorum
longus

(a)   Waldrop

FIGURE 10.22
Muscles of the leg, ankle, and foot: (*a*) anterior view, (*b*) posterior
view, and (*c*) cross section.

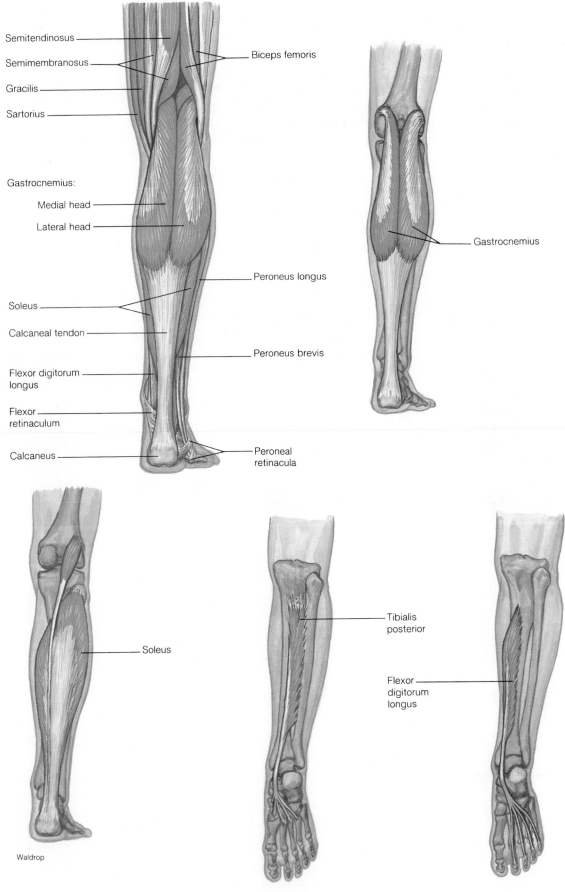

Semitendinosus

Semimembranosus

Gracilis

Sartorius

Biceps femoris

Gastrocnemius:

Medial head

Lateral head

Gastrocnemius

Peroneus longus

Soleus

Calcaneal tendon

Peroneus brevis

Flexor digitorum longus

Flexor retinaculum

Calcaneus

Peroneal retinacula

Soleus

Tibialis posterior

Flexor digitorum longus

Waldrop

(b)

FIGURE 10.22 Continued

TABLE OF MUSCLE ACTIONS

The major muscles of the human body are listed by body region in Table 10.1. The name, general location, origin, insertion, action, and innervation are given for each muscle. In studying muscles, it is important to use all clues from the muscle name to locate and contract the muscle on your own body, to group muscles with similar actions, and to compare them with groups of antagonistic muscles.

CLINICAL TERMS

fibromyositis (fi″bro-mi″o-si′tis)   inflammation of muscle and its accompanying fibrous tissue

fibrosis (fi-bro′sis)   formation of fibrous tissue, often to replace degenerating muscle tissue

fibrositis (fi″bro-si′tis)   inflammation and hyperplasia of fibrous sheaths surrounding skeletal muscle, usually accompanied by stiffness and pain

myalgia (mi-al′je-ah)   pain in muscles

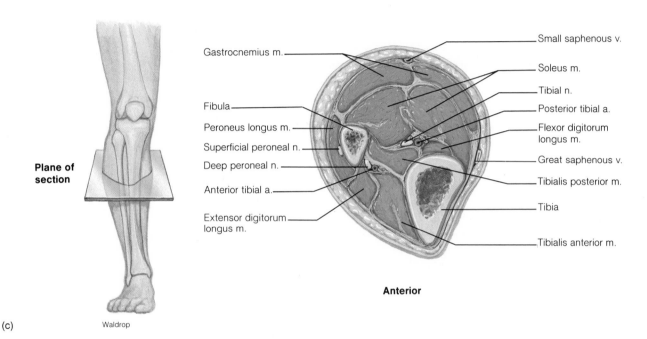

Plane of section

Gastrocnemius m.

Fibula

Peroneus longus m.

Superficial peroneal n.

Deep peroneal n.

Anterior tibial a.

Extensor digitorum longus m.

Small saphenous v.

Soleus m.

Tibial n.

Posterior tibial a.

Flexor digitorum longus m.

Great saphenous v.

Tibialis posterior m.

Tibia

Tibialis anterior m.

Anterior

(c)

Waldrop

FIGURE 10.22 Continued

TABLE 10.1   Muscle Characteristics and Actions

Muscle	Location	Origin	Insertion	Action	Innervation
Muscles of Facial Expression					
Occipitofrontalis	Forehead and scalp	Occipital bone	Muscles of eye orbit	Wrinkles forehead, raises eyebrows	Cranial nerve VII
Orbicularis oculi	Encircles eyelid	—	—	Closes eye	Cranial nerve VII
Orbicularis oris	Encircles mouth	—	—	Puckers lips	Cranial nerve VII
Platysma	Anterior neck	Fascia of deltoid and pectoralis major	Mandible	Draws corners of mouth down	Cranial nerve VII
Buccinator	Cheeks	Maxillae	Skin of sides of mouth	Used in smiling, sucking	Cranial nerve VII
Muscles of Mastication					
Masseter	Side of mandible	Zygomatic arch	Mandible	Closes jaw	Cranial nerve V
Temporal	Lateral skull	Temporal bone	Mandible	Closes jaw	Cranial nerve V
Pterygoids	Medial to ramus of mandible	Pterygoid process of sphenoid	Mandible	Protrudes jaw and moves it from side to side	Cranial nerve V
Muscles of the Tongue					
Genioglossus	Beneath tongue	Mandible	Tongue and hyoid bone	Depresses and protracts tongue	Cranial nerve XII
Styloglossus	Cheek	Styloid process of temporal	Tongue	Elevates and retracts tongue	Cranial nerve XII
Stylohyoid	Behind mandible	Styloid process of temporal	Hyoid bone	Elevates and retracts tongue	Cranial nerve VII
Hyoglossus	In and beneath tongue	Hyoid	Within tongue	Depresses tongue	Cranial nerve XII
Muscles of the Neck					
Sternocleidomastoid	Lateral neck	Sternum and clavicle	Mastoid process of temporal	Flexes and rotates head	Cranial nerve XI and cervical nerves 2, 3
Semispinalis capitis	Back of neck	Upper 6 thoracic and lower 4 cervical vertebrae	Occipital bone	Extends head and bends it laterally	Cervical nerves 1–5
Splenius capitis	Back of neck	Ligamentum nuchae	Mastoid process of temporal	Extends head	Cervical nerves 2–4
		Upper thoracic vertebrae	Occipital bone	Bends and rotates head	
Longissimus capitis	Lateral neck	Upper 6 thoracic and lower 4 cervical vertebrae	Mastoid process of temporal	Extends, bends, and rotates head	Middle and lower cervical nerves
Muscles of the Trunk					
Sacrospinalis	Posterior trunk			Extends spine, maintains posture	Cervical nerve 1 to lumbar nerve 5
Iliocostalis					
Lumborum		Iliac crest and lumbar vertebrae	Lower 6 ribs		
Thoracis		Lower 6 ribs	Upper 6 ribs		
Cervicis		Upper 6 ribs	Cervical vertebrae 4 to 6		
Longissimus thoracis		Iliac crest and lumbar vertebrae	Thoracic vertebrae and ribs		
Cervicis		Upper 6 thoracic vertebrae	Cervical vertebrae 2 to 6		
Capitis		(See neck muscles)			
Quadratus lumborum	Between ilium and last rib	Iliac crest	Twelfth rib	Extends or abducts spine	Thoracic nerve 12 and lumbar nerves 1, 2

Muscle	Location	Origin	Insertion	Action	Innervation
Muscles of Breathing					
Diaphragm	Between abdominal and thoracic cavities	Lower circumference of thorax	Central tendon	Increases volume of thoracic cavity	Phrenic nerves
External intercostals	Superficial, between ribs	Inferior border of rib	Superior border of inferior rib	Elevates ribs	Intercostal nerves
Internal intercostals	Deep, between ribs	Inner surface of rib	Superior border of inferior rib	Depresses ribs	Intercostal nerves
Muscles of the Abdominal Wall					
External oblique	Superficial, lateral abdomen	Lower 8 ribs	Iliac crest, linea alba, and pubis	Compresses abdomen	Intercostal nerves
Internal oblique	Middle layer, lateral abdomen	Iliac crest, fascia of back	Lower ribs, pubis, linea alba	Compresses abdomen	Intercostal nerves
Transverse abdominis	Inner layer, lateral abdomen	Lower ribs, iliac crest, fascia of back	Pubis and linea alba	Compresses abdomen	Intercostal nerves
Rectus abdominis	Medial abdomen	Pubis	Costal cartilage of ribs 5–7	Compresses abdomen and flexes trunk	Intercostal nerves
Muscles of the Pelvic Floor					
Levator ani	Pelvic floor	Pubis, ischium	Coccyx	Supports pelvic organs	Pudendal nerve
Coccygeus	Pelvic floor	Ischium	Coccyx, sacrum	Supports pelvic organs	Pudendal nerve
Muscles of the Pectoral Girdle					
Pectoralis minor	Deep, chest	Ribs 3–5	Coracoid process of scapula	Depresses scapula, rotates shoulder, elevates ribs	Pectoral nerve
Serratus anterior	Lateral thorax	Upper 8 ribs	Vertebral border of scapula	Rotates scapula, elevates ribs	Long thoracic nerve
Trapezius	Superficial, upper back	Occipital bone, ligamentum nuchae, vertebrae C7–T12	Spine of scapula, acromion process of clavicle	Elevates, adducts, or depresses scapula; elevates clavicle	Cranial nerve XI, cervical nerves 3 and 4
Levator scapulae	Posterior neck	Upper cervical vertebrae	Vertebral border of scapula	Elevates scapula	Cervical nerves
Rhomboideus major	Deep, upper back	Vertebrae T2–5	Vertebral border of scapula	Adducts and upwardly rotates scapula	Cervical nerves
Rhomboideus minor	Deep, upper back	C7 and T1 vertebrae	Superior angle of scapula	Adducts scapula	Cervical nerves
Muscles of the Shoulder Joint					
Pectoralis major	Superficial, chest	Clavicle, sternum, cartilage of ribs 2–6	Greater tubercle of humerus	Flexes, adducts, and medially rotates arm	Pectoral nerves
Latissimus dorsi	Superficial, lower back	Vertebrae T6–L5, lower ribs, iliac crest	Humerus	Extends, adducts, and medially rotates arm	Thoracodorsal nerve
Deltoid	Top of shoulder	Clavicle, acromion, and spine of scapula	Deltoid tuberosity of humerus	Abducts arm	Axillary nerve
Supraspinatus	Above spine of scapula	Supraspinous fossa	Greater tubercle of humerus	Abducts arm	Suprascapular nerves
Infraspinatus	Below spine of scapula	Infraspinous fossa	Greater tubercle of humerus	Laterally rotates arm	Suprascapular nerve
Teres major	Between scapula and humerus	Inferior angle of scapula	Lesser tubercle of humerus	Adducts, extends, and medially rotates arm	Lower subscapular nerves
Teres minor	Above teres major	Axillary border of scapula	Greater tubercle of humerus	Laterally rotates arm	Axillary nerve

TABLE 10.1 Continued

Muscle	Location	Origin	Insertion	Action	Innervation
Muscles of the Elbow Joint					
Biceps brachii	Anterior upper arm	Glenoid fossa and coracoid process	Radial tuberosity	Flexes and supinates forearm	Musculocutaneous nerve
Brachialis	Anterior upper arm	Lower anterior humerus	Coronoid process of ulna	Flexes pronated forearm	Musculocutaneous nerve
Brachioradialis	Lateral forearm	Upper lateral humerus	Styloid process of radius	Flexes forearm	Radial nerve
Triceps brachii	Posterior humerus	Scapula and humerus	Olecranon process of ulna	Extends lower arm	Radial nerve
Pronator teres	Anterior upper forearm	Medial epicondyle of humerus	Middle of radius	Pronates and flexes forearm	Median nerve
Supinator	Anterior upper forearm	Lateral epicondyle of humerus	Upper radius	Supinates forearm	Radial nerve
Muscles of the Wrist and Hand					
Flexor carpi radialis	Superficial anterior forearm	Medial epicondyle of humerus	Metacarpals 2 and 3	Flexes and abducts wrist	Median nerve
Flexor carpi ulnaris	Superficial anterior forearm	Medial epicondyle of humerus and distal ulna	Metacarpal 5 and carpals	Flexes and adducts wrist	Ulnar nerve
Extensor carpi radialis longus	Posterior forearm	Lateral epicondyle of humerus	Metacarpal 2	Extends and abducts wrist	Radial nerve
Extensor carpi ulnaris	Posterior forearm	Lateral epicondyle of humerus and ulna	Metacarpal 5	Extends and adducts wrist	Radial nerve
Flexor digitorum profundus	Deep, anterior forearm	Proximal, anteromedial ulna	Base of distal phalanges	Flexes wrist and phalanges	Median and ulnar nerves
Extensor digitorum	Posterior forearm	Lateral epicondyle of humerus	Middle and distal phalanges	Extends wrist and phalanges	Deep radial nerve
Muscles of the Thigh					
Iliopsoas (Iliacus and psoas major)	Iliac fossa and posterior pelvic cavity	Iliac fossa and vertebrae T12 to L5	Lesser trochanter of femur	Flexes thigh (or trunk if femur is fixed)	Femoral nerve and lumbar nerves 2–4
Gluteus maximus	Buttocks	Ilium, sacrum, and coccyx	Gluteal tuberosity of femur	Extends and laterally rotates thigh	Inferior gluteal nerve
Gluteus medius	Buttocks	Lateral surface of ilium	Greater trochanter of femur	Abducts and rotates thigh	Superior gluteal nerve
Gluteus minimus	Buttocks	Lateral surface of ilium	Greater trochanter of femur	Abducts and rotates thigh	Superior gluteal nerve
Tensor fasciae latae	Lateral hip	Anterior superior iliac spine	Fascia lata and tibia	Abducts thigh	Superior gluteal nerve
Adductor brevis	Medial thigh	Pubic bone	Upper linea aspera	Adducts thigh	Obturator nerve
Adductor longus	Medial thigh	Pubic bone	Middle linea aspera	Adducts thigh	Obturator nerve
Adductor magnus	Medial thigh	Pubic bone	Lower linea aspera	Adducts thigh	Obturator nerve
Gracilis	Medial thigh	Pubic bone	Medial head of tibia	Adducts thigh and flexes knee	Obturator nerve

Muscle	Location	Origin	Insertion	Action	Innervation
Muscles of the Knee Joint					
Sartorius	Anterior thigh	Anterior superior iliac spine	Medial head of tibia	Adducts thigh and flexes knee	Femoral nerve
Gracilis	Medial thigh	Pubic bone	Medial head of tibia	Adducts thigh and flexes knee	Obturator nerve
Quadriceps femoris group					
Rectus femoris	Superficial, anterior thigh	Anterior inferior iliac spine	Tibia via patellar tendon	Flexes thigh and extends lower leg	Femoral nerve
Vastus lateralis	Lateral thigh	Linea aspera of femur	Tibia via patellar tendon	Extends lower leg	Femoral nerve
Vastus intermedius	Deep, anterior thigh	Shaft of femur	Tibia via patellar tendon	Extends lower leg	Femoral nerve
Vastus medialis	Medial thigh	Linea aspera of femur	Tibia via patellar tendon	Extends lower leg	Femoral nerve
Hamstring group					
Biceps femoris	Posterior thigh	Ischial tuberosity and linea aspera	Lateral head of fibula	Extends thigh and flexes lower leg	Sciatic nerve
Semitendinosus	Posterior thigh	Ischial tuberosity	Medial shaft of tibia	Extends thigh and flexes lower leg	Sciatic nerve
Semimembranosus	Posterior thigh	Ischial tuberosity	Medial condyle of tibia	Extends thigh and flexes lower leg	Sciatic nerve
Muscles of the Ankle and Foot					
Tibialis anterior	Anterolateral lower leg	Lateral condyle of tibia	Cuneiform and metatarsal 1	Flexes and inverts foot	Deep peroneal nerve
Gastrocnemius	Superficial, calf of leg	Posterior surface of condyles of femur	Calcaneus via Achilles tendon	Flexes lower leg and extends foot	Tibial nerve
Soleus	Deep, calf	Proximal tibia and head of fibula	Calcaneus via Achilles tendon	Extends (plantarflexes) foot	Tibial nerve
Peroneus longus	Lateral leg	Lateral condyle of tibia and head of fibula	Cuneiform and metatarsal 1	Plantarflexes and everts foot	Common peroneal nerve
Peroneus brevis	Lateral leg	Lower lateral shaft of fibula	Metatarsal 5	Plantarflexes and everts foot	Common peroneal nerve
Tibialis posterior	Posterior leg	Posterior surface of tibia and fibula	Tarsals	Extends (plantarflexes) and inverts foot	Tibial nerve

ESSAY

MUSCLE INJURIES HEAD TO FOOT

People often speak of having a Charley horse or shin splints, but do you really know what such injuries entail? If you are like many students, you vaguely associate such injuries with excessive muscle strain from strenuous activity. Let's find out more about these and other muscle injuries, proceeding through the body from head to foot (Figure 10.23).

Some trumpet players cannot help having their cheeks balloon when they play their instruments. Balloon cheeks are due to excessive stretching of buccinator and other cheek muscles from prolonged forcible blowing on the trumpet.

Perhaps you have gotten up in the morning to find that your head is turned slightly to one side and rotating your head is painful or impossible. If so, you have suffered from **torticollis** (tor″tik-ol′is), or wryneck, because of a temporary contracture of the sternocleidomastoid muscle. Your contracture will likely disappear in the next day or so, but many cases of torticollis are congenital (present from birth) and are often due to excessive pulling on the neck during a difficult delivery.

Many people have hernias, or separations between the fibers of muscles, as a result of heavy lifting, obesity, or lack of exercise. Other hernias are congenital or due to unknown causes.

Inguinal hernias (ing′win-al her′ne-az), which account for 90 percent of all hernias, are ruptures in groin area (the weakest part of the abdominal wall in males). Three-fourths of such hernias occur in male infants and children, and only a few occur in females of any age. Protrusion of a loop of intestine through an inguinal hernia, sometimes into the male scrotum, is a serious condition called a **strangulated hernia.** Such hernias usually require immediate surgical repair to prevent intestinal blockage.

Hiatal and umbilical hernias also are relatively common. A **hiatal** (hi-a′tal) **hernia** is a rupture in the diaphragm through which the fundus (superior portion) of the stomach protrudes into the chest cavity. Pain and regurgitation of acid are common symptoms of hiatal hernias. Most hiatal hernias appear in overweight people over 40 and only one infant in 2000 has a congenital hiatal hernia. **Umbilical** (um-bil′ik-al) **hernias,** which usually are congenital, result from incomplete closure of the abdominal wall after the umbilical cord is ligated at birth.

Dupuytren's contracture (de-pwe-trahnz′ kon-trak′tūr), a progressive increase in fibrous tissue in the aponeurosis of the palm of the hand, shortens tendons and causes marked flexion at the metacarpophalangeal joints. Though the cause of this disorder is unknown, predisposition toward it appears to be inherited in some families.

Paralysis and atrophy of muscles that abduct the thumb, called *ape hand,* prevents the thumb from touching the fingers. This condition often develops as a result of nerve damage following wrist slashing in suicide attempts.

A *Charley horse* is a bruising and tearing of muscle fibers in which a **hematoma** (hem″at-o′mah), a local mass of blood that has escaped from tissues, develops. Pain and stiffness are main symptoms. Though such injuries can occur in any muscle, they are most common in the quadriceps femoris and especially common in tackled football players.

Another common sports injury is a *hip pointer,* a bruise over the iliac crest. Hip pointers result from direct trauma and can be accompanied by hematoma in the sartorius muscle, the origin of which is on the anterior superior iliac spine.

Dancers and gymnasts are especially subject to *pulled groin,* a straining, stretching, or tearing of tendons at the origins of the adductor muscles. Practicing high kicks and splits often leads to pulled groin muscles unless stretching is done very gradually. Ossification of these tendons, called *rider's bones,* often occurs in horse riders, whose thighs are adducted for long periods of time.

Pulled hamstrings, which result from violent kicking and hard running, are due to tearing of muscles and tendons, and rupture of blood vessels. A hematoma often develops in the fascia lata. Sudden pulling of a hamstring tendon can cause a runner to drop and writhe with pain. Contractures of hamstring tendons are common complications of knee joint disorders. Such contractures keep the knee partially flexed and the tibia slightly rotated laterally.

Pain and swelling along the anterior leg, or *shin splints,* typically begin after prolonged, vigorous exercise. This condition is due to swelling and reduced blood flow in the tibialis anterior muscle. It is common in northern climates after the first use of heavy snowshoes each winter.

Repeated kicking of a football or soccer ball irritates the anterior distal surface of the tibia. In some cases, *footballer's ankle,* a bony outgrowth protruding from the tibia, develops.

High heel syndrome causes pain, not while people are wearing high heels, but when they wear low heels or go barefoot. Persistent wearing of high heels leads to shortening of calf muscles and their tendons. Attempting to place the foot in a more natural position stretches the shortened structures and causes pain.

Having completed our survey of muscle injuries from head to foot, it seems appropriate to point out that most such injuries can be avoided by using moderation in both the intensity and duration of sports participation. We must eradicate the often expressed view, "no pain, no gain," and replace it with the knowledge that gradual increases in activity can increase muscle strength and endurance. Exercise is essential to good health, but sufficient exercise can be obtained without pushing muscles beyond their physiological limits.

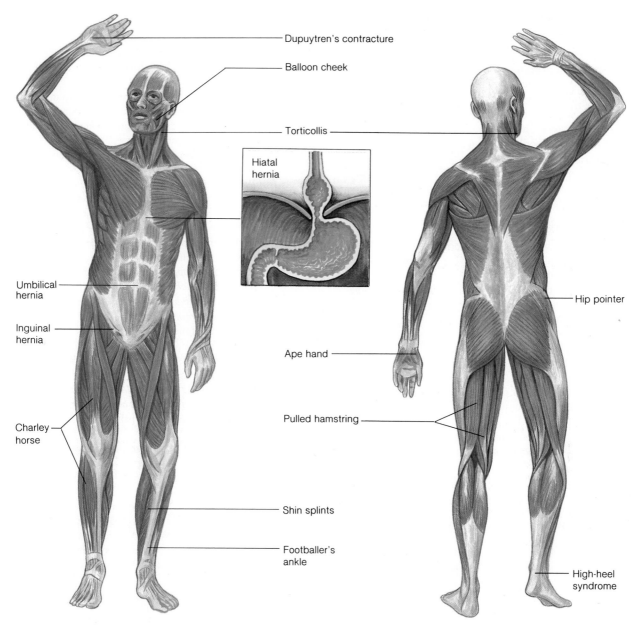

FIGURE 10.23
Location of muscle injuries.

Questions

1. Distinguish the different kinds of hernias.
2. What is a contracture?
3. Relate at least five muscle injuries to particular occupations or sports.
4. What is wrong with the expression, "no pain, no gain"?

CHAPTER SUMMARY

Overview of Muscles
- The many skeletal muscles are grouped by body region.

Development of Skeletal Muscle
- Most skeletal muscles develop from myotomes, each of which is associated mainly with a single spinal nerve.

Properties of Muscles
- The bones and joints form a complex set of levers that are moved by the contraction of muscles.
- Various kinds of levers allow the body or its parts to move.
- Each muscle has its origin at the least movable attachment and its insertion at the most movable attachment.
- Muscle actions describe the kind of movement a muscle causes (refer back to Table 8.2).
- Main actions are performed by agonists (prime movers), aided by synergists, and opposed by antagonists.
- Several anatomical characteristics are used to name muscles.

Muscles of the Axial Skeleton
- Muscles of the head and neck perform functions such as facial expression, mastication, moving the tongue, and moving the head.
- Muscles of the trunk include back muscles that move the back and maintain posture, muscles of breathing, abdominal muscles, and muscles of the pelvic floor.

Muscles of the Appendicular Skeleton
- Muscles of the pectoral girdle move the girdle and hold it to the axial skeleton.
- Muscles of the shoulder cause movements such as abduction, adduction, flexion, extension, and circumduction of the arm.
- Muscles of the elbow flex, extend, and rotate the forearm.
- Muscles of the wrist and hand are quite numerous; they cause various movements of the wrist and move the joints of the fingers and thumb.
- Muscles of the pelvic girdle hold the girdle firmly to the axial skeleton.
- Muscles of the hip cause movements such as abduction, adduction, flexion, extension, and circumduction of the thigh and leg.
- Muscles of the knee flex and extend the leg.
- Muscles of the ankle and foot flex and extend the foot, and perform some rotational movements; they also cause limited movement of the toes.

Table of Muscle Actions
- The actions and other selected attributes of the muscles discussed in this chapter are summarized in Table 10.1.

QUESTIONS AND PROBLEMS

The questions at the end of each chapter are numbered to correspond with the objectives listed at the beginning of the chapter. Italics indicate that a question requires critical thinking skills beyond simple factual recall.

Questions
1. What are the major groups of skeletal muscles?
2. Briefly describe how skeletal muscles develop?
3. *How can principles of levers be used to explain muscle action?*
4. (a) Define the boldface terms pertaining to muscle action.
   (b) What characteristics are used in naming muscles?
5. (a) List the main muscles of facial expression, and give their locations and actions.
   (b) Use the actions of muscles to explain how mastication occurs.
   (c) Which muscles assist in moving the tongue and pushing food to the back of the mouth?
   (d) Which neck muscles are involved in flexion, extension, and rotation?
6. (a) Which muscles are involved in flexing, extending, and rotating the trunk?
   (b) *If a person stopped breathing because of muscle paralysis, which muscles would have been affected?*
   (c) *How do the abdominal muscles function to flex the spine and compress the abdomen?*
   (d) What is the function of the muscles of the pelvic floor?
7. (a) Which muscles hold the pectoral girdle to the axial skeleton?
   (b) *Which muscles and muscle actions are involved in throwing a ball? Opening a door? Writing?*
8. (a) Which muscles hold the pelvic girdle to the axial skeleton?
   (b) *Which muscles and muscle actions are involved in kicking a ball? Walking up a stairway? Standing on tiptoe? Getting out of bed? Jogging?*

Problems
1. Select a specific injury that is frequently sustained by people who engage in your favorite sport. Use your knowledge of the skeletomuscular system and the sport to suggest ways of minimizing the chance of such an injury occurring.
2. Use what you have learned about muscle actions and exercise to devise an exercise program that would improve your physical condition and from which you would get some satisfaction.

The following set of illustrations includes muscle dissections of human cadavers. These photographs will help you visualize the spatial and proportional relationships between the major muscles of actual specimens. The photographs can also serve as the basis for a review of the information you have gained from your study of the human organism.

HUMAN CADAVER MUSCLES

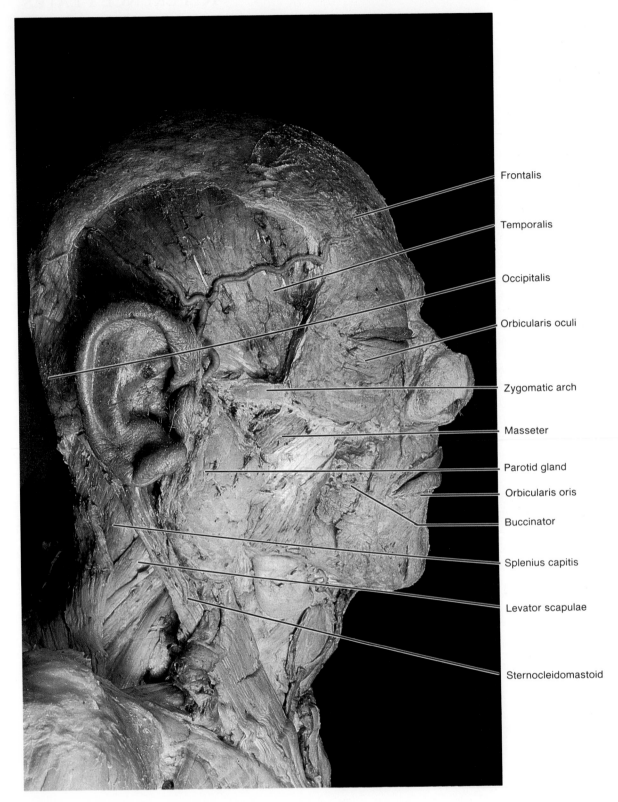

Frontalis

Temporalis

Occipitalis

Orbicularis oculi

Zygomatic arch

Masseter

Parotid gland

Orbicularis oris

Buccinator

Splenius capitis

Levator scapulae

Sternocleidomastoid

PLATE 1
Lateral view of the head.

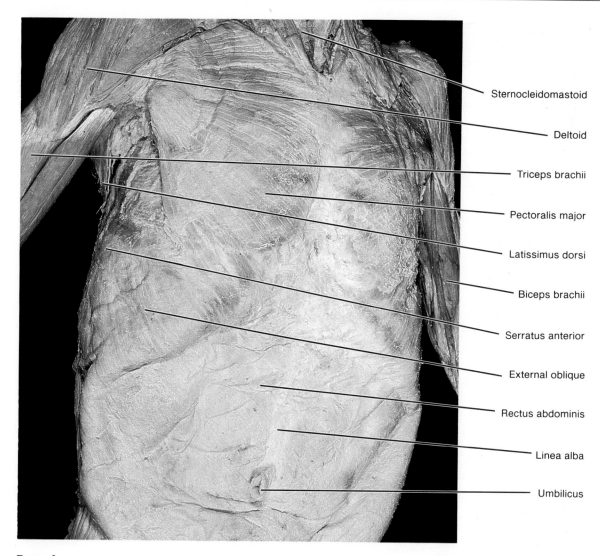

Sternocleidomastoid

Deltoid

Triceps brachii

Pectoralis major

Latissimus dorsi

Biceps brachii

Serratus anterior

External oblique

Rectus abdominis

Linea alba

Umbilicus

PLATE 2
Anterior view of the trunk.

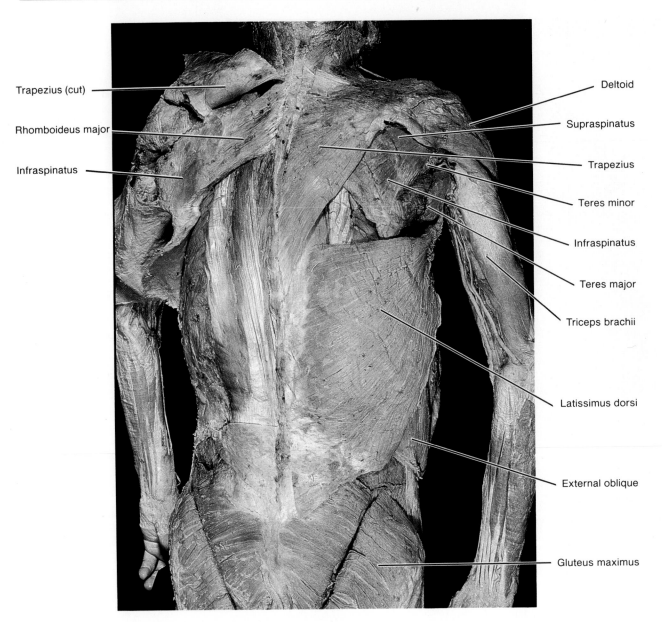

Trapezius (cut)

Rhomboideus major

Infraspinatus

Deltoid

Supraspinatus

Trapezius

Teres minor

Infraspinatus

Teres major

Triceps brachii

Latissimus dorsi

External oblique

Gluteus maximus

PLATE 3
Posterior view of the trunk, with deep thoracic muscles exposed on the left.

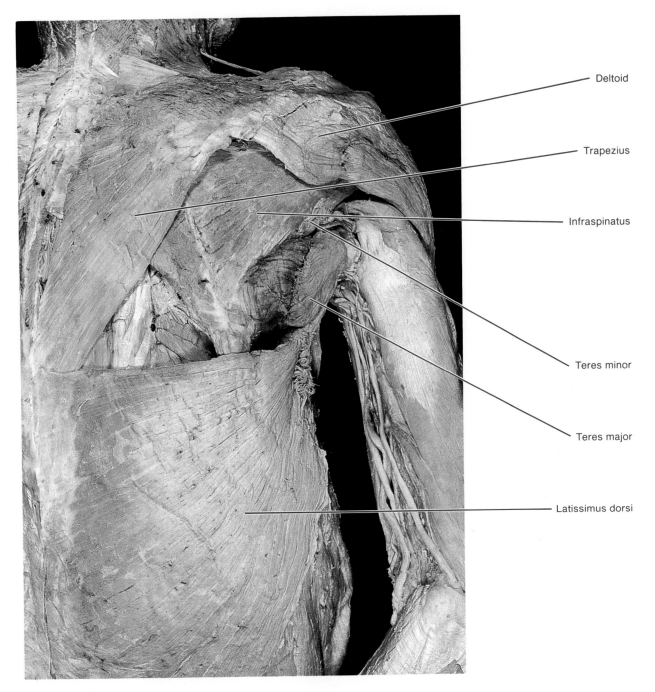

Deltoid

Trapezius

Infraspinatus

Teres minor

Teres major

Latissimus dorsi

PLATE 4
Posterior view of the right thorax and upper arm.

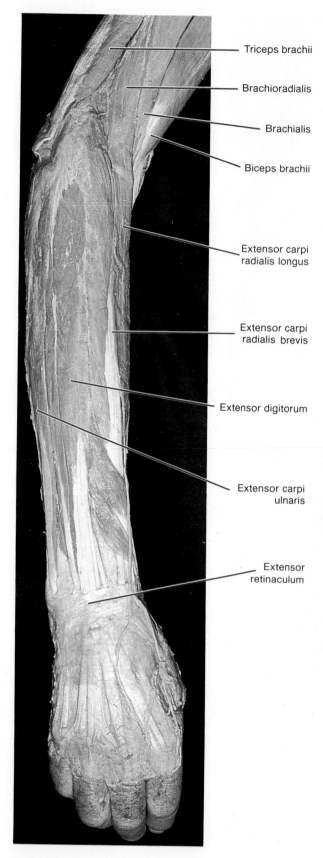

Triceps brachii

Brachioradialis

Brachialis

Biceps brachii

Extensor carpi radialis longus

Extensor carpi radialis brevis

Extensor digitorum

Extensor carpi ulnaris

Extensor retinaculum

PLATE 5
Posterior view of the right forearm and hand.

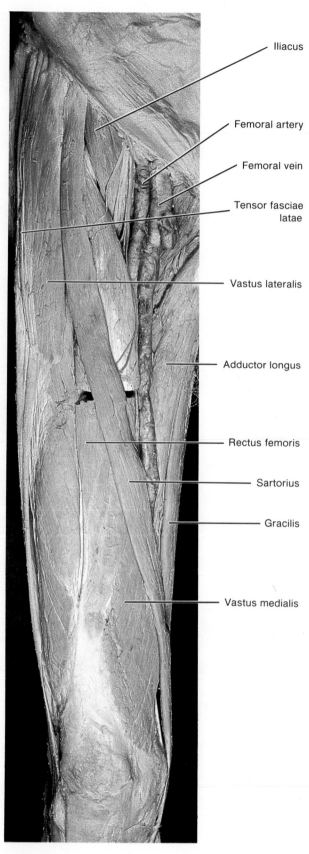

Iliacus

Femoral artery

Femoral vein

Tensor fasciae latae

Vastus lateralis

Adductor longus

Rectus femoris

Sartorius

Gracilis

Vastus medialis

PLATE 6
Anterior view of the right thigh.

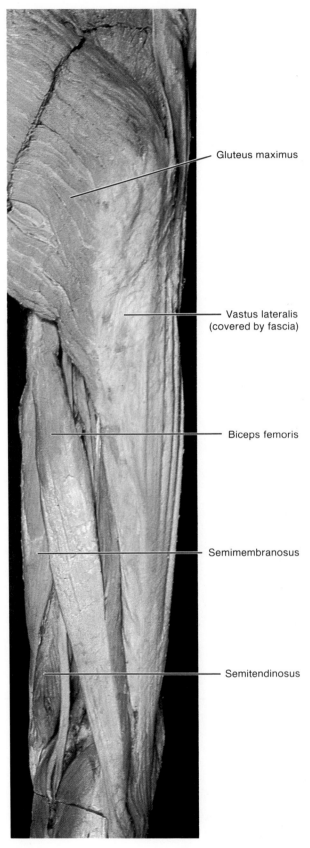

Gluteus maximus

Vastus lateralis
(covered by fascia)

Biceps femoris

Semimembranosus

Semitendinosus

PLATE 7
Posterior view of the right thigh.

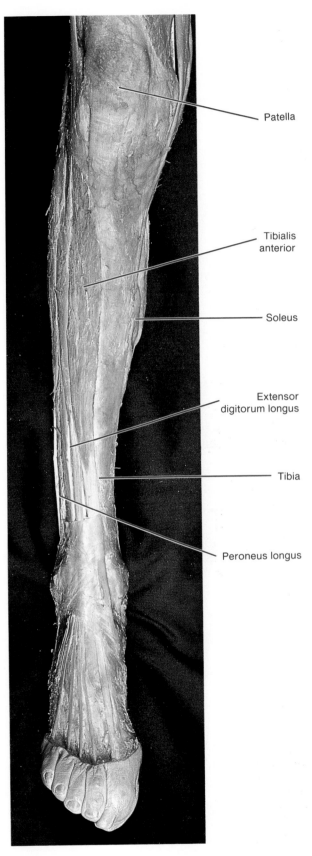

Patella

Tibialis
anterior

Soleus

Extensor
digitorum longus

Tibia

Peroneus longus

PLATE 8
Anterior view of the right leg.

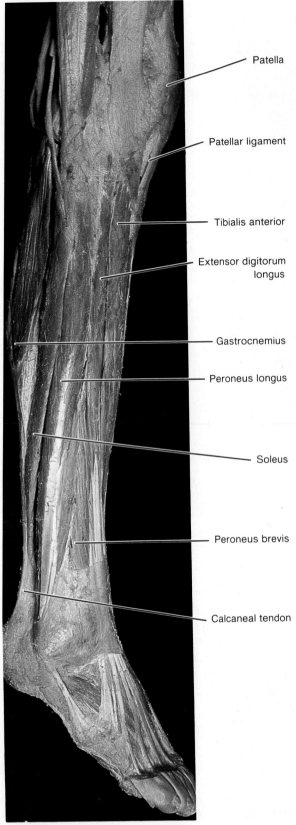

Patella

Patellar ligament

Tibialis anterior

Extensor digitorum
longus

Gastrocnemius

Peroneus longus

Soleus

Peroneus brevis

Calcaneal tendon

PLATE 9
Lateral view of the right leg.

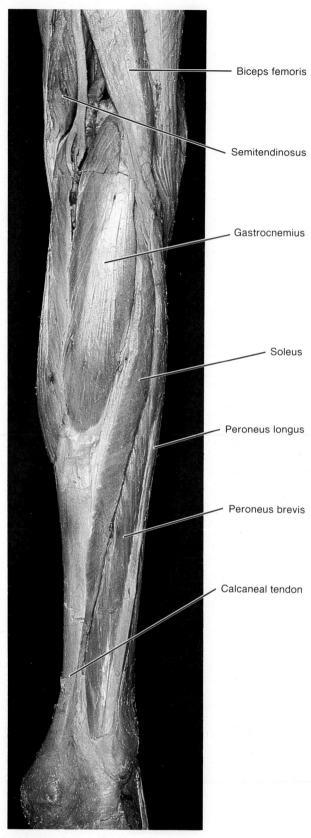

Biceps femoris

Semitendinosus

Gastrocnemius

Soleus

Peroneus longus

Peroneus brevis

Calcaneal tendon

PLATE 10
Posterior view of the right leg.

Exercise is essential for maintaining physical fitness. Many years ago, most people were physically fit because they worked hard every day, and only athletes were concerned with physical fitness. Today many people are working to improve their physical fitness (Figure I2.1). The number of Americans engaging in regular exercise continues to increase—more than 30 million jog, 10 to 20 million swim, and millions more cycle. Unfortunately, the physical education programs in some schools and colleges still emphasize competitive sports where a few make the team and everybody else watches (Figure I2.2). Such programs should teach activities—swimming, tennis, cycling, squash, handball, and the like—to help every student find enjoyable activities for maintaining lifetime fitness. Equally unfortunate is the fact that many adults suffer from back pain and many develop heart disease, in part because of a sedentary lifestyle and a fatty diet.

The key to fitness, it has been said, is to stay active without running yourself ragged—to avoid the extremes of no exercise and dangerously strenuous exercise. Sports physiologists and fitness specialists recognize four categories of physical fitness: cardiorespiratory endurance, body composition, flexibility, and muscle strength. *Cardiorespiratory endurance* concerns the ability of the heart and lungs to support prolonged muscular activity. *Body composition* refers to the proportion of the body composed of fat. *Flexibility* pertains to the ability to move body parts through their full range of motion—describing a large, full circle with each arm and leg, for example. *Muscle strength* refers to the ability to use muscles to exert force.

Aerobic exercise, activity in which the respiratory rate increases but cells still obtain adequate oxygen, is suitable for improving cardiorespiratory endurance and reducing excess body fat. Stretching exercises are needed to increase flexibility. A work-out on an exercise machine or with weights or calisthenics such as push-ups and sit-ups is needed to increase muscle strength. Carrying hand weights while walking or running increases both endurance and strength. Stretching against water resistance while swimming is an excellent fitness activity, except that some weight-bearing exercise also may be needed to reduce the risk of osteoporosis, especially in women.

FIGURE I2.1
People exercise in different ways to maintain physical fitness.

FIGURE I2.2
Only a few people maintain fitness at many sports events.

AN EXERCISE PROGRAM

To achieve fitness, the American College of Sports Medicine recommends the following exercise program, but cautions that a person adopting such a program should first obtain the approval of a physician.

1. Select a rhythmic activity that uses large muscle groups and perform it at aerobic pace. Walking, jogging, swimming, aerobic dancing, skating, and cross-country skiing are examples of suitable activities.
2. Work hard enough to raise the heart rate to 60 to 85 percent of its maximum rate. The maximum human heart rate is about 220 minus the person's age. A 20-year-old student who wishes to exercise at 85 percent of the maximum heart rate would find that value as follows: $220 - 20 = 200$; $0.85 \times 200 = 170$. For a 50-year-old slightly overweight professor exercising at 65 percent of the maximum heart rate, the rate is 110. Some studies suggest that the body burns relatively more fat during exercise at 60 to 75 percent of maximum heart rate than at higher levels.

EXERCISE AND PHYSICAL FITNESS

3. Perform 20 to 60 minutes of continuous activity depending on the intensity of the activity. For example, 20 minutes at 85 percent of maximum heart rate is approximately equivalent to 50 minutes at 60 percent. Working too hard can do more harm than good, so it is important to check the pulse rate five minutes after the exercise period begins. By this time, the heart rate should have dropped below 120 in people under age 50, and below 105 in people age 50 or over.

4. Exercise three to five days per week. It has been estimated that three sessions per week over several months bring the body to about 70 percent of maximum performance and five sessions per week bring it to 97 percent of maximum performance. Exercising more than five times per week does not significantly improve fitness. Exercising less than three times per week is of minimal value, but, of course, any exercise is better than no exercise.

A healthful exercise program includes one-half hour of aerobic activity three times per week and one hour of general activity—housework, yardwork, walking—every day. Aerobic exercise should begin with a five-minute warm-up and end with a five-minute cool-down period of stretching and strengthening exercises.

Effects of Exercise

Much of what is known about the effects of aerobic exercise has come from studies of distance runners. When these runners maintain a 50 to 60 percent heart rate, about half of their energy comes from glycogen and half from free fatty acids. At a heart rate greater than 85 percent, most energy comes from glycogen—at least as long as the glycogen supply lasts. A person with a lean body weighing 70 kg (154 lbs) has about 135,000 calories stored in fat and about 1400 stored in glycogen.

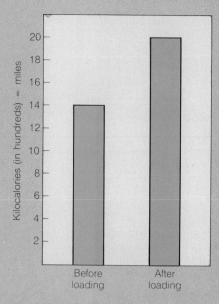

FIGURE I2.3
How carbohydrate loading affects available energy.

Carbohydrate loading, eating large quantities of starchy foods for a few days before a race, can increase the glycogen supply to about 2000 calories. At 100 calories per mile, this provides the runner with energy for an additional six miles at an 85 percent heart rate (Figure I2.3).

Athletes encounter an overwhelming fatigue variously referred to as "the wall," "the bear," or "the bonk." This fuel crisis occurs when glycogen stores are depleted and corresponds to a shift from metabolizing mainly glucose to metabolizing mainly free fatty acids from the degradation of fats stored in adipose tissue. Why this metabolic change should produce such severe fatigue is not known. One possibility is that some time is required for activation of existing enzymes for fatty acid oxidation.

Studies of muscle composition shows that people vary genetically in the proportion of slow-twitch and fast-twitch fibers their muscles contain. Slow-twitch fibers have greater capacity for sustained aerobic metabolism than fast-twitch fibers. The muscles of Olympic-caliber distance runners contain up to 98 percent slow-twitch fibers, so it appears that such people are born with the kind of muscles that respond to training. With or without "Olympic" muscles, people still need exercise to maintain fitness.

The mechanisms by which running and other exercise increase cardiovascular endurance are as follows: During exercise, stimulation by the sympathetic division of the nervous system (Chapter 14) increases the heart rate. When the volume of blood being returned to the heart increases, this stretches muscle fibers and increases volume of blood pumped with each heart contraction. These factors greatly increase the volume of blood leaving the heart per minute; at the same time, blood is being shunted to muscles and away from digestive organs. Sympathetic stimulation also increases the capacity of blood vessels, dilates passageways in the lungs, and increases the respiratory rate. As a result, muscle cells receive more oxygen and more nutrients, and their wastes are removed more rapidly.

Other beneficial effects of exercise have been documented. Exercise lowers blood fats such as acylglycerols and cholesterol, and raises high-density lipoproteins; but after three days without exercise, blood fats shift back to pre-exercise levels. Exercise also increases the responsiveness of cells to insulin (a hormone that causes glucose to enter cells), reduces the likelihood of blood clots in vessels (Chapter 17), and decreases the incidence of heart rhythm disorders (Chapter 18). Finally, exercise has some effects on mood and personality. It appears to counteract anxiety and depression, and to reduce the magnitude of type A (excessive, time-conscious, driving) behavior.

The so-called "runner's high" has yet to be satisfactorily explained. Some physiologists believe that the euphoric feeling reported by many runners is due to release of chemical substances called endorphins from the brain. They cite elevated blood levels of such substances that correspond to the "high." Other physiologists counter that finding endorphins in circulating blood does not necessarily mean they have exerted an effect in brain cells. Also requiring a satisfactory explanation is the observation that some runners suffer withdrawal symptoms if they suddenly stop exercising.

Some researchers note the irony that whereas sedentary people claim to be "allergic" to exercise, a few enthusiastic joggers actually display an allergic reaction to exercise. Though the mechanism of exercise allergy is not fully understood, its existence seems clear from the flushed skin, hives, and breathing difficulty some people experience when they exercise. Such people usually are allergic to aspirin and certain foods, too. They should warm up slowly before engaging in exercise and stop at the first signs of the allergic reaction.

SPORTS INJURIES

The recent increase in sports participation has been accompanied by an increase in sports injuries (Figure I2.4). A runner's body absorbs over 200 tons of force per mile, so it is no surprise that runner's often suffer knee and foot injuries. Swimmers who make 400,000 to 600,000 overhead strokes per day commonly suffer from shoulder pain. In fact, overuse injuries such as these account for nearly half of all sports injuries. In overuse injuries, repetitive trauma overwhelms tissue repair capacity, and inflammation ensues. Tendons, bursa, cartilage, bone, and nerves are most likely to be damaged.

Tendons are likely to be injured when tension is applied quickly or obliquely, when the tendon is already loaded, and when it is weaker than its associated muscle. Tendinitis, inflammation of a tendon, is especially common in the Achilles tendon. Once injured, tendons are slow to heal, partly because they have a relatively poor blood supply. Bursae often become inflamed in the shoulder in swimmers; the elbow, in tennis players; and the knee, in ball players.

FIGURE I2.4
Sports injuries are relatively common.

Cartilage in joint cavities is normally protected from injury by lubricating synovial fluid. Rapid joint movement or movement under a heavy load causes interstitial fluid to leak out of cartilage and reduces secretion of the lubricant. When secretions fail to prevent friction, joint injuries, especially at knees and elbows, often occur. Cartilage, which has no blood supply and receives nutrients by diffusion, is even slower than tendons to heal.

Overuse injuries to bone usually occur as stress fractures from repetitive stresses. Runners compress leg and foot bones as their feet hit the ground, basketball players twist bones as they pivot during quick turns, and gymnasts put stresses on nearly every bone in their bodies. Stress fractures are most common when fatigued muscles are less able to absorb energy. Compared to other sports injuries, stress fractures heal relatively quickly because bone is well supplied with blood vessels. Nerve injury often accompanies trauma to bones and muscles, and myelinated nerves are especially subject to compression damage. Recurrent pounding of the feet on the ground can cause *tarsal tunnel syndrome*, an inflammation of branches of the tibial nerve in the foot. Repetitive wrist movements can cause *carpal tunnel syndrome* (Chapter 13). Weight lifters and rowers suffer radial nerve damage as forearm muscles hypertrophy.

Among teenage and young adult athletes, overtraining is becoming an increasingly severe problem. The main risks are the potential damage to bones and joints in all athletes, and cessation of menstruation in females. Excessive exercise can wear out cartilage, and in growing athletes, damage growth plates. In females, failure to menstruate also means a lack of the female hormone estrogen and the development of brittle bones like those of older women with osteoporosis. Researchers are still trying to determine how much exercise is too much, but easing training during growth spurts—about 11 to 13 in girls and about 13 to 16 in boys—appears to be important.

Treatment of overuse injuries begins with rest. Next, the physician must determine what kind of overuse caused the problem and the athlete must look for ways to reduce the risk of recurrence. Heat, cold, and electrical stimulation, individually or in combination, are used to treat overuse injuries. As healing occurs, the principle of relative rest is applied—the athlete exercises to maintain fitness using activities that do not interfere with healing of the injury.

Will Exercise Help You Live Longer?

Yes, according to a study done by the Institute for Aerobics Research in Dallas, Texas, people who exercise do live longer than sedentary people. More than 10,000 men and more than 3000 women were followed for an average of eight years, but some as long as 20 years. They were assigned to one of five groups according to their level of fitness by a treadmill test from Group 1 (least fit) to Group 5 (most fit). When the numbers and causes of death were compared across groups, the researchers found that the risk of death among the least fit was much higher than among the most fit (Table I2.1). Even a little exercise seems to lower the risk of death, as indicated by the moderate death rates in Groups 2 and 3 compared to the high death rates in Group 1.

TABLE I2.1 Risk of Death and Fitness*

Men	
Level of Fitness	Increased Risk of Death (%)*
Group 1	240
Group 2	40
Group 3	50
Group 4	20
Women	
Level of Fitness	Increased Risk of Death (%)*
Group 1	360
Group 2	140
Group 3	40
Group 4	20

*Expressed as percentage increase over the death rate in the most fit Group 5.

When causes of death were considered exercise was shown to lower the risk of death from both circulatory diseases and cancer. Given the known effects of exercise on the heart and blood vessels, this was not surprising. Relating exercise and a lower risk of death from cancer is more difficult. That exercise speeds the rate at which food passes through the digestive tract might account for fewer cases of colon cancer in people who exercise. In women, exercise reduces the amount of body fat and can reduce the amount of estrogen—a hormone that has been associated with cancer of the breast, ovary, and uterus in older women. More research is needed to fully explain how exercise reduces the risk of death from cancer.

Problems

1. If you are engaged in a fitness program, evaluate it according to the criteria provided here. If you do not have such a program, plan and implement one.
2. Summarize the benefits and risks associated with exercise.
3. Explain in physiological terms how exercise contributes to cardiovascular endurance, body composition, flexibility, and muscle strength.
4. Plan how to minimize the risk of at least three kinds of sports injuries.

CONTROL AND INTEGRATION

OBJECTIVES

1. Briefly describe the organization and general function of the nervous system.
2. Describe the basic properties of neuroglia and relate them to neural function.
3. Describe the basic properties of neurons and relate them to neural function.
4. Explain how neurons display excitability and how excitability contributes to neural function.
5. Describe the kinds of signals generated by neurons.
6. Explain how action potentials are propagated and what factors modify them.
7. Explain how graded potentials differ from action potentials.
8. Explain how signals are relayed across synapses and how such transmission contributes to neural function.
9. Summarize the properties of well-known neurotransmitters and how they control function at the cellular level.
10. Explain how neuroactive peptides and neuropharmacological agents are similar to or different from well-known neurotransmitters.
11. Explain how neuronal pools contribute to the regulation of body functions.
12. Describe how nerve cells respond to injury, and how injuries are repaired.
13. Briefly describe the nature of the following disorders of the nervous system: (a) neuritis, (b) neuralgia, and (c) multiple sclerosis.

11

PHYSIOLOGY OF NERVE TISSUE

ORGANIZATION AND GENERAL FUNCTION

As you read this sentence, millions of neural signals travel through your body. Your nervous system allows you to read words and understand their meaning, while also maintaining your posture and controlling various internal processes. Though neural function is determined by the capabilities of individual neurons, the ability of the nervous system to control the whole body depends on the integrated function of many neurons.

The human nervous system is an extremely complex network of billions of nerve cells of many different kinds. The complexity of this vital regulatory system arises, not so much from anatomical differences among cells, as from similar cells taking on different functions, depending on how they are connected. We can learn much about neural physiology by considering how neurons, the signaling cells, produce their relatively stereotyped signals, and how these cells are connected.

Structurally, the nervous system can be divided into the **central nervous system** and the **peripheral nervous system.** The central nervous system consists of the **brain** and **spinal cord.** The peripheral nervous system consists of the **somatic nervous system** and the **autonomic nervous system.** The somatic nervous system is further divided into the **cranial nerves** and the **spinal nerves,** and the autonomic nervous system is further divided into the **sympathetic division** and the **parasympathetic division** (Figure 11.1a).

FIGURE 11.1
Organization of the nervous system:
(*a*) structural organization, and (*b*) functional organization.

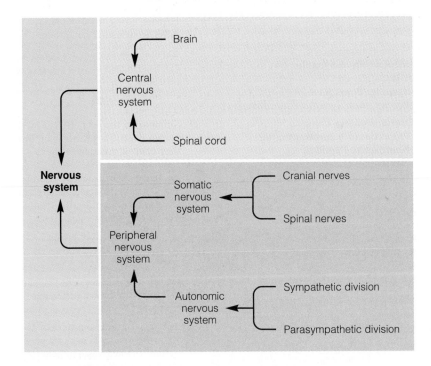

(a)

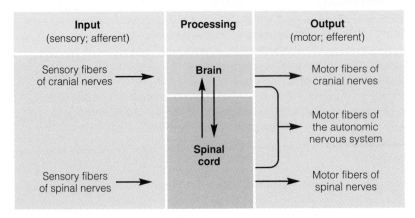

(b)

Functionally, the nervous system is a complex network of interconnected neurons that operates somewhat like a computer (Figure 11.1b). Both the nervous system and a computer have input, processing, and output functions, but the human nervous system is far more intricate and versatile than the most advanced computer. Using the computer analogy, the central nervous system (brain and spinal cord) is the central processing unit. The **afferent** (af′er-ent) **neurons,** which carry signals from receptors via cranial and spinal nerves to the central nervous system, constitute the input component. The **efferent** (ef′er-ent) **neurons,** which carry signals from the central nervous system via cranial and spinal nerves to the effectors (muscles and glands) constitute the output component. Afferent and efferent neurons (the peripheral nervous system) are analogous to input and output devices (the peripheral equipment) of a computer.

Though simplistic, the computer analogy focuses on the important functions of input, processing, and output. To understand neurophysiology, a much more thorough understanding of nervous system components is needed. Thus, we will consider individual functional components of the nervous system in some detail, always keeping in mind that the nervous system functions as an integrated whole.

As we begin our study of neural functions, it is important to recall from Chapter 1 that the body has two control systems—the nervous system and the endocrine system—and that one or both are involved in controlling nearly all body functions. As more is learned about these systems, more relationships are being discovered between them to the extent that describing a single neuroendocrine control system might be more accurate.

CELLS OF NERVE TISSUE

Nerve tissue consists of two general types of cells, **neurons** that transmit signals and **neuroglia** (nu-rog′le-ah) that do not transmit signals. Neurons or their processes (cytoplasmic extensions called axons and dendrites) are found in tracts of the central nervous system (brain and spinal cord), in nerves (bundles of axons outside the central nervous system), and in information processing portions of the brain. Neurons in different locations vary greatly in shape. Because knowing about neuroglia helps us to understand some properties of neurons, we will describe neuroglia first.

Neuroglia

When neuroglia (*glia,* glue) were named, they were erroneously believed to glue neurons together. Neuroglia make up much of the bulk of the nervous system, and in the brain, they outnumber neurons ten to one. They retain the ability to divide throughout life, a capacity that neurons lose by the time nervous system development is complete. Neuroglia are apparently quite metabolically active because they contain significant quantities of mitochondria, endoplasmic reticulum, ribosomes, and lysosomes. Four types of neuroglia—astrocytes, oligodendrocytes, ependymal cells, and microglia—are found in the central nervous system (Figure 11.2). Schwann cells, which make up the bulk of most nerves, and satellite cells are found outside the central nervous system.

Astrocytes have cytoplasmic processes that form bridges between neurons and capillaries, and the astrocytes themselves often are linked by gap junctions that allow small molecules to pass easily between the cells. Astrocytes also may act as phagocytes, initiate immune reactions, and help to synthesize signaling chemicals called **neurotransmitters.** When neurons die, astrocytes proliferate and form "scar tissue" of the central nervous system. This is called **astrocytosis. Oligodendrocytes** (ol″ig-o-den′dro-sītz) lay down a fatty substance called **myelin** (mi′el-in) in spiral layers around axons in the brain and spinal cord, thereby increasing conduction velocity. **Ependymal** (ep-en′dim-al) **cells** form sheets that line cavities in the brain and spinal cord. They may play a role in the formation of cerebrospinal fluid, which fills these cavities. **Microglia** (mi-krog′le-ah) are small phagocytic cells that migrate to the site of an injury in the central nervous system. They are thought to protect neurons against infection by engulfing microorganisms and by clearing away debris from tissue injury. Immature microglia may be capable of developing into other kinds of glial cells.

During embryonic development, **Schwann cells** become associated with an axon and wrap around it several times as they grow (Figure 11.3). Schwann cell membranes, which contain a large amount of myelin, form a multilayer sheath around most axons and a few dendrites. Such processes are said to be **myelinated.** The outer layer of membrane is called the **neurilemma** (nu″ril-em′mah). Gaps between Schwann cells form the **nodes of Ranvier** (rahn-ve-a′) and signals conducted along myelinated axons jump from one node to the next as will be explained later. **Satellite cells** provide support for aggregations of the bodies of neurons.

Neurons

Signals are relayed by neurons, which have a soma (cell body or perikaryon) and processes called axons and dendrites. The **soma** (so'mah) contains much of the cell's cytosol and its organelles—mitochondria, Golgi apparatus, nucleus, nucleolus, and nissl granules. **Nissl** (nis"l) **granules** consist of endoplasmic reticulum and ribosomes, where proteins are synthesized. **Axons** generally transmit signals from the soma to another neuron, and **dendrites** receive signals from other neurons and carry information toward the soma; however, soma, axons, and dendrites occur in many different configurations—all of which receive, process, and transmit information.

Communication occurs between neurons. The simplest form occurs when signals are relayed along the processes of a neuron and from the tip of an axon of one neuron to a dendrite of another neuron. Signal pathways involve at least two kinds of neurons—sensory neurons and motor

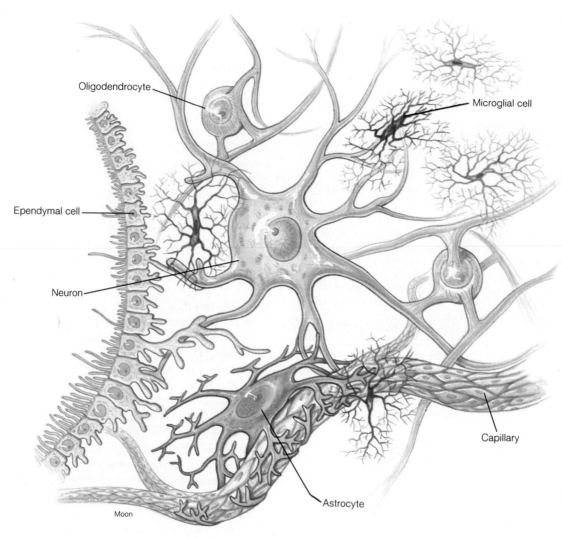

FIGURE 11.2
Neuroglia.

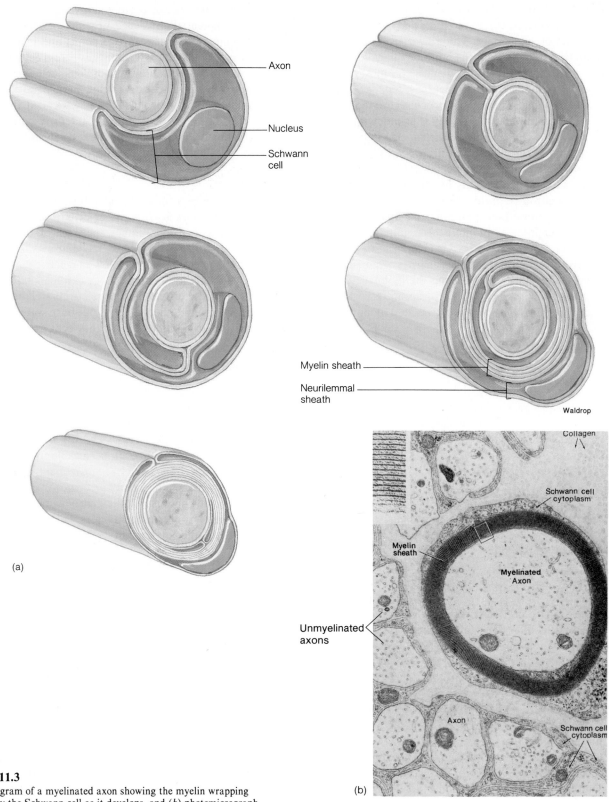

FIGURE 11.3

(a) A diagram of a myelinated axon showing the myelin wrapping created by the Schwann cell as it develops, and (b) photomicrograph of a cross section of a myelinated axon.

neurons (Figure 11.4). **Sensory neurons,** or **afferent neurons,** receive signals from receptors and relay them toward the brain or spinal cord. **Motor neurons,** or **efferent neurons,** relay signals from the brain or spinal cord to effectors such as muscles or glands. In most pathways, **interneurons,** or **association neurons,** found in the spinal cord relay signals from a sensory neuron to a motor neuron.

The typical neural pathway is as follows: A **stimulus** (a detectable change in the environment) initiates signals in receptors, some of which are dendrites of a sensory neuron. Signals travel from dendrites, through the soma, to the axon, and eventually to **axon terminals,** which release a neurotransmitter. The neurotransmitter diffuses across a **synapse** (sin′aps), a narrow cleft between neurons, to dendrites of an interneuron, initiating a signal in it. The interneuron releases a neurotransmitter that sends signals across a synapse to the dendrites of a motor neuron. Finally, the signals travel along the motor neuron to its axon terminals, where a neurotransmitter produces an effect, causing a muscle to contract or a gland to secrete.

Excitability of Neurons

In neurons, **excitability,** the ability to respond to a stimulus, causes a change in a membrane that initiates or helps to initiate a signal. To understand excitability and signals, we need to review some fundamental principles of electricity.

Principles of Electricity

Electrical charges are arbitrarily designated as either positive or negative; among charged particles, like charges repel and unlike charges attract, creating electrical force between charged particles (Chapter 2). The attracting force between particles of unlike charge increases as the quantity of charges increases or as the square of the distance between particles decreases. The repelling force between particles of like charge increases as the quantity of charges increases or as the square of the distance between the particles decreases.

Charged particles behave like magnets. If you bring oppositely charged poles of magnets together they tend to "jump" toward each other; if you bring like-charged poles of magnets together they tend to "jump" apart. Also, strong magnets exert greater force than weak ones.

Electrical force is related to energy and work. **Energy** is the capacity to do work, and **work** is force times the distance over which the force acts. (Work = force × distance.) Energy is needed to hold oppositely charged molecules apart, and they have the potential to do work when they come together. This is like magnets doing work when opposite poles attract each other.

The force between any particles of unlike charge is called a **potential difference,** or simply a **potential.** Potential is measured between two points and is expressed in units called **volts.** Small potentials across cell membranes are measured in thousandths of volts, or **millivolts** (mV). When a potential exists, it causes charged particles to flow if a conductor is present. This flow is called **current.** According to Ohm's Law, current intensity equals **voltage** (potential in volts) divided by the **resistance** of the medium through which the current flows.

current = voltage/resistance

The concepts of voltage and resistance are important in studying neurons (and other cells) because potentials exist across cell membranes (between extracellular and intracellular fluids). However, cellular events differ from electrical events. Electrons flow in electrical current, but ions flow in living systems. Both extracellular and intracellular fluids conduct current relatively well because they contain large numbers of ions, but pure water and membranes resist ion flow because they contain few ions.

Diffusion Potentials

Diffusion of charged particles creates **diffusion potentials** and is important in neuron function. Let us first see how diffusion produces potentials in a nonliving system with solutions of 0.5 molar NaCl and 0.1 molar NaCl separated by a membrane freely permeable to both Na^+ and Cl^- (Figure 11.5a). Each compartment is electrically neutral (has the same number of positive and negative charges); but because of the concentration gradient between the two compartments, ions will diffuse from compartment 1 to compartment 2. Both sodium and chloride ions have hydration shells—water molecules attached to them. Chloride ions have fewer water molecules, weigh less, and diffuse more rapidly than sodium ions.

Both sodium and chloride ions diffuse down their concentration gradients, but the more rapid diffusion of chloride ions leads to a surplus of negative charges in compartment 2 and a surplus of positive ions in compartment 1 (Figure 11.5b). This creates a temporary potential between the two compartments, lasting only until Na^+ diffusion comes to equal Cl^- diffusion. A potential difference created by differential diffusion of charged particles between two solutions is a **diffusion potential.**

As negative ions increase in compartment 2, an electrical gradient is created such that some Cl^- ions are repelled away from compartment 2 and return to compartment 1. Ion movements due to the electrical gradient counteract those due to the concentration gradient until an equilibrium is reached and no further net flow of ions occurs. In a system where the membrane is permeable to

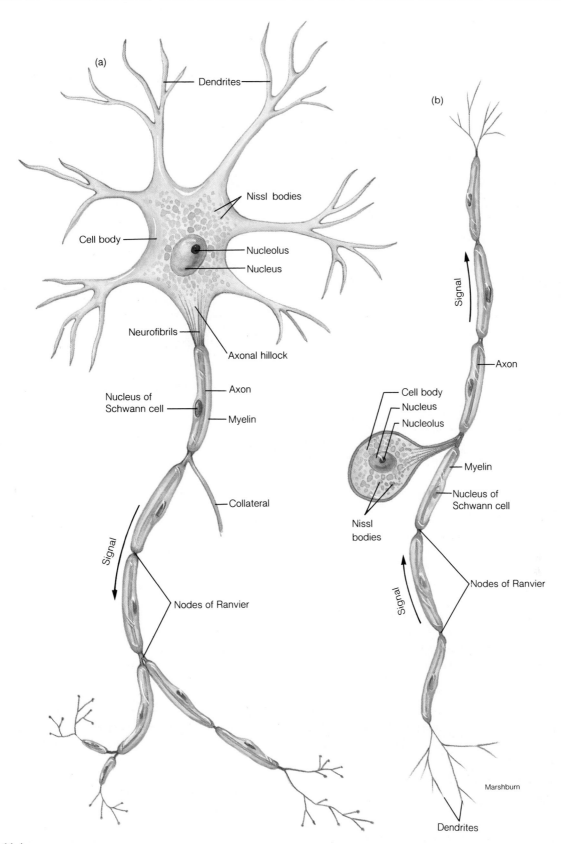

FIGURE 11.4
(*a*) Motor neuron, and (*b*) sensory neuron.

Compartment 1
0.5 M NaCl

Compartment 2
0.1 M NaCl

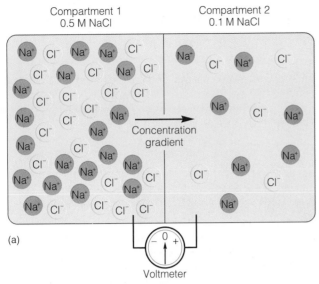

(a)

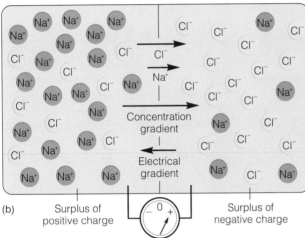

(b)

Surplus of
positive charge

Surplus of
negative charge

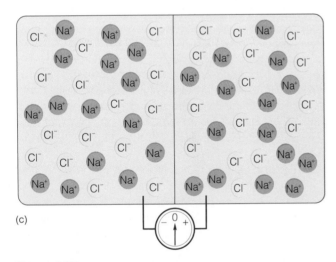

(c)

FIGURE 11.5

Diffusion potential and subsequent equilibrium: (*a*) The concentration gradient causes ions to diffuse from compartment 1 to compartment 2. (*b*) Because Cl$^-$ diffuses more rapidly than Na$^+$, an electrical gradient opposing the concentration gradient develops. (*c*) The system eventually comes to equilibrium when concentrations of each of the ions Na$^+$ and Cl$^-$ are the same in both compartments and no net movement occurs.

both Na$^+$ and Cl$^-$ (Figure 11.5c), equilibrium is reached when the ion concentrations in the two compartments are equal. At equilibrium, the potential is zero.

In contrast, living systems have selectively permeable membranes—membranes that allow some substances to pass through and block others. In such a system, an **equilibrium potential** exists when the system has no net flow of a given ion. In a hypothetical membrane permeable to K$^+$ but impermeable to Na$^+$ and Cl$^-$ (Figure 11.6), K$^+$ diffuses down its concentration gradient until it has established an equal but opposite electrical gradient. The resulting potential is the equilibrium potential—the potential at which no net diffusion occurs.

The asymmetric distribution of ions across a selectively permeable membrane, as shown by K$^+$ in Figure 11.6, creates two opposing forces: (1) a chemical force and (2) an electrostatic force. The chemical force is the ratio of the concentrations of the ions inside and outside a membrane: $W_c = RT \ln[C_o]/[C_i]$ where W_c = chemical force

R = the ideal gas constant (relates pressure and volume of a gas at a given temperature)
T = absolute temperature (temperature in °C + 273°)
ln = natural logarithm (2.3 ln = log$_{10}$)
C$_o$ = ion concentration outside the membrane
C$_i$ = ion concentration inside the membrane.

The electrostatic force is proportional to the product of the charge on the ion and the potential difference: $W = ZFE$ where W_e = electrostatic force

Z = the charge (valence) of the ion
F = Faraday's constant (number of coulombs of electrical charge per mole of electrons)
E = equilibrium potential (potential difference in mV).

These two opposing forces are equal:

$$W_c = W_e, \text{ so } RT \ln[C_o]/[C_i] = ZFE.$$

If we rearrange this equation to solve for E, the equilibrium potential, we have the **Nernst equation**.

$$E = \frac{RT \ln[C_o]}{ZF [C_i]}$$

This equation expresses the equilibrium potential for any ion present on both sides of a membrane. We can simplify the equation by finding RT/ZF for an ion with a 1$^+$ valence at 37° C and converting the natural log to base 10: $E = 61.5 \log[C_o]/[C_i]$. For an ion with a 1$^-$ valence, we can avoid a $-$ sign by reversing the positions of the concentrations: $E = 61.5 \log [C_i]/[C_o]$.

If we apply the Nernst equation to Cl$^-$ using the concentrations in Table 11.1, we find that $E_{Cl^-} = 61.5 \log 9/125 = -70$ mV. Similarly, we find E_{K^+} is -90 mV and E_{Na^+} is = 60 mV. The equilibrium potentials for Cl$^-$, K$^+$, and Na$^+$ help to determine the membrane potential of a resting (unstimulated) neuron.

Compartment 1
0.1 M KCl

Compartment 2
0.1 M NaCl

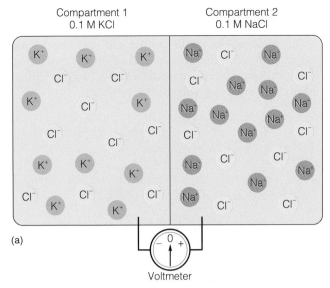

(a)

Voltmeter

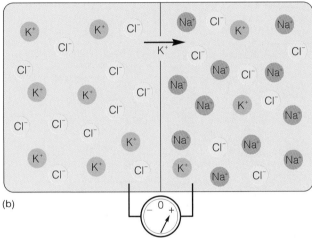

(b)

FIGURE 11.6

An equilibrium potential across a membrane permeable only to K^+: (*a*) Compartment 1 contains KCl and compartment 2 contains NaCl, both in 0.1 molar concentrations. (*b*) Equilibrium is reached when K^+ has moved down its concentration gradient to the point at which the diffusion potential remaining equals the electrical potential created. The potential difference at this point is the equilibrium potential.

TABLE 11.1	Ion Concentrations In and Around a Typical Neuronal Membrane in Millimoles/Liter of Solution at Equilibrium		
Ion	**Inside Cell**	**Outside Cell**	**Potential**
Na^+	15	150	+60
K^+	150	5	−90
Cl^-	9	125	−70

Resting Membrane Potential

Every cell has a membrane potential, but membrane potentials vary from −5 to −100 mV with the inside of the cell negative to the outside. In nerve and muscle cells, this potential varies significantly with stimulation. The potential across a membrane not recently stimulated is called the **resting membrane potential**—about −70 mV in a typical resting neuron. This potential and the ways it can be altered are directly related to neural signals.

Different concentrations of ions inside and outside a membrane give rise to the resting membrane potential. Na^+, K^+, Ca^{2+}, Mg^{2+}, Cl^-, HCO_3^-, PO_4^{3+}, and SO_4^{2-} and charged protein molecules are present on both sides of the membrane in different concentrations. Differences in the Na^+, K^+, and Cl^- concentrations (Table 11.1) account for much of the potential between intracellular and extracellular fluids, which can be measured as shown in Figure 11.7.

Neuron membranes are selectively permeable—freely permeable to K^+ and Cl^-, relatively impermeable to Na^+, and totally impermeable to large, negatively charged intracellular protein molecules. Such selective permeability contributes to the resting membrane potential. According to Table 11.1, K^+ would produce a potential of −90 mV, Na^+ a potential of +60 mV, and Cl^- a potential of −70 mV.

Cl^- diffuses down its concentration gradient into the cell until an equal electrical gradient develops in the opposite direction. Because Cl^- moves passively and has an equilibrium potential equal to the resting membrane potential, it contributes little to generating that potential.

K^+ and Na^+ contribute significantly to the resting membrane potential (Figure 11.8) as diffusion occurs along concentration and electrical gradients. K^+ moves out of the cell along the concentration gradient and into the cell along the electrical gradient. Na^+ moves into the cell along both gradients. Membrane impermeability slightly impedes K^+ movement and greatly impedes Na^+ movement; thus, the net movement of K^+ and Na^+ is determined by the combined effects of concentration and electrical gradients and by selective membrane permeability.

Because the membrane is more permeable to K^+ than to Na^+, its resting potential is nearer the equilibrium potential for K^+ than for Na^+ (−70 mV is nearer to −90 mV than to +60 mV). The resting potential is due mainly, but not entirely, to the high membrane permeability to K^+ and the net diffusion of K^+ out of the cell. At −70 mV, net diffusion of K^+ is out of the neuron, Cl^- has no net diffusion, and Na^+ diffusion is impeded by low membrane permeability.

If no energy were expended to maintain the resting membrane potential, diffusion would create a membrane potential of about -60 mV. Many cells maintain a resting potential of -70 mV by using energy to operate Na^+-K^+ pumps in the membrane, as shown in Figure 11.8 and as previously described in Chapter 3. This pump is **electrogenic;** that is, it contributes to an electrical gradient across the membrane by moving different numbers of the two ions. Typically, for each molecule of ATP hydrolyzed, three Na^+ are moved out of the neuron and only two K^+ are moved in. Positive charges accumulate outside the neuron. It becomes more negative inside, and Na^+ and K^+ diffuse down the electrical gradient into the neuron.

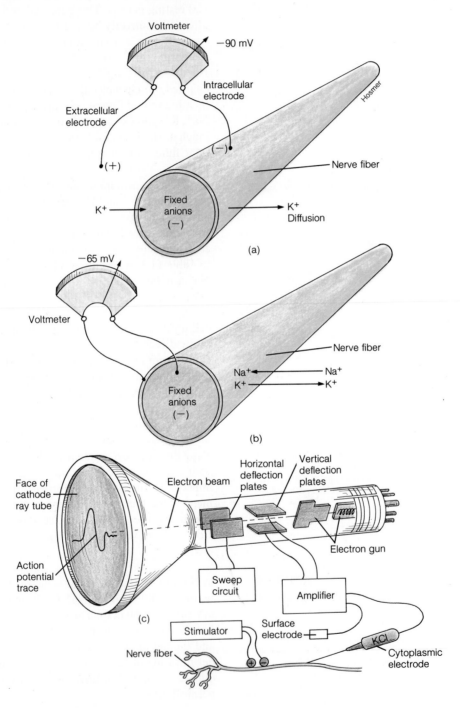

FIGURE 11.7
(a) Equilibrium potential of K^+, (b) equilibrium potential of Na^+, and (c) how an oscilloscope displays potentials.

In summary, several forces interact to maintain a resting membrane potential. K+ concentrations inside and outside a resting neuron remain constant because of the following:

1. Fairly large quantities of K+ diffuse down its concentration gradient out of the neuron.

2. Small quantities of K+ diffuse down the electrical gradient into the neuron.

3. The membrane pump actively transports sufficient K+ into the neuron to maintain constant but different concentrations of K+ on each side of the membrane.

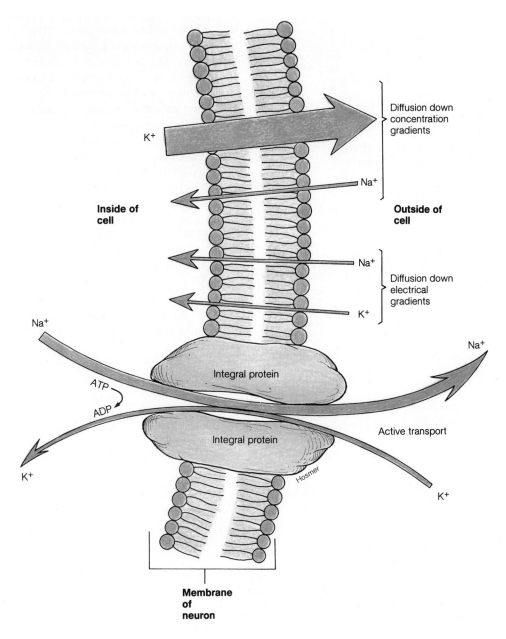

FIGURE 11.8

Summary of movements of K+ and Na+ in maintaining a resting membrane potential (−70 mV): K+(out) due to concentration gradient = K+(in) due to electrical gradient and active transport. Na+(in) due to concentration and electrical gradients = Na+(out) due to active transport.

Na$^+$ concentrations inside and outside a resting neuron remain constant because of the following:

1. Small quantities of Na$^+$ diffuse into the neuron along its concentration gradient and the electrical gradients.
2. The membrane pump actively transports sufficient Na$^+$ out of the neuron to maintain constant but different concentrations of Na$^+$ on each side of the membrane.

NEURONS AS SIGNALING UNITS

In the nervous system, signaling, or relaying information, depends on altering resting membrane potentials to produce local and propagated electrical signals and on signal integration (Figure 11.9). Let's see what happens in signaling and signal integration.

Local signals exert their effects where they are produced by providing information to neurons. They include receptor potentials and synaptic potentials. Stimulating a sensory receptor generates **receptor potentials.** For example, light striking a receptor cell in the retina of the eye causes the receptor cell to generate a receptor potential. One neuron transmitting a signal to another neuron in a neural pathway generates **synaptic potentials.**

Signals called **action potentials** cause a large change in a neuron's potential—over 100 mV from −70 mV to about +35 mV—and are propagated (conducted) the full length of the neuron. An action potential is an **all-or-none** phenomenon: It is either generated and propagated along the entire neuronal membrane at maximum intensity or it fails to occur to any degree. Action potentials arriving at axon terminals lead to the release of a neurotransmitter, which diffuses to a neighboring neuron and generates a synaptic potential.

Signal integration occurs by processing local potentials, which by themselves cannot trigger an action potential. Local potentials are **graded potentials** in that they vary in intensity and can alter the membrane potential by a small degree in either direction. An **excitatory potential** contributes to the development of an action potential; an **inhibitory potential** creates a more negative potential, making an action potential more difficult to generate. A neuron can be simultaneously affected by several potentials of each type. The **trigger zone** of a neuron, located in the axon hillock near the cell body, sums the excitatory and inhibitory input it receives. Only if the sum is sufficiently large and excitatory will the trigger zone respond by generating an action potential. Table 11.2 summarizes the different kinds of potentials.

Changes in membrane potentials can be demonstrated in the laboratory by placing a stimulating electrode at one point in an axon and a recording electrode at another point (Figure 11.10). Applying current from the positive side of a power source causes negative charges to move out of the axon and positive charges to move into it.

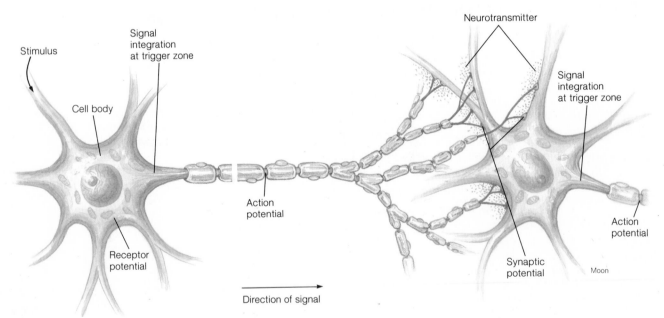

FIGURE 11.9
Summary of sites at which different kinds of signals are generated.

TABLE 11.2   Definitions of Kinds of Potentials

Potential	A potential difference or force between two points	Graded potential	A local potential of small, variable strength, not strong enough to initiate an action potential alone
Diffusion potential	A potential resulting from different rates of diffusion of kinds of charged particles	Excitatory potential	A local potential that contributes to initiating an action potential
Equilibrium potential	A potential having no net flow of charged particles because opposing forces balance each other	Inhibitory potential	A local potential that makes initiating an action potential more difficult
Resting membrane potential	A potential across a membrane that has not been stimulated recently	Receptor potential	A local potential produced when a sensory receptor is stimulated
Local potential	A potential that exerts its effect where it occurs	Synaptic potential	A local potential produced when one neuron transmits a chemical signal to another neuron

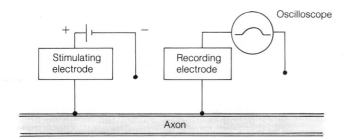

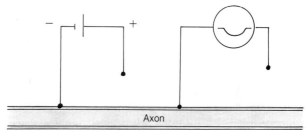

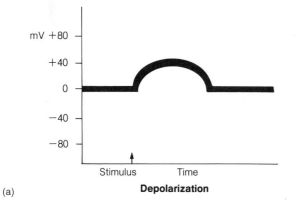

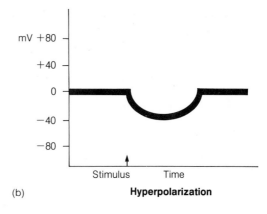

(a)   **Depolarization**

(b)   **Hyperpolarization**

FIGURE 11.10

Effects of stimulating an axon by applying current: (*a*) When the stimulating electrode is connected to the positive side of the power supply, the axon is depolarized (hypopolarized). (*b*) When the stimulating electrode is connected to the negative side of the power supply, the axon is hyperpolarized.

The membrane potential becomes less negative (is less polarized) than the resting potential and is said to be **depolarized.** Conversely, applying current from the negative side of the power source causes positive charges to move out of the axon and negative charges to move into it. The membrane potential becomes more negative (is more polarized) than the resting potential and is said to be **hyperpolarized.** Weak currents behave like graded potentials, altering the membrane potential by only a small amount (up to about 10 mV). Stronger depolarizing currents depolarize the membrane to its **threshold** (usually a potential less negative than -55 mV). When a neuron reaches its threshold, it generates an action potential.

Action Potentials

An **action potential** is sudden large change in the membrane potential that is propagated along the membrane as a **wave of depolarization** (Figure 11.11). Action potentials are due to changes in membrane permeability to Na^+ and K^+ and involve the following events (Figure 11.12). Once a stimulus is received, depolarization occurs gradually until it reaches the threshold and suddenly thereafter until it reaches a maximum of about $+35$ mV. The sudden rise and fall in the membrane potential is the **spike potential.** Within the spike, the positive potential (0 to $+35$ mV) is called **overshoot,** and the return to the resting potential is

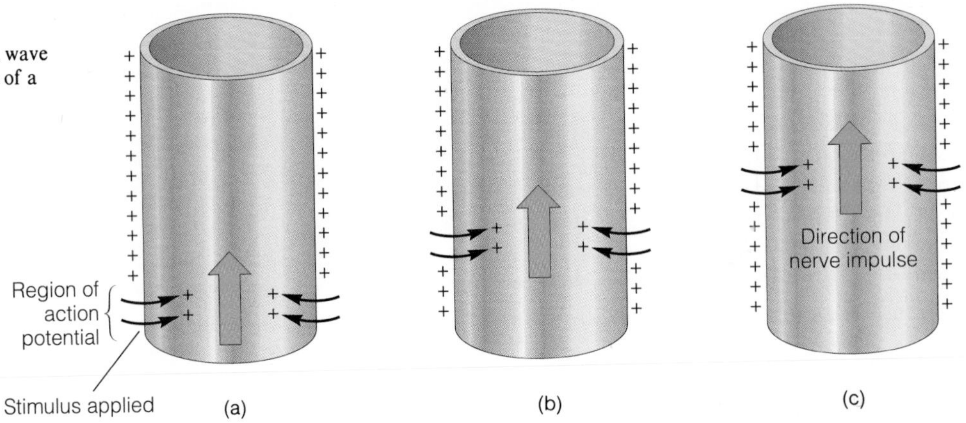

FIGURE 11.11
An action potential is propagated as a wave of depolarization along the membrane of a neuron.

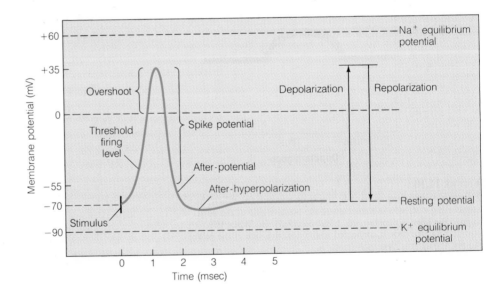

FIGURE 11.12
Events in an action potential.

called **repolarization.** A brief drop below the resting potential, the **after-hyperpolarization,** makes the membrane temporarily more difficult to stimulate.

Changes in membrane permeability account for the events in an action potential. Na^+ diffusing into the neuron along both electrical and chemical gradients starts depolarizing the membrane, making it more permeable to Na^+. When the membrane has depolarized by about 15 mV and reached its threshold, it suddenly becomes highly permeable to Na^+ and rapid depolarization produces a potential of about +35 mV. Increased Na^+ permeability leading to even greater permeability is an example of positive feedback activated by sufficiently strong graded potentials. Weaker graded potentials simply "die out" because they do not cause sufficient depolarization to trigger a sudden increase in Na^+ permeability.

Changes in Na^+ permeability probably involve sodium pores—integral membrane proteins with gating molecules imbedded in them (Figure 11.13). Most sodium pores are closed at the resting potential. Local potentials initiate depolarization, causing gating molecules, which are sensitive to voltage changes, to rotate and open the pores. Such pores, called **voltage-gated ion channels,** are not fully understood but appear to open a few at a time as depolarization begins, and suddenly open many at a time when a certain voltage is reached. This self-accelerating process suddenly and rapidly increases sodium permeability.

When most of the change in membrane potential has occurred, sodium channels begin to close and Na^+ permeability decreases. At the same time, voltage-gated potassium channels open and K^+ permeability increases rapidly. K^+ permeability reaches a peak immediately after the peak spike potential. K^+ leaves the cell much faster than Na^+ enters and the membrane repolarizes. After Na^+ permeability returns to the resting level, continued decreasing K^+ permeability accounts for after-hyperpolarization.

For a brief period during and after a neuron produces an action potential, it is refractory, or resistant, to further stimulation. This **refractory period** (Figure 11.14) includes an **absolute refractory period** when the neuron cannot be stimulated no matter how strong a stimulus is applied and a **relative refractory period** when the neuron can be stimulated by a strong stimulus. The absolute refractory period lasts less than 2 msec during which the membrane remains maximally permeable to Na^+. The neuron cannot be stimulated because it cannot become

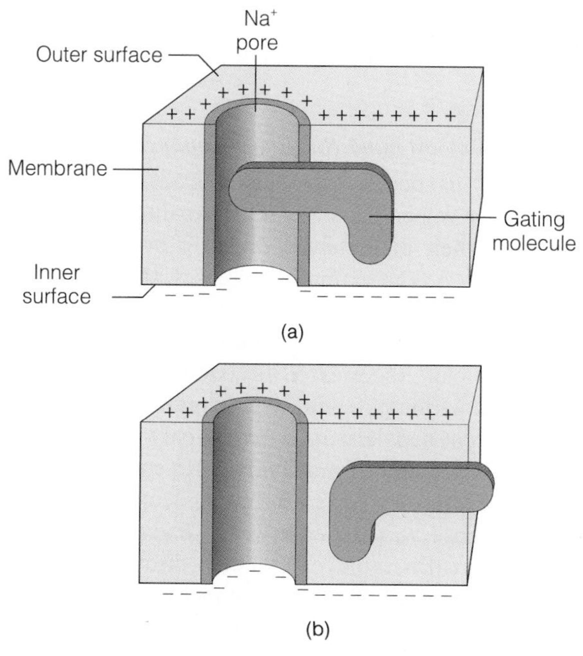

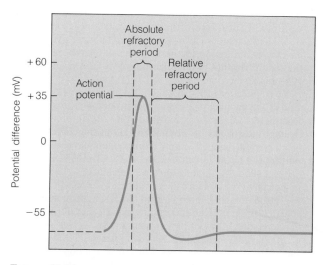

FIGURE 11.14

Refractory periods. A neuron is absolutely refractory during most of the period of the spike potential and relatively refractory until the resting potential is restored.

FIGURE 11.13

A model of the proposed action of a gating molecule: (*a*) gating molecule blocking an Na^+ pore, and (*b*) gating molecule after rotation in the membrane allowing the Na^+ pore to open.

more permeable to Na^+ as it must do to respond. The relative refractory period lasts 5 to 15 msec when Na^+ permeability is decreasing and K^+ permeability is increasing. Na^+ permeability has decreased enough that it can be stimulated to increase if the stimulus is strong enough.

Neurons expend energy to reestablish ion concentrations after action potentials. The sodium-potassium membrane pump continuously transports Na^+ out and K^+ into cells and maintains the gradients that provide energy for action potentials. It also assists in restoring concentrations of these ions to the resting level after an action potential. Because propagation of action potentials changes Na^+ and K^+ concentrations only in small regions of the membrane at any one time, millions of action potentials can form in a short time without significantly affecting the overall ionic concentrations in the neuron or the interstitial fluid.

Changes in membrane permeability during action potentials were first described by Hodgkin and Huxley, working with squid giant axons, which reach diameters of 1 mm. These unmyelinated axons allowed experiments that were impossible in smaller mammalian neurons, and many researchers have since experimented with them. Potentials can be measured on either side of the membrane, and cytoplasm can be removed and replaced with liquids of known composition. It is also possible to electronically "clamp" the voltage, or hold the voltage across the membrane constant. Membrane permeability to ions at various fixed voltages then can be studied. Some procedures developed on giant axons have been adapted to smaller neurons, and have led to the conclusion that basic properties of axons are similar in both squid and humans.

Certain drugs have been found to block a particular kind of ion channel. For example, tetrodotoxin (a poison from puffer fish ovaries) and saxitonin (a toxin produced by microscopic marine animals called dinoflagellates) block Na^+ channels. Tetraethylammonium ion, easily synthesized in the laboratory, blocks K^+ channels. These drugs make it possible to study permeability independent of the movement of the blocked ion.

Tetrodotoxin has been used to block signals in many research projects. One such study showed that during development (in cat embryos) spontaneous signals along neurons are important in establishing normal connections between neurons. Certain connections failed to develop when tetrodotoxin prevented such signals. Other studies focus on finding a sensor that can detect poisons that bind to membrane channels, regardless of which ion the channel normally regulated. Having a generalized sensor would greatly simplify detection of environmental toxins.

Propagation of Action Potentials

Once a neuron is stimulated sufficiently to initiate an action potential, voltage changes cause adjacent portions of the membrane to depolarize to the threshold. The neuron obeys the **all-or-none principle:** it either conducts a full-sized action potential or none at all. Therefore, a neuron can be considered "on" or "off." It remains "off" until it receives a threshold stimulus that turns it "on."

Action potentials are propagated as waves of depolarization and repolarization in both directions along an axonal membrane. This is like plucking a taut string and sending waves along it in both directions. Only action potentials traveling toward axon terminals function as signals. Each wave of depolarization and repolarization takes only about 2 msec and results in the action potential of constant size and rate moving along the entire axon.

Propagation of action potentials and the passage of current both involve electrical phenomena, but propagation along a neuron and conduction along a wire differ in three ways:

1. Action potentials move at speeds up to 100 m/sec, but electrical current moves at the speed of light— 3×10^8 m/sec.
2. Action potentials display the all-or-none phenomenon, but current flow decreases with distance.
3. Charged ions move in propagation of action potentials, whereas electrons move in electrical current.

Conduction Velocity and Saltatory Conduction

Although any particular axon conducts action potentials at a particular velocity, axons vary in conduction velocity according to their diameters. Velocity is proportional to axon diameter. As the diameter increases, the membrane surface increases and internal resistance to the conduction decreases. Other factors being equal, the greater the diameter of an axon, the greater its conduction velocity.

When an action potential is propagated along a myelinated neuron, depolarization occurs only at the nodes, which usually are about 1 mm apart. Such conduction was described as **saltatory** (sal'tat-or"e), or leaping (*saltere,* to leap), by those who first observed it. The signal doesn't really leap, but it is rapidly conducted through extracellular fluids (the path of least resistance) between depolarizations at nodes (Figure 11.15). Conduction through extracellular fluids is faster than along a membrane because ions move more rapidly in fluid. Ions cross the membrane only at the nodes, where most sodium channels are located. Potassium channels are found in the membrane under the myelin. It appears that channels are distributed

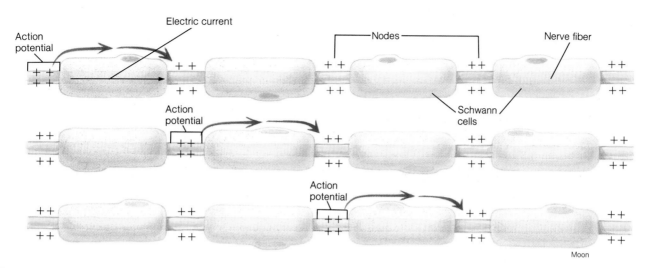

FIGURE 11.15
Saltatory conduction.

in this manner in embryos before myelination begins and the channels probably determine the placement of myelin and nodes.

Saltatory conduction is both faster and more energy efficient than conduction along unmyelinated axons. It occurs at velocities up to 100 m/sec in large, myelinated axons, such as some motor neurons that innervate skeletal muscles. In contrast, small unmyelinated fibers, such as those that supply visceral organs, conduct action potentials at velocities of only 0.5 to 1 meter per second. It requires less energy than conduction along the entire membrane because fewer ATP molecules are hydrolyzed to operate the membrane pump.

Glial cells do not produce signals like neurons do, but they may play a role in signaling. Current can pass readily between glial cells linked by gap junctions. When an action potential travels along a neuron, neighboring glial cells depolarize as K^+ accumulates in interstitial fluids around them. Glial cells may take up K^+ and depolarize in a kind of chain reaction, with the first depolarized cells drawing current through gap junctions and causing adjacent cells to depolarize. Exactly how K^+ uptake and glial cell depolarization affects neurons is not yet known, but it might cause glial cells to release substances such as nerve growth factors and neurotransmitters that neurons need. Glial cells are known to release the neurotransmitter gamma-aminobutyric acid (GABA), and they have voltage-sensitive sodium and potassium channels, which might be involved in some way in maintaining similar channels in neurons. Much is being learned about glial cells now that methods are available to study these small cells.

Graded Potentials

As we have seen, graded potentials, such as receptor potentials and synaptic potentials, affect only part of the membrane of a neuron. For example, a stimulus can depolarize a membrane to -65 mV at one point while it remains at -70 mV at another point. Negatively charged ions then move toward the stimulated region (-65 mV) and positively charged ions move away from it. Conversely, certain stimuli can hyperpolarize a membrane to -75 mV, and this can cause adjacent regions to become slightly hyperpolarized.

Graded potentials cause varying degrees of depolarization or hyperpolarization, but their effects are limited to small regions of the neuronal membrane. Depending on the flow of ions, they can alter the resting potential by as much as 10 mV in either direction. Graded potentials are decremental (decrease with distance) and are not propagated, yet they initiate many action potentials.

Strength of Stimuli

To initiate an action potential, a stimulus must be strong enough to depolarize a membrane to its threshold. A stimulus of minimum strength to generate an action potential is a **threshold stimulus** (Figure 11.16a). Whether graded potentials cause a neuron to reach threshold depends in part on conditions around the neuron. High extracellular Ca^{2+} concentration and certain local anesthetics decrease membrane permeability to Na^+ and make a neuron more difficult to stimulate. Hyperpolarizing the neuron by lowering the extracellular K^+ concentration also makes it more

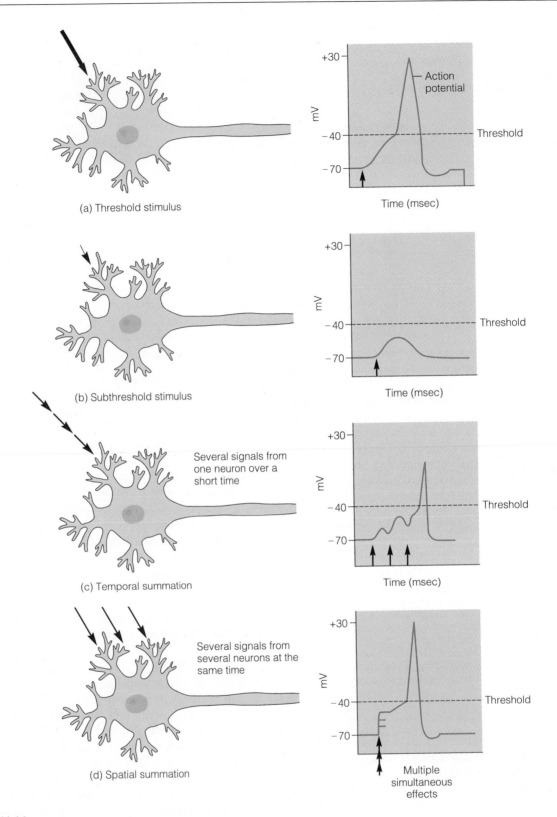

(a) Threshold stimulus

(b) Subthreshold stimulus

(c) Temporal summation

Several signals from one neuron over a short time

(d) Spatial summation

Several signals from several neurons at the same time

FIGURE 11.16

(*a*) A threshold stimulus initiates an action potential, (*b*) but a subthreshold stimulus is too weak to initiate an action potential. (*c*) In temporal summation, each of a series of stimuli bring the potential closer to the threshold. (*d*) In spatial summation, each of several stimuli arriving at the same time contribute to bringing the action potential to the threshold.

difficult to stimulate. Low extracellular Ca^{2+} concentration or the drug veratridine increases membrane permeability to Na^+ and makes a neuron easier to stimulate.

Increasing the strength of the stimulus beyond the threshold level will not increase the size of the action potential. Yet, a strong stimulus can cause a neuron to generate and propagate a series of action potentials. This increases the frequency of action potentials arriving in centers in the brain. As we shall see in later chapters, the brain makes use of variations in the frequency of action potentials in interpreting signals.

Summation

A stimulus too weak to initiate an action potential is called a **subthreshold stimulus** (Figure 11.16b). Such stimuli produce graded potentials. Each subthreshold stimulus depolarizes the membrane slightly and the cumulative effects of several such stimuli can initiate an action potential. These cumulative effects constitute **summation,** which affects the membrane in the region of the axon hillock and can be temporal or spatial. In **temporal summation** (Figure 11.16c), stimuli reach a single neuron in rapid succession over a period of a few milliseconds. In **spatial summation** (Figure 11.16d), stimuli reach adjacent points on the membrane at the same time.

> When your dentist injects a local anesthetic into the gum near a nerve that supplies a tooth, the region quickly becomes numb and you feel no pain while work is done on the tooth. Local anesthetics such as procaine (Novocain) and tetracaine make neurons in the jaw more difficult to stimulate by blocking the action of sodium gating molecules. With neuronal membranes impermeable to sodium, action potentials cannot be generated, and no pain signals can be sent to your brain. If instead, your dentist uses a general (whole body) anesthetic such as nitrous oxide, it blocks pain signals after they have reached the central nervous system.

SYNAPTIC TRANSMISSION

Signals reaching axon terminals are transmitted to another neuron across a synapse. In **chemical synapses,** the signal is relayed by a chemical neurotransmitter. In **electrical synapses,** the signal is relayed by ion flow across a cytoplasmic bridge. Gap junctions in cardiac and smooth muscle are electrical synapses. Most other synapses in the body are chemical synapses.

In chemical synapses, the **presynaptic** (pre″sin-ap′tik) **neuron** releases neurotransmitter and the **postsynaptic** (post″sin-ap′tik) **neuron** receives it. In a typical synapse (Figure 11.17), a **presynaptic knob** of an axon is separated from the **postsynaptic region** of a dendrite, cell body, or axon by a very small space called the **synaptic cleft.**

Mechanism of Synaptic Transmission

Axon terminals continuously synthesize neurotransmitter and store it in **vesicles,** which aggregate near the surface of the presynaptic knobs. When an action potential arrives at a presynaptic knob, it increases the membrane permeability to Ca^{2+} via voltage-sensitive calcium channels, which remain open throughout depolarization. As Ca^{2+} enters the presynaptic knob, it stimulates neurotransmitter release, probably by causing vesicles to fuse with the cell membrane.

Once released, neurotransmitter molecules diffuse across the narrow synaptic cleft and bind with receptor proteins of the postsynaptic membrane. Such binding directly opens gated pores in some membrane, but it may act in other ways. Opening gated channels lets ions flow across the membrane, altering the voltage. The kinds of ions moving and their direction of movement depend on which neurotransmitter acts as the ligand. Stimulatory neurotransmitters cause voltage changes that result in synaptic potentials. The chemical neurotransmitter signal is changed into a graded synaptic potential. This is called **transduction.** Inhibitory neurotransmitters cause opposite voltage changes and interfere with signaling. Finally, the neurotransmitter dissociates from the receptor and is inactivated in the cleft by a postsynaptic membrane enzyme or is returned to the presynaptic knob for reuse. The steps in the transmission of a signal across a synapse are summarized in Figure 11.17.

Synaptic Potentials

The effects of transmission across a synapse depend on the rate at which neurotransmitter molecules reach receptor sites on the postsynaptic membrane and on the interaction of transmitter (ligand) and receptor. Such interactions create graded excitatory or inhibitory postsynaptic potentials (Figure 11.18).

In an excitatory interaction, the transmitter-receptor complex increases the permeability of the postsynaptic membrane, allowing Na^+ to flow into the neuron. For the

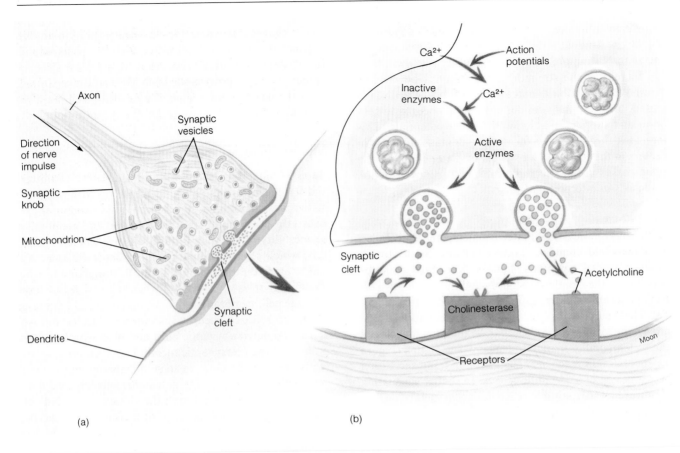

(a)

(b)

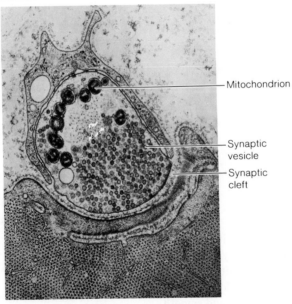

Mitochondrion

Synaptic
vesicle

Synaptic
cleft

FIGURE 11.17

A synapse: (*a*) a general diagram, (*b*) a detailed view of
neurotransmitter release, and (*c*) a transmission electron micrograph,
showing a synaptic knob filled with synaptic vesicles.

neurotransmitter acetylcholine, it also allows K^+ to flow out of the neuron. Such ion movements depolarize the membrane and create an **excitatory postsynaptic potential (EPSP)**. Neurotransmitter from a single presynaptic knob is unlikely to generate an action potential, but it does depolarize the membrane toward the threshold. Additional neurotransmitter, which increases the size of the EPSP can reach the membrane by subsequent releases from the same knob (temporal summation) or from other knobs of the same neuron or other neurons (spatial summation). **Facilitation,** increasing the EPSP, usually leads to an action potential.

In an inhibitory interaction, the transmitter-receptor complex increases postsynaptic membrane permeability to K^+ and/or Cl^-. K^+ flows out or Cl^- flows in to hyperpolarize the membrane. Hyperpolarization creates an **in-hibitory postsynaptic potential (IPSP)**. When a neuron is hyperpolarized, the membrane potential is farther from the threshold than in the resting state and, therefore, requires a stronger than usual stimulus to reach the threshold level. **Inhibition,** increasing the IPSP, makes a neuron more difficult to stimulate.

Some neurons receive excitatory and inhibitory signals from different axon terminals of many other neurons (as many as 10,000 for certain brain neurons). Whether a neuron generates an action potential is determined by the net effect of these events. Both spatial and temporal summation occur, and summation usually is needed to generate action potentials. The various events at the synapses ultimately affect the trigger zone near the axon hillock, where action potentials are most often generated.

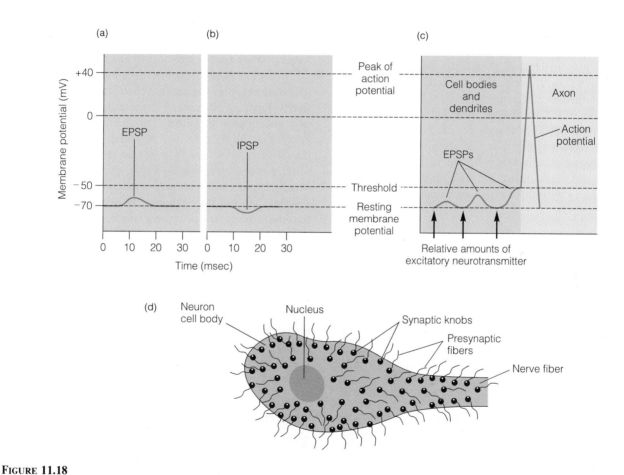

FIGURE 11.18

Alteration of postsynaptic potentials: (*a*) During an EPSP, hypopolarization facilitates the initiation of an action potential. (*b*) During an IPSP, hyperpolarization inhibits initiation of an action potential. (*c*) Several EPSPs can generate an action potential. (*d*) Many synaptic bulbs can impinge on a single neuron and each can elicit an EPSP or IPSP on that neuron.

Properties of Synaptic Transmission

Having considered the basic process of synaptic transmission, we are ready to look at some important properties of the process.

Principle of Forward Conduction

Once generated, an action potential spreads both directions along the membrane, but only the signals reaching axon terminals are effective in conveying information. The release of neurotransmitters from axon terminals, but not from dendrites, maintains one-way signals in the nervous system. This constitutes the **principle of forward conduction.** Signals go from axon hillock to axon terminals within neurons and from axon terminals across synapses to the next neuron.

Synaptic Delay

Transmission across synapses is slower than conduction along neurons because of **synaptic delay,** the time required for the release, diffusion, and action of the transmitter (about 0.5 msec). Because of synaptic delay, the time a signal takes to travel between two points can be used to estimate the number of synapses in the neural pathway. For example, at 100 meters per second (10 cm per msec), a signal can travel 5 cm along a neuron in the same time it takes to cross a synapse. In the brain where synapses are close together, traversing a 5-cm pathway with one synapse would take 1 msec; traversing the same distance over a pathway with three synapses would take 2 msec—twice as long.

Synaptic Fatigue

Repeated stimulation of presynaptic knobs leads to **synaptic fatigue,** depletion of stored transmitter and temporary inability to release it. In this rare phenomenon, the postsynaptic neuron cannot be stimulated for several seconds while the presynaptic axon terminals synthesize neurotransmitter. Synaptic fatigue is the only kind of neural fatigue. Action potential delays are of shorter duration and are due to refractory periods.

Signal Integration

In signal integration (Figure 11.19), the effects at a neuron's various synapses are summed algebraically. Sufficient depolarization in the trigger zone causes an action potential. Synapses farthest from the trigger zone have the least effect, and those nearest the trigger zone have the greatest effect. An action potential is produced if, and only if, summation depolarizes to threshold in the trigger zone.

Summary

The properties of action potentials and synaptic potentials differ significantly (Table 11.3). Action potentials are large, brief, and travel over long distances. Synaptic potentials are small, of variable duration, subject to summation, and travel over short distances. Only action potentials display thresholds, refractory periods, and all-or-none conduction. Action potentials always depolarize, but synaptic potentials can either depolarize or hyperpolarize.

NEUROTRANSMITTERS

Chemical substances called neurotransmitters are released from presynaptic neurons and interact with specific receptor sites on postsynaptic neurons. At least 30 known or putative neurotransmitters have been found. **A putative neurotransmitter** satisfies some but not all criteria for being a neurotransmitter. Many neurotransmitters have been found in brain tissue, and more are likely to be discovered.

Early in embryological development, neurons have the potential to synthesize several chemical transmitters, but as development proceeds, most neurons come to produce only one neurotransmitter. For many years, it was believed that all mature neurons produce a single neurotransmitter, but it now appears that some mature neurons contain two, or occasionally three, chemical transmitters. Much remains to be learned about multiple transmitter neurons, and the **Dale principle of neuronal specificity**—that mature neurons produce a single specific neurotransmitter—still applies to the majority of neurons.

In most neurons, neurotransmitters bind to specific receptors and alter membrane potentials, but more complicated arrangements have been found, especially in the brain. For example, some neurotransmitters act on more than one kind of receptor. Also, it now appears that many neurotransmitters act directly or indirectly on voltage-sensitive ion channels and that effects on ion channels occur over a much longer time span than those of transient synaptic potentials.

Properties of Neurotransmitters

The six neurotransmitters for which some functions are reasonably well established display wide diversity in chemical structure (Figure 11.20). Each has particular properties worthy of special consideration (Table 11.4).

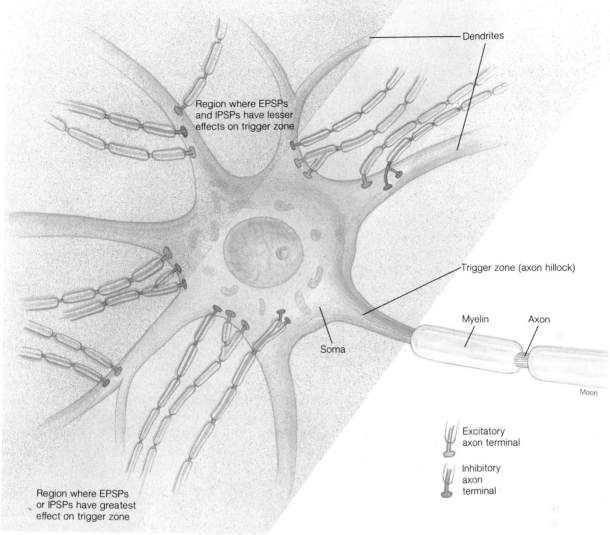

FIGURE 11.19

Signal integration. Action potentials are initiated in the trigger zone if, and only if, summation results in hypopolarization to the threshold level in the trigger zone. Events at synapses closest to the trigger zone have the greatest effect on that zone.

TABLE 11.3   Properties of Action Potentials and Synaptic Potentials

Property	Action Potential	Synaptic Potential	Property	Action Potential	Synaptic Potential
Amplitude	Large (70–110 mV)	Small, but summed to create an action potential	Threshold	Approximately 15 mV less negative than resting membrane potential	None
Duration	Brief (1–10 msec)	Variable (5 msec–20 min)			
Summation	None; amplitude maximal regardless of conditions initiating it	Graded; amplitude depends on conditions initiating it	Refractory period	Consists of absolute and relative refractoriness	None
			Conduction Signal	All-or-none Hypopolarizing	Decremental Hypo- or hyperpolarizing
			Distance	Long	Short

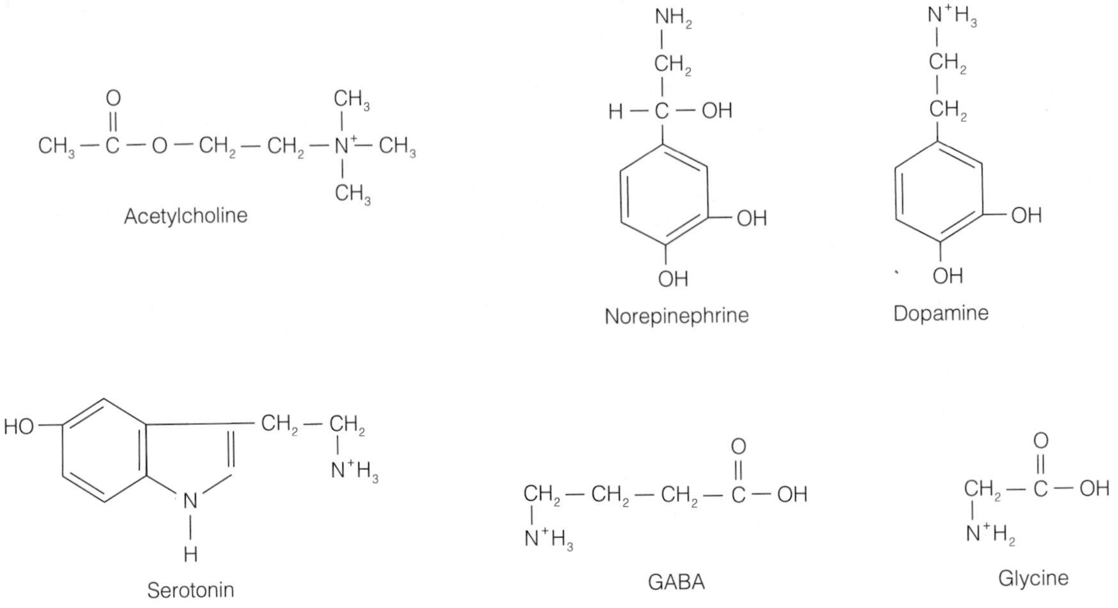

FIGURE 11.20
Neurotransmitters.

TABLE 11.4   Properties of Neurotransmitters

Neurotransmitter	Where Found	Actions
Acetylcholine	Neuromuscular junctions, autonomic nervous system, and brain	Excites muscles, decreases heart rate, and relays various signals in the autonomic nervous system and the brain
Norepinephrine	Sympathetic nervous system and brain	Regulates activity of visceral organs and some brain functions
Dopamine	Brain	Involved in control of certain motor functions
Serotonin	Brain and spinal cord	May be involved in mental functions, circadian rhythms, and sleep and wakefulness
Gamma-aminobutyric acid	Brain and spinal cord	Inhibits various neurons
Glycine	Spinal cord	Inhibits neurons

The neurotransmitter **acetylcholine** (as″e-til-ko′lēn) is released at neuromuscular junctions; that is, junctions between motor neurons and skeletal muscle cells, at certain brain synapses, and at many synapses in the autonomic nervous system. Like other neurotransmitters,

acetylcholine's action depends on its effect on postsynaptic membranes. For example, at receptors on neuromuscular junctions, it increases permeability to Na^+ and K^+ and its excitatory effect leads to muscle contraction. At receptors in a specialized conduction system in the heart, it increases membrane permeability to K^+ but not to Na^+, so it causes hyperpolarization and decreases the rate of firing of parts of the heart's signal conducting system.

Acetylcholine is inactivated by the postsynaptic membrane enzyme **cholinesterase** (ko″lin-es′ter-ās), which catalyzes hydrolysis to acetate and choline. Some acetate and choline molecules are actively transported back to the presynaptic neuron and resynthesized into acetylcholine; the rest enter the bloodstream and are carried to other cells.

Cholinesterase activity is essential to remove acetylcholine from receptors so they can relay new signals. Some flea collars, warfare nerve gases, pesticide strips, and other insecticides contain a **cholinesterase inhibitor** (diisopropyl flurophosphate) that prevents acetylcholine breakdown. Acetylcholine remains bound to receptors and prevents their reexcitation; nerve signals cease, and the organism dies. The drug atropine, which competes with acetylcholine for receptors, can serve as an antidote for poisoning with cholinesterase inhibitors.

Another neurotransmitter **norepinephrine** (nor″ep-ĭ-nef′rin), or noradrenalin, is secreted by many neurons of the sympathetic nervous system and by some brain neurons. Depending on the receptor to which it binds, norepinephrine can be excitatory or inhibitory. This transmitter is important in regulating the activity of visceral (internal) organs and in controlling certain brain functions. Norepinephrine can be inactivated at the postsynaptic neuron by the enzyme monoamine oxidase, reabsorbed back into the presynaptic knob, or inactivated in the liver by the enzyme catecholamine O-methyl transferase (COMT).

The neurotransmitter **dopamine** (do′pah-mēn) is chemically related to norepinephrine and to epinephrine, the predominant secretion of the adrenal medulla. All these substances are **catecholamines** (kat″eh-kol′ah-mēnz), which are synthesized from the amino acid tyrosine. Dopamine is a neurotransmitter of certain brain neurons involved in motor control. The related substance epinephrine appears to act as a neurotransmitter in some neurons of the medulla oblongata, where many vital functions are controlled.

Dopamine deficiency is associated with Parkinson's disease, a disorder characterized by muscle rigidity, tremors, and an uncoordinated, forward-leaning gait. Excesses of dopamine may be involved in mental disorders, such as schizophrenia. These disorders are discussed in Chapter 12.

Serotonin (ser″o-to′nin), a neurotransmitter synthesized from the amino acid tryptophan, is released by certain neurons in the brain and spinal cord. It occurs in higher concentrations in blood platelets and in certain cells of the digestive tract, where it apparently has functions other than neurotransmission. The functions of serotonin as a transmitter are poorly understood, but it may play a role in the regulation of wakefulness and sleep, and other circadian rhythms.

Gamma-aminobutyric acid (GABA), found only in the central nervous system, has an inhibitory effect at its receptors. GABA makes postsynaptic membrane potentials more negative by increasing the membrane's permeability to K^+ and Cl^-; thus, the membrane becomes hyperpolarized.

The simple amino acid **glycine** is a transmitter in certain inhibitory synapses in the spinal cord. Though glycine is probably present in all cells, it acts as a neurotransmitter only when it is released from a neuron.

Neuroactive Peptides

In recent years, about 25 small diffusible molecules called **neuroactive peptides** have been shown to relay signals between neurons. These molecules are stored and released by presynaptic neurons onto adjacent neurons with appropriate receptors. Some serve as hormones outside the brain, but many now appear to play a role in neurotrans-

TABLE 11.5	Neuroactive Peptides	
Substance	**Where Secreted***	**Where Discussed in This Text**
Cholecystokinin-pancreozymin (CCK-PZ)	Digestive tract, maybe cerebral cortex	Chapter 23
Endorphins and enkephalins	Various parts of central nervous system, digestive tract	Chapter 12
Hypothalamic hormones (CRH, GRH, LRH, PRH, somatostatin, TRH)	Hypothalamus, maybe other parts of the brain; somatostatin from digestive tract	Chapters 12 and 16
Oxytocin	Posterior pituitary gland, maybe parts of brain	Chapter 16
Substance P	Digestive tract, some afferent neurons, many other brain neurons	Chapter 23
Vasoactive intestinal peptide (VIP)	Maybe digestive tract and hypothalamus	Chapter 23
Vasopressin	Posterior pituitary gland, maybe other parts of the brain	Chapter 16

*Sites of secretion most clearly established are listed first, though these sites may involve some function other than release of neurotransmitters.

mission, too. Information about selected neuroactive peptides is summarized in Table 11.5.

Neuropharmacologic Agents

In addition to endogenous neurotransmitters normally found in the body, certain exogenous substances from outside the body affect neuron excitability by mimicking or blocking the action of a neurotransmitter. Substances that mimic fit a neurotransmitter receptor so well that a neuron acts as if the neurotransmitter itself were present. Nicotine in small quantities has that effect on some acetylcholine receptors. Substances that block bind with receptors but fit poorly and fail to function. Curare paralyzes skeletal muscles by blocking acetylcholine receptors in this way.

Certain substances increase neuron excitability. Those that affect membrane permeability to Ca^{2+} and maybe other ions include substances found in beverages—theobromine in cocoa, theophylline in tea, and caffeine in coffee. Strychnine and tetanus toxin (a poison made by certain bacteria) increase neuron excitability by inhibiting the action of inhibitory transmitters, such as glycine, and can cause convulsions.

Other substances decrease excitability. Dilantin may stabilize the threshold of neurons against hyperexcitability by promoting Na^+ efflux. It is used to prevent seizures—sudden discharges from groups of neurons in the brain.

Some anesthetics increase the stimulation required to reach the threshold for an action potential. Lipid soluble anesthetics such as ether dissolve in neuronal membranes and increase K^+ permeability, causing hyperpolarization and making membranes less responsive to stimulation. The tranquilizer and muscle relaxant diazepam (Valium) enhances the inhibitory effects of GABA. The hallucinogen lysergic acid diethylamide (LSD) is chemically similar to serotonin and binds to its receptors; however, its effects—hallucinations and bizarre behaviors—are quite different from those of serotonin. A variety of other substances that specifically affect neuromuscular junctions—the synapses between axons and muscle cells—were discussed in Chapter 9.

Neuronal Pools

An aggregation of thousands to millions of interconnected neurons in a complex set of synapses is a **neuronal pool.** How such pools operate depends on the synapses and circuits they contain.

Several kinds of synapses are recognized according to which parts of the two neurons join (Figure 11.21). The numerous **axodendritic** (ak″so-den-drit′ik) **synapses** relay signals from the axon of one neuron to the dendrites of the next neuron. They constitute the long distance signaling

system of neurons. Other synapses operate over shorter distances, especially in the brain. **Axoaxonic** (ak″-so-ak-son′ik) **synapses** transmit from axon to axon, **dendrodendritic** (den″dro-den-drit′ik) **synapses** transmit from dendrite to dendrite, and **axosomatic** (ak″so-so-mat′ik) **synapses** transmit from axon to soma.

Few neural paths are as simple as one axon to one dendrite. As shown in Figure 11.22, several axons can synapse with the dendrites and cell body of a single neuron and create a **convergence** of pathways, or a single axon can synapse with dendrites and cell bodies of several neurons and create a **divergence** of the pathway. Convergent and divergent pathways provide a mechanism for sorting and integrating neural signals. Regardless of the synaptic configuration, the process of chemical transmission is essentially the same.

Complex neuronal pools are found mainly in the brain, where they regulate body functions. It has been estimated that the human brain contains 100 billion neurons. If each neuron had a length from dendrite to axon terminal of only 1 mm, all the neurons in the brain placed end to end would reach around the earth's equator about ten times! Among these billions of neurons, each neuron probably synapses with hundreds to thousands of other neurons. With so many synapses on a given neuron, the need for the many different transmitters found in the brain becomes obvious.

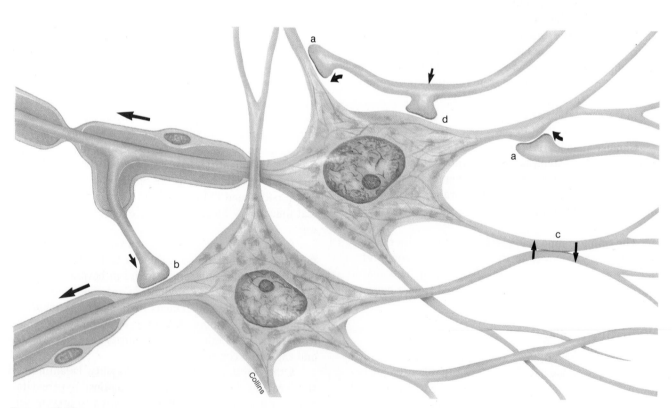

Figure 11.21
Kinds of chemical synapses: (*a*) axodendritic, (*b*) axoaxonic, (*c*) dendrodendritic, and (*d*) axosomatic.

If all the neurons released the same neurotransmitter and the transmitter had the same effect on all receptors, there would be no way to distinguish one signal from another. Confusion, such as having many telephones all with the same telephone number, would result.

Confusion in neuronal pools is reduced by the following:

1. Each transmitter-receptor complex has a specific effect.
2. Each postsynaptic neuron probably has different kinds of receptors and responds to a transmitter bound to a particular receptor in a particular way.
3. The postsynaptic graded response at single receptors fails to generate an action potential; several similar receptors must receive signals to cause a significant effect on the neuron.
4. Some transmitter-receptor interactions are excitatory and others are inhibitory, and the net effect is the algebraic sum of signals reaching the axon hillock.

A single neuron's action as a result of signal integration is the most basic level of integration in the nervous system. In neuronal pools, the same mechanism operates over many neurons simultaneously and signal integration becomes an extremely complex process. In a neuronal pool, presynaptic axon terminals synapse with as many as a thousand postsynaptic neurons; likewise, postsynaptic neurons can have another thousand synaptic connections. Some connections lead to excitation and some to inhibition. The number of terminals discharging on a neuron, whether the discharges are facilatory or inhibitory, and the rate at which neurotransmitter molecules are discharged determine whether an action potential will arise. Sometimes a single neurotransmitter, such as norepinephrine, has an excitatory effect at some synapses and an inhibitory effect at others. Furthermore, spatial and temporal summation can initiate action potentials even though none of the individual presynaptic signals alone could exert such an effect.

REACTIONS OF NERVE CELLS TO INJURY

For years, neurophysiologists believed that in animal development neurons became less able to divide and that dead cells could not be replaced. Studies in certain birds have

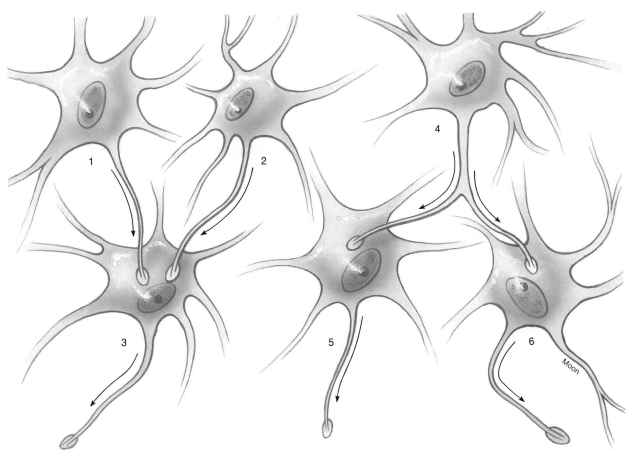

FIGURE 11.22
Axons of neurons 1 and 2 converge on neuron 3, whereas axon branches of neuron 4 diverge to neurons 5 and 6.

shown that new neurons can be produced in adulthood and that astrocytes provide pathways for new cells to migrate to their functional area in the brain. Whether new cells can be produced in response to injury is not yet known; however, progress is being made in stimulating growth of axons in the central nervous system. These developments include making implantable brain electrodes that can stimulate far away muscles, transplanting neural tissue that provides a channel for regrowth, and studying proteins that facilitate or inhibit regeneration.

Though little is known about cell division in neurons, much more has been learned about how injured neurons repair themselves (Figure 11.23). When an axon is cut, the ends retract, seal off, and swell, forming retraction bulbs. Swelling is due to the accumulation of mitochondria and vesicles, which normally move along an uninjured axon. Both ends swell because transport occurs in both directions along an intact axon.

In the distal segment, which has been severed from the cell body, axon terminals and the mitochondria themselves begin to swell within a day after injury. Swollen mitochondria are characteristic of impaired metabolism in dying cells. Within a week, the axon terminals lose contact with their postsynaptic cells. Over a period of two to three months, the entire distal segment—axon, axon terminals, and myelin sheath—degenerates and is phagocytized, though the neurilemma may persist. This is called **wallerian degeneration.**

Degeneration also occurs in a small portion of the proximal segment, usually from the point of injury back toward the cell body to the first branch in the axon. Provided the cell body survives, the proximal segment grows toward its former connection with a neuron or a skeletal muscle cell at 1 to 4 mm per day. If the neurilemma provides a conduit, new growth follows the old path. If no conduit exists, growth is haphazard and reestablishment of original synapses is unlikely.

Within two or three days of injury, the cell body itself reacts. If it dies, all connections between the dead cell and those with which it synapsed are severed. If it survives, it undergoes **chromatolysis** (kro″mat-ol′is-is), a process in which the nissl granules dissolve. Simultaneously, the cell body synthesizes large quantities of RNA and protein, apparently in preparation for regrowth of the axon stub. If the axon makes synaptic connections, chromatolysis ceases; but if it fails to make them, the cell may degenerate completely.

Glial cells can contribute to or interfere with healing. Astrocytes and microglia phagocytize cellular debris and toxic products of degeneration in the central nervous

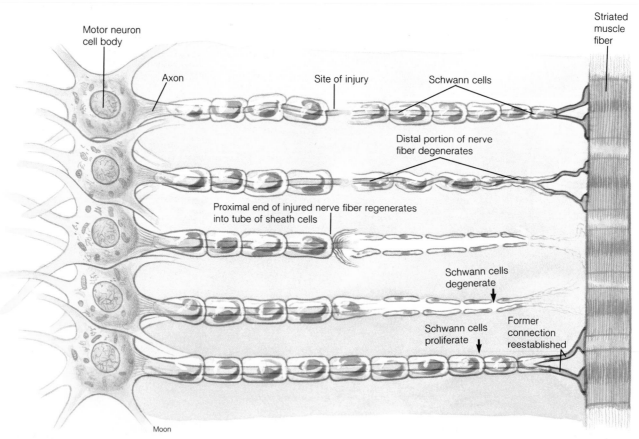

FIGURE 11.23
Degeneration and regeneration of a peripheral axon.

system; Schwann cells do the same in the peripheral nervous system. Various kinds of glial cells can interfere with healing by blocking the path of a regenerating axon, and they can proliferate to become tumors called **gliomas** (gli-o'maz).

Transneuronal degeneration, degeneration of cells associated with a degenerating neuron, occurs after injury at certain sites. When axons that supply skeletal muscle cells degenerate, the muscle cells become paralyzed, lose bulk, and eventually waste away. When axons that extend from the retina to a part of the brain called the thalamus degenerate, thalamic neurons that synapse with them also degenerate. One explanation of transneuronal degeneration is that without neural stimulation, muscle cells and neurons cannot maintain normal metabolism. Frequent changes in membrane potentials may be needed to maintain neurons.

Compared with tissues whose cells can divide, neural tissue has limited capacity for repair. Neurons compensate for this deficiency to some extent by regrowing axons. Surgical techniques are being developed to rejoin cut ends of peripheral nerves before degeneration occurs (Figure 11.24). These techniques involve freezing severed nerve

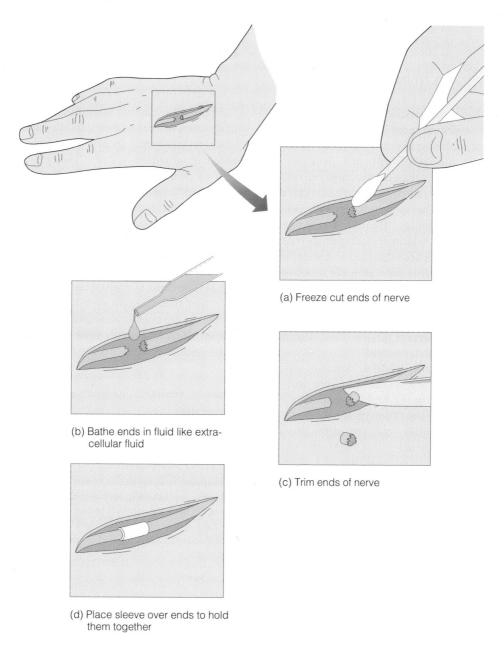

(a) Freeze cut ends of nerve

(b) Bathe ends in fluid like extra-cellular fluid

(c) Trim ends of nerve

(d) Place sleeve over ends to hold them together

FIGURE 11.24
Surgical procedure to repair a cut nerve.

ends and bathing them in a fluid similar to intracellular fluid, trimming the ends, and holding them in close approximation with a thin sleeve that allows regrowth. Infusion of a protein called **nerve growth factor** prevents degeneration of brain cells in some animals, and it may prove useful in preventing or retarding degeneration of neurons in humans with disorders such as Alzheimer's disease. Implanted fetal brain tissue, which can grow and release neurotransmitters in adults, may effectively treat disorders such as Parkinson's disease. Implanted mature nerve tissue cannot conduct signals, but it may provide a channel for axon growth. Other promising experiments in nervous system repair involve implanting electrodes that can transmit electrical signals to cause movement in paralyzed limbs.

CLINICAL APPLICATIONS

More a symptom than a specific disease, **neuritis** (nu-ri'tis) is a general term for disturbances in the peripheral nervous system. Sensory or motor fibers, or both, can be damaged. Pain is nearly always present, and paralysis or unusual sensations such as tingling or numbness can occur. Neuritis can be caused by trauma (blows, fractures), exposure to toxic substances (heavy metals, some drugs), or deficiencies of some B vitamins. Treatment is directed toward the cause, if known. Aspirin and other drugs reduce pain, and weakened muscles are massaged and stretched.

Neuralgia (nu-ral'je-ah) is pain in a circumscribed area innervated by a particular sensory peripheral nerve. It is usually impossible to determine the cause of the pain, and very difficult to relieve it. In severe cases, the sensory nerve supplying the painful area is surgically cut. The area becomes numb, but the pain is relieved.

The neurological disorder **multiple sclerosis** (sklero'sis) is characterized by the loss of myelin from neurons in multiple, scattered sites in the brain and spinal cord, altering conduction in neurons. Though demyelinization is not fully understood, proteins released in the process damage neurons. Eventually, sclerotic (hard) scar tissue replaces myelin and neurons, and interferes with conduction (Figure 11.25).

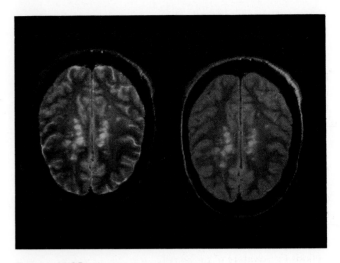

FIGURE 11.25
Sclerotic scar tissue of multiple sclerosis.

Multiple sclerosis most often affects young adults 20 to 40 years of age. At onset, mild symptoms such as slight visual impairment, speech disturbances, and uncoordinated handwriting appear. Though neural damage increases over the course of the disease, remissions often occur and the disease can stabilize for a year or more. Remission probably results from myelin repair by glial cells. Following remissions, new symptoms or previous symptoms in a more incapacitating form may appear. Muscle weakness, lack of muscular coordination, loss of bladder and bowel control, mental impairment, and eventual death result. The cause of multiple sclerosis is unknown, but viral infection that triggers an autoimmune reaction against myelin in genetically susceptible individuals seems likely. Experiments in animals show that administering certain antibodies may allow myelin replacement.

CLINICAL TERMS

amyotrophic (ah-mi''o-tro'fik) lateral sclerosis atrophy of muscles associated with degeneration of neurons in the lateral spinal cord

Guillian-Barré (ge-yan' bar-ra') **syndrome** muscle weakness associated with encephalitis

neuropathy (nu-rop'ah-the) any disorder of neural tissue

shingles an inflammatory disorder caused by herpes zoster virus, which damages cell bodies of dorsal root ganglia

tic douloureux (tik doo-loo-roo') a spasmodic twitching of facial muscles accompanied by pain, also called trigeminal neuralgia

ESSAY
BIOELECTRICITY

by Norman N. Goldstein
Professor of Biological Sciences
California State University at Hayward

Imagine being treated for illness by having the electric torpedo fish applied to your body to jolt the affected part. Greek physicians did just that in the first century A.D. Within the last twenty years, electrotherapy based on scientific principles has yielded new methods for relieving pain, hastening bone fracture repair, healing wounds, correcting scoliosis (a curvature of the spine), and relaxing muscle spasms. Better understanding and broader application of electrotherapy depends on learning more about the intimate relationship between electricity and life functions.

Living things generate bioelectrical fields and live in the electromagnetic fields of the earth. If, along with using microscopes to describe cell anatomy, we were to use voltmeters and analyses of bioelectromagnetic lines of force, we might develop an electrical anatomy. Bridging physical and electrical anatomies with metabolic reactions could yield a more complete physiological picture.

Membranes acting as selective barriers distribute certain ions—K^+, Na^+, Ca^{2+}, Cl^-—asymmetrically. The separation of charge produces a gradient, which, in turn, affects membrane selectivity. The separation and movement of ions confer excitability on sensory, nerve, and muscle cells, and on certain plant cells. Changes in ionic distribution accompany growth, locomotion, and other processes in all cells.

Studies of electrical phenomena in several species have uncovered some surprises. The common aquarium greenery, *Lemma* (duckweed) and *Elodea,* manifest surprisingly large membrane potentials—much greater than those in animal cells. The fungus *Neurospora,* a valuable and long-used tool of the geneticist, has a comparably high membrane potential. The greatly elongated cells of the algae *Nitella* and *Chara,* when stimulated, propagate action potentials similar to those in animal axons, but with very high peaks (175 mV) and durations of several seconds. Although these algal action potentials have no known role in communication, they may play a role in cytoplasmic streaming. Action potentials in the Venus fly trap and sundew plant are more clearly adaptive. When unwary insects bumble over trigger hairs in open trap-like leaves, action potentials in the trigger hairs close the traps. Such action potentials can be recorded along plant stems.

Conspicuously rapid plant movements have fascinated many scientists, yet we know little about the mechanisms involved. When touched or stimulated by heat or ammonia vapor *Mimosa pudica* quickly folds its leaflets and its petioles droop as action potentials of 50 to 100 mV in amplitude and a few seconds in duration pass through them. Though these events involve loss of plant tissue turgor in response to stimuli, the relationship of this loss to action potentials is unclear.

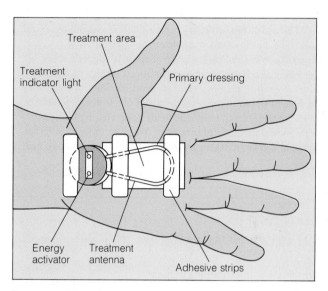

FIGURE 11.26
Electric bandage.

From the perspective of health care professionals, studying the effect of electrical currents on animal tissue growth has been fruitful and has led to some medical applications. Limb regeneration in amphibians appears to be guided by electrical fields spreading as a carpet from the stump of the severed limb. Similar currents have been seen in accidental guillotine-like fingertip amputations in children. If a finger severed beyond the distal joint is kept moist, even with no flap sewn over it, satisfactory regrowth may occur. Even adults with fingertip amputations can benefit from such "conservative" management. Recalcitrant bone fractures have been healed with both direct and pulsating currents. Certain degenerative diseases of bone have been halted by prolonged localized impression of an electrical field. Direct current fields appear to hasten wound healing, and certain devices have received approval for marketing from the Federal Drug Administration (Figure 11.26).

Acute and chronic pain can be controlled by a **transcutaneous electric nerve stimulator** (TENS), an electronic pulse generator, which can be carried in a small pocket. TENS produces a perceptible but not uncomfortable pulse between one or two pairs of electrodes strategically placed on the body surface. The resulting pain relief at joints, muscles, traumatized tissue, or surgical wounds may be due to stimulation of large sensory neurons that blocks signal conduction in small pain neurons. Some very painful syndromes can be attenuated by direct implantation of TENS electrodes into the spinal cord or brain. This effect probably involves release of natural opiate-like substances called endorphins and enkephalins, and activation of pain inhibitory neural circuits.

The use of bioelectricity can be likened to a double-faced mask. The smiling face symbolizes the application of electricity to control biological processes. It suggests the possibility of reliably augmenting natural biological fields to enhance growth, healing, and regeneration. The frowning face symbolizes the worrisome concern that electrotherapy and environmental electromagnetic fields generated by humans might stimulate harmful processes. We need to know whether cancer and other diseases among people who work or live near

microwave paths, radar stations, power line radiation, or radio antennae are related to these sources of non-ionizing radiation. Studies focused on these issues are fraught with measurement problems and have failed to yield definitive conclusions. The possible harm from the prevalence of electric field exposure seems so ominous, however, that federal policy is being considered without awaiting a final verdict from science.

Questions

1. In what ways do living cells respond to electricity?
2. How do living things generate electricity?
3. How can electricity be used therapeutically?
4. What might be some positive and negative effects of environmental electrical phenomena?

CHAPTER SUMMARY

Organization and General Function

- Components of the nervous system are summarized in Figure 11.1.
- In general, the nervous system receives and processes information from the external and internal environment and initiates responses.

Cells of Nerve Tissue

- Neurons are signal conducting cells of the nervous system.
- Neuroglia are nonconducting cells that produce myelin, and protect and possibly nourish neurons.

Excitability of Neurons

- Excitability, the ability to respond to stimuli, depends on electrical events and changes in membrane permeability.
- A potential difference created by the unequal distribution of ions inside and outside a cell polarizes the cell membrane.
- Resting, polarized membranes of neurons typically have a potential of about -70 mV, the inside being negative to the outside.

Neurons as Signaling Units

- When a membrane is stimulated and its permeability altered, sodium ions flow into the cell and the potential approaches the equilibrium potential for sodium.

- Shortly after this event, the membrane permeability to potassium increases, and the potential approaches the equilibrium potential for potassium.
- An action potential is generated when membrane permeabilities are altered as shown in Figures 11.11 and 11.12.
- Action potentials occur when stimuli exceed the threshold level, either occur at full size or not at all (the all-or-none principle), and display absolute or relative refractoriness.
- Conduction velocity of an axon increases with diameter.
- Saltatory conduction along a myelinated axon is rapid because depolarization occurs only at the nodes of Ranvier.
- Graded potentials are local, decremental effects on membrane permeability; they include receptor potentials and synaptic potentials, and are subject to temporal and spatial summation.

Synaptic Transmission

- A synapse is a specialized junction between the axon terminal of a presynaptic neuron and a portion (usually a dendrite) of a postsynaptic neuron.
- Most synapses are chemical; neurotransmitter released from axon terminals diffuses across the synaptic cleft.
- Binding of the neurotransmitter to receptors on the postsynaptic neuron elicits a synaptic potential.
- Most neurotransmitters are inactivated by enzymes.
- Signals at synapses obey the principle of forward conduction going from presynaptic axon to postsynaptic neuron.

- Excitatory postsynaptic potentials (EPSPs) depolarize the postsynaptic neuron, and inhibitory ones (IPSPs) hyperpolarize it.
- The net effects of excitatory and inhibitory influences at the trigger zone determine whether action potentials occur; such effects serve to coordinate and integrate body function.
- Synaptic delay is due to time required for release, diffusion, and action of a neurotransmitter.
- Synaptic fatigue is due to depletion of neurotransmitter.

Neurotransmitters

- Neurotransmitters (Table 11.4) are substances released from presynaptic axons, usually one kind per neuron.
- Neuroactive peptides are summarized in Table 11.5.

Neuronal Pools

- Neuronal pools are complex arrangements of synapses found mainly in the brain that regulate body functions.
- A single axon synapsing with several postsynaptic neurons creates a divergent pathway; several axons synapsing with a single postsynaptic neuron create a convergent pathway.

Reactions of Nerve Cells to Injury

- Injured neurons undergo degeneration of damaged tissue and axonal regeneration if the cell body survives.
- New synapses can be formed, usually with less precise distribution than before injury.

Clinical Applications

- Neuritis is pain or other unusual sensation anywhere in the peripheral nervous system, whereas neuralgia is pain in a specific peripheral nerve.
- Multiple sclerosis is a progressive degenerative disorder due to the loss of myelin from neurons.

QUESTIONS AND PROBLEMS

The questions at the end of each chapter are numbered to correspond with the objectives listed at the beginning of the chapter. Italics indicate that a question requires critical thinking skills beyond simple factual recall.

Questions

1. (a) *How are the components of the nervous system organized?*
   (b) What, in general, does the nervous system do?
2. (a) What are the distinguishing properties of neuroglia?
   (b) *How do these properties relate to neuroglial function?*
3. (a) What are the distinguishing properties of neurons?
   (b) *How do these properties relate to neuron function?*

4. (a) *How do electrical principles relate to neuron function?*
   (b) *What is a resting membrane potential, and how is it produced?*
   (c) *How does the membrane pump affect neural function?*
5. (a) *How do receptor and synaptic potentials differ?*
   (b) What is the trigger zone's role in action potentials?
6. (a) *What happens at the axon membrane in each phase of an action potential?*
   (b) *What factors determine whether an action potential will be generated and how fast it will travel?*
   (c) *How is an action potential propagated along an unmyelinated neuron? Along a myelinated neuron?*
   (d) *How do graded potentials affect neuron membranes?*
7. *How do graded potentials differ from action potentials?*
8. (a) What events take place at a chemical synapse?
   (b) *How do EPSPs and IPSPs generate action potentials?*
   (c) *Explain the principles of forward conduction, synaptic delay, and synaptic fatigue.*
   (d) *What are the similarities and differences in action potentials and synaptic potentials?*
   (e) *How does signal integration occur, and how does it affect neural function?*
9. Define neurotransmitter, and list the properties of the main neurotransmitters.
10. (a) What are neuroactive peptides?
    (b) *How do exogenous substances affect neuronal membrane permeability?*
11. (a) *What is a neuronal pool, and what is its significance?*
    (b) *How do different kinds of synapses and divergent and convergent pathways function in neuronal pools?*
12. (a) How do neurons respond to injury?
    (b) What are the limitations on repair of nerve injuries?
13. (a) Distinguish between neuritis and neuralgia.
    (b) What neural damage produces symptoms of multiple sclerosis, and what are some of its effects?

Problems

1. Predict the effects on a neuron of the closing of all the channels for each of the following ions—Na^+, K^+, and Ca^{2+}.
2. What would happen if myelin were removed from neurons in an arm or a leg? In the brain?
3. Prepare a short report on current research on one of the following topics: (a) voltage-gated ion channels, (b) putative neurotransmitters, or (c) neuroactive peptides.

12

CENTRAL NERVOUS SYSTEM

THE NERVOUS SYSTEM AND ITS DEVELOPMENT

Neurons, which generate and transmit signals, are largely responsible for functions throughout the nervous system. We find out what is going on in our environment by way of signals received in our brains. When we respond, our brains send signals that allow us to carry on various activities, such as talking and moving. We considered the properties of neurons in the last chapter. In this and the next few chapters, we will look at broader functions of major components of the nervous system, beginning in this chapter with the central nervous system—brain and spinal cord. We will consider the somatic nervous system and its role in motor functions in Chapter 13 and the autonomic nervous system in Chapter 14. Finally, we will consider sensory functions in Chapter 15.

As described in Chapter 4, the outer surface of an embryo consists of ectoderm. Early in development, dorsal ectoderm becomes neural ectoderm and differentiates into neurons and neuroglia, except for microglia, which are derived from mesoderm. The meninges, membranes that cover the external surfaces of the brain and spinal cord, are connective tissue derived from mesoderm. Similar sheaths around bundles of fibers in the peripheral nervous system are also derived from mesoderm.

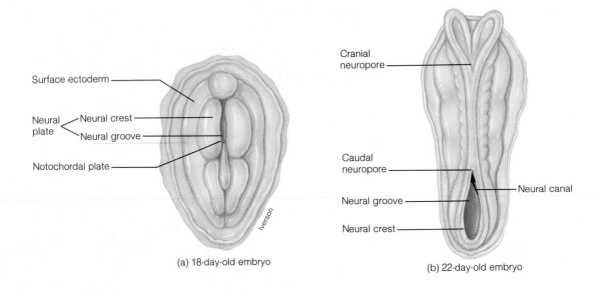

(a) 18-day-old embryo

(b) 22-day-old embryo

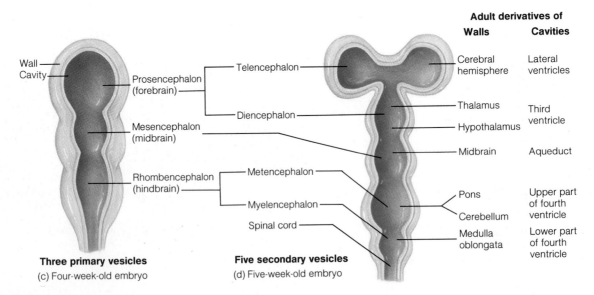

Three primary vesicles
(c) Four-week-old embryo

Five secondary vesicles
(d) Five-week-old embryo

FIGURE 12.1
Development of the nervous system.

Ectodermal cells of the neural plate in the embryo's posterior midline form a neural fold, which deepens and closes over to form a hollow **neural tube.** Cells extending laterally make up the **neural crests** (Figure 12.1a). By the end of the first month of development, the neural tube extends from the head to the tail of the embryo and is completely closed. The long cavity within the neural tube becomes the central canal of the spinal cord and the ventricles of the brain.

The head end of the neural tube enlarges and differentiates into the **forebrain** (prosencephalon), **midbrain** (mesencephalon), and **hindbrain** (rhombencephalon). The forebrain further differentiates into the telencephalon and diencephalon, and the hindbrain further differentiates into the metencephalon and the myelencephalon. Parts of the mature brain derived from each embryonic region are listed in Table 12.1. The remainder of the neural tube becomes the spinal cord.

Portions of the neural tube also give rise to the motor, or efferent, portions of the cranial and spinal nerves, the retina and optic nerves of each eye, and a portion of the pituitary gland (Figure 12.1b). Neural crest cells give rise to sensory, or afferent, portions of the cranial and spinal nerves, the autonomic nervous sytem, and the medulla of the adrenal gland. The fact that parts of glands are derived from the same embryonic tissue as the nervous system illustrates the close relationship between the body's control systems.

The nervous system has **plasticity;** that is, it can be modified. Studies on visual systems in kittens show that if one eye is sutured shut in early months after birth, the kitten has a permanent brain deficit in interpreting signals from that eye. Light stimulation during a critical period (2 to 4 months in kittens) appears to be essential to the development of normal brain processing. One mechanism by which light might alter brain function is through the dephosphorylation of a protein called microtubule associated protein 2 (MAP2), which is part of the cytoskeleton of neurons. In the dark, MAP2 remains phosphorylated and interferes with cytoskeleton development in certain neurons. In the light, MAP2 is dephosphorylated, and links form between elements of the cytoskeleton. Such links apparently contribute to controlling molecular movements associated with synaptic transmission and normal brain processing of light signals.

Normally, the brain and spinal cord are surrounded by the skull and vertebrae, respectively. In about 1 per 1000 live births, the neural tube fails to close, producing a **neural tube defect** (Figure 12.2). If the defect arises during the first month of pregnancy, neural tissue usually remains exposed on the body surface. An opening at the anterior end of the neural tube results in **anencephaly** (an''en-sef'al-e), failure of the brain to develop. An opening at the posterior end results in *spina bifida* (spi'nah bif'id-ah). Spina bifida is often accompanied by *myelomeningocele* (mi''el-o-men-ing'go-sēl) in which part of the spinal cord and its meninges protrude and form a lump on the lower back. Individuals with anencephaly usually are stillborn, but those with spina bifida and myelomeningocele have varying degrees of paralysis, depending on which spinal nerves are affected and how severely their distribution is disturbed.

Though the cause of neural tube defects is not known, the incidence of such defects is especially high in Ireland and Wales where an average of six cases per 1000 live births is reported. Less than 10 percent of cases occur in families with a history of such defects so defective genes are not a major cause of neural tube defects. Many environmental factors have been considered in the search for the cause of these defects; poor maternal nutrition seems to be of some importance. Vitamin supplements given to mothers prior to and after conception may reduce the incidence of neural tube defects. An excess of a fetal blood protein called alpha-fetoprotein in amniotic fluid during the first half of pregnancy suggests a neural tube lesion.

TABLE 12.1 Embryonic Divisions of the Brain and Structures Derived From Them		
	Embryonic Brain Divisions	**Mature Brain Structures**
Forebrain (prosencephalon)	Telencephalon	Cerebral cortex Basal nuclei Olfactory bulbs
	Diencephalon	Thalamus Hypothalamus
Midbrain (mesencephalon)		Cerebral peduncles Corpora quadrigemina Several important nuclei
Hindbrain (rhomb-encephalon)	Metencephalon	Pons, including its nuclei Cerebellum
	Myelencephalon	Medulla oblongata, including its nuclei

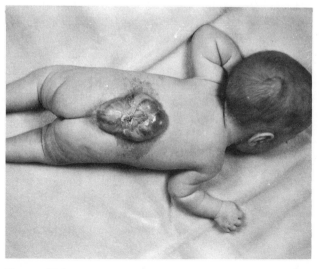

FIGURE 12.2
Neural tube defect.

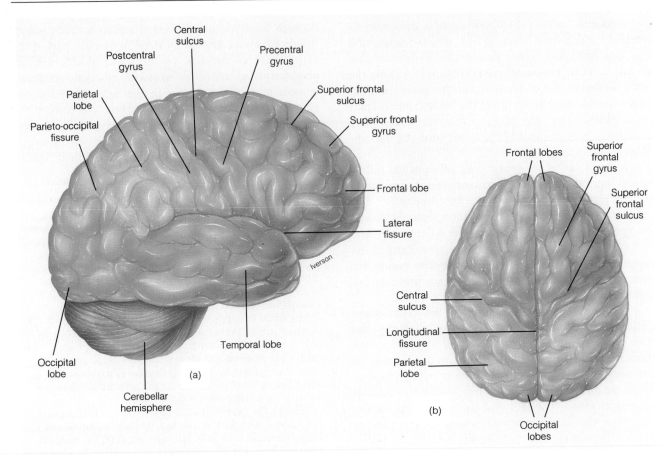

FIGURE 12.3
The external surface of the brain, showing the lobes of the cerebrum, gyri, sulci, and the relationships of the cerebrum to other parts of the brain: (*a*) lateral view, and (*b*) superior view.

EVOLUTION OF THE BRAIN

The vertebrate brain has undergone millions of years of evolution, and the mature human brain contains structures that reflect this evolution. One step along the evolutionary path was the *reptilian brain,* which was somewhat like the brain of modern reptiles with rudiments of all basic vertebrate brain components. Next to evolve was the **limbic** (lim′bik) **system,** which seems to be concerned with emotional experiences and with interpreting the significance of perceived objects and events. For example, monkeys whose limbic system has been partially destroyed will eat nuts and bolts and other nonfood items, presumably because their limbic system cannot distinguish between food and nonfood items. The most recently evolved **neocortex** (ne″o-kor′teks) makes up the largest part of the human cerebrum. Parts of the human neocortex concerned with speech, spatial abilities, memory, and planning do not complete development until about age 10; thus, it is understandable that speech and other intellectual abilities develop over a time period correlated with brain development.

REGIONS OF THE BRAIN

The human brain is an exceedingly complex organ with many components (Figure 12.3 and Table 12.2). By far the largest part of the human brain is the **cerebrum,** which is divided into left and right hemispheres. All conscious processes occur in the cerebrum.

The surface of the cerebrum is greatly folded. The upward folds, or **gyri** (gi′re)—singular, gyrus (ji′rus)—alternate with the downward grooves, or **sulci** (sul′se)—singular, sulcus (sul′kus). Larger sulci called **fissures** divide the cerebrum into two hemispheres and each hemisphere into lobes. The **longitudinal fissure** separates the

TABLE 12.2   Major Structures in the Brain and Their Functions

Structure	Function	Structure	Function
Medulla oblongata	Receives and integrates signals from spinal cord and sends signals to the cerebellum and thalamus Contains centers that regulate heart beat, blood pressure, respiratory rate, coughing, and some other involuntary movements		Provides structural and functional connection between the nervous and endocrine systems by its relationship to the pituitary gland In combination with the limbic system participates in physiological response to emotional experiences
Pons	Relays signals between the medulla and more superior parts of the brain, between the hemispheres of the cerebellum, and between the cerebellum and cerebrum	Cerebellum	Receives sensory signals from the eyes; organs of balance; and receptors in muscles, tendons, and joints Uses sensory information to control and integrate complex voluntary movements
Midbrain	Relays sensory signals between the spinal cord and the thalamus, and motor signals between the cerebral cortex and the pons and spinal cord Controls reflexive movements of the head and eyeballs in response to visual stimuli; controls reflexive movements of the head and trunk in response to auditory stimuli	Cerebrum	Contains areas that receive and process sensory signals (somatic sensory area, visual area, auditory area) and that initiate motor signals for voluntary movements (somatic motor area and speech area) Contains association areas where sensory signals are interpreted, memories are stored, and complex processing occurs Contains tracts of association fibers, commissural fibers, and projection fibers that relay signals between the cerebral cortex and other parts of the nervous system
Thalamus	Contains many nuclei through which it relays all sensory signals (except smell signals) to the cerebral cortex Relays motor signals from the cerebral cortex toward the spinal cord Relays signals to the cerebral cortex that maintain consciousness Processes some crude sensations	Basal nuclei	Help to control muscle tone and thereby help to coordinate voluntary movements
		Limbic system	Contains pleasure and punishment centers, so plays a role in emotional feelings Hippocampus plays a role in determining what memories will be stored
Hypothalamus	Receives sensory signals from internal organs by way of the thalamus and uses these signals to control actions of the autonomic nervous system and pituitary gland, thereby helping to maintain homeostasis	Reticular formation	Contains nuclei involved in wakefulness and sleep

two hemispheres. The **central sulcus** separates the frontal and parietal lobes of each hemisphere and the **lateral fissure** separates the temporal lobe from the remainder of the hemisphere.

The second largest part of the human brain, the **cerebellum** (ser″e-bel′um), is located under the occipital bones posterior and inferior to the cerebrum. It has two hemispheres and small, nearly horizontal surface folds. The cerebellum helps to coordinate and control movements initiated by the cerebrum.

The remainder of the brain is the **brain stem.** Its various regions are responsible for regulating many important processes—heart function, blood pressure, breathing, body temperature, thirst, and appetite—to name a few.

All the major regions of the brain are interconnected by thousands of myelinated tracts and several nuclei. A **tract** is a bundle of axons in the central nervous system. A **nucleus** is an aggregation of nerve cell bodies. In general, tracts consist of **white matter** because they are myelinated, and nuclei (being unmyelinated) consist of **gray matter.** White matter is internal in the cerebrum, and gray matter makes up the cerebral cortex. A gradual transition of intermingled gray and white matter occurs in the brain stem. In the spinal cord, gray matter is internal, and white matter is on the surface. In the peripheral nervous system, bundles of axons are called **nerves,** and an aggregation of cell bodies of afferent neurons is a **ganglion** (gang′gle-on). Many nerves branch from the brain and spinal cord, and each has one or more ganglia associated with it.

Brain Stem

Small, but nonetheless important, regions found in the brain stem include the medulla oblongata, pons, midbrain, thalamus, and hypothalamus (Figure 12.4).

Medulla Oblongata

The **medulla oblongata** (med-ul'ah ob''long-gah'tah), which is continuous with the spinal cord, contains important nuclei (gray matter) and tracts (white matter) that carry signals between the spinal cord and other parts of the brain.

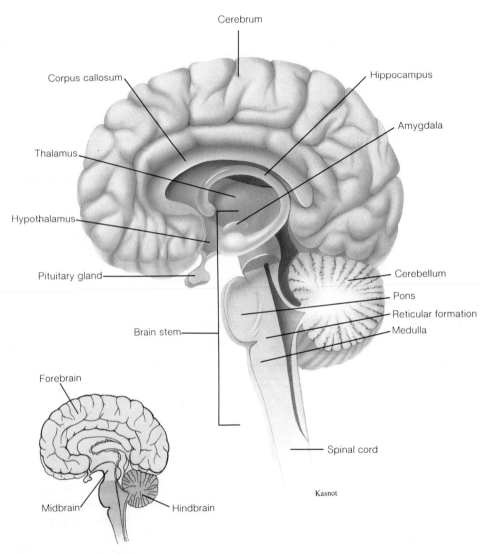

Cerebrum

Corpus callosum

Thalamus

Hypothalamus

Pituitary gland

Brain stem

Forebrain

Midbrain

Hindbrain

Hippocampus

Amygdala

Cerebellum

Pons

Reticular formation

Medulla

Spinal cord

Kasnot

FIGURE 12.4
A sagittal section through the brain showing the cerebrum, the cerebellum, and the components of the brain stem.

Some nuclei of the medulla receive signals from axons in the tracts, integrate them with other information, and send the resulting signals on to the cerebellum or the thalamus. Other nuclei, usually called centers, receive signals from other brain regions, integrate the information, and send out signals that control various internal functions. Signals from the **cardiac center** regulate heart rate. Those from the **vasomotor center** regulate blood vessel diameters. Those from the **respiratory center,** together with some from the pons, regulate the rate and depth of breathing. Still other medullary centers control coughing, sneezing, vomiting, swallowing, and hiccoughing. Cranial nerves IX through XII originate in nuclei in the medulla.

Pons

Lying superior to the medulla in the brain stem is the **pons** (literally, bridge). It is well named because many fibers pass through it carrying signals between brain regions. Some signals from the spinal cord and medulla go to more superior brain regions. Others go between the two hemispheres of the cerebellum or between the cerebellum and the cerebrum. Signals going between the cerebellum and the cerebrum are especially important in the coordination of voluntary movements. Nuclei of cranial nerves V through VIII and several respiratory centers also are located in the pons.

Midbrain

Located superior to the pons, the **midbrain** has on its anterior surface two bundles of fibers called the **cerebral peduncles** (ser′e-bral ped′ung-kl). These fibers carry sensory signals from the spinal cord to the thalamus, and motor signals from the cerebral cortex to the pons and on to the spinal cord.

The posterior surface of the midbrain has two pairs of rounded protrusions collectively called the *corpora quadrigemina* (kor′po-rah kwod″rij-em′in-ah). One pair, the *superior colliculi* (kol-ik′u-li), control head and eyeball movements in response to visual and other stimuli. The other pair, the *inferior colliculi,* control head and trunk movements in response to auditory stimuli. The nuclei of cranial nerves III and IV, which control movements of the eyeballs, also are located in the midbrain.

Other important nuclei include the **substantia nigra** (sub-stan′she-ah ni′grah), a dark area near the cerebral peduncles, and the **red nucleus,** located more inferiorly. As we shall see, these nuclei are connected to the basal nuclei of the cerebrum and are concerned with motor functions.

Thalamus

Forming the junction between the brain stem and cerebrum, the **thalamus** (thal′am-us) has many nuclei. Except for the sense of smell, signals from all major sense organs reach the cerebrum by way of the thalamus, and such signals help to maintain consciousness. The thalamus relays some motor signals from the cerebrum to the spinal cord and processes a few sensory signals such as crude sensations of pain and temperature without relaying them to the cerebral cortex. The thalamus behaves like a telephone switchboard, relaying signals through synapses; and occasionally, it interprets and responds to signals like an operator.

Hypothalamus

Located inferior to the thalamus, the **hypothalamus** (hi″pothal′am-us) is a center for homeostasis. It receives signals from many organs and uses the information to regulate many internal processes by way of the autonomic nervous system (Chapter 14). The hypothalamus connects the nervous and endocrine systems both structurally and functionally. It receives and sends neural signals, and it synthesizes hormones, some traveling in the blood to the anterior pituitary gland and others traveling down axons to the posterior pituitary gland (Chapter 16).

The hypothalamus also receives signals from the limbic system as it receives emotional stimuli. The hypothalamus processes information and relays hormonal signals to the pituitary gland and neural signals to the autonomic nervous system. When you are frightened, the limbic system relays signals to the hypothalamus, which sends various signals to prepare for an emergency. Your blood pressure and heart rate increase and your muscles receive more glucose and oxygen so they can contract forcefully over a longer period of time. When stress is prolonged, the hypothalamus initiates other processes that help your body to withstand the stress (Chapter 16).

Certain hypothalamic centers operate by negative feedback to regulate body temperature and blood osmotic pressure. When your body temperature rises or drops, the hypothalamus initiates changes that bring the temperature back to normal. Likewise, when blood osmotic pressure increases (the blood has become too concentrated), the hypothalamus sends signals that result in the sensation of thirst. When you drink fluids, some of the fluid enters the blood and brings the blood osmotic pressure back to normal. The hypothalamus also contains hunger and satiety centers that regulate food intake (Chapter 23).

CEREBELLUM

The cerebellum (Figure 12.5) consists of lateral hemispheres and a central worm-shaped **vermis** (ver′mis). A cross section through a hemisphere shows a branching treelike arrangement of white matter called the **arbor vitae** (tree of life). You are never aware of the actions of your cerebellum, yet you depend on it for nearly every movement you make. If your cerebellum were not functioning properly, you could not throw or catch a ball, write a word, or even walk in a straight line.

The cerebellum is attached to the brain stem by three pairs of tracts, the *cerebellar peduncles,* each of which carry signals between the cerebellum and an associated brain stem structure. The *superior cerebellar peduncles* are associated with the midbrain, the *middle cerebellar peduncles* with the pons, and the *inferior cerebellar peduncles* with the medulla.

The cerebellum has an outer layer of gray matter called the **cerebellar cortex.** Among the kinds of cortical neurons, the **Purkinje** (pur-kin′je) **cells** are the most thoroughly studied. Their dendrites can have as many as 100,000 synapses, and their axons contribute to the many

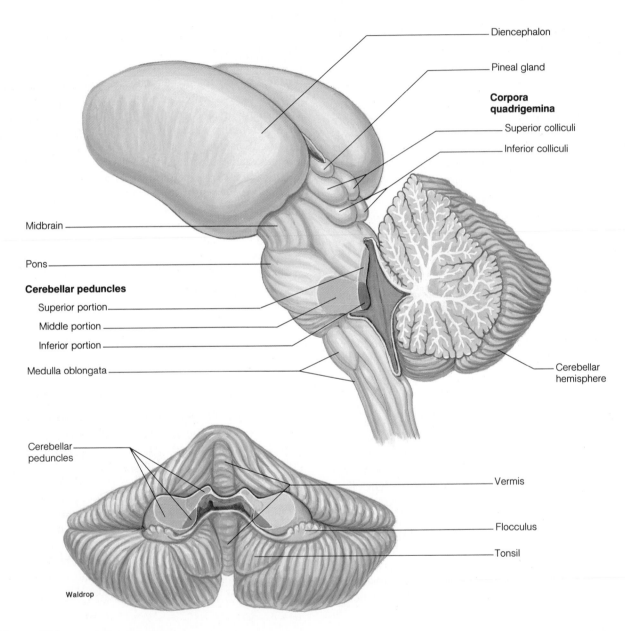

FIGURE 12.5
External components and internal components of the cerebellum and the location of the cerebellar peduncles on the brain stem.

myelinated tracts between the cerebellar cortex to the brain stem. These tracts also contain fibers of neurons from other parts of the brain that carry signals to the cerebellar cortex. Together, signals in these tracts control and integrate muscle contraction.

By controlling muscle contraction, the cerebellum coordinates movements and helps to maintain posture and balance. All its activities take place unconsciously. The cerebellum receives signals from the eyes, organs of balance, tendons, joints, and muscles themselves. It also receives signals from collateral branches of motor neurons responsible for initiating a voluntary muscle action. Most signals from the cerebellum are inhibitory and tend to dampen and smooth movements. They help to maintain muscle tone, integrate contractions of several muscles in complex movements, and refine and coordinate voluntary movements.

Cerebrum

In everyday language, we sometimes use the word *brain* when we really mean the part of the cerebrum that carries out conscious thought processes. Except for some crude

sensations that are interpreted in the thalamus, awareness of any object, event, or thought means that it has been processed somewhere in the cerebrum. Also, any intentional movement is initiated by the cerebrum.

Each hemisphere of the cerebrum has four lobes—the frontal, parietal, occipital, and temporal—located beneath the skull bones of the same name. Furthermore, the cortex of each lobe has certain functional areas, which receive signals from, and send signals to, various other parts of the nervous system. Various tracts and nuclei beneath the cortex are essential to this signalling process.

Tracts

Cerebral white matter beneath the cortex consists of three kinds of tracts, or bundles of myelinated fibers (Figure 12.6). **Association fibers** carry signals between two functional areas within the same hemisphere. **Commissural** (kom-mis'u-ral) **fibers** carry signals between the two hemispheres. The largest commissural tract is the **corpus callosum** (kor'pus kah-lo'sum). **Projection fibers** carry signals between the cerebral cortex and other parts of the brain and spinal cord.

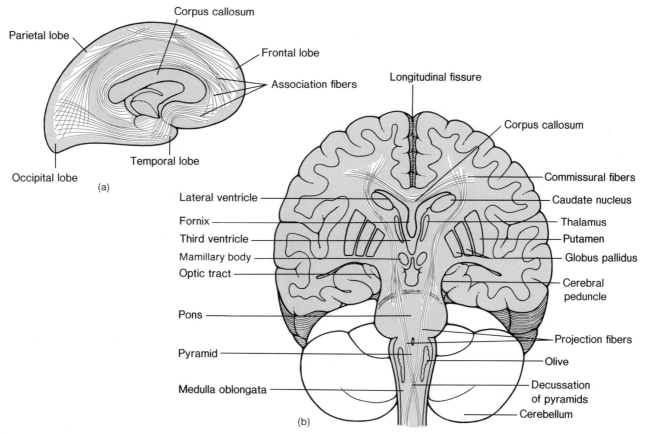

FIGURE 12.6
Some tracts of the cerebrum: (*a*) association fibers and (*b*) commissural and projection fibers.

Basal Nuclei

Deep inside the cerebrum are at least two pairs of **basal nuclei** (Figure 12.7), with one of each pair in each hemisphere. The largest of these aggregates of neuronal cell bodies is the *corpus striatum* (stri-a'tum). Most neuroanatomists agree that it includes the *caudate* (kaw'dāt) *nucleus,* the *globus pallidus* (glo'bus pal'id-us), and the *putamen* (pu-ta'men). The smaller, almond-shaped *amygdaloid* (am-ig'dal-oid) *nucleus* is attached to the tail of the caudate nucleus. These nuclei are connected to each other and to *subthalamic nuclei* such as the substantia nigra and the red nucleus of the midbrain.

The main function of the basal nuclei is to control muscle tone, often by inhibiting contraction. Activities of basal nuclei are coordinated with those of the cerebellum to control gross intentional, but automatic, movements, such as walking. These nuclei also appear to maintain and control muscle tone required for specific fine muscle actions. For example, when you prepare to write something on a piece of paper, you place your arm in a certain position. The basal nuclei maintain appropriate tone in arm muscles so you can use muscles in your hand to write. Destruction of the basal nuclei makes it impossible to maintain the arm in a position that allows the fine muscles of the hand to be used effectively. Damage to basal nuclei can lead to the various kinds of tremors (involuntary movements).

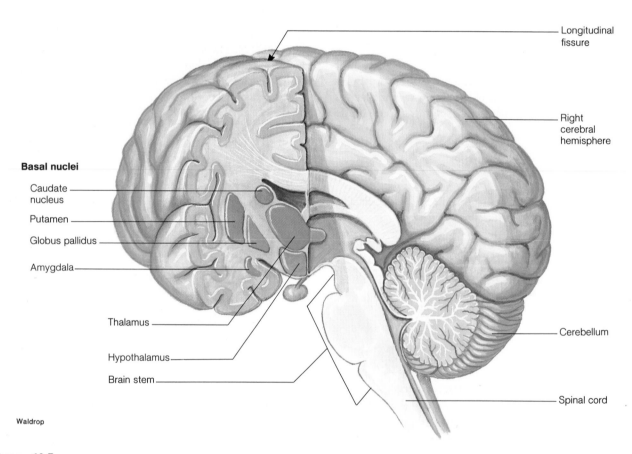

Waldrop

FIGURE 12.7
The location of the basal nuclei of the cerebrum.

Limbic System

The **limbic system** is located mostly in the cerebrum, but extends into the brain stem (Figure 12.8). Its components, found in each hemisphere, consist of two gyri of the cerebral cortex and two nuclei. The gyri are the **hippocampal** (hip″o-kam′pal) **gyrus** and the **cingulate** (sin′gu-lāt) **gyrus.** The nuclei are the **hippocampus** and the **amygdaloid nucleus,** which is also part of the basal nuclei. Some anatomists consider the anterior thalamic nuclei and the hypothalamus to be part of the limbic system, even though these structures are in the brain stem.

The limbic system functions in emotional aspects of behavior and stores some memories. Many emotional experiences including rage, sexual arousal, aggressiveness, fear, and loss of fear have been associated with some part of the limbic system. Experiments with monkeys show that emotions and memory are closely related. Experiences the monkeys remembered best were associated with reward or punishment. The hippocampus seems to consider emotional qualities of experiences in determining what will be remembered. Memory and emotion are discussed in more detail later.

Both pleasant and unpleasant odors elicit emotional feelings. This probably involves relaying signals from smell receptors along tracts to the amygdaloid nucleus and other parts of the limbic system. Such signals are integrated and relayed to the cerebral cortex.

Numerous experiments have used electrodes placed in selected areas of the brains of monkeys or other animals. Some allow an animal to stimulate its own brain by pressing a lever. By noting the effects of electrodes in specific areas, such experiments identified reward and punishment centers. Reward, or pleasure, centers have been found in the hypothalamus, the amygdaloid nuclei, and some other areas in the basal nuclei and thalamus. Some monkeys find stimulating a reward center so satisfying that they will press a lever to receive stimulation up to 7000 times an hour! Punishment centers have been found in the thalamus and the hypothalamus. In some cases, stimulation of a punishment or pain center inhibits a pleasure or reward center, thereby preventing the experiencing of pleasurable feelings.

CEREBRAL CORTEX

The **cerebral cortex,** with its 100 billion neurons of at least ten different kinds, covers an area of one-fourth square meter. Found external to the tracts of the cerebrum, this gray matter is 2 to 5 mm thick and contains six vertical layers of nonmyelinated cells. Horizontal fibers connecting adjacent cells and vertical fibers connecting the cortex with the thalamus and other structures creates a system of connections with as many as 100 trillion synapses.

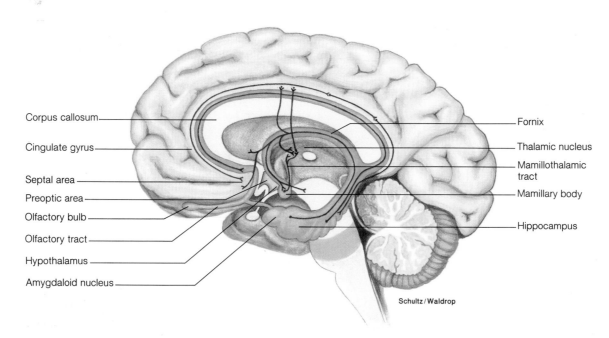

Schultz/Waldrop

FIGURE 12.8
The major components of the limbic system, with some neural pathways shown in black.

Functional Areas

Certain broad functional areas of the cortex have been identified by observing the effects of injuries and by electrical stimulation during surgery. Such stimulation helps the surgeon to avoid disturbing vital brain areas. Cortical stimulation is not painful because the brain lacks pain receptors and a local anesthetic is used for opening the cranial cavity. Patients are alert and can report experiences as different areas of the cortex are stimulated.

The central fissure in each hemisphere, which provides a landmark for locating functional areas of the cortex, lies between the **postcentral gyrus** in the parietal lobe and the **precentral gyrus** in the frontal lobe. The cortex of the postcentral gyrus is the **somatic sensory area.** It receives signals from sensory receptors in the skin and from taste receptors in the mouth. The cortex of the precentral gyrus is the **somatic motor area.** It initiates motor signals to skeletal muscles. Both the sensory and motor areas are spatially arranged with large numbers of neurons allocated to body regions with highly specialized functions such as the mouth and hands (Figure 12.9).

As you listen to someone speak and comprehend what is said, and speak in response, you use several functional areas (Figure 12.10). You use the **visual area** in the occipital lobe to process light signals to see the person speaking to you. You use the **auditory area** in the temporal lobe to process signals from sound receptors. (We will have more to say about cortical areas that process sensory signals in Chapter 15.) When you think about what you are hearing and how you will respond, your **frontal lobe** processes your thoughts and the **parietal lobe** oversees some frontal lobe activities. **Broca's** (bro'kahz) **motor speech area** in the frontal lobe formulates sentences and initiates speech.

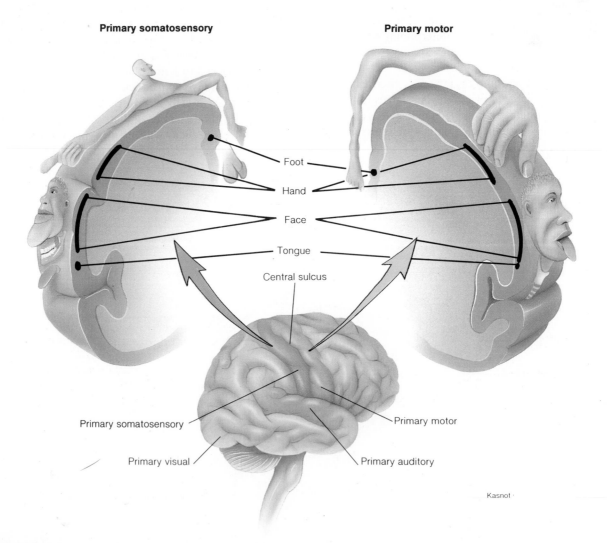

Primary somatosensory

Primary motor

Foot
Hand
Face
Tongue
Central sulcus

Primary somatosensory

Primary visual

Primary motor

Primary auditory

Kasnot

FIGURE 12.9

The locations of the somatic sensory and somatic motor cortices, with sketches of the parts of the body each portion serves. Each of the body parts in the figure is represented in proportion to the area of the cortex with which it is associated.

From T. L. Peele, *Neuroanatomical Basis for Clinical Neurology,* 2d ed. Copyright © 1961 McGraw-Hill, Inc.

Broca's area was named for Paul Broca, a nineteenth century French surgeon and anthropologist who identified a number of functional areas in the cortex. This area, which usually develops only in the left hemisphere, allows words to be formed by coordinating the movements of the lips and tongue. When the area is damaged, a person can make sounds and say simple words such as yes and no, but cannot say longer words or speak in sentences. An adjacent area coordinates respiratory and vocal cord movements so that speech can occur. The somatic motor area (just described) initiates throat and tongue muscle contractions, and the auditory area provides feedback about the sound of the voice.

The **facial recognition area** has connections with the primary visual cortex and the limbic system, a subcortical system concerned with emotions. When one considers that much of our experience involves recognizing and relating to other people, it is not surprising that an area of the cortex would be devoted to facial recognition, or that it would have the connections it has.

Association Areas

From the preceding introduction to functional areas of the cerebrum, we can see that association areas receive and interpret information from corresponding sensory areas. They use information stored in memory to interpret incoming signals.

The **somatic sensory association area** analyzes information from sensory receptors in the skin. If you were to pick up warm popcorn in one hand and cold gravel in the other, you would have no difficulty perceiving the temperature, texture, and relative weight of the substances or which hand held each substance. Furthermore, you could do this without using any other senses. Your somatic sensory association area integrates signals pertaining to touch, pressure, warmth, cold, and pain into patterns and interprets shapes, sizes, and temperatures of objects touched.

The **visual association area** interprets information from sensory receptors in the retina after preliminary processing (Chapter 15). It analyzes information and allows one to perceive shapes, sizes, distances, relationships among objects seen, and even the meaning of written words. Lesions in this area can cause **dyslexia** (dis-lek'se-ah), or word blindness, in which a person has difficulty understanding written words.

The **auditory association area** interprets information from sound receptors (also after preliminary processing). It analyzes information and allows one to perceive patterns of sounds, such as spoken words and music.

The **premotor cortex,** or **motor association area,** is located in the frontal lobe and is connected by association fibers to the somatic motor and somatic sensory cortices, the sensory association areas, the thalamus, and, most importantly, the basal nuclei. The premotor cortex coordinates movements required in skilled activities; certain areas of it control voluntary movements of the eyes, hands, and head.

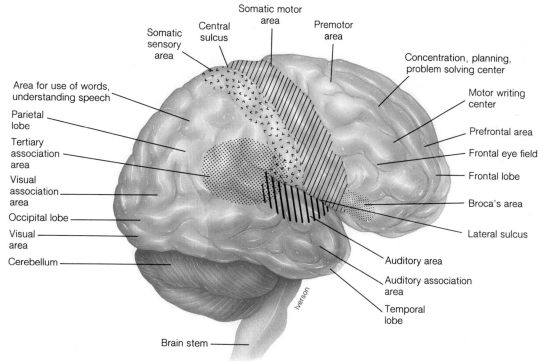

FIGURE 12.10
Other functional areas of the cortex.

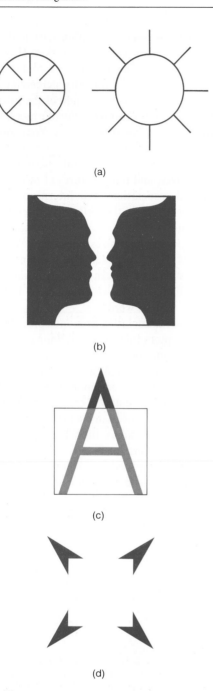

(a)

(b)

(c)

(d)

FIGURE 12.11
Optical illusions. Describe what you see in each part of this figure, then see text for explanations.

The **prefrontal cortex** is much larger in humans than in other animals, and early investigators were partially correct in relating it to intellectual functions. Destruction of the prefrontal cortex, however, leads not so much to impaired intellectual function as to distractability, uninhibited behavior, and sudden mood changes. It functions in sequencing thoughts, elaborating highly abstract thoughts, solving complicated problems, planning for the future, considering consequences of possible actions, and controlling moral behavior.

The **tertiary association area,** also called the general interpretative area, receives and integrates information from other association areas. Its temporal portion is called *Wernicke's* (ver'nik-ēz) *area* in honor of Karl Wernicke, a nineteenth century German physician. Wernicke's area uses memory to extract meaning from patterns of sensory information. Damage to this area prevents comprehension of thoughts expressed in written or spoken sentences, though words have been seen or heard. An affected person would experience something akin to what we experience when we read or hear words spoken in a language we cannot understand. Electrically stimulating this area can elicit such vivid recall that people feel like an event is actually happening over again. The left tertiary association area is functional in almost all right-handed people, but such specialization is less clearcut in left-handed people.

Perception of **optical illusions** illustrates some of the ways association areas may process information. (Study Figure 12.11 before you read on.) In Figure 12.11a, the circle with the lines extending into it probably looked slightly smaller than the one with the lines extending out of it, though both circles are the same size. Your brain compares the circles, but it also considers the lines and the overall pattern, and perceives one circle as larger than the other.

The brain can make only one interpretation at a time of information it receives, but it sometimes supplies "information" not actually provided. In Figure 12.11b, you saw either dark silhouettes or a light column on a dark background. When you perceive one, the other is background. Your brain cannot make two simultaneous interpretations of the same information. In Figure 12.11c, the letter A appeared to be partially overlaid with a transparent screen. Actually, the lower part of the letter A is printed in lighter ink, but your brain uses previous experience to supply the idea of transparency. Finally, in Figure 12.11d, you might describe the figure as a square with an arrow at each corner, but your brain supplied the idea of a square.

Researchers can study specific cortical activities by measuring variations in the blood flow or metabolic rate in cortical areas in live human subjects (Figure 12.12).

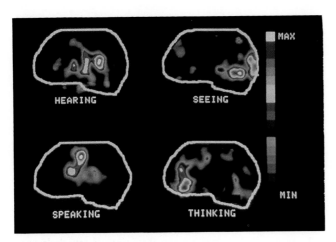

FIGURE 12.12
Activities of specific cortical areas related to blood flow.

Lateralization

Functional differences between left and right sides of the body, or **lateralization,** include left- or right-handedness, a dominant eye or foot, and left hemisphere function of Broca's speech area and the tertiary association area. At birth, both cerebral hemispheres seem to have equal capacities, becoming lateralized gradually in early childhood. Why lateralization occurs is unknown. When very young children sustain damage to one cerebral hemisphere, the other takes over nearly all cortical functions and development proceeds with little impairment. Recent PET and similar studies show cerebral activity in both hemispheres even in processes thought to be lateralized.

If not apparent from handedness or other behavior, cerebral lateralization can be determined by injecting a fast-acting barbiturate sodium amytal into an artery that goes to one cerebral hemisphere. After such an injection, a person is asked to count out loud. If the person stops counting and cannot resume counting, the injection has entered the dominant hemisphere.

The left hemisphere controls speech in almost all right-handed people and in many left-handed ones. In 20 to 40 percent of left-handed people, the right hemisphere controls speech, showing that speech control as well as handedness can be left or right dominant. A few left-handed people have speech control in both hemispheres so lateralization is not essential to normal speech. Observations from autopsies show larger bundles of fibers between hemispheres in left-handers than in right-handers, suggesting that more signals may be relayed between hemispheres in left-handers.

One thing these measurements show is that in resting subjects, the cerebral cortex is anything but resting. During rest, cerebral activity is greatest in the prefrontal area. This finding is consistent with the subjects' reports that they were thinking about problems or planning future activities. Different patterns of cerebral activity occur during reading silently, reading aloud, writing, or speaking.

Positron emission tomography (PET) is a noninvasive technique used to determine the anatomical location and rates of reactions in living organisms (Figure 12.12). PET has three requirements:

1. Substances to be studied must be labeled with a radioactive isotope, such as ^{11}C, ^{15}O, or ^{13}N, that emits positrons (positively charged electrons) when it decays. Positrons combine with electrons to emit gamma rays that can penetrate the skull.
2. A positron tomograph, an array of gamma ray detectors placed around the circumference of the head, provides an image of a cross section of the brain.
3. The reaction studied must be well enough understood that its location and rate can be determined from positron emissions.

PET allows the study of many processes. Positron-emitting oxygen or glucose is used to relate active cortical regions with certain mental tasks or to find diseased brain tissue. Labeled neurotransmitters and drugs are used to determine their sites of action and the nature of their receptors. Such studies not only increase our understanding of normal brain structure and function, they also identify chemical changes associated with anesthetics, seizures, cerebral hemorrhage, brain tumors, psychiatric and degenerative diseases, and drug addiction.

Hand position during writing is lateralized and correlated with dominance (Figure 12.13). People who write with their hand in a noninverted position have cerebral dominance opposite to handedness. This includes a large number of right-handers with left cerebral dominance and a few left-handers with right cerebral dominance. People who write with their hand in an inverted position have same-sided dominance—many left-handers with left cerebral dominance and a very small number of right-handers with right cerebral dominance. Cerebral dominance cannot be determined by writing position in a few people, probably because they lack clearly defined dominance.

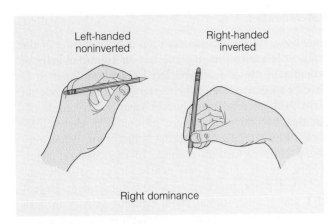

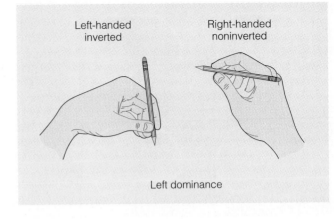

FIGURE 12.13
Writing position of the hand is correlated with cerebral dominance.

Visual processing also is lateralized, as can be shown from experiments on humans with "split brains" (disconnected cerebral hemispheres). Such surgery is sometimes done to reduce seizures in severe cases of epilepsy. People with "split brains" perform tasks of daily living normally, probably because each hemisphere processes signals and directs appropriate responses without interhemisphere communication. Deficits can be detected if only one hemisphere receives a signal. For example, if the word *orange* is presented to the left hemisphere (displayed so the image reaches only the right side of each eye), the person can say *orange*. If the image reaches the right hemisphere, the person cannot say *orange,* but can select an orange from among several objects by touch even when hidden from view.

When different images are presented to noncommunicating hemispheres, the patient can point to a picture of an object that relates to each. For example, when the left hemisphere recognizes a grocery basket, the person might match it with a box of cereal, and when the right hemisphere recognizes a car key the person might match it with a car. Asked to explain the selections, the person might say the cereal belongs in the grocery basket and the car is to be used to get to the grocery store. The left brain seems to have a region that acts as an **interpreter.** It organizes available information into a meaningful pattern. The interpreter, being unaware of the car key, invented a relationship between the car and the cereal and basket.

That emotional expression can be lateralized is clear from studies of normal people with connected hemispheres. To show this, people expressing an emotion were photographed and composite pictures consisting of one side of the face and its mirror image were made (Figure 12.14). Composites of the left side of the face generally show stronger emotion than composites of the right side of the face.

Processing of Language

Though facial expressions and gestures are important, language is the main means of human communication. Language processing, an important higher neural function, illustrates how sensory analysis and motor coordination work together in most higher level functions.

Sensory analysis begins with words reaching our consciousness through visual pathways when we read them or through auditory pathways when we hear them. Signals are processed in the proper association area and relayed to the tertiary association area where their meaning is interpreted.

After words are interpreted, a motor response such as speaking or writing, is often initiated. The first step in a motor response is to formulate the response and place words in an appropriate order. These processes probably occur in the tertiary association area and the sensory association area because people with lesions in these areas have difficulty formulating verbal responses. For a written response, signals go to the premotor and motor cortices,

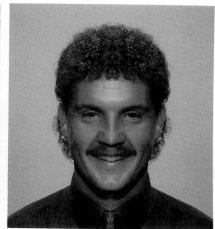

FIGURE 12.14
Lateralization can be seen in facial expressions.

which, in turn, relay them to other parts of the motor system. For a spoken response, signals go to Broca's area where the movements required to pronounce words are initiated. Speaking the words requires activity of the mouth portion of the somatic motor cortex, the premotor cortex, and the auditory cortex, which provides feedback about sounds being made.

WAKEFULNESS AND SLEEP

Falling asleep and awakening follows a daily (circadian) rhythm, with different stages reflected in characteristic brain waves. The **reticular formation,** a diffuse column of nuclei in the core of the brain stem (Figure 12.15), contains several nuclei that participate in regulating the wakefulness-sleep cycle. The nuclei include the *reticular activating system* (RAS), the *locus ceruleus,* the *pontine cells,* and the *raphe nuclei.* Knowledge of how these nuclei function is based, in part, on brain waves and electroencephalograms.

Brain Waves and Electroencephalograms

When electrodes are placed at various sites on the scalp, electrical potential differences between brain sites can be observed and recorded. These differences are the average of the **synaptic potentials** (EPSPs and IPSPs) between sites. **Brain waves,** fluctuations in potential differences, have frequency and amplitude. A record of brain waves over time is called an **electroencephalogram** (e-lek″tro-en-sef′ah-lo-gram″), or EEG.

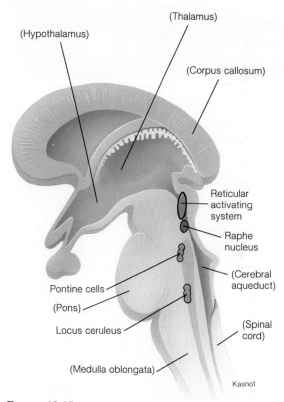

FIGURE 12.15
Some parts of the reticular formation implicated in the regulation of sleep and wakefulness. (Structures with labels in parentheses are included as points of reference and are not directly involved in the regulation of sleep and wakefulness.)

(a)

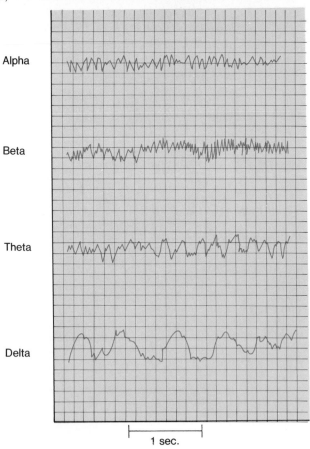

Alpha

Beta

Theta

Delta

|← 1 sec. →|

(b)

FIGURE 12.16
(*a*) A person having an encephalogram made, and (*b*) different kinds of brain waves.

TABLE 12.3 Types of Brain Waves Seen in Electroencephalograms

Wave Type	Frequency (Cycles/Second)	Functional Status
Alpha	8–13	Awake, but relaxed with eyes closed
Beta	14–25	Mentally alert
Theta	4–7	In children; rare in adults, in some brain disorders
Delta	⅓–3½	Deep sleep, infancy, serious brain disorders

Four kinds of brain waves are seen in normal EEGs (Figure 12.16 and Table 12.3). **Alpha waves,** rhythmic waves with a frequency of 8 to 13 cycles per second, occur when a person is awake but relaxed with eyes closed. **Beta waves,** less rhythmic, lower amplitude waves twice as frequent as alpha waves, occur when a person is mentally alert. They are most frequently recorded from the frontal and parietal scalp over association areas where mental activities occur. **Theta waves,** high amplitude waves with a frequency of 4 to 7 cycles per second, are detected in the temporal and parietal regions of the brains of normal children and also in some people of any age with brain disorders. **Delta waves,** high amplitude, low frequency waves (no faster than 3.5 cycles per second) occur during deep sleep.

Information about the brain inferred from brain waves is purely descriptive. Using EEGs to study the brain is like listening to noises outside a factory and trying to determine what is going on inside. Still, EEG patterns aid in the diagnosis of some disorders (Figure 12.17) though they do not explain them. The complete absence of brain waves denotes death even when a patient's respiration and blood circulation are being maintained by life support systems. Also, brain waves are useful in studying stages of sleep.

Wakefulness

Wakefulness is maintained by the reticular formation, particularly the reticular activating system (RAS). Nuclei in the reticular formation receive signals from other parts of the brain stem and from the spinal cord, cerebellum, cerebral cortex, basal nuclei, and limbic system; they send signals to the spinal cord, cerebellum, and cerebral cortex. The RAS relays sensory signals to the cerebral cortex and the cerebral cortex sends back signals that stimulate the RAS. This positive feedback mechanism maintains wakefulness (Figure 12.18).

The RAS also determines which incoming signals are to be relayed to the cortex, but how it makes this selection is unknown. Somehow, it allows one to concentrate on a particular task such as homework while at the same time being aware of other signals such as music or a TV program. The RAS usually relays strong or novel signals but

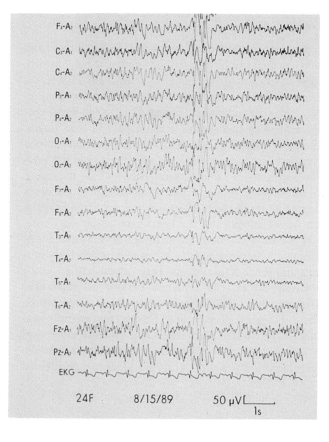

(a)

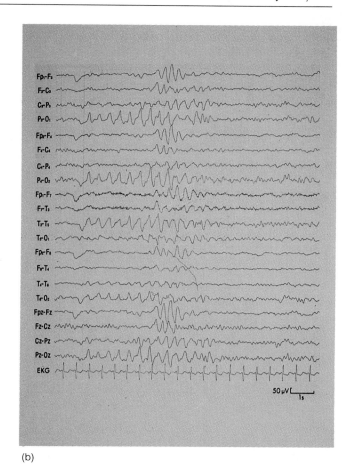

(b)

FIGURE 12.17
Alterations of electroencephalograms related to brain disorders:
(*a*) epileptic seizure; (*b*) generalized diffuse swelling.

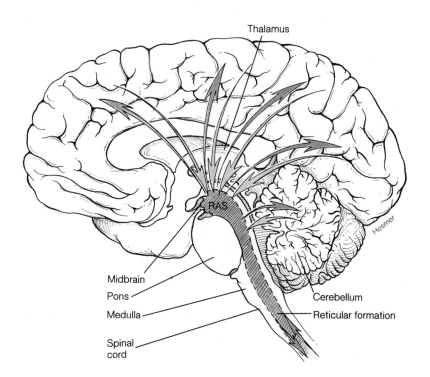

FIGURE 12.18
The reticular formation sends and receives signals from various parts of the brain. When stimulated, the reticular activating system (RAS) sends signals (shown in black) to the cerebral cortex. The cortex, in turn, sends signals (shown in color) back to the RAS. This positive feedback mechanism is a major factor in maintaining wakefulness.

rarely relays weak or repetitive signals in familiar patterns. This mechanism allows a parent to wake suddenly at the sound of a crying baby but to sleep through louder noises. It also explains why the sound of a continuously humming motor or fairly constant traffic noise does not wake us, while a ringing telephone does. Screening out continuous stimuli is a form of habituation. **Habituation,** the ignoring of sensory signals, is controlled largely by the RAS and does not involve adaptation of sensory receptors.

Coma, or unconsciousness, a state from which a person cannot be awakened, was once equated with lack of neural activity, especially in the brain stem. We now know it involves increased activity of brain stem cells that release acetylcholine when the brain has been traumatized. Acetylcholine suppresses RAS activity that normally would maintain wakefulness and serves to shut down injured tissue. Whether this shutdown contributes to healing is not known.

Coma in children is a frequent component of Reye's syndrome, a disorder associated with the use of aspirin in children with viral diseases. How aspirin and the viral disease interact to cause coma and damage to the liver and other organs is not clearly understood.

Sleep

Sleep, a natural temporary absence of wakefulness from which a person can be aroused, is essential to good health, though its exact contribution is not understood. Five stages of sleep have been identified using electroencephalography and other techniques (Figure 12.19). **Non-rapid-eye-movement sleep** (non-REM sleep) occurs in four stages. In Stage 1, a drifting sensation at the onset of sleep, alpha waves decrease in frequency and amplitude, and RAS activity decreases. If aroused, most people claim they were not really asleep. In Stage 2, **sleep spindles,** bursts of brain waves at a rate of about 15 per second, occur. In Stage 3, delta waves appear and the heart rate, respiratory rate, blood pressure, and body temperature decrease. In Stage 4, deep, oblivious sleep, delta waves persist and body processes and RAS activity reach their lowest levels.

Every 80 to 120 minutes, periods of **rapid-eye-movement sleep** (REM sleep) occur. REM sleep also is called paradoxical sleep because EEG waves resemble beta waves of the awake, alert state. The eyeballs move rapidly, and physiological processes increase to waking levels. Dreaming usually occurs during REM sleep, and dreams

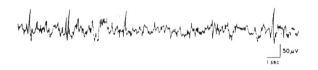
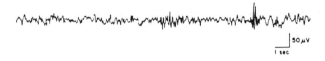

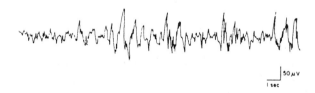

FIGURE 12.19
EEG tracings showing stages of sleep.

that occur near the end of a sleep period tend to have the most intense emotional content—and are the most likely to be remembered.

Four to six cycles of alternating REM and non-REM sleep occur during a typical night. Most REM episodes last from five to 20 minutes but can last as long as 50 minutes. Typically, 90 to 120 minutes of a night's sleep is REM sleep. REM sleep differs from non-REM sleep in several ways as summarized in Table 12.4.

How sleep and wakefulness are regulated is poorly understood and controversial, but the following factors may be involved:

1. During waking hours, serotonin-secreting neurons cause synthesis of polypeptide sleep factors in some brain cells.
2. Serotonin-secreting neurons in the raphe nuclei either initiate both REM and non-REM sleep directly or secrete sleep factors that modify these sleep stages.
3. Norepinephrine-secreting neurons in the locus ceruleus probably act on the somatic motor cortex to suppress skeletal muscle tone during REM sleep. This effect may prevent people from acting out their dreams, and it explains how one can suddenly awaken from a dream momentarily unable to move.
4. Pontine cells become active shortly before and during REM sleep and initiate rapid eye movements. Inhibitory neurons of the locus ceruleus probably suppress pontine cells and eye movements during waking and non-REM sleep, but not during REM sleep.
5. Various neurons secreting norepinephrine, dopamine, or acetylcholine work together to maintain wakefulness and arousal.

One way to assess the functions of sleep is to observe the effects of sleep deprivation. A day or two without sleep reduces attention span and impairs mental ability and memory. Several days without sleep leads to bizarre and psychotic behavior, and an accumulation of stress-related hormones and a serotonin-like substance in the blood, which may cause hallucinations. Awakening sleepers as they enter REM sleep worsens the effects of sleep deprivation. After deprivation, sleep periods are longer with more than the usual time spent in REM sleep. REM sleep may be especially important in restoring normal neural function.

During normal sleep, the effects of experiences on brain cells while a person is awake appear to be sorted, with some stored as memories and others obliterated.

TABLE 12.4 A Comparison of REM and Non-REM Sleep

REM Sleep	Non-REM Sleep
EEG similar to wakeful, alert state	EEG shows damped alpha waves followed by delta waves
Muscle tone is inhibited, except in eye muscles	Muscle tone maintained
Dreaming occurs	No dreaming, but sleeper may report thoughts of recent events if awakened
Right cerebral hemisphere most active	Left cerebral hemisphere most active
Respiratory rate, heart rate, and blood pressure show irregular fluctuations	Respiratory rate, heart rate, blood pressure, and body temperature show slight decreases
Sleeper is relatively difficult to arouse	Sleeper is relatively easy to arouse
Occurs in periods of 5 to 20 minutes but can last as long as 50 minutes	Occupies remainder of cycles of 80 to 120 minutes
Generally follows an initial period of non-REM sleep	Generally constitutes the first period of natural sleep
Accounts for 90 to 120 minutes of an adult's nighttime sleep (15–20 percent); accounts for half of an infant's sleep	Accounts for the remainder of a typical adult's nighttime sleep (80–85 percent); accounts for half of an infant's sleep
When sleeper is deprived of REM sleep, deficit is generally made up the next night, but REM debt takes longer to make up than non-REM debt	When sleeper is deprived of non-REM sleep, deficit is generally made up the next night; non-REM debt is made up more quickly than REM debt

Much remains to be learned about these processes, but they seem to prepare brain cells for mental activity on waking.

LEARNING AND MEMORY

Learning is a change in behavior in response to environmental stimuli. It involves brain plasticity, maturation of neural pathways, and interaction of brain structures with stimuli. For example, learning to walk is automatic when appropriate neural pathways have matured, but learning to use language depends more on brain structure. In fact, linguists believe that, because languages have common elements of grammar and structure, language structure must be determined by constraints of brain structure. **Memory** is the storage of a portion of what is learned in a form that allows recall.

Understanding today's experiments on the neural basis of learning and memory requires at least a modest understanding of kinds of learning (Figure 12.20). **Habituation** is a decreased response to a repeated stimulus with neither beneficial nor harmful effects. For example, during the first weeks you live near a train track, whistling trains wake you every night but after a while they no longer wake you. **Sensitization** is an increased response to a repeated harmful or irritating stimulus. If you have a persistent itchy nose, you rub it gently at first, but the longer it lasts, the more vigorously you rub it.

Classical conditioning, first studied by Pavlov around 1900, requires association of a conditioned stimulus such as a flashing light with an unconditioned stimulus such as the sight of food. The sight of food usually elicits salivation, but a conditioned animal begins to salivate when the light flashes. If a conditioned stimulus (light) is repeatedly presented without the unconditioned stimulus (food), **extinction** occurs—the conditioned response decreases, becomes less probable, or both.

Operant conditioning, or **trial-and-error learning** associates a stimulus and a favorable behavior called an **operant.** An animal with access to a lever may press it either during random activity or by prior learning. If the animal gets food when it presses the lever, lever-pressing is rewarded and increases as the animal learns to associate the lever with food. Similarly, a behavior that prevents a harmful stimulus, such as an electric shock, also is rewarded and increases. **Reinforcement,** rewarding certain behaviors and failing to reward others (but not necessarily punishing), fosters repetition of rewarded behaviors and nonrepetition of unrewarded behaviors. Without reinforcement, extinction occurs; but intermittent reinforcement (reward after the animal pulls the lever a certain number of times) slows extinction. Intermittent rewards in operant conditioning cause a behavior such as a human pulling a slot machine lever to undergo slow extinction!

Humans also learn by sensory learning, imitation, discrimination, abstraction, and insight. **Sensory learning** is based on sensory experience—learning a song from hearing it or learning physiology from reading this book. **Imitation** is copying another person's behavior; it is important in learning language. **Discrimination** involves detecting subtle differences between stimuli, such as differences in the pitch of musical notes or the shapes of letters of the alphabet. **Abstraction** is recognizing qualities or concepts apart from specific objects. The concept of "twoness" has meaning without applying it to specific objects. **Insight,** or **reasoning,** is the ability to respond appropriately to a new situation and to use prior learning to solve a new problem mentally without trial-and-error behavior.

Many physiologists today are interested in the neural basis for learning and memory. They study certain mollusks because the mollusks display learning and have millions of large neurons (instead of trillions of small ones that humans have). And experiments on them are technically and ethically possible.

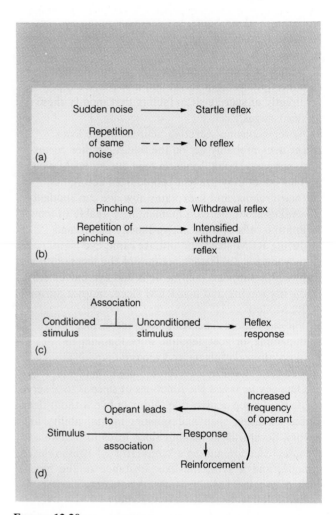

FIGURE 12.20

Simple kinds of learning: (*a*) habituation, (*b*) sensitization in *Aplysia,* a small mollusk with a relatively simple nervous system, (*c*) classical conditioning, and (*d*) operant conditioning.

Aplysia, a large marine snail that has no shell, is frequently used in learning studies. It has a brain in the head end and a mantle cavity containing the gill, and is covered by a mantle shelf that terminates in a siphon. Stimulation of the mantle shelf or siphon by light touch causes withdrawal of the gill, siphon, and mantle shelf into the mantle cavity. This behavior can be modified by habituation, sensitization, and classical conditioning. In habituation, light touch causes sudden withdrawal, but after repeated touching, withdrawal no longer occurs. In sensitization, an electrical shock elicits withdrawal, and repeated shocks lead to more intense withdrawal. Specific neurons have been associated with gill-siphon withdrawal (Figure 12.21). Cellular changes occur at synapses of sensory neurons with interneurons and motor neurons. At some synapses, the size of the synaptic potential is proportional to the amount of transmitter released. Each action potential releases less neurotransmitter during habituation and more during sensitization.

Studies of classical conditioning in another marine snail *Hermissenda* show that the mechanism of conditioning may be quite similar to that of sensitization. As the animal becomes conditioned, certain neurons accumulate Ca^{2+}, which closes K^+ channels and makes the membrane more excitable. Continued stimulation during learning exerts positive feedback: as membrane excitability increases, it leads to more Ca^{2+} accumulation. Positive feedback in learning makes sense when one considers that learning is a change in behavior. Negative feedback maintains a steady state, and positive feedback results in change.

Like learning, memory involves cellular and molecular events. Evidence is accumulating from humans with memory disorders that the hippocampus is a main organ of memory. Studies of the hippocampus in rats and in laboratory cultures suggest two processes that might account for memory. In one process (Figure 12.22), signals crossing

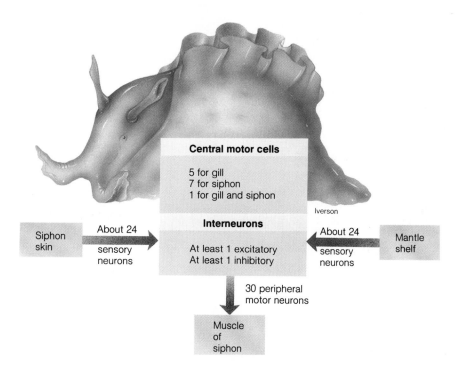

FIGURE 12.21

Schematic diagram of neurons mediating the gill-siphon withdrawal reflex. Sensory neurons synapse directly on interneurons and on motor neurons, and the motor neurons synapse directly on muscle cells of the siphon.

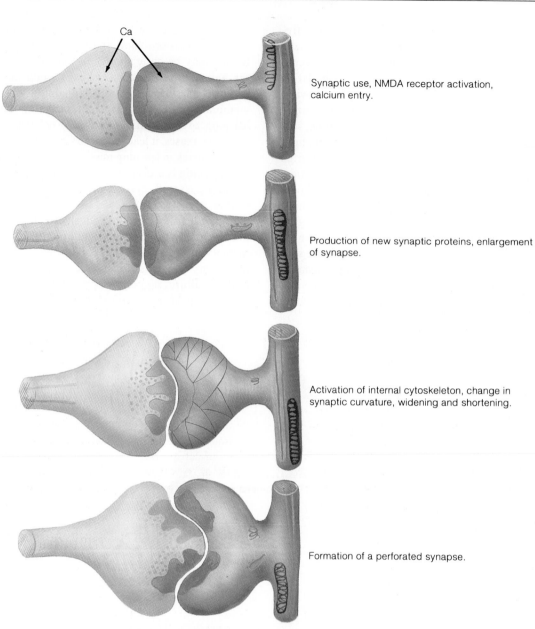

Ca

Synaptic use, NMDA receptor activation, calcium entry.

Production of new synaptic proteins, enlargement of synapse.

Activation of internal cytoskeleton, change in synaptic curvature, widening and shortening.

Formation of a perforated synapse.

Division of the synapse and formation of new dendritic spines and dendritic length.

Paras

FIGURE 12.22

A possible mechanism for information storage involving NMDA receptors and division of synapses.

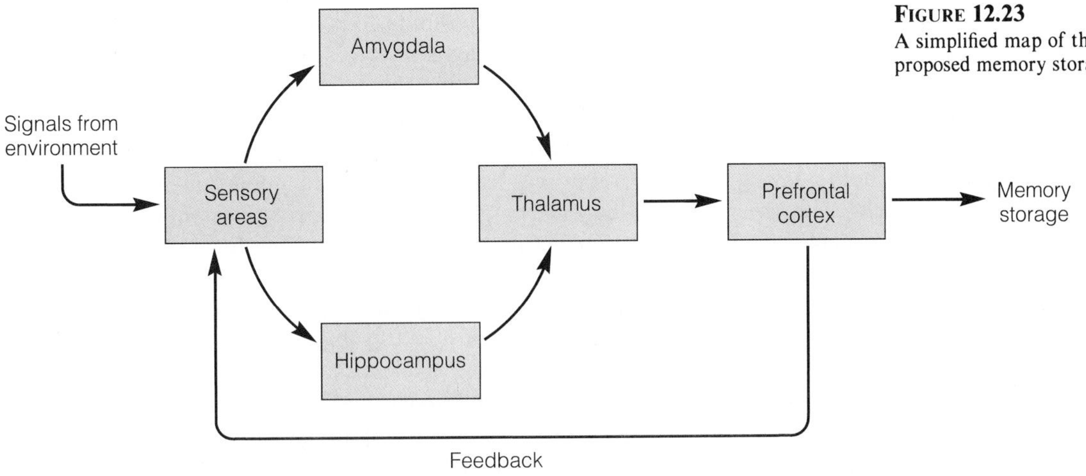

FIGURE 12.23
A simplified map of the flow of signals in two proposed memory storage pathways.

synapses activate certain receptors called N-methyl-D-aspartate (NMDA) receptors, and such activation increases the flow of Ca^{2+} into the neuron. This activates the cytoskeleton and ultimately results in increasing the number of dendrite spines and causing synapses to divide. New dendrite spines and new synapses seem to indicate memory storage. In another process, signals activating NMDA receptors cause a series of biochemical changes that ultimately result in **long-term potentiation** (LTP) of neurons. The persistence of LTP keeps a group of cells ready to fire. When any cell in the group is stimulated, the whole group fires and recalls the same experience that originally produced the LTP. For example, cells in your hippocampus store an original experience such as sitting with friends around a fireplace eating popcorn. Later, when you smell popcorn, you recall the whole fireplace experience.

Memories persist for different periods of time. **Short-term memory** (memory for less than a day) might involve synthesis of the cellular messenger cAMP or diminished degradation of it. Sensitization in mollusks suggests that short-term memory grades into **long-term memory** (memory for days to life). Accumulation of cAMP may account for short-term memory, whereas division of synapses or long-term potentiation may act in long-term memory. Increasing the number of stimuli appears to increase the length of memory. For example, a single stimulus produces memory for several hours, four stimuli produce memory for a day, and four stimuli per day for four days produce memory for several weeks. This seems to apply to humans because students who regularly review what they have learned in class and through reading remember more of what they have learned.

Studies of humans who have experienced brain trauma, deep anesthesia, or brain anoxia (lack of oxygen) show that memory for recent events may be lost but older memories are retained. One explanation is that changes in neurons associated with memory (new synapses or potentiated cells) occur near the cell body first. As a neuron accumulates new information, changes occur farther from the cell body. Degenerative changes, when they occur, begin far from the cell body and move toward it.

The ability to recall previous experience can be disrupted by destroying memory stores (Figure 12.23) or by interfering with the search and select processing of recall. Analogously, the recall of music on a tape or data on a computer disk can be prevented permanently by damage to the tape or disk, or temporarily by the absence of appropriate equipment. Similar events probably occur in humans after head injuries because some memories are permanently lost, but some are regained.

Related to memory storage is the problem of what kinds of memories are stored. Reports of patients receiving brain stimulation during surgery show that some memories must consist of records of experience. These records vary in accuracy as shown by comparing eyewitness reports of an event or the retelling of stories. Such accounts are coherent, but original details are often replaced with plausible substitutions. The brain apparently reconstructs experience from what is remembered by adding inferences to make a coherent story. Physiological events in such reconstructions are not yet understood.

Memories play a special role in problem solving. When humans and some other animals are faced with an unfamiliar situation or a new problem, bits and pieces of information are recalled and used in new ways to mentally construct a solution. Suppose you are told to dig a hole six feet deep with vertical walls (something you have never done before). As you finish, you face the problem of getting out of the hole. One person solved this problem by pushing the shovel blade into the wall of the hole and climbing out on the handle. Though memory stores contained no record of such an experience, the brain made use of various memories to conceive of a shovel as a ladder instead of a digging tool. How the brain constructs solutions to novel problems is not known, but human problem solving behaviors confirm that such mechanisms must exist. The mechanisms also must be capable of enhancement because practice using brainstorming and other creative problem solving techniques increases problem solving ability.

Researchers are identifying signal routes in memory storage (Figure 12.23). Two regions on the inner temporal lobe surfaces of both hemispheres—the hippocampus and the amygdala—appear to be essential storage areas. In monkeys, removing either the hippocampus or the amygdala has little effect on memory storage, but removing both severely impairs storage of new memories. Humans with damage to both areas also fail to store memories. They can enjoy television reruns, old magazines, and old jokes as new experiences, but they cannot remember where they put their clothes or other possessions, or where to find a bathroom.

Finding that memories could be stored if either the hippocampus or the amygdala were intact suggested two pathways for memory storage. Further studies show that both of these structures receive sensory signals and relay them to the thalamus, and that the thalamus relays signals on to the prefrontal cortex. Furthermore, the prefrontal cortex supplies feedback signals to sensory areas of the cortex. Though there is much overlap in function, the two pathways apparently serve different functions. Signals processed by the hippocampus are important in memory of spatial relationships, whereas those processed by the amygdala allow association of memories from different senses—smell, taste, touch, hearing, and vision.

EMOTIONAL BEHAVIOR

Emotion is the affective (subjective feeling) aspect of consciousness. Human emotions include joy, love, fear, anger, pleasure, anxiety, and aggression. They are agreeable or disagreeable, but never neutral. Studies in humans and animals point to the limbic system, especially the hypothalamus, as the mediator of emotional behavior. The amygdala may contribute sensory input to emotions, and allow emotions to influence perception and memory storage. The amygdala also makes endogenous opiates (endorphins and enkephalins), and releases them when it receives signals from the hypothalamus. These substances probably account for strong impressions and lasting memories elicited by emotionally charged events.

A special aspect of emotional behavior is that many odors elicit emotional feelings. The association of odors with feelings is possible because of a direct connection between the limbic system, especially the amygdala, and the olfactory bulb. Signals from smell receptors, but not other receptors, are relayed through the limbic system on their way to the cerebral cortex.

Studies of pleasure began with the accidental discovery that electrical stimulation of certain regions of the rat's limbic system (the septum or the median forebrain bundle) leads to a pleasure response. In a situation in which rats can press a lever to stimulate their own pleasure centers, they choose pressing the lever—up to 2000 times per hour—over obtaining food even when hungry. Electrical stimulation can alleviate severe pain and depression in humans, but using such stimulation to control human behavior raises serious ethical issues.

Studies in rats show that stimulating a pleasure center reinforces learning. Given the opportunity after satisfactory learning performance, rats repeatedly self-stimulate the pleasure center. They later perform the learned tasks better than rats not allowed self-stimulation.

Fear, an emotional response to danger, and anxiety, the same reponse in the absence of danger, probably have the same neurological mechanisms. Much of what is known about fear and anxiety has been learned by studying the effects of certain tranquilizers on the release of neurotransmitters. The benzodiazepine tranquilizers (Valium and Librium) reduce the release of catecholamines (dopamine and norepinephrine) and serotonin at synapses in the brain. They also facilitate transmission at postsynaptic receptors that respond to GABA (gamma-aminobutyric acid) by competing with a protein that often blocks GABA receptors. Because GABA is an inhibitory transmitter, anxiety probably is reduced by inhibiting certain brain neurons.

Extreme anxiety and fear are often accompanied by increases in the metabolic rate and in the breathing rate mediated by the sympathetic nervous system (Chapter 14). When metabolism increases, lactate production also increases; but in people with a disorder called anxiety neurosis, lactate production increases markedly and causes trembling, heart palpitations, and dizziness. Such symptoms may be due to lactate binding to calcium causing a temporary calcium deficiency in muscle and nerve cells.

Aggression includes behaviors ranging from assertiveness to outright violence and is usually accompanied by anger. Both hormones and the hypothalamus apparently play a role in aggression. Among male prison inmates, the degree of violence of criminal offenses is highly correlated with blood testosterone concentrations. Among females committing violent crimes, over half had done so the week before a menstrual period. Electrical stimulation of the anterior and medial hypothalamus causes cats to hiss, arch their backs, and bare their teeth, but fails to cause them to attack. Stimulating the lateral hypothalamus causes them to make ferocious attacks without emotional displays. Whether the same might happen in humans is not known.

MENINGES, VENTRICLES, AND FLUIDS

The brain and spinal cord are protected by membranes called meninges (men-in'jēz). Some spaces between meninges and cavities in the brain and spinal cord contain cerebrospinal fluid. The movements of both this fluid and blood are carefully regulated.

Meninges

The meninges consist of three layers of connective tissue membranes (Figure 12.24). The outermost **dura mater** (du'rah ma'ter), literally "tough mother," is a tough, white fibrous tissue that lines the cranial cavity and the vertebral canal. It terminates in a blind sac at the second sacral vertebra, where spinal nerves in the cauda equina pass through it.

Between the dura mater and the skull and vertebrae is the **epidural** (ep''ĭ-du'ral) **space,** which is filled with fatty tissue and other connective tissues that cushion and support the brain and spinal cord. An epidural anesthetic is injected into this space and diffuses into neural tissue. In certain places in the cranial cavity, the dura mater forms two layers. The inner one extends between brain lobes and helps to hold the brain in place in the cranial cavity. The

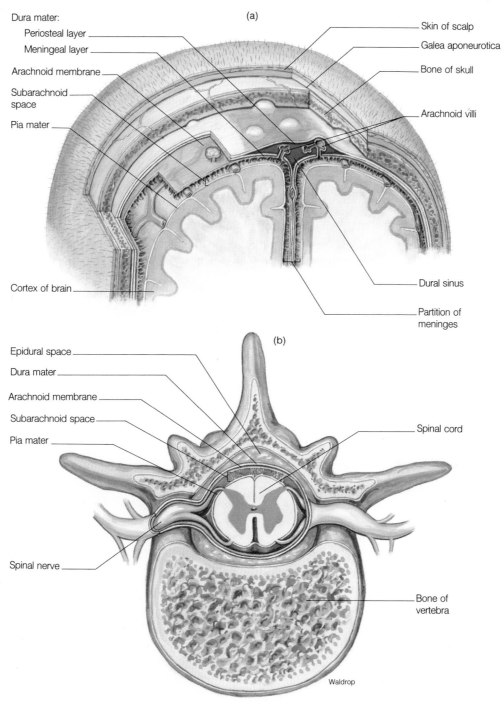

FIGURE 12.24

The meninges and associated structures (*a*) around the brain and (*b*) around the spinal cord.

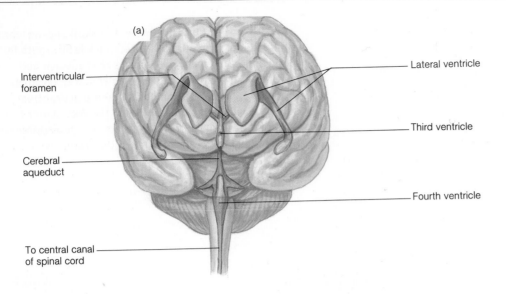

(a)

Interventricular foramen

Lateral ventricle

Third ventricle

Cerebral aqueduct

Fourth ventricle

To central canal of spinal cord

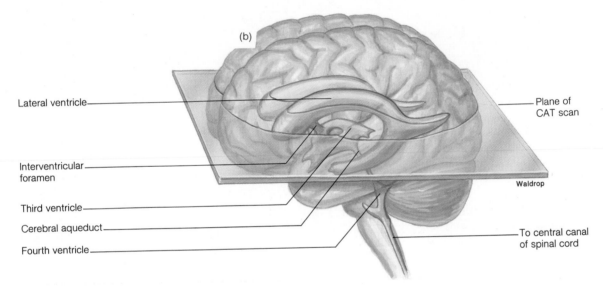

(b)

Lateral ventricle

Plane of CAT scan

Waldrop

Interventricular foramen

Third ventricle

Cerebral aqueduct

Fourth ventricle

To central canal of spinal cord

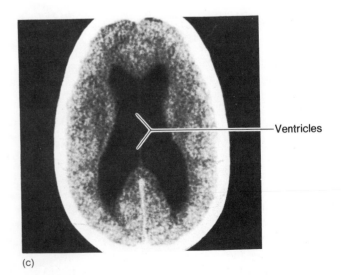

Ventricles

(c)

FIGURE 12.25

The ventricles of the brain: (*a*) anterior view, and (*b*) lateral view, shown as if the brain were transparent. (*c*) Ventricles as seen in a CAT scan.

spaces between the layers form **dural sinuses,** which collect venous blood from brain tissues to be returned to the heart.

The **arachnoid** (ar-ak′noid) **layer** is a thin, cobweb-like membrane. It is separated from the dura mater by the **subdural space,** which contains a small amount of serous fluid. It is separated from the pia mater by the **subarachnoid space,** which contains cerebrospinal fluid. **Arachnoid villi** (or arachnoid granulations) are small folds in the arachnoid layer that form one-way passages for the diffusion of cerebrospinal fluid from the subarachnoid space into the blood in the dural sinuses.

The **pia** (pi′ah) **mater,** literally "tender mother," is a delicate membrane that follows all the contours of the brain and spinal cord. It consists of modified loose connective tissue and is well supplied with blood vessels.

Ventricles

The **ventricles** (ven′trik-lz) of the brain consist of four hollow, fluid-filled cavities inside the brain (Figure 12.25). Each cerebral hemisphere contains a **lateral ventricle,** which is connected to the third ventricle by an **interventricular foramen.** The **third ventricle** is a narrow channel through the thalamus. It is connected by the **cerebral aqueduct** (aqueduct of Sylvius) in the midbrain portion of the brain stem to the **fourth ventricle** in the pons and medulla. The fourth ventricle is continuous with the central canal of the spinal cord. Three openings in the roof of the fourth ventricle, a pair of **lateral apertures** (foramina of Magendie) and a **median aperture** allow cerebrospinal fluid to move into the subarachnoid space.

The roof of each ventricle contains a network of capillaries called a **choroid plexus** (ko′roid plek′sus). These capillaries receive blood from small arteries in the pia mater and secrete cerebrospinal fluid into the ventricles.

Cerebrospinal Fluid

All brain ventricles, the central canal of the spinal cord, and the subarachnoid space around both the brain and spinal cord are filled with cerebrospinal fluid (CSF). CSF is formed as the choroid plexuses actively secrete sodium ions into the ventricles. These positively charged ions draw negatively charged ions, particularly chloride ions, along with them. They maintain an osmotic pressure somewhat greater than that in blood—and water enters the ventricles by osmosis. CSF is subject to changes in pH and in oxygen and carbon dioxide content; such changes initiate signals that help to regulate respiration (Chapter 22). Both the volume of about 150 ml and the pressure in CSF remain nearly constant. It is secreted at a rate of about 840 ml per day and reabsorbed from the arachnoid villi into the dural sinuses at the same rate (Figure 12.26).

CSF acts as a shock absorber for the brain and spinal cord. When a severe blow to the head allows the brain to displace CSF and hit the opposite cranial wall with sufficient force, concussion (tissue damage) occurs opposite the blow. Such tissue damage usually is not permanent, but can cause a massive electrical discharge within the brain. The electrical discharge may be responsible for convulsions and memory loss for recent events in such injuries.

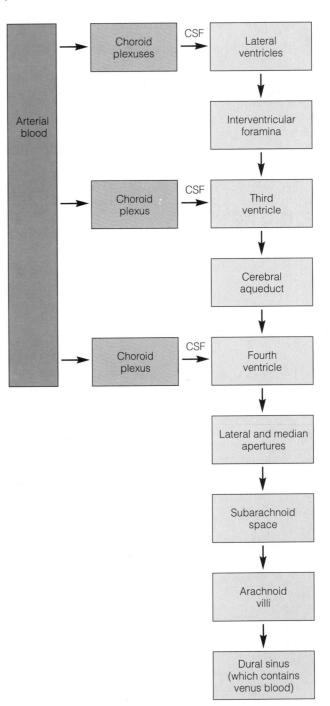

FIGURE 12.26
Flow of cerebrospinal fluid.

FIGURE 12.27
Movement of substances out of and into
cerebrospinal fluid.

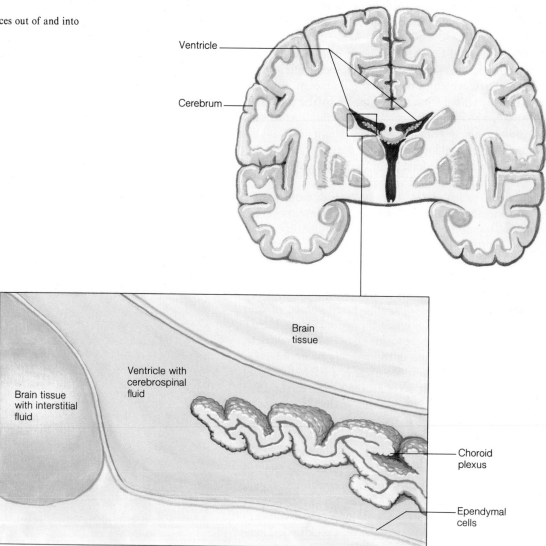

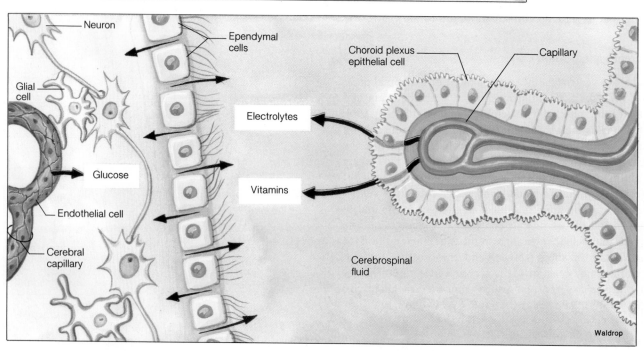

The choroid plexuses actively transport vitamin C, folates, and some other vitamins from blood to CSF. Both the vitamins and electrolytes in CSF move from ependymal cells to interstitial fluids by facilitated diffusion (Figure 12.27). The choroid plexuses, sometimes called the kidneys of the brain, also clear products of neurotransmitters, other metabolic wastes, antibiotics, and other drugs from CSF.

Babies are sometimes born with a high CSF pressure from excess fluid or an obstruction that prevents its outflow into the subarachnoid space. This condition is called **hydrocephalus** (hi-dro-sef'al-us), or water on the brain (Figure 12.28). If the fluid is not drained, pressure causes bone separation and skull enlargement in infants and brain tissue compression in patients of any age. Untreated, compression leads to hypotension (low blood pressure), respiratory failure, and death. Hydrocephalus is treated by inserting a shunt in a ventricle to drain fluid from the brain to the heart or the abdominal cavity.

Blood Supply

Of all the organs in the body, the brain is especially dependent on its blood supply for sufficient glucose, amino acids, and oxygen to maintain normal function. Whereas most other tissues can metabolize fat as well as glucose, the brain normally requires glucose and it can, in fact, absorb as much as 75 percent of the glucose available in the blood.

Blood is carried to the brain by the internal carotid and vertebral arteries. Much of this blood is directed through the base of the brain via the **circle of Willis,** which provides multiple pathways (Figure 12.29) for blood to reach brain cells. (Other brain blood vessels are considered in Chapter 19.) Blood leaves the brain via the dural sinuses, wide tubelike spaces between double layers of dura mater in certain locations over the surface of the brain.

Arteriograms (Figure 12.29b) can be made to determine the degree to which blood circulates through the arteries of an organ. An X-ray opaque dye is injected into a major artery, and X-ray photographs are taken of the artery and its branches as the dye circulates through them. Restricted passage of the dye indicates blockage of a vessel, which can be caused by a blood clot or a deposit of fatty substances in the arterial wall (atherosclerosis, Chapter 20).

Many brain capillaries are specially constructed to form a **blood-brain barrier** (Figure 12.30), which limits entry of substances into brain cells. Compared to other capillaries, these capillaries are modified in three ways: (1) their walls lack pores, (2) cells of the walls overlap and increase the wall thickness, and (3) the capillaries are surrounded by astrocyte foot processes and basement membrane. (The basement membrane consists of glycoprotein from cells of the capillary and fibers from the astrocytes.) By limiting the passage of substances from blood to brain cells, the blood-brain barrier protects brain cells from toxic substances. It also prevents them from receiving medications such as antibiotics, tranquilizers, and antidepressants. This is one reason brain infections and mental illnesses are difficult to treat.

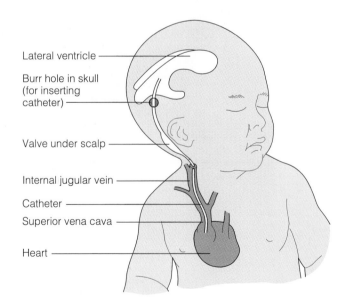

Lateral ventricle

Burr hole in skull (for inserting catheter)

Valve under scalp

Internal jugular vein

Catheter

Superior vena cava

Heart

FIGURE 12.28
Hydrocephalus can be treated by placing a shunt between a cerebral ventricle and the heart or the abdominal cavity.

Anterior
cerebral artery

Middle
cerebral artery

Posterior
communicating
artery

Posterior
cerebral artery

Anterior
communicating artery

Internal carotid artery

Basilar artery

Vertebral artery

Schultz / Waldrop

(a)

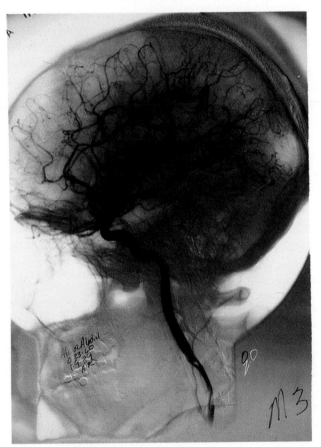

(b)

FIGURE 12.29
(a) A ventral view of the brain, showing the
distribution of arteries in the brain. The
circle of Willis is formed by the anterior
and posterior communicating arteries.
(b) Arteriogram of the cerebrum.

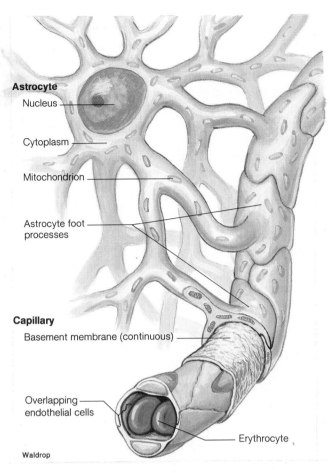

Astrocyte
Nucleus
Cytoplasm
Mitochondrion
Astrocyte foot processes
Capillary
Basement membrane (continuous)
Overlapping endothelial cells
Erythrocyte

Waldrop

FIGURE 12.30
The structure of a capillary in the brain surrounded by an astrocyte foot process, showing how the blood–brain barrier is created.

SPINAL CORD

The **spinal cord** (Figure 12.31) extends from the brain stem to the level of the second lumbar vertebra—a distance of 42 to 45 cm in an adult. Its anteroposterior diameter is less than its left-to-right diameter, and it is enlarged in two regions. The **cervical enlargement,** between the fourth cervical and first thoracic vertebrae, is the region from which nerves supplying the arm arise. The **lumbar enlargement,** between the ninth and twelfth thoracic vertebrae, is the region from which the nerves supplying the legs arise. The lumbar enlargement is in the thoracic, and not the lumbar, region because the spinal cord grows at a slower rate than the vertebral column. By adulthood, the region below the second lumbar vertebra contains only spinal nerves that have branched from the spinal cord at higher levels. These spinal nerves form the **cauda equina** (e-kwi′nah), or horse's tail. The posterior end of the spinal cord, the **conus medullaris** (ko′nus med′u-la″ris), is anchored to vertebrae by connective tissue called the **filum terminale** (fi′lum ter-min-a′le).

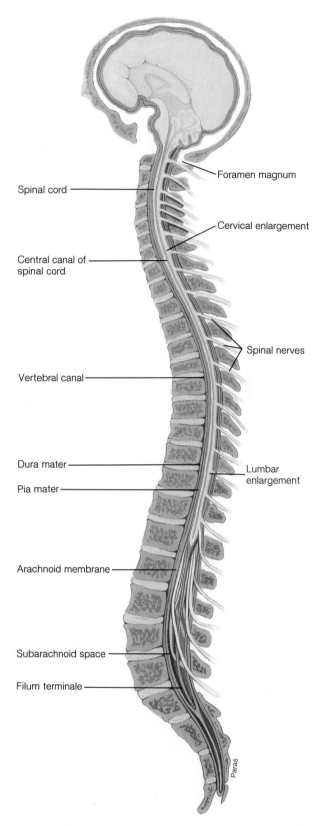

Spinal cord
Foramen magnum
Cervical enlargement
Central canal of spinal cord
Vertebral canal
Spinal nerves
Dura mater
Pia mater
Lumbar enlargement
Arachnoid membrane
Subarachnoid space
Filum terminale

Paras

FIGURE 12.31
Gross anatomy of the spinal cord.

The spinal cord is separated into left and right symmetrical halves by the **posterior median sulcus** and the **anterior median fissure** (Figure 12.32). The hollow **central canal** contains cerebrospinal fluid. The central butterfly-shaped area (gray matter) consists of cell bodies and interneurons. Surrounding the gray matter are bundles of myelinated fibers, which form the white matter of the spinal cord. The cut ends of these fibers are seen in cross section.

In each segment of the spinal cord, a spinal nerve arises from each side of the cord. Each nerve has two roots.

The **dorsal nerve root** consists of bundles of sensory axons (carrying afferent signals) whose cell bodies are in the **dorsal root ganglion**. These axons extend to the **posterior horn** where they synapse with interneurons and neurons of spinal tracts. Interneurons synapse with other interneurons or with motor neurons whose cell bodies are in the **anterior horn.** Aggregations of axons of motor neurons (carrying efferent signals) form the **ventral nerve root.** The **lateral horns** are aggregations of gray matter between the anterior and posterior horns found in the thoracic and upper lumbar regions of the spinal cord.

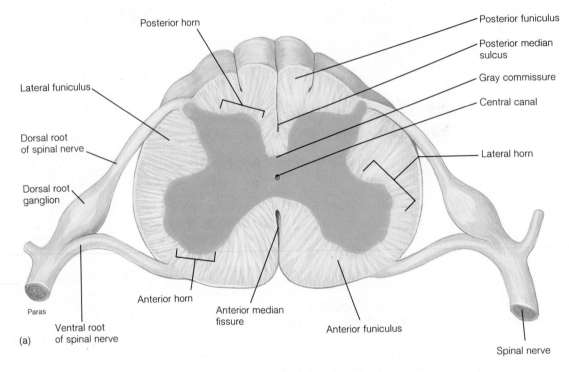

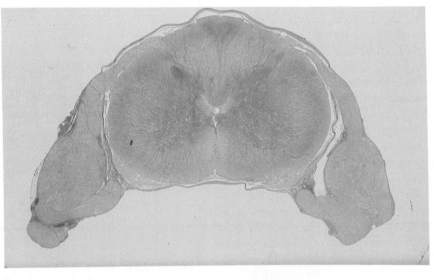

FIGURE 12.32

A cross section of the spinal cord:
(a) diagram and (b) photomicrograph.

SPINAL TRACTS

Within the white matter of the spinal cord are two kinds of fasciculi, or bundles of axons: the **ascending tracts,** which carry sensory signals from spinal nerves to the brain, and **descending tracts,** which carry motor signals from the brain to the spinal nerves (Table 12.5 and Figure 12.33). All spinal tracts are paired and **bilateral**—a member of each pair is found on each side of the spinal cord. Tracts that remain on the same side of the body from their origin to their destination are **ipsilateral** (ip″si-lat′er-al). Tracts that **decussate** (de-kus′āt), or cross from one side of the spinal cord to the other, are **contralateral** (kon″tra-lat′er-al). In decussated fibers, the signal destination is on the opposite side of the body from its origin.

TABLE 12.5 Major Spinal Tracts

Ascending Tracts*	Termination	Characteristics	Descending Tracts†	Origin	Characteristics
Gracile and cuneate fasciculi	Medulla (relayed to cerebral cortex)	Ipsilateral to medulla; brain neurons cross over. Information allows sensing body position, distinguishing closely adjacent touch stimuli, and recognizing sizes, shapes, and textures.	Lateral and anterior corticospinal	Motor area of cerebral cortex	Contralateral; lateral tract crosses over in medulla, anterior crosses over near termination. Referred to collectively as pyramidal tracts. Initiate voluntary movements.
Posterior and anterior spinocerebellar	Cerebellum	Ipsilateral. Information aids in the control of movement and posture.	Rubrospinal	Red nuclei	Contralateral; crosses over in brain stem. Helps to regulate movement and posture.
Lateral spinothalamic	Thalamus (relayed to cerebral cortex)	Contralateral; crosses over at level of origin. Information about pain and temperature.	Reticulospinal	Reticular formation	Contralateral; crosses over in brain stem. Increases motor activity.
Anterior spinothalamic	Thalamus (relayed to cerebral cortex)	Contralateral: crosses over at level of origin. Information about pressure and crude touch, some of which helps to regulate posture and muscle action.	Vestibulospinal	Vestibular nuclei	Ipsilateral. Maintains equilibrium and balance.

*All ascending tracts originate in the posterior horns of the spinal cord.

†All descending tracts terminate in the ventral (anterior) horns of the spinal cord.

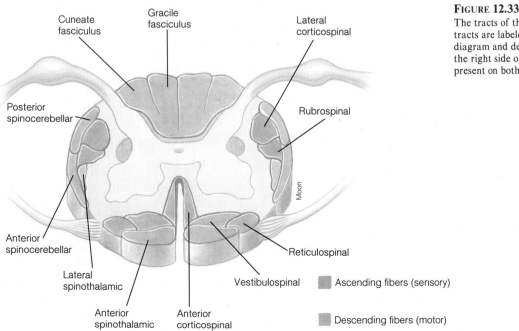

Cuneate fasciculus
Gracile fasciculus
Lateral corticospinal
Posterior spinocerebellar
Rubrospinal
Anterior spinocerebellar
Lateral spinothalamic
Reticulospinal
Vestibulospinal
Anterior spinothalamic
Anterior corticospinal

Moon

■ Ascending fibers (sensory)

■ Descending fibers (motor)

FIGURE 12.33
The tracts of the spinal cord. Ascending tracts are labeled on the left side of the diagram and descending tracts are labeled on the right side of the diagram, though all are present on both sides of the cord.

Most spinal tract names indicate origins and destinations. For example, the ascending spinocerebellar tracts carry signals from synapses with peripheral nerve endings in the spinal cord to the cerebellum. Likewise, descending corticospinal tracts carry signals from the cerebral cortex to synapses with peripheral nerves in the spinal cord. Ascending tracts carry signals from sensory receptors in the skin that detect touch, pressure, and other sensations. They also carry signals from muscles, tendons, and joints that detect tension and body position. All descending tracts carry signals that control movement.

The major ascending spinal tracts (Figure 12.34) are the gracile and cuneate fasciculi (bundles), the spinocerebellar tracts, and the spinothalamic tracts. The **gracile**

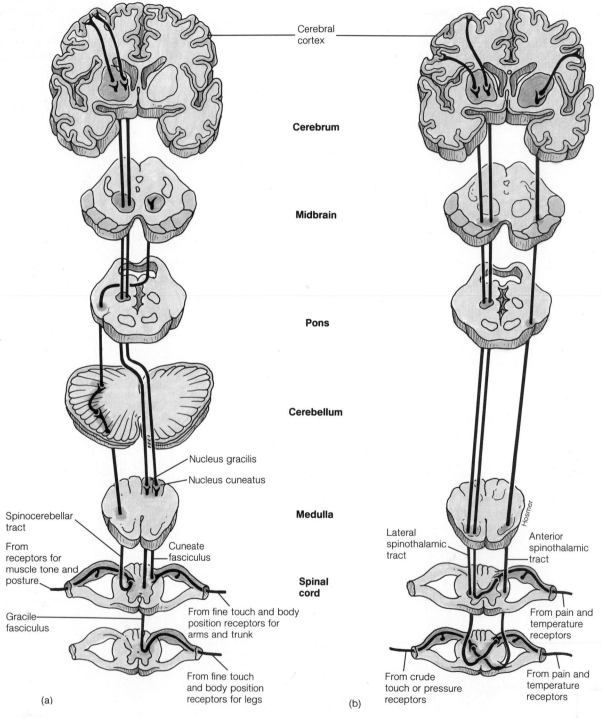

FIGURE 12.34
Pathways of afferent signals through spinal tracts: (*a*) gracile and cuneate fasciculi and spinocerebellar tracts; and (*b*) spinothalamic tracts.

fasciculi (gras'il fas-ik'u-li) and the **cuneate** (ku'ne-āt) **fasciculi** carry signals from skin receptors sensitive to fine touch and from joint receptors that detect body position. The cuneate fasciculus carries signals from the arms and trunk, and the gracile fasciculus carries signals from the legs. These fasciculi terminate in the medulla at the cuneate and gracile nuclei, respectively. There the fibers synapse with neurons whose axons decussate and carry signals to the thalamus. Some signals go to the somatic sensory area, where they allow perception of fine touch or body position.

The **spinocerebellar tracts** receive signals from muscle spindles and relay them to the cerebellar cortex. The cerebellum processes the information and sends new signals to the red nucleus and reticular formation in the midbrain. These signals are integrated with motor signals and used to control movement and posture.

The **lateral spinothalamic tract** receives signals from pain and temperature receptors in the skin and relays them to the thalamus, which in turn relays them to the cerebral cortex. Except for movements made as a result of pain or skin temperature changes, this tract is not involved in regulation of movement. The **anterior spinothalamic tract** sends signals from crude touch and pressure receptors in the skin to the thalamus, which relays them to the cerebral cortex. Information such as pressure on the soles of the feet is used in maintaining posture, so this tract indirectly contributes to regulating motor functions.

The major descending spinal tracts are grouped into the pyramidal tracts and the extrapyramidal tracts. The **pyramidal** (pĭ-ram'id-al) **tracts** (Figure 12.35a), so named because they pass through the pyramid of the medulla, include the lateral corticospinal tracts and the anterior

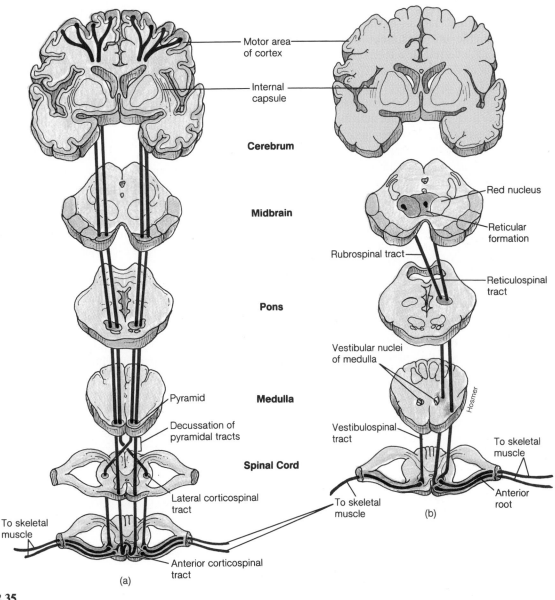

FIGURE 12.35
Pathways of efferent signals through spinal tracts: (*a*) corticospinal (pyramidal) tracts, and (*b*) extrapyramidal tracts.

corticospinal tracts. The **extrapyramidal tracts** (Figure 12.35b) include the rubrospinal tracts and the reticulospinal tracts, which decussate in the brain stem, and the vestibulospinal tracts, which do not decussate.

The **lateral corticospinal tracts** and the **anterior corticospinal tracts** (pyramidal tracts) carry signals from the cortical somatic motor area to the ventral horns at various levels of the spinal cord. A neuron in these tracts is an **upper motor neuron.** It synapses with a **lower motor neuron** whose cell body is located in the ventral horn of the spinal cord and whose axon extends along a peripheral nerve. Damage to upper motor neurons can lead to spastic paralysis (paralysis of muscles in the contracted state). Spastic paralysis occurs because peripheral function remains, but higher order controls are reduced or absent. Damage to lower motor neurons can lead to flaccid paralysis (paralysis of muscles in the relaxed state) because signals fail to reach the muscles from any control level.

Among the extrapyramidal tracts, the **rubrospinal tract** originates in the red nucleus of the midbrain, the **reticulospinal tract** in the reticular formation, and the **vestibulospinal tract** in the vestibular nucleus. All terminate in the ventral horns of the spinal cord. The rubrospinal tract carries signals from the cerebrum that control movement and posture. The reticulospinal tract carries signals from various sources that increase motor activity, but these signals are partly counteracted by other signals. The vestibulospinal tract carries signals from the organs in the inner ear that help to maintain equilibrium.

To show how integrated signals control movement, suppose you are sitting on an uncomfortable chair in a hot, stuffy room. Signals to the somatic sensory area of your cerebrum allow you to perceive your sitting position via gracile and cuneate fasciculi and pressure from the chair via anterior spinothalamic tracts. Your cerebellum receives information for maintaining posture from the spinocerebellar tracts. You decide to get up from your chair, walk to a window, and open it. Motor signals initiated in your cerebral cortex pass through the pyramidal tracts and peripheral nerves, and cause contraction of muscles required for your actions. Simultaneous signals from other parts of the cerebrum and the cerebellum are relayed via extrapyramidal tracts to maintain your balance and coordinate your movements.

CLINICAL APPLICATIONS

Disorders of learning and memory are common and poorly understood human maladies. In the United States, nearly 4 million children and adolescents have learning disabilities such as attentional deficit disorder, and more than 2 million older persons suffer from Alzheimer's disease.

Children with **attention deficit hyperactivity disorder** have a short attention span. They often develop behavioral and emotional problems partly from frustrations in school and social situations. This disorder may be due to improper screening of stimuli by the thalamus so a child's attention shifts from schoolwork to any distracting stimulus. Amphetamines and similar drugs have been used to treat attention disorders, with temporary calming effects and side effects such as suppressed growth. Highly structured, self-paced instruction seems to help these children focus on their schoolwork.

Alzheimer's (altz'hi-merz) **disease** is characterized by memory loss and sometimes marked personality changes. It exists in two forms—the familial form is due to a defective gene and usually begins around age 50, and the sporadic form usually begins much later. The same kinds of brain lesions appear in both forms (Figure 12.36). **Senile plaques** contain large amounts of a substance called amyloid protein and glial cells. **Neurofibrillary** (nu-ro-fi'bril-a-re) **tangles** are disorganized masses of fibers within neurons that also contain amyloid protein. Such abnormalities in the cerebrum lead to memory loss, and in the limbic system, to personality changes. The accumulation of the amyloid protein may be due to an inhibitor of an enzyme that normally breaks down excess amyloid. Whether these abnormalities are the cause or effect of the disease is not known. Researchers are looking for ways to treat or prevent Alzheimer's disease.

Understanding the physiology of emotional disorders is of great practical significance to health professionals and it provides a means of increasing our knowledge of brain function. Evidence of biochemical imbalances is especially useful because it provides a basis for therapy.

Manic-depressive psychosis (man'ik-de-pres'iv si-ko'sis), or affective psychosis, is a severe mental disturbance with great mood swings from manic (overactive) behavior to deep depression (dejection, absence of cheer-

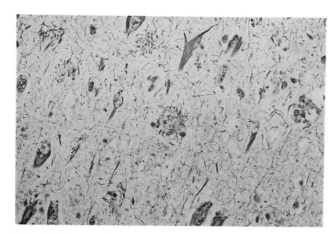

Figure 12.36
A brain lesion from a patient who died of Alzheimer's disease.

fulness and hope). Some patients experience only depression, and great variation exists in the degree of mania and depression among patients and within the same patient at different times. Both genetic and environmental factors probably contribute to the disease, which may involve a deficiency of dopamine or its derivative norepinephrine.

Cells that make monoamine transmitters, such as dopamine and norepinephrine, usually also contain an enzyme monoamine oxidase (MAO), which degrades the transmitters. Drugs called monoamine oxidase inhibitors, which slow degradation and allow accumulation of the neurotransmitters, are used to treat depression. Another group of drugs, the tricyclic antidepressants, inhibit the reuptake of norepinephrine into the presynaptic knobs, thereby extending the length of time it is available to act. The element lithium has been used to treat manic conditions and to stabilize the mood of patients who experience manic-depressive mood changes. Though its mode of action is unknown, it may affect the movement of ions across cell membranes. Given during a manic episode, lithium is retained in the body until after the manic behavior has subsided, but why this occurs is not known.

Seasonal affective disorder (SAD) is a kind of depression accompanied by carbohydrate craving that occurs in winter months. Its incidence increases with latitude from about 1 in 40 in Florida to about 1 in 5 in Alaska. A seemingly unrelated disorder, premenstrual syndrome (PMS), causes the same symptoms in many women during the week before a menstrual period. Some researchers believe serotonin may be implicated in both disorders, but how it exerts its effects is unknown. Both SAD and PMS respond to treatment with bright light. Half an hour of exposure to bright light when symptoms are present helps to alleviate them.

Schizophrenia (skiz″o-fre′ne-ah) is a complex of disturbances in perception, thinking, and feeling, and may be a name for several related diseases. Symptoms include detachment from reality, withdrawal, disrupted communication with other people, and sometimes delusions and hallucinations. Factors implicated in the pathology of schizophrenia include a genetic predisposition and a virus-like agent (found in the cerebrospinal fluid of some normal and some schizophrenic persons).

The brains of schizophrenics have structural and biochemical abnormalities. CAT scans show enlarged brain ventricles and autopsies show temporal lobe and hippocampal atrophy. Some brains contain 50 percent more dopamine than normal—almost all in the limbic system. This finding led to treatment with antipsychotic drugs such as chlorpromazine that block dopamine receptor sites. Though these drugs alleviate symptoms of schizophrenia, they also cause the number of dopamine receptor sites to increase. Dosages must then be increased to achieve the same effects. **Tardive dyskinesia** (tahr′div dis″kin-e′ze-ah), characterized by involuntary movements such as facial twitches, rapid eye blinking, and tongue and lip movements, often develops with long-term use of such drugs. This syndrome probably is due to a dopamine deficiency and resembles parkinsonism (Chapter 13).

CLINICAL TERMS

agraphia (ah-graf′e-ah)   inability to write because of a neural disorder

alexia (ah-lek′se-ah)   inability to read because of a neural disorder

amnesia (am-ne′ze-ah)   loss of or lack of memory

aphasia (ah-fa′ze-ah)   inability to speak or write, or inability to comprehend spoken or written language because of a neural disorder

apraxia (ah-prak′se-ah)   inability to perform purposeful movements, without paralysis of the parts concerned

coma (ko′mah)   a state of unconsciousness from which the individual cannot be aroused

dyslexia (dis-lek′se-ah)   impaired ability to read (especially reversal of letters in words) because of a neural disorder

ESSAY

SUBSTANCES THAT MODIFY BRAIN FUNCTIONS

Hundreds of substances—addictive drugs used for their mind altering properties, therapeutic drugs, and some substances found in foods—are known to modify brain function.

Addictive drugs such as caffeine, nicotine, cocaine, codeine, morphine (and its derivative heroin) are classified chemically as alkaloids and are extracted from plants. Alcohol, another addictive drug, affects the brain and many other organs (Chapter 24).

Archeological evidence indicates that chewing of coca plant leaves (now known to consist of nearly 2 percent cocaine) was practiced in the Andes 5000 years ago. According to an Incan myth, the god Inti created coca to alleviate hunger and thirst. Explorers returning to Europe from South America in the sixteenth century brought glowing reports of the effects of coca leaf chewing. By the late nineteenth century, cocaine had been extracted from leaves, purified, and tested. It served as a local anesthetic when injected subcutaneously and as a central nervous system stimulant when ingested. Because of the stimulant effect, some physicians, including Sigmund Freud, thought it might be used as an antidote to morphine addiction. Freud tested the drug (many times) and described "how a small dose lifted me to the heights in a wonderful fashion." The American surgeon William Halstead became addicted while trying to demonstrate that cocaine acted as a nerve blocker. Well-wishing colleagues treated him with morphine and he became addicted to both drugs.

Recent studies show that cocaine prevents inactivation of neurotransmitters, thereby prolonging their effects and maybe accounting for the euphoric effects of cocaine. Other effects of cocaine include increased heart rate and blood pressure; a slight increase in body temperature; and, sometimes, dilation of pupils. Like cocaine, morphine and other opiates also have been used for centuries. They have three effects—analgesia, sedation, and emotional effects. Each effect is due to the drug binding to a specific type of brain receptor.

Three characteristics of addiction are **tolerance** (the requirement for larger doses for the same effect), **physiological dependence** (the need for a drug to prevent withdrawal symptoms), and **habit-forming nature** (the quality that causes addicts to spend their time and resources seeking more of the drug). Repeated use of opiates, alcohol, and barbiturates lead to addiction; repeated smoking or injection of cocaine is habit-forming and leads to mild physiological dependence, but appears not to produce tolerance.

Though most kinds of addiction are difficult to treat, *methadone* (meth'a-dōn) is effective in treating heroin addiction. Methadone apparently binds to opiate receptors and satisfies the craving for heroin without eliciting its effects. Dosage is calibrated for each addict to maintain function without withdrawal symptoms or narcotization. Methadone binds to and is gradually released from a substance in the blood, so the blood concentration remains available to brain cells for up to 24 hours after a single dose.

Many drugs are used for their therapeutic effects on mental disorders. Some act at dopamine-mediated synapses and affect learning and memory. The tranquilizer reserpine appears to impair learning and memory by preventing dopamine storage in the secretory vesicles of axons. Chlorpromazine acts similarly by blocking dopamine receptor sites for dopamine. Amphetamines in small amounts temporarily facilitate learning and memory by stimulating the release of dopamine at synapses. Endogenous opiates and apomorphine, a substance derived from morphine, also facilitate learning and memory by stimulating binding of dopamine to its receptors. Endorphins and enkephalins also interact with dopamine to produce emotional good feelings. Some day they may be used to treat individuals with impaired learning ability and memory.

Questions

1. What kinds of substances are known to modify brain function?
2. Summarize the history of drug use.
3. What are the characteristics of addiction?
4. How are drugs used to improve brain function?

CHAPTER SUMMARY

The Nervous System and Its Development

- Neural ectoderm of the posterior surface of the embryo develops into a neural tube and neural crests. The neural tube gives rise to the brain and spinal cord, and motor portions of the cranial and spinal nerves. Neural crests give rise to sensory neurons, the autonomic nervous system, and the adrenal medulla.

Evolution of the Brain

- The brain evolved in three major stages: the reptilian brain, the limbic system, and the neocortex.

Regions of the Brain

- The major regions of the brain are the cerebrum, the cerebellum, and brain stem.

Brain Stem

- The medulla oblongata, which is continuous with the spinal cord, controls vital functions such as heart rate, blood pressure, respiration, and certain other functions.
- The pons, located superior to the medulla and adjacent to the cerebellum, transmits signals to and from the cerebrum and cerebellum, and also helps to regulate respiration.
- The midbrain relays signals to other parts of the brain and helps to control head and eye movements.
- The thalamus acts as a relay center for transmitting signals to and from the cerebrum.
- The hypothalamus acts as a center of homeostasis by regulating many internal processes and forming a connection between the nervous and endocrine systems.

Cerebellum

- The cerebellum consists of two lateral hemispheres and a median vermis, all of which lie above the brain stem inferior to the cerebrum.
- The cerebellum coordinates movements and helps to maintain posture and balance.

Cerebrum

- The cerebrum, the largest part of the human brain, consists of two hemispheres with an outer cortex (gray matter) and inner myelinated tracts (white matter). Other components of the cerebrum include the basal nuclei and limbic system.

Cerebral Cortex

- The cerebral cortex with its six layers of interconnected nonmyelinated neurons is organized into functional regions connected by myelinated tracts.
- Functional regions include the somatic sensory and motor areas, primary sensory areas, Broca's motor speech area, the premotor and prefrontal cortices, and the facial recognition area.

- The visual, auditory, and somatic sensory association areas process particular kinds of sensory information, and the tertiary association area integrates information from them.
- Lateralization, the presence of functional differences between the left and right sides of the body, is marked in the cerebrum. The left hemisphere usually controls speech and has an interpreter that integrates signals from the two hemispheres.
- Processing language involves sensory analysis and motor coordination, and uses most cortical areas.

Wakefulness and Sleep

- Electroencephalograms display brain waves and are useful in monitoring sleep and other processes that occur in the brain.
- Sleep stages 1 through 4 represent increasing depth of sleep and REM (rapid-eye-movement) sleep is characterized by processes summarized in Table 12.4.
- Sleep may be regulated by nuclei in the reticular formation, serotonin, norepinephrine, and sleep factors.
- Sleep seems to restore brain neurons to functional levels that prepare it for another period of activity.

Learning and Memory

- Learning is modification of behavior as a result of experience; memory is the process of storing some of what is learned in a recallable form.
- Learning has been studied in mollusks, which have a few large neurons and learn by habituation, sensitization, and operant conditioning.
- Memory probably involves division of synapses and long-term potentiation. In humans, memories appear to be stored in the hippocampus and amygdala.

Emotional Behavior

- Emotions such as pleasure, fear and anxiety, and aggression are mediated by the limbic system, especially the hypothalamus.

Meninges, Ventricles, and Fluids

- The meninges are connective tissue membranes that surround and protect the brain; from outer to inner layers, they are the dura mater, the arachnoid layer, and the pia mater.
- The ventricles of the brain are four hollow cavities filled with cerebrospinal fluid.
- Cerebrospinal fluid, which is secreted by choroid plexuses, acts as a shock absorber and regulates pressure within the ventricles and central canal of the spinal cord.
- Blood reaches the brain by alternate pathways in the circle of Willis. The blood-brain barrier restricts entry of substances into brain tissue.

Spinal Cord

- The spinal cord extends from the base of the brain to the second lumbar vertebra and gives rise to 31 pairs of spinal nerves. It contains an internal area of cell bodies and interneurons and outer region of myelinated spinal tracts.

Spinal Tracts

- Spinal tracts and their functions are summarized in Table 12.5.

Clinical Applications

- Children with attentional deficit disorder have a short attention span and usually are hyperactive.
- Alzheimer's disease causes memory loss and personality changes in elderly patients. It is characterized by deposits of amyloid protein in senile plaques and by neurofibrillary tangles.
- Manic-depressive psychosis displays wide swings in emotions. Seasonal affective disorder is a winter depression. Schizophrenia is a complex of altered perception, thinking, and feeling.

QUESTIONS AND PROBLEMS

The questions at the end of each chapter are numbered to correspond with the objectives listed at the beginning of the chapter. Italics indicate that a question requires critical thinking skills beyond simple factual recall.

Questions

1. List the major steps in nervous system development.
2. List the order in which the components of the brain evolved.
3. What are the main regions of the brain?
4. (a) What are the parts of the brain stem, and where is each located?
   (b) What are the major functions of each of the parts of the brain stem?
   (c) *Why is the hypothalamus appropriately called the center for maintenance of homeostasis?*
   (d) *For each of the parts of the brain stem, what would happen if it failed to function?*
5. (a) Where is the cerebellum, and what are its main parts?
   (b) What is the general function of the cerebellum?
6. (a) Where are the main tracts of the cerebrum located, and what do they do?
   (b) Where are the basal nuclei, and what do they do?
   (c) Where is the limbic system, and what does it do?
7. (a) Where is the cerebral cortex located, and what is its structure?
   (b) Where are each of the major functional areas of the cerebrum, and what does each do?
   (c) *If a blood clot or other obstruction prevented blood flow to a cerebral area, what would happen to cells in that area?*
   (d) *Using what you have learned about association areas, predict the effects of injury to each association area.*

   (e) *What cortical functions occur when someone asks a question and you answer? When you read a paragraph and write a summary of it? When you recognize and speak to an old friend?*
   (f) What is lateralization of function, and how can it be observed in humans?
8. (a) What are the characteristics of the stages of REM and non-REM sleep?
   (b) *How are wakefulness and sleep regulated?*
   (c) What are the physiological effects of sleep? Of sleep deprivation?
9. (a) What are the major findings regarding learning and memory from studies of mollusks? Of mammalian brains?
   (b) *How might research on learning and memory be applied to helping you learn?*
   (c) How is the expression of emotions related to brain centers and neurotransmitters?
10. (a) What is the location, structure, and function of the meninges?
    (b) Where and what are the ventricles of the brain?
11. (a) What is cerebrospinal fluid, how is it formed, and how does it circulate?
    (b) What are the functions of cerebrospinal fluid?
12. *What are the special properties of blood circulation through the brain?*
13. Describe the location and gross anatomy of the spinal cord.
14. (a) Summarize the location and function of each of the spinal tracts.
    (b) *For any two motor and any two sensory tracts, explain how you could determine that a tract had been damaged.*

Problems

1. Suppose a friend asks you for directions to your home. Prepare a brief description of the neural processes that occur as you interpret the request and respond to it.
2. Think of your earliest childhood memory. Explain briefly what processes might have been involved in storing that memory.
3. Research the effects of a specific substance that modifies brain function. If possible, present your findings to your class.
4. Observe a sleeping animal or human child. Record eye movements (if visible through the eyelids) or eyelid movements and changes in position. (Sometimes puppies' legs make running movements during sleep.) Explain your findings in terms of sleep stages.
5. Much knowledge of motor pathways has been learned from the effects of neural damage. Use what you know about neural controls to explain the following: (a) cuneate tract damage in the right cervical area, (b) pyramidal tract damage in the left lumbar area, (c) motor cortex damage in the right lateral region, (d) bilateral vestibular nuclei damage, (e) bilateral basal nuclei damage, and (f) inability of cerebellum to send efferent signals.

OUTLINE	OBJECTIVES
Somatic Nervous System Components	1. List the components of the somatic nervous system, and relate them to the whole nervous system.
Nerve Structure	2. Describe the structure of a nerve.
Cranial Nerves	3. List the twelve cranial nerves in order, and describe the location and function of each.

Spinal Nerves
    Structure and Distribution
    Plexuses
	4. Describe the structure of spinal nerves, their distribution, and their arrangement into plexuses.

Levels of Neural Function	5. Name and define the body's levels of neural function.

Reflexes
    Reflex Arc
    Examples of Reflexes
	6. Describe the properties of a reflex arc, and give examples to explain how reflexes control body functions.

Higher Order Control of Movement
    Program of Action
    Brain Functions in Movement
	7. List the events in motor functions, and explain how they are controlled at local and higher levels.

Postural Reflexes and Locomotion
    Postural Reflexes
    Locomotion
	8. Summarize the mechanisms involved in maintaining posture and balance.
	9. Explain how locomotion is accomplished.

Reaching for a Book—An Integrated Movement	10. Summarize the processes involved in integrated movement.
Clinical Applications	11. Explain how nuclear magnetic resonance imaging and reflexes can be used diagnostically. 12. Briefly describe the following conditions: (a) sciatica, (b) carpal tunnel syndrome, (c) backache, (d) cerebral palsy, (e) Parkinson's disease, and (f) poliomyelitis.
Essay: Spinal Cord Injuries Chapter Summary Questions and Problems	

13

SOMATIC NERVOUS SYSTEM

SOMATIC NERVOUS SYSTEM COMPONENTS

Fortunately, the whole nervous system normally operates in an integrated fashion. Otherwise, people would have trouble receiving information from the environment, interpreting it, and responding to it. The nervous system is very complex, and it can be better understood by looking at some of its components separately. In Chapter 12, we considered the brain and spinal cord—the components of the central nervous system. All other parts of the nervous system constitute the peripheral nervous system. In this chapter, we will consider the somatic nervous system, which controls and integrates the activity of the skeletal muscles. In the next chapter, we will consider the autonomic nervous system, which controls the visceral organs. The neural signals for both of these divisions are carried through the peripheral nerves, which are described in the following paragraphs.

NERVE STRUCTURE

A bundle of nerve fibers, usually myelinated axons, in the peripheral nervous system is called a **nerve** (Figure 13.1). Such axons are surrounded by several connective tissue sheaths. The **endoneurium** surrounds each fiber, the **perineurium** binds groups of fibers together into **fascicles,** or bundles, and the **epineurium** covers the whole nerve. Blood and lymph vessels are often located within the connective tissue sheath between fascicles of large nerves and make up a significant portion of the total nerve diameter. Nerves can be grouped by location as cranial or spinal. **Cranial nerves** carry signals to and from the brain, and **spinal nerves** carry signals to and from the spinal cord. Except for a few cranial nerves, all nerves contain both sensory (afferent) and motor (efferent) fibers and are called **mixed nerves.**

CRANIAL NERVES

Each cranial nerve passes through a foramen in the skull as it runs between the organ it supplies and the brain. The 12 pairs of cranial nerves are identified by name and by Roman numerals I through XII in order of their attachment from the front to the back of the brain (Table 13.1 and Figure 13.2). Any nerve that carries motor signals to muscles also has fibers that carry signals concerning muscle length and tension back to the brain. (Though the nerves are paired, they are discussed in the singular in the following paragraphs.)

TABLE 13.1 Characteristics of Cranial Nerves

Nerve	Type	Pathways	Functions
I Olfactory	Sensory	From nasal epithelium to olfactory bulb	Smell
II Optic	Sensory	From retina of eye to nuclei in thalamus	Sight
III Oculomotor	Motor, proprioceptive*	From midbrain to four eye muscles; from ciliary body to midbrain	Movement of eyeball and eyelid, change in pupil size, eye muscle length and tension
IV Trochlear	Motor, proprioceptive*	From midbrain to superior oblique muscle; from muscle to midbrain	Movement of eyeball, eye muscle length and tension
V Trigeminal	Sensory, motor	From pons to muscles of mastication; from cornea, facial skin, tongue, lips, and teeth to pons	Chewing food, sensations from facial organs
VI Abducens	Motor, proprioceptive*	From pons to lateral rectus muscle; from muscle to pons	Movement of eyeball, eye muscle length and tension
VII Facial	Sensory, motor	From pons to facial muscles; from facial muscles and taste buds to pons	Facial expressions, secretion of saliva and tears, facial muscle length and tension, taste
VIII Vestibulocochlear	Sensory	Organs of hearing and balance to pons	Hearing, balance
IX Glossopharyngeal	Sensory, motor	From medulla to muscles of pharynx, pharyngeal muscles and taste buds to medulla	Swallowing, secretion of saliva, facial muscle length and tension, taste
X Vagus	Sensory, motor	From medulla to viscera, from viscera to medulla	Visceral muscle movement, information from viscera
XI Accessory or spinal accessory	Motor, proprioceptive*	From medulla to pharyngeal and neck muscles; from muscles to medulla	Swallowing, head movements, neck muscle length and tension
XII Hypoglossal	Motor, proprioceptive*	From medulla to muscles of tongue, tongue muscles to medulla	Speech, swallowing, tongue muscle length and tension

*Proprioceptive signals are afferent signals from muscles, tendons, and joints.

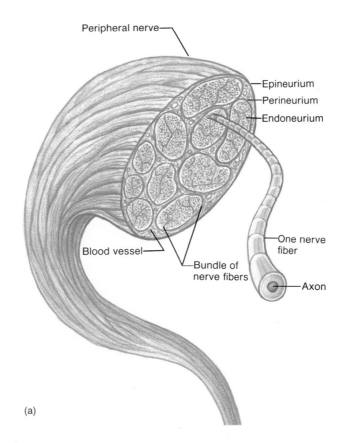

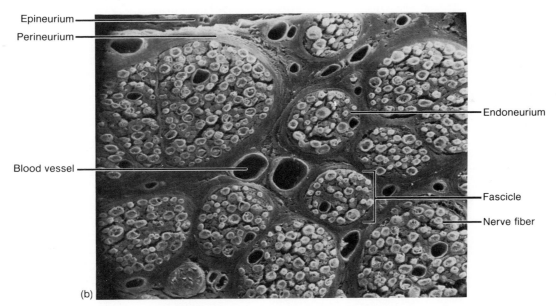

FIGURE 13.1
The structure of a nerve: (*a*) diagram, and (*b*) photomicrograph.

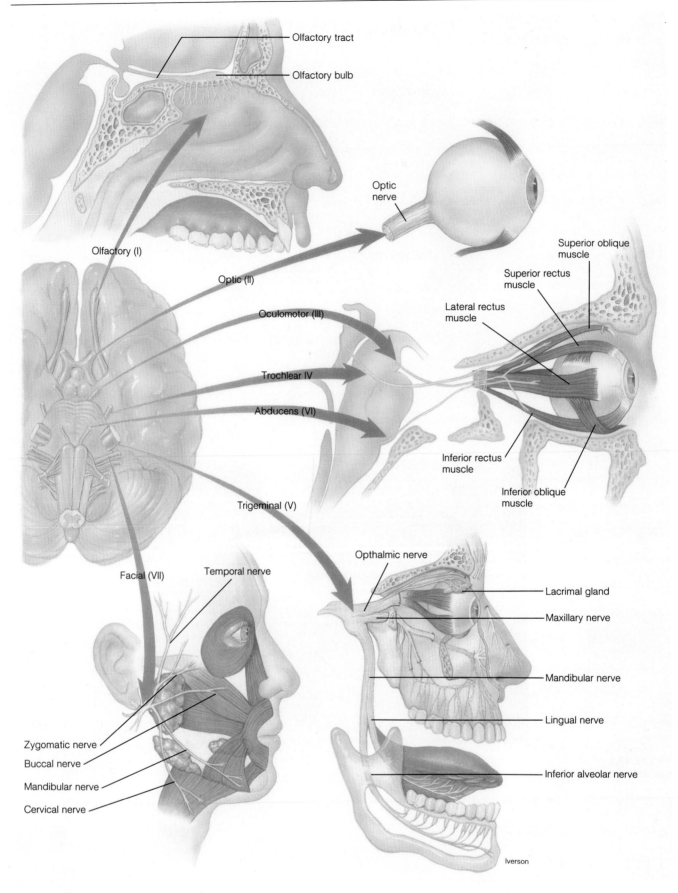

FIGURE 13.2
The cranial nerves.

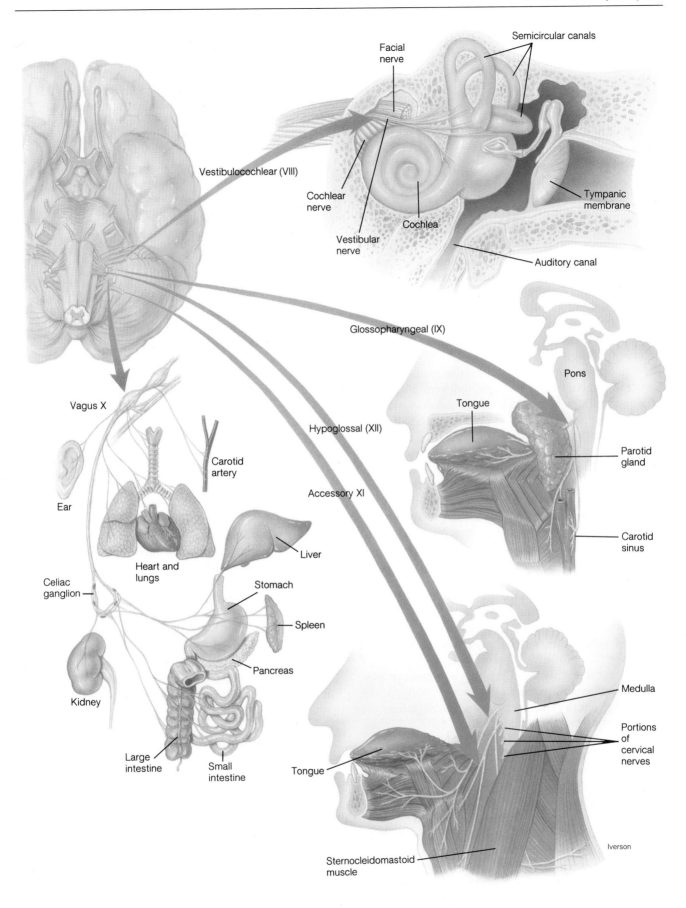

Facial nerve

Semicircular canals

Vestibulocochlear (VIII)

Cochlear nerve

Vestibular nerve

Cochlea

Tympanic membrane

Auditory canal

Glossopharyngeal (IX)

Pons

Tongue

Parotid gland

Hypoglossal (XII)

Carotid sinus

Accessory XI

Vagus X

Carotid artery

Ear

Liver

Heart and lungs

Stomach

Spleen

Celiac ganglion

Pancreas

Kidney

Medulla

Portions of cervical nerves

Large intestine

Small intestine

Tongue

Sternocleidomastoid muscle

Iverson

The **olfactory nerve** (I) is a sensory nerve. Its fibers run from smell receptors in the nasal epithelium through the cribriform plate of the skull and enter the **olfactory bulb,** which lies inferior to the frontal lobes of the cerebrum. In the olfactory bulb, fibers synapse with neurons of the **olfactory tract,** and signals are relayed to the cerebral cortex and limbic system. The pathway for smell is the only sensory pathway that does not pass through the thalamus.

The **optic nerve** (II) also is a sensory nerve. The left and right nerves carry signals from neurons in the retina of an eye to the optic chiasma, where medial fibers of each nerve decussate and join the lateral fibers from the other nerve. Fibers pass through the optic chiasma without synapsing to form **optic tracts** that synapse with neurons in a thalamic nucleus. Thalamic neurons carry signals to the visual area of the cerebral cortex. Some optic nerve fibers go to the superior colliculi of the midbrain and synapse with neurons that carry signals to cranial nerves III, IV, and VI. These nerves control eye movements in accordance with what is being seen.

The **oculomotor nerve** (III) contains motor and proprioceptive fibers. Motor fibers extend from a midbrain nucleus to muscles of the upper eyelid and extrinsic muscles of the eyeball. They control eyelid and eyeball movements. Some fibers carry signals from the autonomic nervous system (Chapter 14) through the ciliary ganglion and control the size of the pupil and the shape of the lens. Proprioceptive fibers carry signals from receptors in the eye muscles that detect length and tension of muscles.

The **trochlear nerve** (IV), the smallest cranial nerve, also has motor and proprioceptive fibers. Motor signals go from a midbrain nucleus to the superior oblique muscle of the eye, and proprioceptive fibers go from the muscle to the midbrain.

The **trigeminal nerve** (V), the largest cranial nerve, contains both sensory and motor fibers. The sensory portion has three large branches, the **ophthalmic, maxillary,** and **mandibular nerves,** which carry signals from the eye, upper jaw, and lower jaw, respectively. The sensory branches come together to form the **semilunar,** or **gasserian, ganglion** and terminate on the ventrolateral surface of the pons. The motor portion arises as a separate root in the pons near the sensory portion. In general, motor fibers parallel sensory fibers. Motor branches of the mandibular nerve control the chewing muscles.

The **abducens nerve** (VI) contains motor and proprioceptive fibers. The motor fibers extend from a nucleus in the pons to the lateral rectus muscle of the eyeball. Proprioceptive signals are relayed back to the pons.

The **facial nerve** (VII) contains sensory fibers from taste buds in the anterior two-thirds of the tongue to a nucleus in the pons. Signals arriving in the pons are relayed to the thalamus and then to the appropriate part of the somatic sensory cortex. It also contains motor fibers that carry signals from the pons to muscles of the scalp and face, the latter controlling facial expression. Other motor fibers, which are part of the autonomic nervous system (Chapter 14), stimulate secretion by salivary and lacrimal (tear) glands.

The mainly sensory **vestibulocochlear nerve** (VIII) has two main branches, both of which go to the inner ear. The **cochlear,** or **auditory,** nerve carries signals from sound receptors in the cochlea. The cell bodies of its fibers are in the **spiral ganglion,** and the axons extend to the thalamus. Signals are relayed from the thalamus to the auditory cortex and cerebellum. The **vestibular nerve** originates in the semicircular canals and other inner ear structures associated with balance. The cell bodies of its fibers are in the **vestibular ganglion,** and the axons extend to nuclei in the thalamus. Signals are relayed from the thalamus to the cerebral cortex and to the cerebellum.

The **glossopharyngeal nerve** (IX) has sensory fibers originating in taste buds in the posterior third of the tongue and terminating in the thalamus. Some sensory fibers carry signals from the carotid sinus, an organ in the wall of the carotid artery in the neck, that helps to regulate blood pressure. It also has motor fibers extending from a nucleus in the medulla to muscles of the pharynx and the parotid salivary gland.

The **vagus nerve** (X) carries both sensory and motor fibers between the brain stem and visceral organs. It is an important component of the autonomic nervous system (Chapter 14).

The **accessory nerve** (XI) carries motor and proprioceptive fibers from the medulla and the most superior portions of the spinal cord. Fibers from the medulla supply voluntary muscles of the soft palate, pharynx, and larynx, including muscles used in swallowing and speaking. Fibers from the spinal portion supply muscles of the neck, shoulders, and rib cage. Proprioceptive signals from these muscles are relayed back to the medulla.

The **hypoglossal nerve** (XII) carries motor and proprioceptive fibers. The motor fibers originate in a nucleus of the medulla and control tongue movements in speech and swallowing. Branches of the first three cervical nerves join with the hypoglossal in controlling parts of the tongue and some neck muscles. Proprioceptive signals from these muscles are relayed back to the medulla.

Spinal Nerves

Each spinal nerve has afferent and efferent fibers. In general, afferents carry signals through the dorsal root to the spinal cord and efferents carry signals from nuclei in the spinal cord through the ventral root. Like many cranial nerves, all spinal nerves are mixed nerves.

Structure and Distribution

The **spinal nerves** are formed by the union of the dorsal and ventral roots a short distance beyond the spinal cord (Figure 13.3). After passing through the intervertebral

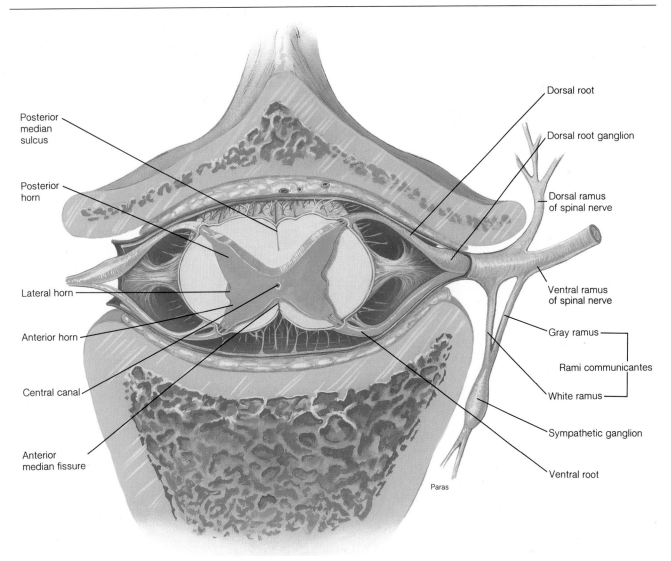

FIGURE 13.3
The distribution of the branches of a spinal nerve.

foramen, each spinal nerve separates into posterior and anterior branches. The **posterior branch** innervates muscles and skin of the posterior part of the body. The **anterior branch** innervates the limbs and the lateral and anterior parts of the body. Spinal nerves in the thoracic and lumbar regions have a **visceral branch,** which innervates internal organs. Most also have a small **meningeal branch,** that innervates vertebral ligaments, meninges, and blood vessels of the spinal cord itself.

The 31 pairs of cranial nerves are named and numbered according to the vertebrae with which they are associated (Figure 13.4)—eight pairs of **cervical nerves** (C1 through C8), 12 pairs of **thoracic nerves** (T1 through T12), five pairs of **lumbar nerves** L1 through L5), and one pair of **coccygeal nerves (Co1).** Though there are only seven cervical vertebrae, there are eight pair of cervical nerves because the atlas has a pair both superior and inferior to it. In other regions, each pair of spinal nerves lies inferior to the vertebra of the same name.

Plexuses

A branching, interconnected network of nerves is called a **plexus.** The anterior branch of each spinal nerve (except T2 through T12) enters a plexus from which fibers are redistributed to nerves in the limbs and other body areas (Figure 13.4). Such plexuses are paired—one on each side of the body.

The **cervical plexus,** formed from branches of nerves C1 through C4 (and some fibers of the accessory and hypoglossal nerves), lies in the neck beneath the sternocleidomastoid muscle. Most nerves branching from the cervical plexus supply the skin and muscles of the neck, scalp, and upper part of the shoulders. The **phrenic nerve** branches from the cervical plexus and innervates the diaphragm, a major muscle of the breathing mechanism. Damage to the spinal cord above the cervical plexus can impair or stop breathing by interfering with signals in the phrenic nerve.

The **brachial plexus** includes branches of nerves C5 through T1 and forms a network near the clavicle. Nerves leading from the brachial plexus branch repeatedly and innervate skin and muscles of the arm. Some of the major branches of the brachial plexus are the **radial nerve** innervating the posterior arm, the **musculocutaneous nerve** innervating the anterior upper arm, and the **median nerve** and **ulnar nerve,** both innervating the anterior forearm. Sharp pain from bumping the "crazy bone" at the elbow is due to stimulation of the ulnar nerve.

The **lumbar plexus** is formed by nerves L1 through L4 (and in some people, part of T12). It lies in the psoas major muscle, a deep muscle between the lumbar vertebrae and the femur. The largest single branch of the lumbar plexus

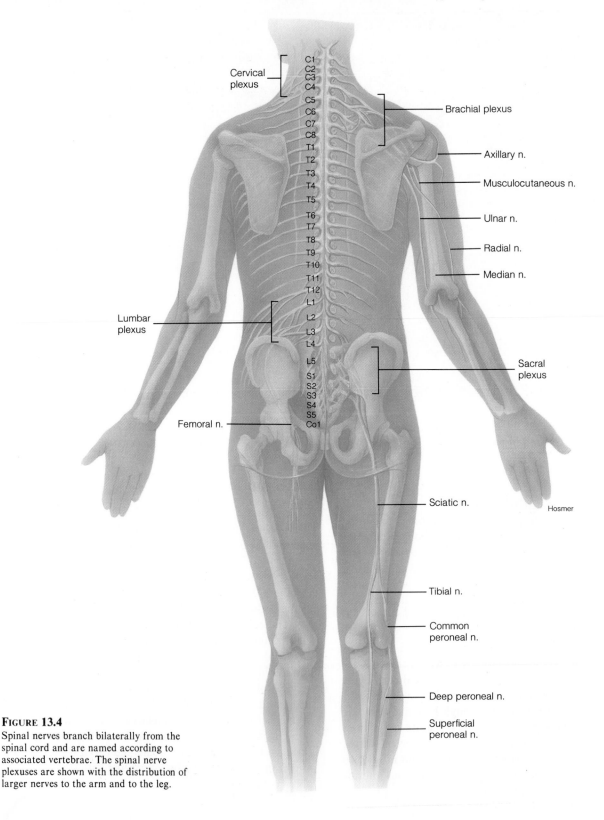

FIGURE 13.4
Spinal nerves branch bilaterally from the spinal cord and are named according to associated vertebrae. The spinal nerve plexuses are shown with the distribution of larger nerves to the arm and to the leg.

is the **femoral nerve,** which innervates muscles and skin of the anterior thigh and leg. Other branches supply the lateral and medial thigh, the anterolateral abdominal wall, and portions of the pelvic organs.

Closely associated with the lumbar plexus is the **sacral plexus,** formed from branches of nerves L4 through S3 (and in some people part of S4). Because of close proximity and overlapping branches, the lumbar and sacral plexuses, are sometimes referred to collectively as the **lumbosacral plexus.** The major nerve of the sacral plexus is the **sciatic nerve,** the largest nerve in the body. It branches into the **common peroneal** and **tibial nerves,** both of which supply the leg. Smaller nerves of the sacral plexus supply the gluteal muscles and portions of the pelvic

organs. Intramuscular injections into the gluteus muscle should be given with care to avoid damage to the sciatic nerve that can result in partial paralysis or loss of some sensation.

Except for nerve C1, sensory fibers of spinal nerves become associated with embryonic dermatomes (segments of mesoderm that form the dermis) and motor fibers become associated with embryonic myotomes. Even in adults, the segments of skin supplied by a particular spinal nerve are called dermatomes (Figure 13.5). Growth of appendages and unequal growth in other areas create dermatomes of unequal size and shape with varying degrees of overlap. Abnormalities in skin sensations in a dermatome can be used to detect spinal nerve damage.

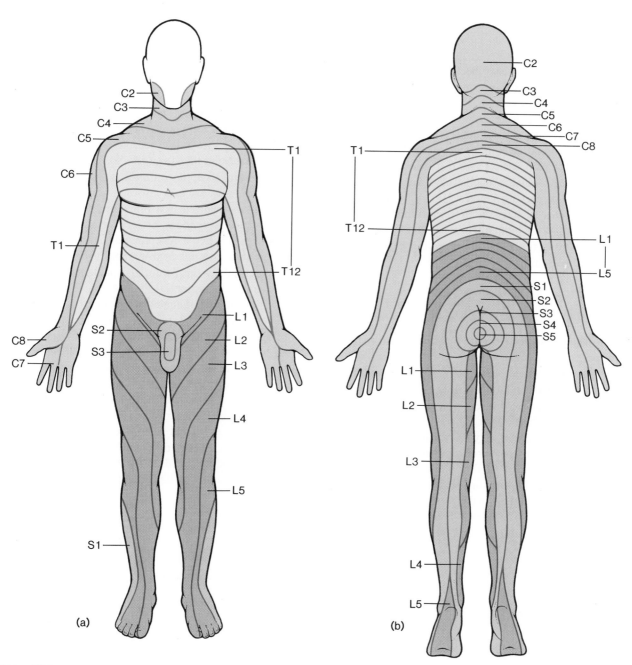

FIGURE 13.5
The arrangement of dermatomes and spinal nerve distribution to them.

LEVELS OF NEURAL FUNCTION

Within the human nervous system, three major levels of function have evolved. They are: (1) the spinal cord level; (2) the lower brain level, involving all parts of the brain except the cerebral cortex; and (3) the higher brain, or cortical, level. Function at the spinal cord level is limited principally to reflexes. A **reflex** involves a stimulus, processing in the central nervous system (usually the spinal cord), and a response. Reflexes are automatic and almost instantaneous.

Functions at the lower brain level include subconscious, usually reflexive, activities such as regulation of respiration, blood pressure, and body temperature; salivation and other responses to food; control of balance; and coordination of movement. Even some emotional responses, such as reactions to pain and pleasure, excitement, anger, and sexual activities must occur at the lower brain level because they have been observed in animals lacking a cerebral cortex. Lower brain activity, therefore, coordinates many vital functions and may contribute to emotional responses as well.

Functions at the cortical level include mainly the storage and interpretation of information including learned patterns of motor responses. These responses can be used voluntarily to control motor functions as diverse as speech, riding a bicycle, or playing a musical instrument. Certain areas of the cortex such as the prefrontal lobes and significant portions of the temporal and parietal lobes are involved in abstract functions—interpreting experiences, thinking, and reasoning.

REFLEXES

Reflexes, which constitute a kind of homeostatic control system, occur mainly at the spinal cord level, but some visceral reflexes involve the brain stem and hypothalamus. In a reflex, a given stimulus usually produces the same general kind of response. For example, when your finger touches a hot object, you jerk it away; however, the intensity of the stimulus can vary as can the response to it. In one instance, you might merely extend your finger and avoid a hot object, but in another you might move your whole body to avoid the stimulus. Reflexes do not require cortical function, but signals to the cortex let us know that a reflex has occurred. The human body displays a great variety of reflexes, which regulate body functions.

Most reflexes serve to protect the body. Blinking the eyes protects them from approaching objects. Dilation or constriction of the pupils regulates how much light enters the eyes. A muscle in the middle ear contracts in response to a sudden loud noise, thereby protecting the hearing mechanism. Stretching of muscles and tension in tendons elicit reflexes that regulate muscle contraction. The withdrawal reflex (jerking away from an injurious object) minimizes tissue damage. Food on the tongue releases saliva, and food touching the palate initiates the swallowing reflex. Changes in blood pressure elicit reflexes that regulate the pressure itself. Because reflexes require only modest processing by the central nervous system, they help the body to react quickly to changes in the external or internal environment. No time is lost in conscious decision making about how to react to the stimulus.

Reflex Arc

The conduction pathway of a typical spinal **reflex arc** (Figure 13.6) proceeds from a receptor through an afferent neuron to a synapse in the spinal cord. From the synapse, signals usually are relayed through one or more **interneurons,** or **association neurons,** in the spinal cord to a motor neuron and then to an effector. For example, stimulating a skin receptor initiates a signal that travels through the reflex arc and causes a muscle to contract.

Receptors involved in a reflex arc can be located in the skin, muscles, tendons, visceral organs, or in or near a sense organ. When a receptor responds to a change in the external or internal environment, it produces a membrane potential. If the potential is sufficiently strong, it initiates an action potential in a neuron. The action potential is conducted along a sensory neuron to a synapse in the dorsal horn of the spinal cord or brain stem. Signals are relayed across synapses by neurotransmitters and converted to action potentials, usually in one or more interneurons. In a few synapses, the signal goes directly to a motor neuron. Interneurons relay signals to a motor neuron. Many reflexes also have interneurons that relay signals across the spinal cord and to adjacent segments of the spinal cord above and below the level of the incoming signal. For some reflexes, other pathways that are not part of the reflex conduct signals to the cerebral cortex, making us aware that the reflex has occurred.

Examples of Reflexes

Reflexes control various motor functions that protect and maintain homeostasis. We will look at a few examples.

The **withdrawal reflex,** or flexor reflex, occurs in response to a painful stimulus (Figure 13.7a). It relays signals to adjacent segments on one side of the spinal cord and causes flexion of joints in the injured limb and its withdrawal from the source of the stimulus.

The **crossed-extensor reflex** (Figure 13.7b) likewise occurs in response to a painful stimulus, but it causes movement that maintains balance by counteracting the withdrawal reflex. In this reflex, interneurons relay signals to several segments on both sides of the spinal cord.

When you move one arm away from a painful stimulus, the crossed-extensor reflex causes you to move the other arm so as to maintain balance.

Proprioceptive reflexes adjust tension in muscles in accordance with muscle stretching, the size of the load, and other factors. When you lift an object, muscle fibers are subject to stretching and must exert tension to overcome the stretching as well as to lift the load. Local receptors in muscles detect stretching and initiate stretch reflexes. Similar receptors in tendons detect tension and initiate tendon reflexes. Afferent neurons from the organs relay signals to the spinal cord that provide negative feedback to muscles and control contraction.

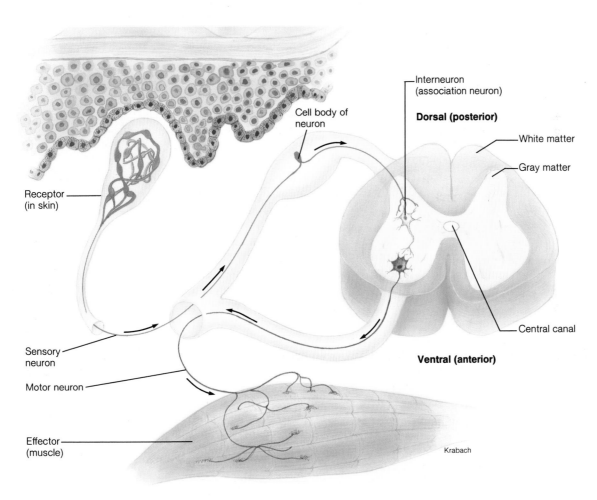

FIGURE 13.6
A generalized reflex arc.

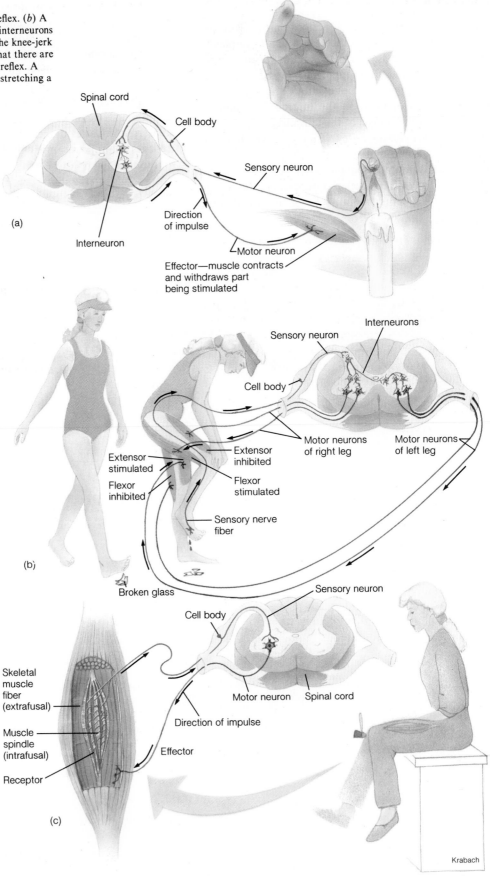

FIGURE 13.7
(a) The withdrawal, or flexor, reflex. (b) A crossed-extensor reflex involves interneurons that cross the spinal cord. (c) The knee-jerk reflex is a stretch reflex. Note that there are no interneurons involved in this reflex. A muscle spindle is stimulated by stretching a muscle.

Spinal cord

Cell body

Sensory neuron

Direction of impulse

Interneuron

Motor neuron

Effector—muscle contracts and withdraws part being stimulated

(a)

Interneurons

Sensory neuron

Cell body

Motor neurons of right leg

Motor neurons of left leg

Extensor stimulated

Extensor inhibited

Flexor inhibited

Flexor stimulated

Sensory nerve fiber

(b)

Broken glass

Sensory neuron

Cell body

Skeletal muscle fiber (extrafusal)

Muscle spindle (intrafusal)

Receptor

Motor neuron   Spinal cord

Direction of impulse

Effector

(c)

Krabach

The **stretch reflex,** as illustrated by the knee-jerk reflex (Figure 13.7c), regulates motor response to increased length (stretching) of a muscle. The components of the stretch reflex are the muscle spindle, afferent fibers that carry signals to the spinal cord, and efferent fibers that carry signals to the muscle. A **muscle spindle** consists of **intrafusal fibers,** specialized muscle fibers that act as receptors to detect stretching. The contractile fibers of a muscle are called **extrafusal fibers.** When intrafusal fibers are stretched, they initiate signals in the reflex arc that cause the extrafusal fibers to contract in proportion to the stretching, thereby maintaining constant tension in a muscle. Stretch receptors are particularly important in maintaining posture.

In some situations, muscles display **clonus,** a sequence of rhythmic contractions (Figure 13.8). For example, when one is standing on tiptoes and muscles are stretched, the stretch reflex causes muscles to contract and the whole foot touches the floor momentarily. Clonus produces an oscillation between tiptoe and foot on the floor positions.

Stretch receptors also help to keep the mouth closed, except when it is voluntarily opened for eating or talking. When the jaw starts to open, stretch reflexes initiate muscle contractions to close it. If a person falls asleep in a sitting position, gravity causes the lower jaw to drop. Because stretch receptors lose tone during sleep, they fail to close the mouth.

When you hold out your hand to receive a heavy object, your contracted biceps muscle holds your forearm parallel with the floor. When the object is placed in your hand, your hand will drop. The degree of drop is decreased by the **load reflex** in which stretching of the muscles supporting the load excites stretch receptors. The stretch reflex increases muscle tension sufficient to support the load.

The **tendon reflex** (Figure 13.9) involves a receptor called the **Golgi tendon organ,** a network of nerve endings among the connective tissue fibers of a tendon. When muscles contract, they exert tension on tendons and stimulate tendon organs. These organs respond, not to stretching, but to tension. Signals from tendon organs go to inhibitory interneurons in the spinal cord and cause inhibitory postsynaptic potentials in motor neurons. This prevents extrafusal fibers from contracting and reduces tension in the muscle. Tendon organs send signals at a rate proportional to the degree of tension in a tendon (and its attached muscle).

Tendon organs vary in their sensitivities. Those that respond to extreme tension help to prevent muscle tearing, such as during weight-lifting. Those that respond to lesser tension regulate tension according to the contractile force needed for a particular action.

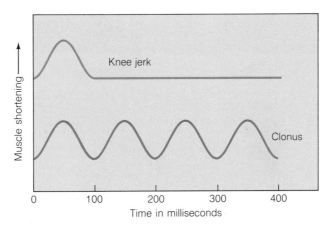

FIGURE 13.8
Compared to a simple reflex, clonus consists of a sequence of rhythmic contractions.

In addition to local control of muscle tension, signals from tendon organs are relayed via spinal cord tracts to the cerebellum and cerebral cortex. Such signals are used by motor control systems to integrate and coordinate tensions in various muscles to accomplish desired movements. An activity such as rowing a boat requires contraction and relaxation of several different sets of muscles. A combination of stretch reflexes, tendon reflexes, and other control mechanisms maintains appropriate tensions for smooth, rhythmic strokes of the oars.

Other receptors called **joint kinesthetic** (ken″es-thet′ik) **receptors** are found in the capsules of synovial joints. They send signals that allow the cerebral cortex to recognize joint positions and joint movements. The cerebellum and other parts of the brain use these signals to coordinate movements.

HIGHER ORDER CONTROL OF MOVEMENT

Most motor functions are far more complex than the reflexes just discussed, though they may incorporate reflexes and other control mechanisms. For example, when you stoop to pick up a pencil, you must coordinate arm and finger movements with a visual image of the pencil and its location, maintain a grasp on the pencil, lift the pencil, and maintain posture and balance throughout the entire activity. You would need to make additional adjustments if the object being picked up were an egg with a cracked shell or a heavy wrench.

Regardless of the complexity of movement—blinking the eyes, picking up a pencil, or playing a piano—the basic functional unit for all movement is the motor unit, a motor

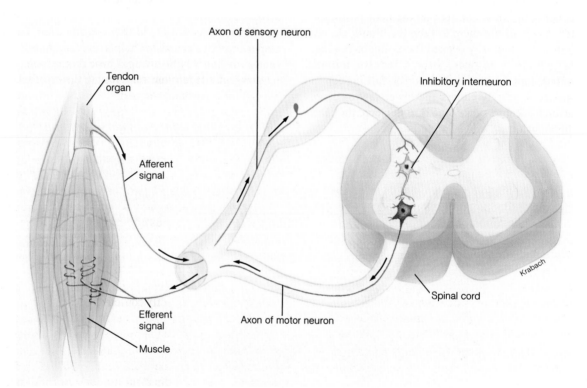

FIGURE 13.9
A tendon reflex. Tension exerted on Golgi tendon organ elicits afferent signals that excite an inhibitory interneuron. This causes an inhibitory postsynaptic potential in the alpha efferent neuron and reduces tension in the muscle.

neuron and the muscle fibers it innervates (Chapter 9). Signals sent by motor neurons to skeletal muscle fibers are always excitatory; however, pathways to the motor neurons contain synapses that receive excitatory and inhibitory signals. The net effect of these signals determines what movement will occur.

Program of Action

Voluntary motor activity begins with the initiation process, the intention to pick up a pencil, for example, but physiologists still do not know exactly where or how this occurs. They do know that certain parts of the cortex that participate in mental activity and parts of the cerebellum display increased electrical activity prior to movement. These areas are very likely involved in early coordination of movement and may be involved in its initiation. They and the basal nuclei all receive information from various receptors about body position prior to and during movement. Information from all these sources is used to create a **program of action**—a plan for how to carry out an activity.

Once a program of action is established, it is relayed to the motor cortex via the thalamus, and the motor cortex initiates a pattern of action potentials. Certain action potentials go via motor axons directly to motor units, but collaterals of those axons send signals to the cerebellum and the basal nuclei. The cerebellum and basal nuclei continuously receive information about body position and ongoing movements (the course of a ball one is trying to catch, for example) and provide signals that are used to constantly revise the program. They compare actual movements (one's movement relative to the ball) with the plan and notify the motor cortex to alter movements to accommodate the new plan. Continuous cycling of information between the motor cortex and the cerebellum and basal nuclei allows coordination of movement even when different movements occur concurrently (playing massive chords on the piano) or in rapid succession (typing at great speed). So, voluntary motor activities—things we consciously choose to do, usually with some purpose in mind—involve processes that occur at all levels of neural function.

Learning plays an important role in voluntary activity. When we start to learn a motor skill, the program that initiates it is not clearly established. A person learning to play the piano or typewrite deliberately presses one key at a time and waits for sensory feedback that the action was correct before striking the next key. After sufficient practice, programs become established and motor skills tend to become automatic.

Regardless of how they are controlled, most movements require muscle contractions to support the body against gravity and to maintain posture and balance while the movement is occurring. For example, the muscles of a ballerina dancing on her toes must simultaneously keep her body in a balanced, upright position and execute the dance steps. Certain muscles of typists and piano players must maintain posture and arm position while other muscles control finger movements.

Brain Functions in Movement

Several brain stem nuclei help to control movement. The **vestibular nuclei** use signals from organs of balance in the inner ear (Chapter 15) to cause reflexes that help to maintain balance. The **red nuclei** are excited mainly by signals from the cerebellum that regulate movement. They send signals to the reticular formation and through the rubrospinal tract to muscles to regulate movement and posture. The **reticular formation** receives and relays signals from many sources—the vestibular nuclei, cerebellum, basal nuclei, cerebral cortex, and hypothalamus. The **basal nuclei** work with the motor cortex and cerebellum in an integrated unit. The basal nuclei generally regulate muscle tone required for precise movements. For example, in handwriting, subconscious signals from one of the basal nuclei stabilize postural muscles and muscles that steady the shoulder and arm, thereby allowing conscious precise movements of the hands and fingers. Output signals from the basal nuclei also reach the motor cortex and provide the main means by which extrapyramidal tracts influence pyramidal tracts (Chapter 12).

Though the cerebellum rarely elicits movement and we never become directly conscious of its activities, it does coordinate muscular activities, especially rapid movements. Loss of cerebellar function results in inability to coordinate movements during running, talking, and rapid finger movements. The main function of the cerebellum is to monitor and adjust motor activities initiated elsewhere. The cerebellum receives signals from the somatic motor cortex, the basal nuclei, and the reticular formation. It also receives signals from the skin, muscle spindles, and tendon organs along the spinocerebellar tracts.

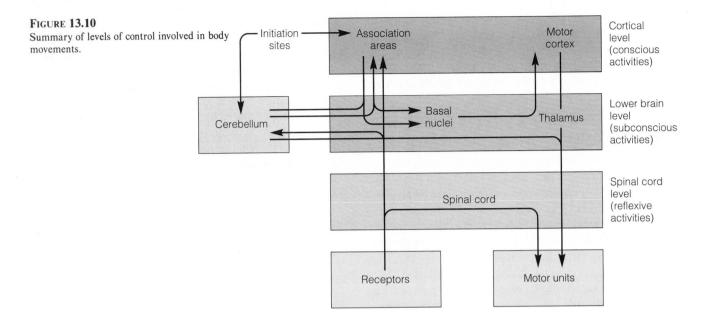

FIGURE 13.10
Summary of levels of control involved in body movements.

These signals travel at velocities in excess of 100 meters per second—the most rapid conduction rate in the central nervous system. The cerebellum continuously processes information and sends signals to all the parts of the brain from which it receives signals. It also continuously receives feedback about all aspects of motor activity and, thereby, plays an important role in adjusting the program of action to changing conditions.

The somatic motor cortex (previously shown in Figure 12.10) initiates signals that lead to voluntary movements. A relatively large cortical area is devoted to controlling muscles of the hands, face, lips, and tongue because they are supplied with many small motor units. Signals that excite the somatic motor cortex come from the sensory, visual, and auditory cortices, the motor association area, cerebellum, basal nuclei, and thalamus. Using the various signals it receives, this part of the brain controls all voluntary movements.

Body movement and its regulation are far more complex than this brief summary indicates. The levels of control of movement are summarized in Figure 13.10.

POSTURAL REFLEXES AND LOCOMOTION

In spite of the sturdy support provided by your skeleton, you cannot move about or even maintain a balanced standing or sitting position without coordinated muscle contractions and postural reflexes. To illustrate how control and communication processes regulate motor functions, we will consider two situations: (1) how these processes maintain posture and balance, and (2) how they function in locomotion.

Postural Reflexes

Having the same components as other reflexes, postural reflexes go through the spinal cord or parts of the brain for processing. Some examples of postural reflexes are the labyrinthine reflexes and righting reflexes.

Labyrinthine reflexes, responses to tilting the head, are initiated in the inner ear (Chapter 15) by the effects of gravity on changes in head position. Tonic labyrinthine reflexes, which go through the medulla, help to keep the head level. **Neck righting reflexes** are initiated by the stretching of neck muscles, as might occur when one begins to doze off while sitting in a chair. Spindles in stretched muscles elicit signals that go through the midbrain and stimulate muscles of the neck, shoulders, and thorax to contract. **Optical righting reflexes** are initiated by visual cues from receptors in the eyes. Signals go through the cerebral cortex and cause muscle contractions that right the head. Postural reflexes not only maintain posture and balance, they also allow the body to respond quickly to regain balance when it is upset. These reflexes occur automatically, though we are usually aware of their effects.

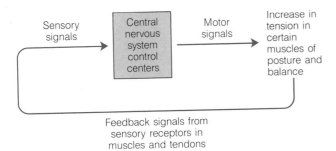

FIGURE 13.11

Summary of control and communication processes in maintaining posture and balance. Postural reflexes are initiated by afferent signals, including feedback signals from muscles. Afferent signals are processed by centers in the central nervous system and motor signals are initiated. Motor signals increase tension in certain muscles of posture and balance.

Various sensory signals help to maintain posture and balance. Signals from the eyes, organs of balance, skin receptors, joint receptors, and muscle spindles, together provide much redundant information about body position and orientation in space. We know that some information is redundant because blind people and people with inner ear disorders still can maintain posture and balance. Available signals are integrated and coordinated in many places in the central nervous system, but the cerebellum is an especially important integrating center. Processes involved in controlling posture and balance are summarized in Figure 13.11.

Astronauts sleep tied in bags attached to the space vehicle wall (Figure 13.12). Unless their arms are tied in the bags, they assume a position over the astronauts heads. If astronauts are asked the position of their arms before they open their eyes, they cannot answer correctly. What does this tell us about the role of gravity in our awareness of the position of our body parts? Does it mean that gravity is necessary to stimulate joint and other receptors to send signals needed to sense limb position?

Locomotion

Human locomotion, or walking erect, requires maintenance of posture while the body is in motion and balanced at intervals on one foot as well as a variety of coordinated muscle actions. Walking is a rhythmic process involving forward movement of first one leg and then the other. Each step is a phasic process. One foot is in the **swing phase** (off

FIGURE 13.12

Astronaut in sleeping bag.

the ground and swinging forward) while the other foot is in the **stance phase** (on the ground with the trunk pivoting about the stationary foot). Swinging arm movements are coordinated with phasic leg movements and assist in maintaining balance.

Walking and other rhythmic activities are directed by an aggregation of stored information called a **neural oscillator,** or **pattern generator.** Exactly where or how this information is stored is not yet known, but it probably is a special kind of program of action. Afferent feedback, mostly to the cerebellum, provides information about both limb activity and reflexes processed in the spinal cord. This information is used to control switching between swing and stance phases, to regulate muscle contractions for each step, and to adjust for changes in the terrain or to avoid small obstacles.

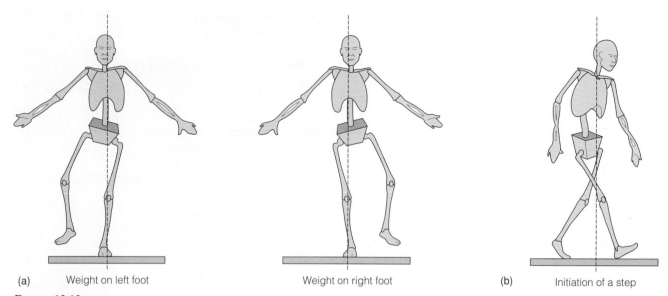

(a) Weight on left foot   Weight on right foot   (b) Initiation of a step

FIGURE 13.13
Shifts in the body's center of gravity (shown by dotted line).

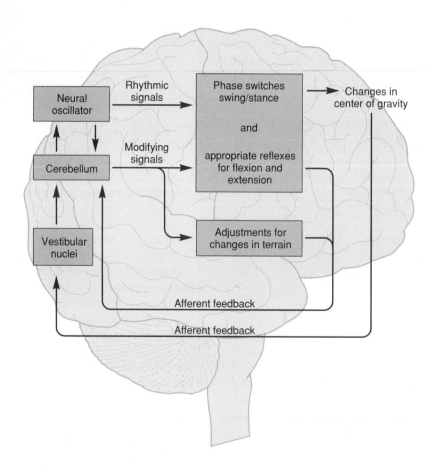

FIGURE 13.14
Summary of events in locomotion. The neural oscillator generates
rhythmic signals that control phase switches and flexion and extension
reflexes. Changes in the center of gravity and responses of receptors in
muscles and tendons provide afferent feedback, which modifies signals
and helps to produce smooth locomotion.

In addition to rhythmic activity and its regulation by afferent feedback, walking requires shifting balance from one foot to the other and a means of creating forward motion. Both occur by shifts in the body's center of gravity and neuromuscular responses to those shifts (Figure 13.13). These processes are controlled mainly by the cerebellum. If you concentrate on walking as you take a few slow steps, you can feel shifts in your body's center of gravity. When your left leg is in the stance phase, the center of gravity shifts to the left of the body's midline and allows your right leg to go through the swing phase. Your body's center of gravity shifts forward, so you must either take a forward step or fall. The many tumbles of toddlers learning to walk are due to their inability to respond properly to shifts in the center of gravity.

In summary, the complex act of walking involves the neural oscillator, alternating contractions of muscles on the left and right sides of the body, coordination via the cerebellum and other brain centers, and shifts in the body's center of gravity (Figure 13.14). Purposeful walking from one point to another is a voluntary activity, and the actual execution and regulation of the process once learned require no conscious thought.

REACHING FOR A BOOK—AN INTEGRATED MOVEMENT

Reaching for a book on a high shelf might begin with the idea that you need the book. You get out of your chair, stand on the toes of one foot, and reach above your head and far to the right. You grasp the spine of the book, pull it from the shelf, and place it on your desk (Figure 13.15). You can do all these things without thinking about anything except identifying the book—and in less time than it has taken you to read about them.

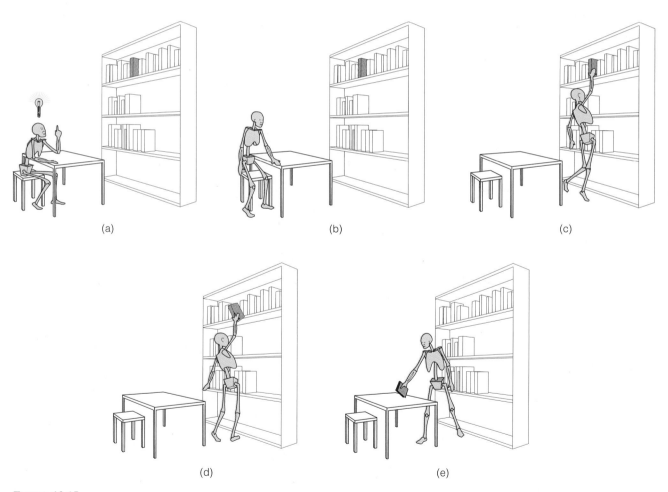

(a)

(b)

(c)

(d)

(e)

FIGURE 13.15
Processes involved in reaching for a book: (*a*) idea and generating a program of action, (*b*) standing, (*c*) balancing on one foot and reaching for the book, (*d*) grasping the book, and (*e*) lifting the book from the shelf onto the desk.

Almost in the same instant that you have the idea to reach for the book, your brain develops and initiates a program of action. This program makes use of information about the location, relative position, and the direction and rate of motion of each moving body part. Signals from your motor cortex cause muscle actions that let you stand up. Sensory cues from your eyes, organs of balance, skin, muscle spindles, and tendon organs, along with adjustments initiated by the cerebellum maintain your equilibrium in a standing position. As you balance on one foot and reach for the book, different motor signals cause appropriate muscles to contract or relax. More sensory cues help you to aim your hand at the proper book and maintain your balance in this precarious position. All the while, your cerebellum compares actual movements with the intended program and adjusts muscle tensions accordingly. As you grasp the book, touch sensations help to adjust the program to provide exactly the amount of tension to slide the book from the shelf. As you lift the book and move it to the desk, the load reflex helps to adjust tension so you can maintain a grasp on the book and support its weight. At the same time, the cerebellum is rapidly adjusting tension in many different muscles so you are not thrown off balance by the weight of the book.

Throughout the entire sequence of movements, various parts of your nervous system continuously receive and send signals that cause appropriate movements. All this is accomplished simply by adjusting muscle tensions.

The **restless leg syndrome** (RLS), first described in 1865 by the English physician Thomas Willis, is characterized as discomfort in both legs of a peculiar restless quality, aching and tension in calf muscles. The affected person has a persistent desire to shift the legs because of the uncomfortable sensations. RLS is intermittent and typically begins while the patient is relaxed or resting in bed. Walking seems to alleviate the symptoms. Many different causes—nutritional deficiencies, chronic bronchitis, reduced venous blood flow, caffeine, various medicines, neurological disorders, anxiety, and depression—have been suggested, but none has been clearly established.

CLINICAL APPLICATIONS

Neurological disorders can affect the brain, spinal cord, and cranial and spinal nerves, structures we have studied in this and the previous chapter. Many disorders can be diagnosed by imaging techniques and by abnormalities in reflexes.

The newest of imaging techniques is nuclear magnetic resonance (NMR), also referred to as magnetic resonance imaging (MRI). To obtain such an image, the patient is placed in a chamber within a very large magnet.

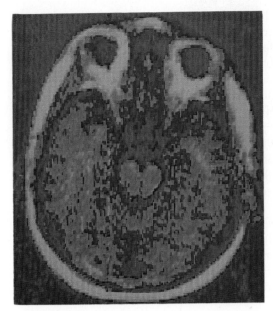

FIGURE 13.16
A magnetic resonance image (MRI) of the human brain.

The strong magnetic field—as much as 40,000 times as strong as the earth's magnetic field—causes hydrogen molecules to spin very rapidly. Energy from spinning molecules is translated by a computer into a visual image (Figure 13.16). MRI is particularly useful for visualizing soft tissue, distinguishing between gray and white matter, for example. It also can be used to study metabolic reactions as they occur in living tissue.

Though MRI provides much clearer images of soft tissues than CAT scans (Chapter 12), it has some disadvantages. Its magnetic force can, at highest power, pull metal objects—loose tooth fillings, pacemakers, joint prostheses, and the like—out of the body, and some evidence suggests that it may actually cause tissue damage. Finally, it is much more costly than other imaging techniques.

Certain reflexes are used clinically to assess neurological function (Figure 13.17). In healthy individuals, each reflex produces a predictable response to a specific stimulus. Abnormal responses help to locate neurological damage or disease.

Stroking the sole of the foot in infants up to 18 months of age normally initiates the **Babinski** (bab-in'ske) **reflex** (spreading of toes away from the sole) and in older persons the **plantar reflex** (curling of toes toward the sole). The Babinski reflex is due to immaturity and incomplete myelinization of corticospinal tracts. When it occurs after maturation of the corticospinal tracts, it indicates damage or disease somewhere in those tracts.

The **ankle-jerk reflex** results in extension of the foot when the Achilles tendon is stimulated. It is weak or absent when peripheral nerves have been damaged and exaggerated when spinal tracts have been damaged. The rate of

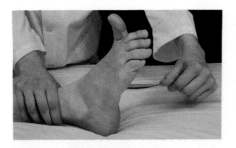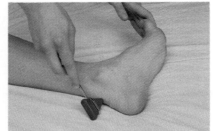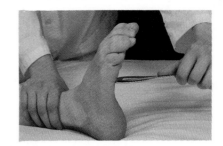

FIGURE 13.17
Some clinically important reflexes.

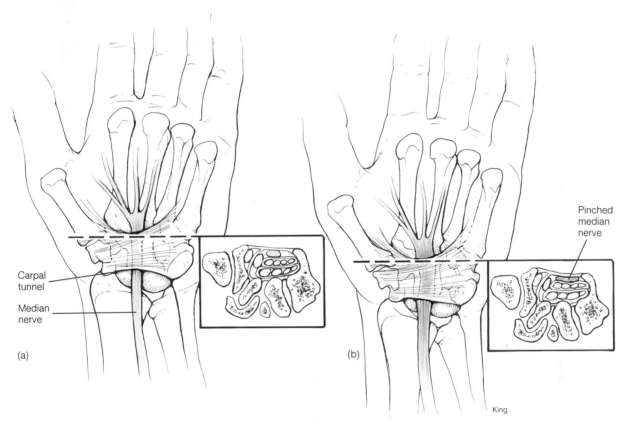

Carpal
tunnel

Median
nerve

(a)

Pinched
median
nerve

(b)

King

FIGURE 13.18
(*a*) A normal carpal tunnel, and (*b*) a carpal tunnel affected with
carpal tunnel syndrome.

relaxation of muscles after the ankle-jerk reflex also is
useful in diagnosing abnormal thyroid function. In hy-
pothyroidism (underactive thyroid gland), relaxation is
slow; and in hyperthyroidism (overactive thyroid gland),
it is accelerated.

Sciatica is an irritation (neuritis) of the sciatic nerve.
The most common cause is a herniated intervertebral disk
in the lumbar area (Chapter 7), which exerts pressure on
the sciatic nerve. It also can be caused by osteoarthritis
or pregnancy (if the fetus causes pressure on the sciatic
nerve). Treatment includes bed rest, exercises, muscle relaxants, physical therapy, and pain medication.

Carpal tunnel syndrome, a painful tingling sensation
in the palm of the hand and palm side of the fingers, can
be caused by pressure on the median nerve as it passes
through the carpal tunnel (Figure 13.18). This condition
is common among people who frequently use their hands
for grasping and twisting movement and is aggravated by
typing and by subjecting hands to prolonged vibrations.
Other distinguishing characteristics of carpal tunnel syndrome are that tingling does not affect the little finger and
that pain is alleviated by extending the wrist.

Over 5 million Americans suffer from **backache** yearly,
and most of them bring the condition on themselves by
lack of exercise. Weak back muscles under the stress of

prolonged sitting are easily strained by almost any unusual movement. A spasm occurs in the strained muscle causing pain in two ways—the muscle swells and restricts its own blood flow, and it contracts around and squeezes a nerve. Moist heat, bed rest, and muscle relaxants can alleviate backaches from muscle strain. But most backaches can be prevented by regular exercise that includes mild stretching of back muscles. Even standing and walking around the office several times a day can help.

A variety of neurological disorders have been associated with damage to the basal nuclei (Figure 13.19). Disorders categorized as **cerebral palsy** (pawl′ze) cause varying degrees of impairment. Affected individuals usually are quite intelligent though they may be unable to walk or talk. A palsy called *athetosis* (ath″et-o′sis), characterized by writhing movements, results from damage to the globus pallidus and sometimes other basal nuclei as well. Such damage appears to disrupt normal feedback circuits among the basal nuclei, thalamus, and cerebral cortex. *Huntington's chorea,* an inherited disorder characterized by continuous, uncontrolled, random movements, results from widespread, diffuse damage to the caudate nucleus and putamen. Neurons that secrete the inhibitory transmitter GABA are reduced in activity or number, and random movements fail to be inhibited.

Parkinson's disease is characterized by muscle rigidity, resting tremors (involuntary trembling movements of body parts when no movements are intended), and inability to initiate movement. It results from widespread damage to the **substantia nigra,** a component of the basal nuclei located in the midbrain region of the brain stem. Damage can spread to the globus pallidus and other nuclei. Though the cause of Parkinson's disease is still unknown, a neurotoxin or an exceedingly small infectious particle (smaller than a virus) could be responsible for some cases of the disease. Its effects are due to decreased release of the neurotransmitter dopamine from neurons of the substantia nigra. The absence of dopamine's inhibitory signals allows acetylcholine from other neurons to exert excitatory effects that produce rigidity and tremors.

Available treatments for Parkinson's disease are less than ideal. L-dopa (L-dihydroxyphenylalanine) alleviates some symptoms of Parkinson's disease. Because it has numerous undesirable side effects, physicians prescribe it mainly for advanced cases. Dopamine has no effect because it cannot cross the blood-brain barrier. L-dopa, which can cross the barrier, is presumed to be converted to dopamine in the brain or to compensate directly for the dopamine deficiency. Transplanting tissue to the deteriorating region of the brain has alleviated symptoms in some patients, but long-term results are not known. Tissue from the substantia nigra of aborted fetuses or a patient's own adrenal medulla, which makes dopamine, has been used, but how these tissues function is not clearly understood. The new drug deprenyl, which inhibits the enzyme monoamine oxidase B, has been shown to delay the need for L-

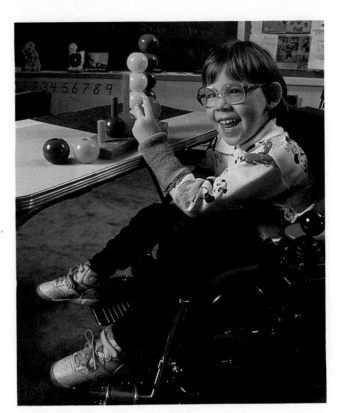

FIGURE 13.19
Disorders resulting from damage to basal nuclei.

dopa by six to eight months in drug trials. Inhibition of the enzyme appears to slow the progress of Parkinson's disease by preventing the formation of a metabolite that kills substantia nigra cells. Surgical destruction of tracts that initiate tremors is being tried.

Poliomyelitis is an acute viral infection that destroys cell bodies of motor neurons in the anterior horn of the spinal cord. Bulbar poliomyelitis affects motor neurons in the brain stem and can damage respiratory centers in the medulla. Symptoms include stiff neck, muscle pain, headache, and fever. As the disease runs its course, muscles served by the dying neurons become paralyzed. Because the disease affects only motor neurons, sensory function remains even in paralyzed limbs. If some motor neurons survive, affected individuals can sometimes regain partial or even total use of limbs through physical therapy. Effective immunization has been available for several decades, but many children and adults still are unprotected from this seriously crippling disease.

Clinical Terms

ataxia (ah-tak′-se-ah)   inability to coordinate muscular activity
paresis (par′-es-is)   slight or incomplete paralysis
vagotomy (va-got′-o-me)   cutting of the vagus nerve
whiplash   an injury caused by sudden hyperextension of the neck, which can lead to spinal nerve damage in the cervical area

Essay
Spinal Cord Injuries

While a spinal cord injury occurs at a specific site, its effects are widespread. Our studies of skeletal, muscular, and neurologic factors in movement in this and earlier chapters have prepared us for considering such injuries.

Spinal cord injuries result from falls and blows, especially in conjunction with automobile and diving accidents. The extent of disability caused by a cord injury is determined by its location and its severity, which can be assessed by studying a myelogram of the injured area (Figure 13.20). Completely severing the cord prevents all sensory and voluntary motor activity below the injury. In the upper lumbar region, it results in **paraplegia** (par″ah-ple′je-ah), loss of sensation and voluntary motor function in the legs. In the midcervical region, it results in **quadriplegia** (kwod″rip-le′je-ah), loss of sensation and voluntary movement in all four limbs. Above spinal segment C3, it causes respiratory arrest.

Partially severing the cord can result in paralysis, loss of sensation, or both below the level of injury in the affected spinal tracts. Precise **hemisection** (severing of the left or right side of the cord) produces a specific pattern of impairment, the **Brown-Séquard** (sa-kār) **syndrome.** All voluntary motor function and touch and pressure sensations are lost below the site of injury on the same side as the injury. Sensations of heat, cold, and pain are lost on the side of the body opposite the injury. Knowledge of spinal tract pathways and decussations make these effects predictable and help neurologists to assess an injury.

Spastic paralysis often develops over a period of weeks or months because the motor neurons to the affected muscles no longer receive inhibitory signals from the brain. In contrast, flaccid paralysis more often occurs in injuries to peripheral nerves because the affected muscles receive no signals at all.

Spinal shock follows severe trauma to the spinal cord. At first, all cord reflexes in the area below the injury cease to function. The person loses bladder and bowel control, and arterial blood pressure drops precipitously in response to a change in body position, such as moving from a horizontal to a vertical position. Loss of reflexes below the injury is ascribed to the lack of facilitating signals from higher centers. If the reflex arcs below the injury are intact, they begin to function again after a period of several days to several months.

The treatment of spinal cord injuries usually involves six to eight weeks in traction (Figure 13.21). Vertebral fractures are reduced by traction or surgery. Electrical stimulation of muscles maintains contractility, and isometric muscle retraining uses undamaged neurons to create new neuromuscular pathways. The anti-inflammatory drug methylprednisolone is being used immediately after spinal cord injury with remarkable reduction in the degree of paralysis.

Until recently, it was thought that the central nervous system had extremely limited regenerative ability; however, experiments in animals have shown that peripheral nerve tissue can be grafted into damaged areas of the cord. Such

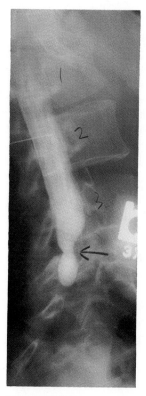

FIGURE 13.20
A myelogram showing a spinal cord injury.

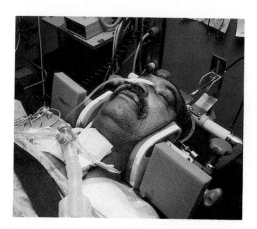

FIGURE 13.21
A patient with a spinal cord injury in traction.

grafts provide sheaths through which some axons of the spinal tracts seem to be able to reestablish connections. Without such sheaths, axon growth is prevented by glial cell invasion. Though many technical problems remain, grafting may eventually offer new hope for individuals who sustain spinal cord injuries.

Complications of spinal cord injury include urinary infections, decubitus ulcers (bed sores), and spinal deformity. Urinary infections can be controlled by antibiotics and by frequent drainage of the bladder (by catheter until urinary function is regained). The risk of decubitus ulcers can be reduced by adequate dietary protein and frequent turning of the injured person to relieve pressure on bony prominences (sacrum, hips, shoulders, and heels). Children are especially prone to spinal deformity because their vertebral columns are still growing. Using a frame that holds a child erect (instead of a wheelchair) helps to prevent deformities during the ambulatory phase of recovery.

Recovery of sensation and movement varies with the location and severity of the injury. In patients with a completely severed spinal cord, fewer than one in six regains any useful function. On the other hand, paraplegics whose injury is in the lower lumbar region can sometimes walk with good braces and crutches if they have strong trunk muscles.

Among patients with partially severed cords, four-fifths of those with thoracolumbar injuries and almost half of those with cervical injuries recover some useful function.

Rehabilitation is an essential part of the treatment of spinal cord injuries. Patients are taught how to use remaining functions to pursue a vocation. Education is important to help the patient and the family adjust to a different lifestyle. Finally, special equipment to help compensate for disabilities often makes the difference between helplessness and independence.

Questions

1. What determines the degree of disability from spinal cord injuries?
2. What are the effects of hemisection?
3. What is spinal shock?
4. What complications often arise during recovery from spinal cord injuries?
5. What progress in the treatment of spinal cord injuries has been made recently?

CHAPTER SUMMARY

Somatic Nervous System Components

- The somatic nervous system consists of spinal and cranial nerves, which carry signals between the central nervous system and skin, muscles, and sense organs. It and the autonomic nervous system make up the peripheral nervous system.

Nerve Structure

- Nerves usually contain both afferent and efferent fibers (axons) arranged in bundles. Each fiber is covered by endoneurium, each bundle by perineurium, and the whole nerve by epineurium.

Cranial Nerves

- The properties of cranial nerves are summarized in Table 13.1.

Spinal Nerves

- Spinal nerves carry signals from sensory receptors to the spinal cord and from the spinal cord to muscles. Their numbers and locations are closely associated with vertebrae.
- Nerve networks called plexuses, located in the cervical, brachial, lumbar, and sacral regions, redistribute signals going to or from the spinal cord.

Levels of Neural Function

- Neural function at the spinal cord level is limited mainly to reflexes. Function at the lower brain level includes subconscious regulation of internal processes. Function at the higher brain level involves storage and interpretation of information and voluntary motor functions.

Reflexes

- A simple reflex arc consists of a receptor, a sensory neuron, usually one or more interneurons, a motor neuron, and an effector.
- Reflexes are instantaneous, stereotyped responses to specific stimuli. Many help to maintain homeostasis by protecting the body or regulating muscle contraction.
- Local control of motor neurons is exerted by stretch reflexes, the load reflex, and the tendon reflex.

Higher Order Control of Movement

- To initiate motor function, a program of action is created; the motor cortex sends signals to motor units, and the cerebellum and basal nuclei continuously adjust those signals. Cycling of information between these structures and the motor cortex allows coordination of movements.
- Most movements result from a combination of voluntary (conscious) and involuntary (subconscious) activities. Learning establishes programs for well-practiced motor skills.

- In the brain stem, vestibular nuclei receive signals from the organs of balance, the reticular formation excites muscles that support the body against gravity, and the red nuclei help to regulate movement and posture.
- Basal nuclei regulate skeletal muscle tone, control gross intentional movements, and relay signals to the motor cortex that influences extrapyramidal tracts.
- The primary motor cortex sends signals via pyramidal fibers that produce voluntary movement. Signals from sensory cortices and the motor association area coordinate and fine tune speech and finger movements.
- The cerebellum continuously receives feedback about all aspects of movement, compares the current status with the program of action, and makes appropriate adjustments.
- Integration of motor control requires information from receptors, accurate muscle control, and coordination of muscle actions.
- The nervous system controls and integrates movement solely by adjusting muscle tension.

Postural Reflexes and Locomotion

- Posture and balance are maintained mainly by reflexes initiated by sensory signals from eyes, organs of balance, and joint and muscle receptors. The cerebellum integrates such information.
- Postural reflexes include labyrinthine, neck righting, and optical righting reflexes.
- Locomotion requires coordinated muscle actions, especially rhythmic alternation of the swing and stance phases of each step; afferent feedback helps to control switching between swing and stance phases and facilitates adjustments for changes in terrain.

Reaching for a Book—An Integrated Movement

- The processes involved in reaching for an object, an example of integrated movement, are summarized in Figure 13.15.

Clinical Applications

- Magnetic resonance imaging provides clear images of sections through the body using electrical signals instead of radiation.
- Abnormal reflexes can indicate neural damage or disease.
- Sciatica is irritation of the sciatic nerve. In carpal tunnel syndrome, a pinched nerve causes pain and tingling in the hand. Most backaches are due to strains in muscles made susceptible by lack of exercise. Cerebral palsy refers to a group of disorders in movement because of cerebral damage. Parkinson's disease is due to degeneration of cells in the substantia nigra. Poliomyelitis is a viral infection that destroys cell bodies of motor neurons.

QUESTIONS AND PROBLEMS

The questions at the end of each chapter are numbered to correspond with the objectives listed at the beginning of the chapter. Italics indicate that a question requires critical thinking skills beyond simple factual recall.

Questions

1. What is the somatic nervous system, and how does it relate to the total nervous system?
2. Describe the structure of a nerve.
3. (a) Describe the distribution and function of each of the 12 cranial nerves.
   (b) *What are the consequences of damage to cranial nerves?*
4. (a) Describe the structure and distribution of the spinal nerves.
   (b) Explain how plexuses redistribute branches of spinal nerves.
5. List and define the three levels of neural function.
6. (a) Describe the general structure and function of a reflex arc.
   (b) Explain the signal pathway and the effect of the withdrawal, crossed-extensor, and stretch reflexes.
7. *(a) What factors are involved in higher order control of movement?*
   (b) How do the following structures control movement: brain stem structures, basal nuclei, cerebral cortex, and cerebellum?
8. *How are posture and balance maintained?*
9. *Describe the processes involved in locomotion.*
10. *Explain how the main components of a complex movement are integrated.*
11. *What diagnostic procedures are available to detect damage or disease in the nervous system?*
12. What is the nature of the disorder in each of the following: (a) sciatica, (b) carpal tunnel syndrome, (c) backache, (d) cerebral palsy, (e) Parkinson's disease, and (f) poliomyelitis?

Problems

1. Prepare a short essay similar to the "reaching for a book" example, describing a voluntary activity you often perform (writing, playing a musical instrument, swimming, or hitting a tennis ball).
2. Find out more about CAT, PET, and MRI procedures—how they differ, and what information they provide.

14

AUTONOMIC NERVOUS SYSTEM

ORGANIZATION AND GENERAL FUNCTIONS

When you are frightened, your heart races and you breathe rapidly. You can run faster and jump higher than at other times. When you are well fed, relaxed, and contented, your heart rate and breathing are slower. Unless you are highly motivated, you may find it difficult to get out of your chair to tackle a big job. These examples, of course, are extremes along a continuum of physiological activity that may occur in your body. Most of the time, however, your bodily functions are maintained within a narrow range between these extremes.

The **autonomic nervous system** helps to maintain conditions in your body within a narrow range regardless of environmental changes. Autonomic means self-controlling, and the system is aptly named in that it regulates many internal processes without our conscious awareness. This system also is called the **visceral nervous system** because it regulates the function of the **viscera,** or internal organs such as the heart, lungs, and intestines. It also regulates the function of sweat glands in the skin and smooth muscle in blood vessel walls throughout the body. Though the autonomic nervous system allows you to respond to an emergency and "unwind" after it is over, you spend only a small part of your life in these states. That the autonomic nervous system can cope with extreme situations is important. That it responds continuously and automatically to small changes is even more important in controlling internal organ functions and maintaining homeostasis.

Autonomic motor neurons carry signals that control visceral organs, but they are activated by afferent signals like all other components of the nervous system. For example, motor signals control the rate and depth of breathing, but sensory signals from the lungs and from sensors that detect changes in blood carbon dioxide provide feedback that adjusts such signals. Other functions—volume of blood pumped by the heart, blood vessel diameters, and sweat production—are similarly regulated. Such regulation usually involves negative feedback.

Autonomic efferent pathways are composed of two neurons: a **preganglionic neuron** and a **postganglionic neuron.** Signals from the preganglionic neuron are transmitted by way of acetylcholine across a synapse in a **ganglion** (an aggregation of cell bodies) to the postganglionic neuron (Figure 14.1). Preganglionic neurons are myelinated; postganglionic neurons are not. The properties of the autonomic nervous system are distinctly different from those of the somatic nervous system, as summarized in Table 14.1.

The autonomic nervous system has two parts: the **sympathetic division** and the **parasympathetic** (par″ah-sim″path-et′ik) **division.** Preganglionic sympathetic neurons originate in the thoracolumbar (thoracic and lumbar) region of the spinal cord, whereas preganglionic parasympathetic neurons originate in the medulla oblongata, midbrain, and sacral region of the spinal cord. These divisions are also called the **thoracolumbar** (tho″rak-o-lum′bar) and **craniosacral** (kra″ne-o-sa′kral) divisions, respectively.

Most organs have **dual innervation,** or **reciprocal innervation;** that is, they are supplied by both sympathetic and parasympathetic fibers (Figure 14.2). An important effect of dual innervation is to maintain organ functions within a narrow range, thereby contributing to homeostasis in internal conditions. For example, your heart rate varies throughout the day, increasing as you engage in activity and decreasing as you relax; yet, your heart rate usually is kept within a range from about 70 to 110 beats per minute, going higher only when you engage in prolonged strenuous activity. Dual innervation allows sympathetic signals to foster activity—increase the heart and respiratory rates, decrease digestive activities, dilate bronchioles, stimulate sweating, and increase the blood glucose concentration. Parasympathetic signals cause changes in the opposite direction—decrease heart and respiratory rates, increase digestive activities, constrict bronchioles, and decrease the blood glucose concentration. The specific rate of an internal function at any time reflects the net effect of sympathetic and parasympathetic signals.

Another effect of dual innervation is to attain optimal function for existing environmental conditions. An example is the adjustment of pupil size to prevailing light. Sympathetic signals dilate pupils and parasympathetic signals constrict them, as described in Chapter 15.

DEVELOPMENT

Before the end of the first month of human embryonic development, **neural crest cells** separate from the neural tube (refer back to Figure 12.1.). These cells migrate laterally

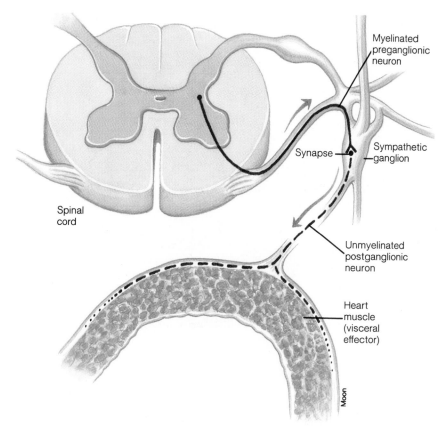

FIGURE 14.1
The arrangement of neurons in an autonomic pathway.

TABLE 14.1	A Comparison of the Properties of the Somatic Motor and Autonomic Divisions of the Peripheral Nervous System

Somatic Motor	Autonomic
Innervates skeletal muscle	Innervates smooth muscle, cardiac muscle, and glands
Pathway contains one efferent neuron	Pathway contains two efferent neurons
Innervation is always excitatory	Innervation can be excitatory or inhibitory
Transmitter is acetylcholine	Transmitter is acetylcholine or norepinephrine
Motor signals lead to voluntary activity	Motor signals lead to involuntary activity

and anteriorly to form two chains of ganglia, one on either side of the spinal cord. Some of these cells, the so-called chromaffin cells, migrate along a different path and form the medulla of the adrenal gland (Chapter 16). Neural crests also give rise to several ganglia in the head region—the ciliary ganglia near the eye, for example. Subsequent growth of the spinal and cranial nerves establishes connections between the autonomic and other peripheral nerves.

NEUROTRANSMITTERS

Efferent autonomic neurons synthesize and secrete neurotransmitters, which must be inactivated to prevent continuous stimulation and to allow repolarization of stimulated neurons. In these respects, autonomic neurons are no different from other neurons. Differences in sympathetic and parasympathetic functions are determined by which neurotransmitter is released and how that transmitter interacts with the receptor to which it binds (Figures 14.3 and 14.4).

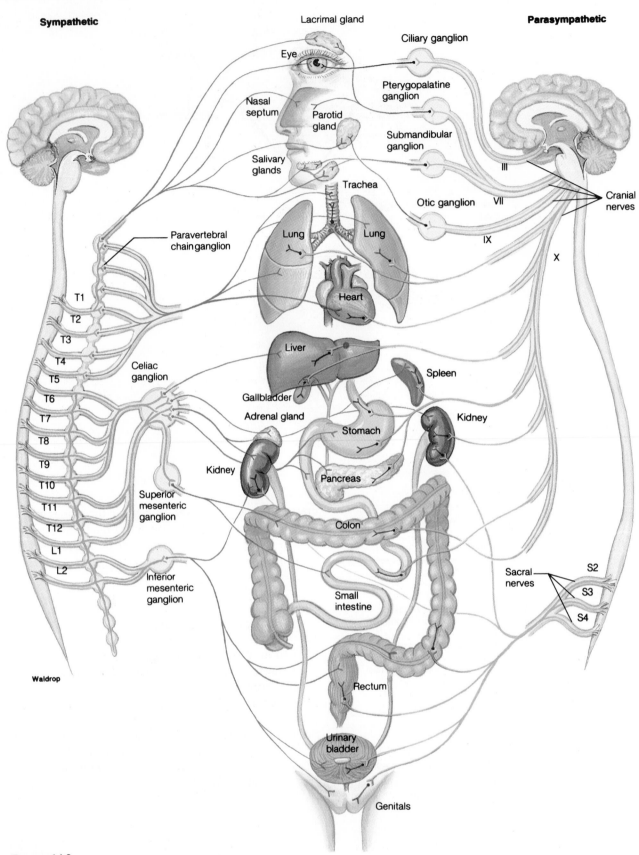

Sympathetic

Parasympathetic

Lacrimal gland

Ciliary ganglion

Eye

Pterygopalatine ganglion

Nasal septum

Parotid gland

Submandibular ganglion

Salivary glands

Trachea

Otic ganglion

III

VII

Cranial nerves

Paravertebral chain ganglion

Lung

Lung

IX

X

T1
T2
T3
T4
T5
T6
T7
T8
T9
T10
T11
T12
L1
L2

Heart

Celiac ganglion

Liver

Spleen

Gallbladder

Kidney

Adrenal gland

Stomach

Kidney

Kidney

Pancreas

Superior mesenteric ganglion

Colon

Inferior mesenteric ganglion

Small intestine

Sacral nerves

S2
S3
S4

Rectum

Urinary bladder

Genitals

Waldrop

FIGURE 14.2
The autonomic nervous system. The sympathetic division is shown in red, and the parasympathetic division is shown in blue. Preganglionic neurons are darker in color than postganglionic neurons.

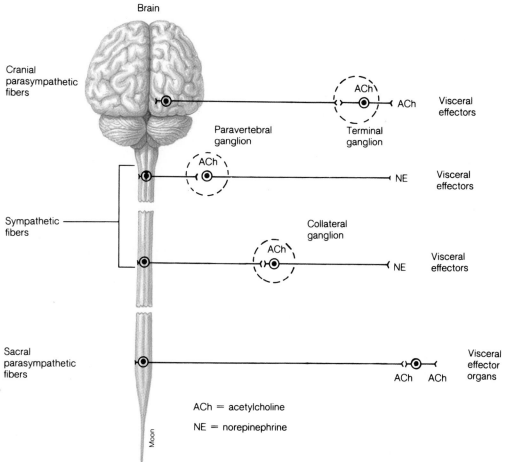

Brain

Cranial
parasympathetic
fibers

ACh

ACh

Visceral
effectors

Paravertebral
ganglion

Terminal
ganglion

ACh

NE

Visceral
effectors

Sympathetic
fibers

Collateral
ganglion

ACh

NE

Visceral
effectors

Sacral
parasympathetic
fibers

Visceral
effector
organs

ACh   ACh

ACh = acetylcholine

NE = norepinephrine

Moon

FIGURE 14.3
The fibers of the autonomic nervous system and their usual
neurotransmitters.

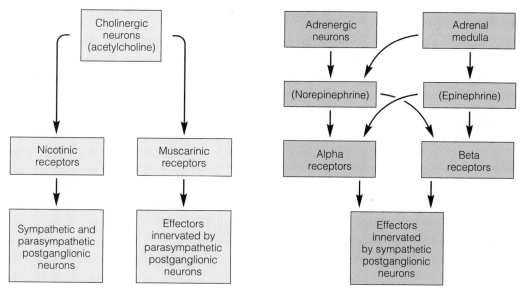

Cholinergic
neurons
(acetylcholine)

Adrenergic
neurons

Adrenal
medulla

(Norepinephrine)

(Epinephrine)

Nicotinic
receptors

Muscarinic
receptors

Alpha
receptors

Beta
receptors

Sympathetic and
parasympathetic
postganglionic
neurons

Effectors
innervated by
parasympathetic
postganglionic
neurons

Effectors
innervated
by sympathetic
postganglionic
neurons

FIGURE 14.4
Summary of neurotransmitters and receptors involved in the function
of the autonomic nervous system.

The autonomic nervous system produces two neuro-transmitters. **Acetylcholine** is released by all presynaptic neurons and by all parasympathetic postganglionic neurons. It also is released by a few sympathetic postganglionic neurons, notably those that innervate sweat glands and those that dilate blood vessels in skeletal muscle. Neurons that release acetylcholine are said to be **cholinergic** (ko''lin-er'jik). **Norepinephrine** (noradrenalin) is released by most sympathetic postganglionic neurons; such neurons are said to be **adrenergic** (ad''ren-er'jik).

Cholinergic Effects

When acetylcholine is released by cholinergic neurons, its effects are determined by the nature of the receptor with which it interacts. Acetylcholine receptors are of two kinds: *nicotinic* (nik''o-tin'ik) *receptors* and *muscarinic* (mus''kar-in'ik) *receptors*. (These receptor names derive from early investigations in which nicotine and muscarine, a toxin from mushrooms, were used to study cholinergic neurons. The actions of acetylcholine are said to be nicotinic when they are like the effects of nicotine or muscarinic when they are like the effects of muscarine.

Nicotinic receptors are found on both sympathetic and parasympathetic postganglionic neurons—the neurons that are stimulated by signals from preganglionic neurons. Such receptors are said to be type 1 nicotinic receptors. (Type 2 nicotinic receptors are found in the cell membranes of skeletal muscles.) The result of the interaction of acetylcholine with nicotinic receptors depends mainly on the acetylcholine concentration. Small amounts stimulate such receptors and large amounts inhibit them.

Muscarinic receptors are found at several sites—on organs innervated by postganglionic parasympathetic neurons (heart and most smooth muscle), on organs innervated by sympathetic cholinergic neurons (sweat glands and some smooth muscle in blood vessels that supply skeletal muscles). Acetylcholine acting on muscarinic receptors reduces the heart rate, increases intestinal motility (movements that mix and propel food), causes sweating, and dilates blood vessels in skeletal muscles.

Adrenergic Effects

Like cholinergic effects, adrenergic effects also are determined by the nature of the receptor with which the neurotransmitter, in this case, norepinephrine, interacts. The action of norepinephrine from neurons is augmented, however, by norepinephrine and epinephrine secreted as hormones from the adrenal medulla. The adrenal medulla (the core of the adrenal gland) develops embryologically from the same tissue as autonomic neurons. When stimulated by adrenergic neurons, the adrenal medulla releases about four times as much epinephrine as norepinephrine. Compared with the rapid and short-lived action of norepinephrine at synapses, the action of epinephrine and norepinephrine released into the blood is slow and long-lasting.

Receptors for epinephrine and norepinephrine are found in most internal organs, especially in smooth muscle, and they respond to these substances regardless of whether they come from neurons or the adrenal medulla. Four types of receptors—$alpha_1$, $alpha_2$, $beta_1$, and $beta_2$ have been identified. Researchers have found certain chemical substances that bind to particular receptors, and they use them to study receptor properties. Epinephrine generally excites both alpha and beta receptors, whereas norepinephrine excites mainly alpha receptors. The specific effects of either substance on an organ depend on the type, quantity, and sensitivity of the organ's receptors.

Many organs have both alpha and beta receptors, and the effects of stimulating one type can be directly opposite to those of stimulating the other type. For example, in arterial smooth muscle, stimulating $alpha_1$ receptors generally causes muscle contraction and blood vessel constriction, whereas stimulating $beta_2$ receptors causes muscle relaxation and blood vessel dilation. Similarly, stimulating $beta_1$ receptors accelerates the breakdown of fat deposited in adipose tissues, and stimulating $alpha_2$ receptors prevents it. Actions are not always opposite. Stimulating either alpha or beta receptors in intestinal smooth muscle reduces contractility, thereby decreasing motility.

Differences in the effects of stimulating various receptors appear to depend on what kind of response occurs in the cell membrane of a stimulated cell. Stimulating beta receptors, either type 1 or type 2, activates the enzyme adenylate cyclase. This, in turn, releases a second messenger molecule called cyclic AMP. (Second messengers receive a message delivered to a cell membrane and relay it to appropriate cell components. See Chapter 16 for more detail.) In contrast, stimulating $alpha_1$ receptors increases the permeability of the cell membrane to calcium ions and stimulating $alpha_2$ receptors inhibits adenylate cyclase and prevents the relase of cyclic AMP. These observations explain how stimulation of alpha or beta receptors can have opposite effects.

Neuropeptides

According to the classic view, autonomic function involves one transmitter-one receptor synapses (Figure 14.5a), but more complex relationships exist. Finding that some autonomic neurons produce effects not attributable to either cholinergic or adrenergic effects led to the discovery of multiple transmitters in a single neuron and neuropeptides that modulate autonomic functions. Some neuropeptides act on either a special presynaptic receptor or a postsynaptic receptor (Figure 14.5b). Sometimes two substances act on their respective receptors in a complex synapse (Figure 14.5c), or one acts as a modulator of the other (Figure 14.5d).

Studying neuropeptides is further complicated in several ways. First, they are present in very small concentrations—less than one-ten thousandth the concentration of conventional neurotransmitters. Immunologic methods (building molecules that will react specifically with them) are needed to locate neuropeptides. Second, most neuropeptides vary among species by one or more amino acids, so the information that investigators learn from animals may not apply to humans. Finally, neuropeptides are found not only in neurons but also in cells of the hypothalamus and various glands.

Neuropeptides have been found in all efferent components of the autonomic system—sensory, preganglionic, postganglionic sympathetic, and postganglionic parasympathetic—and in the adrenal medulla. A few examples will illustrate the kinds of functions neuropeptides perform.

Many sensory neurons contain the neuropeptide *substance P,* an 11–amino acid molecule, first discovered in horse tissues in 1931. Substance P is a powerful *vasodilator* (vas″o-di-la′tor), a substance that relaxes smooth muscle and increases blood vessel diameters. Substance P also plays a role in the body's response to tissue injury. In addition to sending pain signals to the brain, sensory neurons also display a *local axon reflex* (Figure 14.6). They send signals along collateral branches directly to blood vessels without going through the spinal cord or any other neuron. The signals release substance P, which dilates blood vessels and increases their permeability so fluids leak into injured tissue. Other neurons also release substance P. In autonomic ganglia, substance P prolongs depolarization of ganglion cells. It acts at other sites to lower blood pressure or decrease the heart rate.

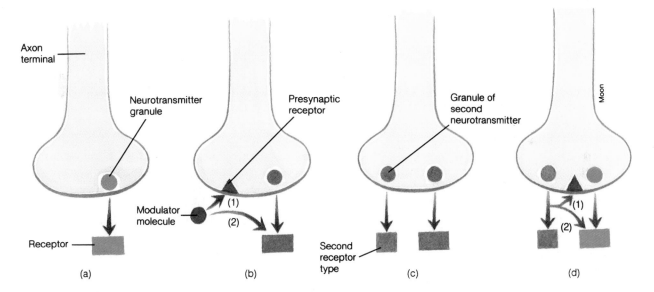

FIGURE 14.5
(*a*) Simple one neurotransmitter–one receptor type synapse.
(*b*) Neuromodulator molecule can exert effects on (*1*) a presynaptic receptor or (*2*) a postsynaptic receptor. (*c*) A two neurotransmitter–two receptor synapse. (*d*) A two neurotransmitter–two receptor synapse in which one transmitter also can act as a neuromodulator on (*1*) a presynaptic receptor or on (*2*) a postsynaptic receptor of the other transmitter.

Autonomic neurons, including those in many local ganglia and in the vagus nerve, release enkephalins. Which kinds of enkephalins are released and where they come from vary among species. Much more research is needed to determine their effects.

Some postganglionic sympathetic neurons release *neuropeptide Y*, which has been found to have two actions—constriction of some blood vessels and inhibition of signals initiated by norepinephrine at organ receptors.

Some postganglionic parasympathetic neurons release *vasoactive intestinal peptide* (VIP). VIP probably modulates the effects of acetylcholine on blood vessel walls and smooth muscle in various internal organs. In salivary glands, it increases secretion and dilates blood vessels, probably by increasing the ability of acetylcholine to bind to muscarinic receptors.

Neuropeptides also affect the adrenal medulla. Substance P prolongs epinephrine and norepinephrine output during times of stress. Enkephalins are stored in adrenal medulla cells in a concentration of about 6 nanomoles (10^{-9} or billionths of a mole) per gram of wet tissue. They are stored as large protein molecules called *polyproteins,* from which enkephalins and other opiatelike peptides can be derived. Adrenal enkephalins may modulate blood pressure by inhibiting release of epinephrine and norepinephrine. They may contribute to stress-induced analgesia seen in victims of serious injuries who feel no pain for some time after the injury.

As we consider sympathetic and parasympathetic functions, keep in mind that many functions initiated by neurotransmitters are very likely modulated by neuropeptides. In another decade, the roles of neuropeptides will be much better understood.

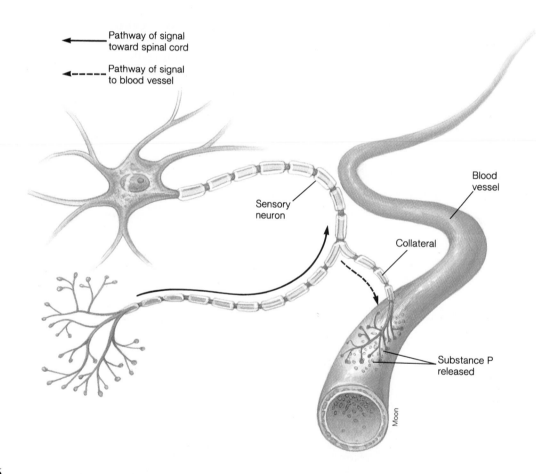

Pathway of signal toward spinal cord

Pathway of signal to blood vessel

Sensory neuron

Blood vessel

Collateral

Substance P released

FIGURE 14.6
In a local axon, reflex signals go along a sensory neuron toward the spinal cord. The reflex signal also goes along a collateral in the opposite (from the normal) direction to a blood vessel where it elicits secretion of substance P. Substance P causes dilation and increased permeability of the blood vessel.

SYMPATHETIC DIVISION

Structure

The sympathetic division of the autonomic nervous system consists of paired paravertebral **sympathetic chain ganglia,** preganglionic neurons, and postganglionic neurons. The cell bodies of preganglionic neurons are in the lateral horns of the spinal cord. Their myelinated axons extend along a spinal nerve and through the *white ramus communicans* (ra'mus kom-mu'nek-anz) to a sympathetic chain ganglion (Figure 14.7). Here, the axon can take one or more of three paths:

1. It can synapse directly with a postganglionic neuron in the chain ganglion.
2. It can extend through the chain ganglion to another ganglion, called a **prevertebral ganglion,** or **collateral ganglion,** and synapse there with a postganglionic neuron.
3. It can extend up or down the chain to another chain ganglion and either synapse directly or go to a collateral ganglion.

The cell bodies of postganglionic neurons are in chain ganglia or collateral ganglia. The unmyelinated axons of some postganglionic neurons extend from cell bodies in chain ganglia along the *gray ramus communicans* to a spinal nerve. These axons terminate in smooth muscle of blood vessels, sweat glands, or small muscles around hair follicles in the skin. Other postganglionic axons extend from cell bodies in chain ganglia or prevertebral ganglia via **splanchnic** (splank'nik) **nerves** to smooth muscle in visceral organs and blood vessels.

Postganglionic fibers (axons) are 30 times as numerous as preganglionic fibers. A single preganglionic fiber can stimulate many postganglionic fibers, causing wide divergence of neural signals. For example, a signal arriving in the celiac ganglion near the stomach can spread to several visceral organs and create a generalized effect. This may account for the severity of what some people describe as a blow to the "solar plexus."

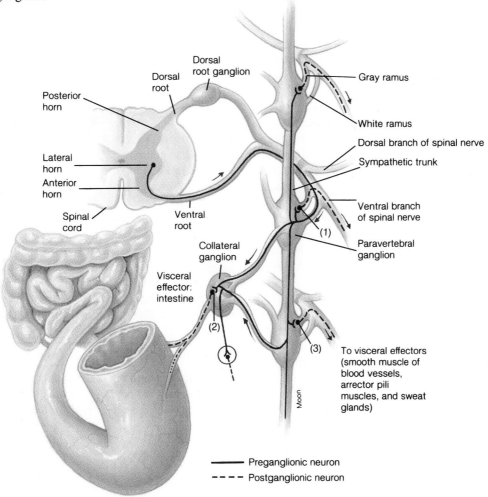

FIGURE 14.7
The arrangement of chain ganglia of the sympathetic nervous system and the nerve pathways through them. Synapses numbered 1, 2, and 3 are described in the text.

Some sympathetic fibers are distributed directly to the adrenal medulla. Axons of preganglionic cells pass without a synapse from the spinal cord, through chain ganglia, along a splanchnic nerve, and through the celiac ganglion to the adrenal medulla. Adrenal medulla cells are modified postganglionic cells, which release epinephrine and norepinephrine into the blood.

Functions

The sympathetic division of the autonomic nervous system works continuously and in conjunction with the parasympathetic division to regulate many body functions (Table 14.2). Such regulation operates largely by negative feedback to maintain homeostasis. Sometimes it operates by positive feedback to accelerate a process such as uterine contraction during labor. Though sympathetic and parasympathetic functions are discussed separately here, they are carefully integrated. Under most physiological conditions, the sympathetic (more than the parasympathetic) division adjusts the level of organ functions according to signals from hypothalamic and other central nervous system centers.

During stress, signals from the hypothalamus and parts of the brain stem travel through the spinal cord to the preganglionic sympathetic neurons. These signals diverge and excite many postganglionic neurons leading to *mass activation*, in which nearly all organs innervated by sympathetic neurons are stimulated. As signals diverge, sympathetic stimulation of organs increases accordingly. Though very strong stimulation only rarely occurs, its effects demonstrate sympathetic functions. Pupils dilate, allowing more light to enter the eyes. The liver and adipose tissues release nutrients into the blood. Sweat glands become more active and get rid of heat from strenuous muscular activity. Blood clots more rapidly if an injury occurs. Digestive tract smooth muscle relaxes and digestive secretions diminish even if there is food in the tract. This is why you may have a "lump in your stomach" when you encounter a stressful situation shortly after a meal. Blood is diverted from the digestive tract and kidneys to the skeletal muscles and heart, where it supplies oxygen and nutrients for strenuous activity. In extreme cases, the kidneys produce a smaller volume of urine, thereby compensating for fluid loss due to sweating (or bleeding, if it occurs). All in all, strong sympathetic stimulation causes the body to mount a massive campaign to meet the demands of a stressful situation.

Lesser sympathetic stimulation produces smaller adjustments. Suppose you are out jogging and start up a hill. Afferent signals relayed from your heart and lungs to nuclei in your brain stem elicit sympathetic signals that increase the volume of blood your heart pumps and the number of breaths you take per minute. When you come to level ground, afferent signals elicit fewer sympathetic signals and your heart and lungs "slow down" a bit. Throughout your run, your sympathetic nervous system adjusts internal organ functions according to demands placed on them.

Panic disorder is characterized by sudden episodes of severe anxiety accompanied by a variety of physical symptoms—shortness of breath, sudden increase in the heart rate, chest pain, dizziness, sweating, trembling, and faintness. It is diagnosed in only about four per thousand people in the general population, but many cases go undiagnosed. The disorder usually begins between ages 15 and 29, and probably is caused by bursts of excessive adrenergic activity initiated by the central nervous system. Why it occurs is not known. Once diagnosed, the disorder can be treated with the drug alprazolam. With careful regulation of the dose and patient education, relapses can be prevented.

PARASYMPATHETIC DIVISION

Structure

Like the sympathetic division, the parasympathetic division consists of ganglia, preganglionic neurons, and postganglionic neurons, but in different locations (refer back to Figure 14.2). Axons of neurons with cell bodies in the brain go through cranial nerves to structures such as salivary glands and eye muscles; those with cell bodies in the sacral spinal cord go through spinal nerves to pelvic organs. These presynaptic axons synapse with a small number of postganglionic neurons in ganglia located on or near the organs they serve. These neurons are usually short and terminate within the tissues of the organ they serve. Parasympathetic signals have less divergence and more precise, focused effects than those in the sympathetic division.

Functions

The parasympathetic division, like the sympathetic, acts continuously by negative feedback to maintain homeostasis. It, too, can act by positive feedback to accelerate a

TABLE 14.2   Autonomic Effects on Various Organs

Organ	Sympathetic Receptor Type	Effects of Stimulation	
		Sympathetic	*Parasympathetic**
Iris	α	Contract dilator muscle and dilate pupil	Contract constrictor and constrict pupil
Ciliary muscle	β_1	Relax muscle (minor effect)	Contract muscle and accommodate lens for near vision
Sweat glands	Cholinergic	Stimulate secretion	None
Lacrimal gland		None	Secrete normal or excessive amount of tears
Salivary glands	α	Reduce amount of saliva	Increase amount of saliva
Heart	β_1	Increase heart rate and stroke volume	Decrease heart rate
Smooth muscle of blood vessels in most internal organs	α_1 β_2	Contract Relax	None
Adrenal medulla		Secrete epinephrine and norepinephrine	None
Liver	β_2	Increase breakdown of glycogen, decrease bile secretion	None
Digestive glands (except salivary)		None	Increase secretion
Stomach			
Motility and tone	β_1	Decrease	Increase
Sphincter	α	Contract	Relax
Intestine			
Motility and tone	α, β_2	Decrease	Increase
Sphincter	α	Contract	Relax
Spleen	α	Contract and discharge stored blood	None
Fat cells	α_2 β_1	Inhibit fat breakdown Stimulate fat breakdown	
Kidney	α	Constrict blood vessels and decrease urine volume	None
Urinary bladder			
Muscle wall	β_2	Relax	Contract
Sphincter	α	Contract	Relax
Uterus	β_2 α	Inhibit contraction if nonpregnant Stimulate contraction if pregnant	Contract (minor effect)
Genital organs	?	Constrict blood vessels of sperm passageways and cause ejaculation	Dilate blood vessels of clitoris and penis and cause erection
Arrector pili muscles	α	Cause erection of hair (goose flesh)	None

*All parasympathetic receptors are muscarinic.

process such as contraction of the urinary bladder. In general, parasympathetic function opposes sympathetic function; if one speeds a process, the other slows it down. This is accomplished by reciprocal innervation, as is demonstrated by the iris of the eye (Figure 14.8). Sympathetic signals cause contraction of radial muscle and dilate the pupil, whereas parasympathetic signals cause contraction of constrictor muscle and constrict the pupil.

Strong parasympathetic stimulation can occur after a satisfying meal under relaxing circumstances. Smooth muscle of the digestive tract contracts rhythmically, mixing and propelling food, and digestive glands secrete copious quantities of digestive juices. The heart rate decreases and the bronchi constrict. Occasionally, people display the effects of parasympathetic stimulation as they attempt to cope with stressful situations, such as fear and depression. In acute fear, they have diarrhea, urinary incontinence, and bronchial constriction—all parasympathetic effects. In depression, they feel powerless and react

by giving up or giving in. This probably is a parasympathetic response when a sympathetic fight or flight response would have been more appropriate.

Autonomic neuropathy (disordered autonomic function) is common in diabetics and widespread through their bodies. Exactly how it relates to diabetes is not known, but it is characterized by a small increase in heart rate and a large increase in blood pressure during strenuous activity. Normal individuals respond to strenuous activity by a modest increase in the heart rate and a small increase in blood pressure. Autonomic neuropathy such as dizziness, nighttime diarrhea, and impotence can be seen in both diabetics and nondiabetics; however, the death rate among diabetics with autonomic neuropathy is much higher than that among normal individuals, and somewhat higher than that among diabetics without autonomic neuropathy. Most such deaths are sudden and unexpected; some are due to kidney failure, but no cause can be found for many others.

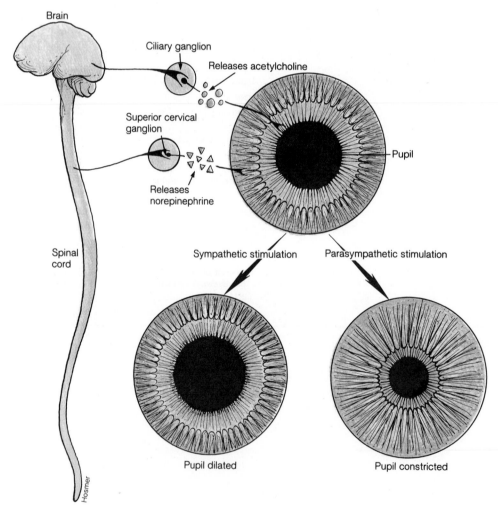

FIGURE 14.8
Reciprocal innervation.

CONTROL OF AUTONOMIC FUNCTIONS

The continuous and reciprocal actions of the sympathetic and parasympathetic divisions on organs maintain these organs at a functional level consistent with homeostasis (Figure 14.9). This constitutes **autonomic tone.** For example, as a person's level of activity changes, sympathetic or parasympathetic signals adjust the heart rate to the activity level. The heart rate at any given time is due to the net effect of the two opposing kinds of signals. The same process controls the size of the pupil and the degree of motility of the digestive tract and urinary bladder.

Some functions are controlled largely by sympathetic or parasympathetic signals, but not both. In most instances, smooth muscle in the walls of blood vessels serving internal organs is controlled by sympathetic signals. Stimulation of alpha$_1$ receptors causes such muscle to contract and stimulation of beta$_2$ receptors causes it to relax. The relative degree of stimulation of each kind of receptor maintains blood vessel diameters appropriate to meet the needs of tissues. Sweat glands and blood vesssels in skeletal muscles also are controlled by sympathetic signals, but these signals are relayed to muscarinic receptors. In contrast, tears result from parasympathetic stimulation.

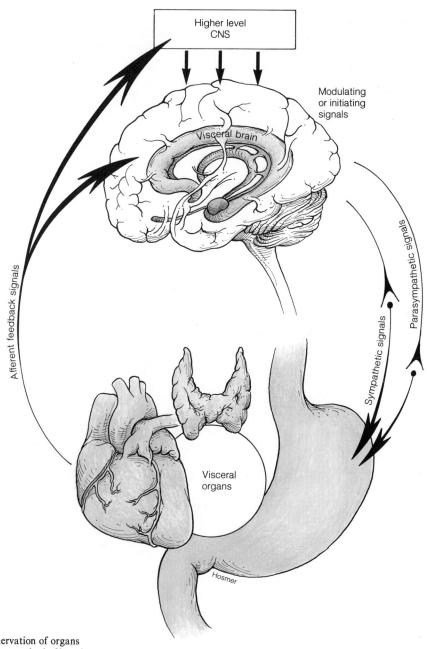

FIGURE 14.9
The net effect of dual innervation of organs by sympathetic and parasympathetic fibers is to maintain homeostasis.

Though tears usually are associated with stress, they generally come after a stressful situation has been met and the parasympathetic division has taken over. Also, parasympathetic stimulation causes dilation of blood vessels in the clitoris and penis, and produces erection. This may be one reason that sexual performance is often impaired when individuals are under stress.

Most autonomic signals are mediated via reflex arcs. Afferent signals are carried by spinal nerves and the spinal cord, or the vagus nerve to a center in the medulla oblongata. Efferent signals from the center travel via pre- and postganglionic neurons to exert their effects on organs. For example, food in the digestive tract stimulates certain receptors in the walls of digestive organs. These receptors initiate afferent signals that ultimately stimulate mainly parasympathetic neurons. (Trace these pathways on Figures 14.2 and 14.7.) Parasympathetic stimulation increases smooth muscle motility and the secretion of enzymes and other fluids in the intestinal tract. If a stressful situation arises before digestion is complete, sympathetic signals exceed parasympathetic signals. This slows digestive processes and accelerates other processes, such as the heart rate and breathing rate, until the needs of the situation are met.

As noted earlier, the word "autonomic" is appropriate to denote that the system functions without conscious control, but it is inappropriate in that the system is not entirely autonomous (self-controlled). Control is imposed by reflexes involving the spinal cord, centers in the medulla oblongata, and other parts of the brain stem often referred to collectively as the **visceral brain** (Figure 14.10). In the visceral brain, the central hypothalamus is interconnected with several parts of the limbic system—the amygdala, hippocampus, parts of the thalamus and basal nuclei, parts of the septum between cerebral hemispheres, and the paraolfactory area (which connects olfactory sensations and emotions). Cortical components of the limbic system are the frontal area, cingulate gyrus, and hippocampal gyrus. These components form a ring of cortical tissue that relays signals between association areas and subcortical limbic components.

The hypothalamus acts as a central relay station in the visceral brain. It receives signals from nearly all components of the limbic system and from the cerebral cortex via the cingulate gyrus and the hippocampus. It also sends signals to the limbic system, thalamus, cerebral cortex, reticular formation, pons, and medulla. These signals help to regulate visceral organs and contribute to motivation

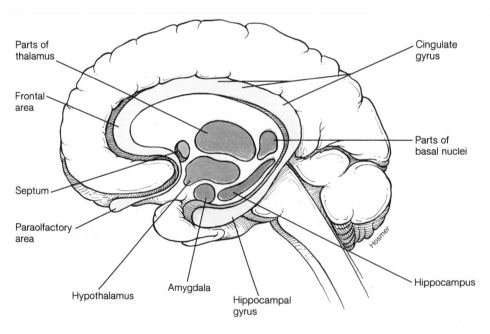

FIGURE 14.10
The visceral brain.

and other emotional behaviors, thereby suggesting a biological basis for certain behaviors. For example, the degree of shyness or spontaneity children display is related to physiological factors. Compared to outgoing children, shy children displayed consistently higher heart rates under stress and secreted more of the stress-related hormone, cortisol, even when not under stress.

Studies of individuals in meditative states have shown that their body processes respond in ways almost opposite to that of the sympathetic response. Meditation appears to produce a generalized parasympathetic response, whereas stress usually produces a generalized sympathetic response. These responses and other characteristics that distinguish sympathetic and parasympathetic functions are summarized in Table 14.3.

During both emotional stress and meditation, the cerebral cortex can exert control over the autonomic nervous system. The emotional stress of viewing a shockingly unpleasant situation or hearing unexpected bad news can cause a person to faint. Sympathetic signals cause sudden dilation of blood vessels, especially in muscles, causing the blood pressure to drop and the blood supply to the head to decrease. Fainting also can result from parasympathetic stimulation, as from a sudden drop in the heart rate.

DRUGS AND AUTONOMIC FUNCTION

Drugs are used in humans to augment or block autonomic nervous system function at two points: (1) at cholinergic effector organs, which have receptors for acetylcholine, and (2) at adrenergic effector organs, which have receptors for norepinephrine. The actions of the drugs at adrenergic effector organs vary, depending on whether they alter norepinephrine synthesis or release, and on which receptors they affect.

At cholinergic effector organs, muscarinic receptors can respond to either parasympathetic or sympathetic stimulation. Both pilocarpine and methacholine excite these receptors. These drugs, called **parasympathomimetics** (par″ah-sim″pah-tho-mi-met′ikz), increase parasympathetic activity. Pilocarpine is used to reduce pressure within the eyeball because it dilates the canal that drains aqueous humor from the anterior chamber (see glaucoma, Chapter 15).

Drugs called **parasympatholytics** (par″ah-sim″pah-tho-lit′ikz) decrease parasympathetic activity. Atropine, a widely used drug, blocks the action of acetylcholine at

TABLE 14.3 Comparison of Sympathetic and Parasympathetic Divisions	
Sympathetic Division	**Parasympathetic Division**
Physiological Characteristics	
Accelerated heart rate	Lowered heart rate
Increased blood pressure	Decreased blood pressure
Increased metabolic rate	Decreased metabolic rate
Increased oxygen consumption and respiratory rate	Decreased oxygen consumption and respiratory rate
Beta waves predominate in EEGs	Alpha waves predominate in EEGs
Increase in lactic acid in blood	Decrease in lactic acid in blood
Decrease in electrical resistance of skin	Increase in electrical resistance of skin
Fight or flight response	Meditation response*
Anatomical Characteristics	
Preganglionic neurons originate in thoracolumbar regions	Preganglionic neurons originate in craniosacral regions
Contains chain ganglia and other visceral ganglia	Has ganglia very close to the organ supplied
Each preganglionic fiber synapses with many postganglionic fibers, so many organs are affected	Each preganglionic fiber synapses with only a few postganglionic fibers so only one organ is usually affected
Action is diffuse	Action is limited to one or a few organs
Chemical Characteristics	
Division is said to be adrenergic	Division is said to be cholinergic
Norepinephrine is the transmitter for most postganglionic fibers	Acetylcholine is the transmitter for postganglionic fibers
Adrenal medulla is stimulated to release epinephrine and norepinephrine	

*Some of the characteristics of the meditation response may be due to suppression of sympathetic function rather than being due to parasympathetic action.

muscarinic receptors by binding specifically to the receptors. It prevents the normal transmitter from reaching the receptor, so the effector organ is not stimulated. Atropine has no effect on nicotinic receptors because it fails to bind to them; thus, atropine suppresses parasympathetic activity specifically. It is used during eye examinations to dilate pupils and to relax the ciliary muscles. Dilating the pupil allows examination of the retina. Relaxing the ciliary muscles allows assessment of vision in unstrained eyes. Atropine stops muscle spasms in the digestive tract and in the ducts of certain organs.

TABLE 14.4   Drugs That Affect Adrenergic Effector Organs

Site of Action	Drug	Action	Effect on Sympathetic Activity	Medical Application
Axon terminals of postganglionic neurons	Ephedrine	Augment release of norepinephrine	Augment	Nasal decongestant, antiasthmatic
	Guanethidine (Ismelin)	Deplete stores and prevent release of norepinephrine	Depress	Lower blood pressure
Alpha receptors	Methoxamine (Vasoxyl)	Excite α receptors, especially in blood vessels	Augment	Increase blood pressure in shock by vasoconstriction without affecting heart
	Phenoxybenzamine (Dibenzyline)	Block α receptors	Depress	Lower blood pressure
Beta receptors	Isoproterenol (Isuprel)	Excite β receptors	Augment	Bronchodilation, increase heart rate in patients with heart block (Chapter 18)
	Propanolol (Inderal)	Block β receptors	Depress	Lower blood pressure by decreasing cardiac output and suppressing renin release (Chapters 14 and 22)
	Practolol	Block β_1 receptors	Depress heart receptors	Not in general use
	Butoxamine	Block β_2 receptors	Depress receptors except in heart	Not in general use

Many drugs are available to augment or suppress adrenergic effects (Table 14.4). Augmenting drugs are called **sympathomimetics** (sim″pah-tho-mi-met′ikz); suppressing drugs are called blocking agents or **sympatholytics** (sim″pah-tho-lit′ikz). Some drugs have highly specific effects—they excite or block a particular type of receptor. Such drugs are exceedingly important in modern medical practice.

Four sympathomimetics currently in use are phenylephrine, clonidine, phentolamine, and isoproterenol. Phenylephrine, which stimulates alpha₁ receptors, is a commonly used bronchodilator. It is found in Chlor-Trimeton, Dimetapp, and some other medicines used to alleviate symptoms of upper respiratory infections and certain allergies. Clonidine, which stimulates alpha₂ receptors, can be used to treat hypertension (high blood pressure). It appears to exert its effects on adrenergic receptors in the central nervous system and to inhibit centers in the medulla that would accelerate the heart rate and constrict blood vessels. Phentolamine stimulates both kinds of alpha receptors, but it is used mainly to diagnose

pheochromocytoma, a disorder in which adrenal medulla tissue proliferates and releases epinephrine and norepinephrine, thereby causing extremely high blood pressure. Phentolamine quickly reduces high blood pressure caused by pheochromocytoma. Isoproterenol (Isuprel), which excites beta receptors, relaxes bronchial spasm in asthma, emphysema, and bronchitis. It also relaxes digestive tract smooth muscle, and it increases the quantity of blood pumped by the heart.

The drug propanolol (Inderal) is a **beta blocker;** that is, its sympatholytic action is due specifically to its ability to bind to beta receptors and prevent them from being stimulated. Propanolol is effective in treating hypertension and some cases of angina pectoris (chest pain due to reduced blood flow in the heart muscle) and cardiac arrhythmias (irregular heartbeats). Its antihypertensive action may be due to its ability to block sympathetic signals in vasomotor centers in the brain or in the heart muscle itself. It also may act by inhibiting the release of a substance called renin from the kidneys, which ordinarily initiates a sequence of reactions that elevate blood pressure.

ESSAY
BIOFEEDBACK

Evidence that functions of the autonomic nervous system could be consciously controlled were for many years limited to reports about Buddhist monks in meditation and practitioners of yoga. Today, some autonomic nervous system functions have been brought under conscious control through operant conditioning with appropriate biofeedback. Without biofeedback training, most people have difficulty perceiving their own autonomic processes, such as blood pressure or heart rate, and some even fail to perceive tension in skeletal muscles. **Biofeedback** provides light, sound, or other observable signals related directly to electronically sensed physiological processes. When presented while a person is trying to control a physiological function, such biofeedback can reinforce learning and help the person to perceive and control such functions. Though much remains to be learned about biofeedback, some of what is known is summarized here.

Numerous experiments with humans have shown that biofeedback can be used to alter processes under autonomic control—the blood circulation to the skin, sweating, heart rate and rhythm, blood pressure, salivation, and possibly gastrointestinal motility. It also can be used to relax skeletal muscle or increase alpha brain waves. Because alpha waves are associated with the wakeful, relaxed state, their increase is presumed to indicate that relaxation has been achieved.

To provide biofeedback, the status of a physiological process monitored by an electronic device is displayed for the individual. For example, an electrode that senses muscle tension is connected to an electrical circuit so that as muscle tension changes the subject sees different lights. A red light might indicate high tension; a yellow one, moderate tension; and a green one, little tension. The subject watches the changes in the lights and attempts to perceive how the muscle feels when the green light is on. He or she tries deliberately to maintain the muscle in a state that keeps the green light on. Sounds also can be used to provide feedback. The pitch of a sound or the frequency of a clicking sound can be made to increase with muscle tension. The subject's task is to keep the pitch or frequency of clicks as low as possible.

Regardless of the physiological process monitored or the signal used, the purpose is to provide information or feedback about changes in the process. The recipient of such information uses it in two ways: (1) to learn how his or her body feels when the process increases or decreases, and (2) to learn to purposely cause the process to remain at the desired level.

People who suffer from recurring headaches due to muscle tension or migraine headaches have been helped to reduce their pain by biofeedback. Feedback helps them to recognize tense muscles and to learn to relax them. The benefit of biofeedback in migraine headaches was discovered when a women suffering from such a headache participated in a biofeedback experiment intended to increase hand temperature. As her hand temperature increased, her headache faded. Subsequently, it has been shown that about one-third of migraine sufferers are helped by this technique.

Biofeedback has been tried in several other situations with some degree of success. Feedback from monitors of both blood pressure and muscle tension enables some people to lower their blood pressure by as much as 25 mm Hg. Feedback of heart rate has been used to treat abnormal heart rates. Feedback of alpha brain waves has been used to treat anxiety and relieve pain. Feedback of other brain waves can help seizure-prone individuals to recognize signs of impending seizures and sometimes to prevent them.

Though biofeedback has been tried only in small numbers of people, its success rate is sufficiently high to warrant further trials. Being a very low-risk treatment, biofeedback can be tried before more hazardous treatments such as drugs or surgery are used.

Questions
1. What is biofeedback?
2. What kinds of conditions can biofeedback be used to treat?
3. How do patients know when they are successful in receiving biofeedback?
4. Do you think medical use of biofeedback will increase or decrease in the future?

CHAPTER SUMMARY

Organization and General Functions
- The autonomic nervous system consists of the sympathetic and the parasympathetic divisions, which provide dual innervation to most visceral organs.
- The general function of the autonomic nervous system is to control the functions of internal organs, thereby helping to maintain homeostasis.

Development
- The autonomic nervous system develops from the neural crests.

Neurotransmitters
- Acetylcholine is releasd by cholinergic neurons, including all preganglionic neurons, all parasympathetic postganglionic neurons, and sympathetic postganglionic neurons that innervate sweat glands and blood vessels in skeletal muscles.
- Norepinephrine is released by adrenergic neurons, including most of the sympathetic postganglionic neurons.

- Acetylcholine receptors on postganglionic neurons are nicotinic receptors; those on effector organs are muscarinic receptors.
- Norepinephrine receptors are classified as alpha$_1$, alpha$_2$, beta$_1$, or beta$_2$ receptors. Effector organs have one or more types of receptors.
- Epinephrine (from the adrenal medulla) excites both alpha and beta receptors, and norepinephrine (from neurons or the adrenal medulla) excites alpha receptors to a greater extent than beta receptors. The predominant receptors on various organs and the effects of sympathetic or parasympathetic stimulation are summarized in Table 14.2.

Neuropeptides

- Neuropeptides are short chains of amino acids, such as substance P, vasoactive intestinal peptide, and enkephalins. They affect autonomic and other neural functions.
- Actions of neuropeptides are shown in Figure 14.5, and the local axon reflex is shown in Figure 14.6.

Sympathetic Division

- The arrangement of neurons in the sympathetic division is summarized in Figure 14.7.
- The sympathetic division interacts with the parasympathetic division to regulate function of internal organs, prepares the body to meet emergencies, and is augmented by secretions from the adrenal medulla.

Parasympathetic Division

- The parasympathetic division has preganglionic neurons extending from cranial or sacral nuclei to ganglia in or near the organs it serves. It also has short postganglionic axons extending from the ganglia to the organs.

Control of Autonomic Functions

- Both the sympathetic and parasympathetic divisions act continuously and produce autonomic tone in most organs, but some processes are controlled by sympathetic signals (diameter of internal blood vessels) or parasympathetic signals (tears).
- Autonomic signals are controlled by reflexes; centers in the medulla oblongata; the hypothalamus; other parts of the visceral brain; and, during stress, the cerebral cortex.

Drugs and Autonomic Function

- Drugs can affect the autonomic nervous system at cholinergic or adrenergic effector organs. Such effects can be to block or to augment autonomic effects.
- Drugs that affect adrenergic effector organs are most important medically.

Questions and Problems

The questions at the end of each chapter are numbered to correspond with the objectives listed at the beginning of the chapter. Italics indicate that a question requires critical thinking skills beyond simple factual recall.

Questions

1. (a) How does the autonomic nervous system differ from the somatic motor system?
   (b) What is the main function of the autonomic nervous system?
   (c) *How might removal of thoracic ganglia on one side of the body affect autonomic functions?*
2. From what tissue does the autonomic nervous system develop?
3. (a) What neurotransmitters are released at various sites in the autonomic nervous system?
   (b) How do cholinergic and adrenergic receptors differ?
   (c) *What would be the effect of blocking cholinergic receptors? Of blocking adrenergic receptors?*
4. (a) What are neuropeptides, and by what mechanisms do they act?
   (b) What specific effects are known for substance P, vasoactive intestinal peptide, and enkephalins?
5. *List the major effects of sympathetic stimulation on the various systems of the body, and explain how neural signals produce such stimulation.*
6. *List the major effects of parasympathetic stimulation on the various systems of the body, and explain how neural signals produce such stimulation.*
7. (a) *What is autonomic tone, and how is it maintained?*
   (b) *What is meant by dual innervation, and how is it involved in maintaining homeostasis?*
8. *How do drugs affect the autonomic nervous system?*

Problems

1. Suppose you are working in a hospital emergency room and are called upon to give cardiopulmonary resuscitation for an extended period of time. Describe the role of your autonomic nervous system in helping you respond to the emergency.
2. After the emergency is over, you get a chance to relax for a few minutes. Describe the role of the autonomic nervous system in your "recovery" from dealing with the emergency.
3. Several days later in a practice session, you attempt to perform cardiopulmonary resuscitation on a dummy for as long a time as you can. You discover that you become exhausted in half the time that you were able to perform in the emergency. Explain this observation.

OBJECTIVES

1. Explain the general principles of receptor functions.

2. Describe the properties and mechanisms of somatic senses including pain and its control.

3. Describe the similarities and differences in the mechanisms of taste and smell.

4. Define the properties of light pertinent to vision.

5. Describe the external eye structure; and explain how the eyes are protected, how they move, and how tears form.

6. Describe the internal eye structure; and explain how light travels through the eye and how accommodation occurs.

7. Relate the structure and function of rods to vision in dim light.

8. Relate the structure and function of cones to color vision.

9. Describe the path of signals from rods and cones to the cerebrum, and explain the effects of binocular vision.

10. Briefly describe how defects in vision occur.

11. Define the properties of sound pertinent to hearing.

12. Describe the structure of the external, middle, and internal ear; and explain how vibrations are transmitted through it.

13. Summarize the mechanisms involved in sound perception.

14. Briefly describe how hearing disorders occur.

15. Distinguish between static and dynamic equilibrium, and explain how each is maintained.

16. Briefly describe how disturbances in balance occur.

15

SENSORY ORGANS

RECEPTORS IN GENERAL

Walking in the woods on a crisp autumn day (Figure 15.1) allows us to discover how valuable our senses really are. The air feels refreshingly cool, but the sun's direct rays are warm. Bright red, orange, and yellow leaves provide a sharp contrast with the green needles of evergreen trees. Fallen leaves rustle as we walk through them. The smell of wood smoke beckons us to a neighbor's fireplace—and perhaps to a cup of hot, spicy cider. For experiences like this, we depend on our sensory organs—the subject of this chapter.

Receptors are specialized cells that respond to stimuli (changes in the external or internal environment) by initiating and transmitting signals. Signals generated in response to stimuli provide the only means by which the central nervous system receives information about conditions inside or outside the body. Signals that reach consciousness inform us about our environment, our body position, and even our aches and pains. Those that do not reach consciousness go to the spinal cord or subcortical areas of the brain to control such things as muscle tension and blood pressure. Some receptors send signals that provide feedback to control a process.

In discussing sensory organs, it is important to distinguish between sensations and perceptions. Signals that reach the cerebral cortex create conscious **sensations.** Each of the main kinds of sensation—touch, temperature, pain, taste, smell, sight, sound, and balance—is a **modality** of sensation. Interpreting sensations according to what is remembered from previous experience is **perception.** Hearing music is a sensation; recognizing a familiar tune is perception. Similarly, observing printed words as black shapes on a white paper is a sensation; attaching meaning to the words is perception. Once we have perceived the meaning of a sensation, we can respond to it by movement, speech, or thought. The sequence of events—detection of a change by a receptor, conduction of an afferent signal, sensation and perception by the cerebral cortex, and conduction of an efferent signal to muscles—occurs whenever we consciously respond to a change in our environment.

We may think we perceive our world directly and precisely, but that is not the case. First, our receptors do not detect all that is going on in our environment. Our sound and smell receptors are less sensitive than those of some other animals. Dogs, for example, can hear higher pitched sounds and detect smaller concentrations of odorous substances than humans. We and most other animals lack receptors to detect magnetic fields and radio waves. Second,

FIGURE 15.1
How many senses do you use to enjoy walking in the woods on a crisp autumn day?

what our receptors detect can be lost or distorted (as in a bad telephone connection) before it reaches the cerebrum. Finally, the cerebral cortex interprets sensations according to its stored information regardless of how limited the information may be. Reading a book to a baby is futile because the baby's cerebral cortex lacks sufficient stored information to interpret what it hears. In spite of these limitations, signals from sensory receptors provide our only means of perceiving our environment. They are, in fact, responsible for maintaining consciousness itself.

Types of Receptors

Receptors are classified by their locations and by the types of stimuli they can detect. By location, receptors fall into three categories (Table 15.1): **Exteroceptors** (eks″ter-o-sep′torz), located near the body surface, detect changes in the external environment. They include receptors for touch, temperature, light, sound, balance, taste, smell, and pain from external stimuli. **Proprioceptors** (pro″pre-o-sep′torz) are located in muscles, tendons, and joints (Chapter 13). Muscle spindles and Golgi tendon organs detect stimuli that help to regulate muscle contractions. Joint kinesthetic receptors detect position and movement at joints. **Visceroceptors** (vis″er-o-sep′torz), also called enteroceptors, are located in internal organs where they detect stretching of blood vessel walls, digestive tract movements, and similar changes. Signals from visceroceptors are relayed to the brain stem where they regulate internal processes.

The type of stimulus a receptor detects is of particular physiological significance. **Chemoreceptors** (ke″mo-re-sep′torz) respond to changes in concentrations of chemical substances. They include receptors for taste, smell, and detection of chemical changes in the blood. **Mechanoreceptors** (mek-an″o-re-sep′torz) respond to mechanical stimuli such as pressure or movement of fluids. They include receptors for hearing, balance, touch, pressure, and muscle length and tension. **Thermoreceptors** (ther″mo-re-sep′torz) detect warmth and cold. **Photoreceptors** (fo″to-re-sep′torz) respond to light and are found only in the retina of the eye. **Nociceptors** (no″se-sep′torz), or pain receptors, detect tissue damage. Though stimulating nociceptors produces the sensation of pain, applying a very strong stimulus to many kinds of receptors can cause pain.

Properties of Receptors

Regardless of location or type of stimuli detected, receptors have several common properties. These properties are very important in understanding sensory functions.

Transduction

A primary property of receptors is **transduction,** the conversion of energy from one form to another. You already know about many transducers. Light bulbs transduce electrical energy to light, microphones transduce sound to electrical energy, and gasoline engines transduce chemical energy to mechanical energy. Sensory receptors are transducers that convert some form of energy into a neural signal called a **generator potential.** For example, receptors in the eye respond to light or changes in light intensity, those in the ear respond to sound waves, and those in taste buds respond to certain chemical substances. Each transduces the stimulus into a generator potential.

TABLE 15.1   Kinds of Receptors

Nature of Stimuli	Location		
	Exteroceptor	*Proprioceptor*	*Visceroceptor*
Chemoreceptor	Taste, smell		Concentrations of substances in blood
Mechanoreceptor	Touch, pressure, hearing, balance	Muscle tension and stretching, joint position	Stretching of blood vessel walls and other internal structures
Thermoreceptor	Warmth, cold		
Photoreceptor	Sight		
Nociceptor	Pain	Pain	Pain

To cause a receptor to respond, a stimulus must be of the proper type and of sufficient strength to elicit a response. Such a stimulus depolarizes the receptor membrane and leads to a generator potential, which initiates an action potential in an afferent neuron. The properties of generator potentials and action potentials are compared in Table 15.2.

The Law of Specific Nerve Energies

Regardless of the modality of a stimulus, adequately stimulated receptors initiate action potentials. This is the **law of specific nerve energies.** If all receptors initiate action potentials, how does the nervous system recognize different kinds of stimuli? The answer is that signals from each kind of receptor are relayed to specific sites in the central nervous system. For example, when signals from the retina of the eye are relayed to the cerebral cortex, they produce the sensation of light. This is true even if the retina is stimulated electrically instead of by light rays. If pathways could be altered so that signals from sound receptors went to the visual cortex and those from light receptors to the auditory cortex, we would see thunder and hear lightning. The quality of a sensation depends on what part of the brain receives the signal and not on the nature of the signal itself.

Detection of Stimulus Intensity

Our sense organs allow us to detect variations in the brightness of light, the loudness of sounds, or the severity of pain. This ability to detect **stimulus intensity** is a third property of receptors. Though receptors function solely by producing generator potentials, stimulus intensity can be detected by the number of receptors stimulated or by the amplitude of the generator potential in a receptor. To produce a generator potential, a stimulus must exceed the receptor threshold, the minimum strength that elicits a response. This is called the **law of adequate stimulus.** Increasing either the strength or frequency of a stimulus can increase the amplitude of the generator potential in a receptor and also can affect more receptors. Whenever the generator potential reaches threshold, it initiates an action potential and is then annihilated. Both the rate at which new generator potentials develop and the number of receptors producing generator potentials affect the frequency of action potentials. The rate at which action potentials reach the central nervous system varies, and such variations are used to interpret the intensity of the original stimulus.

TABLE 15.2 A Comparison of Generator Potentials and Action Potentials

Generator Potentials	Action Potentials
Stimuli, which can be additive, elicit receptor potentials. Potential that excites axon is generator potential	Elicited by generator potentials and not directly by stimuli
Can produce graded response; stronger or more frequent stimuli increase the amplitude of the generator potential	Does not produce graded response; once threshold for action potential has been reached, an increase in the strength or frequency of stimuli does not affect the size of the action potential but can increase its frequency
Has a highly variable duration, but exceeds 1 to 2 msec; dependent on duration of stimulus	Has a nearly constant duration between 1 and 2 msec; independent of stimulus
Has no refractory period (period before next signal can occur)	Has a refractory period of about 1 msec
Receptor potential loses amplitude as it passes along receptor cell (where receptor is separate cell); remaining generator potential is transferred to axon	Maintains constant amplitude as it passes along axon

Differential Sensitivity

A fourth property of receptors is **differential sensitivity.** Within the normal range of stimuli strengths, each type of receptor has a low threshold for and responds to its kind of stimulus. It has a high threshold for and fails to respond to other stimuli. Receptors that respond to a salty taste do not respond to light, sound, or the smell of baking bread. Unusually strong stimuli sometimes stimulate receptors that normally would not respond to them. For example, receptors in the retina normally respond only to light, but a strong sharp blow to the eye or head creates the sensation of light and we "see stars."

Adaptation

Adaptation is the reduction or loss of response to a constant strength, or long-lasting stimulus. It is a fifth property of receptors. How adaptation occurs is not known, but receptors are known to vary in their rate of adaptation. Furthermore, failure of receptors to adapt can be a protective device. Pain receptors, proprioceptors, and chemoreceptors that detect changes in blood oxygen or carbon dioxide are slow to adapt. They elicit a high frequency of action potentials as long as they are stimulated

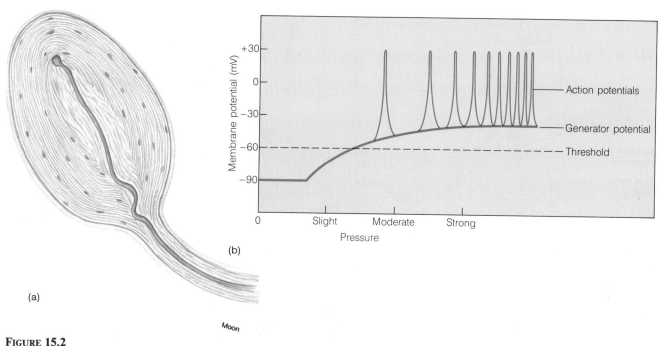

Moon

FIGURE 15.2

(a) A Pacinian corpuscle consists of a single nonmyelinated terminal of a sensory neuron imbedded in concentric layers of connective tissue. (b) As pressure on a Pacinian corpuscle is increased, the generator potential increases proportionately. After pressure reaches threshold level, action potentials are produced with increasing frequency in proportion to the increased generator potential.

and provide information about potentially dangerous situations. Continuous pain from an injured limb, for example, keeps us from using it and protects against further injury. Pressure and touch receptors adapt rapidly and the information they provide is rarely important for survival.

Nature of Generator Potentials

Generator potentials have been carefully studied by placing microelectrodes in **Pacinian** (pah-sin'e-an) **corpuscles,** large receptors deep in the dermis and mesenteries that detect pressure and vibrations. A Pacinian corpuscle consists of a single nonmyelinated axon terminal embedded in a capsule of concentric layers of connective tissue (Figure 15.2a). Pressure applied to the outermost layers deforms the nerve ending membrane, increasing its permeability and causing local depolarization. Such depolarization constitutes a generator potential.

Varying the pressure applied to Pacinian corpuscles produces graded potentials (Figure 15.2b). A small amount of pressure produces a generator potential like an excitatory postsynaptic potential, and several subthreshold stimuli may be needed to elicit an action potential in the Pacinian corpuscle axon. Threshold pressure causes depolarization and produces a generator potential that elicits an action potential. Above normal threshold pressure can increase the frequency of action potentials.

Other experiments show that the axon and axon terminal must be intact for a Pacinian corpuscle to produce a generator potential. Removal of the capsule does not prevent generator potentials but severing the axon from its terminal does. Applying local anesthetics to the axon does not prevent generator potentials, but it does prevent action potentials at the first node of Ranvier where they normally start.

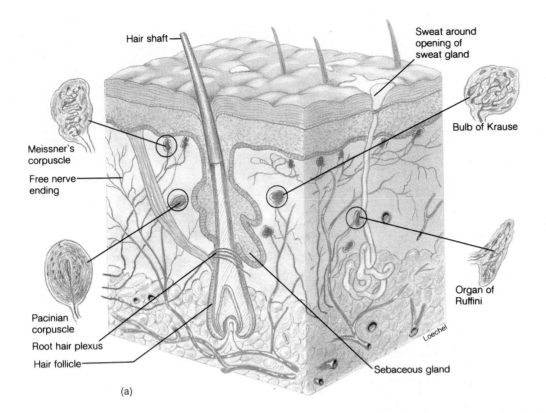

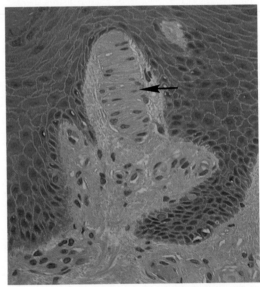

FIGURE 15.3
(*a*) Sensory receptors found in the skin, and (*b*) photomicrograph of a
Meissner's corpuscle.

SOMATIC SENSES

Skin Receptors

Five sensory modalities attributed to skin receptors are
pain, touch, pressure, warmth, and cold (Figure 15.3). **Free
sensory terminals,** unencapsulated and unmyelinated receptors, are found between epidermal cells and associated
with hairs. They generate signals interpreted as pain and
touch (slight pressure). **Meissner's corpuscles,** encapsulated terminals in the upper dermis, also respond to touch.
An abundance of these rapidly adapting receptors in fingertips and lips allows one to detect the exact point at

which the skin is touched (Figure 15.4). These corpuscles also allow us to distinguish textures and to feel low frequency vibrations (up to 60 cycles per second). Pacinian corpuscles respond to strong pressure and to vibrations at frequencies between 60 and 500 cycles per second. Pain receptors, Meissner's corpuscles, and Pacinian corpuscles also are found in deep tissues and joints where they respond to pressure, stretch, or any mechanical deformation.

Biologists do not agree about what kinds of receptors detect warmth and cold, but they do agree that such sensations exist. Most researchers believe temperature receptors are located mainly in the upper dermis. **Cold receptors** usually are myelinated and **warmth receptors, cold-pain receptors,** and **warmth-pain receptors** probably lack myelin. Cold receptors appear to be three or four times as numerous as warmth receptors. They are most numerous in lips, moderately numerous in fingertips, and rare in broad skin surfaces. Temperature is perceived according to the pattern of receptors stimulated and the strength of the stimuli. The variation in signal generation with temperature change allows these receptors to detect rates of change in temperature. We use this ability when we first turn on a hot water tap and sense how fast the water is getting hot.

Signals from various skin receptors and proprioceptors in muscles travel along somatic nerves and spinal tracts to the cerebrum (see Figure 15.5 and refer back to Figure 12.34). Because of the precise spatial correspondence between receptor sites and cortical sites to which the signals are relayed, or projected, we can perceive both the nature of the stimulus and the site at which it occurred (refer back to Figure 12.9). Most signals travel rapidly along large-diameter fibers and elicit sensations of fine touch, vibrations, and pressure. Those in the lateral spinothalamic tracts travel slowly along small-diameter fibers and elicit sensations of crude touch, pressure, warmth, cold, pain, tickle, and itch. Once thought to involve mild stimulation of pain receptors, tickle and itch are now thought to involve small bare nerve endings.

Humans experience a complex assortment of somatic sensations from skin receptors. Though sight, sound, taste, and smell are often thought of as the main sensory modalities, modalities detected by skin receptors are exceedingly important components of sensory perception. Skin receptors detect potentially dangerous environmental conditions and allow us to perceive fine distinctions among tactile and pressure sensations.

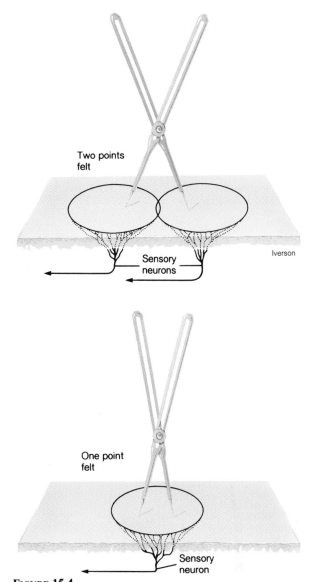

Iverson

FIGURE 15.4
If two points of touch are far enough apart to stimulate two neurons, they are felt separately. If they stimulate only one neuron, they cannot be distinguished.

Pain

Pain is ordinarily a protective sensation. Free terminals that respond to painful stimuli adapt slowly, if at all, providing persistent signals that tissue damage is occurring. Sometimes pain persists in the absence of an identifiable cause. Such chronic pain is difficult to relieve and serves no protective function.

Pain is classified as pricking, burning, or aching. We feel **pricking pain** from tissue irritation. Such signals travel to the somatic sensory area by small myelinated fibers. We feel **burning pain,** the most excruciating of all pain, when skin is burned and for a long time afterward. Cuts and abrasions also elicit burning pain. We feel **aching pain** deep below the body's surface, usually over a wide

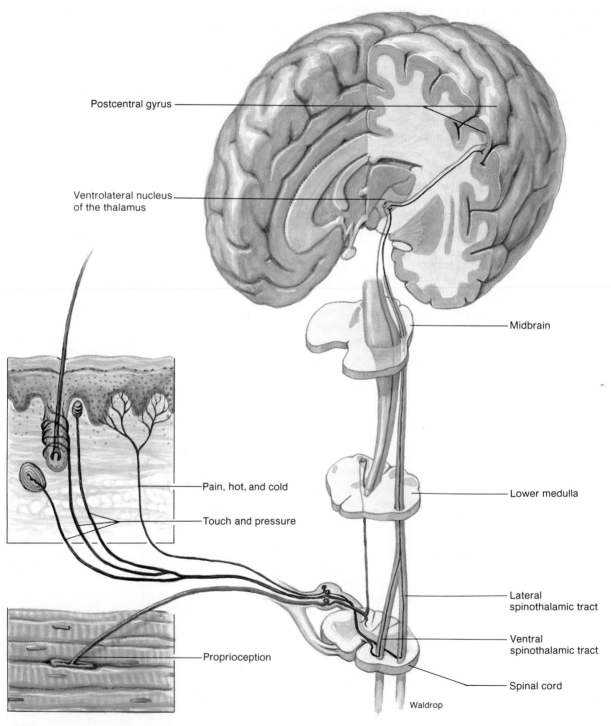

Postcentral gyrus

Ventrolateral nucleus of the thalamus

Pain, hot, and cold

Touch and pressure

Proprioception

Midbrain

Lower medulla

Lateral spinothalamic tract

Ventral spinothalamic tract

Spinal cord

Waldrop

FIGURE 15.5

Pathways from skin receptors and proprioceptors to the somatic sensory area of the cerebrum.

area. Both burning and aching pain signals reach the reticular activating system by small nonmyelinated, slow-conducting fibers. Easily relayed to the cerebral cortex, such signals elicit sensations that can rouse a person from sleep.

Most pain receptors are located in the skin, but some are found in muscles, joint capsules, ligaments, teeth, periosteum, the cornea, visceral organs, and arterial walls. Visceral receptors are relatively sparse, but they can signal severe pain if sufficient numbers are stimulated. Pain from digestive tract ulcers or abdominal surgery is severe visceral pain.

Fast pain receptors respond more quickly than slow ones and are associated with temperature and pressure receptors. **Slow pain receptors** are believed to be stimulated by **kinins** (ki'ninz), polypeptides released from injured tissues. Kinins appear to stimulate receptor terminals as long as they are being released from the tissue. Tissue **ischemia** (is-ke'me-ah), or reduced blood flow, can cause enough tissue damage to release kinins.

Visceral pain can be referred, or perceived as coming from a site different from the site of tissue injury. For example, **referred pain** from a "heart attack" can be felt in the shoulder and arm. Sites of this and other referred pain are illustrated in Figure 15.6. Pain referral appears to depend on visceral and somatic pain fibers converging with the same fibers in spinal cord tracts (Figure 15.7). When visceral pain fibers are intensely stimulated, the cerebral cortex can interpret the signals as coming from the skin where numerous receptors make intense signals more likely. Also visceral referred pain is felt in a somatic area whose nerves join the same region of the spinal cord as those from the painful viscera. The sites at which visceral nerves join the spinal cord vary greatly, so different patients report pain from the same visceral site at different somatic sites. Sometimes amputees experience **phantom pain,** painful sensations in limbs they no longer have. Such pain is due to signals in nerves that once supplied the amputated limb.

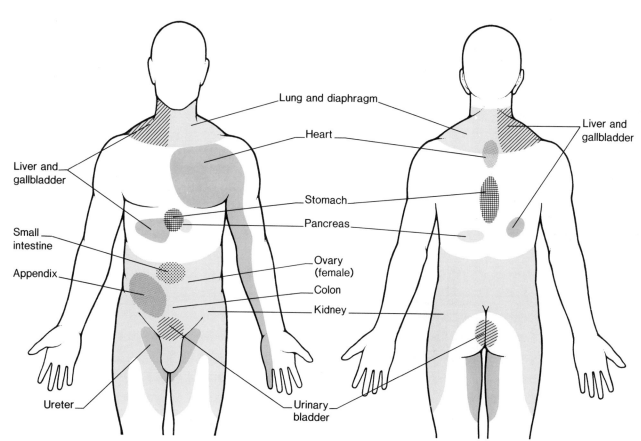

FIGURE 15.6
Sites of referred pain.

Headaches have many causes. Tension headaches are due to muscle spasms in the scalp and neck, though the pain may be referred to inside the head or as a tight band around the head. Infections of the sinuses and meninges and certain eye disorders also can cause headaches. Stimulation of pain receptors in the brain is not a cause of headaches because the brain contains no such receptors. **Migraine headaches,** once thought to be caused by muscle spasms in the walls of arteries of the brain and scalp, are now believed to be caused by fluctuations in the neurotransmitter serotonin or other upsets in the brain's chemistry. Caffeine, food allergies, and many other phenomena can trigger migraine headaches. About one in five humans has suffered from migraine headaches, some several times a week. A few people have chronic daily migraine attacks from barbiturates or ergotamine, the drugs used to treat the headaches. As they have more and more headaches, victims become dependent on the drugs. Between doses they develop withdrawal symptoms that include headache. A vicious cycle of headache, medication, and another headache develops. Non-narcotic drugs—dihydroergotamine and the tranquilizer chlorpromazine—relieve migraine pain without causing physical dependency.

Though the sensitivity of pain receptors varies little, people show great variation in pain perception. Among human volunteers exposed to gradually increasing skin temperature in a specific area, all first perceived pain between 43.5° and 47° C—most at precisely 45° C, the temperature at which tissue damage first occurs. However, at painful temperatures, the volunteers varied significantly in the degree of pain they reported. Such variations involve the brain's perception of pain and not the response of the receptors.

Two related theories, the gating theory and the endorphin theory, purport to explain individual reactions to pain and its control. According to the **gating theory,** both pain signals and nonpain signals (such as those from mechanoreceptors) are relayed to the brain simultaneously. Certain synapses act as gates that admit nonpain signals, but fail to admit pain signals. Variation in the effectiveness of the gating mechanism helps to explain different reactions of individuals to pain. It also may explain the reason some people can reduce their pain by rubbing a painful area or by directing their attention to absorbing activities.

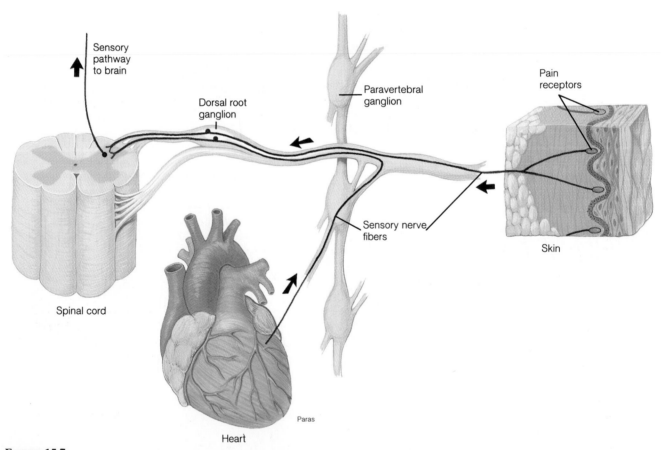

FIGURE 15.7
Pain signals originating in the heart travel the same pathway through the spinal nerve root as signals originating in the skin of the shoulder and arm. They follow common pathways to the cerebral cortex, where their source can be misinterpreted as the skin.

According to the **endorphin** (en-dor'fin) **theory,** pain stimuli cause certain neurons to release polypeptides called **endorphins** and smaller molecules called **enkephalins** (en-kef'al-inz). These molecules appear to suppress pain perception by binding to receptors in the brain, as will be explained in the essay at the end of this chapter.

CHEMICAL SENSES

Taste and smell are chemical senses because their receptors respond to chemical substances. Both are rapidly adapting senses.

Sense of Taste

Receptors for the sense of **taste,** or **gustatory** (gus'tat-o''re) **sense** are found in **taste buds** (Figure 15.8). Most taste buds are located in papilla of the tongue but some are found on the soft palate and portions of the pharynx. Chemical substances reach a taste bud through an opening called a **taste pore.** A taste bud contains several **taste (gustatory) cells.** Each taste cell has receptors at the unattached end, which consist of folds in the cell membrane called microvilli, and a synapse with an afferent neuron at its base. Certain dissolved substances entering taste pores stimulate taste hairs to produce a generator potential, which elicits an action potential in the afferent neuron.

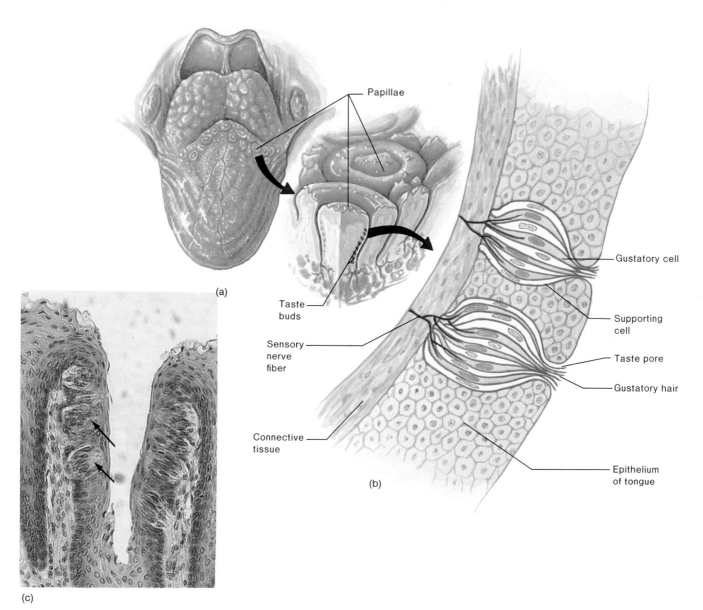

Papillae

(a)

Taste buds

Sensory nerve fiber

Connective tissue

(b)

Gustatory cell

Supporting cell

Taste pore

Gustatory hair

Epithelium of tongue

(c)

FIGURE 15.8

(*a*) Most taste receptors are located on the surface of the tongue in taste buds. (*b*) The structure of a taste bud. (*c*) Photomicrograph of taste buds.

These neurons enter cranial nerves VII or IX and conduct signals to the thalamus, from which the signals are relayed to the somatic sensory area of the cerebral cortex and interpreted (refer back to Figure 13.2).

Only four tastes—sweet, sour, salty, and bitter—are recognized. All other sensations that we call tastes are combinations of basic tastes or are detected as smells. That food tastes "flat" when your nose is congested shows that your perception of the flavor of food is greatly dependent on your sense of smell.

Though all taste buds are slightly sensitive to each of the basic tastes, some are particularly sensitive to one taste (Figure 15.9). Those on the tip of the tongue are most sensitive to sugars that the brain interprets as sweet. Those on the sides of the tongue detect mainly acids (substances that release H^+) that the brain interprets as sour. Those on the tip and sides of the tongue are most sensitive to salts, especially Cl^- in salts such as NaCl and KCl. Those at the back of the tongue and along the pharynx detect bitter substances such as quinine, some inorganic salts, caffeine, nicotine, and urea. Mucus in the pharynx during a cold stimulates pharyngeal taste buds, allowing us to taste it. In low concentrations, the artificial sweetener saccharine is detected by taste buds on the tip of the tongue and interpreted as sweet; and in higher concentrations, it is detected by taste buds at the back of the tongue and interpreted as bitter. Bitter substances, which are often toxic, can elicit signals that prevent swallowing a substance.

Sense of Smell

The human sense of **smell,** or **olfactory sense,** depends on a relatively small number of receptors and is quite limited compared to many animals; yet, humans can smell chemical substances such as methylmercaptan at a concentration of 0.0000004 mg per liter of air. Methylmercaptan is added to odorless natural gas to warn of gas leaks.

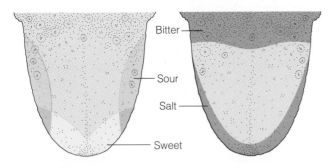

FIGURE 15.9
Taste buds in certain areas of the tongue are particularly sensitive to one of the four basic tastes.

Smell receptors called **olfactory hairs** are microvilli of **olfactory cells** in the epithelium of the nasal passages (refer back to Figure 13.2). Axons extend from the base of olfactory cells through openings in the horizontal plate of the ethmoid bone and come together to form the olfactory nerve (cranial nerve I).

Among the many chemical substances in the environment, only about one per thousand has an odor; that is, can be detected by human olfactory receptors. To be detected, substances must be soluble in both water and lipid, and must be volatile enough to reach the receptors during inhalation.

A possible explanation of the mechanism of smell is that substances bind with receptors on olfactory hairs and activate the enzyme adenylate cyclase. The enzyme makes cAMP that probably alters membrane permeability and creates a generator potential. The generator potential initiates action potentials in an axon of the olfactory nerve. Such signals are relayed via the **olfactory tracts** directly to the medial aspect of the frontal lobes of the cerebrum where they are interpreted. They do not go through the thalamus as do other sensory signals. Computer simulations of smell suggest that cortical cells first look for similarities to known smells and then for differences within groups of similar smells. As smells are identified, they probably are fixed in memory by making certain cells odor-specific.

Some signals from olfactory receptors go to the limbic system where they contribute to emotional aspects of smell. For example, we experience pleasant feelings when we smell a fragrant flower or good food and unpleasant feelings when we smell mercaptans or rotten food.

The senses of taste and smell are compared in Table 15.3.

TABLE 15.3	Comparison of the Senses of Taste and Smell	
Characteristic	**Taste**	**Smell**
Receptor sensitivity	Four primary tastes	Numerous categories of smells
Adaptation	Rapid	Rapid
Reduction with aging	Small because taste cells are continually replaced	Significant because olfactory cells that die are not replaced
Signals from receptors	Travel via cranial nerves VII and IX; pass through thalamus and then to cerebrum	Travel via cranial nerve I to olfactory bulb; pass through olfactory tract to limbic system and cerebrum

Vision

Properties of Light

Light rays travel through air at 300,000 kilometers per second, but they travel through transparent liquids and solids, such as those of the eyes, much more slowly. The speed of light rays in air relative to their speed in another medium is the refractive index of the medium. For example, if light travels through a kind of glass at 250,000 km/sec, the refractive index of the glass is 1.2 (300,000/250,000). The refractive index of air itself is 1.0.

Light rays from distant objects are parallel as they travel through air. When they strike glass at an angle perpendicular to the rays themselves, their velocity is decreased but they continue on their original paths. When light rays strike a glass surface at an angle, however, they are bent as they cross the interface between air and glass at both surfaces (Figure 15.10). The bending of light rays

striking a surface at an angle is called **refraction.** Increasing the refractive index or the angle at which the light enters a new medium causes more refraction. The same phenomenon occurs between air and water, displacing where objects appear to be under water. We may miss when we reach for an underwater object because it is not quite where it seems to be.

These properties of light help to explain events at refractive surfaces of the eye and of corrective lenses. When parallel light rays strike a convex surface, those perpendicular to the center of the surface pass straight through, but those striking the curved surface of the lens at an angle are refracted (Figure 15.11a). The light rays converge at the **focal point** behind the convex lens. The distance from the lens to the focal point is the **focal length.** It is determined by the degree of curvature of the lens and by the distance between the source of light and the lens. Refraction as light passes through a concave surface (Figure 15.11b) causes rays that strike it at an angle to diverge.

The External Eye

The human adult eye, or eyeball, is a hollow spherical structure about 2.5 cm in diameter. Less than 20 percent of the surface of the eyeball is visible between the eyelids, the remainder being surrounded by the bony eye orbit.

In addition to the bony orbit, the eyes are protected by eyelashes, eyelids, conjunctiva, and a lacrimal apparatus (Figure 15.12). Each eyelid, or **palpebra** (pal'pĕ-brah), consists of a skin fold and associated muscles. Contraction of the orbicularis oculi closes the eye and contraction of the levator palpebra superioris raises the upper eyelid. Each eyelid also contains a tarsal plate and tarsal glands. The **tarsal plate** of dense connective tissue helps maintain the shape of the eyelid. **Tarsal,** or *meibomian* (mi-bo'me-an), **glands** secrete a lubricant that helps to keep the eyelids from sticking together.

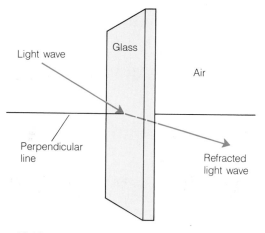

FIGURE 15.10
When parallel light rays strike a piece of glass at an angle other than 90°, the rays are bent as they strike the glass and their velocity is decreased.

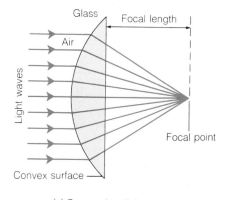

(a) Converging light waves

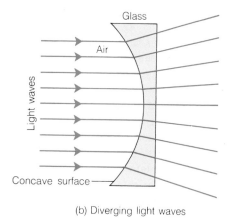

(b) Diverging light waves

FIGURE 15.11
(*a*) When parallel light rays strike a convex lens, those that strike the center (perpendicular to the lens surface) are not bent, but those that strike the lens at other angles converge (are bent toward the center of the lens toward a focal point). (*b*) When parallel light rays strike a

concave lens, those that strike the center are not bent, but those that strike the lens at an angle diverge (bend away from the center of the lens).

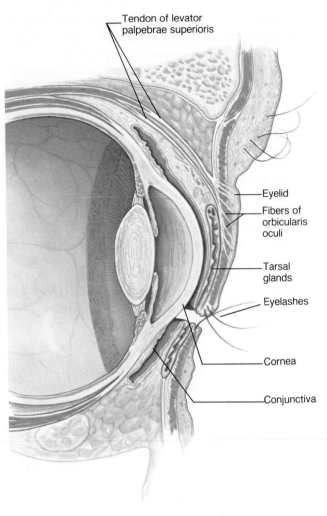

(a)

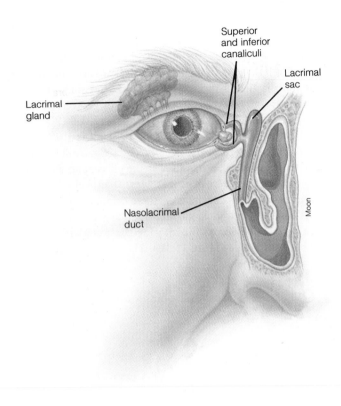

(b)

FIGURE 15.12
(*a*) A vertical section through the eyeball showing external structures.
(*b*) The lacrimal apparatus.

The inner surface of each eyelid and the exposed surface of the eyeball are covered with a sheet of mucus-secreting stratified squamous epithelium, the **conjunctiva** (kon″junk-ti′vah). A fold in this membrane between the eyelid and the eyeball helps to keep foreign objects from sliding around or under the eyeball. Blinking, which occurs about 10 times per minute, spreads fluid over the conjunctiva and keeps the eyeball surface moist. A blink reflex closes the eye when an object is moving toward it.

The **lacrimal** (lak′rim-al) **apparatus** consists of a lacrimal gland, located superior and lateral to the eyeball, and ducts that drain secretions (tears) from the eyeball into the nasal cavity. **Tears** contain mucus from conjunctival glands, which serves to lubricate the eyeball, and **lysozyme** (li′so-zīm), an enzyme that can destroy bacteria.

Eye Movements

Vision is enhanced by mechanisms that direct the eyes toward objects to be viewed, scan those objects, and follow the path of moving objects. Eye movements are controlled by three pairs of muscles associated with each eyeball (Table 15.4 and Figure 15.13). Each pair of muscles is reciprocally innervated—when one contracts the other relaxes.

TABLE 15.4 Extrinsic Eye Muscles
Muscles of the Eyeball
Medial rectus—medial rotation
Lateral rectus—lateral rotation
Superior rectus—superior and medial rotation
Inferior rectus—inferior and medial rotation
Superior oblique—inferior and lateral rotation
Inferior oblique—superior and lateral rotation
Muscles of the Eyelids
Levator palpebrae superioris—opens eye
Orbicularis oculi—closes eye

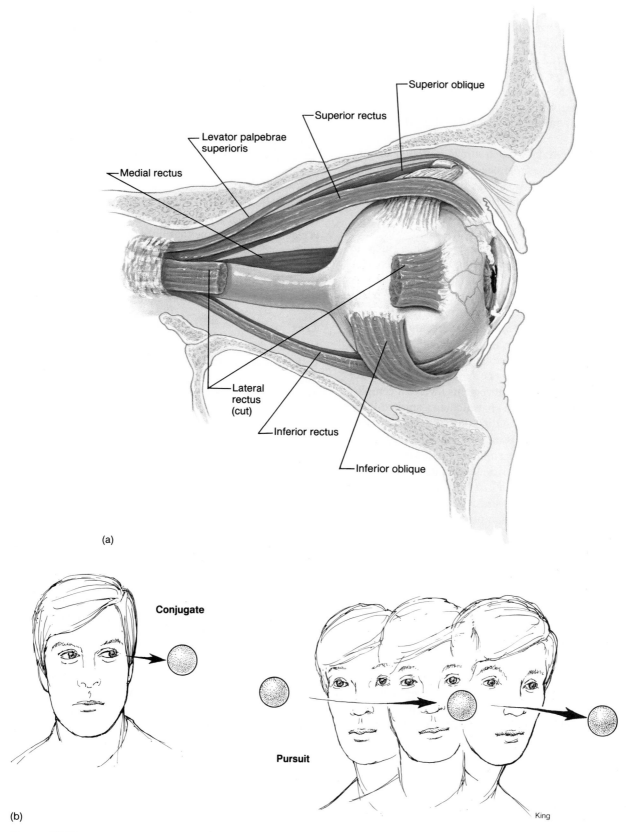

(a)

(b)

FIGURE 15.13
(*a*) Extrinsic eye muscles and (*b*) eye movements.

Conjugate movements are simultaneous movements of both eyes in the same direction—left, right, up, down. They are directly controlled by several nuclei in the brain, but these nuclei are strongly influenced by the cerebral cortex. Although eye movements are coordinated subconsciously, they can be voluntarily initiated to direct the eyes toward particular objects.

Fixation movements, movements that fix the eyes on specific parts of the visual field, are controlled by two mechanisms. Voluntary fixation allows a person to consciously move the eyes from one object and fix them on another. Involuntary fixation locks the eyes on an object and prevents the focus from drifting until it is voluntarily moved.

Saccadic (sah-kād'ik) **movements** occur when the visual scene is moving before the eyes (as when riding in a car) or when the eyes are scanning a visual scene (as when reading). The eyes fix on first one and then another part of the visual field. Saccades (jumps of the eyes) occur several times per second, but are so rapid that the eyes are fixed on a particular site about 90 percent of the time. Because the brain suppresses images during the saccades, we are usually unaware of these movements.

Pursuit movements allow one's eyes to follow the path of a moving object. Pursuit movements involve a complex cortical process. Suppose you are watching a spirited Ping-Pong game with the players hitting the ball across the table many times before one misses. At first, your eye movements are jerky and you have difficulty following the ball. After a few seconds, your eye movements are smoother and you can follow the path of the ball almost exactly. Your cerebral cortex makes subconscious, automatic calculations and generates a plan of movement so you can keep your eyes on the ball.

Regulation of eye movements by the vestibular apparatus, which maintains balance as explained later, allows the eyes to remain focused on a particular object in spite of rapid movements of the body or head. For example, when irregular head movements occur as one rides over a bumpy road, the vestibular apparatus causes the eyes to move rapidly and automatically in the opposite direction. This process occurs in thousandths of a second—much more rapidly than saccadic movements. Dancers and gymnasts learn to assist this process by fixing their eyes on a particular object and refixing on the same object as they spin or rotate. This helps to avoid dizziness and disorientation.

The Internal Eye

The eyeball consists of three tissue layers and contains the lens and two fluid-filled chambers (Figure 15.14).

Fibrous Tunic

The outer **fibrous tunic** of the eye consists of tough white sclera and a transparent anterior cornea. The **cornea** (kor'ne-ah) admits light, and its convex surface plays a major role in focusing light rays as they enter the eye. Lacking blood vessels, the cornea obtains oxygen and nutrients by diffusion from adjacent fluids and tissues. It also obtains some oxygen from air. Corneas were transplanted successfully before other organs because their lack of blood vessels prevents tissue rejection (Chapter 21).

Vascular Tunic

The middle **vascular tunic,** or **uvea** (u've-ah), consists of the choroid, ciliary body, and the iris. The **choroid** (ko'roid) is a delicate membrane that lines the inner scleral surface and extends to the ciliary body. The **ciliary** (sil'e-er"e) **body** includes a **ciliary process,** which secretes fluid into the anterior part of the eye, and **ciliary muscles,** which control the shape of the lens.

Lens

Suspended from the ciliary body across the hollow eyeball is the **lens,** a transparent, biconvex elastic structure, consisting of epithelial cells that contain large fibers of clear cytoplasm. A capsule surrounding the lens consists of layers of protein molecules secreted by the epithelial cells. The lens changes shape from moment to moment depending on the distance from the retina to objects being viewed, thereby focusing light from an object. Changes in lens shape and surface contours throughout life account for some changes in vision.

Iris

The **iris** is a doughnut-shaped, colored structure with an opening called the pupil. Its entire perimeter is attached to the ciliary body and it contains two sets of smooth muscle fibers, a circular constrictor set near the pupil and a radial dilator set near the perimeter (refer back to Figure 14.8). These muscles, which are under autonomic control, determine the diameter of the pupil and, thus, the amount of light entering the eye. In dim light, the radial muscles contract, dilate (enlarge) the pupil, and allow more light to enter. In bright light, the circular muscles contract, constrict the pupil, and allow less light to enter.

Changes in pupil size are usually related to light intensity, but can be altered by emotional state. For example, pupils dilate when people look at things that please them and constrict when they look at things they find distasteful. A perceptive salesperson can identify potential buyers from their pupils.

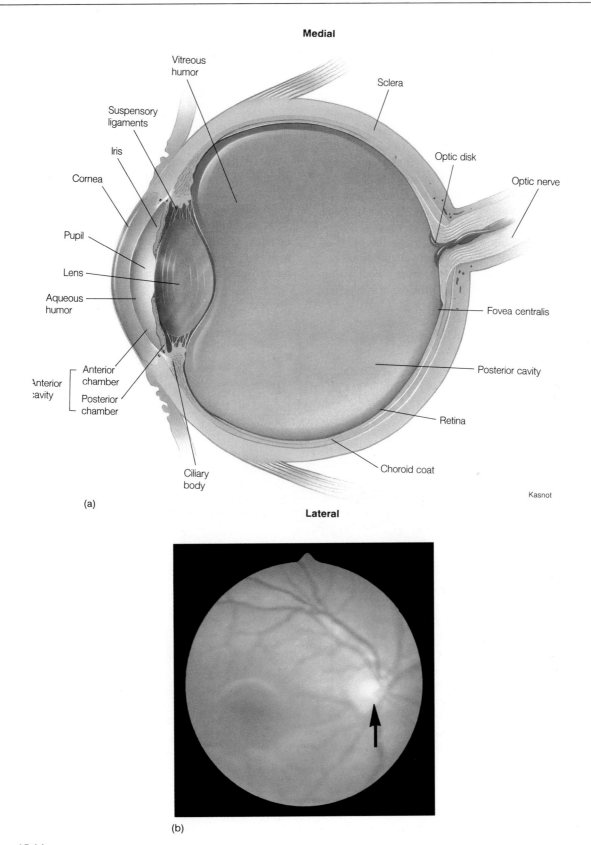

FIGURE 15.14
(a) A horizontal section through the eyeball showing layers of the wall and internal structures. (b) Photograph of the inside of the eye. Arrow points to the optic disk.

Anterior Cavity

Between the cornea and lens, light rays pass through the **anterior cavity** across which the iris (colored part of the eye) is suspended. This cavity is filled with **aqueous humor,** a watery liquid continuously secreted by epithelial cells in the ciliary process. Aqueous humor is formed by active transport of Na^+, movement of Cl^- and probably HCO_3^- down an electrical gradient, and water moving by osmosis. Amino acids, vitamin C, and glucose enter aqueous humor by active transport or facilitated diffusion.

Aqueous humor helps to maintain eyeball shape and provides nutrients to the lens and cornea. Light rays bend at its interfaces with the cornea and lens. Aqueous humor drains out of the anterior chamber through the **canal of Schlemm** at the same rate it is secreted—about 30 ml per day (Figure 15.15). Pressure in the chamber is kept within a small range of 11 to 22 mm Hg. If the canal is blocked or constricted, intraocular pressure increases causing **glaucoma** (glaw-ko'mah). Unless drainage is increased by drugs or surgery, excessive pressure impairs blood circulation and causes blindness by destroying receptor cells in the retina.

Posterior Cavity

After refraction by the lens, light passes through the **posterior cavity** before striking the retina. This cavity is filled with jellylike **vitreous** (vit're-us) **humor,** which refracts light only slightly. It helps to maintain the eyeball shape and keeps the retina smoothly pressed against the choroid. Vitreous humor is formed only during embryonic development. Until synthetic vitreous fluid recently became available, it was very important to prevent loss of vitreous humor from a punctured eyeball.

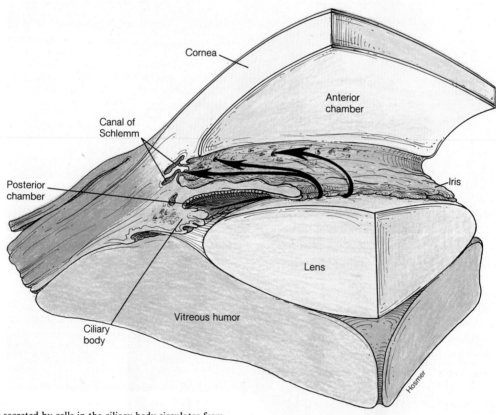

FIGURE 15.15
Aqueous humor secreted by cells in the ciliary body circulates from the posterior chamber through the pupil to the anterior chamber, from which it drains into the canal of Schlemm.

Retina

The inner **retina** (ret'in-ah), or **nervous tunic,** contains light receptor cells. Light rays focus most directly on the **fovea centralis** (fo've-ah sen-tra'lis), a slight depression in the retina where vision is sharpest. Medial to the fovea is the **optic disk (blind spot),** where blood vessels and nerve fibers leave the eyeball and no receptors are present. The nerve fibers form the optic nerve (cranial nerve II).

As light rays pass through the eye, they are refracted at each interface between two media—air to cornea, cornea to aqueous humor, aqueous humor to lens, and lens to vitreous humor. The cornea and lens have the greatest effect on focusing light rays at the focal point on the retina. The pupil regulates the amount of light entering the eye.

Accommodation

The ability of the eye to focus on near objects, called **accommodation,** involves lens, iris, and certain eye muscles. Parallel light rays from a distant object (farther than 6 m from the eyes) are very nearly perpendicular to the corneal surface. After passing through the cornea, they require little bending by the lens to converge to a focal point on the retina. The more divergent light rays from close objects (between 10 cm and 6 m) strike the cornea at an angle. They require greater bending by the lens to converge to a focal point. When the eyes focus on a near object, the ciliary muscles contract, regulate the refractive power of the lens, and bring the object into focus. Humans cannot focus on objects closer than 10 cm because this mechanism reaches its maximum convergence at that distance.

Ciliary muscles control lens shape (Figure 15.16). **Meridional fibers** extend from the choroid to the **suspensory ligament,** which is attached to the entire perimeter of the lens capsule. When these fibers contract, they pull the choroid layer forward, reducing tension on the suspensory ligament and, thus, on the lens. **Circular fibers** form a sphincter around the suspensory ligament. When they contract, they decrease the diameter of the suspensory ligament and also reduce tension on the lens. Under lessened tension, the natural elasticity of the lens allows it to become more spherical and its refractive power increases. When the ciliary muscles relax, tension in the suspensory ligament increases and the ligament exerts force on the lens capsule. This flattens the lens and decreases its refractive power.

During accommodation, other processes help to create a sharp image of a near object. The pupils constrict as the distance of the object from the eyes decreases. This shuts out the most divergent rays and compensates for the decreased depth of field, or how distant an object is. (A pinhole camera demonstrates this phenomenon. By shutting

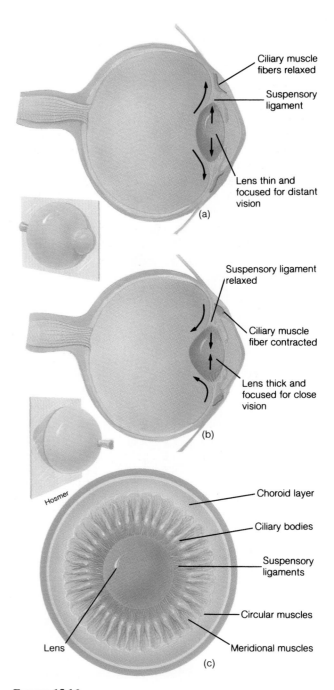

FIGURE 15.16
The lens is attached to the ciliary body by the lens capsule and the suspensory ligament. (*a*) When the ciliary muscles are relaxed, tension on the suspensory ligament increases and the lens becomes thinner and focuses on distant objects. (*b*) During accommodation when the ciliary muscles contract, they decrease the tension on the suspensory ligament and allow the lens to become thicker and focus on close objects.
(*c*) Both meridional and circular fibers contract during accommodation.

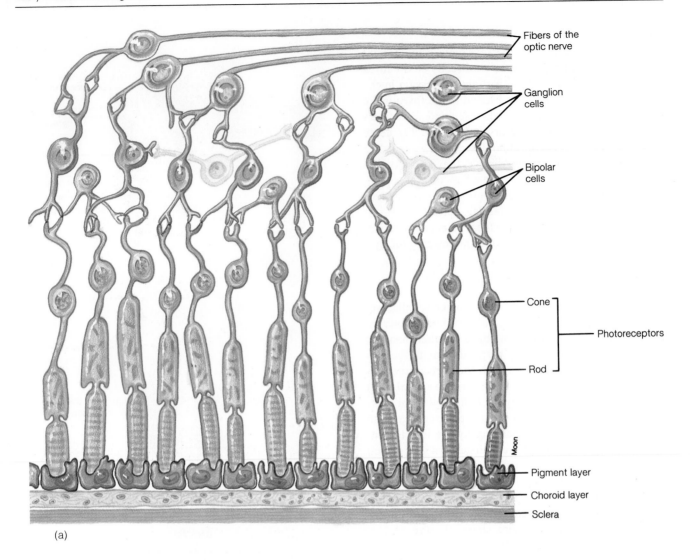

(a)

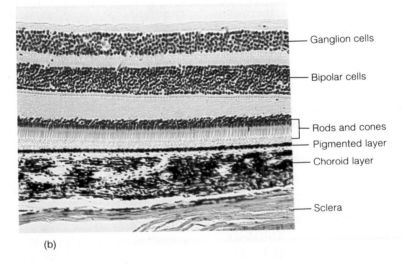

(b)

FIGURE 15.17
Light waves pass through the eye, including the transparent neurons
in the retina, where they stimulate the rods and cones. Electrical
signals are then conducted back through the retina by way of the
bipolar cell to ganglion cells, whose axons are the fibers of the optic
nerve. (*a*) A drawing, and (*b*) a photomicrograph of a section through
the retina.

out all divergent rays and allowing only a point of light to strike the film, the camera is always in focus.) Also, the eyeballs rotate medially as certain external eye muscles contract. This directs both eyes toward the near object.

> Most animals lack modifications of the ciliary muscles that allow humans to see close objects sharply. Even in humans, focusing on close objects strains those muscles. When you are using your eyes for long periods to see close objects, look away periodically to allow your ciliary muscles to relax.

Retinal Cells

The human retina has several layers of cells concerned with detecting light and relaying signals to the optic nerve (Figure 15.17). Each retina contains two kinds of receptor cells—about 125 million rods and about 5.5 million cones. Cones are most numerous in the fovea and gradually decrease in number from the fovea outward. Rods are absent from the fovea, but are present elsewhere in the retina. Both rods and cones contain pigment molecules that change shape when they absorb light waves. This change in shape initiates a series of chemical reactions and produces a generator potential.

In addition to the receptor cells, the retina also contains **bipolar cells** and **ganglion cells** that carry signals from the receptors to the fibers of the optic nerve. Considerable convergence occurs between receptor cells and bipolar cells, and between bipolar and ganglion cells. This convergence usually involves other retinal cells called horizontal cells and amacrine cells.

Light reaching the retina passes through the transparent neurons and strikes the pigmented layer of the retina. Melanin in the pigmented and choroid layers absorbs light, preventing reflection inside the eyeball and allowing more precise image formation. Albinos lack sharpness of vision because they are genetically deficient in melanin.

Properties of Rods and Cones

Rods and cones are similar in structure (Figure 15.18). A rod has a long outer segment, an inner segment, and a synaptic body. The outer segment contains a light-sensitive pigment **rhodopsin** (ro-dop'sin), stored in disks of intracellular membranes. The inner segment contains organelles, especially mitochondria. The synaptic body transfers signals from the rod to bipolar cells. Cones are similar to rods, except that they are shorter and thicker, and contain one of three color-sensitive pigments. The significance of these pigments is discussed with color vision. Rods and

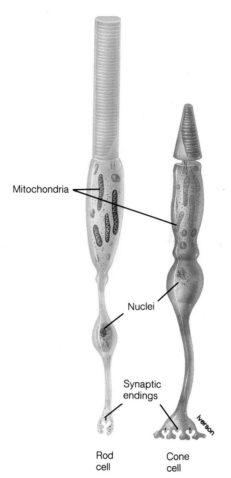

Figure 15.18
The internal structure of a rod and a cone.

cones stimulated by light hyperpolarize instead of depolarizing, as do most other receptors. Hyperpolarization, which results from decreased permeability of receptor cell membranes to Na^+, lasts as long as stimulation persists.

Vision in Dim Light

Rods, being much more light-sensitive than cones, allow vision in dim light, but they do not detect colors. In dim light, we see black, shades of gray, and white according to the light intensities detected by the rods.

Rhodopsin consists of the enzyme **opsin** (op'sin) and **retinal** (ret'in-al), a carotenoid pigment derived from vitamin A. The tips of rods are imbedded in the pigmented epithelium, which stores vitamin A. Rhodopsin is continuously synthesized in the inner segment, migrates to the base of the outer segment, and is incorporated into membrane disks. New disks produced hourly replace those used up at the tip of the outer segment.

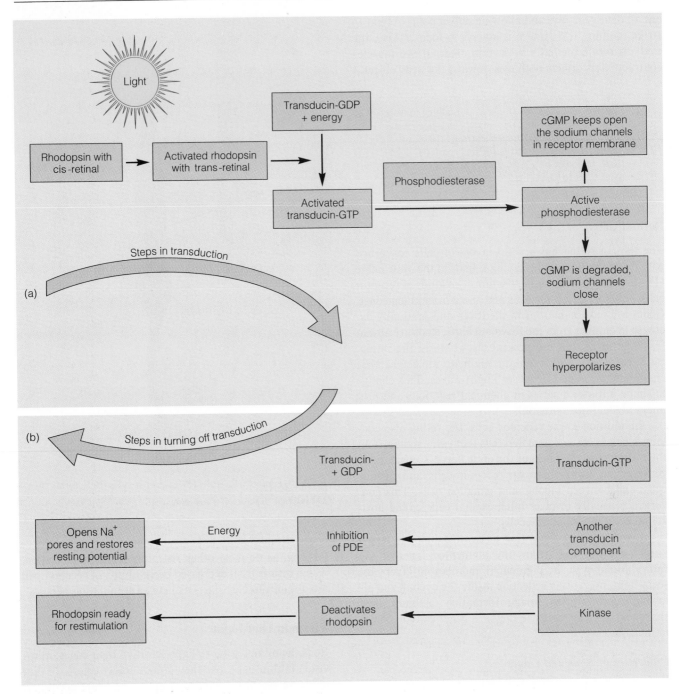

FIGURE 15.19

(*a*) Events in transduction of light energy to a neural signal.

(*b*) Events in turning off transduction.

Eliciting signals from light receptors is a photochemical process (Figure 15.19a). Inactive rhodopsin contains the isomer *cis* retinal. A single **photon** (the smallest unit of light) can activate a molecule of rhodopsin, converting retinal to its *trans* isomer and activating the enzyme **transducin** (trans-du′sin). Inactive transducin has guanosine diphosphate (GDP) bound to it, and during activation GDP is converted to guanosine triphosphate (GTP). Transducin-GTP behaves as an intracellular transmitter and moves from the disks to the cell membrane, where it activates the enzyme phosphodiesterase (PDE). In an unstimulated cell membrane, cyclic guanosine monophosphate (cGMP) maintains a resting membrane potential by keeping sodium pores open. In light, PDE hydrolyzes cGMP, thereby allowing sodium pores to close and the cell membrane to hyperpolarize. Hyperpolarization spreads along the membrane to the synaptic terminals where it initiates a signal in a bipolar cell.

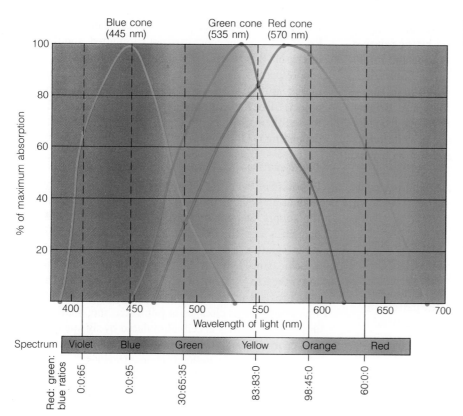

FIGURE 15.20
Light absorption curves for cone pigments.

Amplification, which occurs in two reactions, accounts for the fact that a single photon of light can elicit a neural signal. First, each molecule of rhodopsin activates many molecules of transducin by converting GDP to GTP. Second, the powerful enzyme PDE can hydrolyze over 4000 molecules of cGMP per second.

Deactivation of transducin and PDE turns off receptors and makes membranes ready for restimulation (Figure 15.19b). A chemical timer component of transducin initiates deactivation by converting GTP to GDP. This reaction also makes energy available for future reactions. Another component of transducin inhibits PDE, thereby allowing sodium pores to open and restoring the resting membrane potential. Finally, a kinase deactivates rhodopsin by adding phosphates to amino acids in opsin. This rhodopsin forms a complex with another protein, **arrestin** (ah-res′tin), that prevents it from binding to transducin until it is again stimulated by light.

Color Vision

Humans perceive light of wavelengths 400 to 700 nanometers (nm) as a spectrum of color. Properties of pigments in colored objects determine how objects absorb or reflect light and, thus, what color they appear to have. For color to be perceived, certain wavelengths must reach light-sensitive pigments in cones. Cones are named by the pigment they contain. Blue cones have the pigment **cyanolabe** (si-an′o-lāb″), green cones have **chlorolabe** (klor′o-lāb), and red cones have **erythrolabe** (ĕ-rith′ro-lāb). All these pigments contain retinal, which serves the same function as it does in rods. When cone pigments are excited by light, they initiate signals probably in much the same way as rods do. Because cones require a higher intensity of light to become excited than rods, they are not active in dim light. This is the reason that even colored objects appear as shades of gray in dim light.

Each pigment absorbs light of particular wavelengths (Figure 15.20). Studies with monochromatic light (light of a single wavelength) demonstrate that maximal (100 percent) light absorption occurs in blue cones at 445 nm, in green cones at 535 nm, and in red cones at 570 nm. These studies also allow calculation of ratios (the percent of maximum absorption at which each cone operates) for any particular wavelength. The ratios are arbitrarily stated in the order red:green:blue. For example, light at 490 nm produces the ratio 30:65:35. The brain interprets the sensation produced by this ratio as green. Light and pigments are physical entities, but color exists only in our perceptions.

We perceive colors by comparisons of relative numbers of signals from different kinds of cones in cerebral visual areas. Visual areas also consider the contrast between an object and its background. We describe colors according to three qualities—hue, saturation, and brightness. **Hue,** the name of a color, is detected by the relative degree of light absorption in each kind of cone and information relayed to the visual association area. **Saturation,** the concentration of a color, is detected by the intensity of sensations, using the same mechanism as hue. **Brightness** is the degree to which a color stands out from its background. For example, a red object on a white background appears brighter than the same object on a pink background.

The interpretation of color is complex and not completely understood, but we can summarize what is known as follows:

1. The visual cortex receives signals about both objects and background.
2. It also receives signals that provide coded information about hue, saturation, and brightness. Together, these signals relay a complex pattern of information about the object and its background to the visual association area. (If the modern multiplex hypothesis is correct, these signals consist of compound waves.)
3. The visual association area integrates and interprets signals; the result is perception of all the complex and varied patterns of color we can see.

Color affects mood as advertisers and interior decorators know. The colors of packages help determine what products we buy even though we may not consciously make our choices by package color. Pink rooms in psychiatric hospitals discourage antisocial behavior. Blues and greens give living rooms a restful quality. And mood affects what color clothing we choose as is demonstrated by what we wear at the beach, in the office, or at a funeral.

Color vision does have some limitations. Because at least two cones must be stimulated for the signaling mechanism to operate, very tiny objects that reflect light to only one cone cannot be seen. Also because the fovea lacks blue cones, small objects that reflect light solely to fovea can be seen only if they stimulate red or green cones. Finally, because blue cones are least numerous throughout the retina, differences perceived in shades of blue are less precise than among shades of red or green.

How We Perceive White and Black

When an object reflects light containing many wavelengths in the visible range, the object looks white because the light stimulates all kinds of cones equally and no color is perceived. The sensation that we call white also arises when light of complementary colors such as red and green excite two kinds of cones simultaneously and equally. The sensation that we perceive as black is the absence of light. Receptors (rods or cones) are not stimulated, and the brain interprets this absence of stimuli as the color black. Blind people perceive black because receptors cannot be stimulated. But how can people with normal vision see black objects? Black objects absorb light of all wavelengths, and we see them outlined by other objects—white or colored depending on the wavelengths those objects reflect.

Light and Dark Adaptation, and Night Blindness

Exposure of the retina to strong light over a period of time, such as an hour on the beach or in a wheat field, exhausts the supply of light-sensitive pigments. Light sensitivity decreases and the eyes are said to be *light adapted.* In contrast, after a period of darkness, such as while watching a movie, stored vitamin A is converted to retinal and used to synthesize pigments. Light sensitivity increases and the eyes are said to be *dark adapted.* Light or dark adaptation also involves changes in pupil size and small changes in signal intensities in cells stimulated by signals from rods and cones. These latter changes occur much more rapidly than do changes in pigment concentrations.

Extreme light- or dark-adaptation accounts for a millionfold changes in the sensitivity of receptors to light. Farmhands have difficulty seeing when they come from a field into a dark barn. Moviegoers have difficulty seeing when they come out of a theater onto a brightly lit street. These difficulties are short-lived, because light sensitivity is constantly adjusted to changes in illumination. Such adjustments allow receptors to detect contrasts between light and dark areas, and provide sharp vision under many illumination levels.

Night blindness is a rare severe vitamin A deficiency, which leads to deficiencies in retinal, rhodopsin, and sometimes light-sensitive cone pigments. Dim light fails to stimulate any receptors, but both rods and cones respond to daylight in spite of their pigment deficiencies. Because the liver stores vitamin A, the diet must be greatly deficient in vitamin A for several months for night blindness to develop. An intravenous injection of vitamin A cures night blindness in minutes because receptors quickly use it to synthesize rhodopsin.

Central Visual Pathways

Signals initiated by rods and cones are relayed through bipolar cells to ganglion cells. They travel along axons of ganglion cells, which project through the optic disk, and become myelinated fibers of the optic nerve. Each optic nerve contains about one million such fibers. Through convergence, they relay signals from more than 130 million receptors. About 80 percent of optic nerve fibers convey signals from receptors in the fovea centralis where convergence is minimal and vision is sharpest.

Optic nerve fibers go to the **lateral geniculate** (jen-ik'-u-lāt) **nuclei** of the thalamus. Those from the lateral side of each retina go directly to the ipsilateral nucleus, and those from the medial side of each retina cross over at the **optic chiasma** (ki-as'mah) and go to the contralateral nucleus (Figure 15.21). Some fibers from each retina go to midbrain nuclei (superior colliculi), where they convey information used to coordinate head and eye movements.

Fibers reaching geniculate nuclei synapse with cells that process signals and relay them to the **primary visual cortex.** Both the lateral geniculate nuclei and the primary visual cortex reflect a spatial map of the retina. The primary visual cortex detects lines and sharpness of borders in images projected upon it. For example, when you look at a solid object, the primary visual cortex detects the outline of the object. The greater the contrast between the object and its background, the more vivid is the object's outline in the primary visual cortex. The primary visual cortex also detects the orientation of objects and determines the length of lines. It relays signals to the **visual association area** where they are interpreted (Chapter 12).

The *receptive field hypothesis,* proposed by Nobel prize winners David Hubel and Torsten Wiesel in the 1960s, has been the most widely accepted theory of visual image processing. This hypothesis proposes that the retina consists of numerous receptive fields in which each different kind of image, such as a square, circle, or triangle, stimulates one or more neurons in the brain. From such stimuli, the cerebrum accumulates information about shapes represented in each of many receptive fields. Recently, Barry Richmond and Lance Optican have formulated the *multiplex filter hypothesis* to explain visual processing. This hypothesis proposes that visual systems add together signals and send them as compound waves that are split into separate signals at their destinations. Such signals are somewhat like frequency-modulated (FM) radio signals. More research is needed to fully develop and test this hypothesis, but waves simulated by computers as predicted by the hypothesis have very closely resembled actual waves detected in visual systems. An important component of current research is to determine the neural code for visual signals—exactly what signals the brain uses to make the multiplex signal.

Our brains use signals from light receptors to construct images, but they also construct mental images of things previously seen and even things imagined. In certain studies, people have been asked to form mental images of objects previously shown to them. Retinal and brain activity was measured while they formed the mental images. Though no retinal activity was observed, mental activity was the same as that observed when the images were actually seen. Furthermore, mental images and previous knowledge help to identify features of objects actually seen.

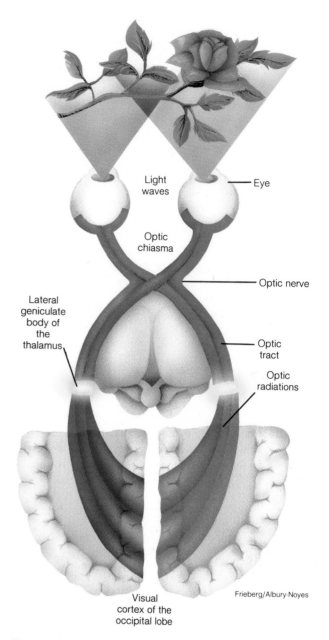

FIGURE 15.21
Pathway of signals from the retina through the optic chiasm to the brain showing binocular vision.

In humans, both eyes can be focused on the same visual field to produce **binocular vision** (Figure 15.21). Signals from the left side of a visual field (detected by receptors in the right half of each retina) are relayed to the right primary visual cortex. Similarly, those from the right visual field are relayed to the left cortex. Each side of the visual cortex relays a slightly different perspective to the visual association area, where differences between the perspectives are used to interpret distance, depth, and height and width of objects in the total visual field. Cues

FIGURE 15.22
The focal length of a lens: (*a*) parallel rays from a distant light source have a relatively short focal length, (*b*) divergent rays from a nearby light source have a longer focal length because part of the power of the lens is used to compensate for divergence.

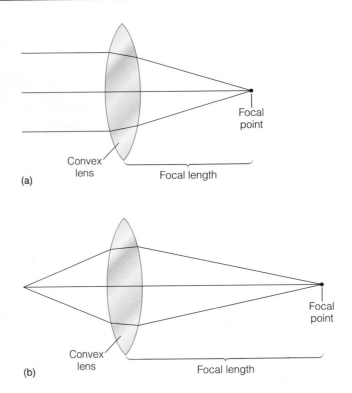

from binocular vision help to maintain balance and to regulate posture and body movements. They help car drivers to estimate how far away other cars are, and how fast they are traveling.

> Many animals have eyes on the sides of their heads—one seeing to the left and one to the right. These animals lack binocular vision but they can maintain vigilance against predators over a large proportion of their environment. Binocular vision in humans probably evolved from arboreal ancestors in which being able to judge distances, especially heights, had survival value.

Defects in Vision

Human vision defects result from abnormalities in eyeball shape, curvature of corneal and lens surfaces, lens elasticity, or lens transparency. Some of these defects can be counteracted by corrective lenses. To understand how corrective lenses work, we need to know a little more about optics. We already know about focal points and focal lengths of lenses and that increasing surface curvature of a lens increases its refraction.

A convex lens focusing parallel light from a distant source (Figure 15.22a) has a relatively short focal length. In contrast, divergent rays from a nearby point source, passing through the same lens, have a longer focal length (Figure 15.22b). Some of the refractive power of the lens is used to compensate for divergence.

Another property of lenses is that they invert images (Figure 15.23). To see that this is so, think of a person as a series of point sources (head, shoulders, trunk, legs, and feet). An image of each point source is formed behind the lens at its focal point. Consequently, the image is upside down. Similarly, if we consider left and right arms and hands as another series of points, we can see that the image also is reversed left to right.

The refractive power of a lens is measured in diopters (Figure 15.24). A **diopter** (di-op′ter) is defined as 1 m divided by the focal length of a convex lens. A 1 diopter lens converges parallel light rays to a focal point 1 m from the lens. A 2 diopter lens has a focal length of 0.5 m and a 4 diopter lens has a focal length of 0.25 m. The power of a concave lens to cause divergence is determined by finding the power of a compensating convex lens. If a 1 diopter convex lens exactly compensates, then the concave lens has a refractive power of −1 diopter. Concave and convex lenses of the same power aligned so light rays pass through both neutralize each other. Such a system has a zero refractive power and the light rays leaving it will be parallel just as they were when they entered it.

Refractive Disorders

The normal eyeball is spheroidal, and in normal vision, or **emmetropia** (em-et-ro′pe-ah), light rays focus exactly at the surface of the retina. Either lengthening or shortening

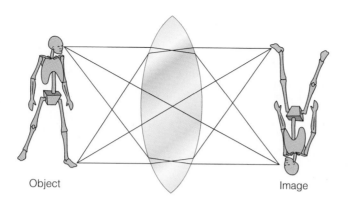

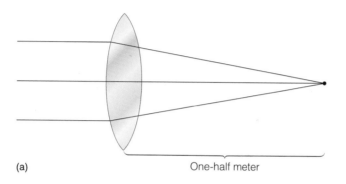

FIGURE 15.23

A convex lens forms an inverted image of an object at its focal point. If we treat the head and feet of the figure on the left as point sources, we can see how the inversion occurs. If we were to treat the left and right hands as point sources and use a horizontal section through the lens, we could demonstrate left-right reversal of the image.

Object

Image

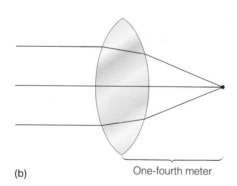

FIGURE 15.24

Refractive power of lenses: (a) +2 diopters, (b) +4 diopters, and (c) −4 diopters.

(a)

One-half meter

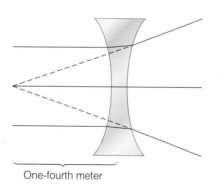

(b)

One-fourth meter

One-fourth meter

(c)

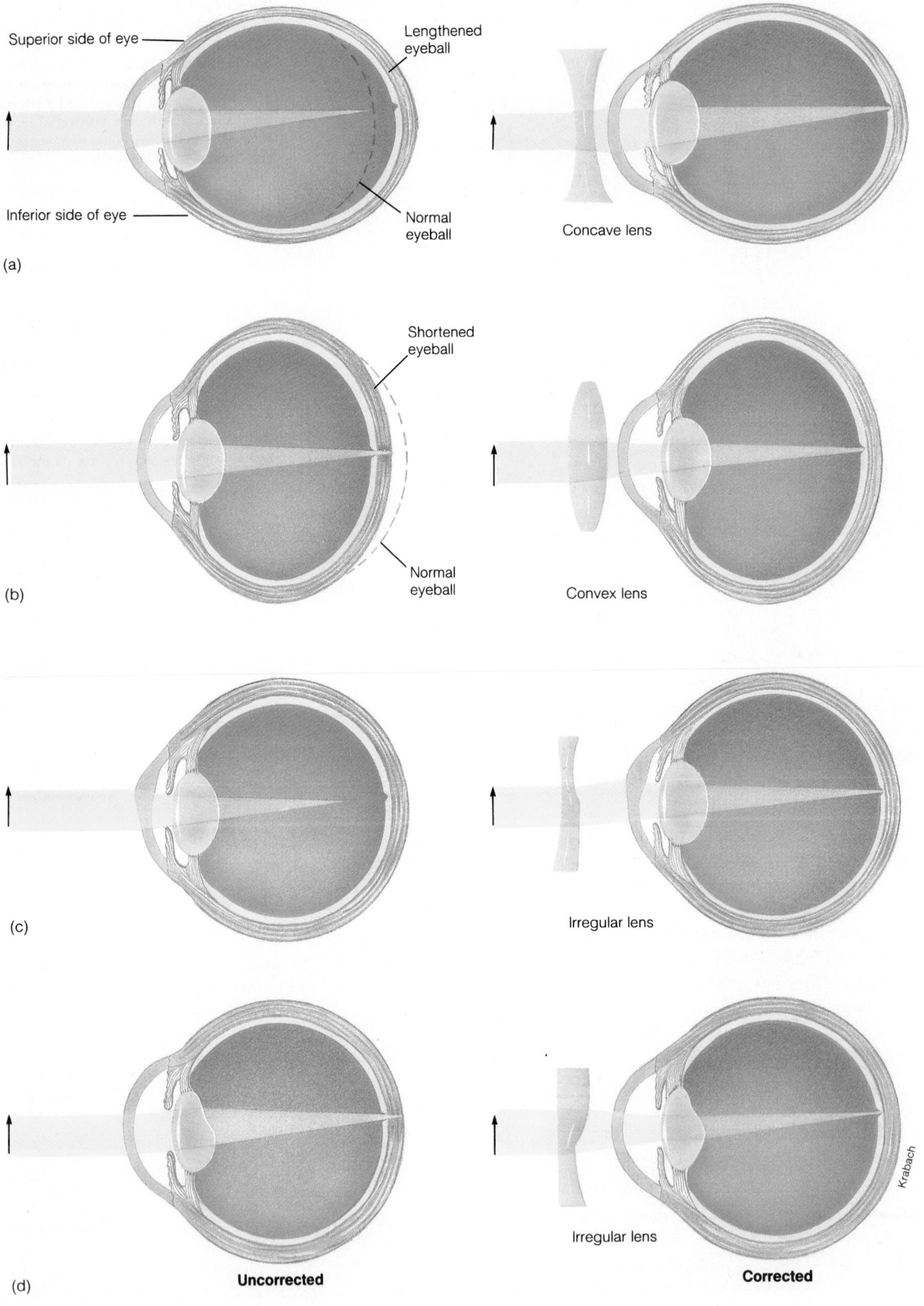

Superior side of eye

Lengthened eyeball

Inferior side of eye

Normal eyeball

(a)

Concave lens

Shortened eyeball

Normal eyeball

(b)

Convex lens

(c)

Irregular lens

(d)

Uncorrected

Irregular lens

Corrected

FIGURE 15.25

How corrective lenses change the bending of light waves in
(a) myopia, (b) hypermetropia, and (c) astigmatism due to an
irregular corneal surface, and (d) astigmatism due to an irregular lens
surface.

the anterior-posterior axis of the eyeball alters its focal length and irregular refractive surfaces distort images (Figure 15.25). In **myopia** (mi-o'pe-ah), the eyeball is elongated and light rays converge to a focal point in front of the retina. Images from distant objects are more distorted than those from near objects, so affected individuals are said to be nearsighted. A corrective concave lens can cause light rays to diverge and focus on the retina.

In **hypermetropia** (hi''per-me-tro'pe-ah), the eyeball is shortened and light rays converge toward a focal point behind the retina. Images from near objects are more distorted than those from distant objects, so affected individuals are said to be farsighted. A corrective convex lens increases convergence so light rays focus on the retina.

Light rays striking an irregular surface of the cornea or lens are refracted unevenly. Some focus on the retina and others in front of or behind it. This condition, called **astigmatism** (as-tig'mat-izm), produces blurred images. It can be alleviated by corrective lenses with compensating irregular curvature.

Light rays pass through a normal transparent lens with little scattering (random alterations in refraction). In a **cataract** (kat'ah-rakt), alteration of lens proteins cause lens cells to become opaque. Such opaque areas scatter light so it fails to come to a single focal point. Cataracts are common in the elderly, but they can be present at birth. They also are common in diabetic individuals, possibly because high blood glucose increases osmotic pressure in the lens tissue and damages lens proteins. An opaque lens can be removed and replaced with a plastic lens implant, usually as outpatient surgery.

With aging, the lens loses elasticity and ability to accommodate, and near vision is impaired. This condition, called **presbyopia** (pres''be-o'pe-ah), is treated with bifocal lenses that increase convergence of light from near objects. Trifocal or multifocal lenses improve mid-range vision. Contact lens wearers can use a lens for distance vision in one eye and one for near vision in the other eye. After a period of adaptation, this arrangement has proven satisfactory in about 80 percent of people who have tried them.

Color Blindness

About 9 percent of human males are colorblind due to a defective gene on their single X chromosome. The most common form, which accounts for over half of all cases, is due to a deficiency of green-cone pigment. Three other forms of color blindness, each accounting for about 15 percent of all cases, are (1) a deficiency of red-cone pigment, (2) an absence of red cones, or (3) an absence of green cones. A small number of color blindness cases are caused by a deficiency of blue pigment, an absence of blue cones, or an absence of more than one kind of cone. Females are colorblind only if they inherit the same defective gene from each parent.

HEARING

Hearing involves conduction of mechanical vibrations from sound waves through the ear to receptors. Understanding hearing requires some knowledge of the properties of sound.

Properties of Sound

Sound energy is transmitted through the atmosphere by vibration of gas molecules. A disturbance called a sound wave consists of alternating regions of **compression,** where gas molecules are pushed together and pressure is relatively high, and **rarefaction,** where gas molecules are spread far apart and pressure is relatively low (Figure 15.26). For example, a vibrating tuning fork produces just

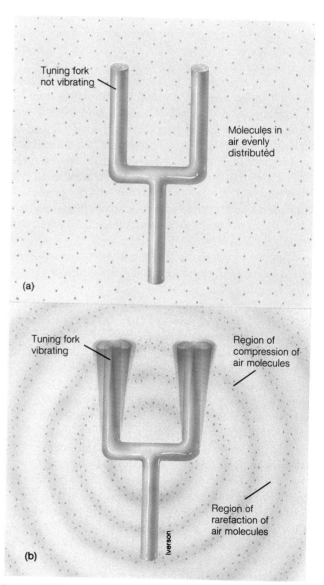

FIGURE 15.26
(*a*) Gas molecules in air are evenly distributed when not disturbed by sound waves. (*b*) They become distributed in regions of compression and rarefaction when disturbed by sound waves.

FIGURE 15.27
Frequency and amplitude of sound waves:
(*a*) moderate amplitude and frequency of
about 20 hertz, (*b*) higher amplitude and
frequency of about 40 hertz, and (*c*) lower
amplitude and frequency of about 80 hertz.

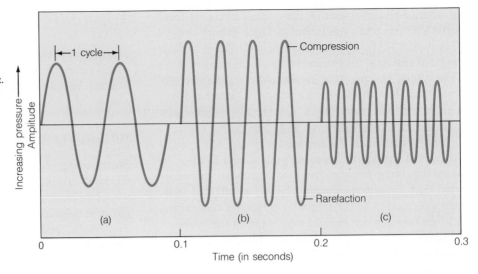

such regions of compression and rarefaction. Pressure differences in these regions cause gas molecules to move. The resulting sound waves travel for miles as long as moving gas molecules cause other gas molecules to move.

Sound waves vary in frequency and amplitude (Figure 15.27). The **frequency** of a sound wave is the number of vibration cycles per second. A single sound wave is one vibration or one cycle of compression and rarefaction. If this cycle is repeated 50 times per second, the sound wave has a frequency of 50 cycles per second, or 50 hertz. (A hertz equals one cycle per second.) The **amplitude** of a sound wave is the pressure difference between regions of compression and rarefaction. It is proportional to the maximum displacement of air molecules as the wave moves through the air.

Qualities of sound—pitch, intensity, and timbre—are related to frequency and amplitude. **Pitch,** the highness or lowness of a tone, is determined by the frequency of vibrations. The normal human ear is sensitive to frequencies from low-pitched sounds (20 hertz) to high-pitched sounds (20,000 hertz). A piano produces sounds from about 30 to 4000 hertz. Among singers, male voices range between 60 and 400 hertz, and female voices range between 150 and 1000 hertz. Humans can detect the full range of pitches noted above if the sounds are relatively loud, but can detect only pitches between 500 and 5000 hertz in barely audible sounds.

Intensity, or loudness, is related to amplitude and is measured in decibels (db). A **decibel** (des′ib-el) is an arbitrary unit of a special logarithmic scale, in which the logarithm of sound intensity is multiplied by ten. Compared to a 20 db sound, a 30 db sound is ten times as loud, a 40 db sound is 100 times as loud, and a 50 db sound is 1000 times as loud. The decibel levels of selected sounds are shown in Figure 15.28. Extremely loud sounds cause pain and can permanently damage receptor cells.

Timbre (tam′br), or tone quality, is produced by higher frequency vibrations (harmonics or overtones) superimposed on a basic pitch. When several different musical instruments sound the same note (basic pitch), each

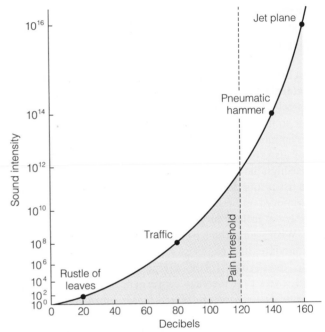

FIGURE 15.28
Decibel levels of selected sounds. Though individuals vary in pain threshold for sound, most experience pain at about 120 decibels.

produces vibrations at a specific frequency associated with the note and additional higher frequency vibrations. Because each instrument—piano, trumpet, violin, or oboe—produces a different pattern of higher frequency vibrations, it has a distinctive timbre.

Structure of the Ear

The **ear** (Figure 15.29) has external, middle, and inner regions. Sound vibrations are conducted in different media in each region—air in the outer ear, bone in the middle ear, and fluid in the inner ear.

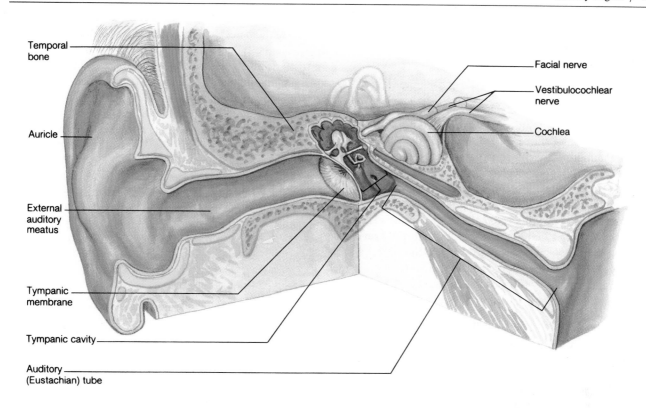

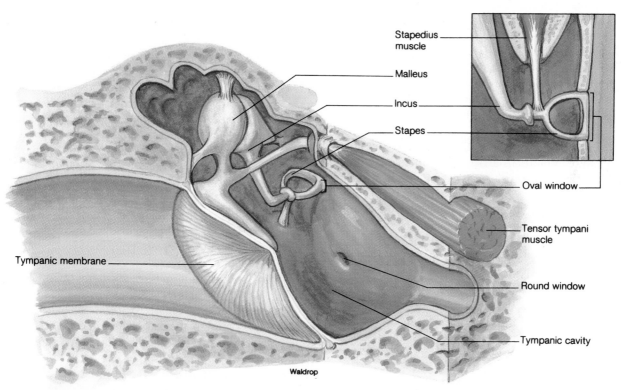

Waldrop

FIGURE 15.29

The structure of the ear with a detailed view of the middle ear
showing bones, membranes, and muscles associated with the
attenuation reflex.

The External Ear

The outer **external ear,** includes the half-moon shaped **auricle,** or **pinna** (pin′nah), and a canal called the **external auditory meatus.** The auricle consists of elastic cartilage covered with skin that contains sweat and sebaceous glands. Both the **helix** (rim) of the auricle and its fleshy lobe vary in size and shape. The canal is lined with skin containing numerous touch and pain receptors, hairs, and **ceruminous** (sĕ-roo′min-us) **glands** that secrete **cerumen** (sĕ-roo′men), or earwax. Foreign objects can become entrapped in the hairs and wax before reaching the eardrum. Sound waves striking the pinna travel through the air along the 2.5 cm canal to the translucent, fibrous **tympanic** (tim-pan′ik) **membrane** (eardrum). Sound waves striking the tympanic membrane cause it to vibrate with amplitude and frequency proportional to the sound waves themselves. The tympanic membrane separates the external and middle ear.

The Middle Ear

Found in a cavity in the temporal bone, the **middle ear** (Figure 15.29), contains small bones and membrane-covered openings (tympanic membrane, oval window, and round window) associated with hearing. From the tympanic membrane, vibrations are transferred through the bones from the **malleus** (mal′e-us) to the **incus** (ing′kus) to the **stapes** (sta′pēz). The **footplate** of the stapes conducts vibrations to the membrane of the **oval window.** As vibrations are transferred across the oval window to fluid in the inner ear, the **round window** bulges into the middle ear.

Conduction through bone decreases amplitude but increases pressure. Vibrations arrive at the footplate with an amplitude three-fourths the original amplitude at the malleus, but at a pressure 1.33 times that at the tympanic membrane. The surface area of the oval window is about one-seventeenth that of the tympanic membrane, so pressure at the oval window membrane is multiplied 17 times. Together, these factors increase vibration pressure about 22 times (1.33×17).

An **attenuation reflex** is activated by very loud sounds being conducted through the bones of the middle ear. In this reflex, sound signals relayed to the midbrain elicit signals that cause contraction of small muscles, the *tensor tympani* attached to the malleus and the *stapedius* (stape′de-us) attached to the stapes. Such contractions withdraw bones from the eardrum and oval window membrane and make them nearly immovable. This reduces bone conduction and can decrease the perceived sound intensity by 30 to 40 decibels. Because the reflex takes 40 to 80 msec to cause muscle contraction, it fails to protect the inner ear from very sudden loud noises, but it does reduce vibration intensity. It is most effective against low frequency sounds, so it also masks low frequency background noises and allows a person to concentrate on higher frequency (above 1000 hertz) sounds.

The middle ear also is connected with the pharynx by the **auditory tube** (Eustachian tube) and to the mastoid cells of the temporal bone by the **tympanic antrum** (an′trum). The auditory tube equalizes pressure between the middle ear cavity and atmospheric pressure in the pharynx.

The Inner Ear

When vibrations cross the oval window membrane, they are transferred to fluid in the **cochlea** (kok′le-ah), the part of the inner ear associated with hearing (Figure 15.30). The cochlea contains a bony supporting structure, the **modiolus** (mo-di′o-lus), and three coiled fluid-filled passageways, the **scala vestibuli** (ska′lah ves-tib′u-le), the **scala tympani** (tim-pan′e), and the **cochlear duct.** The scala vestibuli and the scala tympani join at the apex (helicotrema), and fluid called **perilymph** (per′i-limf) flows through both passageways. The **cochlear duct,** which is separated from the scala tympani by the **basilar membrane** and from the scala vestibuli by the **vestibular membrane,** is filled with a fluid called **endolymph** (en′do-limf). Vibrations first reach the perilymph and are transferred across membranes to the endolymph. Fluid has far greater inertia (resistance to movement) than air, but the large pressure increase at the oval window is sufficient to cause vibrations or pressure waves in these fluids. As the waves pass through the fluids, they cause the round window to bulge slightly toward the middle ear.

Pressure waves exert force on the fluid and deflect the basilar membrane into the scala tympani. Movements of the basilar membrane excite receptors in the **organ of Corti** (kor′te) and encode information about the qualities of sound.

The organ of Corti (Figure 15.30), which extends the full length (34 mm) of the cochlear duct, contains about 16,000 receptors called **hair cells** and many supporting cells. Each hair cell has at its apex about 100 stereocilia most of which are attached to the **tectorial** (tek-to′re-al) **membrane.** At its base, which lies on the basilar membrane, it has a fiber of the cochlear nerve fiber projecting from it.

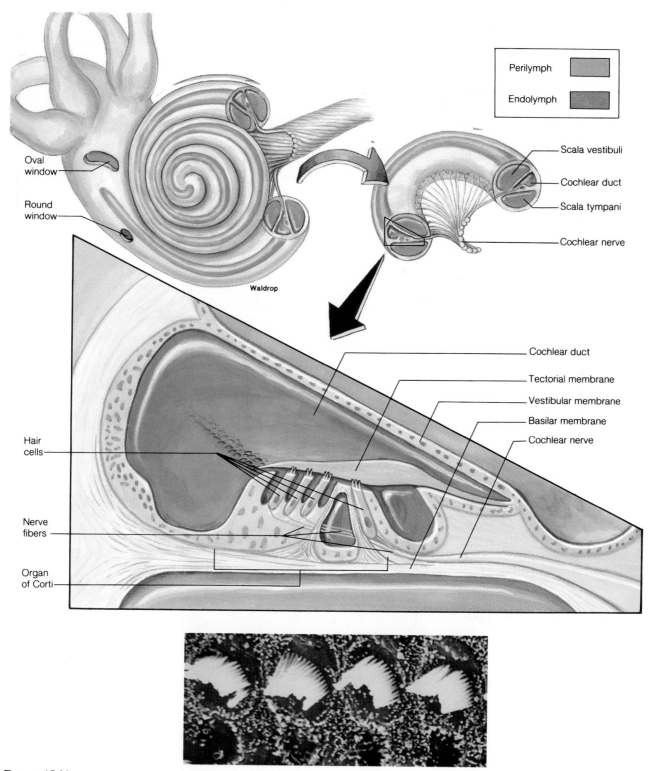

FIGURE 15.30
The structure of the cochlea showing a detailed view of the organ of
Corti and a scanning electron photomicrograph of hair cells.

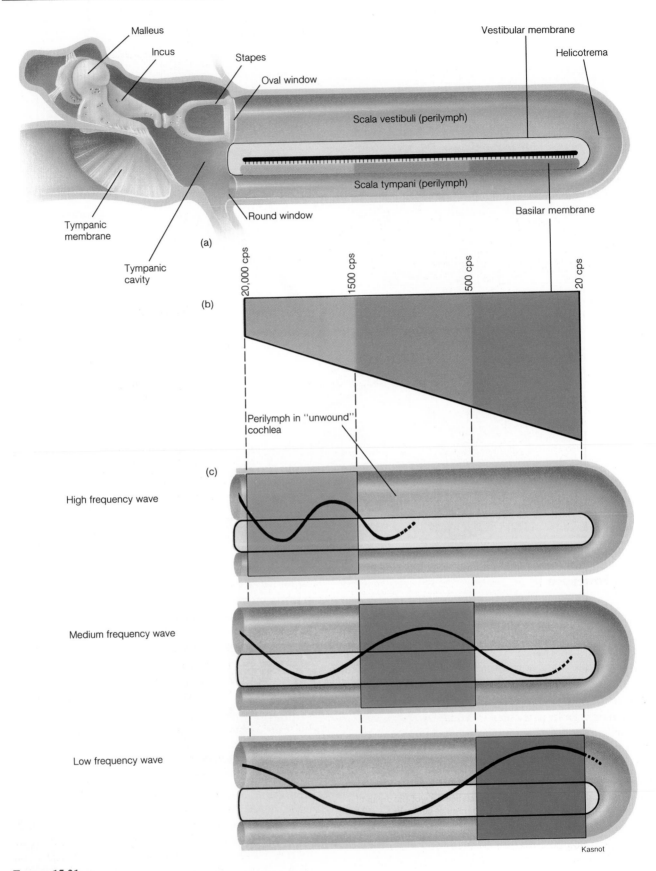

FIGURE 15.31
(a) An "unwound" cochlea showing location of passageways,
(b) sensitivities of hair cells by location, and (c) path of sound waves
of different frequencies through cochlea.

Kasnot

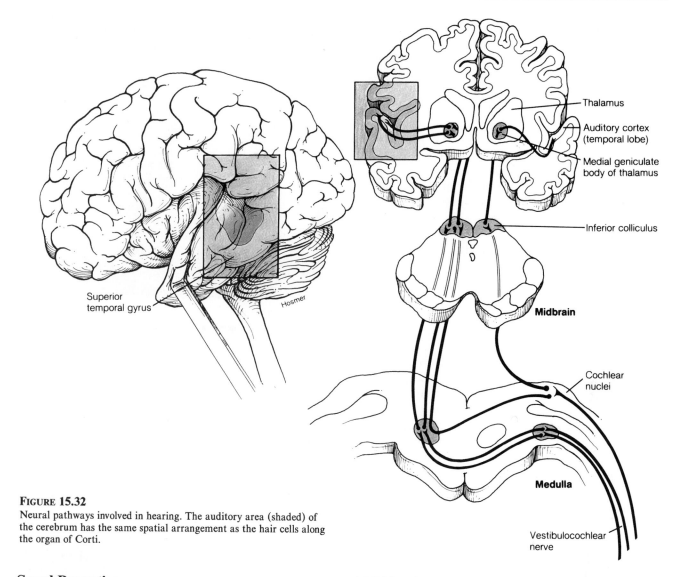

FIGURE 15.32
Neural pathways involved in hearing. The auditory area (shaded) of
the cerebrum has the same spatial arrangement as the hair cells along
the organ of Corti.

Sound Perception

Pressure waves in perilymph travel toward the helicotrema with wave forms characteristic of their frequency.
Each wave deflects a particular region of the basilar membrane (Figure 15.31) according to the membrane's ability
to respond to certain pressure waves. Regions nearest the
stapes are deflected by high-pitched sounds and those
nearest the apex by low-pitched sounds. Such deflections
excite hair cells in particular locations by twisting their
stereocilia. The twisting opens ion channels at the base of
stereocilia, transducing mechanical pressures to generator potentials. These generator potentials encode precise
information. For example, if music is played into an animal's ear and electrical signals from the hair cells are fed
into an audio amplifier, the music can be played back with
great fidelity. Generator potentials travel to hair cell bases
and elicit action potentials in cochlear nerve fibers.

Each sound frequency stimulates a group of hair cells
to generate signals in their nerve fibers, at the same time
inhibiting hair cells on either side of them. This **lateral
inhibition** increases the precision with which specific
pitches are heard. Signals from specific hair cells arrive
in particular locations in the auditory cortex, and the signal
location is used to interpret the pitch of the sound.

Intensity also is detected by basilar membrane deflections and hair cell responses to them. Louder sounds
cause greater membrane deflection, which increases both
the number of hair cells stimulated and the frequency of
action potentials they generate. The auditory cortex uses
the number and frequency of signals to interpret sound
intensity.

Timbre is detected from the pattern of signals from
simultaneously detecting pitch in several regions. Signals
denoting the basic pitch have a greater intensity than those
denoting the harmonics.

All signals from each ear are relayed from the organ
of Corti via the vestibulocochlear nerve to the brain
(Figure 15.32). The signals pass through the **spiral ganglion** to the **cochlear nuclei** in the medulla, where they
synapse with other neurons. Some relay signals to the same

side of the brain and others relay them to the opposite side. Signals from both cochleas travel along both sides of the brain stem including through the **medial geniculate nuclei** to the auditory cortex of the temporal lobe. The auditory cortex corresponds spatially to regions of the organ of Corti that are sensitive to different pitches (Figure 15.32). Some signals also travel to the reticular activating system and, thereby, activate a larger part of the cerebral cortex. Finally, information from the auditory cortex is interpreted in the auditory association area.

Think for a moment of the variety of pitches, intensities, and timbres produced by a full symphony orchestra. You can tell which instruments are playing, recognize many different pitches simultaneously, and detect variations in the intensities of the sounds. As this example shows, a vast complex of information is encoded by the organ of Corti, processed in the auditory cortex, and decoded in the auditory association area.

Hearing Disorders

Because the auditory tube allows microorganisms to reach the middle ear, **otitis media** (o-ti′tis me′de-ah), inflammation of the middle ear is a common disorder. Before antibiotics became available, such infections often spread through the tympanic antrum to the mastoid cells. Now the infections usually are treated before they spread. Many children are especially susceptible to otitis media. An operation called a **myringotomy** (mir″in-got′o-me), the insertion of a small tube in an incision in the tympanic membrane (Figure 15.33), helps to prevent fluid build-up in the middle ear. The tubes are eventually sloughed from the ear without additional surgery.

Deafness, or loss of hearing, can be caused by defective conduction or by nerve damage, and can be partial or complete. Studies of the relative effects of different sensory losses have shown that people have more difficulty adjusting to profound (complete) deafness than to most any other severe sensory deficit.

In the last two decades, the incidence of hearing loss in young people has greatly increased. Much hearing loss is due to continuous bombardment of the hearing mechanism with excessively loud music from radios, concerts, and discotheques.

Any factor that impairs transmission of vibrations to the cochlea can cause **conduction deafness.** These factors include earwax accumulation, otosclerosis (fusion of the stapes to the oval window via spongy bone growth as shown in Figure 15.34), and tympanic membrane damage. Earwax is treated by simple removal. Otosclerosis requires surgery to separate the bones and replace the stapes with a prosthesis that can transmit vibrations to the oval window. Though some damage can be repaired by the tympanic membrane itself, extensive damage leaves scar

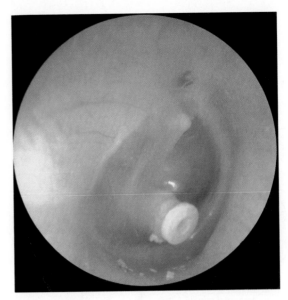

FIGURE 15.33
A tympanic membrane after insertion of a tube.

tissue that greatly decreases conduction. The tympanic membrane can be replaced or reinforced surgically, using fascia from the temporalis muscle. Deafness, especially that from unrepaired tympanic membrane damage, can be treated with a hearing aid that amplifies sound and conducts vibrations through the ear or through skull bones. A problem with hearing aids is that they amplify both nearby sounds and background noise.

> Why does a recording of your own voice sound different than the voice you hear when you speak? The sound of your own voice is conducted partly by skull bones and partly by air. Because of this, your voice sounds different when you speak and when you hear a recording from which vibrations are conducted entirely by air.

Nerve deafness from damage to receptors or nerve fibers can be caused by excessive exposure to loud noises, tumors, brain damage, or congenital defects. Also, some loss of cochlear nerve function usually accompanies aging. Electronic devices can be implanted in the cochlea to respond to sound and stimulate any remaining nerve fibers. This enables the nerve deaf to perceive rhythms of speech and music, and to detect environmental sounds such as sirens and ringing phones. Such a device and its surgical implantation cost more than $10,000. Though the device has limited effectiveness, several thousand patients have received it.

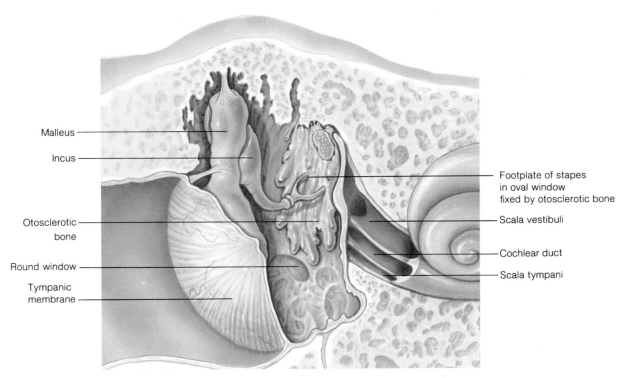

Malleus

Incus

Otosclerotic
bone

Round window

Tympanic
membrane

Footplate of stapes
in oval window
fixed by otosclerotic bone

Scala vestibuli

Cochlear duct

Scala tympani

FIGURE 15.34
Otosclerotic bone in the middle ear.

BALANCE

The organs of balance, located in the inner ear near the cochlea, consist of the saccule, utricle, and semicircular canals (Figure 15.35). The **utricle** (u′tre-k′l), and probably the **saccule** (sak′ūl), are concerned with **static equilibrium** and provide information about stationary head position. The **semicircular canals,** in contrast, are concerned with **dynamic equilibrium** and provide information about changes in head position.

Static Equilibrium

Within each utricle is a **macula** (mak′u-lah) containing hair cells with their stereocilia imbedded in the gelatinous substance. Also associated with the gelatinous substance are particles of calcium carbonate called **otoliths** (o′to-lithz), or ear stones. Otoliths make the gelatinous substance heavier than the endolymph that fills the remainder of the cavity within the macula. A similar structure in the saccule may operate in the same way, but it appears to have little or no function in humans.

Hair cells of the macula respond to gravity (Figure 15.35). When the head is in an upright position, gravity causes the gelatinous mass and otoliths to press upon the hair cells. When the head position is changed, the gelatinous mass shifts position and the stereocilia are bent. Such bending causes ion channels to open and creates generator

potentials in the affected hair cells. The generator potential releases a neurotransmitter that diffuses across a synapse and initiates an action potential in a fiber of the vestibular nerve (part of cranial nerve VIII).

Signals from the vestibular nerve are relayed to the vestibular nuclei and to motor areas in the medulla and cerebellum, where they elicit positional reflexes (Chapter 13). These signals coordinate body movements with head position. Some signals also go to the cerebral cortex, which interprets them as different head positions.

The macula also responds to linear acceleration as when a person starts to run. Movement of otoliths elicits signals interpreted as a feeling of falling backward. The runner then leans forward at an angle that exactly compensates for the backward movement of the otoliths.

Dynamic Equilibrium

In each inner ear, a set of three **semicircular canals** is arranged so each canal occupies a plane roughly perpendicular to the other two, somewhat like the arrangement of the floor and the two adjacent side walls in the corner of a room (Figure 15.36). Any kind of movement will disturb endolymph in at least one canal. The semicircular canals detect movement better than the utricles because of their geometric arrangement and because of their greater responsiveness to endolymph movements.

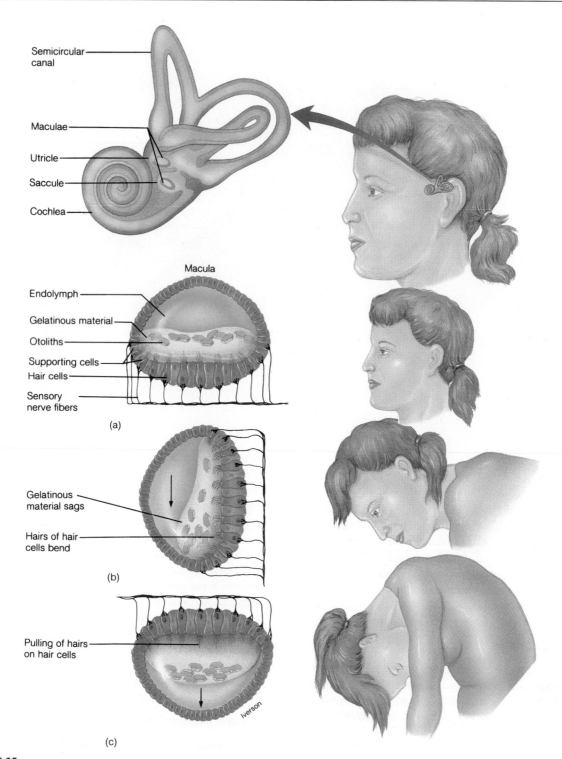

Semicircular canal

Maculae

Utricle

Saccule

Cochlea

Macula

Endolymph

Gelatinous material

Otoliths

Supporting cells

Hair cells

Sensory nerve fibers

(a)

Gelatinous material sags

Hairs of hair cells bend

(b)

Pulling of hairs on hair cells

Iverson

(c)

FIGURE 15.35

Organs of equilibrium include the utricle, probably the saccule, and the semicircular canals, all of which are located in the inner ear. Orientation of macula in static equilibrium: (*a*) with head upright, (*b*) with head tilted, and (*c*) with head down.

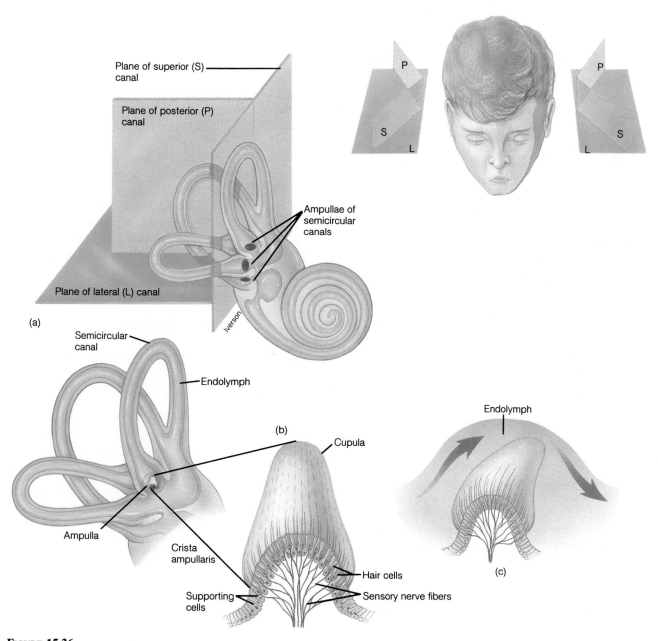

(a)

Plane of superior (S) canal

Plane of posterior (P) canal

Plane of lateral (L) canal

Ampullae of semicircular canals

Iverson

Semicircular canal

Endolymph

(b)

Ampulla

Crista ampullaris

Supporting cells

Cupula

Hair cells

Sensory nerve fibers

Endolymph

(c)

FIGURE 15.36

Dynamic equilibrium. (*a*) Semicircular canals are oriented in mutually perpendicular planes on both sides of the head. (*b*) Each canal is filled with endolymph and has an ampulla, which consists of hair cells embedded in a cupula. (*c*) Head movements cause endolymph to flow, deflecting each cupula, and stimulating hair cells. The direction and speed of movements can be sensed by the rate and degree of deflection of each cupula.

Each semicircular canal has an enlarged area with a spherical **ampulla** (am-pul′lah). Inside each ampulla is a **crista** (kris′tah) made up of hair cells and supporting cells. The hair cells have their stereocilia imbedded in a gelatinous mass called the **cupula** (ku′pu-lah) and their bases connected to branches of the vestibular nerve. When the hair cells are bent by endolymph moving in the canal, they send signals to the nerve fibers. Because the hair cells of the canal in the plane of movement respond maximally, the brain can use signals from them to determine the direction and rate of movement. Turning the head displaces fluid toward an ampulla on one side and away from an ampulla on the other side. Nodding the head stimulates hair cells bilaterally in one pair of canals and deflects fluid away from other hair cells.

When the head is set in motion, a slight lag occurs before the endolymph begins to move. You can demonstrate this kind of lag by quickly picking up a full glass of water with a sweeping motion. Some water lags behind and spills out of the glass. The lag in endolymph movement causes the cupula to move in the opposite direction of the head movement, bends stereocilia, and creates generator potentials. The remainder of the signaling process is the same as in the utricle. Signals arriving in the cerebellum elicit motor signals that maintain balance and coordination during starting, stopping, and turning. These signals and the movements they elicit are called **acceleratory reflexes.** Some signals also go to the cerebral cortex where their interpretation creates awareness of motion.

In addition to detecting specific movements, receptors in the semicircular canals also detect changes in rate of movement. After the initial lag, endolymph flows at the same rate as the head is moving and the cupula is no longer displaced. As long as the rate of movement is constant, the number of signals generated by hair cells in the cristae remains at a minimum. This explains the effects of sudden acceleration and deceleration of an elevator. It also explains why you feel a strong sensation of motion when your car first begins to move, but are barely aware of motion after the car has traveled at a constant speed for a period of time. When the rate or direction of movement changes, hair cells generate many sensory signals and you quickly become aware of a change in movement.

Signals from receptors for both static and dynamic equilibrium are relayed to the cerebellum and the vestibular nuclei. Signals from the eyes and from receptors in joints, tendons, muscles, and skin also are relayed to the vestibular nuclei. These and other signals are used to control muscle contraction and thereby equilibrium and balance.

Disturbances in Balance

Motion sickness, a temporary disturbance in semicircular canal function, results from repetitive changes in the rate and direction of movement, and from conflicting vestibular and visual signals. For example, when riding in a boat, your head position constantly shifts as you try to look at various objects. Fixing your eyes on a single point helps to prevent motion sickness.

Meniere's (men″e-ārz′) **disease,** an accumulation of endolymph caused by irritation or damage to the inner ear, can affect the semicircular canals or the cochlea. Irritation of the semicircular canals causes dizziness and loss of equilibrium. Irritation of the cochlea causes **tinnitus** (tin-i′tus), or ringing in the ears, and can lead to deafness. No permanent cure exists for this disease, but drugs can be used to control the dizziness.

CLINICAL TERMS

ametropia (am-et-ro′pe-ah) any imperfection in the refraction of light in the eye

blepharitis (blef″ar-i′tis) inflammation of the eyelids

detached retina separation of the retina from the underlying pigmented layer, which leads to blindness unless diagnosed early and reattached with a laser

diplopia (dip-lo′pe-ah) seeing double images of single objects

hemianopia (hem″e-an-o′pe-ah) blindness in half the field of vision of one or both eyes

iritis (i-ri′tis) inflammation of the iris

keratitis (ker″at-i′tis) inflammation of the cornea

labyrinthitis (lab″ir-in-thi′tis) inflammation of the labyrinth of the inner ear

mastoiditis (mas″toid-i′tis) inflammation of the antrum and cells of the mastoid process of the temporal bone

nystagmus (nis-tag′mus) an involuntary rapid movement of the eyeball in any direction

presbycusis (pres″bĕ-ku′sis) lessening of hearing acuity with age

ptosis (to′sis) drooping of the upper eyelids from nerve or muscle damage

retinitis pigmentosa (ret″in-i′tis pig″men-to′sah) a disorder caused by the inability of choroid cells to remove debris from sensory cells; eventually results in blindness

retinoblastoma (ret″in-o-blas-to′mah) a tumor derived from retinal cells

trachoma (trak-o′mah) an infectious inflammation of the conjunctiva

vertigo (ver′ti-go) a sensation in which the surroundings appear to revolve around the individual, or the feeling that the individual is revolving in space

ESSAY
ENDORPHINS AND ENKEPHALINS

Opiates have relieved pain—and created addicts—for centuries. During the search for nonaddictive substitutes, the discovery of opiate receptor sites on brain cells raised an interesting question: How can such sites exist if opiates are not natural body substances? The first clue came from beta-lipotropin, a protein made and degraded in the anterior pituitary gland to **endorphins** and **enkephalins** (Figure 15.37). Two enkephalins, three endorphins, and **dynorphin** (dīn-or′fin), a polypeptide that includes an enkephalin, have been identified and shown to bind to opiate receptors. These substances sometimes are referred to as endogenous (from inside the body) opiates. Further confirmation of the actions of endogenous opiates came from work with **naloxone** (nal-oks′ōn), or Narcan, which appears to bind to the same receptors as endogenous opiates. Naloxone is used to treat drug overdoses because it replaces opiates bound to receptors and counteracts their effects.

Endogenous opiates play important roles in analgesia (pain relief). Electrical stimulation of certain regions of brain stem gray matter produces analgesia, possibly because it releases endogenous opiates. Studies in animals provide additional evidence. (1) Morphine and brain stimulation, both at nonanalgesic levels, together produce analgesia. (2) Tolerance for opiates (the need for larger doses to achieve the same effect) develops not only from repetitive doses of morphine but also from repetitive brain stimulation. (3) Naloxone partially antagonizes the effects of stimulation.

The observation that naloxone fails to completely antagonize the effects of stimulation raised the question of whether both opiate and nonopiate analgesic systems exist. The answer seems to be "yes," at least in rats under special conditions, where foot shocks administered at 3-minute intervals produce analgesia not blocked by naloxone. Foot shocks administered by 30-minute intervals under the same conditions produced analgesia that *was* blocked by naloxone.

Conditioning experiments with rats show that they can learn to activate endogenous opiates to inhibit pain perception. When the rats have learned that a certain cue will be followed by a shock, they display analgesia by the time the shock is delivered. Such analgesia can be blocked by naloxone.

Certain limited observations on humans receiving acupuncture and other forms of stimulation seem to suggest that both opiate and nonopiate analgesic systems are present. For example, stimulation near a pain site usually produces analgesia not blocked by naloxone and probably involves a nonopiate system. However, stimulation of sites far from a pain site, such as those used in acupuncture, produce analgesia that can be blocked by naloxone and that probably involve endogenous opiates.

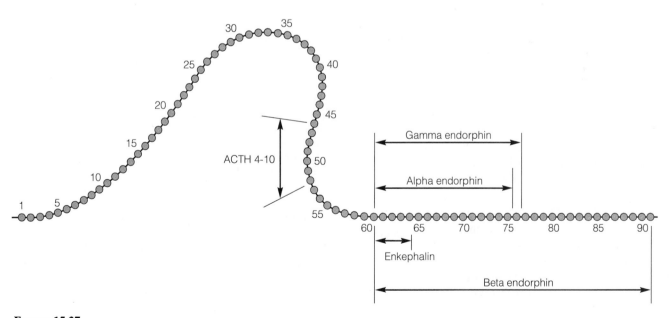

FIGURE 15.37
A diagram of a molecule of beta-lipotropin, showing psychoactive components. Each of the 91 elements of this protein is a specific amino acid.

In studies of the central nervous system peptides, cholecystokinin (CCK), already known as a gastrointestinal hormone, was found in the brain. CCK acts as an opiate antagonist, but it also provides negative feedback to restore normal pain sensitivity after the release of endogenous opiates.

How endorphins and enkephalins produce their effects is unclear, but postsynaptic inhibition (hyperpolarization of postsynaptic neurons) seems not to be involved. One proposed mechanism for enkephalin action is **presynaptic inhibition**: released at a synapse, it binds to presynaptic terminals of adjacent neurons and inhibits their release of an excitatory neurotransmitter.

Efforts to use endorphins and enkephalins in pain therapy so far have met with little success. It may be that they remain active for such a short time that they must be released continuously to maintain analgesia. Or they may be rapidly counteracted by CCK. Various effects of endogenous opiates have been reported (Table 15.5). So, the search for nonaddictive analgesics continues and may eventually lead to pain relief without the risk of addiction.

Questions

1. What events led to the discovery of endorphins and enkephalins?
2. How are opiates thought to relieve pain?
3. Of what clinical value is naloxone?
4. What is presynaptic inhibition?

TABLE 15.5   Effects Reported for Selected Endogenous Opiates

Substance	Reported Effects
Alpha-endorphin	Analgesic and tranquilizing effects in rats
Beta-endorphin	Powerful analgesic effect in mice and rats; in small sample of humans, countering of depression and anxiety, and reduced hallucinations in mental disturbances; being studied for alleviation of pain in terminal cancer and for reducing withdrawal symptoms in opiate addiction
Gamma-endorphin	Increased irritability, violent behavior, and increased sensitivity to pain in rats.
Enkephalins	Appear to be the body's natural analgesic molecules; can cause long-lasting pain relief in rats. May improve memory and induce pleasure in humans. May initiate epilepsy in rats.

CHAPTER SUMMARY

Receptors in General

- Receptors are specialized cells that detect changes in the external or internal environment.
- Signals from receptors reaching the cerebral cortex create conscious sensations; those reaching subconscious levels regulate many body functions.
- Receptors are classified by location as exteroceptors, proprioceptors, and visceroceptors, and by the stimuli they detect as chemoreceptors, mechanoreceptors, thermoreceptors, photoreceptors, and nociceptors.
- Properties of receptors include transduction, detection of stimulus intensity, selective response to one modality of sensation, production of action potentials, and varying degrees of adaptation.

Somatic Senses

- Sensory modalities attributed to skin receptors include pain, touch, pressure, warmth, and cold; signals are relayed to the somatic sensory area of the cerebral cortex.
- Some pain receptors respond to kinins released from injured tissues; referred pain is perceived as coming from a site other than the injury and probably involves convergence of signals.
- Theories of pain control include the gating theory and the endorphin theory.

Chemical Senses

- The chemical senses include taste and smell.
- Properties of these senses are summarized in Table 15.3.

Vision

- When light rays strike a convex lens, they converge; and when they strike a concave lens, they diverge.
- Eye structures that serve protective functions include bony eye sockets, eyelashes, eyelids, and the tear apparatus.
- Eye movements include conjugate, fixation, saccadic, and pursuit movements.
- The eyeball has an outer fibrous tunic made up of the sclera and cornea; a vascular tunic made up of the choroid, ciliary body, and iris; and a nervous tunic, or retina.
- A lens suspended across the hollow eyeball separates the anterior cavity filled with aqueous humor and the posterior cavity filled with vitreous humor.
- The cornea and lens cause light to focus on the retina, the iris regulates the amount of light reaching the retina, aqueous humor nourishes the cornea and lens, and both aqueous and vitreous humors help maintain the shape of the eyeball.
- The retina contains rods that respond to dim light and cones that detect color in bright light; both rods and cones hyperpolarize when stimulated.
- Light energy activates rhodopsin and leads to activation of a chain of enzymes that amplify hyperpolarization of the cell membrane. Dim light activates rhodopsin, but brighter light is required to activate cone pigments that allow color vision. Color is perceived by the brain comparing cone signals to detect hue, saturation, and brightness.
- Axons of ganglion cells carry signals to the lateral geniculate nuclei; visual signals are relayed (possibly as summed waves) and processed sequentially in the spatially organized primary visual cortex and the visual association area.
- Binocular vision allows distance and depth perception.
- Light-adapted eyes have little light-sensitive pigment and lessened sensitivity; dark-adapted eyes have more pigment and increased sensitivity.
- Night blindness results from a deficiency of vitamin A; hence, a deficiency in light-sensitive pigments.
- Defects in refraction include myopia, hypermetropia, astigmatism, presbyopia, and cataracts.
- Color blindness is caused by defective genes that lead to defective cones.

Hearing

- Hearing involves the conduction of sound waves, their conversion to pressure waves, and the response of receptors to mechanical vibrations. Sound waves are air disturbances with regions of compression and rarefaction that vary in frequency and amplitude.
- Pitch is determined by the frequency of vibrations, intensity by the amplitude of vibrations, and timbre by higher frequency vibrations superimposed on a basic pitch.
- The external ear includes the pinna and external auditory meatus. It is protected by hairs and wax.
- The middle ear contains three bones and membrane-covered openings for transmitting vibrations; it also has openings that connect with the pharynx and mastoid cells. It is protected by the attenuation reflex.

- Sound waves are conducted through air to the tympanic membrane, amplified by bones and membranes of the middle ear, and transmitted as pressure waves that deflect the basilar membrane.
- Movement of the basilar membrane in the inner ear excites hair cells in the organ of Corti, creating generator potentials that elicit action potentials.
- Signals are relayed to the medial geniculate nuclei on both sides of the brain, on to the auditory cortex, and finally interpreted in the auditory association area.
- Different pitches are detected by selective excitation of hair cells in certain regions of the organ of Corti; different intensities are detected by the degree of basilar membrane deflection; timbre involves simultaneous stimulation of different areas of the organ of Corti at different intensities.
- Otitis media is a middle ear infection. Deafness can involve blocking of conduction or damage to receptors or nerve fibers.

Balance

- In static equilibrium, a gelatinous mass in the macula of the utricle responds to gravity to detect different head positions; signals from hair cells control positional reflexes and coordinate body movements with the position of the head.
- In dynamic equilibrium, hair cells in semicircular canals detect angular head movements. Signals relayed to the cerebellum help to maintain balance and coordination during starting, stopping, and turning movements via acceleratory reflexes.
- Motion sickness results from disturbances in the semicircular canals. Meniere's disease is due to an accumulation of endolymph in the semicircular canals or cochlea that can produce dizziness, tinnitus, and even deafness.

QUESTIONS AND PROBLEMS

The questions at the end of each chapter are numbered to correspond with the objectives listed at the beginning of the chapter. Italics indicate that a question requires critical thinking skills beyond simple factual recall.

Questions

1. (a) How are receptors classified by location and by modality?
   (b) *What is the difference between sensation and perception?*
   (c) *For any three receptor types, show how each demonstrates all the general properties of a receptor.*
2. (a) What sensory modalities are detected by skin receptors, and how are their signals relayed to the brain?
   (b) What are the properties of pain?
   (c) *How might pain control be achieved?*
3. (a) *What are the similarities and differences between the senses of taste and smell?*
   (b) Why are the senses of taste and smell called chemical senses?

4. Define refraction, focal point, and focal length.
5. (a) What natural mechanisms protect the eyes?
   (b) What different kinds of eye movements are possible, and what functions do they perform?
6. (a) *Trace the pathway of light through the eye, and explain what happens at each interface between two media.*
   (b) *How does the eye accommodate to focus on near objects?*
7. (a) *How does dim light stimulate receptors, and what reactions lead to rods relaying signals?*
   (b) *How are the receptors prepared for restimulation?*
8. *How does vision in bright light differ from vision in dim light?*
9. Trace the path of signals from light receptors to visual areas of the brain.
10. (a) *Using properties of light and lenses, explain how lenses can be used to correct faulty vision.*
    (b) What causes color blindness?
11. Define frequency, amplitude, pitch, intensity, and timbre.
12. (a) *How are sound waves relayed through the ear to receptors?*
    (b) How are the ears protected?
13. (a) *How are the three properties of sound detected?*
    (b) *How is information about sound relayed to the brain, and how is it interpreted?*

14. (a) What is otitis media?
    (b) How do conduction and nerve deafness differ, and how are they treated?
15. (a) *How do static and dynamic equilibrium differ?*
    (b) *How is each maintained?*
16. Describe two disorders of the organs of equilibrium.

Problems

1. What is the focal distance of a lens of $+3$ diopters? Of a lens of $+5$ diopters? What lens would neutralize the power of a lens of -4 diopters?
2. Determine the pitch of sounds with frequencies of 256 and 512 hertz. (You will need an outside reference to do this.) Then, use a piano to determine the range of your own voice in hertz.
3. Suppose you have a mild headache, but get interested in your studies and forget it. How might this happen?
4. Do some library research on (*a*) the advantages of contact lenses for correcting some kinds of visual defects or (*b*) the implantation of electronic devices to allow people suffering from nerve deafness to hear.
5. For some activity such as dancing, diving, or playing tennis or Ping-Pong, use what you have learned about sensory receptors and the nervous system to explain how afferent signals control your movements in that activity.

16

THE ENDOCRINE SYSTEM

ORGANIZATION AND GENERAL FUNCTION

Perhaps you have a friend or relative who must limit dietary sugars because of diabetes. Maybe you have seen an adult with a growth on the neck called a goiter or a child who is not nearly as tall as other children of the same age. You must have heard of teenage boys and young men who are so anxious to develop strong, heavy muscles that they take anabolic steroids.

All these conditions are related to chemical messenger molecules called hormones that circulate continuously in your blood. Diabetes is caused by a lack of activity of the hormone insulin from the pancreas. Goiter is due to defective thyroid gland activity. Some children fail to grow at a normal rate because they lack growth hormone from the pituitary gland.

Though the testes normally release steroid hormones that foster the development of male sexual characteristics including muscular development, some athletes take anabolic (metabolism stimulating) steroids to increase muscular development. Unfortunately, anabolic steroids have numerous adverse effects. Short-term effects include inducing acne and impairing sexual performance. Long-term effects include elevated blood cholesterol and increased risk of heart attack; severe mental disorders; infertility; premature termination of bone growth in adolescence; and, sometimes, liver cancer.

To qualify as a hormone (hor'mōn), a chemical substance must be synthesized and secreted by a cell, travel in the blood, and bind to specific receptors of **target cells,** where it exerts regulatory effects. Not all regulatory substances are hormones. Some are neurotransmitters or metabolic products such as carbon dioxide. As we have seen, neurotransmitters act at synapses. Though they are released from cells and act by binding to receptors, they are not usually considered hormones because they do not travel in the blood. Norepinephrine released from a synapse is a neurotransmitter; secreted by cells of the adrenal medulla, it is a hormone. Carbon dioxide, a waste product of cellular metabolism, travels in the blood and is detected by cells in certain brain centers that regulate breathing. It is not considered a hormone because most cells release it.

Hormone secreting, or endocrine, cells are found in ductless glands that make up the **endocrine system** and in many other tissues. Secretions of endocrine glands (*endo,* within) travel through the body in the blood. In contrast, secretions of exocrine glands (*exo,* outside) like sweat glands and sebaceous glands are released onto the skin surface, and secretions of digestive glands go into the lumen of the digestive tract.

Like the nervous system, the endocrine system is a control system, but it differs significantly from the nervous system. Hormonal signals travel more slowly than neural signals, but they remain available in the blood, continue to bind to receptors, and exert their effects over a longer period of time. The functions of both the endocrine and nervous systems are closely related, and they work together to regulate body functions.

The glands of the endocrine system (Figure 16.1) include the pituitary (hypophysis), thyroid, parathyroid, adrenals, and the islets of Langerhans of the pancreas. The

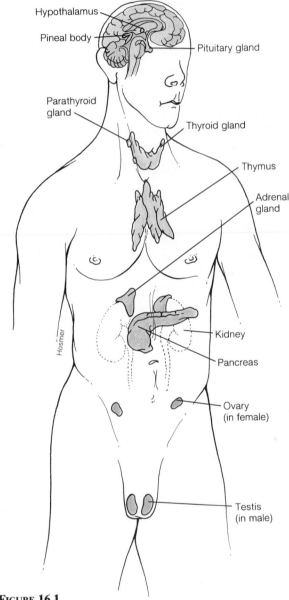

FIGURE 16.1
The glands of the endocrine system are located in various parts of the body. In addition to discrete glands, many organs contain diffuse endocrine structures.

hypothalamus, gonads, thymus, pineal gland, heart, kidneys, and the placenta also have endocrine functions. In addition, the **diffuse endocrine system** (DES) consists of endocrine cells scattered through certain epithelial tissues, such as the digestive tract lining.

Some physiologists consider *cachectin* (kak-ek′tin), or *tumor necrosis* (ne-kro′sis) *factor,* to be a hormone. It is secreted by macrophages, cells of the body's defense system. Cachectin can cause sudden death when large numbers of bacteria are present in the blood. Smaller amounts of cachectin in patients with less virulent infections, injuries, or cancer cause **cachexia** (kak-ek′se-ah)—weight loss, tissue wasting, and loss of appetite.

DEVELOPMENT

Some endocrine glands are derived from each of the embryonic germ layers (Table 16.1). The hypophysis arises from two ectodermal sources (Figure 16.2). The **adenohypophysis** (ad″en-o-hi-pof′is-is), or anterior pituitary

TABLE 16.1   Embryonic Origin of the Endocrine Glands and the Chemical Nature of Their Secretions

Gland	Embryonic Origin	Chemical Nature of Secretion
Adenohypophysis	Ectoderm	Proteins or polypeptides
Neurohypophysis	Ectoderm	Polypeptides
Adrenal medulla	Ectoderm	Amines derived from one amino acid
Thyroid	Endoderm	Modified amino acids
Parathyroids	Endoderm	Polypeptides
Pancreas	Endoderm	Proteins or polypeptides
Adrenal cortex	Mesoderm	Steroids
Gonads	Mesoderm	Steroids

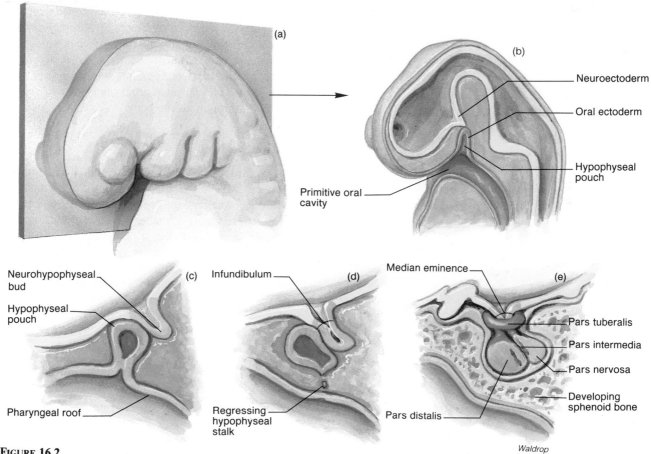

FIGURE 16.2

The pituitary gland is formed from Rathke's pouch, an ectodermal structure in the roof of the mouth and neural ectoderm. Rathke's pouch separates from the mouth and comes to lie adjacent to the neural portion of the pituitary. The anterior and intermediate lobes of the pituitary are derived from Rathke's pouch, and the posterior lobe is derived from neural ectoderm: (*a*) 4 weeks, (*b*) 5 weeks, (*c*) 6 weeks, (*d*) 8 weeks, (*e*) completed development.

gland, develops from ectoderm called **Rathke's** (rahth'kez) **pouch** near the roof of the mouth. The **neurohypophysis** (nu"ro-hi-pof'is-is), or posterior pituitary, develops from neural ectoderm in the floor of the brain. An intermediate lobe of the pituitary gland exists in rudimentary form in humans. It releases melanocyte-stimulating hormone that stimulates pigment production in melanocytes in human skin and iris cells.

The thyroid and adrenal glands also have a dual origin. Thyroid follicular cells are derived from endoderm, whereas parafollicular cells come from neural crest ectoderm. The adrenal medulla is derived from neural crests and the adrenal cortex from mesoderm.

Other glands come from single germ layers. The parathyroid glands and the pancreas are derived from pouches of endoderm in the digestive tract lining. The gonads are derived from mesoderm.

The chemical nature of hormones is related to the embryonic origin of the glands that secrete them (Table 16.1). Ectodermal and endodermal glands secrete hormones made of amino acids, including modified amino acids, polypeptides, and proteins. Mesodermal glands secrete steroids.

Early studies of endocrine gland tissue were made with light microscopes and simple stains. Today, transmission and scanning electron microscopes allow more precise observations and special stains detect hormones within the cells that synthesize them. Hormone effects have been studied by surgically removing glands from animals and subsequently transplanting glands or administering extracts from them to try to restore normal function. Over the years, techniques for purifying and identifying hormones have improved, but measuring hormones in body fluids has been difficult because of their small concentrations—as small as 0.000001 mg/ml. Bioassays and radioimmunoassays now allow such measurements. **Bioassay** (bi"o-as'a) uses a laboratory animal lacking the ability to produce a particular hormone, such as growth hormone. The animal's response to an extract of unknown concentration is compared to its response to a known quantity of the hormone. **Radioimmunoassay** (ra"de-o-im"u-no-as'a) uses a radioactive hormone and antibodies that bind specifically with the hormone. (Antibodies are substances the body makes in response to foreign substances.) Such a system, carefully calibrated, can be used to assay very small quantities of a natural hormone.

PROPERTIES OF HORMONES

Prohormones

Some hormones are synthesized as inactive molecules called **prohormones** and later converted to their active form. For example, insulin is synthesized as **preinsulin** on ribosomes, transported through the endoplasmic reticulum and converted to **proinsulin** by the removal of a small chain of amino acids. Proinsulin is stored in the Golgi apparatus. When proinsulin is secreted, a segment called **C-peptide** is cleaved from it to make active insulin.

Transport of Hormones

Most small hormone molecules are transported by proteins. Specific carrier proteins are available for some hormones, but many are carried by nonspecific globulins or albumins. For example, most thyroid hormone molecules in the blood are carried by *thyroxine-binding globulin,* but these hormones also can be carried by albumin. Hydrocortisone from the adrenal cortex is carried by a specific protein *transcortin* (trans-kor'tin). Progesterone from the ovary or placenta also can be carried on transcortin or on albumin. On entering the blood, these molecules bind loosely to protein. When some hormone molecules leave the blood and move toward cells, more molecules are released from the carriers. This mechanism prevents too rapid a release of a hormone from the blood and helps to distribute it throughout the body.

Negative Feedback

Negative feedback controls the secretion of nearly every hormone. As the blood concentration of a hormone increases above the amount needed, the increased concentration tends to suppress further secretion of the hormone. Then, the blood concentration of the hormone begins to decrease as some of it binds to receptors. As the blood concentration decreases, suppression also decreases, and hormone secretion begins again. This self-regulation of the blood concentration of a hormone is negative feedback.

Hormonal control systems are similar to neural control systems in some respects, but they also have some special properties. Hormonal control systems are compared with neural control systems in Table 16.2.

TABLE 16.2	A Comparison of Neural and Hormonal Control Systems	
Characteristic	**Neural Control Systems**	**Hormonal Control Systems**
Nature of message	Action potential and neurotransmitter	Hormone
Mode of transmission	Conduction on neuron and transmission at synapses	Through blood
Effects	Change membrane potential of affected cells; carry afferent and efferent signals; involved in perception, consciousness, learning, memory, emotion, and control of movements	Change rates of cellular metabolic activities of target cells
Time for effects to occur	Usually less than one second	Usually several minutes
Duration of effects	Milliseconds to seconds (except memory storage)	Minutes to hours, sometimes even days

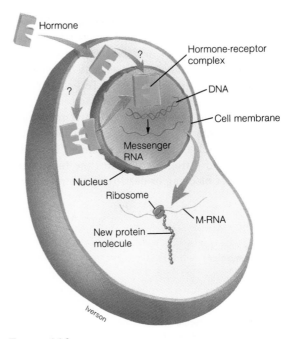

FIGURE 16.3

Steps in the action of steroid and thyroid hormones: The hormone diffuses through the cell membrane and across the cytoplasm to the nuclear membrane. It binds to a receptor on the membrane or in the nucleus and activates a gene. Gene activation causes transcription of messenger RNA and synthesis of a protein.

MECHANISMS OF HORMONE ACTION

Though endocrine glands can release hormones spontaneously, certain stimuli—blood concentrations of some substances, neural signals, or hypothalamic releasing or inhibiting hormones—can increase or decrease the rate of secretion. Once in the blood, a hormone can reach almost any cell in the body; however, each hormone affects only its target cells, the cells that have receptors for it. Nearly all cells have receptors for thyroid and growth hormones but only intestinal epithelium, bone, and certain kidney cells have receptors for parathyroid hormone.

Hormones exert their effects at the cellular level by changing the rate of certain cellular processes. For example, the pancreatic hormone insulin increases the rate at which glucose enters cells (except those of the liver and the brain). How hormones produce their physiological effects is not fully understood, but it is possible to describe, in a general way, some mechanisms by which hormones act.

Steroid and Thyroid Hormone Actions

Steroids, being relatively small molecules synthesized from cholesterol, are lipid soluble and easily pass through cell membranes. They diffuse into all cells and apparently diffuse out of nontarget cells without having any effect. In target cells, steroids bind to protein receptors. Some researchers believe the receptors are located on the nuclear membrane, and others think they are inside the nucleus, perhaps associated with DNA or are segments of DNA itself. The mechanism of action of thyroid hormone and vitamin D now appears to be the same as that of steroid hormones. All bind to intracellular receptors and exert their effects by activating or inactivating a segment of DNA usually 50 to 100 genes long. Activated DNA directs mRNA transcription and the mRNA moves to the ribosome where it directs protein synthesis (Chapter 3). We will call this the **protein synthesis mechanism** of hormone action (Figure 16.3). Hormones that act by this mechanism show their effects relatively slowly because of the time required for protein synthesis.

This understanding of steroid hormone action allows a new interpretation of some endocrine abnormalities. Instead of being due to excesses or deficiencies of the hormones themselves, some abnormalities may be due to variations in the activity of the receptor. Testicular feminization, in which male secondary sex characteristics fail to develop, may be due to deficient or defective receptors for the male sex hormones. Similarly, cells of hormone-dependent tumors, such as certain breast tumors, have estrogen receptors, and estrogen stimulates the tumors to grow. If a means of blocking or destroying the receptors could be found, the tumor would lack stimulation and cease to grow. Someday, contraceptives that act by blocking receptors for reproductive hormones may be available.

Protein and Polypeptide Hormone Actions

Many protein and polypeptide hormones, such as those from the pituitary and parathyroid glands, and glucagon and insulin from the pancreas act by the cyclic adenosine monophosphate mechanism, or the **cAMP mechanism** (Figure 16.4). In the cAMP mechanism, a hormone (the first messenger) binds to a receptor on the outer surface

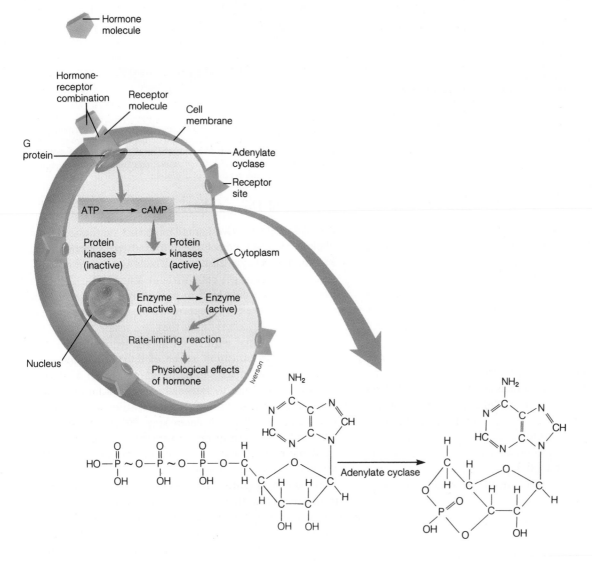

FIGURE 16.4

Steps in the action of some polypeptide hormones: The hormone binds to a receptor on the cell membrane and thereby activates the enzyme adenylate cyclase. Prostaglandins sometimes amplify the activity of adenylate cyclase. The second messenger cAMP is formed; cAMP acts directly or indirectly by activating a kinase to produce the physiological effects of the hormone.

of the cell membrane. Each receptor has an associated G protein. Some G proteins are excitatory and others are inhibitory. An excitatory G protein activates the enzyme **adenylate cyclase** (ad-en'il-āt si'klās) in the cell membrane. An inhibitory G protein inhibits it. Active adenylate cyclase converts ATP to cAMP, and cAMP acts as a **second messenger.**

One role of the second messenger cAMP can be to activate an enzyme called a protein kinase. **Protein kinase** (ki'nās), using phosphate and energy from ATP, phosphorylates another enzyme and, thereby, activates it. When an enzyme is activated, it catalyzes a specific reaction—often a rate-limiting reaction in a metabolic pathway. If the rate-limiting reaction occurs, all the reactions in the pathway occur and these reactions constitute the physiological effects of the hormone. Many effects of hormones (and some neurotransmitters) are mediated via enzyme phosphorylation.

In certain cell membranes, cyclic nucleotide *cGMP* (cyclic guanosine monophosphate) is produced from GTP (guanosine triphosphate) by the enzyme *guanylate* (gwan'il-āt) *cyclase*. The cGMP functions like cAMP in hormone action. Atrial natriuretic hormone from the heart acts by the cGMP mechanism.

An enzyme called **hormone-sensitive lipase** is activated by phosphorylation in response to any of several hormones including epinephrine, glucagon, thyroid hormones, and the pituitary hormone ACTH. It is inhibited by insulin. Activation of hormone-sensitive lipase occurs as follows:

1. The hormone binding to a receptor activates adenylate cyclase.
2. The messenger cAMP is released and activates a protein kinase.
3. The protein kinase phosphorylates (activates) the hormone-sensitive lipase.
4. The lipase initiates the breakdown of fats in adipose tissues.

Some hormone actions accelerate processes and some inhibit them. Such actions constitute a complex set of on–off switches that regulate cell function. Cells respond to different hormones according to the kinds of receptors they have. Because specific events follow the binding of a hormone to its receptor, each hormone produces particular physiological effects.

In some instances, neurotransmitters have actions similar to hormones. For example, both the hormone glucagon and the neurotransmitter norepinephrine can trigger a series of chemical reactions that degrade glycogen. The fact that both hormone and neurotransmitter effects can be mediated through a similar mechanism illustrates the close relationship between the body's two major control systems.

Calcium and Calmodulin

Like neural signals and hormones, calcium ions play a significant regulatory role in a wide array of physiological processes, such as blood clotting, bone formation, movements within cells, release of neurotransmitters, and muscle contraction and relaxation. Calcium is now recognized as an important second messenger in cells. It binds to *calcium-binding proteins*—proteins similar to hormone receptors located in the cytosol.

The calcium-binding protein **calmodulin** (kal-mod'u-lin) binds or releases calcium ions, depending on conditions within the cell. In unstimulated mammalian cells, the concentration of free Ca^{2+} in the cytosol is exceedingly small—10^{-7} moles per liter. A neural or hormonal stimulus to a cell can cause an influx of Ca^{2+} into the cytosol from outside the cell or from within the endoplasmic reticulum, raising the concentration to 10^{-6} moles per liter or higher. The increase in the Ca^{2+} concentration causes formation of a Ca^{2+}-calmodulin complex. The Ca^{2+}-calmodulin complex can activate enzymes such as adenylate cyclase or phosphorylase kinase. Other calcium-binding proteins include troponin in skeletal and cardiac muscle, and a vitamin D-dependent calcium-binding protein that binds Ca^{2+} from food in the small intestine and facilitates its absorption.

Recent studies suggest that mechanisms involving calcium and those involving cAMP often work together to create a complex cellular-regulating system. Ca^{2+} regulates enzymes that synthesize and degrade cAMP, and cAMP regulates the movement of Ca^{2+} into and out of cells. Both cAMP and Ca^{2+} control intracellular events by controlling protein kinases.

Properties of Receptors

Receptors are as necessary as hormones for hormone actions, so the number of receptors affects hormone activity. Cells can react to high-hormone concentrations by reducing their number of active receptors. How receptors become inactive is not fully understood, but it appears that membrane receptors can sink below the surface and that both membrane and cellular receptors can change their molecular shape.

Some hormones affect not only the number of their own receptors, but also those for other hormones. For example, excess thyroid hormones increases epinephrine receptors on heart cells in proportion to the degree of excess thyroid hormones. The effect is to increase the heart rates, presumably because greater numbers of epinephrine receptors are stimulated. When hyperthyroid patients receive epinephrine their heart rates increase more than those of normal individuals receiving the same amount of epinephrine.

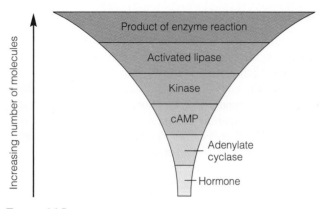

Increasing number of molecules

Product of enzyme reaction

Activated lipase

Kinase

cAMP

Adenylate cyclase

Hormone

FIGURE 16.5
Many hormones cause a cascade effect called enzyme amplification, where the number of active molecules present at each step is much larger than in the previous step.

Enzyme Amplification

The ability of a very small amount of a hormone to exert a significant effect is explained by **enzyme amplification** (Figure 16.5), a cascade effect which occurs in the cAMP mechanism as follows: A single hormone molecule binding to a single receptor site activates several molecules of adenylate cyclase. Each adenylate cyclase molecule makes many molecules of cAMP, and each molecule of cAMP activates many molecules of the appropriate protein kinase. Each protein kinase molecule, in turn, activates many molecules of a particular enzyme, such as hormone-sensitive lipase. Each lipase molecule breaks down many lipid molecules.

Steroid hormones display similar amplification as one molecule of messenger RNA directs the synthesis of many molecules of a particular protein (usually an enzyme), and each enzyme molecule produces many product molecules.

Inactivation of Hormones

To serve as regulators, hormones, like neurotransmitters, must be inactivated once they have exerted their effects. Most hormones are inactivated or degraded by liver enzymes and excreted. Thyroid hormones are excreted in bile, whereas steroid hormones are excreted in urine. Certain end products of steroid hormones appear in urine in amounts proportional to the amount secreted; thus, end-product concentrations in urine often can be used clinically to estimate the amount of the hormone being secreted. Such tests are used to identify athletes who are using anabolic steroids. Epinephrine and norepinephrine (whether secreted as a hormone or a neurotransmitter) are degraded by the enzymes monoamine oxidase and catechol-o-methyltransferase. End products of the degradation are excreted in the urine, where they, too, can be measured. Peptide and protein hormones are inactivated by proteolytic enzymes. The protein insulin cannot be given by mouth because enzymes digest it before it can be absorbed into the blood.

PROSTAGLANDINS

Discovered in the 1930s and named for a presumed association with the prostate gland, **prostaglandins** (pros"tah-glan'dinz) are now known to be important chemical regulators of cellular function. They are derived from arachidonic acid (a 20-carbon unsaturated fatty acid) and contain a 5-carbon ring and two side chains (Figure 16.6). Prostaglandins are synthesized and secreted as needed. They act by binding to extracellular receptors either near the cells that made them or after traveling in the blood. They are extremely potent—some cause smooth muscle fibers to contract at a concentration of only 1 nanogram per milliliter.

The physiological effects of prostaglandins are extremely varied and can be antagonistic. Some augment the action of other regulatory molecules. Much remains to be learned about prostaglandin actions but when a hormone alters the cAMP concentration, prostaglandins augment that effect. When insulin increases cAMP and causes cells to take up glucose, prostaglandins further increase the cAMP concentration. In fat cells where insulin decreases the concentration of cAMP and prevents the release of fatty acids, prostaglandins further decrease cAMP. Some prostaglandins cause fever during infections, and aspirin reduces fever by inhibiting prostaglandins.

Two groups of prostaglandins (PGs), designated by their chemical structure as PGEs and PGFs, usually have antagonistic actions. For example, PGEs dilate blood vessels and bronchial passages, whereas PGFs constrict them; thus, the balance between these two types of prostaglandins helps to maintain homeostasis in certain physiological processes. Both PGEs and PGFs, however, cause uterine smooth muscle to contract, and both have been used to induce abortions. In fact, during the first six months of pregnancy, the uterus is more sensitive to prostaglandins than to oxytocin (a hormone that stimulates uterine contraction during labor). Though better agents are available to induce abortion during the first trimester (one-third of the term, or three months) of pregnancy, PGEs and PGFs are used during the second trimester. PGEs also increase blood vessel permeability, inhibit aggregation of blood platelets, and act as sedatives.

PITUITARY GLAND

A small structure lying beneath the hypothalamus, the **pituitary gland,** or hypophysis (Figure 16.7), consists mainly of the **anterior pituitary** (adenohypophysis) and the **posterior pituitary** (neurohypophysis). Both are regulated by the hypothalamus, which provides anatomical and physiological connections between the nervous and endocrine systems.

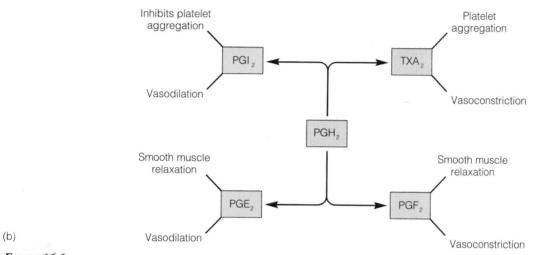

(a) Prostaglandin E₁(PGE₁) Prostaglandin E₂(PGE₂)

(b)

FIGURE 16.6
(*a*) In naming prostaglandins (PG), a letter such as E indicates the kind of ring it contains, and the subscript indicates the kind of side chains it has. (*b*) Some actions of prostaglandins.

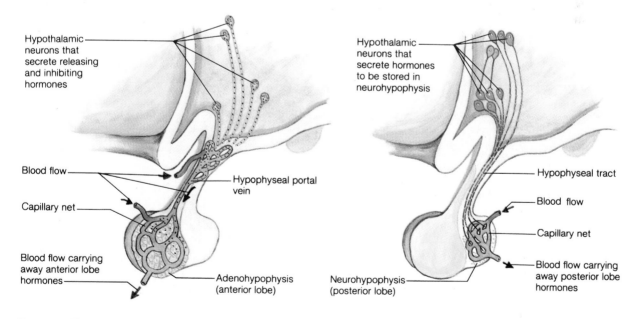

FIGURE 16.7
The pituitary gland (lateral view) includes the adenohypophysis (anterior pituitary) and the neurohypophysis (posterior pituitary).

The adenohypophysis is connected with the hypothalamus via the **hypophyseal portal system,** venules which carry blood between capillaries of the hypothalamus and those of the adenohypophysis. Releasing and inhibiting hormones from the hypothalamus travel through these vessels and regulate adenohypophysis function.

The neurohypophysis and hypothalamus are connected by the **hypophyseal tract,** a nerve tract of axons whose cells bodies are in the hypothalamus. These cells synthesize hormones that travel down the axons and are stored in the neurohypophysis, making the neurohypophysis a storage area rather than a gland.

Hormones of the Adenohypophysis

Several hormones of significance in humans are synthesized and secreted from the adenohypophysis (Figure 16.8). They are growth hormone, prolactin, follicle-stimulating hormone (FSH), luteinizing hormone (LH), thyroid-stimulating hormone (TSH), and adrenocorticotropic hormone (ACTH). Beta-lipotropin, whose action is less clearly understood, is synthesized and released from the adenohypophysis. Also discussed here is melanocyte-stimulating hormone (MSH) from the poorly developed intermediate lobe of the pituitary gland. Among these

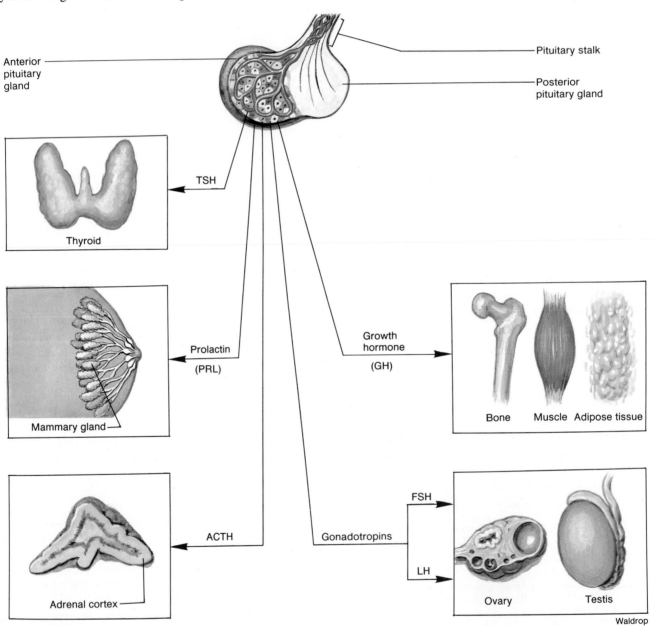

Waldrop

FIGURE 16.8
The hormones of the adenohypophysis and their effects.

hormones, all of which are polypeptides, only growth hormone has a general effect on nearly all cells. Several are **tropic** (tro'pik) **hormones**—hormones that act on other glands. They include the gonadotropins, FSH and LH, and the chemically related TSH. Tropic ACTH is synthesized along with beta-lipotropin and MSH as a single polypeptide.

In general, a tropic hormone from gland A stimulates gland B to release its secretion. The hormone from gland B acts on its target cells, and it also acts by negative feedback to inhibit hormone secretion from gland A. This mechanism serves to maintain the concentration of the hormone from gland B fluctuating within a narrow range around its set point (Figure 16.9).

Growth Hormone

Somatotropin (so''mat-o-tro'pin), also called **growth hormone** (GH), acts directly on some cells to stimulate growth, and acts indirectly on others to release proteins, such as insulin-like growth factor I. Growth hormone also causes cartilage and bone cells to divide and osteoblasts to secrete collagen. It promotes Ca^{2+} absorption from the intestine. Thyroid hormones and insulin work with growth hormone to promote collagen synthesis.

Growth hormone and growth factors are essential to stimulate growth to adult size and to maintain adult size thereafter. Together, they increase the rate at which cells take up amino acids and use them to synthesize proteins. They also stimulate free fatty acid release from fat cells and glycogen breakdown in the liver, which releases glucose into the blood. These functions provide adequate nutrients between meals and during stress.

Much bone and muscle loss associated with reduced strength in aging may be due to declining growth hormone secretion after age 50. At age 30, the body mass is about 10 percent bone, 30 percent muscle, and 20 percent fatty tissue. By age 75, it is about 8 percent bone, 15 percent muscle, and 40 percent fatty tissue.

Growth hormone has been synthesized in the laboratory using recombinant DNA. The human gene for synthesis of the hormone is combined with bacterial DNA and the bacterial cells then produce human growth hormone, which can be isolated from other metabolic products and purified for therapeutic use in humans.

Researchers have long been puzzled by the observation that African pygmies and some American children who fail to grow at a normal rate have normal amounts of growth hormone in their blood. Recent findings indicate that a genetic defect that produces deficient or defective receptors for growth hormone may account for the inability of growth hormone to stimulate growth.

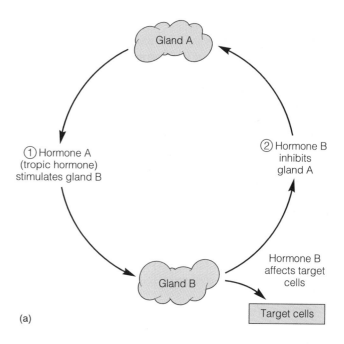

(a)

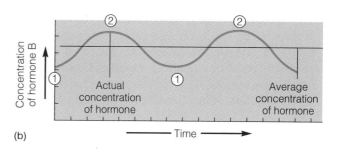

(b)

FIGURE 16.9
(a) Negative feedback from a tropic gland to the gland that produces the tropic hormone. (b) Fluctuation of the blood concentration of a hormone around its set point.

Prolactin

The protein hormone **prolactin** (pro-lak'tin), or lactogenic hormone, stimulates milk secretion in mammary glands previously prepared for milk production by other hormones—estradiol, progesterone, corticosteroids, and insulin (Chapter 29).

Follicle-Stimulating Hormone

The glycoprotein **follicle-stimulating hormone** (FSH) stimulates the maturation of an ovum (egg) each month during a female's reproductive years. It also stimulates maturation of sperm in males. When FSH is released in females, it initiates maturation of 10 to 20 groups of cells called follicles. One of these follicles dominates the others, matures, and produces estrogen hormones. The estrogens act by negative feedback to suppress further release of FSH (Chapter 29).

Luteinizing Hormone

Another glycoprotein hormone involved in reproduction is **luteinizing** (loo'te-in-īz''ing) **hormone** (LH). In females, it stimulates ovulation (release of an ovum) and causes follicular cells to produce progesterone, which among other things stimulates mammary gland development. In males, LH stimulates development of interstitial cells (Leydig's cells) of the testes and causes them to produce testosterone (Chapter 29).

Thyroid-Stimulating Hormone

The glycoprotein called **thyroid-stimulating hormone** (TSH), or thyrotropin, stimulates the thyroid gland to synthesize and secrete its hormones, triiodothyronine (T_3) and thyroxine (T_4). When these hormones reach sufficient concentration in the blood, they act by negative feedback to suppress further release of TSH.

Adrenocorticotropic Hormone

Adrenocorticotropic (ad-re''no-kor''te-ko-trop'ik) **hormone** (ACTH), or corticotropin, is a polypeptide that acts on the cortex of the adrenal gland to regulate synthesis and secretion of several of its hormones, especially the glucocorticoids. Glucocorticoids, in turn, stimulate the release of fatty acids and glucose into the blood and help the body to resist stress and inflammation. One of the glucocorticoids, cortisol, is especially important in regulating ACTH secretion. When the adrenal cortex has been sufficiently stimulated by ACTH, cortisol acts, by negative feedback, to suppress further release of ACTH. The role of ACTH in stress is discussed in this chapter's essay.

Beta-Lipotropin

Although not proven to have a specific effect in humans, **beta-lipotropin** (lip''o-trop'in) has been isolated from the anterior pituitary glands of various animals. Endorphins and enkephalins can be made from beta-lipotropin in the anterior pituitary and in the brain directly (Chapter 15). More research is needed to understand when and why the brain and the anterior pituitary synthesize and secrete these substances.

Melanocyte-Stimulating Hormone

The intermediate lobe of the pituitary gland, a rudimentary structure in humans, secretes very small quantities of **melanocyte-stimulating** (mel'an'o-sīt) **hormone** (MSH). MSH is identical to the first 13 amino acids of ACTH, and MSH effects in humans are probably masked by the greater effect of ACTH. Patients with Cushing's disease, a disorder caused by excess ACTH, sometimes develop excessive skin pigment. This hyperpigmentation is due to an excess of MSH produced along with the excess ACTH.

TABLE 16.3 Hypothalamic Hormones that Regulate Secretion of Hormones from the Adenohypophysis

Hormone	Abbreviation	Function
Thyrotropin-releasing hormone	TRH	Stimulates release of TSH
Corticotropin-releasing hormone	CRH	Stimulates release of ACTH
Gonadotropin-releasing hormone	GnRH	Stimulates release of FSH and LH, depending on concentration
Growth-hormone-releasing hormone	GHRH	Stimulates release of GH
Growth-hormone-inhibiting hormone (somatostatin)	GHIH	Inhibits release of GH
Prolactin-releasing hormone	PRH	Stimulates release of prolactin
Prolactin-inhibiting hormone	PIH	Inhibits release of prolactin

Regulation of the Adenohypophyseal Hormones

Several adenohypophyseal hormones are regulated by small peptide hormones from neurosecretory cells of the hypothalamus. They are released into the hypothalamic capillaries and travel through the hypophyseal portal vein to the adenohypophysis. Each regulatory hormone stimulates or inhibits release of an anterior pituitary hormone (Table 16.3). **Somatostatin** (so''mat-o-stat'in), or growth-hormone-inhibiting hormone (GHIH), also stimulates the production of glucose from noncarbohydrate sources when blood sugar falls below the normal concentration.

Hormonal regulation operates at several levels. Hypothalamic hormones regulate hormones of the anterior pituitary. Tropic hormones from the anterior pituitary regulate other glands such as the thyroid, adrenal cortex, and gonads. Hormones from these glands act by negative feedback to inhibit the release of both tropic and hypothalamic hormones (Figure 16.10). Complex relationships between hormones of the gonads, adenohypophysis, and hypothalamus help to regulate reproductive functions (Chapter 29). Neural signals elicited by sexual arousal, stress, anxiety, trauma, variations in the light-dark cycle, and the sucking of a breast-fed infant, also regulate hypothalamic hormones.

Certain hypothalamic hormones such as TRH appear to have wide distribution and multiple functions. TRH was characterized as a tripeptide in 1969, and a few years later it was shown to stimulate the release of TSH from the anterior pituitary by activating adenylate cyclase. Since then, it has been shown to release certain other pituitary

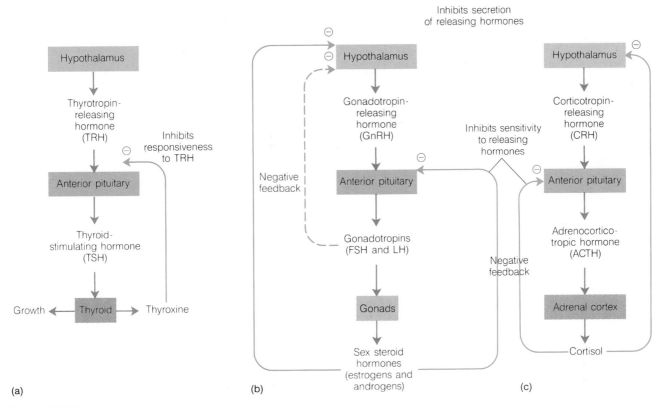

FIGURE 16.10
Examples of negative feedback in the regulation of function of the anterior pituitary gland and glands affected by tropic hormones: (a) thyroid, (b) gonads, and (c) adrenal cortex.

hormones, such as prolactin and in psychiatric and metabolic disorders, growth hormone as well. In addition to the hypothalamus, it has been found in other parts of the brain—hippocampus, amygdala, pons, medulla, and cerebral cortex. TRH is present in high concentrations in brain tissue in Huntington's chorea, schizophrenia, and sometimes in Alzheimer's disease and depression. When administered experimentally to animals, TRH elicits many effects like those resulting from stimulation of cholinergic receptors (Chapter 14). Whether TRH stimulates cholinergic neurons or cholinergic receptors directly is not known. TRH also has been found outside the brain—in the digestive tract, placenta, and prostate gland—but its functions in these sites are unknown.

Hormones of the Neurohypophysis

Two chemically similar peptide hormones, oxytocin and antidiuretic hormone (Figure 16.11), are called **neurosecretions** because they are synthesized in hypothalamic neurons and are stored in the neurohypophysis (refer back to Figure 16.7). When action potentials cause their release from axons, they enter the blood and act as hormones.

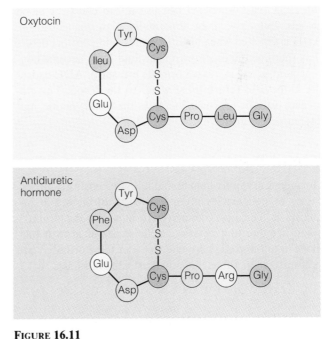

FIGURE 16.11
The amino acid composition of polypeptide hormones oxytocin and antidiuretic hormone is quite similar.

Oxytocin

The hormone **oxytocin** (ok″se-to′sin) stimulates contraction of smooth muscle in the uterus and the contractile cells around mammary gland ducts. Distention of the cervix and vagina or stimulation of the nipple cause oxytocin release; thus the hormone is released in large quantities during sexual intercourse, labor, and lactation. Oxytocin can be used to induce labor in near-term pregnancy, but it is not known whether oxytocin plays a significant role in the natural onset of labor.

Oxytocin release during lactation is reasonably well understood. During breast-feeding, an infant's sucking of the nipple sends neural signals, which are relayed to the neurohypophysis and cause oxytocin release. Within 30 to 60 seconds, oxytocin causes smooth muscle contraction in mammary gland ducts and milk is ejected. The release of milk by the sucking of an infant is an example of neural-hormone **positive feedback** (Figure 16.12). Positive feedback, which empties body cavities but is otherwise unusual in a living organism, accelerates a process. For control by positive feedback to be effective, there must be a mechanism outside the feedback loop to stop the process, otherwise the process would accelerate indefinitely. In this example, the infant's sucking accelerates milk flow by positive feedback and milk flow stops when the sucking ceases or the breast is empty.

Antidiuretic Hormone

Antidiuretic (an″tĭ-di″u-ret′ik) **hormone** (ADH) prevents excess water loss in urine. This hormone is also called **vasopressin** (vas″o-pres′in) because it can constrict blood vessels, but only in an abnormally high concentration. ADH is secreted when hypothalamic osmoreceptors detect that the blood and cerebrospinal fluid have become concentrated (exert increased osmotic pressure). ADH causes the kidneys to return more water to the blood. This decreases osmotic pressure, increases the blood volume, and decreases the urine volume (Chapter 27).

Pituitary Disorders

Hormones normally help to maintain homeostasis, but excesses or deficiencies often lead to disturbances in homeostasis. The most common disturbances in the pituitary gland are due to excesses or deficiencies of growth hormone or antidiuretic hormone. Some of these disorders are summarized in Table 16.4 and illustrated in Figure 16.13.

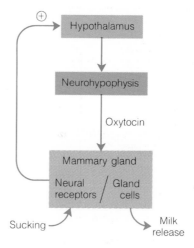

FIGURE 16.12

A neural-hormonal positive feedback loop. The sucking of an infant stimulates receptors in the nipple of the mammary gland. Neural signals travel to the hypothalamus, where they stimulate the release of oxytocin from the neurohypophysis. Oxytocin travels in the blood to the mammary gland and stimulates the release of milk.

Medical uses of ADH illustrate some important aspects of **hormone therapy.** The natural range of blood concentrations of a hormone is referred to as its **physiological level,** but hormones sometimes are given in amounts that raise blood concentrations above the physiological level. The higher concentration is referred to as the **pharmacologic level.** Vasopressin at a pharmacologic level will cause rapid and extreme vasoconstriction. More often, hormone therapy supplies a hormone to overcome a deficiency and maintain a physiological level. This is called **replacement therapy.** For example, diabetes insipidus is due to a deficiency of ADH and results in excessive excretion of dilute urine. ADH administered at replacement levels controls the symptoms of the disease without unduly lowering the blood osmotic pressure or unduly constricting blood vessels. This hormone is given by injection or nasal spray because it would be digested if given by mouth.

GHRH and GH, both of which can now be made in the laboratory, are used to induce growth in children of small stature for their age. Not all children respond to such treatment. It appears that those who have a hypothalamic defect and those with "sluggish" pituitary glands respond as the GHRH is replaced. Those whose anterior pituitary cannot produce growth hormone even when GHRH is given often respond to GH directly.

TABLE 16.4   Some Pituitary Disorders

Disorder	Possible Cause	Hormone	Excess or Deficiency	Effects
Pituitary dwarfism	Destruction or congenital deficiency of GH-producing cells	Growth hormone	Deficiency	Small, but well-proportioned body; sexual immaturity
Giantism	Pituitary tumor before adult size is reached	Growth hormone	Excess	Large, well-proportioned body
Acromegaly	Pituitary tumor after adult size is reached	Growth hormone	Excess	Disproportionate increase in thickness of bones of face, hands, and feet
Panhypopituitarism	Tumor or thrombus	All	Deficiency	Depressed thyroid, adrenocortical, and gonad function
Diabetes insipidus	Damage to the hypothalamus	ADH	Deficiency	Excessive excretion of dilute urine
High ADH blood level*	Excessive stimulation of ADH-secreting neurons, or pituitary tumor	ADH	Excess	Excessively dilute blood and low sodium concentration in plasma

*Also known as SIADH (syndrome of inappropriate ADH secretion).

FIGURE 16.13
Pituitary disorders.

THYROID GLAND

Located inferior to the thyroid cartilage of the larynx, the **thyroid gland** consists of left and right lobes connected by an isthmus. The gland contains follicles of cuboidal epithelium filled with colloid (see Figures 2.34 and 16.14). **Colloid** consists of a protein thyroglobulin into which thyroid hormone molecules have been incorporated. Both thyroglobulin and hormone molecules are made by epithelial cells. The thyroid gland can store enough hormone in the colloid to supply the body for about two months. Some of the cells around the follicles, the parafollicular C-cells, produce the hormone calcitonin.

Synthesis, Storage, and Release of Hormones

Thyroid hormones are amino acid derivatives synthesized by iodination of the amino acid tyrosine (Figure 16.15). The thyroid gland removes iodide ions from the blood and converts them to molecular iodine. It then replaces certain hydrogen atoms of tyrosine molecules with iodine. Precursor molecules called monoiodotyrosine (MIT) and diiodotyrosine (DIT) are synthesized first, and then the two precursor molecules are coupled to make the active thyroid hormones *triiodothyronine* (tri″i-o″do-thi′ron-ēn), or T_3, and *thyroxine* (T_4). Completed hormones are stored attached to thyroglobulin within follicles.

When the gland is stimulated to secrete hormones, T_3 and T_4 are separated from the thyroglobulin and released into the blood where they are carried by a thyroxine-binding globulin or an albumin. Far more T_4 than T_3 is released, and T_4 binds more tightly to the blood proteins; however, T_3 has much greater hormone activity and most of the T_4 is converted to T_3 by the endoplasmic reticulum of liver, kidney, and other cells. Nearly all cells are target cells for thyroid hormones. In cells that convert T_4 to T_3, some T_3 goes to the nucleus and some is released into the blood plasma and transported to other cells.

Thyroid hormones increase oxygen consumption and metabolism in most tissues, and they stimulate protein synthesis. They accelerate oxygen consumption and metabolism by increasing the concentration and activity of mitochondrial enzymes. The higher metabolic rate produces more ATP for use by the Na^+–K^+ pump and other cellular activities. The presence of T_3 in the nucleus of a cell and binding of T_3 to nuclear receptors initiates protein synthesis in a manner similar to that of steroid hormones.

The primary regulator of thyroid hormones is TSH (thyroid-stimulating hormone) from the anterior pituitary. When blood hormone concentrations decrease, TSH is released and travels to the thyroid gland where it stimulates synthesis and release of the hormones. As blood

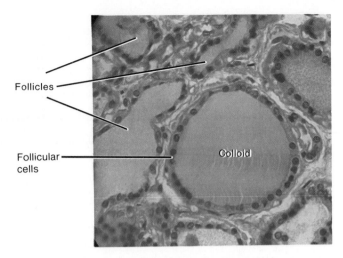

FIGURE 16.14
Microscopic anatomy of the thyroid gland. A scan of thyroid gland 24 hours after ingestion of radioactive iodine was shown in Figure 2.34.

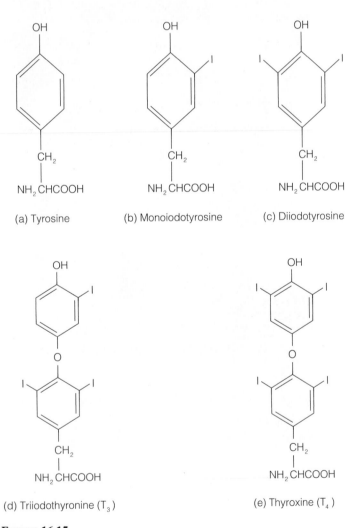

(a) Tyrosine (b) Monoiodotyrosine (c) Diiodotyrosine

(d) Triiodothyronine (T_3) (e) Thyroxine (T_4)

FIGURE 16.15
Thyroid hormones and their precursors: (*a*) tyrosine, (*b*) monoiodotyrosine, (*c*) diiodotyrosine, (*d*) triiodothyronine (T_3), and (*e*) thyroxine (tetraiodothyronine or T_4).

concentrations of the hormones increase, they suppress TSH by negative feedback (refer back to Figure 16.10). TRH can stimulate TSH release, and it can be inhibited by thyroid hormones. TRH is most effective in adapting the body to the environment. Living in a cold environment stimulates TRH release and living in a warm climate suppresses it. In pregnancy, a placental TSH-like hormone increases thyroid hormone release and accelerates the metabolic rate sufficiently to maintain growth and development of the fetus.

Calcitonin (kal″sit-o′nin) is secreted from the C-cells of the thyroid gland when the blood calcium concentration is high. Its actions are indirect in that it blocks the actions of parathyroid hormone, which is discussed in succeeding paragraphs.

Thyroid Disorders

Disorders of the thyroid gland (Table 16.5 and Figure 16.16) typically affect multiple organ systems because the metabolic rate of all cells is affected. Several thyroid function tests are available to diagnose these disorders. A very reliable test measures concentrations of T_3 and T_4 in the blood. Another uses radioactive iodine and measures thyroid gland activity. The more active the gland, the more radioactive iodine it incorporates and, thus, the higher the radioactivity of the gland. Most thyroid disorders are treatable by replacing deficient hormones or iodine or by using drugs, surgery, or radioactive substances to reduce hormone production. The same radioactive iodine used in diagnosis can be used in higher doses to destroy thyroid gland cells and reduce overproduction of the hormones.

TABLE 16.5 Some Thyroid Disorders

Disorder	Possible Cause	Excess or Deficiency	Effects
Cretinism*	Absence of thyroid function in infant	Deficiency	Mental retardation, impaired growth, low body temperature, and abnormal bone formation
Myxedema	Decreased thyroid function in adult	Deficiency	Tissue edema, mental and physical sluggishness, and weight gain
Simple or endemic goiter*	Deficiency of iodine in diet	Deficiency	Enlargement of thyroid gland because lack of hormone allows TSH to be released
Graves' disease*	Autoimmune reaction with antibodies to thyroid tissue produced	Excess†	Weight loss, slightly elevated temperature, excitability, exophthalmia (protrusion of eyeballs because of fluid accumulation behind them), and diffuse goiter because of many widely dispersed centers of excess hormone synthesis throughout the gland

*See Figure 16.16.

†Excess is created by antibody called long-acting-thyroid-stimulator (LATS) that binds to TSH receptors and causes prolonged stimulation.

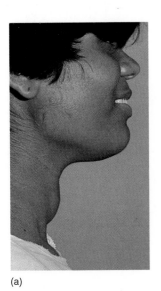
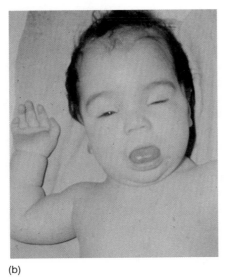
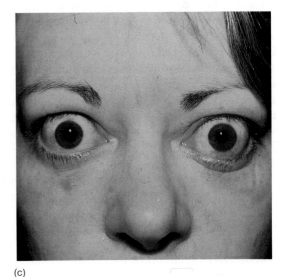

(a)   (b)   (c)

FIGURE 16.16
Thyroid disorders: (a) goiter, (b) cretinism, and (c) Graves' disease.

Parathyroid Glands

The **parathyroid** (par″ah-thi′roid) **glands** consist of four groups of cells encapsulated with connective tissue—two on the posterior surface of each lobe of the thyroid gland (Figure 16.17). Epithelial cells within the gland synthesize and secrete **parathormone** (par″ah-thor′mōn), or parathyroid hormone (PTH), a large polypeptide that helps to regulate the metabolism of calcium and certain other minerals.

With respect to calcium metabolism, PTH and calcitonin have antagonistic effects. Their actions, along with those of vitamin D, maintain calcium homeostasis (see Figure 6.7). When blood calcium decreases, the parathyroids release PTH. It increases calcium absorption from the intestine by activating vitamin D, which stimulates synthesis of a calcium carrier protein in the small intestine. PTH also decreases calcium excretion by the kidneys and can remove calcium from bones if necessary to maintain a normal blood calcium concentration.

When blood calcium increases, calcium acts by negative feedback and suppresses PTH secretion. At the same time, it stimulates calcitonin secretion by thyroid C-cells. Calcitonin lowers blood calcium by moving it to bones, causing kidneys to excrete it, and suppressing its absorption. Calcitonin has far greater effects in children than in adults because children have more active osteoclasts. These bone-degrading cells release up to 5 grams of calcium into extracellular fluids daily in children, but only 0.8 grams in adults. Calcitonin acts for one to two hours as a calcium regulator until it is overridden by PTH. Neither PTH nor calcitonin are regulated by the pituitary gland.

Parathyroid disorders can lead to severe physiological disturbances. **Hypoparathyroidism** (hi″po-par″ah-thi′roid-izm), a PTH deficiency, usually is due to gland injury, sometimes in conjunction with thyroid surgery. Low blood calcium decreases the threshold of motor neurons and allows them to depolarize more easily. Muscle spasms of **hypocalcemic tetany** (hi″po-kal-se′mik tet′an-e) can occur, especially in the larynx, arms, and legs. If the laryngeal muscle spasms block the airway, death by suffocation can occur. If hyperventilation (rapid breathing) occurs, it raises the pH of body fluids and further decreases neuron thresholds, so tetany can occur even when blood calcium is only slightly lowered.

Hyperparathyroidism (hi″per-par″ah-thi′roid-izm), an excess of PTH, usually is caused by a parathyroid tumor. Excess PTH removes large amounts of calcium from bones (leading to fractures) and causes calcium to precipitate in kidney tubules (producing kidney stones). The presence of high calcium kidney stones suggests a parathyroid tumor.

Adrenal Glands

Located superior to each kidney, the paired **adrenal** (ad-re′nal) **glands** have a small inner medulla and a thick outer cortex (Figure 16.18). The **adrenal medulla,** being derived from the same tissue as sympathetic neurons, secretes hormones in response to sympathetic stimulation (Chapter 14). The **adrenal cortex** has three zones of cells, which together secrete more than 25 different corticosteroid hormones. Though cells in each zone can produce nearly all of these hormones, the zones are specialized to synthesize and secrete particular hormones. The outer **zona glomerulosa** (zo′nah glom-er-u-lo′sa), which lies directly under the fibrous connective tissue capsule, produces mineralocorticoids that regulate mineral concentrations in body fluids. The middle **zona fasciculata** (fas-sik-u-lat′ah) produces glucocorticoids, which help to regulate metabolism of carbohydrates and other nutrients. The inner **zona reticularis** (rĕ-tik″u-lar′is), adjacent to the medulla, produces adrenal sex hormones, as well as moderate amounts of glucocorticoids.

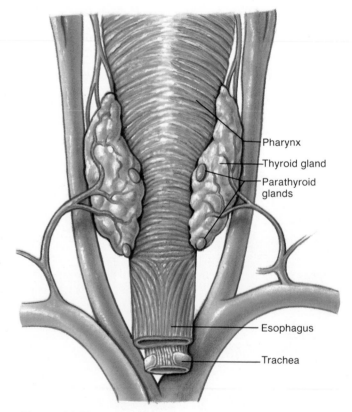

Pharynx

Thyroid gland

Parathyroid glands

Esophagus

Trachea

Figure 16.17
The parathyroid glands are located on the posterior surface of the thyroid gland, typically two in each lobe.

Hormones of the Adrenal Medulla

The hormones of the adrenal medulla, **epinephrine** (adrenaline) and **norepinephrine** (noradrenaline), are synthesized from the amino acid tyrosine via dopamine. Vitamin C is required for activity of the enzyme that converts dopamine to norepinephrine. About four-fifths of the norepinephrine is converted to epinephrine by the addition of a methyl group. Compared with the brief action of norepinephrine released at synapses, it and epinephrine released as hormones can act for up to three minutes—the time required for enzymes in liver and other tissues to inactivate them.

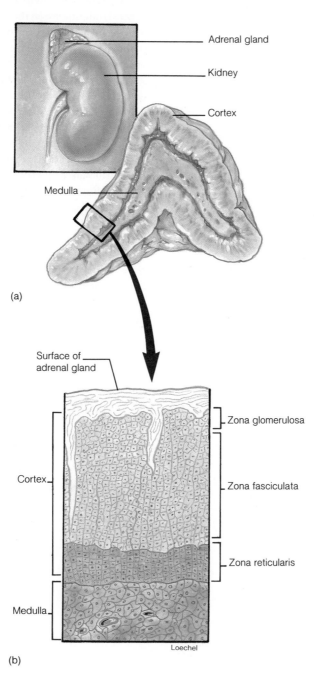

(a)

(b)

FIGURE 16.18
The adrenal glands consist of (*a*) a medullary portion, and (*b*) three cortical zones, surrounded by a tough fibrous capsule.

Hormones of the Adrenal Cortex

Hormones of the adrenal cortex are corticosteroids synthesized from cholesterol. They fall in three groups—**mineralocorticoids, glucocorticoids,** and **androgens** (Figure 16.19). Small chemical differences in these hormones account for their vastly different functions.

The major mineralocorticoid **aldosterone** (al-dos'ter-ōn) regulates the sodium and potassium content of body fluids (Figure 16.20). Aldosterone is secreted when sodium decreases or potassium increases in the blood. It acts on the Na^+/K^+ pump in kidney tubule cells and causes sodium to be returned to the blood and potassium to be excreted. As the sodium concentration in the blood increases, water follows it by osmosis, and the blood volume also increases. The overall effects of aldosterone are to increase blood sodium and blood volume, and to decrease blood potassium.

The **glucocorticoids** (gloo-ko-kor'ti-koidz) help to regulate, not only carbohydrate metabolism as the name suggests, but also protein and fat metabolism. Though glucocorticoids regulate specific metabolic processes, their effects are far-reaching because altering one metabolic process usually alters others, too. Glucocorticoids of the human adrenal cortex include the potent **cortisol** (kor'tisol), or hydrocortisone, and the less potent **corticosterone** (kor''tik-os'ter-ōn). These hormones are secreted after several hours without food when liver stores of glucose have been depleted. They stimulate synthesis of glucose from noncarbohydrate sources, cause fat cells to break down fats and release fatty acids into the blood, and cause many cells to break down proteins and release amino acids into the blood (Chapter 24). Increasing blood fatty acids causes many cells to use more fatty acids and less glucose to meet their energy needs.

Cortisol and corticosterone counteract some effects of inflammation and help the body to respond to stress. In general, inflammation increases capillary permeability and causes lysosomes of injured cells to release enzymes. How cortisol counteracts inflammation is not well understood, but it seems to reduce permeability of capillaries and make lysosomal membranes more stable. Cortisol helps the body to respond to stress by mobilizing fatty acids and other nutrients that provide energy to meet the demands of a stressful situation. It also stimulates the adrenal medulla to release epinephrine and norepinephrine. Cortisol is exceedingly important in maintaining homeostasis through its continuous regulation of metabolism and through its special effects during inflammation and stress.

Adrenal sex hormones, or gonadocorticoids, are mostly **androgens** (an'dro-jenz), chemically similar to testosterone and convertible to it in some tissues. These hormones stimulate pubic hair development in females, but they normally have no effects in males. Adrenal tumors can produce large amounts of androgens that can cause masculinization of a female or precocious sexual development in a young male.

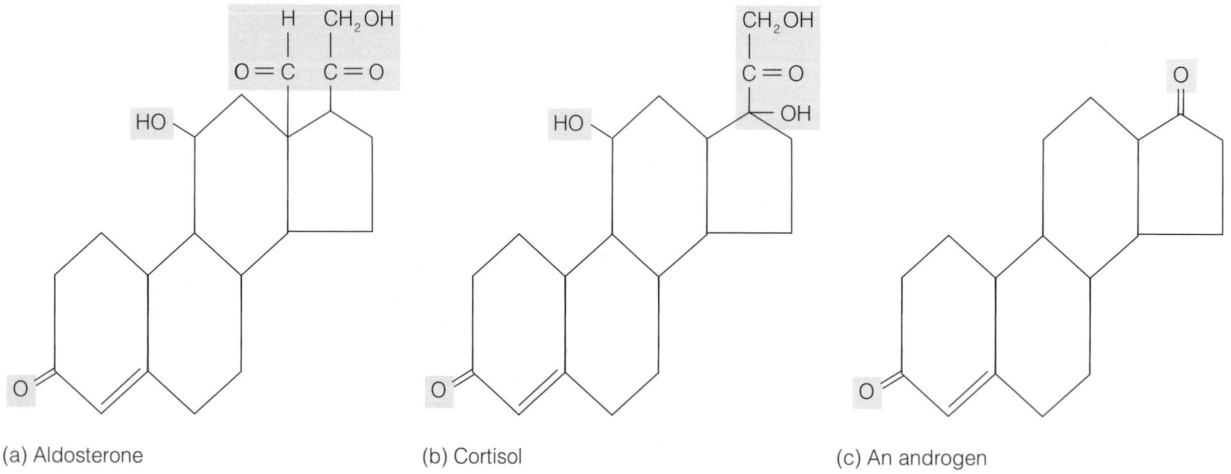

(a) Aldosterone   (b) Cortisol   (c) An androgen

FIGURE 16.19
Note the small differences in the structure of the steroid hormones secreted by the adrenal cortex: (*a*) aldosterone, (*b*) cortisol, and (*c*) an androgen.

In addition to their other functions, several steroid hormones—cortisol, aldosterone, testosterone, and estradiol—work together to regulate bone metabolism. Cortisol can cause breakdown of collagen and increase blood calcium. Aldosterone causes other minerals to be returned to the blood from kidney tubules and makes them available for deposition in bone. Testosterone promotes protein synthesis during bone development, calcium retention, and calcium deposition in bones. Estradiol promotes protein synthesis and increased bone density by calcium accumulation.

Regulation of Hormones of the Adrenal Cortex

Aldosterone secretion is regulated by the blood potassium concentration; a blood peptide **angiotensin** (an''je-o-ten'sin); ACTH; and, to a lesser degree, blood sodium concentration (Chapter 27).

Cortisol and corticosterone secretions are controlled mainly by ACTH as already shown in Figure 16.10. ACTH also stimulates aldosterone release, but only cortisol exerts significant feedback on ACTH. A high cortisol level suppresses release of both ACTH and CRH. The adrenal cortex receives less stimulation and cortisol secretion decreases. The resulting low cortisol level fails to suppress CRH and ACTH, and they increase secretion of cortisol and other adrenocortical hormones.

Adrenal Disorders

No disorders have been traced to deficient adrenal medulla hormones, but norepinephrine deficiency in the brain can lead to affective disorders (Chapter 12). Pheochromocytoma, a tumor consisting of epinephrine-secreting cells, causes symptoms suggestive of excessive sympathetic stimulation, but especially high blood pressure. One

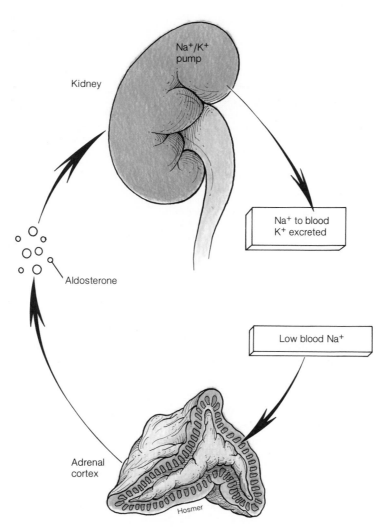

FIGURE 16.20
Aldosterone regulates sodium, potassium, and water in body fluids.

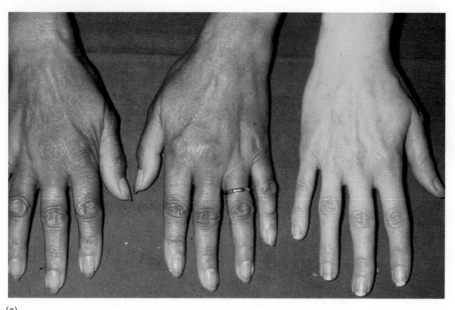

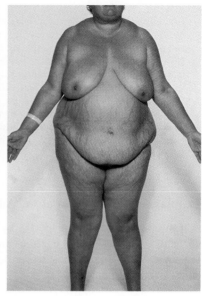

(a)

(b)

FIGURE 16.21
Adrenocorticotrophic disorders: (*a*) Addison's disease, and
(*b*) Cushing's disease.

TABLE 16.6   Some Adrenocortical Disorders

Disorder	Possible Causes	Excess or Deficiency	Effects
Addison's disease	Loss of function of whole cortex	Deficiency of both cortisol and aldosterone; excess of ACTH by loss of negative feedback	Muscle weakness and fatigue due to lack of stimulation of metabolism, high blood potassium and low blood sodium, bronzing of skin because excess ACTH has a melanocyte-stimulating effect
Hyperaldosteronism	Tumor of the adrenal cortex	Excess aldosterone	Low blood potassium and high blood sodium leads to hypertension and accumulation of fluids in tissues
Cushing's disease	Tumor of adrenal cortex or of anterior pituitary that produces ACTH	Excess cortisol	Moon-shaped face, trunk obesity, and diabetes caused by cortisol counteracting effects of insulin

means of diagnosing pheochromocytoma is to give a dose of phentolamine, a drug that blocks alpha adrenergic receptors. A sharp decrease in diastolic blood pressure of more than 25 mm Hg (Chapter 18) provides a positive diagnosis.

Adrenocortical disorders involve excesses or deficiencies of aldosterone, cortisol, or both. These disorders (Figure 16.21 and Table 16.6) are treated by supplying hormones to compensate for deficiencies and by removing tumors that cause excesses of the hormones.

Cortisol and some of its derivatives are widely used to counteract inflammation in a variety of chronic diseases. These drugs have significant effects on the brain and on the mood of patients, and can cause Cushing's disease and depression. Some steroids act as anesthetics and others produce euphoria, probably by altering membrane excitability of nerve cells. Much remains to be learned about the action of steroids on brain cells, but mood changes associated with stress, the menstrual cycle, and pregnancy almost certainly are caused by steroids.

ENDOCRINE FUNCTIONS OF THE PANCREAS

The **pancreas,** which lies inferior to the stomach in a bend of the duodenum, is both an exocrine and an endocrine gland (Figure 16.22). Its exocrine secretions aid in digestion as discussed in Chapter 23. The endocrine cells of the pancreas are arranged in clusters called **islets of Langerhans** (lahn′ger-hanz). About three-fourths of these cells are **B-cells,** which produce the hormone **insulin** (in′su-lin).

TABLE 16.7 Summary of Endocrine Functions

Glands	Hormones	Target Cells	Major Effects	Negative Feedback	Disorders
Adenohypophysis	Growth hormone (GH) (somatotropin)	Most cells	Growth, maintenance of adult size, protein synthesis, release of fats and glucose into blood	Blood nutrient level	Dwarfism, giantism, acromegaly
	Prolactin	Mammary glands	Secretion of milk		
	Follicle-stimulating hormone (FSH)	Ovaries	Maturation of ova and production of estrogen in female	Estrogen	
		Testes	Maturation of sperm in male	Inhibin	
	Luteinizing hormone (LH) (Interstitial cell stimulating hormone in male)	Ovaries	Release of ova and production of progesterone	Progesterone	
		Testes	Development of interstitial cells and production of testosterone	Testosterone	
	Adrenocorticotropic hormone (ACTH)	Adrenal cortex	Release of hormones (other than mineralocorticoids) from adrenal cortex	Cortisol	
	Thyroid-stimulating hormone (TSH) (thyrotropin)	Thyroid gland	Synthesis and release of T_3 and T_4	T_3 and T_4	
Neurohypophysis	Oxytocin	Smooth muscle of uterus and mammary ducts	Cause uterine contraction and release of milk		
	Antidiuretic hormone (ADH) (vasopressin)	Kidney tubules	Water reabsorption	Blood osmotic pressure	Diabetes insipidus
		Smooth muscle of blood vessels	Constrict blood vessels and raise blood pressure	Blood pressure	
Thyroid	Thyroxine (T_4) and triiodothyronine (T_3)	Most cells	Increase oxygen consumption and metabolic rate; increase protein synthesis	T_3 and T_4 suppress TSH secretion	Cretinism, myxedema, goiter, Graves' disease
	Calcitonin	Osteoblasts in bone	Cause bones to take up calcium	Low blood calcium	
Parathyroids	Parathormone	Intestine	Increase absorption of calcium	High blood calcium	Hyperparathyroidism, hypoparathyroidism
		Kidneys	Decrease excretion of calcium		
		Osteoclasts in bone	Cause reabsorption of calcium from bone		
Adrenal medulla	Epinephrine and norepinephrine	Typically, tissues affected by sympathetic stimulation	Generally similar to sympathetic stimulation but of longer duration		Tumors of adrenal medulla
Adrenal cortex	Aldosterone	Kidneys	Increase sodium reabsorption, increase potassium secretion, increase extracellular volume	High blood sodium or low blood potassium	Hyperaldosteronism, Addison's disease, Cushing's syndrome, adrenogenital syndrome
	Cortisol	Liver	Breakdown of fat and protein for energy, gluconeogenesis	Cortisol suppresses ACTH	
		Other tissues	Antiinflammatory effect		
	Androgens	Sex organs	Masculinization of female, precocious puberty in males, but normally has little effect		

TABLE 16.7   Continued

Glands	Hormones	Target Cells	Major Effects	Negative Feedback	Disorders
Pancreas	Glucagon	Mainly liver	Increase breakdown of glycogen to glucose, gluconeogenesis	High blood glucose	
	Insulin	Most tissues except brain	Increase take-up of glucose and amino acids, increase glycogen storage, decrease gluconeogenesis	Low blood glucose	Diabetes mellitus, hypoglycemia
	Somatostatin (growth-hormone-inhibiting hormone)	Pancreas and intestine	Suppress release of pancreatic and digestive hormones		
	Pancreatic polypeptide	Pancreas and intestine	Suppress release of pancreatic and digestive hormones		
Ovaries	Estradiol	Follicles	Maturation of ova	Estradiol suppresses GnRH and FSH (See Ch. 29)	
		Uterus	Development of lining		
		Mammary glands	Development of duct system		
		Tissues displaying secondary sex characteristics	Development of characteristics at puberty		
	Progesterone	Uterine lining	Development of glandular lining, development of placenta, maintenance of pregnancy	Progesterone suppresses GnRH and LH (See Ch. 29)	
		Mammary glands	Development of secretory tissue		
	Relaxin	Pubic symphysis	Soften ligaments		
Testes	Testosterone	Sperm-forming cells	Maturation of sperm	Testosterone suppresses GnRH and LH	
		Tissues displaying secondary sex characteristics	Development of characteristics at puberty		
		Many tissues	Growth		
Placenta	Human chorionic gonadotropin (RCG)	Maternal ovary	Production of estrogen and progesterone		
	Chorionic growth hormone prolactin	Fetus, maternal breasts	Effects similar to pituitary counterparts		
	Estrogens and progesterone	Fetus, maternal tissues	Effects similar to pituitary counterparts		
Thymus	Thymosins	Lymphoid tissue	Development of cells concerned with immunity		
Pineal gland	Melatonin and serotonin	Various tissues(?)	Regulate cyclic phenomena(?)		
Blood	Erythropoietin	Bone marrow	Development of erythrocytes		
Digestive tract	Gastrin, secretin, cholecystokinin (pancreozymin)	See Chapter 23			

CLINICAL TERMS

Hashimoto's thyroiditis   an inflammation of the thyroid gland, probably caused by an autoimmune reaction

hirsutism (her'sūt-izm)   abnormal hairiness

neuroblastoma (nu''ro-blas-to'mah)   a malignant tumor consisting of neural tissue, often tissue of the sympathetic division growing in the adrenal medulla

pituitary cachexia (kak-ek'se-ah)   a general ill health, weakness, and wasting caused by pituitary failure; Simmonds' disease

Waterhouse-Friderichsen syndrome   a sudden, severe disease caused by a meningococcus and characterized by fever, coma, and hemorrhage, especially in the adrenal glands

STRESS

The concept of physiological stress was first developed by Hans Selye at McGill University. In 1935, while studying hormones injected into rats, Selye discovered that regardless of what he injected, all animals responded in the same way with enlarged adrenal glands, bleeding ulcers, and atrophy of lymphatic tissues. On the basis of his observations, Selye defined several terms: **Stress** is a condition produced by a variety of injurious agents; **stressors** are agents that produce stress. The **general adaptation syndrome** refers to a group of changes that appear in animals under stress. Selye chose his words carefully. By *general,* he meant that the response was the same for many different stressors. By *adaptation,* he meant that the animal's response made it possible for the body to cope with the stress—to adapt to it. By *syndrome,* he referred to a set of signs and symptoms that occurred together.

We now know the severe stress Selye observed was due to stressors stimulating excess ACTH secretion, which caused excess cortisol secretion. Stressors can be pleasant or unpleasant. Winning a race can be as stressful as losing; being around people all the time can be as stressful as being isolated; and the birth of a baby can be as stressful as the death of a loved one. Milder stressors produce less severe effects without excess cortisol secretion, and the body's homeostatic mechanisms usually can prevent damage.

Selye distinguishes between productive stress that "keeps us on our toes" and harmful stress that causes tissue damage.

The former, he called **eustress;** the latter, he called **distress.** Distress is continual stress that requires constant readjustment. Caring for a seriously ill relative, working at a frustrating and unsatisfying job, or even continuous boredom can be a distressing situation. Mental or emotional distress is more likely to lead to health problems than physical distress from muscular work. What constitutes distress differs among individuals and in the same individual at different times. A job that distresses one person may bring out another's best performance. An event we find tolerable one day may be almost intolerable another day. How much sleep we have had, what we have eaten, our heredity, our relationships with other people, our past experiences, and many other factors determine whether stimuli will produce manageable stress or distress.

According to Selye, the body's first response to stress is the **alarm reaction** where stimuli from the brain and sense organs cause the hypothalamus to release CRH, which causes the anterior pituitary to release ACTH. Since Selye's work, others have shown that the hypothalamus also releases epinephrine, oxytocin, and vasopressin, all of which cause ACTH release. ACTH stimulates the adrenal cortex to release increased amounts of cortisol and other glucocorticoids (Figure 16.27). These hormones produce several effects— decreased allergic reaction, decreased resistance to disease, and increased fat metabolism. Hydrocortisone (cortisol)

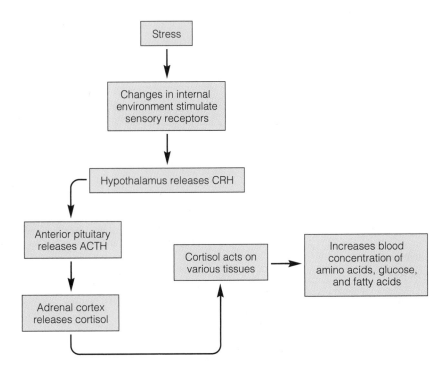

FIGURE 16.27
Mechanism for the release of cortisol in the alarm reaction and the nature of its effects.

treatments for inflammation, allergy, or autoimmune diseases take advantage of the body's alarm reaction. Prolonged treatment can cause the same ill effects as the exhaustion stage of stress.

In addition to affecting the adrenal cortex, a stressor also stimulates the sympathetic nervous system (Figure 16.28) and produces the "fight-or-flight" reaction—increased blood pressure and blood volume, decreased digestion with distribution of blood away from the digestive tract to muscles and active organs, and increased blood glucose concentration. These effects are prolonged because sympathetic stimulation (and maybe stimulation by cortisol) causes the adrenal medulla to release excess hormones.

Depending on the results of the alarm reaction, the body either develops resistance to the stress or becomes exhausted. By adjusting to a stressor through the alarm reaction, the body reaches the **resistance stage** and no further response to the stressor occurs. Cortisol production and sympathetic stimulation return to normal. Normal cortisol levels inhibit CRH release by negative feedback, but new stressors can override this mechanism and release CRH and other hormones. Our bodies have experienced the alarm reaction and resistance stage many times. But when the body fails to reach the resistance stage and stress continues, it reaches the **exhaustion stage,** and ACTH and cortisol secretion exceeds that in the alarm reaction. Stress-related disorders such as gastrointestinal ulcers, migraine headaches, disturbed heart rhythms, and some mental illnesses typically appear in the exhaustion stage. In spite of high cortisol secretion, the body is unable to cope with the stressor. Unless those effects can be reversed, the ultimate result is death.

Selye's pioneering work encouraged others to study stress. It is now believed that in emotional stress, the limbic system mediates the stressor response to the hypothalamus. The hypothalamus stimulates the posterior pituitary and the cortex and medulla of the adrenal gland. It also releases antidiuretic hormone, which conserves body fluids and counteracts fluid loss from sweating or bleeding. Whether cortisol helps organisms to resist stress as Selye thought, or is merely a consequence of it, is still not clear. Whatever future studies show about stress, Selye will always deserve credit for initiating work on this important problem.

Questions

1. How did Selye define the general adaptation syndrome, and how did he distinguish between eustress and distress?
2. What happens in the alarm reaction?
3. How do the resistance and exhaustion stages of stress differ?
4. What advances have been made in understanding stress since Selye's studies?

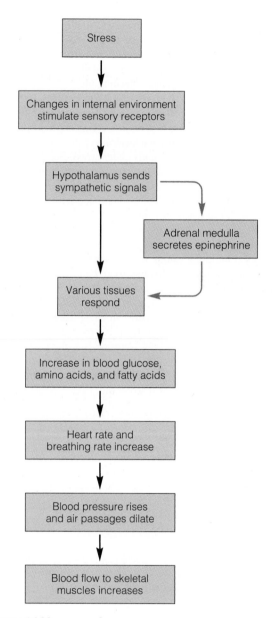

FIGURE 16.28
Mechanism for the stimulation of the sympathetic nervous system and the adrenal medulla in the alarm reaction, and the nature of its effects.

CHAPTER SUMMARY

Organization and General Function
- The general function of endocrine tissues (shown in Figure 16.2) is to produce hormones.
- Hormones are chemical control molecules secreted from endocrine cells that travel in blood to receptors on target cells.

Development
- Endocrine glands that develop from ectoderm or endoderm secrete hormones made of amino acids. Those that develop from mesoderm secrete steroids.

Properties of Hormones
- Some hormones are synthesized in inactive forms called prohormones and later activated.
- Hormone secretion is generally regulated by negative feedback.
- Most hormones are transported in the blood on protein to which they bind loosely.

Mechanisms of Hormone Action
- Hormones exert their effects at the cellular level by changing the rate of certain cellular processes.
- The mechanism of action of steroid and thyroid hormones is summarized in Figure 16.3; that of many polypeptide hormones in Figure 16.4.
- Cyclic nucleotides are important second messengers in the action of polypeptide hormones, and they, in turn, may be affected by prostaglandins.
- Hormones are inactivated—epinephrine and norepinephrine by enzymes in target cells, and most other hormones by liver enzymes before being excreted through bile or urine.

Prostaglandins
- Prostaglandins are synthesized in cell membranes as needed, affect only nearby cells, and are quickly inactivated.
- Some occur in antagonistic pairs that constrict or dilate blood vessels or passageways; others increase blood vessel permeability or elevate body temperature.

Pituitary Gland
- The pituitary gland (hypophysis) consists of the anterior adenohypophysis and the posterior neurohypophysis; pituitary and other hormones and their functions are summarized in Table 16.7.
- The hypothalamus produces releasing and inhibiting factors that regulate the anterior pituitary as shown in Table 16.3; it also secretes the hormones that are stored in the neurohypophysis.

Thyroid Gland
- Thyroid hormones, mainly T_3, increase the rate of cellular metabolism and stimulate protein synthesis.
- TSH stimulates thyroid hormone secretion and thyroid hormones provide negative feedback that suppresses further secretion.

Parathyroid Glands
- Low blood calcium stimulates release of parathormone and high blood calcium acts by negative feedback to suppress its release.
- High blood calcium stimulates release of calcitonin and low blood calcium acts by negative feedback to suppress its release.

Adrenal Glands
- The adrenal glands are two glands in one—a medulla and a cortex; the adrenal medulla augments sympathetic stimulation.
- ACTH stimulates the adrenal cortex to secrete its hormones; cortisol acts by negative feedback to suppress ACTH, which, in turn, reduces secretion of adrenocortical hormones.

Endocrine Functions of the Pancreas
- The islets of Langerhans of the pancreas contain several kinds of hormone-secreting cells—mainly insulin, which lowers blood glucose, and glucagon, which raises it.
- When blood glucose is low, glucagon is secreted and insulin is inhibited; when it is high, insulin is secreted and glucagon is inhibited.

Endocrine Functions of the Gonads
- The gonads, or ovaries and testes, synthesize and secrete steroid hormones as summarized in Table 16.7.
- Hypothalamic and anterior pituitary hormones stimulate secretion of these hormones, and the gonadal hormones act by negative feedback on the hypothalamus and anterior pituitary to regulate secretions as will be explained in Chapter 29.

Endocrine Functions of Other Structures
- The placenta produces several hormones as summarized in Table 16.7.
- The thymus produces thymosins that contribute to the development of immunity.
- The pineal gland produces melatonin and serotonin, and probably regulates cyclic phenomena.
- Various hormones from diffuse endocrine tissues of the digestive tract regulate digestive and metabolic processes.
- Other hormones include natriuretic hormone that increases sodium excretion and relaxes smooth muscle in blood vessels, erythropoietin that stimulates the development of erythrocytes, and vitamin D that stimulates synthesis of a calcium carrier protein.

QUESTIONS AND PROBLEMS

The questions at the end of each chapter are numbered to correspond with the objectives listed at the beginning of the chapter. Italics indicate that a question requires critical thinking skills beyond simple factual recall.

Questions

1. (a) What is a hormone?
   (b) What is the general function of the endocrine system, and what are its components?
2. (a) Which endocrine glands are derived from two embryonic sources, and how do they develop?
   (b) *How are germ layers and hormones related?*
3. (a) What is a prohormone?
   (b) How are hormones transported?
   (c) *Briefly explain the role of negative feedback in controlling hormone secretion.*
   (d) *How do neural and hormonal control systems differ, and how are they alike?*
4. (a) *Compare and contrast the mechanisms of action of steroid and thyroid hormones with that of polypeptide hormones.*
   (b) *How do calcium and calmodulin function as regulators of cell functions?*
5. (a) *How do properties of receptors affect hormone actions?*
   (b) How does enzyme amplification affect the body's response to certain hormones?
   (c) How are hormones inactivated?
6. What are the main properties of prostaglandins, and what do they do?
7. (a) What are the functions of the hormones secreted by the anterior pituitary gland?
   (b) *How are anterior pituitary hormones regulated?*
   (c) *What are some effects of excesses and deficiencies of anterior pituitary hormones?*
8. (a) What are the functions of the hormones released from the posterior pituitary gland?
   (b) Where do posterior pituitary hormones come from, and how are these hormones regulated?
9. *What hormone excesses and deficiencies are known to lead to pituitary disorders?*
10. How are the hormones of the thyroid gland synthesized, stored, and released?
11. (a) What are the effects of thyroid hormones?
    (b) *How are thyroid hormones regulated?*
12. *What are the effects of excess or deficient thyroid hormones?*
13. (a) *What is the function of parathormone, and how is it regulated?*
    (b) *What is the source of calcitonin, and how is its function related to that of parathormone?*
14. (a) What hormones does the adrenal medulla secrete?
    (b) *What do adrenal medulla hormones do, and how are they regulated?*
    (c) What hormones does the adrenal cortex secrete?
    (d) *What do adrenocortical hormones do, and how are they regulated?*
15. *What are some effects of excesses or deficiencies of adrenal hormones?*
16. (a) What are the major pancreatic hormones?
    (b) *What do pancreatic hormones do, and how are they regulated?*
17. What are the major functions of hormones of the gonads?
18. What other human hormones have been identified, and what are their main functions?

Problems

1. How would the following symptoms narrow down a possible cause of an endocrine disorder: (a) bronze-colored skin, (b) extreme thirst, and (c) patient complaining of being cold? For each of the above, explain what defect in endocrine function is likely to be present.
2. According to the results of some studies in rats, insulin and vasopressin can be given orally if they are coated with an impervious substance. Why is the coating necessary?
3. Research the views of physiologists on the role of cortisol in stress, and participate in a class debate on the topic.
4. Select any three endocrine disorders, and explain how failure of regulatory mechanisms contribute to the disorder.
5. Describe the actions and regulation of hormones that affect the following target tissues: (a) all cells, (b) liver, (c) bone, (d) ovaries, (e) testes, (f) kidneys, and (g) smooth muscle.
6. What endocrine disorders would you look for in an individual complaining of fatigue?

CIRCADIAN RHYTHMS

Daily cyclic variations in physiological conditions are called **circadian** (sir-ka′de-an) **rhythms** (*circa,* about; *dian,* day). A **rhythm,** or cycle, is a sequence of events that occurs again and again in the same order and over the same time interval. Both the human body temperature and sleep-wake cycles have a period of about 24 hours and a frequency of 1/24 hours. The **period** of a rhythm is the time required for one cycle. The **frequency** of a rhythm is how many cycles occur in a given period. Most cycles have a *range* (maximum to minimum values) and a *mean,* or average value. Normal human body temperature has a range of about 1° C (36.5° to 37.5° C) and a mean of about 37° C.

Rhythms undergo advances or delays called **phase shifts.** Going to or from daylight savings time constitutes a small phase shift in your sleep-wake rhythm— a *phase advance* in the spring and a *phase delay* in the fall. Jet travel subjects people to large phase shifts, advances with eastward travel and delays with westward travel.

A **biological clock** is a regulator of circadian and other rhythms. Such a regulator must relay time-related information so it influences physiological processes, and it must be subject to resetting to synchronize rhythms. The human body has no organ that can be called its biological clock, but several structures help to regulate a cycle, and one probably serves as a "pacemaker" to coordinate the others.

Claims have been made that biorhythms based on birth date persist throughout life, producing "critical days" on which a person is accident prone or capable of above-average performance. Studies that purport to demonstrate such cycles generally lack adequate controls and have failed to find any physiological basis for the cycles. Cycles that use zodiac or other astrological signs are no more valid than those that use birth dates.

Early studies of human circadian rhythms used deep caves or polar regions during summer periods of continuous light to isolate the subjects from time cues. Special isolation units now provide environments without time cues and with more of the "comforts of home." The first step in studying rhythms is to determine whether a rhythm exists by plotting data on a graph (Figure I3.1). If a rhythm is found, the next step is to determine whether it is **exogenous** (eks-oj′en-us), timed by environmental factors such as day-night variations, or **endogenous** (en-doj′en-us), timed by some mechanism within the organism.

Identifying cycles as exogenous, endogenous, or some combination can be quite difficult. Humans in a normal day-night environment display a 24-hour sleep-wake cycle. Isolated from light-dark cues, most display a longer (approximately 25-hour) sleep-wake cycle. Such a rhythm is **free-running**— not modified by environmental factors.

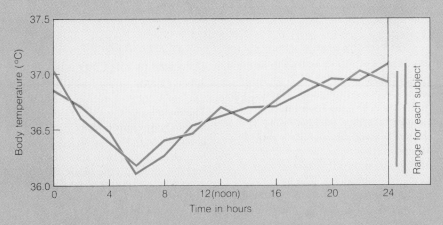

FIGURE I3.1
Hypothetical data on body temperature of two human volunteers over a 24-hour period.

It appears that an endogenous biological clock controls the sleep-wake cycle, but that the cycle is synchronized with the 24-hour day-night cycle by environmental cues. Synchronization of the clock by environmental factors is called **entrainment** (en-trān'ment). Environmental factors that can entrain the clock are called **zeitgeber** (tsīt'ga-ber), or time givers. Zeitgeber include daily variations in environmental temperature, social cues, and especially daily light-dark cycles.

To study the effects of abrupt time shifts, lights in an isolation unit are manipulated to create an artificial day length—perhaps six hours shorter than normal like the experience of flying across the Atlantic Ocean—followed by several 24-hour days. Most rhythms gradually shift to the new 24-hour day, but they vary in the time required to synchronize with the new day. After a six-hour advance, exogenous rhythms synchronize in a day or two, but endogenous rhythms take longer. Such studies have demonstrated certain qualities of an endogenous rhythm:

1. The rhythm persists in the absence of environmental cues.
2. The period of the rhythm is not exactly 24 hours if zeitgeber is excluded, but remains near 24 hours even when a different day length is imposed.
3. The rhythm is slow to adapt to time shifts.

Rhythms that do not meet these criteria are probably exogenous or controlled by a combination of endogenous and exogenous factors.

In the late 1960s, speculations about biological clocks focused on hypothalamic cells. Today, the **suprachiasmatic** (soo″prah-ki-as-mat′ik) **nuclei** (SCN) **pacemaker,** above the optic chiasm in the hypothalamus, is thought to coordinate circadian rhythms. Light stimuli can be relayed to and entrain the SCN pacemaker, but much remains to be learned about how this occurs. According to the circadian regulation model (Figure I3.2), when light stimulates the retina, signals are relayed to the SCN pacemaker and account for the major effects of light on circadian rhythms. Raphe nuclei, concerned with sleep-wakefulness, release serotonin, which relays signals to the lateral geniculate nucleus and the SCN pacemaker. Such signals may

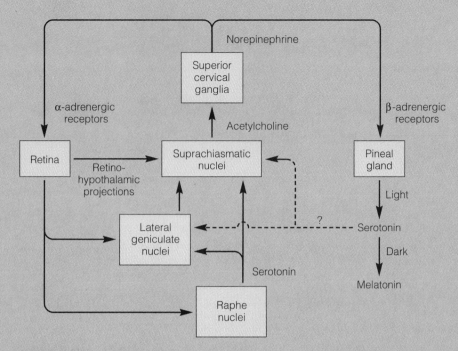

FIGURE I3.2

An anatomical model of circadian regulation in mammals. The suprachiasmatic nuclei (SCN) act as the pacemaker. The SCN are influenced directly and indirectly by light striking the retina. Signals from the SCN reach both the retina and the pineal gland. The pineal gland displays a circadian rhythm of serotonin secretion in light and melatonin secretion in darkness.

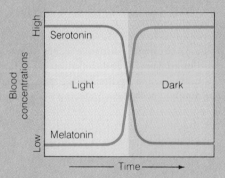

FIGURE I3.3

Synthesis and secretion of serotonin and melatonin by the pineal gland has a circadian rhythm driven by the brain pacemaker but influenced by the light-dark cycle of the environment.

coordinate sleep-wake and night-day cycles. Signals from the SCN go to the retina and the pineal gland. Those going to the retina have no known effect on circadian rhythms, but those going to the pineal gland stimulate synthesis of an enzyme that makes serotonin and inhibits an enzyme that converts serotonin to melatonin. The pineal gland thereby displays a circadian rhythm: it synthesizes and secretes serotonin in light, and converts serotonin to melatonin and secretes it in darkness (Figure I3.3). Melatonin receptors have been detected in the SCN, providing further evidence that both are involved in regulating circadian rhythms.

EXAMPLES OF CIRCADIAN RHYTHMS

The existence of circadian rhythm in human body temperature has been known since the mid-nineteenth century, but rhythms have been studied only in recent decades. In the 1960s, sleep-wake rhythms were studied in volunteers living in caves, who let the researchers know by telephone when they awoke and when they retired. Such experiments showed that free-running sleep-wake cycles usually have a period of slightly more than 24 hours (Figure I3.4a).

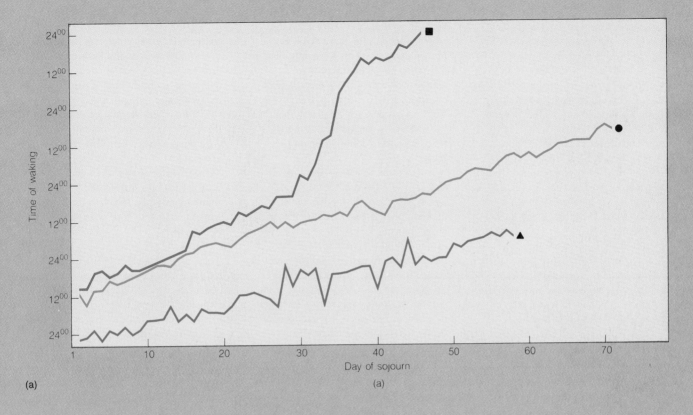

(a)

(a)

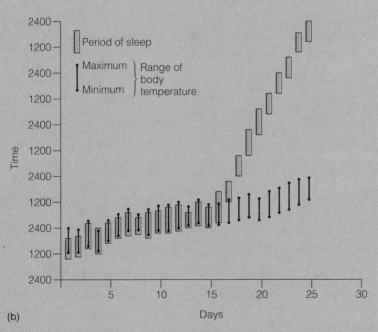

(b)

FIGURE I3.4

(*a*) Time of waking over successive days in three subjects isolated from time cues in caves. Triangle, Suffre; circle, Workman; square, Lafferty (latter half of sojourn). (From Conroy and Mills, 1970, Fig. 7.2) Note that rhythm suddenly becomes erratic for one subject. (*b*) Sleep periods and body temperature rhythms are correlated for first 15 days of the above study, but then become desynchronized.

From these and other observations, the sleep-wake rhythm seems to be an endogenous rhythm. The "Monday morning blues" probably is a disruption of the sleep-wake cycle. After a weekend of staying up late and sleeping late, some people find it difficult to get going on Monday morning.

Measurements of deep body (usually rectal) temperature show a temperature rhythm with a period of about 25 hours.

This suggests an endogenous rhythm, but it might be synchronized with other rhythms and controlled by a higher level regulatory mechanism. In a classic experiment (Figure I3.4b), a subject living in isolation displayed synchronous temperature and sleep-wake rhythms with an average period of 25.7 hours for 14 days. On day 15, the rhythms desynchronized spontaneously. The sleep-wake rhythm shifted to a much

longer period (average 33.4 hours) while the body temperature rhythm retained nearly the same period (average 25.1 hours); thus, normally synchronized rhythms can become desynchronized, and temperature is an endogenous rhythm. That the body temperature rhythm takes much longer than the sleep schedule to adjust after jet travel provides further evidence of its endogenous nature.

Many hormones—the CRH-ACTH-cortisol system, growth hormone, TSH and others—have circadian rhythms. Cortisol and ACTH rhythms seem to confirm negative feedback of cortisol on ACTH. Growth hormone (GH) has a strong rhythm closely related to sleep with the peak blood concentration right after sleep onset (22:00 to 23:00 on the 24-hour clock). Sleep appears to release GH and the GH rhythm follows the sleep rhythm in free-running experiments even when sleep is desynchronized with other rhythms. TSH secretion displays a circadian rhythm, reaching a maximum with the onset of sleep, declining throughout the sleep period, and remaining at a minimum throughout the waking period. In contrast, T_3 and T_4 levels vary throughout the day and night.

Observations of the effects of time zone changes and shift work help researchers understand circadian rhythms. Such schedule changes also have important effects on human health and performance.

Time zone transitions occur during travel to the east or west, as noted earlier. With ground travel, zeitgeber entrain rhythms to the new environment almost as fast as time zones are traversed. With air travel, zeitgeber entrain exogenous rhythms fairly rapidly, but they entrain endogenous rhythms much more slowly. Rhythms thereby become desynchronized, and we say we are suffering from "jet lag." Comparisons of flights of the same duration to the east, west, or south show that all flights produce fatigue, decreased performance, and altered temperature and other rhythms. Normal rhythms are restored by the second postflight day after southward flights, but take four to eight days after eastward or westward flights through several time zones. Body temperature lags after eastward flights and advances after westward flights, as expected from the time change and the delay typical of resetting of endogenous rhythms.

Length of time spent in different time zones also affects rhythms, as is shown by the effects of flying through eleven times zones and either returning home immediately or after a five-day stay. If the travelers returned home immediately, rhythms were appropriate for the home environment—no adaptation to the distant environment occurred. If they stayed five days, most rhythms adapted (or nearly adapted) to the distant environment, and the travelers had to adapt again on their return home.

Adaptation to eastward travel is slower than adaptation to westward travel regardless of whether the trip is made toward or away from home or in daylight or darkness. This differential rate of adaptation may involve the way in which the human oscillator system responds to delays or advances in zeitgeber. Because free-running cycles tend to be slightly more than 24 hours, it appears to be easier for endogenous cycles to entrain to a longer than 24-hour day during westward travel than to a shorter than 24-hour day during eastward travel.

Rhythms vary in time needed to adapt to a new environment. Strongly endogenous rhythms such as body temperature are slowest to adapt and strongly exogenous rhythms such as sleep-wakefulness are fastest to adapt. The rate of adaptation also varies. When many time zones are involved, the initial rate of adaptation is faster, but full adjustment takes longer. As the rhythms approach those for the new environment, the rate of adaptation decreases.

Many travelers use what is known about the effects of travel over time zones. For short trips, they maintain home time if possible. For longer trips, they recognize that adaptation takes time and avoid commitments that require performance at old sleep times— late afternoon after westward trips and early morning after eastward trips. On trips over many time zones, they use all local zeitgeber—bright outdoor light, and meals and other activities on the new time schedule—to accelerate adaptation. If possible, they preadapt for a few days before leaving home by getting up earlier each day for eastward travel or going to bed later each day for westward travel.

Shift work is necessary to maintain continuous essential services—medical, fire, and police—and to keep industrial processes and expensive equipment operating at optimum efficiency. Though many workers recognize the necessity of shifts, they find adjusting to shifts, especially night shifts, difficult. The most significant evidence for disorders associated with shift work was collected during World War II when many people were forced into shift work. Subsequently, self-selection has let many workers unable to cope with shifts find other jobs.

Most shift workers get too little sleep when they must sleep in the daytime— they are easily wakened and often cannot get back to sleep. Body temperature is out of phase with the sleep-wake rhythm for several days after each shift change. Lack of sleep and out-of-phase body temperature both contribute to fatigue. When workers switch from day to night shifts, the body temperature rhythm adapts in about seven days; but when they switch from night to day shifts, it adapts in about three days. This difference may be due to zeitgeber entraining rhythms to the normal day-night rhythm.

Rapidly rotating shift work—two days on each shift and two days off—has a lesser effect on body temperature rhythms than slower rotating shift work. Changing from morning (06:00–14:00) to evening shift (14:00–22:00) had little effect, but changing from evening to night (22:00–06:00) altered the temperature rhythm for a day or two. Rotating shifts at three- to four-day intervals produced the greatest number of health disturbances, especially gastrointestinal complaints. Weekly rotations produced moderate numbers of complaints, and rotations after two or more weeks produced the least complaints. When workers who had worked all three shifts were changed to two shifts (morning and evening) they showed significant improvement on several measures—better social life, fewer sleep and mood complaints, and fewer gastrointestinal disturbances.

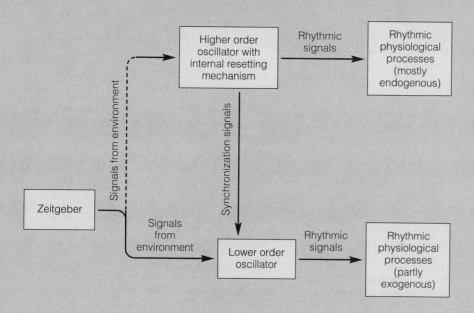

A model of circadian regulation. Oscillators are coupled in a hierarchy, and both communicate by sending rhythmic signals to control physiological processes. The higher order oscillator also sends signals that synchronize rhythms to the lower order oscillator. Both oscillators receive signals from the environment. The lower one is reset by zeitgeber, but the higher one is only slightly affected by zeitgeber because it has its own internal resetting mechanism. Note that this model does not involve feedback.

MECHANISMS OF CIRCADIAN RHYTHMS

Various studies have focused on models of circadian regulation. One such model (Figure I3.5) proposes the SCN pacemaker as a lower order controller of relatively easily entrained circadian rhythms with its parts controlling different rhythms. This model also proposes a higher order controller, possibly located in the reticular formation, that regulates less easily entrained rhythms and integrates all rhythms. How controllers of rhythms exert their effects is not known. Negative feedback has been suggested, but such mechanisms oscillate with nearly fixed amplitude about a set point and have a variable period. Circadian rhythms have variable amplitude and a fixed period, so negative feedback would not likely control them.

The effects of zeitgeber on rhythms are variable and depend on the strengths of both zeitgebers and any oscillators that may be involved. Rhythms controlled by weak oscillators are most easily entrained by a zeitgeber, as is illustrated by the effects of the light-dark cycle on the sleep-wake rhythm. Each body rhythm probably is influenced, not only by an internal controller, but also by zeitgeber interacting with controllers in complex ways.

Social factors can act as human zeitgeber as shown by pairs of subjects in an isolation unit. They displayed identical temperatures and sleep-wake rhythms, though they had not been told to synchronize their activities. In one pair, the sleep-wake rhythm became desynchronized from the temperature rhythm in exactly the same way in both members—a sleep-wake rhythm of 31.5 hours and a temperature rhythm of 25.1 hours.

CLINICAL IMPLICATIONS OF CIRCADIAN RHYTHMS

Knowledge of human circadian rhythms has great potential for improving clinical practice. For example, the efficacy of drugs has been related to circadian rhythms in animals. When information is available about drug rhythms in humans, it may be possible to improve therapy by giving different doses at different times of day.

Identifying more human circadian rhythms could improve accuracy of diagnoses and make treatments more effective. We already use morning temperature to determine the end of the febrile phase of an illness because body temperature is typically lowest in the morning. We could use cortisol secretion rhythms to diagnose and monitor disorders such as Cushing's disease (hypersecretion) and Addison's disease (hyposecretion). Fetal movements display circadian rhythms with maximal movement in the evening. Diagnosis of fetal distress by measuring fetal movement should be done at other times of the day.

One treatment based on a circadian rhythm—bright light to treat SAD (Chapter 12)—is already in use. The light apparently causes a phase shift in melatonin production. Depression that responds to this treatment may be due to a disorder of circadian rhythms.

Considering circadian rhythms in treatment is important to get the greatest benefit with the least adverse effects from drugs. Rhythms in the effects of selected drugs and other agents are summarized in Table I3.1. Differences between maximal and minimal effects can be extreme, and times of maximal and minimal effects usually are not 12 hours apart as one might expect.

| TABLE I3.1 | Susceptibility Rhythms for Selected Drugs and Other Agents |

Drug	Animal in Which Tested	Means of Evaluating Effects	Time of Day Given For: Maximum Effect	Minimum Effect
Insulin (10 IU/day up to 9 days)	Hamsters	Percent mortality	100%–22:00	5%–16:00
Ethanol (0.8 ml 25% solution)	Mice	Percent mortality	60%–20:00	10%–24:00
X radiation (410–450 R)	Mice	Percent mortality	85%–04:00	15%–24:00
Bacterial endotoxin (5 mg/kg)	Mice	Percent mortality	85%–16:00	5%–24:00
Amphetamine (26 mg/kg)	Rats	Percent mortality	80%–03:00	5%–06:00
Strychnine (2 mg/kg)	Rats	Percent mortality	85%–18:00	35%–12:00
Nicotine (0.057 mg/kg)	Rats	Percent mortality	80%–18:00	10%–14:00
Phenobarbital— long acting (190 mg/kg)	Rats	Percent mortality	100%–12:00	0%–24:00
Phenobarbital-Na—short acting (90 mg/kg)	Rats	Percent mortality	80%–18:00	30%–14:00
Oxazolone, an allergen (0.07 ml of 10% solution to surface of ear)	Rats	Immune response after second administration as change in ear thickness	150 µm–10:00	20 µm–16:00
Histamine (300 µg subcutaneously)	Guinea pigs	Area of erethyma on second administration on third administration	1.7 cm²–10:00 / 2.2 cm²–10:00	1.2 cm²–16:00 / 1.2 cm²–16:00
Oxygen (100%)	Rats	Minutes exposure to produce grand mal seizure	30 min–20:00 to 02:00	65 min–07:00 to 10:00
Hydroxyurea, a teratogen (750 mg/kg)	Rats	Number of teratogenic abnormalities	0–12:00	2.5–24:00
Morphine (8 mg/kg)	Rats	Percent showing analgesia by tail pinch test	90%–24:00	40%–08:00
Pentobarbital-Na (35 mg/kg)	Rats	Durations of anesthesia	90 min–18:00 to 22:00	55 min–08:00
Histamine	Humans	Area of erythema	7 cm²–07:00	2.5 cm²–19:00
Lidocaine (4 mg intradermal) (40 mg around tooth)	Humans	Duration of anesthesia	80 min–14:00 / 30 min–14:00	30 min–24:00 to 08:00 / 12 min–24:00 to 08:00
Ethanol (0.75 g/kg)	Humans	Percent decrease from maximum blood level in 2 hours	47%–17:00	15%–13:00
Arabinosylcytosine, a chemotherapeutic drug (200 mg/kg)	Mice	Percent mortality	75%–24:00	10%–14:00

Data from L.E. Scheving, "Chronotoxicology in General and Experimental Chronotherapeutics of Cancer" in *Chronobiology: Principles and Applications to Shifts in Schedules*, 1980.

Certain drugs used to prevent rejection of transplanted organs suppress the body's immune system, including its ability to fight infection. These drugs also decrease the release of leukocytes (white blood cells) into the blood, letting the number in circulation reach dangerously low levels. Because the number of circulating leukocytes varies rhythmically from a maximum just before midnight to a minimum just before noon, such variations should be considered in assessing the effects of the drugs.

A new branch of clinical practice, **chrononcology** (kron-on-kol'o-je), applies knowledge of rhythms to the treatment of cancer. Two important considerations in designing cancer therapy are rhythms of cell division in malignant and normal cells, and rhythms of the patient's tolerance for chemotherapeutic agents. For many kinds of malignant and normal cells, it is possible to predict approximately when most cells of a particular kind will be synthesizing DNA. Chemotherapeutic drugs and radiation have their maximum effects when most malignant cells are synthesizing DNA. If most normal cells synthesize DNA at a different time, damage to them can be minimized. Responses to chemotherapeutic agents also are rhythmic. An agent can be highly toxic at one time of day and only moderately toxic at another.

Chronotherapy (kron-o-ther'a-pe), the timing of treatments with rhythms, is more effective than **homeostatic therapy,** giving constant doses at regular intervals. In studies of over 1000 mice inoculated with certain leukemic cells, chronotherapy was vastly superior to homeostatic therapy. Among untreated mice, none lived more than 10 days. Among mice receiving homeostatic treatment with arabinosylcytosine (ARA-C), 50 percent lived at least 10 days and 18 percent lived more than 60 days. Among mice receiving ARA-C chronotherapy, 90 percent lived at least 10 days and 37 percent lived more than 60 days. Chronotherapy is now used in some human cancers with good results.

The patient usually receives several drugs according to elaborate schedules based on cellular rhythms and on the patient's tolerance rhythms.

Clinical implications of circadian rhythms extend to medical personnel, who display the same reduced performance at night as other shift workers. Fortunately, the stimulation of an emergency tends to counteract lower performance. Though this benefits patients, shift work may still impair the health of personnel.

A bold proposal related to clinical use of circadian rhythms is that every person should have a profile of rhythms made while in a healthy state. This profile, which would contain a period and range of all measurable rhythms, could be used to determine optimal treatment schedules if illness occurs. It would allow detection of rhythm abnormalities indicative of certain diseases, too. Some evidence already exists that specific rhythm changes can be related to particular metabolic, sleep, respiratory, renal, psychiatric, and endocrine disorders. Examples include cortisol and ACTH levels in Cushing's and Addison's diseases, plasma tryptophan and tyrosine levels in depression, luteinizing hormone levels in anorexia nervosa, and abnormal mitotic rhythms in cancer.

Problems

1. Summarize the current theories regarding mechanisms of circadian rhythms.
2. Design an experiment using available methods to determine whether a rhythm is mainly endogenous or mainly exogenous.
3. Design and, if possible, carry out an experiment to determine whether your mental performance displays a circadian rhythmicity.
4. Prepare a list of activities to help travelers adjust to time-zone transitions when traveling east and when traveling west.
5. Survey recent literature to determine whether new theories regarding circadian rhythms have been proposed.

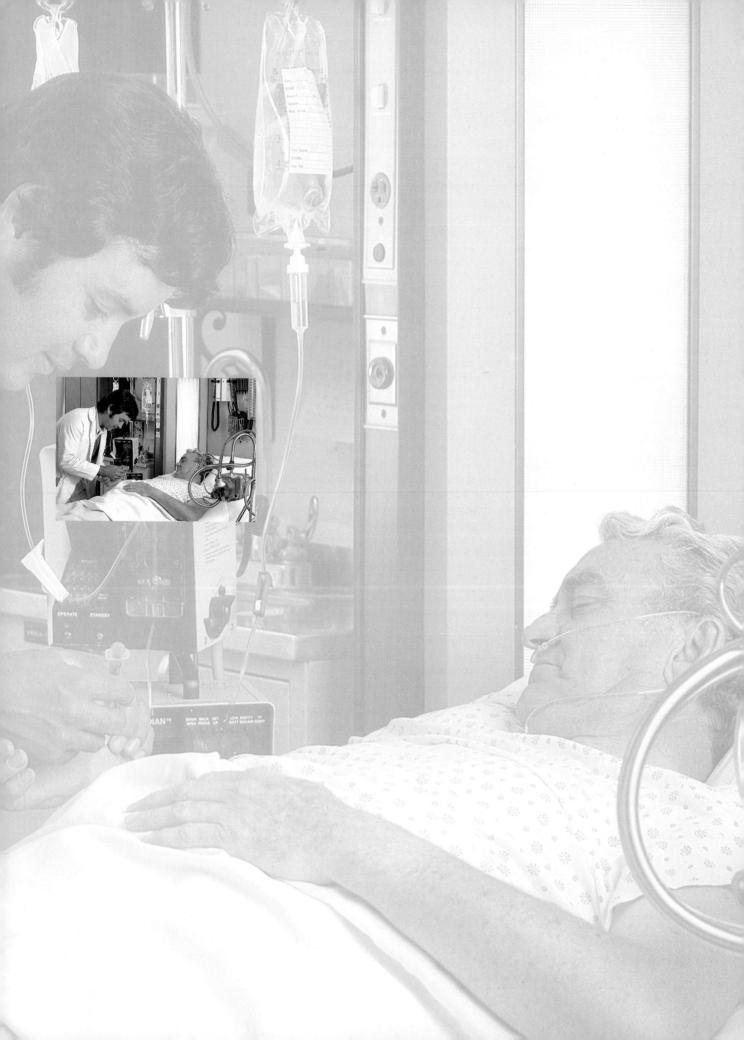

HOMEOSTATIC SYSTEMS

OBJECTIVES

1. Describe the role of blood in the circulatory system and the general composition and functions of blood.

2. Describe the composition of plasma and the functions of its components.

3. List the formed elements of blood, and explain how they arise.

4. Describe the properties and functions of erythocytes, hemoglobin, and iron metabolism.
5. Explain how the number of erythrocytes is regulated.
6. Summarize disorders associated with erythrocytes.

7. Describe the properties and functions of leukocytes.
8. Summarize the disorders of leukocytes.

9. Describe the properties and functions of platelets.

10. Explain the process of hemostasis: how it is started, stopped, and prevented.
11. Explain how disorders of hemostasis cause abnormal clotting or prevent clotting.

12. Describe the composition and function of lymph.

17

BLOOD AND LYMPH

BLOOD: A CIRCULATORY SYSTEM COMPONENT

Blood is pumped by the heart through an extensive system of closed vessels that reach nearly every body cell. These components—blood, heart, and blood vessels—make up the circulatory system (Figure 17.1). Lymph and lymph vessels are closely associated with these components. Together blood and lymph deliver nutrients to cells, remove wastes from them, and contribute to overall body–water balance.

Health scientists, athletes, nutritionists, and almost all of us are interested in blood—it carries substances that keep cells healthy, and healthy cells are essential to a healthy body. Health scientists are particularly interested in blood because many diagnostic tests are performed on blood.

Blood consists of 55 to 60 percent **plasma** (plaz'mah), the liquid portion of the blood, and 40 to 45 percent **formed elements,** including red blood cells called **erythrocytes** (er-ith'ro-sītz), white blood cells called **leukocytes** (loo'ko-sītz), and cell fragments called **platelets** (Table 17.1 and Figure 17.2). The presence of formed elements and plasma proteins make blood about five times as viscous as water. It has a pH of 7.35 to 7.45, depending on how much carbon dioxide it is carrying.

Blood makes up about 8 percent of the body weight in a lean person, but the amount varies according to body size and the amount of fat stored in the body. Small, thin women have only about 3 liters of blood, and larger women, including those whose bodies contain large amounts of fat have about 4 liters. Small men have about 4 liters of blood, but most men have from 5 to 6 liters.

Blood performs several important functions (Table 17.2). It transports nutrients and oxygen to the cells, and carbon dioxide and other waste products away from them,

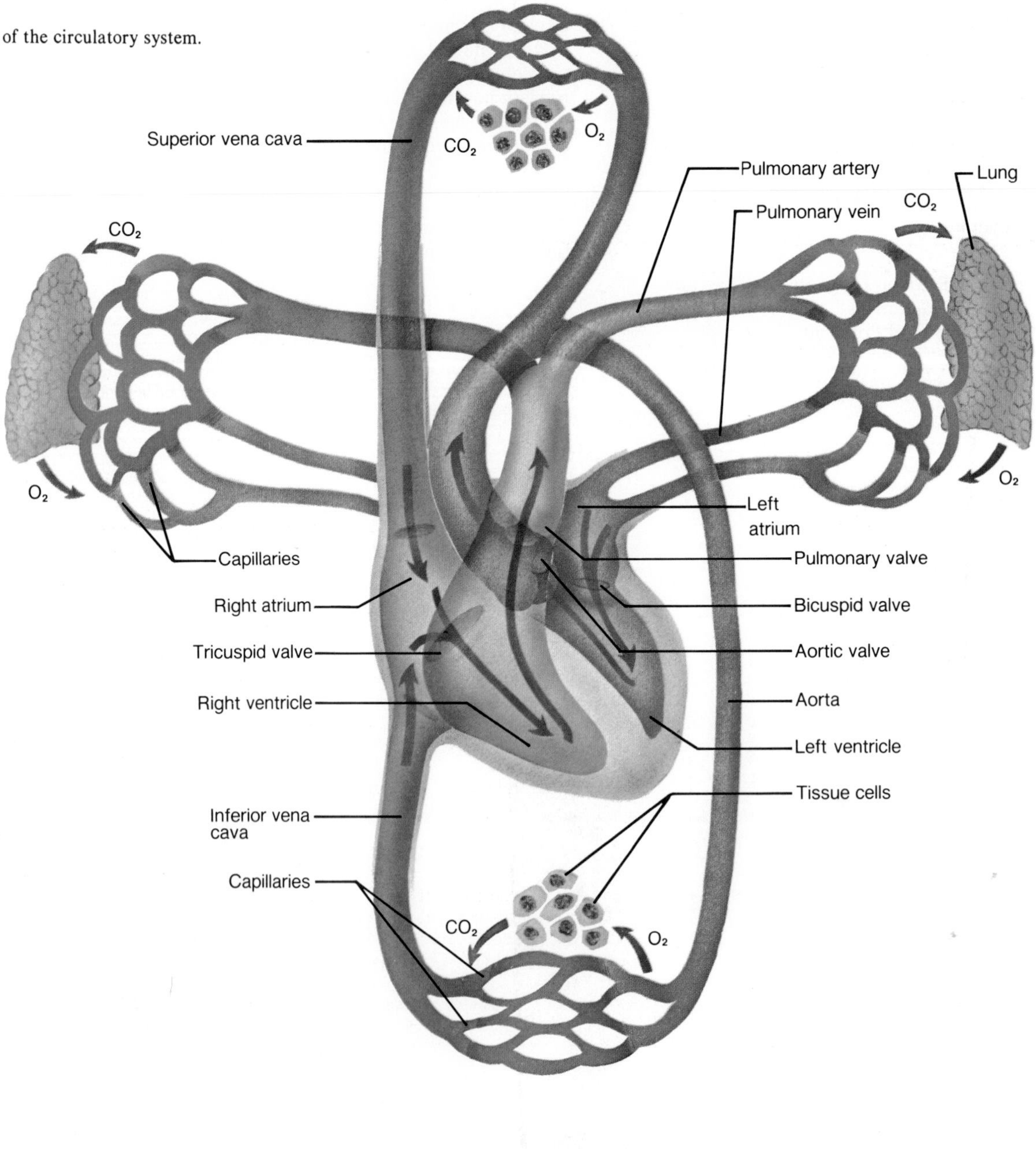

FIGURE 17.1
The general plan of the circulatory system.

helping to assure that cells can carry out their metabolic processes. Blood also transports hormones that control various body functions. Certain leukocytes phagocytize foreign matter and others contribute to immunity (Chapter 21). Clotting factors prevent blood loss after tissue injury. Homeostatic mechanisms maintain the concentrations of blood components within a small range in spite of constant addition and removal of substances. They also maintain cellular water, body fluid pH, body temperature, and the concentrations of many substances in a narrow range.

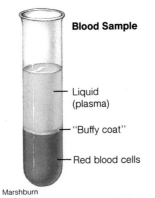

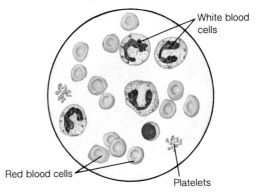

FIGURE 17.2
Components of blood.

TABLE 17.1	Normal Numbers of Formed Elements in Human Blood
Element	**Normal Numbers per Microliter***
Erythrocytes	
Adult male	4.6 to 6.2 million
Adult female	4.2 to 5.4 million
Infant and child	4.5 to 5.0 million
Neonate	5.0 to 5.1 million
Adult at 4000 meters (altitude)	6.0 to 8.0 million
Leukocytes	5000 to 9000
Percent of all leukocytes	
Neutrophils	50%–70%
Eosinophils	1%–4%
Basophils	0.1%
Monocytes	2%–8%
Lymphocytes	20%–40%
Platelets	250,000 to 300,000

*1 microliter = 1/1,000,000 liter = 1 cubic millimeter.

TABLE 17.2	The Functions of Blood	
Functions	**Examples**	
Transport		
Nutrients	Glucose, amino acids, vitamins, and minerals are carried to nearly all cells.	
Oxygen	Oxygen is transported from the lungs to all cells of the body by erythrocytes.	
Carbon dioxide	Carbon dioxide is transported from cells to the lungs.	
Waste products	Urea, uric acid, and other waste products are carried from cells to the liver, lungs, and kidneys.	
Hormones	Hormones are carried from endocrine cells to target cells.	
Protection		
Invasion of microorganisms	Leukocytes engulf and destroy some microorganisms. Antibodies produced by certain leukocytes react with specific antigens, such as those on microorganisms.	
Damage from toxic substances	Some toxic substances are carried to the liver, where they are rendered nontoxic, and then to the kidneys, where they are excreted. Some toxic substances are inactivated by antibodies called antitoxins.	
Blood loss	The clotting mechanism seals off blood vessels. Substances released from platelets initiate clotting after tissue injury.	
Regulation		
Water content of cells	Centers in the hypothalamus detect changes in the blood osmotic pressure and indirectly cause the blood to gain or lose water, thereby maintaining nearly constant water content of cells.	
pH of body fluids	Certain neural centers detect carbon dioxide concentration and regulate the pH of body fluids indirectly. (An excess of carbon dioxide causes acid to accumulate.) Buffers in blood help to keep blood pH nearly constant.	
Body temperature	Temperature sensors in the hypothalamus detect changes in body temperature and act to increase or decrease blood flow to the skin, thereby increasing or decreasing heat loss, respectively.	

TABLE 17.3   Constituents of Plasma

Constituent	Amount/Concentration*
Water	90% of plasma
Electrolytes (inorganic)	<1% of plasma
Na^+	142 mEq/l (142 mmol/l)
K^+	4 mEq/l (4 mmol/l)
Ca^{2+}	10 mEq/l (5 mmol/l)
Mg^{2+}	3 mEq/l (1.5 mmol/l)
Cl^-	103 mEq/l (103 mmol/l)
HCO_3^-	27 mEq/l (27 mmol/l)
HPO_4^{2-}	4 mEq/l (2 mmol/l)
SO_4^{2-}	1 mEq/l (0.5 mmol/l)
Gases	<1% of plasma
CO_2 (mostly as HCO_3^-)	60 ml/100 ml
O_2	0.2 ml/100 ml
N_2	0.9 ml/100 ml
Nutrients	<1% of plasma
Glucose and other carbohydrates	100 mg/100 ml
Amino acids	40 mg/100 ml
Lipids	500 mg/100 ml
Cholesterol	150–250 mg/100 ml
Vitamins	Traces
Trace elements	Traces
Waste products	<1% of plasma
Urea	<20 mg/100 ml
Creatinine	<1 mg/100 ml
Uric acid	5 mg/100 ml
Bilirubin	0.2–1.2 mg/100 ml
Proteins	6% of plasma (2.5 mmol/l)
Albumins	4.5 g/100 ml
Globulins	2.5 g/100 ml
Fibrinogen	0.3 g/100 ml
Hormones	Very small amounts
pH	7.35–7.45

*Concentrations for some substances are expressed in both milliequivalents (mEq) and millimoles (mmol). One millimole is one-thousandth of a gram molecular weight of a substance. For substances that have a valence of 1, mEq and mmol are equal; for substances that have a valence of 2, 2 mEq equal 1 mmol.

PLASMA

Plasma, the fluid that remains when the formed elements are removed from blood, is more than 90 percent water (by weight) and contains proteins, electrolytes (inorganic ions), gases, nutrients, waste products, hormones, and blood clotting factors (Table 17.3). Cholesterol and other blood lipids are transported on various blood proteins (Chapter 24). If clotting factors are removed from plasma, the fluid that remains is called **serum** (se'rum). Each liter of plasma contains 60 to 80 grams of protein, mostly made in the liver and composed of about 55 percent albumin, 38 percent globulin, and 7 percent fibrinogen, an important protein in the blood clotting mechanism.

Albumins (al-bu'minz) are large globular proteins negatively charged at normal blood pH. They transport calcium ions and relatively water-insoluble blood components and drugs such as aspirin, barbiturates, and digitalis (a heart drug). Albumins exert osmotic pressure that helps hold water in the blood.

Globulins (glob'u-linz), proteins larger than albumins, vary in size and function. *Alpha globulins* include *haptoglobulin* that transports hemoglobin from erythrocytes, *ceruloplasmin* (se-roo''lo-plaz'min) that transports copper, *prothrombin* (pro-throm'bin) and other blood clotting factors, and some lipid transport proteins. *Beta globulins* include *transferrin* (trans-fer'rin), which transports iron, and some lipid transport proteins. *Gamma globulins,* also called immunoglobulins or antibodies, are concerned with immunity (Chapter 21). They are synthesized by white blood cells called **lymphocytes** (lim'fo-sītz).

FORMED ELEMENTS AND THEIR ORIGINS

As already noted, formed elements of the blood include erythrocytes, leukocytes, and platelets. The term **formed elements** is used because platelets have form and are important elements of blood, but they are not cells. Of these elements, oxygen-carrying erythrocytes are by far the most abundant; they make up about 40 percent of the total blood volume in females and about 45 percent in males. Leukocytes are classified as granular if their cytoplasm contains easily stained granules or agranular if it does not.

Formed elements arise by **hemopoiesis** (he''mo-poi-e'sis) or **hematopoiesis** (he-mat''o-poi-e'sis), shown in Figure 17.3. Hemopoiesis begins in stem cells in the yolk sac of an embryo. These cells divide and their progeny migrate to and form colonies in the liver, spleen, lymph nodes, thymus, and later bone marrow. By adulthood, hemopoiesis occurs only in **lymphoid** (lim'foid) **tissue** such as lymph nodes where lymphocytes form, and in **myeloid** (mi'el-oid) **tissue** such as bone marrow where all other elements form. Bone marrow is a fatty tissue found in hollow cavities in bones.

Erythropoiesis (er-ith''ro-poi-e'sis), the production of erythrocytes, occurs in **red bone marrow,** the color of which is due to the presence of developing erythrocytes. Before about age five, all bones contain red marrow. After that age, marrow in long bones, such as the femur and tibia, becomes increasingly fatty and loses its ability to produce erythrocytes—and its red color. This tissue, now called **yellow bone marrow,** can be induced to form erythrocytes again if the need for new cells is great, as after **hemorrhage** (severe blood loss). Marrow in the sternum, vertebrae, and ribs produces erythrocytes throughout life, though it becomes less productive with age.

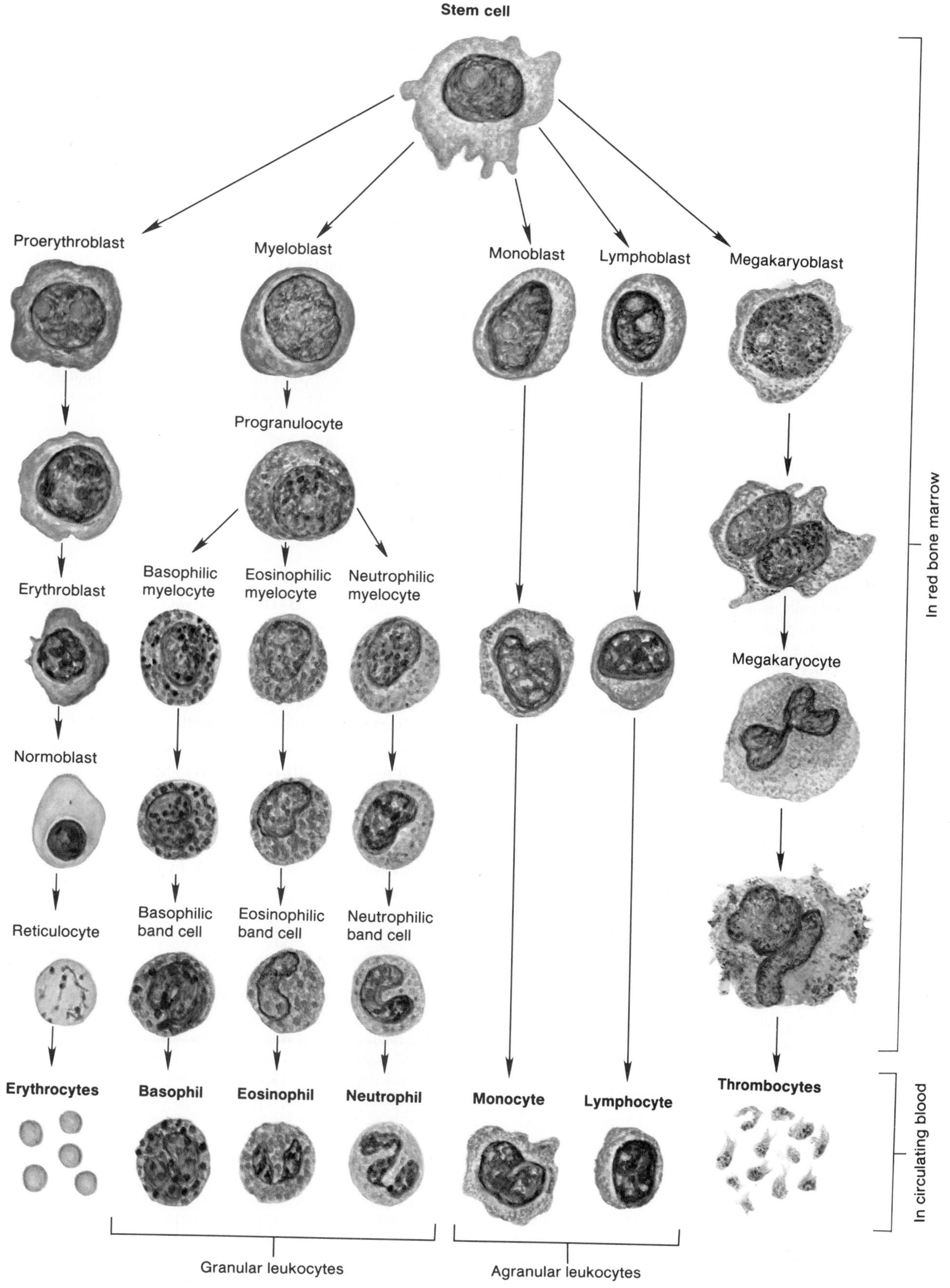

Stem cell

Proerythroblast

Myeloblast

Monoblast

Lymphoblast

Megakaryoblast

Progranulocyte

Erythroblast

Basophilic
myelocyte

Eosinophilic
myelocyte

Neutrophilic
myelocyte

Megakaryocyte

Normoblast

Reticulocyte

Basophilic
band cell

Eosinophilic
band cell

Neutrophilic
band cell

Erythrocytes   **Basophil**   **Eosinophil**   **Neutrophil**   **Monocyte**   **Lymphocyte**   **Thrombocytes**

Granular leukocytes

Agranular leukocytes

In red bone marrow

In circulating blood

FIGURE 17.3

All formed elements of blood are derived from undifferentiated stem
cells. Some of the intermediate cells in the differentiation of each kind
of blood element are shown here.

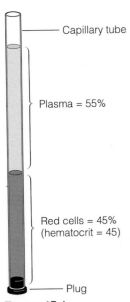

- Capillary tube

Plasma = 55%

Red cells = 45%
(hematocrit = 45)

- Plug

FIGURE 17.4
A hematocrit.

The **hematocrit** (hem-at′o-krit), the percent of blood volume composed of red blood cells, is determined by centrifuging the sample until the cells are packed in the bottom of the tube (Figure 17.4). Leukocytes and platelets, which are not included in the hematocrit, form a grayish upper layer called the **buffy coat**. This routine laboratory study provides an estimate of the oxygen-carrying capacity of the blood.

ERYTHROCYTES

Properties of Erythrocytes

Erythrocytes (Figure 17.5) are biconcave disks 7 to 8 μm in diameter and about 2 μm thick at the edges. This biconcave shape increases the cell's surface area, compared to a sphere of the same volume, and facilitates gas diffusion into and out of the cell. Circulating erythrocytes lack nuclei, mitochondria, microtubules, and the machinery for synthesizing protein. In the absence of mitochondria, erythrocytes obtain energy from glucose without using oxygen. During their lifespan of about 120 days, erythrocytes travel a total distance of about 300 miles.

The erythrocyte plasma membrane is rendered flexible, yet durable, by a meshlike network of proteins beneath the membrane. These proteins include actin filaments, actin-binding proteins, **spectrin** (spek′trin), and *ankyrin* (an-ki′rin), which crosslinks spectrin molecules. These proteins enable erythrocytes to withstand elongation as they pass through narrow capillaries and to regain a normal shape in larger blood vessels. Erythrocytes deficient in spectrin are not easily deformed and have difficulty passing through small capillaries, especially those in the spleen. Such cells become trapped and are destroyed.

Hemoglobin

Each erythrocyte contains about 280 million molecules of **hemoglobin** (he′mo-glo″bin), a pigment that is bright red when oxygenated. It gives erythrocytes the name "red blood cells." Hemoglobin (Figure 17.6) consists of the protein **globin** with associated nonprotein pigment units called **heme** (hēm). Globin contains four polypeptides—two identical alpha chains and two identical beta chains. Imbedded in each chain is a molecule of heme, a chemically complex molecule called a porphyrin. Each heme contains an iron ion that can bind an oxygen molecule (O_2), so each hemoglobin molecule can carry four O_2 molecules.

One might wonder what the advantage is of hemoglobin being enclosed in erythrocytes rather than free in the plasma. Observing what happens when many erythrocytes rupture and release hemoglobin into the plasma provides one important reason. Free hemoglobin molecules become trapped in passageways within the kidneys and interfere with blood filtration. Erythrocytes do not interfere because they are too large to enter the filtration process.

Hemoglobin in erythrocytes increases the oxygen-carrying capacity of blood to about 70 times the amount that can dissolve in plasma. Hemoglobin also transports carbon dioxide, though much carbon dioxide is carried dissolved in plasma. Because carbon dioxide binds to globin, hemoglobin can carry both oxygen and carbon dioxide simultaneously. Under normal physiological conditions, however, erythrocytes traveling from the lungs to other tissues carry large quantities of oxygen and almost no carbon dioxide. In the tissues, they lose oxygen and acquire carbon dioxide, but they still carry more oxygen than carbon dioxide on their way back to the lungs (Chapter 22).

Carbon monoxide (CO) is a deadly gas that binds to iron 200 times more tightly than oxygen (O_2). Bound to hemoglobin, it interferes with oxygen transport, but it kills a person by binding to cytochrome oxidase, a heme-containing enzyme that cells need to obtain energy. Because it is a colorless, odorless gas, people can be overcome by CO unknowingly. A first sign of CO poisoning is extremely red skin because hemoglobin carrying CO is brighter red than oxygenated hemoglobin. This is followed by marked euphoria, then sleepiness, coma, and death. Removing a victim from the CO source is not sufficient treatment because CO remains bound to iron. Transfusion of erythrocytes can increase the oxygen-carrying capacity of the blood and pure oxygen can be given under pressure to replace some CO. A common cause of CO poisoning is a faulty gas heater, which releases CO when it fails to completely burn fuel to CO_2 and water. CO from combustion engines contributes to air pollution, and CO in the air provides an index of the pollution level.

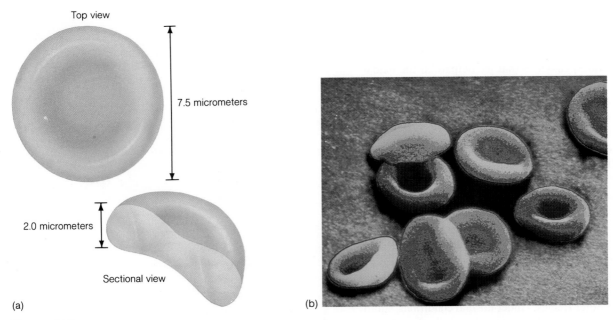

Top view

7.5 micrometers

2.0 micrometers

Sectional view

(a)

(b)

FIGURE 17.5
Erythrocytes: (*a*) diagram, and (*b*) as seen with a scanning electron microscope.

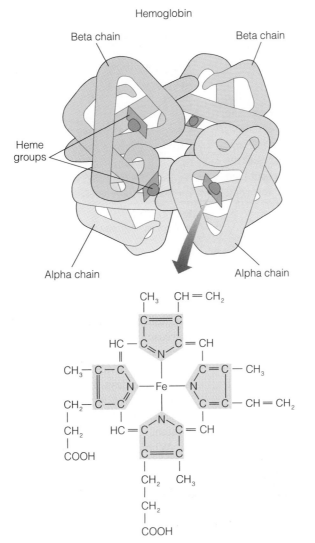

Hemoglobin

Beta chain

Beta chain

Heme groups

Alpha chain

Alpha chain

FIGURE 17.6
The three-dimensional structure of a hemoglobin molecule. The molecular structure of heme is enlarged. Note the iron atom in the center of the molecule.

Iron Metabolism

In addition to hemoglobin synthesis, iron is essential for the synthesis of myoglobin, an oxygen carrier in muscle, and cytochromes, enzymes that participate in the oxidation of nutrients for energy. Iron metabolism (Figure 17.7) begins with oxidation of Fe^{3+} in some foods to Fe^{2+}, and the binding of Fe^{2+} to the protein **gastroferrin** (gas″tro-fer′in) in the stomach. Gastroferrin ferries iron to the small intestine, where it is transferred to a carrier protein in the intestinal lining cells and absorbed into the blood. Once absorbed, iron is oxidized to Fe^{3+} and combines loosely with a plasma protein **transferrin.** Of the body's 4 grams of iron, over two-thirds is in hemoglobin and most of the rest is stored in the liver in **ferritin** (fer′it-in). Ferritin consists of a protein **apoferritin** (ap″o-fer′it-in) with iron loosely bound to it.

When the plasma iron concentration drops, iron is released from ferritin and carried by transferrin to bone marrow and other tissues. Under normal conditions, small amounts of iron (about 0.6 mg) are lost daily through the feces. Bleeding due to injury or excessive menstrual flow can seriously deplete iron stores. When excess iron accumulates in the body, it is deposited in cells in insoluble granules of **hemosiderin** (hem″o-sid′er-in). Such deposits result from many blood transfusions and not from excessive dietary iron. Dietary iron can be increased by eating iron-enriched cereals, green leafy vegetables, and red meats.

Iron absorption appears to be regulated mainly by the amount of iron already available in the body, though the details of regulatory processes are not fully understood. When apoferritin is saturated with iron, it cannot accept iron from transferrin and transferrin, too, becomes saturated. When transferrin is saturated, it cannot accept iron from the intestinal mucosa and iron fails to be absorbed until some transferrin becomes available.

Regulation of Erythrocyte Numbers

Erythrocytes are constantly dying and being replaced at a rate of about 2.5 million per second (Figure 17.8). Because replacement occurs at the same rate as destruction, the total erythrocyte number and volume in the blood remains remarkably constant. Iron and vitamins such as vitamin B_{12} and folic acid are necessary for the production of new erythrocytes, so adequate amounts of these substances must be available in the diet to maintain normal erythrocyte production.

Tissue oxygenation is the major regulator of erythrocyte production. Conditions that reduce the quantity of oxygen reaching the tissues stimulate erythrocyte production. These conditions include circulatory disorders that reduce blood flow to the tissues, lung disorders that impair blood oxygenation, disorders that reduce erythrocytes and hemoglobin, and living at high altitudes where the atmospheric oxygen concentration is low.

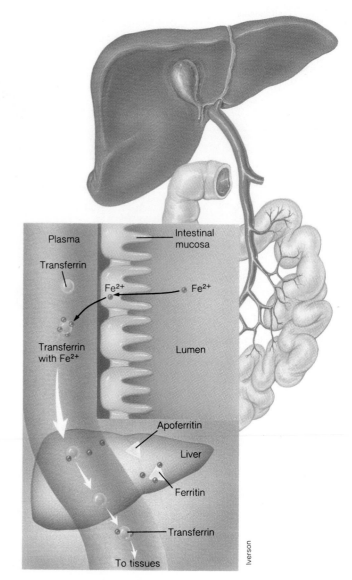

FIGURE 17.7
Absorption, transport, and storage of iron.

Red bone marrow is stimulated to produce erythrocytes, not by lowered blood oxygen directly, but by a hormone **erythropoietin** (er-ith″ro-poi′et-in), which is synthesized by the kidney and secreted when blood oxygen decreases. The liver may produce small quantities of erythropoietin. The hormone induces stem cells in red bone marrow to divide and differentiate into erythrocytes. Within five days, the number of circulating erythrocytes increases. Men have more hemoglobin and erythrocytes than women because male hormones stimulate their production.

When a person donates a pint (475 ml) of blood, about 10 percent of the total blood volume is removed. Through a variety of homeostatic mechanisms, the body replaces lost volume in a matter of hours, plasma proteins in one to three days, and lost cells and platelets in three to four weeks.

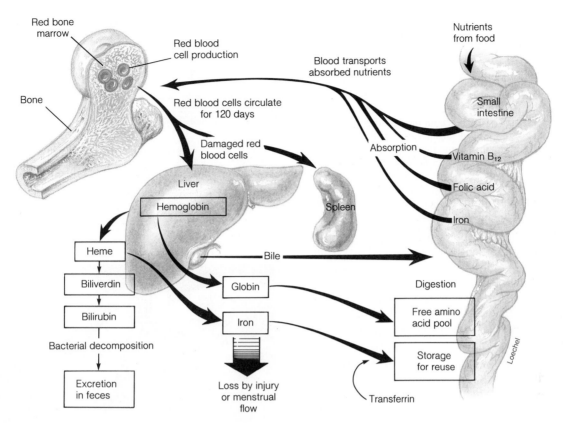

FIGURE 17.8
Formation and destruction of erythrocytes, including degradation of hemoglobin.

The continuous destruction of erythrocytes is not as wasteful a process as it might first appear (Figure 17.8). Phagocytic cells in the liver, spleen, and bone marrow engulf and digest cells that have lost their ability to deform. The phagocytes separate hemoglobin into heme and globin, and digest the globin to amino acids, which are used to make new proteins. They further separate heme into iron and porphyrin. Transferrin carries the iron to red bone marrow where it is reused in making erythrocytes and to the liver where it is stored as ferritin. The phagocytic cells degrade the porphyrin to the green pigment **biliverdin** (bil″i-ver′din) and then to the yellow-orange pigment **bilirubin** (bil″i-roo′bin). Bilirubin is secreted in bile and excreted in the feces.

Disorders of Erythroctyes

Anemia, a decrease in hemoglobin or in erythrocyte numbers, has many causes. All anemias cause fatigue and other symptoms related to the reduced oxygen-carrying capacity of the blood.

Nutritional anemia is due to a dietary deficiency of iron or vitamins required to make normal erythrocytes or a failure to absorb such nutrients from the intestine (Chapter 26). Erythrocytes are reduced in number and contain decreased amounts of hemoglobin. Nutritional anemias can be treated by providing an adequate diet including excesses of poorly absorbed nutrients.

Hemorrhagic (hem″or-aj′ik) **anemia** results from excessive blood loss—severe injury, bleeding ulcers, or excessive menstrual flow. Sudden excessive blood loss can be fatal unless blood is replaced by transfusions. The loss of small amounts of blood over time causes chronic fatigue. Finding and treating the cause of bleeding is the most appropriate treatment.

Hemolytic (he″mo-lit′ik) **anemia** results from **hemolysis** (he-mol′is-is), the rupture of the erythrocyte membranes, as can occur in streptococcal infections or immune reactions against a person's own red blood cells. Hemolysis releases the contents of erythrocytes into the plasma. It can lead to **jaundice** (jawn′dis), a yellowing of the skin due to bilirubin accumulation. Treatment is directed toward removing or alleviating the cause. Transfusions are used only in severe cases. The parasitic disease malaria causes hemolytic anemia by stimulating cells to release a substance called **tumor necrotic factor** (TNF), which rapidly lyses red blood cells. When released in cancer patients, TNF cuts off the blood supply to tumors. Now recognized to be the same as cachectin, it also causes wasting of other tissues.

Jaundice is a common disorder in newborns, especially premature ones, as erythrocytes are broken down faster than the immature liver can get rid of bilirubin. Erythrocyte numbers normally decrease from about 6 million per microliter in a fetus to about 5 million per microliter by the time an infant is born. This process is greatly accelerated in premature infants, where the blood bilirubin can exceed 20 times the normal concentration. Excess bilirubin in the blood causes **kernicterus** (ker-nik'ter-us), the degeneration of brain neurons, and permanent brain damage. Jaundiced infants are exposed to ultraviolet light, which accelerates bilirubin breakdown. A new treatment uses a synthetic porphyrin that contains tin instead of iron to bind to the enzyme that normally converts hemoglobin to bilirubin. More hemoglobin is excreted directly and less is converted to bilirubin, so jaundice is decreased and the risk of brain damage lessened. The treatment must be monitored carefully because too much hemoglobin excretion can cause kidney damage.

Aplastic (a-plas'tik) **anemia** is due to destruction of bone marrow by leukemia, X ray, and other forms of radiation, or certain drugs such as chloramphenicol. The term *aplastic* means not to form, so aplastic anemia results from failure of erythrocytes to form. In leukemia, malignant cells dividing out of control destroy normal hemopoietic cells. Bone marrow transplants from genetically compatible individuals are sometimes used to treat this disorder.

Pernicious (per-nish'us) **anemia** is a shortage of erythrocytes because certain stomach lining cells fail to produce a substance called **intrinsic factor,** which is required for vitamin B_{12} absorption. The resulting vitamin deficiency leads to abnormally large, irregularly shaped erythrocytes with especially fragile cell membranes. Pernicious anemia is treated with vitamin B_{12} injections to compensate for the failure to absorb it.

Space anemia, a 10 to 15 percent decrease in the mass of erythrocytes, occurs in astronauts during a space flight. Though associated with weightlessness, how weightlessness decreases erythrocyte mass is unknown. The condition corrects itself after four to six weeks back on earth with no apparent harm. Whether harm might result from extended space visits is unknown.

Sickle cell anemia occurs in individuals who have inherited a certain recessive gene for hemoglobin synthesis from each parent (Chapter 30). Their hemoglobin differs from normal hemoglobin by only a single amino acid in the beta polypeptides, and such hemoglobin alters erythrocyte function. Under conditions of reduced oxygen, the hemoglobin molecules stick to one another and form fibrils that deform the cells into sickle shapes; hence, the name of the disease (Figure 17.9). The sickled cells aggregate in blood vessels and block circulation. Pain in bones

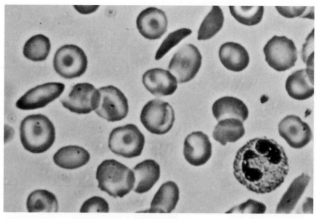

(a)

(b)

FIGURE 17.9
Sickled erythrocytes: (*a*) by light microscope, and (*b*) by scanning electron microscope.

and joints due to occluded vessels is a common complaint of victims of this disease. Individuals who have a single gene for sickle cell anemia are said to have "sickle cell trait." Though generally not affected by the disease, they can transmit the defective gene to their offspring.

The sickle cell gene is most common in subtropical Africa, the Mediterranean region, and parts of Asia where the parasitic disease malaria is, or once was, common. Having one sickle cell gene increases a person's resistance to malaria by causing erythrocytes to sickle when the parasites enter them. The sickled cells die. They and the parasites are removed from circulation. In regions where the entire population was exposed to malaria, many people lacking the gene died of malaria and many with two sickle cell genes died of sickle cell anemia; hence, many survivors carry one sickle cell gene.

Thalassemia refers to several inherited hemolytic anemias in which hemoglobin synthesis is defective. It is common among people of the Mediterranean area and

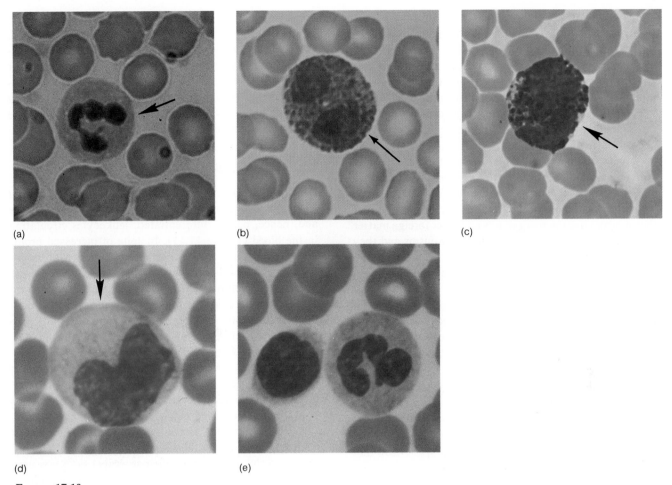

FIGURE 17.10
(*a*) Neutrophil, (*b*) eosinophil, (*c*) basophil, (*d*) monocyte, and
(*e*) lymphocyte.

among those whose ancestors came from that area. The disease can be mild in individuals who have one defective gene but severe in those who have two defective genes. It can be treated with blood transfusions.

In contrast to anemia, **polycythemia** is a condition in which abnormally large numbers of erythrocytes are produced. Hematocrits as high as 80 percent with correspondingly reduced plasma volume are seen. The extremely viscous blood flows so slowly that tissues fail to receive enough oxygen in spite of the many erythrocytes. Primary polycythemia is a malignancy of erythrocyte precursors and leads to the presence of many immature red cells. Secondary polycythemia can result from smoking, chronic heart or lung diseases, or prolonged strenuous activity at high altitude. The last of these is an apparently normal, appropriate response to tissue hypoxia.

LEUKOCYTES

Properties of Leukocytes

Leukocytes, the largest of blood cells, are classified as granular or agranular. **Granular leukocytes** include **neutrophils** (nu′tro-filz), **basophils** (ba′so-filz), and **eosinophils** (e″o-sin′o-filz); and **agranular leukocytes** include **monocytes** (mon′o-sītz) and **lymphocytes** (Figure 17.10). Because mature granular leukocytes have multilobed nuclei, they are sometimes referred to as **polymorphonuclear** (pol″e-mor″fo-nu′kle-ar) **leukocytes** (PMNLs). Hormones similar to erythropoietin are now known to stimulate leukocyte production. Differentiation of stem cells into each kind of blood cell is controlled by a different hormone.

Granular Leukocytes

Neutrophils, the most numerous of all leukocytes, migrate to any site of injury and guard the body against invasion by microorganisms and foreign substances. They are produced and released into the blood in greater numbers after injury or infection, and they migrate from the blood to affected tissues. Their actions in tissue constitute part of the **inflammatory process.** By a process called **chemotaxis** (ke″mo-tak′sis), injured cells release chemicals that attract neutrophils from capillaries toward the site of an injury. By another process called **diapedesis** (di″ah-pĕ-de′sis), neutrophils elongate and squeeze through spaces between cells in the wall of a capillary. At the site of injury, they phagocytize microorganisms and other foreign matter. In most instances, they kill the microorganisms and they, themselves, die. Even when neutrophils do not phagocytize foreign substances, they live only a few days after they are released from bone marrow.

Eosinophils, which increase in number during allergic reactions, probably have several functions—detoxifying foreign substances, turning off inflammatory reactions, and breaking down products of immune reactions. They may also be phagocytic.

Basophils, though not phagocytic, synthesize and secrete histamine and heparin, both of which participate in the body's defense mechanisms. **Histamine** (his′tah-mēn) initiates the inflammatory response and **heparin** (hep′ar-in) inhibits blood clotting, especially where clots might occlude vessels.

Certain cells called **mast cells** are prominent in connective tissue around capillaries. When such tissue is injured, the mast cells release histamine and heparin. This led to the idea that mast cells might be basophils that have migrated to the tissues. Most investigators now believe that mast cells and basophils arise from different kinds of cells and become more alike as they develop.

Agranular Leukocytes

Monocytes are especially adept at phagocytizing large particles of debris. When they enter tissues, they are transformed into macrophages, which have two important functions. They digest debris left from neutrophils that have died after ingesting smaller particles, and they stimulate lymphocytes to participate in immunologic reactions.

Some lymphocytes circulate in the blood but most are found in lymphoid tissues, such as lymph nodes, spleen, and tonsils. They function in immunity (Chapter 21).

Lymphocytes and monocytes enter blood along with lymph that has collected in lymph vessels. From the blood, these cells move by diapedesis to the tissues and are returned to the lymph vessels. Because of this recycling, it is difficult to estimate their longevity. Among lymphocytes labeled with radioactive tracers, a few have been found in blood 100 to 300 days after labeling and some researchers believe they circulate for years.

Disorders of Leukocytes

Leukemia (loo-ke′me-ah), a malignancy of blood-forming tissues, is characterized by large numbers of immature and, hence, poorly differentiated, leukocytes circulating in the blood (Figure 17.11). In spite of increased proliferation, the leukocytes are unable to function normally. Lymphocytes proliferate in **lymphoid leukemia,** whereas neutrophils proliferate in **myeloid leukemia.** Either form can be **acute** (rapidly developing, but of short duration) or **chronic** (slowly developing and of long duration). Untreated acute leukemia can lead to death in a few months, but chronic leukemia allows survival for years. The diversion of bone marrow to producing malignant cells rather than normal cells leads to anemia, decreased platelets, bleeding, and a tendency to develop bacterial infections because of the reduced numbers of normal neutrophils.

Prior to the 1960s, acute lymphoid leukemia, the most common form of childhood malignancy, was almost invariably fatal. Since then, chemotherapeutic drugs used in combination have proven effective in producing remissions, during which the individuals are symptom-free. Follow-ups of children treated before 1975 showed that nearly half remained completely free of the disease for two and one-half years, at which time they stopped receiving therapy. Among these disease-free children, 80 percent had no relapse in four years after therapy was stopped, and none of the 80 percent that remained healthy for four years has had a later relapse. Today, bone marrow transplants are sometimes done after radiation or chemotherapy, and even higher remission rates are achieved.

Infectious mononucleosis (mon″o-nu″kle-o′sis), called the "kissing disease" because it is transmitted by close contact, is caused by the Epstein-Barr virus. Though this virus infects nearly everyone and persists in lymphocytes for life, it causes disease only in susceptible children and young adults. Symptoms include fever, fatigue, a severe sore throat, and enlarged lymph nodes. Infectious

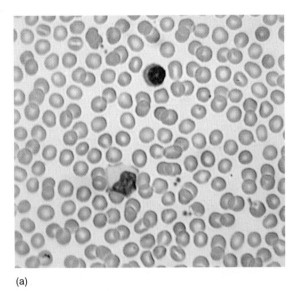

(a)

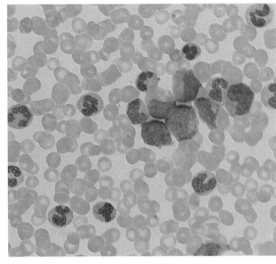

(b)

FIGURE 17.11
Blood smears from (*a*) a normal person and (*b*) a person with leukemia.

mononucleosis is mentioned here because it greatly increases the number of circulating leukocytes and the proportion of lymphocytes among them. The disease is treated by alleviating symptoms and watching for complications, which are rare but may include rupture of the spleen, neurological disorders, hepatitis, and immune reactions against the body's own erythrocytes or platelets. Recovery takes several weeks and is usually complete.

Chronic mononucleosis—is it real and, if so, what causes it? As early as 1985, some researchers suggested that fever and persistent fatigue not associated with any specific illness might be a chronic effect of Epstein-Barr virus (EBV) infection in susceptible people. Studies of patients with chronic fatigue showed EBV in some but not all, and found other latent viruses such as herpes and measles viruses in many. Other researchers think the condition is a kind of depression unrelated to any infection. The disease has been renamed *chronic fatigue syndrome,* and a list of persistent or recurrent symptoms has been established to make diagnosis more precise. These actions will allow researchers to study clear-cut cases and perhaps identify the cause.

PLATELETS

Platelets, also called **thrombocytes** (throm′bo-sītz), are fragments of megakaryocytes only 2 to 4 μm in diameter. They are replaced every 5 to 10 days. Platelets contain enzymes in mitochondria and can oxidatively metabolize glucose for energy. Their main functions are to plug leaks in small blood vessels and to release substances essential for the initiation of blood clotting.

HEMOSTASIS

The arrest of bleeding is called **hemostasis** (he″mo-sta′sis). Blood vessels carry blood between cells in multicellular organisms as long as they remain intact. When, and only when, a vessel is damaged, blood loss must be stopped without stopping blood flow. In hemostasis, following an injury to a small blood vessel, three processes—blood vessel spasm, platelet plug formation, and clotting of blood, around the injury—occur (Figure 17.12). Related to hemostasis is the dissolving of blood clots after they have arrested bleeding.

Blood Vessel Spasm

When a blood vessel is injured, the smooth muscle in its walls contracts and reduces blood flow to the injured area. The contraction apparently is stimulated by the injury and possibly by nerve signals from pain receptors in the injured tissue. The contraction can block blood flow completely by pressing cut surfaces together and allowing sticky endothelial cells of the vessel lining to adhere to one another. Contraction lasts up to 30 minutes in small veins and mild injuries, and longer in larger vessels or more severe injures.

Formation of a Platelet Plug

Platelets normally move freely through blood vessels. If the vessel lining is disrupted and underlying collagen is exposed, platelets adhere to the collagen and to each other to form a **platelet plug** at the site of the injury. Cytoplasmic vacuoles in adhering platelets release substances such as ADP and serotonin, which increase platelet stickiness. This causes circulating platelets to adhere to those already attached to the collagen. In other words, stickiness produces more stickiness by positive feedback and clumps of platelets rapidly increase in size (Figure 17.13). Platelets also release epinephrine, and epinephrine and serotonin cause further blood vessel constriction.

A pair of prostaglandins also contribute to controlling blood flow and bleeding. Platelets release a prostaglandin called thromboxane A_2 (TXA_2), which acts for only about 30 seconds to cause platelet aggregation and vasoconstriction. Arterial lining cells release an antagonistic prostaglandin called PGI_2 (prostacyclin), which acts for about two minutes to prevent platelet aggregation and cause vasodilation. TXA_2 may go too far and contribute to emboli (clots in blood vessels), atherosclerosis, and heart attacks. PGI_2 seems to protect against these disorders, and it probably reduces the likelihood of metastasis of cancer cells by preventing them from adhering to membrane surfaces.

Blood Clotting

Bleeding from small injuries and breaks in capillaries can be stopped by vessel spasms and platelet adhesion, but more extensive injuries require blood clotting. Blood clotting, which can be initiated by extrinsic or intrinsic mechanisms, involves a series of steps ending in the formation of a blood clot (Figure 17.14).

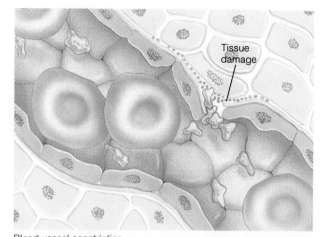

Blood vessel constriction

Platelet aggregation

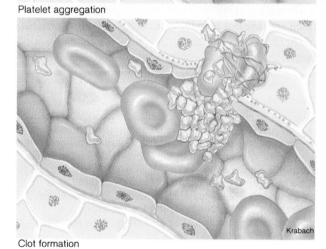

Clot formation

FIGURE 17.12
Summary of the events involved in hemostasis.

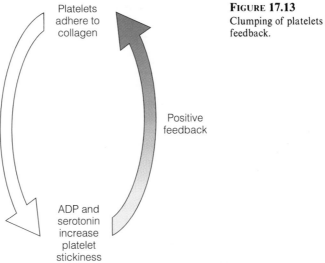

FIGURE 17.13
Clumping of platelets involves positive feedback.

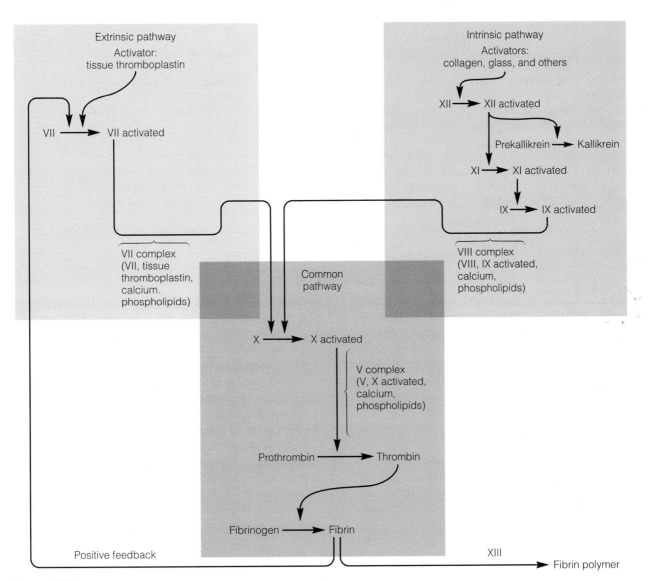

FIGURE 17.14
The mechanisms of blood coagulation showing the extrinsic pathway, the intrinsic pathway, and the final common pathway.

TABLE 17.4   Blood Coagulation Factors

Factor Number	Name	Nature and Origin	Function
I	Fibrinogen	Protein synthesized in liver	Precursor to fibrin
II	Prothrombin	Enzyme synthesized in liver in presence of vitamin K	Precursor to thrombin
III	Tissue thromboplastin	Lipoprotein released from tissues	Activates factor VII
IV*	Calcium ions	Inorganic ion present in plasma	Necessary for reactions in most stages of process
V	Proaccelerin or labile factor	Protein synthesized in liver	Required for extrinsic and intrinsic mechanisms
VI	Number no longer used; substance now shown to be activated factor V		
VII	Proconvertin	Enzyme synthesized in liver in presence of vitamin K	Necessary for extrinsic mechanism, activates factor X
VIII	Antihemophilic factor	Globulin synthesized in liver; absent in inherited disorder hemophilia A	Necessary for intrinsic mechanism
IX	Plasma thromboplastin (or Christmas factor)	Enzyme synthesized in liver; absent in inherited disorder hemophilia B	Necessary for intrinsic mechanism
X	Stuart-Power factor	Enzyme synthesized in liver in presence of vitamin K	Necessary for extrinsic and intrinsic mechanism
XI	Plasma thromboplastin antecedent	Enzyme synthesized in liver	Necessary for intrinsic mechanism
XII	Hageman factor	Enzyme synthesized in liver	Necessary for intrinsic mechanism and to activate plasmin
XIII	Fibrin stabilizing factor	Enzyme found in platelets and plasma	Crosslinks filaments and makes fibrin polymer
Prekallikrein		Enzyme activated by small amount of factor XII	Activates more factor XII; accelerates cascade
High-molecular weight kininogen		Cofactor in plasma	Assists kallikrein in activating factor XII
Pf$_1$	Platelet accelerator	Platelets; same as factor V	Accelerates action of platelets
Pf$_2$	Thrombin accelerator	Platelets	Accelerates thrombin formation
Pf$_3$	Platelet thromboplastic factor	Platelets	Necessary for intrinsic mechanism
Pf$_4$	Platelet factor 4	Platelets	Binds heparin, a natural anticoagulant, during clotting

*IV no longer used; referred to simply as Ca^{2+}.

Numerous inactive blood clotting factors circulate in blood plasma at all times. They include protease enzymes (that activate other proteins), cofactors, and enzymes and substrates for clotting reactions. Several factors are synthesized in the liver with the aid of vitamin K. Without vitamin K, these protein factors lack an important carboxyl group and function poorly in clotting reactions. Clotting factors are described in Table 17.4.

Mechanisms of Initiating Clotting

Tissue damage external to blood vessels themselves activates the **extrinsic mechanism** of blood clotting. In addition to causing blood vessel spasm and platelet plug formation, tissue damage causes injured cells to release a membrane lipoprotein called **tissue thromboplastin** (throm"bo-plas'tin). With the help of calcium ions and other factors, tissue thromboplastin activates factor VII. Active factor VII, in turn, activates factor X, and the blood clotting process is underway.

Blood coming in contact with damaged tissue, such as a negatively charged collagen surface within a blood vessel wall, activates the **intrinsic mechanism** of blood clotting. Steps in this mechanism are as follows: Factor XII converts prekallikrein to *kallikrein* (kal"i-kre'in). Kallikrein and kininogen from platelets accelerate the activation of factor XII, and they work with factor XII to activate factor XI. Active factor XI activates factor IX; active factor IX, in the presence of factor VIII and calcium ions, activates factor X.

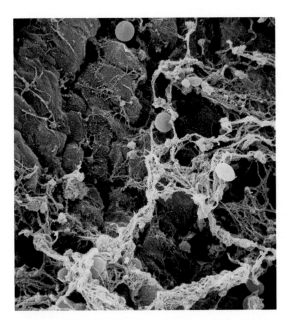

FIGURE 17.15
A mesh of fibrin with entrapped platelets and cells.

Crossovers occur between the extrinsic and the intrinsic pathways. Activated factor XII from the intrinsic system activates factor VII of the extrinsic system. Also, factor VII activated in the extrinsic pathway activates factor IX of the intrinsic pathway. Such crossovers prevent uncontrolled bleeding in some clotting factor deficiencies by activating whichever clotting mechanism is functional.

Formation of a Clot

Once factor X is activated by either mechanism, the pathways converge. Factor X converts the blood protein **prothrombin** to the active enzyme **thrombin.** Thrombin acts on another blood protein **fibrinogen,** converting it to **fibrin.** Polymerization of fibrin occurs by the overlapping of molecules and formation of crosslinks between them (Figure 17.15). Crosslinking is established by an enzyme called *fibrin stabilizing factor,* or factor XIII. Polymerized fibrin is a fibrous material that forms a loose meshwork over the injured area. Blood cells become trapped in the meshwork, reinforce the platelet plug, and close off the opening; thus, a blood clot consists of fibrin, blood cells, and platelets.

The sequence of reactions in blood clotting constitutes a cascade—each step in the process causes production of more molecules of the product of the next step.

Also, once thrombin is formed, it activates more factor VII, thereby acting as a positive feedback mechanism to accelerate the clotting process.

> People get dietary vitamin K from green leafy vegetables, but many receive a fast-acting, synthetic vitamin K, menadione, before surgery. This allows those with deficient intake or absorption of the vitamin, or liver disease to make optimum use of any clotting factors the liver is able to synthesize. Newborns undergoing surgery also receive vitamin K because their livers are immature and may not make enough of some clotting factors.

Inhibition of Clotting

In addition to initiating clotting, mechanisms also exist to prevent clotting unless a vessel has been damaged or to inhibit it after bleeding has been arrested. These mechanisms include heparin activating a protein called *antithrombin* (an"ti-throm'bin). Giving heparin to people prone to form thrombi makes use of this mechanism. Clotting also is limited to some degree by clotting factors being used up in reactions and by PGI_2 (prostacyclin), inhibiting platelet aggregation. The competing actions of PGI_2 and thromboxane help to regulate the clotting process.

> A cell surface thrombin-binding protein called *thrombomodulin* (throm-bo-mod'u-lin) from blood vessel linings also inhibits clotting by converting thrombin to a substance called protein C activator. *Protein C activator* activates *protein C,* an anticoagulant that inactivates factors V and VIII. Protein C production appears to depend on the presence of a dominant gene. Infants lacking the gene often succumb to uncontrolled clotting in blood vessels—most obvious in small vessels visible through the skin—unless they are treated with protein C concentrates.

Dissolution of Clots

Blood clots are temporary structures that seal off a damaged area until healing can take place. After half an hour or more, the clot retracts and becomes smaller and more dense, probably by the action of platelets trapped in the clot. As the fibrin filaments gather around the platelet aggregation, cytoplasmic processes of platelets attach to fibrin and pull the fibers closer together. This is called **clot retraction.** When it occurs in a test tube, one can observe that fluid is squeezed from the clot.

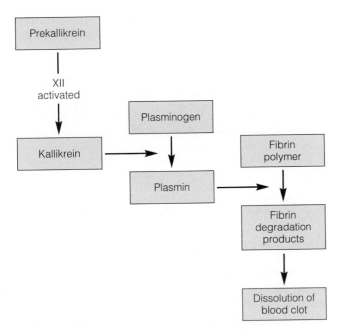

FIGURE 17.16
The mechanism by which blood clots are dissolved.

At the same time a clot is forming, the inactive enzyme **plasminogen** (plaz-min'o-jen) is deposited in the clot (Figure 17.16). When plasminogen is activated, it becomes **plasmin** (plaz'min) and begins to digest the clot. Several factors participate in activating plasmin. Activated factor XII not only participates indirectly in the activation of thrombin, it also is broken down to form proteolytic fragments called *prekallikrein activators.*

The protein *tissue plasminogen activator* (tPA) appears to be released from the epithelial linings of blood vessels and other tissues in response to injury and other stimuli. tPA causes digestion of fibrin specifically by activating plasminogen.

In addition to the above mechanisms, thrombin also activates plasmin and plasmin breaks active factor XII into prekallikrein activators, thereby accelerating clot digestion by positive feedback. Much of the plasmin is incorporated into the clot where it degrades both fibrinogen and fibrin. Plasmin, being a slow-acting proteolytic enzyme, gradually dissolves away the clot while tissue repair is taking place. The products of the action of plasmin are called **fibrin degradation products** (FDP), or fibrin split products. Maintaining an intact circulatory system through which blood flows freely requires a continuous balance between hemostasis and fibrinolysis.

Anticoagulants

Anticoagulants (an''tĭ-ko-ag'u-lantz) are substances that reduce the clotting ability of blood by interfering with reactions in the clotting mechanism. In the body, **heparin,** a complex polysaccharide produced by both mast cells and basophils, reduces the blood's ability to clot. It interferes with activation of thrombin and certain other clotting factors. **Coumarin** (koo'mar-in), an anticoagulant derived from sweet clover, competes with vitamin K and interferes with the synthesis of factors V and VII in the liver. Similar drugs such as *dicumarol* (di-koo'mar-ol) and *warfarin* (war-far'in) also reduce the likelihood of blood clots forming within blood vessels. Coumarin derivatives take up to two days to act, but their effects last longer than heparin and they can be taken orally. In the laboratory, anticoagulants, including heparin, are used to prevent blood samples from clotting before tests can be performed. Sodium citrate and ammonium oxalate bind to calcium ions so they cannot facilitate reactions in the clotting process.

At one time, when physicians believed some diseases were caused by "bad blood," leeches were used to remove blood from patients. Leeches attach by suckers and penetrate the skin, releasing *hirudin* (hĭ-roo'din) from glands near the suckers. As hirudin enters human blood, it inactivates thrombin and prevents blood from clotting while the leech is feeding. Leeches are used today to keep capillary blood from clotting after surgery to reattach fingers.

Vipers often kill their prey by injecting them with anticoagulants. *Arvin* (ar'vin), an anticoagulant in Malayan viper venom, decomposes fibrinogen faster than it can be synthesized. It may prove useful in dissolving clots, especially in blood vessels of the retina.

Disorders of Hemostasis

A **thrombus** (throm'bus) is a blood clot attached to the wall of a blood vessel where it formed. An **embolus** (em'bo-lus) is any foreign object traveling in the bloodstream. Emboli include thrombi that have broken away from the site of formation, clumps of bacteria, and air bubbles. Several theories have been proposed to explain the causes of clot formation in blood vessels. One theory is that damaged vessel linings initiate the clotting process. Platelets adhering to exposed collagen fibers would activate the intrinsic clotting mechanism. Other theories are that thrombus formation is initiated by inflammation from infection or toxic substances or by sluggish blood flow. Thrombi are relatively common in leg veins of inactive patients. In most hospitals, patients wear elastic stockings that maintain pressure on leg veins, thereby helping to maintain blood flow.

Once a clot has formed, it is potentially dangerous. If it remains attached as a thrombus, it can occlude a vessel and reduce blood supply to surrounding tissues. When this happens in a coronary (heart) blood vessel, it is called a

coronary thrombosis (kor'ŏ-na-re throm-bo'sis) and is a common cause of heart attacks. If it breaks loose as an embolus, it can travel to a smaller vessel and occlude it. A **pulmonary embolus,** a clot formed in some other vessel (such as a leg vein) that has lodged in a vessel carrying blood from the heart to the lungs, is especially dangerous because it can severely impair both blood flow from the heart and gas exchange in the lungs. Emboli or thrombi in the brain can cause a cerebrovascular accident, or stroke.

Treatment for thrombi and emboli often involves the use of substances to dissolve the clots and thereby restore circulation to a tissue before severe ischemia and cell death results. **Streptokinase** (strep''to-ki'nās), an enzyme made by certain hemolytic streptococci, digests most any kind of protein. Given intravenously, it digests blood clots in coronary vessels especially if given within a few hours of the onset of symptoms. Tissue plasminogen activator appears to act faster and more specifically than streptokinase in dissolving blood clots. Its amino acid sequence is known and it can be made by genetically engineered bacteria.

Hemophilia (he''mo-fil'e-ah) refers to a group of inherited diseases, each caused by a different genetic defect that prevents the synthesis of a coagulation factor. The most common type, hemophilia A, which accounts for about 80 percent of all cases, is due to the absence of factor VIII. The next most common type, Christmas disease or hemophilia B, is due to the absence of factor IX (Christmas factor, named for the patient in whom it was identified). Both are inherited as sex-linked recessive characteristics and affect males almost exclusively. An autosomal recessive disorder, von Willebrand's disease, leads to low factor VIII activity. The factors most often missing in hemophilia function in the intrinsic mechanism, and symptoms such as ease of bruising and joint pain result from internal bleeding. Small breaks in blood vessels in joints where the vessels are continually subjected to stress are not repaired as they are when the intrinsic clotting mechanism is operating normally. Though the extrinsic mechanism usually is functional, it is not activated by internal vessel damage.

Thrombocytopenia (throm''bo-si''to-pe'ne-ah), a condition of too few platelets, leads to decreased constriction of ruptured vessels, deficient platelet plug formation, and limited clot retraction. Individuals suffering from thrombocytopenia often have small subcutaneous hemorrhages, and minor bumps cause large bruises.

In **disseminated intravascular coagulation** (DIC), many small clots are disseminated throughout the body and hemorrhages at other sites occur simultaneously. It occurs in the presence of some bacterial toxins, complications of childbirth, burns, and various kinds of trauma.

Though the reasons for clot formation are not well understood, such clots block small blood vessels and impede blood flow. They also deplete the supply of clotting factors so hemorrhage occurs at some sites while clots are present at others. Ultimately, treatment should be directed toward alleviating the cause, but immediate treatment may be necessary to avert death. Whether to treat the patient for hemorrhage or clotting is a matter of some controversy.

LYMPH

Lymph (limf) is interstitial fluid that has drained into lymph vessels and is being returned to the blood. Lymph closely resembles blood plasma except that it contains more or less of some constituents. Lymph formed in the liver contains large amounts of protein such as clotting factors and transports them to the blood. Lymph from the small intestine contains large amounts of newly digested lipids, which are absorbed into lymph vessels. Lymph from most other tissues contains less protein than blood plasma because few protein molecules pass from blood into tissue fluids. Lymph normally lacks erythrocytes and platelets, but contains large numbers of lymphocytes from lymph nodes and other lymphatic tissues. It also contains small numbers of other kinds of leukocytes that entered tissue fluids by diapedesis.

CLINICAL TERMS

agranulocytosis (ah-gran''u-lo-si-to'sis) an acute disease with marked deficiency of leukocytes and mucous membrane lesions

eosinophilia (e''o-sin''o-fil'e-ah) the presence of excessive numbers of eosinophils often associated with allergies and parasitic infections

hemolysis (he-mol'is-is) the rupture of erythrocytes, which releases hemoglobin into the plasma

leukopenia (loo''ko-pe'ne-ah) reduction in the number of circulating leukocytes

macrocytosis (mak''ro-si-to'sis) the presence of abnormally large erythrocytes

microcytosis (mi''kro-si-to'sis) the presence of abnormally small erythrocytes

neutrophilia (nu''tro-fil'e-ah) the presence of excessive numbers of neutrophils, often associated with a bacterial infection

prothrombin time a laboratory test to determine how much prothrombin is available in the blood and, therefore, how rapidly it will clot

purpura (pur'pu-rah) the presence of purplish patches on skin and mucous membranes due to ruptured subcutaneous blood vessels

septicemia (sep''tis-e'me-ah) the presence of pathogenic bacteria in the blood

ESSAY
BLOOD IN DIAGNOSIS AND THERAPY

Blood plays an exceedingly important role in the diagnosis of human disease in two ways. First, because it circulates through the body, it reflects the physiological status of the body's tissues. Second, it can be obtained for study with a minimum of discomfort and inconvenience to the patient. Chemical tests detect abnormalities in nearly every system of the body. From years of accumulated observations, the normal range of concentrations is known for an enormous number of substances found in the blood. Some frequently studied substances and disorders associated with their excesses or deficiencies are listed in Table 17.5.

Cell counts also identify disorders. Excess white blood cells often indicate bacterial infection. A **differential white cell count,** in which percentages of each type of leukocyte are determined, can help to diagnose leukemia, mononucleosis, and various bacterial and viral infections, parasitic infections, and other diseases. The size and shape of red cells themselves can suggest sickle cell anemia and other anemias. When cells are of normal size, the hematocrit provides a reasonably accurate red cell count. Platelet counts also can be done, and a reduced platelet count can indicate aplastic bone marrow.

Blood is useful, not only in diagnosis, but also in therapy (Figure 17.17). Its transport capacity is employed to distribute medications to cells. One reason medications are given around the clock is to try to maintain nearly constant blood concentrations. Extensive research is underway to develop drug delivery systems that provide continuous medication. Such systems include: (1) application of drug-saturated bandages to the skin, (2) surgical implantation of a drug pellet, (3) placement of a reservoir that can be refilled with a hypodermic syringe, and (4) a pump filled with a fluorocarbon propellant that maintains constant pressure and, thus, constant and continuous release of the drug. Even such systems fail to compensate for circadian rhythms in the rate of metabolism of some drugs. Eventually, it may become possible to regulate drug concentrations in accordance with these rhythms.

Blood components—erythrocytes, platelets, protein fractions, heparin, or other molecules—have therapeutic value. They can provide proper replacement for losses in diseases and injuries. Individuals with platelet deficiencies can be treated with platelet transfusion, leaving remaining parts of the same blood for use in treating others. Certain parts can be used for replacing clotting factors that are missing in individuals with hemophilia, or for supplying ready-made antibodies to help a person resist or combat an infection. Blood plasma is rarely

TABLE 17.5	Blood Components and Their Use in Diagnosis and Management of Disease
Blood Components	**Used in Diagnosis and Management of:**
Glucose	Diabetes mellitus, hypoglycemia
Cholesterol, lipids, and lipoproteins	Heart and blood vessel diseases
Proteins	Kidney disease, malnutrition, impaired amino acid absorption
Enzymes	Tissue damage following a heart attack or associated with diseases of the liver or other organs
Uric acid	Gout, kidney disease
Bilirubin	Excessive breakdown of erythrocytes, liver disease, blockage of bile ducts
Blood urea nitrogen	Kidney failure
Electrolytes (Na^+, K^+, Cl^-, etc.)	Kidney function, metabolic disorders
Blood gases (CO_2 and O_2)	Lung function
pH	Acidosis, alkalosis
Hormones	Endocrine disorders

used today because of the risk of transmitting hepatitis, but albumin can be given to increase the blood volume following hemorrhage.

Many people avoid donating blood and fear receiving it because of the possibility of acquiring AIDS, hepatitis, or another virus. Infection of blood donors is prevented by never reusing equipment that comes in contact with either a donor or a recipient, and discarding blood suspected to contain viruses.

Experiments to develop synthetic blood are underway. One such blood consists of an emulsion of **perfluorocarbons** (per-floo-ro-kar'bonz)—hydrocarbons with some hydrogen atoms replaced by fluorine atoms. Perfluorocarbons carry about one-fourth the oxygen carried by real blood (5 ml instead of 20 ml of oxygen per liter). They supply cells with oxygen because they have a very low viscosity and circulate much more rapidly at normal blood pressure. A mixture of 30 percent perfluorocarbon and 70 percent blood can maintain cells almost indefinitely—and with no harmful side effects in tests with healthy volunteers. Synthetic blood may provide an answer to the problem of treating individuals whose religious or other beliefs prevent them from accepting blood or those for whom the appropriate type of blood is not available.

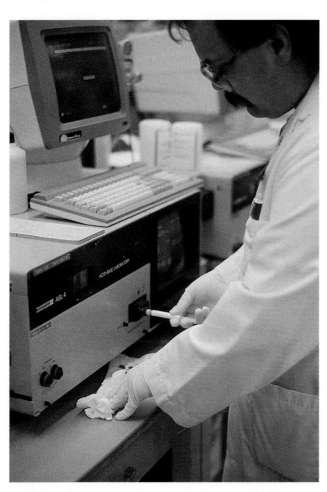

FIGURE 17.17
Modern hospital laboratories perform many different kinds of tests on blood.

Questions

1. Why is blood important in diagnosis and therapy?
2. What is a differential white cell count, and what is its significance?
3. When you donate a pint of blood, how can it be used to treat several patients?
4. What are perfluorocarbons, and of what medical use are they?

CHAPTER SUMMARY

Blood: A Circulatory System Component

- The circulatory system consists of the heart, a closed system of blood vessels, and blood.
- Blood consists of formed elements, erythrocytes, leukocytes, and platelets, in fluid plasma.
- The functions of blood include transport, protection, clotting, and helping to regulate fluids and body temperature.
- Properties of blood are summarized in Tables 17.1 and 17.2.

Plasma

- Plasma carries many substances including gases, electrolytes, nutrients, hormones, and wastes—some dissolved and others transported on particular plasma proteins.
- Plasma proteins help to maintain blood osmotic pressure and participate in blood clotting and immunity.

Formed Elements and Their Origins

- Formed elements include erythrocytes; granular basophils, eosinophils, and neutrophils; agranular lymphocytes and monocytes; and platelets.
- All formed elements are derived from stem cells by hemopoiesis either in bone marrow or lymphatic tissues; erythropoiesis is the production of erythrocytes.

Erythrocytes

- Erythrocytes are biconcave disks that lack nuclei and live only about 120 days; they contain hemoglobin that carries oxygen and some carbon dioxide.
- As erythrocytes die, their hemoglobin is degraded, the iron and amino acids are reused, and the heme is converted to bilirubin and subsequently excreted.
- When blood oxygen falls below normal, the kidneys release erythropoietin, a hormone that stimulates erythropoiesis.
- Disorders of erythrocytes include several kinds of anemia (a deficiency of erythrocytes or hemoglobin) and polycythemia (an excess of erythrocytes).

Leukocytes

- Among the five kinds of leukocytes, neutrophils and monocytes act as phagocytes, basophils release histamine and heparin, and eosinophils may detoxify and turn off immune reactions; lymphocytes are primarily concerned with immunity.
- Leukemia, a group of malignancies that cause excess production of leukocytes, can involve either lymphoid or myeloid tissue and can be acute or chronic.
- Infectious mononucleosis is an infection of lymphocytes caused by Epstein-Barr virus.

Platelets

- Platelets are fragments of megakaryocytes.
- They aggregate to seal injured blood vessels and contribute to the clotting mechanism.

Hemostasis

- Hemostasis is the arrest of bleeding and involves blood vessel spasm, platelet plug formation, and clotting of blood.
- Clotting can be initiated by extrinsic or intrinsic factors and proceeds as summarized in Figure 17.14.
- Clots are degraded by the mechanism described in Figure 17.16.
- Anticoagulants can be used in the body and in the laboratory to prevent clotting.
- Bleeding disorders include several hemophilias (absent clotting factors caused by genetic defects) and thrombocytopenia (a platelet deficiency).
- Abnormal clots occur where collagen is exposed, and include fixed thrombi and moving emboli, both of which can block vessels.
- Disseminated intravascular coagulation is a disorder in which multiple clots and hemorrhage occur simultaneously.

Lymph

- Lymph is interstitial fluid that has drained into lymph vessels; its composition is very similar to plasma.
- Lymph carries fluid from tissues back to blood.

QUESTIONS AND PROBLEMS

The questions at the end of each chapter are numbered to correspond with the objectives listed at the beginning of the chapter. Italics indicate that a question requires critical thinking skills beyond simple factual recall.

Questions

1. (a) *What use might be made of knowing normal values for blood constituents?*
   (b) What are the major functions of blood?
2. (a) What are the major components of plasma?
   (b) How does serum differ from plasma?
3. (a) What kinds of formed elements are found in blood?
   (b) What are the origins of formed elements?
4. (a) How are erythrocytes produced?
   (b) *How is erythrocyte production regulated?*
5. (a) How are components of erythrocytes reused?
   (b) *What mechanisms rid the body of nonreusable erythrocyte components?*
   (c) *How is iron absorbed, transported, stored, and excreted?*
6. (a) *What factors control the number of erythrocytes in blood?*
   (b) *What disorders are associated with excesses, deficiencies, or defects of erythrocytes?*
7. (a) What are the characteristics and relative numbers of the different kinds of leukocytes?
   (b) *What are the major functions of the different kinds of leukocytes?*
8. *What kinds of disorders are associated with leukocytes?*
9. What are platelets, and what are their functions?
10. (a) What is hemostasis, and what factors contribute to it?
    (b) *Explain the similarities and differences in the extrinsic and intrinsic blood clotting mechanisms.*
    (c) *How are clots dissolved?*
    (d) *How is clotting prevented when it is not needed?*
11. *What are the effects of disorders in hemostasis?*
12. *How does lymph differ from blood in composition and function?*

Problems

1. How can tissues be oxygen deficient in both leukemia and polycythemia?
2. What disorder might cause an adult female to have 3.5 million erythrocytes and 12,000 leukocytes per microliter of blood?
3. What disorder might cause an adult male to have blood proteins making up no more than 4 percent of the plasma volume, a blood glucose level of 140 mg/100 ml, and electrolytes in excess of 1.5 percent of the plasma volume?
4. In giving first aid to an injured person, pressure is often applied to a bleeding wound. Use your knowledge of physiology to explain how pressure helps to stop bleeding.

OUTLINE

OBJECTIVES

1. Describe the general plan of the heart, its location, and its general function in the circulatory system.
2. Describe the embryonic development of the heart.
3. Describe the properties of each tissue layer in the heart and pericardium.
4. Describe the anatomical plan of the heart, and trace the flow of blood through its chambers.
5. Describe the circulation of blood through the heart wall itself and the innervation of the heart muscle.
6. Explain the nature of action potentials in cardiac contractile fibers and how they are initiated and regulated.
7. Relate electrocardiogram components to heart function.
8. Describe the cardiac cycle, and explain its significance.
9. Define cardiac output, and describe the factors affecting it.
10. Relate electrocardiogram abnormalities to the diagnosis of heart disorders.
11. Explain how congenital heart defects disrupt heart function.

18

HEART

ORGANIZATION AND GENERAL FUNCTIONS

Make a clenched fist. Slowly count to 70, opening and closing your fist with each count. Your heart is about the size of your clenched fist, and it contracts and relaxes about 70 times per minute when the body is at rest throughout adulthood and more frequently in childhood. Though your hand muscles would ache after a few minutes of contracting 70 times per minute, your heart is able to contract continually at that rate without fatigue.

The **heart** is a cone-shaped, hollow muscular organ about 12 cm long from base to apex, 9 cm wide at its widest point, and 6 cm thick. Inside, the heart has four chambers separated by septa and valves. (A septum is a wall that separates two chambers.) The heart functions as two side-by-side pumps forcing blood through the blood vessels. Skeletal landmarks that can be palpated (felt) through the skin such as the ribs and sternum help to define the location of the heart (Figure 18.1). The bulk of the heart lies beneath the sternum and between the lungs in the **mediastinum** (me"de-as-ti'num). The heart itself is enclosed in a tough, loose-fitting membrane, the **pericardial** (per"i-kar'de-al) **sac.**

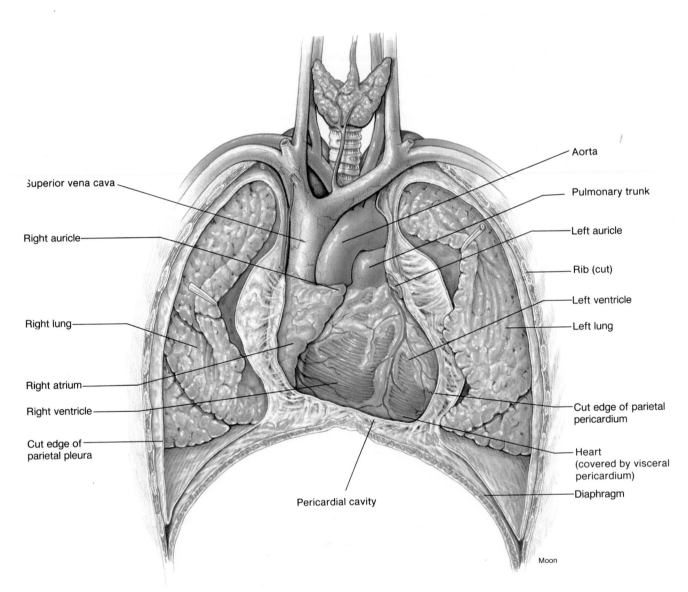

FIGURE 18.1
Location of the heart and pericardial sac.

The body's major blood vessels carry blood to or from the heart, and are connected to smaller vessels to form a closed system. These vessels form three major circulatory pathways—the **pulmonary circulation,** serving respiratory portions of the lungs; the **coronary circulation,** serving the heart wall itself; and the **systemic circulation,** serving all other parts of the body.

DEVELOPMENT

The heart is the first major organ to become functional during embryonic development (Figure 18.2). Between the third and fifth weeks of development, the heart consists of a simple, pulsating tube that receives blood from veins and pumps it out into arteries. During the fifth week, the heart grows rapidly and becomes S-shaped. It then twists around so that the venous end, which receives blood from the venae cavae, comes to lie posterior to and slightly superior to the arterial end. Septa grow inside the tube, dividing it into two tubes. Each tube develops into a pump, one destined to supply the lungs and the other to supply the remainder of the body. The single tube exiting the heart divides into the aorta and the pulmonary trunk at about the time that the heart chambers form. The arterial end becomes separated into two ventricles and the venous end becomes separated into two atria.

FIGURE 18.2
The development of the heart.

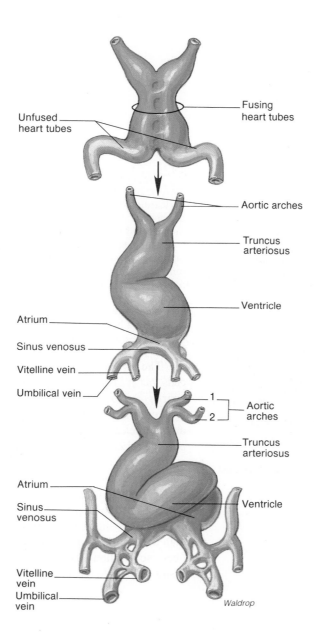

Unfused heart tubes

Fusing heart tubes

Aortic arches

Truncus arteriosus

Ventricle

Atrium

Sinus venosus

Vitelline vein

Umbilical vein

1
2
Aortic arches

Truncus arteriosus

Atrium

Sinus venosus

Ventricle

Vitelline vein

Umbilical vein

Waldrop

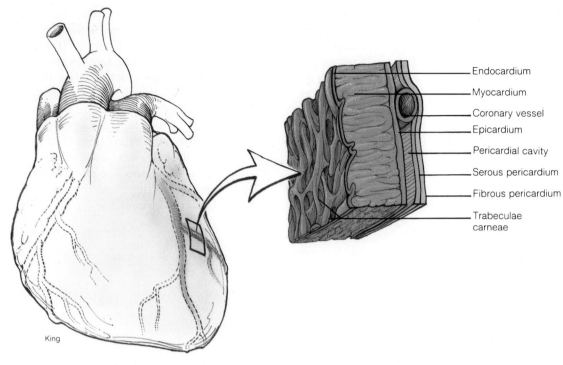

FIGURE 18.3
The tissue layers of the heart and pericardium.

ANATOMY OF THE HEART

The heart is a complex organ with thick walls, specialized surrounding tissues, chambers, and valves through which blood is pumped. It has its own blood supply and innervation.

Heart Wall and Surrounding Tissues

The heart wall consists of a thin lining called the **endocardium** (en″do-kar′de-um), a thick muscular **myocardium** (mi″o-kar′de-um), and an outer **epicardium** (ep″i-kar′de-um), or **visceral pericardium.** Connected to the epicardium is a sac called the **parietal pericardium** that surrounds the heart and the pericardial cavity (Figure 18.3). Both the visceral and parietal pericardium secrete a serous (watery) fluid that lubricates the membrane surfaces and reduces friction when the heart contracts and relaxes within the **pericardial cavity.**

Pericardial abnormalities include pericarditis and cardiac tamponade. **Pericarditis** (per″i-kar-di′tis), an inflammation of the pericardial membranes due to infection or trauma, can cause the membranes to adhere and interfere with the heart's pumping movements. In **cardiac tamponade** (kar′de-ak tam″pon-ād′), the pericardial cavity becomes engorged with fluid or blood. This fluid reduces the heart's pumping efficiency by preventing complete filling of the heart chambers between contractions. Severe cardiac tamponade, as from the rupture of a heart chamber, leads to death in a few minutes!

The myocardium consists mainly of contractile muscle cells, but it also has specialized muscle cells that form the heart's conduction system. Like skeletal muscle fibers (Chapter 10), contractile cardiac muscle fibers contain myofilaments composed of actin and myosin, sarcoplasmic reticulum, and T tubules. In contrast to the long, multinucleated fibers of skeletal muscle, the fibers of cardiac muscle consist of uninucleated cells that undergo extensive branching and that are electrically coupled by gap junctions at the **intercalated** (in-ter′kal-āt-ed) **disks.** The electrical resistance across intercalated disks is much less than that across the external membranes of cardiac muscle fibers, so ions flow freely across them and action potentials easily propagate from one myocardial cell to the next.

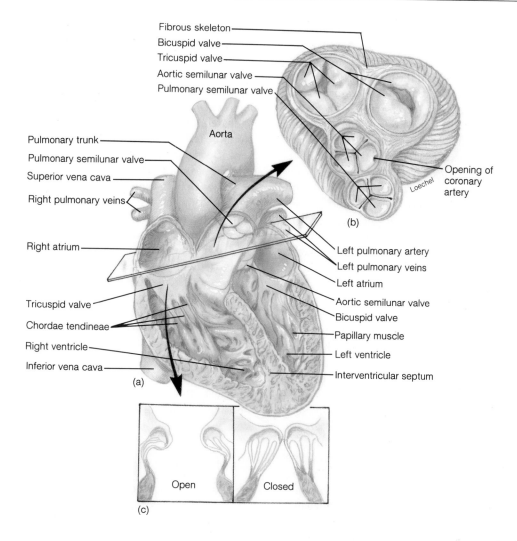

Fibrous skeleton
Bicuspid valve
Tricuspid valve
Aortic semilunar valve
Pulmonary semilunar valve

Aorta

Pulmonary trunk
Pulmonary semilunar valve
Superior vena cava
Right pulmonary veins

Right atrium

Tricuspid valve
Chordae tendineae
Right ventricle
Inferior vena cava

(a)

Opening of coronary artery

Loechel

(b)

Left pulmonary artery
Left pulmonary veins
Left atrium
Aortic semilunar valve
Bicuspid valve
Papillary muscle
Left ventricle
Interventricular septum

Open   Closed

(c)

FIGURE 18.4
(*a*) The internal anatomy of the heart in cross section, viewed anteriorly. (*b*) The valves of the heart and their supporting fibrous rings, as seen in a cutaway superior view after removal of the atria. (*c*) Changes in papillary muscles and chordae tendinae when valves are open and closed.

The myocardium consists of two functional units. One forms the **atria** (a'tre-ah), chambers of the heart that receive blood. (One such chamber is an **atrium** [a'tre-um].) The other forms the **ventricles** (ven'trik-lz), chambers that eject blood. Each atrium has a flaplike appendage called an **auricle** (aw'rik-l). The atria and ventricles are separated by valves and rings of fibrous endocardial tissue. These rings separate the electrical activity of the atria from that of the ventricles. The ease of movement of action potentials and the extensive branching of cardiac fibers causes each functional unit—the atria or the ventricles—to contract together. In contrast to skeletal muscle where single fibers obey the all-or-none law and contract fully or not at all, each functional unit of cardiac muscle obeys the all-or-none principle. Contractions of the units are coordinated by the specialized myocardial conductive cells.

Atrial contractile cells contain large numbers of secretory granules. Recently, these granules have been found to consist of a polypeptide called **atrial natriuretic hormone.** This hormone plays an important role in regulating blood pressure and blood volume (Chapter 27), and sodium, water, and potassium excretion (Chapter 28).

The endocardium consists of a thin layer of simple squamous epithelium that lines the chambers of the heart and the rings of fibrous tissue that surround the support the heart valves. This lining is continuous with the endothelium that lines all blood vessels and forms the entire wall of capillaries.

Chambers and Valves

The heart has four chambers, a left atrium and ventricle, and a right atrium and ventricle, separated by the muscular **interatrial** (in″ter-a'tre-al) **septum** and the **interventricular** (in″ter-ven-trik'u-lar) **septum** (Figure 18.4). Each side functions as a separate pump.

Four valves, supported by connective tissue rings, passively control the direction of blood flow through the heart (Figures 18.4a and 18.4b). Valves, which are opened and closed by pressure changes in the blood, prevent backflow of blood to the chamber from which it came.

The **atrioventricular** (a"tre-o-ven-trik'u-lar), or **AV, valves** are the **tricuspid** (tri-kus'pid) **valve** on the right and the **mitral** (mi'tral), or **bicuspid, valve** on the left. They prevent backflow to the atria when the ventricles contract. Their cusps (flaps) are connected to the ventricular walls by fibrous **chordae tendineae** (kor'de ten-din'e) and **papillary** (pap'il-er"e) **muscles** (Figure 18.4c). These muscles contract with the ventricles, pull the cusps toward the ventricles, and oppose valve closure. As pressure in contracting ventricles closes the valves, the papillary muscles prevent valve cusps from bending backward into the atria. Weakening of papillary muscles allowing the cusps to bend into the atria is called **valve prolapse.**

Pressure from blood ejected as ventricles contract opens the pulmonary and aortic **semilunar** (sem"e-lun'ar) **valves.** Arterial pressure closes them, preventing backflow into relaxed ventricles. Leaflets of the semilunar valves are smaller than cusps of the AV valves and lack chordae tendinae and papillary muscles.

The turbulence (agitated, irregular motion) in blood caused by heart valves closing makes sounds that are audible with a stethoscope as a "lub-dub." The lub (first sound) occurs as the AV valves close and the dub (second sound) occurs as the semilunar valves close. Abnormal sounds associated with valves are called **heart murmurs.** In **aortic stenosis** (a-or'tik sten-o'sis), a condition in which the aortic semilunar valve opening is abnormally small, blood flow from the left ventricle to the aorta is restricted. Pressure increases inside the ventricle, and over time, the ventricle wall thickens as it works to eject blood through the constricted opening. The turbulence thereby created is heard as a heart murmur. In other heart murmurs, **incompetent valves** fail to close completely and blood regurgitates (flows back into the chamber from which it came). Regurgitation causes a blowing or swishing sound and reduces net output of the chamber.

People with streptococcal infections should receive prompt antibiotic treatment to avoid developing rheumatic fever (heart inflammation, increased heart rate, and fever). Repeated bouts of rheumatic fever can lead to rheumatic heart disease, in which the bicuspid valve is nearly always damaged. Infection of the endocardium and heart valves is common among drug addicts who probably inadvertently inject streptococci and other pathogenic bacteria into their blood when they use dirty needles. The reason pathogens attack heart valves and streptococci attack the mitral valve, in particular, is not known.

Blood moves in an orderly sequence through the chambers and valves of the heart (Figure 18.5). Unoxygenated blood from the systemic circulation enters the right atrium and passes through the tricuspid valve to the right ventricle. It is pumped through the pulmonary semilunar valve into the pulmonary circulation where it is oxygenated and returned to the left atrium. Blood passes from the left atrium through the bicuspid valve to the left ventricle and is pumped through the aortic semilunar valve into the systemic circulation.

Though both ventricles contract at the same time, the left one contracts with greater force than the right. The wall of the left ventricle contains more muscle and is noticeably thicker than the wall of the right ventricle. Pressure of blood ejected into the systemic circulation via the aorta is about five times as great as in that entering the pulmonary circulation. This greater aortic pressure provides sufficient force to propel blood through the entire systemic circulation. Though pressures differ, the volume of blood pumped by each side of the heart is the same.

The left and right heart pumps work together. Both atria contract simultaneously, forcing blood into their respective ventricles. Then, the ventricles contract simultaneously, forcing blood into the arteries. Valves alternate opening and closing when the atria contract and when the ventricles contract (Figure 18.6).

Blood Supply and Innervation

In addition to the aorta, pulmonary vessels, and venae cavae, major blood vessels supplying the heart wall itself can be seen on the heart's surface in anterior and posterior views (Figure 18.7). Both the **right and left coronary arteries** branch from the aorta very near the heart. These arteries, in turn, branch many times into smaller arteries, and eventually into capillaries that carry blood to all the cells of the myocardium. These arteries form **anastomoses** (an-as"to-mo'ses), or connections between branches, that allow blood from one branch to enter the capillary network of another branch. Anastomoses provide alternate pathways should small vessels become blocked. Blood from capillaries drains via smaller veins to the **great cardiac vein,** then into the **coronary sinus,** and is returned to the right atrium. Especially in the wall of the right ventricle, some blood enters smaller veins or wide, thin-walled sinusoids, and flows from them directly to the right ventricular chamber.

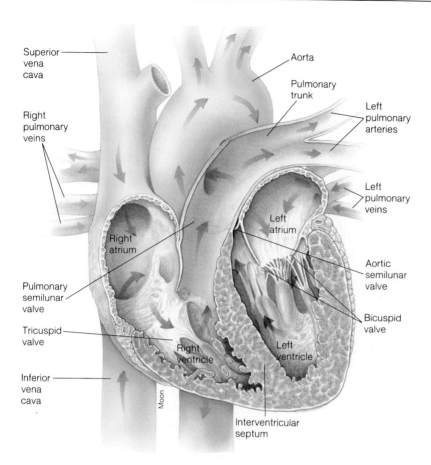

Superior vena cava

Right pulmonary veins

Right atrium

Pulmonary semilunar valve

Tricuspid valve

Inferior vena cava

Aorta

Pulmonary trunk

Left pulmonary arteries

Left pulmonary veins

Left atrium

Aortic semilunar valve

Bicuspid valve

Right ventricle

Left ventricle

Interventricular septum

Moon

FIGURE 18.5
The pathway of blood through the chambers and valves of the heart.

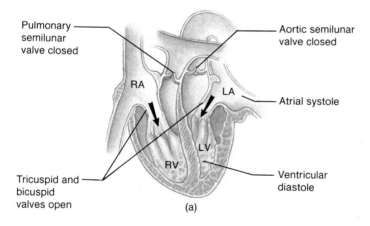

Pulmonary semilunar valve closed

RA

LA

Aortic semilunar valve closed

Atrial systole

Tricuspid and bicuspid valves open

RV

LV

Ventricular diastole

(a)

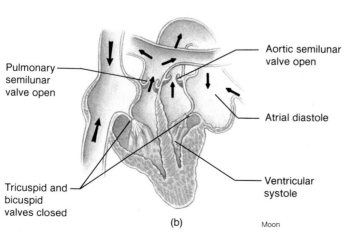

Pulmonary semilunar valve open

Aortic semilunar valve open

Atrial diastole

Tricuspid and bicuspid valves closed

Ventricular systole

(b)

Moon

FIGURE 18.6
(*a*) When the atria contract, the tricuspid and bicuspid valves open and blood empties into the ventricles. The semilunar valves remain closed. (*b*) When the ventricles contract, the semilunar valves open and the tricuspid and bicuspid valves close.

FIGURE 18.7
External anatomy of the heart (*a*) anterior view, and (*b*) posterior view.

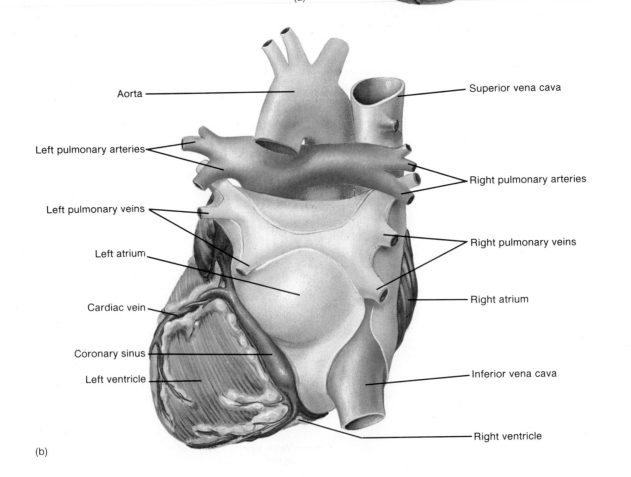

Right pulmonary arteries

Right pulmonary veins

Aorta

Superior vena cava

Right atrium

Right coronary artery

Right ventricle

Inferior vena cava

Ligamentum arteriosum

Pulmonary trunk

Left pulmonary arteries

Left pulmonary veins

Left atrium

Left coronary artery

Circumflex artery

Anterior interventricular artery

Left ventricle

Cardiac vein

Apex

(a)

Aorta

Left pulmonary arteries

Left pulmonary veins

Left atrium

Cardiac vein

Coronary sinus

Left ventricle

Superior vena cava

Right pulmonary arteries

Right pulmonary veins

Right atrium

Inferior vena cava

Right ventricle

(b)

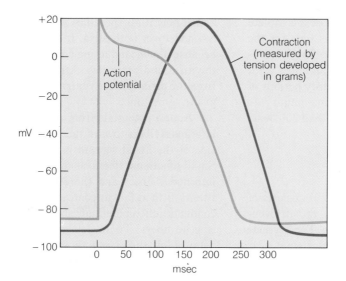

FIGURE 18.8
Relationship between the action potential and muscle contraction in the heart.

Contractions greatly reduce blood flow in myocardial vessels, so cells receive nutrients and oxygen and get rid of wastes mainly between contractions when the heart is relaxed. The period of relaxation is sufficient to maintain the cells if the arteries are of normal diameter; however, if an artery becomes narrowed by atherosclerosis (Chapter 20), or blocked by a blood clot, the cells supplied by that vessel are deprived of nutrients.

One final attribute of heart anatomy—its innervation—is important in understanding heart function. The atria are well supplied with both sympathetic and parasympathetic nerve fibers, but the ventricles are supplied almost solely with sympathetic fibers. Neural signals affect both the rate and strength of contractions. In general, sympathetic stimulation increases both the rate and strength of contractions, and parasympathetic stimulation decreases the rate at which the conduction system spontaneously initiates contractions.

INITIATION AND REGULATION OF THE HEARTBEAT

In a normal heart, the two atria contract simultaneously and then the two ventricles contract simultaneously, with fibrous rings around the valves preventing the spread of contraction waves from atria to ventricles. The heart has a **conduction system,** consisting of specialized myocardial fibers, that stimulates contractile fibers and coordinates atrial and ventricular contraction. This system acts by initiating and propagating action potentials.

Action Potentials in Cardiac Contractile Fibers

Action potentials in cardiac contractile fibers have some special properties (Figure 18.8). The resting membrane potential of normal cardiac muscle is about −90 mV. When an action potential is generated, the membrane potential suddenly changes to about +20 mV. It maintains a plateau of polarization for 0.15 second in atrial muscle and from 0.25 to 0.30 second in ventricular muscle before it repolarizes. Cardiac muscle has certain properties not found in skeletal muscle that account for this plateau. Sodium ions flowing across the membrane into cardiac muscle cells cause the sudden spike of depolarization. Then, calcium ions flowing into the cells maintain the plateau and prevent repolarization. When calcium ions cease to flow into muscle cells, potassium ions flow out and repolarization occurs.

Properties of action potentials are closely related to the refractory periods of muscle fibers. Like skeletal muscles, cardiac muscle also has absolute and relative refractory periods. In skeletal muscle, the absolute refractory period is short (1 to 2 msec); whereas, in cardiac muscle, it is extremely long (250 to 300 msec in ventricular muscle and 150 msec in atrial muscle). Furthermore, the total cardiac refractory period corresponds closely to the plateau in the action potential. During most of this time, no stimulus can cause a contraction, but at the end of the plateau, a stimulus sometimes causes an abnormal or premature contraction. Because the refractory period of the atria is shorter than that of the ventricles, the atria are capable of contracting more rapidly than the ventricles.

The long cardiac refractory period enforces complete relaxation between contractions and allows time for the chambers of the heart to refill and blood to flow in coronary vessels. It prevents tetany by setting an upper limit on the heart rate. Because only 200 refractory periods of 300 msec each can occur in a minute, the maximum rate of ventricular contraction is somewhat less than 200 beats per minute.

The Conduction System

The heart's **conduction system** (Figure 18.9) consists of the sinoatrial (SA) node, internodal fibers, the atrioventricular (AV) node, and the Purkinje system. The **SA node** is located in the posterior wall of the right atrium near the superior vena cava. The inherently self-excitatory SA node fibers depolarize automatically and rhythmically. Because of "leaky membranes," they have an unstable membrane potential of -55 to -60 mV instead of the -90 mV found in contractile fibers. This unstable potential, called the **pacemaker potential,** undergoes gradual depolarization as Na^+ and Ca^{2+} leak into the fibers eventually reaching the threshold for an action potential.

Once the threshold voltage of -45 to -50 mV is reached, the SA nodal fibers generate an action potential. Na^+ moves rapidly into the fibers and causes sudden depolarization. The membranes remain depolarized for 0.1 second, after which the permeability of the membrane to K^+ increases and that to Na^+ decreases. Repolarization occurs as K^+ moves out of the fibers. When the membrane has repolarized, its permeability to K^+ again decreases while its permeability to Na^+ and Ca^{2+} increases, and the next pacemaker potential begins to develop. Another cycle in the automatic rhythmicity of the SA nodal fibers is initiated and this cycle is repeated continuously throughout life.

Because the SA node generates action potentials at a greater frequency than other parts of the conduction system, it is the **pacemaker.** As pacemaker, it sets the contraction rate and synchronizes the actions of the atria and ventricles.

Action potentials generated by SA node depolarization spread along atrial contractile cells to the sites where they fuse with contractile fibers. Cell-to-cell conduction is facilitated by gap junctions at the intercalated disks, and action potentials spread as waves of contraction from the SA node over both atria. It does not reach the ventricles because of nonconducting connective tissue rings between the atria and ventricles. Atrial contraction pushes blood from the atria into the ventricles on both sides of the heart at the same time.

Action potentials from the SA node also pass along **internodal fibers** toward the AV node, located on the right side of the atrial septum just superior to the ventricles. Small **junctional fibers** conduct action potentials between internodal fibers and the AV node slowly—about one-twenty-fifth as fast as contractile cells. The AV node itself conducts action potentials about one-fourth as fast as contractile fibers. These delays, which constitute **AV node delay,** allow the atria to empty into the ventricles before the ventricles receive a signal to contract.

Once fibers of the AV node are excited, they conduct action potentials to the **Purkinje system.** This system of rapidly conducting fibers originates as the **atrioventricular bundle,** which passes through the fibrous connective tissue around the valves to the ventricular septum. Here, the bundle divides into the left and right **bundle branches,** which lie just beneath the endocardium on each side of the septum. **Purkinje fibers** extend from the bundle branches to the apex (inferior tip) of the heart and upward into the lateral myocardium. Their ends fuse with contractile fibers to which they conduct action potentials.

Once action potentials from the SA node reach the Purkinje system, they initiate a rapidly spreading wave of depolarization through the ventricular myocardium from the apex upward. This causes a corresponding wave of contraction that pushes blood out of the ventricles. The entire ventricular myocardium depolarizes before any of it repolarizes.

Though the SA node normally sets the heart rate, it is not the only tissue that can spontaneously generate action potentials. Each portion of the conduction system has this capacity—and its own inherent rate of depolarization. If the SA node sends no pacemaker potentials, the AV node spontaneously generates action potentials 40 to 60 times per minute. In the absence of other stimuli, the Purkinje fibers generate action potentials 15 to 40 times per minute.

In abnormal situations, such as reduced blood flow or overstimulation (as by caffeine), clusters of contractile fibers themselves sometimes spontaneously generate action

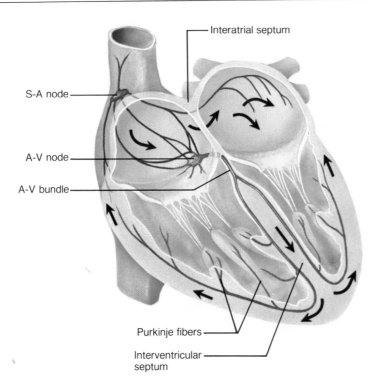

S-A node

A-V node

A-V bundle

Interatrial septum

Purkinje fibers

Interventricular septum

FIGURE 18.9
The conduction system of the heart.

potentials. These fibers constitute an **ectopic** (ek-top′ik) **focus** (plural, **foci**). (Ectopic means located at a site other than the normal one.) Ectopic foci can override the SA node and disrupt its regulatory activity, causing premature contractions of the ventricles or becoming the pacemaker by establishing a rhythm faster than that of the SA node.

If the SA node fails to set the pace for heart contractions, the AV node takes over, but it sends signals at its much slower rate. People with SA node failure are continuously fatigued and lethargic unless they receive an **artificial pacemaker,** which reestablishes a normal heart rate. This electrical stimulator is implanted under the skin and electrodes extend to the ventricles. Artificial pacemakers stimulate the ventricles rather than the atria because strong, regular ventricular contractions are needed to maintain normal blood flow. As we shall see, atrial contractions are not essential to heart function. The stimulator is operated by a battery that must be replaced about once every 3 to 4 years, or once every 15 years for the new atomic battery.

PUMPING ACTION OF THE HEART

The heart's pumping action occurs as a repetitive cycle of contraction and relaxation regulated by the conduction system. This cycle is called the **cardiac cycle.** Certain events in the cardiac cycle can be recorded as an electrocardiogram.

The Electrocardiogram

A graphic record of the electrical activity of cardiac muscle is an **electrocardiogram** (e-lek″tro-kar′de-o-gram″), or ECG. ECGs are used to study cardiac function and to identify disorders of the myocardium or conduction system. Negatively charged surfaces of depolarized cells in one region of the heart are sufficiently different from positively charged surfaces of resting cells in another region that current flows between them. Such currents can be

FIGURE 18.10
(a) A patient having an electrocardiogram made, (b) a portion of the electrocardiogram, and (c) an enlarged normal cycle.

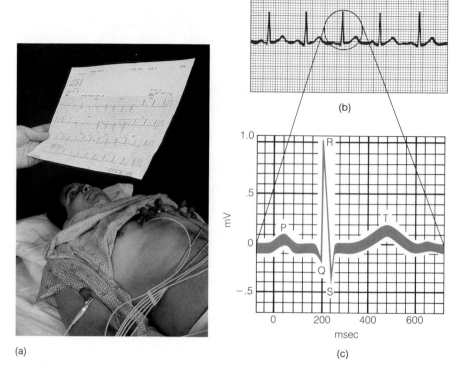

(a)

(b)

(c)

detected painlessly by electrodes placed on the skin (Figure 18.10a). Wires from the electrodes carry signals to an instrument called an **electrocardiograph** (Figure 18.10b). Heart specialists, or **cardiologists,** use different electrode positions to measure current between different regions of the heart. Much study is needed to understand electrocardiography, but experienced cardiologists often can identify the site and degree of heart damage from ECGs.

A normal electrocardiogram (Figure 18.10c) consists of a P wave, a QRS complex, and a T wave. (The letters used to name waves were arbitrarily selected and do not stand for any particular words.) A set of these waves is produced each time an action potential passes through the heart—once every 0.8 second for a heart rate of 72 beats per minute. The **P wave** follows the spontaneous initiation of an action potential in the SA node and represents atrial depolarization. The **QRS complex** follows the excitation of the Purkinje system and represents ventricular depolarization. The QRS complex has greater amplitude than the P wave because the ventricles have more muscle mass than the atria. The **T wave** represents repolarization of the ventricles. The relatively weak repolarization of the atria is obscured by the stronger electrical signals from depolarization of the ventricles.

Cardiac Cycle

The heart, like any pump, has a pumping (emptying) phase and a filling phase. Pumping occurs during contraction, or **systole** (sis'to-le); filling occurs during relaxation, or **diastole** (di-as'to-le). As noted earlier, the left ventricle exerts much greater force than the right, though both pump the same blood volume. The **cardiac cycle** (Figure 18.11) is a cycle of contraction and relaxation of the entire cardiac muscle. At a resting heart rate (72 beats per minute), each cycle takes 0.8 second, and descriptions here are based on that rate. Changes in the heart rate due to exercise and other factors, also change the duration of various phases of the cycle. Five phases are recognized.

Phase 1—Atrial Contraction

At the time an action potential is initiated in the SA node, the heart is relaxed, the AV valves (tricuspid and mitral) are open, and the ventricles are filling with blood. Blood pressures in the venae cavae and pulmonary veins that are slightly higher than those in the ventricles force blood through the atria and into the ventricles until they are more than three-fourths full. The action potential from the SA node causes atrial depolarization (P wave) and **atrial contraction** (systole), which lasts 0.10 second, increases pressure, and forces more blood into the ventricles. Atrial contraction is not essential because the ventricles become nearly full without it.

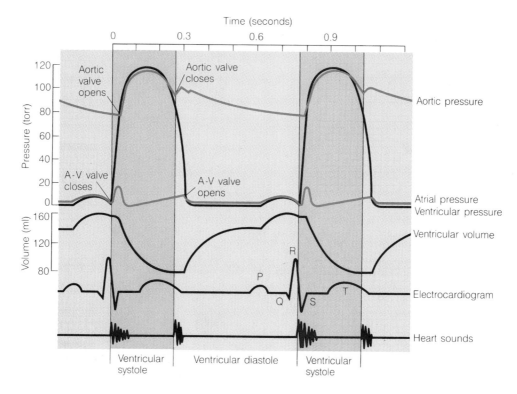

FIGURE 18.11
The cardiac cycle. Pressures refer to the left side of the heart and aorta. The pressures in the right side of the heart and the pulmonary trunk are lower than the ones shown here. (See text for explanation of figure.)

Phase 2—Isovolumetric Ventricular Contraction

About 0.16 second after SA node depolarization, a signal has passed through the AV node to the Purkinje system and started depolarization of the ventricles (QRS complex). Ventricular systole begins. The first 0.05 second of systole is **isovolumetric ventricular contraction** after the AV valves have closed and before the semilunar valves have opened. The volume of blood in the ventricles does not change though the ventricles contract and pressure increases. Because the AV valves are slightly pushed backward into the atria, atrial pressures increase by a small amount during this phase. Closing of the AV valves creates turbulence in the ventricular blood detectable with a stethoscope as the "lub" of "lub-dub." At the end of this phase, aortic and pulmonary artery pressures are at their lowest—known as **diastolic** (di''as-tol'ik) **pressure.**

Phase 3—Rapid Ventricular Ejection

When the pressures in the ventricles exceed the pressures in the pulmonary trunk and the aorta, the semilunar valves are forced open and blood is ejected from the ventricles. This phase, **rapid ventricular ejection,** lasts about 0.25 second—the time the semilunar valves remain open. Ventricular pressures continue to rise during the first part of this phase because the ventricles contract faster than blood leaves them. As blood is ejected, arterial pressures rise and remain elevated because arterial elasticity increases the pressure on arterial blood. During this phase, each ventricle ejects a volume of 70 to 90 ml of blood. The phase ends when ventricular pressures fall below arterial pressures and the semilunar valves close. Closing of these valves creates the "dub" of the "lub-dub" and signifies the end of systole.

Phase 4—Isovolumetric Ventricular Relaxation

For about 0.05 second after the semilunar valves close and before the AV valves reopen, the ventricles relax and ventricular blood volume remains constant at its lowest level of about 50 ml. This is called **isovolumetric ventricular relaxation.**

Phase 5—Rapid Ventricular Filling

While both the ventricles and the atria are relaxed, blood flows passively from the venae cavae and pulmonary veins into the atria until the atrial pressures exceed the ventricular pressures. The AV valves open and blood enters the ventricles. This **rapid ventricular filling** lasts about 0.35 seconds and is most rapid during the first 0.10 second. Toward the end of the phase, a new action potential in the SA node marks the beginning of another cardiac cycle. AV delay prevents the ventricles from receiving the action potential until ventricular filling is complete.

Relationships between signals in the conduction system and events in the cardiac cycle are summarized in Figure 18.12.

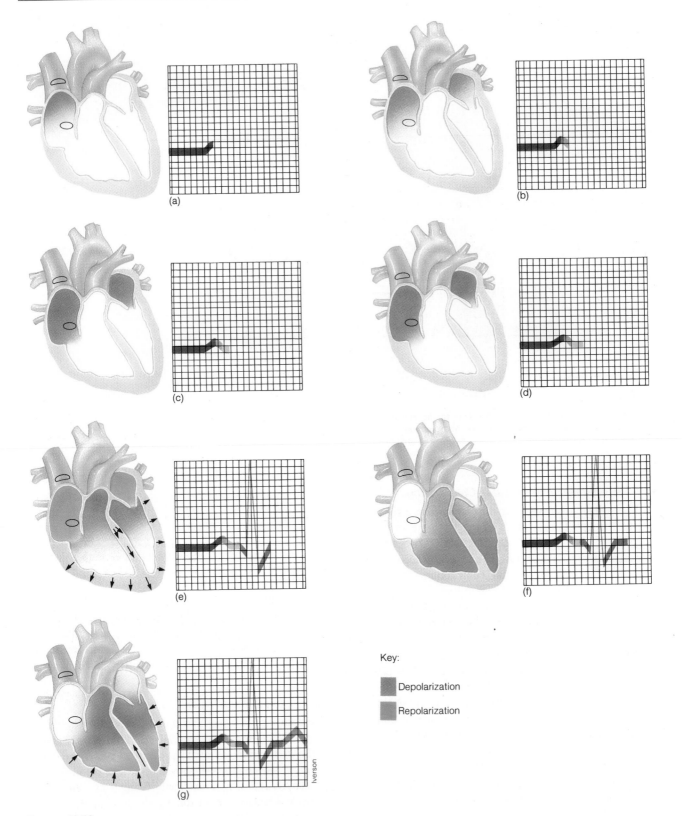

Key:

■ Depolarization

■ Repolarization

FIGURE 18.12
Relationships between the conduction system and the cardiac cycle.

CARDIAC OUTPUT AND FACTORS THAT AFFECT IT

Cardiac output is the volume of blood pumped from a ventricle per minute and normally is the same for each ventricle. It is the product of the **heart rate** (heartbeats per minute) times the **stroke volume** (blood volume ejected by one ventricle during a contraction), as shown here.

heart rate × stroke volume = cardiac output

For example, in a resting adult when the heart beats 72 times per minute and pumps 70 ml with each stroke or beat, the cardiac output is 5040 ml per minute. As the body's total blood volume is 4 to 6 l, a volume approximately equal to that total is pumped through each side of the heart every minute. During strenuous exercise, the heart rate reaches 150 beats per minute and the stroke volume reaches 120 ml, bringing cardiac output to 18 l per minute. In trained athletes, cardiac output can go to 35 l per minute—seven times the body's total blood volume. Conversely, cardiac output decreases after excessive blood loss and in other circulatory disorders. Any factor that alters heart rate or stroke volume alters cardiac output. Intrinsic (within the organ) factors affect mainly stroke volume, and extrinsic (outside the organ) factors affect both heart rate and stroke volume.

Intrinsic Control Factors

The heart intrinsically adapts to changes in the **end-diastolic volume** (EDV), or **preload,** the volume of blood in the heart when it has finished filling. Increasing EDV stretches myocardial fibers, and stretching increases tension up to the physiological limits of the fibers. (As stretch becomes extreme, tension decreases.) Increased tension causes fibers to contract with greater force. This increase in **contractility,** or the force of contraction, is intrinsic to the muscle fibers themselves and is independent of innervation or other extrinsic factors. Because of stretching, EDV also increases the stroke volume. Conversely, decreasing EDV reduces stretching and muscle tension, thereby decreasing stroke volume and contractility.

EDV, the main intrinsic regulator of stroke volume and contractility, is affected by several factors:

1. **Venous return** When venous return decreases because of lowered blood volume, dilation of veins, or other factors, ventricular filling decreases. When venous return increases because of constriction of veins or elevated blood volume, ventricular filling increases.
2. **Atrial contraction** Stronger atrial contractions increase ventricular filling, and weaker ones decrease ventricular filling.

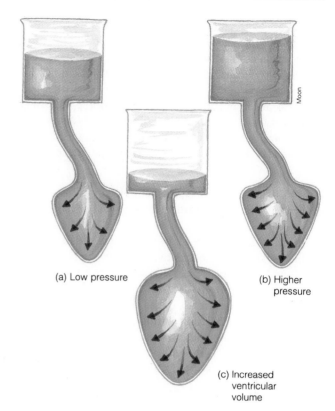

FIGURE 18.13
Operation of the Frank-Starling law of the heart: (*a*) Blood entering the ventricle under low pressure causes little stretching. (*b*) Blood entering under high pressure causes greater stretching. (*c*) Increasing ventricular end-diastolic volume increases stroke volume.

3. **Negative intrathoracic pressure** Normal negative pressure (vacuum) within the chest aids venous return by "sucking" blood into the heart. Less negative pressure or positive pressure decreases venous return.
4. **Negative intrapericardial pressure** Normal negative pressure within the pericardial sac aids ventricular filling. Accumulation of fluid in the pericardial cavity (cardiac tamponade) limits ventricular filling.

The **Frank-Starling law of the heart** states that within physiological limits the heart pumps all the blood that enters it. Named for Otto Frank of Germany and Ernest Starling of England who studied cardiac physiology in the late nineteenth century, this law can be demonstrated by varying the volume of fluid perfused (pumped through) a recently removed animal heart (Figure 18.13). This and the above discussion show that muscle fiber length regulates cardiac output. Such regulation ensures that the heart ejects as much blood as it receives and that the left and right sides of the heart pump the same volume of blood.

Historically, cardiologists have used the work of Frank and Starling for nearly a century because it accounts for alterations in cardiac output from diseased hearts. It also describes how transplanted hearts, which lack innervation, regulate cardiac output. Recent studies show that the law is less useful in studies of hearts within the body and at their optimum functional level. Some studies show that heart expansion after a contraction regulates filling. Others show that the heart stores energy from systolic contraction:

1. As the heart contracts, it forces blood upward and propels itself downward, stretching elastic blood vessels and connective tissue attached to it. As the heart relaxes, it moves upward toward the incoming blood. This is analogous to increasing the rate of filling by moving a glass upward under a stream of water running from a faucet.
2. As the heart contracts, it also stores energy within its own tissues by compressing elastic elements and muscle fibers. As these tissues expand during relaxation, they create a negative pressure in the heart chambers that pulls blood into the heart.

A second kind of intrinsic regulation pertains to heart muscle metabolism. Stretching heart muscle fibers accelerates their metabolic activities, thereby increasing their contractile strength. This returns the fibers to near their original lengths and, at the same time, increases cardiac output.

One remarkable feature of the intrinsic regulation of stroke volume is that normal arterial pressure, or **afterload,** has almost no effect on cardiac output. If, however, systemic arterial pressure exceeds 170 torr, cardiac output decreases significantly as the heart is forced to pump blood against the resistance of a large afterload (Chapter 20).

As mentioned earlier, intrinsic regulatory mechanisms exert their main effects on stroke volume, but stretching muscle fibers can affect heart rate. When the walls of the right atrium are stretched by increased venous return, the heart rate increases by 10 to 30 percent, thereby increasing cardiac output.

Extrinsic Control Factors

Extrinsic control of the heart is accomplished by sympathetic and parasympathetic reflexes that directly affect heart rate and contractility. Altering contractility can change stroke volume. To understand sympathetic and parasympathetic effects, it is important to recall that opposing effects continuously maintain autonomic tone. The net effect of sympathetic and parasympathetic signals at any time, regulates the rate and strength of contractions. These autonomic effects have special names. Effects on heart rate are **chronotropic** (kron″o-trop′ik) **actions** (*chrono*, time). Effects on contractility are **inotropic** (in″o-trop′ik) **actions** (*ino*, fiber).

Chronotropic actions affect the conduction system, especially the SA node. Sympathetic stimulation, such as epinephrine application directly to an animal heart, increases the heart rate (a positive chronotropic effect). Severing connections between sympathetic nerve fibers and an animal heart decreases the heart rate. Conversely, parasympathetic stimulation, such as acetylcholine application directly to an animal heart, decreases the heart rate (a negative chronotropic effect), and severing the connections between these nerves and the heart increases the heart rate. Severing connections between all autonomic nerves and the heart increases the heart rate to 100 beats per minute; thus, the intrinsic SA node rate of about 100 beats per minute must be brought down to the resting rate of 72 beats per minute by parasympathetic influence. Digitalis and its more commonly used derivative digoxin also have negative chronotropic effects.

How chronotropic actions occur is not entirely clear, but acetylcholine from parasympathetic neurons seems to hyperpolarize the SA node by increasing membrane permeability to potassium. Norepinephrine from sympathetic neurons and epinephrine from the adrenal medulla apparently have an opposite action. They decrease membrane permeability to potassium and make depolarization easier.

In inotropic actions, sympathetic stimulation increases contractility (a positive inotropic effect) and parasympathetic stimulation decreases contractility (a negative inotropic effect). Sympathetic signals are far more numerous and, thus, more effective than parasympathetic signals.

Sympathetic stimulation has an additional effect on the heart. At a given EDV, norepinephrine from sympathetic neurons and epinephrine from the adrenal medulla increase contractility, not by stretching fibers, but by increasing force at muscle fiber cross-bridges. These substances stimulate cardiac beta-adrenergic receptors, which release cAMP, and cAMP increases the rate at which calcium ions are released from the sarcoplasmic reticulum. Muscle force is stronger and develops more rapidly. This allows the myocardium to remain relaxed and available for filling for a larger portion of the cardiac cycle. Sympathetic signals, thus, cause the heart to eject a greater blood volume with each contraction and allow it to fill more completely between contractions.

Several factors modify inotropic effects. Nicotine, caffeine, theophylline in tea, and theobromine in chocolate have positive inotropic effects. Nicotine stimulates norepinephrine release, which increases contractility. The other substances have positive inotropic effects because they inhibit cAMP breakdown. Digitalis from the foxglove plant also exerts a positive inotropic effect. It inhibits the Na^+/K^+ pump in muscle cell membranes, allowing intracellular Na^+ to increase. This apparently releases Ca^{2+}, which binds to troponin and facilitates contraction.

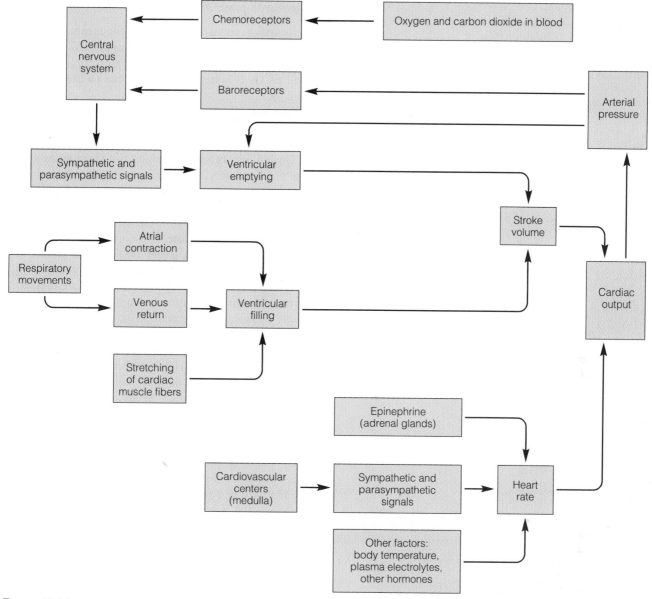

FIGURE 18.14
Mechanism of regulating cardiac output.

Quinidine, local anesthetics, and drugs such as pro-panolol, which block beta-adrenergic receptors, exert inotropic effects. Quinidine, a drug related to quinine and likewise obtained from the bark of cinchona tree, decreases excitability and conduction velocity of the heart. Local anesthetics and beta-blockers have similar effects. Even products of normal metabolism—carbon dioxide and acid—can exert negative inotropic effects, especially if the oxygen concentration is reduced. These metabolic effects occur when blood vessels that supply heart muscle cells are blocked, as occurs in a "heart attack."

Intrinsic and extrinsic factors that regulate cardiac output are summarized in Figure 18.14. Cardiac output also is affected by vasodilation, vasoconstriction, and the actions of the kidneys, aldosterone, renin, angiotensin, and antidiuretic and atrial natriuretic hormones as explained in later chapters.

CLINICAL APPLICATIONS

Disorders considered here—arrhythmias and congenital defects—affect mainly the heart. Disorders involving blood circulation to the heart—heart attacks and heart failure—are considered in Chapter 20.

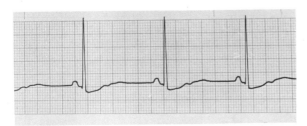

(a)

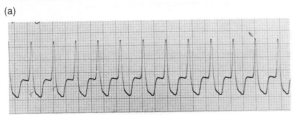

(b)

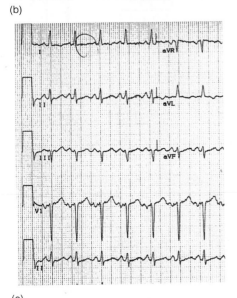

(c)

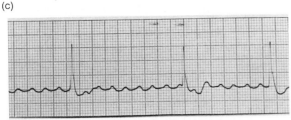

(d)

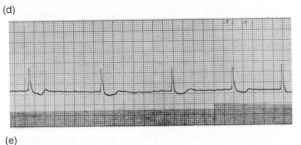

(e)

FIGURE 18.15

Abnormal electrocardiograms: (*a*) sinus bradycardia, (*b*) sinus tachycardia, (*c*) premature ventricular contractions, (*d*) flutter, and (*e*) fibrillation.

Electrocardiogram Abnormalities

Electrocardiograms are important components in the diagnosis of heart disorders. **Arrhythmias** (ah-rith′me-ahz), disturbances in the heart rate, account for many abnormal ECGs (Figure 18.15). They can be produced by ectopic foci.

Sinus arrhythmia can be a normal event related to breathing. During deep breathing, nerve signals that regulate breathing also affect brain centers that control heart rate. During inspiration (breathing in), these signals increase the heart rate, and during expiration (breathing out), they decrease it. More marked sinus arrhythmias can indicate abnormal SA node function.

Sinus bradycardia (brad″e-kar′de-ah) is a heart rate less than 60 beats per minute. It is normal in trained athletes whose hearts eject a larger than normal volume of blood per contraction and adequately supply tissues in fewer beats per minute. Low body temperature and excessive parasympathetic stimulation slow the heart rate, probably by initiating reflexes that act on the SA node to reduce the rate of contraction.

Sinus tachycardia (tak″e-kar′de-ah) is a rapid heart rate, usually in excess of 100 beats per minute. It occurs normally during exercise and also can be caused by high body temperature (fever) or by excessive sympathetic stimulation.

Premature ventricular contractions usually are caused by ventricular ectopic foci that initiate action potentials independent of the SA node. The premature contractions are interposed between normal waves initiated by the SA node. They are seen from time to time in most adults over age 30 and usually are not associated with heart disease.

Heart block arrhythmias occur when the conduction system fails to transmit some signals. **Partial heart block** exists when some action potentials from the SA node fail to pass through the AV node. All action potentials from the SA node produce a P wave, but some fail to cause ventricular depolarization and a QRS complex. **Complete heart block** occurs when no action potentials from the SA node pass through the AV node. The SA node establishes one rhythm and the AV node and Purkinje system another, so no relationship exists between P waves and QRS complexes. **Bundle branch block** occurs when no action potentials pass through either the left, right, or both bundle branches. Heart block can decrease the heart's pumping efficiency by disrupting coordination between atrial and ventricular contractions.

Flutter is regular, very rapid beating of the atria or the ventricles. Because the chambers cannot fully fill between contractions, flutter is an extremely inefficient heart rate and very little blood is ejected from the heart.

In **fibrillation** (fib-ril-a′shun), which often follows flutter, the myocardium contracts very rapidly in an uncoordinated fashion. Fibrillation can be caused by **circus**

rhythms, electrical signals being continuously recycled. Various myocardial regions beat at their own independent rates and very little blood is ejected. Atrial fibrillation is not immediately life threatening because the ventricles fill adequately without atrial contraction; however, it should be treated with digoxin or quinidine to prevent heart failure or formation of thrombi or emboli. Ventricular fibrillation is immediately life threatening unless it can be stopped by **defibrillation.** An instrument called a **defibrillator** passes a strong electrical current through the heart, momentarily placing the entire myocardium in a single absolute refractory period. Defibrillation is used as an emergency life-saving technique where the only hope for the patient's survival is that after defibrillation, the SA node will spontaneously reestablish a normal sinus rhythm.

Congenital Heart Defects

Originating during embryonic development, **congenital** (kon-jen'it-al) **heart defects** can seriously impair heart function. Many such defects are caused by infections such as **rubella** (roo-bel'ah), also called German measles, or medications used by the mother during the first two months of pregnancy when the embryonic heart is developing. Infants with heart defects often have other defects because agents that damage the heart can damage other organs that are vulnerable at the same time in development.

Heart defects fall in three groups according to their effects on function: (1) **stenosis** (constriction) of a channel that impairs blood flow, (2) a **left-to-right shunt,** which allows blood to bypass the systemic circulation, and (3) a **right-to-left shunt,** which allows blood to bypass the pulmonary circulation. All such defects impair the patient's ability to withstand exertion and can be life threatening.

Coarctation (ko''ark-ta'shun) **of the aorta** (Figure 18.16a), a stenosis of the aorta, causes the left ventricle to work excessively hard to force blood through the aorta. It causes enlargement of the left ventricle.

Left-to-right shunting can be caused by a patent (open) ductus arteriosus, an atrial septal defect, or a ventricular septal defect. In the fetus, the ductus arteriosus (Figure 18.16b) allows blood to flow from the pulmonary trunk to the aorta, bypassing the lungs. At birth, when the lungs expand, higher pressure in the aorta reverses blood flow in the ductus arteriosus, shunting it from left to right. The ductus normally closes within a few hours to a few days. In *patent* (open) *ductus arteriosus,* the ductus fails to close and blood is recycled through the lungs. After several years if the ductus is not surgically closed, the entire heart becomes greatly enlarged. This enlargement is due to the increased pumping needed to maintain systemic blood flow while pumping two to three times the normal amount of blood through the lungs.

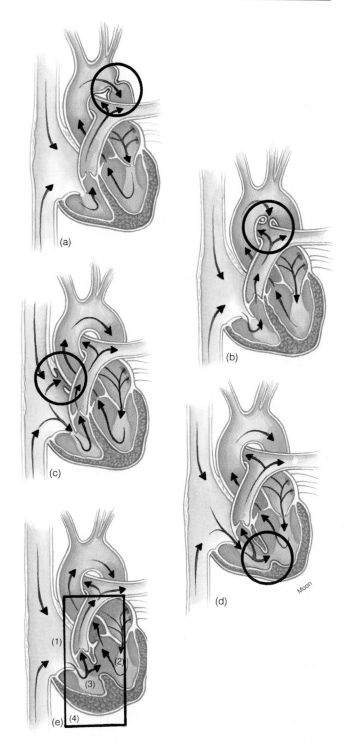

FIGURE 18.16
Some common congenital defects of the heart: (*a*) coarctation (narrowing) of the aorta, (*b*) patent ductus arteriosus, (*c*) atrial septal defect, (*d*) ventricular septal defect, and (*e*) tetralogy of Fallot. In *a–d,* the affected area is encircled. In *e,* the four affected areas are indicated by number: (*1*) aorta originating from both ventricles, (*2*) stenosed pulmonary artery, (*3*) interventricular septal defect, and (*4*) greatly enlarged right ventricle.

An **atrial septal defect,** an opening in the interatrial septum (Figure 18.16c), is often caused by failure of the *foramen ovale* to close. Like the ductus arteriosus, the foramen ovale allows blood to bypass the lungs in a fetus. At birth a flap of tissue blocks the foramen ovale and pressure in the left atrium keeps it closed until fibrous tissue grows to seal it. If the foramen ovale fails to close or if a larger defect exists in the interatrial septum, significant left-to-right shunting occurs. If not repaired by open-heart surgery, the heart enlarges and fluid accumulates in lung tissue because of high pulmonary pressure.

A **ventricular septal defect,** an opening in the interventricular septum (Figure 18.16d), allows blood to flow from the left to the right ventricle. If the defect is not repaired by open-heart surgery, the walls of the right ventricle become as thick as those of the left ventricle. Much of the heart's pumping action is devoted to pumping blood through the opening in the septum and recycling it through the pulmonary circulation.

The **tetralogy of Fallot** (tet-ral′o-je of fal-o′)—four simultaneous heart defects—creates a right-to-left shunt (Figure 18.16e). The defects include (1) the aorta originating from the right, or both, ventricles, (2) pulmonary artery stenosis, (3) an interventricular septal defect, and (4) an enlarged right ventricle. As much as three-fourths of the blood may be pumped from the right ventricle into the aorta without passing through the lungs and without being oxygenated. An infant with tetralogy of Fallot is sometimes called a "blue baby" because blood is poorly oxygenated and the skin appears blue. Surgical procedures can correct these multiple defects.

CLINICAL TERMS

angina pectoris (an-jin′-ah pek′tor-is)   severe chest pain due to anoxia of the myocardium

antiarrhythmic (an″ti-ah-rith′mik) **drug**   a substance that helps to restore a normal heart rhythm or helps to prevent arrhythmias

asystole (ah-sis′to-le)   inability of the heart to perform a complete contraction

auscultation (aws″kul-ta′shun)   the act of listening for sounds within the body

cardiomegaly (kar″de-o-meg′al-e)   enlargement, or hypertrophy, of the heart

congestive heart failure   inability of the heart to pump an adequate amount of blood because of accumulation of fluids around the heart or in the lungs

cor pulmonale (kor pul″mo-nal′e)   failure of the right side of the heart to pump sufficient blood because of obstructed blood flow in the lungs

echocardiography (ek″o-kar″de-og′raf-e)   the use of high frequency soundwaves to study heart function and diagnose disorders

endocarditis (en″do-kar-di′tis)   an inflammation of the heart lining and valves

palpitation (pal″pi-ta′shun)   a sudden increase in the heart rate that is felt by the individual

REPLACEMENT AND ASSISTANCE FOR DISEASED HEARTS

Over 40 million Americans have some form of heart or blood vessel disease, and the combined costs of treatment and lost income exceed 50 billion dollars annually. About 4 million people, 10 percent of those with cardiovascular diseases, have coronary artery disease. As many as 160,000 Americans die each year waiting for heart transplants.

Scientists have been interested in heart replacement for more than a century. In 1880, Henry Martin studied heart physiology by removing intact heart and lungs from animals and using a mechanical device to perfuse blood through them. Other physiologists built and experimented with pumps to replace heart action during surgery. A 1928 machine was a prototype for today's heart-lung machine that keeps tissues alive during heart and lung surgery.

The first human heart transplant was done in 1967 by the South African surgeon Christian Barnard and his surgical team. They used a healthy heart from a victim of another disorder to replace a diseased heart. During the next ten years, about 360 transplantations were done, mostly at Stanford University Medical Center. Survival rates were poor for early transplants, but have improved with recent procedures—two-thirds of the patients live more than a year and more than half are alive five years after surgery.

In spite of improved procedures, heart transplantation has not come into widespread use for three reasons:

1. Too few hearts are available to meet the need. Transplantable hearts come mainly from healthy accident victims, who are far less numerous than the patients who could benefit from them.
2. The problem of the recipient's body rejecting the transplanted organ is a serious one, as explained in Chapter 21.
3. Heart transplantation will remain very expensive, even if other problems were solved.

In recent years, interest in the totally artificial heart has developed because it could be mass produced from biologically inert materials. Such a heart was implanted in a patient already declared dead to demonstrate that it would effectively perfuse tissues. Several artificial hearts have been implanted either to maintain lives until donors could be found or as permanent treatments.

The heart most often implanted was the Jarvik-7 heart developed by Robert Jarvik and co-workers at the University of Utah. The Jarvik-7 heart, designed to accommodate human anatomical and physiological requirements, had two polyurethane ventricles and four tilting disk valves connected to an electric air compressor that causes ventricular contraction similar to normal heart action. Because of its vigorous pumping, the Jarvik-7 heart created pathology as

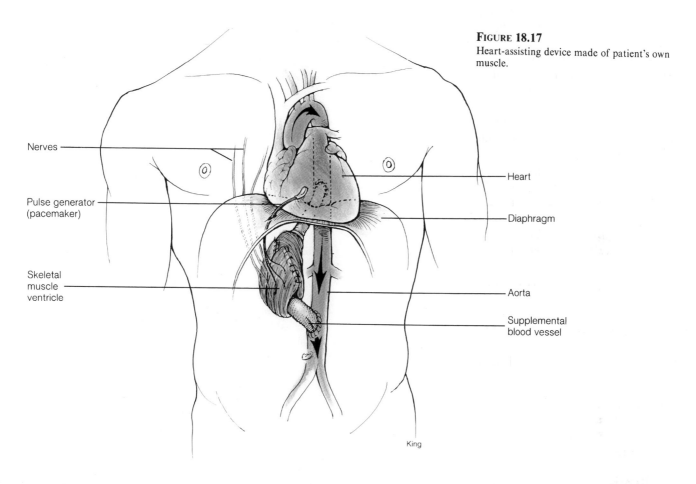

FIGURE 18.17
Heart-assisting device made of patient's own muscle.

Nerves

Pulse generator
(pacemaker)

Skeletal
muscle
ventricle

Heart

Diaphragm

Aorta

Supplemental
blood vessel

King

life-threatening as that caused by the diseased heart it replaced. More heart replacement will likely await improved design.

Even if the design problems are solved so that implants in humans are feasible, ethical problems will remain. The cost of manufacturing and implanting an artificial heart now exceeds $100,000—far more than most potential recipients can afford. Can and will taxpayers and insurance premium payers accept the burden of providing such a heart for all who need them? If not, who will choose which individuals will receive them, and how will these choices be made? Should such ethical problems have been resolved before allocating $10 million in government funds per year for several years to heart replacement research?

Most physiologists have considered skeletal muscle too easily fatigued to function continuously as cardiac muscle, but recent studies have proven differently. Skeletal muscle used to construct ventricles and electrically stimulated, pumped

continuously in free-running beagles for up to eight weeks. In a modification of this technique, a muscular bag is constructed from the patient's back muscle. It is aligned parallel to the aorta and its innervation maintained (Figure 18.17). Such a device, so far tested only in animals, can increase cardiac performance by 25 to 50 percent. Muscle from a person's own body is readily available, not rejected as foreign, and is less expensive to arrange than transplantation. It is suitable to assist, but not replace, ailing hearts.

Questions

1. What is the magnitude of the problem of heart disease?
2. How have heart transplantation procedures changed over the years?
3. How do the problems associated with an artificial heart differ from those associated with transplanting a natural heart?
4. What is the likely future of heart replacement?

CHAPTER SUMMARY

Organization and General Functions

- The heart, located in the thorax, is a double pump that forces blood through vessels to all parts of the body.

Development

- The heart first differentiates as a simple tube, then becomes curved and separated into chambers with septa and valves.

Anatomy of the Heart

- The heart wall has an outer epicardium, a middle myocardium, and an inner endocardium; it is surrounded by a pericardial sac and a fluid-filled pericardial cavity.
- The myocardium consists of many contractile cells and some specialized conductive cells; atrial and ventricular contractile cells are separated by nonconductive endocardial tissue.
- The right side of the heart pumps blood to the pulmonary circulation; the left side pumps it to the systemic circulation.
- The atrioventricular valves prevent backflow into the atria during ventricular contraction; the semilunar valves prevent backflow from the arteries while the ventricles fill.
- Blood goes via the coronary arteries to the myocardium and is returned to the right atrium via the coronary sinus.
- Atria are well supplied by sympathetic and parasympathetic nerve fibers; ventricles have mainly sympathetic fibers.

Initiation and Regulation of the Heartbeat

- The orderly sequential contraction of atria and then the ventricles is maintained by the conduction system acting on contractile cells.
- Action potentials from the conduction system cause atria and then ventricles to contract synchronously; plateaus in action potentials are due to many calcium ions entering cells.
- A refractory period corresponding to the plateau prevents cardiac muscle from contracting without first relaxing completely.
- In the heart's conduction system, SA node fibers depolarize automatically and rhythmically, and send action potentials to the atria and to the AV node; signals to the AV node are delayed so the atria contract before the ventricles.
- Once action potentials reach the AV node, they pass through the AV bundle, bundle branches, and Purkinje fibers to contractile cells of the ventricles; the ventricles then contract in a wave from the apex upward.
- Each part of the conduction system has an inherent rate of depolarization; when contractile cells abnormally generate action potentials, they disrupt the normal SA node pacemaker activity.

Pumping Action of the Heart

- An electrocardiogram (ECG) is a graphic record of electrical activity in cardiac muscle that detects potential differences between parts of the heart from electrodes placed on the skin.
- A normal ECG has a P wave associated with atrial depolarization, a QRS complex associated with ventricular depolarization, and T wave associated with ventricular repolarization.
- The cardiac cycle consists of one cycle of contraction (systole) and relaxation (diastole) of the heart.
- When the ventricles contract, each pumps the same volume of blood, but the thicker wall of the left ventricle contracts with greater force.
- Phases of the cardiac cycle are: (1) atrial contraction, when the ventricles fill with blood; (2) isovolumetric ventricular contraction, when valves are closed and ventricular pressures increase; (3) rapid ventricular ejection, when semilunar valves are open and blood is forcefully pumped from the ventricles; (4) isovolumetric ventricular relaxation, when valves are closed and ventricular pressures are falling; and (5) rapid ventricular filling, when the ventricles are relaxed, the AV valves are open, and blood flows passively from the arteries through the atria into the ventricles.

Cardiac Output and Factors That Affect It

- Cardiac output is the product of the heart rate times the stroke volume, the volume of blood ejected with each ventricular contraction.
- Intrinsic control of cardiac output occurs as the volume of the blood being returned to the heart alters the tension in the fibers of the myocardium.
- According to the Frank-Starling law of the heart, within physiological limits, the heart pumps all the blood that enters it.
- Extrinsic reflex control is accomplished by sympathetic and parasympathetic signals. Sympathetic signals increase heart rate and parasympathetic signals decrease it, probably by altering potassium permeability of cell membranes. Such signals also increase myocardial contractility by making more Ca^{2+} bind with troponin.

Clinical Applications

- ECGs provide information about normal heart physiology and about abnormalities, especially arrhythmias including tachycardia, bradycardia, premature ventricular contractions, flutter, and fibrillation.
- Congenital heart defects, which usually are caused by infections or exposure to toxic substances during the first two months of pregnancy, include stenosis, left-to-right shunting, and right-to-left shunting.

QUESTIONS AND PROBLEMS

The questions at the end of each chapter are numbered to correspond with the objectives listed at the beginning of the chapter. Italics indicate that a question requires critical thinking skills beyond simple factual recall.

Questions

1. Describe the shape, size, and location of the heart.
2. What changes occur as the heart develops from a simple tube?
3. *(a) What are the main characteristics of the tissues of the heart and the pericardium?*
   (b) What are the functions of each heart layer?
4. (a) Describe the structure and arrangement of the chambers of the heart.
   (b) How do heart valves prevent backflow of blood?
   (c) In what sequence does blood pass through the chambers and valves of the heart?
5. *(a) How does blood circulate through the myocardium?*
   (b) Why is a special coronary circulation needed?
   (c) How is the heart innervated?
6. *(a) What are the roles of movements of ions in action potentials and refractory periods in myocardial muscle?*
   (b) What is the advantage of the heart's long refractory period?
   (c) What is the pathway of signals in the conduction system?
   (d) What is the significance of delaying action potentials moving toward the ventricles?
7. What physiological events are represented by the waves recorded in an electrocardiogram?
8. *(a) What are the significant events that occur in each of the five phases of the cardiac cycle?*
   (b) How does each event relate to normal heart function?
   (c) For each of the events in the cardiac cycle, what would be the effects of greatly slowing that event?
9. (a) What two factors determine cardiac output?
   (b) How can cardiac output be modified normally and in disease processes?
   (c) What is the Frank-Starling law of the heart, and by what mechanism does it operate?
   (d) How are sympathetic and parasympathetic signals involved in regulating heart rate and stroke volume?
10. *Briefly describe abnormalities seen in electrocardiograms, and relate them to heart physiology.*
11. *What are the physiological effects of the three major groups of congenital heart defects?*

Problems

1. For each of the following, draw an ECG and explain the abnormal physiology: (a) sinus bradycardia, (b) right bundle branch block, (c) fibrillation, and (d) complete heart block.
2. Calculate the cardiac output for each of the following: (a) 60 beats per minute, each pumping 75 ml of blood; (b) 130 beats per minute, each pumping 55 ml of blood; and (c) 190 beats per minute, each pumping 5 ml of blood. Suggest at least one physiological state that might result in each of the above cardiac outputs.
3. What would happen if the volume of blood moving from the heart to the aorta were greater than, or less than, that moving from the heart to the pulmonary trunk?

OBJECTIVES

1. Review the general plan of the circulatory system, and describe the role of blood vessels in it.

2. Describe briefly the development of the blood vessels.

3. Describe the structure of each type of blood vessel, and explain how structure is related to function.

4. Locate and identify the major blood vessels of the human body.

5. List the special characteristics of blood vessels of the brain, digestive tract and liver, spleen, kidneys, muscles and skin, and placenta and fetus.

6. Describe the organization and general function of the lymphatic system.

7. Describe the tissues of the lymphatic system.

8. Locate and identify the vessels and nodes of the lymphatic system.

9. Explain the functions of the lymphatic system, and describe how lymph moves through the system.

10. Describe briefly the causes, effects, and treatment of aneurysms.

19

BLOOD VESSELS AND LYMPHATICS

ORGANIZATION AND GENERAL FUNCTIONS OF BLOOD VESSELS

Blood flows continuously throughout your body within a complex array of blood vessels. It moves against gravity from your heart to your head and from your toes back to your heart. When you exercise, certain vessels widen to send more blood to active muscles. After a meal, other vessels widen to send more blood to digestive organs. Together, blood vessels carry nutrients to and remove wastes from all body cells.

The body's blood vessels form several closed circuits. The coronary circuit serves the heart (Chapter 18), the pulmonary circuit serves the respiratory portion of the lungs, and the systemic circuit serves all other cells of the body (refer back to Figure 17.1). Each circuit includes arteries that transport blood from the heart, arterioles that branch from arteries, capillaries that branch from arterioles, venules that receive blood from capillaries, and veins that receive blood from venules and return it to the heart.

In certain locations, vessels called **portal** (por'tal) **veins** carry blood between two sets of capillaries without going through the heart. In humans, a large hepatic portal vein carries blood from capillaries of the intestine and other abdominal organs to the liver, and another portal vein carries blood from the hypothalamus to the anterior pituitary gland (Chapter 16).

> Reptiles and birds have a renal portal system. Blood from capillaries in the tail region passes through a portal vein and enters capillaries of the kidneys. Blood from the kidneys enters veins that carry it to the heart.

DEVELOPMENT

Blood vessels, which are derived from mesoderm, begin to develop as cords of cells during the third week of embryonic life (Figure 19.1). Soon a lumen forms within each cord, creating a tube, and each tube becomes a blood vessel. In larger blood vessels, the original tube becomes a unicellular lining called the endothelium with connective tissue and smooth muscle layers around it. Eventually, blood vessels extend throughout all tissues of the embryo as thousands of branches form by budding. Although there is a typical arrangement of the body's major blood vessels, many variations are seen in the distribution of the smaller vessels. These variations are due to the somewhat random pattern of budding that arises as the embryo develops.

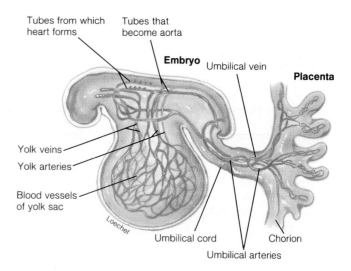

FIGURE 19.1
The structure of the circulatory system of the human embryo at about 3 weeks.

Extensive early development of blood vessels in the human embryo is essential to survival because the human ovum contains very little stored food material. The embryo quickly establishes a circulatory system having close contact with the maternal blood of the placenta. Throughout development, it exchanges nutrients and wastes across placental membranes between the fetal and maternal blood. Even though human embryos have almost no yolk, they develop a yolk sac with many blood vessels like vertebrate embryos that have a large yolk containing stored nutrients.

During embryonic development, chemical substances called **angiogenic factors** stimulate **angiogenesis** (an"je-o-jen'es-is), the growth of blood vessels. Later in life, these substances accelerate wound healing, and in pregnant women, placental development. When blood vessels, especially those in the heart or brain, become blocked, angiogenic factors stimulate growth of blood vessels that form alternate pathways. Angiogenic factors also can stimulate abnormal blood vessel growth. In the retina, such growth causes blindness; in arthritic joints, it destroys cartilage; and in tumors, it enhances the blood supply and allows rampant growth.

TISSUES OF BLOOD VESSELS

Various kinds of blood vessels—arteries, capillaries, and veins—can be distinguished by tissue layers (Figure 19.2). Arterioles are similar to arteries, and venules are similar to veins.

Tunica intima

Connective tissue

Elastic tissue

Tunica media

Tunica adventitia

Serosa

Artery

Arteriole

Capillary bed

Capillary bed

Venule

Vein

Tunica intima

Direction of blood flow

Elastic tissue

Tunica media

Tunica adventitia

Hosmer

FIGURE 19.2

Various types of blood vessels and their walls.

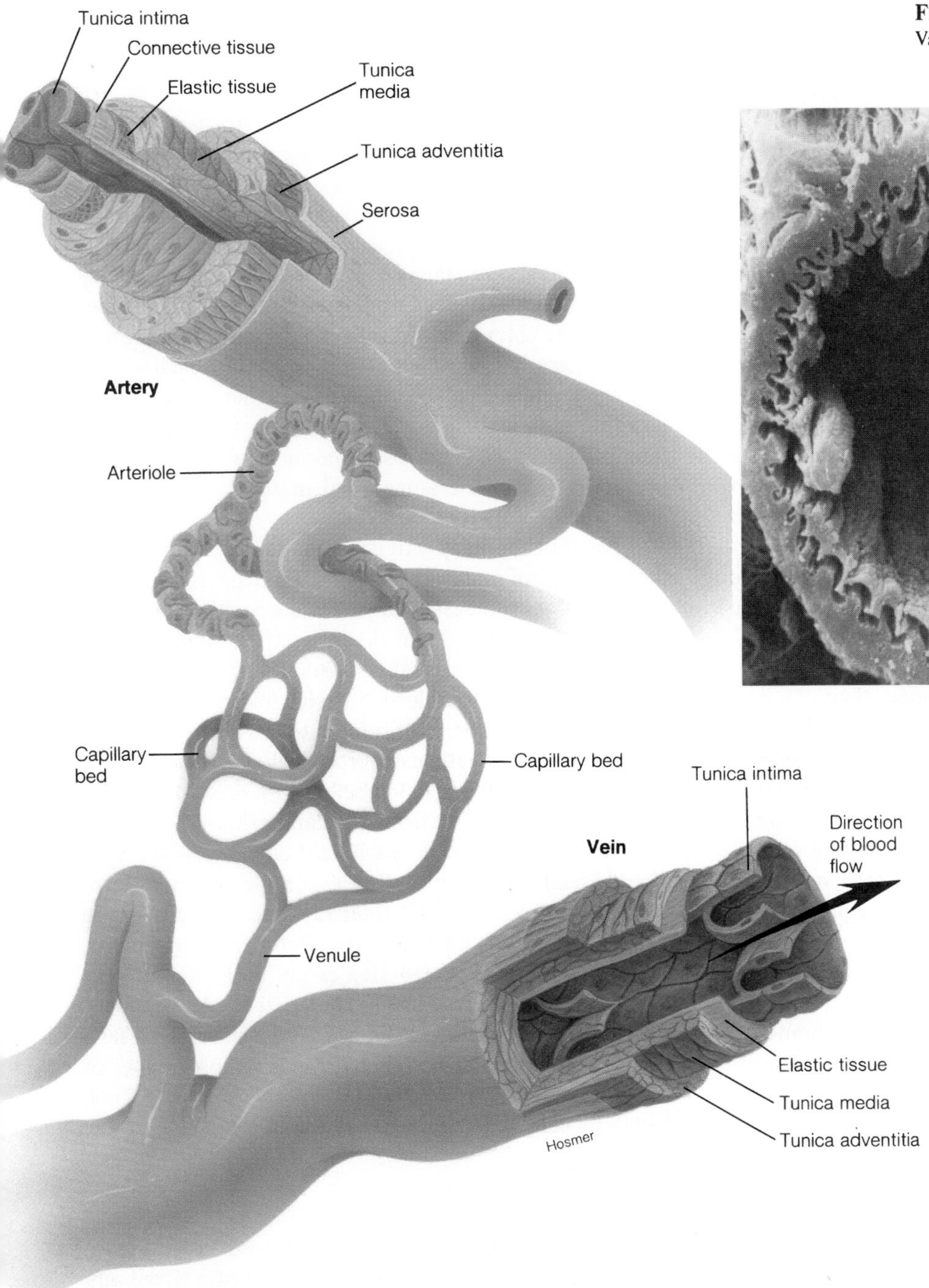

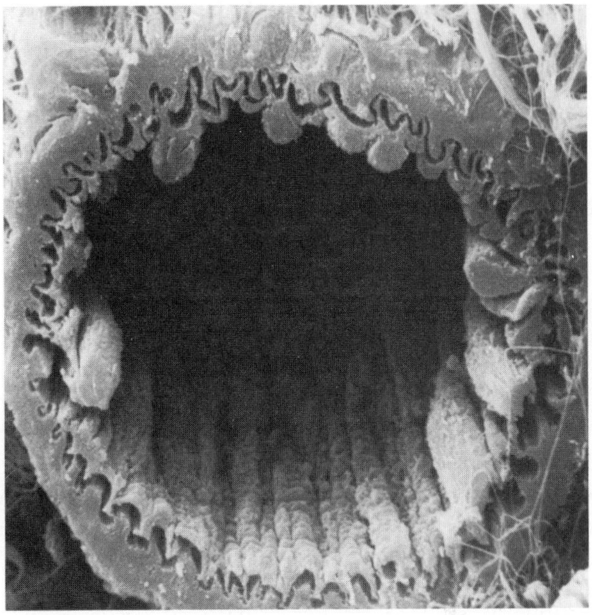

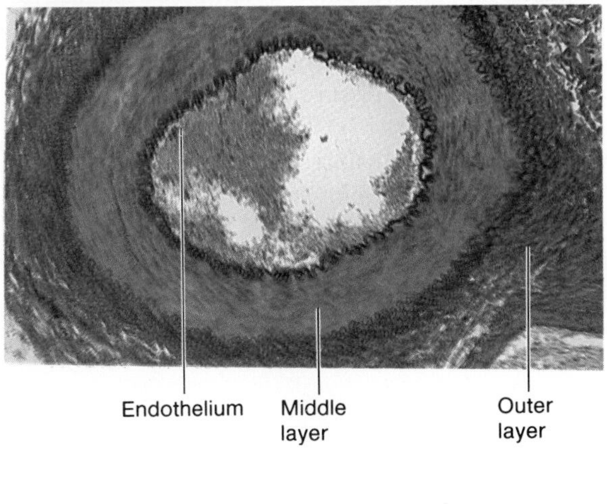

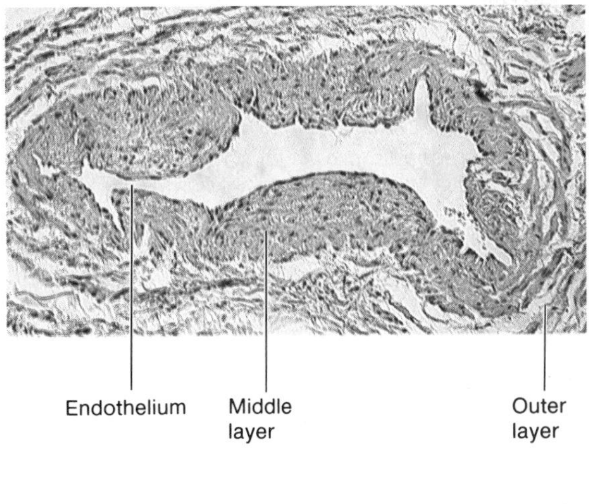

Endothelium   Middle layer   Outer layer

Endothelium   Middle layer   Outer layer

Arteries and Arterioles

Arteries, having the thickest walls of all blood vessels, consist of three tissue coats (tunics) surrounding the lumen in concentric circles. The inner **tunica interna** (too′nik-ah in-ter′nah) or **intima** (in′tim-ah), has a layer of simple squamous cells called the **endothelium** (en″do-the′le-um) and a thin layer of areolar connective tissue with elastic fibers. The thick middle **tunica media** (me′de-ah) consists of elastic tissue and smooth muscle. The outer **tunica externa** (eks-ter′nah), or **adventitia** (ad″ven-tish′e-ah), is made of white fibrous connective tissue with a few elastic fibers and smooth muscle cells. The walls of large arteries are themselves nourished by blood vessels called **vasa vasorum** (va′sah vas-o′rum), a vessel of a vessel.

The structure of an artery is closely related to its function. The largest arteries receive blood from the heart under high, pulsating pressure. Their thick walls help them to withstand high pressure. Their many elastic fibers, particularly in the tunica media, allow the walls to stretch to accommodate the sudden increases in pressure and spring back to their former diameter as pressure decreases. Elastic arteries stretch and spring back with each heartbeat over and over throughout a lifetime.

In arteries farther from the heart, the tunica media contains relatively more smooth muscle and less elastic tissue. Muscle contraction decreases the lumen diameter, causing **vasoconstriction** (vas″o-kon-strik′shun). Muscle relaxation increases the lumen diameter, causing **vasodilation** (vas″o-di-la′shun). The degree of vasoconstriction or vasodilation of muscular arteries helps to regulate blood flow to various tissues.

Arteries branch and rebranch many times as they traverse the tissues. Most tissues receive blood from more than one arterial branch, and sometimes the branches rejoin, or **anastomose,** within a tissue. Anastomoses provide alternate pathways for blood to reach any given tissue and help to assure that a tissue will receive blood even when a vessel supplying the tissue is blocked. These alternate pathways are referred to collectively as **collateral circulation.**

Arterioles, like the small arteries from which they branch, carry blood away from the heart to the smallest of all vessels, the capillaries. Wall thickness decreases gradually, so it is difficult to determine where a small artery becomes an arteriole. In many tissues, the junctions of arterioles with capillaries are marked by the presence of **precapillary sphincters,** short vessel segments with particularly thick smooth muscle. This muscle allows the arteriole to control the entry of blood into the capillaries.

Capillaries

Thin-walled **capillaries** have inside diameters varying from 4 to 12 μm and averaging 8 μm—the right size for red blood cells to pass through single file. To pass through smaller capillaries, red blood cells are temporarily distorted into elongated oval shapes. Capillary walls consist of a single layer of endothelial cells continuous with the endothelial lining of the arterioles and venules. Each arteriole branches into many capillaries, forming a **capillary network** (Figure 19.3). Capillary sphincters in arterioles regulate the entry of blood into capillaries. Capillary networks generally have at least one **thoroughfare channel** that lacks sphincters. Such channels assure that enough blood flows through a tissue at all times to sustain slowly metabolizing cells. When a tissue becomes more active, the precapillary sphincters open and increase blood flow in proportion to activity.

The thin capillary wall is well suited to its function—the exchange of materials between blood and tissue fluids. Because of its thinness, many substances diffuse through the capillary wall and into the interstitial fluids that surround cells. Likewise, many substances from interstitial fluids diffuse into the blood. Loosely joined cells in capillary walls allow some substances to pass between them. Even white blood cells can force their way between endothelial cells, but red blood cells and large protein molecules normally do not leave the capillaries.

Certain tissues have specialized capillaries. We have already seen that brain capillaries are specialized to form the blood-brain barrier (Chapter 12). In the kidneys where large quantities of substances are filtered out of the blood, capillaries are **fenestrated** (fen′es-tra″ted); that is, perforated with thousands of small holes (*fenestra,* window). The liver and some other tissues have **sinusoids** (si′nusoidz), large diameter vessels lacking a complete layer of endothelium. Flow is sluggish in sinusoids and phagocytic Kupffer's cells ingest foreign material.

Veins and Venules

Like arteries, **veins** have three coats (refer back to Figure 19.2), but they have lesser amounts of elastic tissue and smooth muscle and greater amounts of fibrous connective tissue than arteries. This wall structure is adequate in veins because they do not receive blood under high pressure as arteries do. In fact, the pressure in veins is insufficient to return blood to the heart.

When a person is in an upright position, gravity moves blood toward the heart in veins superior to it, but opposes movement in veins inferior to the heart. Movement of blood through veins depends on muscle contractions and valves. Skeletal muscle contractions help to push blood forward in veins. **Valves** (Figure 19.4) allow blood to flow toward

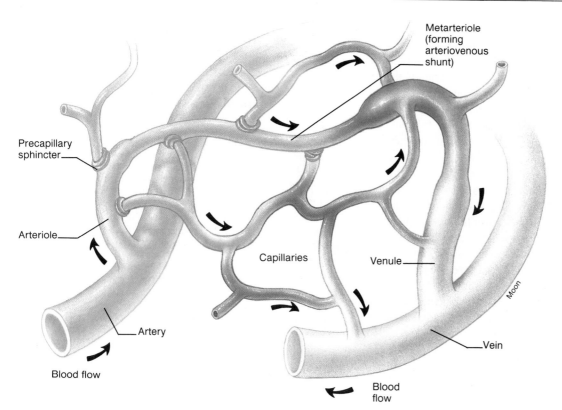

FIGURE 19.3
A capillary network, showing the relationship of arterioles, precapillary sphincters, capillaries (including a thoroughfare channel), and venules.

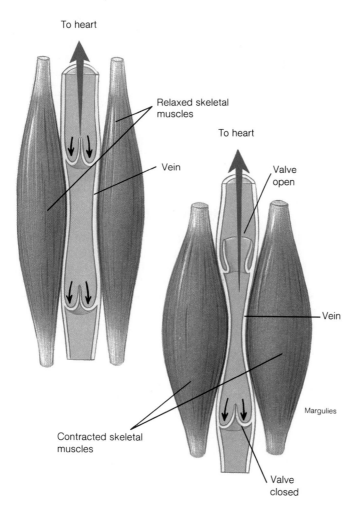

FIGURE 19.4
The location and structure of valves in a vein.

the heart, but prevent it from flowing backward. Contractions during exhalation create negative pressure near the heart and help to move blood through the thorax. Calf muscle contractions during walking press on veins and help push blood through them. Blood that is pushed upward by muscle contractions is prevented from falling back by valves.

Standing motionless for long periods of time allows blood to pool in the leg veins because of the lack of muscle contractions. People who must be on their feet should walk around, stand first on one foot and then the other, or periodically bend their knees to help keep blood moving in leg veins. Wearing support stockings also aids circulation.

The walls of leg veins can be stretched by prolonged periods of standing and by the stresses of pregnancy. The valves no longer close the lumen of stretched veins and backflow occurs. Backflow causes additional stretching and eventually the walls are greatly weakened and the veins distended. Such distended veins are called **varicose** (var'ik-ōs) **veins,** and those in the anal canal are called **hemorrhoids.**

Like arteries, veins vary in diameter. Tiny **venules,** which collect blood from the capillaries, converge to form small veins, and small veins converge to form large veins. Some veins, called **sinuses,** have very large lumens and thin walls consisting mainly of endothelium. Venous sinuses are found in several locations—the liver, spleen, brain, and heart. For example, dural sinuses beneath the dura mater drain blood from the brain, and the coronary sinus in the heart drains blood from the myocardium back into a heart chamber.

Structure and function are closely related in blood vessels. The structure of arteries allows them to withstand high pressure. Similarly, the structure of arterioles allows them to regulate blood flow to capillaries, that of capillaries allows for exchange of substances, and that of veins prevents backflow of blood.

DISTRIBUTION OF BLOOD VESSELS

Systemic Arteries

The **aorta** is the largest artery of the systemic circulation. It and its major branches are depicted in Table 19.1 and Figure 19.5. More detailed diagrams of the distribution

TABLE 19.1   The Branches of the Aorta

Artery	Region of Body Served
Ascending Aorta	
Coronary (right and left)	Heart muscle
Arch of the Aorta	
Brachiocephalic (right)	
Right common carotid	Right side of head and neck
Right subclavian	Right shoulder and arm
Left common carotid	Left side of head and neck
Left subclavian	Left shoulder and arm
Thoracic Aorta	
Intercostals	Intercostal and chest muscles
Superior phrenics	Diaphragm
Bronchials	Nonrespiratory portion of bronchi and lungs
Esophageals	Esophagus
Abdominal Aorta	
Inferior phrenics	Diaphragm
Celiac	Liver, stomach, spleen, and pancreas
Superior mesenteric	Small intestine, ascending and transverse colon
Suprarenals	Adrenal (suprarenal) glands
Renals	Kidneys
Gonadals (ovarian or testicular/spermatic)	Ovaries or testes
Inferior mesenteric	Transverse, descending, and sigmoid colon; rectum
Common iliacs	
Internal iliacs	Urinary bladder, muscles of buttocks, uterus, or prostate gland
External iliacs	Hip and leg

of certain arteries are shown in Figures 19.6 through 19.10, and summarized in Tables 19.2 through 19.7. Figure 19.11 shows how vessels appear in arteriograms.

Systemic Veins

In general, veins follow the same pathways as the arteries serving a particular tissue, but the blood in them flows in the opposite direction. The **superior** and **inferior venae cavae** are the largest veins of the systemic circulation. They and the major veins that drain into them are shown in Figures 19.12 to 19.17, and summarized in Tables 19.8 through 19.13.

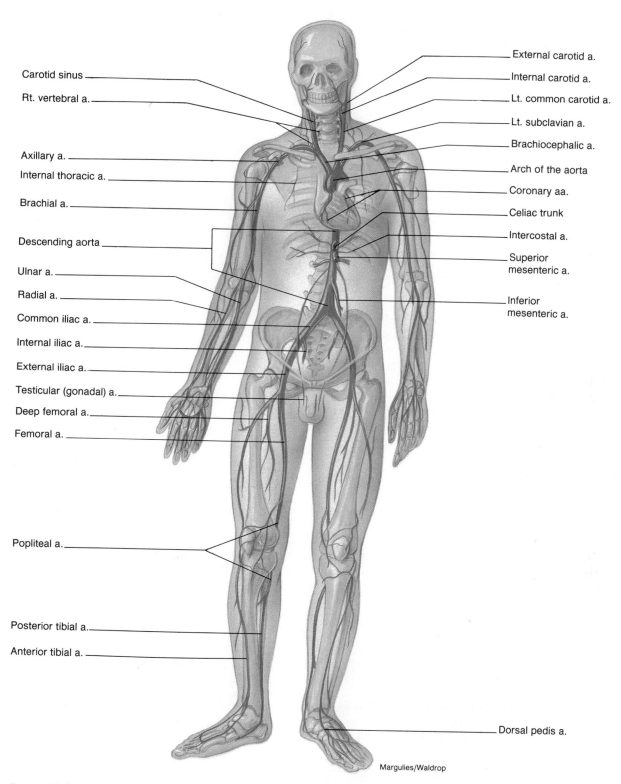

Carotid sinus

Rt. vertebral a.

Axillary a.

Internal thoracic a.

Brachial a.

Descending aorta

Ulnar a.

Radial a.

Common iliac a.

Internal iliac a.

External iliac a.

Testicular (gonadal) a.

Deep femoral a.

Femoral a.

Popliteal a.

Posterior tibial a.

Anterior tibial a.

External carotid a.

Internal carotid a.

Lt. common carotid a.

Lt. subclavian a.

Brachiocephalic a.

Arch of the aorta

Coronary aa.

Celiac trunk

Intercostal a.

Superior mesenteric a.

Inferior mesenteric a.

Dorsal pedis a.

Margulies/Waldrop

FIGURE 19.5

An anterior view of the aorta and major arteries.

FIGURE 19.6
Arteries (*a*) of the head and neck, and (*b*) of
the brain.

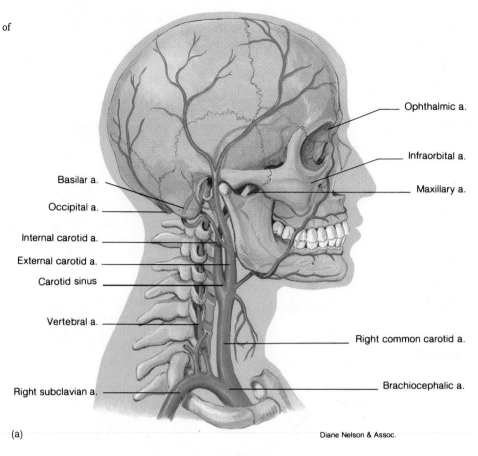

Ophthalmic a.

Infraorbital a.

Maxillary a.

Basilar a.

Occipital a.

Internal carotid a.

External carotid a.

Carotid sinus

Vertebral a.

Right common carotid a.

Right subclavian a.

Brachiocephalic a.

(a)

Diane Nelson & Assoc.

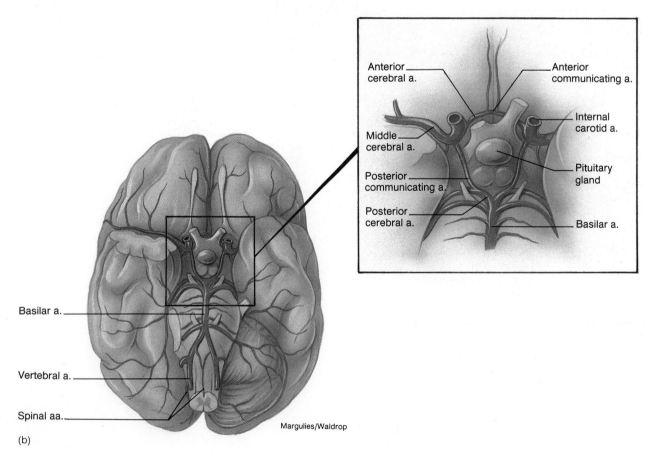

Anterior
cerebral a.

Anterior
communicating a.

Middle
cerebral a.

Internal
carotid a.

Posterior
communicating a.

Pituitary
gland

Posterior
cerebral a.

Basilar a.

Basilar a.

Vertebral a.

Spinal aa.

Margulies/Waldrop

(b)

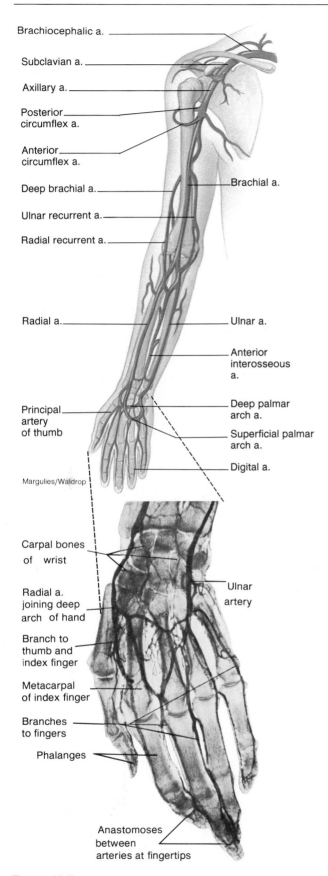

Brachiocephalic a.

Subclavian a.

Axillary a.

Posterior circumflex a.

Anterior circumflex a.

Deep brachial a.

Ulnar recurrent a.

Radial recurrent a.

Radial a.

Principal artery of thumb

Margulies/Waldrop

Brachial a.

Ulnar a.

Anterior interosseous a.

Deep palmar arch a.

Superficial palmar arch a.

Digital a.

Carpal bones of wrist

Radial a. joining deep arch of hand

Branch to thumb and index finger

Metacarpal of index finger

Branches to fingers

Phalanges

Ulnar artery

Anastomoses between arteries at fingertips

FIGURE 19.7
Arteries of the shoulder, arm, and hand.

TABLE 19.2	Arteries of the Head, Neck, and Brain
Artery	**Region of Body Served**
Common carotid	
External carotid	Thyroid gland; salivary glands; tongue; neck, throat, scalp, and facial muscles; jaws and teeth.
Internal carotid	Anterior and middle cerebral arteries supply the cerebrum; other branches supply pituitary, eyeball and eye muscles, lacrimal gland, and nasal cavity.
Vertebral	Spinal cord and vertebrae of the neck. Left and right vertebrals join to form the basilar artery, which supplies the occipital lobe of the cerebrum and the cerebellum.

TABLE 19.3	Arteries of the Shoulder and Arm
Artery	**Region of Body Served**
Subclavian	
Dorsoscapular	Muscles and skin of shoulder and upper back
Internal thoracic (mammary)	Muscles and skin of chest and upper abdomen; breast, membranes of the thoracic cavity, pericardium
Axillary	Continuation of subclavian beyond first rib; muscles of the chest and shoulder, head of humerus and shoulder joint
Brachial	Continuation of axillary beyond shoulder joint; branches supply humerus, muscles and skin of upper arm; frequently used for blood pressure measurements; branches into radial and ulnar arteries distal to the elbow
Radial	Branches supply elbow and muscles of forearm on the thumb side; bones and joints of the wrist
Ulnar	Branches supply elbow and muscles of forearm on ulnar side; bones and joints of the wrist
Palmar arches	Anastomoses of branches of radial and ulnar arteries; branches from the palmar arches supply the digits; create extensive collateral blood supply to hand

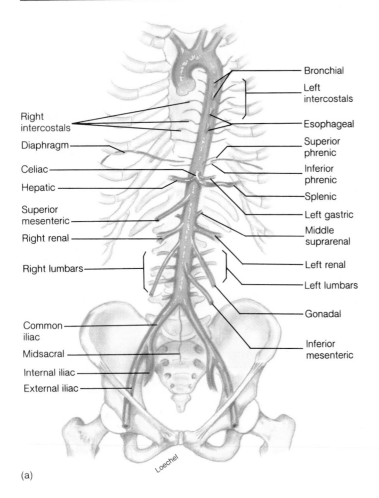

(a)

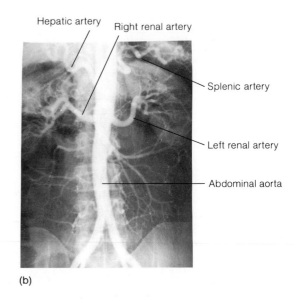

(b)

FIGURE 19.8
The major arteries branching from the descending aorta: (*a*) a diagram, and (*b*) an arteriogram.

TABLE 19.4	Arteries of the Descending Aorta
Artery	**Region of Body Served**
Bronchial	Airways, nonrespiratory portion of lungs
Intercostals	Spinal cord, skin and muscles of the back and chest, including intercostal muscles, breasts
Esophageal	Esophagus
Superior phrenic	Diaphragm
Inferior phrenic	Diaphragm
Celiac (see Table 19.5)	
Superior mesenteric (see Table 19.6)	
Middle suprarenal	Adrenal gland
Renal	Kidneys and ureters; a pair of very large arteries responsible for transporting large quantities of blood through kidneys for waste removal and electrolyte balance
Gonadal (ovarian or testicular, sometimes called spermatic)	Gonads
Inferior mesenteric (see Table 19.6)	
Lumbar	Muscles and skin of lumbar region; lumbar vertebrae
Common iliac (see Table 19.7)	
Midsacral	Sacral vertebrae and rectum

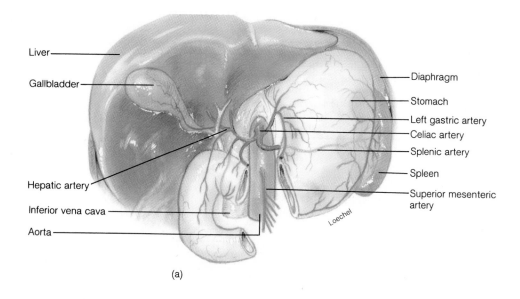

(a)

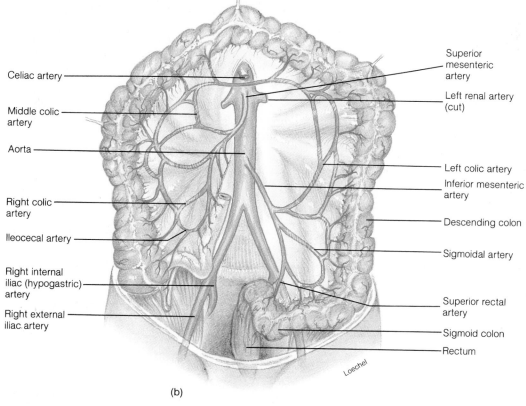

(b)

FIGURE 19.9

(a) The branches of the celiac artery, and (b) the branches of the superior and inferior mesenteric arteries.

TABLE 19.5	Branches of the Celiac Artery
Artery	**Region of Body Served**
Left gastric	Stomach and esophagus
Right gastric	Stomach
Hepatic	Liver, gallbladder, and bile duct; stomach, duodenum, and pancreas
Splenic	Spleen, stomach, pancreas, and omentum

TABLE 19.6	Branches of the Superior and Inferior Mesenteric Arteries
Artery	**Region of Body Served**
Superior Mesenteric	
Intestinal	Many branches to the walls of the small intestine
Inferior pancreatoduodenal	Pancreas and duodenum
Ileocecal	Ileum and cecum
Right colic	Ascending colon
Middle colic	Transverse colon
Inferior Mesenteric	
Left colic	Descending colon
Sigmoid	Descending and sigmoid colon
Superior rectal	Rectum

FIGURE 19.10
The arteries of the pelvic area and leg.

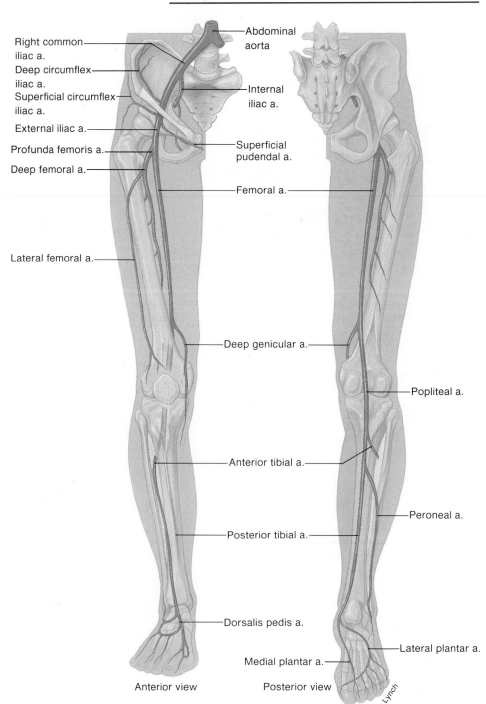

Anterior view   Posterior view

TABLE 19.7   The Arteries of the Pelvic Area and Leg

Artery	Region of Body Served	Artery	Region of Body Served
Common iliac	Forms left and right branch from abdominal aorta near the sacrum	Popliteal	Continuation of femoral artery as it passes into the popliteal fossa behind the knee; branches to form the anterior and posterior tibial arteries
Internal iliac (hypogastric)	Pelvic organs, skin and muscles of pelvic area		
External iliac	Becomes the femoral artery as it passes to the area of the thigh	Anterior tibial	Skin and muscles of anterior leg and dorsum (upper surface) of foot
Femoral	Muscles of the upper thigh; forms many anastomoses with external iliac and with arteries of the lower leg	Posterior tibial	Skin and muscles of posterior leg, ankle joint, heel, and sole of foot
		Dorsalis pedis, dorsal arch, and plantar arch	Anastomosing vessels that serve the midportion of the foot and give rise to branches that serve the toes

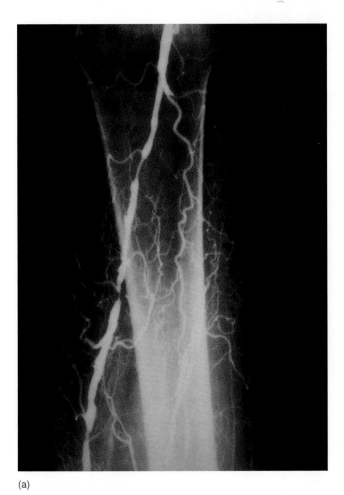

(a)

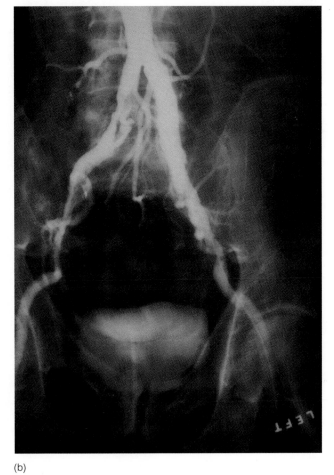

(b)

FIGURE 19.11
Arteriograms of (*a*) the leg, and (*b*) the pelvis and urinary tract.

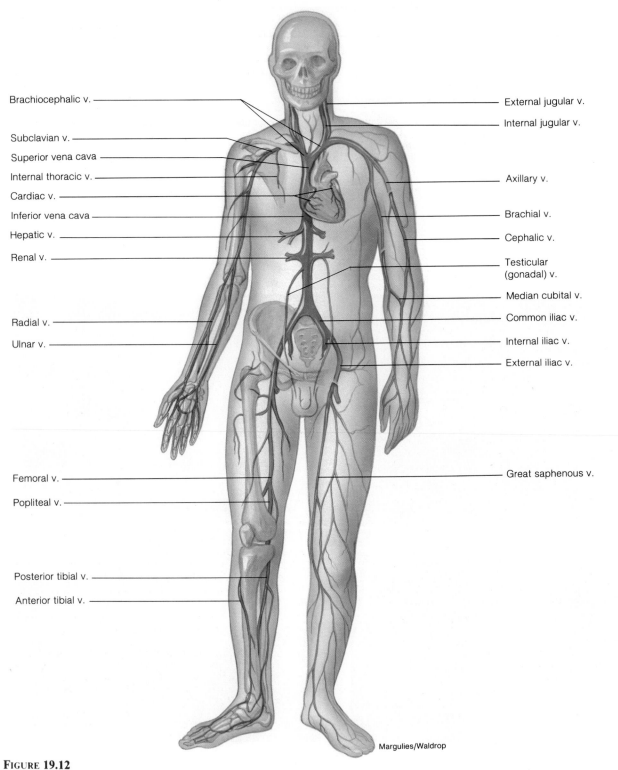

Brachiocephalic v.

External jugular v.

Internal jugular v.

Subclavian v.

Superior vena cava

Internal thoracic v.

Axillary v.

Cardiac v.

Inferior vena cava

Brachial v.

Hepatic v.

Cephalic v.

Renal v.

Testicular
(gonadal) v.

Median cubital v.

Radial v.

Common iliac v.

Ulnar v.

Internal iliac v.

External iliac v.

Femoral v.

Great saphenous v.

Popliteal v.

Posterior tibial v.

Anterior tibial v.

Margulies/Waldrop

FIGURE 19.12
The major veins of the human body in anterior view.

Venous sinuses

Superior
ophthalmic v.

FIGURE 19.13
Veins of the head, neck, and brain.

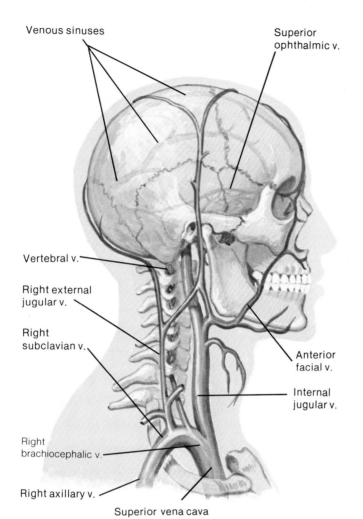

Vertebral v.

Right external
jugular v.

Right
subclavian v.

Anterior
facial v.

Internal
jugular v.

Right
brachiocephalic v.

Right axillary v.

Superior vena cava

TABLE 19.8	Veins Draining into the Superior and Inferior Venae Cavae
Vein	**Region of Body Served**
Superior Vena Cava	
Azygos	Thorax
Brachiocephalic (innominate)	Head, neck, and arms
Internal jugular	Inside of skull and brain
External jugular	Skin and muscles of scalp and face
Subclavian	Shoulder and arm
Inferior Vena Cava	
Hepatic	Liver
Suprarenal	Adrenal gland
Renal	Kidneys
Right gonadal (ovarian or testicular/spermatic)	Gonad
Common iliac	Pelvic area and legs

TABLE 19.9	Veins Draining the Head, Neck, and Brain
Vein	**Region of Body Served**
Brachiocephalic (left and right)	Receives blood from all other veins of head, neck, and brain
Internal jugular	Receives blood from dural sinuses inside cranial cavity
External jugular	Receives blood from salivary (parotid) glands, skin and muscles of the face, scalp, and neck
Vertebral	Receives blood from the cerebellum

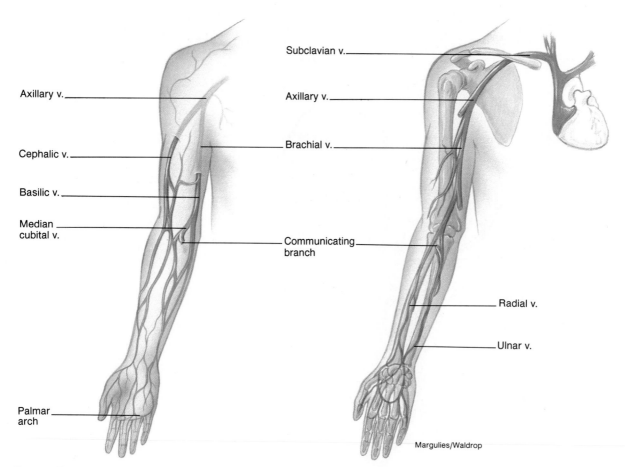

Axillary v.

Cephalic v.

Basilic v.

Median cubital v.

Palmar arch

Subclavian v.

Axillary v.

Brachial v.

Communicating branch

Radial v.

Ulnar v.

Margulies/Waldrop

FIGURE 19.14
Veins of the shoulder, arm, and hand.

TABLE 19.10	Veins Draining the Shoulder and Arm
Vein	**Region of Body Served**
Subclavian	Drains into the brachiocephalic; receives the axillary vein
Axillary	Drains the cephalic, brachial, and basilic veins
Cephalic	Drains the radial side of upper arm and forearm
Brachial	Drains the skin and muscles of the upper arm
Basilic	Drains the medial side of the lower arm; receives the median cubital and median antebrachial
Dorsal arch	Anastomoses of larger veins in the hand; also drains the digits

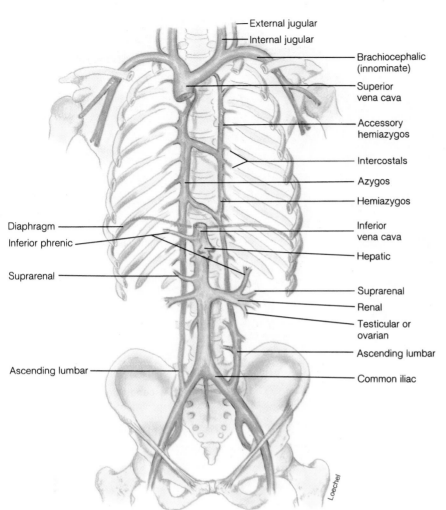

External jugular

Internal jugular

Brachiocephalic
(innominate)

Superior
vena cava

Accessory
hemiazygos

Intercostals

Azygos

Hemiazygos

Inferior
vena cava

Hepatic

Suprarenal

Renal

Testicular or
ovarian

Ascending lumbar

Common iliac

Diaphragm

Inferior phrenic

Suprarenal

Ascending lumbar

Loechel

FIGURE 19.15
Veins of the thorax and abdomen.

TABLE 19.11   Veins Draining the Thorax and Abdomen

Vein	Region of Body Served	Vein	Region of Body Served
Superior Vena Cava		*Inferior Vena Cava*	
Azygos	Drains into superior vena cava; serves as a collateral vessel for draining thoracic and abdominal skin and muscles	Right inferior phrenic	Diaphragm
		Hepatic	Drains liver, which receives all blood from organs of the digestive tract (see Table 19.12)
		Right suprarenal	Adrenal gland
Brachiocephalic		Right renal	Kidney
Hemiazygos	Drains into brachiocephalic; also serves as a collateral vessel for draining thoracic and abdominal skin and muscles	Right ovarian or testicular	Gonad
		Left renal	Receives branches from left inferior phrenic, left suprarenal, and left ovarian or testicular, in addition to draining kidney
Ascending lumbar	Drains into hemiazygos; drains blood from lumbar skin and muscles	Common iliac	Receives blood from pelvic area and leg; also receives left ascending lumbar, which forms a collateral pathway for the return of blood to the inferior vena cava

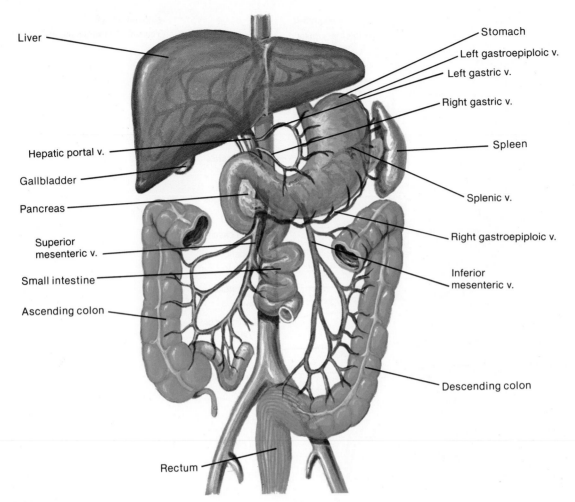

FIGURE 19.16
Veins of the hepatic portal system.

TABLE 19.12 Veins of the Hepatic Portal System

Vein	Region of Body Served
Hepatic portal vein	Receives blood from organs of the digestive system; transports blood to the liver, where it is circulated through hepatic sinusoids and carried to the hepatic veins
Splenic	Drains blood from the spleen and stomach to the hepatic portal vein
Superior mesenteric	Drains blood from the small intestine and part of the large intestine to the hepatic portal vein
Inferior mesenteric	Drains blood from part of the large intestine to the splenic vein

TABLE 19.13 Veins of the Pelvis and Leg

Vein	Region of Body Served
Common iliac	Drains into inferior vena cava
Internal iliac (hypogastric)	Drains pelvic organs and skin and muscles of the pelvic area
External iliac	Drains all of the veins of the leg into the common iliac
Femoral	Drains the deep muscles and bones of the leg; becomes the popliteal behind the knee; drains the peroneal vein, the anterior tibial, and the posterior tibial
Great saphenous	Drains the medial superficial skin and muscles of the leg
Dorsalis pedis, dorsal arch, and plantar arch	Establish collateral circulation through the foot and drain the digits

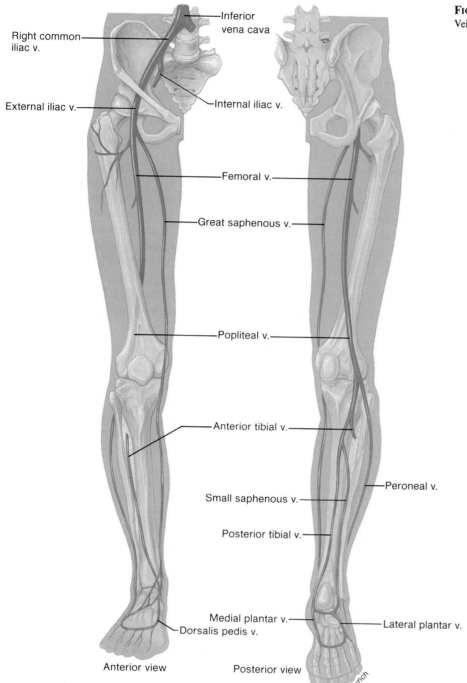

Right common
iliac v.

Inferior
vena cava

External iliac v.

Internal iliac v.

Femoral v.

Great saphenous v.

Popliteal v.

Anterior tibial v.

Peroneal v.

Small saphenous v.

Posterior tibial v.

Medial plantar v.
Dorsalis pedis v.

Lateral plantar v.

Anterior view

Posterior view

Lynch

FIGURE 19.17
Veins of the pelvic area and leg.

Pulmonary Vessels

Blood from the right ventricle goes to the lungs and is returned to the left atrium via the **pulmonary circulation** (refer back to Figure 17.1). Capillaries of the pulmonary circulation are in the respiratory membranes of the lungs where oxygen diffuses into the blood and carbon dioxide diffuses out of it. From the right ventricle, blood enters the **pulmonary trunk** and passes on to the right and left **pulmonary arteries,** which carry blood to the right and left lungs, respectively. These arteries branch into lobar arteries, smaller arteries and arterioles, and, finally, into capillaries. After passing through the capillaries, the blood enters small venules. Venules converge to small veins and, finally, into two **pulmonary veins** in each lung. All four pulmonary veins return blood to the left atrium.

In the pulmonary circulation, arteries carry unoxygenated blood from the heart to the respiratory portion of the lungs, and veins carry oxygenated blood from the lungs to the heart. In all other parts of the body (except in certain fetal vessels) arteries carry oxygenated blood and veins carry unoxygenated blood.

Special Circulatory Pathways

In addition to the general characteristics of blood vessels and the distribution of major blood vessels already discussed, circulatory pathways in certain organs and tissues have special characteristics.

Brain

Special anatomic characteristics of the brain and skull affect blood circulation to and from the brain (Figures 19.6 and 19.12). The **carotid arteries** carry blood along the neck before branching. The **external carotid arteries** serve mainly the scalp whereas the **internal carotid arteries** serve the brain. They empty into the **circle of Willis** at the base of the brain, which provides alternate pathways via several cerebral arteries for blood to enter the brain. If one pathway is obstructed, blood can flow through others to reach nearly all brain cells. Exchange of substances between capillaries and interstitial fluids is controlled by modification of capillary walls that form the blood-brain barrier (Chapter 12). The skull forms a noncompressible chamber around the brain and its blood vessels. Veins inside the skull are not compressed by increased external pressure as are veins in other parts of the body.

Digestive Tract and Liver

Blood enters digestive organs via the several abdominal arteries (Figure 19.9). Branches of the **celiac artery** serve the liver, stomach, spleen, and pancreas (Table 19.5). Branches of the **superior mesenteric artery** serve the small intestine and part of the large intestine, and branches of the **inferior mesenteric artery** serve the remainder of the large intestine. Some blood goes directly to the liver, but most goes first to the capillaries of the digestive tract or spleen. From these capillaries, it enters veins that drain into the **hepatic portal vein** (Figure 19.16). Blood entering the hepatic portal vein from capillaries of the small intestine is laden with nutrients for several hours after a meal. The nutrients are carried to the liver, where some are stored and others are processed before they are delivered to the tissues.

The liver receives an extremely large proportion of the cardiac output—about 28 percent under resting conditions. It serves as a blood reservoir, holding up to 500 ml of blood in its thin-walled **sinusoids.** In times of stress, sympathetic signals can cause the liver capsule to contract, forcing as much as 300 ml of blood out of the sinusoids into the circulation.

Spleen

The spleen is a lymphatic tissue, which receives blood from the celiac artery (Figure 19.9) and returns it to the hepatic portal vein (Figure 19.16). Like the liver, the spleen contains many large sinusoids filled with slowly moving blood. Cells along the walls of sinusoids phagocytize worn out red blood cells. When the body is under stress, sympathetic signals cause the capsule to constrict and push as much as 250 ml of blood into the systemic circulation. Because this blood contains unusually large numbers of erythrocytes, its release increases the volume of circulating erythrocytes by 3 to 4 percent.

Kidneys

The kidneys receive blood from the large **renal arteries** (Figure 19.8) and return it to the **renal veins** (Figure 19.15). Certain kidney capillaries are fenestrated and otherwise specialized to control the size of particles leaving the blood and to remove wastes from the blood (Chapter 27). They also adjust the blood volume and the concentrations of various blood constituents (Chapter 28). Under resting conditions, the kidneys receive about 23 percent of the cardiac output, thereby assuring that the blood composition is continuously adjusted and maintained within the normal range.

Muscles and Skin

The muscles and skin receive blood from many arteries, among which are the **right brachiocephalic, left common carotid, left subclavian, intercostal,** and **common iliac arteries,** all of which branch from the aorta. These vessels branch into smaller arteries that serve muscles and skin in various parts of the body (Tables 19.2, 19.3, and 19.7).

The volume of blood flowing through skeletal muscles varies greatly in proportion to physical activity. Each 100 g of skeletal muscle tissue receives 4 to 7 ml of blood per minute during rest and 50 to 75 ml during strenuous exercise—more than a tenfold increase.

Constriction and dilation of skin arterioles is important in regulating body temperature. When the body temperature rises, heat is lost as arterioles dilate and allow more blood (as much as 2500 ml/min) to pass through capillaries near the skin surface. When the body temperature falls, heat is conserved as arterioles constrict and limit the amount of blood passing through the capillaries. Arterioles of the skin and muscles often dilate simultaneously because exercise produces large amounts of heat.

Placenta and Fetus

Blood circulation through a developing fetus differs from that after birth in several ways (Figure 19.18). Deoxygenated fetal blood reaches the placenta through a pair of **umbilical arteries,** and oxygenated blood enters the fetus through a single **umbilical vein.** The umbilical vein branches into two vessels near the liver, with one joining the hepatic portal vein and the other joining the inferior vena cava as the **ductus venosus** (ve-no'sus). Most of the blood from the placenta and the fetal digestive organs passes through the ductus venosus to the inferior vena cava, bypassing the liver.

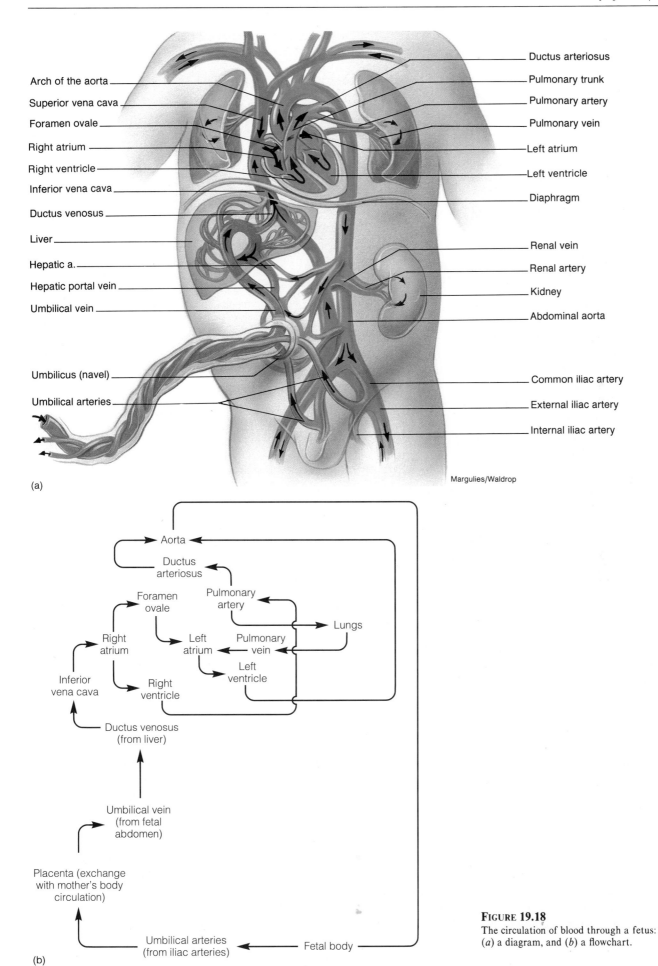

Arch of the aorta

Superior vena cava

Foramen ovale

Right atrium

Right ventricle

Inferior vena cava

Ductus venosus

Liver

Hepatic a.

Hepatic portal vein

Umbilical vein

Umbilicus (navel)

Umbilical arteries

Ductus arteriosus

Pulmonary trunk

Pulmonary artery

Pulmonary vein

Left atrium

Left ventricle

Diaphragm

Renal vein

Renal artery

Kidney

Abdominal aorta

Common iliac artery

External iliac artery

Internal iliac artery

Margulies/Waldrop

(a)

Aorta

Ductus arteriosus

Foramen ovale

Pulmonary artery

Lungs

Right atrium

Left atrium

Pulmonary vein

Inferior vena cava

Right ventricle

Left ventricle

Ductus venosus (from liver)

Umbilical vein (from fetal abdomen)

Placenta (exchange with mother's body circulation)

Umbilical arteries (from iliac arteries)

Fetal body

(b)

FIGURE 19.18
The circulation of blood through a fetus:
(a) a diagram, and (b) a flowchart.

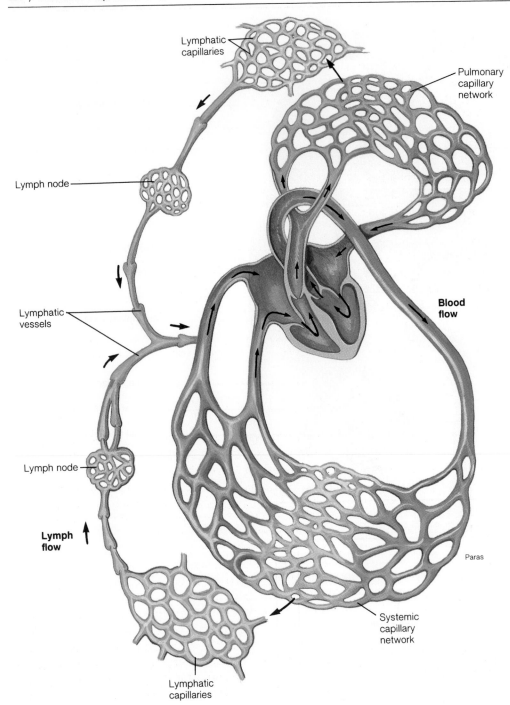

Lymphatic
capillaries

Pulmonary
capillary
network

Lymph node

Lymphatic
vessels

**Blood
flow**

Lymph node

**Lymph
flow**

Paras

Systemic
capillary
network

Lymphatic
capillaries

FIGURE 19.19
A schematic diagram showing the relationship of the lymphatic
system to the blood circulatory system.

Of blood entering the right atrium, some goes to the
right ventricle and some goes through the foramen ovale
to the left atrium. Only blood going to the right ventricle
passes through the pulmonary trunk. Much of the blood
reaching the pulmonary trunk passes through the ductus
arteriosus to the aorta, bypassing the pulmonary circulation.

Circulation of fetal blood through the placenta supplies the fetus with nutrients and oxygen, and rids it of
wastes and carbon dioxide. The placenta serves the functions of the digestive, respiratory, and excretory systems
of the fetus. Because of the exchange of substances between maternal and fetal blood, the mother's body carries
out numerous functions for the fetus. Her liver processes
nutrients and makes them ready for absorption across the
placenta. It also detoxifies toxic substances, thereby minimizing their effect on the fetus. Premature infants sometimes have jaundice, a yellowing of the skin and mucous
membranes, due to bilirubin in the blood and tissues from
unmetabolized hemoglobin of worn out red blood cells.
Though liver function is limited, fetal circulation must be
sufficient to supply nutrients and oxygen to liver cells.

The umbilical vessels, ductus venosus, foramen ovale, and ductus arteriosus function only in the fetus. They allow for circulation of blood through the placenta and limit it in the fetal liver and lungs. Shortly after birth, these special fetal circulatory pathways cease to function. Tying the umbilical cord blocks circulation through umbilical vessels. Within a few hours after birth, the muscular ductus venosus contracts, blocking passage of blood through it. As the lungs expand when breathing begins, blood vessels in the lungs dilate. Increased blood flow to the lungs causes left atrial pressure to exceed right atrial pressure. A flap of tissue on the left side of the interatrial septum occludes the foramen ovale and the higher left than right atrial pressure holds it in place. Over a few months, the flap fuses with the septum and permanently closes the foramen. Though the ductus arteriosus remains open for a few hours to a few days, high aortic pressure pushes blood toward the pulmonary trunk. Eventually, the ductus arteriosus atrophies, severing the last connection between pulmonary and systemic circuits.

LYMPHATIC ORGANIZATION AND FUNCTION

The **lymphatic system** is closely related to the circulatory system. Though this system lacks a heart, it has lymphatic vessels, lymph, and a variety of lymphatic tissues. Interstitial fluid becomes lymph as it is collected by the lymphatic vessels. Lymphatic tissues include the lymph nodes, thymus gland, tonsils, and spleen. The appendix and Peyer's patches of the intestine also contain lymphatic tissue. Though the body's lymphatic tissues together weigh less than 2 kg, they contain 2 trillion cells.

An important function of lymphatic circulation is to collect plasma proteins and fluid that has moved from blood to interstices (spaces) and return them to the blood. Because lymph capillaries are more permeable to protein molecules than blood capillaries, most protein molecules in interstitial fluid are returned to the blood via lymph vessels. Excess interstitial fluid also enters the lymph capillaries. In a 24-hour period, one-fourth to one-half of the body's total amount of plasma proteins and 2 to 4 l of interstitial fluid are returned to the blood via lymphatics. Continuous return of proteins and fluid helps to maintain body fluid balance and normal circulatory functions. The relationship of the lymphatic system to the blood circulatory system is shown in Figure 19.19.

Certain cells and tissues of the lymphatic system defend the body against infectious agents and other foreign substances. Some of its cells phagocytize the microorganisms, and others produce antibodies, proteins that react with molecules called antigens on the microbial surfaces. Antibodies inactivate microbes, kill them, or increase their chances of being phagocytized (Chapter 21).

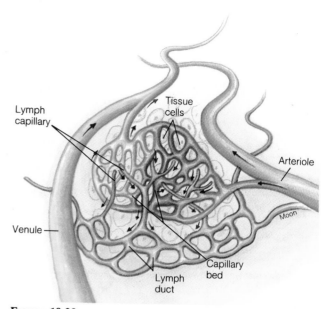

FIGURE 19.20
The location and structure of lymph capillaries, collecting ducts, and lymphatics.

LYMPHATIC TISSUES

Most **lymphatic tissues** consist of a network of reticular fibers containing many lymphocytes. Of these tissues, the most numerous are lymph nodes scattered throughout the body. They are connected to lymphatic vessels and have lymph flowing through them. As lymph passes through the lymph nodes, microorganisms and other foreign substances are removed from it.

The lymphatic vessels begin as saclike **lymph capillaries** (Figure 19.20). These capillaries are about 50 μm in diameter (larger than blood capillaries) with endothelial walls like those of blood capillaries. They are distributed throughout the body and pass near almost all body cells. Lymph capillaries drain into larger vessels called **collecting ducts,** which, in turn, drain into still larger vessels called **lymphatics.** The lymphatics are similar to veins except that their walls are thinner and contain more valves than veins. The walls of some collecting ducts and all lymphatics contain smooth muscle.

Lymph nodes, situated along the lymphatics, are bean-shaped organs varying from 1 to 25 mm in diameter. Each lymph node consists of a network of reticular connective tissue and is covered by a fibrous connective tissue capsule. **Afferent lymphatic vessels** carry lymph to them and **efferent lymphatic vessels** carry lymph from them (Figure 19.21). Within a node, lymph flows through large, thin-walled **lymph sinuses,** the walls of which are lined with phagocytic cells. Interspersed in the reticular network are clusters of cells called **germinal centers.** These cells give rise to lymphocytes, which function in immunity (Chapter 21).

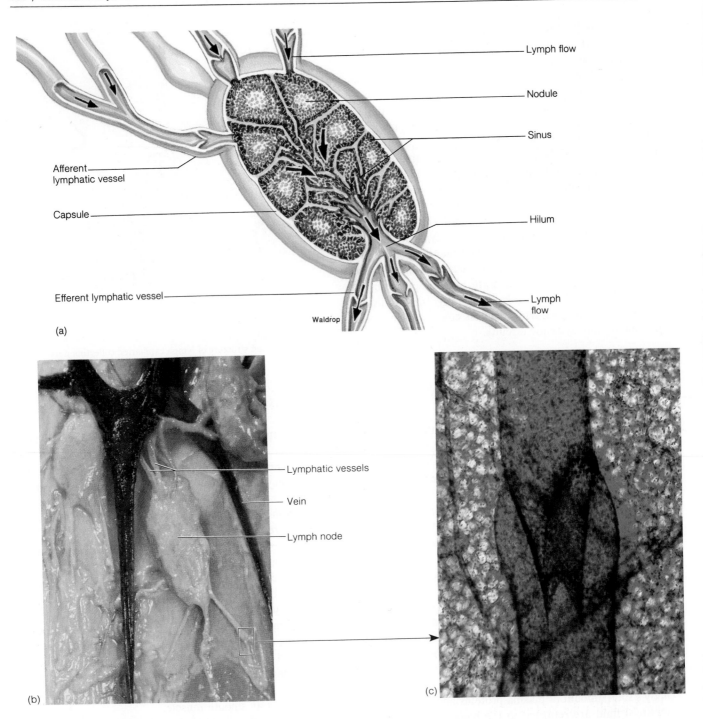

Lymph flow

Nodule

Sinus

Afferent lymphatic vessel

Capsule

Hilum

Efferent lymphatic vessel

Lymph flow

Waldrop

(a)

Lymphatic vessels

Vein

Lymph node

(b)

(c)

FIGURE 19.21
The structure of a lymph node: (*a*) diagram, (*b*) photomicrograph, and (*c*) valve in lymph vessel.

Lymph nodes are subject to several disorders. Though they help to fight infection, they themselves can become infected. An infected lymph node contains a pus-filled abscess surrounded by inflamed tissue. Lymph nodes also become inflamed from allergic reactions and can become cancerous as in leukemia, Hodgkin's lymphoma, and other lymphomas. Cells that break from tumors in other sites can be transported through the lymph to lymph nodes, where they can multiply as secondary sites of malignant growth. This is called metastasis, as noted in Chapter 3.

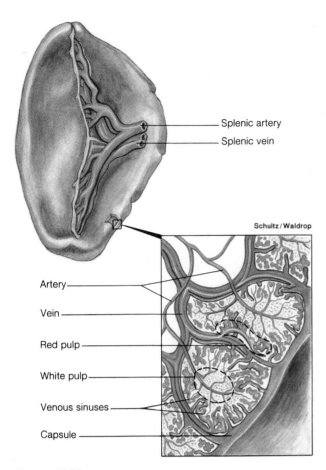

Schultz/Waldrop

Artery

Vein

Red pulp

White pulp

Venous sinuses

Capsule

FIGURE 19.22
The structure of the spleen.

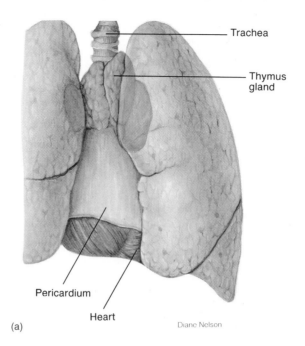

Trachea

Thymus gland

Pericardium

Heart

(a)

Diane Nelson

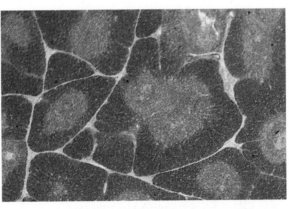

(b)

FIGURE 19.23
(*a*) The location of the thymus gland, and (*b*) photomicrograph of thymus tissue.

The **spleen,** the largest of the lymphatic tissue masses, is located in the upper left quadrant of the abdominal cavity. Internally, it is divided into lobules. Like a lymph node, it has an inner reticular network and a fibrous connective tissue capsule (Figure 19.22). The reticular network consists of **red pulp** and **white pulp.** Blood from the splenic artery enters venous sinuses of the red pulp, where dead erythrocytes are phagocytized. The white pulp consists of lymphatic tissue arranged around blood vessels. Cells called B-lymphocytes are produced by the spleen and most lymphatic tissues.

The **thymus gland** (Figure 19.23), a two-lobed structure located in the mediastinum, has many internal lobules and is covered by a thin fibrous capsule. The thymus regresses during adolescence, and is much smaller in adults than in children. Each lobule contains epithelial cells and lymphocytes. The thymus produces large numbers of cells called T-lymphocytes, which play a role in immunity. It also produces **thymosins,** which stimulate T-lymphocyte development. Oxytocin and substances called neurophysins, which transport hormones, have recently been isolated from the thymus. Their actions are being studied.

Tonsils are masses of lymphatic tissue associated with the respiratory tract (Chapter 22). They phagocytize foreign substances from the blood. If overloaded with microorganisms, they can become inflamed and enlarged. Chronically infected, enlarged tonsils make breathing difficult and can be removed surgically in a *tonsillectomy.* Because tonsils are now known to be important in defending the body against infection, fewer tonsillectomies are performed today than in the past.

FIGURE 19.24
The location of the major lymphatic vessels and lymph nodes. The shaded area shows the portion of the body not drained by the thoracic duct.

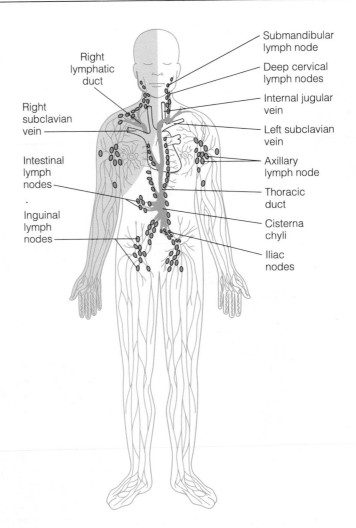

LYMPHATIC SYSTEM ANATOMY

Small lymph vessels that drain the lymph capillaries empty into larger vessels (Figure 19.24). The **thoracic duct** arises from an enlarged area in a lymph vessel, called the **cisterna chyli** (sis-ter′nah ki′li), which drains intestinal and iliac lymph nodes. The thoracic duct also drains lymph from several small lymph vessels and empties into the left subclavian vein. Other small vessels drain lymph from the right arm, shoulder, thorax, and neck and empty directly into the right subclavian vein.

Groups of lymph nodes interspersed along the lymphatics are particularly concentrated in certain regions. The **deep cervical lymph nodes** are located along the internal jugular veins. The **submandibular lymph nodes** are located along the lower border of the mandible. Both of these groups of lymph nodes can become enlarged and painful to touch if they become inflamed as they defend the body against an infection. The **axillary lymph nodes** are located in the armpit and chest area. Lymph vessels associated with these nodes drain fluid from the mammary glands and, thus, can be responsible for the spread of malignant cells from a breast tumor. The **inguinal lymph nodes** are located in the groin area and drain lymph from the legs and genital area. In addition, many small individual lymph nodes are scattered throughout the body.

LYMPHATIC SYSTEM PHYSIOLOGY

Movement of lymph through the lymphatic vessels is produced not by a heart but by rhythmic contractions of the lymph vessels themselves and by contractions of adjacent skeletal muscles. Such contractions, including those associated with breathing, apply pressure to the lymph capillaries and smaller lymphatic vessels, pushing lymph toward the main lymphatic ducts. Valves in the lymphatic vessels, similar to those in veins, prevent backflow. Interstitial fluid not returned to the blood accumulates in the tissues. This condition is called **edema** (ĕ-de′mah).

Structure and function are closely related in lymphatics. Lymph capillaries are structurally specialized to pick up fluid not returned to blood capillaries. Cells in lymphatic tissues are specialized to carry out phagocytosis or contribute to immunity.

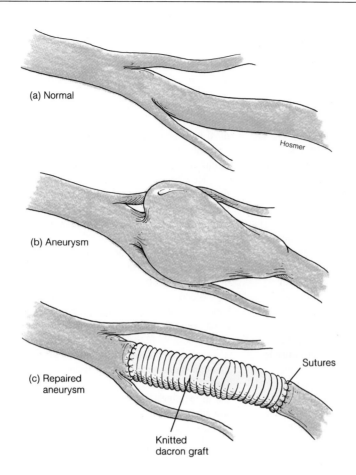

(a) Normal

Hosmer

(b) Aneurysm

(c) Repaired aneurysm

Sutures

Knitted dacron graft

FIGURE 19.25
(*a*) A normal artery, (*b*) an aneurysm, and (*c*) a repaired aneurysm. Aneurysms often, but by no means always, occur near branch points. A knitted dacron graft is sutured to the ends of the healthy vessels from inside the aneurysm and the damaged tissue is removed.

CLINICAL APPLICATIONS

An **aneurysm** (an'u-rizm) is a balloonlike localized dilation of an arterial wall. Normally, elastic tissue and smooth muscle enable arterial walls to withstand the pulsating pressure in blood leaving the heart. If a segment of the wall weakens, it expands under pressure and forms an aneurysm (Figure 19.25). Should an aneurysm develop in a major vessel, it must be surgically repaired to prevent vessel rupture and death from hemorrhage.

Several factors can cause or contribute to the formation of aneurysms. They include: (1) atherosclerosis, explained in the Essay that follows; (2) abnormally large amounts of enzymes that digest elastin and collagen in arterial walls; (3) genetic factors that lead to the development of unusually large, weak arteries; (4) turbulent flow, especially near branch points; and (5) decreased blood flow in blood vessels serving the arterial wall itself. Whatever the cause of an aneurysm, its presence indicates a failure of the vessel wall to withstand prevailing pressures.

CLINICAL TERMS

cyanosis (si"an-o'sis) bluishness of skin, usually due to insufficient oxygenation of blood

elephantiasis (el"ef-an-ti'as-is) an inflammation and blockage of lymph ducts due to a worm infection

hematoma (hem"at-o'mah) a solid mass of clotted blood in a tissue caused by leakage of blood vessels

lymphadenectomy (lim-fad"en-ek'to-me) the surgical removal of lymph nodes

lymphedema (lim"fe-de'mah) collection of fluid in soft tissues because of an excess of lymph

lymphoma (lim-fo'mah) any tumor consisting of lymphatic tissue

Raynaud's (ra-nōz') **disease** painful ischemia, usually of the hands or feet, due to marked vasoconstriction on exposure to cold

shunt to turn aside or divert; passage between the sides of the heart or between two blood vessels, as in an arteriovenous shunt

splenectomy (sple-nek'to-me) surgical removal of the spleen

splenomegaly (sple"no-meg'al-e) enlargement of the spleen

ATHEROSCLEROSIS

Narrowing of arteries by **plaque** deposits and roughening of walls are hallmarks of **atherosclerosis** (ath″er-o-skler-o′sis), shown in Figure 19.26. Blood clots form on rough surfaces and block already narrowed arteries. Factors such as high dietary cholesterol, high blood levels of low density lipoproteins (LDLs), and defective cells that take in too much cholesterol are thought to contribute to atherosclerosis. Cigarette smoking, hypertension, and diabetes mellitus also have been reported to add to the risk of atherosclerosis. A first sign of atherosclerosis is a fatty streak in an arterial wall. The streak results from blood monocytes migrating into tissue beneath endothelium, thereby becoming macrophages and filling with cholesterol and LDLs.

Normal cells have LDL receptors that stop functioning when cells have taken in enough cholesterol. In plaque, the macrophages, endothelial cells, and underlying smooth muscle cells can have special LDL receptors that continue to move cholesterol into the cells long after the cells' needs are met. Cholesterol accumulates because human cells lack enzymes to digest it. Dying cholesterol-laden cells may release toxins that trigger adjacent cells to take in more cholesterol, spreading plaque formation and cell death along an artery. (Cholesterol metabolism is considered in more detail in Chapter 24.)

Along with injured cells and cholesterol, plaque contains collagen, connective tissue, and fibrinogen. In coronary arteries, plaque contributes to myocardial infarctions (heart attacks). In cerebral arteries, it contributes to cerebrovascular accidents (strokes), as discussed in Chapter 20.

Much evidence for the role of LDLs in atherosclerosis comes from studies of humans with familial hypercholesterolemia (FH) and animals with a similar disorder. FH is caused by a dominant gene that cannot direct LDL receptor synthesis. Normal cells have two recessive genes (*hh*) both of which direct LDL receptor synthesis. Cells with one defective gene (*Hh*) make about half the normal number of LDL receptors and those with two defective genes (*HH*) make no LDL receptors. In most human populations, about one in 500 people has one defective gene and one per million has two. The former have twice the normal blood LDLs, and

most have heart attacks by age 35. The latter have more than six times the normal blood LDLs and most have heart attacks in childhood—some as early as age two.

How blood LDLs increase in people not affected by FH is not clear, but animal fat may be involved. Eating animal fat, which contains large amounts of cholesterol, seems to decrease LDL receptors, whereas avoiding animal fat seems to increase them. Though animal fat intake is highly correlated with blood LDL concentration, only about half the people with high blood LDLs have atherosclerosis. Researchers are looking for other factors that contribute to atherosclerosis.

Blood cholesterol can be lowered somewhat by diet and exercise, and to a greater degree by various drugs. A diet low in cholesterol and containing several servings per day of foods high in soluble fiber (oat bran, apples, beans) appears to lower blood cholesterol. This fiber apparently acts by the same mechanism as certain drugs used to treat high cholesterol. Both the fiber and the drugs bind to bile salts (steroids that aid in fat digestion) and cause their excretion. The liver must use cholesterol to continuously replace bile salts, so less cholesterol is available to enter cells.

The idea that fish oils containing *omega-3 fatty acids* may help to control blood cholesterol is based on observations that Eskimos and some other cultural groups have low blood cholesterol and a low incidence of atherosclerosis in spite of eating diets high in animal fats. Omega-3 fatty acids, which have a double bond between the third and fourth carbons from the methyl end of the molecule, appear to lower blood lipids by increasing membrane fluidity so cells take up and use more lipids.

Questions

1. What is atherosclerosis, and what vessels does it most often affect?
2. How is familial hypercholesterolemia related to atherosclerosis?
3. How can normal individuals use diet to decrease their risk of atherosclerosis?

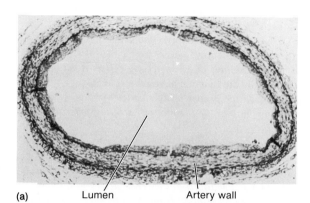

(a)   Lumen   Artery wall

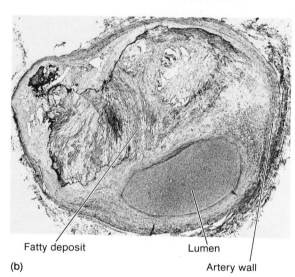

Fatty deposit   Lumen

(b)   Artery wall

FIGURE 19.26
Cross sections of arteries: (*a*) normal, and (*b*) atherosclerotic.

CHAPTER SUMMARY

Organization and General Functions of Blood Vessels
- The circulatory system consists of the heart, blood, and blood vessels in the coronary, pulmonary, and systemic circuits. The main function of the circulatory system is to transport substances to and from all cells of the body.

Development
- Blood vessels begin development as cords of cells that form hollow tubes, branch extensively by budding, and become distributed to all tissues of the body.

Tissues of Blood Vessels
- Arteries, which carry blood from the heart, consist of endothelium called the tunica interna, elastic connective tissue and smooth muscle forming the tunica media, and fibrous connective tissue in the tunica externa. The vasa vasorum supply blood to the walls of large blood vessels.
- Arterioles, small blood vessels that carry blood from arteries to capillaries, have precapillary sphincters at their junctions with most capillaries.
- Capillaries are 4 to 12 μm in diameter and consist of endothelium, which allows exchange of substances with tissues. Some capillaries are fenestrated or form sinusoids.
- Veins have the same three layers as arteries, but contain more connective tissue and less smooth muscle and elastic tissue. They also contain valves, which prevent backflow, and can form sinuses. Skeletal muscle contractions help to maintain venous blood flow.
- Venules are small vessels that connect capillaries and veins.

Distribution of Blood Vessels
- The arteries of the body are shown in Figures 19.5 through 19.10 and are described in Tables 19.1 through 19.7.
- The veins of the body are shown in Figures 19.12 through 19.17 and are described in Tables 19.8 through 19.13.

Special Circulatory Pathways
- Brain tissue is especially sensitive to decreases in glucose and oxygen. Anastomoses provide alternate pathways and a blood-brain barrier prevents many substances from entering brain tissue. The noncompressible skull prevents vessel compression.
- The digestive tract and liver receive oxygenated blood from branches of the aorta. Blood containing nutrients from digestive organs goes to the liver via the hepatic portal vein.
- The spleen has large sinusoids that can release blood into circulation when the body is under stress. Phagocytic cells in the linings of sinusoids destroy worn out erythrocytes.
- The kidneys filter wastes from the blood and adjust concentrations of various substances in the blood.

- Blood flow in muscles increases during exercise and decreases during rest. Dilation of blood vessels during exercise is caused mainly by changes in muscle metabolism.
- Skin blood flow changes directly with body temperature and plays an important role in temperature regulation.
- In the fetus, the umbilical vein carries blood from the placenta to the fetus and the umbilical arteries carry blood from the fetus to the placenta. The foramen ovale and the ductus arteriosus allow blood to bypass the pulmonary circulation before birth and close at birth.

Lymphatic Organization and Function

- The lymphatic system consists of lymph, lymphatic vessels, and lymphatic tissues. It returns proteins and tissue fluids to the blood and defends the body against infections.

Lymphatic Tissues

- Lymphatic vessels include saclike, blind-ended, thin-walled capillaries, which are larger and more permeable than blood capillaries; collecting ducts, which collect lymph from the capillaries; and lymphatics, which return lymph to the blood.
- Lymph nodes consist of reticular connective tissue and lymph sinuses that receive lymph from afferent lymphatic vessels and release it into efferent lymphatic vessels.
- The spleen consists of red pulp that phagocytizes worn out erythrocytes and white pulp that produces lymphocytes.
- The thymus, located in the mediastinum, is composed of epithelial cells and lymphocytes, and produces T-lymphocytes, the hormone thymosin, and probably several other hormones.
- The tonsils, imbedded in mucous membranes of the respiratory tract, phagocytize foreign substances in lymph.

Lymphatic System Anatomy

- Lymphatics and lymph nodes are shown in Figure 19.24.

Lymphatic System Physiology

- Lymphatics help to maintain fluid balance and have cells that phagocytize foreign materials and contribute to immunity.

Clinical Applications

- Aneurysms, dilations in weakened arterial walls, can be caused by atherosclerosis and several other contributing factors. Massive hemorrhage occurs if an aneurysm ruptures before the dilation can be surgically repaired.

QUESTIONS AND PROBLEMS

The questions at the end of each chapter are numbered to correspond with the objectives listed at the beginning of the chapter. Italics indicate that a question requires critical thinking skills beyond simple factual recall.

Questions

1. (a) How is the circulatory system organized?
   (b) What are the functions of blood vessels?
   (c) How is the function of an organ affected if the blood supply to it is decreased?
2. (a) How do blood vessels grow to reach all cells?
   (b) Why does the human embryo require a blood supply early in development?
3. (a) How are walls of arteries and veins similar, and how are they different?
   (b) How do blood vessel structures reflect functions?
   (c) What is the function of vasa vasorum?
   (d) How do the properties of elasticity and contractility contribute to the functions of arteries, arterioles, and veins?
   (e) What is the function of valves in veins, and why are valves not needed in arteries?
   (f) How do sinuses and sinusoids differ?
4. *Trace blood flow (a) from the celiac artery to the external jugular vein, (b) from the internal carotid artery to the renal vein, and (c) from the ulnar artery to the great saphenous vein.*
5. (a) *What are the special characteristics of blood circulation in the (i) brain, (ii) digestive tract, (iii) liver, (iv) spleen, and (v) kidneys?*
   (b) *What factors affect blood volume in muscles and skin?*
   (c) *Trace the flow of blood from the placenta through a developing fetus and back to the placenta.*
6. (a) What are the components of the lymphatic system?
   (b) What are the functions of the lymphatic system?
7. (a) Compare lymph vessels to blood vessels.
   (b) What are the main characteristics of each of the types of lymphatic tissue?
8. (a) Trace the flow of lymph from each of the following locations back to the circulatory system: (i) legs, (ii) intestine, (iii) right arm, and (iv) left side of head.
   (b) Where are concentrations of lymph nodes located?
9. *(a) What forces cause lymph to move through vessels?*
   (b) What do lymph nodes do under normal conditions?
   (c) What might infected lymph nodes do?
10. *What is an aneurysm, and what are its causes and effects?*

Problems

1. Describe in as much detail as possible what happens in various blood vessels in your body when you engage in strenuous exercise.
2. Combine what you have learned in this and other chapters to explain all of the changes that occur in the circulatory system of an infant at or shortly after birth.
3. Use what you have learned about the lymphatic system to explain why tonsils become enlarged and inflamed, and why cancer cells are often found in lymph nodes.

20

CIRCULATION

PRINCIPLES OF CIRCULATION

When we step into the shower, we expect to regulate water flow by manipulating faucet handles. Our expectations usually are met because modern plumbing systems maintain pressure in pipes so water flows when a valve is opened. Blood flow in the circulatory system follows the same principles as fluid flow through a pipe.

The principles governing the circulation of blood pertain to flow, pressure, pressure gradients, and resistance. **Flow** is the volume of fluid moving through a tube in a given period of time. **Pressure** is the force per unit of a tube's cross-sectional area, tending to push fluid through the tube. A **pressure gradient** is a difference between pressures at any two points in a tube. **Resistance** is the force that opposes flow. Fluid flows down a pressure gradient from a region of higher pressure to a region of lower pressure. Flow is proportional to the size of the pressure gradient provided resistance remains the same. If resistance increases, flow decreases, and *vice versa*.

Flow can be described by the following equation.

$$F = \frac{P}{R} : \text{flow (ml/min)} = \frac{\text{pressure gradient}}{\text{resistance}}$$

Students familiar with electricity will see that the equation for the flow of blood is the same as Ohm's law, which describes the flow of current: current = change in potential/resistance.

Resistance results mainly from friction between moving fluid and vessel walls. The main factor in resistance is the blood vessel's cross-sectional area (size of the lumen), which is proportional to the vessel's radius (Figure 20.1a). As blood flows along the walls of a vessel, friction offers resistance and impedes flow, but friction is lacking in the lumen and flow is unimpeded. This is called **laminar flow** (Figure 20.1b). Resistance (R) decreases as the fourth power of the vessel radius (r) increases: $R = 1/r^4$. Provided pressure and other factors are held constant, large vessels have much less resistance than small vessels because relatively more fluid moves through the lumen without encountering resistance from friction with vessel walls.

Increasing the length of a vessel increases the quantity of vessel wall and, thus, the resistance. Other factors being equal, the resistance of a vessel is directly proportional to its length (Figure 20.2).

Viscosity, a fluid's own internal resistance, is due to friction between molecules and particles within the fluid itself. Given a constant pressure gradient, a viscous fluid such as molasses flows much more slowly than a less viscous fluid such as water. Because blood contains cells and large protein molecules, it is about five times as viscous as water. Normally, blood viscosity remains nearly constant, but conditions such as dehydration (loss of body water) or excesses of blood cells (as in polycythemia and leukemia) increase blood viscosity.

Pressure in blood vessels is generated by ventricular contractions pushing blood into arteries, and pressure gradients contribute to flow from regions of higher pressure (arteries) to regions of lower pressure (capillaries or veins). Pressure gradients, like resistance, are related to cross-sectional areas of vessels. Large arteries branch many times before the branches reach the size of capillaries, as shown by the hypothetical vessel in Figure 20.3. The cross-sectional area of the large vessel is greater than any single branch, but only about half the total cross-sectional area of all eight branches. The pressure is correspondingly less in the branches than in the large vessel. Actual blood vessels branch extensively and total cross-sectional areas of all the vessels of a certain type vary considerably (Table 20.1 and Figure 20.4a).

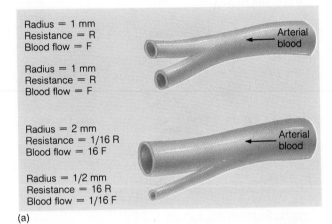

Radius = 1 mm
Resistance = R
Blood flow = F

Arterial blood

Radius = 1 mm
Resistance = R
Blood flow = F

Radius = 2 mm
Resistance = 1/16 R
Blood flow = 16 F

Arterial blood

Radius = 1/2 mm
Resistance = 16 R
Blood flow = 1/16 F

(a)

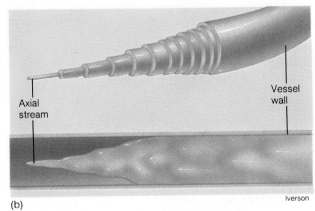

Axial stream

Vessel wall

(b)

Iverson

FIGURE 20.1
(*a*) The effect of the internal radius of a vessel on resistance when pressures and other factors are held constant. (*b*) The property of laminar flow allows blood to move faster in the center than along the wall of the vessel.

Pressure gradients maintain continuous blood flow (Figure 20.4b). For a given vessel, flow (F) is the product of the velocity (V), or the distance blood travels in a given time, and the vessel's cross-sectional area (A): $F = VA$. In other words, flow is how much fluid moves and velocity is how fast it moves. Rearranging the equation to $V = F/A$, we can see that velocity is the volume of flow divided by a vessel's cross-sectional area.

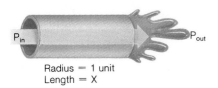

Radius = 1 unit
Length = X

Radius = 1 unit
Length = 2X

Iverson

FIGURE 20.2
The effect of the length of a vessel on resistance: When P_{in} (pressure entering vessel) is the same for both vessels, P_{out} (pressure leaving a vessel) is less in the longer vessel.

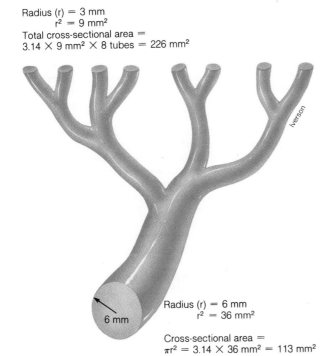

Radius (r) = 3 mm
r^2 = 9 mm²
Total cross-sectional area =
3.14 × 9 mm² × 8 tubes = 226 mm²

Iverson

Radius (r) = 6 mm
r^2 = 36 mm²

6 mm

Cross-sectional area =
πr^2 = 3.14 × 36 mm² = 113 mm²

FIGURE 20.3
Cross-sectional area of blood vessels increases as the vessels branch.

TABLE 20.1	Estimates of Cross-sectional Area of Different Classes of Human Blood Vessels	
Class of Vessel	**Total Cross-sectional Area (mm²)**	**Typical Mean Pressure (torr)**
Aorta	250	100
Small arteries	2,000	90
Arterioles	4,000	60
Capillaries	250,000	20
Venules	25,000	5
Small veins	8,000	2
Venae cavae	800	0

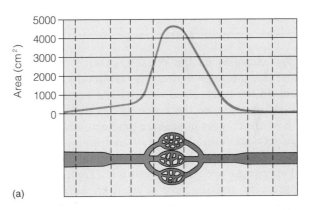

(a)

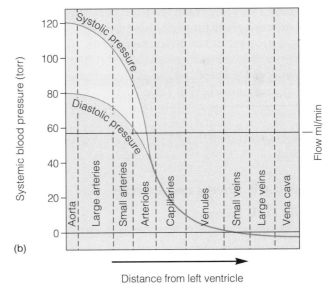

(b)

Distance from left ventricle

FIGURE 20.4
(a) Cross-sectional area, and (b) pressure and flow related to different kinds of vessels.

TABLE 20.2 Summary of the Principles of Circulation

	Definitions	Concepts
Flow	Volume of fluid moving through a tube in a given period of time	Constant total flow is maintained throughout the circulatory system, though there are many variations in pressure, resistance, velocity, and cross-sectional area. In any particular vessel, flow is directly proportional to pressure and inversely proportional to resistance.
Pressure	Force per unit area tending to push a fluid through a tube	Because of the rhythmic contractions of the heart and resistance encountered during flow, pressure is high and pulsating in the arteries and arterioles, lower in the capillaries, and lowest in the venules and veins.
Resistance	Force tending to oppose the flow of a fluid through a tube	Resistance is affected by the radius and length of vessels. Resistance is inversely proportional to the fourth power of the radius; thus, large vessels have much less resistance than small ones. Other factors being equal, resistance is proportional to the length of the vessel. When two vessels have the same input pressure, the output pressure is lower in the longer vessel.
Viscosity	A fluid's own internal resistance to flow	Viscosity impedes flow, other factors being equal.
Velocity	Distance traveled in a given time; the speed of a moving fluid	Total cross-sectional area is lowest in arteries, greatest in capillaries, and intermediate in veins; it is a major factor in determining velocity. Velocity is high and pulsating in arteries, low in arterioles nearest capillaries, lowest in capillaries, and gradually increases in venules and veins.

Flow increases as the pressure gradient or cross-sectional area increases. It decreases as viscosity or length increase. In the body, blood viscosity and vessel lengths are nearly constant, so the major factors that can alter blood flow are changes in pressure gradients and changes in cross-sectional area. Because of compensating variations in cross-sectional area and velocity, total flow across the whole circulatory system is constant (Figure 20.4b). High pressure arteries have a relatively small cross-sectional area, but very large pressure gradients. Capillaries have a very large total cross-sectional area and very small pressure gradients. Veins have a small total cross-sectional area and small pressure gradients. Flow is assisted by muscle contractions and valves (Chapter 19).

Terms and concepts related to the principles of circulation are summarized in Table 20.2.

CIRCULATION IN DIFFERENT VESSEL TYPES

Circulation in Arteries and Arterioles

Arterial Blood Pressure

When we speak of blood pressure, we usually mean arterial pressure, or more precisely, the systolic and diastolic arterial pressures in the systemic circuit. **Systolic pressure** is the maximum pressure developed in arteries, and **diastolic pressure** is the minimum pressure in the arteries during the cardiac cycle. Because the left ventricle contracts with greater force than the right, both the systolic and diastolic pressures in the aorta are about five times those in the pulmonary arteries.

One way of expressing blood pressure is to equate it to the height in millimeters of a column of mercury (mm Hg) in an evacuated tube. (The tube must be evacuated to eliminate the effect of atmospheric pressure.) The trend today is to express pressures in **torr,** a unit equal to 1 mm

Hg. The normal systemic pressures in young adult males are approximately 120 torr systolic and 80 torr diastolic pressure. Corresponding pressures for young adult females are slightly lower. Normal variations in pressure with age and sex are summarized in Table 20.3. Venous pressures are expressed as single measurements because they do not have systolic and diastolic fluctuations. At the venae cavae, venous pressure is less than 5 torr.

To measure human systemic arterial blood pressure, one needs a **sphygmomanometer** (sfig″mo-man-om′et-er), an inflatable cuff with a pressure meter, and a **stethoscope** (steth′o-skōp), a device to direct body sounds to the ears (Figure 20.5). The sphygmomanometer cuff is placed around the arm over the brachial artery and inflated to a pressure greater than the systolic pressure, blocking blood flow through the artery. The stethoscope bell is placed over the artery's inner aspect (antecubital fossa) at the elbow. Air is then slowly released from the cuff, reducing pressure on the artery. Soon blood spurts through the compressed artery under the cuff and its turbulence creates sounds, called **Korotkoff's** (ko-rot′kofs) **sounds,** audible through the stethoscope. The pressure shown on the sphygmomanometer when the first sound is heard is the systolic blood pressure. As more air is released from the cuff, the sounds gradually become muffled. The pressure when the last muffled Korotkoff's sound is heard is the diastolic pressure. When the cuff pressure is lower than the diastolic pressure, no sounds are heard because blood flow through an open artery makes no sound.

Systolic and diastolic arterial pressures can be altered by a number of factors such as exercise, reflexes that alter resistance, and disorders such as hemorrhage and atherosclerosis. These and other factors will be discussed later in this chapter.

Pulse and Pulse Pressure

Pulse is the stretching and recoil of an artery with pressure changes. **Pulse pressure** is the difference between the systolic and diastolic pressures. When the heart ejects

TABLE 20.3 Normal Arterial Blood Pressures at Selected Ages*

Age	Systolic		Diastolic		Age	Systolic		Diastolic	
	Male	*Female*	*Male*	*Female*		*Male*	*Female*	*Male*	*Female*
1 day	70†				16 years	118	116	73	72
3 days	72†				17 years	121	116	74	72
9 days	73†				18 years	120	116	74	72
3 weeks	77†				19 years	122	115	75	71
3 months	86†				20–24 years	123	116	76	72
6–12 months	89	93	60	62	25–29 years	125	117	78	74
1 year	96	95	66	65	30–34 years	126	120	79	75
2 years	99	92	64	60	35–39 years	127	124	80	78
3 years	100	100	67	64	40–44 years	129	127	81	80
4 years	99	99	65	66	45–49 years	130	131	82	82
5 years	92	92	62	62	50–54 years	135	137	83	84
6 years	94	94	64	64	55–59 years	138	139	84	84
7 years	97	97	65	66	60–64 years	142	144	85	85
8 years	100	100	67	68	65–69 years	143	154	83	85
9 years	101	101	68	69	70–74 years	145	159	82	85
10 years	103	103	69	70	75–79 years	146	158	81	84
11 years	104	104	70	71	80–84 years	145	157	82	83
12 years	106	106	71	72	85–89 years	145	154	79	82
13 years	108	108	72	73	90–94 years	145	150	78	79
14 years	110	110	73	74	95–106 years	145	149	78	81
15 years	112	112	75	76					

*Mean arterial blood pressure; derived from various studies.
†Value for both males and females.
© 1970 CIBA-GEIGY Limited, Basle, Switzerland. Reprinted by permission.

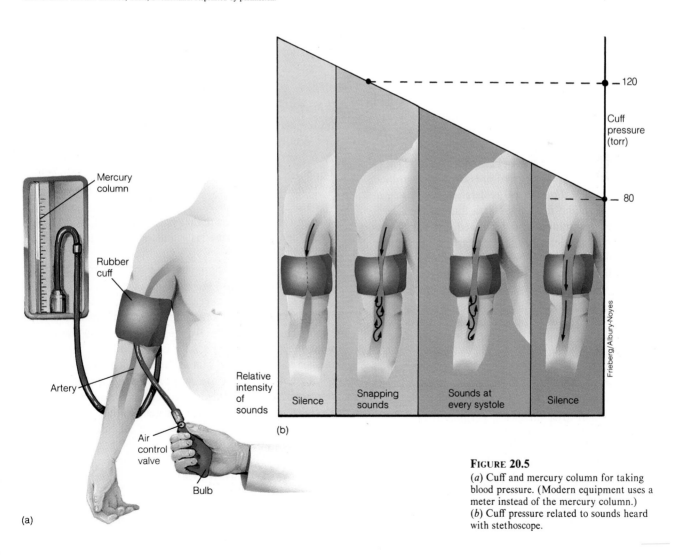

FIGURE 20.5

(*a*) Cuff and mercury column for taking blood pressure. (Modern equipment uses a meter instead of the mercury column.)
(*b*) Cuff pressure related to sounds heard with stethoscope.

blood, it initiates a wave of stretching that passes along arterial branches as the blood flows through them. The pulse rate is normally the same as the heart rate because each artery pulses every time the heart beats. Apical pulse can be assessed at the apex of the heart, using a stethoscope. Pulse is usually taken on the radial artery in the wrist, but can be taken on any artery sufficiently near the body surface to be felt (palpated). Try to feel a pulse in your brachial, temporal, carotid, femoral, popliteal, and posterior tibial arteries.

Resting blood pressure displays circadian rhythm—from 100 torr systolic and 40 torr diastolic at about the middle of the sleep period to 130 torr systolic and 80 torr diastolic 12 to 13 hours later. The rhythm is not due to daily activity patterns because it persists during continuous bed rest.

Pulse rate also has a circadian rhythm, varying from about 50 beats per minute at the middle of the sleep period to about 80 beats per minute at the middle of the wakeful period. How this rhythm is produced is not known, but it entrains to environmental cues because subjects living in a laboratory on an artificial 28-hour day display a 28-hour pulse cycle.

Mean Arterial Pressure

Because arterial pressure varies between systolic and diastolic pressure over the cardiac cycle, physiologists developed the idea of **mean arterial pressure** (MAP), which is approximately equal to the diastolic pressure plus one-third the pulse pressure. One-third rather than one-half of the pulse pressure is used because for much of the cardiac cycle the pressure is nearer the diastolic than the systolic pressure. For typical resting pressures of 122/80 (systolic/diastolic), MAP is 94 ($122 - 80 = 42$; $42/3 = 14$; $80 + 14 = 94$). Mean arterial pressure is determined by cardiac output and **peripheral resistance,** resistance in peripheral vessels, and it changes with these factors.

Peripheral resistance is determined by blood viscosity, friction, vessel length, and the degree of dilation or constriction of the vessels. As we have seen when the heart ejects blood into an artery, the elastic walls stretch and the stretched walls exert force that adds to the force from ventricular contraction driving blood toward arterioles. Smooth muscle in the walls of arteries and, especially, in arterioles regulates vessel diameters and contributes to peripheral resistance (Figure 20.6). Consequently, both arteries and arterioles are called **resistance vessels.** When smooth muscle contracts and cross-sectional area decreases, peripheral resistance increases. When it relaxes and cross-sectional area increases, peripheral resistance decreases.

Regulation of blood vessel diameters controls the distribution of blood to various organs. When tissues increase in metabolic activity, their arterioles dilate, increasing the blood volume they receive. Skeletal muscle receives 10 times as much blood during strenuous exercise

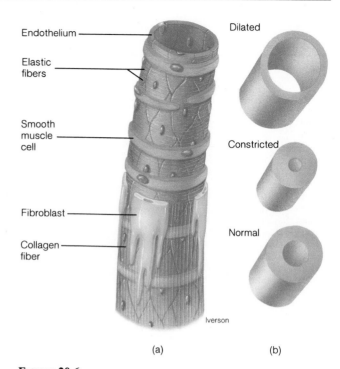

FIGURE 20.6
The structure of the wall of (*a*) an arteriole is related to (*b*) its ability to constrict or dilate.

as during rest, while arterioles in the digestive organs constrict, allowing only half the normal blood volume in those organs. Such changes are controlled by autoregulation and extrinsic factors.

Autoregulation

Autoregulation, regulation by tissues of their own blood flow, occurs in kidneys, skeletal muscle, cardiac muscle, brain, liver, and mesenteries (membranes that support abdominal organs). In autoregulation, flow is maintained by compensating changes in which pressures and resistance increase simultaneously or decrease simultaneously. Though autoregulation makes only small changes in blood flow, it operates continuously and maintains appropriate local blood flow.

Autoregulation occurs by two mechanisms. By the **myogenic** (mi″o-jen′ik) **mechanism,** as pressure in a vessel increases, stretching increases. Smooth muscle contracts in response to the stretching and maintains muscle tension proportional to pressure. By the **metabolic mechanism,** as metabolic products, especially carbon dioxide, accumulate, arteriolar muscle relaxes and blood flow increases. As metabolic products are swept away, muscle tension increases and blood flow decreases. Other factors that dilate arterioles include decreased oxygen, higher tissue temperature as metabolic processes give off heat, increased blood potassium, increased cellular ADP as energy is released from ATP, and lowered pH from acidic metabolites including carbon dioxide. (Carbon dioxide combines with water and ionizes, giving off hydrogen ions.) Reversal of any of the above conditions constricts arterioles and decreases blood flow.

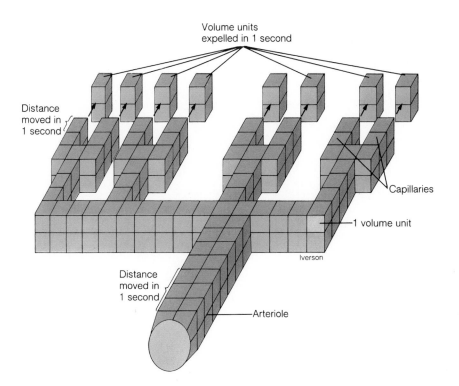

Volume units
expelled in 1 second

Distance
moved in
1 second

Capillaries

1 volume unit

Iverson

Distance
moved in
1 second

Arteriole

FIGURE 20.7
Relationships between cross-sectional area and velocity of flow in arterioles and capillaries.

Extrinsic Regulation

Signals from outside of regulated tissues produce **extrinsic regulation.** They include sympathetic (and occasionally parasympathetic) signals, certain hormones, and substances called kinins. Extrinsic regulation, which supplements autoregulation, maintains adequate blood flow to various tissues in accordance with metabolic needs. It is very important in maintaining blood flow to vital organs, such as the brain and heart.

Most arteriolar smooth muscle is well supplied with sympathetic motor neurons that release norepinephrine, and has few, if any, parasympathetic motor neurons (with exceptions noted below). Norepinephrine interacts with alpha-adrenergic receptors in the arteriolar smooth muscle, causing it to contract and the vessels to constrict. The degree of constriction is determined by the degree of sympathetic stimulation (sympathetic tone) in the muscle. Sympathetic stimulation, though continuous, varies in degree. Maximal stimulation causes arterioles to constrict maximally, and minimal stimulation causes them to dilate maximally.

Exceptions to sympathetic stimulation via norepinephrine include the following:

1. Certain skeletal muscle arterioles are innervated by sympathetic neurons that release acetylcholine. Such arterioles dilate during stress and exercise.
2. Arterioles in external genital organs are well supplied with parasympathetic neurons. Parasympathetic signals play an important role in sexual arousal (Chapter 29).

Certain hormones and other chemical substances act as **vasoconstrictors** (constrictors of blood vessels) or **vasodilators** (dilators of blood vessels). Other factors aside, vasoconstrictors raise blood pressure and vasodilators lower it simply by changing the total blood vessel volume. Epinephrine from the adrenal medulla acts in skeletal muscle and the liver as a vasodilator at low concentration and a vasoconstrictor at high concentration. Angiotensin II (from the plasma protein angiotensinogen) causes vasoconstriction and elevates blood pressure. The recently identified atrial natriuretic hormone counters the effects of angiotensin II. These substances also are important in regulating body fluids (Chapter 28). Following tissue injury, plasma proteins called kininogens are activated to **kinins,** which cause vasodilation and lower blood pressure. They also play a role in the inflammatory reaction (Chapter 21).

Circulation in Capillaries

Capillaries form networks, or capillary beds, between arterioles and venules, and blood flow in these networks is regulated mainly by precapillary sphincters (Chapter 19). Capillaries typically hold only about 5 percent of the blood volume, much of it in thoroughfare channels. Increasing the metabolic activity of a tissue causes precapillary sphincters to open and allows more blood to move through the tissue.

Capillaries have high resistance, mainly because of their very small diameter. This resistance greatly reduces blood velocity in the capillaries, but it does not impede overall flow because of the very large number of capillaries (Figure 20.7).

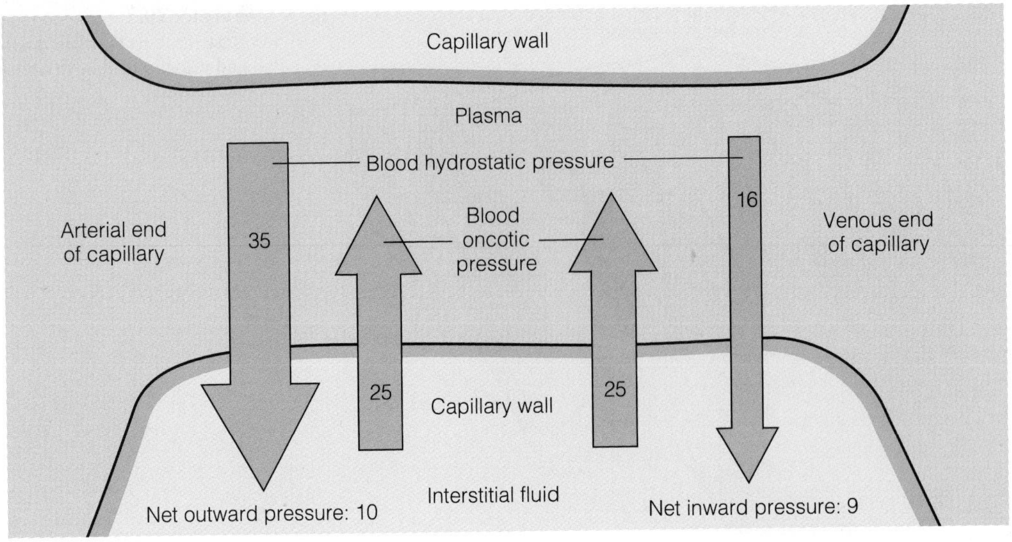

FIGURE 20.8
Pressures in torr moving fluids out of and into capillaries. Arrows indicate direction of movement and their thickness indicates the relative pressure.

Capillaries are sometimes called **exchange vessels** because all exchanges between blood and tissue fluids take place across their walls. Because such exchanges are so essential to life, one might claim that the rest of the circulatory system exists to get blood to and from the capillaries. Nutrients enter capillaries in the digestive tract, mainly from the small intestine. Oxygen enters and carbon dioxide leaves lung capillaries. Wastes are filtered out of the blood across capillaries in the kidneys.

Most capillaries are only about 1 mm long, and the single layer of thin cells in their walls allows materials to move across or pass between endothelial cells. Capillaries come within 20 to 30 μm (1 or 2 cell widths) of nearly all body cells, so substances from capillaries easily diffuse across interstitial fluid to cell membranes, and wastes from cells likewise diffuse to capillaries. Substances move across capillary walls by diffusion, osmosis, and filtration. These processes were described in Chapter 3.

Substances leaving capillaries move down gradients. Oxygen dissolved in plasma diffuses down a concentration gradient out of capillaries into interstitial fluid and, eventually, into cells. As oxygen leaves the plasma, hemoglobin releases more oxygen, and the gradient is maintained. Carbon dioxide moves down its gradient in the opposite direction—from cells, through interstitial fluid, and into capillaries. Glucose diffuses from its higher concentration in the blood to interstitial fluid, but requires special active transport mechanisms to enter cells (Chapter 24). Fat-soluble nutrients dissolve in lipids in capillary walls and diffuse into interstitial fluid. Ions and many small water soluble molecules pass through junctions between capillary wall cells to enter interstitial fluid. Filtration pressure, or blood hydrostatic pressure (the pressure in a fluid) maintained by the heart's pumping action, augments diffusion through capillary walls.

According to one explanation of capillary function, water leaves the blood by filtration and returns by osmosis. Osmosis results from **oncotic** (ong-kot′ik) **pressure,** the colloid osmotic pressure created mainly by large plasma protein molecules. Filtration and oncotic pressures in capillaries vary in skin, muscle, kidney, and other tissues according to blood pressure and the composition of blood and interstitial fluids.

To understand how filtration and oncotic pressures might operate to maintain water balance, let us look at the effects of some typical capillary pressures (Figure 20.8). Filtration pressure varies from about 35 torr at the arterial end to about 16 torr at the venous end of a capillary. Oncotic pressures remain nearly constant at about 30 torr inside the capillary and 5 torr in interstitial fluids, thereby creating a net oncotic pressure of 25 torr.

Movement of substances across capillary walls depends on net pressures. The arterial end has a net outward pressure of about 10 torr (35 torr blood hydrostatic pressure less 25 torr net blood oncotic pressure). The venous end has a net inward pressure of about 9 torr (25 torr net blood oncotic pressure less 16 torr blood hydrostatic pressure). Because of pressure gradients—10 torr pushing substances out of the capillary and 9 torr drawing substances into the capillary—about nine-tenths of the fluid forced out at the arterial end is returned at the venous end. The remaining tenth, about 2 l per day, drains from interstitial spaces to lymph capillaries and enters lymph vessels, through which it is eventually returned to the blood.

Another explanation of capillary function asserts that whenever a capillary is open, filtration pressure exceeds oncotic pressure throughout its length. This would mean that water leaves open capillaries, but can be returned by

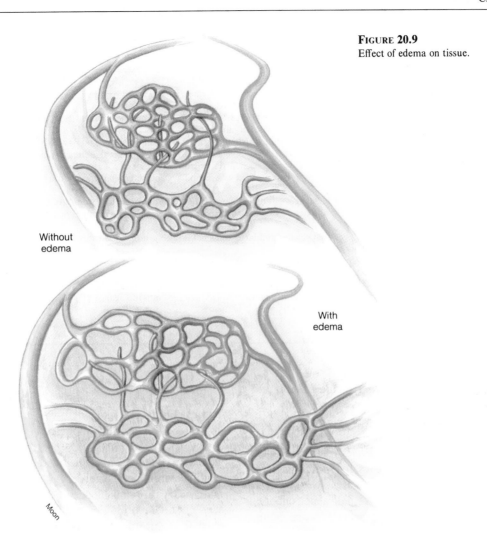

Figure 20.9
Effect of edema on tissue.

Without
edema

With
edema

Moon

oncotic pressure only when the capillary is closed by capillary sphincter contraction. Even if this latter mechanism proves to describe capillary function more accurately, the volumes of fluid moving into and out of capillaries would be about the same as for the former mechanism.

Circulation in Venules and Veins

Veins and venules, the **capacitance** (kap-as′it-ans) **vessels,** contain about 60 percent of the total blood volume. From venules that drain capillaries to veins near the heart, both pressure and total cross-sectional area decrease, and velocity increases. As blood flows into narrower channels, velocity increases. This is analogous to the increase in the current as a river flows from a region with a wide, flat, riverbed to a narrow rocky gorge. Hydrostatic pressure drops from about 15 torr where blood enters venules to no more than 5 torr in large veins near the heart. Valves in veins and skeletal muscle contractions contribute to venous circulation (Chapter 19) as do venous smooth muscle, breathing movements, and heart contractions. Increasing the venous blood volume stretches veins and causes smooth muscle in their walls to contract and resist further stretching.

The pressure at the entrance to the right atrium, called **central venous pressure,** averages about 4 torr. After ventricles contract and eject blood, atrial volumes increase and atrial pressures decrease, sucking blood into the atria (Chapter 18). If ventricular contractions are slow or ineffective or if blood volume increases, this mechanism becomes less effective and the central venous pressure increases. A larger blood volume returns to the heart than the heart can easily accommodate.

Lymphatics and Edema

Failure of the lymphatic system to return sufficient lymph to the blood results in **edema** (Figure 20.9), the accumulation of abnormal amounts of interstitial fluid. Inadequate lymph drainage occurs after removal of lymph nodes or when lymph vessels are blocked. For example, in a mastectomy (surgical removal of a breast), axillary (armpit) lymph nodes are also removed. Lymph node removal causes edema by disrupting lymph drainage from the arm, but the edema usually is temporary. In filariasis, parasitic worms block lymph vessels, causing accumulation of several liters of fluid and massive swelling, especially in the legs and scrotum.

TABLE 20.4 Factors That Affect Circulation

Vessels	Factors
Arteries	Elasticity and compliance minimize variation in blood pressure.
Arterioles	Smooth muscle responds to changes in the degree of sympathetic stimulation by constricting or dilating, and helps to assure a continuous supply of blood to vital organs.
	Metabolites accumulate in blood, causing vasodilation and thereby increasing blood flow to metabolically active tissues.
	Epinephrine and norepinephrine have modest effects on blood flow. Under special circumstances, angiotensin II causes vasoconstriction and kinins cause vasodilation; atrial natriuretic hormone counters the effects of angiotensin II.
Capillaries	Entry of blood into most capillaries is controlled by precapillary sphincters, but thoroughfare channels assure minimal blood flow to all tissues.
	Filtration pressure pushes substances out of capillaries. Some substances pass through spaces between cells in capillary walls. Filtration pressure and net oncotic pressure are primarily responsible for fluid balance between capillaries and interstitial fluid.
Venules and veins	The pressure gradient between venules and large veins helps to return blood to the heart.
	Valves prevent backflow.
	Contraction of smooth muscle in venous walls helps to maintain pressure.
	Contraction of skeletal muscles around veins pushes blood toward the heart.
	Breathing movements squeeze blood from abdominal veins toward the heart.
	Ventricular contraction reduces pressure in atria and sucks blood into the heart.
Lymph vessels	Rhythmic contraction of lymph vessels propels lymph.
	Valves prevent backflow.
	Contraction of skeletal muscles and breathing movements also help to propel lymph.

Other factors that cause edema include increased capillary permeability, high filtration pressure, or low net oncotic pressure. Histamines and kinins released in tissue injury increase capillary permeability, but they have only local effects. Filtration pressure increases when arterioles are dilated and blood enters capillaries under greater than normal pressure or when heart contractions weaken and backpressure develops in veins. Too small a net oncotic pressure can result from deficient plasma proteins or an excess of osmotic substances in interstitial fluid. Diseased livers can fail to make enough plasma proteins and diseased kidneys can allow them to be removed from the blood. Such disorders in organ functions often lead to **ascites** (as-si'tēz), the accumulation of interstitial fluid in the abdominal cavity. NaCl accumulation in interstitial fluid, which often occurs with reduction in plasma proteins, draws water into interstitial fluid by osmosis.

Circulation is affected by a variety of properties of blood and lymph vessels as summarized in Table 20.4.

CONTROL OF CIRCULATION

Properties of vessels themselves and local regulatory factors control circulation, as we have seen, but more generalized control is exerted by centers in the brain and by signals from baroreceptors and chemoreceptors.

Control Centers in the Brain

The **cardiovascular** (kar″de-o-vas′ku-lar) **center** (Figure 20.10), a diffuse group of neurons in the medulla, receives signals from other parts of the central nervous system and from baroreceptors and chemoreceptors. It responds by sending sympathetic or parasympathetic signals that regulate heart rate and blood pressure. The cardiovascular center is divided functionally (but not anatomically) into the cardioinhibitory area, the vasomotor pressor area, and the vasomotor depressor area. The **cardioinhibitory** (kar″de-o-in-hib′it-or-e) **area** receives signals from baroreceptors when arterial pressure rises, and sends signals that decrease the heart rate.

Whether a specific cardioacceleratory area exists is not clear, but fear, anxiety, and emotional excitement can directly initiate sympathetic signals that increase arterial pressure by constricting arterioles and increase venous pressure and venous return by constricting veins. These factors increase heart rate and stroke volume and, therefore, greatly increase cardiac output.

The **vasomotor pressor area** and the **vasomotor depressor area** act reciprocally to maintain **vasomotor tone,** the degree of vessel constriction. An increase in signals to the pressor area or a decrease in signals to the depressor area causes arterioles to constrict and raise arterial pressure. The opposite actions lower arterial pressure. This mechanism also constricts veins and increases the blood volume returning to the heart, but venous constriction is far less significant than arteriolar constriction in regulating arterial pressure.

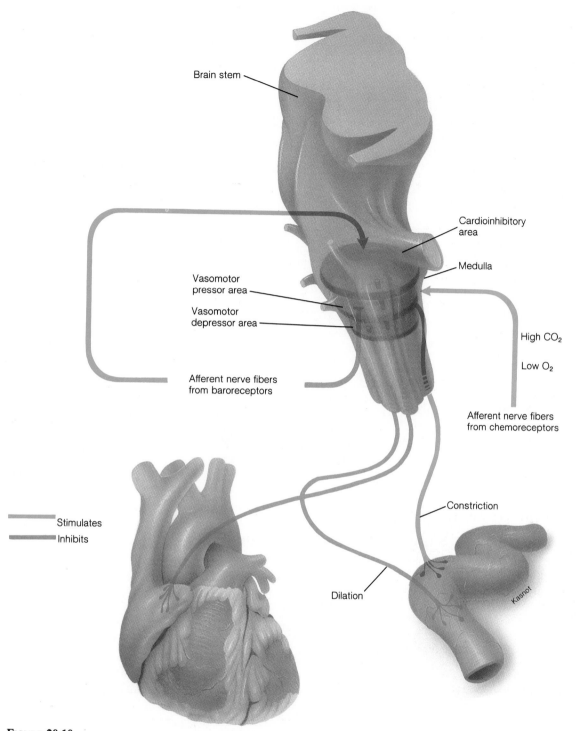

FIGURE 20.10
The cardiovascular center: its stimuli and responses.

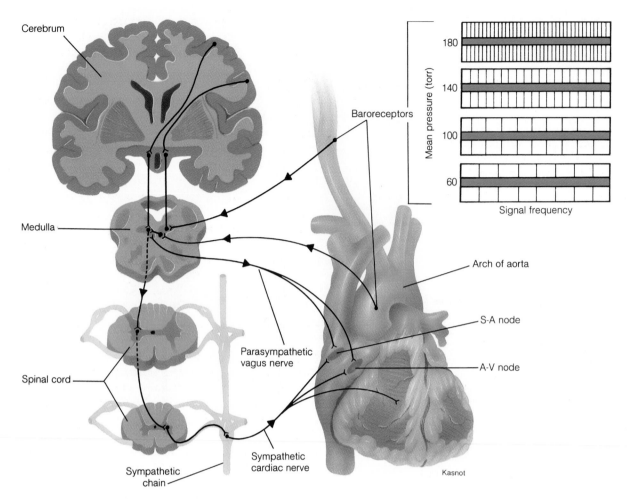

FIGURE 20.11
Role of baroreceptors in regulating heart function and blood pressure.

Pressor or depressor areas can be stimulated by changes in blood pressure, blood carbon dioxide or oxygen, body temperature, and emotional stimuli. Many such signals are relayed from receptors that detect changes in pressures or chemical composition of blood.

Receiving a sudden emotional shock can suppress sympathetic activity, causing blood vessels to dilate and arterial pressure and venous return to decrease. These changes decrease the heart rate and stroke volume and, thus, decrease cardiac output. Blood flow to the brain drops precipitously and a person faints, especially if he or she was standing. Fainting initiates a compensatory process that increases blood flow to the brain. Attempting to keep the person in an upright position works against compensation.

Baroreceptors

Pressure receptors called **baroreceptors** (Figure 20.11) respond to arterial wall stretching. Primary baroreceptors are found in the **carotid** (kar-ot′id) **sinus** of each common carotid artery and in the aortic arch. Other baroreceptors probably are distributed throughout the arteries. Baroreceptors send signals continuously at a rate proportional to the arterial pressure. The signals are relayed to various parts of the cardiovascular center where they affect both vasomotor tone and heart rate. When arterial pressure is high, frequent signals act on the cardioinhibitory area to inhibit vasoconstriction and slow the heart rate. When arterial pressure is low, signals are so infrequent that they fail to excite the cardioinhibitory area and allow vasoconstriction and an increased heart rate.

Baroreceptors adapt to high or low arterial pressure in a few days, after which they maintain new abnormal pressures instead of normal ones. Consequently, in patients with hypertension (high blood pressure), baroreceptors actually help to maintain abnormally high pressures.

Trained emergency medical personnel sometimes take advantage of the carotid sinuses' response to pressure by massaging them to slow the heart rate during tachycardia (rapid heart rate). Untrained individuals should not attempt carotid massage; improperly performed, it can stop the heart completely.

Chemoreceptors

In addition to baroreceptors, the carotid arteries and aorta also contain **chemoreceptors,** which respond to changes in blood concentrations of oxygen, carbon dioxide, and acid (hydrogen ions). Signals from chemoreceptors are important in regulating breathing (Chapter 22), but they also affect blood pressure. Too little oxygen or too much carbon dioxide or acid increases blood pressure and heart rate. If the rate and depth of breathing also increase, blood picks up more oxygen and gets rid of more carbon dioxide (and acid) as it passes through the pulmonary circuit.

INTEGRATED FUNCTION: SYSTEMIC ARTERIAL PRESSURE REGULATION

Systemic arterial pressure is an important variable in integrated circulatory function. Regulating it maintains systolic, diastolic, and mean arterial pressures nearly constant even when cardiac output and other factors vary significantly. Though blood continuously flows to all tissues, the volume going to different tissues and organs varies somewhat when the body is at rest and even more with exercise, changes in body position, and circulatory disorders.

Among the various factors affecting blood volume in a tissue, the cross-sectional area serving the tissue is especially important. As noted earlier, when vessel length remains constant (as it does in the body), cross-sectional area determines resistance. When pressure is constant, flow increases as resistance decreases and vice versa. Dilation and constriction of vessels, mainly arterioles, which change cross-sectional area, are the main means of regulating

blood flow to various tissues. For example, during exercise, resistance decreases in arterioles of muscle, increases in those of the kidneys and digestive organs, and remains constant in those of the brain. The body's ability to adapt to change by varying cardiac output and the distribution of blood to various tissues while maintaining constant pressure is an important example of homeostasis.

Arterial pressure is regulated by short-term and long-term factors. Short-term factors act within seconds to hours and partially correct abnormal arterial pressure. Long-term factors take longer to act but maintain arterial pressure at a nearly constant level over days, weeks, or even years.

Short-term regulation involves mainly reflexes triggered by signals from baroreceptors and chemoreceptors responding to changes in pressure in major arteries. The **CNS ischemic** (is-kem′ik) **response,** another short-term regulator, operates when blood flow to the vasomotor center itself decreases. **Ischemia** (decreased blood flow) leads to a lack of nutrients and an accumulation of wastes in a tissue. In the CNS ischemic response, excess carbon dioxide probably stimulates the vasomotor center, causing a quick rise in arterial pressure. Short-term regulators rapidly adapt to pressure changes and lose their effect.

Long-term regulation occurs mainly by fluid regulation (Chapter 28). When receptors in the kidneys detect a decrease in arterial pressure, they elicit signals that release aldosterone or activate angiotensin II. These substances cause the kidneys to return more fluid to the blood, thereby increasing blood volume. As blood volume increases, so does arterial pressure. If arterial pressure goes too high, the atria release atrial natriuretic hormone, which causes more fluid to be excreted. These mechanisms operate effectively over long periods of time to maintain arterial pressure in the normal range.

Regulatory systems that maintain mean arterial pressure within a narrow, tolerable range involve negative feedback. When the pressure varies outside the normal range, signals to the cardiovascular center cause alteration in vessel diameters that brings the pressure back into the normal range (Figure 20.12a). In certain disorders, such as the sudden loss of a large volume of blood for which the body cannot compensate, positive feedback develops (Figure 20.12b). The low blood volume decreases the heart's pumping efficiency and cardiac output decreases. As cardiac output decreases, less blood flows through the coronary arteries to the heart muscle, the pumping efficiency decreases further, and a vicious cycle is set up. Unless negative feedback can override positive feedback, death ensues.

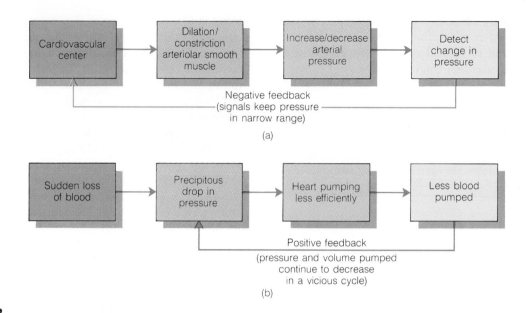

FIGURE 20.12

(*a*) A negative feedback system maintains mean arterial pressure at nearly a constant level. (*b*) A positive feedback loop leads to progressively lowered pumping efficiency of the heart following the loss of a large volume of blood.

PHYSIOLOGY OF CIRCULATION IN CERTAIN TISSUES

Circulation is continuous and flow is constant over the whole circulatory system, but various tissues contain different amounts of blood. Those amounts change with a shift from rest to exercise (Figure 20.13). Such variations are achieved mainly by adjusting the cross-sectional area of arterioles.

Heart

Under most conditions relatively high resistance in the coronary circuit slows blood flow, increasing exchanges between blood and cells. When the cells need more nutrients and oxygen, they release adenosine, excess K^+, and prostaglandins, all of which dilate blood vessels. When cells are well supplied with nutrients and oxygen, they release the peptide *endothelin,* which constricts coronary blood vessels, possibly by opening calcium channels in the endothelial cells. A normal concentration of endothelin probably contributes to normal blood pressure, but an excess might lead to hypertension. This idea is supported by studies of sarafotoxins, snake venoms with 21 of 25 amino acids in common with endothelin. Sarafotoxins kill mice in a few minutes by excessively elevating blood pressure.

Brain

The brain contains about 15 percent of the body's blood regardless of whether a person is engaging in exercise, mental activity, or sleep, but certain areas of the brain receive more blood than others. Gray matter receives about six times as much as white matter. Brain blood flow in children exceeds 100 ml/min per 100 g of tissue—about twice that in adults.

Several factors contribute to the constancy of cerebral blood flow. Nearly constant intracranial pressure prevents blood vessel compression and ischemia. Should ischemia occur, the CNS ischemic response elevates arterial pressure throughout the body, including the cerebrum. In severe ischemia, arterial pressure rises to dangerously high levels as cardiac output reaches a maximum. The very homeostatic mechanism that regulates cerebral circulation can, under extreme conditions, create pressures high enough to damage tissues. In severe brain edema, intracranial pressure exceeds arterial pressure, stopping blood flow wherever it occurs and causing cells to die from lack of oxygen and nutrients.

Autoregulation and local metabolic effects have special importance in the brain. Autoregulation in cerebral arterioles distributes blood to active areas of the cerebrum during problem solving and other higher level processes (see Figure 12.12). Carbon dioxide and, thus, H^+ are more potent vasodilators in the cerebrum than in other tissues, probably because H^+ acts directly on cerebral vessels.

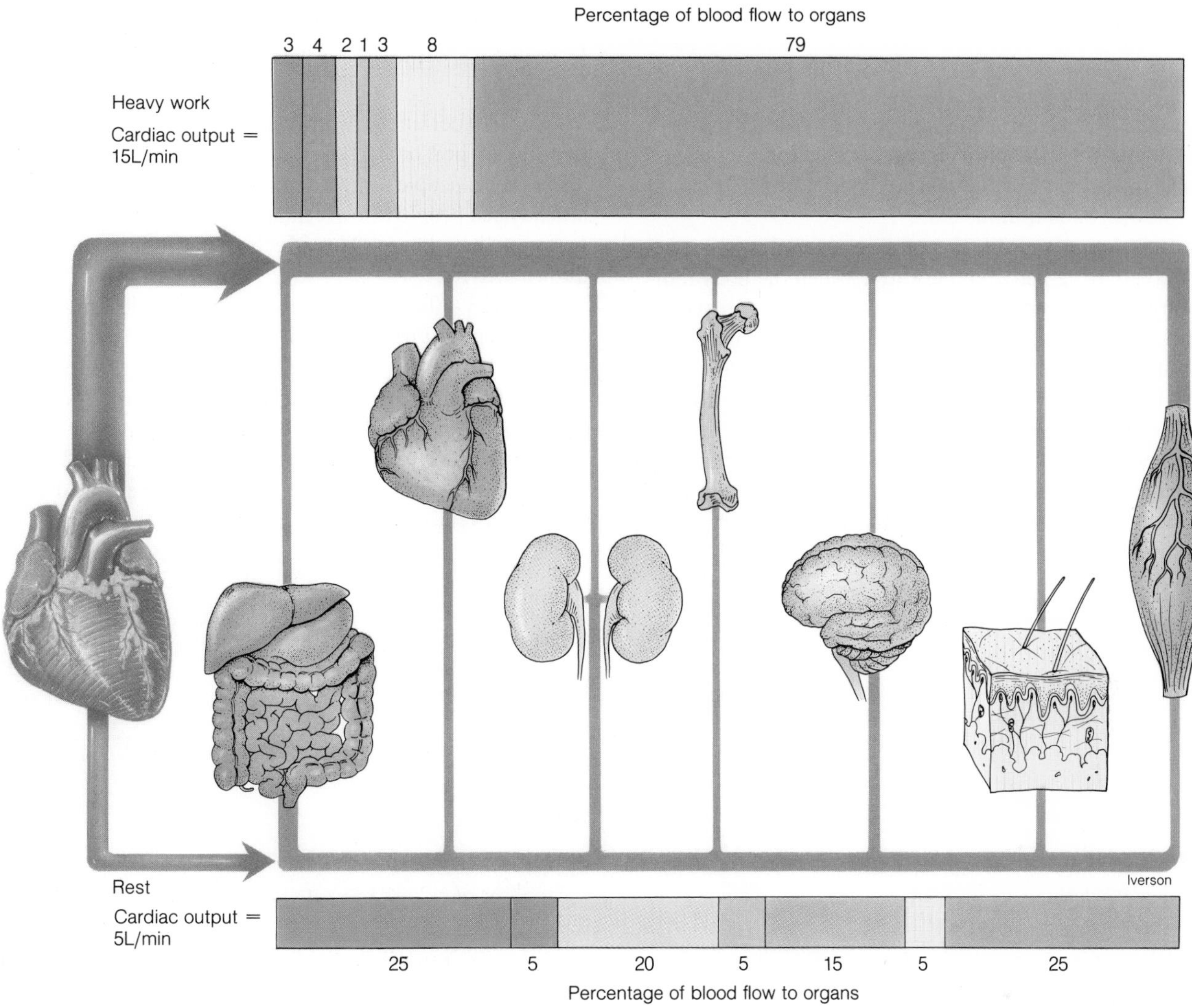

Percentage of blood flow to organs

3 4 2 1 3   8   79

Heavy work

Cardiac output =
15L/min

Iverson

Rest

Cardiac output =
5L/min

25   5   20   5   15   5   25

Percentage of blood flow to organs

FIGURE 20.13
Effects of exercise on distribution of blood to various organs. The
areas indicated in the graphs are proportional to blood flow.

From P. Astrand and K. Rodahl, Textbook of Work Physiology, 2d ed. Copyright © 1977
McGraw-Hill, Inc., New York, NY. Reproduced with permission of McGraw-Hill, Inc.

Brain tissue is especially sensitive to shortages of glucose and oxygen and to toxic ammonia. Using glucose as
its primary energy source, the brain metabolizes from 20
to 60 percent of the available glucose. Lacking glucose,
the brain can use ketone bodies for energy, but not large
fatty acids. The brain requires oxygen to metabolize glucose and more than five minutes of oxygen deprivation
permanently damages brain cells. Cells minimize release
of ammonia by attaching waste amino groups from the
breakdown of proteins to the amino acid glutamic acid to
form glutamine. This reaction is reversed in the kidneys
where amino groups are excreted in the urine.

Digestive Tract and Liver

The liver receives as much as 28 percent of cardiac output
under resting conditions—one-fifth from the hepatic artery
and four-fifths from the hepatic portal vein. Pressure drops
precipitously as these vessels join liver sinusoids. During
stress, sympathetic signals can constrict presinusoidal vessels and dilate postsinusoidal ones, emptying as much as
300 ml of blood from liver sinusoids into the circulation.

Muscles and Skin

Dilation of blood vessels in muscles during exercise is due mainly to accelerated muscle metabolism and the accumulation of carbon dioxide and lactic acid. Changes in sympathetic tone, blood oxygen concentration, and vascular compression during contraction also help regulate blood flow in skeletal muscles.

Constriction and dilation of skin arterioles is important in regulating body temperature. When the body temperature rises, signals from the sympathetic nervous system cause arterioles to dilate and allow more blood (as much as 2500 ml/min) to pass through capillaries near the skin surface. When body temperature falls, heat is conserved as arterioles constrict and shunt blood through arteriovenous anastomoses (Figure 20.14), especially in the head, fingers, and toes. Under very cold conditions, skin may freeze to conserve heat to keep the heart functioning.

EFFECTS OF CHANGES IN POSITION ON CIRCULATION

We've all experienced light-headedness when we suddenly hop out of bed or "go over the top" on a roller coaster. We've seen people's faces turn red when they stoop over and advised people who are about to faint to put their heads between their knees. The effects of position changes are due to gravity.

Gravity affects pressures in both arteries and veins. In a 160 cm tall person who is standing, the mean arterial pressure (MAP) at the heart level is 100 torr. Because of gravity, the MAP in a head artery is 62 torr and the MAP in a foot artery is 180 torr. Similar, but smaller, differences exist in veins. In contrast, gravity has little effect on a person lying down because all blood vessels are nearly at heart level.

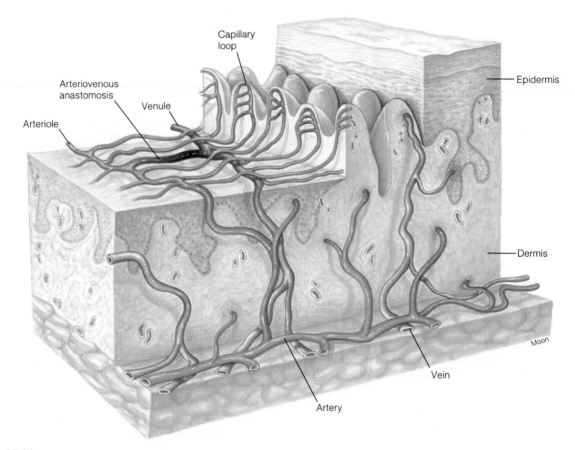

FIGURE 20.14
Distribution of blood vessels in skin and the location of arteriovenous shunts.

When a person shifts from a lying to a standing position, gravity causes blood to collect in veins below the heart, dilates veins, and decreases venous return. This reduces stroke volume, which decreases arterial pressure. Compensation (Figure 20.15) begins with a slowing of inhibitory signals from baroreceptors to the cardiovascular center. With less inhibition, the heart rate increases and blood vessels constrict, increasing total peripheral resistance and venous return. Both stroke volume and heart rate increase, significantly increasing cardiac output. Increasing both cardiac output and peripheral resistance usually returns mean arterial pressure to normal in a few seconds. When compensation is delayed, as it is in many elderly people, the condition is called **postural hypotension** (hi-po-ten′shun). It can cause the person to faint.

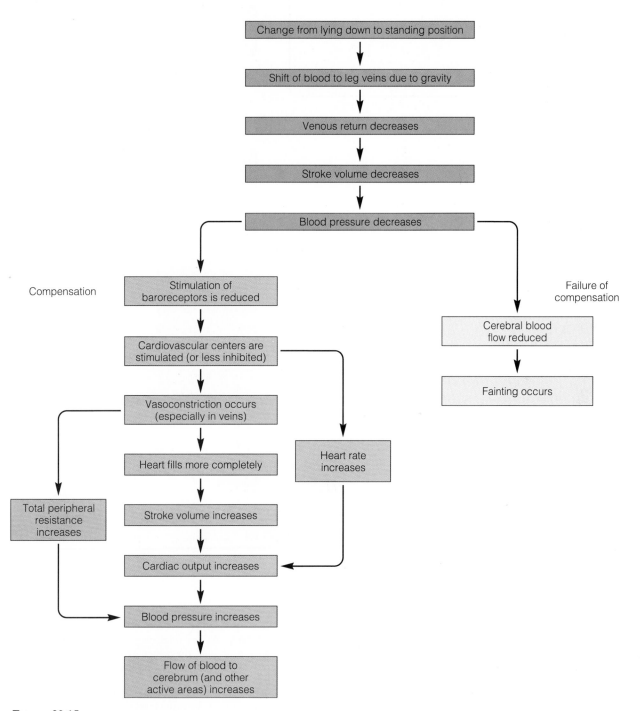

FIGURE 20.15
Effects of and compensation for moving from a lying to a standing position.

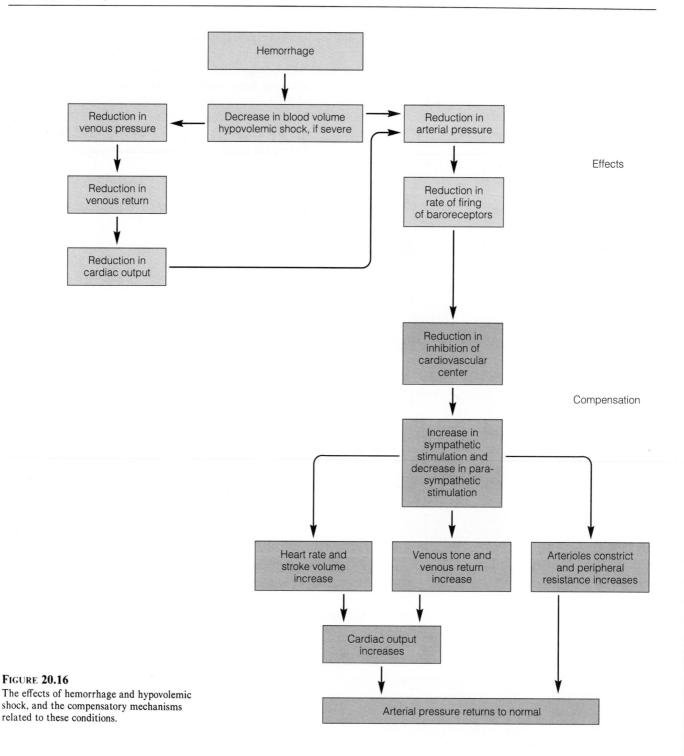

FIGURE 20.16
The effects of hemorrhage and hypovolemic shock, and the compensatory mechanisms related to these conditions.

Acceleratory forces in airplanes and space capsules are stronger and more physiologically disruptive than gravity. Such forces are measured in G units (1 G equals the force of gravity). When airplanes make sharp turns at high speed, the pilot and passengers experience an acceleratory force proportional to the square of the velocity and increasingly strong as the radius of the turning circle decreases. Astronauts are subjected to similar forces as the booster rockets fire. These forces can reach 10 G. At any force over 4 G, blood pressure in the head drops and the person experiences a "blackout" of vision in a few seconds and faints shortly thereafter. Pilots of commercial airplanes avoid high speed, sharp turns. Astronauts use reclining seats to keep their heads at the same level as the rest of their bodies during blast-off, thereby minimizing the effects of acceleration.

CLINICAL APPLICATIONS

Several human ills—high blood pressure, heart attacks, and strokes—are circulatory disorders. Developing and maintaining physical fitness (Interlude 2) can help to prevent these disorders.

Hemorrhage and Hypovolemic Shock

Shock is a condition in which the blood flow is totally inadequate to meet the metabolic needs of cells. **Hemorrhage,** or excessive blood loss, can cause **hypovolemic** (hi″po-vo-le′mik) **shock.** Because of blood loss, the blood volume decreases (becomes hypovolemic); blood pressure and blood flow are correspondingly lowered. A person in shock is ashen pale, has a rapid weak pulse, and cool, moist skin. Pallor is due to reduced blood flow near the skin surface. The rapid weak pulse represents frequent heartbeats with small stroke volume—the body's response to the drop in systemic arterial pressure. The cool, moist skin results from activities of the sympathetic nervous system. When the skin is cyanotic (bluish), blood oxygen has been depleted and flow is too depressed to maintain adequate circulation through the lungs.

Hemorrhage elicits compensatory mechanisms (Figure 20.16). Low blood pressure slows baroreceptor firing, which results in sympathetic signals that increase heart rate and stroke volume and constrict arterioles; therefore, cardiac output and mean arterial pressure increase. If compensation is adequate, arterial pressure returns to normal. If not, a blood volume expander, such as albumin or the polysaccharide dextran, is needed. Shock can result from allergic reactions (Chapter 21) and other disorders, but compensation is the same in all cases.

Cerebrovascular Accident

The sudden interruption of blood flow to a portion of the brain because of a blocked or ruptured blood vessel is called a **cerebrovascular** (ser″ĕ-bro-vas′ku-lar) **accident** (CVA), or **stroke.** A blood vessel gets blocked from a lodged embolus and/or atherosclerosis. A rupture usually occurs in a weakened wall of an aneurysm. In a CVA, ischemic brain cells die in a few minutes and paralysis, loss of feeling, or loss of speech often follow, depending on the location and degree of brain damage. Effects of a CVA are often seen on the opposite side of the body from the brain damage because of decussation (Chapter 12). One explanation for cell death is that damaged cells secrete large quantities of glutamate, which diffuses to and damages other cells probably by causing a sudden Ca^{2+} influx. These damaged cells likewise release glutamate, further widening the region of cell damage.

Many CVA victims have a history of **transient ischemic attacks**—short periods of headache, confusion, and other diffuse symptoms, due to reduced cerebral blood flow

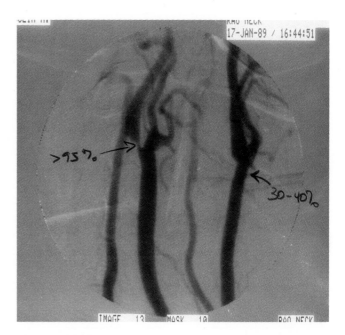

FIGURE 20.17
Arteriogram showing occlusion of an internal carotid artery.

and cerebral artery spasms. Such attacks warn that a CVA is likely, but a CVA may be avoided by locating plaque deposits in carotid arteries in **arteriograms** (ar-te′re-o-gramz″), shown in Figure 20.17, and surgically opening the arteries and removing the plaque.

Recovery after a CVA has depended in part on how well **collateral circulation,** blood vessel branches that bypass blocked or damaged ones, supplies surviving neurons. It also depends on how well those neurons take over functions previously performed by neurons that have been destroyed. Today, clot digesters such as tPA (Chapter 17) are being used as emergency treatment for stroke victims as well as for heart attack victims.

Hypertension

High arterial pressure, or **hypertension** (hi″per-ten′shun), is defined as a resting arterial pressure that exceeds 140/90 torr (systolic/diastolic) for a prolonged period of time. Many patients have "borderline hypertension"—diastolic pressure of 90 to 95 torr and systolic pressure of 140 to 160 torr. Diastolic pressure above 95 torr exerts a continuous pressure that can damage blood vessels and can lead to cardiovascular disorders. Arterial pressure increases with age (Table 20.3), especially as arteries lose their elasticity, but this does not account for the high incidence of hypertension in people who are middle-aged and older.

Hypertension can be primary, secondary, or malignant. About 90 percent of cases are **primary,** or **essential, hypertension,** the cause of which is unknown. Elderly patients with essential hypertension typically have several conditions associated with aging—less elastic arteries, less

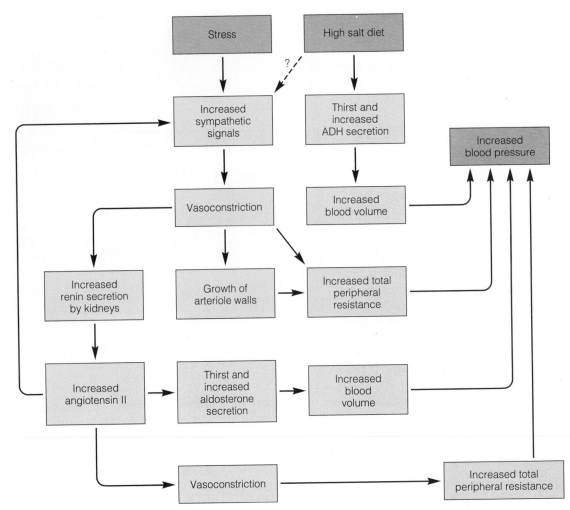

FIGURE 20.18
A possible mechanism of hypertension.

responsive baroreceptors, and greater peripheral resistance—that elevate blood pressure. Nearly 10 percent of cases are **secondary hypertension.** They result from a primary disease, such as hypersecretion of hormones that elevates blood pressure, narrowing of arteries, polycythemia, toxemia of pregnancy (Chapter 29), and the use of birth control pills. Once pressure becomes elevated, baroreceptors adapt to maintain the higher pressure. A few untreated cases progress to **malignant hypertension** in which *diastolic* pressure exceeds 120 torr. In this condition, retinal hemorrhages can cause blindness, and cardiovascular and renal functions can deteriorate to the point of death within two years.

Genetic predisposition, obesity, and stress can contribute to hypertension. Evidence for a genetic factor comes from observations that children of hypertensive parents are twice as likely to develop hypertension as those of normotensive parents. Obesity increases the total quantity of tissue to which the heart must deliver blood and, thus, increases peripheral resistance, blood volume, and

cardiac output. When cardiac output and peripheral resistance increase simultaneously, arterial pressure increases greatly. Stress increases sympathetic stimulation, which, in turn, elevates arterial pressure.

Stress and a high-salt diet may work together to produce hypertension (Figure 20.18). Excessive sympathetic stimulation, which can be due to genetically abnormal receptors, causes vasoconstriction and increased peripheral resistance. The kidneys secrete renin, which produces a chain of events that increases blood volume (Chapter 28) and further increases peripheral resistance. Eating salty foods increases the burden on overworked kidneys, but eating foods high in potassium seems to decrease that burden. Though high arterial pressure can increase sodium and water excretion slightly, the damage it does to blood vessels far outweighs its benefit. Diets of mainly processed foods (with high sodium and low potassium) common in the United States are particularly hazardous to one-fifth of the population who are susceptible to hypertension.

By increasing the workload on the heart and by damaging arteries through continuous excessive pressure, hy-

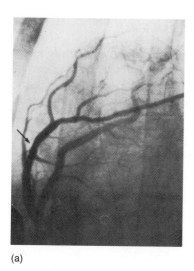 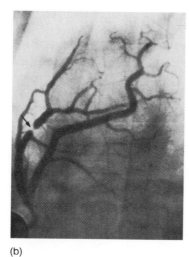 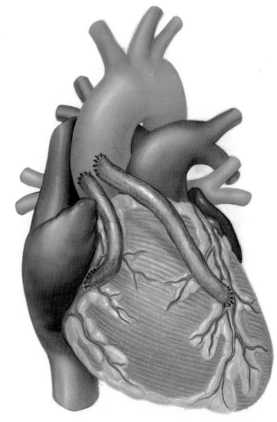

(a)   (b)

FIGURE 20.19
(a) Arteriogram showing occlusion of a coronary artery, and (b) the results of coronary bypass surgery.

pertension can lead to heart failure, kidney failure, and blockage or aneurysms of the aorta and cerebral arteries. The left ventricle enlarges as it pumps against greater resistance. The enlarged ventricle needs more nutrients and oxygen, and is especially vulnerable to cell death if the blood supply decreases. Stretching of myocardial fibers increases contraction strength up to a point, but can eventually lead to heart failure.

Recognition of complications of hypertension has provided impetus for improving treatment. Modes of treatment are related to physiological factors that affect blood pressure (BP)—heart rate (HR), stroke volume (SV), and total peripheral resistance (TPR). The patient exercises to lower HR, restricts sodium intake and takes diuretics (drugs that increase water excretion) to decrease SV, and may receive relaxation therapy and drugs to block sympathetic stimulation or relax smooth muscle. A new category of antihypertensives, the ACE inhibitors, lower blood pressure by inhibiting **a**ngiotensin-**c**onverting **e**nzyme, thereby preventing angiotensin from raising blood pressure (Chapter 27). The lowered death rate from cardiovascular diseases since the 1960s in the United States may be partly due to successful treatment of hypertension.

Recent observations that certain drugs used to treat hypertension raise blood cholesterol may alter that treatment. Diuretics, such as the widely used thiazides, raise cholesterol by about 7 percent, LDLs by about 15 percent, and triglycerides by about 22 percent. Beta adrenergics lower HDLs. The present recommendation is to monitor blood lipids, changing medication if a lipid increases by 20 percent or if HDLs decrease.

Ischemic Heart Disease and Myocardial Infarction

As atherosclerosis in coronary arteries worsens, many patients suffer from **ischemic heart disease,** reduced blood flow to myocardial cells. Ischemia and **angina pectoris** (an'jin-ah pek'tor-is), severe chest and referred arm pain, occur when exercise demands more of the myocardium than its blood supply can accommodate. Nitroglycerine, a drug quickly absorbed from beneath the tongue, alleviates angina by dilating blood vessels—increasing blood flow in coronary vessels, lowering peripheral resistance, and decreasing the workload on the heart.

Ischemic heart disease can be treated by coronary bypass surgery, in which segments of veins taken from the patient's legs are sewn to coronary arteries so blood flows around plaque-filled segments of those arteries (Figure 20.19). Successful surgery of this type greatly improves the myocardial blood supply.

A **myocardial infarction** (MI), damage to myocardial cells because of a lack of blood supply, often occurs suddenly when a blood clot completely blocks an already atherosclerotic artery. Arrhythmias arising from infarcted tissue are a common cause of sudden death. Symptoms that lead to an MI are often referred to as a heart attack.

In an MI, the size of the infarcted area depends on how much of the myocardium is deprived of blood. Collateral circulation may keep most cells alive in a small infarction and all damage will eventually be repaired. Three zones of injury—central dead cells, a region of nonfunctional cells, and outer ischemic cells—develop in a large infarction (Figure 20.20). During the healing process, fibrous tissue replaces all cells in the central zone and some in the nonfunctional zone. Most cells in the outer zone and many in the nonfunctional zone survive and recover function. Older patients with a history of angina may survive an MI that would kill a younger patient because they have developed collateral circulation as their coronary arteries gradually narrowed.

The degree of tissue damage from an MI can be assessed by measuring blood concentration of certain enzymes released from damaged myocardial cells. These enzymes include creatine kinase (CK), lactic acid dehydrogenase (LDH), and serum glutamic-oxaloacetic transaminase (SGOT). CK transfers a phosphate group from phosphocreatine to ATP, generating energy-rich ATP (Chapter 9). LDH removes hydrogen from lactic acid, and SGOT transfers amino groups from glutamic acid to oxaloacetic acid (Chapter 24).

In addition to affecting heart muscle, MIs also affect circulation. Impaired pumping decreases cardiac output and serious hypotension (low blood pressure) results. Sometimes, and for no known reason, parasympathetic

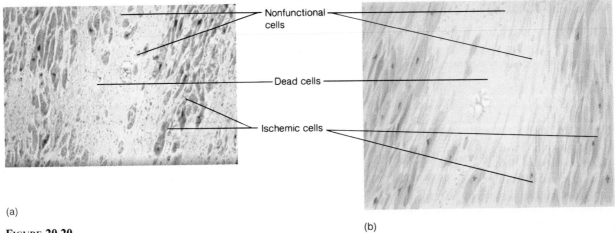

(a)

(b)

FIGURE 20.20
(a) Photomicrograph, and (b) diagram of tissue affected by a myocardial infarction.

signals greatly exceed sympathetic ones. This further lowers blood pressure and cardiac output, but it also spares the muscle tissue further damage. Overworking a seriously damaged heart can lead to complete heart failure as described below.

Treatment for MI now begins as soon as medical assistance arrives with clot-dissolving drugs such as tPA (Chapter 17). Still important during recovery are complete bed rest to reduce the workload on the heart and oxygen to assure that blood reaching damaged cells will contain as much oxygen as possible. In the last decade, **angioplasty,** the use of a "balloon" to open atherosclerotic arteries by pressing plaque tightly against the walls, has been used after many MIs. Recent studies indicate that only patients with recurrent symptoms need angioplasty and that those who are recovering without further symptoms gain no benefit from it.

The search for ways to improve MI treatment continues. Because *liposomes* (intracellular membrane-bound particles filled with lipids) accumulate in ischemic tissues, they could be made in the laboratory, laden with drugs, and used to carry the drugs to injured cells. *Prostacyclin* (pros"tah-si'klin), a natural potent prostaglandin, lowers peripheral resistance and reduces arterial pressure without changing the heart rate. It also inhibits platelet aggregation that might cause clots and reduces enzyme release from injured cells. A single *aspirin* tablet every other day appears to reduce the chance of an MI in men over 50 and to help prevent a second MI in those who have had one. Though one effect of aspirin is to inhibit prostaglandins, it also appears to interfere with certain steps in blood clotting.

Heart Failure

The inability of the heart to adequately supply blood to tissues is called **heart failure.** Patients can recover from acute (sudden, severe) heart failure following an MI as the myocardium heals and pumps sufficient blood to all tissues. Chronic (slowly developing, worsening) heart failure eventually leads to death as the heart gradually weakens and pumps less and less blood. Chronic heart failure can be caused by hypertension, atherosclerosis, defective heart valves, or congenital lesions.

Heart failure can be left-sided or right-sided. When the left side fails, blood backs up in the left atrium and pulmonary veins, fluid accumulates in lung tissue, breathing is impaired, and little oxygen enters the blood. The overworked heart that needs more oxygen gets even less. When the right side fails, blood backs up in the liver and systemic veins and peripheral resistance increases, placing a greater load on the heart. Any heart failure decreases cardiac output, lowers renal blood flow, and reduces excretion of wastes and excess fluid. If heart failure on one side continues, the other side will also fail. Heart failure sometimes is called **congestive heart failure** because of congestion due to tissue fluid accumulation and severe edema.

When heart failure is caused by a heart defect, it can often be corrected surgically. When it is caused by irreparable damage, diuretics can help remove fluid from edematous tissue and reduce peripheral resistance, lessening the load on the heart. Digitalis strengthens heart contractions, increases cardiac output, and helps to correct arrhythmias that sometimes accompany heart failure. Unfortunately, digitalis is toxic, and the difference between an effective and a toxic dose is small.

CLINICAL TERMS

antihypertensive (an"ti-hi"per-ten'siv) **drug** a substance that helps to reduce blood pressure

endarterectomy (end"ar-ter-ek'to-me) the removal of the inner wall of an artery to reduce occlusion of the vessel

phlebothrombosis (fleb"o-throm-bo'sis) inflammation of a vein caused by a thrombus; such a thrombus is easily detached to become an embolus

phlebotomy (fleb-ot'o-me) the opening of a vein to remove blood

thrombectomy (throm-bek'to-me) the removal of a venous thrombus

thrombophlebitis (throm"bo-fle-bi'tis) inflammation of a vein that results in the formation of a thrombus

vasodilator (vas"o-di'la-tor) **drug** a substance that increases the diameter of blood vessels

ESSAY

RISK FACTORS IN HEART AND BLOOD VESSEL DISEASE

Nearly a million Americans die of heart and blood vessel diseases each year—more than from all other causes including cancer and accidents. Another 27 million have cardiovascular disease. Hypertension is the most common and most potent factor in the development of atherosclerosis. Combinations such as hypertension, obesity, and diabetes mellitus almost always lead to cardiovascular disease. Much work has been done to identify risk factors in cardiovascular disease, especially those factors that people can control. Though they cannot control risk factors such as age, sex, and heredity, they can control smoking, exercise, and diet.

Age, Sex, and Heredity

Up to age 55, women are less prone to MI than men, and estrogens may be responsible for the difference. Degeneration of arterial smooth muscle cells and thickening of arterial walls by plaque apparently increase with age. These changes are seen in animal experiments even when the animals are maintained on a low-fat, cholesterol-free diet. Some families have a higher incidence of heart and blood vessel diseases than others because of life-style and genetic factors, such as the one for familial hypercholesterolemia.

Smoking, Exercise, and Diet

Smoking, especially cigarette smoking, has been correlated with both cancer and cardiovascular disease. Nicotine packs a one-two punch—it accelerates the heart rate, increasing the tissue demand for oxygen, and it constricts coronary arteries, decreasing the availability of oxygen when most needed. Many Americans have stopped smoking; others have shifted to low-tar, low-nicotine cigarettes, but they still increase their risk of cardiovascular disease from nicotine and carbon monoxide in smoke. Because cigarette smoking is so dangerous, so prevalent, and its effects so reversible, stopping it ranks first among measures that could lower the risk of cardiovascular disease.

Exercise appears to reduce the risk of cardiovascular disease by strengthening the heart and altering blood lipids. Regular exercise that keeps the heart rate in the training range increases contraction strength and stroke volume. This allows the heart to provide an adequate blood supply in fewer contractions than before the exercise program was begun. Such strengthening of the heart probably reduces the risk of cardiovascular disease, but proof of its effectiveness is needed. Exercise also seems to lower LDLs and raise HDLs in the blood, but exactly how this occurs is not understood.

Diet is important in controlling body weight and blood lipids. Excessive body weight contributes to cardiovascular disease by placing a constant strain on the heart to pump blood through more tissue. Animal fats, especially cholesterol, also have been implicated in cardiovascular disease (Chapter 26).

Progress

Various campaigns over the past two decades have reduced the incidence of cardiovascular disease. A larger proportion of cases of hypertension are detected and treated. Many smokers have "kicked the habit" and cigarette advertising has been reduced. Many people exercise regularly and some have lost weight. All these factors lower blood pressure, and exercise and diet can lower blood cholesterol, too. Keeping blood pressure and blood cholesterol within normal limits minimizes plaque deposition in arteries. These findings should help health professionals to motivate patients to follow their recommendations.

Questions

1. Which risk factors can people change?
2. Evaluate your risk of cardiovascular disease on fixed and changeable factors.
3. What can you do now to lower your risk?
4. If you have family members who could lower their risk, how might you encourage them to do so?

CHAPTER SUMMARY

Principles of Circulation

• The principles of circulation are summarized in Table 20.2.

Circulation in Different Vessel Types

• Blood pressure is created by the heart and pushes blood through vessels. Systolic pressure occurs during blood ejection and diastolic pressure occurs during ventricular relaxation. Pulse pressure is the difference between systolic and diastolic pressures. Mean arterial pressure is the average pressure driving blood through the arteries.

Control of Circulation

• Circulation is regulated by factors listed in Table 20.4 and by brain centers, baroreceptors, and chemoreceptors. The cardiovascular center produces cardioinhibition, vasomotor constriction, and vasomotor dilation. The vasomotor area acts reciprocally to maintain vasomotor tone according to signals from receptors. Baroreceptors send signals at a rate proportional to the blood pressure, and such signals reduce cardiac output and cause vasodilation. Chemoreceptors increase blood pressure when blood carbon dioxide or acid increases.

Integrated Function: Systemic Arterial Pressure Regulation

- Regulation of systemic arterial pressure is an important homeostatic mechanism that maintains blood flow to all tissues even as cardiac output varies.
- Mean arterial pressure is equal to the product of cardiac output and total peripheral resistance.
- Short-term regulation of arterial pressure occurs mainly by reflexes, whereas long-term regulation involves a renal-body fluid-pressure control mechanism.

Physiology of Circulation in Certain Tissues

- Coronary blood flow is regulated according to the needs of cells by various neural and chemical factors.
- The brain must have an adequate supply of oxygen and glucose; various parts of the brain use these substances at different rates, according to the brain's activity.
- Blood flow to muscles and skin increases greatly during exercise, supplying energy for muscle contraction and helping to regulate the body temperature.

Effects of Changes in Position on Circulation

- Changing from a lying to a standing position reduces venous return and blood pressure; baroreceptors respond and cause an increase in cardiac output and blood pressure.

Clinical Applications

- Disturbances in circulation are accompanied by disturbances in homeostasis: Hemorrhage can cause hypovolemic shock. A cerebrovascular accident can cause brain damage and paralysis. Hypertension increases the work load on the heart and the likelihood of infarction. Ischemic heart disease impairs heart function and can lead to a myocardial infarction, which causes further impairment. Heart failure leads to severely decreased cardiac output and fluid accumulation in the tissues.

QUESTIONS AND PROBLEMS

The questions at the end of each chapter are numbered to correspond with the objectives listed at the beginning of the chapter. Italics indicate that a question requires critical thinking skills beyond simple factual recall.

Questions

1. *(a) What factors affect pressure and resistance?*
   (b) How do changes in pressure and resistance affect blood flow?
   (c) How can blood flow be constant over the entire circulatory system while pressures vary so greatly?
2. (a) Describe the events in the brachial artery when blood pressure is being measured.
   (b) What is pulse, and what is pulse pressure?
   (c) If systolic pressure is 125 torr and diastolic pressure is 77 torr, what is the mean arterial pressure?
   (d) How is circulation regulated in arteries?

3. *How do autoregulation and extrinsic regulation in arterioles differ?*
4. *(a) How is circulation regulated in capillaries?*
   (b) How do pressure differences contribute to exchanges between capillaries and tissues?
   (c) Use hydrostatic and oncotic pressures to explain the movement of substances out of and into capillaries.
   (d) Apply what you know about capillary exchange to explain the effects of excess tissue fluid and elevated or lowered blood hydrostatic pressure.
5. *How is circulation regulated in venules and veins?*
6. *Explain at least three causes of edema.*
7. *How would the following affect heart rate and circulation: (a) inactivity of cardioinhibitory center, (b) nonresponsiveness of baroreceptors, and (c) blockage of lymph vessels?*
8. *(a) How is systemic arterial pressure regulated?*
   (b) In what ways is the regulation of arterial pressure a good example of a homeostatic mechanism?
9. *How might the body's ability to direct blood to different organs and tissues maintain homeostasis during (a) strenuous exercise, (b) excessive cold, (c) excessive heat, and (d) the digestion of a heavy meal?*
10. (a) What happens to circulation when a person faints?
    (b) What compensatory events occur after fainting?
    (c) How does gravity affect pressure in blood vessels?
    (d) How does changing from a lying to a standing position affect pressure in blood vessels?
    (e) What is postural hypotension and how does the body normally prevent it?
11. *How do the following disorders disturb homeostasis: (a) hypovolemic shock, (b) cerebrovascular accident, (c) hypertension, (d) myocardial infarction, and (e) heart failure?*

Problems

1. Suppose you are working with a patient who suffers from hypertension. How would you teach the patient the importance of continuing to follow medical advice?
2. If the heart is pumping 5 l of blood per minute, what is the total flow through each type of blood vessel?
3. What is significant about a central venous pressure of 15 torr?
4. Research (a) enzyme release from damaged heart cells in myocardial infarction or (b) drug use in treating hypertension.
5. Apply what you have learned about the circulatory system to explain why arrhythmias might follow a myocardial infarction.
6. Suppose you are working with a person who has difficulty breathing, a low urinary output, and tissue edema. Why would you or would you not order an ECG? Blood enzyme studies? What would you expect to learn from these tests?

OBJECTIVES

1. Briefly define basic concepts about nonspecific and specific body defenses.
2. Describe the steps in inflammation and healing.
3. Explain how interferon and complement serve as nonspecific defenses.
4. Describe the components of the immune system, and explain what is meant by the dual roles of specific immunity.
5. Summarize the properties of antigens, histocompatibility proteins, and antibodies (immunoglobulins).
6. Describe the characteristics of humoral immunity.
7. Describe the characteristics of cell-mediated immunity.
8. Briefly explain how killer cells kill.
9. Summarize the general properties of specific immunity.
10. Relate immunization to different kinds of immunity.
11. Briefly summarize the characteristics of hypersensitivity and autoimmune disorders.
12. Distinguish between inherited and acquired immunodeficiencies.

21

BODY DEFENSE MECHANISMS

BASIC CONCEPTS OF BODY DEFENSE

When you have a cold, you treat the symptoms expecting that no matter how miserable you are, soon you will be well again—at least until the next cold. Your body fights colds and other infections (diseases caused by microorganisms invading the body) with both nonspecific and specific defense mechanisms.

In a general sense, **immunity** (*immune,* safe or free of burden) refers to all physiological mechanisms that enable an organism to recognize and defend against infectious agents. **Nonspecific immunity** defends the body against any infectious agent. **Specific immunity** defends against particular infectious agents. It arises from functions of the **immune system,** which consists of lymphocytes, the thymus gland, and other lymphatic tissues. **Immunology** is the study of specific immunity and how the immune system responds to specific infectious agents.

Nonspecific defenses function regardless of which disease agent threatens the body. The body's first such defense is the skin. Intact skin is a mechanical barrier to microbes and foreign substances, and skin glands secrete substances that are toxic to microorganisms. Second, mucous membranes lining the digestive, respiratory, and urinary tracts also provide a mechanical barrier. Glands in these membranes secrete mucus, which traps foreign particles, and cilia on some membranes help to sweep foreign substances out of tracts. When microorganisms or foreign substances penetrate skin and membrane barriers, they initiate the inflammatory process. Phagocytic cells migrate to the invading substances and attempt to engulf them. Slightly acidic tissue fluids and lysosomes from damaged cells help to destroy invading substances.

Specific defenses produce immunity to particular diseases. Such defenses respond to an **antigen** (an'ti-jen), a molecule foreign to the body and usually associated with a disease agent—the chicken pox virus or the diphtheria bacterium. An antigen typically causes the immune system to produce specific **antibodies** (an'ti-bod"ez), molecules that inactivate the antigen. Specific immunity ordinarily develops while a person has a disease, such as chicken pox or mumps, and keeps the person from having the disease again. Humans do not develop lasting immunity to colds, partly because there are many immunologically different viruses that can cause colds and immunity to any one of them typically lasts only a few years.

Sometimes specific defenses react inappropriately or not at all, and allergies, reactions to blood transfusions, immune reactions to one's own tissues, and immunodeficiency diseases result. The same defenses that rid the body of disease agents, reject and destroy transplanted organs, cancer cells, and other substances that the immune system identifies as foreign.

NONSPECIFIC DEFENSES

The body's main nonspecific defense is inflammation. Certain special substances—interferon and complement—also act as nonspecific defenses.

Inflammation

Do you remember what happened the last time you had a severe cut? You probably experienced the five "cardinal signs" of **inflammation**—swelling, redness, heat, pain, and loss of function. Inflammation follows tissue damage from mechanical injury, microorganisms, harmful chemicals, heat and electrical burns, ultraviolet light, and allergies.

Steps in the Inflammatory Process

Inflammation is the first step in combatting infection (Figure 21.1). When bacteria or other foreign materials enter the skin, some antibodies against them may be available from previous immune reactions (described later). Most microbes are engulfed by polymorphonuclear leukocytes, which release lysosomal enzymes. These enzymes destroy microbes and activate complement, which, in turn, causes mast cells to release **histamine.** Histamine diffuses to and dilates capillaries and venules and makes their walls more permeable. Dilation increases the blood flow and delivers clotting factors and nutrients to the injured area and removes wastes from it. The extra blood makes the skin red and hot. Increasing permeability allows clotting factors and excess fluid to seep from the blood toward the injured cells, where the fluid causes swelling. Histamine also causes the runny nose and red, watery eyes of hay fever and the itching of insect bites.

Tissue injury activates small peptides called **kinins,** found in blood in inactive form, and activates clotting factors if there is bleeding. Kinins further increase blood flow and vessel permeability and attract phagocytes to the injured tissue. Clotting factors stop bleeding by forming a clot where blood vessels are severed. The clot dries and becomes a **scab.** Clots also form in tissue fluids, walling off the injured area from the rest of the body.

Inflamed tissues release **leukocytosis-promoting factor** (LP factor), which causes leukocytosis, an increase in the number of blood leukocytes. Many leukocytes leave capillaries by **diapedesis** (di"ah-ped-e'sis), or passing between cells in capillary walls, and move toward chemical attractants in injured tissues. Leukocytes, especially avidly phagocytic neutrophils, congregate in tissue fluids in this manner.

Pain after tissue injury may be due to activation of **bradykinin** and other kinins, which also dilate blood vessels. In one experiment, volunteers received an injection of bradykinin or saline, and neither they nor those giving the injections knew which substance was used. Those receiving bradykinin reported more pain than those receiving saline. How bradykinin stimulates pain receptors in the skin is unknown, but **prostaglandins** seem to intensify its effect. Pain protects a body part by minimizing its use during healing. Aspirin relieves pain by interfering

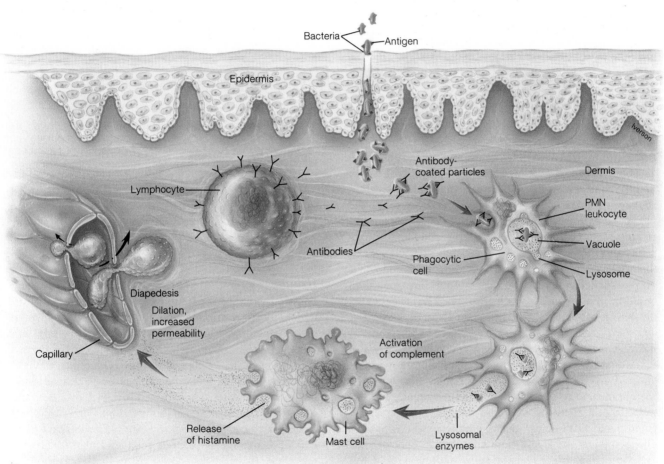

FIGURE 21.1
Steps in the processes of inflammation and healing.

with prostaglandin synthesis, but no pain relievers have been found to interfere with the bradykinin release or action.

Anti-inflammatory drugs act in several ways. Antihistamines, which lessen cold and allergy symptoms, block histamine receptors on cell membranes. Keeping histamine from reaching such receptors prevents symptoms such as runny nose, watery eyes, and itching. Like aspirin, corticosteroids used to treat severe inflammation block prostaglandin synthesis but they can have undesirable side effects. Nonsteroidal anti-inflammatory drugs (NSAIDs) also inhibit prostaglandin synthesis, but they probably also prevent neutrophils from releasing substances that destroy cartilage in rheumatoid arthritis and prevent phagocytes from releasing lysosomal enzymes and superoxides that damage tissue.

Phagocytosis

Phagocytosis (Chapter 3) is an important component of most inflammatory reactions, and phagocytes are attracted to injured tissues by kinins and other substances as noted above. The phagocytes of human tissues are mainly neutrophils, monocytes, and macrophages. **Macrophages** (mak′ro-fāj-ez), monocytes that have left the blood and entered the tissues, have different names in different tissues (Table 21.1). Neutrophils phagocytize bacteria and other small particles, and monocytes and macrophages phagocytize parts of dead cells and other larger particles—even neutrophils that have engulfed bacteria. Eosinophils, which also may be phagocytic, defend against certain parasitic infections.

Phagocytes reach foreign substances by **chemotaxis,** movement along a concentration gradient toward a chemical attractant. The phagocytic cell membrane forms a vacuole around the object and lysosomes fuse with the membrane and release enzymes into the vacuole. The enzymes digest microbes into small molecules (amino acids, sugars, and fatty acids) that the phagocyte can use for energy to find and digest more microbes. Phagocytes use large amounts of oxygen to form microbe-killing hydrogen peroxide, superoxide (O_2^-), and hypochlorite. Hypochlorite is the ingredient in household bleach that kills microorganisms. Some attractants act in blood, causing neutrophil membranes to stiffen and trapping these phagocytes in inflamed lungs and other organs.

TABLE 21.1   Names of Macrophages in Various Tissues

Name of Macrophage	Tissue
Alveolar macrophage (dust cell)	Lung
Histiocyte	Connective tissue
Kupffer's cell	Liver
Microglial cell	Neural tissue
Osteoclast	Bone
Sinusoidal lining cell	Spleen

During phagocytosis, many phagocytes themselves die as do cells of injured tissues. The accumulation of dead phagocytes, materials they have ingested, and tissue debris forms the white or yellow fluid called **pus.** Pus continues to form until the infection or tissue damage has been brought under control.

Infectious microbes battle phagocytes in various ways. Bacteria that cause tuberculosis make a waxy capsule and some that cause pneumonia are coated with protein or polysaccharide. Some bacteria and most viruses are not destroyed by phagocytosis, probably because they inactivate lysosomal enzymes. Viruses profit by being phagocytized—they use cell nutrients to multiply, safe from other defenses. Whether microbes cause disease depends on who wins the battle—the phagocyte or the microbe.

Healing

During the entire inflammatory reaction, healing is also underway. It is stimulated by a variety of polypeptide growth factors including growth hormone, insulin, and factors that cause particular tissues such as epidermis, cartilage, and nerve tissues to grow. Once inflammation has subsided and tissue debris has been cleared away, healing accelerates. Capillaries grow as blood clots dissolve. New epidermis replaces what was destroyed both on skin surfaces and in organ linings. **Fibroblasts** (fi′bro-blastz) proliferate and secrete collagen and other fibers to form **scar tissue,** a connective tissue that replaces nerve and muscle cells that cannot divide. Scar tissue neither contracts nor carries signals, but it forms a durable "patch" that allows normal tissue to function. Large scars can be disfiguring so the edges of deep cuts are often stitched together to reduce the quantity of scar tissue formed.

Several factors affect the healing process. The tissues of young people heal more rapidly than those of older people. The cells of young people divide more quickly, their bodies are generally in a better nutritional state, and their blood circulation is more efficient. Good circulation is important because blood contains many factors that facilitate healing. Certain vitamins also are important—vitamin A for epithelial cell division, vitamin C for collagen synthesis, and vitamin K for blood clotting. Vitamin E also may promote healing and prevent formation of excessively large, disfiguring scars. Growth hormone stimulates cell division, but it also causes macrophages to produce more than twice as much bacteria-killing superoxide.

Fever

Fever acts as a nonspecific defense against microorganisms (Chapter 25). It makes a person feel ill enough to rest, stimulates certain immune responses, and increases the rate of chemical reactions including those that inactivate or kill disease-causing microbes. Physicians who know the benefits of fever let it run its course unless it goes above 38.5° C (101° F) orally, or 39° C (102° F) rectally, except in patients with severe heart disease, fluid imbalances, or a risk of convulsions. They caution parents not to give aspirin to young children with a viral disease because it may increase the risk of **Reye's** (rīz) **syndrome.** This syndrome increases blood ammonia levels and causes vomiting, liver damage, deep coma, and cerebral edema. Unless aggressively treated, such edema causes permanent, usually fatal, brain damage.

Peritonitis, an inflammation of the peritoneum and abdominal cavity, is still a common infection in spite of the availability of antibiotics. Microorganisms usually reach the peritoneal cavity during surgery. They are spread to the lymph and blood by contraction and relaxation of the diaphragm, which sucks them through small openings in its epithelial lining into lymph vessels. Macrophages, various white blood cells, and mast cells in the abdominal cavity produce inflammation as they do in other parts of the body.

Blood usually does not clot in the abdominal cavity during surgery because cells of the peritoneum itself release plasminogen activators. After surgical or other trauma and in ischemia and infection, plasminogen activator decreases and thromboplastin increases. This enhances clotting and leads to fibrin adhesions (sticking together of organ surfaces).

In spite of the body's defenses, peritonitis is difficult to treat and poses a great risk of spreading throughout the body. For these reasons, surgical teams should take every precaution to prevent such infections.

Interferon

When viruses infect cells, they take over the cell's DNA, causing it to make viral nucleic acids and proteins. These products are used to assemble new viruses. As early as the 1930s, infection with one virus was known to prevent infection by another virus at least for a time. Several decades later, virus-infected cells were shown to release a small protein now called **interferon** (in″ter-fe′ron) that prevented infection of adjacent cells. Efforts to purify interferon led to the discovery of related molecules among different species and even among different tissues in the same species. Among human cells, leukocytes make *alpha interferon,* fibroblasts make *beta interferon,* and lymphocytes (T cells) make *gamma interferon.* Fever combats viral infections by increasing interferon production.

Once released, interferon binds to surface receptors on adjacent cells, where a single molecule stimulates a cell to make many molecules of a substance called **antiviral protein.** When viruses enter such cells, antiviral protein prevents the virus from making either nucleic acid or protein. The actions of interferon and antiviral protein are summarized in Figure 21.2.

The ability to make interferon has led to a search for therapeutic uses. It was first made in the laboratory in human fibroblast and white blood cell cultures. Improvements in recombinant DNA techniques in bacteria have made it much more plentiful and much less expensive. For certain bone cancers, after most malignant tissue is surgically removed or destroyed by radiation, interferon reduces the incidence of metastasis (spread of cancer cells). It also is beneficial in kidney cell cancer and melanoma (a skin cancer that is difficult to treat), but not in lung, breast, or colon cancer. Interferon delays the onset of influenza symptoms and sometimes lessens the severity of genital herpes. It inhibits cold viruses, but causes irritation and nasal bleeding.

Complement

The **complement system,** known simply as **complement,** is a set of over 20 proteins that circulate in the plasma in inactive form and account for about 10 percent (by weight) of plasma proteins. Complement is so named because it complements, or completes, some immunologic reactions. It also enhances phagocytosis, produces inflammation, and breaks down some microorganisms. It exerts these nonspecific effects on any microbe and acts long before specific immune reactions can occur.

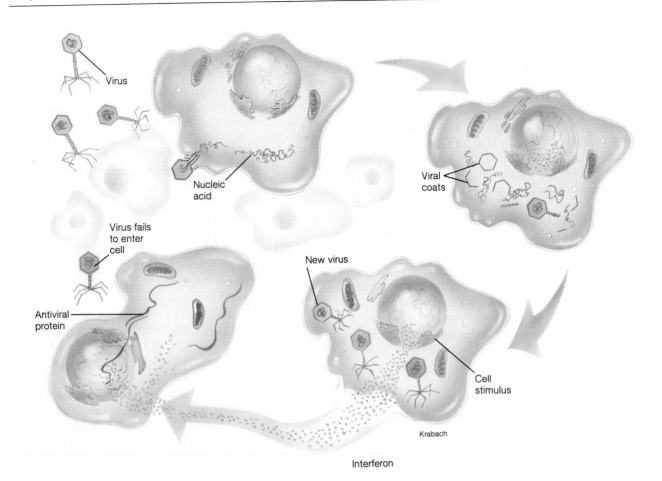

Virus

Nucleic acid

Viral coats

Virus fails to enter cell

New virus

Antiviral protein

Cell stimulus

Krabach

Interferon

FIGURE 21.2
Interferon and antiviral protein.

Two complement reaction pathways have been identified (Figure 21.3). Complement proteins C1 through C9 (C stands for complement) of the complement system participate in the **classic pathway.** Properdin, factor B, and factor D replace C1, C2, and C4 in the **properdin** (pro′perdin) **pathway.** Because C3 and C5 through C9 are common to both pathways, the effects of the complement system are the same regardless of the pathway by which C3 is produced.

Various reaction products contribute to inflammation by increasing blood vessel permeability, stimulating chemotaxis, and causing mast cells to release histamine. Some bind to the surface of microorganisms already coated with antibodies from specific immune reactions, causing them to be phagocytized. The C5 through C9 reaction sequence, known as the **membrane attack complex,** directly damages cell membranes so the cell's contents leak out (Figure 21.4). This process is called **immune cytolysis** (si-tol′is-is).

Advantages of complement as a defense mechanism are that a very small amount of activating substance can start the cascade and the cascade occurs very rapidly. A disadvantage is that a deficiency of one protein impairs the whole process.

SPECIFIC DEFENSES—IMMUNITY

As an infant, you probably received vaccines—substances that made you immune to diseases such as diphtheria, tetanus, whooping cough, poliomyelitis (polio), measles, German measles, and mumps. Your parents probably developed immunity by having some of these diseases. Either receiving a vaccine or having a disease can confer specific immunity to that disease.

Immune System Components

Functions of the immune system are carried out by **B lymphocytes (B cells)** and **T lymphocytes (T cells),** which originate in bone marrow from stem cells and differentiate as

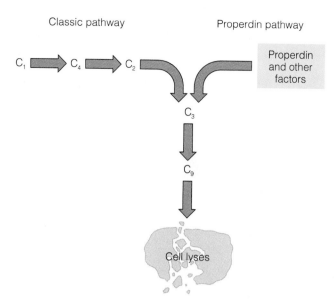

FIGURE 21.3
Pathways of complement activation.

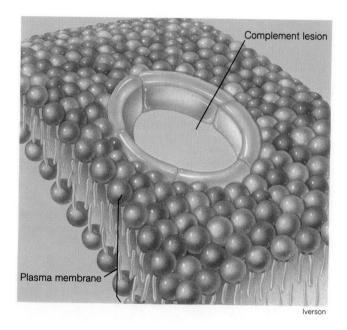

FIGURE 21.4
Immune cytolysis.

shown in Figure 21.5. Characteristics of these cells are summarized in Table 21.2.

Differentiation of B lymphocytes was first studied in birds where it occurs in the **bursa of Fabricius** (fab-ris′e-us). Though humans have no such bursa, they do have B cells and must have bursal-equivalent tissue—probably bone marrow or lymphoid tissues in the digestive tract. B cells are found in all lymphoid tissues—lymph nodes, spleen, tonsils, adenoids, appendix, and Peyer's patches of the small intestine. They also make up about one-fourth of blood lymphocytes.

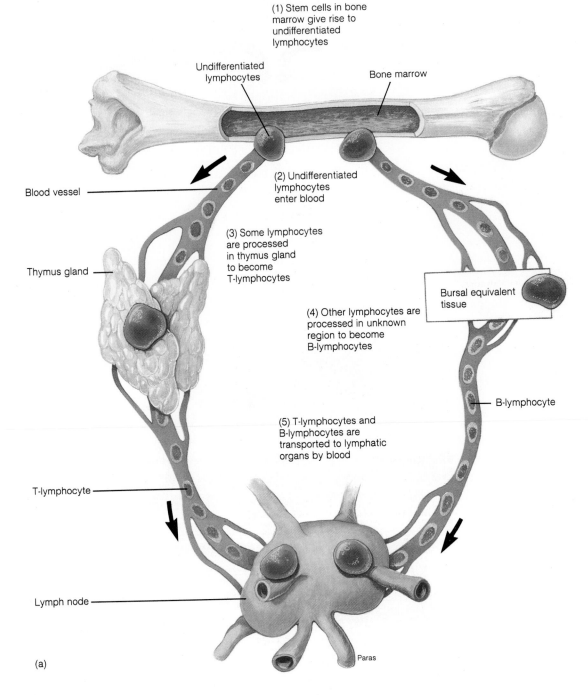

(1) Stem cells in bone
marrow give rise to
undifferentiated
lymphocytes

Undifferentiated
lymphocytes

Bone marrow

Blood vessel

(2) Undifferentiated
lymphocytes
enter blood

(3) Some lymphocytes
are processed
in thymus gland
to become
T-lymphocytes

Thymus gland

Bursal equivalent
tissue

(4) Other lymphocytes are
processed in unknown
region to become
B-lymphocytes

B-lymphocyte

(5) T-lymphocytes and
B-lymphocytes are
transported to lymphatic
organs by blood

T-lymphocyte

Lymph node

Paras

(a)

FIGURE 21.5
Differentiation of lymphocytes: (*a*) sites at which differentiation
occurs, and (*b*) steps in differentiation.

TABLE 21.2   Characteristics of B Cells and T Cells

	B cells	T cells
Site of production	Bursal-equivalent tissues	Thymus or elsewhere under the influence of thymic hormones
Type of immunity	Humoral	Cell-mediated and assist humoral
Subpopulations	Plasma cells and memory cells	Cytotoxic, helper, suppressor, delayed hypersensitivity, and memory cells
Antigen receptors	Yes	Yes

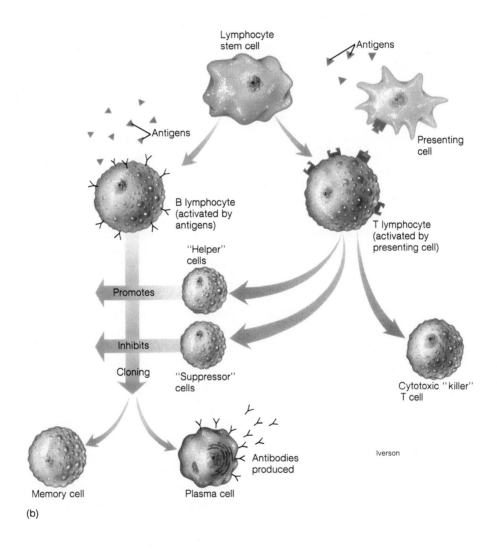

(b)

	B cells	T cells
Surface antibodies	Yes	No
Secretory product	Antibodies	Lymphokines
Percent of leukocytes		
Peripheral blood	15–30	55–75
Lymph nodes	20	75
Bone marrow	75	10
Thymus	10	75

T lymphocytes differentiate in the **thymus gland,** which begins to process lymphocytes around the time of birth and releases them into the blood as T cells. It also secretes **thymosin,** which stimulates lymph nodes and other tissues to make lymphocytes, and other hormones that cause lymphocytes to become T cells. After puberty, T cells are thought to form in bone marrow. They are found wherever B cells are found and make up about three-fourths of blood lymphocytes. T cells further differentiate into (1) cytotoxic (killer) T cells, (2) delayed type hypersensitivity T cells, (3) helper T cells, and (4) suppressor T cells, all of which are found in lymphatic tissues and blood.

Dual Nature of the Immune System

Specific immunity can be humoral or cell-mediated, and many foreign substances trigger both kinds of responses. **Humoral** (hu'mor-al) **immunity** results from antibodies mainly in blood. When stimulated by an antigen, B lymphocytes divide to form plasma cells, which, in turn, release antibodies. Humoral immunity destroys bacterial toxins, bacteria, and viruses (before they enter cells). **Cell-mediated immunity** results from T cell actions at the cellular level, especially where antigens are inside cells or are otherwise inaccessible to antibodies. It kills virus-infected cells and probably destroys fungi, other eukaryotic parasites, cancer, and foreign tissues, including transplanted organs.

> Describing blood-borne immunity as *humoral* comes from a time when blood, bile, and other body fluids were called humors and people thought illnesses were due to imbalances in humors. Our current use of the expressions "good humor" or "bad humor" is derived from this mistaken idea about body fluids.

Immune System Molecules

Immunity is a specific defense, in part, because of the nature of the molecules that participate in it. Molecules of special importance include antigens, major histocompatibility complex proteins, and antibodies.

Antigens

Immunity develops in response to **antigens,** substances that the immune system identifies as foreign. Most antigens are large complex proteins with molecular weights greater than 10,000. Some are polysaccharides, and a few are glycoproteins (carbohydrate and protein) or nucleic acids.

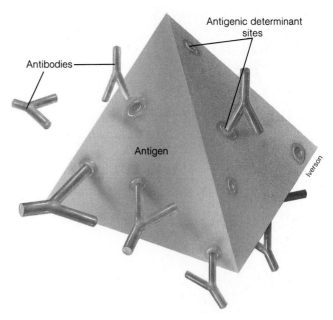

FIGURE 21.6
Antigenic determinants.

Every antigen has at least one **antigenic determinant,** a site to which antibodies can bind, and large antigens can have several such sites. Certain small molecules called **haptens** (hap'tenz) bind to a large protein molecule where they become antigenic determinants (Figure 21.6). Neither the hapten nor the protein alone is antigenic, but together they are. When penicillin molecules (haptens) bind to protein molecules, they can act as antigens. The body's response to such antigens can be a severe allergic reaction.

Major Histocompatibility Complex Proteins

Cells have many naturally occurring surface proteins that can behave as antigens. Only identical twins have identical cell surface proteins, but human cells have proteins not found in other species and family members have more common proteins than unrelated individuals. Of these proteins, **major histocompatibility** (his"to-kom-pat'ib-il'it-e) **complex** (MHC) **proteins** account for differences in surface antigens.

MHC proteins on cell surfaces are important in immune reactions. Macrophages and certain other cells can present antigens to B or T cells. Such a cell is called an **antigen presenting cell** (APC). When APCs ingest and digest microbes, they process microbial antigens, combine them with MHC proteins, and insert the MHC-microbial antigen complex in their membranes. APCs can display

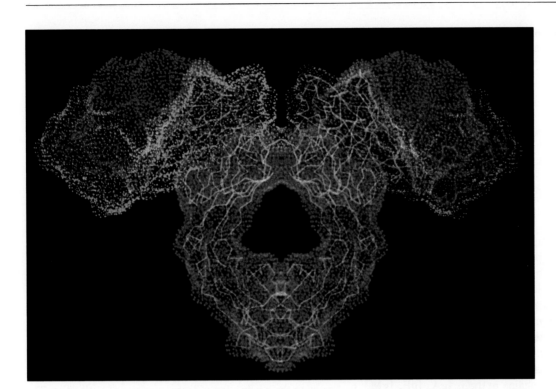

FIGURE 21.7
Antibody structure.

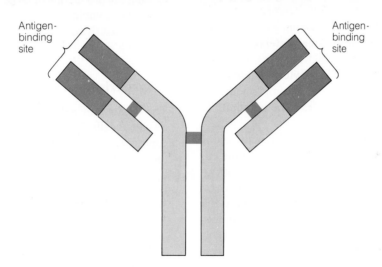

Class I or Class II proteins. Cytotoxic T cells also have Class I MHC proteins and helper T cells have Class II MHC proteins. Immune reactions occur when matching MHC proteins combine as an APC presents an antigen.

Antibodies (Immunoglobulins)

One way the immune system responds to a foreign substance is by making **antibodies,** proteins that can bind specifically to the antigenic determinant of an antigen.

Antibodies also are called **immunoglobulins** (im″u-no-glob′u-linz), or Ig. An antibody consists of two pairs of polypeptide chains—short **light** (L) **chains** and longer **heavy** (H) **chains** (Figure 21.7)—held together by disulfide bonds. Each antibody has a pair of identical variable regions, or antigen-binding sites, at the adjacent ends of the L and H chains. The variable regions of one kind of antibody are different from those of any other kind of antibody. H chain tails have another binding site that can attach to a cell membrane.

Properties	Classes of Immunoglobulins				
	IgG	*IgA*	*IgE*	*IgD*	*IgM*
Number of units	1	1 or 2	1	1	5
Percent of total blood antibodies	75–85	5–15	0.5	0.2	5–10
Activation of complement	Yes	Yes	No	Yes	Yes
Crosses placenta	Yes	No	No	No	No
Binds to phagocytes	Yes	Yes	No	?	No
Binds to lymphocytes	Yes	Yes	Yes	?	Yes
Binds to mast cells and basophils	No	No	Yes	?	No

TABLE 21.3 Properties of Immunoglobulins

How a cell has enough genetic information to make billions of different antibodies, each with a unique variable region, has puzzled biologists. One way might be that many short pieces (like letters of the alphabet) of DNA each specify a small segment of a variable region, while a few longer pieces specify nonvariable regions. Then antibodies might be assembled from different assortments of segments, like words are assembled from letters.

Five immunoglobulin classes have been defined according to distinctive H chains as IgG, IgA, IgE, IgM, and IgD (Table 21.3).

The most important blood immunoglobulin, **IgG**, makes up about 20 percent of plasma proteins. Variable regions of IgGs attach to microbial antigens and their tissue-binding sites attach to phagocytes, thereby fostering engulfment. Most IgG activates complement and crosses the placenta, where it provides some immunity to a newborn baby.

IgA is found in tears, milk, saliva, and mucus, and is present on all membranes that line tracts open to the environment (digestive, respiratory, and genitourinary). Having a molecular structure that resists proteolytic enzyme action, IgA binds microbial antigens before the microbes invade tissues. It activates complement, but does not cross the placenta. Being abundant in **colostrum** (kol-os'trum), the fluid secreted by mammary glands for a few days after delivery, IgA provides an infant with some immunity.

Among other Igs, **IgE** binds to receptors on basophils and mast cells in body fluids and skin. Simultaneous binding of IgE to antigens, such as pollens, drugs, or foods, causes mast cells to release histamine and other substances that elicit allergy symptoms. Made by both B cells and plasma cells, **IgM** consists of 5 units connected by H chains with 10 antigen-binding sites. It binds antigens and clumps microbes early in infections and activates complement. **IgD** is found mainly on B cell membranes, rarely secreted, and has no known function.

HUMORAL IMMUNITY

Humoral immunity is most effective against bacterial infections, but can act in blood to neutralize toxins and viruses. Its ability to protect the body depends mainly on recognizing antigens of disease-causing microbes while they are in blood or body fluids. Humoral immunity is mediated by B lymphocytes that respond to a specific antigen, such as on the membrane of a disease organism, by causing the production of large numbers of antibodies. The antibodies combine with an antigen like a key fits in a lock, thereby inactivating the antigen and, in most instances, initiating events that destroy the organisms.

Characteristics of Humoral Immunity

The events that give rise to humoral immunity typically involve B cells, T cells, and macrophages. Each B cell has copies of the antibody that it is programmed to make, extending from its membrane surface. It recognizes an antigen by a molecular shape that can bind to the antibody. In the case of a large antigen with numerous antigenic determinants, B cell antibodies bind the antigen immediately. Antigens with fewer antigenic determinants, found on most infectious organisms, require processing by macrophages before they can bind to B cells.

During processing, macrophages typically phagocytize and degrade the organism and insert some of its antigens into their own cell membranes adjacent to their MHC proteins. Helper T cells with the same MHC proteins and receptors for the foreign antigen bind to the macrophage. Together, the macrophage and helper T cell present the processed antigen to the B cell. The macrophage also releases **lymphokines** (lim'fo-kīnz), chemical substances that trigger certain immunologic reactions. The presentation together with the lymphokines activates the B cell. These events are summarized in Figure 21.8. Presenting the processed antigen to a B cell initiates humoral immunity. Presenting it to a cytotoxic T cell, initiates cell-mediated immunity as we shall see later.

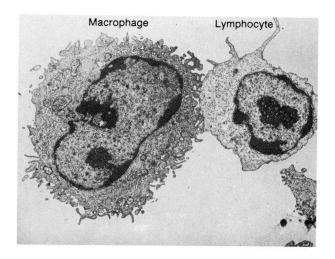

Macrophage   Lymphocyte

FIGURE 21.8
How macrophages (or other antigen-presenting cells) process and present antigens to helper T cells.

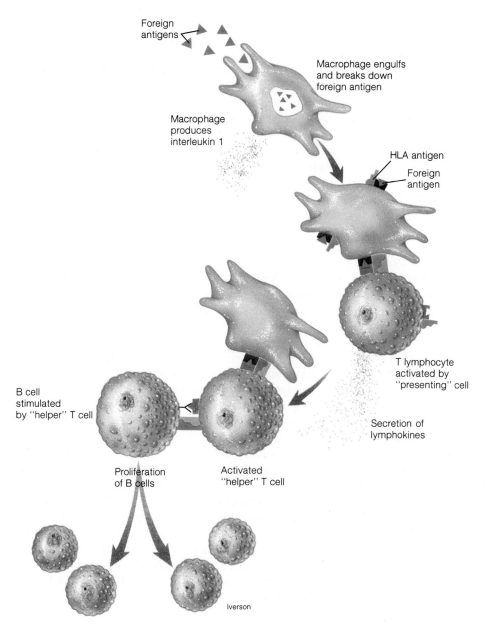

Foreign antigens

Macrophage engulfs and breaks down foreign antigen

Macrophage produces interleukin 1

HLA antigen

Foreign antigen

T lymphocyte activated by "presenting" cell

Secretion of lymphokines

B cell stimulated by "helper" T cell

Proliferation of B cells

Activated "helper" T cell

Iverson

FIGURE 21.9
Proliferation of activated B cells.

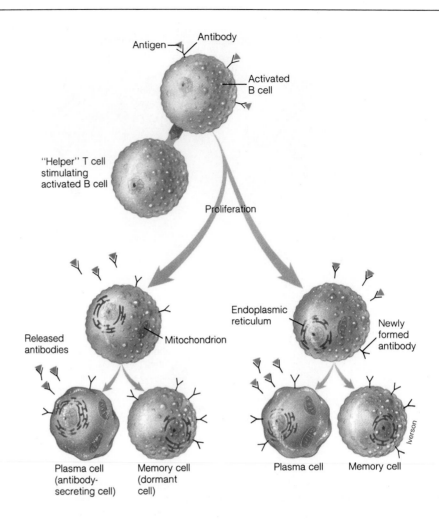

Activated B cells divide repeatedly to form a **clone** (klōn), a group of cells identical to the parent cell—in this case, large lymphocytes all with the same antibody on their surfaces (Figure 21.9). Most of the lymphocytes become **plasma cells,** which synthesize and release antibodies like those on their membranes at a rate of up to 2000 per second. As plasma cells die, B cells divide to replace them so antibodies are released as long as the antigen is present. Some of the lymphocytes become **memory cells,** which remain dormant until activated by later exposure to the antigen.

Suppressor T cells can block helper T cells, and they can terminate antigen-antibody reactions. Helper and suppressor T cells secrete helper and suppressor factors, respectively. Whether B cells are helped or suppressed seems to depend on the relative amounts of the factors present at a reaction site.

The formation of antigen-antibody complexes is the main means by which humoral immunity inactivates infectious agents. During the first week of an infection,

helper T cells foster growth and differentiation of plasma cells. This process subsides as suppressor T cells inhibit further antibody production.

Primary and Secondary Responses

In humoral immunity, recognition of an antigen initiates the **primary response.** B cells release IgM directly and form plasma cells that release IgG. Blood antibodies increase over a period of 1 to 10 weeks with IgM more abundant at first and IgG more abundant later. Memory cells that were formed during the primary response survive for years in lymphoid tissues ready to recognize a particular antigen.

Recognition of an antigen by memory cells initiates a **secondary response.** Some memory cells divide to produce B cells and plasma cells, and others remain as memory cells. Less IgM is produced, and it is produced over a shorter period; and more IgG is produced sooner than in the primary response. The primary and secondary responses are compared in Figure 21.10. As we shall see later,

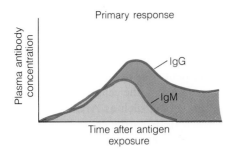

Primary response

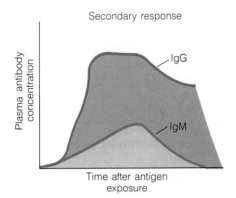

Secondary response

FIGURE 21.10
Primary and secondary responses to an antigen in humoral immunity.

vaccines used to immunize people against specific diseases cause a primary response without causing disease. Subsequent secondary responses prevent the person from having the disease.

Monoclonal (mon″o-klo′nal) **antibodies** consist of a clone of identical antibodies made in the laboratory by **hybridomas** (hi″brid-o′maz), cells containing genetic information from myeloma cells (malignant cells of the immune system) and lymphocytes (Figure 21.11). Hybridomas divide repeatedly and make large quantities of an antibody.

First made in 1975 by researchers Cesar Milstein and Georges Kohler, the ability of monoclonal antibodies to identify specific antigens has practical applications. Certain antibodies detect pregnancy only 10 days after conception, and others quickly diagnose several serious viral diseases, now including AIDS. They may soon identify cancer cells before they grow into tumors and deliver drugs or radioactive substances to destroy the cells.

Molecules called *catalytic antibodies* now exist, at least in the laboratory. Like other antibodies, they react with an antigen; but as an enzyme, they also chemically rearrange the antigen. Catalytic antibodies may someday inactivate toxins, cut through the protein coats of viruses, or cure genetic disorders by repairing defective DNA.

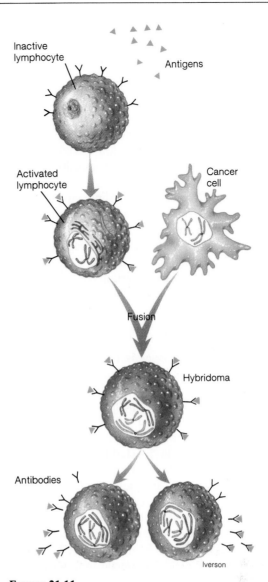

FIGURE 21.11
How monoclonal antibodies can be made: inactive lymphocytes, as from a patient's blood sample, are activated by exposure to an antigen. The active lymphocytes are fused with cancer cells to form hybridomas. The hybridomas divide rapidly and give rise to many cells that produce monoclonal antibodies.

CELL-MEDIATED IMMUNITY

In cell-mediated immunity, T cells interact directly with cells displaying foreign antigens. Such interactions rid the body of cells infected by viruses and certain other pathogens, and reject tumor cells. Sometimes, they also cause allergic reactions and rejection of intentionally transplanted tissues.

Characteristics of Cell-Mediated Immunity

Several types of T cells and various lymphokines participate in cell-mediated immunity. The cell-mediated immune reaction often begins with a macrophage processing an antigen, typically from a virus or malignant cell. The macrophage places the antigen on its surface adjacent to MHC class I proteins. Binding of a cytotoxic T cell with the same MHC protein and a receptor for the antigen to the macrophage initiates a cell-mediated immune reaction.

Finding out how T cells recognize antigens has posed a problem for immunologists. It now appears that a T cell responds to an antigen fragment 10 to 20 amino acids long located in a groove in an MHC protein. Recognition is by chemical properties, not by shape as in B cells, and particular MHC proteins present certain amino acid sequences. Learning more about which fragments trigger immune responses will improve our understanding of immune reactions and help with the practical matter of making vaccines.

Binding with a macrophage initiates T cell division and differentiation, including memory cell formation (Figure 21.12). Each cell is sensitized to the antigen that initiated the process, and each type has a different function. Some cells act directly and others release lymphokines. Macrophages that have processed an antigen secrete the lymphokine, **interleukin 1** (IL-1), which activates **helper T cells.** Helper T cells, in turn, secrete lymphokines such as **interleukin 2** (IL-2) and gamma interferon. IL-1 and IL-2 activate **suppressor T cells, delayed type hypersensitivity T cells,** and **cytotoxic (killer) T cells.** Interferon and interleukins together cause null cells to become **natural killer cells.** As in humoral immunity, some T memory cells form and persist, ready to respond to familiar antigens.

Each type of T cell has a different function. Helper and suppressor T cells modify humoral immunity as described earlier. Delayed-type hypersensitivity T cells participate in a kind of allergic reaction to be described later, and they release various lymphokines. Among those lymphokines, *macrophage chemotactic factor* helps macrophages find microbes, *macrophage activating factor* stimulates phagocytes to engulf microbes, *migration inhibiting factor* prevents macrophages from leaving infection sites, and *macrophage aggregation factor* causes macrophages to congregate at such sites. Cytotoxic T cells kill mainly virus-infected cells, and natural killer cells kill mainly malignant and transplanted cells. Natural killer cells must bind to a cell that contains a foreign antigen before they can act, but they need not have had previous exposure to the foreign antigen.

When microbes evade humoral immunity and invade cells, they cause persistent infections unless a cell-mediated immune reaction can destroy the cells and the organisms in them. A T cell infection, such as AIDS, is especially devastating because it destroys the very cells that might have combatted the infection.

How Killer Cells Kill

Both cytotoxic and natural killer cells make a lethal protein and fire it at target cells. They contain granules of a lethal protein **perforin** (per'fo-rin), which they release when they bind to a target cell. Perforin bores holes in the target cell membranes so essential molecules leak out and the cells die. By killing infected cells while they are few in number, cytotoxic T cells prevent the spread of infection, but at the expense of destroying host cells. Natural killer cells similarly destroy malignant cells before they can multiply. Both kinds of killer cells can act repeatedly to destroy many infected cells.

Perforin does not damage healthy cells because cells that can release it do so only after binding to infected or malignant cells. Why perforin does not attack membranes of cells that make it is not known, but such cells may produce a protein that inactivates perforin.

Some disease-causing agents, especially parasites and fungi, have their own lethal proteins that help them invade and kill human cells. Learning more about these proteins may lead to ways of treating infections by blocking such proteins. Conversely, learning how to enhance the action of the body's own lethal proteins might increase the chance that malignant cells and AIDS infected cells would be killed while few in number.

That skin functions in nonspecific defenses is well known, but its role in specific immunity has been only recently recognized. Three kinds of cells—*Langerhans cells, Grandstein* (grand'stin) *cells,* and *keratinocytes* (ker-at-in'o-sītz)—function in specific immune reactions in the skin. When an antigen evades surface barriers and enters the epidermis, Langerhans or Grandstein cells can bind it, process it, and present it to a T cell. Langerhans cells present antigens to helper T cells and Grandstein cells present them to suppressor T cells. Langerhans cells, being more active than Grandstein cells, usually cause helper T cells to inactivate antigens. Grandstein cells probably keep Langerhans cell activities in check. Keratinocytes, long known to contain keratin, also produce lymphokines, which stimulate proliferation of T cells that attack antigens in the skin.

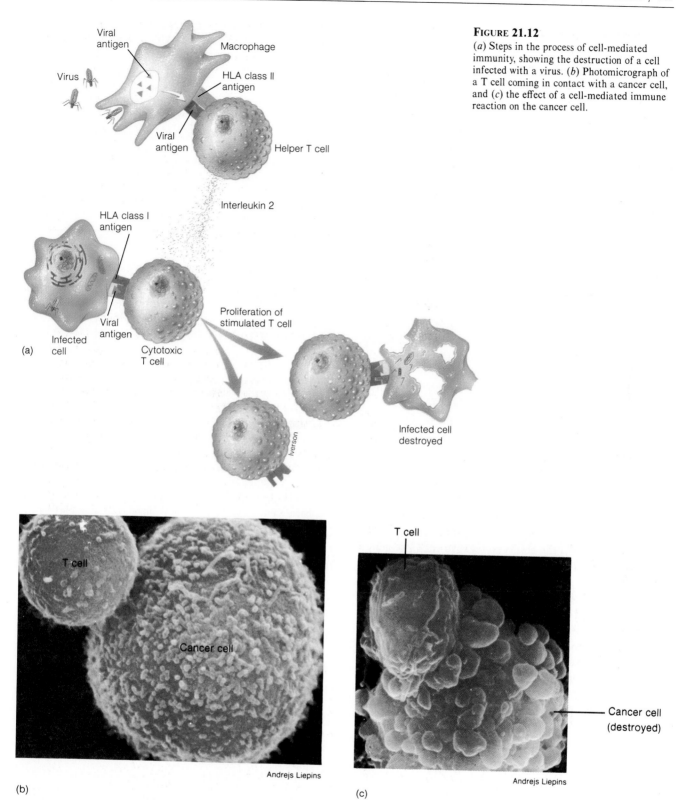

(a)

(b)

(c)

Andrejs Liepins

Andrejs Liepins

T cell

Cancer cell

T cell

Cancer cell
(destroyed)

Viral
antigen

Macrophage

Virus

HLA class II
antigen

Viral
antigen

Helper T cell

Interleukin 2

HLA class I
antigen

Proliferation of
stimulated T cell

Infected
cell

Viral
antigen

Cytotoxic
T cell

Infected cell
destroyed

FIGURE 21.12

(*a*) Steps in the process of cell-mediated immunity, showing the destruction of a cell infected with a virus. (*b*) Photomicrograph of a T cell coming in contact with a cancer cell, and (*c*) the effect of a cell-mediated immune reaction on the cancer cell.

GENERAL PROPERTIES OF IMMUNITY

Immunity, whether humoral or cell-mediated, has the following properties: (1) recognition of self versus nonself, (2) specificity, (3) heterogeneity, and (4) memory (Table 21.4).

The immune system must distinguish between **self** (part of the body) and **nonself** (foreign substances). Otherwise, it might destroy as many normal cells as foreign ones. Immunologists are still studying how immune cells make this distinction. According to the **clonal selection theory,** proposed by Frank Macfarlane Burnet in 1957, each lymphocyte in an embryo can detect only one substance, and lymphocytes responding to a normal body protein are quickly killed. This leads to **tolerance,** or failure to destroy self. Lymphocytes that detect foreign substances remain functional.

Distinguishing between self and nonself is now believed to be a lifelong process carried out by many kinds of APCs and T cells "educated" in the thymus to recognize an antigen. As a part of their education, the T cells learn to destroy foreign antigens and to ignore body proteins. APCs regularly insert MHC proteins and antigen fragments on their surfaces. Some antigen fragments are derived from normal cellular proteins whereas others are foreign. T cells with the same MHC protein as the APC temporarily bind to it. They combine sequentially and at random with many different APCs, creating a kind of immune surveillance in which a cell-mediated reaction occurs only when a foreign antigen is detected.

By age two or three, when the immune system is mature, it not only distinguishes self and nonself, it also displays **specificity,** the ability to react to each antigen in a particular way. As we have seen, B cells recognize antigens by shape, and T cells recognize them by chemical properties of antigen fragments. This arrangement helps to assure that by one mechanism or the other, the immune system can rid the body of foreign substances without destroying the body's own proteins.

Whereas specificity refers to the immune system's ability to attack particular antigens, **heterogeneity** refers to a diverse assortment of B and T cells each capable of responding to a different antigen. Given such a lymphocyte population, it is likely that one or more will attack an antigen regardless of whether it has ever been present in the body.

TABLE 21.4   Attributes of Specific Immunity

Attribute	Description
Recognition of self and nonself	Ability to distinguish between substances naturally present in the body and substances foreign to the body
Specificity	Ability to react differently to each foreign substance
Heterogeneity	Presence of a large number of different lymphocytes, each capable of recognizing and responding to a different antigen
Memory	The ability to recognize a previously encountered antigen

In the immune system, **memory** is the ability to recognize an antigen to which it has responded previously. It allows a rapid and specific immune response to a microbe that has caused a previous infection. Original memory cells or their progeny persist for years or decades ever ready to react.

The emerging science of **psychoneuroimmunology** is concerned with discovering relationships among the body's regulatory systems, the immune system, and behavior. For example, neuroendocrine regulation of immune functions is becoming increasingly recognized. Oxytocin and neurophysin have been identified in human thymus and can be synthesized there under neural stimulation. ACTH, once thought to be synthesized only in the anterior pituitary gland, has been found in cultures of white blood cells. Relatedness of chemical substances—hormones, neurotransmitters, neuropeptides, and lymphokines—has led to the general name **information molecules** for all of them.

Studying the effects of information molecules, in turn, has led to relating mind and health. The immune system acts like a sense organ, detecting infectious organisms. White blood cells make certain hormones, which carry messages to the brain. Endorphins and other neuropeptides affect pain perception, mood, and certain aspects of behavior. Stressful events (death of a loved one, divorce, and the like) are correlated with significant alterations—suppression or enhancement—of immune function. Some patients who respond with anger and denial to the diagnosis of a fatal disease such as cancer or AIDS have an enhanced immune response. Some biologists believe we have seen only "the tip of the iceberg." They anticipate new discoveries in psychoneuroimmunology and see exciting therapeutic possibilities.

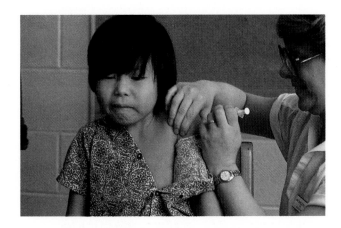

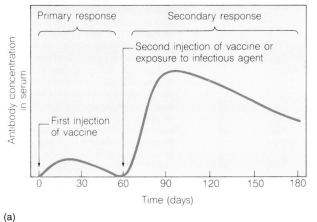

(a)

(b)

FIGURE 21.13
Ways of acquiring immunity: (*a*) artificial active immunity can be acquired from receiving a vaccine. The first vaccine produces a modest response, but a second exposure to the same antigen in vaccine or an infectious agent produces a much stronger response. (*b*) Artificial passive immunity can be acquired from breast milk and especially colostrum.

IMMUNITY AND IMMUNIZATION

Innate immunity, or **genetic immunity,** is determined by genetic factors. It includes immunity common to all members of a species, but it also accounts for individual differences. People who seem never to have an infection probably have especially effective immune systems, just as other people have particularly good eyesight, hearing, or physical stamina. **Acquired immunity** is obtained in some manner other than by heredity. It can be active or passive and naturally or artificially acquired (Figure 21.13).

Active immunity is created when the person's own immune system defends against an infectious agent and memory cells persist to act when the same antigens are encountered again. It can last from a few weeks to a lifetime, depending on how long memory cells function. **Naturally acquired active immunity** follows exposure to an infectious agent. **Artificially acquired active immunity** follows **immunization** with a vaccine, and the immune response is the same as that which occurs during a disease. Vaccines made with live, inactivated organisms generally confer longer lasting immunity than those made with dead organisms or toxins, but they also create a small risk that

TABLE 21.5	Recommended Immunizations for Normal Infants and Children in the United States	
Diseases	**Vaccine**	**Dosage Schedule**
Diphtheria	Toxoid	2, 4, and 6 months, 1½ and 4 to 6 years
Tetanus	Toxoid	Same as diphtheria, administered in DPT vaccine
Pertussis	Killed bacteria	2, 4, and 6 months, 1½ years
Poliomyelitis	Live viruses (Types I, II, and III)	2, 4, and 6 months, 1½ and 4 to 6 years
Measles	Live virus	15 months
Mumps	Live virus	15 months
Rubella	Live virus	15 months

Data from Centers for Disease Control.

TABLE 21.6	Characteristics of Types of Immunity
Type of Immunity	**Characteristics**
Innate	Inherited; present from birth; lasts throughout life; due to some inborn ability to resist infections that affect other organisms
Acquired	Not inherited; can be natural or artificial and active or passive
Naturally acquired active	Develops from a specific immune response to a foreign antigen on an infectious organism; occurs through having a disease; lasts as long as antibodies or memory cells persist
Naturally acquired passive	Results from mother's antibodies crossing the placenta or being transmitted in colostrum or milk; does not stimulate recipient to make antibodies; lasts only as long as original antibodies persist
Artificially acquired active	Develops from a specific immune response to a foreign antigen in a vaccine; stimulates immune system just as having a disease does; lasts as long as antibodies or memory cells persist
Artificially acquired passive	Results from ready-made antibodies being introduced into the body, such as in gamma globulin or other serum product; does not stimulate recipient to make antibodies; lasts only as long as original antibodies persist

the vaccine recipient will become infected. Booster shots, which elicit a secondary response, are often needed if a vaccine does not confer lifetime immunity. Immunizations recommended for normal infants and children in the United States are listed in Table 21.5.

Passive immunity is created by introducing ready-made antibodies into the body without stimulating the immune system to make antibodies. **Naturally acquired passive immunity** occurs when a mother's antibodies cross the placenta to a fetus or pass in colostrum to a newborn infant. The fetus receives IgG, which circulates in the blood, whereas the infant receives IgA, which remains in the gut and provides local immunity. **Artificially acquired passive immunity** occurs when antibodies made by others are transferred to a new host in gamma globulin or another serum product. Ready-made antibodies are destroyed by the immune system, so they confer only temporary immunity.

Types of immunity are summarized in Table 21.6.

CLINICAL APPLICATIONS

Most **immunologic disorders** are due to excessive or inappropriate immune reactions called **hypersensitivity** (hi″per-sen′sit-iv″it-e) or inadequate reactions called **immunodeficiency** (im″u-no-def-ish′en-se). Both harm, rather than protect, the body—hypersensitivity being too much of a good thing and immunodeficiency too little of a good thing.

Hypersensitivity

Immunologists recognize four types of hypersensitivity. **Immediate (Type I) hypersensitivity** includes common allergies and is mediated by IgEs. **Cytotoxic (Type II) hypersensitivity** occurs in transfusion reactions and is elicited

by cell surface antigens. **Immune complex (Type III) hypersensitivity** such as rheumatoid arthritis is due to deposition of antigen-antibody complexes and is often also an **autoimmune** (aw″to-im-ūn′) **disorder. Cell-mediated (Type IV) hypersensitivity,** or delayed hypersensitivity, such as poison ivy is mediated by delayed type hypersensitivity T cells.

Immediate (Type I) Hypersensitivity

Allergy (al′er-je) is immediate hypersensitivity to an **allergen** (al′er-jen), an ordinarily innocuous substance such as pollen, household dust containing mites and their fecal pellets, molds, antibiotics and other drugs, vaccines, and certain foods. Food allergies can develop in infants because the digestive tract remains permeable to protein molecules at least 3 months after birth. Whether such allergies develop appears to depend on how well IgA antibodies bind potential allergens before they are absorbed.

Allergies include atopy and anaphylaxis. **Atopy** (at′o-pe), which means "out of place," refers to localized allergic reactions such as skin redness, swelling, and itching. Inhaled allergens cause runny nose, watery eyes, and bronchial constriction. Ingested allergens inflame mucous membranes of the digestive tract. The common atopy hay fever plagues 20 million Americans. It can be distinguished from the common cold by the increased numbers of eosinophils in nasal secretions.

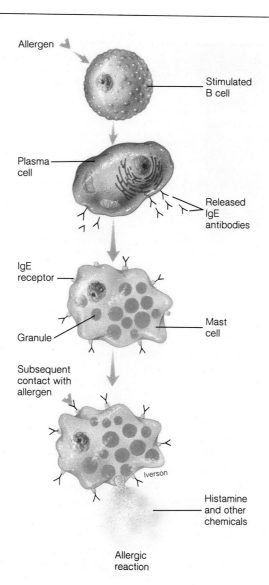

FIGURE 21.14
Immediate hypersensitivity.

Generalized **anaphylaxis** (an″af-il-aks′is) begins with sudden skin reddening, intense itching, and hives, especially on the face, chest, and palms of hands. It can progress to respiratory anaphylaxis or anaphylactic shock. Allergic asthma often causes respiratory anaphylaxis, in which the airways become so severely constricted and filled with mucus that suffocation is likely. Insect venom allergies can cause anaphylactic shock, in which blood vessels dilate and become more permeable, so blood pressure drops precipitously. Anaphylaxis can be life threatening unless immediate treatment (usually with epinephrine) to dilate respiratory passageways and increase blood pressure is available.

Allergic reactions occur after sensitization to an allergen (Figure 21.14). In **sensitization,** B cells make IgE against an allergen, but why some people become sensitized is not known. In sensitized people, IgE attaches by its H chain end to mast cell and basophil receptors, leaving allergen binding regions exposed. When the allergen binds, sensitized mast cells and basophils release histamine, prostaglandins, and other mediators of allergic reactions. Histamine dilates capillaries and makes them more permeable, causing redness and swelling. It also causes itching. Both histamine and prostaglandins contract bronchial smooth muscle, making breathing difficult. **Slow-reacting substance of anaphylaxis** (SRS-A), first found in mast cells, consists of substances 100 to 1000 times as potent as histamines and prostaglandins in eliciting allergic reactions.

A person who does not know exactly what substances cause allergic reactions can have diagnostic tests, in which very small quantities of different allergens are injected into the skin. Redness and swelling at particular sites indicate to which substances the person is allergic (Figure 21.15a).

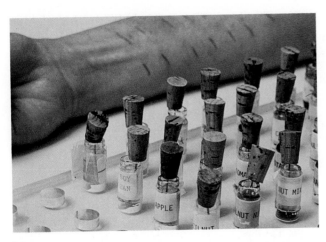
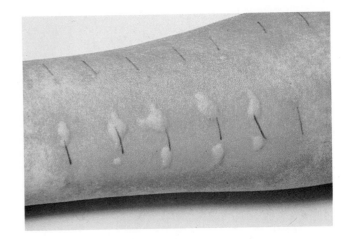

(a)

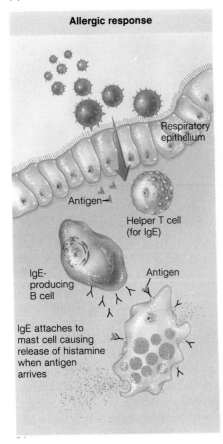
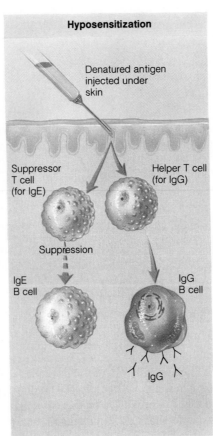
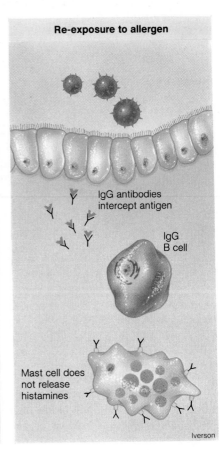

Allergic response

Respiratory epithelium

Antigen

Helper T cell (for IgE)

IgE-producing B cell

Antigen

IgE attaches to mast cell causing release of histamine when antigen arrives

Hyposensitization

Denatured antigen injected under skin

Suppressor T cell (for IgE)

Helper T cell (for IgG)

Suppression

IgE B cell

IgG B cell

IgG

Re-exposure to allergen

IgG antibodies intercept antigen

IgG B cell

Mast cell does not release histamines

Iverson

(b)

FIGURE 21.15
(a) Diagnosis (includes both pHs), and (b) treatment of allergy.

Once the allergen is identified, the allergy can be treated by **hyposensitization** (allergy shots), in which the patient receives gradually increasing doses of an allergen (Figure 21.15b). Hyposensitization is thought to elicit IgG called **blocking antibodies** that bind with the allergen before it reacts with IgE. Intended to cure an allergy by preventing release of mediators of allergic symptoms, this treatment is not always effective and can itself cause anaphylactic shock. Antihistamines alleviate symptoms caused by histamine, and corticosteroid hormones suppress the inflammatory response.

TABLE 21.7   Characteristics of ABO Blood Types

Blood Types	Genes	Agglutinogens on Cells	Agglutinins in Plasma	Plasma Agglutinates Blood Types	Cells Agglutinated by Plasma of Blood Types
A	AA, AO	A	b	B, AB	B, O
B	BB, BO	B	a	A, AB	A, O
AB	AB	AB	None	None	A, B, O
O	OO	None	a, b	A, B, AB	None

Cytotoxic (Type II) Hypersensitivity Reactions

Cytotoxic reactions are elicited by erythrocyte and other cell surface antigens. Examples include transfusion reactions and hemolytic disease of the newborn.

Transfusion Reactions   Erythrocyte membranes have surface proteins called **agglutinogens** (ag″loo-tin′o-jenz) that behave as antigens and determine a person's blood type. When antibodies called **agglutinins** (ag-loo′tin-inz) react with the antigens, as in a mismatched blood transfusion, the cells agglutinate, or clump. Genetically determined erythrocyte antigens A and B produce **ABO blood groups.** Each person has two genes from among three possibilities *A, B,* and *O. A* and *B* direct the synthesis of antigens A and B, respectively, but *O* makes neither (Table 21.7).

Antibodies, or agglutinins, for the A and B antigens are called anti-a and anti-b, respectively. Since no O antigen exists, there is no corresponding anti-o antibody. Plasma normally does not contain antibodies that can react with its antigens, but it contains antibodies against missing AB antigens (Figure 21.16).

A **transfusion reaction** occurs when a patient's blood contains matching antigens and antibodies. It is most likely when donor cell antigens bind to recipient antibodies because the large recipient blood volume (about 5 l) compared to the donor blood volume (about 0.5 l) allows great numbers of antibodies to agglutinate donor cells. The recipient's IgM antibodies cause a Type II hypersensitivity reaction against the foreign antigen. Complement is activated and erythrocytes are hemolyzed (ruptured) and agglutinated (clumped).

Symptoms of a transfusion reaction include fever, low blood pressure, and nausea, but the most serious is kidney failure. Hemolyzed erythrocytes release a substance that constricts blood vessels, especially those in the kidney. Clumps of cells block small blood vessels, interfering with the kidney's filtering of wastes from the blood. Large quantities of hemoglobin and erythrocyte fragments directly damage renal tissues.

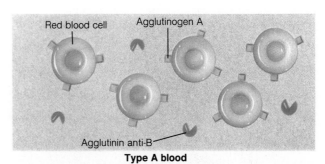

Type A blood

Type B blood

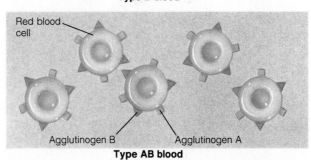

Type AB blood

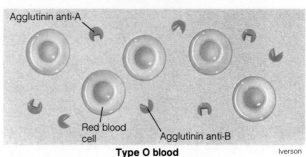

Type O blood

Iverson

FIGURE 21.16
The antigens and antibodies present in types A, B, AB, and O blood groups.

Because donor antigens are most likely to cause transfusion reactions, individuals with type O blood (which lacks A and B antigens) have been called "universal donors." Likewise, because recipient antibodies most often participate in transfusion reactions, people with type AB blood (which lacks anti-a and anti-b antibodies) have been called "universal recipients." Applying the concept of universal donors and recipients is not practical because human blood contains other antigens such as Rh, M, N, Kell, Duffy, and Lewis, which, in rare instances, cause transfusion reactions. Today, blood and blood components are carefully cross-matched before transfusion. **Cross-matching** involves mixing recipient cells with donor plasma and donor cells with recipient plasma (Figure 21.17). If no agglutination occurs in either mixture, the transfusion can proceed safely.

Hemolytic Disease of the Newborn   Blood with Rh antigens on erythrocytes is Rh-positive, and blood lacking them is Rh-negative. Anti-Rh antibodies form only after an Rh-negative person is sensitized with Rh-positive blood. Sensitization typically occurs when an Rh-negative woman carries an Rh-positive fetus that inherited its father's Rh antigen. The fetal Rh antigen leaks across the placenta during delivery, miscarriage, or abortion, and sensitizes

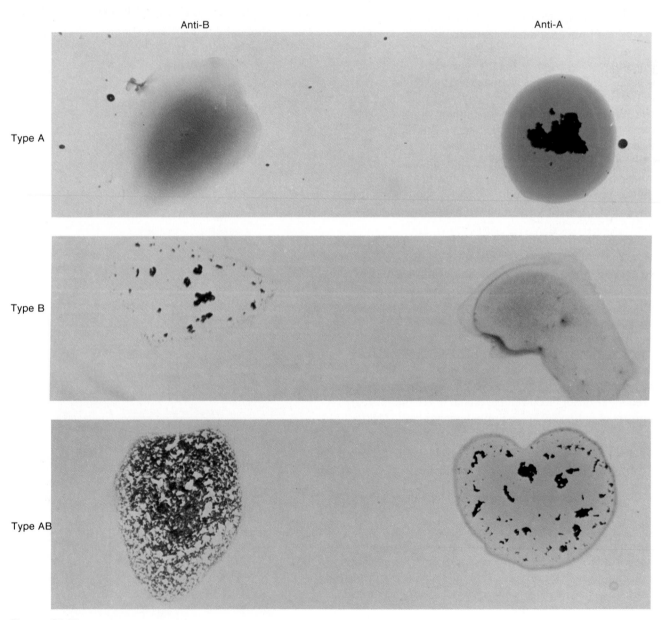

FIGURE 21.17
Effects of mixing different types of blood with known samples of antibodies. This laboratory test resembles cross-matching.

the mother (Figure 21.18a–c). The first fetus usually is not damaged because sensitization rarely occurs during pregnancy.

During a second or subsequent pregnancy with an Rh-positive fetus, the mother's sensitized immune system makes anti-Rh antibodies that cross the placenta and cause **hemolytic disease of the newborn** (Figure 21.18d–e). Fetal erythrocytes agglutinate, complement is activated, and erythrocytes are destroyed. The baby's liver and spleen enlarge as they rid the body of large numbers of damaged erythrocytes. Such babies have **jaundice** because of high blood bilirubin from erythrocyte breakdown.

Hemolytic disease of the newborn can be prevented by injecting anti-Rh antibodies into Rh-negative mothers within 72 hours after delivery. The antibodies are presumed to bind Rh antigens in the mother's blood before they sensitize her immune system. All Rh-negative women should receive such treatment after delivery, miscarriage, or abortion in case the fetus may have been Rh-positive. Before treatment became available, hemolytic disease occurred in about 0.5 percent of all pregnancies, and 12 percent of affected fetuses were stillborn.

In rare instances, a sufficient quantity of some fetal protein or erythrocyte cell surface antigen crosses the placenta, causing the mother's immune system to make antibodies against fetal tissues. If such antibodies reach the fetus, they can damage it, and may cause stillbirths and spontaneous abortions. Maternal proteins reaching the fetus do not elicit antibodies because of the immaturity of the fetal immune system.

Immune Complex (Type III) Hypersensitivity

Immune complex disorders—glomerulonephritis, rheumatoid arthritis, and systemic lupus erythematosus—result from sensitization to self, production of antibodies against cellular proteins or other substances, and deposition of antigen-antibody complexes. Such disorders are a kind of autoimmunity (discussed in more detail later). Once sensitization occurs, antibodies can bind to live cells or parts of damaged cells in blood vessel walls and other tissues. Kupffer's cells in the liver phagocytize large complexes if they are coated with both antibodies and complement, but small complexes and those lacking antibodies or complement persist.

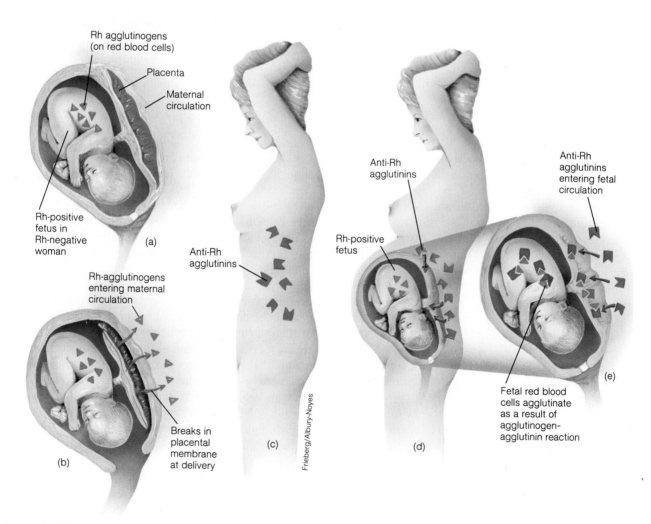

FIGURE 21.18

Hemolytic disease of the newborn: (a–c) sensitization, and (d–e) effect on second fetus.

Cell-Mediated (Type IV) Hypersensitivity

Cell-mediated hypersensitivity, unlike hypersensitivities that elicit antibodies, involves delayed-type hypersensitivity T cells. Reactions take more than 12 hours to develop and cause tissue destruction over several days. They are elicited by environmental substances (such as poison ivy), infectious agents, transplanted tissues, and the body's own malignant cells. On first exposure, the antigen binds to macrophages and sensitizes delayed-type hypersensitivity T cells. When the same antigen appears again, T cells release lymphokines that cause macrophages to ingest antigens and kill microorganisms, keep the macrophages at the site, and cause skin redness and swelling.

Contact dermatitis, for example, occurs after sensitization to poison ivy oils, rubber, certain metals, dyes, cosmetics, some plastics, and topical medications. Poison ivy, common in the United States, is acquired by direct contact with plant parts and occasionally by inhaling smoke from burning plants. Sensitivity can develop at any age, even after years of contact without reacting. Once a person is sensitized, an exceedingly small quantity of oil causes a reaction. Scratching lesions does not spread the oil, but it leaves skin vulnerable to infection.

Photosensitivity dermatitis requires both sunlight and a photosensitizing agent such as the artificial sweetener calcium cyclamate, aminobenzoic acid in sunscreens, sulfonamides (sulfa drugs), or even a fig plant. Once a person is sensitized, dermatitis can reappear with each exposure to sun.

Autoimmune Disorders

Hypersensitivities that elicit **autoantibodies** (aw″to-an′tib-od″ez), antibodies against self, are called **autoimmune disorders.** Cell destruction occurs by hypersensitivity reactions (usually Types II or III), and the disorders range from those affecting a single organ or tissue to those affecting many parts of the body (Table 21.8).

Several factors contribute to autoimmune disorders. Genetic factors may lead directly to disease or predispose toward it. Mutations can give rise to cells that make autoantibodies. Circulation of previously sequestered embryonic antigens might elicit antibodies. Sympathetic nervous system damage decreases the numbers of suppressor T cells and might account for some autoimmune diseases. An important factor is **antigenic mimicry** in which certain foreign and self substances are so similar that receptors fail to distinguish between them.

Myasthenia gravis (mi″as-the′ne-ah gra′vis) is an autoimmune disorder of skeletal muscle (Chapter 9). Two kinds of antibodies have been found—those that block calcium channels and prevent release of acetylcholine, and those that block acetylcholine receptors directly.

TABLE 21.8 The Spectrum of Autoimmune Disorders

Disorder	Organ(s) Affected	Antibody Made against:
Organ-Specific Disorders		
Autoimmune hemolytic anemia	Erythrocytes	Erythrocytes
Glomerulonephritis	Kidneys	Streptococci*
Graves' disease	Thyroid gland	TSH
Hashimoto's thyroiditis	Thyroid gland	Thyroglobulin
Juvenile diabetes	Pancreas	Pancreatic beta cells and insulin
Myasthenia gravis	Skeletal muscles	Acetylcholine
Pernicious anemia	Stomach	Vitamin B_{12} binding site
Rheumatic fever	Heart	Streptococci*
Ulcerative colitis	Colon	Colon
Systemic (Disseminated) Disorders		
Goodpasture's syndrome	Basement membranes	Basement membrane
Rheumatoid arthritis	Basement membranes and other tissues	Nuclei, gamma globulins, and Epstein-Barr virus
Scleroderma	Connective tissues	Nuclei
Systemic lupus erythamatosus	Many tissues	DNA, lymphocytes, erythroplatelets, neurons
Vasculitis	Blood vessels	Circulating immune complexes

*Streptococci resemble certain human tissue proteins.

Rheumatoid arthritis exerts its most obvious effects on joints, but also affects other tissues (Figure 21.19). It develops in the prime of life (usually age 30 to 40) and often causes crippling damage to joints of the hands and feet. Rheumatoid arthritis begins with joint inflammation as phagocytes release lysozymes that partially digest certain IgG antibodies, causing them to become antigenic. B cells make IgM antibodies to the antigens and cause more joint inflammation. IgM appears in joint fluid and subcutaneous nodules (lumps beneath the skin). It also appears in blood as **rheumatoid factor** in patients and some of their asymptomatic relatives. How relatives can have rheumatoid factor without the disease has not been explained. Symptoms are treated with nonsteroidal anti-inflammatory drugs, and physical therapy is used to keep joints movable. Severely damaged joints can be replaced surgically. The disease cannot be cured.

Multiple sclerosis causes severe central nervous system damage (Chapter 12), and autoimmune reactions probably play a role in that damage. Researchers using an animal model of the disease, allergic encephalomyelitis in

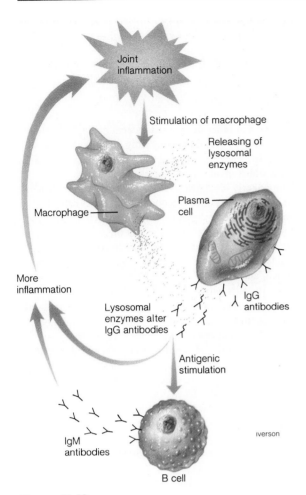

Figure 21.19
Rheumatoid arthritis.

rats, have developed a vaccine that protects the rats. This vaccine contains "good T cells" with receptors that react with "bad T cells" before they can destroy myelin. Whether a similar vaccine will be effective in humans remains to be seen.

The first evidence of **systemic lupus erythematosus** (lu'pus er-ith-e'mah-to'sus), or SLE, may be a butterfly-shaped rash over the nose that someone thought resembled a wolf bite (*erythematose*, red; *lupus*, wolf). This autoimmune disease occurs four times as often in women as in men, usually appears during the reproductive years, and eventually affects the whole body. Autoantibodies are mainly against DNA components, but can be against blood cells, neurons, and other tissues. Immune complexes cause inflammation and impair function wherever they are deposited—in the kidneys, skin, blood vessels, joints, and the brain. Most lupus patients die from kidney failure. Finding **LE cells,** neutrophils filled with damaged cells in a tissue,

provides a positive diagnosis. No cure is available, but drugs can suppress inflammation, fever, and further immune reactions.

Immunodeficiencies

Immunodeficiencies can be primary or secondary. **Primary immunodeficiencies** are genetic or developmental disorders in which B or T cells are absent or defective. **Secondary immunodeficiencies** are due to damage to developed T or B cells from malignancies, malnutrition, drugs that suppress the immune system, or infections such as AIDS (Interlude 4).

Agammaglobulinemia (a-gam"ah-glob"u-lin-e'me-ah), the first immunodeficiency disease discovered, is a sex-linked, inherited absence of B cells seen mostly in male infants. (Sex-linked inheritance is explained in Chapter 30.) After losing maternal antibodies by about nine months of age, such infants develop severe infections because they cannot produce antibodies. They are treated with antibodies in gamma globulin and antibiotics.

DiGeorge syndrome, a T cell deficiency caused by impaired development of the thymus gland, usually is due to a teratogenic agent (something that interferes with embryological development). Cell-mediated immunity is impaired in proportion to thymus damage, but humoral immunity protects against many diseases.

Severe combined immunodeficiency disease (SCID) is particularly debilitating because a genetic defect in stem cells causes extreme deficiency of both B and T cells. Keeping such a child in a germfree environment has been the only treatment, but new ways are being developed. A bone marrow transplant from a compatible donor (usually a brother or sister) can be effective. Transplantation of liver tissue from a fetus of less than twelve weeks gestation provides stem cells that sometimes persist in the recipient and produce normal T and B cells.

CLINICAL TERMS

Goodpasture's syndrome   a systemic autoimmune disease caused by immunologic reactions to basement membrane antigens

rheumatic fever   heart inflammation following streptococcal infection in which antibodies to streptococci bind to antigens in heart cells

scleroderma (skler-o-der'mah)   an autoimmune reaction to nuclear materials that affects mainly connective tissues

serum sickness   a response to foreign, usually intentionally injected protein

vasculitis (vas"ku-li'tis)   inflammation of blood vessels due to circulating immune complexes

---

IMMUNOLOGY OF ORGAN TRANSPLANTS AND CANCER

Transplantation, moving **graft tissue** from one site to another, began with skin grafts among animals of the same species. Grafts that at first appeared healthy became inflamed and dropped off the recipient's body within a few weeks. This reaction, originally thought to be due to infection, is now known to be immunologic **transplant rejection.** Antigens in graft tissue elicit host immunologic reactions that destroy the graft. Such **host-versus-graft disease** is the most common cause of transplant rejection in humans. In the much less common **graft-versus-host disease,** antigens in host tissue cause graft cells to immunologically destroy host tissue. This can occur in immunodeficient patients receiving bone marrow transplants.

Transplant rejection displays specificity and memory, and is initiated mainly by MHC proteins. If antigens on a donor kidney, heart, or other transplantable organ differ from the recipient's antigens, as they likely would in randomly chosen donors and recipients, recipient T cells destroy donor tissue. Such reactions are minimized by tissue typing and matching of proteins on donor and recipient tissues before a transplant is performed. Most tissue typing deals with MHC proteins on leukocytes, and the proteins are called **human leukocyte antigens** (HLAs).

HLAs are determined by genes designated as HLA-A, HLA-B, HLA-C, and HLA-D (Figure 21.20). HLA-D has an associated HLA-DR gene. Information in a gene can code for one of many different antigens—35 B antigens have been identified. D and DR antigens cause the strongest rejection reactions, and some HLA antigens may contribute to autoimmune and other disorders. HLA tissue typing is so precise that it has replaced blood typing in paternity suits and criminal investigations.

Rejection can occur in minutes or months, but does not occur in grafts, such as corneal transplants, that lack blood vessels. Damage is done by cytotoxic T cells or macrophages stimulated by lymphokines. Acute graft rejection (Type II hypersensitivity) occurs rapidly, but only in previously sensitized recipients. Slower rejections are mainly cell-mediated (Type IV) and occur in two to five days in previously sensitized hosts and in seven days to three months without prior sensitization.

Rejection reactions can be minimized by **immunosuppression** (im″u-no-sup-resh′un), decreased responsiveness of the immune system. Ideally, immunosuppression should be specific, causing tolerance of transplanted tissue and maintaining intolerance to infectious agents. In this respect, radiation and cytotoxic drugs are less than ideal. **Radiation** of lymphoid tissues suppresses immune and other functions of lymphoid tissues. **Cytotoxic drugs** (methotrexate and other drugs used to treat cancer) interfere with DNA synthesis and damage rapidly dividing cells more than other cells. They have a somewhat selective effect on B and T cells, which divide rapidly after sensitization.

A more nearly ideal drug, *cyclosporine* (si″klo-spo′rin) *A,* temporarily suppresses T cells without affecting B cells. It helps prevent transplant rejection without reducing resistance to infections, but it may increase the risk of cancer. Cyclosporine eye drops may soon be used to treat an autoimmune kind of "dry eye," a loss of tearing that causes constant pain and gradual loss of sight, especially in rheumatoid arthritis patients.

In some ways, **cancer cells** behave much like foreign graft tissue because they have cell-surface antigens not found on normal cells. Such antigens make an ideal target for immunologic destruction. According to the **theory of immune surveillance,** T cells continuously seek and destroy these cells with abnormal antigens before they develop into cancers. We develop cancer only if the T cells fail in their surveillance. And they might fail if the cancer cell antigens stimulate B cells to make innocuous antibodies that bind to the antigens without damaging the cells. These so-called *enhancement antibodies* block T cell access to the antigens and enhance cell growth.

The ability of immune reactions to destroy cancer cells has led researchers to develop immunotoxins and cancer vaccines. An **immunotoxin** is a monoclonal antibody with an anticancer drug or radioactive substance attached. The antibody binds to a cancer cell antigen and the attached substance destroys the cell. Vaccines can contain a patient's own cancer antigens or antigens frequently found on cancer cells. The many different antigens of cancer cells (some of which also are found on normal cells) and the ability of some malignant cells to release substances that inhibit division of T cells make immunotoxin and vaccine production difficult. Great care is needed to assure that these agents will destroy malignant cells but not normal ones.

Questions

1. How does host-versus-graft disease differ from graft-versus-host disease?
2. Find out more about how HLA antigens and DNA fragments are used in paternity suits and criminal investigations.
3. How does rejection of organ transplants occur, and how can it be prevented?
4. How are cancer cells similar to graft cells?
5. How do the body's methods of fighting cancer compare with medical methods?

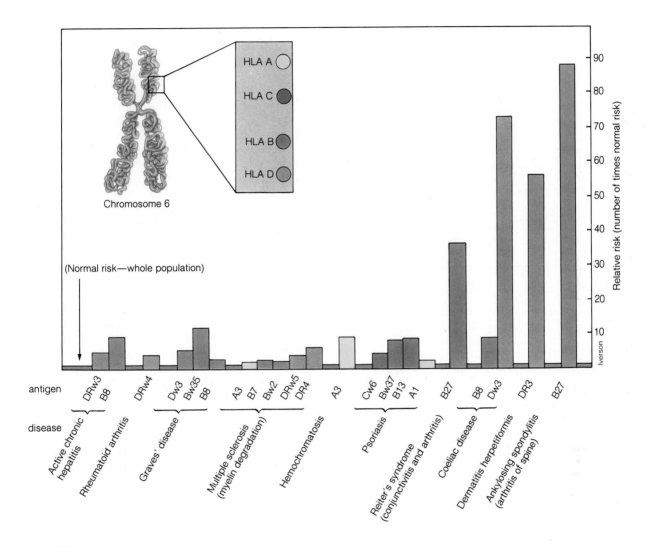

FIGURE 21.20
Correlation of HLA and diseases.

CHAPTER SUMMARY

Basic Concepts of Body Defense

- The human body fights infections with nonspecific and specific defenses.

Nonspecific Defenses

- Inflammation follows tissue damages, starts repair, and is mediated by histamine and other substances from injured cells.
- Phagocytes (neutrophils, monocytes, and macrophages) remove infectious agents and debris from injured tissues.
- During healing, which is stimulated by growth factors, fibroblasts secrete fibers that form scar tissue and new epidermis replaces what was destroyed.
- Interferon released from virus-infected cells stimulates adjacent healthy cells to make an antiviral protein.
- Complement enhances phagocytosis, lyses cells directly, and participates in a variety of immune reactions.

Immunity—Specific Defenses

- Specific immunity arises from functions of the immune system and can be humoral or cell-mediated.
- Components of the immune system include B and T lymphocytes, bursal equivalent tissues, thymus, and other lymphoid tissues.
- Antigens are substances recognized as foreign and antibodies are substances made in response to antigens.
- MHC proteins are natural cell surface antigens, some of which assist in immunologic reactions.
- Immunoglobulins (antibodies) contain polypeptide chains with unique variable regions. Five types have been identified.

Humoral Immunity

- B cells sensitized to a foreign antigen form a clone of plasma cells and some memory cells. Plasma cells release large numbers of antibodies. Helper and suppressor T cells modify humoral immune reactions.
- A primary immune response on initial exposure to a foreign substance releases large numbers of IgM followed by IgG. A secondary response on subsequent exposure releases smaller numbers of IgM and more rapidly produced, more numerous IgG.

Cell-Mediated Immunity

- Cell-mediated immunity makes use of several kinds of T cells and lymphokines they secrete.
- Macrophages process foreign antigens and present them to T cells, which then become sensitized.
- Cytotoxic and natural killer T cells secrete lethal protein perforin that kills cells; helper and suppressor T cells modify immune reactions, and delayed hypersensitivity T cells participate in certain hypersensitivity reactions.

General Properties of Immunity

- Properties of immunity include recognition of self and nonself, specificity, heterogeneity, and memory.

Immunity and Immunization

- Immunity can be innate, acquired naturally or artificially, and active or passive (Table 21.6).
- Immunization with vaccines produces artificially acquired active immunity.

Clinical Applications

- Immunologic disorders include hypersensitivities and immunodeficiencies.
- Immediate (Type I) hypersensitivity (allergies and anaphylaxis) results from IgE made in response to allergens.
- Cytotoxic (Type II) hypersensitivity, which is elicited by cell-surface antigens after sensitization, destroys cells as in transfusion reactions and hemolytic disease of the newborn.
- In immune complex (Type III) hypersensitivity, antigen-antibody complexes are deposited and damage tissues such as in glomerulonephritis and certain autoimmune diseases.
- Cell-mediated (Type IV) hypersensitivity, mediated by delayed hypersensitivity T cells, takes more than 12 hours to develop and causes tissue damage over several days, as in poison ivy and rejection of transplanted organs.
- Autoimmune disorders result from hypersensitivity to self and may involve antigenic mimicry; they include myasthenia gravis, rheumatoid arthritis, and systemic lupus erythematosus.
- Immunodeficiencies include inherited diseases (agammaglobulinemia, diGeorge syndrome, and severe combined immunodeficiency) and acquired diseases such as AIDS.

QUESTIONS AND PROBLEMS

The questions at the end of each chapter are numbered to correspond with the objectives listed at the beginning of the chapter. Italics indicate that a question requires critical thinking skills beyond simple factual recall.

Questions

1. (a) How are specific and nonspecific defenses alike?
   (b) *What basic concepts are required to understand immunology?*
2. (a) How is inflammation initiated?
   (b) How do phagocytes contribute to inflammation?
   (c) How does healing occur?
3. (a) What is interferon, and how does it defend the body?
   (b) What is complement, and how does it defend the body?
4. (a) What cells and tissues make up the immune system?
   (b) *Why is specific immunity said to have a dual role?*
5. (a) What are the properties of antigens and MHC proteins?
   (b) What are the properties of immunoglobulins, and how do the types of immunoglobulins differ?
6. (a) What events give rise to antibodies in humoral immunity?
   (b) *How do antibodies fight infections?*
   (c) *How do primary and secondary responses differ?*

7. (a) What are the general functions of the different kinds of T cells?

    (b) *How do macrophages and lymphokines participate in cell-mediated immunity?*

8. *How do killer cells kill?*

9. (a) *How is the immune system believed to distinguish between self and nonself?*

    (b) Define specificity, heterogeneity, and memory as they apply to the immune system.

10. *What kind of immunity might prevent you from getting distemper from your dog, hepatitis after direct contact with a friend having the disease, measles, a second case of the same disease, or an infection right after you were born?*

11. (a) *What are the basic properties of each of the four types of hypersensitivity?*

    (b) *What are the distinguishing characteristics of a transfusion reaction, glomerulonephritis, allergy, poison ivy, hemolytic disease of the newborn, and anaphylactic shock?*

    (c) *Summarize the immunologic factors in rheumatoid arthritis, myasthenia gravis, and systemic lupus erythematosus.*

12. What are the causes and effects of inherited immunodeficiencies?

Problems

1. Describe the events that take place from the time you skin a knee until the scab finally comes off.

2. If you worked in a family health clinic, what could you do to assure that all infants and young children in the neighborhood receive appropriate immunizations.

3. Prepare a research report about one of the following topics: (a) immunoglobulins, (b) cell-mediated immunity, (c) an autoimmune disease, (d) an immunodeficiency disease, (e) transplant rejection, or (f) immunological defense against cancer.

INTERLUDE 4

AIDS

In Zaire on Christmas Eve 1976, Danish physician Grethe Rask continued her two-year fight against an unrelenting fatigue as she attempted to prepare a holiday meal. Her colleagues saw that she was beset by more than fatigue—weight loss, swollen lymph nodes, difficulty breathing, and one infection after another. Her doctors knew her body lacked T cells, but they found no evidence of lymph cancer. This talented and dedicated surgeon died less than a year later at age 47—her lungs filled with fluid and her mysterious disease undiagnosed.

AIDS AGENT FOUND AS DISEASE SPREADS

New cases of the mysterious disease appeared sporadically, but with increasing frequency, and deaths were duly recorded. By 1982, the disease was named **acquired immune deficiency syndrome** (AIDS). The incidence of the disease escalated while its cause remained unknown. By 1985, the human immunodeficiency virus type 1 (HIV-1), which infects T cells (Figure I4.1) was found to cause AIDS—but not before more than 10,000 cases had appeared in the United States.

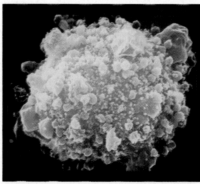

(a)

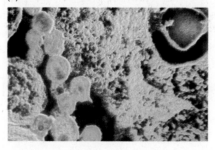

(b)

FIGURE I4.1
(*a*) False color scanning electron micrograph of a lymphocyte infected with the AIDS virus, and (*b*) a false color transmission electron micrograph showing mature AIDS viruses (orange) between T lymphocytes.

Knowing the cause has not slowed the pace of the AIDS epidemic. The number of cases has increased astronomically, both in the United States and many other countries. By the end of 1989, an estimated 1.5 million Americans were infected with HIV, though many had not shown disease symptoms and over half did not know they were infected. The Centers for Disease Control project that by the end of 1992, 365,000 Americans will have been diagnosed with AIDS and 263,000 will have died from it. Most cases have been among homosexual or bisexual males, intravenous drug abusers, and the sexual partners and infants of these groups. Some people became infected from blood or blood products, but careful screening of blood since 1985 has minimized this risk.

The virus is transmitted in semen, blood, and other body fluids and by unsterilized needles among drug abusers. Anal intercourse readily transmits the disease because the rectum is more easily abraded than the vagina and abrasions allow the virus to enter the blood. Infected women, who may be prostitutes, drug abusers, or partners of bisexual men or drug users, can transmit the virus to their infants during pregnancy, at delivery, and by breast-feeding. In Africa, AIDS transmission is mainly heterosexual—and usually among those people with multiple partners.

INCUBATION PERIOD VARIABLE

The AIDS incubation period is difficult to determine because most patients don't know when they were infected. Estimated incubation periods are 2.0 years for children under five, 7.8 years for homosexual men, and 8.2 years for adults with blood-transmitted infections. Diagnosis is made by finding antibodies, which may not be detectable for several months after infection. New highly sensitive viral culture tests found HIV in one-third of a group of homosexual men even though none of them had HIV antibodies and some had been infected as long as 36 months before the tests. This high incidence of HIV presence without antibodies raises serious questions about relying on antibody detection for diagnosis and about how many people really are infected.

How AIDS Affects the Body

AIDS specifically damages lymphocytes called T4 cells, which include helper T cells. Disease symptoms appear only after the virus starts replicating and begins to damage T4 cells. The AIDS virus, which has its genetic information stored in RNA, begins replicating by the virus's own enzymes converting viral RNA to DNA. Then, enzymes in the host's lymphocytes use the DNA to direct the synthesis of more viruses until nearly all T4 cells are killed. As T4 cells decrease in numbers, they fail to stimulate B cells to make antibodies and suppressor T cells now have relatively greater influence and depress other immune functions. Lymphokine secretion decreases so macrophages and cytotoxic T cells are not activated. T4 cells that survive lack surface antigen receptors and cannot elicit immune reactions.

Early AIDS symptoms resemble many other diseases—fever, sweats, nausea, headache, sore throat, loss of appetite, muscle and joint pain, swollen lymph nodes, rash, and reduced blood lymphocytes and platelets. Weight loss and muscle wasting also occur. As AIDS progresses, severe opportunistic infections (caused by agents that the body usually resists), malignancies, and other complications appear (Figure I4.2). Opportunistic infections include pneumocystic pneumonia, severe diarrhea, and encephalitis. Many patients develop a malignancy called Kaposi's sarcoma, and some develop a mental disorder called AIDS dementia.

Individuals with AIDS antibodies and mild symptoms are sometimes said to have **AIDS related complex** (ARC). Whether ARC is a separate disease or a precursor to AIDS is not entirely clear, but long-term studies of ARC patients show that in most patients, the disease progresses to AIDS.

Diagnosis of AIDS

The widely used ELISA (**e**nzyme-**l**inked **i**mmuno**s**orbent **a**ssay) test detects HIV-1 antibodies (Figure I4.3) and was developed to screen the nation's blood supply. Though a sensitive and accurate test, a confirmatory test usually is done before a diagnosis is made. AIDS antibodies indicate that a person has been infected though the infection may not be active. A quick (two-minute) monoclonal anti-AIDS antibody test is now available.

Some scientists think all people with HIV-1 antibodies eventually develop AIDS, but **seroreversion** (se''ro-re-ver'shun), the loss of antibodies once present and the absence of symptoms, has been observed in a few individuals. Most have DNA made from HIV-1 RNA in some body cells. A few have no detectable evidence of AIDS infection, but the virus may be hiding in cells that were not tested. One kind of seroreversion occurs when the immune system is so severely damaged that it cannot produce antibodies and HIV-1 antibodies eventually disappear.

Less than Perfect Treatment

The first AIDS drug, AZT (3'azido-3'thymidine or zidovudine), which became available in 1987, inhibits replication of the HIV virus and slows the progress of the disease. It fails to eradicate the virus, and so it does not cure the disease. The Food and Drug Administration has now approved AZT for use in children. AZT also delays the onset of symptoms in HIV-positive individuals and some are receiving the drug. The drug probenecid, now used to treat gout, was developed to increase the length of time penicillin stays in the blood. It appears to help keep AZT in the blood and increases AZT's effect on the AIDS virus but it also increases AZT's toxicity. AZT causes anemia because of its toxicity to bone marrow, costs thousands of dollars per year, and is ineffective in some patients.

Dideoxyinosine (DDI) stimulates immune function and blocks HIV replication in HIV-infected patients. Many patients who become unable to tolerate the side effects of AZT switch to DDI.

Other drugs are under study. For example, the antibiotic fusidic acid inhibits HIV-1 viruses in cultures, and if proven effective in humans, it might cost only one thousand dollars per year. Pressure from desperate AIDS patients caused the Food and Drug Administration to allow importation of small quantities of drugs not licensed in the United States for personal use. Controlling opportunistic infections, treating malignancies, and alleviating symptoms also are important aspects of AIDS treatment. Regrettably, no patient has yet recovered from AIDS.

Antibodylike molecules called CD4 have been created by genetic engineering to help combat AIDS. CD4 molecules appear to bind to HIV viruses and they also activate complement.

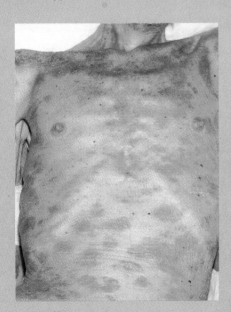

Figure I4.2
This AIDS patient shows marked emaciation and extensive seborrhoeic dermatitis.

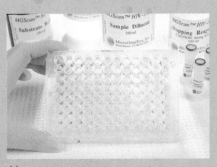

(a)

(b)

Figure I4.3
AIDS testing makes use of (*a*) the ELISA test and (*b*) the Western blot test.

Cytotoxic T cells can recognize and ingest CD4 and HIV viruses bound to it; thus, CD4 shows promise as a possible treatment for AIDS.

RISK TO MEDICAL PERSONNEL

Nurses, physicians, laboratory technicians, and other medical personnel who work with AIDS patients must use care to avoid contact with their body secretions. Wearing gowns and gloves and disposing of needles in puncture proof containers are important precautions.

WILL THERE BE AN AIDS VACCINE?

Efforts to develop an AIDS vaccine have not yet succeeded, partly because the AIDS virus mutates rapidly. Before immunization against existing strains could occur, new strains with previously unknown antigens would be causing the infections. Researchers hope to find an antigenic viral surface structure not subject to rapid change and use it to make a vaccine. Anger at being infected with AIDS and maintaining fighting spirit seems to increase the ability of the immune system to combat rather than succumb to AIDS. Though such emotions may delay the progression of AIDS, their effects on the immune system are not sufficient to prevent or stop the disease. Lacking any other means of preventing or curing AIDS, educating people to avoid exposure to it is the only means of prevention.

Problems

1. Trace the early history of AIDS.
2. Which cells does HIV attack, and how does it reach them?
3. What are the prospects for curing and for treating AIDS?
4. How can you protect yourself from exposure to AIDS?
5. How can the spread of AIDS be prevented?

OUTLINE

OBJECTIVES

1. Distinguish between ventilation and gas exchange.
2. Describe the overall plan of the respiratory system, and explain its general functions.
3. Describe the general embryological development of the respiratory system.
4. Identify the tissues of the respiratory tract, and explain how the epithelial tissues are specialized to perform their particular functions.
5. Locate and identify structures in the nasal cavity, pharynx, and larynx; and explain their respiratory functions.
6. Describe the overall structure and location of the lungs and surrounding membranes, including blood supply and innervation.
7. Locate and identify the anatomical structures of the bronchial tree, and explain their functions.
8. Describe the mechanics of inspiration and expiration and the factors that affect these processes.
9. Define the volumes associated with ventilation, and relate them to pulmonary function tests.
10. Relate the concept of partial pressure to gas exchange.
11. Explain how gas exchange occurs in the lungs.
12. Explain how gas exchange occurs in other tissues.
13. Describe how oxygen is transported in the blood.
14. Describe how carbon dioxide is transported in the blood.
15. Explain how blood gases relate to respiration.
16. Summarize the relationships between the transport and exchange of gases.
17. Describe how neural signals regulate depth and frequency of breathing.
18. Describe how chemical factors affect breathing.
19. Explain how respiration is regulated at high altitude, under water, and during anesthesia.
20. Summarize the effects of exercise on circulatory and respiratory functions.
21. Describe the alteration in function in the following conditions: (a) asthma, (b) emphysema, (c) hyaline membrane disease, (d) smoking, (e) bronchitis, (f) pneumonia, and (g) hypoxia.

22

RESPIRATORY SYSTEM

ORGANIZATION AND GENERAL FUNCTIONS

You would live no more than a few minutes without oxygen and only a little longer if your cells had no way to rid themselves of carbon dioxide. A continuous supply of oxygen and respiratory processes to deliver it to cells are essential to life.

Your **respiratory system,** which moves gases between the environment and your blood, carries out two major functions, ventilation and gas exchange. **Ventilation** is the mechanical process of conveying gases into and out of the lungs. **Gas exchange** is the movement of gases across membranes in lungs and other tissues. Blood plasma and hemoglobin in erythrocytes contribute to respiratory functions by transporting gases.

The respiratory system (Figure 22.1) consists of a conducting portion and a respiratory portion. The **conducting portion,** or **upper respiratory system,** includes the nasal cavity, pharynx, larynx, trachea, bronchi, and non-respiratory bronchioles, through which gases enter and leave the system during breathing. The **respiratory portion,** or **lower respiratory system,** includes the respiratory bronchioles and alveoli, which are thin-walled membranes richly supplied with capillaries where gas exchange occurs. The bronchi and all structures branching from them are enclosed within paired organs called **lungs.** The circulatory system supports respiratory functions. Blood vessels transport gases between the lungs and tissues.

DEVELOPMENT

The primitive embryonic gut is an endodermal tube that grows at both ends. Anteriorly, **pharyngeal** (far-in′je-al) **pouches** grow outward from the walls of the tube in humans as in all vertebrate embryos. Though in fish these pouches form the gills, in humans they form the auditory tube, fossae in which the tonsils lie, the parathyroid glands, and the thymus gland. A diverticulum (pocket) gives rise to the thyroid gland (Figure 22.2).

A **laryngotracheal** (lar-ing″o-tra′ke-al) **bud** grows posteriorly from the pharynx, forming the larynx and trachea. It then branches to form the left and right bronchi and all of their branches (Figure 22.3).

FIGURE 22.1
The structure of the respiratory system.

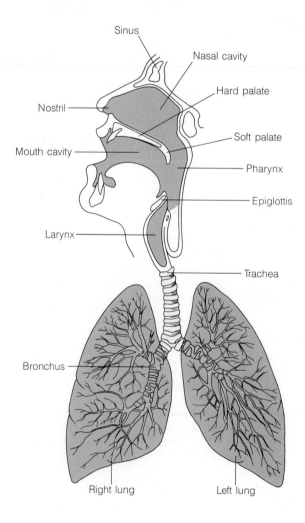

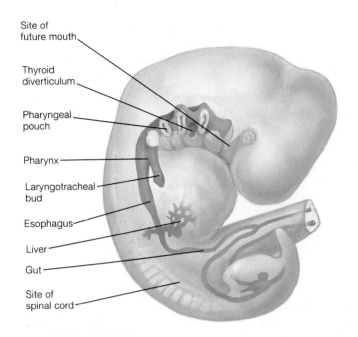

FIGURE 22.2
A side view of an embryo early in the second month of development showing the pharynx and laryngotracheal buds that contribute to the development of the respiratory system.

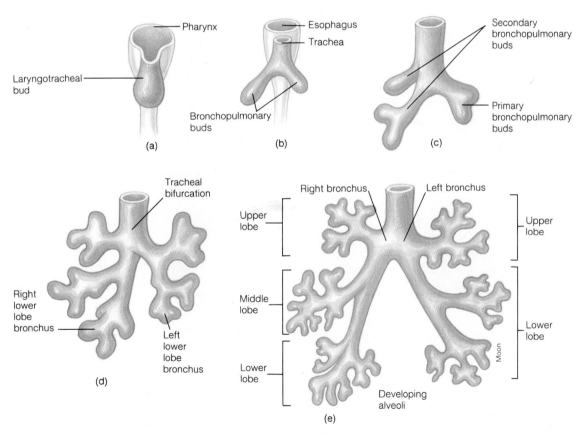

FIGURE 22.3
Development of bronchioles and alveoli: (a–b) occur during the first month of development, (c–e) occur during the second month.

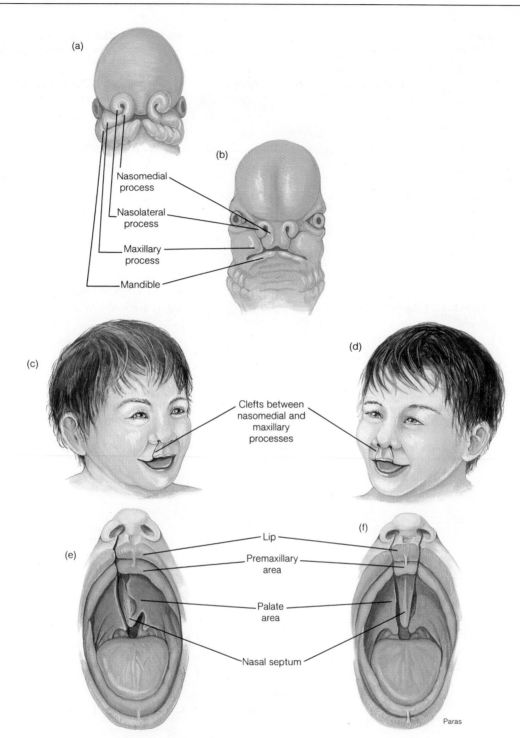

FIGURE 22.4

Development of face, including nose, lip, and palate: (*a*) location of medial and lateral processes at 5½ weeks, (*b*) normal fusion of processes at 7 weeks, (*c*) unilateral cleft lip, (*d*) bilateral cleft lip, (*e*) unilateral cleft lip and palate, and (*f*) bilateral cleft lip and palate.

The entire respiratory system is lined with epithelium of endodermal origin. Up to about the twenty-fourth week of development, the alveoli consist of cuboidal epithelium, which by the time of birth become squamous and thin enough to allow gas exchange. Respiratory difficulties in premature infants often result from alveoli that are too thick for adequate gas exchange.

In the head area during the second month of development, several lateral processes grow medially and meet with the nasomedial process to form the nose, lips, and palate (Figure 22.4). Incomplete fusion causes cleft lip and cleft palate. Cleft lip can be repaired in early infancy and cleft palate at about 18 months of age.

TISSUES

Most epithelium lining respiratory passageways contains goblet cells and is covered with cilia (Figure 22.5). The goblet cells secrete mucus that coats the surface and traps dust, smoke particles, and other debris that enter passageways with air. Cilia in bronchial tubes beat toward the pharynx, moving mucus and the debris trapped in it into the pharynx, from which they can be expectorated or swallowed. This mechanism is called the **mucociliary** (mu″ko-sil′e-ar″e) **escalator.** The alveoli and smallest bronchioles consist of continuously moist nonciliated squamous epithelium across which gas exchange takes place.

Other tissues of the respiratory system include the cartilage that supports all but the smaller respiratory passages, the smooth muscle of the bronchioles, and the connective tissues that support and protect the tract. The characteristics of these tissues were discussed in Chapter 4.

RESPIRATORY SYSTEM ANATOMY

Air normally enters the respiratory system at the nasal cavity. It moves through the pharynx, larynx, trachea, bronchi, and bronchioles before it reaches the alveoli.

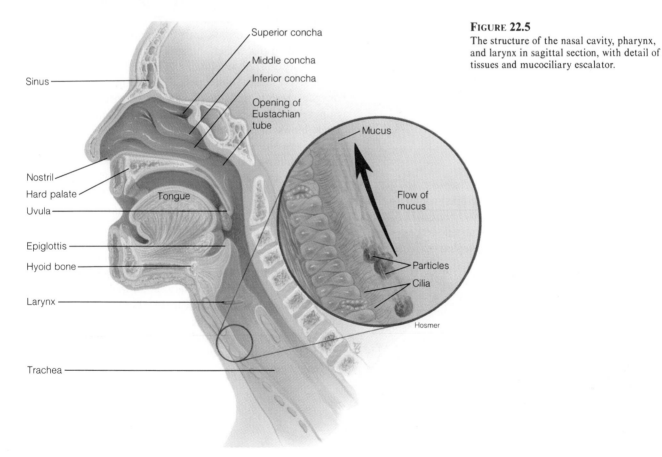

FIGURE 22.5
The structure of the nasal cavity, pharynx, and larynx in sagittal section, with detail of tissues and mucociliary escalator.

Sinus

Superior concha
Middle concha
Inferior concha
Opening of Eustachian tube

Nostril
Hard palate
Tongue
Uvula
Epiglottis
Hyoid bone
Larynx

Trachea

Mucus
Flow of mucus
Particles
Cilia

Hosmer

Nasal Cavity

Beginning at the **external nares,** or nostrils, the **nasal cavity** (Figure 22.5) extends through the nose deep into the facial bones of the skull to the **internal nares.** The nose consists of bone and cartilage covered with skin. The nasal cavity is separated from the mouth by the **hard palate** and **soft palate.** It is divided into left and right halves by the **nasal septum,** which is composed of the vomer and the perpendicular plate of the ethmoid bone. Extending from the roof of the nasal cavity are three pairs of **nasal conchae.** Olfactory receptor cells are located in the superior portion of the membranes that cover the conchae. The portion of the nasal cavity in front of the conchae is the **vestibule.**

As air passes through the nasal cavity, it is warmed, moistened, and filtered. Mucus traps large particles of debris, preventing them from passing beyond the pharynx. Some small particles are exhaled, but others are inhaled deep into the lungs where they can become lodged in smaller passageways. As air swirls through the nasal cavity, airborne molecules of odorous substances are swept into contact with moist membranes containing olfactory receptors. These substances dissolve in the fluid and then stimulate receptor cells (Chapter 15).

The sinuses (hollow cavities within certain skull bones) open from the nasal cavity. They are lined with mucous membranes and can become infected by microorganisms in the nasal cavity. The resulting inflammation, **sinusitis** (si″nus-i′tis), causes the membrane to swell and produce excess mucus. Swelling of membranes and blockage of passageways in the nasal cavity and sinuses cause pain and probably alter the quality of the voice.

Pharynx

From the nasal cavity, the **pharynx** (far′inks) conducts air to the larynx and food to the esophagus (Figure 22.5). It is divided into three areas: the nasopharynx, oropharynx, and laryngopharynx. The superior **nasopharynx** (na″zo-far′inks), which lies next to the internal nares, is connected to each middle ear chamber by the auditory tube. Superior and posterior to the auditory tubes are the **pharyngeal tonsils.** When these tonsils become enlarged, usually because of chronic infection, they are called **adenoids** (ad′en-oidz). The nasopharynx, a passageway for the respiratory tract only, joins the **oropharynx** (o″ro-far′inks),

or throat. Near the **fauces** (faw′sēz), the boundary between the mouth and the oropharynx, are the **palatine** (pal′ah-tīn) **tonsils,** and on the dorsal side of the root of the tongue are the **lingual** (ling′gwal) **tonsils.** All tonsils are masses of lymphatic tissue (Chapter 19). The oropharynx and the inferior **laryngopharynx** (lar-ing″go-far′inks) serve both the respiratory and digestive tracts.

Larynx

The voice box, or **larynx** (lar′inks), is a rigid structure between the laryngopharynx and the trachea (Figure 22.6a). It contains the vocal cords, which produce sound when they vibrate. Rigidity of the larynx is maintained by muscles, ligaments, and nine cartilages—three unpaired and three paired. The large unpaired **thyroid cartilage** forms the anterior surface of the larynx. Inferior to it is the ring-shaped **cricoid** (krik′oid) **cartilage,** and superior to it is the **epiglottis** (ep″ĭ-glot′is). The paired cartilages are the **arytenoid** (ar″et-e′noid), **corniculate** (kor-nik′u-lāt), and **cuneiform** (ku-ne′if-orm).

During swallowing, the body of the larynx is pulled toward the hyoid bone. This movement shields the **glottis** (glot′is), the opening into the larynx from the esophagus, and allows food to pass into the esophagus without entering the larynx. When this mechanism fails, food blocks the airway and causes choking. The victim cannot breathe or talk. Cyanosis (bluishness of the skin due to lack of oxygen) results and death will occur within about five minutes unless the obstructing material is removed. Quick back blows or sudden compression of the victim's abdomen are often effective remedies because they force air out of the lungs and dislodge the object.

Inside the larynx, supported by the cartilages and stretched across the glottis are the **vocal cords** (Figure 22.6b). The outer **false vocal cords** are nonfunctional, but the inner **true vocal cords** vibrate and produce sound when air passes through the glottis. During puberty, testosterone causes the male larynx to enlarge and the vocal cords to lengthen. These changes produce the male **Adam's apple** (enlarged thyroid cartilage) and lower the pitch of the voice. Pitch is regulated in speaking and singing by changing the tension on the vocal cords—the greater the tension, the higher the pitch. This is accomplished by intrinsic (within the organ) laryngeal muscles moving the arytenoid cartilages.

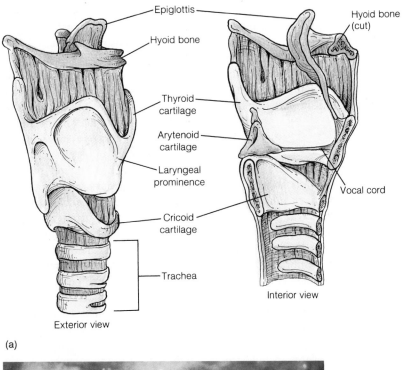

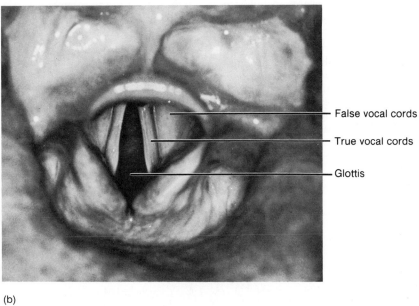

FIGURE 22.6
(*a*) The larynx, and (*b*) vocal cords.

Phonation (fo-na'shun), the production of speech sounds, requires vibration of the vocal cords and movements of the tongue and lips. These processes are controlled by centers in the brain. Though other animals make sounds and sometimes use signs, speech is a human ability. Some anthropologists believe the configuration of the human larynx and hyoid bone account, in part, for that ability.

Lungs and Thoracic Cavity

The lungs lie in the **thoracic cavity,** which is bounded by the rib cage and muscular **diaphragm.** Each lung is covered with a serous membrane called **visceral pleura** and occupies a **pleural cavity** lined by serous membranes called **parietal pleura** (Figure 22.7). Because the lungs completely fill the pleural cavities and the membranes touch, the pleural cavities are only potential spaces. The pleurae constantly secrete and reabsorb a watery fluid that keeps the membranes moist; thus, they slide over each other during breathing movements, but adhere like two sheets of moist glass. Adherence of pleura also is facilitated by the **intrapleural** (or intrathoracic) pressures, which are about 4 torr *less* than atmospheric pressure.

Pleurisy (ploor'is-e), an inflammation of the pleurae, increases friction between membranes and makes breathing painful. **Pneumothorax** (nu''mo-tho'raks) (nu''mo-tho'raks), a condition in which air enters a pleural cavity (usually after a chest injury), causes severe pain and dyspnea (labored breathing). This condition is treated by using a chest tube to suck air out and recreate a negative pressure that allows the lung to expand.

Each lung has a superior apex and flat, inferior base. The **costal** (kos'tal) **surfaces** of the lungs lie next to the ribs and the **medial surfaces** lie next to the mediastinum. The larger right lung has three lobes; the smaller left lung has only two lobes and a concavity, the **cardiac notch,** where the heart lies. The **lobes** of the lungs are separated by fissures (Figure 22.7). Each lobe is covered by an elastic connective tissue membrane and is subdivided into many **lobules,** each having its own respiratory passageways, arteriole, venule, and lymphatic vessel.

The lungs contain both pulmonary and systemic blood vessels. Pulmonary arteries carry blood to alveolar capillaries, where gas exchange occurs, and the pulmonary veins drain blood from these capillaries. Bronchial arteries carry blood to the nonrespiratory tissues and bronchial veins drain them. Some anastomoses exist between the pulmonary and systemic circuits.

The lungs have both somatic and autonomic nerves. Among somatic nerves, which control breathing movements, the phrenic nerve supplies the diaphragm and the thoracic spinal nerves supply the intercostal muscles. Sympathetic and parasympathetic nerves regulate smooth muscle tone in the bronchioles.

Healthy lungs always contain air, but diseased lungs may contain excessive amounts of fluid. Sometimes the condition of lungs is important in determining the cause of death. For example, the lungs of live-born infants will float, whereas those of stillborn infants will not.

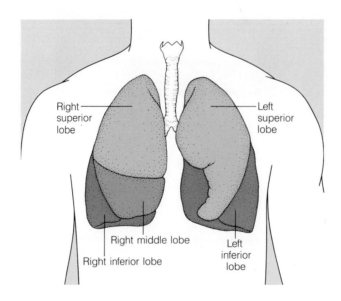

FIGURE 22.7
Lobes of the lungs.

Trachea

The windpipe, or **trachea** (tra'ke-ah), is a tube supported by cartilage that carries air between the larynx and the primary bronchi of the lungs (Figure 22.8). It is 2 to 3 cm in diameter and 12 to 13 cm long, with walls composed of an inner ciliated mucous membrane, a middle intermittent hyaline cartilage layer, and an outer connective tissue layer. C-shaped rings of cartilage extend around the trachea except for a short distance on the posterior side, where smooth muscle supports their ends. The cartilage rings keep the airway open, and the smooth muscle allows the trachea to be pushed forward and to be slightly collapsed as food passes through the esophagus.

Bronchi and Bronchioles

The **bronchi** (brong'ki) branch from the trachea as the limbs branch from a tree trunk to form the **bronchial** (brong'ke-al) **tree** (Figure 22.9). One **primary bronchus** (brong'kus) goes to each lung; the right bronchus is wider in diameter and lies in a more vertical position than the left bronchus. Each bronchus branches many times—the first several branches forming conducting bronchi, bronchioles, and terminal bronchioles—and the last forming respiratory bronchioles, alveolar ducts, and alveolar sacs where gas exchange occurs. This complex branching arrangement greatly increases the surface area exposed to air flowing into and out of the lungs. Air flow is fairly rapid in primary bronchi, but it slows as the total cross-sectional area increases with each branching.

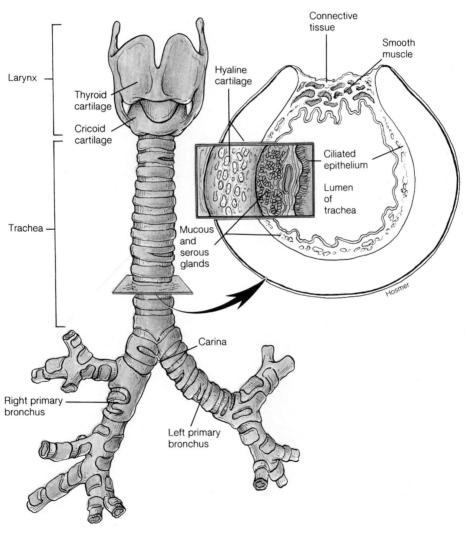

FIGURE 22.8
The trachea, with cross section, and bronchi.

Large bronchi have cartilage rings similar to those in the trachea, but the amount of cartilage decreases with tube diameter so smaller bronchioles lack cartilage. Smooth muscle innervated by the autonomic nervous system partly surrounds tubes that have cartilage and completely surrounds those that lack it. This muscle is important in regulating the diameter of respiratory passageways. When it constricts passageways excessively, as in **bronchial asthma,** the patient has great difficulty breathing.

The epithelium changes as passageways branch. Large bronchi have pseudostratified, ciliated columnar epithelium, but in the smaller ones, it is cuboidal and can be ciliated or nonciliated.

Respiratory Structures

The respiratory portion of the lungs consists of **respiratory bronchioles** (brong'ke-ōlz), **alveolar** (al-ve'o-lar) **ducts, alveolar sacs,** and **alveoli** (al-ve'o-li), all located at the ends of terminal nonrespiratory bronchioles (Figure 22.9b and c) and all capable of gas exchange. Some alveoli project from respiratory bronchioles, but most are arranged in clusters around an alveolar duct. Together, the human lungs contain about 300 million alveoli. Each cluster of alveoli is surrounded by elastic connective tissue and is supplied with capillaries of the pulmonary circuit and lymphatic vessels, which drain interstitial fluid from around the alveoli.

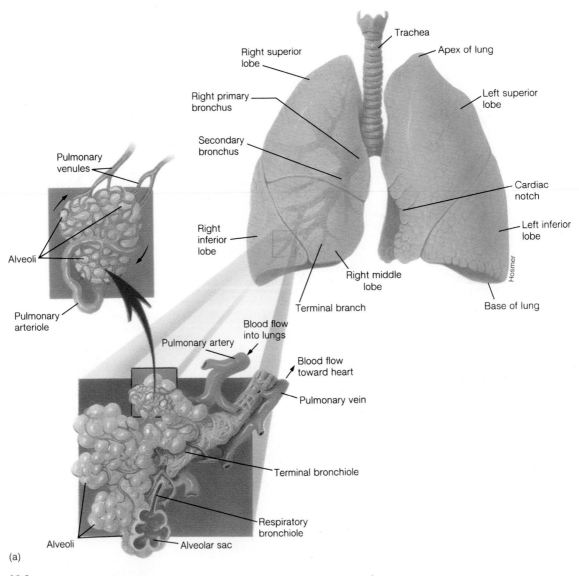

FIGURE 22.9

(*a*) External and internal structure of the lungs. (*b*) Plastic cast of conducting portions of lungs. (*c*) Photomicrograph of lung tissue.

The total surface area of the alveoli amounts to about one square meter per kilogram body weight. So, a person weighing 68 kg (150 lbs) would have 68 m² of alveolar surface area—about half the surface area of a tennis court. This surface is well supplied with capillaries, and alveolar and capillary walls are extremely thin—less than 1 μm in thickness.

VENTILATION

Breathing, or ventilation, provides a continuous supply of air to the respiratory membranes of the alveoli. It occurs in two phases: **inspiration,** breathing in atmospheric air, and **expiration,** breathing out a mixture of gases that differs from air as a result of gas exchange.

Mechanisms of Inspiration and Expiration

During inspiration (Figure 22.10a), the diaphragm and the external intercostal muscles contract. When the dome-shaped diaphragm contracts, it is drawn downward toward the abdomen by about 1.5 cm during quiet inspiration and by as much as 7 cm during deep inspiration. The external intercostal muscles pull the ribs upward and outward like lifting the handle on a bucket. This increases the antero-posterior dimension of the thoracic cavity. During quiet respiration, the diaphragm accounts for about three-fourths and the intercostals for about one-fourth of the thoracic volume change. The shading in Figure 22.10b shows how the thoracic cavity changes shape during inspiration.

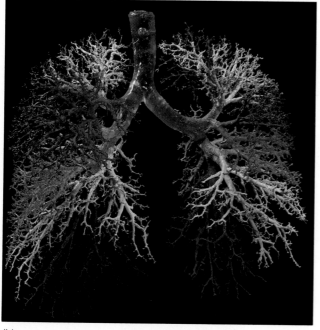

(b)

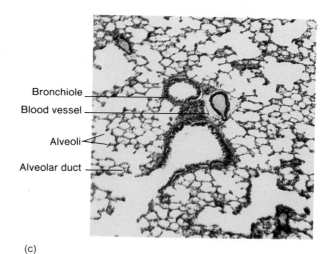

Bronchiole

Blood vessel

Alveoli

Alveolar duct

(c)

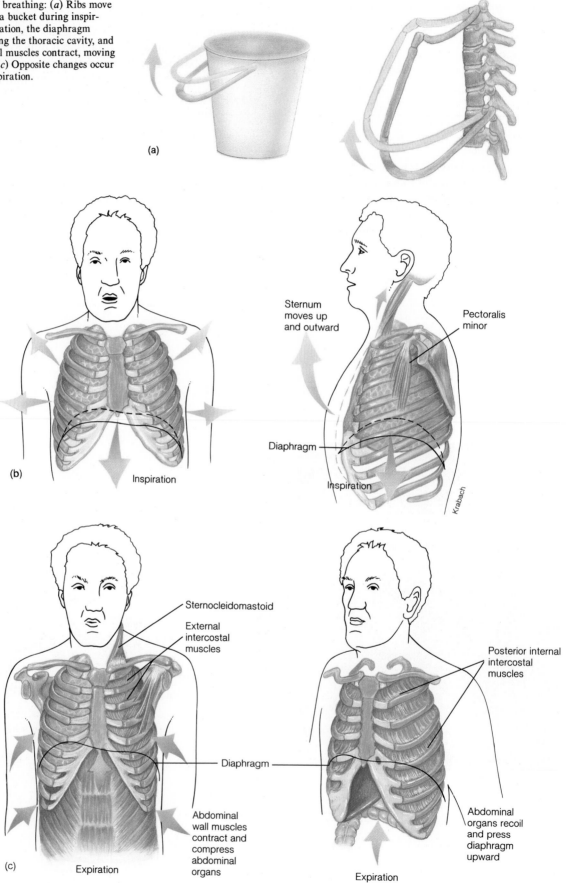

FIGURE 22.10
The mechanism of breathing: (*a*) Ribs move like the handle on a bucket during inspiration. (*b*) In inspiration, the diaphragm contracts, elongating the thoracic cavity, and external intercostal muscles contract, moving the ribs outward. (*c*) Opposite changes occur during forceful expiration.

During expiration, the diaphragm and the external intercostal muscles relax and intrathoracic volume passively returns to the pre-inspiratory volume. During forceful expiration, however, the internal intercostal muscles contract and pull the ribs inward and downward and decrease the anteroposterior dimension of the thoracic cavity. The abdominal muscles also contract and push the diaphragm toward the lungs while depressing the ribs and further decrease the thoracic cavity volume. The shading in Figure 22.10c shows how the thoracic cavity changes shape during expiration.

Inspiration and expiration result from pressure gradients along airways associated with changes in the intrathoracic volume. Pressures and volumes follow **Boyle's law:** the pressure exerted by a gas is inversely proportional to its volume, provided the temperature remains constant. In other words, decreasing the volume of a given quantity of gas increases the pressure it exerts, and *vice versa.* Such

changes create gradients along airways that convey gases back and forth between the atmosphere and the alveoli. This is analogous to moving air in and out of a syringe.

Inspiration requires energy to contract muscles and is an active process. For air to flow into the lungs, pressure inside the lungs must be less than atmospheric pressure (Figure 22.11). Such a gradient is created by the diaphragm and external intercostal muscles contracting and expanding the intrathoracic volume. This expands lung volume because the pleura adhere to each other. **Intra-alveolar** (in-trah-al-ve′o-lar) **pressure** becomes slightly less than atmospheric pressure as the same amount of gas occupies a larger volume. Air moves along the pressure gradient into the lungs until intra-alveolar pressure equals atmospheric pressure. In the moment between inspiration and expiration, alveolar pressure equals atmospheric pressure (760 torr at sea level) and no gradient exists (Figure 22.11).

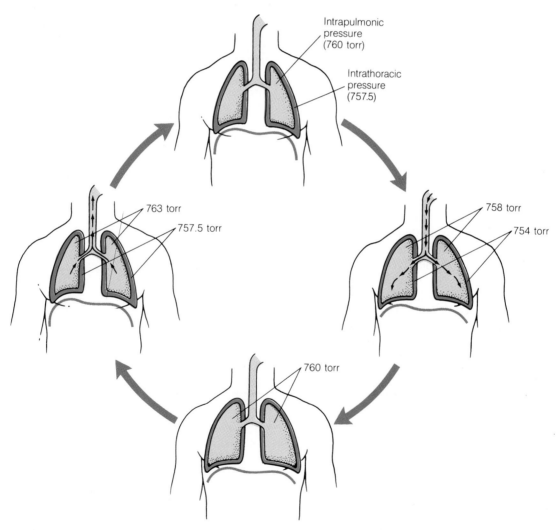

FIGURE 22.11

Changes in pressures during the respiratory cycle (atmospheric pressure = 760 torr).

Expiration, like inspiration, results from a pressure gradient. As the diaphragm and external intercostal muscles relax, the volume of the thoracic cavity decreases. This raises the intra-alveolar pressure to about 1 torr above atmospheric pressure as the same amount of gas occupies a smaller volume. Air moves along the pressure gradient out of the lungs until the intra-alveolar pressure again equals atmospheric pressure (Figure 22.11).

Compliance, or stretchability, of the lungs and the chest wall contributes to inspiration. Compliance is the volume increase of the lungs for each unit increase in intra-alveolar pressure. Elastic fibers give lung tissue compliance, whereas muscle fiber elasticity gives the chest wall compliance.

During expiration, compliance is reversed as the lungs and chest wall display **elastic recoil;** and the volume of both the lungs and the thoracic cavity decreases. Thoracic volume decreases as muscles relax and return to their original length, and lung volume decreases as elastic fibers in lung tissue recoil and surface tension tends to draw alveolar surfaces together.

Surface tension on alveolar membranes is created by two forces: (1) Water molecules are drawn together by hydrogen bonds between them, creating a thin layer over the membrane. (2) The thin water layer contracts to its minimum surface area and this draws the surfaces toward each other. We can see the effects of surface tension by placing a few drops of water on the inner surface of a rubber balloon and noticing that the walls of the balloon adhere to each other. Applying the **law of LaPlace** to alveoli, pressure created by surface tension is proportional to the surface tension and inversely proportional to the radius of the alveolus. If surface tension is constant, small alveoli are under greater pressure than large ones.

Though surface tension on alveolar membranes contributes to decreasing intra-alveolar volume, this process could go too far and cause the alveoli to collapse completely. Excessive collapse of alveoli is prevented by a mixture of phospholipids called **surfactant** (sur-fak'tant), or surface active agent, which is secreted by Type II alveolar cells and coats alveolar and other respiratory surfaces (Figure 22.12). (Type I alveolar cells make up the alveolar walls, themselves.) Surfactant acts like a detergent to lower surface tension by breaking up the water layer. This prevents collapse of the alveoli while preserving sufficient moist membrane for gas exchange. We can see the effect of surfactant by placing a drop of detergent in the wet balloon and noting how easily the balloon walls separate.

Factors Affecting Inspiration and Expiration

Like blood flow, airflow into and out of the lungs is proportional to the pressure gradient and inversely proportional to resistance. Let's look at the effects of these factors.

Pressure Gradients

As we have seen, pressure gradients move air into and out of the lungs. Providing that resistance does not change, the size of the pressure gradient determines the airflow. For example, when one "takes a deep breath," the muscles of inspiration contract more forcefully than in resting breathing. The intrathoracic volume increases in proportion to the strength of contraction, and a greater volume of air moves into the lungs. Likewise, when one exhales forcefully, the muscles of exhalation contract and decrease the intrathoracic volume below that which occurs in resting breathing. The degree of volume reduction is still proportional to the strength of the contraction, and a greater volume of gases moves out of the lungs.

Airway Resistance

Airflow into and out of the lungs also depends on airway resistance, which is due to the same factors as resistance in blood vessels. Airway resistance is directly proportional to the length of the passageway and inversely proportional to the fourth power of the radius of the passageway. Large bronchi offer little resistance, but small ducts offer much resistance.

Resistance in small bronchioles is determined by autonomic neural signals and chemical factors. Sympathetic signals decrease resistance by reducing smooth muscle contraction and dilating bronchioles. Parasympathetic signals do just the opposite. They increase resis-

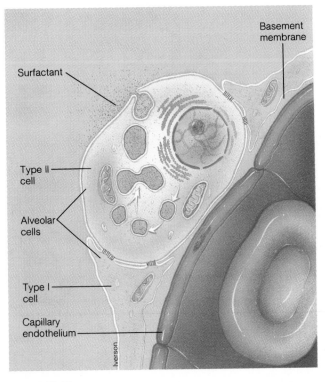

FIGURE 22.12
Type I alveolar cells make up the thin walls, whereas Type II cells secrete surfactant.

tance by increasing muscle contraction and constricting bronchioles. Chemical factors also have opposing effects. Epinephrine increases airflow by relaxing smooth muscle. Histamine and other substances from injured tissue or allergic reactions decrease airflow by increasing muscle contraction and mucus production.

Resistance in alveoli is somewhat less than might be expected from their small diameters because of the very low velocity of airflow. Turbulence and laminar flow, which contribute significantly to resistance where velocity is high, produce much less resistance in alveoli.

Blood Flow

Both the quantity of blood flowing through alveolar capillaries and the concentrations of oxygen and hydrogen ions in larger pulmonary vessels affect airflow. These factors help to maintain a balance between blood flow and airflow.

The amount of carbon dioxide reaching alveoli is proportional to the local capillary blood flow, and carbon dioxide relaxes smooth muscle and increases blood flow. Reduced blood flow decreases carbon dioxide and allows smooth muscle to contract. Variations in blood flow to clusters of alveoli help to keep local airflow in balance with the amount of carbon dioxide entering the alveoli (Figure 22.13a).

A similar mechanism counterbalances the response of pulmonary vessels to variations in the blood concentrations of oxygen and hydrogen ions (Figure 22.13b). When

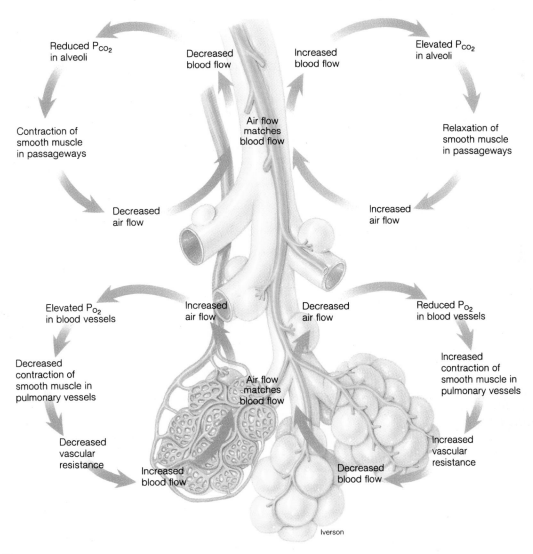

FIGURE 22.13
Homeostatic mechanisms that adjust (*a*) airflow to match blood flow, and (*b*) blood flow to match airflow at respiratory membranes.

airflow to a cluster of alveoli increases, the blood receives more oxygen and increased oxygen relaxes smooth muscle in pulmonary vessels. Vascular resistance decreases, and blood flow increases in proportion to the increased airflow. Conversely, when airflow to a cluster of alveoli decreases, the blood receives less oxygen. Vascular resistance increases, and blood flow decreases in proportion to the decreased airflow.

Smooth muscle of pulmonary vessels also is sensitive to variations in H^+ concentration. H^+ is released when carbon dioxide combines with water to form carbonic acid, some of which ionizes. Increased H^+ causes contraction of vascular smooth muscle and increases resistance, whereas decreased H^+ relaxes muscle and decreases resistance. Increased resistance impedes blood flow, allowing more carbon dioxide (and indirectly H^+) to be removed from blood in pulmonary capillaries.

Counterbalance of Airflow and Blood Flow

The above mechanisms work together to maintain a precise match between blood flow and airflow in alveoli. Carbon dioxide adjusts airflow to match blood flow by altering airway resistance. Oxygen and H^+ adjusts blood flow to match airflow by altering vascular resistance. Oxygen and H^+ have opposite effects in pulmonary and systemic circulation. In pulmonary vessels, they increase movement of oxygen into blood and carbon dioxide into alveoli. In systemic vessels, they increase movement of oxygen into tissues and carbon dioxide into blood.

Air Volumes and Pulmonary Function

Air volumes moved during ventilation are used to describe and assess respiratory function (Figure 22.14). The study of air volumes is called **spirometry** (spi-rom'et-re), and air is captured and measured with a **spirometer** (spi-rom'et-er).

Tidal volume, the volume of air moved into or out of the lungs with each resting-level inspiration or expiration, is about 500 ml. Tidal volume includes alveolar volume and physiological dead space volume. **Alveolar volume** (about 350 ml) consists of air that reaches respiratory surfaces and participates in gas exchange. **Physiological dead space volume** (about 150 ml) includes (1) **anatomical dead space volume,** the volume of nonrespiratory airways, and (2) air in alveoli that fails to reach respiratory surfaces. Physiological dead space volume increases in diseases that damage alveoli.

Air volumes larger than tidal volume can be moved into or out of lungs by forcible inhalation or exhalation. Most people can forcibly inhale 2000 to 3000 ml of air beyond tidal volume by taking one very deep breath. This is the **inspiratory reserve volume.** Similarly, most people can forcibly exhale about 1000 ml of air after exhaling the tidal volume. This is the **expiratory reserve volume.** After the expiratory reserve volume is exhaled, about 1200 ml of air, the **residual volume,** remain in the alveoli. The sum of all volumes mentioned above is the **total lung capacity**—a volume of 5000 to 6000 ml, depending on body size.

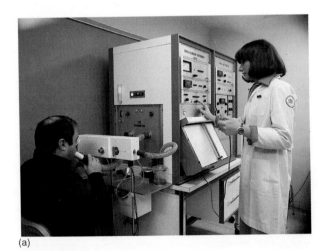

(a)

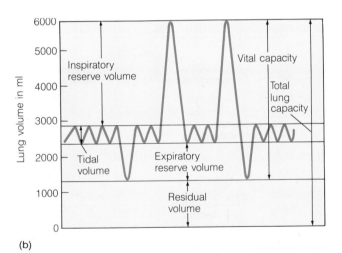

(b)

FIGURE 22.14
(a) Patient using a modern spirometer to measure respiratory performance. (b) Respiratory volumes.

Combinations of some of these volumes are meaningful. **Vital capacity,** the maximum volume that can be moved into and out of the lungs in one breath, equals total lung capacity minus the residual volume. **Inspiratory capacity,** the total volume that can be inhaled after a normal exhalation, includes tidal volume plus the inspiratory reserve volume. **Functional residual capacity** (FRC), the sum of expiratory reserve volume and residual volume, represents resting lung volume. The main work of breathing occurs when volume is increasing or decreasing from the FRC.

Minute respiratory volume (MRV), the product of tidal volume and respiratory rate, is an important measure of respiratory function. A person with a 500 ml tidal volume and a respiratory rate of 12 breaths per minute, has an MRV of 6 l. A large, young male forcing himself to breathe as deeply and rapidly as possible (voluntary hyperventilation) might reach an MRV of 150 l for a few seconds and sustain an MRV of about 100 l for a few minutes of strenuous exercise. Such hyperventilation increases blood oxygen uptake and carbon dioxide removal. Abnormally low MRV usually is related to impaired airflow by constriction of passageways or blockage with mucus.

Other abnormalities in specific volumes sometimes can be used to diagnose or evaluate respiratory impairment. For example, **forced expiratory volume** (FEV), the percent of the vital capacity that can be exhaled per second, is measured for the first few seconds of exhalation. A person who can exhale less than three-fourths of the vital capacity in the first second probably has emphysema or some other disease that obstructs airflow.

GAS EXCHANGE

Gas exchange depends on diffusion of gases across membranes between alveoli and blood, and between blood and the cells of various tissues. How much and how fast a gas diffuses depends on the partial pressures of the gases.

Partial Pressures of Gases

The earth's atmosphere is a mixture of gases—nearly four-fifths nitrogen, about one-fifth oxygen, and small amounts of carbon dioxide and other gases. According to **Dalton's law,** each gas in a mixture exerts a pressure, its **partial pressure,** independently of other gases. The total atmospheric pressure is the sum of the partial pressures of gases in the atmosphere. According to **Henry's law,** gases dissolve in blood and tissue fluids in proportion to their partial pressures and their solubility in the liquid.

Dry atmospheric air at sea-level conditions is 20.9 percent oxygen and 0.04 percent carbon dioxide, and exerts a total pressure of 760 torr. From this information, we can determine that the partial pressure of oxygen is 158 torr (20.9 percent of 760) and the partial pressure of carbon dioxide is 0.3 torr (0.04 percent of 760). In addition, nitrogen exerts a pressure of about 596 torr and water vapor about 6 torr (more in humid air). Nitrogen is breathed in and out, and diffuses into and out of the blood, but is not directly involved in normal respiration.

The different partial pressures of oxygen and carbon dioxide in alveoli, blood, and tissues, create diffusion gradients where net diffusion is from higher to lower partial pressure. Diffusion itself changes partial pressures and gradients. Net diffusion is greater and faster in large gradients than in small ones.

Gas Exchange in the Lungs

As air enters alveoli, it mixes with gas remaining in the alveoli after the last expiration. Less than 10 percent of the total lung volume is replaced with each respiration. Gas exchange occurs continuously across respiratory membranes, with alveolar gas losing oxygen to the blood and gaining carbon dioxide from it. Gas in respiratory passageways at the end of inspiration never reaches the respiratory membranes and remains unchanged. The composition of inhaled, alveolar, and exhaled gases is summarized in Table 22.1.

Gases exchanged between the alveoli and the blood first dissolve in water, where they continue to exert partial pressures, and then cross several barriers—alveolar epithelium, basement membrane, and capillary endothelium (refer back to Figure 22.12) in a distance of less than 1 μm.

TABLE 22.1	Composition of Inhaled, Alveolar, and Exhaled Gases (expressed as partial pressures in torr)*		
Gas	**Inhaled**	**Alveolar**	**Exhaled**
Oxygen	158.0	100.0	116.0
Carbon dioxide	0.3	40.0	26.8
Nitrogen	596.0	573.0†	570.2
Water vapor	5.7	47.0	47.0
Total	760.0	760.0	760.0

*Partial pressures are given for sea-level conditions. Above sea level, the barometric pressure decreases and so does the partial pressure of the gases.

†The decrease in the partial pressure of nitrogen is due mainly to the relative increase in water vapor and not to the removal of nitrogen in the gas exchange process.

To follow the course of diffusion in the lungs, let us consider the average partial pressures of the gases in the alveoli and in the blood (Figure 22.15). (We refer to average pressures because pressures are constantly changing.) The average partial pressure of oxygen, P_{O_2}, is 100 torr in the alveoli and 40 torr in blood entering alveolar capillaries. Net oxygen diffusion is out of the alveoli and into the blood. The partial pressure of carbon dioxide, P_{CO_2}, is 40 torr in the alveoli and 45 torr in blood entering alveolar capillaries. Net carbon dioxide diffusion is out of the blood and into the alveoli. As diffusion proceeds, blood oxygen increases to a maximum and blood carbon dioxide decreases to a minimum as blood leaves the alveolar capillaries.

Gas exchange in the lungs is affected by both the partial pressures of gases and the nature of pulmonary circulation. Compared to the systemic circulation, the pulmonary circulation is a low pressure, low resistance pathway with a small capillary filtration pressure. When

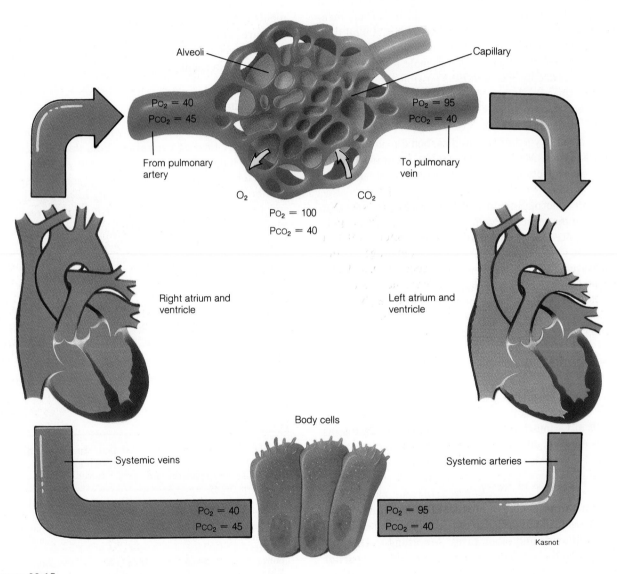

FIGURE 22.15
The diffusion of gases in external and internal respiration, as partial pressures of each gas (in torr).

TABLE 22.2	Average Partial Pressures (in torr) of Oxygen and Carbon Dioxide				
Gas	Atmosphere	Alveoli	Arterial Blood	Cell	Venous Blood
P_{O_2}	158	100	95	25	40
P_{CO_2}	0.3	40	45	46	45

the body is at rest, blood flows through a pulmonary capillary in about 0.75 seconds, but oxygen diffuses so rapidly that the partial pressure in the capillary equals that in the alveoli in 0.25 seconds. Therefore, blood flow can become as much as three times as fast during exercise and still allow P_{O_2} in capillaries to equal that in alveoli. Equalization of P_{CO_2} between blood and alveoli also occurs rapidly. Both oxygenation of blood and removal of carbon dioxide from it are limited by blood flow (perfusion) and not by diffusion.

From the foregoing discussion, we might expect the partial pressures of gases in blood entering the systemic circulation blood to equal those in alveoli. This is not the case because of anastomoses between pulmonary and systemic circuits. Some blood reaching the lungs via the pulmonary arteries drains into the systemic intercostal and azygos veins, and some blood reaching the lungs via the bronchial arteries drains into the pulmonary veins. Thus, blood in systemic veins leaving the lungs has been partially oxygenated and some carbon dioxide has been removed. Blood in pulmonary veins has been more thoroughly oxygenated and has more carbon dioxide removed. In systemic arterial blood leaving the heart, P_{O_2} is about 95 torr and P_{CO_2} is about 40 torr.

Gas Exchange in the Tissues

In blood reaching systemic capillaries, partial pressures of gases create gradients that cause net oxygen diffusion from the blood into interstitial fluid and, eventually, into cells and net carbon dioxide diffusion from cells via interstitial fluid into blood. P_{O_2} in the capillaries is about 95 torr and in cells about 25 torr. P_{CO_2} in capillaries is about 40 torr and in cells about 45 torr. These pressures change with cellular metabolism, and maximum pressure gradients occur between capillaries and rapidly metabolizing muscle cells during strenuous exercise. Such cells use more oxygen and release more carbon dioxide than slowly metabolizing cells.

The body's overall metabolic rate also affects the partial pressures of blood gases. At an average metabolic rate, P_{O_2} is about 40 torr and P_{CO_2} is about 46 torr by the time the blood returns to the lungs. But during strenuous exercise, venous blood contains less oxygen and more carbon dioxide than usual. Partial pressures of gases as a result of gas exchange in the lungs and in the tissues are summarized in Table 22.2.

Transport of Gases

Once oxygen enters the blood, it is transported to the cells of all body tissues. As oxygen is used and carbon dioxide is released in cellular respiration, carbon dioxide enters the blood and is transported to the lungs. Special mechanisms are involved in the transport of both oxygen and carbon dioxide.

Transport of Oxygen

Oxygen diffusing into the blood first dissolves in the plasma, but quickly diffuses into erythrocytes and combines with hemoglobin, forming **oxyhemoglobin.** As some oxygen moves to hemoglobin, more can dissolve in plasma. As much as 99 percent of blood oxygen is bound to hemoglobin in erythrocytes, but the quantity varies, depending on how much hemoglobin the blood contains. Because of hemoglobin, blood typically carries about seventy times as much oxygen (200 ml/l) as can dissolve in plasma (3 ml/l). In tissue capillaries, this process is reversed. As oxygen diffuses into interstitial fluid, lowering the plasma P_{O_2}, hemoglobin continuously releases more oxygen into the plasma.

When completely saturated, each hemoglobin molecule carries four molecules of oxygen. Oxygen combines with and is released from hemoglobin, as shown in the following equation.

$$\text{H-Hb} \; + \; \text{O}_2 \; \leftrightarrow \; \text{HbO}_2 \; + \; \text{H}^+$$

deoxygenated   oxygen   oxyhemoglobin
hemoglobin

This reaction goes to the right in the lungs during oxygen loading and to the left in the tissues during oxygen unloading. When P_{O_2} is high and P_{CO_2} is low, as in the alveolar capillaries, hemoglobin readily combines with available oxygen. When hemoglobin reaches the tissues where P_{CO_2} is higher and P_{O_2} is lower, hemoglobin readily releases oxygen.

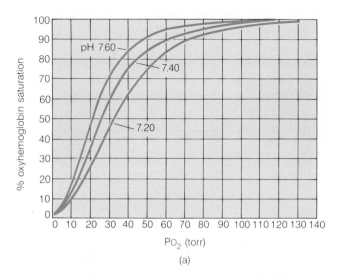

(a)

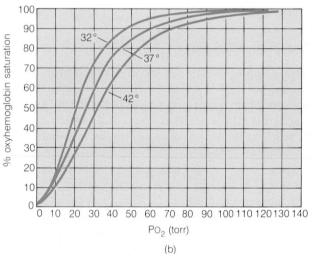

(b)

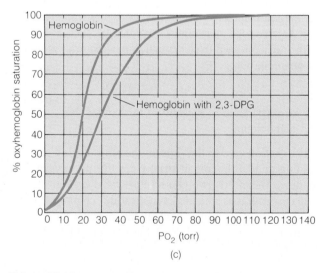

(c)

Hemoglobin is 100 percent saturated with oxygen when every hemoglobin molecule has bound four oxygen molecules. The degree of saturation increases as the Po_2 increases, markedly up to Po_2 of 50 torr and to a lesser degree at higher partial pressures. Changes in pH and temperature alter the molecular shape of hemoglobin and its affinity for oxygen (Figures 22.16a and b). Lowering the pH or raising the temperature decreases affinity; conversely raising the pH or lowering the temperature increases affinity. The former occurs in metabolically active tissues and the latter in resting tissues.

Metabolically active cells such as contracting muscles produce lactic acid and give off heat. Blood passing through such tissues becomes more acidic and warmer, and its hemoglobin releases more oxygen. Erythrocytes contain the enzyme carbonic anhydrase, which causes carbon dioxide entering the cells to combine with water. Some of the resulting carbonic acid ionizes.

$$CO_2 + H_2O \leftrightarrow H_2CO_3 \leftrightarrow H^+ + HCO_3-$$

As the above reaction goes to the right, H^+ lowers the blood pH and hemoglobin loses more oxygen. All these processes increase oxygen availability in plasma near metabolically active cells. They also act locally to assure that hemoglobin releases oxygen in proportion to the metabolic activity of tissues.

Finally, 2,3-diphosphoglycerate (DPG), a metabolic product of erythrocytes, binds to hemoglobin and causes it to release oxygen (Figure 22.16). This process occurs continuously, assuring some oxygen release regardless of the local pH or temperature.

Transport of Carbon Dioxide

Carbon dioxide is transported in three ways—dissolved in plasma, on globin, and as bicarbonate (Figure 22.17 and Table 22.3). Being about twenty times more water-soluble than oxygen, much carbon dioxide can be carried in plasma without a carrier like hemoglobin. This high solubility also facilitates carbon dioxide's diffusion from cells across interstitial fluid into blood. Some carbon dioxide enters erythrocytes and binds to globin, forming **carbaminohemoglobin** (kar-bam''in-o-hem''o-glo'bin). Carbonic anhydrase causes the remaining carbon dioxide to combine with water to form carbonic acid (see previous equation). As some carbonic acid ionizes, H^+ bind to negative charges on globin and bicarbonate ions diffuse into plasma, some binding to plasma proteins.

FIGURE 22.16

(a) The oxyhemoglobin dissociation curves for pH. These curves illustrate two principles: (1) the percent saturation of hemoglobin increases with Po_2, rapidly up to Po_2 of 50 torr and more slowly at higher Po_2; and (2) pH affects the percent saturation at any Po_2. The curve labeled pH 7.4 is the normal dissociation curve. At pH (7.6), hemoglobin is more saturated at any given Po_2 and at pH (7.2), it is less saturated at any given Po_2. (b) The oxyhemoglobin dissociation curves for temperature (Celsius). These curves illustrate the relationship between Po_2 and percent saturation of hemoglobin, and show the effect of temperature on hemoglobin saturation. The curve labeled 37° is the normal curve. At 42° C hemoglobin is less saturated at any Po_2, and at 32° C, hemoglobin is more saturated at any Po_2. (c) Effects of 2,3-DPG on hemoglobin saturation.

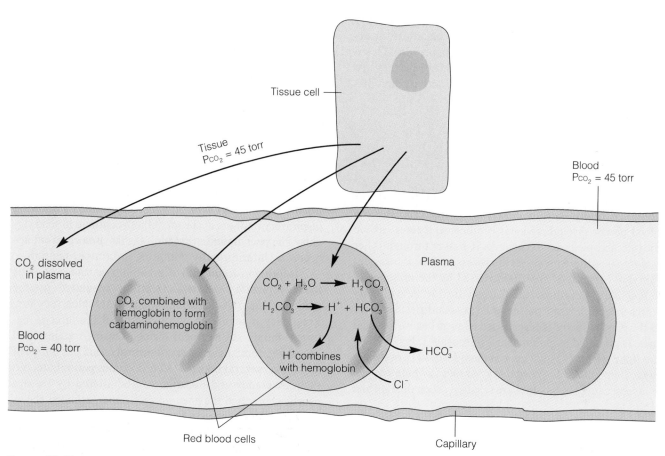

Tissue cell

Tissue
P_{CO_2} = 45 torr

Blood
P_{CO_2} = 45 torr

CO_2 dissolved
in plasma

Plasma

CO_2 combined with
hemoglobin to form
carbaminohemoglobin

$CO_2 + H_2O \longrightarrow H_2CO_3$

$H_2CO_3 \longrightarrow H^+ + HCO_3^-$

Blood
P_{CO_2} = 40 torr

H^+ combines
with hemoglobin

HCO_3^-

Cl^-

Red blood cells

Capillary

FIGURE 22.17
Mechanisms of carbon dioxide transport.

TABLE 22.3	Transport of Carbon Dioxide
Mechanism	**Percent**
Dissolved gas	5–7
Bicarbonate ions	70–72
Associated with hemoglobin and plasma proteins	22–23
Total	100

Exposed amino acids of globin in desaturated (deoxygenated) hemoglobin bind both hydrogen ions and carbon dioxide. As oxygen binds to heme (Chapter 17), hemoglobin can carry oxygen and carbon dioxide simultaneously. Because of the pressure gradients at gas exchange sites, it carries mostly oxygen from the alveoli to other tissues, and mostly carbon dioxide back to the alveoli.

About 70 percent of bicarbonate ions formed in erythrocytes diffuse into the plasma, creating an electrical gradient across erythrocyte membranes. Electrical balance between erythrocytes and plasma is restored by **chloride shift,** the diffusion of chloride ions into erythrocytes.

When blood reaches alveolar capillaries, carbon dioxide, bicarbonate, and hydrogen ion movements are reversed. Carbon dioxide diffuses from plasma into the alveoli. Carbonic acid reforms and carbonic anhydrase breaks it into carbon dioxide and water. Hemoglobin also releases hydrogen ions and carbon dioxide and is made ready to carry more oxygen. Carbon dioxide diffusion continues until the P_{CO_2} in the blood and alveoli are equal.

Blood Gas Measurements

In the laboratory, electrodes sensitive to particular gases are used to measure their partial pressures in plasma. Each such electrode generates a current proportional to the partial pressure of the gas it senses. Blood gases usually are measured on arterial blood to assess lung function. If perfusion, airflow, and gas exchange are normal, the partial pressures of the gases in arterial blood are normal—P_{O_2} about 95 torr and P_{CO_2} about 40 torr. Low P_{O_2} indicates impaired function. Adding oxygen to the air a person breathes, as is done for hospitalized patients with heart or lung disorders, increases the rate of delivery of oxygen to tissues. High P_{CO_2} indicates either impaired lung function or a metabolic disorder that increases blood acidity.

Lung function is very important in correcting blood pH in response to **acidosis** (as″id-o′sis), low blood pH, or **alkalosis** (al″kah-lo′sis), high blood pH (Chapter 28). Gases usually are not measured in venous blood because they are highly variable, depending on metabolic activity, and difficult to interpret.

SUMMARY OF TRANSPORT AND EXCHANGE OF GASES

Though we have considered transport and exchange of gases separately, they are closely integrated processes. As blood passes through alveolar capillaries, it loses carbon dioxide and picks up oxygen. As it passes through other body tissues, these processes are reversed. Blood's oxygen-carrying capacity is greatly increased by the presence of hemoglobin in erythrocytes. Carbon dioxide, being highly water-soluble, is transported in large quantities dissolved in the plasma. Some is transported as bicarbonate ions and some is attached to globin. Gas transport is augmented by the **Bohr effect**: (1) high carbon dioxide and H^+ in tissues cause hemoglobin to release oxygen and (2) high oxygen in alveoli causes hemoglobin to release carbon dioxide and H^+.

REGULATION OF RESPIRATION

Breathing normally occurs at a regular, rhythmic rate, but both its rate and depth can be altered by neural and chemical signals.

Neural Regulation

Breathing muscles must receive signals from motor neurons to contract. Neurons must, in turn, receive signals from centers in the cerebrum for conscious control or centers in the pons and medulla for automatic control of breathing. Signals from the cerebral center allow you to vary your rate and depth of breathing and even to "hold your breath" for a short time. If you hold your breath long enough, the automatic mechanism overrides the conscious one and forces you to resume breathing. In rare instances, the automatic breathing mechanism fails and a person must make a conscious effort to breathe.

Damage to the automatic breathing mechanism is called *Ondine's* (on′dīnz) *curse* from the French fairy tale about the water sprite Ondine, whose lover became unfaithful. As punishment, she placed a curse on him that required him to purposefully perform automatic body processes, such as breathing. Eventually, he became exhausted, slept, and stopped breathing.

Centers in the pons and medulla send signals that simultaneously stimulate inspiratory muscles and inhibit expiratory muscles or vice versa. This phenomenon is called **reciprocal innervation.** Respiratory control areas are believed to consist of areas of fairly widely scattered neurons that work together (Figure 22.18a). The **rhythmicity area,** which maintains resting level breathing, includes some self-excitatory neurons that initiate inspiration and some neurons that allow expiration by preventing inspiration. Respiratory centers in the pons include the median apneustic area and bilateral pneumotaxic areas. (*Apneusis* is the arrest of breathing during inspiration.) The **apneustic** (ap-nu′stik) **area** prolongs inspiration by allowing inspiratory neurons of the medulla to continue to send signals. The **pneumotaxic** (nu″mo-taks′ik) **areas** prevent apneusis by intermittently inhibiting inspiratory neurons.

Depth and frequency of breathing depend on signals from inspiratory neurons. Depth is regulated by the number of neurons firing and the rate at which they fire. Frequency is regulated by the length of the interval between bursts of neuronal firing. During resting-level breathing, these neurons spontaneously send out bursts of signals for several seconds, 12 to 15 times per minute, but their frequency and duration can be altered by the apneustic center and the pneumotaxic center, as noted above.

Stretch receptors in tissues of the lungs help prevent overinflation of the lungs during deep breathing by sending inhibitory signals proportional to the degree of stretching to the apneustic center. Inhibition of the apneustic center prevents further inhalation. This response, called the **Hering-Breuer reflex,** limits the amount of air a person can inhale in a very deep breath. It is initiated when the volume of air inhaled exceeds that of a normal inspiration by about 50 percent.

Chemical Regulation

When the partial pressures of carbon dioxide or oxygen or the concentration of hydrogen ions in the blood changes, chemical regulation of breathing can occur. **Chemoreceptors** that detect such changes are found in **carotid bodies** in the walls of carotid arteries and in **aortic bodies** in the walls of the aortic arch (Figure 22.18b). Chemoreceptors are stimulated by an increase in P_{CO_2} or H^+ in arterial blood, and, under some conditions, by a decrease in P_{O_2}. When stimulated, they send signals to inspiratory neurons and increase the frequency of breathing. This increases *both* the rate at which oxygen enters the blood and the rate at which carbon dioxide leaves it.

Chemical regulation of respiration also occurs in a **chemosensitive area** near the ventral surface of the medulla. This area responds directly and quickly to increased P_{CO_2} or H^+ in cerebrospinal fluid. It also responds to increased blood P_{CO_2} as carbon dioxide easily crosses the blood-brain barrier and enters the chemosensitive area, where it combines with water to form carbonic acid. Some carbonic acid then ionizes and releases H^+, which directly stimulates the chemosensitive area.

Carbon dioxide may have a greater effect in cerebrospinal fluid than in blood because cerebrospinal fluid lacks the buffers (substances that resist pH change) found in

blood. Carbon dioxide that reaches cerebrospinal fluid quickly stimulates the chemosensitive area, whereas carbon dioxide that remains in the blood may be counteracted by buffers.

As we have seen, in normal breathing, a cycle of inhalation and exhalation takes four to five seconds. Certain abnormal conditions produce periodic breathing, the most

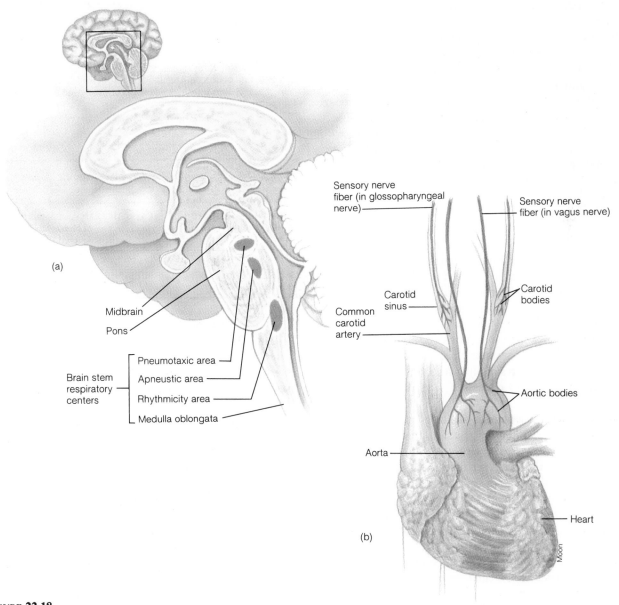

FIGURE 22.18
(*a*) A schematic diagram of neural structures that regulate respiration. (*b*) The locations of carotid and aortic bodies.

common being **Cheyne-Stokes** (chān'stōks) **breathing** (Figure 22.19) in which breathing waxes and wanes every one to three minutes. To explain Cheyne-Stokes breathing, let us assume that a period of rapid, deep breathing has just occurred. The P_{CO_2} of blood in the lungs is lower than usual, and it fails to initiate breathing when it reaches the chemosensitive area. While breathing is inhibited, the P_{CO_2} of the blood in the lungs becomes excessive. When this blood reaches the chemosensitive area, it initiates a cycle of rapid, deep breathing (overbreathing). Thus, Cheyne-Stokes breathing is caused by larger than normal oscillations in P_{CO_2}. It often appears in patients with heart failure or brain stem damage and in those near death.

The abnormality of Cheyne-Stokes breathing emphasizes the importance of homeostatic mechanisms that minimize wide oscillations of a physiological condition. Whereas large oscillations produced by marked variations in P_{CO_2} are seen in Cheyne-Stokes breathing, normal controls respond to small changes in P_{CO_2} and breathing is of normal depth and frequency.

Regulation of Respiration under Special Circumstances

A variety of conditions significantly affect the regulation of respiration. High altitude, underwater conditions, and anesthesia will be discussed here. Cardiopulmonary resuscitation (CPR) will be described because of its great importance in emergency situations.

High Altitude

As altitude increases, barometric pressure decreases and so do the partial pressures of all gases in the atmosphere. Decreased P_{O_2} leads to decreased saturation of hemoglobin with oxygen (Table 22.4). Oxygen deprivation becomes increasingly severe as altitude increases, and consciousness is lost when oxygen saturation falls below 50 percent. Increasing the oxygen content of inspired air increases the quantity of oxygen taken up by hemoglobin, but at very high altitudes, pressurized cabins or suits are needed to keep oxygen bound to hemoglobin. Airplanes that fly at high altitudes and space vehicles have equipment to pressurize their cabins to near sea-level conditions.

In the natural atmosphere at high altitudes humans display great capacities to adapt to low P_{O_2} (Figure 22.20). Immediate and long-term compensatory mechanisms come into play, especially at altitudes above 8000 feet. The respiratory rate increases immediately as one ascends to a high altitude and chemoreceptors are activated by low P_{O_2}. Removal of CO_2 temporarily limits the increase in respiratory rate to no more than 60 percent above normal, but

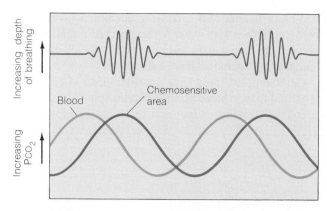

FIGURE 22.19

Cheyne-Stokes breathing. Oscillations in P_{CO_2} lead to alternating spells of rapid, deep breathing and spells of little or no breathing.

TABLE 22.4	Effects of Altitude on Barometric Pressure, P_{O_2}, and Oxygen Saturation of Hemoglobin

Altitude (feet)	Barometric Pressure (torr)	P_{O_2} in Air (torr)	Oxygen Saturation of Hemoglobin in Arterial Blood (%)
0	760	159	97
10,000	525	110	90
20,000	350	73	70

a rate that maintains oxygenated hemoglobin eventually becomes established. This rate can be seven times normal. Because carbon dioxide removal is accelerated, the blood pH becomes slightly more basic.

Observations of experienced mountain climbers ascending and remaining briefly at the summit of Mt. Everest (over 29,000 ft) showed some startling results. First, some climbers made the ascent without auxiliary oxygen supplies, though many physiologists thought this feat impossible. Second, some physiological changes were more extreme than believed consistent with consciousness—alveolar P_{CO_2} of 7.5 torr, blood pH of 7.7, and P_{O_2} below 30 torr.

Other compensatory mechanisms operate in mountain climbers and in people living at less extreme altitudes (8000 to 10,000 ft) for several months. Cardiac output increases during the first few days, but gradually returns to normal. After two or three weeks, both hemoglobin and total blood volume begin to rise until after several months, hemoglobin content increases by about 50 percent and blood volume by about 25 percent, filling the many new

FIGURE 22.20
Mountain climbers should ascend slowly to allow their bodies to adjust to lower partial pressure of oxygen in the atmosphere. Above 10,000 feet, they should hyperventilate (take frequent deep breaths) to increase availability of oxygen at gas exchange membranes.

capillaries that develop at high altitude and increasing the blood's oxygen-carrying capacity. Drinking extra water helps to increase blood volume, and headaches that some people have while acclimating to high altitude may be due, in part, to lack of adequate water.

The affinity of hemoglobin for oxygen decreases with altitude because phosphates including DPG produced in erythrocyte metabolism bind to the hemoglobin. Up to an altitude of about 15,000 ft, this fosters oxygen unloading in the tissues. At higher altitudes, it interferes with oxygen loading in the lungs and decreases the availability of oxygen to the tissues accordingly.

Increases in the numbers of mitochondria in cells also have been observed, especially in people who have always lived at altitudes above 13,000 ft. Larger numbers of mitochondria increase the amounts of oxidative enzymes in cells and, thereby, increase their capacity to use available oxygen.

Underwater Conditions

Pressure on a person submerged in water is greater than sea-level atmospheric pressure and doubles for every 33 feet of water depth. As pressure increases, nitrogen diffuses into the blood in proportion to the pressure and elevated blood nitrogen causes **nitrogen narcosis** (nar-ko'sis), a depression of the central nervous system including the respiratory centers. (Drugs, such as alcohol, barbiturates, and heroin also cause narcosis.) In narcosis, the respiratory centers fail to increase the depth and frequency of breathing, even when P_{CO_2} increases and blood pH decreases. Nitrogen narcosis can start with euphoria—sometimes called "rapture of the deep." Impairment of nerve transmission and unconsciousness follow, probably because nitrogen dissolves lipids in cell membranes and acts as an anesthetic.

When a diver who has been in deep water for an hour or so ascends too rapidly, nitrogen dissolved in body fluids comes out of solution and forms bubbles, causing **decompression sickness.** This disorder is also called "caisson disease" because it affects workers in protective caissons under water or in tunnels. It is called "the bends" because sufferers literally become bent with excruciating pain. The bubbles obstruct blood flow and lead to hypoxia, and can disrupt cells and cause permanent nerve damage. Decompression sickness is preventable by carefully controlling the ascent according to the time and depth of a dive so nitrogen diffuses from the blood slowly and bubbles do not form. A study of "bends" victims found that many had a patent foramen ovale, which allowed nitrogen bubbles to bypass the lungs (where they are sometimes trapped) and become distributed in other tissues.

Breath-hold diving is far more limited among humans than among many other air-breathing animals. Such dives usually do not exceed one minute or 5 m and are limited by the risk of **hypothermia** (hi''po-therm'e-ah), the lowering of the internal body temperature below 35° C (Chapter 25). In spite of this generalization, some victims of near-drowning accidents have survived up to an hour's submersion in ice cold water. One explanation is that rapid chilling quickly leads to unconsciousness, preventing the use of oxygen in a panicky struggle and reducing metabolic needs for oxygen.

Head-out water immersion has several physiological effects. It increases heart stroke volume, thereby increasing cardiac output; causes diuresis (increased urine volume); and redistributes blood flow toward the liver, fatty tissues, skin, and gut. These effects make head-out water immersion a possibly beneficial treatment for heart transplant patients and patients with congestive heart failure or liver disease.

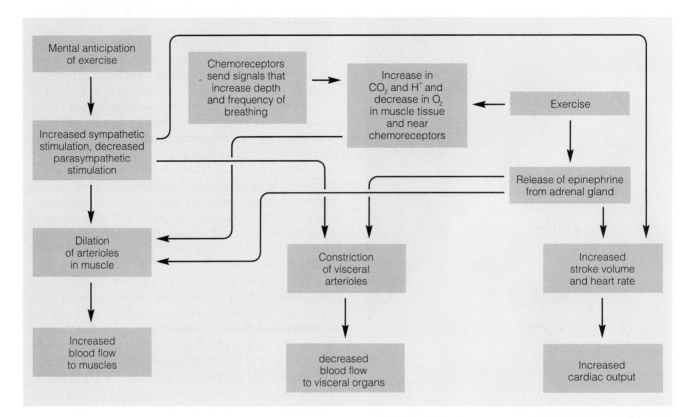

(a)

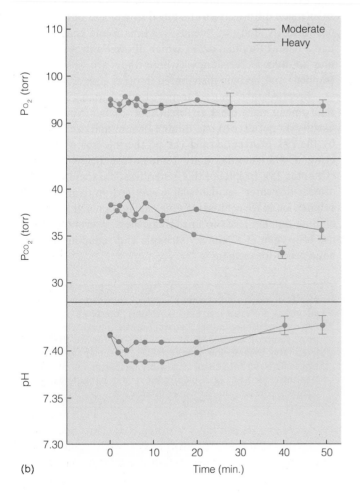

(b)

FIGURE 22.21

(*a*) A flowchart of the respiratory events in the cardiovascular system in response to exercise; (*b*) changes in blood gases during moderate and heavy exercise.

Anesthesia

Of the various substances that produce anesthesia (loss of sensation), those with the greatest effect on respiration are **inhalation anesthetics** (an″es-thet′ikz), or gases administered by inhalation. Such anesthetics—nitrous oxide, cyclopropane, and ethylene—diffuse from alveoli to the blood and are transported to the brain. They cross the blood-brain barrier and impair transmission of signals across synapses.

Cardiopulmonary Resuscitation

If a person stops breathing and his or her heart stops beating, **cardiopulmonary resuscitation** (kar″de-o-pul′mon-er-e re-sus″it-a′shun), or CPR, can temporarily maintain blood circulation and airflow. The three main steps in CPR are: (1) to make sure the airway to the lungs is open, (2) to provide air by mouth-to-mouth breathing, and (3) to keep the blood circulating by compressing the heart. Everyone should become proficient at CPR, but *no one should attempt it without adequate training.* Improperly performed, CPR can do great harm. Many colleges and several organizations such as the American Red Cross and the American Heart Association offer courses in CPR.

EFFECTS OF EXERCISE ON CIRCULATORY AND RESPIRATORY FUNCTIONS

Exercise increases CO_2 and H^+ in blood passing through muscles, thereby dilating arterioles in muscles. It also releases epinephrine, which increases cardiac output and constricts visceral arterioles. When chemoreceptors detect elevated CO_2 or H^+, they send signals that increase the depth and frequency of breathing. As cardiac output increases, the blood volume in the pulmonary circulation and alveolar capillaries increases in proportion to the increase in the systemic circulation. The combination of increased alveolar blood flow and increased depth and frequency of breathing helps to maintain blood gases and pH in a normal range (Figure 22.21).

CLINICAL APPLICATIONS

Respiratory disorders alter breathing, gas exchange, or both, and therefore disrupt homeostasis. Asthma, emphysema, and bronchitis are chronic obstructive disorders.

Asthma (az′mah) can be inherited or caused by allergens in the respiratory tract that cause histamine release. Histamine constricts bronchioles and severely reduces airflow. Though inhalation tends to push bronchioles open, exhalation tends to compress already constricted bronchioles. Epinephrine and sympathomimetic drugs are often used to treat asthma.

(a)

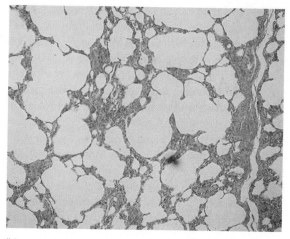

(b)

FIGURE 22.22
The lung of (*a*) a normal person, and (*b*) a person with emphysema.

In **emphysema** (em″fis-e′mah), many septa between alveoli are destroyed and much elastic tissue is replaced by connective tissue (Figure 22.22). Surface area for gas exchange is greatly reduced and expiration is especially difficult because of reduced elasticity. The lungs remain inflated and the individual must work to exhale. Eventually, alveolar capillaries are damaged by compression from inflated alveoli. Pressure increases in the pulmonary arteries and the right ventricle must work harder to force blood through the pulmonary circulation.

Most cases of emphysema are caused by smoking and by the inhalation of noxious substances over a period of time. Some are caused by an inherited deficiency of the enzyme, alpha-1 antitrypsin (AAT). This enzyme normally degrades elastase, an enzyme that helps to clean up wound debris, but in excess it damages alveoli. Smoking increases the amount of elastase in the lungs and inactivates AAT. Emphysema cannot be cured but the symptoms can be alleviated by removing the irritating substances, performing breathing exercises, and using drugs to dilate the air passageways. AAT, made by genetic engineering, is now available to treat emphysema.

An inflammation of the bronchi, **bronchitis** (brong-ki'tis), is caused by infection or another form of irritation. The linings of the bronchi swell and produce excess mucus, and cilia fail to remove it. In chronic bronchitis, fluid can accumulate in bronchial tissues and further narrow the air passageways. Treatment includes antibiotics, the removal of irritating agents, and the breathing of humidified air to relieve coughing.

Pneumonia (nu-mo'ne-ah) is characterized by acutely inflamed alveoli unable to exchange gases and sometimes filled with fluid and dead white blood cells. Pneumonia can be caused by infection with bacteria, fungi, or viruses; chemical irritants; or other foreign substances. It can affect one lobe of a lung *(lobar pneumonia)*, both lungs *(double pneumonia)*, or both alveoli and bronchi *(bronchial pneumonia)*. Antibiotics usually are effective against bacterial pneumonias, and oxygen can be given to increase its partial pressure and, thus, its diffusion rate (Figure 22.23).

Hyaline membrane disease (HMD), also called **respiratory distress syndrome of the newborn,** is caused by a deficiency in surfactant. Premature infants born before the twenty-eighth week of development are most susceptible to HMD because their lungs do not produce sufficient surfactant. The disease gets its name from the fact that after a few hours of labored breathing, the lungs fill with a high-protein fluid that has a hyaline, or glassy, appearance on autopsy. One of the major problems in treating the disease is to prevent **atelectasis** (at"el-ek'tas-is), the collapse of the alveoli. Currently, the most effective means of doing this is to provide gases to the airway under positive pressure so that alveoli remain expanded during exhalation. An experimental treatment of blowing surfactant into the lungs is being tested in a few high-risk infants. Even with the best available treatment, many infants succumb to HMD.

Smoking or the accidental inhalation of smoke or other toxic substances causes abnormalities in respiratory passageways that can lead to **lung cancer.** Smoke in the respiratory tract causes deterioration of cilia and enlargement of goblet cells in epithelia in the trachea and bronchi. Goblet cells produce excess mucus that cilia fail to remove. This process leads to smoker's cough. Carbon monoxide in smoke binds to hemoglobin and reduces its oxygen-carrying capacity. Other chemicals in smoke induce malignant changes in epithelial cells, and such cells can invade underlying tissues. Some investigators believe that only four marijuana cigarettes do the same damage as a pack of tobacco cigarettes. Furthermore, some cigarettes have filters that contain crocidolite (a kind of asbestos). A study of workers in factories where the filters are manufactured showed them to have eight times the incidence of cancer and 15 times the incidence of asbestosis as the general population. Finally, smoking allows carcinogens to bind to DNA—the event most likely to contribute to lung cancer. These effects can be prevented by not smoking and avoiding exposure to smoke and other noxious substances.

Hypoxia (hi-pok'se-ah), an insufficiency of oxygen in body tissues, can be caused by carbon monoxide poisoning, hypoventilation, and blood and circulatory disorders. When carbon monoxide is present in air, it enters alveoli and diffuses into the blood. It binds to hemoglobin and to the enzyme cytochrome oxidase. Hypoventilation (reduction in the minute respiratory volume) can be due to temporary airway obstruction as in asthma, destruction of alveoli as in pneumonia, or neuromuscular disorders as breathing muscle paralysis in poliomyelitis. Heart defects that allow mixing of arterial and venous blood (Chapter 18), and anemias also lead to hypoxia. Living at a high altitude leads to temporary hypoxia, but the body compensates for this by enlargement of the lungs and right ventricle, and increased erythrocyte production. More erythrocytes lead to more DPG release. This unloads more oxygen in the tissues, but at very high altitudes interferes with oxygen loading in the lungs.

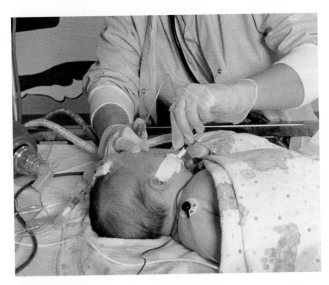

FIGURE 22.23
Nurse attends to critically ill child being kept alive on a respirator.

CLINICAL TERMS

anoxia (an-oks′e-ah)   deficiency of oxygen

apnea (ap-ne′ah)   cessation of breathing

asphyxia (as-fiks′e-ah)   suffocation

atelectasis (at″el-ek′tas-is)   partial collapse of the alveoli, imperfect expansion of the alveoli

dyspnea (disp-ne′ah)   difficult or labored breathing

hypercapnia (hi″per-kap′ne-ah)   an excess of carbon dioxide in the blood

hypoxemia (hi″pok-se′me-ah)   a deficiency of oxygen in the blood

laryngectomy (lar″in-jek′to-me)   removal of the larynx

rale   an abnormal respiratory sound usually heard through a stethoscope

rhinitis (ri-ni′tis)   an inflammation of the nasal mucous membrane; runny nose

tracheostomy (tra″ke-os′to-me)   cutting of the trachea, usually to provide an artificial airway

tuberculosis (too-ber″ku-lo′sis)   a bacterial infection usually affecting the lungs

ESSAY

SUDDEN INFANT DEATH

Parents sometimes put a perfectly normal-appearing infant to bed at night and in the morning, discover that the infant is dead. Cessation of breathing during sleep in infants, usually those less than one year old, is called **sudden infant death syndrome** (SIDS). About 7000 babies—one in 500—die of this malady each year in the United States. SIDS is the leading cause of infant death in this and many other countries.

Though the causes of SIDS are not yet completely understood, much has been learned about its anatomic and physiological effects. Autopsies of victims show thickening of muscle in the smaller branches of pulmonary arteries in about 60 percent of the cases. This may be due to an attempt to compensate for underventilation; it actually causes arterial constriction. Thickening of the right ventricle from increased pulmonary load also is seen.

Underventilation lowers arterial oxygen and causes other changes seen in over half of the SIDS victims studied. These changes include: (1) increased epinephrine release, (2) elevated erythrocyte production, (3) growth-lag characteristic of oxygen deprivation, and (4) abnormal retention of brown fat. Brown fat normally surrounds vital organs at birth and helps neonates to maintain body temperature. Brown fat cells contain large numbers of mitochondria and require oxygen to release heat. Persistence of brown fat for a longer period in SIDS infants than in normal infants may be due to oxygen deficit. Excessive epinephrine, in turn, may be due to stress created by oxygen deficit, and epinephrine may actually stimulate brown fat retention. (Brown fat reappears in adults whose adrenal glands produce excessive epinephrine.) Oxygen deficit stimulates erythropoietin release and erythrocyte production. Finally, growth is retarded in SIDS infants as in animals kept in low-oxygen environments.

That many SIDS infants are hypoxemic (have low blood oxygen) is clear, but *why* they become hypoxemic is not so clear. Respiratory centers fail to respond normally to excess carbon dioxide, but why this happens is still not understood. Glial cell proliferation and delayed myelination (and

therefore maturation) of brain stem neurons may be involved. Copper deficiency at birth has recently been suggested as a possible cause of delayed myelination. Finding high levels of hypoxanthine and adenosine (breakdown products of AMP) and elevated fetal hemoglobin in victims of SIDS suggests chronic hypoxia as a cause. Breakdown of AMP indicates cells lack oxygen to capture energy in ATP, and adenosine itself is known to inhibit respiration. The mothers of such infants often have chronic low oxygen from anemia or asthma.

In the search for causes of SIDS, genetic factors and events during pregnancy and delivery have been investigated. Because the SIDS incidence in families is no higher than in the general population, some researchers think genetic factors probably are not involved. Because the incidence of SIDS is higher in infants with type B blood than in infants with other blood types, other researchers think that a genetic factor may be involved. Several events during pregnancy are correlated with SIDS. They include bacterial infection of amniotic fluid, smoking, the use of barbiturates, and other oxygen-lowering conditions. SIDS also is correlated with crowded housing and a deficiency of the vitamin biotin combined with environmental stress. One or more of these factors are present in only about one-third of infants whose deaths are attributed to SIDS, so further research is needed to understand what roles (if any) they play in the disorder.

Predicting which infants will suffer respiratory arrest is difficult, but infants who have experienced respiratory difficulties are often placed on an electronic monitor that sounds an alarm when the infant stops breathing. Parents are trained in resuscitation techniques to handle such emergencies. One study showed that the monitor and the parents' efforts saved all but four of over 150 infants who experienced respiratory difficulty while connected to it.

Questions

1. What abnormalities are seen in SIDS victims on autopsy?
2. What are some possible causes of SIDS?
3. How does a monitor help to save infants from SIDS?

CHAPTER SUMMARY

Organization and General Functions

- The respiratory system consists of a set of passageways that carries gases to and from the respiratory membrane; gases are exchanged between the blood and lungs across the respiratory membranes.

Development

- The respiratory system develops from portions of the pharynx and from the laryngotracheal bud.

Tissues

- The nasal cavity, pharynx, and bronchi are lined with ciliated epithelium in which the cilia beat to move mucus and debris toward the pharynx.
- The epithelium of the alveoli and respiratory bronchioles is squamous epithelium, which allows gas exchange.
- Cartilage, smooth muscle, and connective tissue are found in most of the walls of passageways in the lungs.

Respiratory System Anatomy

- The nasal cavity, which contains nasal conchae and olfactory receptors, warms, filters, and moistens air, and its mucus and cilia help to remove foreign particles.
- The pharynx, divided into the nasopharynx, oropharynx, and laryngopharynx, forms the passageway between the nasal cavity and the larynx and contains tonsils.
- The larynx, between the laryngopharynx and the trachea, is supported by cartilages and contains the vocal cords. The epiglottis helps to prevent food from entering the glottis.
- The lungs, which are covered by visceral pleura, lie in the pleural cavities, with their costal surfaces next to the ribs and their medial surfaces near the mediastinum. The larger right lung has three lobes, the smaller left lung has two, and each lobe is divided into lobules covered by elastic tissue.
- The conducting portion of the bronchial tree consists of the trachea, bronchi, and nonrespiratory bronchioles. It is supported by cartilage rings or plates so it remains open for passage of gases.
- The respiratory portion of the bronchial tree consists of respiratory bronchioles, which branch into alveolar ducts, alveolar sacs, and alveoli. It has thin membranes through which gases diffuse between the lungs and blood.
- The diaphragm is innervated by the phrenic nerve, and the intercostal muscles are innervated by thoracic spinal nerves; these nerves relay signals that regulate breathing. Autonomic nerve signals maintain bronchiolar smooth muscle tone.

Ventilation

- Ventilation (inspiration and expiration) provides a continuous supply of air to the respiratory membranes. In inspiration, the volume of the thoracic cavity increases, creating a pressure gradient that draws air into the lungs. In expiration, the volume of the thoracic cavity decreases, creating a pressure gradient that causes air to flow out of the lungs.
- The total lung capacity includes tidal volume, inspiratory reserve, expiratory volume, and residual volume. Tidal volume is further divided into alveolar volume and dead space volume.

Gas Exchange

- Gas exchange depends on pressure gradients of oxygen and carbon dioxide on the two sides of a membrane. In the lungs, net oxygen diffusion is toward the blood and net carbon dioxide diffusion is out of the blood. In other tissues, net oxygen diffusion is out of the blood and net carbon dioxide diffusion is toward the blood.

Transport of Gases

- Only a small amount of oxygen in blood is dissolved in plasma with up to 99 percent being carried on hemoglobin molecules in erythrocytes.
- Most carbon dioxide in the blood is carried as bicarbonate ions formed in erythrocytes and released into the plasma. Some is dissolved in the plasma or bound to hemoglobin and certain plasma proteins.

Summary of Transport and Exchange of Gases

- Oxygen diffuses from air in alveoli into blood in the lungs and out of blood to other tissues. Carbon dioxide diffuses from other tissues into blood and out of blood into alveoli.

Regulation of Respiration

- Resting rhythmic breathing is maintained mainly by the self-excitatory capacity of the inspiratory neurons of the respiratory center.
- Depth and frequency of breathing are altered by the apneustic center, which prolongs inspiration, and by the pneumotaxic center, which facilitates expiration. Excessive inflation of the lungs inhibits the apneustic center via the Hering-Breuer reflex.
- Chemoreceptors and the chemosensitive area of the medulla respond to changes in P_{CO_2}, hydrogen ion concentration, and P_{O_2}. They send signals to the inspiratory neurons to regulate breathing.
- Regulation of respiration is altered by high altitude, underwater conditions, and inhalation anesthetics.

Effects of Exercise on Circulatory and Respiratory Functions

- The effects of exercise on circulatory and respiratory functions are summarized in Figure 22.21.

Clinical Applications

- Asthma, emphysema, and bronchitis are chronic obstructive disorders, whereas pneumonia interferes with gas exchange.
- Hyaline membrane disease, caused by a deficiency of surfactant in premature infants, involves the collapse of alveoli and, therefore, impaired gas exchange.
- Smoking damages epithelial cells, increases mucus, and can lead to malignant changes in cells; tobacco smoke contains carbon monoxide, which impairs oxygen transport.
- Hypoxia is an insufficiency of oxygen in the tissues of the body.

QUESTIONS AND PROBLEMS

The questions at the end of each chapter are numbered to correspond with the objectives listed at the beginning of the chapter. Italics indicate that a question requires critical thinking skills beyond simple factual recall.

Questions

1. *How do ventilation and gas exchange differ?*
2. (a) What are the main features of the respiratory system?
   (b) What are the main functions of the respiratory system?
3. Briefly describe the embryological development of the respiratory system.
4. *What adaptations of tissues contribute to the functions of the respiratory system?*
5. Name the structures within the following organs and give their functions: (a) nasal cavity, (b) pharynx, and (c) larynx.
6. *(a) Which structures associated with the lungs contribute to pressure changes, and which contribute to movement of gases between the respiratory membranes and the atmosphere?*
   (b) How does the circulatory system serve the lungs?
   (c) What neural signals does the respiratory system receive?
7. (a) Describe the structure and function of the conducting portion of the bronchial tree.
   (b) Describe the structure and function of the respiratory portion of the bronchial tree.
   (c) What criteria must a structure meet to be classified as a respiratory structure?
8. *(a) How is air caused to enter the lungs?*
   (b) How is air caused to leave the lungs?
   (c) What are the roles of compliance, elastic recoil, and surfactant in ventilation?
   (d) How is balance with respect to airflow and pulmonary blood flow maintained?
9. *(a) What different air volumes are concerned with ventilation?*
   (b) How can abnormalities in air volumes be used to detect functional disorders?
10. *How can partial pressures and Boyle's law help to explain gas exchange?*
11. Where and how does gas exchange occur in the lungs?
12. *How does gas exchange in other tissues differ from that in the lungs?*
13. *What factors affect the transport of oxygen in the blood, and how is oxygen released from the blood?*
14. *By what mechanisms is carbon dioxide transported in the blood?*
15. *How can blood gas measurements be used to assess respiratory functions?*
16. *In what ways do the respiratory and circulatory systems work together in transport and exchange of gases?*
17. (a) *How is resting level breathing maintained?*
    (b) *How are depth and frequency of breathing altered in accordance with physiological conditions?*
18. *How are chemical signals involved in regulating breathing?*
19. (a) *How is regulation of respiration altered at high altitude and under water?*
    (b) What are the effects of inhalation anesthetics?
20. *How does exercise improve cardiopulmonary fitness?*
21. What kinds of alterations in function occur in asthma, emphysema, hyaline membrane disease, smoking, bronchitis, pneumonia, and hypoxia?

Problems

1. Given the following respiratory volumes, calculate the total lung capacity: (a) 2500 ml inspiratory capacity, 500 ml tidal volume, and 2100 ml functional residual capacity; and (b) 2000 ml inspiratory reserve, 550 ml tidal volume, 1000 ml expiratory reserve, and 2000 ml functional residual capacity.
2. Find the minute respiratory volume given a breathing rate of 15 breaths per minute, an inspiratory reserve of 2000 ml, and a tidal volume of 500 ml.
3. Suppose partial pressures of gases on the two sides of a membrane are 90 torr for oxygen and 50 torr for carbon dioxide on side A of a membrane, and 50 torr for oxygen and 52 torr for carbon dioxide on side B of a membrane. In which direction would each gas diffuse? Under what conditions might such pressures exist in the lungs?
4. Suppose the pressure in the lungs is 758 torr. At standard atmospheric pressure, will air move into or out of the lungs? Will it move faster or slower than normal?
5. Estimate the saturation of hemoglobin with oxygen under the following conditions: (a) P_{O_2} = 50 torr and blood pH = 7.5; (b) P_{O_2} = 60 torr and body temperature = 40° C; and (c) P_{O_2} = 40 torr, blood pH = 7.3, and body temperature = 41° C.
6. How could you reassure the parents of a child who frequently holds his breath to frighten them?

OUTLINE

OBJECTIVES

DIGESTIVE SYSTEM

ORGANIZATION AND GENERAL FUNCTIONS

Before your body can benefit from a hamburger, green salad, or any other food you eat, the food must be broken down into small molecules. And those molecules must get into the blood where they can be transported to your body's cells. Breaking down food and getting it to the blood are functions of the digestive system.

The **digestive system** (Figure 23.1) receives food in large bites and converts it into small molecules as it passes through the mouth, pharynx, esophagus, stomach, and intestines. It also moves small molecules into blood or lymph. Accessory organs such as the salivary glands, liver, and pancreas aid in these processes, which are precisely controlled by a variety of homeostatic mechanisms.

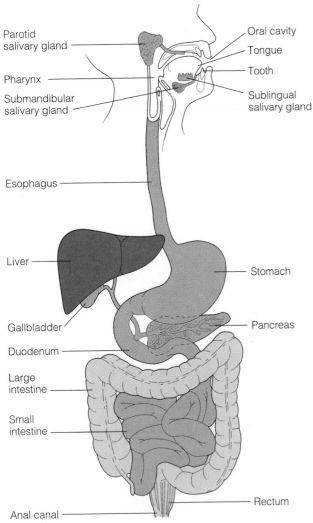

Parotid salivary gland

Pharynx

Submandibular salivary gland

Esophagus

Liver

Gallbladder

Duodenum

Large intestine

Small intestine

Anal canal

Oral cavity

Tongue

Tooth

Sublingual salivary gland

Stomach

Pancreas

Rectum

FIGURE 23.1
Organs of the digestive system.

DEVELOPMENT

Shortly after the three germ layers appear (Chapter 4), the primitive gut cavity forms a tube that is closed off at both ends (Figure 23.2). Surface ectoderm meets gut endoderm, creating two-layered membranes across the ends of the tube. The oral membrane ruptures at about 24 days to form the **stomodeum** (sto''mo-de'um), which later differentiates into the mouth. The cloacal membrane ruptures at about seven weeks to form the anal opening. As the tube elongates and differentiates, outgrowths form the liver and pancreas. By the eighth week, all the organs of the digestive tract are present in at least rudimentary form.

Certain digestive tract defects are seen at birth. In normal development, cells proliferate into the lumen of the esophagus and are later eroded away. Without this erosion, the esophagus remains blocked, causing **esophageal atresia** (ĕ-sof''ah-ge'al ah-tre'ze-ah). This condition is often accompanied by a tracheoesophageal fistule (tra''ke-o-e-sof'ah-je-al fis'tu-lah), which allows aspiration of food into the lungs. In **pyloric** (pi-lo'rik) **stenosis,** muscle that controls movement of food into the intestine is overdeveloped. As the muscle contracts, an affected infant vomits forcibly. So little food enters the small intestine that the infant becomes malnourished. **Imperforate anus** results from the failure of the cloacal membrane to rupture. These conditions can be corrected surgically.

TISSUES AND THEIR FUNCTIONS

Tissues

Throughout most of its length, the digestive tract has four basic layers, with various modifications to carry out particular functions. From the outer surface inward to the lumen, the layers (Figure 23.3) are as follows: (1) The outer membranous **serosa** (se-ro'sah) secretes a serous lubricating fluid. (2) A layer of smooth muscle, the **muscularis** (mus''ku-la'ris), composed of outer longitudinal fibers and inner circular fibers, mixes and propels food through the tract. (3) The **submucosa** (sub''mu-ko'sah) is loose connective tissue richly supplied with blood and lymphatic vessels. Both the muscularis and the submucosa are supplied with nerve endings. The **myenteric** (mi''en-ter'ik) **plexus** (plexus of Auerbach) supplies the muscularis, and the **submucosal plexus** (plexus of Meissner) supplies the submucosa. (4) The **mucosa** (mu-ko'sah) itself has three layers: (a) The thin outer **muscularis mucosae** contains smooth muscle. (b) The middle **lamina propria** (lam'in-ah pro'pre-ah) consists of loose connective tissue, blood vessels, glands, and some lymphoid tissue. (c) The inner **epithelium** lines the digestive tract and secretes mucus.

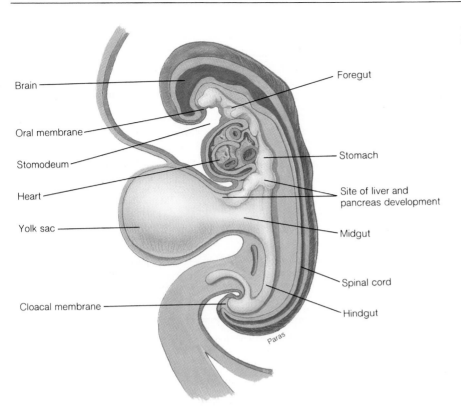

FIGURE 23.2
The digestive tract at the beginning of the fourth week of development.

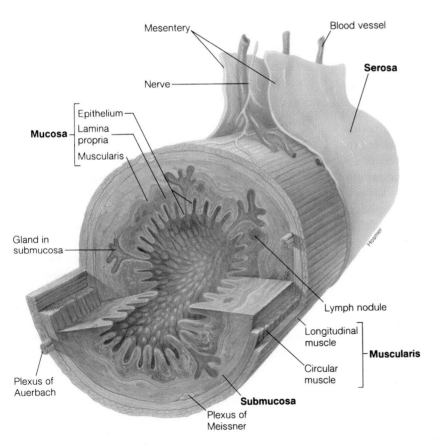

FIGURE 23.3
Layers of the digestive tract.

TABLE 23.1   Modifications of Regions of the Digestive Tract

Region	Serosa	Muscularis	Submucosa	Mucosa
Mouth and Pharynx	(Do not follow basic plan; consist of outer skin, middle skeletal muscles, and inner stratified, squamous, nonkeratinized epithelium)			
Esophagus	Adventitia, a nonserous connective tissue	Upper third skeletal muscle, middle third mixed skeletal and smooth muscle, and lower third smooth muscle	*	*
Stomach	*	Additional oblique layer of muscle internal to circular layer	*	Rugae apparent in empty stomach; many glands extend into lamina propria
Small intestine	*	*	*	Many glands extend into lamina propria; plicae circularis, villi, microvilli increase surface area; blood capillaries and lacteals abundant in lamina propria
Large intestine	*	Longitudinal muscle bands, the taeniae coli, form haustra	*	Few glands; no modifications to increase surface area
Rectum	Terminal portion surrounded by skeletal muscle	Internal smooth muscle sphincter and external skeletal muscle sphincter	*	Terminal portion contains anal columns

*No major modification of basic plan

Along the 4.5 m (15 ft) length of the living adult digestive tract, various areas are specialized to carry out specific functions. These specializations are summarized in Table 23.1.

In addition to the tissues of the digestive tract wall, a set of continuous membranes supports and protects the tract and abdominal cavity (Figure 23.4). These membranes include the **visceral peritoneum** (per″it-o-ne′um), the same membrane as the serosa, the **parietal peritoneum,** and the **mesenteries** (mes′en-ter″ēz) and **omenta** (o-men′tah). Each membrane has one or more layers of epithelial cells supported by connective tissue. The parietal peritoneum lines the entire abdominal cavity. At the posterior midline, left and right membranes come together to form a double membrane called a mesentery. Each abdominal organ is suspended by this mesentery. As the sheets separate to surround an organ, they become visceral peritoneum. Mesenteries forming a four-layer, folded apron over organs is an omentum—the **lesser omentum** (o-men′tum) between the liver and stomach, and the **greater omentum** between the stomach and small intestine.

Mesenteries are well supplied with blood vessels that carry blood to and from digestive organs. Omenta serve as fat storage depots and contain phagocytic cells that help to fight abdominal infections. *Peritonitis* (per″it-o-ni′tis), an inflammation of these membranes, can be due to infection or chemical irritation.

General Functions

The digestive system has four functions—motility, secretion, digestion, and absorption.

By **motility,** muscle contractions of the digestive tract mix food and digestive juices and propel the mixture through the tract. Food in the lumen stimulates mechanical and chemical receptors in the tract wall, which send signals via plexuses to smooth muscle in the tract. Such signals alter muscle tension and the intensity and rate of rhythmic contractions. The relay of contractions along the tract propels foodstuffs. **Intrinsic** signals can regulate gut motility locally, and **extrinsic** autonomic signals, which act mainly on the myenteric plexus, modulate their effects. In general, sympathetic signals decrease gut motility and parasympathetic signals increase it.

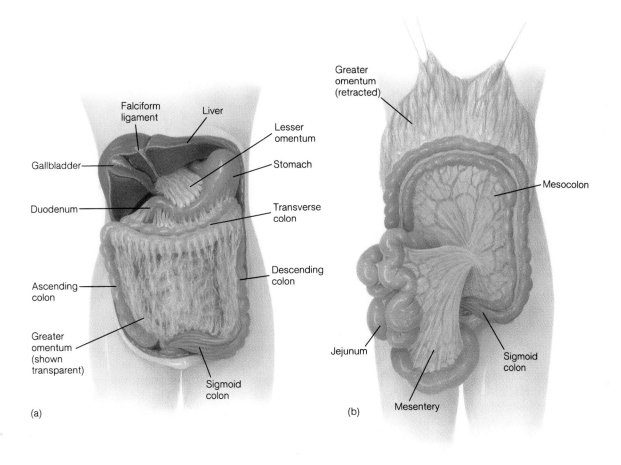

(a)

(b)

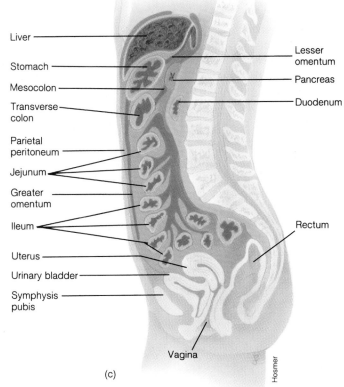

(c)

FIGURE 23.4

Mesenteries and omenta: (*a*) surface view, (*b*) deep view, and (*c*) lateral view.

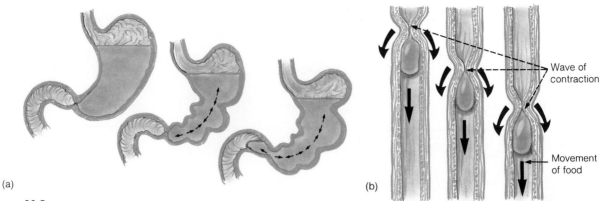

(a)

(b)

Wave of
contraction

Movement
of food

FIGURE 23.5
(*a*) Mixing, and (*b*) peristalsis.

Gut motility can be local or spread over most of the digestive tract (Figure 23.5). **Local contractions,** which originate and act in many different, limited regions, mix food with digestive juices. Waves of contraction called **peristalsis** (per″is-tal′sis) spread from the esophagus to the anus, and propel the contents of the lumen toward the anus. Waves of contraction are initiated by distention in a region of the tract, which stimulates wall receptors. Signals from the receptors cause contraction a few centimeters above the distention and relaxation below it. As the contents of the distended area are pushed toward the anus, new areas of distention arise and the cycle is repeated many times to propel foodstuffs through the entire tract.

Secretion includes the release of mucus, enzymes, and other digestive secretions. Mucus lubricates and protects. Enzymes and other digestive secretions are normally released in response to the presence of food in the tract, with the kind and amount of food determining the kinds and amounts of secretions.

Digestion, or breaking of large food molecules into smaller ones, occurs in the lumen of the digestive tract—mostly in the small intestine. Enzymes break molecules, usually by hydrolysis (Chapter 2), into their components—amino acids, monosaccharides, fatty acids, and other substances.

Absorption, the process of moving nutrients across the mucosa into blood or lymph, takes place mainly in the small intestine. Structural modifications there increase the surface area and facilitate the process.

MOUTH

Initial processing of food occurs in the **mouth** and involves the tongue, salivary glands, and teeth. The **oral cavity,** buccal cavity, or mouth (Figure 23.6), is bounded by the **teeth**; the **gingiva** (jin-ji′vah), or gums; the hard and soft palates; and the oropharynx. The border between the mouth and the oropharynx is the **fauces** (faw′sēz). Outside the oral cavity is the **oral vestibule,** the space between the gums and the lips. Both the oral cavity and the vestibule are lined with mucous membranes.

The thick, muscular **tongue,** attached to the floor of the mouth by the **frenulum** (fren′u-lum), helps to move food during chewing and swallowing. Its upper surface has rough projections called **papillae,** which are sensitive to touch. Some papillae contain taste buds (Chapter 15).

Salivary Glands and Digestion

Three pairs of **salivary glands** secrete saliva into the mouth through ducts (Figure 23.7). The large **parotid** (par-ot′id) **glands,** located near the ears, release their secretions into **parotid ducts.** Each duct passes over a buccinator muscle and opens into the oral vestibule near the upper second molars. The small **sublingual** (sub-ling′gwal) **glands** beneath the tongue have ducts that open into the floor of the oral cavity. The medium-sized **submandibular** (sub″mandib′u-lar) **glands** in the oral cavity near the angle of each mandible release secretions through **submandibular ducts.**

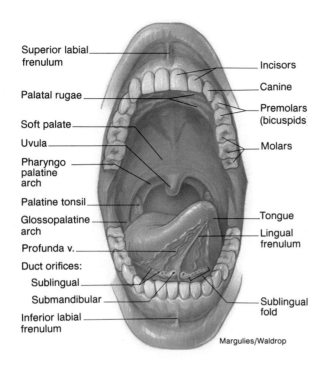

Superior labial frenulum

Palatal rugae

Soft palate

Uvula

Pharyngo palatine arch

Palatine tonsil

Glossopalatine arch

Profunda v.

Duct orifices:

Sublingual

Submandibular

Inferior labial frenulum

Incisors

Canine

Premolars (bicuspids)

Molars

Tongue

Lingual frenulum

Sublingual fold

Margulies/Waldrop

FIGURE 23.6
Anterior view of an open mouth.

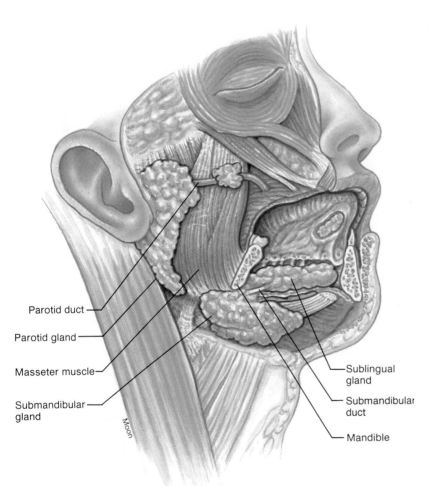

Parotid duct

Parotid gland

Masseter muscle

Submandibular gland

Sublingual gland

Submandibular duct

Mandible

Moon

FIGURE 23.7
Location of salivary glands.

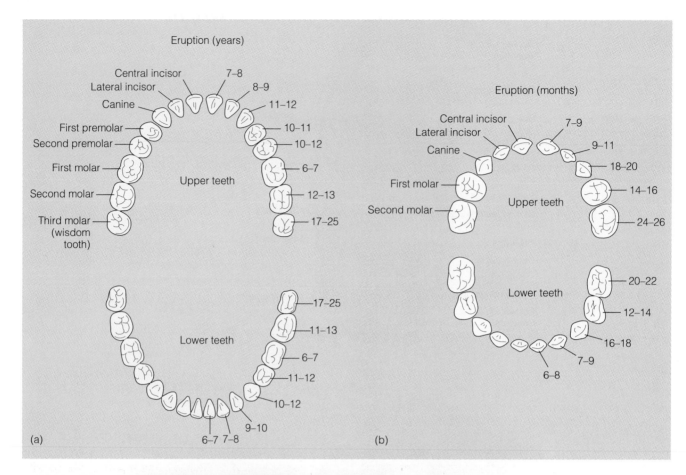

Eruption (years)

Central incisor — 7–8
Lateral incisor — 8–9
Canine — 11–12
First premolar — 10–11
Second premolar — 10–12
First molar — 6–7
Second molar — 12–13
Third molar (wisdom tooth) — 17–25

Upper teeth

Lower teeth
17–25
11–13
6–7
11–12
10–12
9–10
6–7 7–8

(a)

Eruption (months)

Central incisor — 7–9
Lateral incisor — 9–11
Canine — 18–20
First molar — 14–16
Second molar — 24–26

Upper teeth

Lower teeth
20–22
12–14
16–18
7–9
6–8

(b)

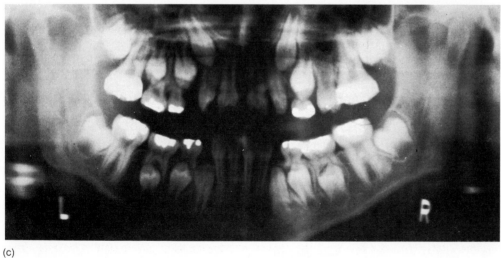

(c)

FIGURE 23.8
(a) Permanent teeth, (b) deciduous teeth, and (c) X ray of developing and erupting teeth.

These ducts open into the oral cavity near the lower central incisors. *Mumps* is a viral infection of the salivary glands. It is also called *parotitis* (par″ot-i′tis) because the parotid gland usually is the one infected.

Salivary glands secrete over 1 l of saliva daily. The parotids secrete mainly the starch-digesting enzyme **salivary amylase** (am′il-ās), whereas the others secrete both the enzyme and **mucin** (mu′sin), a glycoprotein that lubricates membranes and binds food particles together. Amylase and mucin account for only 0.5 percent of the saliva volume, the rest being mainly water and electrolytes. Electrolytes in saliva are subsequently reabsorbed in the intestine and continuously recycled. Saliva also contains a small amount of fat-digesting **lingual lipase** (li′pās) and thiocyanate (SCN^-), which may destroy some bacteria in food.

Salivary amylase, the first enzyme to act on food, begins starch digestion in the mouth. To demonstrate this, chew a piece of bread thoroughly and avoid swallowing it as long as you can. Eventually, the bread will begin to taste sweet as disaccharides are released from the starch. Food usually is swallowed before salivary amylase action is complete.

Teeth and Chewing

Each human adult jaw normally contains 16 teeth—eight in each symmetrical left and right side—making a full adult complement of 32 teeth (Figure 23.8a). From the midline of the upper or lower jaw to the angle of the mandible are two **incisors** (in-si′zerz); one **cuspid** (kus′pid), or canine; two **bicuspids** (bi-kus′pidz), and three **molars** (mo′larz). The posteriormost molars are sometimes called "wisdom teeth." Each tooth is adapted for a particular function. Incisors bite, canines tear, and bicuspids and molars grind.

In the human infant, the first **deciduous** (de-sid′u-us) **teeth** (milk teeth) erupt through the gums at about six months of age and more teeth erupt periodically until the child has a set of 20 teeth at about age two (Figure 23.8b). Each quadrant of the mouth contains two incisors, one cuspid, and two primary molars.

At about age six, the first permanent molars erupt posterior to the deciduous teeth, and the deciduous teeth begin to be replaced by the permanent teeth. As each permanent tooth completes its development beneath the gum, it moves toward the surface. If it is beneath a deciduous tooth, it exerts pressure and erodes roots until the deciduous tooth falls out.

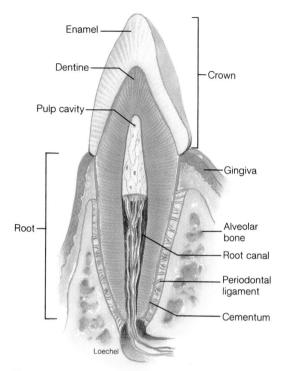

FIGURE 23.9
A longitudinal section through a tooth.

Structurally, a tooth has a **crown** above the gum and a **root** below it. The crown is covered with **enamel,** the hardest substance in the body, and the root covered with **cementum** (se-men′tum), a substance similar to bone (Figure 23.9). Enamel withstands the grinding of one tooth against another during chewing. The **neck** of a tooth is a slight constriction at the gumline that separates the root and crown. Under the hard outer layer of both the crown and the root is **dentin** (den′tin), another bonelike substance that forms the bulk of the tooth. The central **pulp cavity** extends down into the roots as the **root canals.** Blood and lymph vessels and nerves enter the pulp cavity through the **apical** (ap′ik-al) **foramina** at the tips of the roots. **Pulp** consists of connective tissue in which vessels and nerves are imbedded. Each tooth is held in an **alveolus** (tooth socket) by fibers of a **periodontal** (per″e-o-don′tal) **ligament** between the cementum and the socket.

The presence of a **bolus** (bo′lus), or mass, of food in the mouth initiates chewing movements. The bolus first inhibits chewing muscles and the lower jaw drops. The drop initiates a stretch reflex and causes jaw muscles to contract and press on the bolus. The rhythmic stretching and contracting movements break food apart, create small particles, and increase surface area. Chewing breaks up cellulose in cell walls of foods derived from plants and releases nutrients from the cells. When a bolus is thoroughly chewed, the tongue voluntarily pushes it into the pharynx and swallowing begins.

Though absorption is not a primary function of the mouth, some absorption can occur through the thin membrane beneath the tongue. Medications such as nitroglycerine (used to alleviate angina) can be administered by this route.

PHARYNX AND ESOPHAGUS

Extending from the soft palate to the larynx and esophagus is the **pharynx** (Chapter 22). In addition to serving as an airway, the pharynx also serves as a food passageway between the oral cavity and esophagus. The **esophagus** (ĕ-sof′ag-us) is a muscular tube about 25 cm long lying posterior to the trachea. It has a **pharyngoesophageal sphincter** (far-ing″go-e-sof′ah-je″al sfingk′ter) near the pharynx and a **gastroesophageal** (gas″tro-e-sof′ah-je″al) **sphincter** near the stomach. Esophageal tissues vary from the general plan in that: (1) the lining is stratified squamous epithelium, (2) muscle ranges from mostly skeletal near the pharynx to mostly smooth near the stomach, and (3) above the diaphragm, a connective tissue **adventitia** (ad″ven-tish′e-ah) replaces the serosa.

Once the tongue pushes a bolus into the pharynx, **deglutition** (deg″loo-tish′un), or swallowing, is essentially involuntary and automatic. The bolus activates swallowing receptors along the pharyngeal surface that send signals to the swallowing center in the brain stem. Signals from the swallowing center initiate pharyngeal peristaltic contractions, during which elevation of the soft palate closes the nasal passages and the epiglottis keeps food from entering the glottis. The wave of peristalsis opens the pharyngoesophageal sphincter and propels the bolus into the esophagus. Except during swallowing, this sphincter remains closed and keeps air from entering the esophagus during breathing. Swallowing a mouthful of food takes only one or two seconds, and breathing is temporarily arrested during that time.

The esophagus functions solely to conduct food to the stomach. The primary peristaltic wave that propels a bolus into the esophagus continues throughout the esophagus, pushing the bolus all the way to the stomach. Muscle relaxation preceding a peristaltic contraction opens the gastroesophageal sphincter. Passage of a bolus through the esophagus takes less than 10 seconds and usually requires only peristaltic contractions. If the esophagus becomes distended, esophageal contractions also help to move the bolus. Except when food is entering the stomach (or during vomiting), the gastroesophageal sphincter remains closed and prevents regurgitation of acidic stomach contents.

Secretion in the pharynx is limited to mucus from goblet cells in its epithelial lining. In the esophagus, it is limited to mucus secreted from tubuloalveolar glands in the lamina propria and submucosa. No digestion or absorption occurs but salivary amylase continues to act on carbohydrates.

STOMACH

Structure

The **stomach,** a hollow, J-shaped organ that can store as much as one liter of food for four to six hours, is connected to the esophagus at one end and the small intestine at the other. It has a greater and a lesser curvature and several anatomical regions adjacent to its central body (Figure 23.10a). The **cardiac region** is nearest the esophagus and the **fundus** (fun′dus) extends superiorly from it. The narrow **pylorus** (pi-lo′rus) terminates in the **pyloric** (pi-lor′ik) **sphincter,** which separates the stomach from the small intestine. Internally, the wall is folded into numerous ridges called **rugae** (roo′je).

The stomach mucosa consists of simple, columnar epithelium interrupted by millions of **gastric** (gas′trik) **pits** that open into the gastric glands (Figure 23.10b). Nearest to the surface are **mucous neck cells,** which with mucosal goblet cells, synthesize and secrete enough mucus to coat the entire gastric mucosa. Deeper in the gland are the **chief cells, parietal cells,** and **argentaffin** (ar-jen′taf-in) **cells.** Chief cells secrete **pepsinogen** (pep-sin′o-jen), an inactive form of the enzyme **pepsin** (pep′sin) and small amounts of **gastric amylase** and **gastric lipase.** The amylase is inactive because of acidity, and the lipase acts only on butterfat. Parietal cells secrete **hydrochloric** (hi″dro-klor′ik) **acid** (HCl) and **intrinsic factor,** a polysaccharide that binds to vitamin B_{12} and facilitates its absorption in the small intestine. Argentaffin cells secrete serotonin and histamine, but the reason these substances are released into the stomach is not known. The pylorus contains **G-cells** that secret the hormone **gastrin** into the blood and pyloric glands that secrete a particularly viscous alkaline mucus.

Motility

When food is present in the stomach, gentle mixing contractions agitate gastric secretions and food particles next to the mucosa. Stronger peristaltic contractions propel the mixture toward the pylorus. Stomach distention causes stretch receptors to send signals via the vagus nerve to the medulla. The medulla then sends signals, also via the vagus nerve, that initiate secretion, mixing, and peristaltic contractions. Increasing distention causes stretch receptors to send more signals and such signals accelerate stomach emptying. As the food and digestive juice mixture reaches the pylorus, mixing contractions become stronger, and by the time it reaches the pyloric sphincter, it is a semifluid mixture called **chyme** (kĭm). Each time the pyloric sphincter opens, as it does before a peristaltic contraction, several milliliters of chyme enter the small intestine.

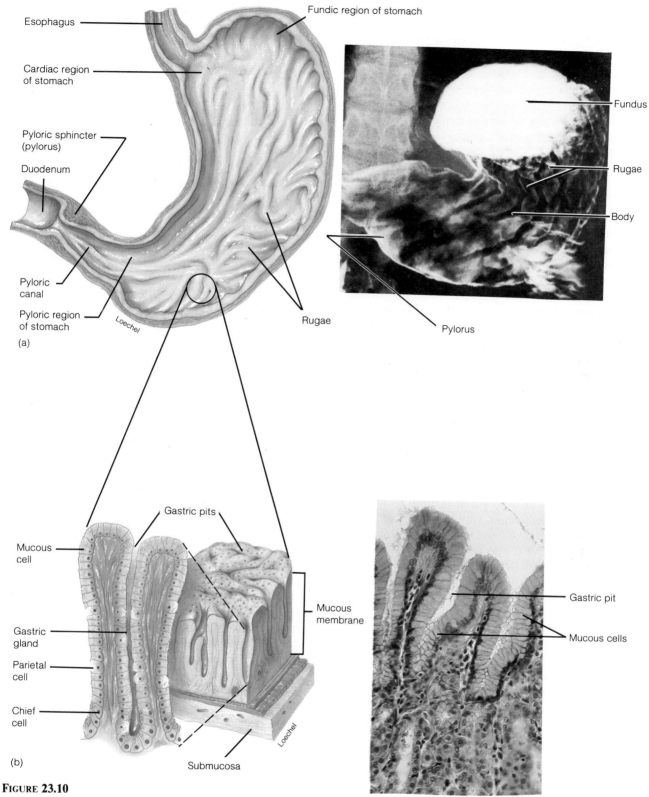

Esophagus

Fundic region of stomach

Cardiac region of stomach

Pyloric sphincter (pylorus)

Duodenum

Pyloric canal

Pyloric region of stomach

Loechel

Rugae

(a)

Fundus

Rugae

Body

Pylorus

Gastric pits

Mucous cell

Gastric gland

Parietal cell

Chief cell

Mucous membrane

Loechel

(b)

Submucosa

Gastric pit

Mucous cells

FIGURE 23.10

(*a*) Macroscopic anatomy of the stomach in diagram and X ray.
(*b*) Microscopic anatomy of the stomach in diagram and
photomicrograph.

Enteric hormones and the enterogastric reflex also
regulate stomach motility. Proteins and certain other substances in the stomach initiate release of the hormone **gas-**
trin (gas′trin). Though gastrin affects mainly secretion, it
also stimulates peristalsis. Certain other enteric hormones

reduce stomach motility. The **enterogastric** (en"ter-o-gas'trik) **reflex** slows peristalsis when the duodenum is distended or irritated, or when chyme is fat laden, very acidic (below pH 4), excessively hypertonic or hypotonic. It maintains chyme flow into the small intestine at a rate no faster than the chyme can be processed. This reflex is initiated by signals from duodenal receptors relayed back to the stomach. Strong signals can stop chyme flow until events in the duodenum have digested fat, buffered acid, or adjusted tonicity.

Secretions and Their Regulation

In the stomach, acid and enzyme secretion creates conditions that might allow digestion of the stomach itself if not properly regulated. HCl synthesis and secretion probably involve the following steps (Figure 23.11): Water in the parietal cells ionizes to form H^+ and OH^-, and Cl^- and CO_2 diffuse from interstitial fluids into the parietal cells. Carbonic anhydrase causes CO_2 to combine with H_2O to form carbonic acid (H_2CO_3). This acid partially ionizes and releases bicarbonate (HCO_3^-) and H^+. The HCO_3^- diffuses into the interstitial fluid, where it replaces negative charges lost by Cl^- entering cells. H^+ from carbonic acid combines with OH^- from water, reforming some water molecules. The Cl^- and the remaining H^+ are actively transported separately across parietal cell membranes to small passageways called canaliculi (little canals) made of folds in the membrane.

Gastrin stimulates the release of H^+ and Cl^- from the canaliculi into the lumen of the stomach. Only a small quantity of water is released with the ionized HCl, so the acid reaching the lumen has a very low pH of 0.8. Once it enters the lumen, HCl mainly lowers the pH sufficiently to activate pepsin, but it also kills microorganisms that have reached the stomach. Pepsinogen is activated to pepsin, a proteolytic (protein-digesting) enzyme, by acidity of pH 2 and by previously activated pepsin.

Though the whole digestive tract lining is coated with mucus, stomach mucus is very important in preventing damage by acid and pepsin. Especially thick, sticky, alkaline mucus provides special protection in the pyloric region where the most acid chyme accumulates. In addition, mucosal cells are tightly joined in a solid barrier that prevents acid seepage from the lumen back through the mucosa.

Gastric secretions are regulated by a multiplicity of interacting factors. **Cephalic** (sef-al'ik) **factors** originate in the brain. Food in the mouth or the sight or smell of food stimulates centers in the limbic system and hypothalamus. Such stimulation sends signals via the vagus nerve to release gastric secretions. **Gastric factors** are primarily local; that is, they originate and act in the stomach itself. Food in the stomach stimulates mechanical and chemical receptors in the stomach wall to send signals via the submucosal plexus to parietal cells, causing them to secrete HCl. **Secretagogues** (se-kre'ta-gogz), substances

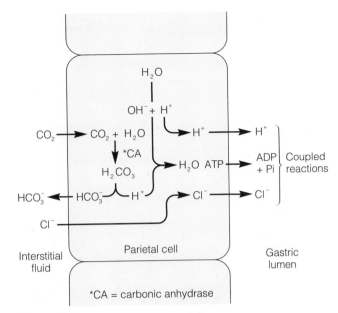

FIGURE 23.11
The formation of hydrochloric acid in the parietal cells of the gastric mucosa.

such as partially digested proteins, alcohol, and caffeine that stimulate secretion, act locally and specifically to release gastrin. Gastrin, being a hormone, is secreted into the blood, through which it is carried to gastric glands. Gastrin stimulates HCl secretion, until stomach acidity below pH 2 inhibits gastrin secretion by negative feedback. **Intestinal factors** such as the enterogastric reflex act by feedback from the small intestine on the stomach.

Digestion and Absorption

In the stomach, both digestion and absorption are limited. Hydrolysis of proteins to shorter chains called polypeptides by the action of pepsin and HCl is the main digestive action. A little butterfat and starch may be broken down by **gastric lipase** and **gastric amylase,** respectively. Absorption is severely limited by the thick mucous that prevents HCl and pepsin from digesting the mucosa. Only alcohol, aspirin, and some lipid-soluble drugs can cross the mucus and be absorbed in the stomach.

Processes that occur in the stomach are summarized in Table 23.2.

LIVER AND PANCREAS

Because functions of the liver and pancreas are essential to normal small intestine functions, we will digress to consider those organs.

Structure of the Liver

The **liver** is a large organ weighing about 1.4 kilograms and occupying the upper right quadrant of the abdominal cavity (Figure 23.12a). Its main right and left lobes are

TABLE 23.2   Processes that Occur in the Stomach

Motility	Mixing contractions break up food particles and mix them with digestive juices. Peristaltic contractions propel food and digestive juices toward the pyloric sphincter. Peristaltic contractions also open the pyloric sphincter and push chyme into the small intestine.	Digestion	Pepsin and HCl digest proteins to polypeptides. Gastric amylase may digest small quantities of starch. Gastric lipase digests butterfat.
Secretion	Parietal cells secrete HCl and intrinsic factor. Chief cells secrete pepsinogen, which is activated to pepsin by HCl or active pepsin, and small amounts of gastric amylase and gastric lipase.	Absorption	Absorption is limited by the thick mucous barrier. The main substances absorbed in the stomach are alcohol, aspirin, and fat-soluble drugs.

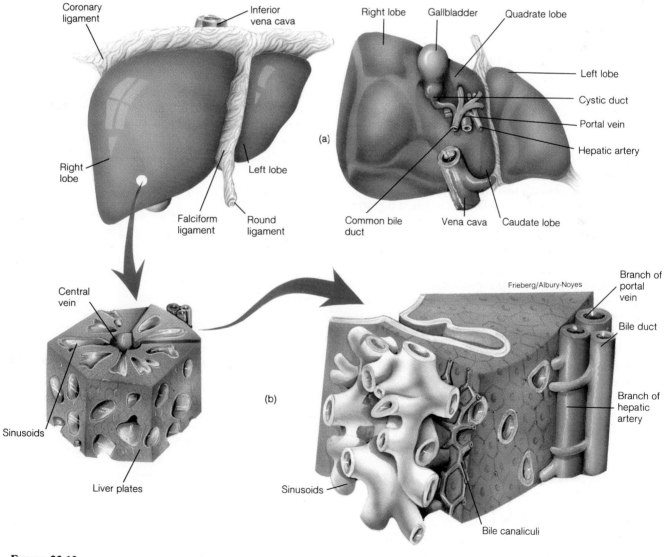

FIGURE 23.12

(*a*) Macroscopic structure of the liver, and (*b*) microscopic structure of the liver.

separated by the **falciform** (fal′sif-orm) **ligament,** a membrane that is continuous with the peritoneum. The **round ligament,** a fibrous cord of connective tissue between the falciform ligament and the umbilicus, represents the remains of the fetal umbilical vein. Seen from the inferior side, the right lobe of the liver has two smaller lobes, the posterior *caudate lobe* and the inferior *quadrate lobe.* Also visible on the inferior surface are the gallbladder, common bile duct, hepatic portal vein, hepatic artery, and the inferior vena cava.

The liver has numerous functional units called **lobules** (Figure 23.12b). Each cylindrical lobule is about 2 mm in diameter and is completely surrounded by connective tissue. From the **central vein** in the center of a lobule, cords of liver cells radiate out in all directions. In the spaces between these cords of cells are **sinusoids,** wide thin-walled blood vessels lined with endothelial cells and **Kupffer's** (koop′ferz) **cells.** Kupffer's cells are phagocytes that remove microorganisms and foreign substances from the blood. Around the periphery of each lobule are five to seven **portal triads,** clusters of three vessels—a branch of the hepatic portal vein, a branch of the hepatic artery, and a bile duct.

Blood enters the liver from both the hepatic artery and the hepatic portal vein. Blood in the hepatic artery, which comes from the aorta, carries oxygen and lipids it has received from the lymphatics. Blood in the hepatic portal vein, which comes from the capillaries of the small intestine and other digestive organs, carries monosaccharides, amino acids, and other nutrients. Blood from branches of both vessels passes along the triads and through the sinusoids to the central vein of a lobule. Blood from the central veins drains into a hepatic vein.

Liver cells secrete small quantities of **bile,** a digestive secretion, into minute canaliculi that carry it to bile ducts in the triads. These ducts converge to form the left and right **hepatic** (he-pat′ik) **ducts,** which, in turn, form the **common hepatic duct.** The common hepatic duct joins the **cystic** (sis′tik) **duct** from the gallbladder to form the **common bile duct.** Outside the liver, the common bile duct joins the pancreatic duct near the duodenum. When the **sphincter of Oddi** (od′e) is closed, some bile flows along the cystic duct to the gallbladder. Pathways of blood and bile inside the liver are shown in Figure 23.13a, and pathways of bile outside the liver are shown in Figure 23.13b.

The small pear-shaped **gallbladder,** attached to the ventral surface of the liver, has an epithelial lining and a muscular wall. Water is reabsorbed from bile in the gallbladder until the bile becomes 5 to 12 times as concentrated as when secreted. The gallbladder releases concentrated bile after meals.

Digestive Functions of the Liver

The main digestive function of the liver is to produce bile, a fluid consisting of water, bile acids, cholesterol, bile pigments, certain other lipids, and electrolytes. The liver uses cholesterol to make the primary **bile acids, cholic** (ko′lik) **acid** and **chenodeoxycholic** (ke″no-de-ok″se-kol′ik) **acid.** **Bile salts** are bile acids that have lost a hydrogen ion and gained a sodium or potassium ion, and the terms can be used interchangeably. Bile acids combine with either the amino acid **glycine** (gli′sēn) or an amino acid derivative **taurine** (taw′rēn) to form **conjugated bile acids,** which travel in bile to the small intestine where they emulsify (break up) fat globules. In the intestine, most bile salts are deconjugated (undergo loss of glycine or taurine) and recycled to the liver for reuse.

Bile contains several substances with no digestive function. The bile pigments **biliverdin** and **bilirubin,** hemoglobin breakdown products, are merely transported in the bile to the small intestine where they can be excreted. Lipids found in bile include cholesterol, fatty acids, and the phospholipid **lecithin** (les′ith-in). Lecithin and bile salts help to keep cholesterol in solution. If cholesterol precipitates, it forms gallstones. Electrolytes in bile are similar to those in plasma.

Other Liver Functions

Along with aiding fat digestion, the liver carries out many other functions (Table 23.3). It helps to regulate metabolism of carbohydrates and fats, synthesizes proteins, removes wastes, and stores vitamins. As these processes require many enzymes, it is not surprising that liver cells contain more enzymes than most other cells.

TABLE 23.3 Functions of the Liver

Carbohydrate Metabolism
Glycogenesis—the removal of excess glucose from the blood and the storage of glucose in glycogen
Glycogenolysis—the conversion of glycogen back to glucose for release into the blood when the glucose concentration drops
Gluconeogenesis—the synthesis of glucose from noncarbohydrate nutrients as occurs when the glucose and glycogen supplies are depleted

Fat Metabolism
Synthesis of cholesterol, the precursor of all steroids in the body
Synthesis of phospholipids used in growth and repair of cells
Synthesis of lipoproteins, carrier molecules by which lipids are transported to other cells for energy and to adipose cells for storage

Protein Metabolism
Synthesis of plasma proteins
Deamination of amino acids and the production of ammonia

Other Synthetic Functions
Synthesis of urea, a nitrogenous waste product that is formed, in part, from ammonia released by deamination
Synthesis of several factors that are essential for the clotting of blood
Detoxification of toxic substances, usually by the addition of some chemical substance rendering the toxic substances less harmful

Decomposition of Hemoglobin from Worn-out Erythrocytes

Storage Functions
Storage of a year's supply of vitamin A, several months' supply of vitamin D, and several years' supply of vitamin B$_{12}$
Combination of iron with the liver protein **apoferritin** for later use in the synthesis of hemoglobin

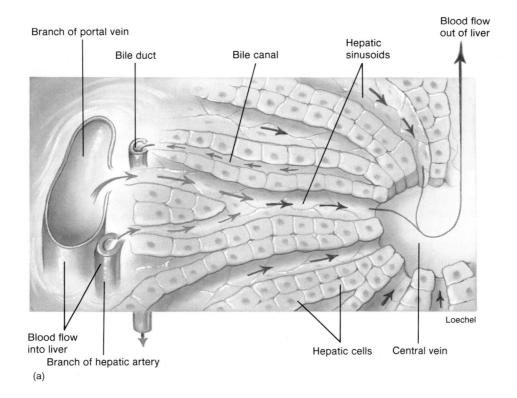

Branch of portal vein

Bile duct

Bile canal

Hepatic sinusoids

Blood flow out of liver

Blood flow into liver

Branch of hepatic artery

Hepatic cells

Central vein

Loechel

(a)

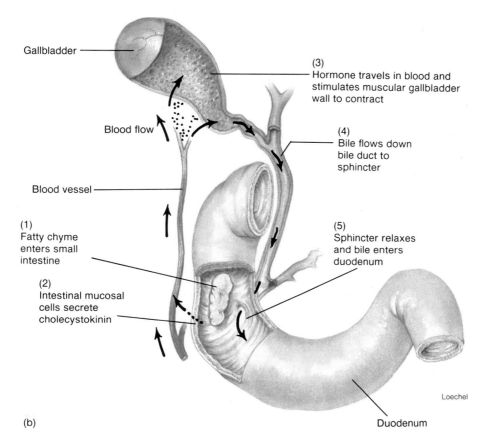

Gallblader

(3) Hormone travels in blood and stimulates muscular gallbladder wall to contract

Blood flow

(4) Bile flows down bile duct to sphincter

Blood vessel

(1) Fatty chyme enters small intestine

(5) Sphincter relaxes and bile enters duodenum

(2) Intestinal mucosal cells secrete cholecystokinin

Loechel

(b)

Duodenum

FIGURE 23.13
(*a*) Blood and bile flow in opposite directions in liver lobules. (*b*) Bile is stored in the gallbladder until cholecystokinin stimulates its secretion through ducts to the duodenum.

The liver helps to maintain a normal blood glucose concentration by storing excess glucose as glycogen, such as after a meal, releasing it as the blood glucose drops, and making glucose from other nutrients if needed (Chapter 24). It makes cholesterol (sometimes even if not needed) and blood proteins—albumins, globulins, blood clotting factors, and transport proteins for lipids, some hormones, and minerals such as iron and copper.

The liver rids the body of wastes and toxic substances. It converts toxic ammonia from amino acids to much less toxic urea, which is transported to the kidney for excretion. It also renders a variety of toxic substances less toxic by modifying their molecular structure. (See the essay in Chapter 24).

Finally, the liver stores several vitamins. It can store a year's supply of vitamin A, several months' supply of vitamin D, and several years' supply of vitamin B_{12}.

A functional liver is essential for life. Unfortunately, toxic substances, especially large amounts of alcohol over a period of years can severely damage it. Among the estimated 7 million alcoholics in the United States, nearly a third will suffer severe liver damage within 5 to 20 years (Chapter 25). Liver damage begins with accumulation of triacylglycerols and free fatty acids in the liver, to make up from 3 to 40 percent of the organ weight. Liver cells die, and fibrous connective tissue infiltrates, blocking blood flow through the liver and making it orange and leathery—like a football. This condition is called **cirrhosis** (sir-ro′sis) **of the liver.** (Cirrhosis means orange-yellow.)

After liver damage, bile duct obstruction, or red blood cell destruction (as in a transfusion reaction), the liver may be unable to degrade heme as fast as it is released from red blood cells. The bile pigment bilirubin accumulates in tissues, causing **jaundice,** or yellowing of the skin. If toxic substances are avoided and the patient begins to eat a high-protein, low-fat diet before damage is severe, the liver can repair itself by cell division in which new cells replace dead ones. Such large organ regeneration is unusual in higher organisms such as humans, and it can mean the difference between life and death.

Structure of the Pancreas

The **pancreas** (pan′kre-as) is a soft, spongy organ about 2.5 cm wide and 12 to 15 cm long located inferior to the stomach (Figure 23.14). Its head fits into a curve of the duodenum, and its body and tail extend laterally. A **pancreatic** (pan″kre-at′ik) **duct** (duct of Wirsung) collects pancreatic juice from smaller ducts within the pancreas and carries it to the duodenum. Usually, the pancreatic duct joints the common bile duct, and the two enter the duodenum together about 10 cm below the stomach at the **ampulla of Vater** (am-pul′lah of fat′er). Variations in pancreatic ducts are common and usually have no significant effects. Some people have separate pancreatic and bile

ducts entering the duodenum and others have a second pancreatic duct, the **duct of Santorini** (sahn″to-re′ne), emptying into the duodenum about 2 cm nearer the stomach. A few have both these modifications.

Digestive secretions from the pancreas include enzymes from cells in **acini** (as′in-i) and water and bicarbonate ions from tubule cells. Tiny ducts carrying enzymes from acini converge into larger tubules that add water and bicarbonate before the secretion goes to the duodenum.

Digestive Functions of the Pancreas

Acinar cells of the pancreas secrete three proteolytic enzymes, **trypsin** (trip′sin), **chymotrypsin** (ki″mo-trip′sin), and **carboxypeptidase** (kar-bok″se-pep′tid-ās) as inactive **trypsinogen** (trip-sin′o-jen), **chymotrypsinogen** (ki″mo-trip-sin′o-jen), and **procarboxypeptidase** (pro″kar-bok″se-pep′tid-ās), respectively. Activation of these enzymes in the small intestine as they mix with chyme prevents them from digesting pancreatic or intestinal tissues. Acinar cells also secrete **pancreatic amylase, pancreatic lipase, cholesteryl esterase** (ko-les′ter-il es′ter-ās), **ribonuclease** (ri″bo-nu′kle-ās), and **deoxyribonuclease** (de-ok″se-ri″bo-nu′kle-ās). All pancreatic enzymes act in the small intestine. Bicarbonate buffers the acid in chyme and water dilutes hyperosmotic chyme.

SMALL INTESTINE

Structure

The **small intestine,** a tubular organ averaging about 2.5 cm in diameter, extends from the pylorus to the large intestine, and its loops and coils fill a large part of the abdominal cavity (refer back to Figure 23.1). During life, the small intestine is about 3 meters long, though when longitudinal muscle relaxes after death, it can be twice as long.

The small intestine has three regions. The **duodenum** (du″o-de′num), the 25 cm nearest the stomach, receives chyme from the stomach and secretions from the liver and pancreas. The middle **jejunum** (jĕ-joo′num) is about 120 cm long, and the remainder, the **ileum** (il′e-um), is about 160 cm long. Both the jejunum and the ileum are suspended by mesentery. Lymph notes are found throughout the intestine and its mesenteries.

The layers of the small intestine closely follow the general plan of the digestive tract. **Intestinal glands,** or *crypts of Lieberkühn* (kriptz of le′ber-kun), in the mucosa secrete digestive enzymes. Submucosal **Brunner's glands** in the duodenum and goblet cells throughout the intestine secrete lubricating mucus. The submucosa of the ileum contains lymphatic tissues called **Peyer's patches.**

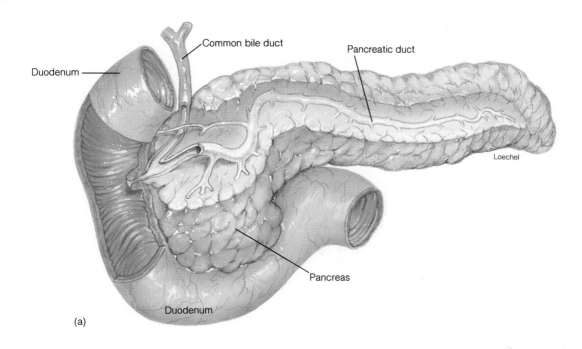

(a)

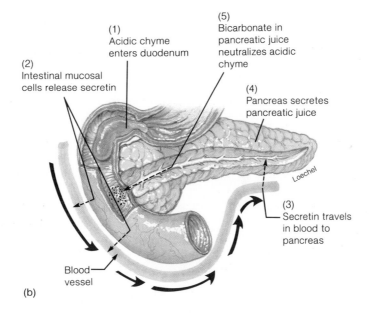

(b)

FIGURE 23.14
(*a*) The pancreas and its ducts, and (*b*) secretin and pancreozymin stimulation of bicarbonate and enzymes secretion respectively.

Several structures increase the surface area of the small intestine. The **plicae circulares** (pli′ke sir-ku-la′ris), circular folds of the submucosa are found throughout the small intestine, but are most pronounced in the jejunum. These folds increase surface area by about three times. Fingerlike projections of the mucosa called **villi** (vil′i) increase the surface area by about 10 times (Figure 23.15).

Each villus has surface epithelial cells with about 600 folds in the cell membrane called **microvilli** (mi″kro-vil′i). The microvilli, also called a *brush border,* increase the surface area by another 20 times. Surface area is exceedingly important for efficient absorption, and the total absorptive area of the small intestine is about 250 m² —the area of a 10- by 25-m swimming pool or one-tenth the area of a football field.

FIGURE 23.15
Location and structure of villi.

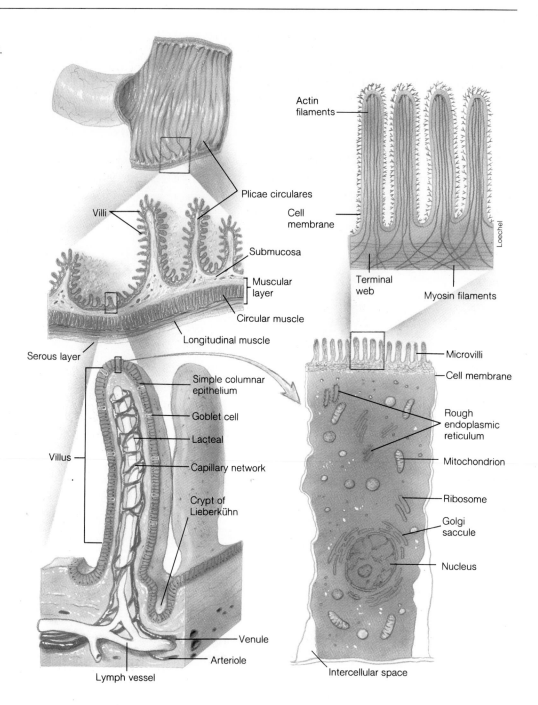

Motility

Small intestine motility serves both to propel and mix chyme. Propelling peristaltic contractions are initiated by intestinal wall stretching. Stretch receptors relay signals that activate the **myenteric reflex,** which affects both circular and longitudinal smooth muscle of the muscularis layer. Circular muscle contracts behind the stretched region and longitudinal muscle contracts temporarily shortening the region. Chyme is pushed toward the large intestine at about 1 cm per minute.

Within each villus, imbedded in a core of connective tissue, are blood capillaries, a lymph vessel called a **lacteal** (lak′te-al), and a nerve fiber. The close proximity of blood and lymph vessels to the epithelium facilitates absorption. Most nutrients travel from the digestive tract to the blood, but fats enter lacteals and later drain with lymph into the blood.

Segmentation contractions, local ring-like chopping contractions occurring at regular intervals along the small intestine, mix the chyme and increase its exposure to enzymes in the digestive juices. As one set of segmentation contractions subside, another set appears at intervals between the first set.

Contractions of the muscularis mucosae, which has fibers extending into the villi, move villi intermittently and agitate fluids around them. Such agitation increases the rate of absorption. **Villikinin** (vil″ĭ-ki′nin), a mucosal hormone, stimulates movement of villi. Microvilli themselves move when their actin and myosin filaments slide over one another.

Chyme distending the terminal ileum opens the **ileocecal** (il″e-o-se′kal) **valve** and peristalsis pushes the chyme into the large intestine. The **gastroileal** (gas″tro-il′e-al) **reflex,** which moves chyme through the ileocecal valve, is initiated by food in the stomach. Its main function is to force chyme from a previous meal out of the small intestine and make way for a new meal.

Secretion

The liver, pancreas, and the intestinal epithelium itself secrete substances into the intestinal lumen. Bile from the liver (via the gallbladder) and enzymes, bicarbonate, and water from the pancreas reach the duodenum via ducts. Brunner's glands and mucosal goblet cells secrete mucus. Cells in the crypts of Lieberkühn secrete various enzymes and about 2 l of fluid per day—more if chyme is excessively acidic or hypertonic. This fluid, which has a neutral pH and is similar to interstitial fluid, dilutes chyme and provides a watery medium for digestion and absorption. Because most of the fluid is reabsorbed, net fluid movement is from the lumen toward the blood and lymph.

Intestinal Enzymes

Epithelial cells in the mucosa and in the crypts of Lieberkühn synthesize several enzymes. Some enzymes act within the cells as nutrients are being absorbed, and others act at membrane surfaces or in the lumen as they are released from sloughed mucosal cells. **Enterokinase** (en″ter-o-ki′nās) converts trypsinogen to active trypsin. **Dipeptidase** (di-pep′tid-ās), and **aminopeptidase** (am″in-o-pep′tid-ās) break peptide bonds, and **maltase** (mawl′tās), **lactase** (lak′tās), and **sucrase** (soo′krās) break glycosidic bonds.

Regulation of Secretion

Secretion in the small intestine is regulated mainly by chyme, which stimulates Brunner's glands (submucosal glands in the duodenum) to secrete mucus and the crypts of Lieberkühn to secrete fluid. The hormone VIP (vasoactive intestinal peptide) stimulates secretion of electrolytes, which causes water to enter the lumen by osmosis. The water helps dissolve nutrients.

Digestion

More digestion takes place in the small intestine than in all other parts of the digestive tract together. Pancreatic enzymes, with the assistance of bile salts, digest most nutrients. Before digestion can begin in the small intestine, acidic chyme must be buffered by bicarbonate ions from the pancreas because enzymes that act here require a nearly neutral pH. Raising the pH stops the action of pepsin, but allows many other enzymes to act.

Digestion of Carbohydrates

Dietary carbohydrates enter the small intestine as starch, partially digested starch and disaccharides, and undigestible cellulose and other kinds of fiber. Pancreatic amylase completes digestion of starch to the disaccharide maltose. Maltase hydrolyzes maltose to two molecules of glucose. Lactase hydrolyzes lactose (milk sugar) to glucose and galactose, and sucrase hydrolyzes sucrose (table sugar) to glucose and fructose. Cellulose and other kinds of fiber remain undigested because the human body has no enzymes that can attack them. Fiber has no nutrient value, but it provides roughage that facilitates movement of unabsorbed material through the large intestine. Digestion of carbohydrates produces monosaccharides (Figure 23.16).

Complete digestion of dietary carbohydrate supplies cells with glucose that can be used for energy or can be converted to fat. The laboratory discovery of an amylase inactivator in kidney beans led to marketing of the substance as an aid in weight loss. Preventing complete starch digestion, it was reasoned, would prevent absorption of much dietary carbohydrate and lead to weight loss. Careful studies show that, unfortunately, the amylase inhibitor fails to function in the human intestine.

Digestion of Proteins

After partial digestion by pepsin in the stomach, proteins enter the small intestine as polypeptides. Alkaline conditions stop the action of pepsin, and pancreatic proteolytic enzymes (Figure 23.17) take over. Trypsinogen is activated to trypsin by intestinal enterokinase, after which trypsin activates other proteolytic enzymes. This chain of events helps to assure that enzymes are mixed with chyme before they are activated and minimizes the risk that they will digest pancreatic or intestinal tissue. Trypsin and chymotrypsin digest large polypeptides to smaller ones by breaking peptide bonds at various locations along the chain. They release individual amino acids only if they happen to act at the end of a chain. Carboxypeptidase releases single amino acids from the carboxyl end of a peptide.

Mucosal enzymes also digest proteins. Dipeptidase breaks dipeptides into amino acids and aminopeptidase releases single amino acids from the amino end of a chain. Proteins are completely digested to amino acids (Figure 23.18).

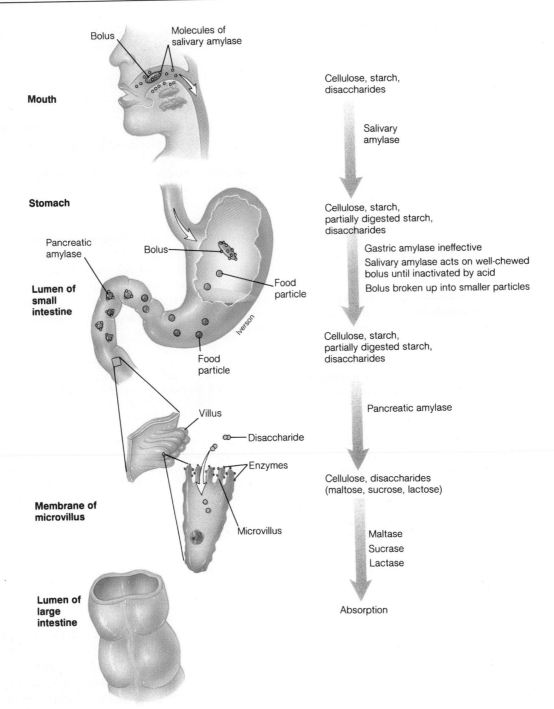

FIGURE 23.16
Digestion of carbohydrates.

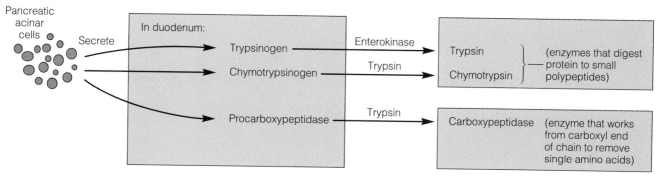

FIGURE 23.17
Pancreatic proteolytic enzymes.

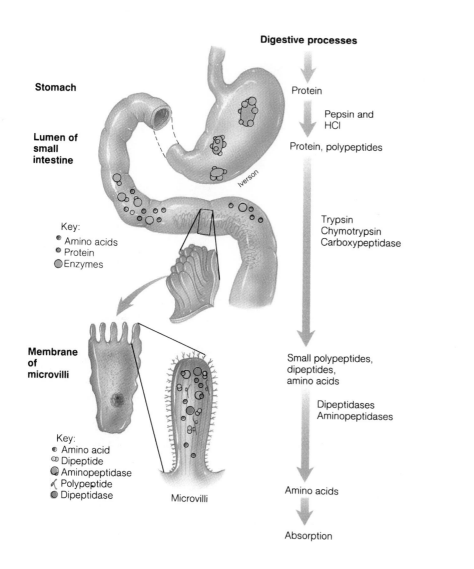

FIGURE 23.18
Digestion of proteins.

Digestive processes

Protein

Pepsin and HCl

Protein, polypeptides

Trypsin
Chymotrypsin
Carboxypeptidase

Small polypeptides,
dipeptides,
amino acids

Dipeptidases
Aminopeptidases

Amino acids

Absorption

Stomach

Lumen of small intestine

Key:
Amino acids
Protein
Enzymes

Membrane of microvilli

Key:
Amino acid
Dipeptide
Aminopeptidase
Polypeptide
Dipeptidase

Microvilli

FIGURE 23.19
Digestion of lipids.

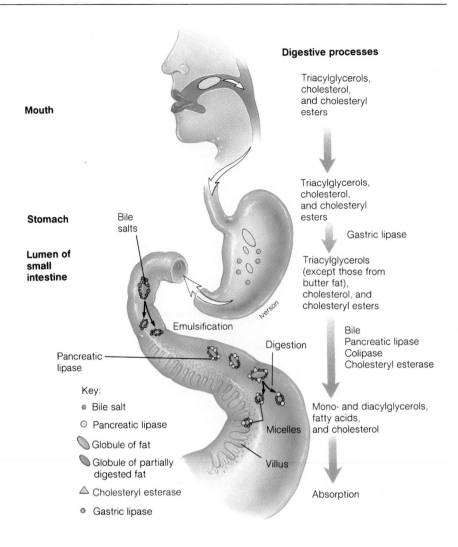

Digestive processes

Triacylglycerols, cholesterol, and cholesteryl esters

Triacylglycerols, cholesterol, and cholesteryl esters

Gastric lipase

Triacylglycerols (except those from butter fat), cholesterol, and cholesteryl esters

Bile
Pancreatic lipase
Colipase
Cholesteryl esterase

Mono- and diacylglycerols, fatty acids, and cholesterol

Absorption

Mouth

Stomach

Lumen of small intestine

Bile salts

Emulsification

Pancreatic lipase

Digestion

Micelles

Villus

Iverson

Key:
- Bile salt
- Pancreatic lipase
- Globule of fat
- Globule of partially digested fat
- Cholesteryl esterase
- Gastric lipase

Digestion of Lipids

Lipids enter the small intestine undigested, except for some butterfat digested in the stomach. They include triacylglycerols, cholesterol, and cholesteryl esters (cholesterol with a fatty acid attached to its alcohol group). Before digestion can begin, lipids must be emulsified.

Emulsification (e-mul″sĕ-fĭ-ka′shun), the mixing of lipids with watery digestive juices, is accomplished by bile salts, which have water-soluble and fat-soluble parts. The fat-soluble part combines with a fat globule and the water-soluble part projects into the watery medium. This reduces surface tension on the fat globule and keeps it suspended in the water. Agitation by gentle mixing contractions then breaks large fat globules into smaller ones. Lecithins in bile also help to break up fat globules. Fragmenting large globules into many tiny ones greatly increases the total surface area exposed to digestive enzymes. Pancreatic lipase, with the assistance of the protein **colipase** (ko-li′pāz), breaks ester linkages in triacylglycerols and yields glycerol, free fatty acids, and mono- and diacylglycerols.

The resulting small particles called **micelles** (mi-selz′) contain bile salts, cholesterol, triacylglycerols, and products of digestion such as mono- and diacylglycerols, free fatty acids, and glycerol. Polar molecules with their polar ends toward the watery medium keep micelles suspended as they migrate to microvilli. Fatty acids, glycerol, and acylglycerols are absorbed and bile salts are released for reuse. Dietary cholesterol and cholesteryl esters are emulsified along with other lipids. The esters are hydrolyzed by the pancreatic enzyme cholesteryl esterase. Both cholesterol and fatty acids are ferried in micelles to microvilli where they are absorbed. Lipid digestion is summarized in Figure 23.19.

Digestion of Nucleic Acids

Foods, which consist mostly of previously living cells, contain nucleic acids that are digested along with other nutrients. Pancreatic ribonuclease and deoxyribonuclease break RNA and DNA, respectively, into nucleotides. Intestinal phosphatases split phosphate from nucleotides, forming nucleosides. Intestinal nucleosidases split nucleosides into purines, pyrimidines, ribose, and deoxyribose, most of which are absorbed and used by cells to make new nucleotides.

Absorption

In an average-size adult, about 1 kg of food and 7.5 l of gastrointestinal secretions enter the digestive tract daily. Most of this volume is absorbed in the small intestine—300 to 400 g of monosaccharides, 50 to 100 g of amino acids, about 100 g of lipids, 50 g of various ions, and 8 l of water.

Absorption takes place across mucosal cells of the intestinal villi. Immediately adjacent to the mucosal cell membranes is a thin **unstirred water layer** through which substances diffuse to reach the mucosal membrane. Substances that reach the membrane pass through it by a variety of mechanisms, including simple diffusion, facilitated diffusion, and active transport. After crossing the surface membrane of mucosal cells, substances also must pass through the inner cell membrane, the interstitial fluid, and the single layer of cells in the walls of blood or lymph capillaries. Various nutrients make this journey by different mechanisms.

Absorption of Monosaccharides

Digested carbohydrates and sugars from nucleic acids are absorbed as monosaccharides. Smaller five-carbon monosaccharides probably are absorbed by simple diffusion, but absorption of six-carbon sugars is more complex.

Glucose moves across the surface membrane of mucosal cells by **co-transport;** that is, on a carrier molecule that also carries sodium ions (Figure 23.20a). Energy from ATP actively transports Na$^+$ and a glucose molecule attached to the carrier moves with it. Because of cotransport, the rate of glucose transport is proportional to Na$^+$ concentration in the intestinal lumen, at least until the carrier molecules become saturated and transport at their maximum rate. After being transported into the mucosal cell, glucose moves by diffusion out the basal side of the cell, across the interstitial fluid, and into a blood capillary. Na$^+$ moves to the interstitial fluid, probably by the sodium-potassium pump, but few enter the blood. Galactose is transported by the same mechanism and on the same carrier.

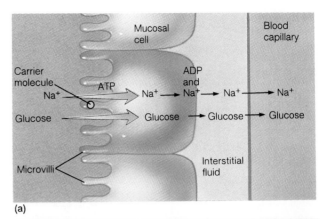

(a)

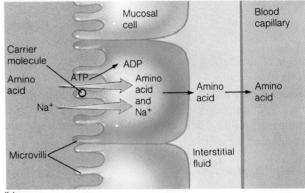

(b)

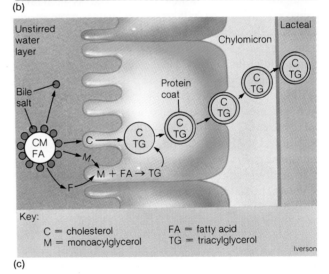

Key:

C = cholesterol	FA = fatty acid
M = monoacylglycerol	TG = triacylglycerol

Iverson

(c)

FIGURE 23.20
(*a*) The co-transport of glucose and sodium ions during absorption.
(*b*) The transport of amino acids during absorption. (*c*) The absorption of lipids.

FIGURE 23.21
Enterohepatic circulation of bile salts.

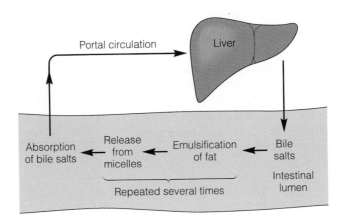

In contrast, fructose is transported by a different carrier and moves by facilitated diffusion. Neither ATP nor sodium ions are involved in fructose absorption. Most fructose in mucosal cells is phosphorylated and converted to glucose, so intestinal capillary blood contains mostly glucose and very little fructose.

Absorption of Amino Acids

Of digested proteins, which are absorbed as individual amino acid molecules, half come from dietary proteins and half from the breakdown of proteins within the digestive tract. The latter consist of about equal amounts of enzymes and proteins from sloughed epithelial cells. Amino acids are transported across mucosal surface membranes on different carrier molecules that also transport Na^+. As with glucose transport, amino acid transport requires energy from ATP, but the carrier molecule does not move Na^+ unless it also has an amino acid attached. Each of four kinds of carrier molecules transports a specific group of structurally similar amino acids. Amino acids move out of mucosal cells to the blood in the same way as glucose molecules. Amino acid transport is summarized in Figure 23.20b.

Absorption of the Products of Lipid Digestion

Products of lipid digestion—free fatty acids, cholesterol, glycerol, and mono- and diacylglycerols—travel in micelles across the unstirred water layer to the mucosal membrane. The processes by which these molecules enter mucosal cells are not well understood, but the following events appear to occur. Micelles probably disintegrate near the membrane releasing a collection of lipid-soluble molecules, which then diffuse through membrane lipids and enter mucosal cells. Some micelles may actually fuse with mucosal cell membranes and enter the cell intact. Inside the mucosal cells, free fatty acids recombine with mono- and diacylglycerols to reform triacylglycerols. A small globule of triacylglycerols and cholesterol is coated with protein to make a particle called a **chylomicron** (ki″lo-mi′kron). Chylomicrons are extruded from the mucosal cells into the interstitial fluids, and then to lacteals (Figure 23.20c). Chylomicrons circulate through lymphatic vessels and enter the blood at the left subclavian vein.

Bile salts released when micelles disintegrate remain in the intestinal lumen and emulsify more fat globules. As bile salts are reused, they are swept toward the ileum, where most are actively transported into mucosal cells and diffuse into the blood. Less than 5 percent remain in the intestine to be excreted with the feces. Bile salts returned to the liver are reexcreted into bile. This **enterohepatic** (en″ter-o-hep-at′ik) **circulation** (Figure 23.21) is so efficient that some bile salt molecules make the trip five times a day.

Absorption of the Products of Nucleic Acid Digestion

Nucleic acid digestion releases five-carbon sugars (ribose and deoxyribose), purines, pyrimidines, and phosphates. The sugars and phosphates probably are absorbed by simple diffusion, and the purines and pyrimidines by active transport.

Absorption of Water, Minerals, and Electrolytes

Large quantities of water from ingested fluids and digestive juices pass through the small intestine daily, and 90 percent of this water is absorbed. Water moves freely in

TABLE 23.4	Processes That Occur in the Small Intestine
Motility	Peristaltic contractions propel chyme toward the ileocecal valve.
	Segmentation contractions exert a chopping effect and mix the chyme with digestive juices.
	Contractions of the muscularis mucosa move the villi and agitate fluid around them, thereby increasing absorption.
	Contractile proteins in the microvilli cause them to move; this movement also increases absorption.
Secretion	The small intestine receives bile from the liver and a variety of digestive enzymes, bicarbonate, and fluid from the pancreas.
	Brunner's glands and goblet cells secrete mucus.
	Cells in the crypts of Lieberkühn secrete large quantities of a fluid similar to interstitial fluid.
	Epithelial cells of the mucosa produce enzymes that act within the cells or are released into the lumen.
Digestion	Carbohydrates are digested to monosaccharides; pancreatic amylase digests starch to maltose; maltase, lactase, and sucrase digest specific disaccharides to monosaccharides.
	Proteins are digested to individual amino acids; trypsin and chymotrypsin break long chains that remain from the action of pepsin to shorter chains; carboxypeptidase and aminopeptidase remove amino acids from the carboxyl and amino ends of peptides, respectively; and dipeptidase breaks dipeptides into amino acids.
	Lipids are emulsified by bile salts. Pancreatic lipase, assisted by colipase, digests triacylglycerols to glycerol, free fatty acids, and mono- and diacylglycerols. Cholesteryl esterase removes fatty acids from cholesteryl esters.
	Nucleic acids are digested to nucleotides by ribonuclease and deoxyribonuclease; phosphatases and nucleosides complete digestion of nucleotides into sugars, phosphates, purines, and pyrimidines.
Absorption	Carbohydrates are absorbed as monosaccharides. Glucose and galactose are absorbed by co-transport, fructose is absorbed by facilitated diffusion, and five-carbon sugars probably are absorbed by simple diffusion.
	Proteins are absorbed as amino acids. Four different carrier molecules each carry different types of amino acids in the process of co-transport.
	Products of lipid digestion probably are absorbed by passive diffusion through the lipid part of the mucosal membrane. In the mucosal cells, these products are packages in protein coated chylomicrons.
	Products of nucleic acid digestion are absorbed by two processes: sugars probably by diffusion and phosphates, purines, and pyrimidines by active transport.
	Water moves freely in both directions across the mucosa, but net movement is toward the blood.
	Various minerals and electrolytes are absorbed by simple diffusion or active transport.
	Most water-soluble vitamins are rapidly absorbed; fat-soluble vitamins also are easily absorbed if pancreatic lipase and bile salts are sufficient.

both directions across the mucosal membrane in the small and large intestines, but the net movement is out of the intestine into the blood.

Some sodium and potassium ions diffuse in both directions across the mucosal membrane. Sodium also is actively transported, as we have seen, and potassium is actively secreted into the intestinal lumen in the jejunum, ileum, and colon.

Calcium and iron are actively absorbed in the small intestine. Calcium ions are absorbed on a carrier protein, and the availability of the carrier molecule is enhanced by vitamin D and parathyroid hormone. Iron ions are absorbed on gastroferrin; its absorption rate is regulated by the amount of iron already stored in the ferritin in the intestinal mucosa and certain other tissues. Transport and storage proteins help to assure adequate supplies of certain key elements—calcium for maintaining bone strength and normal muscle and nerve function, and iron for making new erythrocytes.

Chloride ions move to equalize charges. They passively follow active absorption of sodium ions, or they can be actively transported into the lumen of the ileum and the colon when H^+ from bacterial action accumulates in the lumen.

Absorption of Vitamins

Most vitamins released by digestion are absorbed in the upper part of the small intestine. Typically, water-soluble vitamins (B-complex and C) are rapidly absorbed, but vitamin B_{12} binds to intrinsic factor in the stomach and the complex is absorbed across the mucosa of the ileum. Fat-soluble vitamins (A, D, E, and K) are absorbed with other fats, provided pancreatic lipase and bile salts are available.

Processes that occur in the small intestine are summarized in Table 23.4.

LARGE INTESTINE AND RECTUM

Structure

The **large intestine,** or **colon** (ko′lon), is about 1.5 meters in length and extends from the ileocecal valve to the rectum (Figure 23.22). Its diameter varies but is always larger than that of the small intestine. The **cecum** (se′kum) is a pouchlike structure near the ileocecal valve, to which the **vermiform** (ver′mi-form) **appendix,** a slightly coiled, blind tube about 8 cm long, is attached. The colon has four regions—the **ascending, transverse, descending,** and **sigmoid** (sig′moid) **colon.** The **right colic** (hepatic) **flexure** (flek′sher) marks the boundary between the ascending and transverse colon; the **left colic** (splenic) **flexure** marks the boundary between the transverse and descending colon. The sigmoid colon begins near the left iliac crest and extends in an S-shaped path to the rectum. Mesenteries support the colon.

The appendix is a well-developed lymphoid organ and probably defends against infection, especially in infants and children. It atrophies and is replaced by connective tissue in old age.

The colon has three bands of longitudinal muscle, the **taeniae coli** (te′ne-e ko′le), that contract and draw the rest of the wall into small pouches called **haustra** (haws′trah). It also has fat deposits called **epiploic** (ep″i-plo′ik) **appendages.**

The **rectum** (rek′tum) comprises the last 20 cm of the digestive tract and terminates in the 2 cm long **anal canal** and an opening called the **anus.** The rectum is firmly attached to the sacrum by the peritoneum and it lacks haustra. Inside the anal canal are **anal columns,** folds in the lining that are well supplied with blood vessels. **Hemorrhoids** are enlarged, inflamed anal, or hemorrhoidal, veins. The anus has an **internal anal sphincter** comprised of smooth muscle and an **external anal sphincter** composed of skeletal muscle.

Motility

In the large intestine, motility is sluggish. As chyme passes through the ileocecal valve into the cecum and on to the ascending colon, the haustra become distended. In a process called **haustral** (hos′tral) **churning,** distention stimulates contraction of the muscle strips of one haustra, which pushes the chyme ahead into the next haustra. Strong propulsive movements called **mass movements** occur in the large intestine only a few times a day, usually after meals. How contractions of the colon are regulated is not well understood, but intrinsic innervation and certain hormones such as CCK-PZ probably are involved. The contractions propel waste material into the transverse and descending colon and into the rectum.

Chyme that has reached the descending colon is called feces. **Feces** (fe′sēz) are about three-fourths water and one-fourth solids. Bacteria and undigested roughage each account for approximately one-third of the solid volume; the remainder is composed of inorganic material and undigested nutrients. Within the undigested materials are remnants of enzymes and epithelial cells sloughed from the intestinal lining. The brown color of feces is due to products of bacteria breaking down bilirubin.

Feces entering the rectum stimulate the **defecation** (def″e-ka′shun) **reflex** and are excreted through **defecation.** Peristaltic waves pass along the descending colon, sigmoid colon, and rectum. Relaxation that precedes the peristaltic wave dilates the anal canal and opens the internal anal sphincter. A second stronger reflex sends signals to the spinal cord and back via parasympathetic fibers, intensifying peristalsis near the anus. Voluntary control of the external anal sphincter provides a way to allow or inhibit defecation. Control of this sphincter is learned, as any parent of a toddler knows.

Studies in which volunteers were fed small beads with a meal have been used to estimate the time required for substances to traverse the digestive tract. About 70 percent of the beads appeared in the feces within 72 hours of the meal, but as much as a week was required to recover all of the beads.

Other Functions

In the large intestine, secretions are limited to mucus from goblet cells in the epithelium and crypts of Lieberkühn and bicarbonate from the mucosal epithelium. The bicarbonate helps to neutralize acids produced by bacterial action. No enzymes are secreted in the large intestine.

Except in newborns and people who have undergone prolonged antibiotic therapy, the large intestine contains numerous bacteria, called intestinal flora. They enter the digestive tract early in life with food and maintain themselves by metabolizing nutrients that remain in the chyme. Any digestion in the large intestine is performed by the bacteria, which release byproducts of their metabolic processes, especially amino acids and the vitamins K, B_{12}, thiamine, and riboflavin. Small amounts of these substances are absorbed in the large intestine.

From the liter of chyme entering the large intestine daily, 100 to 200 ml of water, a few electrolytes, and some vitamins are absorbed. When water is present, the large intestine can absorb up to 5 l per day. It also efficiently absorbs sodium ions without cotransport if stimulated by aldosterone. Because of the absorptive capacities of the colon, fluids and medications are frequently instilled rectally, especially in children.

INTEGRATED DIGESTIVE FUNCTIONS

Having considered how motility, secretion, digestion, and absorption occur, we will now look at overall function and its regulation as it pertains to the stomach, liver, pancreas, and intestine. Finally, we will consider how enteric hormones integrate digestive functions.

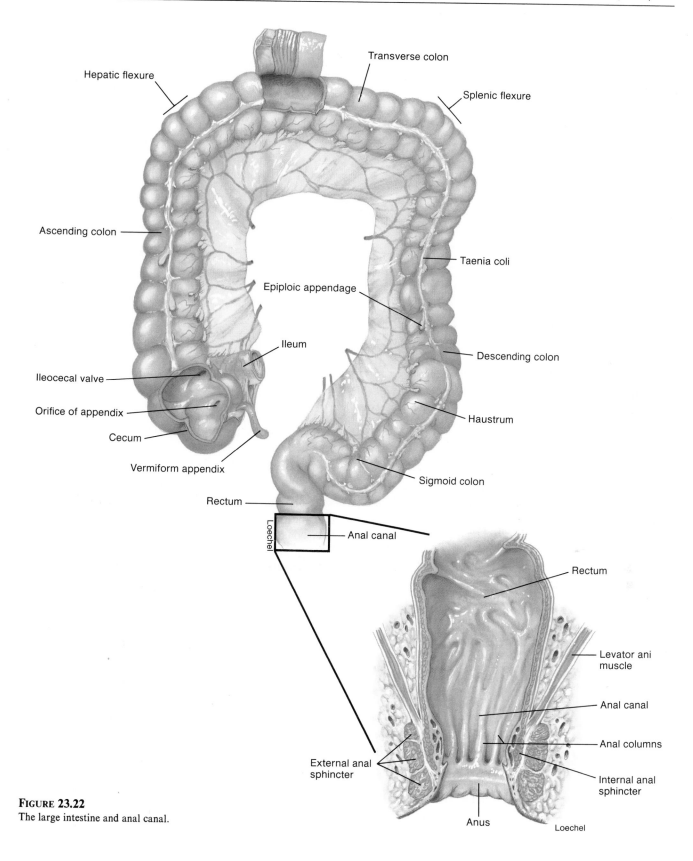

Hepatic flexure

Transverse colon

Splenic flexure

Ascending colon

Taenia coli

Epiploic appendage

Ileum

Descending colon

Ileocecal valve

Orifice of appendix

Cecum

Haustrum

Vermiform appendix

Sigmoid colon

Rectum

Loechel

Anal canal

Rectum

Levator ani muscle

Anal canal

Anal columns

External anal sphincter

Internal anal sphincter

Anus

Loechel

FIGURE 23.22
The large intestine and anal canal.

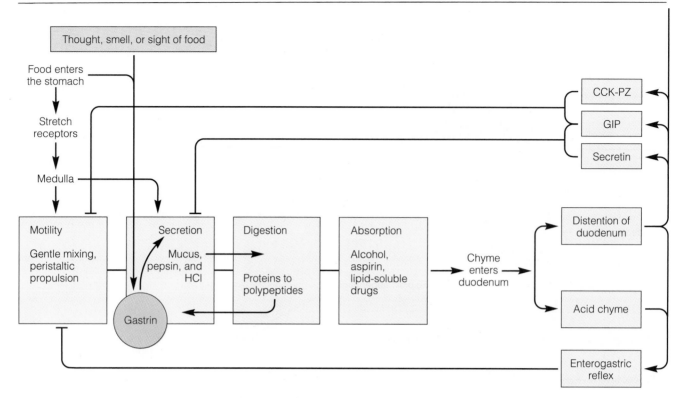

FIGURE 23.23
Processes that occur in the stomach and their regulation.

Regulation of Stomach Function

Food entering the stomach initiates both motility and secretion of mucus, pepsinogen, and gastrin; gastrin stimulates secretion of HCl. As digestion proceeds, accumulation of polypeptides suppresses HCl secretion by negative feedback. Absorption is limited to molecules that can diffuse through mucus on the stomach lining. When acid chyme reaches the small intestine, it initiates neural and hormonal signals that reduce stomach motility and secretions (Figure 23.23).

Regulation of Liver and Pancreatic Function

Functions of the liver and pancreas are regulated by neural and hormonal stimuli. The same parasympathetic stimuli that act on the stomach also stimulate enzyme synthesis in the pancreas; however, the hormones secretin and CCK-PZ are responsible for releasing bile and pancreatic secretions. Both hormones are secreted by the duodenal epithelium in response to the presence of chyme, and they reach the liver and pancreas via the blood.

Secretin stimulates the liver to produce and secrete bile and the pancreatic tubular cells to secrete water and bicarbonate ions. Acidity is a particularly strong stimulator of secretin release—the more acid the chyme, the more secretin is released. As bicarbonate reaches the duodenum, it neutralizes the acidity, and as acidity is reduced so is secretin release.

CCK-PZ has two important effects, which were discovered independently. The responsible hormones were named, but later, it was learned that the same substance produces both effects. As cholecystokinin, the hormone stimulates bile release from the gallbladder. (The name aptly describes this action: *chol,* bile; *cyst,* bladder; and *kinin,* to move.) When it reaches the gallbladder, the hormone causes the muscular wall and duct to contract rhythmically. These peristaltic waves push bile along the bile duct and open the sphincter between the duct and the duodenum. As pancreozymin, the hormone stimulates the secretion of pancreatic enzymes. (Again, the name aptly describes the action: *pancreo,* pancreas; and *zymin,* enzyme producer.) Both secretin and CCK-PZ potentiate or augment the other's effects, and their combined effects are greater than those of either alone.

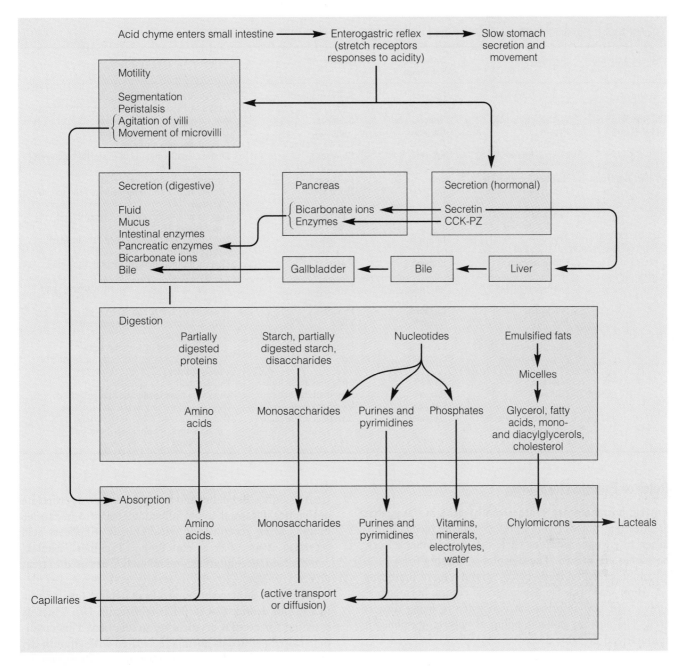

FIGURE 23.24
Processes that occur in the liver, pancreas, and small intestine and
their regulation.

Regulation of Small Intestine Function

Acid chyme entering the small intestine initiates both motility and release of secretin and CCK-PZ. The hormones, in turn, cause the pancreas to release bicarbonate ions, fluid, and enzymes, and the gallbladder to release bile. Bicarbonate ions neutralize the chyme, bile emulsifies fats, and the enzymes digest various nutrients to small molecules. Intestinal enzymes complete the digestive process. Amino acids, monosaccharides, glycerol, fatty acids,

mono- and diacylglycerols, cholesterol, nucleic acid components, vitamins, minerals, electrolytes, and water are made available for absorption. Motility of villi and microvilli facilitate absorption. Most substances are absorbed by active transport or diffusion and enter the blood. Products of fat digestion are absorbed, packaged into chylomicrons, and enter lacteals.

Processes that occur in the liver, pancreas, and small intestine are summarized in Figure 23.24. Actions of all substances involved in digestion are summarized in Table 23.5.

TABLE 23.5   Summary of Substances Involved in Digestion

Site of Action	Substance	Nature of Substance	Source	Action
Mouth	Salivary amylase	Enzyme	Salivary glands	Digests starch to intermediate products (and maltose, if time permits)
Stomach	Hydrochloric acid	Mineral acid	Stomach	Activates pepsin, kills bacteria
	Pepsin	Enzyme	Stomach	Digests proteins to polypeptides
	Rennin	Enzyme	Child's stomach	Coagulates milk proteins
	Gastric lipase	Enzyme	Stomach	Digests small amounts of butterfat to glycerol and fatty acids
Small Intestine	Sodium bicarbonate	Salt	Pancreas	Neutralizes hydrochloric acid
	Pancreatic amylase	Enzyme	Pancreas	Digests starch and intermediate products to maltose
	Pancreatic lipases	Enzymes	Pancreas	Digests fats to fatty acids, glycerol, and cholesterol
	Trypsin	Enzyme	Pancreas	Digests proteins and polypeptides to small polypeptides
	Chymotrypsin	Enzyme	Pancreas	Digests proteins and polypeptides to small polypeptides
	Carboxypeptidases	Enzymes	Pancreas	Remove amino acids from the carboxyl end of polypeptides
	Nucleases	Enzymes	Pancreas	Break down DNA and RNA
	Bile salts	Steroids	Liver	Help to form emulsions and micelles
	Enterokinase	Enzyme	Intestinal epithelium	Activates trypsin
	Dipeptidases	Enzymes	Intestinal epithelium	Break dipeptides into amino acids
	Aminopeptidase	Enzyme	Intestinal epithelium	Remove amino acids from the amino end of polypeptides
	Maltase	Enzyme	Intestinal epithelium	Digests maltose to glucose
	Lactase	Enzyme	Intestinal epithelium	Digests lactose to glucose and galactose
	Sucrase	Enzyme	Intestinal epithelium	Digests sucrose to glucose and fructose
	Nucleotide-digesting enzymes	Enzymes	Intestinal epithelium	Break DNA and RNA into purines, pyrimidines, phosphate, ribose, and deoxyribose

Roles of Enteric Hormones

Enteric hormones are synthesized by cells in the gastric or intestinal mucosa and secreted into the blood. In addition to gastrin, secretin, CCK-PZ, many other enteric hormones are known. These peptides, ranging from 11 to 43 amino acids, typically act by stimulating or inhibiting cAMP or Ca^{2+} release. Research on enteric hormones is active, and new information is constantly being discovered. The current understanding of enteric hormones is summarized in Table 23.6.

CLINICAL APPLICATIONS

The human digestive tract is subject to many disorders that can disrupt homeostasis in one way or another. We shall consider the causes and effects of a few of the more common disorders.

Gastritis (gas-tri′tis), inflammation of the gastric mucosa, is a common ailment caused by irritation, alcohol, infections, or excessive gastric secretions. Persistent inflammation and excessive secretions can destroy the mucus and begin to digest the stomach wall itself, causing a **peptic ulcer.** Similar damage occurs as a **duodenal** (du″o-de′nal) **ulcer** when inadequately buffered gastric secretions damage the intestinal wall. Gastritis is

treated with a bland diet and antacids, and an appropriate antibiotic, if caused by an infection. Because histamine binding to H_2 receptors on parietal cells stimulates acid secretion, drugs such as cimetidine (Tagamet), Zantac, and Pepcid that inhibit histamine binding are used to treat ulcers.

Stomach damage or removal can deplete the intrinsic factor and decrease absorption of vitamin B_{12}, which is essential for normal erythrocyte production. This condition, called **pernicious anemia,** is treated with vitamin B_{12} injections because the vitamin given by mouth will not be absorbed. When pernicious anemia is caused by gastritis, treating the gastritis often cures the anemia.

Vomiting, or ejection of stomach contents through the mouth, can be caused by irritating substances in the stomach or by unpleasant emotional or sensory experiences, including motion sickness. Sensory signals via the vagus nerve or other sensory pathways act on the vomiting center in the medulla. They initiate a reflex that simultaneously relaxes the gastroesophageal sphincter and contracts the diaphragm. Vomiting can lead to significant acid loss and alkalosis, an increase in the blood pH (Chapter 28). It is treated with drugs such as diphenyloxylate (Lomotil) to reduce digestive tract motility or dimenhydrinate (Dramamine) to counteract overstimulation of the organs of balance during travel. Prolonged vomiting can require hospitalization to correct alkalosis.

TABLE 23.6 Current Understanding of Enteric Hormones

Hormone	Site of Production	Main Site of Action	Main Functions
Gastrin	G-cells of the pyloric portion of stomach	Stomach	Stimulates secretion of acid and motility of upper part of stomach.
Secretin	Mucosal glands of upper small intestine	Pancreas, stomach	Increases pancreatic secretion of bicarbonate; decreases gastric section; augments action of CCK-PZ.
Cholecystokinin-pancreozymin (CCK-PZ)	Mucosal glands of upper small intestine	Pancreas, gallbladder	Causes contraction of gallbladder; stimulates secretion of pancreatic enzymes; augments action of secretin; inhibits gastric emptying.
Glucagon	Duodenum	Pancreas	Stimulates release of insulin from pancreas when glucose concentration is high in intestine.
Gastric inhibitory peptide (GIP)	Duodenum and jejunum	Stomach, pancreas	Inhibits gastric secretion and motility; stimulates insulin secretion.
Vasoactive intestinal peptide (VIP)	Small intestine	Small intestine, stomach, blood vessels	Stimulates intestinal secretion of electrolytes (and thus water); inhibits acid secretion; dilates peripheral blood vessels.
Motilin	Duodenum	Stomach	Stimulates acid secretion, increases motility.
Substance P	Neurons and endocrine-type cells of gastrointestinal tract	Small intestine	Increases motility of small intestine (not proven to enter blood); may increase mucus secretion.
Bombesin (gastrin-releasing peptide)	Neurons? of gastrointestinal tract?	Stomach, small intestine, gallbladder	Increases gastrin secretion; increases motility of small intestine and gallbladder.
Somatostatin (growth-hormone-inhibiting hormone)	Pyloric mucosa?	?	Same hormone from hypothalamus has several effects on gastrointestinal tract, but effects of locally produced hormone not yet clear; probably inhibits gastrin secretion.
Chymodenin	Duodenum?	Pancreas	Stimulates chymotrypsin release from pancreas.
Bulbogastrone	Duodenum?	Stomach	Inhibits HCl secretion.
Urogastrone	Various	Stomach	Inhibits HCl secretion.
Villikinin	Intestinal mucosa	Villi	Stimulates movement of villi and increases lymph flow.
Enkephalins	Various	Cholinergic neurons	Suppresses release of acetylcholine?
Neurotensin	N-cells of villi	?	Modulates neural signals?

Hepatitis (hep″ah-ti′tis), liver inflammation, can be caused by viruses, bacteria, protozoa, and toxic substances. Outbreaks of hepatitis often can be traced to poor sanitation in food handling or the sharing of contaminated needles among drug abusers. Liver function is impaired as cells die, and the patient suffers from jaundice, lack of appetite, and extreme weakness. Hepatitis is treated with rest; low-fat, nutritious food; and antibiotics for bacterial infections.

Blockage of the pancreatic duct, infection, and excessive alcohol can cause **pancreatitis** (pan″kre-at-i′tis), inflammation of the pancreas. If secretions accumulating in the pancreas inactivate a trypsin inhibitor, proteolytic enzymes can digest an entire pancreas in a matter of hours. Deficient intestinal bicarbonate from impaired tubule secretion can disturb the body's acid-base balance. Acute pancreatitis is treated with pain medication, lactated Ringer's solution to restore acid-base balance, and antibiotics for bacterial infections.

In **celiac** (se′le-ak) **disease,** or **sprue** (sproo), absorption is impaired because the epithelium of villi fails to be continuously replaced as it normally is. In susceptible people, celiac disease is caused by ingestion of **gluten** (gloo′ten) in wheat and rye products. A substance in gluten called **gliadin** (gli′ad-in) disrupts epithelial development. In severe cases, villi become blunted or disappear completely over large regions of the small intestine. Absorption is greatly impaired and tissue wasting and vitamin deficiencies occur. If diagnosed before tissue damage is severe, a gluten-free diet prevents further damage and allows intestinal epithelium to regenerate. A similar disease known as **tropical sprue,** usually seen in people who have spent time in the tropics, is thought to be due to a bacterial agent because it responds to antibiotic treatment.

Humans vary widely in the normal frequency of defecation from after each meal to only once every three days. **Constipation** is the passage of dry, hardened feces at less than a person's normal frequency. Constipation often results from too little dietary fiber or from too frequent inhibition of the voluntary defecation reflex. Repeated inhibition reduces muscle tone and the strength of reflexes. Fiber increases peristaltic movements and causes feces to be removed before they become dehydrated into a hard stool. Though laxatives relieve constipation, their regular use also reduces natural reflexes.

FIGURE 23.25
Appendix: (*a*) normal, and (*b*) inflamed.

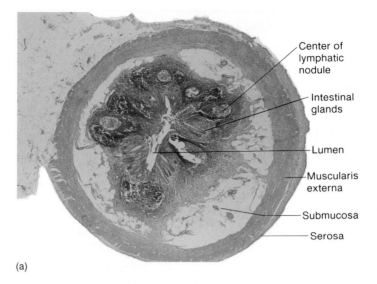

Center of
lymphatic
nodule

Intestinal
glands

Lumen

Muscularis
externa

Submucosa

Serosa

(a)

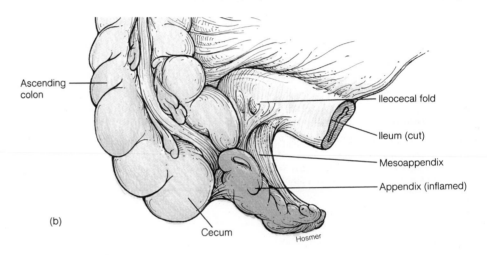

Ascending
colon

Ileocecal fold

Ileum (cut)

Mesoappendix

Appendix (inflamed)

(b)

Cecum

Hosmer

An infection or inflammation of the mucosa anywhere in the intestinal tract can cause excessively frequent bowel movements, or **diarrhea.** Both motility and secretions increase. In severe cases, electrolyte and fluid loss can be life threatening. Diarrhea is treated by restricting food and giving medications (Lomotil) to reduce motility, and by replacing lost fluids and electrolytes.

Appendicitis, acute infection of the appendix (Figure 23.25), causes cramping pains and sometimes vomiting. Pain is typically sharpest at McBurney's point (one-third of the distance from the anterior superior iliac spine to the umbilicus). Swelling can cause the appendix itself to rupture. Every effort is made to remove an infected appendix before it ruptures and spreads infection throughout the peritoneal cavity.

In **ulcerative colitis** (ul″ser-a′tiv ko-li′tis), the mucosa of the colon becomes ulcerated (damaged and eroded) and strong frequent peristaltic contractions occur. The cause of this disease is unknown, but possibilities include bacterial action, abrasion, and proteolytic digestion of mucosal cells. A temporary **ileostomy** (il″e-os′to-me), a surgically created external opening from the intestine, is often performed to divert feces and give the colon an opportunity to heal. If healing is not satisfactory, a permanent ileostomy may be necessary.

Obstructions can occur anywhere along the intestinal tract. Causes include congenital defects, malignancies, fibrotic scar tissue from healed ulcers, and spasms and paralysis of muscles. The effects of an obstruction depend on its location. Pyloric obstruction results in vomiting of acidic stomach contents and alkaline blood pH. Obstruction in the upper small intestine leads to vomiting fluids with neutral pH, and in the lower small intestine to vomiting alkaline fluids and acidic blood pH. Obstructions of the sigmoid colon rarely cause vomiting. Surgery usually is required to remove an obstruction.

CLINICAL TERMS

achalasia (ak-al-a′ze-ah)   failure of the esophageal sphincter to open, and dilation and hypertrophy of the esophagus because of damage to nerve ganglia supplying the esophagus

achlorhydria (ah″klor-hi′dre-ah)   absence of acid secretions in the stomach

aphagia (ah-fa′je-ah)   inability to swallow

cholecystitis (ko″le-sis-ti′tis)   inflammation of the gallbladder

cholelithiasis (ko″le-lith-i′as-is)   presence of gallstones

colonoscopy (ko-lon-os′ko-pe)   examination of the colon with a lighted instrument

colostomy (ko-los′to-me)   a surgical procedure to form an artificial opening into the colon

Crohn's disease   chronic, regional inflammation of the small or large intestine

diverticulitis (di″ver-tik-u-li′tis)   formation of small inflamed pouches in the lining of the colon

dumping syndrome   an excessive increase in the rate at which contents of the stomach enter the small intestine, often caused by vagotomy or gastric surgery

flatus (fla′tus)   presence of gas in the stomach or intestines

Hirshsprung's disease   constriction of the lower colon and dilation of the portion above the constriction due to congenital absence of nerve ganglia that control peristalsis of the colon

pyorrhea (pi″o-re′ah)   an inflammation of the dental periosteum

ESSAY

CYSTIC FIBROSIS

A common fatal genetic disease, **cystic fibrosis** (CF) occurs in one in 2000 Caucasian infants born in the United States, but in only one in 17,000 African Americans. CF is inherited as an autosomal recessive characteristic (Chapter 30), which affects mainly the epithelium of the lungs, pancreas, and sweat glands. The CF gene has been isolated and sequenced, and the defect appears to decrease permeability of chloride ion channels of certain epithelial cells.

CF is characterized by excessive secretion of thick mucus that blocks ducts especially in the lungs and pancreas. Symptoms can be mild to severe, depending on how strongly the genes that cause it are expressed. In the lungs, airways can be sufficiently blocked to make breathing difficult, and trapped mucus creates ideal conditions for chronic lung infections. In the pancreas, blocked ducts prevent enzymes from reaching the small intestine. Pancreatic lipase usually is most deficient, so fat digestion is limited and deficiencies of fat soluble vitamins lead to night blindness (vitamin A), rickets (vitamin D), neurological degeneration (vitamin E), and slowed blood clotting (vitamin K).

Vomiting and the failure of a newborn to pass meconium (waste that accumulates in the colon prior to birth) suggest that a newborn may have cystic fibrosis. Affected infants fail to thrive, have little appetite, and grow slowly because limited calories are used for the excessive work of breathing. Both infants and children have frequent bulky stools, edema, and a persistent, noisy cough. Adolescents and adults—now that victims live to adulthood—often develop diabetes, hemoptysis (coughing up blood), pancreatitis, cirrhosis of the liver, jaundice, and various lung complications. Female patients may be infertile because of a cervical plug of thick mucus.

Diagnosis of CF has depended on detecting a high salt (NaCl) concentration in sweat due to increased Cl^- secretion and the inability to reabsorb it. The high negative charge from Cl^- in the ducts of sweat glands draws Na^+ into the ducts and water follows by osmosis. Now, DNA analysis of fetal leukocytes allows prenatal diagnosis, and similar analysis in prospective parents makes it possible to detect carriers of the defective gene.

CF is treated with nutritional supplements and pulmonary therapy. Nutritional supplements include pancreatic enzymes and vitamins, and if the patient fails to maintain normal weight, high-calorie supplements. Pulmonary therapy includes postural drainage (head below the base of the lungs), chest percussion (gentle blows to the chest and back with a cupped hand to help loosen mucus), antibiotics to control infections, and bronchodilators to keep the airway open. The use of mist tents does not thin mucus, and their use is no longer recommended. Some years ago, children with CF nearly always died in childhood or adolescence. Though no cure is available, management of symptoms has improved sufficiently over the past few decades that some patients are living to adulthood, and a few even have children.

Questions

1. What are the symptoms of cystic fibrosis?
2. How is this disease acquired?
3. What treatments are available for CF?
4. How can CF be prevented?

Chapter Summary

Organization and General Functions

- The general functions of the digestive system are motility, secretion, digestion, and absorption.
- The major organs of the digestive system are the mouth, pharynx, esophagus, stomach, small intestine, large intestine, rectum, and anus.
- The accessory organs of the digestive system are the salivary glands, liver, and pancreas.

Development

- The digestive system develops from the primitive gut into recognizable organs by the eighth week of development.
- Common congenital defects include esophageal atresia, pyloric stenosis, and imperforate anus.

Tissues and Their Functions

- The basic layers of the digestive tract from the outside inward are: (1) the serosa (visceral peritoneum), which is continuous with the mesenteries; (2) the muscularis, which consists of longitudinal and circular muscle fibers; (3) the submucosa, which consists of connective tissue, blood vessels, and lymph vessels; and (4) the mucosa, which contains muscle, connective tissue, blood and lymph vessels, glands, and an innermost epithelium.
- The functions of the digestive tract are carried out by (1) muscles that move and mix food substances, (2) glands that synthesize and secrete digestive juices and mucus, (3) secretions that digest large particles into small molecules, and (4) a mucosa across which small molecules can be absorbed.

Mouth

- The mouth (oral cavity) contains the tongue and the teeth, and receives saliva from the salivary glands.
- The salivary glands are located below the ears and below the floor of the mouth; their secretions are carried to the mouth by ducts. They secrete salivary amylase, which digests starch, and mucus.
- Teeth are composed of an enamel, cementum, dentin, and a pulp cavity. The numbers and types of deciduous and permanent teeth are shown in Figure 23.8.

Pharynx and Esophagus

- Food passes from the mouth through the pharynx and esophagus to the stomach.
- Motility in the pharynx and esophagus produces deglutition (swallowing) by automatic rhythmic contractions of pharyngeal and esophageal muscle.

Stomach

- The stomach is a hollow bag with a sphincter at each end and rugae in the inner wall. It contains gastric glands with mucous neck cells, chief cells that secrete pepsin, parietal cells that secrete HCl and intrinsic factor, and argentaffin cells that secrete serotonin and histamine.
- Motility in the stomach is produced by peristaltic contractions that mix and propel partially digested food toward the small intestine.
- Digestion consists mainly of the breaking of proteins into polypeptides, but gastric lipase also digests butterfat.
- Absorption is limited to substances such as aspirin, alcohol, and a few drugs that can penetrate mucus.

Liver and Pancreas

- The liver is a large, lobed organ in the upper right quadrant of the abdominal cavity. Each of its many lobules has a central vein that carries blood to the hepatic vein, branches of the hepatic artery that carry oxygen and lipids, branches of the hepatic portal vein that carry other nutrients to the liver, bile ducts that carry bile from liver cells to the hepatic ducts, and sinusoids that contain Kupffer's cells.
- Bile contains water, bile salts, cholesterol, certain other lipids, bile pigments, and electrolytes.
- The gallbladder, which has an epithelial lining and a muscular wall, stores and concentrates bile and releases it as needed for the emulsification of fats.
- The pancreas is a soft spongy organ connected to the duodenum by a duct. It has acinar cells that secrete enzymes, and tubule cells that secrete bicarbonate and water. Pancreatic enzymes can digest nearly every substance in the human diet. Bicarbonate neutralizes acid and water dilutes chyme.

Small Intestine

- The small intestine includes the duodenum, jejunum, and ileum. Mucosal glands secrete mucus and enzymes, and the plicae circularis, villi, and microvilli greatly increase the internal surface area. Blood and lymph vessels receive absorbed nutrients.
- Motility includes peristalsis, segmentation contractions, and movements of villi.
- Enzymes, in addition to those from the liver and pancreas, include enterokinase, aminopeptidase, dipeptidase, maltase, lactase, and sucrase, some of which act within cells rather than in the lumen.
- Digestion breaks all nutrients in the human diet into absorbable molecules.
- In absorption, most sugars and amino acids are actively transported across the mucosal membrane and diffuse out of mucosal cells across interstitial fluids to blood capillaries. Products of fat digestion diffuse into mucosal cells, where triacylglycerols are resynthesized and packaged in chylomicrons. Water, vitamins, minerals, and electrolytes also are absorbed.

Large Intestine and Rectum

- The large intestine includes the cecum and the ascending, transverse, descending, and sigmoid colon. The colon has longitudinal bands of muscle that draw it into haustra.
- The rectum includes the anal canal and has folds in the inner lining.
- Motility consists of haustral churning, mass movements, and periodic defecation.
- Other functions include secretion of mucus and bicarbonate, digestion carried out by intestinal flora, and absorption of a little water, electrolytes, and nutrients from bacterial metabolism.

Integrated Digestive Functions

- Regulation of processes in the stomach is summarized in Figure 23.23.
- Regulation of processes in the small intestine is summarized in Figure 23.24.
- The roles of enteric hormones in regulating digestive functions are summarized in Table 23.6.

Clinical Applications

- Disturbances in motility occur in vomiting, constipation, and irritable colon syndrome.
- Disturbances in secretion occur in gastritis, ulcers, cirrhosis of the liver, hepatitis, pancreatitis, and diarrhea.
- Disturbances in digestion occur in gallstones and pancreatitis.
- Disturbances in absorption occur in pernicious anemia, sprue (celiac disease), and ulcerative colitis.

QUESTIONS AND PROBLEMS

The questions at the end of each chapter are numbered to correspond with the objectives listed at the beginning of the chapter. Italics indicate that a question requires critical thinking skills beyond simple factual recall.

Questions

1. (a) What are the general functions of the digestive system?
   (b) What are the major organs and accessory glands of the digestive tract?
2. (a) Briefly describe how the digestive system forms.
   (b) What are some common congenital defects of the digestive system?
3. (a) What are the main characteristics of the layers of the digestive tract?
   (b) *How are nerve plexuses associated with tract layers?*
   (c) *How do the layers of the tract contribute to the four major functions of the digestive system?*
4. What important structures are associated with the mouth?

5. *What digestive functions occur in the mouth?*
6. How does the structure of the pharynx differ from that of the esophagus?
7. (a) What is deglutition, and how does it occur?
   (b) What other functions do the pharynx and esophagus perform?
8. (a) Describe the macroscopic structure of the stomach.
   (b) What microscopic structures of the stomach contribute to digestive functions?
9. *What events occurring in the stomach contribute to (a) motility, (b) secretion, (c) digestion, and (d) absorption?*
10. *How are gastric secretions regulated?*
11. (a) Describe the macroscopic structure of the liver.
    (b) What microscopic liver structures contribute to digestion?
    (c) *How does blood flow through the liver?*
12. *How does the liver contribute to digestion?*
13. (a) Describe the macroscopic structure of the pancreas.
    (b) *Relate the microscopic structure of the pancreas to its secretions.*
    (c) *What would happen if proteolytic enzymes were released in active form?*
14. *How does the pancreas contribute to digestion?*
15. (a) Describe the macroscopic structure of the small intestine.
    (b) What microscopic structures of the small intestine contribute to digestion?
16. *How does motility occur in the small intestine?*
17. (a) What secretions are produced in the small intestine?
    (b) What are the actions of various secretions found in the small intestine?
18. *Summarize the steps in digestion that occur in the small intestine.*
19. (a) *How are carbohydrates absorbed?*
    (b) *How are lipids absorbed?*
    (c) *How are amino acids absorbed?*
    (d) *How are other nutrients absorbed?*
20. Describe the structure of the large intestine and rectum.
21. *How does the large intestine contribute to (a) motility, (b) secretion, (c) digestion, and (d) absorption?*
22. *How are digestive processes coordinated?*
23. *What are the properties and actions of enteric hormones?*
24. *How do each of the clinical applications disturb homeostasis?*

Problems

1. Given that a glass of milk contains lactose, protein, butterfat, vitamins, calcium, and phosphorus, explain in detail what happens to it as it passes through the digestive tract.
2. Repeat the above explanation for a meal consisting of a greasy hamburger on a bun.
3. Survey current information on enteric hormones and update Table 23.6.

OUTLINE

OBJECTIVES

1. Summarize the general principles of nutrient metabolism.
2. Briefly describe carbohydrate transport, glycogenesis, and glycogenolysis.
3. List the major components of glucose metabolism.
4. Describe the main events in glycolysis, and explain its significance.
5. Describe the main events in the Krebs cycle, and explain its significance.
6. Explain how electron transport and oxidative phosphorylation occur, and summarize energy capture from glucose metabolism.
7. Briefly state the significance of the pentose phosphate pathway.
8. Briefly describe lipid transport, fatty acid degradation, fatty acid synthesis, and cholesterol metabolism.
9. Explain where and how lipoproteins are formed and degraded.
10. Define nitrogen balance, amino acid pool, transamination, and deamination.
11. Briefly describe processes and significance of the urea cycle, the use of amino acids for energy, and gluconeogenesis.
12. Briefly summarize nucleotide metabolism.
13. Summarize the overall process of metabolism, and list some ways it is controlled.
14. Explain how lactose intolerance, gallstones, and gout disrupt homeostasis.

24

METABOLISM

OVERVIEW OF NUTRIENT METABOLISM

In the last chapter, we saw how the food we eat is broken down into small molecules and absorbed into the blood. In this chapter, we will see how cells use nutrient molecules to capture energy for cellular activities and to synthesize cellular components.

Metabolism refers to the sum total of chemical reactions in the body. It includes **catabolic** (kat''ah-bol'ik), or **degradative,** reactions in which nutrients are used to provide energy for cells and **anabolic** (an-ah-bol'ik), or **synthetic,** reactions in which the substance of the cell is produced. Like reactions in the digestive tract, reactions in cells also are catalyzed by enzymes. Three kinds of nutrients—carbohydrates, lipids, and proteins—are the substrates, or reactants, of most catabolic reactions. Their breakdown products serve as building blocks for anabolic reactions. In many metabolic pathways, several molecules called **intermediates** appear between the initial reactant and the final product (Figure 24.1). Water participates in reactions directly or as a reaction medium. Vitamins are used to make coenzymes and minerals act as cofactors in many enzyme reactions.

Glucose provides energy by the following summary reaction:

glucose + oxygen → carbon dioxide + water + energy.

Outside the body, this reaction could occur by simple burning, but inside cells, it takes place in a series of specific enzyme-controlled reactions. In certain reactions, energy is captured in high energy bonds in ATP or similar nucleotides (Chapter 3). Energy stored in high energy bonds can be released later in a controlled manner for cells to use in their many activities.

Certain coenzymes transport electrons through oxidation reactions that ultimately capture energy in ATP. (Recall from Chapter 2 that oxidation is the addition of oxygen or the removal of electrons or hydrogen.) Coenzymes work with mitochondrial enzymes to accept electrons or whole hydrogen atoms as they are removed from substrates. FAD (flavin adenine dinucleotide) and NAD (nicotinamide adenine dinucleotide) illustrate this.

$$FAD + 2H \rightarrow FADH_2$$

$$NAD^+ + 2H \rightarrow NADH + H^+$$

oxidized   reduced

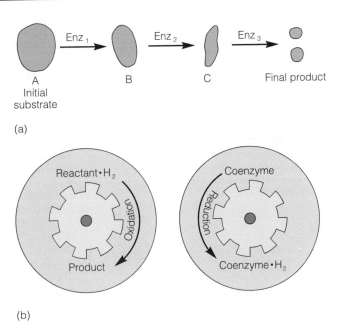

FIGURE 24.1
(*a*) Metabolic pathways often involve a series of enzyme-controlled reactions in which the product of one reaction becomes the reactant of the next. (*b*) Many metabolic reactions require coenzymes.

CARBOHYDRATE METABOLISM

Carbohydrates make up a large part of a normal diet and much of this carbohydrate is glucose. We will see how carbohydrates are transported, how glycogen is stored and released, and how glucose is broken down for energy.

Carbohydrate Transport

The simple sugars glucose, fructose, and galactose are absorbed and carried in the blood to the liver where most of the fructose and galactose is converted to glucose. The liver releases glucose to maintain a normal blood level and stores any excess as glycogen. Insulin facilitates the entry of glucose into cells; without it, cells can starve while surrounded by fluids rich in glucose.

Once in a cell, glucose undergoes **phosphorylation** (fos''fōr-il-a'shun), in which ATP is used to add a phosphate group to its end (sixth) carbon to form glucose-6-phosphate (G-6-P). Fructose and galactose are phosphorylated to form F-6-P and gal-6-P. Phosphorylated sugars are interconvertible. Phosphorylation keeps sugars from crossing the membrane again and enables them to undergo other reactions. Because sugars are phosphorylated as they enter cells, unphosphorylated sugars move down concentration gradients to enter cells.

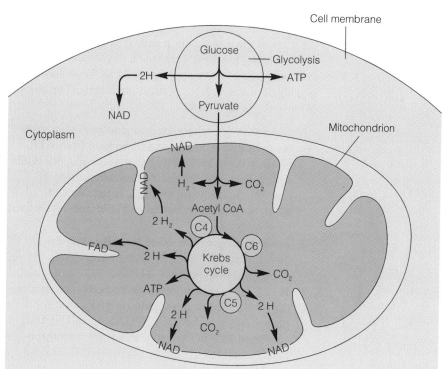

FIGURE 24.2
An overview of glucose metabolism.

Glycogenesis and Glycogenolysis

Synthesis of glycogen from glucose, or **glycogenesis** (gli″ko-jen′es-is), occurs when blood glucose is above normal. Breakdown of glycogen to glucose, **glycogenolysis** (gli″ko-jen-ol′is-is), occurs when blood glucose is below normal. Glycogen can be made from phosphorylated six-carbon sugars, using energy from uridine triphosphate (UTP). Removing glucose from glycogen requires no energy, but an inorganic phosphate group is added to form G-1-P. When the phosphate is removed, free glucose becomes available. These processes, catalyzed by different sets of enzymes, occur mainly in the liver.

Insulin fosters glycogenesis when glucose is present in excess, whereas epinephrine and glucagon foster glycogenolysis when cells need more glucose. These and other hormones precisely regulate the amount of glucose available to cells (Chapter 25). The physiological advantage of such precise regulation is to provide all body cells with a source of energy as they need it.

Muscle cells and most other cells store glycogen that they can break down for energy. Liver cells can contain up to 8 percent glycogen. Storing glucose in large glycogen molecules helps to maintain osmotic balance in cells. If cells stored all glucose as individual molecules, intracellular osmotic pressure would draw enough water into cells to rupture their membranes.

Overview of Glucose Catabolism

Catabolism of glucose for energy requires three processes: glycolysis, which occurs in the cytosol, and the Krebs cycle and electron transport, which occur in mitochondria (Figure 24.2). Glycolysis breaks glucose down to pyruvate and does not require oxygen. The Krebs cycle and electron transport break pyruvate down to carbon dioxide and water, and do require oxygen. All of these pathways involve a series of chemical reactions, and each reaction requires a specific enzyme. Many intermediates are acids that release H^+ and become ions in the aqueous reaction medium. Such ions have names ending in "ate." For example, pyruvic acid becomes pyruvate and acetic acid becomes acetate.

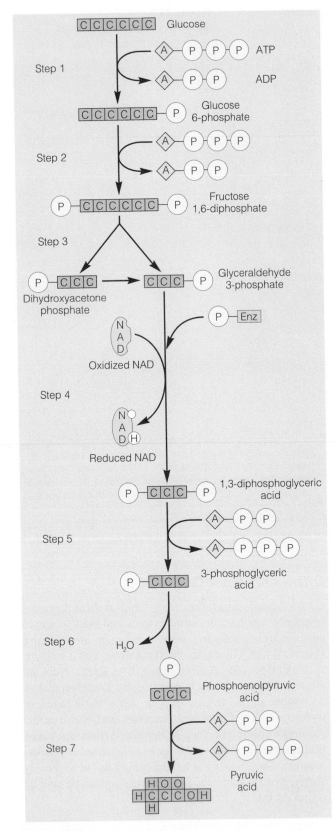

FIGURE 24.3

Glycolysis. The energy from two molecules of ATP are required to raise the energy level of glucose so that it can undergo the remainder of the reactions. The phosphorylated 6-carbon is broken into two 3-carbon molecules. Each 3-carbon molecule loses hydrogen to NAD and release sufficient energy to form two molecules of ATP. The end product of glycolysis is two molecules of pyruvate.

Glycolysis

As soon as glucose enters a cell, it undergoes **glycolysis** (gli-kol'is-is), in which a molecule of glucose is split into two molecules of pyruvate (Figure 24.3). Without oxygen, glycolysis proceeds from pyruvate to lactate. The steps in glycolysis are as follows:

1. Glucose is phosphorylated to glucose-6-phosphate (G-6-P), using energy and phosphate from ATP. (This is the same reaction that keeps glucose inside cells.) The G-6-P molecule rearranges to form fructose-6-phosphate (F-6-P). (Fructose and galactose can be converted to F-6-P and catabolized by glycolysis.)
2. F-6-P is phosphorylated to form fructose-1,6-diphosphate (F-1, 6-diP), using energy and phosphate from another ATP.
3. F-1, 6-diP is split into two three-carbon molecules, dihydroxyacetone phosphate (DHAP) and 3-phosphoglyceraldehyde (3-PGA). Each DHAP is converted to 3-PGA. From this point, two three-carbon molecules must be catabolized to account for one glucose molecule.
4. Inorganic phosphate from the cytosol is added to each 3-PGA molecule. Two hydrogen atoms are transferred from it to NAD^+ to form $NADH + H^+$ and 1,3-diphosphoglycerate (1,3-DPG).
5. Phosphate and energy from 1, 3-DPG are added to ADP to make ATP and 3-phosphoglycerate, which becomes 2-phosphoglycerate.
6. Water is removed from 2-phosphoglycerate to form phosphoenolpyruvate (PEP).
7. PEP is converted to pyruvate, and phosphate and energy are added to ADP to form ATP. The fate of pyruvate depends on whether oxygen is available.

In muscles during strenuous exercise when glycolysis is very rapid and pyruvate is made faster than mitochondria can use it, some pyruvate is converted to lactate by the addition of hydrogen from $NAD^+ + H^+$ (Figure 24.4a). Lactic acid released from skeletal muscles goes to the liver where it is used to make glucose. The glucose goes to muscles where it is metabolized to lactic acid. This process is called the **Cori cycle** (Figure 24.4b).

Moving hydrogen from $NADH + H^+$ to pyruvate forms lactate and frees NAD^+ to remove more hydrogen. This keeps glycolysis operating, so cells have some energy even when oxygen is not available. When oxygen becomes available, hydrogen temporarily stored in lactate is transferred back to NAD^+.

With respect to energy, each three-carbon molecule formed from glucose generates two ATP molecules for a total of four ATP per six-carbon glucose molecule. Because energy from two ATP was used in phosphorylation, the net energy yield from glycolysis in the absence of oxygen is two ATP per glucose molecule.

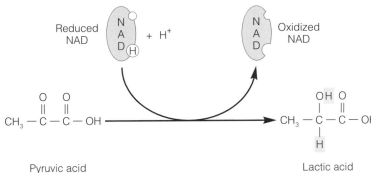

FIGURE 24.4
(*a*) The interconversion between pyruvate and lactate. The reaction goes to the right when oxygen is in short supply and to the left when the oxygen supply is adequate. (*b*) The Cori cycle.

(a)   Pyruvic acid   Lactic acid

Skeletal muscles   **Liver**

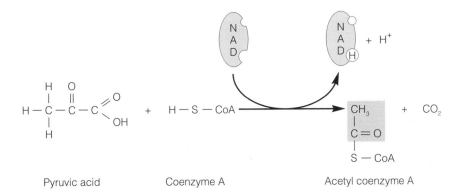

(b)

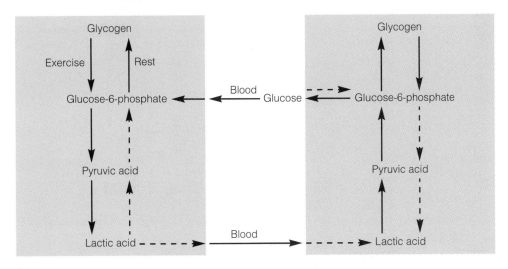

Pyruvic acid   Coenzyme A   Acetyl coenzyme A

FIGURE 24.5
Pyruvate is converted to acetyl-CoA. In this complex reaction, CO_2 is released and hydrogen is transferred to NAD^+.

Pyruvate to Acetyl-CoA

When oxygen is available, pyruvate enters mitochondria and is converted to **acetyl-coenzyme A** (as″et-il-ko-en′zīm a), or acetyl-CoA, by a complex reaction (Figure 24.5) in which pyruvate loses carbon dioxide, releases two hydrogen atoms to NAD^+, and acquires energy as it combines with coenzyme A. Acetyl-CoA now contains sufficient energy to enter the Krebs cycle.

Krebs Cycle

The **Krebs cycle** is a sequence of chemical reactions in which acetyl-CoA is oxidized to carbon dioxide and hydrogen atoms or hydrogen ions and electrons. It also is called the **citric** (sit′rik) **acid cycle** (its first product being citric acid) or the **tricarboxylic** (tri-kar-box-il′ik) **acid cycle** because citric and some other acids have three carboxyl groups. Hydrogen atoms and electrons removed from

FIGURE 24.6
The Krebs cycle.

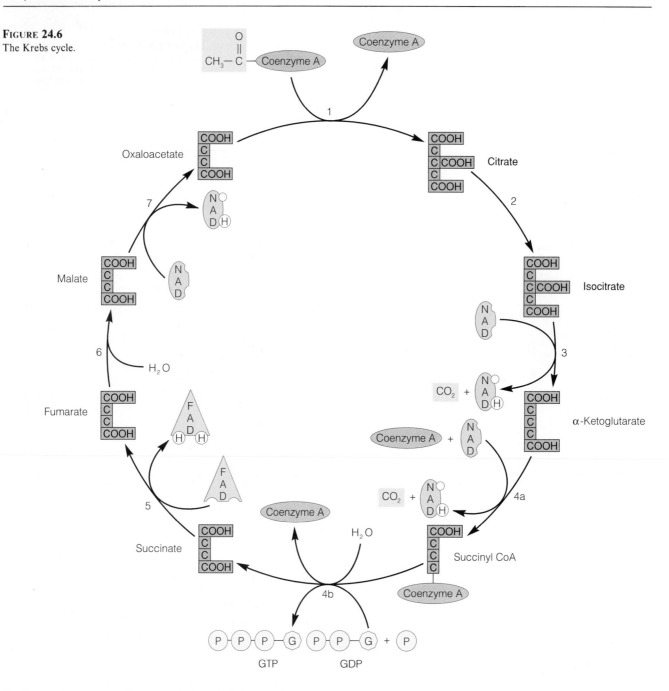

Krebs cycle intermediates undergo electron transport during which energy is captured in ATP. The Krebs cycle and electron transport occur simultaneously and only in the presence of oxygen.

Enzymes that control Krebs cycle reactions are located in the matrix of mitochondria, and molecules are passed from one enzyme to the next as they go through the cycle. The cycle begins with acetyl-CoA combining with oxaloacetate. When the cycle is completed, oxaloacetate becomes available to combine with another acetyl-CoA and goes through the cycle again and again.

The Krebs cycle (Figure 24.6) can be summarized as follows:

1. Acetyl-CoA combines with oxaloacetate, water is added, and CoA is removed to form citrate.
2. Citrate rearranges to form isocitrate.
3. Isocitrate loses carbon dioxide and two hydrogen atoms, and NAD^+ is reduced to $NADH + H^+$.
4. Alpha-ketoglutarate loses carbon dioxide and hydrogen, and gains water, forming $NADH + H^+$ and succinate. During this reaction, energy is transferred to a high-energy bond in guanosine triphosphate (GTP), a nucleotide similar to ATP.
5. Succinate loses two more hydrogen atoms, forming $FADH_2$ and fumarate.

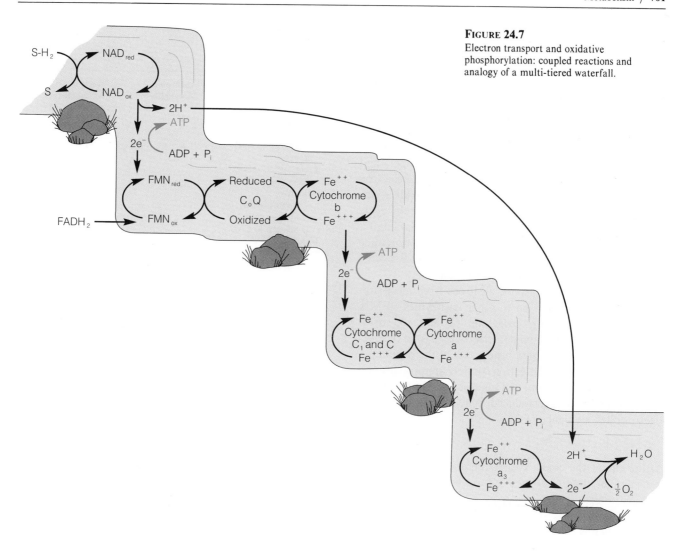

FIGURE 24.7
Electron transport and oxidative phosphorylation: coupled reactions and analogy of a multi-tiered waterfall.

6. Fumarate gains water to form malate.
7. Malate loses two hydrogen atoms, forming NADH + H$^+$ and oxaloacetate, which can combine with acetyl-CoA to start the cycle over again.

Significant events in the Krebs cycle include the following:

1. Two molecules of carbon dioxide are released, thereby disposing of the two carbons in an acetyl group.
2. Four pairs of electrons (or hydrogen atoms) are transferred to coenzymes for later oxidation.
3. Energy is captured in GTP directly from the substrate. Energy in GTP can be transferred to ATP.

$$GTP + ADP \rightarrow GDP + ATP$$

Electron Transport

So far, we have encountered six reactions in glucose metabolism that transfer electrons to coenzymes—from 3-PGA during glycolysis, from pyruvate, and from four intermediates of the Krebs cycle—citrate, alpha-ketoglutarate, succinate, and malate. To metabolize a whole glucose molecule, each reaction takes place twice—once for each three-carbon molecule into which glucose was split. Thus, 12 pairs of hydrogen atoms enter electron transport for each glucose molecule metabolized.

How can 24 hydrogen atoms (12 pairs) be removed from one glucose molecule when the molecule contains only 12 hydrogen atoms? Study the Krebs cycle, and try to answer this question before you read on.

Did you find that water is added in three reactions—the formation of citrate, succinate, and malate? Because two turns of the Krebs cycle are required to metabolize one glucose molecule, six water molecules and, therefore, 12 hydrogens are added to those already present in glucose. The water used in these reactions is called **metabolic water.**

Electron transport (Figure 24.7) involves the transfer of electrons from a substrate through a series of carriers and the capture of energy in ATP at certain transfers.

Electrons and hydrogen ions ultimately combine with oxygen to form water. During each transfer, one substance is oxidized as another is reduced. For example, a substrate with hydrogen (substrate H_2) loses hydrogen (is oxidized) at the same time coenzyme NAD^+ gains electrons (is reduced) and H^+ is released. Substrate-H_2 refers to any substrate that can be oxidized—3-PGA, pyruvate, and Krebs cycle intermediates.

Oxygen must be available for electron transport and the Krebs cycle to occur. Enzymes that catalyze electron transport are located on the cristae of mitochondria, adjacent to those in the matrix that catalyze Krebs cycle reactions. As electrons are removed during oxidation of a substrate, they are used in a coupled reaction to reduce coenzymes in the cristae. Each reaction requires a specific enzyme and most also require a coenzyme.

During electron transport, electrons from most substrates go to NAD and on to other coenzymes—FMN (flavin mononucleotide), coenzyme Q, and the cytochromes b, c1 and c, a and a3. Electrons that go to FAD enter the electron transport system, not at the beginning, but at coenzyme Q. All electron carriers are proteins and most contain an iron atom and some sulfur. Oxidized iron has a valence of 3^+ and reduced iron (after it accepts an electron) has a valance of 2^+.

$$Fe^{3+} + e^- \rightarrow Fe^{2+}$$

At the end of the chain, two H^+ combine with two electrons and an atom of oxygen to form water.

$$2e^- + 2H^+ + \tfrac{1}{2}O_2 \rightarrow H_2O$$

Thus, oxygen is the final hydrogen acceptor. All 12 pairs of hydrogen atoms from the metabolism of one molecule of glucose go through the electron transport process.

$$12H_2 + 6O_2 \rightarrow 12H_2O$$

Oxidative phosphorylation is the addition of phosphate and energy to ADP to form ATP. It occurs at three sites along the electron transport chain where sufficient energy is available to make a high-energy bond. These sites are at the following electron transfers: (1) from FMN to cytochrome b, (2) from cytochrome b to cytochromes c1 and c, and (3) from cytochrome a to cytochrome a3. Energy capture by oxidative phosphorylation during electron transport is analogous to water flowing down a multi-tiered waterfall. Electron transfers that release little energy are represented as small cascades, and those that provide sufficient energy to make a high-energy bond are represented as large cascades.

Oxidative phosphorylation is not fully understood, but the **chemiosmotic** (kem″e-o-os-mot′ik) **theory** for which Peter Mitchell received a Nobel Prize in 1978 offers this explanation. During oxidation, H^+ from the matrix concentrates in the intermembranous space of a mitochondrion, creating a **chemiosmotic gradient.** This gradient of

both electrical charge and osmotic pressure drives phosphorylation reactions. The movement of a pair of protons down the gradient (back across the membrane to the matrix) makes energy available to form a high-energy bond.

Pairs of electrons transferred to NAD^+ go through the entire electron transport process and yield three molecules of ATP by oxidative phosphorylation. Pairs of electrons transferred to FAD enter electron transport after the first phosphorylation and yield only two molecules of ATP. In other words, electrons on NAD^+ go down the entire waterfall, those on FAD go down the second and third large cascades.

Oxidative phosphorylation is significant because it makes energy available in a form that cells can use. We can move body parts or our whole body because muscles exert force on bones as individual muscle cells expend energy from ATP in the contraction process.

Summary of Energy Capture

How much energy is captured from a glucose molecule depends on whether it is metabolized with or without oxygen. Under anaerobic (without oxygen) conditions, glucose is metabolized to lactic acid in so-called **fast glycolysis** with a net yield of only two ATPs. Under aerobic (with oxygen) conditions, glucose is completely metabolized by **slow glycolysis,** the Krebs cycle, and electron transport to carbon dioxide and water, and yields 38 ATPs. Glycolysis yields a net of two substrate level ATPs and the Krebs cycle yields two more substrate level ATPs, one per cycle. Most ATP comes from electron transport of 12 pairs of electrons—10 pairs entering at NAD produce 30 ATPs (three per pair) and two pairs entering at FAD produce four ATPs (two per pair). Anaerobic and aerobic metabolism of glucose is summarized in Table 24.1.

TABLE 24.1 Energy Yields in ATPs for Glucose by Anaerobic and Aerobic Metabolism

Process	Anaerobic Metabolism	Aerobic Metabolism
Glycolysis (substrate level)	4	4
Less energy to start process	−2	−2
Krebs cycle—2 acetyl-CoAs (substrate level)		2
Oxidative phosphorylation		
Hydrogen to NAD		30
Hydrogen to FAD		4
Total	2	38

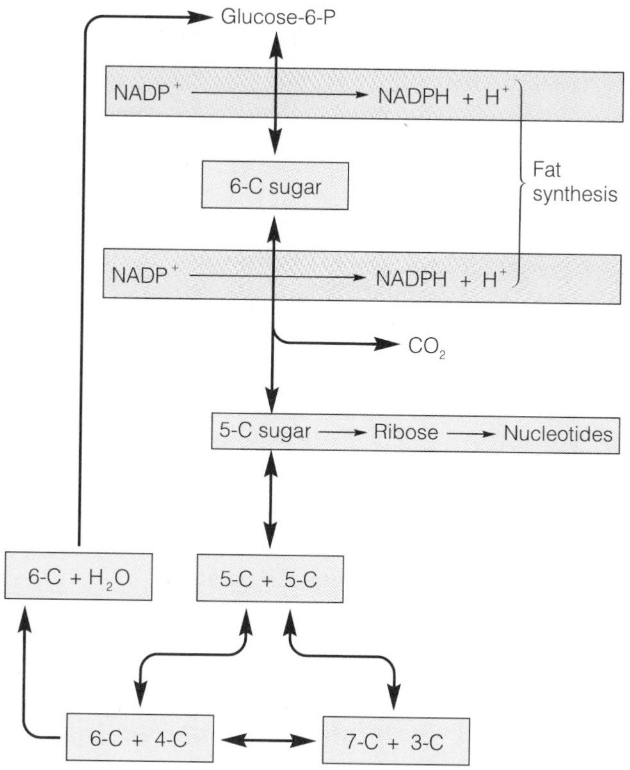

FIGURE 24.8
Some of the reactions of the pentose
phosphate pathway.

Pentose Phosphate Pathway

Though most carbohydrates are catabolized via glycolysis, the Krebs cycle, and electron transport, sometimes as much as 30 percent of the body's glucose is metabolized by the **pentose phosphate pathway** (Figure 24.8). This pathway does not require oxygen and operates in the cytosol of cells—mainly in the liver, fatty tissues, and red blood cells. Though it does not produce ATP directly, the pentose phosphate pathway is significant because it reduces the coenzyme $NADP^+$ to $NADPH + H^+$ required for fatty acid synthesis and five-carbon sugars required for the synthesis of nucleic acids and many structural proteins such as collagen.

LIPID METABOLISM

Most human dietary lipids are acylglycerols (formerly called triglycerides), but cholesterol and fat-soluble vitamins also are plentiful. Half of the energy in foods that Americans eat comes from lipids, and as much as half of the dietary carbohydrate is converted to lipids—later used for energy or stored as fat. So, lipid metabolism is an extremely important part of total metabolism.

Lipid Transport

After absorption, acylglycerols, phospholipids, cholesterol, and other lipids are packaged with protein in **chylomicrons** and then are transported in lymph and blood to the liver. By the time chylomicrons enter the blood after a meal, blood glucose and insulin already have increased. Insulin activates **lipoprotein lipase,** an enzyme bound to the membranes of blood vessel lining cells. As chylomicrons pass through blood vessels, lipoprotein lipase digests their acylglycerols and phospholipids to smaller molecules such as fatty acids and glycerol that can enter cells.

Lipid Storage

Lipids are stored mainly in adipose (fat) cells where they can make up 95 percent of the cell volume. When insulin causes glucose to enter fat cells, it is metabolized by glycolysis to three-carbon molecules, which are converted to glycerol. Fatty acids that diffuse into fat cells combine with glycerol to form acylglycerols. Lipids are excellent energy storage molecules because they are insoluble in watery cell fluids and contain much hydrogen (which yields large amounts of energy in oxidative phosphorylation). Lipids in subcutaneous tissues insulate the body and delay heat loss when the environment is cooler than body temperature. Fat deposits contain mostly the same fatty acids that occur in foods—stearic acid (a saturated 18-carbon molecule), oleic acid (an unsaturated 18-carbon molecule), and palmitic acid (a saturated 16-carbon molecule).

Lipid Catabolism

Acylglycerols, the main lipids used for energy, are hydrolyzed to glycerol and three fatty acids (Chapter 3). Glycerol is converted to 3-PGA and can be metabolized by glycolysis or the pentose phosphate pathway.

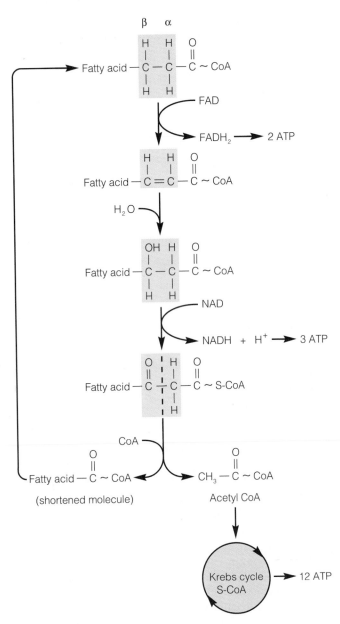

FIGURE 24.9
Beta oxidation.

TABLE 24.2   Energy Yields in ATPs for Saturated and Unsaturated Fatty Acids

Process	Saturated Fatty Acid (Stearic) (No Double Bonds)	Unsaturated Fatty Acid (Linoleic) (2 Double Bonds)
Beta oxidation	40	36
Less ATP used to start process	−1	−1
Krebs cycle and biological oxidation	108	108
Total	147	143

Fatty acid-CoA undergoes beta oxidation in the mitochondrial matrix by the following steps:

1. Fatty acid-CoA loses two hydrogen atoms to FAD, and a double bond forms between the alpha and beta carbon atoms.
2. Reduced fatty acid-CoA gains water at its double bond and immediately loses hydrogen atoms. Oxygen forms a ketone group at the beta carbon, and electrons are transferred to NAD^+.
3. The bond between the alpha and beta carbons breaks releasing acetyl-CoA, which goes to the Krebs cycle.
4. Coenzyme A binds to the beta carbon, making a new fatty acid-CoA shorter by two carbons.

Each new shorter fatty acid-CoA undergoes beta oxidation until finally a four-carbon fatty acid forms two molecules of acetyl-CoA. Thus, an 18-carbon fatty acid goes through beta oxidation eight times and yields nine acetyl-CoA molecules. (Eight, and not nine, trips through beta oxidation are required because the last trip produces two acetyl-CoAs.) Most fatty acids have an even number of carbons, so beta oxidation yields two-carbon acetyl groups almost exclusively.

Fatty acids contain relatively more hydrogen and less oxygen (are more reduced) than sugars, so they yield more energy on oxidation. Each beta oxidation produces five ATPs by electron transport of hydrogens (two ATPs from hydrogens on FAD and three ATPs from those on NAD^+). For an 18-carbon fatty acid, eight beta oxidations yield 40 ATPs less one ATP required to start the process for a net of 39 ATPs. In addition, nine acetyl-CoA molecules go through the Krebs cycle and their hydrogens go through electron transport with 12 ATPs being captured from each acetyl-CoA for another 108 ATPs. Of course, beta oxidation, like the Krebs cycle, is coupled to electron transport and requires oxygen, NAD^+, and FAD. In unsaturated fatty acids, double bonds already exist at some beta oxidation sites, and hydrogens are not transferred to FAD. The energy yield from beta oxidation decreases by two ATPs per double bond. Energy yields from fatty acids are shown in Table 24.2.

Fatty acids from fat hydrolysis are broken down into two-carbon units by **beta oxidation** (Figure 24.9), so named because oxidation occurs at beta carbons (the second carbon from the carboxyl group). Beta oxidation takes place in the mitochondrial matrix, but fatty acids require processing before they can enter the mitochondria. In the cytosol, a fatty acid combines with coenzyme A, using energy from ATP to form a molecule called fatty acid-coa, which then combines with carnitine, a seven-carbon nitrogenous substance, to form fatty acid-carnitine and free coenzyme A. Carnitine ferries the fatty acid across the inner mitochondrial membrane. Coenzyme A replaces carnitine, and carnitine is released to transport more fatty acid molecules.

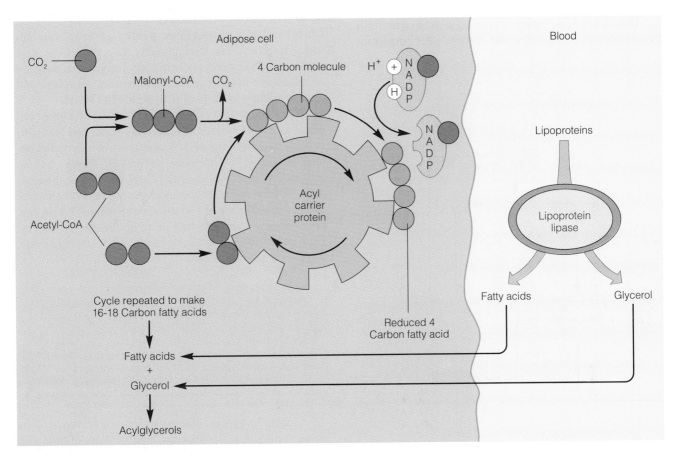

The top portion shows chemical structures:

$CH_3 - \overset{O}{\underset{||}{C}} - CH_2 - \overset{O}{\underset{||}{C}} - OH \longrightarrow CH_3 - \overset{OH}{\underset{|}{CH}} - CH_2 - \overset{O}{\underset{||}{C}} - OH$

Acetoacetic acid   Beta-hydroxybutyric acid

$CH_3 - \overset{O}{\underset{||}{C}} - CH_3$

Acetone

FIGURE 24.10
Ketone bodies are products of incomplete fatty acid metabolism.

FIGURE 24.11
Synthesis of fatty acids and fats.

Liver cells rapidly metabolize fatty acids and can make more acetyl-CoA than they need. They use excess acetyl-coA to make **ketone** (ke′tōn) **bodies**—acetoacetic acid, beta-hydroxybutyric acid, and acetone (Figure 24.10). Sometimes cells make ketone bodies from incompletely metabolized fatty acids. Ketone bodies diffuse freely out of the liver, and acetone, being volatile, is exhaled in the lungs. Ketone bodies that enter cells are converted back to acetyl-CoA and used for energy.

When cells obtain energy mainly from fats, as in starvation or untreated diabetes, **ketosis** (ke-to′sis), the accumulation of ketone bodies in blood, occurs. Ketone bodies, being acidic, cause **acidosis,** an acidic blood pH. Severe acidosis leads to coma and death.

Lipid Synthesis

In addition to making ketone bodies, the fat cells (and the liver) can use acetyl-CoA to make fatty acids by **beta reduction.** Though beta reduction is the reverse of beta oxidation, it is controlled by different enzymes and occurs on a large cytoplasmic molecule called **acyl** (as′il) **carrier protein** instead of in mitochondria. In addition to acetyl-CoA, fatty acid synthesis requires reduced NADP from the pentose phosphate pathway and bicarbonate ions (HCO_3^-) from CO_2. As excess acetyl-CoA is used to make fatty acids or lipoprotein lipase releases them from chylomicrons, the fatty acids move to adipose cells and combine with glycerol to make acylglycerols (Figure 24.11).

Regulation of Lipid Metabolism

Most cells synthesize fatty acids when they have more than enough glucose and acetyl-CoA. They use them for energy when they lack glucose. When acetyl-CoA is plentiful, some of it is made into malonyl-CoA, an intermediate in the pathway for fatty acid synthesis. Malonyl-CoA binds to carnitine and inhibits the fatty acid breakdown. Conversely, when glucose is in short supply, fatty acid-carnitine inhibits the conversion of pyruvate to acetyl-CoA. Thus, a cell synthesizes fatty acids or oxidizes them, but not both at the same time (Figure 24.12).

In addition to providing energy, lipids form structural components of membranes and certain other important molecules. Cells can make all the lipids they need provided *linoleic acid* is available. Linoleic acid is an **essential fatty acid**; that is, it is essential in the diet because human cells cannot make it. It is needed to synthesize membranes and prostaglandins, and it may be needed for the transport and metabolism of cholesterol.

Cholesterol Metabolism and Lipoproteins

Cholesterol, present in any diet containing animal fats, is absorbed and packaged in chylomicrons like other lipids. Most cells make a little cholesterol. Liver cells make a lot. Cholesterol gives membranes fluidity at body temperature. The more cholesterol a membrane contains, the more fluid it is. Cholesterol is completely insoluble in water—an advantage in keeping membranes from dissolving, but a disadvantage in keeping it suspended in blood.

Cholesterol is needed to make and maintain cell membranes and synthesize steroid hormones and bile salts. It helps to waterproof skin. The derivative 7-dehydrocholesterol is used to synthesize vitamin D. Though human cells can make cholesterol, they lack enzymes to break it down. The body's main means of getting rid of cholesterol is through excretion of bile salts in the feces. This process removes only about one-half gram of cholesterol per day. Even when more cholesterol is lost than is taken in in foods, the liver easily replaces it.

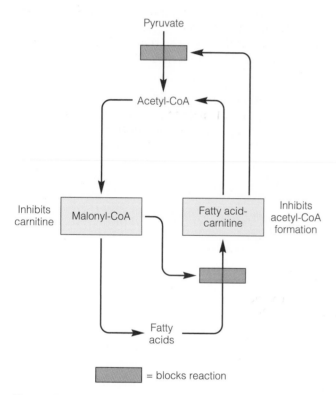

FIGURE 24.12

Regulation of synthesis and breakdown of fatty acids. When fatty acids are being synthesized, malonyl-CoA inhibits the formation of fatty acid-carnitine. When fatty acids are being oxidized, fatty acid-carnitine inhibits the formation of acetyl-CoA from pyruvate. Thus, a cell can synthesize or oxidize fatty acids, but it cannot do both at the same time.

> Cholesterol was first isolated from gallstones over a century ago. The genetic disorder familial hypercholesterolemia (high blood cholesterol) was identified and associated with heart attacks in young patients in 1938. Today the importance of cholesterol research can hardly be underestimated. No less than 13 Nobel Prizes have been awarded for such studies.

Lipids, being insoluble in water, are transported in aqueous plasma as **lipoproteins**, combinations of lipid and protein. They include chylomicrons, very low density lipoproteins (VLDLs), low density lipoproteins (LDLs), and high density lipoproteins (HDLs). Each lipoprotein particle consists of a core of lipids such as acylglycerols and cholesteryl esters surrounded by a lipoprotein membrane made of a protein called apolipoprotein, phospholipids, and cholesterol. (Cholesteryl esters consist of cholesterol and a fatty acid attached by an ester bond.) Lipoprotein composition is summarized in Table 24.3.

TABLE 24.3   Lipid and Protein Composition (in percent) of Lipoproteins

Lipoprotein	Protein	Acylglycerols	Cholesterol	Cholesteryl esters	Phospholipids
Chylomicrons	2%	84%	2%	5%	7%
VLDLs	9	54	7	12	18
LDLs	21	11	8	37	22
HDLs	50	4	2	20	24

Data from R. Montgomery, et al., *Biochemistry: A case-oriented approach*, 1983.

Lipoproteins circulate in the blood and participate in the following processes (Figure 24.13).

1. Dietary fat leaves the intestine in chylomicrons.
2. In the blood, lipoprotein lipase partly degrades chylomicrons to fatty acids, monacylglycerols, and glycerol, which can enter adipose and other tissues.
3. In the liver, chylomicron remnants are incorporated into HDLs.
4. HDLs, made in the liver and intestine, appear to activate lipoprotein lipase and to take up cholesterol. They transport cholesteryl esters and help to prevent cholesterol deposition in coronary and other blood vessels.
5. VLDLs are made in the liver mostly from acylglycerols and cholesterol synthesized there.
6. Lipoprotein lipase releases fatty acids and glycerol from VLDLs and a little surface cholesterol moves to cell membranes, leaving VLDL remnants.

7. Some VLDL remnants bind to liver cell receptors and the remainder become cholesterol-laden LDLs. LDLs also bind to receptors, enter cells by endocytosis, and are degraded by lysosomal enzymes. Receptors are returned to the cell membrane. (Familial hypercholesterolemia is due to defective LDL receptors.)
8. HDLs incorporate some LDLs and VLDL remnants.

As more is learned about lipoproteins, it is becoming increasingly clear that the risk of coronary artery blockage increases when total cholesterol and LDL cholesterol are too high, when HDL cholesterol is too low, or when the ratio of total cholesterol to HDL cholesterol is greater than 4.5. Excessive lowering of blood cholesterol, however, is not necessarily beneficial. A high incidence of cerebral hemorrhages (ruptured cerebral blood vessels) has been reported among people with total cholesterol under 150 mg/dl. The current interpretation of blood cholesterol values is summarized in Table 24.4.

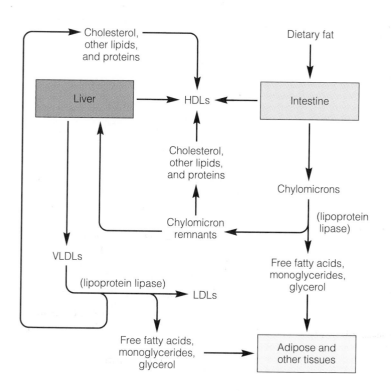

FIGURE 24.13
Lipoprotein metabolism.

TABLE 24.4 Interpretation of Cholesterol Levels

Kind of Cholesterol	Concentrations (in mg/dl)			
	Low	*Normal*	*Borderline*	*High*
Total cholesterol	<150	150–200	200–240	>240
LDL cholesterol		<130	130–160	>160
HDL cholesterol	<35	35 or more		

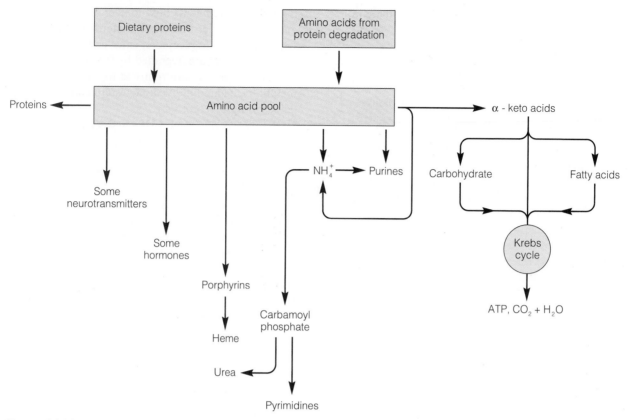

FIGURE 24.14
Metabolism of amino acids from the amino acid pool.

PROTEIN METABOLISM

For several hours after a meal, amino acids from protein digestion are absorbed into the blood. Once in the blood, they are actively transported into cells in 5 to 10 minutes.

Nitrogen Balance

Unlike carbohydrates and lipids, proteins are not stored in cells. Furthermore, enzymes and other cell proteins are regularly degraded and replaced in **protein turnover,** a process that occurs continuously at an overall rate of 350 to 400 g of protein per day in an adult. Rates in tissues vary according to metabolic activity—high in liver, moderate in muscle, and low in cartilage.

When proteins are turned over, many of their amino acids are reused in protein synthesis, but some are degraded and their nitrogen excreted. Nitrogen lost in this way is replaced from dietary proteins to maintain **nitrogen balance.** If nitrogen intake exceeds nitrogen loss, the body is in **positive nitrogen balance,** as occurs in growth, pregnancy, or tissue repair after injury. If nitrogen loss exceeds nitrogen intake, the body is in **negative nitrogen balance,** as occurs in starvation, certain wasting diseases, and weight-loss diets that contain too little protein.

Amino Acid Pool

Of 20 amino acids normally found in proteins, eight **essential amino acids**—leucine, isoleucine, lysine, methionine, threonine, phenylalanine, tryptophan, and valine—must be in the diet because human cells lack the enzymes to synthesize them. Human cells synthesize only small amounts of arginine and histidine, so they, too, should be included in the diet, especially during periods of rapid growth. Given adequate amounts of essential amino acids, cells can make any other amino acids they need.

Amino acids from the diet and from protein turnover form an **amino acid pool,** an assortment of amino acids in cells. Cells use amino acids to make proteins (including enzymes), nucleic acids, and other substances (Figure 24.14). Liver cells make blood proteins such as albumins and blood clotting factors. Connective tissue cells make collagen and elastin, and muscle cells make large quantities of actin and myosin. Amino acids also are used to make neurotransmitters in neurons, hormones in glands, and hemoglobin in erythrocytes.

Protein synthesis is an important component—maybe the most important component—of protein metabolism. Because of its significance in cell function, it was discussed in Chapter 3, but should be reviewed at this time.

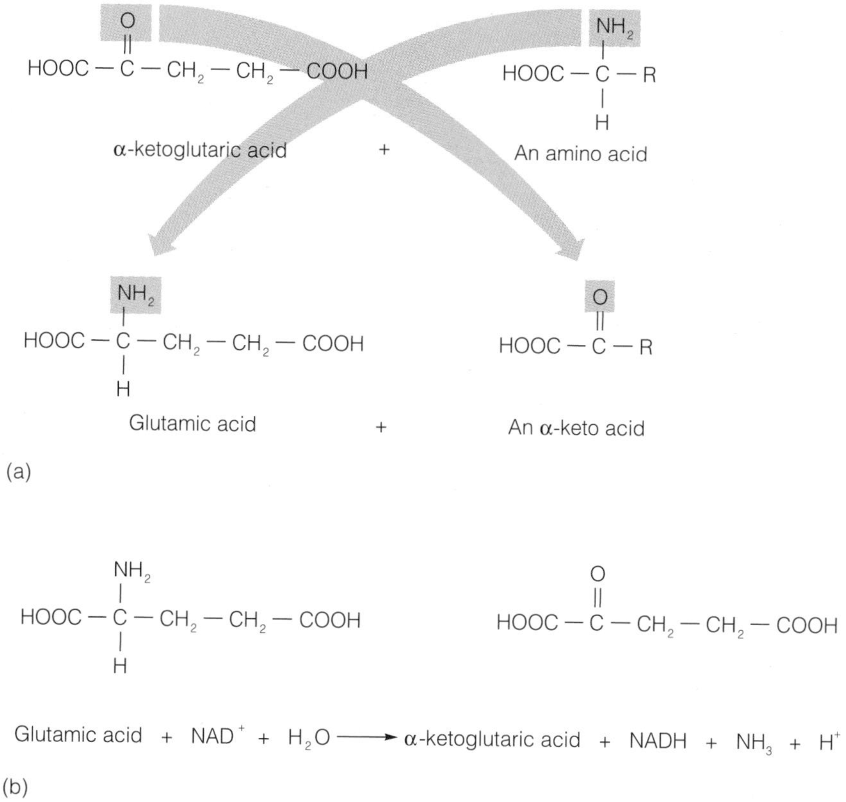

$$\text{HOOC} - \overset{\overset{\displaystyle O}{\|}}{C} - CH_2 - CH_2 - COOH$$

α-ketoglutaric acid   +   An amino acid

$$\text{HOOC} - \overset{\overset{\displaystyle NH_2}{|}}{\underset{\underset{\displaystyle H}{|}}{C}} - R$$

$$\text{HOOC} - \overset{\overset{\displaystyle NH_2}{|}}{\underset{\underset{\displaystyle H}{|}}{C}} - CH_2 - CH_2 - COOH$$

Glutamic acid   +   An α-keto acid

$$\text{HOOC} - \overset{\overset{\displaystyle O}{\|}}{C} - R$$

(a)

$$\text{HOOC} - \overset{\overset{\displaystyle NH_2}{|}}{\underset{\underset{\displaystyle H}{|}}{C}} - CH_2 - CH_2 - COOH$$

$$\text{HOOC} - \overset{\overset{\displaystyle O}{\|}}{C} - CH_2 - CH_2 - COOH$$

Glutamic acid + NAD$^+$ + H$_2$O $\longrightarrow$ α-ketoglutaric acid + NADH + NH$_3$ + H$^+$

(b)

FIGURE 24.15
(*a*) Transamination, and (*b*) deamination.

Use of Amino Acids for Energy

Before they can be used for energy, amino acids undergo transamination or deamination (Figure 24.15). **Transamination** (trans-am-in-a′shun) is the exchange of functional groups between an amino acid and a keto acid (an acid that has a ketone group). Transamination does not change the total number of amino acid molecules in a cell, and it never makes an essential amino acid. It can increase the supply of one amino acid at the expense of a more abundant one, allowing a cell to make a protein it otherwise might not make. In one kind of **deamination** (de-am″in-a′shun), an amino group and two hydrogen atoms are removed from an amino acid, yielding an alpha-keto acid, ammonia, and NADH + H$^+$. Deamination, which decreases a cell's amino acids, occurs when amino acids are used for energy or for the synthesis of glucose or fatty acids.

Nearly all tissues produce ammonia from deamination of amino acids, and the ammonia must be removed before it builds to toxic levels. Ammonia is transported as ammonium ions and attached to the amino acids aspartate and glutamine. When it reaches the liver, it is incorporated into urea, a much less toxic waste, by the following steps in the **urea cycle** (Figure 24.16):

1. An ammonium ion combines with carbon dioxide using energy and phosphate from ATP to form carbamoyl phosphate.
2. Carbamoyl phosphate combines with ornithine to form citrulline. (The amino acids ornithine and citrulline participate in the urea cycle, but are not found in proteins.)
3. An amino group from aspartate is added to citrulline using energy from ATP to form argininosuccinate.
4. Argininosuccinate breaks down to arginine and fumarate. (Fumarate can enter the Krebs cycle.)
5. Arginine combines with water to form ornithine and urea. Ornithine repeatedly combines with carbamoyl phosphate and is recycled. Relatively nontoxic urea travels in the blood and is excreted by the kidneys.

FIGURE 24.16
The urea cycle.

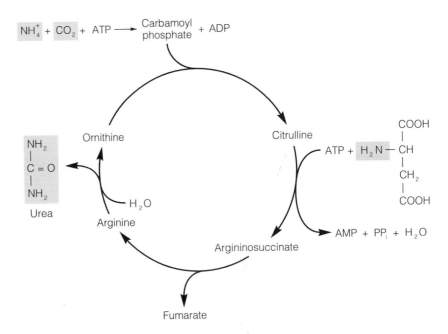

The **carbon skeletons** of amino acids after deamination, many of which are Krebs cycle intermediates such as pyruvate, oxaloacetate, and alpha-ketoglutarate, can be metabolized for energy. This occurs in starvation as carbohydrate and lipid stores are depleted, in protein turnover as small amounts of protein are degraded and some amino acids deaminated, and in positive nitrogen balance when protein intake is excessive. Compared to carbohydrates and lipids, proteins provide the least efficient source of metabolic energy.

Gluconeogenesis

Gluconeogenesis (glu-ko-ne-o-jen′e-sis) is the synthesis of glucose from noncarbohydrate sources, such as amino acids. It occurs between meals when glycogen is depleted and helps keep the blood glucose level in a normal range. Gluconeogenesis provides glucose to brain cells and red blood cells that cannot use other nutrients. It accelerates during starvation when up to 60 percent of the tissue proteins can be degraded before death ensues.

Gluconeogenesis occurs mainly in the liver and kidney cells, which contain the appropriate enzymes. Carbon skeletons are metabolized in the Krebs cycle and by cytosol enzymes to form pyruvate, and energy from GTP is used to form phosphoenolpyruvate (PEP). Energy in PEP drives glycolysis in reverse to make glucose.

NUCLEOTIDE METABOLISM

Though no nucleic acid components are essential nutrients, they are absorbed as nucleotides and nucleosides, and most are degraded to purines, pyrimidines, ribose, deoxyribose, and phosphates. Of these products, phosphates, sugars, and the purine adenine can be used to make nucleic acids. Most other purines and pyrimidines are catabolized for energy and resynthesized when cells make nucleic acids and nucleotides. Most cells can both synthesize and degrade purines and pyrimidines.

Nucleotide metabolism is carefully regulated to allow nucleic acid synthesis in preparation for cell division and protein synthesis as needed. This regulation also allows synthesis of nucleotides such as ATP, GTP, other energy-storing molecules; FAD, NAD, and other electron transport molecules; and control molecules such as cAMP and cGMP.

REGULATION OF CELLULAR LEVEL METABOLISM

Metabolism is regulated within cells mainly by controlling the activity of enzymes, which are localized in particular organelles or other specific sites. For example, adenylate cyclase is found in plasma membranes, enzymes that transfer energy to ATP are bound to the cristae of mitochondria, those that synthesize proteins are bound to ribosomes, and those that synthesize or degrade fats are bound to the smooth endoplasmic reticulum. Enzyme concentrations are regulated according to cellular needs by controlling protein synthesis. Enzyme activity is regulated by several mechanisms, including product inhibition, action of ligands on allosteric enzymes, feedback inhibition, and actions of isozymes.

Product inhibition occurs when accumulation of the product of a reaction suppresses the reaction. In a series of reactions, suppressing the final reaction allows accumulation of the product of the previous reaction and so on back through the series.

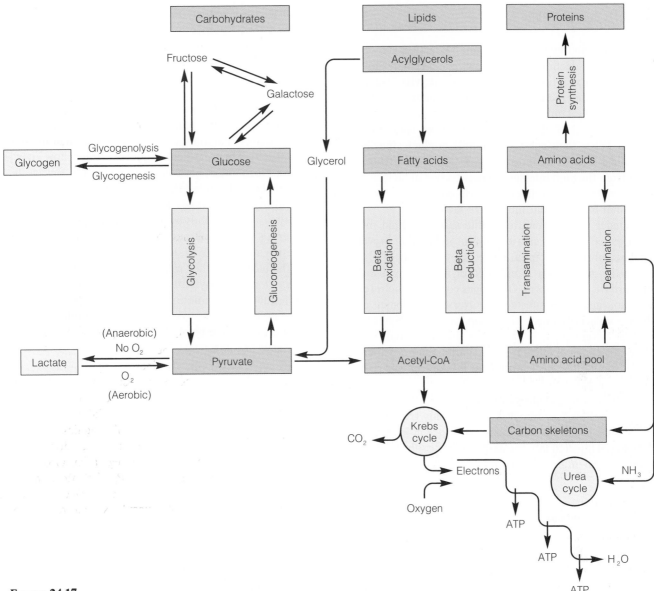

FIGURE 24.17
Summary of metabolism of major nutrients.

Allosteric (al″o-ster′ik) **enzymes** are subject to regulation through a binding site, an **allosteric site,** in addition to the active site where the substrate binds. (Allosteric means other shape.) Ligands that bind to an allosteric site change the shape of the molecule, including its active site. Such reversible changes turn an enzyme on or off, depending on whether the ligand occupies the allosteric site.

Feedback inhibition, or end-product inhibition, may or may not involve an allosteric enzyme. As noted in Chapter 2, it regulates the synthesis of various substances in living cells. The inhibiting product attaches to an allosteric site or otherwise inhibits an enzyme when it is plentiful and fails to inhibit when it is in short supply. Because feedback inhibition acts quickly and directly, the cell wastes no energy making unneeded product or even a special inhibitor; it simply uses the product of a reaction as its inhibitor.

Certain biochemical reactions are catalyzed by **isozymes,** multiple forms of the same enzyme. Though these forms catalyze the same reaction, they can have different effects. For example, the lactic dehydrogenase (LDH) isozyme in heart muscle typically oxidizes lactate to pyruvate, whereas the isozyme in skeletal muscle typically reduces pyruvate to lactate, particularly during exercise. Because of chemical differences, isozymes are subject to allosteric regulation, as when a ligand inhibits one isozyme and has no effect on another. Such selective inhibition serves to regulate cellular metabolism.

Metabolism of major nutrients is summarized in Figure 24.17.

CLINICAL APPLICATIONS

In humans (and most other mammals), lactase activity is high at birth, declines during growth, and remains low in adulthood. Exceptionally low lactase activity causes a condition called **lactose intolerance,** the inability to digest lactose. In the United States, 20 percent of white adults and 70 percent of black adults suffer from lactose intolerance. When they ingest milk or milk products, they have osmotic diarrhea and intestinal gas. Osmotic diarrhea results from the high intestinal concentration of lactose that draws water into the lumen. Gas is produced by certain intestinal bacteria that metabolize lactose.

Under normal conditions, bile salts and lipid molecules called lecithins create a solvent in bile that prevents cholesterol from precipitating. When this solvent fails, or when the gallbladder is inflamed or infected, **gallstones** (Figure 24.18) can form. Gallstones consist of cholesterol, calcium, and other substances. Overweight women over 40 are especially likely to develop gallstones. Symptoms include severe pain, impaired fat digestion, and, sometimes, blockage of ducts from the gallbladder or pancreas to the small intestine. Limiting fat intake seems to reduce the risk of gallstones.

At one time, **cholecystectomy** (ko″le-sis-tek′to-me), surgical gallbladder removal, was the only treatment for gallstones, but drugs and shock waves are now used, too. Chenodiol injected into the gallbladder dissolves stones in patients who might not withstand surgery. It is not widely used because it is toxic to the liver. Extracorporeal lithotripsy (stone fragmenting by shock waves from outside the body) requires a high current, underwater spark discharge focused on a gallstone. For this treatment, the patient must be immersed in water and the treatment area anesthetized. A device is being developed that will fragment gallstones without immersion or anesthetic.

Gout is a severe, painful arthritis often caused by a genetic defect that allows deposition of uric acid in joints (Chapter 9). The treatment of gout illustrates how knowledge of biochemical pathways can be used to thwart disease processes. Limiting the quantity of purines in the diet by limiting meats, especially organ meats, reduces the quantity of amino acids available to synthesize uric acid. The drugs probenecid and sulfinpyrazone increase urate excretion and prevent urate reabsorption from the kidneys into the blood, but they can foster the formation of urate kidney stones. Phenylbutazone increases urate excretion and reduces inflammation. Allopurinol specifically inhibits xanthine oxidase, so xanthine, a soluble precursor of urate, is excreted before urate can form and precipitate in kidneys or joints.

CLINICAL TERMS

ammonemia (am″o-ne′me-ah) accumulation of ammonia in the blood

ketonuria (ke″to-nu′re-ah) accumulation of ketone bodies in urine

FIGURE 24.18
Gallstones in a gallbladder.

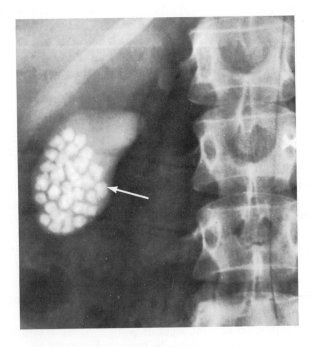

ESSAY

METABOLIC EFFECTS OF DAMAGED AND DEFICIENT ENZYMES

Certain metabolic disturbances result from damaged or deficient enzymes. Agents that damage enzymes usually are environmental toxins such as heavy metals, cyanide ions, carbon monoxide, insecticides, and some medicines. In contrast, enzyme deficiencies usually are due to abnormal genes that fail to synthesize an enzyme or synthesize a defective one.

Heavy metals denature proteins, including enzymes, by distorting the molecular shape. Such distortion is especially detrimental to enzymes because it changes the shape of the active site so the enzyme cannot form a complex with its substrate.

Cyanide ions and carbon monoxide bind with high affinity to molecules such as cytochromes and hemoglobin. Both specifically block oxidative phosphorylation at the last site in the electron transport chain (the cytochrome a-a₃ complex), thereby decreasing the amount of energy captured in ATP. Carbon monoxide also binds to hemoglobin much more tightly than oxygen. Hemoglobin molecules carrying carbon monoxide cannot also carry oxygen, so oxygen transport to cells is severely compromised.

The insecticide rotenone, barbiturates, and some tranquilizers block oxidative phosphorylation at the first site (NAD to FMN) and, thereby, decrease the amount of energy captured in ATP. In addition, certain insecticides specifically inhibit cholinesterase, an enzyme that inactivates acetylcholine. When the enzyme is inhibited, acetylcholine remains active and causes violent spasms. In respiratory muscles, such spasms can be fatal.

Enzyme deficiencies can interfere with the metabolism of all types of nutrients—carbohydrates, lipids, and proteins. These effects include inability to metabolize stored glycogen, alterations in the blood lipoprotein concentrations, and accumulation of toxic levels of some amino acids.

Glycogen storage diseases are due to deficiencies in enzymes that break down glycogen. Affected cells metabolize glucose and store it as glycogen, but cannot break it down. Excessive glycogen deposits interfere with liver and kidney functions. The inability to release glucose from glycogen increases fat metabolism, which can lead to ketosis, and causes low blood glucose, which can lead to muscle weakness.

Hyperlipidemias are excesses of lipoproteins or other blood lipids. One such disorder, an absence of lipoprotein lipase, prevents chylomicron digestion. It becomes apparent as early as the first week after birth and can be treated only with a fat-restricted diet.

Hypolipidemias (deficiencies of blood lipids) usually are caused by a defective enzyme that prevents synthesis of a protein normally found in lipoproteins. Patients deficient in proteins needed to make chylomicrons, VLDLs, or LDLs often have only minor symptoms, but those deficient in proteins to make HDLs usually have excessive cholesterol deposits in their tissues.

Several disorders result from deficiencies in urea cycle enzymes. Hyperammonemia, high blood ammonia concentration, is due to a lack of the enzyme that makes carbamoyl phosphate or the one that makes citrulline. Citrullinuria (accumulation of citrulline in the urine) is caused by a lack of the enzyme that acts on citrulline. Arginosuccinic aciduria (accumulation of arginosuccinic acid in the urine) is caused by a lack of the enzyme that acts on the arginosuccinic acid. Patients with these disorders must limit their protein intake to minimize the amount of waste nitrogen to be removed by the urea cycle.

Several disorders result from a deficiency of an enzyme required to metabolize a specific amino acid. Normally, the essential amino acid phenylalanine is used to make proteins or converted to tyrosine and used to make the pigment melanin and certain hormones and neurotransmitters. In patients with phenylketonuria, the enzyme to convert phenylalanine to tyrosine is deficient, so phenylalanine is converted to other metabolites such as phenylketones. Accumulation of phenylketones in the blood causes irreversible mental retardation. Restricting the diet from early infancy to the small quantity of phenylalanine needed for protein synthesis and supplying tyrosine for synthesis of other products can prevent mental retardation.

Questions

1. What kinds of substances damage enzymes?
2. What causes enzyme deficiencies?
3. How can disorders associated with damaged or deficient enzymes be treated?
4. Can such disorders be prevented? If so, how?

Chapter Summary

Overview of Nutrient Metabolism

- Metabolism includes all anabolic (synthetic) and catabolic (breakdown) reactions that occur in body cells.
- Each reaction is catalyzed by a specific enzyme and some reactions also require a coenzyme.

Carbohydrate Metabolism

- Carbohydrates enter cells as glucose and other simple sugars by facilitated diffusion and active transport.
- Sugars are used for energy or stored as glycogen by glycogenesis.
- Glycogen can be broken down as needed by glycogenolysis to maintain a nearly constant blood glucose concentration.
- Glucose is metabolized for energy by glycolysis, the Krebs cycle, and electron transport.
- Glycolysis occurs in the cytosol and converts glucose to pyruvic acid, and, under anaerobic conditions, to lactic acid.
- The Krebs cycle occurs in the matrix of mitochondria, requires oxygen, and releases carbon dioxide and hydrogen atoms.
- Electron transport occurs in the cristae of mitochondria and requires oxygen. At some electron transfers, oxidative phosphorylation captures energy in ATP.
- A molecule of glucose yields 38 ATP when metabolized with oxygen, and a net of two ATP without oxygen.
- The pentose phosphate pathway occurs in the cytosol, converts glucose to five-carbon molecules, and produces $NADPH + H^+$.

Lipid Metabolism

- Lipids are transported in chylomicrons, digested in the blood by lipoprotein lipase to fatty acids and glycerol that diffuse into cells.
- Fatty acids are metabolized for energy by beta oxidation, the Krebs cycle, and electron transport.
- Fatty acids can be synthesized from excess acetyl-coA by beta reduction, used to make acylglycerols, and stored in fat tissues.
- Fatty acid degradation and synthesis are regulated so they do not occur simultaneously.
- Cholesterol can be synthesized in cells and is used to make bile acids, hormones, and vitamin D.
- Lipoproteins play important roles in the transport of cholesterol and other lipids.

Protein Metabolism

- Nitrogen balance, the ratio of nitrogen intake to excretion, is positive during growth and negative during starvation.
- Amino acids enter cells by active transport, but are not stored in any significant quantity.
- Essential amino acids must be obtained from the diet because human cells lack the enzymes to synthesize them.
- Cells use amino acids from cellular pools to synthesize proteins, nitrogenous bases, and a variety of other substances.
- Amino acids can undergo transamination and deamination; deaminated amino acids can be used for energy or to synthesize glucose or fatty acids.
- Amino groups from deamination are incorporated into urea in the urea cycle.

Nucleotide Metabolism

- Dietary nucleotides usually are degraded to purines, pyrimidines, sugars, and phosphate, and most are catabolized. New purines and pyrimidines are synthesized from amino acids and other substances and used to make nucleotides and nucleic acids.

Regulation of Cellular Level Metabolism

- Most metabolic reactions are controlled at the cellular level. Control mechanisms include product inhibition, actions of ligands on allosteric enzymes, feedback inhibition, and the actions of isoenzymes.

Clinical Applications

- Lactose intolerance is a metabolic disorder of carbohydrate metabolism. Gallstones are a disorder of lipid metabolism. Gout is a disorder of protein and nucleic acid metabolism.

Questions and Problems

The questions at the end of each chapter are numbered to correspond with the objectives listed at the beginning of the chapter. Italics indicate that a question requires critical thinking skills beyond simple factual recall.

Questions

1. *Write a concise paragraph that gives an overview of cellular metabolism.*
2. (a) How are components of carbohydrates transported to cells?
   (b) *What are the differences between glycogenesis and glycogenolysis?*
3. (a) What are the major components of carbohydrate metabolism?
   (b) *Where do glycolysis, the Krebs cycle, and biological oxidation occur?*
4. (a) What are the main steps in glycolysis?
   (b) *What are the products and significance of glycolysis?*
   (c) *How does the metabolism of glucose differ under aerobic and anaerobic conditions?*

5. (a) What are the main steps in the Krebs cycle?

   (b) What are the products and significance of the Krebs cycle?

6. *(a) Distinguish between electron transport and oxidative phosphorylation.*

   (b) What are the main steps in electron transport?

   (c) What are the products and significance of oxidative phosphorylation?

7. *What is the significance of the pentose phosphate pathway?*

8. (a) How are lipids transported in blood, and how are they stored in fat cells?

   (b) How do the processes by which fatty acids are degraded and synthesized differ?

   (c) How are these processes regulated?

   (d) How is cholesterol metabolism related to lipoproteins?

9. (a) Where and how are the different kinds of lipoproteins formed?

   (b) How are lipoproteins degraded?

10. (a) What is nitrogen balance?

   (b) What conditions affect nitrogen balance?

   (c) Define essential amino acid and amino acid pool.

   (d) Distinguish between deamination and transamination.

11. *(a) How are amino acids metabolized for energy, and what happens to the nitrogen they contain?*

   (b) What is gluconeogenesis, and what is its significance?

12. Briefly describe how nucleotides are metabolized.

13. (a) What are the major pathways in anabolism and catabolism?

   (b) How are various nutrients metabolized for energy, and what are the important products of these reactions?

   (c) What regulatory processes control the activity of cellular enzymes?

14. *How do the metabolic disorders described in this chapter disrupt homeostasis?*

Problems

1. How many ATPs are produced by anaerobic metabolism of five glucose molecules? By aerobic metabolism of five glucose molecules?

2. How many ATPs are produced by the complete oxidation of a 16-carbon saturated fatty acid? By an 18-carbon fatty acid with three double bonds?

3. Draw a flowchart showing the probable fate of the nutrients in your last meal.

4. Draw a flowchart to illustrate differences in metabolism under starvation conditions, during the intake of a low-carbohydrate diet, and during the intake of a high-protein diet.

25

REGULATION
OF
METABOLISM

OVERVIEW OF METABOLIC REGULATION

Have you wondered how your body adjusts to changes in the quantity of nutrients in the blood? How you experience hunger or satiety? Or how your body temperature is regulated? We can answer these questions as we study how metabolism is regulated.

As we have seen, many metabolic processes occur at the cellular level and depend on the digestion and delivery of nutrients to cells. In cells, nutrients are catabolized for energy or used to synthesize proteins and other substances that cells need for maintenance, growth, and repair.

Regulatory processes help to supply nutrients to cells according to their metabolic needs. They (1) coordinate nutrient processing with absorption; (2) control hunger and satiety; (3) stimulate protein synthesis as needed for tissue maintenance, growth, and repair; (4) maintain body temperature; and (5) maintain a metabolic rate that will support the above processes.

ABSORPTIVE VS. POSTABSORPTIVE METABOLISM

Almost as soon as food enters the small intestine, absorption begins. **Absorptive metabolism** refers to metabolic events during absorption (for about four hours after a meal). **Postabsorptive metabolism** refers to metabolic events after absorption is complete (for about two hours before a meal and during sleep).

Events in the Absorptive State

In the absorptive state, monosaccharides and amino acids travel via the portal vein directly to the liver. Products of fat digestion travel via lymph vessels and blood to the liver. Blood entering the liver, therefore, is nutrient laden.

During absorption, all cells receive sufficient nutrients, and excess nutrients are processed in a variety of ways (Figure 25.1). Lipoprotein lipase digests triacylglycerols in chylomicrons, and glycerol and fatty acids enter adipose cells, where they reform triacylglycerols and are stored. Such storage conserves excess nutrients during absorptive metabolism when most cells obtain energy from glucose.

Blood glucose during absorption usually exceeds that needed to maintain a normal blood glucose concentration of 70 to 100 mg/dl. Excess glucose is normally converted to other metabolites—polymerized into glycogen or used to make triacylglycerols in liver and adipose cells.

Most amino acids are delivered to cells and used to make proteins, and only limited quantities remain in cellular amino acid pools. Excess amino acids are deaminated mainly in the liver and are used for energy or to synthesize fatty acids. Their amino groups are excreted in urea.

In summary, absorptive metabolism supplies cells with nutrients for energy and synthesis, maintains a normal blood glucose concentration, and conserves energy by storing glycogen and triacylglycerols.

Events in the Postabsorptive State

The postabsorptive state is characterized by the controlled release of nutrients from glycogen and triacylglycerols and, if necessary, the eventual use of proteins for energy (Figure 25.2).

The foremost requirement during postabsorptive metabolism is to maintain the blood glucose concentration in the normal range. When no new glucose is being absorbed, liver glycogen formed only a few minutes to a few hours earlier is degraded and glucose is released into the blood as needed to maintain a normal blood glucose concentration.

Breakdown of triacylglycerols in fat deposits provides a second source of nutrients. Fatty acids go to cells in many tissues and are metabolized for energy; glycerol goes mainly to the liver where it is used to synthesize glucose. Metabolism of fatty acids by cells that can use them, called **glucose sparing,** conserves glucose for neurons and erythrocytes. Also, using glycerol to synthesize glucose augments the glucose supply.

If no glucose is absorbed for several days (starvation conditions), fatty acid metabolism increases and numerous ketone bodies are produced. Under such conditions, neurons can metabolize ketone bodies. Some cellular proteins, especially from muscle cells, are degraded to amino acids. After deamination, their carbon skeletons can be used for energy or to make glucose.

In summary, postabsorptive metabolism supplies cells with nutrients for energy and synthesis, maintains a normal blood glucose concentration, and uses energy from molecules stored in absorptive metabolism.

Regulation of Absorptive and Postabsorptive Metabolism

Complex mechanisms simultaneously regulate absorptive and postabsorptive metabolism and control shifts from one to the other. Blood glucose is regulated by three negative

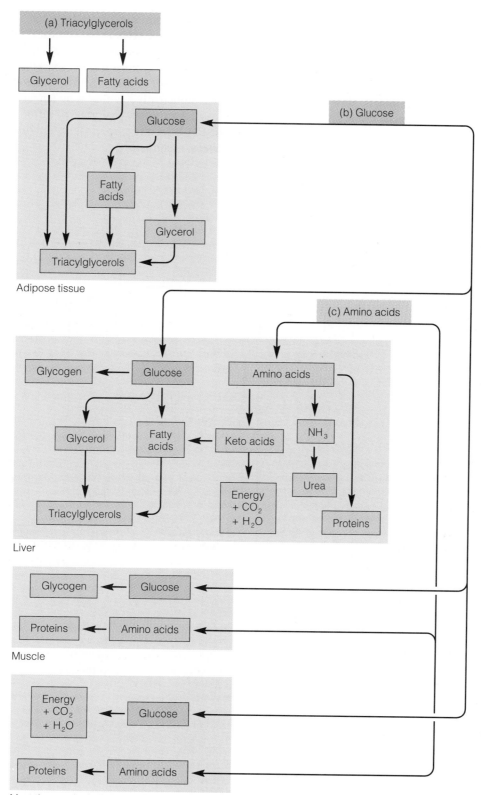

FIGURE 25.1

Metabolic events in the absorptive state. (*a*) Triacylglycerols are digested to glycerol and fatty acids. They are used in adipose tissues to reform triacylglycerols and are stored in fat deposits. (*b*) In addition to being used to maintain blood glucose concentration, glucose is converted to glycogen mainly in the liver, but also in muscle cells. When glucose is present in great abundance, it can be used to synthesize triacylglycerols in the liver and adipose tissues. (*c*) Most amino acids are used by cells to synthesize proteins. Excess amino acids cannot be stored directly, but they can be deaminated, and their carbon skeletons converted to fatty acids, and eventually incorporated into triacylglycerols.

feedback mechanisms—insulin-glucagon, epinephrine-hypothalamic, and growth hormone-hypothalamic mechanisms. Blood glucose normally rises to 120 to 140 mg/dl as glucose is being absorbed for one to two hours after a meal. As these mechanisms exert their effects, it returns to normal within two hours after glucose absorption ceases.

In the **insulin-glucagon mechanism** (Figure 25.3a), insulin and glucagon are released from the pancreas according to negative feedback from the blood glucose concentration. High blood glucose (during absorptive

metabolism) initiates insulin secretion. Insulin causes glucose and amino acids to enter most cells, inhibits breakdown and fosters synthesis of triacylglycerols, and increases glycogen synthesis, especially in the liver. As a result of these actions of insulin, the blood glucose becomes sufficiently low that it no longer stimulates insulin release. A two-hour postprandial (after a meal) blood glucose test is used to assess insulin activity and to screen for diabetes mellitus.

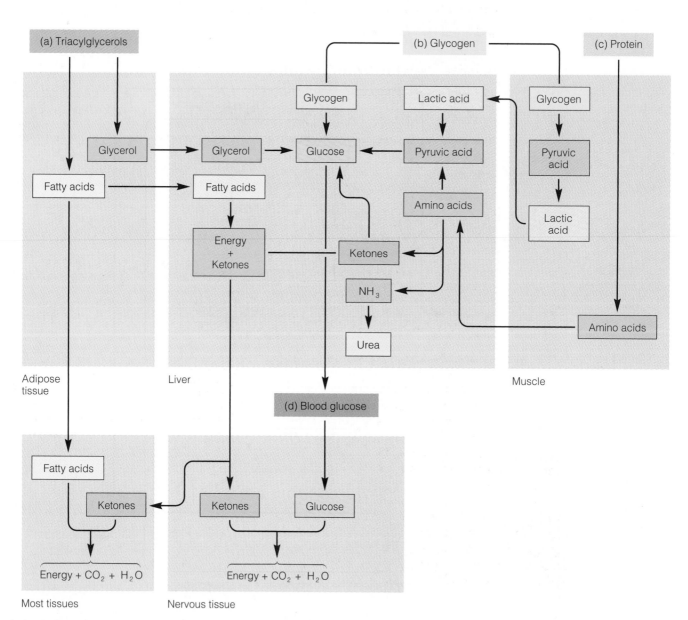

FIGURE 25.2
Metabolic events in the postabsorptive state. (*a*) Triacylglycerols are broken down to fatty acids, which are metabolized for energy, and glycerol, which is converted to glucose in the liver and used for energy. (*b*) Glycogen from the liver and muscles is metabolized to lactic acid or pyruvic acid. These molecules are converted to glucose in the liver. (*c*) When the need for glucose is great, proteins are broken down to

amino acids and transported to the liver, where they are deaminated and their keto-acid derivatives used to synthesize glucose. (*d*) The focal event in the postabsorptive state is to maintain a normal blood glucose concentration and a constant supply of glucose for nervous tissue, even though nervous tissue eventually metabolizes ketones under starvation conditions.

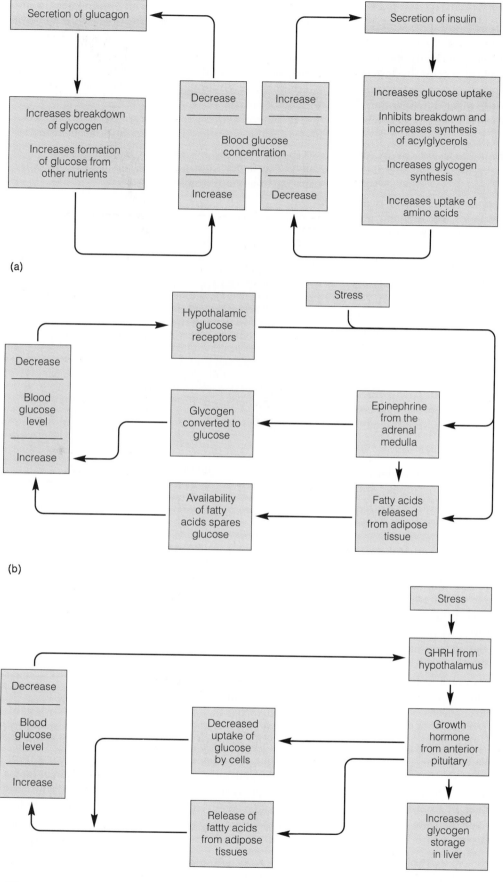

(a)

(b)

(c)

FIGURE 25.3

(*a*) The insulin-glucagon mechanism. When the blood glucose concentration increases, insulin is produced, and its action reduces the blood glucose concentration. When the blood glucose concentration is reduced, glucagon (instead of insulin) is produced, and the action of glucagon increases the blood glucose concentration. This system of negative feedback shown by the "figure-8" in the diagram represents one of the homeostatic mechanisms that helps to regulate the blood glucose concentration. (*b*) The epinephrine-hypothalamic mechanism. A decrease in blood glucose stimulates glucose receptors in the hypothalamus to send nerve signals to adipose tissue and to the adrenal gland. Nerve signals resulting from stress have the same effect. Once stimulated, the adrenal gland releases epinephrine, which stimulates the breakdown of glycogen to glucose in muscles. Fatty acids released from adipose tissue provide an alternative energy source and supplement glucose, thereby increasing the blood glucose concentration. (*c*) The growth hormone-hypothalamic mechanism. When the blood glucose decreases or the body is under stress, the hypothalamus releases GHRH, which, in turn, causes the anterior pituitary to release growth hormone. Growth hormone decreases glucose uptake and increases fatty acid release, thereby increasing the blood glucose concentration. The increase in glycogen storage preserves a supply of glycogen, which can be released later to maintain a stable blood glucose concentration over a longer period of time.

As the blood glucose drops during postabsorptive metabolism, the lack of glucose initiates glucagon secretion. Glucagon increases the breakdown of glycogen and triacylglycerols, and increases glucose synthesis by gluconeogenesis. These effects elevate blood glucose by making glucose available and by supplying glucose-sparing fatty acids. More glucagon and cortisol, which amplifies the effects of glucagon, are secreted during strenuous exercise, starvation, and other stresses.

Glucagon secretion seems to be regulated by the amount of glucose entering glucagon-secreting cells of the pancreas. Insulin facilitates entry of glucose into those cells and suppresses glucagon secretion. When insulin is not being released, glucose fails to enter the cells and glucagon is secreted. In individuals unable to produce insulin, glucagon is secreted even when blood glucose is high. Without insulin, the glucagon-secreting cells fail to get the message that glucose is present and glucagon is not needed.

Blood amino acids also affect the insulin-glucagon mechanism. High concentrations of blood amino acids stimulate the release of insulin *and* glucagon. Having both hormones in the blood tends to stabilize blood glucose, so a high-protein, low-carbohydrate diet can dampen oscillations in the blood glucose concentration. Even without a high amino acid level, both hormones are secreted continuously at a basal level and are present in small amounts in the blood. They exert regulatory effects when they are secreted above the basal rate.

Under normal circumstances, the insulin-glucagon mechanism lowers blood glucose in the absorptive state and elevates it in the postabsorptive state. During any kind of exertion, neural signals to the adrenal medulla initiate epinephrine secretion.

The hormone epinephrine and sympathetic signals initiated by the hypothalamus make up the epinephrine-hypothalamic mechanism (Figure 25.3b). This mechanism can be activated by stress (Chapter 16) or by a decrease in the blood glucose level, which stimulates hypothalamic glucose receptors to relay signals to the adrenal medulla. Stimulation of the adrenal medulla initiates epinephrine secretion. With respect to glucose metabolism, epinephrine behaves much like glucagon. It causes breakdown of glycogen and triacylglycerols and increases gluconeogenesis. This mechanism operates in the postabsorptive state and during stress. It is turned off in the absorptive state and in nonstressful situations by negative feedback that shows an adequate blood glucose level.

Once thought to act only during growth, it is now known that growth hormone affects metabolism throughout life. Stress, starvation, or excitement create postabsorptive-like conditions and activate the **growth hormone-hypothalamic mechanism** (Figure 25.3c). The hypothalamus secretes GHRH, which stimulates the anterior pituitary gland to secrete growth hormone. Growth hormone increases liver glycogen, decreases glucose uptake by cells, increases amino acid uptake and protein synthesis, and releases fatty acids from adipose tissue. These effects increase glycogen *and* blood glucose directly and by glucose sparing.

Prolonged hypersecretion of growth hormone accelerates insulin secretion as homeostatic mechanisms attempt to counteract the excess of growth hormone. This can lead to diabetes mellitus by exhausting the pancreas's capacity to produce insulin.

Though most daily variation in blood glucose results from shifts between absorptive and postabsorptive states, a circadian rhythm in blood glucose regulation exists. When volunteers received three meals of equal energy value at normal meal times, their blood glucose levels changed according to their absorptive and postabsorptive states; yet, the blood glucose took longer to return to the postabsorptive level with each meal. When other healthy volunteers received high carbohydrate meals at two-hour intervals from 9 A.M. to 11 P.M., their blood glucose rose after each meal, but reached a peak at about 10 P.M. These studies suggest that the efficiency of blood glucose regulation is highest early in the day and lowest in the evening.

So far, rhythms have not been observed for protein or lipid metabolism. Obtaining such evidence is difficult because these nutrients are not regulated like glucose; however, the appearance of gastrointestinal disorders with changes in routines seems to indicate that some, as yet unidentified, rhythm has been disturbed.

HUNGER AND SATIETY

Two centers in the hypothalamus, the **feeding center** and the **satiety center,** appear to regulate appetite for food under normal circumstances. Stimulating the feeding center leads to eating, whereas stimulating the satiety center suppresses eating, probably by inhibiting the feeding center. Animals with experimentally damaged satiety centers gain weight for a period of time, but their food intake eventually levels off. The feeding and satiety centers then appear to maintain the new higher body weight.

After many hours without food, humans can experience painful gnawing stomach contractions called hunger pangs, but other signs of hunger usually appear earlier. The blood glucose level seems to be the most direct factor in the regulation of hunger and satiety, but insulin concentration in cerebrospinal fluid and CCK-PZ concentration in the brain also may be involved.

In experiments with animals, feeding increases when the blood glucose concentration drops and stops when the blood glucose rises. As glucose rises, electrical activity increases in the satiety center and decreases in the feeding center. Cells in the satiety center concentrate glucose, and the amount of glucose the cells contain may determine their activity in inhibiting the feeding center.

Though CCK-PZ is involved in the regulation of digestive functions in the gut, this hormone also is found in brain tissue. Among mice, those of normal weight have a higher level of brain CCK-PZ than obese ones. This suggests that CCK-PZ may regulate feeding: higher concentrations inhibit it and lower concentrations allow it.

Receptor sites in the hypothalamus that bind amphetamines are known to exist, and they help to explain the appetite-suppressing effect of amphetamines. Why the sites might be present, what natural substances might bind to them, or the actions of such substances are not known. Depriving animals of food for a few days appears to decrease the number of amphetamine-binding sites, suggesting that the sites do have to do with appetite suppression. The less food an animal has had, the fewer the sites and presumably the greater the appetite. It also may explain the very short-term effect of amphetamines on appetite suppression. A practical aspect of finding amphetamine-binding sites is that it may be possible to develop drugs that act specifically on these sites and suppress appetite without producing the undesirable side effects of amphetamines.

Certain other factors have some effect on food intake. Blood amino acids have a similar but weaker effect than glucose. Over time, an increase in the quantity of adipose tissue tends to decrease feeding. Cold environments increase food intake and warm ones decrease it. Digestive tract distention suppresses hunger as does food intake itself, even when the food is removed from the digestive tract before it can be absorbed. This can occur experimentally in animals by placing a tube from the esophagus to the outside of the body or in humans by self-induced vomiting, as occurs after binge eating in bulimia (discussed later in this chapter).

REGULATION OF PROTEIN METABOLISM

Regulation of absorptive and postabsorptive metabolism helps to regulate protein metabolism. Absorptive processes foster amino acid uptake by cells, whereas postabsorptive processes allow protein breakdown and use of amino acids to make glucose, if needed. Steroid and thyroid hormones control synthesis of specific proteins. They can increase the supply of enzymes, plasma proteins, and other proteins as needed, and they maintain nitrogen balance while fostering growth and tissue maintenance and repair.

Negative nitrogen balance (Chapter 24) can be severely debilitating. Cells cannot replace enzymes to make other proteins they need. A lack of blood clotting proteins leads to excessive bleeding after injuries. Too little globin reduces the numbers of new erythrocytes and leads to anemia. Lymphocytes cannot make enough antibodies, so disease resistance is lowered. Tissue injuries are slow to heal. Growth is impaired in children. Finally, phagocytic cells ingest and degrade plasma proteins. This process makes amino acids available to repair essential cellular structures, but it further depletes plasma proteins. Blood oncotic pressure decreases and tissue edema results. Such edema accounts for the potbellies of starving children.

Normal to positive nitrogen balance provides adequate supplies of amino acids. As newly absorbed amino acids reach the liver, many are used to make plasma proteins. The liver also synthesizes nonessential amino acids by transamination and degrades them by deamination (Chapter 24). Cells use amino acids to synthesize enzymes and maintain cellular structures. Certain cells make special proteins such as antibodies and some hormones.

Unlike carbohydrates deposited in glycogen and lipids deposited in fatty tissue, the body has no protein deposits. When dietary protein is abundant, most cells, and liver cells in particular, synthesize more protein than is essential for normal function. Then, if dietary protein becomes inadequate, the excess protein can be degraded and the amino acids reused or released into the blood for uptake by cells that need them. This mechanism temporarily avoids the effects of dietary protein deficiency. Protein homeostasis is summarized in Figure 25.4.

The various processes of protein metabolism just described occur continuously to maintain body tissues. In two other processes, wound healing and growth, protein synthesis accelerates and more dietary protein is needed to sustain them.

As wounds heal, various types of cells are replaced in varying degrees and by different processes. Epithelial tissues are continually sloughed and replaced, even in the absence of injury. Such tissues are completely replaced in wound healing. Sheetlike epithelial cells divide rapidly and spread laterally to form new layers like the original layers. The liver, pancreas, spleen, and most other glands repair themselves if the basic framework remains intact. Kidney tubules and connective tissues such as bone, cartilage, and the dermis of the skin also repair themselves by some of their cells dividing to replace injured or destroyed portions. Connective tissue forms scars as it replaces other tissues that cannot replace themselves.

The rate of tissue repair in healing depends primarily on an adequate blood supply to the injured area. In general, repair is more rapid in younger than in older people

FIGURE 25.4

Protein homeostasis. Amino acids enter the blood from the intestine. The blood carries amino acids to the liver and to all other tissues. The liver carries out a variety of processes in protein metabolism. Cells in most other tissues use amino acids mainly to synthesize proteins. Small excesses of proteins can be degraded and their amino acids used to make needed proteins when the diet is temporarily deficient in protein.

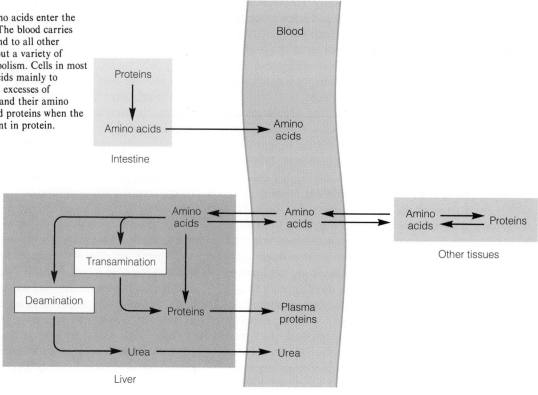

because blood circulation is more adequate in the young. It is also more rapid in healthy people than in those with chronic illness. Repair is slow in immobilized body parts—broken limbs, for example—at any age because immobilization impairs circulation. Another factor in healing is the amino acid supply, which if deficient impairs protein synthesis.

Like wound healing, growth requires adequate supplies of amino acids to create all of the proteins in new cells, including the enzymes to make nonprotein cellular components. Growth occurs at different rates during fetal development, infancy, childhood, and adolescence.

Growth hormone stimulates protein synthesis by facilitating the transport of amino acids into cells, and by accelerating certain steps in the transcription and translation processes of protein synthesis. Mobilized fatty acids provide energy for protein synthesis and raw materials, and energy for lipid synthesis. Growth occurs by synthesis of protein and lipid components of new cells.

Growth hormone itself is regulated by GHRH and GHIH from the hypothalamus (Figure 25.5). What releases GHRH is not entirely clear, but once released it causes the anterior pituitary gland to release growth hormone.

Growth also is stimulated by insulin, androgens, and thyroid hormones. Insulin increases protein synthesis by accelerating amino acid transport and transcription-translation and by making glucose, and therefore energy, available to cells. Androgens act with growth hormone to

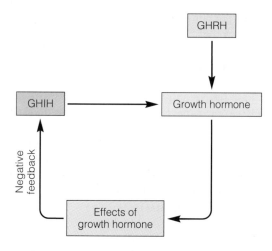

FIGURE 25.5
Growth hormone secretion is regulated by negative feedback.

stimulate protein synthesis in muscles where they produce male musculature. Synthetic androgens (anabolic steroids) are often abused for this purpose (Chapter 9). Thyroid hormones increase the overall metabolic rate, thereby increasing both protein synthesis and protein breakdown. They must be present for growth hormone to act and for normal brain development and normal brain function at all ages.

In contrast to growth stimulation hormones, glucocorticoids in pharmacological concentrations tend to inhibit growth. They mobilize amino acids from tissues

TABLE 25.1   Calories per Hour for Various Types of Activity*

Activity	kcal/hr	Activity	kcal/hr
Sitting at rest	100	Walking 4 mph (some hills)	400
Standing relaxed	105	Moderate recreation (badminton, ballet, calisthenics, golf	
Dressing and undressing	115	without cart, racquetball, table tennis, tennis (doubles),	
Mild activities while sitting (playing cards, writing)	120	volleyball)	400
Moderate activities while sitting (playing piano, typing)	140	Strenuous activities (sawing wood, shoveling snow, digging holes)	450
Light activities		Walking 5 mph	500
Light house and yard work	150	Strenuous recreation (bicycling 10 mph with hills, cross-country	
Painting, power lawn mowing, clerking in store	160	or downhill skiing, jogging 5 mph, swimming 25 yds/min	
Walking 2 mph	200	sidestroke, tennis (singles), water-skiing)	600
Sexual intercourse	280	Rowing 20 strokes/min	800
Light recreation (archery, bicycling 5.5 mph, bowling, canoeing		Walking upstairs	1100
3 mph, moderate dancing, swimming 20 yds/min crawl)	300	Maximal activity (untrained person)	1400
Moderate activities (digging and weeding, heavy housework)	300		

*For a person weighing 70 kg.

except the liver and increase the uptake of amino acids in liver cells. The liver uses some amino acids for protein synthesis, but it deaminates many for use in gluconeogenesis. Prostaglandins also slow growth, probably by interfering with the effect of growth hormone on fat deposits.

REGULATION OF OVERALL METABOLISM

Overall metabolism is regulated so that sufficient energy is available to keep all physiological processes operating, to maintain body temperature, and to allow physical activity.

Metabolic Rate

The overall **metabolic rate,** how fast the body uses energy, is proportional to the rate at which it releases heat. For example, when energy is transferred to ATP during metabolism, about 40 percent of the energy in molecules is captured in high-energy bonds and the remainder is released as heat. The metabolic rate can be determined by measuring heat released from the body in a body-sized chamber, but measuring oxygen use is less cumbersome. Using data from studies that relate heat release to oxygen use and an estimate of body surface area, heat release can be calculated and the metabolic rate determined.

Energy changes in chemical reactions are measured in units called calories. One **calorie** (c) is the amount of energy required to raise the temperature of one gram of water one degree C (from 14° to 15° C). A **kilocalorie** (kcal), which equals 1000 calories, is more convenient to describe human energy metabolism.

Humans use energy for metabolic processes and voluntary activities. Energy used for metabolic processes such as breathing, heartbeat, and maintaining body temperature is called **basal metabolism,** and the rate at which such energy is expended is the **basal metabolic rate** (BMR). The total metabolic rate is the sum of BMR and energy used for voluntary physical activities.

A major factor in determining the basal metabolic rate is lean body mass—total body mass less body fat. BMR can be estimated in relation to body surface area (from which heat is lost). Measured in kilocalories per square meter of body surface per hour, BMR is higher in males than females and higher in the young than in the elderly. People may tend to gain weight as they age because they do not reduce their food intake as their metabolic rate decreases. Some typical values in $kcal/m^2/hour$ are 40 for males in their twenties, 35 for males in their sixties, 36 for females in their twenties, and 32 for females in their sixties. We can determine the daily calories needed for basal metabolism by multiplying the above quantities by body surface area, which varies from 1.4 to 2.0 square m, and by 24 hours per day. Typical values range from 1200 (32 kcal $\times$ 1.4 square m $\times$ 24 hours) for a small, older woman to 1900 (40 kcal $\times$ 2.0 square m $\times$ 24 hours) for a large, young man. Multiplying one's weight in pounds by 10 gives a simple estimate of daily kilocalories for basal metabolism.

Factors That Affect Metabolic Rates

The main factor that increases the metabolic rate above the basal rate is physical activity. Maximal muscular activity for even a short time elevates heat production as much as 100 times that under resting conditions. Swinging a 2-lb weight in each hand while walking more than doubles the calories used. Table 25.1 shows the approximate number of kilocalories that a 70-kg person expends per hour in different activities. In general, physical activity increases the total daily caloric requirements by 40 to 50 percent for a sedentary person (typist, seamstress), 60 to 70 percent for a moderately active person (salesperson, nurse) to 80 to 100 percent for a very active person (mail carrier, bricklayer).

Certain hormones and sympathetic signals can alter the BMR. Excesses of glucocorticoid, thyroid, or growth hormones increase the basal metabolic rate by directly increasing the rate of cellular metabolism. Sympathetic signals also increase the basal metabolic rate, mainly by increasing glycogenolysis in liver and muscle. In newborn infants, such signals liberate heat from **brown fat** around internal organs. Because oxidation and phosphorylation are uncoupled in brown fat metabolism, little energy is captured in ATP and much is released as heat.

Sleep can decrease the metabolic rate 10 to 15 percent below the BMR measured while awake but resting. This change is thought to be due to decreased sympathetic activity and decreased skeletal muscle tone.

Other factors that alter metabolic rates include pregnancy, nutritional status, climate, and body temperature. During the latter half of pregnancy, progesterone and adrenocortical and thyroid hormones increase the metabolic rate as much as 15 percent. This provides energy for fetal growth and for the mother to carry the fetus. Pregnant women sometimes complain of being too hot because body temperature increases with the metabolic rate. Food deprivation can cause cells to metabolize available nutrients more slowly, dropping the metabolic rate 20 to 30 percent below normal. Conversely, too much food appears not to accelerate the basal metabolic rate—it merely increases fat storage. A tropical climate lowers the metabolic rate by 10 to 20 percent, whereas a cold climate raises it by 10 to 20 percent. The effects of climate appear to be mediated by the thyroid gland, which secretes less hormone after adaptation to a hot climate and more after adaptation to a cold climate. Body temperature changes alter the metabolic rate mainly by altering enzyme activity.

> While smokers do not necessarily eat less or exercise more than nonsmokers, they do often gain weight when they quit smoking. Studies have shown that nicotine increases the metabolic rate so that calories are used less efficiently for a given level of activity. Walking an extra mile a day probably compensates for the change in metabolic rate. Rather than using this as an excuse to smoke, one should consider that a weight gain of over 50 lbs would be needed to equal the health risks of smoking.

Respiratory Quotient

The ratio of carbon dioxide released to oxygen consumed during metabolism is the **respiratory quotient** (R). R varies, depending on what is being metabolized. For carbohydrates, R is 1.00 because for every O_2 molecule used, a molecule of CO_2 is released. For triacylglycerols, which contain more hydrogen, R is 0.71 because O_2 is used faster than CO_2 is given off. For amino acids, R is 0.83. During the metabolism of a mixed nutrient meal, at first R is very close to 1.00 as carbohydrate molecules reach cells. It approaches 0.71 when cells metabolize fats, as in starvation or untreated diabetes. R can be used to estimate proportions of nutrients being metabolized at any given time.

BODY TEMPERATURE AND ITS REGULATION

Normal Temperature and Its Variations

The body's temperature-regulating mechanism is remarkably effective in maintaining a nearly constant **core body temperature** (temperature deep inside the body) of a healthy person at rest. This temperature varies only about 1° C above or below the normal body temperature over a wide range of environmental temperatures. Normal body temperature measured orally is said to be 37° C (98.6° F), but rectal temperature is about 0.5° C higher—close to the core body temperature. Many people have a normal body temperature that is higher or lower by 0.5° to 1° C. During strenuous exercise, the core temperature rises as much as 3° C. When this occurs, the regulating system quickly releases heat by initiating sweating and dilating skin blood vessels. The highest temperature consistent with life is 41° to 42° C (106° to 108° F). At these temperatures, convulsions occur, and at 45° C (113° F) proteins are denatured, enzymes fail to function, and death ensues.

Heat Production and Heat Loss

The body produces heat in several ways. When it uses energy from ATP, about 40 percent goes to contract muscles or to do other cell work, but 60 percent is released as heat. Sympathetic stimulation and epinephrine release during stress increase the metabolic rate, thereby increasing heat release. Increased thyroid hormone secretion also increases the metabolic rate and releases more heat.

Depending on the temperature of the environment, heat can be lost in a variety of ways. Large amounts of heat are lost from the body by **radiation.** Heat is radiated as infrared rays from any object with a temperature above the environmental temperature, and the human body is no exception. Objects in the environment also radiate heat, and how much heat the body gains or loses depends on the relative temperatures of the body and its environment. The body loses heat to a cooler environment and gains heat from a warmer one, the degree of loss or gain being proportional to the temperature difference.

Significant heat loss also occurs by **evaporation,** especially in a dry environment. A summer breeze is refreshing because it speeds evaporation of water from skin surfaces. Water loss from the lungs and skin accounts for 300 to 400 calories of heat loss per day, and much more heat is lost during profuse sweating. Finally, heat loss can occur by the transfer of kinetic energy to cooler objects by **conduction** from one object or medium to another. For example, when you sit on a cold chair, heat is conducted to the chair and the seat becomes warm. When you swim, your body loses large amounts of heat to the water.

The skin and subcutaneous tissues insulate the body against heat loss. Subcutaneous adipose tissue is an especially good insulator because it conducts heat only one-third as fast as other tissues; thus, the core temperature can be maintained even when the skin temperature drops.

Regulation of Body Temperature

Body temperature is regulated by heat sensitive neurons in the hypothalamus, heat and cold receptors in the skin, and the signals these cells send to other organs and tissues. To see how body temperature is regulated, let us consider first what happens when body temperature begins to rise, and second what happens when it begins to fall (Figure 25.6).

Compensation for Rising Temperature

As the core body temperature rises, heat sensitive hypothalamic neurons fire more rapidly and initiate signals that dilate skin blood vessels, increase sweating, and inhibit heat production. The direct cause of skin blood vessel dilation is inhibition of sympathetic centers in the hypothalamus, which, in turn, inhibits vasoconstriction and allows skin blood vessels to dilate. Up to 30 percent of the cardiac output flows through the skin. Though heat is continuously transferred from blood vessels to the skin surface, blood vessel dilation greatly increases heat transfer and makes the skin red and hot.

Sweating increases as sympathetic neurons send more signals to sweat-secreting cells. These sympathetic neurons release acetylcholine (not the usual norepinephrine).

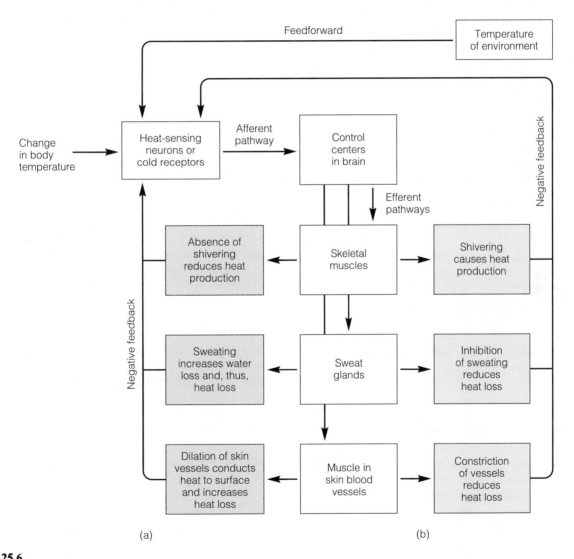

(a) (b)

FIGURE 25.6

Regulation of body temperature. (*a*) Mechanism to decrease body temperature when it rises above normal. (*b*) Mechanism to anticipate decrease and to increase body temperature should it drop below normal.

Stimulating sweat glands allows up to 1.5 l per hour of sweat to reach the skin surface. Individuals having a deficiency or absence of sweat glands lose heat less rapidly and are prone to heat stroke (described below). Unless sweating is profuse, sweat contains fixed concentrations of sodium chloride and other electrolytes. During profuse sweating, as in a person not acclimated to heat, sweat contains enough sodium chloride to disturb the body's electrolyte balance. After a few weeks in a hot climate, aldosterone secretion increases and stimulates cells in sweat ducts to reabsorb more sodium chloride.

Heat production is inhibited by signals from the heat sensitive neurons. Such signals inhibit both shivering and sympathetic stimulation that would accelerate cellular metabolism.

Compensation for Falling Temperature

As the body temperature falls, heat sensitive neurons sense a core temperature decrease; they fire too slowly to relay the signals associated with rising temperature. Signals from cold receptors in the spinal cord and especially in the skin (Chapter 15) both conserve heat and augment heat production. They reduce heat loss by decreasing sympathetic signals that cause sweating and by increasing sympathetic signals that constrict skin blood vessels and cause piloerection. **Piloerection** (pi″lo-e-rek′shun), standing hairs on end, traps warm air over the skin of furry animals, but merely produces "gooseflesh" in humans.

Signals from cold receptors increase heat production by causing shivering and by increasing the metabolic rate. They stimulate the primary motor center for shivering, which relays signals that initiate the muscle contractions we call shivering. Shivering probably involves feedback oscillation in the muscle spindle stretch reflex (Chapter 9) and can increase heat production to four or five times the normal level. Signals from cold receptors also increase sympathetic stimulation and epinephrine release, both of which increase cellular metabolism. This process, called **chemical thermogenesis** (ther″mo-jen′es-is), immediately increases the metabolic rate and heat production of many body cells.

Cooling of heat sensitive neurons initiates secretion of a series of hormones—thyrotropin-releasing hormone, thyroid-stimulating hormone, and, ultimately, thyroid hormones, which increase the metabolic rate (Chapter 16). This mechanism, though it has no immediate effect, is important in adjustment to cold climates.

In humans, behavioral responses are far more important than physiological mechanisms in maintaining body temperature in extreme environments. Humans sense being uncomfortably warm or cold, and adjust clothing and heating or air conditioning before the physiological mechanisms begin functioning. Without clothing and heating, humans could not survive outside the tropical and semitropical regions of the world.

CLINICAL APPLICATIONS

Temperature-related Disorders

Various abnormal conditions disrupt temperature regulation causing **hyperthermia** (hi″per-ther′me-ah), increased body temperature as in fever and heat stroke, and **hypothermia** (hi″po-ther′me-ah), decreased body temperature as in frostbite and generalized hypothermia.

Fever

The body temperature rise called **fever** is due to the resetting of the operating set point for temperature to a higher level, and not to a loss of regulatory ability. This resetting is done by **pyrogens** (pi′ro-jens), proteins released from leukocytes in inflammation and in bacterial infections. How prostaglandins contribute to fever is not known, but their concentration in cerebrospinal fluid rises during fever. Aspirin appears to reduce fever by inhibiting prostaglandin synthesis.

When the body's thermostat has a high set point, the body behaves as if it were hypothermic until the temperature reaches the new set point. Peripheral blood vessels constrict and the person shivers and covers up with blankets. When the thermostat returns to the normal set point—the fever "breaks"—the body behaves as if it were hyperthermic until the temperature returns to the normal set point. Peripheral blood vessels dilate and the person sweats, removes blankets, and drinks cool liquids.

Fever may combat infections, but it can have harmful effects, too. By increasing the metabolic rate, fever speeds up inflammatory and immune responses. It damages tissues by interfering with the actions of many hormones and can cause brain hemorrhage and convulsions, especially in small children.

Heat Stroke

People exposed to extreme heat and high humidity often suffer **environmental heat stroke.** High humidity reduces evaporation and amplifies the effects of heat. During strenuous exercise when the body produces more heat than it can dissipate to the environment, people can suffer **exertional heat stroke.** This condition occurs in athletes and military personnel undergoing physical training even at air temperatures no greater than 27° C (80° F) when humidity is no greater than 60 percent. If the core temperature exceeds 41° C, heat stroke is likely. Heat sensitive neurons fail to send regulatory signals. Symptoms begin with dizziness and abdominal distress, and rapidly proceed to loss of consciousness from brain damage. If brain cells become so hot that vital centers cease operating, death occurs.

To treat heat stroke, the body temperature must be lowered as rapidly as possible. This is done by wrapping the patient in cool sheets or continuously sponging with cool water, using cooling (hypothermia) blankets, operating an electric fan in the room to increase evaporation, and giving intravenous fluids to replace fluids lost.

Malignant hyperthermia, a very rapid rise to a life-threatening temperature, can be caused by certain snake venoms and anesthetics such as halothane in genetically susceptible individuals. The hyperthermia is due not to resetting the thermostat but to the uncoupling of oxidation and phosphorylation reactions. Mitochondria use oxygen rapidly, but release all energy as heat instead of capturing some in ATP.

Frostbite

When exposed to windchill below freezing, **frostbite** of body parts, especially fingers, toes, and earlobes, can occur. If frostbitten tissue is thawed with moderately warm water (38° C to 42° C for 15 to 20 minutes) permanent damage is unlikely. Pain medication and elevation of extremities to minimize edema also are recommended. Too rapid thawing increases tissue damage and can lead to gangrene, with dying cells invaded by microorganisms.

Generalized Hypothermia

A core body temperature below 35° C from exposure to a cool environment is called **generalized hypothermia.** A moderately cool temperature on a damp, windy day can lead to hypothermia as many rain-soaked hikers have learned. Elderly persons with poor blood circulation living in poorly heated homes are especially susceptible to hypothermia, but healthy adults exposed to sudden wet cold can suffer from it, too. Lean, muscular people are more susceptible than fatter, better insulated people. To avoid cold injury, one must keep warm, keep dry, and keep moving.

Hypothermia begins with numbness, paleness, slurred speech, and violent shivering. Dizziness, drowsiness, and incoherence follow. An especially dangerous sign is the victim refusing treatment. Failure of companions to recognize hypothermia has led to unnecessary deaths of otherwise healthy individuals.

The physiological effects of hypothermia are due mainly to a depressed metabolic rate with accompanying decreased respiratory and heart rates. Tissues become ischemic. Capillaries become excessively permeable and water seeps from them, lowering blood volume and blood pressure. Eventually, the heart and lungs function too slowly to support life.

Hypothermia is treated by raising the core body temperature as rapidly as possible without causing tissue damage. From studies of volunteers subjected to cold water immersion, researchers noted that core temperature continues to drop after the volunteers come out of the water. The volunteers recovered best when they walked around drinking hot beverages until their temperatures started to rise and then had a hot shower.

TABLE 25.2	Differences between Insulin-dependent and Non-insulin-dependent Diabetes	
Characteristic	**Insulin-dependent (Juvenile-onset)**	**Non-insulin-dependent (Maturity-onset)**
Age at onset	Usually under 20	Usually over 40
Percent of all diabetics	About 10%	About 90%
Time of year of onset	Fall and winter	No seasonal trend
Appearance of symptoms	Sudden and usually acute	Gradual and slow to develop
Metabolic acidosis	Frequent	Rare
Obesity at onset	Uncommon	Common
B-cells	Decreased	Variable
Insulin	Decreased	Variable
Inflammation of pancreas	Present at onset	Absent
Family history of diabetes	Less common	More common
Association with HLA antigens	Yes	No

Hypothermia can be induced by giving a sedative and cooling the body with cooling blankets or ice to maintain the temperature below 32° C. In heart surgery, the heart is stopped and metabolism slowed, but circulation is maintained mechanically. After surgery is completed, the heart is restarted with electrical shock and no serious physiological effects of hypothermia are seen.

Occasionally, a child survives under ice for 30 minutes or more. Physiological processes are so slow that heartbeats and breathing appear only as temperature rises. One researcher warns that patients are not dead until they are "warm dead."

Diabetes Mellitus

Sugar diabetes, or **diabetes mellitus,** is characterized by excessively high blood glucose and occurs in two forms. In **insulin-dependent diabetes mellitus** (IDDM), the B-cells of the pancreas produce little or no insulin. In **non-insulin-dependent diabetes mellitus** (NIDDM), cells have reduced insulin sensitivity, probably because of altered receptors. Other characteristics of these diseases are summarized in Table 25.2. Over 11 million Americans have diabetes, but only 6.5 million know they have the disease and less than 1 million have IDDM and take insulin daily.

The major effects of diabetes are decreased entrance of glucose into most cells, increased fat metabolism, and depletion of tissue proteins. (Insulin is not required for glucose to enter liver or brain cells, except for certain hypothalamic cells.)

Diabetic symptoms are closely related to the metabolic disorders (Figure 25.7). Cells are starved for glucose and this causes some patients to eat excessively. Because urine contains glucose, more water stays in the urine and patients excrete a large volume. Because fluid loss is excessive, tissues become dehydrated and patients are thirsty;

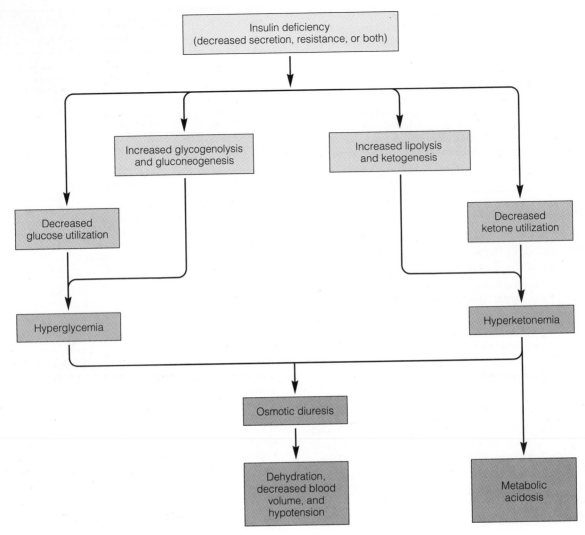

FIGURE 25.7
Physical effects of diabetes.

thus, diabetes is sometimes characterized by the three "polys"—**polyphagia** (pol″e-fa′je-ah), or excessive eating; **polyuria** (pol″e-u′re-ah), or excessive urination; and **polydipsia** (pol″e-dip′se-ah), or excessive thirst.

The long-term effects of diabetes are severely debilitating. Blindness occurs 25 times as often, kidney disease 17 times as often, and gangrene five times as often in diabetics as in nondiabetics. Heart disease and other complications of atherosclerosis are much more common in NIDDM than in IDDM.

Several theories have been proposed to explain effects of diabetes. One is that thickening of capillary basement membranes impairs peripheral circulation and may account for kidney and heart disease and gangrene. Another theory is that intermediates of glucose metabolism, such as sorbitol, accumulate in tissues. Blindness could be caused by lens damage from such deposits or by damage to retinal capillaries. A third theory proposes that excess glucose binds chemically to proteins and that these chemical complexes somehow lead to tissue damage.

NIDDM

Diabetes appearing after age 40 in genetically susceptible individuals, usually is NIDDM. Susceptibility seems to be inherited as an autosomal recessive characteristic. In NIDDM, the pancreas makes insulin—sometimes too much insulin—but cells are insulin resistant because they have too few functional insulin receptors. Obesity greatly increases the risk of NIDDM, probably by decreasing functional insulin receptors. Weight loss seems to overcome insulin resistance to some degree by increasing receptor numbers to near normal. These findings suggest that receptor numbers may be controlled by a negative feedback mechanism. As food intake increases and the blood glucose rises, insulin causes cells to take up more glucose than they can use. Excess cellular glucose somehow inactivates or destroys insulin receptors, thereby protecting cells from excess glucose. When carbohydrate intake remains high, the mechanism seems to prevent nearly all glucose from entering cells. In addition to age, obesity, and

TABLE 25.3   Guidelines for a Diabetic Diet

Constituent	Percent of Total Diet	Comments
Carbohydrate*	55 to 60	Most foods should contain complex carbohydrates and fiber, with little food containing simple sugars.
Protein*	12 to 20	Reduce to 0.8g/kg to minimize the risk of kidney disease.
Lipid	<30	Most foods should contain polyunsaturated fatty acids and little saturated fatty acids and cholesterol.

*Oat bran and legumes, which contain both carbohydrate and protein, are high in soluble fiber. Soluble fiber improves blood glucose control.

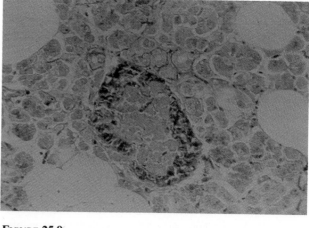

FIGURE 25.8
Damaged beta cells in a patient with IDDM.

genetic propensity, some studies show that for women, each pregnancy may increase the risk of NIDDM by about 10 percent.

Weight loss is the single most important factor in controlling NIDDM, and a sugar-restricted diet also is recommended (Table 25.3). If these treatments fail to maintain blood glucose in a normal range, they are supplemented with **oral hypoglycemics** (hi''po-gli-sem'iks). Tolbutamide stimulates insulin activity, whereas glipizide and glyburide increase insulin receptor sensitivity. Insulin injections are rarely needed.

A controversial theory relates insulin resistance (probably due to defective receptors) and atherosclerosis. Insulin resistance is seen in NIDDM patients, obese individuals, and in some people who are neither diabetic nor overweight. Blood glucose remains high and intracellular glucose low no matter how much insulin the pancreas secretes. According to the theory, this leads to a cluster of factors—high blood pressure, high blood triglycerides, low HDLs—that increase the risk of heart disease.

IDDM

The more severe IDDM, which requires insulin therapy, results from damage to insulin-secreting B-cells (Figure 25.8). IDDM usually begins before age 20 with sudden, acute symptoms, inflammation of the pancreas, and B-cell destruction. IDDM seems to be caused by an autoimmune reaction, possibly with insulin serving as the antigen. Immunosuppressive therapy aimed at producing insulin tolerance may someday prevent B-cell destruction in newly diagnosed IDDM.

The incidence of IDDM is greatest in people who have B8 and B15 HLA antigens, and certain other HLAs seem to confer IDDM resistance. Recently, genetic susceptibility to IDDM has been linked to an error in a single amino acid in an HLA antigen, with people having the genetic error being 100 times more likely to develop IDDM. Among those with the defect, any of several viruses—Coxsackie and those that cause mumps and rubella—can trigger the disease.

Maintaining blood glucose in a normal range is much more difficult in IDDM than in NIDDM. When insulin is deficient, ketoacidosis often occurs as the body metabolizes fats and accumulates ketone bodies. Untreated ketoacidosis can progress to coma, respiratory and cardiac failure, and death. Ketoacidosis is treated with insulin and careful regulation of body fluids and electrolytes. Injecting too much insulin or failing to eat after an injection, can cause **insulin shock.** The blood glucose level falls so far below normal that brain cells receive too little glucose to maintain consciousness. IDDM is treated with a low-sugar diet and daily insulin injections. Insulin cannot be given by mouth because it would be digested by proteolytic enzymes.

Insulin from pigs and cattle has been used to treat humans for many years, but human insulin is now available. It is made by chemically changing pig insulin to human insulin or by designing genetically engineered bacteria to synthesize it.

Obesity

A few people weigh more than weight charts recommend because they have heavy muscles or a large skeleton, but most have excess body fat. **Obesity** is defined as a body weight more than 20 percent above the ideal weight due to excess fat. From laboratory calculations, it appears that weight should increase at the rate of 1 pound for each excess of 3500 kcal eaten and decrease similarly for excess calories used. Any nutrients can be metabolized to acetyl-CoA and used to synthesize fat. Stored fat remains as long as energy from food equals energy used.

The causes of obesity are not well understood, but behavioral, physiological, genetic, and developmental factors are being studied. Behavioral factors include feelings, learned behaviors, and exercise habits. Some people overeat when they are anxious, others when they are bored. Children learn to overeat to satisfy oversolicitous parents. Many people become obese when they fail to reduce food intake as their metabolic rate decreases. Sedentary people may gain weight because they get too little exercise.

Malfunction of hunger-satiety centers is an important physiological factor, and can be due to a hypothalamic tumor, a deficiency in CCK-PZ, or an abnormally high operating set point for body weight. People with a high set point have difficulty losing weight and must maintain stringent long-term control of eating behavior to keep from regaining it. Conversely, people with a low set point must constantly force themselves to eat to gain weight and to maintain weight gain.

Other physiological factors include decreased activity of sodium-potassium pumps and depressed autonomic function. People whose cells have fewer than normal sodium-potassium pumps have a lower basal metabolic rate. They gain weight on a diet that merely maintains weight in a person with a normal number of sodium-potassium pumps. Finally, depressed sympathetic function leads to lower energy use in stressful situations and depressed parasympathetic activity conserves energy by slowing the heart rate and other physiological processes.

In the rare genetic obesity, an enzyme needed to break down stored fat is absent or deficient. Any nutrients converted to fat are stored and cannot be broken down easily.

The metabolism of obese people may differ from that of lean people though the fat in their bodies is qualitatively no different. Some obese people metabolize glucose more slowly than normal. On reducing diets, they apparently metabolize fats more slowly and completely because their blood has fewer ketone bodies than that of normal people on the same diet.

Complications of obesity include dissatisfaction with one's appearance and various physical problems—osteoarthritis, varicose veins, hernias, flat feet, and clumsiness that often leads to accidental injury. Obese people can be insulin resistant and predisposed toward diabetes, hypercholesterolemia, gallstones, and hypertension. Fat deposits increase the effort required for breathing, and carbon dioxide retention can lead to sleepiness and reduced activity. All in all, life expectancy is greatly reduced by obesity.

Treatment of obesity is largely unsuccessful and likely to remain so until causes are better understood. A combination of reduced caloric intake and increased physical activity is a standard treatment for obesity (Figure 25.9). One of the consequences of severe caloric restriction is that the metabolic rate also decreases. Physical activity is particularly important for weight loss because it seems to suppress appetite and helps to maintain a normal metabolic rate. Surgical procedures such as gastric bypass and vertical banded gastroplasty are used in extreme cases. Amphetamines should not be used because they are addictive and they suppress appetite for only a few weeks.

Dieters often experience the yo-yo effect—they gain and lose the same 20 lbs over and over again. Losing weight becomes more difficult each time. One reason may be that reducing food intake tends to increase metabolic efficiency. Another may be that throughout human history, people who had a store of fat were more likely to survive during food shortages, and we are their descendants. To avoid this problem, dieters who lose weight should make every effort to keep it off.

Anorexia Nervosa

Usually seen in adolescent and young adult females, **anorexia nervosa** (an″o-rek′se-ah ner-vo′sah) is a persistent desire to lose weight even when weight is already below normal (Figure 25.10). Patients have a faulty body image and see themselves as fat no matter how much weight they lose. Symptoms are those of starvation—emaciation, hair loss, and cessation of menstruation in women—but the starvation is self-induced. **Bulimia** (bu-lim′e-ah), or binge eating, is often followed by self-induced vomiting. Patients with bulimia often fail to feel full after a meal, and they have only about half the blood CCK-PZ of normal individuals. Bulimia sometimes precedes anorexia, but the relationship between the disorders is not clear.

Victims of anorexia nervosa become dehydrated from excreting an unusually high volume of urine. Administering saline solution fails to increase ADH blood levels as it normally does, so water is not reabsorbed even when blood osmotic pressure is high. ADH in cerebrospinal fluid of anorexics is high. These ADH abnormalities may relate to the development of anorexia nervosa.

What causes anorexia nervosa is unknown, but psychological factors and hypothalamic disorders are being studied. Current treatment makes use of psychotherapy and gavage (stomach tube) feeding. A typical patient receives 3 l per day of fluid containing carbohydrate, hydrolyzed protein, fat, electrolytes, and vitamins, which will cause a weight gain of about 1 lb. Temporary gavage feeding quickly increases weight and restores nutritional balance, but unless the patient's body image and eating habits can be altered, anorexia nervosa can be fatal.

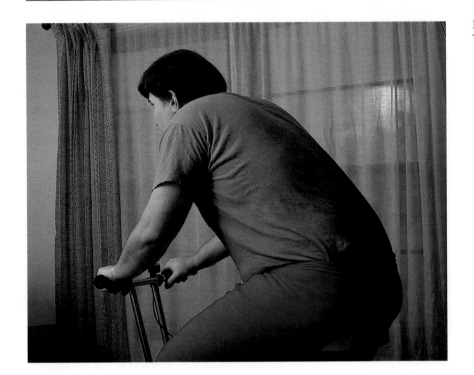

FIGURE 25.9
Treatment for obesity.

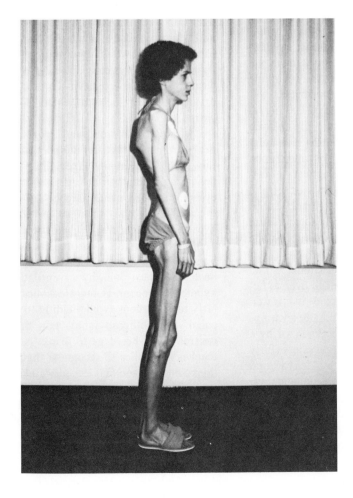

FIGURE 25.10
A patient suffering from anorexia nervosa.

Alcoholism

Physiological, psychological, or social impairment directly associated with persistent and excessive use of alcohol is called **alcoholism.** Both genetic and environmental factors appear to contribute to alcoholism, but exactly how is unclear. In the United States, where 10 percent of the drinkers drink 50 percent of the alcohol, over 10 million people are severe alcoholics and another 7 million are alcohol abusers. Alcoholism is the third leading cause of death in the 25 to 65 age group.

Two types of alcoholism are recognized. Type I, which includes most alcoholics, develops after age 25, usually in stressful environments. Type II occurs in sons of men who became alcoholics near or before age 25, and who also display aggressive behavior. Such individuals have a distinctive brain wave that persists even during abstinence. A genetic factor in alcoholism may be present on the same chromosome as the gene for MN blood antigens and the enzyme alcohol dehydrogenase, which oxidizes alcohol.

Studies of volunteers with a family history of alcoholism and control subjects lacking such a history show differences in alcohol metabolism. In one study, acetaldehyde, which is produced by the action of alcohol dehydrogenase on alcohol, rose much more sharply in the subjects with a family history of alcoholism than in the control subjects. Acetaldehyde production, thus, may be genetically determined and suitable for screening people at risk of becoming alcoholics. Family and other environmental stresses seem to contribute to alcoholism in genetically susceptible individuals. Habitual drinking exacerbates the stresses and creates a vicious cycle.

Blood alcohol concentrations in excess of 100 mg/dl (defined as intoxication by law in most states) affects brain centers and produces diminished coordination and judgment. Higher levels lead to speech impairment and stupor. At blood levels above 400 mg/dl, alcohol appears to disrupt cell membranes. Coma and hypothermia occur and can lead to death. Because alcohol dilates peripheral blood vessels and allows heat loss, it lowers body temperature and, contrary to a popular notion, does not help a person to keep warm while hunting or skiing.

Alcohol is quickly absorbed from the stomach and small intestine, and is oxidized mainly in the liver to acetaldehyde, acetate, and, finally, carbon dioxide and water. When much alcohol is consumed, acetaldehyde accumulates in the liver. Some alcohol may be metabolized by a nonoxidative pathway to form fatty acid ethyl esters (fatty acid combined with alcohol). Hydrogen, acetaldehyde, and fatty acid ethyl esters cause detrimental physiological effects (Figure 25.11). Oxidation of hydrogen provides energy and spares fats; their accumulation leads to high blood lipids, fatty liver, and ketosis. Acetaldehyde damages mainly liver tissue, and fatty acid ethyl esters damage the pancreas and liver. Alcohol also damages brain and muscle. In the brain, it impairs release of neurotransmitters, especially norepinephrine, and alters GABA receptors. In heart and skeletal muscles, it decreases mitochondrial enzymes, impairs both Ca^{2+} transport and sodium-potassium pumps, causes electrolyte imbalances, and decreases muscle contractility. In the myocardium, it alters conduction and may cause arrhythmias.

Acetaldehyde specifically impairs vitamin absorption and decreases vitamin activation in the liver, worsening dietary vitamin deficiencies. It also causes hepatitis (inflammation of the liver) and cirrhosis (destruction of liver cells and replacement with fibrous connective tissue). Connective tissue infiltration obstructs blood flow, interferes with liver functions, and creates back pressure in the hepatic portal vein. Liver cells that die are not replaced because the malnourished state of most alcoholics severely limits regeneration. Impaired urea synthesis leads to hepatic coma—loss of consciousness because toxic ammonia impairs brain cells.

Decreased plasma protein synthesis impairs blood clotting, decreases blood oncotic pressure, and causes tissue edema and decreased blood volume. As blood volume decreases, kidney blood flow decreases, and aldosterone is released. Na^+ is returned to the blood, and K^+ is excreted, creating electrolyte imbalances. Jaundice occurs when the liver fails to conjugate bilirubin and excrete it in the bile, and when connective tissue blocks bile flow to the liver. Finally, hematemesis (vomiting of blood) occurs with the rupture of esophageal varices, dilated esophageal veins weakened by hypertension. Once bleeding begins, hemorrhage is likely because blood clotting is severely impaired.

Most alcohol treatment programs emphasize abstinence from alcohol, peer support, and a diet high in protein and vitamins, especially thiamine. Thiamine deficiency is responsible for uncoordinated locomotion and eye muscles seen in many alcoholics. A nutritious diet helps the alcoholic's body to overcome deficiencies and allows liver cells to regenerate. The drug disulfiram (Antabuse) inhibits the reaction that breaks down acetaldehyde. If a person drinks alcohol after taking Antabuse, vomiting, headache, chest pain, hyperventilation, tachycardia, and hypotension occur. Some alcoholics take the drug regularly to help them avoid yielding to the temptation of alcohol. Psychotherapy may help the alcoholic learn to control his or her urge to drink. Alcoholics Anonymous teaches alcoholics to recognize that hunger, anger, loneliness, and fatigue make them vulnerable to drinking and helps them find safe havens—places offering emotional satisfaction and no alcohol. All available treatments require patient cooperation, and alcoholics often deny their problem and refuse treatment.

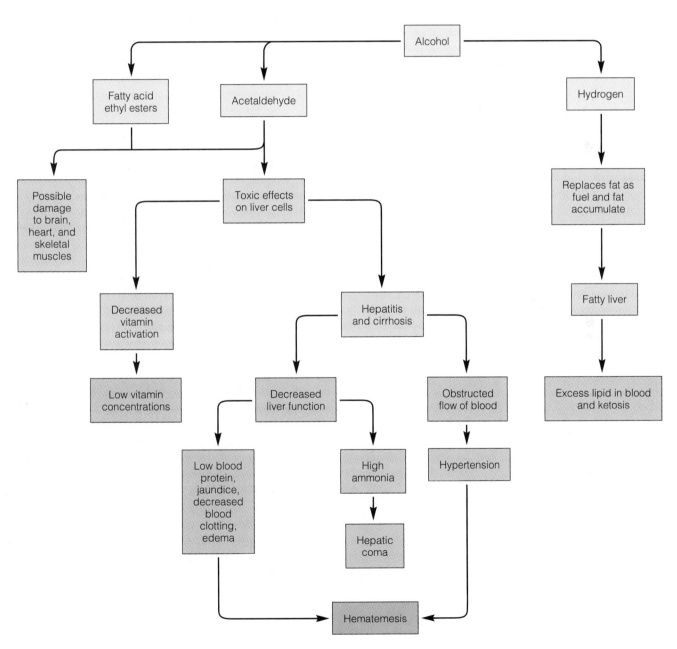

FIGURE 25.11
Some of the effects of excessive alcohol consumption.

ESSAY

DETOXIFICATION—A FIRST DEFENSE AGAINST FOREIGN SUBSTANCES

Rendering poisonous substances harmless can be a matter of life or death. Substances foreign to our bodies enter when we breathe polluted air, consume food containing pesticides or additives; or use tobacco, alcohol, or other legal or illegal drugs. Any foreign substance is potentially toxic (poisonous); its effect depends on how much is present and how well the body can detoxify it.

Most chemical reactions that **detoxify,** or render foreign substances harmless, are carried out by liver cell enzymes. The enzymes ordinarily inactivate hormones and other normal body substances, but they can detoxify many foreign substances, too. Detoxification reactions include oxidation, reduction, hydrolysis, and conjugation (Figure 25.12). Such reactions usually convert fat-soluble substances to water-soluble substances that can be excreted in the kidneys. Many such reactions are inducible; that is, the presence of the potentially toxic substance induces activity of the enzymes that detoxify it.

Combinations of drugs can have unpredictable and undesirable results when they are detoxified by the same enzyme. The drug tolbutamide, used to increase insulin activity, is metabolized and inactivated more slowly when the anti-inflammatory agent phenylbutazone, coumarin anticoagulants, or the antibiotic chloramphenicol is present. In addition to drugs, insecticides, dyes, and other foreign substances can be detoxified in the liver.

When the liver is subjected to large amounts of toxic substances, as in an alcoholic, it produces more detoxifying enzymes—at least until the liver itself becomes impaired. When an alcoholic suddenly abstains from alcohol, the liver makes an overabundance of detoxifying enzymes and sometimes destroys medications before they can have their effect. Individual differences exist in what substances

and how much of them can be detoxified, and good general health and nutritional status maximize this capability. In the elderly, detoxification can be impaired because the body, including the liver, is less efficient. Infants are extremely sensitive to foreign substances because their immature livers fail to make sufficient quantities of detoxifying enzymes.

Fetuses are particularly vulnerable to toxic substances that cross the placenta. The mother may suffer no adverse effects, but the fetus may be deformed, as was tragically demonstrated in the 1950s by the absence of limbs in infants whose mothers took the tranquilizer and anti-nausea drug thalidomide. When the mother uses alcohol or narcotics, the fetus becomes dependent and suffers withdrawal symptoms at birth and usually has other birth defects including mental retardation. Large numbers of infants are hospitalized in intensive care nurseries for alcohol and drug withdrawal. Many also are infected with AIDS.

Even though our livers detoxify foreign substances, it is unwise to expose ourselves unnecessarily to them. New drugs, food additives, insecticides, and other manufactured substances should be evaluated carefully before they are marketed; and strong laws preventing environmental pollution should be enforced.

Questions

1. What is detoxification?
2. How does the body detoxify substances?
3. Why is it important for one physician to know all of the medications a patient is taking?
4. Why are the very young and very old most susceptible to damage from toxic substances?

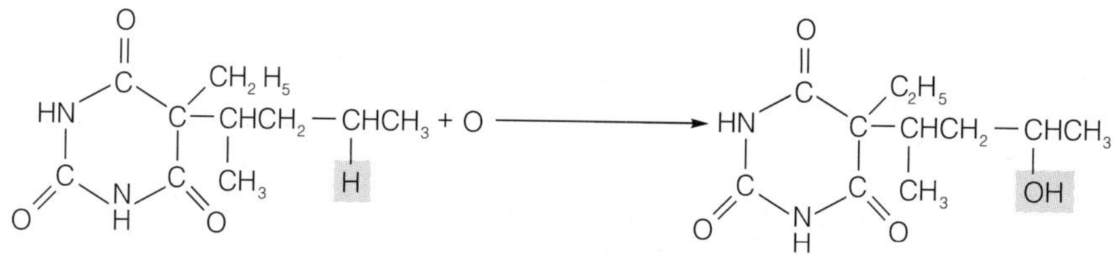

(a)

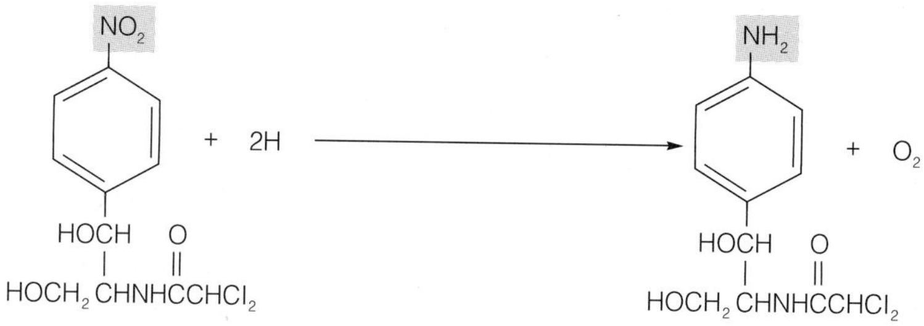

(b)

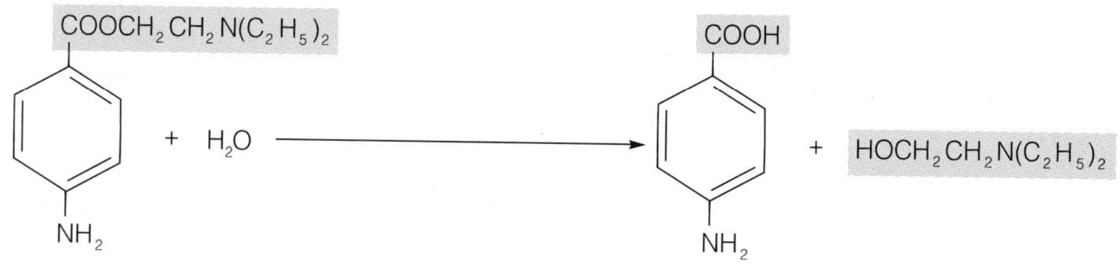

(c)

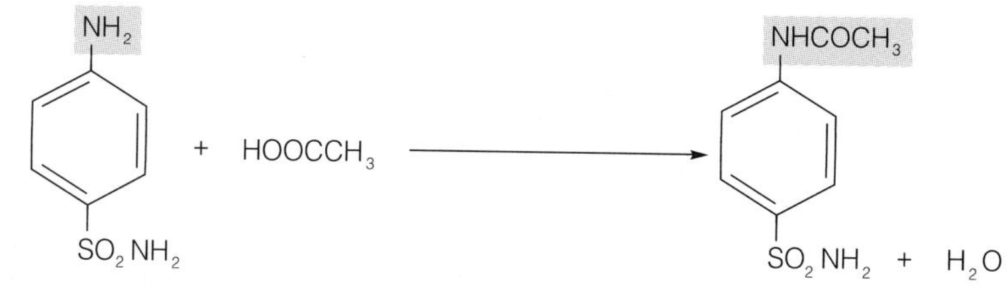

(d)

FIGURE 25.12
The inactivation of foreign substances: (*a*) oxidation of pentobarbital, (*b*) reduction of chloramphenicol, (*c*) hydrolysis of procaine, and (*d*) conjugation of sulfanilamide.

CHAPTER SUMMARY

Overview of Metabolic Regulation

• In cells, catabolism of nutrients captures energy in ATP and anabolism allows maintenance, growth, or repair. A variety of regulatory mechanisms control the rate at which each of these processes occurs.

• Absorptive metabolism is summarized in Figure 25.1, and postabsorptive metabolism is summarized in Figure 25.2.

Absorptive vs. Postabsorptive Metabolism

• Processes that regulate metabolism are summarized in Figures 25.3, 25.4, and 25.5.

Hunger and Satiety

• Hunger and satiety centers in the hypothalamus normally help to control food intake.

Regulation of Protein Metabolism

• Protein metabolism is regulated to maintain nitrogen balance by processes that regulate absorptive and postabsorptive metabolism, but also by growth hormone and thyroid hormones.

Regulation of Overall Metabolism

• Metabolism includes basal metabolism, which accomplishes physiological processes, and energy for physical activity. Factors affecting metabolism include age, sex, pregnancy, nutritional status, climate, body temperature changes, sleep, hormones, and sympathetic signals.

• The respiratory quotient is the ratio of CO_2 released to O_2 used.

Body Temperature and Its Regulation

• The human body maintains a core body temperature within a narrow range around 37° C by regulating heat gain and heat loss. Heat is gained from catabolic reactions that can be accelerated by stress, sympathetic stimulation, and thyroid hormones. It is lost by evaporation, radiation, and conduction.

• In regulating body temperature, heat sensitive neurons in the hypothalamus detect a rise and relay signals that dilate skin blood vessels, increase sweating, and inhibit heat production. Similarly, cold-sensitive cells in the skin and spinal cord relay signals that decrease sweating, constrict skin blood vessels, and initiate shivering.

Clinical Applications

• Hyperthermia includes fever, heat stroke, and malignant hyperthermia. Hypothermia includes frostbite and generalized hypothermia. Treatment usually attempts to restore normal temperature as quickly as possible without further tissue injury.

• Diabetes mellitus, high blood glucose, is caused by damage to the pancreas and a lack of insulin (IDDM) in young patients, or insulin inactivity (NIDDM) in older, usually obese, patients. Ketosis, dehydration, and hunger are common symptoms.

• Obesity, excess body fat, occurs when the food intake exceeds that used for energy but causes are poorly understood.

• Anorexia nervosa, a persistent desire to lose weight, is thought to involve psychological or hypothalamic disorders.

• In alcoholism, hydrogen, acetaldehyde, and fatty acid ethyl esters interfere with normal metabolism and damage the liver, heart, and brain.

QUESTIONS AND PROBLEMS

The questions at the end of each chapter are numbered to correspond with the objectives listed at the beginning of the chapter. Italics indicate that a question requires critical thinking skills beyond simple factual recall.

Questions

1. *What are the major processes involved in metabolism?*
2. *How do absorptive and postabsorptive metabolism differ, and what events occur in each?*
3. *How do the following mechanisms regulate shifts between absorptive and postabsorptive metabolism: (a) insulin-glucagon mechanism, (b) epinephrine-hypothalamic mechanism, and (c) growth hormone-hypothalamic mechanism?*
4. How are hunger and satiety regulated?
5. What factors regulate protein metabolism?
6. (a) *How does basal metabolism differ from total metabolism?*
   (b) *What factors affect metabolic rates?*
7. *What is the significance of the respiratory quotient?*
8. (a) What normal variations occur in body temperature?
   (b) *How is heat produced in the body, and how is it lost?*
9. *What regulatory mechanisms come into play when the body temperature begins to rise? When it begins to fall?*
10. (a) *What are the causes of hyperthermia and hypothermia?*
    (b) How is each kind of body temperature disorder treated?
11. *How is regulation disturbed in (a) non-insulin-dependent diabetes, (b) insulin-dependent diabetes, (c) obesity, (d) anorexia nervosa, and (e) alcoholism?*

Problems

1. What combination of types of nutrients might be metabolized to obtain a respiratory quotient of 0.9?
2. Survey the current literature to determine whether new information on appetite control is available. If so, prepare a report that applies that information to the control of obesity.
3. Prepare a report on any other metabolic disorder.

OBJECTIVES

1. State the basic principles of good nutrition.
2. Explain the body's needs for carbohydrate, protein, and lipid.
3. Explain why the body also needs vitamins, minerals, and water.
4. Explain the relationship between food intake and energy use.

5. Evaluate the guidelines for deciding what to eat.

6. Summarize the specific cellular functions of vitamins and minerals.

7. Relate diet to malnutrition and other nutritional problems.
8. Relate diet to hypertension and other circulatory disorders.
9. Describe the sources and effects of nonnutrient substances in food.
10. Relate diet to cancer and behavioral disorders.

11. Define therapeutic diet, and describe how such diets are used.

26

NUTRITION

Principles of Nutrition

What we eat or don't eat can be a life-or-death matter. Everyday, more than 40,000 people worldwide die from the lack of a nutritious diet while unknown numbers of Americans die from circulatory diseases and cancers, many of which were brought on by years of ingesting too many calories and too much fat and refined sugar. Up to 70 percent of cancers, fatal or treatable, have been linked to diet. Among Americans, 80 million are overweight (more than 10 percent above ideal weight) and nearly 40 million are obese (more than 20 percent above ideal weight). Many suffer from a myriad of health problems associated with obesity (Chapter 25). Diet may be the most important health factor that people can control. Learning what makes up a healthy diet and choosing such a diet is essential to good health.

The science of **nutrition** deals with providing all nutrients—carbohydrates, proteins, fats, vitamins, minerals, and water—for maintaining health through food ingestion. Carbohydrates and fats provide calories to meet the body's energy demands. Carbohydrates can be converted to glycogen and stored in the body. Certain high-carbohydrate foods provide fiber (roughage), which helps move foodstuffs through the digestive tract. Fats also serve as energy storage molecules, and some fats are essential for synthesis of cell membranes, prostaglandins, and other substances. Dietary protein provides amino acids for the synthesis of new proteins and other cell components, and is very important during growth, repair, pregnancy, and lactation. How the body uses carbohydrates, proteins, and fats was described in Chapter 24. Vitamins, minerals, and water also are essential to maintain health. Most vitamins and minerals assist enzymes in catalyzing important reactions. Water, which makes up about 60 percent of the body weight, participates in nearly every chemical reaction as reactant, product, or reaction medium.

For most people, maintaining good nutrition is not simply a matter of ingesting essential nutrients and sufficient calories. Eating is a social process, too. Age, sex, and religious and social customs contribute significantly to what people think is a suitable diet. Certain foods such as bread, fish, and pork have religious significance for some people, and other foods are associated with holidays—such as a Thanksgiving turkey. Health care workers helping people change eating habits should consider these factors in addition to biological nutritional needs.

General Nutritional Requirements

Nutritional requirements are determined at the cellular and molecular levels according to energy expenditure, protein synthesis, enzyme activity, and electrolyte balance.

The Carbohydrate Component

No specific carbohydrates are required in the human diet, but at least 20 percent of total calories should come from carbohydrate, and 50 percent is recommended. Some carbohydrate is needed to provide Krebs cycle intermediates (oxaloacetate, for example) even when the body is obtaining most of its energy from fats. Metabolism of fat without carbohydrate, as occurs in untreated diabetes, can lead to ketosis and acidosis (Chapter 24).

Many Americans eat too little complex carbohydrate (polysaccharide, such as starch), too little fiber (cellulose and other indigestible plant parts), and too much refined sugar. Sucrose (table sugar) contributes to dental caries. Microbes metabolize sugar and deposit dental plaque in which they become entrapped, and produce acid that attacks tooth surfaces.

Dietary fiber is any plant material that humans cannot digest. Two major kinds are insoluble cellulose and soluble pectins. Fiber comes from grains, fruits, and vegetables. Cellulose, found in wheat, celery, and many other vegetables, increases fecal bulk and decreases the time required for material to traverse the intestine. Some of it is digested by intestinal microorganisms, which release metabolic products such as short chain fatty acids, water, carbon dioxide, and other gases. Soluble fiber, found in fruits, oat bran, and peas and beans, delays gastric emptying, slows glucose absorption (making blood sugar easier to control in diabetics), and lowers serum cholesterol and triglycerides.

Either too little or too much fiber can be detrimental. A lack of fiber can contribute to diverticulitis, inflammation of diverticuli (little pouches) in the colon, and colon cancer. Fiber helps to prevent diverticulitis by keeping the colon distended so diverticuli do not form and by reducing colon transit time. Reducing transit time also helps prevent colon cancer, probably by preventing accumulation of carcinogenic products of bacterial metabolism. Fiber also lowers blood pressure in some people. Excess fiber can decrease the absorption of elements such as calcium, iron, copper, and zinc. Many nutritionists are currently studying fiber, but which and how much fiber humans need has not been established. We will consider the effects of different kinds of fiber in more detail later.

The Essential Protein Component

Dietary protein is digested to amino acids and used to make new proteins and other nitrogenous molecules—porphyrins for heme in hemoglobin and nucleotides such as DNA, RNA, and ATP. Of the 20 amino acids used in these synthetic reactions, the **essential amino acids** must be present in the diet because the cells cannot synthesize them in adequate amounts. Of the essential amino acids, eight are needed throughout life, and two are needed during periods of rapid growth (Table 26.1). Though cells can make

TABLE 26.1	Essential Amino Acids

Amino Acids Required throughout Life

Leucine	Phenylalanine
Isoleucine	Threonine
Lysine	Tryptophan
Methionine	Valine

Amino Acids Required during Periods of Rapid Growth

Arginine	Histidine

small quantities of arginine and histidine, they cannot make these amino acids fast enough to support rapid protein synthesis.

Adequate amounts of all amino acids used in protein synthesis must be present at the same time for synthesis to occur. As mRNA codons are read, the appropriate amino acid attached to its tRNA must be immediately available for a polypeptide chain to be lengthened (Chapter 3). Otherwise, the protein cannot be completed. Dietary proteins have high **biological value** when they supply amino acids in the proportions needed for the synthesis of human proteins. The biological value of egg albumin is 100 (the highest possible value) because its amino acid content is nearly the same as the average for human cells. Most meat proteins have a biological value of about 70 and those in cereals and grain have a biological value of about 40. Because plant proteins are deficient in certain human essential amino acids, they must be eaten in particular combinations in the same meal to attain high biological value for humans (see essay at the end of this chapter).

Proteins normally eaten by humans have an average biological value of about 70, a fact used in setting the recommended daily protein allowance of about 0.8 g/kg body weight. Males require slightly more protein per kilogram body weight than females because their bodies contain relatively more muscle mass. Exercise increases the need for calories, protein, and vitamin B_6, which is required for protein metabolism.

Lipid Is Needed but Not Too Much

The **essential fatty acid** linoleic acid is the only specific lipid required in the diet. An essential fatty acid is one that must be present in the diet because the body cannot synthesize it. If cells have sufficient linoleic acid, they can synthesize other unsaturated fatty acids to maintain cell membranes and make prostaglandins and leukotrienes. Because linoleic and other unsaturated fatty acids are abundant in most vegetable fats, a tablespoon of margarine daily satisfies this nutritional need.

Omega-3 fatty acids are highly unsaturated, long-chain fatty acids with a double bond starting at the third carbon from the methyl end. Abundant in fish fats and green, leafy vegetables, they may have important functions though they are not yet recognized as essential nutrients. Omega-3 fatty acids may improve sharpness of vision, lower blood cholesterol and triglycerides, and retard platelet aggregation in atherosclerotic arteries. Fish fats may reduce the risk of atherosclerosis in another way. They contain free radicals (atoms with highly excited electrons) that inhibit a growth factor needed for arterial smooth muscle proliferation.

Most Americans eat too much rather than too little lipid. Excesses of animal fats are particularly detrimental because they contain large quantities of saturated fats and cholesterol, which increase the risk of heart disease, as is discussed later.

Why Vitamins and Minerals

Vitamins are small organic molecules required in the diet because human cells fail to make them or make too little of them to maintain health. Many vitamins are necessary for synthesis of coenzymes such as NAD and FAD or for other essential molecules such as collagen and visual pigments. When first discovered, vitamins were designated by letters A, B, and so forth. As multiple factors were found in the original vitamin B, they were given numerical subscripts. Though vitamins now have names based on their chemical properties, letters are still used. B vitamins and vitamin C are water soluble, whereas vitamins A, D, E, and K are fat soluble. **Minerals** are essential inorganic substances, usually ions. They serve as enzyme cofactors, maintain membrane excitability, and form part of substances such as bone, heme, and thyroid hormones. Functions and requirements for vitamins and minerals are discussed in more detail later.

Advertisers tell us to take vitamins and nutritional supplements every day. Do we really need them? People in good health who eat a variety of foods in a balanced diet should need no nutritional supplements. People with nutritional deficiencies because of a disease or a poor diet may be helped by carefully chosen supplements. Some supplements such as fat-soluble vitamins can be harmful, so it's a good idea to get a physician's advice before taking nutritional supplements of any kind.

The Need for Water

Water maintains adequate hydration of body cells and normal volumes of extracellular fluids. Most metabolic reactions take place in water, and many actually involve water molecules. Dietary water comes from beverages and foods. Watermelon, fruits used to make juices, and squash, cucumbers, and leafy vegetables contain much water. In addition, **metabolic water** is released in cells when hydrogen is transferred to oxygen in oxidation. The body needs enough water, dietary and metabolic, to maintain **water balance** in which intake equals output (Chapter 28).

Balancing Food Intake and Energy Use

Cells can catabolize most nutrients for energy, but they capture only about 40 percent of that energy in ATP and release the remainder as heat. Even so, cellular energy capture is fairly efficient compared to engines that can use only 35 percent of the energy in the fossil fuels they burn. Foods differ in energy content as can be measured by various laboratory techniques. Cells derive four kcal per gram of protein or carbohydrate and about nine kcal per gram of fat ingested, therefore, adding fat raises the energy content of a diet much more than adding carbohydrates or proteins.

Laboratory measurements of the energy in foods are somewhat higher than that obtained by the body for two reasons:

1. Some food does not reach cells because it is not digested or not absorbed. People vary considerably in digestive system efficiency from low in those who can "eat anything and not gain a pound," to high in those in whom "everything they eat turns to fat." A high-fiber, low-fat diet moves through the digestive tract faster than a high-fat, low-fiber diet. Faster transit usually means that a smaller proportion of nutrients are absorbed.

2. On most ordinary diets, the body releases about 10 percent more heat during digestion and metabolism than at other times. This heat, called **specific dynamic action,** is highest for high-protein diets because more energy is needed to digest proteins than other foods.

To maintain a constant body weight, energy obtained from food should equal energy expended for basal metabolism and physical activity. If energy intake exceeds energy use, weight gain occurs. If energy use exceeds energy intake, weight loss will occur. Ideal or desirable weight can be defined as the weight for a given height associated with the lowest mortality (death) rate, based on statistics such as those of the Metropolitan Life Insurance Company (Table 26.2). Weights above or below the ideal range are associated with higher mortality; and the greater the deviation, the greater the risk. Another way to determine normal body weight is the **body mass index** (Figure 26.1).

TABLE 26.2	Desirable Weights for Men and Women, Metropolitan Life Insurance Co., 1983		
Height	**Weight**		
	Small Frame	*Medium Frame*	*Large Frame*
Men*			
5' 2''	128–134	131–141	138–150
5' 3''	130–136	133–143	140–153
5' 4''	132–138	135–145	142–156
5' 5''	134–140	137–148	144–160
5' 6''	136–142	139–151	146–164
5' 7''	138–145	142–154	149–168
5' 8''	140–148	145–157	152–172
5' 9''	142–151	148–160	155–176
5' 10''	144–154	151–163	158–180
5' 11''	146–157	154–166	161–184
6' 0''	149–160	157–170	164–188
6' 1''	152–164	160–174	168–192
6' 2''	155–168	164–178	172–197
6' 3''	158–172	167–182	176–202
6' 4''	162–176	171–187	181–207
Women†			
4' 10''	102–111	109–121	118–131
4' 11''	103–113	111–123	120–134
5' 0''	104–115	113–126	122–137
5' 1''	106–118	115–129	125–140
5' 2''	108–121	118–132	128–143
5' 3''	111–124	121–135	131–147
5' 4''	114–127	124–138	134–151
5' 5''	117–130	127–141	137–155
5' 6''	120–133	130–144	140–159
5' 7''	123–136	133–147	143–163
5' 8''	126–139	136–150	146–167
5' 9''	129–142	139–153	149–170
5' 10''	132–145	142–156	152–173
5' 11''	135–148	145–159	155–176
6' 0''	138–151	148–162	158–179

Courtesy Metropolitan Life Insurance Company.

*Weights at ages 25 to 59 based on lowest mortality. Weight in pounds according to frame (in indoor clothing weighing 5 pounds, shoes with 1'' heels).

†Weights at ages 25 to 59 based on lowest mortality. Weight in pounds according to frame (in indoor clothing weighing 3 pounds, shoes with 1'' heels).

DECIDING WHAT TO EAT

Guidelines are available to help people select a nutritious diet without needing to know the physiological actions of nutrients. One guideline recommends numbers of servings of foods from the basic four food groups. Companies that process and package foods provide label information showing what percentage of recommended dietary allowances a food contains. Over a decade ago, the United States Congress established dietary goals intended to improve eating habits among Americans. Since then, various organizations and agencies have published dietary guidelines.

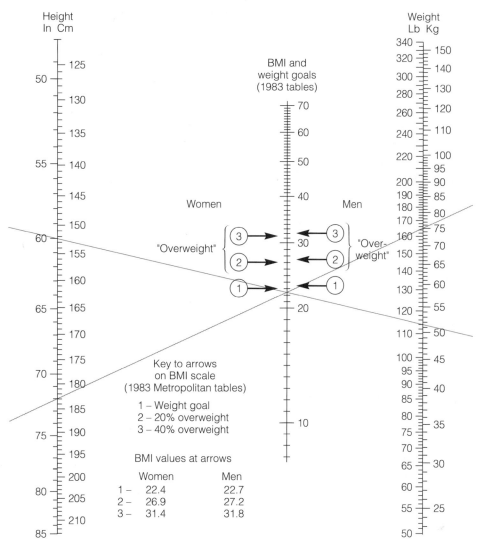

FIGURE 26.1
Body mass index, which is calculated by dividing weight in kilograms by the square of height in meters (kg/m²). To find your body mass index, place a ruler across the figure so that it connects your height on the left scale with your weight on the right scale. The body mass index for ideal weight is 22.4 for women and 22.7 for men—slightly higher because of heavier muscles. The body mass index at which overweight becomes obesity (20 percent overweight) is 27.3 for women and 27.8 for men. For example, for a woman 60 inches tall, 114 pounds is an ideal weight. For a man 72 inches tall, 165 pounds is an ideal weight. Obesity as a metabolic disorder was discussed in Chapter 25.

Basic Four Food Groups

Eating the proper number of servings from each of the **basic four food groups**—milk, meat, fruit and vegetable, and bread and cereal (Table 26.3)—each day provides reasonable assurance of an adequate diet. Eating many different foods over a period of time helps to assure that all vitamins and minerals will be included in the diet. The specific nutritional needs satisfied by each group are summarized in the last column of the table.

Recommended Dietary Allowances

At present, **recommended dietary allowances (RDAs),** the amount of a nutrient thought to satisfy the needs of 97.5 percent of the human population, are the best available standards for human nutritional needs (Table 26.4). Unfortunately, RDAs have not been established for all known essential nutrients, and a diet that satisfies all RDAs may not be nutritious. For example, a processed cereal with

TABLE 26.3 The Basic Four Food Groups and Their Contributions to Nutrition

Group	Recommended Daily Amounts	Nutritional Needs Satisfied
Milk group	Serving: 8 oz milk or 1½–2 oz cheese 2 servings for adults 3 servings for children and pregnant women 4 servings for lactating mothers	Each serving provides 8 g protein, 3 g fat, and about 140 calories. Contains vitamins A and riboflavin, and provides an especially good supply of calcium, phosphorus, potassium, and magnesium. Low-fat dairy products reduce saturated fat and cholesterol in the diet.
Meat group	Serving: 3 oz lean meat, fish, or poultry, or 1 c cooked legume, 2 eggs, or 4 T peanut butter 2 servings	Each serving provides 15–25 g protein, 10–40 g fat, and 150–400 calories. Contains phosphorus, iron, sodium, potassium, riboflavin and other B vitamins. Legumes are very low in fat. Peanut butter is high in fat but also high in niacin. Eggs provide some vitamin A, but are high in cholesterol.

TABLE 26.4 Food and Nutrition Board, National Academy of Sciences—National Research Council Recommended Dietary Allowances,[a] Revised 1989
Designed for the maintenance of good nutrition of practically all healthy people in the United States

Category	Age (years) or Condition	Weight[b] (kg)	(lb)	Height[b] (cm)	(in)	Protein (g)	Fat-Soluble Vitamins Vitamin A (μg RE)[c]	Vitamin D (μg)[d]	Vitamin E (mg α-TE)[e]	Vitamin K (μg)	Water-Soluble Vitamins Vitamin C (mg)	Thiamine (mg)
Infants	0.0–0.5	6	13	60	24	13	375	7.5	3	5	30	0.3
	0.5–1.0	9	20	71	28	14	375	10	4	10	35	0.4
Children	1–3	13	29	90	35	16	400	10	6	15	40	0.7
	4–6	20	44	112	44	24	500	10	7	20	45	0.9
	7–10	28	62	132	52	28	700	10	7	30	45	1.0
Males	11–14	45	99	157	62	45	1000	10	10	45	50	1.3
	15–18	66	145	176	69	59	1000	10	10	65	60	1.5
	19–24	72	160	177	70	58	1000	10	10	70	60	1.5
	25–50	79	174	176	70	63	1000	5	10	80	60	1.5
	51 +	77	170	173	68	63	1000	5	10	80	60	1.2
Females	11–14	46	101	157	62	46	800	10	8	45	50	1.1
	15–18	55	120	163	64	44	800	10	8	55	60	1.1
	19–24	58	128	164	65	46	800	10	8	60	60	1.1
	25–50	63	138	163	64	50	800	5	8	65	60	1.1
	51 +	65	143	160	63	50	800	5	8	65	60	1.0
Pregnant						60	800	10	10	65	70	1.5
Lactating	1st 6 months					65	1300	10	12	65	95	1.6
	2nd 6 months					62	1200	10	11	65	90	1.6

Recommended Dietary Allowances, © 1989, by the National Academy of Sciences, National Academy Press, Washington, DC.

[a]The allowances, expressed as average daily intakes over time, are intended to provide for individual variations among most normal persons as they live in the United States under usual environmental stresses. Diets should be based on a variety of common foods in order to provide other nutrients for which human requirements have been less well defined. See text for detailed discussion of allowances and nutrients not tabulated.

[b]Weights and heights of Reference Adults are actual medians for the U.S. population of the designated age, as reported by NHANES II. The median weights and heights of those under 19 years of age were taken from Hamill et al., (1979) (see pages 16–17). The use of these figures does not imply that the height-to-weight ratios are ideal.

Group	Recommended Daily Amounts	Nutritional Needs Satisfied
Fruit and vegetable group	Serving: 1/2 cup 4 servings, including one good source of vitamin A and one of vitamin C	Foods that provide vitamin A: spinach and other green, leafy vegetables; carrots and other yellow vegetables; and fruits. Foods that provide vitamin C: citrus fruits and juices, and tomatoes. Fruits and vegetables also provide some B vitamins and vitamin K, as well as some minerals. Raw foods in this group have larger amounts of vitamins than cooked foods. All fruits and vegetables provide fiber.
Bread and cereal group	Serving: 1 slice of bread or 1 oz of prepared cereal or 1/2 cup of cooked cereal, rice, pasta, or noodles 4 servings	Each serving provides energy, some protein, and, if made of whole grain, a good supply of B vitamins. Eating food from this group with a legume increases the protein value of both foods. All breads and cereals provide fiber.

Water-Soluble Vitamins					Minerals						
Riboflavin (mg)	Niacin (mg NE)[f]	Vitamin B$_6$ (mg)	Folate (µg)	Vitamin B$_{12}$ (µg)	Calcium (mg)	Phosphorus (mg)	Magnesium (mg)	Iron (mg)	Zinc (mg)	Iodine (µg)	Selenium (µg)
0.4	5	0.3	25	0.3	400	300	40	6	5	40	10
0.5	6	0.6	35	0.5	600	500	60	10	5	50	15
0.8	9	1.0	50	0.7	800	800	80	10	10	70	20
1.1	12	1.1	75	1.0	800	800	120	10	10	90	20
1.2	13	1.4	100	1.4	800	800	170	10	10	120	30
1.5	17	1.7	150	2.0	1200	1200	270	12	15	150	40
1.8	20	2.0	200	2.0	1200	1200	400	12	15	150	50
1.7	19	2.0	200	2.0	1200	1200	350	10	15	150	70
1.7	19	2.0	200	2.0	800	800	350	10	15	150	70
1.4	15	2.0	200	2.0	800	800	350	10	15	150	70
1.3	15	1.4	150	2.0	1200	1200	280	15	12	150	45
1.3	15	1.5	180	2.0	1200	1200	300	15	12	150	50
1.3	15	1.6	180	2.0	1200	1200	280	15	12	150	55
1.3	15	1.6	180	2.0	800	800	280	15	12	150	55
1.2	13	1.6	180	2.0	800	800	280	10	12	150	55
1.6	17	2.2	400	2.2	1200	1200	320	30	15	175	65
1.8	20	2.1	280	2.6	1200	1200	355	15	19	200	75
1.7	20	2.1	260	2.6	1200	1200	340	15	16	200	75

[c]Retinol equivalents. 1 retinol equivalent = 1 µg retinol or 6 µg β-carotene. See text for calculation of vitamin A activity of diets as retinol equivalents.

[d]As cholecalciferol. 10 µg cholecalciferol = 400 IU of vitamin D.

[e]α-Tocopherol equivalents. 1 mg d-α tocopherol = 1 α-TE. See text for variation in allowances and calculation of vitamin E activity of the diet as a α-tocopherol equivalents.

[f]1 NE (niacin equivalent) is equal to 1 mg of niacin or 60 mg of dietary tryptophan.

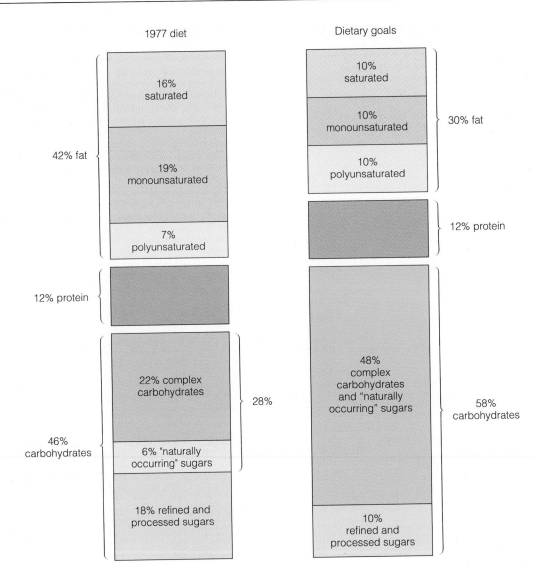

1977 diet

42% fat
- 16% saturated
- 19% monounsaturated
- 7% polyunsaturated

12% protein

46% carbohydrates
- 22% complex carbohydrates
- 6% "naturally occurring" sugars
- 18% refined and processed sugars

28% (complex carbohydrates + naturally occurring sugars)

Dietary goals

30% fat
- 10% saturated
- 10% monounsaturated
- 10% polyunsaturated

12% protein

58% carbohydrates
- 48% complex carbohydrates and "naturally occurring" sugars
- 10% refined and processed sugars

FIGURE 26.2
A typical American diet prior to establishment of dietary goals is compared with a diet consistent with those goals. (From *Dietary Goals for the United States*, U.S. Government Printing Office)

vitamins added to meet RDAs may contain too much refined sugar and other additives. In spite of these limitations, RDAs remain the *only* available scientifically determined measures of human nutritional needs. Special RDAs are provided for infants, children, and pregnant and lactating women. During the last half of pregnancy, a woman generally needs 10 to 15 percent more calories than she did before she became pregnant, and during breastfeeding, she needs about 750 kcal per day to make 850 ml of milk.

Dietary Guidelines

The typical diet in the United States has changed significantly since 1900. People are eating more meat, poultry, fish, dairy products, fats and oils, sugars, and salt, and less grain products, eggs, fruits, and vegetables. Because of these changes and their health implications, the United States Senate published *Dietary Goals for the United States* in 1977. The typical American diet at the time of the report is contrasted with the diet that would satisfy the goals in Figure 26.2. These dietary goals attempt to minimize diseases such as heart disease, stroke, atherosclerosis, hypertension, some kinds of cancer, obesity, diabetes, osteoporosis, tooth decay, and diverticulitis.

The National Research Council found that in 1988, the American diet averaged 36 percent fat, 45 percent carbohydrate, and 19 percent protein. Men averaged 1.75 times and women 1.4 times the protein RDA; and men averaged 435 mg and women 304 mg cholesterol daily. Dietary guidelines also have been prepared by the American Cancer Society, the American Heart Association, and the United States Department of Agriculture (listed here):

1. Eat a variety of foods.
2. Maintain desirable weight.
3. Avoid too much fat, saturated fat, and cholesterol.
4. Eat foods with adequate starch and fiber.
5. Avoid too much sugar.
6. Avoid too much sodium.
7. If you drink alcoholic beverages, do so in moderation.

More specific recommendations from the National Research Council and the Surgeon General are as follows: Eat fish and skinless poultry instead of meat several days per week. To minimize saturated fats and cholesterol, trim fat from red meats when they are used, use low-fat dairy products, and reduce egg yolks in the diet. Have five or more servings of fruits and vegetables (including peas and beans) and six or more servings of starches (including whole-grain foods) per day. Avoid alcoholic beverages if you are, or plan to become, pregnant. To lose weight or prevent weight gain, limit fats, sugars, and other high-calorie foods and engage in regular sustained physical activity.

The Surgeon General's Report also makes recommendations for people with special needs. Fluoride should be obtained from public water supply or dental care products. People especially vulnerable to dental caries should severely limit use of high-sugar foods. Adolescent girls and women should increase their intake of foods high in calcium, especially low-fat dairy products. Children, adolescents, and women of childbearing age should eat lean meats, fish, and iron-enriched cereals to satisfy needs for iron. Finally, people with disorders such as lactose intolerance, phenylketonuria, or food allergies that prevent their following guidelines should make a special effort to substitute nutritious foods for the foods they cannot eat.

WHAT VITAMINS AND MINERALS DO IN CELLS

Actions, sources, and effects of excesses and deficiencies of vitamins and minerals are summarized in Table 26.5. **Water-soluble vitamins** dissolve in body fluids, and the kidneys excrete excesses. **Fat-soluble vitamins** are absorbed with other lipids and stored mainly in the liver. Excesses can be toxic.

Water-Soluble Vitamins

The water-soluble vitamins include a group of vitamins called the B-complex vitamins and vitamin C. The B-complex vitamins will be discussed first.

Thiamine

Also known as vitamin B_1, **thiamine** (thi'am-in) is water-soluble and is destroyed by temperatures above 100° C. Boiling, frying, or pressure cooking foods reduces their thiamine content. Activated thiamine (with a pyrophosphate group) is a coenzyme for enzymes that oxidize and remove carbon dioxide from a molecule. The enzyme that converts pyruvate to acetyl-CoA is an example.

Riboflavin

Originally named vitamin B_2, **riboflavin** (ri''bo-fla'vin) is slightly water soluble and is not destroyed by heat at most cooking temperatures. It consists of a three-ring structure (flavin) with attached ribose. Riboflavin is an essential component of hydrogen carriers such as flavin adenine dinucleotide (FAD).

Niacin

Water-soluble, heat-resistant **niacin** (ni'as-in), or nicotinic acid, is readily converted to nicotinamide, and used to make coenzymes such as nicotinamide adenine dinucleotide (NAD). NAD accepts hydrogen removed from glyceraldehyde-3-phosphate in glycolysis, pyruvate, and several Krebs cycle intermediates. Cells synthesize some niacin from tryptophan if they have pyridoxine (vitamin B_6), so the need for dietary niacin depends on how much tryptophan and pyridoxine food contains. Niacin deficiency causes pellagra.

Pantothenic Acid

Water-soluble, heat-stable **pantothenic** (pan''to-then'ik) **acid** is required for coenzyme A synthesis. Coenzyme A, in turn, is needed for fatty acid oxidation, to allow acetate to enter the Krebs cycle, and to make acetylcholine.

Pyridoxine

Substances with vitamin B_6 activity include **pyridoxine** (pēr''ĭ-dok'sēn), **pyridoxal** (pēr''ĭ-dok'sal), and **pyridoxamine** (pēr''ĭ-doks'ah-mēn), all of which are water soluble and heat stable. Already phosphorylated in food or phosphorylated in cells, these molecules are required for transamination, decarboxylation, the transfer of sulfur between amino acids, and the synthesis of prophyrins and various neurotransmitters. They also help convert the amino acid tryptophan to niacin.

TABLE 26.5   Vitamins and Minerals

	Biochemical Actions	Dietary Sources	Effects of Excess	Effects of Deficiency
Water-Soluble Vitamins				
B_1 (thiamine)	Part of decarboxylation enzyme in Krebs cycle	Organ meats, whole grains, and legumes	None known	Arrests metabolism at pyruvate; beriberi (neurological impairment, heart failure)
B_2 (riboflavin)	Part of FAD and FMN	Dairy products, eggs, and whole grains	None known	Sensitivity to light, eye lesions, and cracks in corners of mouth
Niacin	Part of NAD and NADP	Whole grains, meats and legumes; can be formed from tryptophan	Flushing of skin in sensitive people	Pellagra (skin and digestive lesions, mental disorders)
Pantothenic acid	Part of coenzyme A	Widely distributed	None known	Fatigue, nervous and motor impairment (rare in humans)
B_6 (pyridoxine)	Coenzyme for metabolism of amino acids and fats	Whole grains, meats, and vegetables	None known	Dermatitis, nervous disorders, kidney stones, impaired immunity
Folacin (folic acid)	Coenzyme for metabolism of amino acids and nucleic acids	Meats, legumes, green vegetables, and wheat	None known	Impairs production of erythrocytes, intestinal disturbances
B_{12} (cyanocobalamin)	Coenzyme in nucleic acid metabolism	Meat, eggs, and dairy products	None known	Pernicious anemia, nervous disorders
Biotin	Coenzyme in fat and glycogen synthesis, amino acid metabolism	Egg yolks, legumes, vegetables, and meats	None known	Dermatitis, muscle pains, weakness, and depression
Choline (may not be a vitamin)	Forms phospholipids and acetylcholine	Egg yolk, liver, grains, and legumes	None known	None known in humans
C (ascorbic acid)	Acts in synthesis of collagen and matrix of connective tissue	Citrus fruits	Possibly kidney stones	Scurvy (degeneration of teeth, skin, and blood vessels)
Fat-Soluble Vitamins				
A (carotene)	Forms visual pigments; maintains epithelia	Green and yellow vegetables, fruits, milk, and egg yolks	Headache, loss of appetite, elevated blood calcium, and peeling of skin	Night blindness, excess keratin in tissues of eye
D (calciferol)	Acts in absorption of calcium and bone growth	Fish oils, liver, and fortified dairy products	Kidney damage, vomiting, diarrhea, and weight loss	Bone softness and deformity (rickets in children and osteomalacia in adults)
E (tocopherol)	Maintains integrity of erythrocytes	Green, leafy vegetables; seeds; and oils	None known	Anemia because of fragility of erythrocytes
K (phylloquinone)	Acts in synthesis of prothrombin	Liver; green, leafy vegetables; gut bacteria	May cause jaundice in high doses	Failure of blood coagulation, hemorrhage

Folacin

The vitamin **folacin** (fōl′ah-sin), or folic acid, and certain related compounds are slightly water soluble and are rapidly destroyed by heat. Folacins transfer single carbon groups (such as $-CH_3$, $-CH_2-$, and $-CHO$) in several metabolic processes, including the synthesis of purines, choline, the amino acid serine, and methionine tRNA.

Methionine tRNA is needed to start a polypeptide chain in protein synthesis. Because folacin participates in several processes associated with cell division, folacin deficiency interferes with such processes as red blood cell production, normal growth, and division of malignant cells. Discovering that aminopterin and methotrexate competitively inhibit folacin has led to their use as chemotherapeutic agents.

	Biochemical Actions	Dietary Sources	Effects of Excess	Effects of Deficiency
Minerals				
Calcium	Bone formation, muscle contraction, nerve impulse transmission, blood clotting	Eggs, fish, dairy products, and legumes	Renal damage	Tetany, softening of bones, hemorrhage
Phosphorus	Bone formation, buffers, phosphates in lipids and nucleotides	Dairy products, legumes, meats, and grains	Hypocalcemia	Probably loss of minerals in bones
Magnesium	Cofactor for enzymes, regulates nerve and muscle function	Green vegetables, meat, and milk	Respiratory depression if renal excretion depressed	Tetany
Sodium	Excitability of cells, maintenance of ionic and osmotic balance	Table salt and most foods	Hypertension and edema	Dehydration, renal failure, and cramps in muscles
Potassium	Excitability of membranes in nerve and muscle function	Most foods	Heart arrhythmias	Alteration in muscle contraction and ECG patterns
Sulfur	Part of certain amino acids and other important compounds	Foods containing proteins	None known	None known; deficiency may never have been observed
Chlorine	Osmotic and acid-base balance	Table salt and most foods	Edema	Cramps in muscles and alkalosis
Iron	Part of heme of hemoglobin and cytochromes	Liver, eggs, nuts, legumes, and raisins	Hemochromatosis	Lack of hemoglobin in erythrocytes, anemia
Copper	Acts in hemoglobin formation, mitochondrial function, and melanin synthesis	Liver and meats	Tachycardia, hypertension, and coma	Anemia
Cobalt	Acts in hemoglobin formation	Meats	Cardiomyopathy	Anemia
Iodine	Part of thyroxine	Fish, iodized salt	None	Cretinism, goiter
Manganese	Cofactor in enzymes of oxidative and mucopolysaccharide metabolism	Leafy vegetables and whole grains	Muscle weakness, nervous disturbances	Decrease in rate of cellular respiration
Zinc	Part of insulin and some enzymes in nucleic acid metabolism	Many foods	None known	Growth inhibition, testicular atrophy, skin lesions
Fluorine	Suppresses action of oral bacteria, component of bones and tooth enamel	Milk, dentrifices	Mottling of teeth	Dental caries
Molybdenum	Purine metabolism, oxidation of aldehydes	Most foods and water	None	Goutlike symptoms
Chromium	Mediate insulin effects at cell membranes	Unknown	None	Impaired glucose metabolism
Selenium	Part of glutathione peroxidase, prevents peroxide formation	Most foods and water	Toxic in animals	Muscular pain, cardiomyopathy

Cyanocobalamin

Vitamin B_{12}, or **cyanocobalamin** (si″an-o-ko-bal′am-in), is water soluble and heat stable. The most chemically complex vitamin, it contains pyrrole rings (similar to those in heme) and an atom of cobalt. Cyanocobalamin transfers hydrogen atoms or methyl groups in various reactions, including one that activates folacin. It is obtained mainly from foods derived from animals, so strict vegetarians can develop a cyanocobalamin deficiency. Intestinal microorganisms make the vitamin, but in too small quantities to meet human needs. For absorption, cyanocobalamin must be bound to a transport protein called intrinsic factor, without which pernicious anemia results (Chapter 23). Like folacin, cyanocobalamin is essential normal cell division.

Biotin

Sulfur-containing **biotin** (bi'o-tin) is heat stable and slightly water soluble. It helps to transfer carbon dioxide to make malonyl-CoA in fatty acid synthesis or oxaloacetate from pyruvate. Biotin is inactivated by avidin, an egg-white protein. Because biotin is present in many foods, a deficiency is unlikely unless the diet contains large quantities of raw egg white.

Choline, Inositol, and Lipoic Acid

The vitamin-like molecules *choline* (ko'lin), *inositol* (in-o'si-tol), and *lipoic* (lip-o'ik) *acid* serve as coenzymes in various reactions. They are not true vitamins because they can be made in human cells. For example, when folacin, cyanocobalamin, and methyl groups are available, cells make enough choline to support metabolism. If any of those substances is deficient, choline is likewise deficient. Choline is used to make the neurotransmitter acetylcholine and is a component of the lipid portion of cell membranes. Both inositol and choline form part of phosphoglycerides that are essential parts of cell membranes. Lipoic acid is a coenzyme in the reaction that converts pyruvate to acetyl-CoA. Deficiencies of any of these substances are unlikely because they are found in many foods.

Ascorbic Acid

Humans, other primates, and guinea pigs require **ascorbic** (ah-skor'bik) **acid,** or vitamin C, because they lack an enzyme for its synthesis. Most animals make this water-soluble, heat labile, carbohydrate-like molecule from glucose. In humans, ascorbic acid appears to be needed for adding hydroxyl ($-OH$) groups to certain amino acids and steroids. Because hydroxylated proline and lysine are required for collagen synthesis, an ascorbic acid deficiency leads to defective collagen in bone and other connective tissues. This causes slow wound healing and capillary fragility.

Vitamin C has several other functions. Being an antioxidant, it can counteract oxidation of important molecules within cells. This vitamin also promotes iron absorption. Finally, vitamin C is required for the synthesis of the hormone thyroxine. Factors that increase a person's need for vitamin C include stress, oral contraceptives, smoking, and diabetes.

Large doses of ascorbic acid have been prescribed to prevent colds, but without evidence for effectiveness or a proposed mechanism of action. Large excesses of ascorbic acid can cause kidney stones and mobilize calcium from bones. After taking large doses of vitamin C, some people suffer symptoms of vitamin C deficiency (scurvy) when they stop.

Fat-Soluble Vitamins

The fat soluble vitamins A, D, E, and K are found in the lipid part of foods. They are absorbed, transported in lymph and blood, and enter cells like other lipids.

Vitamin A

Several different **carotene** (kar'o-tēn) molecules have **vitamin A** activity when converted to retinol in the liver. Retinol is used to synthesize visual pigments, rhodopsin in rods and similar pigments in the cones (Chapter 15). Vitamin A also helps to maintain epithelium cells in skin and other organs and the structure and permeability of cell membranes. How the vitamin is involved in these processes is unclear. Smoking may increase a person's need for vitamin A. Deficiency of vitamin A leads to nightblindness.

Vitamin D

Steroids called **calciferols** (kal-sif'er-ols) have **vitamin D** activity. Most human vitamin D comes from activation of *cholecalciferol* (ko''le-kal-sif'er-ol), or vitamin D_3, by ultraviolet light striking the skin. *Ergocalciferol* (er''go-kal-sif'er-ol), or vitamin D_2, also is active in humans. Spending time in bright sunlight reduces the need for vitamin D in the diet. Vitamin D fosters calcium and phosphorus absorption and helps to deposit them in bone (Chapter 6). Fortifying milk with vitamin D helps to assure normal bone growth in children. Childhood vitamin D deficiency leads to rickets.

Vitamin E

Several **tocopherols** (to-kof'er-ols) have **vitamin E** activity. Being fat soluble antioxidants, vitamin E molecules function in the cell membrane to prevent oxidation of polyunsaturated fatty acids in the membrane itself and protect various substances inside the cell including vitamin A from oxidation. The trace element selenium and sulfur-containing amino acids appear to work with vitamin E in preventing oxidation. When vitamin E is deficient, unsaturated fatty acids can be oxidized to peroxides that damage cells and lyse red blood cells. Exposure to smog and a diet high in polyunsaturated fats can increase a person's need for vitamin E.

Vitamin K

Three molecules have **vitamin K** activity in humans. *Vitamin K_1* is found in green, leafy vegetables. *Vitamin K_2* is produced by intestinal microorganisms. And *menadione* (men''ad-i'ōn), or **vitamin K_3,** is made in the laboratory. The liver needs vitamin K to make prothrombin and other blood clotting factors. Vitamin K antagonists slows blood clotting in patients subject to thrombus formation (Chapter 17).

Minerals

Inorganic body components called **minerals** are found mainly in the skeleton and as electrolytes in intracellular and extracellular fluids. *Calcium* and *phosphorus* account for as much as 1 percent of the body weight. The National Institutes of Health have recently recommended that women of any age consume 1000 mg of calcium per day (1500 mg after menopause) to lower the risk of osteoporosis. Milk and milk products contain the highest amounts of calcium of all foods, and four to six servings of low-fat products are needed to meet these requirements. People who cannot tolerate milk products will find tofu (soybean curd) and lime-treated corn tortillas good sources of calcium. Consuming glucose with calcium increases calcium absorption. Smoking, large amounts of alcohol, more than twice the protein needed, or more than 35 grams of fiber can decrease calcium absorption. High caffeine intake—more than 5 cups of coffee—may increase calcium excretion.

Other minerals present in significant quantities are *sodium, potassium, magnesium, sulfur, chlorine,* and *iron.* Minerals present in very limited quantities, the so-called **trace elements,** include copper, cobalt, iodine, manganese, zinc, fluorine, molybdenum, chromium, and selenium. All these minerals are essential in the human diet (Table 26.5).

Certain minerals are worthy of further discussion. *Copper* has recently been reported to be 100 times as effective in lowering plasma cholesterol as clofibrate, a drug commonly used for that purpose. *Magnesium* activates more than 300 enzymes and a severe deficiency can be life threatening. Patients taking glycosides for heart disorders, some antibiotics, and some diuretics are especially susceptible to magnesium deficiency. Such a deficiency can be treated with a magnesium-containing antacid. *Selenium* (sĕ-le′ne-um) and vitamin E protect erythrocytes against hemolysis (rupture) as long as glucose is present. **Selenosis** (se″len-o′sis), caused by too much selenium, occurs in animals grazing on plants grown in high-selenium soil, but human selenosis is rare. This trace element also may help to prevent cancer, heart disease, and sudden infant death syndrome. Selenium and other trace elements may be needed in large amounts by people subjected to smog or large amounts of polyunsaturated fats and by the elderly.

Other trace elements—nickel, silicon, tin, and vanadium—may be needed for normal metabolism, but their functions are not very well understood. Finally, lead, mercury, barium, aluminum, boron, cadmium, and strontium have been identified in human ash. These elements, which probably accumulate as pollutants, have no known functions, and some have proven harmful. Heavy metals such as lead and mercury denature proteins and interfere with oxidative metabolism. Aluminum may contribute to the development of anemia and Alzheimer's disease (Chapters 12 and 30). Minerals as electrolytes are discussed in more detail in Chapter 28.

> Vitamins and minerals can interact as is shown by their actions in bone. Vitamin A facilitates synthesis of chondroitin sulfate, a substance that gives bone plasticity. Vitamin C helps to form cross-linkages between collagen molecules; cross-linkages add tensile strength to bone. Vitamin D, after activation by kidney and liver enzymes, travels in the blood like a hormone. In the intestinal mucosa, it induces synthesis of a calcium carrier protein that transports Ca^{2+} into mucosal cells. Though vitamin D is necessary to maintain adequate blood Ca^{2+} for bone deposition, an excess can destroy bone by stimulating osteoclasts to remove Ca^{2+} from bone. Vitamin D synthesis is stimulated by parathormone and inhibited by high blood Ca^{2+} concentration.

SPECIAL NUTRITIONAL PROBLEMS

Nutritional problems can occur from excesses or deficiencies of certain nutrients and by accidental or intentional ingestion of nonnutrient substances. Most such problems disturb metabolism or interfere with other physiological processes.

Malnutrition

Twenty million Americans, nearly 10 percent of the population, suffer from hunger, whereas 34 million are obese. Though malnutrition (bad nutrition) includes overeating, we have considered obesity in Chapter 25 and will limit this discussion to problems with insufficient nutrients to maintain normal metabolism. Two types of **malnutrition** are recognized, and both are common among children: (1) protein deficiency called **kwashiorkor** (kwash-e-or′kor) and (2) near-starvation called **marasmus** (mah-raz′mus). Protein deficiency may be far less important in malnutrition than was once thought. Though infants and young children need more protein per kilogram of body weight than adults, they do not need a high-protein diet. Less than 7 percent of calories in breast milk come from protein, and many cereals contain more protein than breast milk.

Diets where malnutrition is common usually are not protein deficient, but they may lack essential amino acids. Kwashiorkor, "the disease that comes when a second child is born," is such a deficiency. Recently weaned children obtain barely enough essential amino acids from the adult diet when they are well. If they get sick and eat less food, they can develop amino acid deficiency, which impairs protein synthesis.

Protein deficiency decreases serum proteins, which reduces blood osmotic pressure. Fluid fails to return to the blood, and edema and **ascites** (as-i′tēz), fluid accumulation in the abdominal cavity, develop. Ascites accounts for the potbelly of children suffering from kwashiorkor. Other proteins—blood clotting factors, enzymes to make pigments—are in short supply, and erythrocytes, skin cells, and other cells that normally divide rapidly become depleted. Blood is slow to clot, wounds are slow to heal, hair becomes gray, and anemia and skin lesions develop. If protein deficiency occurs in early childhood, protein synthesis and nerve cell division are impaired, causing slow growth and mental retardation.

Marasmus in children and adults is due to inadequate caloric intake. All essential nutrients—essential amino acids, linoleic acid, vitamins, and minerals—are likely to be deficient. Victims of marasmus suffer from all the symptoms of protein deficiency, especially muscle wasting, anemia, and susceptibility to microbial and parasitic infections. Infections further decrease the amount of nutrients available to body cells. Some people with nutritional deficiencies, especially iron or zinc deficiency, develop a curious condition called **pica** (pi′kah), a craving for nonfood substances. Women and children living in poverty are most susceptible to pica. They sometimes eat wall plaster, paste, or large quantities of ice, though these substances do not alleviate the deficiency.

Vitamin-Responsive Metabolic Disorders

Certain inherited metabolic disorders are caused by an absent or defective gene. Some of them respond to excessively large therapeutic doses of specific vitamins (Table 26.6). Vitamins might work to treat such disorders by stimulating synthesis of an enzyme at a greater rate than would occur without the vitamin or by stimulating synthesis of a coenzyme that increases the rate of the enzyme reaction. Excessively large doses of a vitamin might be therapeutic when they overcome impaired digestion, absorption, transport, or activation of a vitamin. Increasing the amount of the vitamin ingested increases the chances that an adequate amount of the vitamin will reach the cells.

Certain effects of alcoholism and cystic fibrosis are to some extent vitamin responsive. Increasing vitamin intake in alcoholics overcomes dietary deficiencies and allows greater vitamin activation by the damaged liver. Increasing vitamin intake in cystic fibrosis overcomes the deficiencies, especially of fat-soluble vitamins that are not released from foods because of pancreatic enzyme deficiencies.

Restricted diets can cause malnutrition and vitamin deficiencies at any age and any economic level. Vegetarianism leads to vitamin B_{12} deficiency. The so-called macrobiotic diet (consisting mainly of unpolished rice) can lead to protein deficiency and to various vitamin deficiencies. Liquid protein diets or diets limited to only a few foods also lead to deficiencies. Such deficiencies can be prevented by following dietary guidelines listed earlier.

Minerals and Hypertension

High sodium intake has been implicated in hypertension, and moderate potassium offers some protection against it. Excessive potassium, however, can dangerously alter heart contractility, and calcium may have a more significant protective effect. People with normal blood pressure can consume up to 6 grams of salt with no effect. Many physicians recommend that hypertensive patients consume no more than 2 g of salt per day—about the lowest possible sodium intake on a nonhospital diet. They believe that low-sodium intake reduces the amount of medication needed to control blood pressure.

Before antihypertensive drugs became available, a diet of rice, fruit, sugar, and vitamins (the Kemper diet) was sometimes used to treat hypertension. Many patients on the diet showed a drop in blood pressure. This diet *is* low in sodium, but also high in potassium and low in calories. Eaten in moderation, it led to weight reduction, which also tends to lower blood pressure.

Comparisons of eating habits and the incidence of hypertension in different countries and cultures show that hypertension is a complex process involving many factors. Most people in nonindustrialized areas eat a low-sodium, high-potassium diet consisting mainly of natural nonprocessed foods, and they have a low incidence of hypertension. Such people tend to maintain physical fitness (by working hard to make a living) and to lose rather than to gain weight with age.

Calcium might protect against hypertension. In one study, among adults consuming 450 to 600 mg calcium/day, 12 percent had hypertension, whereas among those consuming 1200 to 1400 mg calcium/day only 6 percent had hypertension. Given the many cases of hypertension (and osteoporosis in older women) and the low cost of calcium supplements, perhaps a supplement is warranted for adults who fail to consume more than 1 g calcium/day.

Severe hypertension is related to the ability of blood vessels to constrict and their inability to dilate. Dietary fish oils rich in omega-3 fatty acids have improved blood vessel dilation in patients with hypertension.

TABLE 26.6 Vitamin-Responsive Inherited Metabolic Disorders

Vitamin	Biochemical Defect	Disorder	Manner of Inheritance*	Therapeutic Dose/Day
B₁ (thiamine	Pyruvate decarboxylase deficiency	Pyruvicacidemia	Unknown	5–20 mg
B₆ (pyridoxine)	Glutamate decarboxylase deficiency	Infantile convulsions	Autosomal recessive	10–50mg
	Unknown	Hypochromic anemia	Sex-linked recessive	>10mg
B₁₂ (cyanocobalamin)	Intrinsic factor deficiency, inactive intrinsic factor, transport deficiency	Megaloblastic anemia	All autosomal recessives	5–100 μg
Folic acid	Impaired intestinal folate absorption	Megaloblastic anemia	Unknown	<0.05 mg
	A folate reductase enzyme deficiency	Mental retardation, schizophrenic psychosis	Autosomal recessive	>10 mg
Biotin	Propionyl-CoA carboxylase deficiency	Propionicacidemia ketoacidosis and retardation	Autosomal recessive	10 mg
Niacin	Impaired intestinal and renal transport of tryptophan	Hartnup disease, cerebellar ataxia	Autosomal recessive	>40 mg
D-calciferol	Unknown	Hypophosphatemic rickets	Six-linked dominant	>100,000 units (>4 g chole-calciferol)
	Deficiency of an enzyme in calciferol synthesis	Vitamin D-dependent rickets	Autosomal recessive	>25,000 units (>1 g chole-calciferol)

*See Chapter 30 for an explanation of inheritance.

Fats, Fiber, and Circulatory Disease

Even before USDA and other guidelines became available, researchers had related coronary artery disease to lipoproteins. LDLs contain large amounts of cholesterol that can be deposited in arteries, whereas HDLs carry cholesterol back to the liver and seem to protect against blood vessel diseases (Chapter 18). It now appears that dietary saturated fatty acids contribute more to elevated blood cholesterol than dietary cholesterol itself.

Many researchers are studying blood lipids and their roles in human disease. In one study, monkeys were fed a diet containing 4.5 times as much polyunsaturated fat (PUFA) as saturated fat for three to seven years. The monkeys showed a significant decrease in both HDLs and LDLs, possibly because the unsaturated fats activated enzymes that catabolize lipoproteins.

More recent studies have shown that oat bran, legumes, and other foods containing water-soluble fiber significantly reduce blood cholesterol. Such fiber binds to bile acids in the small intestine. This decreases emulsification and absorption of fats and causes bile salts—and, therefore, cholesterol, to be excreted. As colon bacteria digest soluble fiber, they release short chain fatty acids, which may be absorbed and act in the liver to decrease cholesterol synthesis.

A study comparing nibbling (17 snacks/day) with the same food as eaten in three meals, showed that the nibblers had lower total and LDL cholesterol and lower serum insulin. Insulin stimulates lipid synthesis and proliferation of smooth muscle in blood vessels. Having many small meals minimizes insulin secretion, and the lower insulin reduces the risk of blood vessel disease from both lipid synthesis and smooth muscle proliferation.

The National Cholesterol Education Program makes many of the same dietary recommendations as other guidelines—total fat should be less than 30 percent and saturated fat should be less than 10 percent of calories, and cholesterol should be less than 300 mg/day. If this Step One Diet fails to lower cholesterol, the Step Two Diet, which reduces saturated fat to less than 7 percent of calories and cholesterol to less than 200 mg/day is recommended.

Chitosan, a fiber found in mushrooms, becomes positively charged in the stomach. It attracts and entraps charged micelles in the small intestine. It is so effective in lowering blood cholesterol, that it may soon be licensed for use as a substitute for the drug cholestyramine, which has undesirable side effects.

Nonnutrient Substances in Food

In addition to major nutrients, some commonly used foods and beverages contain nonnutritive substances, such as caffeine, food additives, and pollutants. "Junk foods" and alcohol satisfy the appetite, but have little nutritional value.

Caffeine and Other Xanthines

A group of purines called **xanthines** (zan'thēnz) are ingested as *caffeine* in coffee, tea, and some carbonated beverages, *theophylline* (the-of'ĭ-lin) in tea, and *theobromine* (the"o-bro'min) in cocoa. Xanthines inhibit the breakdown of cyclic-AMP, so any metabolic processes stimulated by cAMP are accelerated and prolonged. Glycogenesis increases, elevating the blood glucose concentration. Lipolysis increases, elevating the free fatty acid concentration in blood. Xanthines increase sympathetic stimulation, which increases insulin secretion, thereby reducing glycogenolysis. Xanthines relax smooth muscle in bronchioles and pulmonary arterioles, increasing both airflow and blood flow. They stimulate myocardial contractility and relax arterial smooth muscle, increasing cardiac output.

Caffeine has both good and bad attributes. The good attributes appear when a person drinks 2 or 3 cups of coffee per day, and more is not better. Caffeine elevates mood, probably by binding to the same brain receptors as adenosine. Adenosine tells the brain to "slow down," and caffeine overrides this effect. Caffeine also improves mental performance, counteracts asthma symptoms, and accelerates calorie use by as much as 10 percent, especially in obese people. The bad attributes of caffeine are that it reduces fertility, may contribute to a heart attack by increasing platelet stickiness, disrupts sleep, and is addictive.

Junk Foods

Candy, carbonated beverages, potato chips, and any other snack foods that contain mainly refined sugar or fat and few vitamins and minerals are commonly known as **junk foods.** Eaten in excess they lead to nutritional deficiencies and undesirable weight gain. Foods and beverages sweetened with saccharine or aspartame pose other hazards. Saccharine has been implicated as a carcinogen in animal studies, and aspartame has been reported to interfere with neurotransmitter actions in the brain. In a study on a small number of people, those receiving aspartame sweetened foods lost more weight than those receiving unsweetened foods. In another small study, people receiving sucrose or aspartame-sweetened foods reported the same degree of satiety regardless of which sweetener was used.

Food Additives

At least 2600 different chemical substances are added to processed foods sold in the United States today. A food additive is something intentionally added to food, usually to improve its appearance, flavor, nutritional value, or the length of time it can remain edible on a grocery shelf. Nutritional supplements can enrich or fortify foods. **Enrichment** is the addition of nutrients to foods, such as breads and cereals, to return them to the same nutritional value that they had before processing. **Fortification** is the addition of nutrients to foods, such as milk and other dairy products, that cause them to have greater than normal nutritional value. Food additives and their properties are summarized in Table 26.7.

Food Pollutants

Contaminants that accidentally enter foods—pesticide residues, heavy metals, microorganisms and their toxins, and insects—are common food pollutants. Pesticides like DDT (dichlorodiphenyl-tricholoroethane) and PCBs (polychlorinated biphenyls) have been used to kill crop-damaging insects. Another pesticide, EDB (ethylenedibromide), used to control insects in grain stores, has been found in grain products. Years after their use has ceased, their residues remain in the environment because living things lack enzymes to degrade them. When animals or humans eat crops treated with pesticides or grown in soil containing pesticides, they ingest pesticide residues too.

Pesticides undergo **biological magnification**—they increase in concentration as they pass along the food chain. As animals eat fat-soluble pesticides on contaminated plants, fatty tissues accumulate pesticide residues. Humans ingest large amounts of pesticide residues from eating animals that have eaten pesticide treated plants. Pesticides can cause neurological disorders, liver damage, and cancer. The hazard of pesticides can be reduced by removing fats from meats and by eating only low-fat dairy products.

Dioxins, a group of chlorinated organic compounds, are being found in small but significant amounts in paper products. These substances, which form during chlorine bleaching of wood pulp in paper manufacturing, are carcinogens. They tend to migrate from paper containers such as milk cartons and meat trays into foods or from coffee filters into coffee. When ingested, they accumulate in tissues. How much dioxin is needed to cause cells to become malignant is not known.

TABLE 26.7   Food Additives

Type	Function	Examples	Typical Uses
Preservatives	To retard spoilage by preventing action of microorganisms	Heating and sealing, dehydration, freezing, pasteurization, salt, sugar, sodium nitrate and nitrite, sodium and calcium propionates, sodium benzoate, sulfur dioxide	Breads, cheeses, meats, jellies, syrups, fruits, and vegetables
Antioxidants	To retard spoilage by preventing oxidation	Lecithin, butylated hydroxyanisole (BHA), butylated hydroxytoluene (BHT)	Oils, cereals, potato chips, and other snack foods
Flavorings and flavor enhancers	To add or enhance flavor	Herbs, spices, synthetic flavorings, monosodium glutamate (over 1100 substances)	Nearly all processed foods
Dyes	To add color appeal or mask undesirable colors	Natural dyes or synthetic coal tar dyes	Soft drinks, cereals, cheeses, ice cream, nearly all other processed foods
Acids and alkalies	To add or mask tartness; to mask undesirable flavors	Phosphoric and citric acids, sodium bicarbonate	Soft drinks, juices, wines, olives, salad dressings
Emulsifiers and stabilizers	To prevent separation of components and provide smooth texture	Lecithin, propylene glycol, acylglycerols, polysorbates, vegetable gums, cellulose, gelatin, seaweed extracts	Soft drinks, ice cream, salad dressings, cheese spreads, cake mixes, margarine, candies
Chelating agents	To tie up trace metals that might cause oxidation, to prevent clouding, and to improve texture	Ethylenediaminetetraacetic acid, sodium phosphate, chlorophyll	Artificial fruit drinks, salad dressings, soft drinks, beer, cheeses, canned and frozen foods

Data from G. T. Miller, Jr., *Living in the Environment.*

Heavy metals such as mercury also are found as food contaminants, especially in ocean fish and sometimes in grains. Methylmercury, a particularly toxic metal, accumulates in the tissues of swordfish, tuna, and other ocean fishes. This mercury comes mainly from natural ocean sediments, and not from human activities. Methylmercury also is used to prevent fungal growth in grain. Ordinarily applied to seed grain, it is ingested only when humans eat the grain instead of planting it. Mercury poisoning causes severe neurological damage and even death, if sufficient quantities are ingested.

Cadmium, another heavy metal, has recently been recognized as a health hazard. Cadmium usually enters the body from water contaminated with industrial wastes. Large amounts can cause disintegration of the skeleton, but small amounts—0.1 part per million—contribute to heart disease and hypertension.

Finally, many food products are contaminated with microorganisms and insect body parts—the very things additives are intended to control. The microorganisms pose a threat to health only if they are capable of causing human infections and are virulent enough to do so, or if they have produced toxins that affect humans. Bacteria that cause food poisoning and viruses that cause hepatitis are examples. *Gonyaulaux,* an alga responsible for "red tides," produces *saxitoxin* (sak″sit-ok′sin), which blocks sodium channels and causes severe neurologic damage. Insect parts, though aesthetically unattractive, may pose a less serious health hazard than most additives used to control them.

Health food stores frequently tout a particular dietary supplement with little or no scientific evidence of its nutritional value. Such practices can be harmful, as in the case of L-tryptophan. Though L-tryptophan is an amino acid found in all proteins and a naturally occurring substance, used as a nutritional supplement it causes a rare blood disease called **eosinophilia.** Symptoms of this disease are aches, fever, painful swollen joints, and sometimes rashes.

The popularity of raw seafood bars has created a new kind of "food additive" problem. One woman was rushed to surgery suffering from symptoms resembling appendicitis. Her surgeon was ready to remove the appendix when a red worm 1.5 in long crawled onto the surgical drape.

Diet and Cancer

How might diet cause or prevent cancer? Researchers interested in that question have identified substances that may cause cancer and others that help to prevent it.

Numerous substances in plants—safrole in root beer, piperine in black pepper, hydrazines in mushrooms, solanine in potatoes, quinones in rhubarb, and theobromine in cocoa—have been reported to cause cancer in laboratory animals. Other carcinogens include alcohol, aflatoxin from molds sometimes found in peanut butter, and nitrosamines formed in the body from natural nitrates and nitrites and those added to foods such as bacon and hot dogs.

Some carcinogens act by forming highly reactive oxygen radicals that damage DNA. These radicals are counteracted by antioxidants such as vitamins A, C, and E and selenium in foods. Some researchers believe 2500 μg RE (retinol equivalents) of vitamin A, 1 g of vitamin C, and 200 IU (international units) of vitamin E daily may reduce the risk of cancer, but they warn that excesses of these nutrients have severe toxic effects.

Some cancers may be related to a kind of biological time warp—a conflict between our bodies that function as they have for centuries and our twentieth century diet and sedentary life-style. To improve the situation, the National Research Council recommended the following dietary changes to reduce the risk of cancer: limiting dietary fats to 30 percent of calories; reducing pickled, salt-cured, and smoked foods; increasing fruits and vegetables; and limiting alcohol. Vegetables from the cabbage family—broccoli, brussels sprouts, cauliflower—and high fiber foods are recommended to reduce the risk of cancer.

High-fat diets are correlated with gastrointestinal, breast, and prostate cancers. Fats should be reduced to minimize the cancer risk from excess bile acids and steroids, which are linked to malignant tumors in laboratory animals. Pickled, salt-cured, and smoked foods should be limited because they can become contaminated during processing with chemicals known to cause cancer in animals. Fruits and vegetables, which contain vitamins A and C, seem to protect against cancer. Vitamin A seems to protect epithelial cells, especially in the lungs. Vitamin C may protect stomach cells by blocking the formation of nitrosamine, a known carcinogen. Alcohol should be limited because of the high incidence of mouth, laryngeal, and esophageal cancer among users, especially those who smoke. Cabbage and similar vegetables probably contain substances that enhance the activity of enzymes that destroy carcinogens. Insoluble fiber decreases transit time of feces and neutralizes soluble fiber-bile acid combinations.

Diet and Behavior

A controversial theory about melatonin, serotonin, and the pineal gland relates three cyclic behaviors, all of which display degrees of carbohydrate craving, weight gain, depression, and excessive sleeping. **Carbohydrate craving obesity** (CCO) is related to daily rhythms where carbohydrate craving occurs especially in the late afternoon and evening. **Premenstrual syndrome** (PMS) is related to monthly rhythms, and **seasonal affective disorder** (SAD) is related to seasonal changes in yearly rhythms. SAD may affect as much as one-fourth of the population in northern latitudes (northern United States and Canada), but is rare near the equator.

As we saw in Interlude 3, the pineal gland secretes melatonin at night and serotonin in daylight when signals elicited by light inactivate the enzyme that converts serotonin to melatonin. The theory that relates carbohydrate craving to serotonin is as follows: When carbohydrate is eaten, insulin secretion increases, causing most amino acids to move into cells. This leaves a high blood concentration of the amino acid tryptophan, which is not affected by insulin. Much of the tryptophan crosses the blood-brain barrier and is used to make the neurotransmitter serotonin. When serotonin is released at brain synapses, it is believed to induce both sleep and satiety with respect to carbohydrates. Some researchers report quite different responses to carbohydrate meals: Normal people with no carbohydrate craving get sleepy, but victims of CCO, being deficient in serotonin, experience carbohydrate satiety and a sense of well-being without becoming sleepy.

CLINICAL APPLICATIONS: THERAPEUTIC DIETS

Current knowledge of nutrition allows the planning and administration of many special diets to treat a variety of physiological disturbances. Some examples of such diets, called therapeutic diets, are summarized in Table 26.8.

A safe and effective weight loss diet (Table 26.9) must supply all daily nutritional requirements and establish long-term changes in eating habits and physical activity. Such a diet would contain the recommended servings from each of the basic four food groups, with choices limited to low-calorie foods. Weight-loss programs such as Time-Calorie Displacement and Weight Watchers meet the above criteria. The Time-Calorie Displacement program assigns small numbers to foods with few calories and high bulk that must be eaten slowly and large numbers to foods with many calories and little bulk that can be eaten rapidly. On this scale, vegetables are 10; fruits, 15; breads and cereals, 50; meats, 75; sweets, 150; and fats, 225. Weight Watchers defines six food groups and requires two to three servings of fruit; at least three servings of vegetables; 3 teaspoons of fat; five to seven ounces of lean fish, poultry, or meat; two to four servings of breads and cereals; and two to three servings of low-fat milk or milk products.

TABLE 26.8 Therapeutic Diets

Kind of Diet	Uses
Low-calorie	Weight reduction, hypertension and other cardiovascular diseases, hypothyroidism
Low-carbohydrate	Minimize hyperglycemia in diabetes, weight control
High-protein and low-carbohydrate	Prevent excessive changes in blood glucose concentration in hypoglycemia
Low-fat	Liver, gallbladder, and pancreatic diseases in which fat digestion and absorption is impaired
Low-fat and cholesterol	Decrease blood cholesterol concentration
Low-protein	Renal and liver diseases in which ammonia accumulates in blood; gout in which uric acid accumulates in blood
Gluten-free diet	Celiac disease
Restricted phenylalanine	Prevent mental retardation in phenylketonuria
Restricted purine	Lower blood uric acid in gout
High-protein	Overcome protein deficiency, provide protein for tissue repair following injury or surgery, kidney disorders in which protein is excreted, various wasting diseases
Lactose-free diet	Prevent discomfort in lactose intolerance
High-sodium	Addison's disease
Low-sodium	Congestive heart failure, renal diseases, and other disorders that lead to edema and ascites; toxemia of pregnancy
High-calcium and phosphorus	Rickets, osteomalacia, dental caries, lead poisoning
High-iron	Nutritional and hemorrhagic anemias
Soft diet	Lack of teeth, inflammation or surgery affecting the mouth
Tube-feeding	Anorexia nervosa, injuries or surgery affecting the mouth, esophagus, or stomach
Bland diet	Inflammation of gut
Restricted residue	Inflammation of gut
High-residue/fiber	Constipation due to impaired intestinal motility

TABLE 26.9 A Sample Low-Calorie Diet

Food Group	Servings	Food Choices and Amounts per Serving
Milk	2–3	8 oz skim or 1% fat milk 4 oz low-fat, plain yogurt 6 oz buttermilk 4 oz low-calorie milk pudding
Meats	2–3	3 oz lean beef, pork, lamb, or veal 3 oz chicken or turkey (skin removed) 4 oz fish or shellfish 1 egg 8 oz tofu ⅓ c cottage cheese
Fruits	1 citrus;	½ c citrus juice, 1 orange, or ½ grapefruit small apple, peach, ½ banana
	1 other	½ c melon or berries
Vegetables	Unlimited; at least 3	Green, leafy vegetables; green beans; and any other vegetables No more than ½ c daily of peas, other beans, corn, or beets
Bread and cereal	2–4	1 slice any type bread 2 slices reduced-calorie bread ½ c cooked cereal ¾ c ready-to-eat cereal ½ c any pasta, potato, or rice

Do all cooking without added fat or sugar.

Eggs, cheeses (except cottage cheese), and beef, pork, and lamb are high in cholesterol. Have only one of these foods per day.

Eat some vegetables raw.

Use up to 4 tablespoons of low-calorie salad dressings, 4 teaspoons of low-calorie margarine, or 2 of each per day.

Avoid presweetened cereals.

Use artificial sweeteners and artificially sweetened gelatin, sodas, coffee, and tea in moderation—a total of 8 servings per day from all categories.

Drink at least 6 8-oz glasses of water per day.

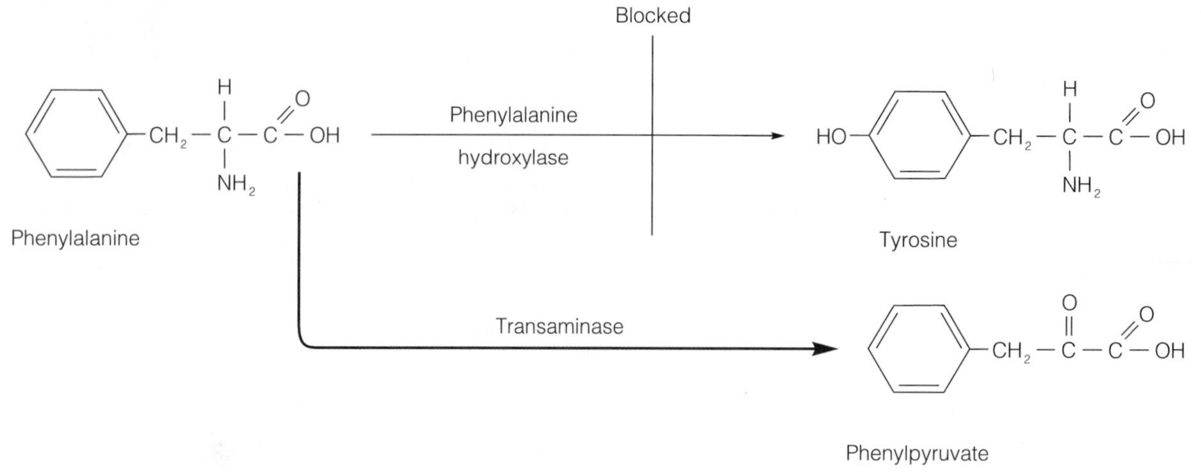

FIGURE 26.3
Disordered metabolism of phenylalanine in PKU.

Other methods of weight loss include drugs, surgery, and very low-calorie, low-carbohydrate, and other fad diets. Drugs usually are not helpful. Most are amphetamine derivatives that only temporarily suppress appetite and can be addictive. Surgery is reserved for the severely obese who are 100 pounds or 100 percent above ideal weight with serious medical conditions. It is painful, expensive, not without risk, and not always successful. Very low-calorie diets—Last Chance, Liquid Protein, and Cambridge Diets—allow only 300 to 600 calories per day and are deficient in many nutrients. Though these diets are intended to be protein-sparing, they can lead to sudden heart attacks, arrhythmias, and depression. Low-carbohydrate diets—Air Force, Calories Don't Count, Drinking Man's, and Dr. Stillman's Diets—cause the body to metabolize mainly fats and allow ketones to accumulate in the blood and urine. Quick initial weight loss is primarily fluid loss by diuresis. Such a diet increases blood LDLs and can cause nausea, fatigue, and possibly ketoacidosis. Fad diets that emphasize specific foods—Rice and Kelp, B$_6$—are dangerous because they lack many nutrients.

The disease **phenylketonuria** (fen''il-ke''to-nu're-ah), or PKU, caused by a single missing enzyme, is treated with a therapeutic diet. Normally the amino acid phenylalanine is metabolized to the amino acid tyrosine; but in PKU, this reaction fails to occur because of a missing enzyme. As a result, some phenylalanine is converted to a ketone called phenylpyruvate and excreted in the urine (Figure 26.3)—hence the name. PKU occurs most often among Caucasians, where 1 in 50 carries a gene that fails to make the needed enzyme and 1 in 10,000 infants are affected. To have this disease, a child must receive a copy of the gene from each parent (Chapter 30). Untreated infants become mentally retarded, probably by phenylalanine accumulating in the blood, entering brain tissue, and preventing synthesis of serotonin or other molecules.

PKU is treated by limiting dietary phenylalanine, which prevents or minimizes brain damage if begun in the first few weeks of life. Infants with the disease are fed a special formula that contains only enough phenylalanine to allow normal protein synthesis. As solid food is added to the diet, proteins must be severely limited because they contain phenylalanine. Most physicians recommend that the low-phenylalanine diet be continued until brain development is virtually complete.

The artificial sweetener aspartame (Nutrasweet) contains phenylalanine, and products containing it have a warning to PKU patients, especially pregnant women. Though a woman usually is not adversely affected, a fetus can be damaged unless the mother follows a low-phenylalanine diet throughout pregnancy.

CLINICAL TERMS

health food   a term that usually means a natural food or an organic food, but that misleadingly implies an unusual health-giving power

natural food   a food grown without fertilizers or pesticides and one that has not undergone any special processing

nutrient density   the quantities of vitamins, minerals, and other essential nutrients relative to the number of calories a food supplies

organic food   a term sometimes used to mean a natural food, but which misleads because all foods are organic (carbon-containing) substances

primary deficiency   a deficiency caused by inadequate intake of a nutrient

secondary deficiency   a deficiency resulting from impaired absorption or excessive excretion of a nutrient

subclinical deficiency   a deficiency not yet apparent

GETTING PROTEINS FROM PLANTS

The American idea that burgers and other red meats are dietary essentials is not substantiated by nutritional studies. Protein is essential in the human diet, but it need not come from meat. What humans really need is an adequate supply of essential amino acids and enough other amino acids to allow protein synthesis. This need can be met by plant proteins if they are eaten in the right combinations at the same meal.

Most plants contain **incomplete proteins**; that is, they have too little of some amino acids—lysine, isoleucine, tryptophan, and the sulfur amino acids, cysteine and methionine—to meet human needs. Various plants lack different amino acids and dairy products easily supplement these deficiencies (Table 26.10). By learning which plant foods supply certain essential amino acids and where dairy products are needed, complementary combinations, such as beans and rice or cereal and milk, that supply all essential amino acids can be made (Table 26.11 and Figure 26.4).

TABLE 26.10	Availability of Selected Essential Amino Acids in Plant Foods			
Food	**Tryptophan**	**Isoleucine**	**Lysine**	**Sulfur-Containing Amino Acids**
Soybeans	+ + +	+ +	+ + +	+
Peas, beans, and other legumes	+	+ +	+ + +	+
Nuts and seeds	+ + +	+	+ + +	+
Grains and whole-grain foods	+ +	+	+	+ +
Corn	0	+	+	+ +
Green vegetables	+ +	+	+ +	+
Dairy products	+ +	+ + +	+ + +	+ +
Eggs	+ + +	+ + +	+ + +	+ + +
Seafoods	+ +	+ +	+ + +	+ +
Meat and poultry	+ +	+ +	+ + +	+ +

+ + + = excellent source, + + = good source, + = poor source, 0 = very poor source

TABLE 26.11	Some Sample Vegetable Protein Combinations and Some Vegetable-Dairy Protein Combinations and Their Protein Content
Combination	**Protein Content**
1 oz cheese and 4 slices whole-grain bread or 1½ c macaroni	15 g
1 c beans and 2 c milk or ⅔ c grated cheese	15 g
⅓ c beans and ½ c sesame seeds	15 g
1 potato and ⅓ c cheese or 1 c milk	12 g
1 c rice and ¾ c soybeans	32 g
1 c rice and ⅓ c sesame seeds	18 g
1 c rice and 1⅓ c milk	30 g
½ c peanut butter and ¾ c milk	36 g

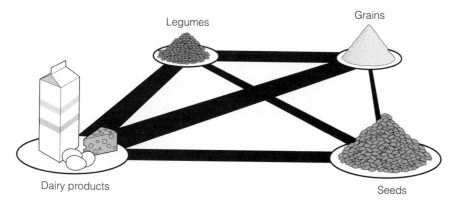

FIGURE 26.4

Summary of complementary protein relationships. Thickness of bars indicates relative value of protein combination.

Because human tissues are similar to tissues of other animals, meat and other animal products provide amino acids in proportions near those needed by humans. Even so, our bodies can use only about 90 percent of the amino acids in an egg, 80 percent of those in milk, and 70 percent of those in meat, poultry, and fish. These percentages, called the **net protein utilization,** represent the proportion of amino acids that can be converted to human proteins. Net protein utilization is directly proportional to biological value: Higher biological value means more amino acids will be used in protein synthesis. Plant foods generally have lower net protein utilization—from 70 percent for rice to 40 percent for kidney beans. To meet protein needs, larger servings of plant than animal foods must be eaten. For example, 4 oz of cooked hamburger provides 30 g of protein, 67 percent or 20 g of which is usable by the body. To obtain 20 g of usable plant protein would require a combination of 8 oz of rice and 6 oz of soybeans with 32 g of protein, 61 percent usable.

Eating less meat has nutritional, ecological, and economic advantages. Nutritionally, increased plant food intake greatly increases vitamins, minerals, and fiber in the diet. Consuming plant proteins also minimizes the risk of biologically magnified pesticide residues in meats. Because plants lack cholesterol and their fats are mostly unsaturated, eating them lowers cholesterol and saturated fat consumption. Finally, eating raw, unprocessed plant foods avoids some hazards of food additives and pollutants.

Ecologically, about 20 lbs of plant proteins are needed to make 1 pound of edible beef protein. Raising animals for human food requires a large grazing area. That same land, if suitable for growing grain, could feed many more people than the animals would feed. Grazing cattle often are grain-fed for a few weeks before they are slaughtered—a practice that uses still more plant materials and puts more saturated fats in the meat.

Economically, meat costs more than rice, beans, and other grains. Even allowing for the larger quantities that must be consumed to provide adequate protein, the cost of plant protein for a meal is much less than that of animal protein.

A strictly vegetarian diet is not recommended because plant foods are deficient in vitamin B_{12} and the amino acids already noted. Small amounts of dairy products supply vitamin B_{12} and increase the biological value of proteins. Meat is not essential to a nutritious human diet.

Questions

1. Which amino acids are most likely to be deficient in plant proteins?
2. How can foods from plants be combined to provide proteins of high biological value?
3. What are some advantages of limiting one's intake of meat?
4. Why is a strictly vegetarian diet not recommended?

CHAPTER SUMMARY

Principles of Nutrition

- Nutrition deals with providing carbohydrates, proteins, fats, vitamins, minerals, and water needed for health.

General Nutritional Requirements

- Carbohydrates meet energy and fiber needs.
- Proteins provide essential and other amino acids for maintenance, growth, and repair of human tissues.
- Fats allow energy storage, and linoleic acid is needed to synthesize prostaglandins and cell membrane components.

- Vitamins are required for the synthesis of certain coenzymes, collagen, and visual pigments. Minerals serve as cofactors for enzymes, in bone formation and membrane excitability, and in heme and thyroid hormone synthesis.
- Water maintains body fluid volumes and participates in or provides a medium for chemical reactions.

Deciding What to Eat

- Dietary guidelines are provided by the basic four food groups (Table 26.3), recommended dietary allowances (Table 26.4), and dietary goals (Figure 26.2).

What Vitamins and Minerals Do in Cells

- The actions, sources, and effects of excesses and deficiencies of vitamins and minerals are summarized in Table 26.5.

Special Nutritional Problems

- The importance of good nutrition is emphasized by considering the effects of nutritional problems. Malnutrition includes kwashiorkor, a protein deficiency, and marasmus, a protein-calorie deficiency. Some genetically determined metabolic disorders respond to large doses of specific vitamins.
- Sodium can worsen hypertension, whereas potassium and calcium may help prevent it. Cholesterol and saturated fatty acids contribute to circulatory disorders.
- Foods commonly consumed by humans contain a variety of nonnutrient substances. Xanthines accelerate glycogenolysis and lipolysis, and stimulate the nervous system. Junk foods have little nutritional value. Food additives and food pollutants can contribute to various disorders.
- Diet can play a role in causing or preventing cancer. Oxidizing radicals and other carcinogens in foods can be counteracted by antioxidants such as vitamins C and E. Reducing fat, salt-cured and smoked foods, and alcohol, and increasing fiber, vegetables (especially those of the cabbage family), and fruits have been recommended to reduce cancer risk.
- Diet may affect cyclic behaviors possibly by affecting melatonin and serotonin secretion by the pineal gland.

Clinical Applications: Therapeutic Diets

- Current knowledge of nutrition can be applied to designing therapeutic diets (Table 26.8). Weight-loss diets should provide all daily nutritional requirements and cause long-term changes in eating habits.

QUESTIONS AND PROBLEMS

The questions at the end of each chapter are numbered to correspond with the objectives listed at the beginning of the chapter. Italics indicate that a question requires critical thinking skills beyond simple factual recall.

Questions

1. (a) Define nutrition, and name six kinds of nutrients.
   (b) *What are the principles of good nutrition?*
2. (a) What nutritional needs do dietary carbohydrates meet?
   (b) Why are essential amino acids needed?
   (c) What nutritional needs do lipids meet, and what are the functions of linoleic acid?
3. (a) Why, in general, are vitamins and minerals required in the human diet?
   (b) What functions does water serve?
4. (a) What factors determine a person's energy requirements?
   (b) *How is the specific dynamic action of food related to energy requirements?*
5. *What are the advantages and disadvantages of guidelines for planning a nutritious diet?*
6. *What vitamin and mineral deficiencies might lead to (a) neurological disorders, (b) skin disorders, (c) reduced wound healing, (d) anemia, (e) impaired blood clotting, (f) functional abnormalities in membranes, and (g) impaired vision? Relate disorders to biochemical actions of vitamins or minerals.*
7. (a) What are the main effects of malnutrition?
   (b) *How do kwashiorkor and marasmus differ?*
   (c) What are the main attributes of vitamin-responsive metabolic disorders?
8. *How might diet contribute to or prevent (a) hypertension and (b) atherosclerosis?*
9. *What nutritional problems are associated with (a) xanthines, (b) pesticides, (c) food additives, and (d) food pollutants?*
10. *In what ways might diet relate to cancer and to behavior?*
11. (a) *How might a weight-loss diet be a therapeutic diet?*
    (b) *What other conditions can be treated with diets?*

Problems

1. Record everything you eat for a day. Compare your diet with the recommendations from the basic four food groups, and plan a menu for a day that corrects any deficiencies in your diet.
2. Compare your diet with dietary goals, and plan a menu that corrects any deficiencies in your diet. (You will need a computer program or reference book to determine the amounts of protein, carbohydrate, saturated and unsaturated fat in different foods to do this.)
3. What excesses or deficiencies might develop on these diets: (a) milk-free, (b) lacking animal protein, (c) lacking whole-grain cereals, (d) lacking citrus fruits, (e) consisting mainly of carrots and other yellow vegetables, and (f) junk food?
4. Do some library research and prepare a report on: (a) hazards of food additives or pollutants, (b) how diet might cause or prevent cancer, or (c) how dietary substances affect behavior.
5. Apply your knowledge of nutrition and obesity to plan a diet and exercise program that an obese person might be likely to follow. (Make reasonable assumptions about the preferences and habits of the person.)
6. Plan a menu for a week that makes use of vegetable proteins and meets the requirements for the basic four food groups.

OUTLINE

OBJECTIVES

27

Organization and General Functions

1. Describe the plan and major functions of the urinary system.

Development

2. Briefly describe the development of the urinary system.

Anatomy of the Kidney
  Gross Anatomy
  Blood Circulation
  Nephrons

3. Locate and identify the anatomical structures of the kidneys, describe the circulation of blood through the kidneys, and distinguish between cortical and juxtamedullary nephrons.

URINARY SYSTEM

Glomerular Filtration
  Filtration Pressures
  Filtration Rate

4. Explain where and how filtration occurs in a nephron, and summarize its effects.

Tubular Reabsorption
  Substances Reabsorbed
  Active Transport in Reabsorption
  The Concept of Renal Threshold

5. Explain where and how reabsorption occurs in a nephron, and summarize its effects.

Tubular Secretion

6. Explain where and how secretion occurs in a nephron, and summarize its effects.

The Countercurrent Mechanism
  Countercurrent Multipliers
  Countercurrent Exchangers

7. Describe the actions and significance of countercurrent multipliers and countercurrent exchangers.

Regulation of Urine Formation
  Excretion of Urea and Other Wastes
  Regulation of Fluid Volume Excreted

8. Explain how urea is excreted, and list other wastes.
9. Describe how the volume of excreted fluid is regulated.

How Kidneys Regulate Blood Flow
  Juxtaglomerular Apparatus
  Renin-Angiotensin Mechanism
  Local Regulation
  Atrial Peptides

10. Explain how the juxtaglomerular apparatus, the renin-angiotensin mechanism, local regulation, and atrial peptides contribute to regulating blood flow.

Measurement of Kidney Function

11. Describe how kidney function can be assessed.

Anatomy of Ureters, Bladder, and Urethra

12. Describe the anatomy of the ureters, bladder, and urethra, including their innervation.

Characteristics of Urine

13. Describe the characteristics of urine.

Micturition and Its Control

14. Explain how micturition is controlled.

Clinical Applications
  Cystitis
  Pyelonephritis
  Glomerulonephritis
  Tubular Disorders
  Disorders of Micturition
  Renal Failure and Dialysis

15. Relate the following clinical conditions to disruption of homeostasis: (a) cystitis, (b) pyelonephritis, (c) glomerulonephritis, (d) tubular disorders, and (e) disorders of micturition.
16. Describe renal failure, and distinguish between hemodialysis and peritoneal dialysis as treatments for it.

Essay: The Role of Urine in Diagnosis
Chapter Summary
Questions and Problems

FIGURE 27.1
The urinary system.

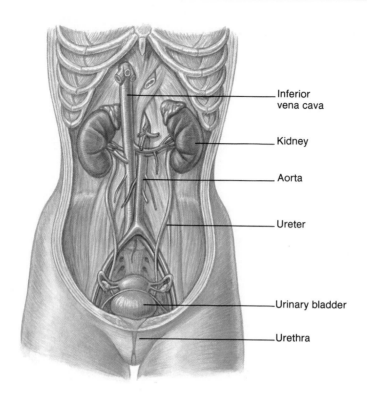

Inferior
vena cava

Kidney

Aorta

Ureter

Urinary bladder

Urethra

ORGANIZATION AND GENERAL FUNCTIONS

If the body fails to make urine or its release is blocked, toxic wastes accumulate in the blood and fluid balance is severely impaired. Unless such conditions are treated, the patient will die in a few days.

The urinary system consists of the kidneys, ureters, urinary bladder, and urethra (Figure 27.1). The kidneys regulate body fluids and remove wastes, and the ureters, urinary bladder, and urethra convey wastes out of the body. **Excretion** is the process of discharging waste materials, and the kidneys are the body's primary excretory organs because they continuously and selectively extract wastes and excess substances from the blood. The kidneys also play an important role in **fluid regulation,** the maintenance of normal fluid volumes; thus, the kidneys are extremely important in maintaining homeostasis of the body's internal environment.

DEVELOPMENT

In the first few weeks of development, urogenital ridges develop from mesoderm of the inner body wall on the left and right of the posterior lumbar region. The superior end of each ridge becomes a rudimentary kidney called a **pronephros** (pro-nef'ros). Ducts from these kidneys grow inferiorly to the cloaca, a common sac at the posterior end of both the urogenital system and the digestive system in the embryo. By the end of the fourth week of development, each pronephros has degenerated, and is being replaced by a **mesonephros** (mes″o-nef'ros), or second kidney, at a more inferior site on the urogenital ridge. This is followed quickly by the development of a third and final kidney, the **metanephros** (met″ah-nef'ros), at a still more inferior site. By the end of the second month of development, the mesonephros also degenerates, except for parts that are appropriated by the reproductive system (Chapter 29).

The **ureteric** (u″ret-er'ik) **bud** gives rise to the ureter and parts of the metanephros, including the renal pelvis, calyces, and collecting ducts, as described later. The metanephric portion of the urogenital ridge gives rise to nephrons, the functional units of kidneys. The bladder forms from part of the cloaca and opens to the exterior through the urethra. The allantois, an embryonic structure of little significance in mammals, collects wastes during the embryological development of reptiles and birds. Some of these structures are shown in Figure 27.2.

Of all the body systems, the urinary system is most susceptible to developmental defects. Kidneys can be displaced, undersized, or filled with cysts. Some people are born with only one kidney, with more than two kidneys, with partial kidneys, or with a "horseshoe" kidney, which results from the fusion of left and right kidneys across the midline of the body. Many people have abnormalities in the arrangement of the ureters. Unless urinary function is impaired, abnormalities can go undetected.

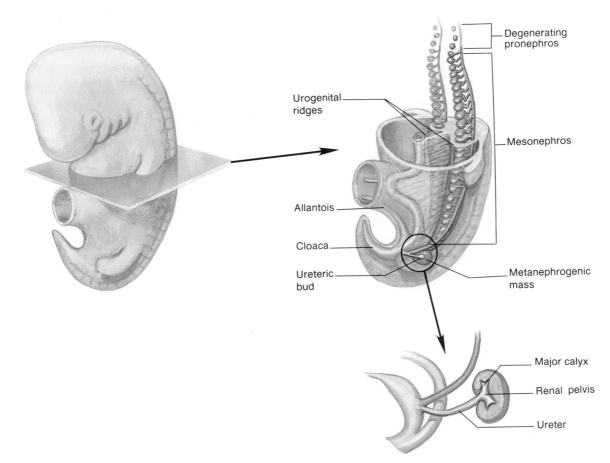

FIGURE 27.2
Development of the human kidneys at 5 weeks.

ANATOMY OF THE KIDNEY

The bean-shaped kidneys are located near lumbar verte-brae 1 through 3, and are **retroperitoneal** (re"tro-per"it-o-ne'al), or behind the parietal peritoneum next to the abdominal wall. The right kidney is pushed downward be-cause the liver occupies such a large area on the right side of the abdominal cavity.

Gross Anatomy

In an adult, each kidney is about 11 cm long, 5 to 7 cm wide, and about 2.5 cm thick. It is covered by the **renal capsule,** a layer of fibrous connective tissue. Around the capsule is a layer of fat, the **adipose capsule,** and an outer fibrous membrane, the **renal fascia.** All these structures protect the kidney, and the fascia anchors it to the abdominal wall. Blood and lymph vessels and nerves enter and leave the kidney, and the ureter attaches at the medial **hilum** (hi'lum) of each kidney (Figure 27.3a).

Internally, a kidney has an outer **cortex** and an inner **medulla** separated by the **juxtamedullary** (juks"tah-med'ul-ar-e) **zone.** The **nephrons** (nef'ronz), the func-tional units of the kidneys, are found mainly in the cortex, but parts of some nephrons extend into the medulla. Also

in the medulla are **collecting ducts,** which collect urine from the nephrons. Cone-shaped aggregations of col-lecting ducts form the **pyramids** of the medulla. The cortex forms **renal columns** extending between the pyramids. All the ducts of a pyramid terminate in a **papilla** and drain into a **minor calyx** (ka'liks), and several minor calyces drain into a **major calyx.** Two or three major calyces drain into the **renal pelvis,** which, in turn, drains into the **ureter** (u-re'ter).

Blood Circulation

Blood enters each kidney through a **renal artery,** which is much larger than arteries serving other organs of com-parable size. While the kidneys are nourished by the blood they receive, they also remove wastes and adjust quan-tities of substances in the blood. At any moment, about one-fourth of the circulating blood volume is in the kid-neys. Each renal artery branches into **interlobar arteries,** which pass between the pyramids and further branch into **arcuate arteries** in the juxtamedullary zone. Anastomoses among these vessels assure that blood will circulate to functional areas of the kidney even when some areas have been damaged. **Interlobular arteries** branch from arcuate arteries, extend into the cortex, and send an arteriole to

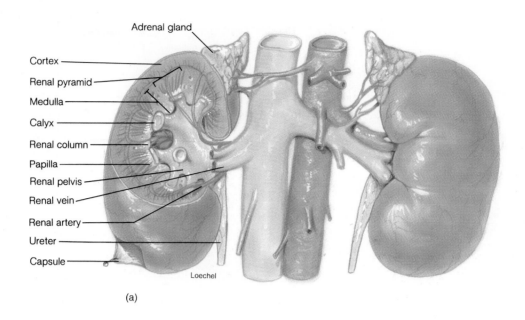

Adrenal gland

Cortex
Renal pyramid
Medulla
Calyx
Renal column
Papilla
Renal pelvis
Renal vein
Renal artery
Ureter
Capsule

Loechel

(a)

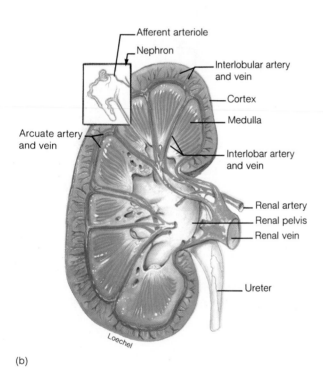

Afferent arteriole
Nephron
Interlobular artery and vein
Cortex
Medulla
Interlobar artery and vein
Arcuate artery and vein
Renal artery
Renal pelvis
Renal vein
Ureter

Loechel

(b)

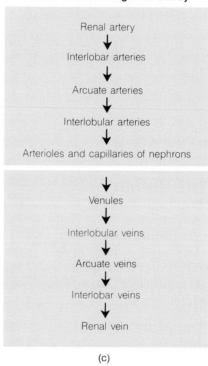

Path of blood through the kidney

Renal artery
↓
Interlobar arteries
↓
Arcuate arteries
↓
Interlobular arteries
↓
Arterioles and capillaries of nephrons

↓
Venules
↓
Interlobular veins
↓
Arcuate veins
↓
Interlobar veins
↓
Renal vein

(c)

FIGURE 27.3

(a) Gross anatomy of the kidneys, (b) the location of kidney blood vessels, and (c) a box showing the path of blood through the kidney.

each nephron. Blood leaving the capillaries of the nephron enters venules and then flows through veins that parallel the arteries just described and that have the same names (Figure 27.3b).

Each kidney's major blood vessels are supplied with autonomic nerves. Sympathetic and parasympathetic signals adjust small blood vessel diameters and, thus, blood volume flowing through the kidneys.

Nephrons

Each of the million **nephrons** in a human kidney consists of a tubule and its associated capillaries. **Cortical nephrons** are located solely in the cortex, whereas **juxtamedullary nephrons** extend into the medulla (Figure 27.4a). In humans, only about 12 percent of nephrons are juxtamedullary.

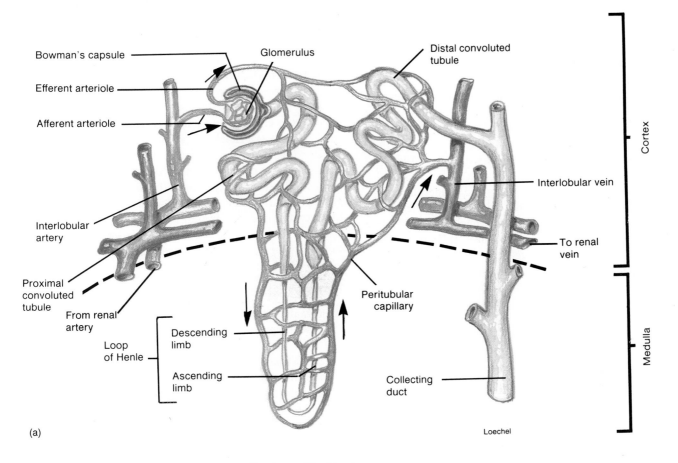

Cortex

Medulla

Bowman's capsule

Glomerulus

Distal convoluted tubule

Efferent arteriole

Afferent arteriole

Interlobular vein

Interlobular artery

To renal vein

Proximal convoluted tubule

From renal artery

Peritubular capillary

Loop of Henle

Descending limb

Ascending limb

Collecting duct

Loechel

(a)

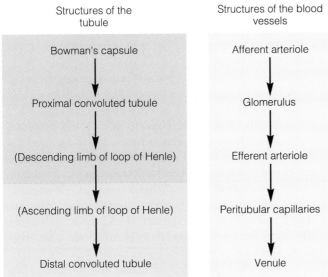

Structures of the tubule

Bowman's capsule

↓

Proximal convoluted tubule

↓

(Descending limb of loop of Henle)

↓

(Ascending limb of loop of Henle)

↓

Distal convoluted tubule

Structures of the blood vessels

Afferent arteriole

↓

Glomerulus

↓

Efferent arteriole

↓

Peritubular capillaries

↓

Venule

(b)

FIGURE 27.4
(*a*) The structure and location of cortical and juxtaglomerular nephrons and (*b*) a box showing relationships between tubule and blood vessel components of a nephron.

All nephrons have a long, coiled, hollow tube attached at one end to a collecting duct with the other end modified into a cup-like **glomerular** (glo-mer′u-lar) **capsule** (Bowman's capsule). Inside the cup is a tuft of capillaries called a **glomerulus** (glo-mer′u-lus). The capsule itself has an inner layer of epithelial cells called **podocytes** (pod′o-sītz) adjacent to the glomerulus and an outer layer of squamous epithelial cells. Substances filtering out of blood in the glomerulus pass through filtration slits in the podocytes and accumulate in the space between the two layers of the capsule.

Portions of the tubule between the glomerular capsule and a collecting duct are differentiated into the **proximal convoluted tubule,** the descending and ascending limbs of the **loop of Henle,** and the **distal convoluted tubule.** The loop of Henle is short in cortical nephrons and long in juxtaglomerular nephrons. These structures form a continuous tube through which kidney filtrate passes and is processed to become urine. Each collecting duct receives urine from several nephrons.

The capillaries surrounding the tubule also are part of the nephron, and in addition to the glomerulus, each nephron has **peritubular** (per-e-tu′bu-lar) **capillaries** (Figure 27.4b). Many **afferent arterioles** branch from each interlobular artery and each goes to a glomerulus, where many substances are filtered out of the blood. From a glomerulus, blood enters an **efferent arteriole** and is carried to the peritubular capillaries. Peritubular capillaries surround the tubule and supply its parts in the following order: proximal convoluted tubule, distal convoluted tubule, and the loop of Henle. The capillaries surrounding the loop of Henle are called **vasa recta** (va′sah rek′tah). Many substances move between the fluid in the tubule and the blood in the peritubular capillaries. Blood leaving the peritubular capillaries enters venules that lead to the interlobular veins.

GLOMERULAR FILTRATION

As blood flows through glomerular capillaries, plasma enters glomerular capsules by filtration (Figure 27.5). Water and low molecular weight substances pass through the endothelium, basement membrane, and filtration slits between podocytes, accumulating between the two layers of the capsule. The sieve-like endothelial layer holds back blood cells, the basement membrane holds back large protein molecules, and filtration slits hold back smaller ones. The few small protein molecules that reach the filtrate are reabsorbed and digested by kidney tubule cells. Consequently, urine contains no protein unless glomeruli are damaged or tubules have impaired reabsorptive capacity.

Filtration Pressures

Filtration depends not only on the above anatomical properties but also on relative pressures in glomeruli and capsules. Hydrostatic pressure in blood entering glomeruli is about 55 torr—about twice that in most capillaries. This pressure is derived partly from the heart's pumping and partly from efferent arterioles having smaller diameters than afferent arterioles. Blood oncotic pressure of about 30 torr and hydrostatic pressure of about 15 torr within the glomerular capsule together exert a 45 torr pressure against filtration; thus, the **net filtration pressure** pushing substances out of a glomerulus is about 10 torr—about 10 times that in most capillaries. Fluid leaving the glomeruli is **kidney filtrate,** sometimes called an **ultrafiltrate** because it is formed under pressure.

Filtration Rate

In normal human kidneys, plasma moving from glomeruli to capsular spaces forms glomerular filtrate at 115 to 125 ml/min—the **glomerular filtration rate** (GFR). GFR, which is proportional to body size and decreases with age, is determined mainly by net filtration pressure. A volume equal to the total adult plasma volume of about 3 l, enters the glomerular capsules once every 25 minutes. Over a 24-hour period, glomerular filtration produces about 180 l of filtrate, from which only about 1 l of urine will be formed. The other 179 liters, including much of the substances dissolved in it, are returned to the blood by reabsorption in other parts of the kidney tubules. Except for proteins, dissolved substances have the same concentration in the filtrate as in the plasma.

GFR is decreased by sympathetic signals that occur when brain receptors detect stress and in cardiovascular shock when baroreceptors detect a drop in blood pressure (Chapter 20). Such signals constrict afferent arterioles and divert blood from the kidneys to the heart and skeletal muscles during stress. In shock, decreasing the GFR conserves water and minimizes the decrease in blood volume.

In the absence of sympathetic signals, GFR remains nearly constant even when the mean arterial pressure (MAP) varies from 70 to 180 torr. (MAP is normally about 100 torr.) Such GFR regulation is called **renal autoregulation.** Within this physiological range, afferent arterioles dilate when the MAP decreases and constrict when the MAP increases in response to chemical substances released from the arterioles themselves. At extreme pressures outside this range, the arterioles fail to maintain a constant GFR. At a MAP of 200 torr, filtrate volume is about seven times normal, whereas at a MAP less than 45 torr, glomerular capillaries collapse and filtration does not occur.

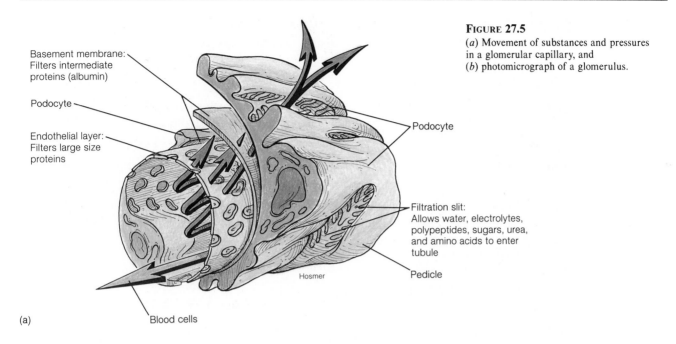

Basement membrane:
Filters intermediate
proteins (albumin)

Podocyte

Endothelial layer:
Filters large size
proteins

Blood cells

Podocyte

Filtration slit:
Allows water, electrolytes,
polypeptides, sugars, urea,
and amino acids to enter
tubule

Pedicle

Hosmer

(a)

FIGURE 27.5
(a) Movement of substances and pressures
in a glomerular capillary, and
(b) photomicrograph of a glomerulus.

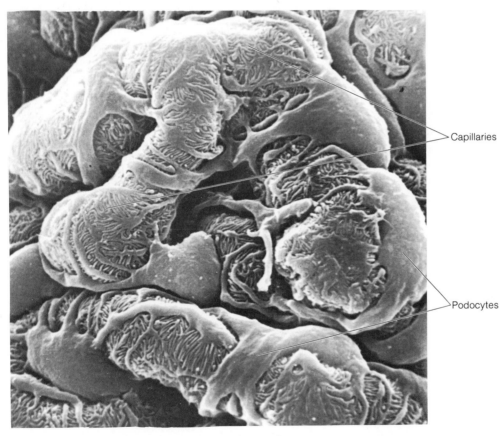

Capillaries

Podocytes

(b)

Tubular Reabsorption

Kidney filtrate flows from the glomerular capsule into the proximal convoluted tubule and on through the remaining segments of the renal tubule to a collecting duct. During this passage, water and solutes are returned to the blood by tubular reabsorption.

Substances Reabsorbed

Water, waste urea, and ions such as Na^+, K^+, and Cl^- can be reabsorbed from all parts of kidney tubules and collecting ducts, but most reabsorption takes place in the proximal tubules. There, metabolically active epithelial cells have a brush border composed of many tiny microvilli that greatly increase the absorptive surface area (Figure 27.6). Active transport of solutes out of the filtrate creates a small osmotic gradient and water leaves the filtrate by osmosis in the same proportion as the solutes. At the end of the proximal tubules, 60 to 70 percent of the filtrate's water and solutes have been reabsorbed.

Active Transport in Reabsorption

Though some Na^+ reabsorption occurs by diffusion, much occurs by active transport and uses energy from ATP. To reach the blood plasma from the tubule lumen, Na^+ diffuses down electrical and concentration gradients across the brush border membrane into the epithelial cells. Na^+ is then actively transported into **peritubular (interstitial) fluid** as K^+ is transported into the cells by a sodium-potassium pump. Carrier molecules, ATPase, and energy from ATP carry out this process, which accounts for more than 5 percent of basal metabolism. Once in peritubular fluid, Na^+ diffuses through peritubular capillaries into blood at a rate proportional to the concentration gradient.

Glucose is reabsorbed by **co-transport** like that in the digestive tract in which energy expended by the sodium-potassium pump indirectly transports glucose along with Na^+. Glucose and Na^+ bind to the same carrier molecule in the brush border membrane. Na^+ enters the cell by moving down its electrical and concentration gradients and carries glucose with it.

Amino acids, Cl^-, certain other ions, the waste uric acid, and creatine (a product of muscle metabolism) also are actively reabsorbed. Four carrier molecules each transport a particular group of amino acid with similar shapes and charges.

The Concept of Renal Threshold

The reabsorption rate of any actively transported solute is the product of the solute's plasma concentration and the GFR. For a plasma concentration of 80 mg/dl of glucose and GFR of 125 ml/min, reabsorption is directly proportional to filtration and occurs at a rate of 100 mg/min. The theoretical maximum reabsorption rate, determined by the availability of ATP, ATPase, and carriers, can be much higher—up to 375 mg/min for glucose and sufficient to remove all but traces of glucose from the filtrate even when the plasma concentration exceeds 300 mg/dl. The plasma concentration at which no more of a solute can be returned to the blood and is retained in the urine is called the **renal threshold.** According to these data, the

Figure 27.6
Proximal tubule epithelium with brush border and other structures associated with reabsorption.

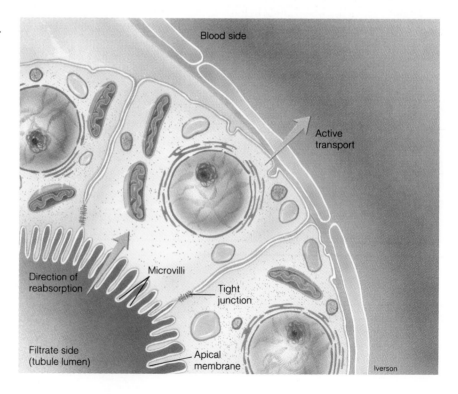

theoretical renal threshold for glucose is about 300 mg/dl, but the actual threshold is nearer 200 mg/dl because not all kidney tubules are working to maximum capacity to bind and transport glucose.

In untreated diabetes mellitus, the glucose concentration in both plasma and glomerular filtrate can become very high—sometimes over 600 mg/dl. Glucose not reabsorbed is excreted. Glucose in the kidney filtrate increases the osmotic gradient and holds water in the filtrate. This reduces water reabsorption and increases urine volume, causing dehydration and thirst.

TUBULAR SECRETION

Whereas large quantities of water and nearly all solutes are reabsorbed from the kidney filtrate, only small quantities of certain plasma solutes are secreted into it. Tubular secretion occurs mainly in the distal tubules, which are impermeable to water. Secretion is a fine-tuning mechanism—it makes final adjustments in the chemical composition of plasma beyond what is accomplished by glomerular filtration and tubular reabsorption.

Like other active transport processes, tubular secretion requires carriers, ATPase, and energy from ATP. H^+, often combined with ammonia, is secreted as a part of the acid-base regulating mechanism. K^+ is secreted and Na^+ is reabsorbed under the influence of aldosterone, which stimulates synthesis of proteins (carriers and enzymes) for active transport in the distal tubule epithelium. Na^+ reabsorption creates an osmotic gradient that conserves water. Penicillin, certain other drugs, and some toxic substances are removed from the blood by tubular secretion. These processes are discussed in Chapter 28.

Filtration, reabsorption, and secretion of plasma constituents are summarized in Figure 27.7 and Table 27.1. In addition to the 60 to 70 percent of solutes and water reabsorbed in proximal tubules, another 20 percent are reabsorbed in the loops of Henle, and the remainder are reabsorbed in distal tubules and collecting ducts.

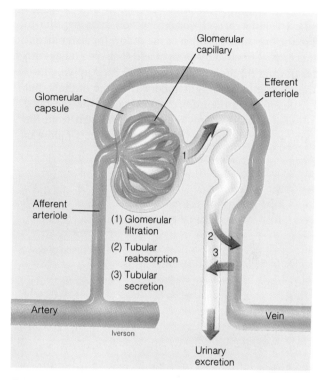

FIGURE 27.7
The three processes by which substances enter or leave the renal tubule are glomerular filtration, tubular reabsorption, and tubular secretion.

TABLE 27.1	Renal Handling of Various Plasma Constituents in a Normal Adult Human on an Average Diet					
Substance	**Per 24 Hours**				**Percentage Reabsorbed**	**Location**
	Filtered	*Reabsorbed*	*Secreted*	*Excreted*		
Na^+ (mEq)	26,000	25,850		150	99.4	P,L,D,C
K^+ (mEq)	600	560*	50*	90	93.3	P,L,D,C
Cl^- (mEq)	18,000	17,850		150	99.2	P,L,D,C
HCO_3^- (mEq)	4,900	4,900		0	100	P,D
Urea (mmol)	870	460†		410	53	P,L,D,C
Creatinine (mmol)	12	1‡	1‡	12	. . .	. . .
Uric acid (mmol)	50	49	4	5	98	P
Glucose (mmol)	800	800		0	100	P
Total solute (mOsm)	54,000	53,400	100	700	87	P,L,D,C
Water (ml)	180,000	179,000		1000	99.4	P, L, D, C

Reprinted with permission from Ganong, W. F.: *Review of Medical Physiology,* 14th Edition, 1989, by Appleton & Lange, Norwalk, CT and San Mateo, CA.

Note: P, proximal tubules; L, loops of Henle; D, distal tubules; C, collecting ducts.

*K^+ is both reabsorbed and secreted.

†Urea diffuses into as well as out of some portions of the nephron.

‡Variable secretion and probable reabsorption of creatinine in humans.

The Countercurrent Mechanism

The **countercurrent mechanism** concentrates urine by increasing osmolarity in peritubular fluids around collecting ducts. It involves the loops of Henle as **countercurrent multipliers** and the vasa recta (capillaries around the loops of Henle) as **countercurrent exchangers** (refer back to Figure 27.4).

In a **countercurrent system,** outflow runs parallel to and in the opposite direction of the inflow for a significant distance, as illustrated by straight and U-shaped tubes with water flowing through them (Figure 27.8). Applying heat to the middle of a straight tube warms the outflow, but has little effect on the inflow. Applying heat to the bend in a U-shaped tube warms the outflow and heat from the outflow warms the inflow; thus, outflow affects inflow in a countercurrent system, whether the system is a heated U-shaped tube or the system that moves Na^+, Cl^-, and water in a loop of Henle.

Countercurrent Multipliers

The loops of Henle of kidney tubules are **countercurrent multipliers.** The loop shape allows passive countercurrents to develop, and active transport of substances out of the system's outflow portion multiplies the countercurrent. Countercurrent multiplication greatly increases the filtrate concentrations of Na^+ and Cl^- at the tips of the loops. More significantly, it greatly increases the osmolarity of peritubular fluids, thereby drawing water from filtrate in collecting ducts and concentrating the urine. How countercurrent multiplication occurs is not fully understood, but a likely possibility is summarized here.

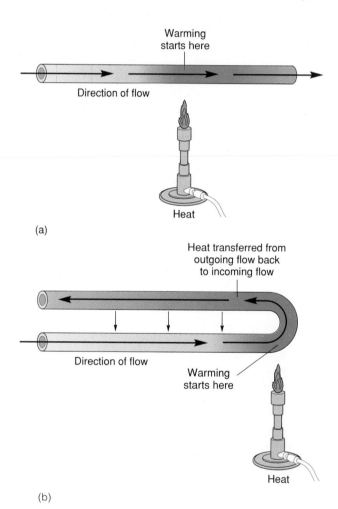

Figure 27.8
(*a*) Heat is transferred from the straight tube to the environment from the point of heat application in the direction of the outflow.
(*b*) Heat is transferred from the bent tube to the environment as in (*a*) *and* from the outflow toward the inflow. The latter constitutes a countercurrent system.

The loop of Henle multiplies Na^+ and Cl^- concentrations because of properties of its walls (Figure 27.9a). The descending (inflow) limb is highly permeable to water and relatively impermeable to solutes. The thin ascending (outflow) limb is highly permeable to Na^+, moderately permeable to urea, and relatively impermeable to water. The thick ascending (outflow) limb is relatively impermeable to both water and solutes, and most ion movements are controlled by carrier molecules. Tubule cell membranes facing the lumen contain a special carrier that simultaneously actively transports 1 Na^+, 1 K^+, and 2 Cl^- from the lumen into cells, though K^+ diffuses back into the lumen. Outer membranes of these cells contain sodium-potassium pumps that actively transport Na^+ out of the cell and K^+ into it and another carrier that actively transports both K^+ and Cl^- out of the cell. Cl^- also diffuses out of the cell.

These properties of a loop of Henle account for countercurrent multiplication as follows (Figure 27.9b and c):

1. The thick ascending (outflow) loop actively transports Na^+ and Cl^- into peritubular fluid, making it hyperosmotic as the filtrate becomes hyposmotic.
2. Water moves by osmosis out of the descending (inflow) loop into the more hyperosmotic peritubular fluid.
3. Countercurrent multiplication of outflow in step 1 affects filtrate composition in step 2, so that filtrate reaching the tip of the loop attains maximum osmolarity as do peritubular fluids around the tip. This effect is an example of positive feedback.
4. Continuous active transport (step 1) maintains the osmotic gradient that draws water out of distal tubules and collecting ducts, thereby concentrating the urine. Water from the filtrate is returned to the blood by diffusing into peritubular capillaries.

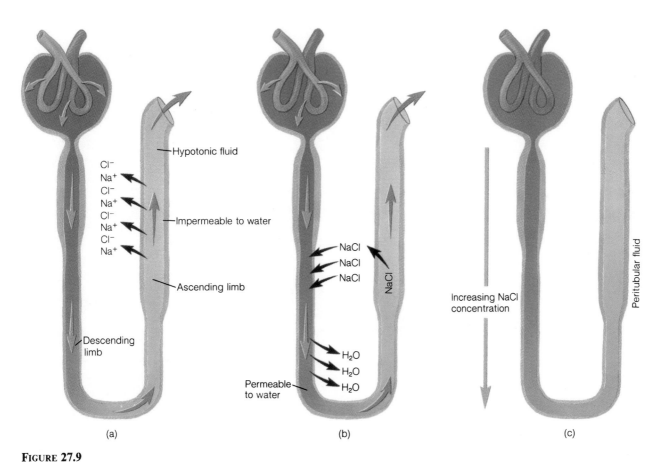

(a) (b) (c)

FIGURE 27.9

(a) The thin descending limb of a loop of Henle is highly permeable to water, whereas the thick ascending limb is impermeable except to substances it actively transports such as NaCl. (b) Each loop of Henle is a countercurrent system in which outflow runs in the opposite direction as inflow and NaCl is continuously recycled. (c) The main effect of the countercurrent system is to concentrate NaCl in peritubular fluid.

Countercurrent Exchangers

When countercurrent multipliers create high peritubular osmolarity, **countercurrent exchangers** maintain it by passively regulating Na^+ and Cl^- in the vasa recta (Figure 27.10). If Na^+, Cl^-, and urea drawn out of collecting ducts left the kidneys with blood, the high peritubular osmolarity needed to concentrate urine would not last long. As blood enters the vasa recta, Na^+, Cl^-, and urea diffuse into the capillaries and water diffuses into the peritubular fluids. As blood leaves the vasa recta, the process is reversed—solutes diffuse out of blood and water returns to it. Blood in the vasa recta has solute levels near those in peritubular fluids, but blood leaving the vasa recta has solute levels appropriate for outside the kidney. These exchanges maintain osmotic gradients created by the countercurrent mechanism without permanently affecting blood solute levels.

High osmolarity in peritubular fluids around the loop of Henle also is fostered by the low-volume, sluggish blood flow in the vasa recta. Most blood entering the kidney is returned to the renal vein through cortical capillaries and venules; only 1 to 2 percent of it passes through the vasa recta.

In summary, one might say that countercurrent multipliers create an osmotic gradient between extracellular fluids and filtrate, and countercurrent exchangers prevent solute loss that could reduce the gradient as blood flows through the medulla (Figure 27.11). The significant effect of the countercurrent mechanism is to create an osmotic gradient that draws water out of the filtrate and concentrates the urine. Other events in the distal tubules and collecting ducts also regulate urine volume and composition, including diffusion of urea out of the filtrate and antidiuretic hormone actions, as we shall soon see.

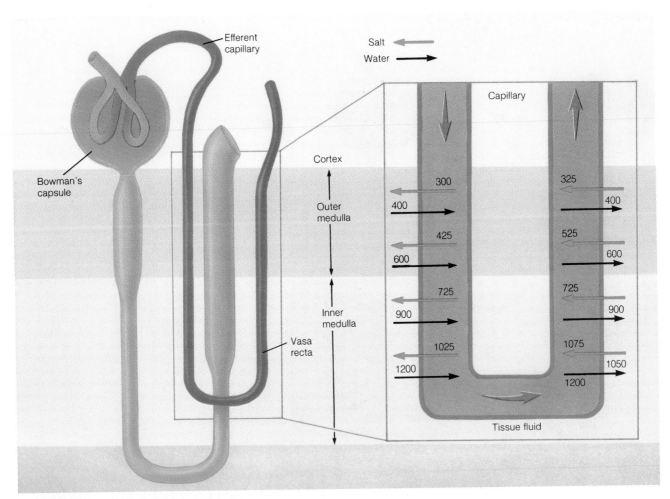

FIGURE 27.10

The vasa recta acquires NaCl as it passes through the medulla but preserves the high peritubular NaCl and normal blood NaCl concentration by losing NaCl as it leaves the medullary region.

REGULATION OF URINE FORMATION

Various regulatory mechanisms in urine formation allow the kidneys to respond quickly to changes in the composition and volume of body fluids and, thereby, maintain fluid homeostasis. To do this, the kidneys excrete nitrogenous wastes, mainly urea, regardless of urine volume, conserve water when fluid volume is low, and excrete excess water when fluid volume is high. The kidneys also regulate blood concentrations of electrolytes, and they assist in maintaining acid-base balance (Chapter 28).

Excretion of Urea and Other Wastes

Urea synthesis (in the urea cycle, Chapter 24) and excretion are the body's main means of getting rid of amino groups from protein catabolism. On a moderate-protein diet, the body produces 25 to 30 g of urea per day—more on a high-protein diet and less on a low-protein diet. Urea is normally excreted at the same rate it is synthesized. If excretion is impaired, as in some kidney disorders, toxic levels can accumulate in body fluids.

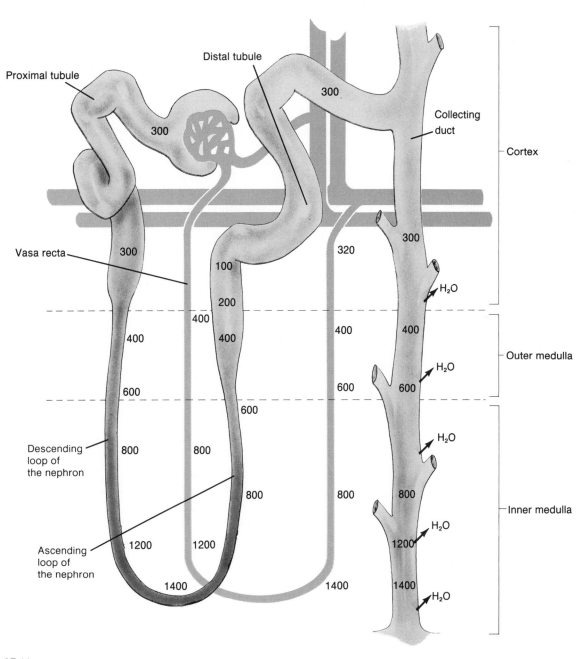

FIGURE 27.11

The countercurrent mechanism operates in the loop of Henle and the countercurrent exchanger in the vasa recta (singular, vas rectum). Their overall function is to concentrate NaCl in extracellular fluids, causing water to diffuse out of the collecting ducts and concentrate the urine.

The rate of urea excretion is determined by the product of plasma urea concentration and GFR—the amount of urea entering the filtrate. For example, at a normal plasma urea concentration of 15 to 35 mg/100 ml and GFR of 125 ml/min, urea enters the filtrate at 20 to 40 mg/min. Kidney tubules are somewhat permeable to urea, and under normal physiological conditions, about 40 percent of urea diffuses back into the blood and 60 percent remains in the filtrate and is excreted. If the urine volume is small and flow through tubules is slow, or the plasma urea level is high, more urea leaves the filtrate and only 10 to 20 percent is excreted. Conversely, if urine volume is large and flow through tubules is relatively rapid, or the plasma urea level is low, less urea leaves the filtrate and up to 70 percent is excreted. Some urea that leaves the filtrate diffuses back into the blood, but a portion remains in peritubular fluids where it contributes to osmolarity, as noted earlier. As a practical matter, high-protein intake leads to concentrated urine because it increases peritubular osmolarity, whereas a high-fluid intake leads to dilute urine because rapid flow allows less time for urea to contribute to osmolarity.

Among nitrogen-containing wastes in human urine, urea accounts for about 85 percent. Another 4.5 percent is creatinine, a product of muscle metabolism excreted at a constant rate; about 3 percent is ammonia; and 2 percent is uric acid from the catabolism of nucleic acids. The remainder consists of very small quantities of various amino acids and traces of other nitrogenous substances.

Regulation of Fluid Volume Excreted

If the entire glomerular filtrate were excreted as urine without becoming more concentrated, body fluids would be depleted in less than an hour! To remove excess solutes while maintaining normal blood volume, the kidneys must concentrate the urine. We have already seen how the countercurrent mechanism and urea accumulation help to concentrate urine, and have noted that aldosterone returns Na$^+$ to the blood with water following by osmosis. Antidiuretic hormone is another extremely important factor.

Antidiuretic Hormone

Interstitial fluid becomes hyperosmotic when blood is hyperosmotic because of too much solute or too little water. Such interstitial fluid causes hypothalamic osmoreceptors to respond by stimulating secretion of **antidiuretic hormone** (ADH). ADH activates an enzyme by the cAMP mechanism, and the enzyme phosphorylates a protein that makes the cell membrane much more permeable to water and a little more permeable to urea. ADH acts on luminal membranes of epithelial cells mainly in collecting ducts and to a lesser degree in distal convoluted tubules. Much water and a little urea move passively into the peritubular fluids. Most of the water then diffuses into the blood, lowering blood osmotic pressure, maintaining blood volume, and preventing dehydration. Lowering blood osmotic pressure prevents further stimulation of osmoreceptors, thereby completing the loop in a negative feedback mechanism (Figure 27.12).

ADH indirectly removes urea from the filtrate. As water is reabsorbed from the collecting ducts, the urea concentration in the ducts increases. When urea becomes more concentrated in the filtrate than in peritubular fluid, its net diffusion is toward the peritubular fluid where it increases osmolarity.

An excess of ADH temporarily increases reabsorption and decreases urine volume. Sustained excessive ADH secretion increases blood volume and arterial pressure, and the higher pressure increases filtration, counteracting the ADH.

When the blood is hyposmotic, such as after one drinks a large quantity of fluid, a less concentrated urine is excreted because osmoreceptors receive no stimulation and ADH is not secreted. In the absence of ADH, cells in distal

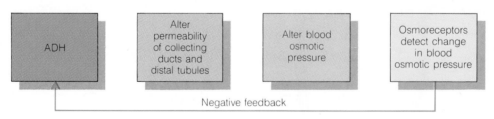

FIGURE 27.12
The role of ADH in regulating blood osmotic pressure and, to some extent, blood hydrostatic pressure.

tubules and collecting ducts remain impermeable and water is excreted.

In ADH deficiency, or **diabetes insipidus** (not to be confused with diabetes mellitus), distal tubules and collecting ducts remain impermeable, and large volumes of fluid—up to 15 l per day—are excreted. The patient is always thirsty, but usually cannot drink enough fluid to compensate for the water loss. This condition is treated by administering ADH either by nasal spray (absorbed through membranes) or injection.

Nephrogenic (nef″ro-jen′ik) **diabetes insipidus,** possibly inherited as a sex-linked characteristic, is due to a poorly understood tubular defect that prevents an ADH response. The kidneys fail to concentrate urine despite high blood ADH, and giving ADH does not help. Patients must cope with the disease by increasing fluid intake and decreasing salt intake.

Diuresis

We have already seen that increased urine production, or **diuresis** (di″u-re′sis), can be caused by blood hyposmolarity and by ADH deficiency. **Osmotic diuresis** is caused by filtrate hyperosmolarity, in which a high solute concentration (such as glucose in untreated diabetes) holds fluid in the filtrate. Osmotic diuresis can raise urine volume to four or five times normal, severely dehydrating tissues and disturbing electrolyte balance.

Diuretics (di″u-ret′iks) include alcohol, caffeine, and certain drugs. The drugs can be classified according to their sites and mechanisms of action (Table 27.2). Diuretics are commonly used to prevent fluid accumulation in hypertension and conditions that decrease cardiac output; however, they can have undesirable side effects and should be used with caution.

HOW KIDNEYS REGULATE BLOOD FLOW

Coregulation of blood flow and kidney function involves the juxtaglomerular apparatus, the renin-angiotensin mechanism, and a local mechanism. Atrial peptides also may be involved.

TABLE 27.2 Properties of Diuretics

Class of Diuretic	Site and Mechanism of Action	Examples
Osmotic diuretics	Increase osmotic pressure in kidney filtrate and cause water to be excreted instead of reabsorbed.	Mannitol, sorbitol
Carbonic anhydrase inhibitors	Inhibit the enzyme carbonic anhydrase in tubule cells and reduce formation of carbonic acid. This prevents secretion of H^+ into filtrate and diffusion of HCO_3^- into the blood and allows increased excretion of Na^+, K^+, and HCO_3^-. Water follows these ions by osmosis. Excess HCO_3^- excretion can cause acidosis.	Acetozolamide (Diamox)
Thiazides	Inhibit reabsorption of Na^+ and Cl^- mainly in distal tubules and reduce osmotic pressure in interstitial fluids. This reduces the kidneys' ability to concentrate urine. It also increases excretion of various electrolytes and can lead to K^+ depletion.	Chlorothiazide (Diuril), hydrochlorothiazide
ATPase inhibitors	Inhibit ATPase that normally provides energy for Na/K pumps in the thick ascending loop of Henle (and proximal and distal tubules). This reduces Na^+ reabsorption and greatly increases Na^+ and water excretion. It can cause HCO_3^- retention and lead to contraction alkalosis—decreased volume and increased pH of body fluids.	Furosemide (Lasix), ethacrynic acid (Edecrin)
Potassium sparing natriuretics	Inhibit Na^+-K^+ exchange in distal tubules and collecting ducts by inhibiting aldosterone (spironolactone) or by inhibiting Na^+ reabsorption (triamterene).	Spironolactone (Aldactone), triamterene (Dyrenium)
Vasopressin (ADH) inhibitors	Inhibit action of vasopressin (ADH) on collecting ducts.	Water, ethyl alcohol
Xanthines	Probably decrease tubular reabsorption of Na^+ and increase glomerular filtration rate.	Caffeine, theophylline

Juxtaglomerular Apparatus

A specialized structure called the **juxtaglomerular apparatus** consists of juxtaglomerular cells and the macula densa (Figure 27.13). **Juxtaglomerular cells,** modified smooth muscle cells in afferent arteriole walls that touch macula densa cells, are filled with an inactive form of the enzyme **renin** (ren'in). The **macula densa** (mak'u-lah den'sah) consists of secretory cells in distal tubule walls. When macula densa cells detect a decrease in blood pressure or GFR, they probably secrete a messenger substance (as yet unidentified) onto the juxtaglomerular cells, causing them to activate and release renin. Decreases in Na^+ in distal tubules and sympathetic signals also can release renin.

Renin-Angiotensin Mechanism

Renin release activates the **renin-angiotensin** (an"je-o-ten'sin) **mechanism** (Figure 27.14), which regulates blood pressure and protects against sodium depletion. Atrial peptides may deal with sodium excesses. Activated renin acts on the plasma protein **angiotensinogen** (an"je-o-ten-sin'o-jen) and releases *angiotensin I*. Almost as rapidly as it is formed, angiotensin I is converted to *angiotensin II* by *plasma converting enzyme,* acting mainly in small blood vessels of the lungs. Angiotensin II remains active in the blood for only a few minutes until degraded by *angiotensinase* (an"je-o-ten'sin-ās).

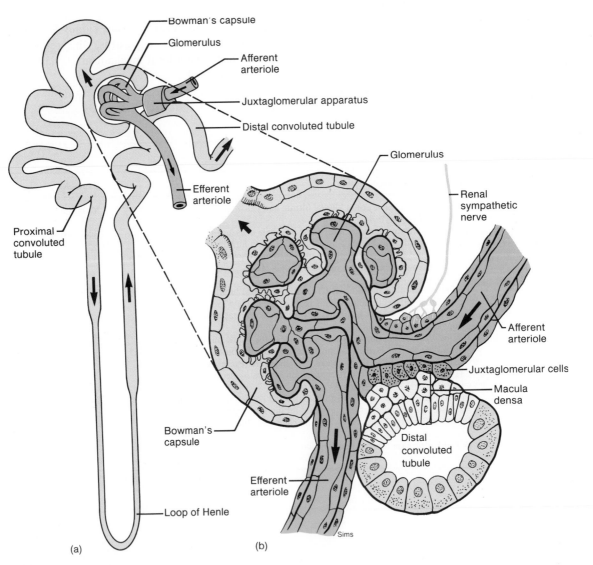

FIGURE 27.13
(*a*) Juxtaglomerular apparatus, and (*b*) enlarged view showing its relationship to the glomerulus and glomerular capsule.

Angiotensin II, a powerful vasoconstrictor, causes rapid constriction of arterioles in the kidneys and elsewhere in the body and mild constriction of venules and veins. Blood vessel constriction increases peripheral resistance and, thereby, elevates blood pressure. Because angiotensin II has a greater effect on efferent than on afferent arterioles, it increases glomerular pressure and, thus, GFR. Angiotensin II also decreases sodium and water excretion by stimulating aldosterone secretion.

Until recently, beta blockers, drugs that block signals that would ordinarily excite beta andrenergic receptors (Chapter 14), have been widely used to treat hypertension. Now drugs that inhibit the renin-angiotensin mechanism are being developed. Some suppress renin release and others inhibit the converting enzyme. Such drugs are useful to treat **renal hypertension**—excessive renin-angiotensin activity usually due to renal artery atherosclerosis—and may work in other kinds of hypertension, too. These drugs may be more effective and cause fewer undesirable side effects than currently available drugs.

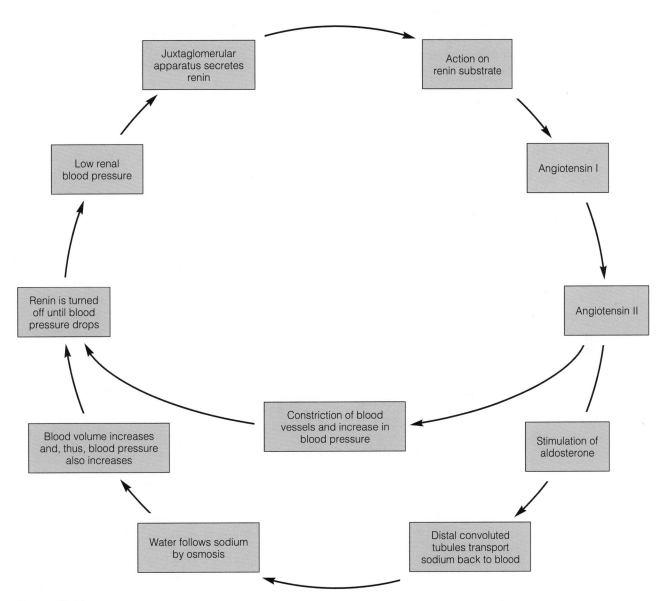

FIGURE 27.14
The effects of the renin-angiotensin system on blood pressure. A decrease in blood pressure turns this system on, and an increase in blood pressure turns it off.

Local Regulation

Glomerular filtration and tubular reabsorption normally are closely coordinated, filtration occurring at about 125 ml/min and reabsorption at about 124 ml/min. Urine is formed at the rate of 1 ml/min. When one rate changes, the other changes in the same direction and *almost* proportionately. Such changes continuously adjust urine volume and maintain fluid homeostasis.

When glomerular filtration decreases and larger quantities of solutes including Cl⁻ have been reabsorbed by the time the filtrate reaches the distal tubules, local regulation occurs. Macula densa cells respond to the lowered Cl⁻ concentration, sending signals that dilate afferent arterioles. This dilation increases glomerular blood flow and, thereby, increases the glomerular filtration. Conversely, when glomerular filtration increases, macula densa cells fail to respond because the Cl⁻ concentration is high. In the absence of signals from the macula densa, afferent arterioles constrict, reducing both blood flow and glomerular filtration.

Though the above mechanism is effective in the short term, decreases in blood pressure lasting more than 20 minutes activate the renin-angiotensin mechanism, and it temporarily reduces renal blood flow. Blood flow is reduced until pressure increases, and GFR is maintained in a normal range; thus, wastes are filtered from the blood over a wide range of blood flow rates. The effects of various regulatory factors are summarized in Figure 27.15.

Atrial Peptides

Several similar small **atrial peptides** are released from atrial muscle tissues when the blood pressure increases and the heart's atria becomes distended. Atrial peptides are

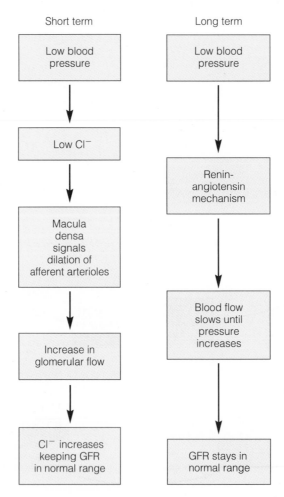

FIGURE 27.15

A summary of local regulatory mechanisms.

also called atrial natriuretic hormone (ANH) or **atrial natriuretic factor** (ANF) because in early experiments they were observed to cause **natriuresis,** or sodium excretion. Natriuresis is significant only under special conditions such as extremely high sodium loads, and the peptides have other actions such as decreasing blood pressure and cardiac output and possibly being brain neurotransmitters.

The mild natriuresis caused by atrial peptides is probably due to a slight alteration in Na$^+$ transport in collecting ducts, but atrial peptides have other renal effects. They reduce reabsorption by inhibiting the effects of angiotensin II and may increase water excretion by inhibiting antidiuretic hormone. They also relax renal arteries and arterioles, allowing glomerular filtration and filtrate volume to increase, thereby increasing urine volume and the quantity of Na$^+$ in it. Finally, they may increase potassium excretion.

MEASUREMENT OF KIDNEY FUNCTION

An important measure of kidney function is **clearance,** the volume of blood that can be cleared, or freed, of a substance per minute. Clearance of a substance is proportional to the sum of the amounts filtered and secreted less the amount reabsorbed. As a substance is cleared from the blood, the amount excreted (E) in mg/min is the concentration in the urine (U) in mg/ml times the rate at which urine volume (V) forms in ml/min.

$$E = U \times V$$

Also, the amount filtered (F) in mg/min is the concentration of the substance in plasma (P) in mg/ml times GFR in ml/min.

$$F = GFR \times P$$

If the substance is neither reabsorbed nor secreted, the amount filtered (F) equals the amount excreted (E), and GFR $\times$ P = U $\times$ V. Rearranging this equation to solve for GFR, we get

$$GFR = \frac{U \times V}{P}$$

In this case, GFR equals the **renal clearance rate,** or clearance in ml/min.

These formulas can be applied to the clearance of *inulin* (in'u-lin), a fructose polysaccharide made by onion and garlic plants. Suppose a person is given inulin intravenously until the plasma concentration (P) reaches 1 mg/ml. Later, the urine inulin concentration (U) is 125 mg/ml and that urine volume (V) forms at a rate of 1 ml/min. Clearance (C) = 125 ml/min.

$$C = \frac{125 \text{ mg/ml} \times 1 \text{ ml/min}}{1 \text{ mg/ml}} = 125 \text{ ml/min}$$

Clearance is equal to GFR because inulin is filtered but neither secreted nor reabsorbed. For the inulin concentration to reach 125 mg/ml, filtrate must form at the normal rate of 125 ml/min. In patients with glomerulonephritis, in which glomeruli are damaged or destroyed, clearance studies show significant decreases in GFR.

Clearance of other substances can be used to study other kidney functions. *Para-aminohippuric* (par"ah-am"in-o-hip-u'rik) *acid* (PAH) clearance estimates blood flow through the kidneys because PAH is filtered and secreted but not reabsorbed. In fact, 91 percent of plasma PAH is cleared in one pass through a kidney. Suppose we administer PAH to a person, obtaining plasma PAH (P) of 0.01 mg/ml. We later find that urine PAH (U) is 4.4 mg/ml and urine volume (V) forms at 1.3 ml/min.

$$C = \frac{4.4 \text{ mg/ml} \times 1.3 \text{ ml/min}}{0.01 \text{ mg/ml}} = 572 \text{ ml/min}$$

If this calculation of clearance accounts for 91 percent of plasma, the total plasma volume is 629 ml/min (572 ml/min divided by 0.91). As plasma makes up about 55 percent of the total blood volume, blood flows through the kidneys at 1144 ml/min (629 ml/min divided by 0.55).

Clearance also can be used to determine what proportion of the nitrogenous waste urea is actually cleared from the blood. Recall that some urea is reabsorbed and some is trapped in peritubular fluids. Suppose we determine that a person has a plasma urea (P) of 0.2 mg/ml and a urine urea (U) of 12 mg/ml when urine volume is formed at 1.3 ml/min.

$$C = \frac{12 \text{ mg/ml} \times 1.3 \text{ ml/min}}{0.2 \text{ mg/ml}} = 78 \text{ ml/min}$$

We see that urea clearance occurs at a rate equal to removing all the urea from 78 ml of plasma per minute, and we have determined a GFR of 125 ml/min. Then we can reason that 78/125 or 62 percent of the urea is being removed from the blood.

The plasma concentration of **creatinine** (kre-at'in-in), a product of muscle metabolism, provides a quick estimate of kidney function. Plasma creatinine is reported with other blood constituents in routine blood analyses. Muscles release creatinine at a constant rate regardless of activity, and the plasma creatinine concentration is nearly constant throughout life. Creatinine clearance also remains constant as long as kidney function is normal. It is filtered and secreted, but it is not reabsorbed. An increase in plasma creatinine suggests that the glomerular filtration rate has decreased and warns that the patient may have kidney disease.

FIGURE 27.16
A low-power photomicrograph of a ureter, showing the layers of its wall. Note the inner transitional epithelium, the longitudinal and circular smooth muscle layers, and the fibrous outer layer.

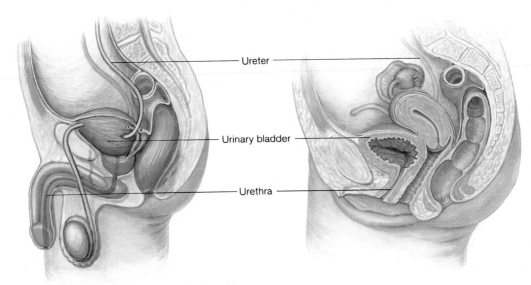

Ureter

Urinary bladder

Urethra

FIGURE 27.17
The location of the urinary bladder (*a*) in a male, and (*b*) in a female.

ANATOMY OF URETERS, BLADDER, AND URETHRA

In addition to kidneys, the urinary system includes paired ureters, a urinary bladder, and a urethra, as already shown in Figure 27.1.

Ureters, which lie behind the peritoneum, originate at a renal pelvis near the hilum of each kidney. They collect urine from the kidney and transport it to the bladder. Ureters are about 25 cm long and have their widest diameter (about 1.5 cm) near the bladder. A ureter wall has three layers—an inner mucous membrane of transitional epithelium, a middle muscular layer, and an outer fibrous layer (Figure 27.16). The mucous membrane has several layers of cells and is continuous with the epithelium of the collecting ducts it drains. The muscular layer contains circular and longitudinal smooth muscle fibers that by alternate contraction propel fluid. The outer fibrous layer protects the ureter and holds it in place.

The **urinary bladder** (Figure 27.17) is a hollow bag in the pelvic cavity inferior to the peritoneum, posterior to the pubic bone, and anterior to the vagina and uterus in females and to the rectum in males. The bladder has an

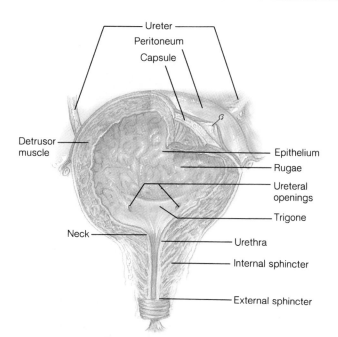

FIGURE 27.18
The internal structure and wall of the urinary bladder.

internal triangular anatomic landmark, the **trigone** (tri′gŏn), bounded by the posterolateral openings through which the ureters enter and the anteromedial opening through which the urethra leaves the bladder. The ureters enter the bladder at an angle, forming a flap-like valve that prevents urine backflow. The easily stretched bladder wall consists of inner transitional epithelium, a middle muscular layer, and outer connective tissue (Figure 27.18). Spherical epithelial cells of an empty bladder flatten as the bladder fills (refer to Chapter 4). Connective tissue consists of peritoneum on the superior surface and fibrous tissue elsewhere.

The muscular layer of the bladder, the *detrusor* (de-troo′sor) *muscle,* consists of inner and outer longitudinal fibers with a middle layer of circular fibers all innervated by parasympathetic fibers. Interspersed among the muscle fibers are stretch receptors that respond to bladder distention. Where the bladder and urethra join, some of the circular smooth muscle is modified to form the **internal sphincter.** Inferior to the internal sphincter is an **external sphincter** made of skeletal muscle.

The **urethra** (u-rě′thrah) conveys urine from the urinary bladder out of the body. In the female, it is only about 4 cm long, but in the male, it passes through the penis and is about 20 cm long. The urethra terminates in the **external urethral** (u-re′thral) **orifice** in the glans penis of males and anterior to the vagina in females.

In both males and females, the urethral lining consists of a mucous membrane continuous with other linings of the urinary tract. In males, a membrane under the mucous membrane attaches the urethra to the surrounding tissues. In females, a venous plexus under the mucous membrane is surrounded by deeper smooth muscle.

CHARACTERISTICS OF URINE

The physical characteristics of urine depend to a large extent on what a person eats and drinks. A transparent, light yellow or amber color is due to the pigment urochrome, a breakdown product of hemoglobin, but color also is affected by foods, such as beets, vitamin supplements, and some medicines. A dark red color can indicate bleeding somewhere along the urinary tract. Oxidation of solutes causes urine to become turbid (cloudy) on standing. Normal urine pH ranges from 5.0 to 7.8, depending on dietary acids and bases. Fruits and proteins, which release amino acids, increase urine acidity and vegetables increase alkalinity. The specific gravity (relative weight compared to the same volume of water) of urine varies from 1.008 for dilute urine to 1.030 for concentrated urine, depending on the concentration of solutes dissolved in it. A 100 ml urine sample weighing 101.2 g has a specific gravity of 1.012. Water has a specific gravity of 1.000.

At least 95 percent of the urine volume is water, the remainder being metabolic waste products—urea, uric acid, ammonia, and creatinine—and excess electrolytes—especially Na^+ and Cl^-—removed from the blood. Glucose is not normally found in urine.

Urine volume is proportional to fluid intake and is affected by nephron function. When filtration pressure drops, nephrons produce less urine, and toxic wastes and excess fluid accumulate in the blood. **Anuria** (ah-nu′re-ah), the failure to produce urine, is a life-threatening condition. **Oliguria** (ol″ig-u′re-ah), the production of a small urine volume (*oligo,* little), is slightly less serious. At the other extreme, **polyuria** (pol″e-u′re-ah), the production of a large urine volume (*poly,* much) usually because of an ADH deficiency, leads to dehydration.

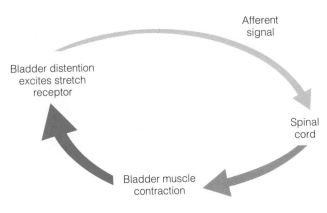

FIGURE 27.19
The micturition reflex.

MICTURITION AND ITS CONTROL

Though the bladder has a capacity of about 750 ml, it rarely fills completely. **Micturition** (mik''tu-rish'un), the release of urine, is controlled by the **micturition reflex** (Figure 27.19). A half-filled bladder is sufficiently distended to stimulate stretch receptors, which send signals to the spinal cord that initiate motor signals in parasympathetic neurons, causing the bladder muscle to contract. Once initiated, the reflex is self-perpetuating—the bladder contracts, further stimulating stretch receptors and causing more contraction. This positive feedback mechanism initiates and sustains bladder contraction until emptying occurs.

Signals from stretch receptors not only initiate motor signals to the bladder, they initiate signals to both facilitatory and inhibitory centers in the brain stem and cerebral cortex. Cortical centers inhibit micturition until the time and place are appropriate. Then, they allow parasympathetic signals to relax the voluntary external sphincter, the main control of urine release. Constant stretching has already relaxed the internal sphincter. By remaining contracted except during micturition, the external sphincter prevents constant dribbling of urine.

CLINICAL APPLICATIONS

Cystitis

Bladder inflammation, or **cystitis** (sis-ti'tis), can be caused by an infection spreading from the urethra, or in males, by pressure from an enlarged prostate gland. Nerve damage, or abnormalities that let the bladder sag, can lead to incomplete emptying and urine stasis (standing) and bacterial growth can cause chronic cystitis. Symptoms include frequent, painful, burning urination, and pressure in the pelvic area. The incidence of cystitis is higher in females than in males because the shorter female urethra allows infectious organisms to reach the bladder more easily.

Pyelonephritis

Inflammation of the renal pelvis and the medulla of the kidney, **pyelonephritis** (pi''el-o-nef-ri'tis), usually is caused by bacteria entering the kidney from the ureters. Preventing fecal material from reaching the urinary tract helps to prevent such infections, and women should wipe from front to back after voiding or defecating to avoid contaminating the urethral area. Partial urethral obstruction slows urine flow and allows bacterial growth and backflow into the renal pelvis. Although pyelonephritis can affect the kidney cortex, it usually affects mainly the countercurrent mechanism. Affected individuals have reasonably normal renal function except for their inability to concentrate urine. Signs and symptoms include frequent and painful urination, fever, and lumbar pain. Prompt antibiotic treatment can prevent permanent kidney damage.

Glomerulonephritis

A common filtration disorder, **glomerulonephritis** (glomer''u-lo-nef-ri'tis), is due to an allergic reaction often initiated by toxins from streptococci. Glomeruli become inflamed and engorged with blood, and antigen-antibody complexes are deposited in capillaries. Damage to the endothelium greatly increases capillary permeability so that proteins and red blood cells leave the blood. If the disease becomes chronic, some endothelium is replaced by impermeable fibrous tissue. As a result, some glomeruli are permeable to all blood components and others are permeable to none. Though the remaining healthy nephrons can compensate for damaged ones by filtering larger quantities of blood, a patient with many nonfunctional glomeruli needs dialysis or a kidney transplant.

Tubular Disorders

Impaired tubular functions, or **tubular disorders,** usually are inherited; they include vitamin D-resistant rickets and renal glycosuria.

Bowed legs and knock-knees in a child learning to walk suggest a vitamin D deficiency called **rickets.** The child has received too little calcium for proper mineralization of the skeleton. In most such cases, giving vitamin D will increase calcium absorption and, if bones are not already deformed, cure the disease. **Vitamin D-resistant rickets** (rickets that does not respond to normal doses of vitamin D) occurs in several forms. A form inherited as a sex-linked characteristic is probably due to a deficient transport protein that impairs phosphate reabsorption. A form inherited as an autosomal recessive is due to failure of mitochondria in kidney tubule cells to activate vitamin D. The first cannot be corrected by vitamin D in any amount, but the second responds to large doses of activated vitamin D. A form is caused by a faulty vitamin D receptor, but whether it is inherited and how to treat it are not yet known.

Renal glycosuria, glucose in the urine because of a tubular disorder, causes a person to lose 5 to 100 g of glucose daily. Blood glucose tends to be low rather than high as it is in diabetes mellitus. Two types of renal glucosuria have been identified and both are probably inherited. In type A, glucose carrier molecules are deficient in number, and in type B, they have reduced affinity for glucose. In both types, glucose is excreted because the renal tubules fail to return it to the blood. Renal glycosuria is a **benign disease;** that is, it requires no treatment. People who have it simply eat enough carbohydrate to compensate for excreted glucose and drink enough fluid to compensate for fluid excreted with the glucose.

Disorders of Micturition

Urine retention, the failure to release urine, is caused by nerve damage or obstruction. Neural signals may be absent or unable to initiate micturition. Abnormal tissue within the urethra or to pressure from an enlarged prostate gland can obstruct urine flow. The bladder fills with urine, micturition fails to occur, and urine continuously dribbles from the bladder.

Incontinence, the inability to control micturition, can be due to neural immaturity before age two, or after that age, to nerve or bladder damage, emotional stress, or irritating substances in the urine.

Renal Failure and Dialysis

The inability of the kidneys to perform their functions, **renal failure,** can be acute or chronic. **Acute renal failure** often follows vascular damage associated with glomerulonephritis or tubule obstruction with the filtrate leaking back into peritubular capillaries. Tissue edema occurs because sodium and water excretion are impaired. Hypertension, waste retention, tissue ischemia, and acidosis soon develop. If renal shutdown is complete, death occurs in one to two weeks without treatment. With appropriate treatment and tissue repair, nearly normal function may be regained. **Chronic renal failure** follows recurrent infections, trauma, and congenital defects such that remaining nephrons are too few to perform normal kidney functions.

The functions of the kidneys can be performed by **dialysis** (di-al'is-is), the transfer of wastes to dialysis fluids, in people suffering from renal failure until or instead of a kidney transplant. **Hemodialysis** (he''mo-di-al'is-is) draws wastes from the blood into dialysis fluid as the blood passes through selectively permeable tubes outside the body. **Peritoneal dialysis** draws wastes into sterile dialysis fluid introduced into the abdominal cavity.

Blood from an artery of a patient undergoing hemodialysis (Figure 27.20) is pumped through a tube within a fluid-filled dialyzer. The dialysis fluid has the same concentrations of electrolytes and nutrients as normal plasma, but lacks waste products. Pores in the tube prevent blood cells and proteins from entering the dialysis fluid but allow smaller molecules and ions to diffuse into the fluid. Waste substances such as urea and ammonia freely diffuse into the dialysis fluid, but only excesses of glucose, amino acids, and electrolytes do so. Nutrients placed in dialysis fluid diffuse into the blood. Following dialysis, blood is returned to the body. Patients who have a chronic need for dialysis often have surgically created subcutaneous fistulas to avoid damage to blood vessels regularly used for dialysis.

Though hemodialysis solves the problem of waste removal in patients with failed kidneys, it creates other problems. Removing blood from the body presents a risk of infection, which can be minimized by sterile techniques. Most patients must limit fluid intake even when thirsty to prevent edema. Because the patient receives dialysis treatment only a few times per week, wastes that accumulate between treatments can impair mental functions. Both the wastes and dependence on a machine can be stressful.

Peritoneal dialysis has many advantages over hemodialysis. Patients have a surgically implanted abdominal valve through which sterile dialysis fluid (1–2 l) drains from a bag into the peritoneal cavity. Waste materials diffuse from abdominal capillaries into the fluid for about two hours, then the fluid is drained back into the bag. Patients can perform this procedure up to four times a day even during work hours. They drink fluids freely, carry on ordinary activities, and do not suffer from waste accumulation. Peritoneal dialysis is much less expensive than hemodialysis and avoids problems associated with removing blood from the body. The main risk is peritonitis, which can be reduced by strict sterile techniques.

CLINICAL TERMS

albuminuria (al''bu-min-u're-ah)   the presence of albumin in the urine

azotemia (az''o-te'me-ah)   the presence of excess urea and other nitrogenous waste products in the urine

cystoscope (sis'to-skōp'')   an instrument for examining the inside of the bladder

enuresis (en''u-re'sis)   involuntary excretion of urine

gout (gowt)   a metabolic disorder in which an excess of uric acid accumulates in the body and is deposited in the joints

nephrosis (nef-ro'sis)   glomerular injury involving proteinuria

nephrotic (nef-rot'ik) **syndrome**   marked proteinuria resulting in edema and a reduction in serum albumin

proteinuria (pro''te-in-u're-ah)   the presence of protein in the urine

renal calculus (kal'ku-lus)   an abnormal aggregation of mineral salts or other substances in the kidney; kidney stone

uremia (u-re'me-ah)   the presence of urinary waste products in the blood

ureteritis (u''re-ter-i'tis)   inflammation of a ureter

urethritis (u''reth-ri'tis)   inflammation of the urethra

Wilm's tumor   an embryonic malignancy of the kidney

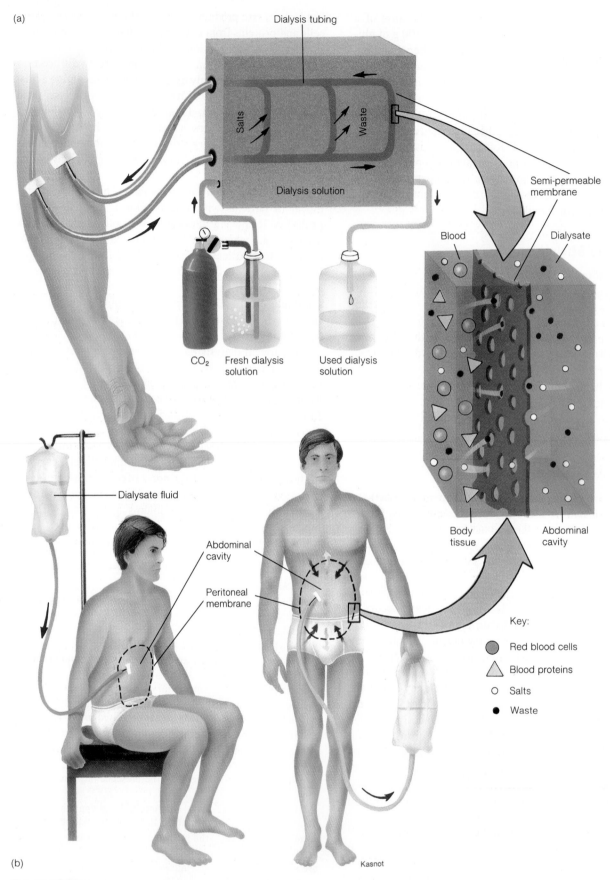

(a)

Dialysis tubing

Salts

Waste

Dialysis solution

Semi-permeable membrane

Blood

Dialysate

CO₂

Fresh dialysis solution

Used dialysis solution

Body tissue

Abdominal cavity

Dialysate fluid

Abdominal cavity

Peritoneal membrane

Kasnot

Key:

⬤ Red blood cells

△ Blood proteins

○ Salts

● Waste

(b)

FIGURE 27.20
(a) Hemodialysis, and (b) peritoneal dialysis.

THE ROLE OF URINE IN DIAGNOSIS

Most of us have provided urine samples when we visited a physician. Such a sample provides valuable information about one's physiological status because it contains all kinds of substances removed from the blood—metabolic products, hormones, acids, bases, electrolytes, and nitrogenous waste materials. A standard urinalysis includes a specific gravity determination; the use of dipsticks to measure pH and to detect glucose, ketones, proteins, and blood, if present; and microscopic examination of urinary sediment. Urine also is used for drug testing (Figure 27.21).

Specific gravity of an early morning specimen below 1.025 indicates the possibility of diabetes insipidus or distal renal tubule disease. Low pH (below 4.6) can indicate acidosis and high pH (above 8.0) can indicate alkalosis. Disorders associated with these and other characteristics of urine are listed in Table 27.3.

Glucose is often found in the urine of untreated diabetics, though it can be absent in an early morning specimen if the nighttime blood glucose does not exceed the renal threshold. In people with a very low renal glucose threshold, glucose appears in urine even when blood glucose is normal.

Protein in urine is associated with glomerulonephritis and a variety of other diseases distinguished by other diagnostic tests. In glomerulonephritis, the urine also contains blood and casts (hardened masses of epithelial cells, fat cells, and red and white blood cells) visible by microscopy. The blood of a glomerulonephritis patient also contains excess urea and creatinine. Victims of multiple myeloma (a kind of cancer) have Bence-Jones protein, an abnormal piece of an immunoglobulin made by the malignant cells, in their blood and urine.

Ketones from excessive fat metabolism appear in the urine of patients with acidosis, untreated diabetes, or other metabolic disorders. Low-carbohydrate, weight-reduction diets also lead to excessive fat metabolism and an undesirable ketonuria.

Bilirubin in the urine indicates that the liver cannot metabolize heme properly. Urobilinogen, derived from bilirubin, is normally excreted in the feces and appears in urine only when present in excess because of liver or blood disease. It sometimes appears in the urine of runners without any known disease.

Renal calculi (kidney stones), which can form anywhere along the urinary tract, consist of calcium phosphate, calcium oxalate, uric acid, or cysteine. Excessive parathyroid hormone secretion can cause calcium-containing calculi.

Hormones or their metabolic end products also are present in urine and can be used to diagnose endocrine disorders, especially those involving steroid hormones. Chorionic gonadotropin, produced by the placenta very early in pregnancy, appears in urine and forms the basis of some common pregnancy tests.

Health can be equated with homeostasis, and kidney function is exceedingly important in maintaining homeostasis. Excreted substances provide much information about homeostatic imbalances that health professionals can use in diagnosis and treatment.

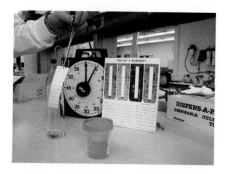

FIGURE 27.21
Technicians perform various tests on urine, including drug tests being done here.

TABLE 27.3 Some Abnormal Characteristics of Urine and the Conditions Associated with Them

Urine Abnormality	Associated Conditions
Red blood cells	Glomerulonephritis, congenital defects, malignancies, chronic infection, sickle cell anemia, drugs such as phenacetin and sulfonamide, bacterial endocarditis
Hemoglobin	Renal hypertension, hemolytic anemia, transfusion reactions, allergic reactions, severe burns, eclampsia (a complication of pregnancy)
Excessive acidity	Acidosis, diarrhea, dehydration, starvation, emphysema
Excessive alkalinity	Renal failure, bacterial infections, vomiting, alkalosis, low-carbohydrate diets
Protein	Glomerulonephritis, pyelonephritis, hypertension, congestive heart failure, bacterial endocarditis, gout, toxemia of pregnancy, potassium depletion, high fever
Glucose	Diabetes mellitus, pancreatitis, lowered renal threshold, coronary thrombosis, hyperthyroidism, shock, pain, excitement
Ketones	Acidosis, vomiting, diarrhea, glycogen-storage disease, starvation, low-carbohydrate diets, hyperthyroidism, eclampsia, trauma, chloroform or ether anesthesia
Bilirubin	Hepatitis, obstructive jaundice, cirrhosis of the liver, carcinoma of the pancreas, noxious fumes, chlorpromazine hepatitis
Urobilinogen	Some of same conditions as bilirubin, hemolytic anemias, pernicious anemia, thalassemia, hepatitis associated with infectious mononucleosis

Questions

1. What kinds of substances does urine normally contain?
2. What diseases can be tentatively diagnosed by changes in the concentration of substances in urine?
3. Name at least five substances or objects present in urine only under abnormal conditions, and describe the condition associated with each.
4. What is meant by the statement that health can be equated with homeostasis?

CHAPTER SUMMARY

Organization and General Functions

- The urinary system consists of the kidneys, ureters, urinary bladder, and urethra.
- The functions of the urinary system are to regulate the composition and volume of the blood and to remove from the blood wastes and substances that are present in excess.

Development

- The urinary system develops mainly from mesoderm of the urogenital ridge. Two pairs of kidneys develop and degenerate, and metanephric kidneys become the functional kidneys.
- The ureters and parts of the kidneys develop from the ureteric bud, and the bladder and urethra develop from the cloaca.

Anatomy of the Kidney

- The kidneys are near lumbar vertebrae behind the peritoneum. Each bean-shaped kidney has a ureter attached at a medial hilum, where blood vessels and nerves enter and leave. Internally, the kidney consists of cortical and juxtamedullary nephrons, collecting ducts located in the pyramids, minor and major calyces, and a renal pelvis.
- Each nephron consists of a glomerular capsule, proximal convoluted tubule, loop of Henle, and distal convoluted tubule; the loop of Henle is long and extends into the medulla in juxtamedullary nephrons.

Glomerular Filtration

- When blood passes through glomeruli, water and substances dissolved in plasma, such as electrolytes, glucose, urea, amino acids, and small polypeptides, enter the glomerular filtrate by filtration. The glomerular filtration rate is determined mainly by net filtration pressure.

Tubular Reabsorption

- Except for wastes and excess electrolytes, most water and solutes from plasma are reabsorbed from the tubule back into the plasma. Glucose, amino acids, and Na^+ are actively transported to the plasma. Cl^- and most electrolytes diffuse according to concentration and electrical gradients. Water moves by osmosis.

Tubular Secretion

- H^+ and K^+ are actively secreted into the distal tubule lumen, as explained in more detail in Chapter 28.

The Countercurrent Mechanism

- In a countercurrent system, outflow runs parallel to and in the opposite direction of the inflow.
- Active transport of Na^+ and Cl^- from the thick segments of loops of Henle creates high osmolarity in peritubular fluid, causing water to move out of collecting ducts by osmosis. Outflow of Na^+ and Cl^- affect the inflow portion of loops. Active transport produces the multiplier effect.
- A countercurrent exchanger preserves high osmolarity in peritubular fluids by solutes diffusing into blood as it enters the vasa recta and out of it as it leaves the vasa recta.

Regulation of Urine Formation

- Urea, a major nitrogenous waste, is filtered from plasma but some is reabsorbed and some is trapped in peritubular fluid.
- Other nitrogenous wastes include creatinine, ammonia, uric acid, and traces of amino acids.
- Antidiuretic hormone (ADH), which is secreted when body fluids become hyperosmotic, increases the permeability of collecting ducts and distal tubules. Much water and some urea are reabsorbed until fluid osmolarity returns to normal.
- Diabetes insipidus is an ADH deficiency in which patients can lose up to 15 l/day of urine unless their deficiency is compensated by inhaling ADH.
- Diuresis, increased urine secretion, can be caused by ADH deficiency (above), hyposmolarity of body fluids, hyperosmolarity of kidney filtrate (osmotic diuresis), alcohol, and caffeine. Drugs called diuretics are used to control edema in hypertension and other disorders.

How Kidneys Regulate Blood Flow

- The juxtaglomerular apparatus includes juxtaglomerular cells, smooth muscle cells in afferent arterioles that secrete renin, and the macula densa, secretory cells in distal tubule cells that respond to a decrease in the glomerular filtration rate by stimulating renin release.
- In the renin-angiotensin mechanism, renin acts on angiotensinogen to produce angiotensin I, which is activated to angiotensin II by plasma-converting enzyme. Angiotensin II causes vasoconstriction, which raises blood pressure directly, and stimulates aldosterone secretion, which raises blood pressure by increasing blood volume.
- Local regulation maintains balance between glomerular filtration and reabsorption.
- Atrial peptides are believed to increase water secretion and relax renal arteries and arterioles, and modestly increase Na^+ secretion.

Measurement of Kidney Function

- Clearance, or renal clearance rate, is a common determination in assessing kidney function. Clearance of inulin, which is neither reabsorbed nor secreted, is equal to the glomerular filtration rate. Clearance of PAH, which is both filtered and secreted, allows estimation of renal blood flow.
- Knowing the glomerular filtration rate and the renal blood flow rate, one can determine from urea clearance what proportion of urea is being excreted.
- Because the plasma concentration and the excretion rate of creatinine normally remain constant, finding an increase in plasma creatinine suggests a kidney disorder.

Anatomy of Ureters, Bladder, and Urethra

- The ureters are tubes that consist of inner transitional epithelium, smooth muscle, and outer fibrous connective tissue.
- The urinary bladder is a hollow bag that consists of the same basic layers; the thick detrusor muscle is innervated by parasympathetic fibers.
- The urethra is a tube lined with epithelium with underlying structures that attach it to adjacent tissues.

Characteristics of Urine
- The physical characteristics of urine depend on fluid intake and kind of food eaten.
- Urine contains at least 95 percent water and the remainder consists of solids such as urea, uric acid, creatinine, and electrolytes.

Micturition and Its Control
- Micturition is the release of urine that has been stored in the urinary bladder. It is controlled by the micturition reflex.

Clinical Applications
- Cystitis is an inflammation of the bladder often caused by a bacterial infection.
- In pyelonephritis, inflammation, usually from a bacterial infection, impairs function of the loops of Henle.
- Glomerulonephritis is an inflammation due to advancing pyelonephritis, trauma, congenital defects, or allergic reactions. It causes some glomeruli to become excessively permeable and others impermeable.
- Tubular disorders, which often impair reabsorption, include vitamin D-resistant rickets and renal glycosuria.
- Disorders of micturition include urine retention (failure of the bladder to drain completely), and incontinence (inability to control urine release).
- Renal failure is the inability of the kidneys to perform their functions. Hemodialysis and continuous ambulatory peritoneal dialysis can be used to compensate for kidney failure.

QUESTIONS AND PROBLEMS

The questions at the end of each chapter are numbered to correspond with the objectives listed at the beginning of the chapter. Italics indicate that a question requires critical thinking skills beyond simple factual recall.

Questions
1. (a) What are the major organs of the urinary system?
   (b) What are the major functions of the urinary system?
2. (a) From what embryonic tissue is the urinary system formed?
   (b) *What is the significance of the three types of kidneys found in the embryo?*
3. (a) Describe the location and size of the kidney.
   (b) *Relate the gross anatomical structures inside the kidney to the flow of urine through it.*
   (c) *Trace the flow of blood through the kidney.*
   (d) Describe the structure of a juxtamedullary nephron, and explain how a cortical nephron differs from it.
4. (a) *Where does filtration occur, and what forces produce it?*
   (b) *What determines which substances leave the blood during filtration and which are held back?*
   (c) *How do various factors affect the glomerular filtration rate?*
5. (a) *Where does reabsorption occur, and what produces it?*
   (b) *What determines which substances are returned to the blood during reabsorption?*

   (c) *What kinds of defects impair reabsorption, and how do they exert their effects?*
6. *Where and how does secretion occur, and what are its effects?*
7. (a) What is a countercurrent system?
   (b) *How does multiplication occur in the countercurrent multiplier?*
   (c) *What do countercurrent multipliers accomplish?*
   (d) *What do countercurrent exchangers accomplish?*
8. (a) *How and to what extent is urea excreted?*
   (b) What other nitrogenous wastes are excreted by humans?
9. (a) *How does antidiuretic hormone regulate urine volume?*
   (b) What is diabetes insipidus, and how is it treated?
   (c) *Under what conditions can diuresis occur in humans?*
10. (a) What are the components of the juxtaglomerular apparatus, and what are their functions?
    (b) *What steps occur in the renin-angiotensin mechanism, and what are its effects?*
    (c) *How does local regulation affect kidney functions?*
    (d) *How do atrial peptides act on the kidneys?*
11. (a) *What kind of information about kidney function is provided by clearance of inulin, PAH, or urea?*
    (b) *What might be the significance of an increase in plasma creatinine?*
12. (a) How are the walls of the ureters, bladder, and urethra similar, and how are they different?
    (b) *How do the internal and external sphincters differ?*
13. What are the physical and chemical characteristics of urine?
14. *What is micturition, and how is it controlled?*
15. (a) *How do pyelonephritis and glomerulonephritis differ?*
    (b) *What are the consequences of tubular defects?*
    (c) *What are the consequences of micturition disorders?*
16. (a) *What is renal failure?*
    (b) *How can dialysis compensate for renal failure?*
    (c) *What are the advantages and disadvantages of hemodialysis and peritoneal dialysis?*

Problems
1. If the concentration of a substance is 1 mg/ml in blood and 20 mg/ml in urine and urine is produced at 1 ml/min, what is the clearance of the substance? (Assume the substance is neither reabsorbed nor secreted.)
2. A patient suspected of having kidney disease was given inulin to a plasma level of 1 mg/ml. The patient produced 90 ml of urine in one hour and the urine contained an inulin concentration of 50 mg/ml. What physiological disorder(s) may be present?
3. How would you explain to the mother of a toddler the reason the child may not be toilet trained before the age of two?
4. List at least ten substances whose concentration in the blood is regulated in some way by the kidneys. Explain the role of the kidneys in regulating each of these substances.
5. Why would a patient in chronic renal failure likely be anemic?

OBJECTIVES

1. List the kinds of mechanisms that help the body maintain homeostasis under a wide range of conditions.

2. Describe the characteristics of the body's fluid compartments including the electrolytes they contain.

3. Explain how the body regulates water balance.

4. Explain how sodium and potassium, calcium and phosphate, and other electrolytes are regulated.

5. Describe the properties of buffers.

6. Explain the mechanisms of maintaining acid-base balance.

7. Describe the effects of disturbances in fluid volumes.

8. Define acidosis and alkalosis, and distinguish between respiratory and nonrespiratory forms of these disorders.

9. Describe the causes and consequences of acid-base imbalances.

28

FLUID, ELECTROLYTE, AND ACID-BASE BALANCE

Physiology of Fluids, Electrolytes, Acids, and Bases

The human body has an immense capacity to maintain internal homeostasis regardless of environmental and dietary changes. It can adapt to sudden sweating after you fly from cold, snowy New York to hot, sunny Buenos Aires. It maintains nearly constant internal conditions even if you eat a whole bag of salty potato chips or drink several glasses of acidic lemonade. These capacities depend on important regulatory mechanisms.

Since considering the principles of homeostasis (Chapter 1), we have looked at many regulatory mechanisms involving cell membranes, neural signals, and hormones. Here we will see how such mechanisms maintain homeostasis of fluids, electrolytes, acids, and bases.

Regulation of Body Fluids

Body fluids are more than 90 percent water and contain a variety of solutes (dissolved substances). Such fluids are found in cells, between cells of most tissues, and in blood and lymph. With respect to water itself, the bodies of adult males of normal weight contain 55 to 60 percent water and those of females have relatively less water and more fat. Because fatty tissue contains almost no water, extremely obese bodies may be no more than 40 percent water and extremely lean ones as much as 75 percent water. Total body water declines with age from about 70 percent at birth to 60 percent at six months and 40 percent in old age.

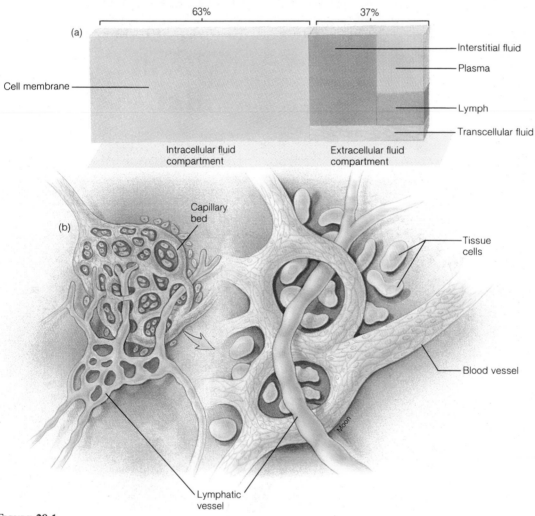

Figure 28.1

(a) Volumes of fluid compartments, and (b) relationships between compartments.

Fluid Compartments

Of the total body fluid, about two-thirds is **intracellular fluid.** The remaining third, **extracellular fluid,** consists mostly of **interstitial fluid** and **plasma.** Small quantities of extracellular fluid also are found in the body as lymph, cerebrospinal fluid, and fluid in joint cavities and inside the eyeballs. Typical volumes of the body's three major fluid compartments—the intracellular compartment, the interstitial fluid compartment, and the plasma compartment—are shown in Figure 28.1.

Electrolytes in Fluids

All body fluids contain **electrolytes** (e-lek′tro-lītz), substances that ionize and can conduct electricity, but their concentrations differ by compartments (Figure 28.2). Electrolyte concentrations are small and are expressed in "milli" units, milliequivalents (mEq) or milliosmols (mOsm). For electrolytes with a 1+ or 1− charge, milliequivalents equal milliosmols. A milliequivalent equals the number of charges in a millimole (mmol) (1/1000 mole) of hydrogen ions. A milliosmol is 1/1000 mole of a substance times the number of particles it forms in solution.

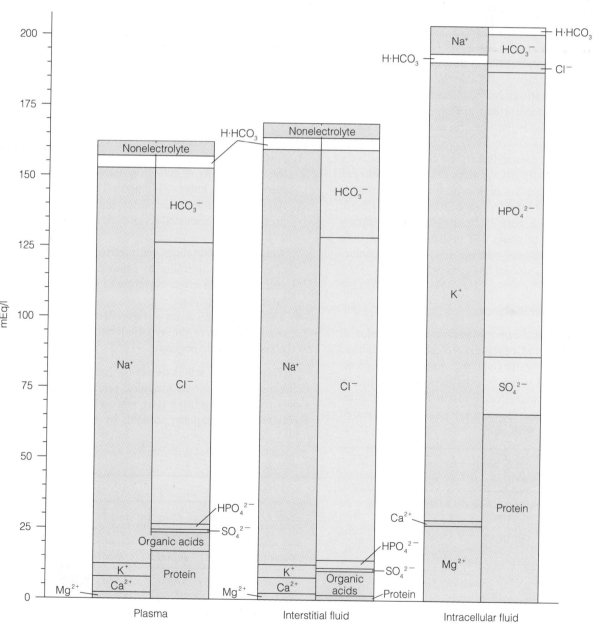

FIGURE 28.2
Composition of plasma, interstitial fluids, and intracellular fluids.

Some sample calculations will help to clarify the distinction between milliequivalents and milliosmols. Sodium, for example, has an atomic weight of 23: 1 mole = 23 g and 1 mmol = 0.023 g. This quantity of sodium ions is also 1 mEq and 1 osmol (Osm). It has the same number of charges as 1 mmol of hydrogen ions and each ion forms one particle in solution. For calcium, with an atomic weight of 40, 1 mmol is 0.040 g. Because calcium ions have two charges (Ca^{2+}), 1 mmol of calcium equals 2 mEq but only 1 mOsm. It has twice as many charges as a millimole of hydrogen ions, but the same number of particles. Given that NaCl has a molecular weight of 58 grams, a 1 mmol/l solution contains 0.058 g/l. NaCl ionizes completely so the solution contains 1 mEq/l Na^+ and 1 mEq/l Cl^-. It exerts 2 mOsm/l because each molecule produces two particles when it ionizes in solution.

Electrolyte concentrations in fluids of different compartments vary significantly (Table 28.1). Of these differences, high interstitial Na^+ concentrations and high intracellular K^+ concentrations are especially important in membrane excitability, particularly in nerve and muscle cells.

Within any one compartment, the cations (positively charged ions) normally equal anions (negatively charged ions)—about 155 mEq/l of each in plasma and interstitial fluids, and about 200 mEq/l of each in intracellular fluid, as already shown in Figure 28.2. All fluid compartments have an osmolarity of about 282 mOsm/l—higher than electrolyte concentrations because uncharged (nonelectrolyte) substances also contribute to osmolarity.

Regulation of Water Balance

Water is essential for the operation of metabolic cycles. It is ingested in food and beverages, released in metabolic reactions (metabolic water), and lost through urine, feces, sweat, and as water vapor in exhaled air. Water loss in exhaled air and in sweat during rest is referred to as **insensible water loss** because such loss cannot be measured directly, but is nonetheless important clinically. When water intake equals water output, the body is said to be in **water balance.** The amounts of water gained and lost by a typical adult on a normal diet in a temperate climate are summarized in Table 28.2.

Under normal physiological conditions, osmotic gradients never persist across cell membranes. Whenever small local differences arise between extracellular and intracellular osmotic pressures, water moves to equalize the pressures in less than a minute. When more extensive disturbances occur, total body equilibrium may not be reestablished so quickly. For example, suppose that a person loses a large volume of water and electrolytes in sweat while working hard outdoors on a hot day. Even when the person drinks sufficient electrolyte replacement fluid, such as Gatorade®,* it takes four to six hours for the fluid to be absorbed and distributed through fluid compartments to reestablish total body equilibrium.

Though osmosis is directly responsible for maintaining equilibrium between fluid compartments, at least four mechanisms regulate the amounts of fluid entering or leaving the body.

1. Antidiuretic hormone conserves water by increasing the permeability of collecting ducts and distal tubules of the kidneys (Chapter 27).
2. Atrial peptides increase water and sodium loss, probably by various effects on the kidneys (Chapter 27).
3. Aldosterone stimulates sodium reabsorption, causing water to follow by osmosis. It also stimulates potassium excretion as will be explained later in this chapter.
4. The thirst mechanism initiates drinking when the osmoreceptors in the hypothalamus detect increased plasma osmotic pressure.

Thirst is a conscious desire for liquid that results in drinking. The hypothalamic **thirst center** has **osmoreceptors** (oz″mo-re-cep′torz) that respond to increased osmolarity of tissue liquids, probably by detecting increases

*Gatorade® is distributed by Stokely-Van Camp, Inc., Chicago, IL.

TABLE 28.1 Variations in Electrolyte Concentrations in Body Fluids

Differences between Extracellular and Intracellular Fluids

Extracellular Fluid	Intracellular Fluid
High in Na^+ and Cl^-	Low in Na^+ and Cl^-
Low in K^+ and HPO_4^{2-}	High in K^+ and HPO_4^{2-}
Relatively high in HCO_3^-	Relatively low in HCO_3^-
Relatively low in Mg^{2+}	Relatively high in Mg^{2+}
Relatively low in protein	High in protein

Differences between Plasma and Interstitial Fluids

Plasma	Interstitial Fluid
Relatively high in protein	Very low in protein
High in Cl^-	Extremely high in Cl^-

TABLE 28.2 Typical Water Input and Output in Normal Water Balance

Water Input	ml/day	Water Output	ml/day
Liquids imbibed	1000	Urine	1300
Water from solid food	1000	Feces	100
Metabolic water	500	Sweat	650
		Water vapor in exhaled air	450
Total input	2500	Total output	2500

in sodium. When these liquids have high osmotic pressure (are too concentrated), osmoreceptors send signals to the cerebrum that cause a person to drink liquids. They also initiate release of antidiuretic hormone, which decreases urine volume and conserves liquid.

Curiously, humans experience relief of thirst immediately after drinking—before osmoreceptors detect added liquid and even before it is absorbed from the digestive tract. How thirst is so suddenly quenched is not fully understood, but the presence of liquid in the stomach seems to be partly responsible. If thirst continued for the 30 minutes it takes for liquids to return equilibrium, too much liquid would be taken in and the body would become overhydrated. The immediate quenching of thirst reduces the range of variation in osmotic pressure.

REGULATION OF SPECIFIC ELECTROLYTES

Having considered the volumes and compositions of fluids themselves, we can now focus on mechanisms that regulate the concentrations of their electrolytes—sodium, potassium, calcium, magnesium, phosphates, chloride, and bicarbonate (Table 28.3). Concentrations in glomerular filtrate and urine are included in the table because excretion of urine is the body's primary means of adjusting electrolyte concentrations. Bicarbonate is included because it is important in acid-base balance, as we shall see.

TABLE 28.3	Selected Electrolytes in Body Fluids (mEq/l)				
Electrolyte	Intracellular Fluid	Interstitial Fluid	Plasma	Glomerular Filtrate	Urine
Na^+	14	142	146	146	128
K^+	140	4	5	5	60
Ca^{2+}	0	4	4	4	5
$HPO_4^{2-}/$ $H_2PO_4^-$	15	2	2	2	50
Cl^-	4	108	103	103	134
Mg^{2+}	62	2	2	2	15
HCO_3^-	10	29	28	28	14

Sodium and Potassium

Among body electrolytes, sodium is most abundant in extracellular fluids and potassium is most abundant in intracellular fluids. Both are found in large quantities in various tissues (Figure 28.3) More than half the body's sodium is deposited in bone and other connective tissues, whereas 95 percent of the total body potassium is confined to intracellular fluids. The imbalance of sodium and potassium concentrations across membranes is maintained by sodium-potassium pumps in cell membranes.

Sodium salts account for more than 90 percent of body fluid osmolarity, so sodium regulation largely determines body fluid volumes. Potassium, an abundant intracellular ion, also must be regulated to maintain normal cell volumes. Both must be precisely regulated for normal nerve and muscle membrane excitability.

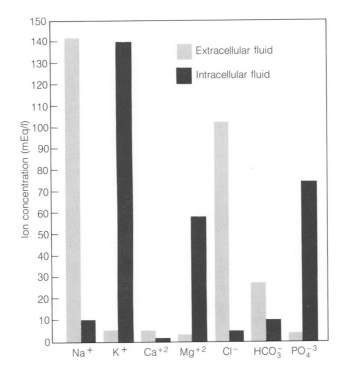

FIGURE 28.3
Distribution of ions in body fluids.

Sodium and potassium levels and fluid volumes are regulated mainly by **aldosterone,** which causes sodium reabsorption and potassium secretion in kidney tubules. Aldosterone acts in the kidneys by combining with a receptor protein in cells of the distal convoluted tubules and collecting ducts. The resulting aldosterone-receptor complex activates DNA, which, in turn, initiates synthesis of proteins needed for the sodium-potassium transport process. Whether proteins are carriers, enzymes, or both is not entirely clear, but from aldosterone experiments on animals, it is clear that the sodium-potassium transport rate increases only after a time lag of about 45 minutes.

Proteins synthesized from the above process act at the membrane surface of sensitive cells (Figure 28.4). Active transport of sodium out of cells creates very low intracellular sodium concentration and a negative potential (-70

mV). The low cellular sodium and the very negative potential both draw sodium from the tubule into the cell. At the same time, aldosterone causes active transport of sodium to interstitial fluid from which it diffuses into plasma; it also causes transport of potassium from interstitial fluid into tubule cells.

Extracellular potassium initiates and maintains aldosterone secretion in proportion to the potassium concentration; thus, both potassium and aldosterone are homeostatically controlled by negative feedback (Figure 28.5). When blood pressure or fluid volume drop, the renin-angiotensin system is activated and angiotensin stimulates aldosterone release (Chapter 27). Sympathetic stimulation and ACTH release, which cause cortisol secretion, also can cause aldosterone secretion. Disorders such as renal hypotension, shock, and heart failure elevate renin, angiotensin, and aldosterone.

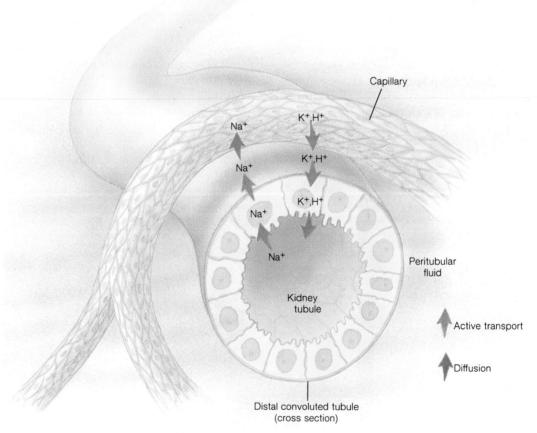

FIGURE 28.4
Movement of Na^+, K^+, and H^+ between distal convoluted tubule and capillary.

Severely restricting dietary sodium intake increases aldosterone release, which, in turn, increases sodium reabsorption and partially defeats the effect of dietary restriction. Possible mechanisms for this effect are as follows:

1. Reducing sodium intake reduces extracellular volume, which, in turn, reduces cardiac output and renal blood flow. Reduced renal blood flow activates the renin-angiotensin system and stimulates aldosterone release.
2. Reducing sodium intake lowers potassium excretion and excess potassium increases aldosterone release.

Failure of the body's regulatory mechanisms to maintain normal sodium or potassium levels can lead to the disorders summarized in Table 28.4.

Calcium and Phosphate

Though calcium and phosphate are found in body fluids as electrolytes, as much as 99 percent of the total body content of these minerals is deposited in bone as the crystalline salt **hydroxyapatite** (hi-drok″se-ap′ah-tīt). Having considered deposition and reabsorption of bone minerals in Chapters 6 and 16, we will limit this discussion to the roles of calcium and phosphate as electrolytes.

Total plasma calcium is maintained between 8.5 and 10.5 mg/dl, distributed in three forms: (1) 40 percent bound to plasma proteins and not ionizable at normal blood pH, (2) 50 percent Ca^{2+}, and (3) 10 percent ionizable inorganic salts such as calcium citrate and calcium phosphate. Though protein-bound calcium remains in plasma,

TABLE 28.4   Some Causes and Effects of Sodium and Potassium Imbalances

Imbalance	Causes	Effects
Hyponatremia (plasma Na^+ less than 136 mEq/l)	Sodium loss Kidney disease, Addison's disease, vomiting, diarrhea, excessive sweating, use of diuretics Water excess Excessive water intake, excessive aldosterone release, oliguria from renal failure	Decreased osmolarity of extracellular fluids leads to excess excretion of fluid until normal osmolarity is restored; can lead to dehydration and shock. Decreased response of nerve and muscle cells to stimulation because of decreased sodium gradient. Water intoxication leads to dilution of intracellular fluids, edema, and decreased concentration of nearly all plasma constituents.
Hypernatremia (plasma Na^+ greater than 150 mEq/l)	Sodium accumulation Intake of excessively salty fluids or foods Water deficit Deficient water intake, increased insensible water loss, excessive use of diuretics, osmotic diuresis because of high concentration of wastes in urine	Increased osmolarity of extracellular fluids and fluid retention; fever due to increased metabolic rate to supply ATP for Na^+-K^+ pump; hyperexcitability of nerve and muscle cells because depolarization is faster and requires only a weak stimulus. Dehydration, water retention to the extent possible, oliguria, shrinkage of cells.
Hypokalemia (plasma K^+ less than 3.5 mEq/l)	Use of diuretics, diarrhea, vomiting, Cushing's syndrome, renal disease	Hyperpolarization of membranes leads to decreased excitability of nerve and muscle cells; this leads to lethargy, muscle weakness, and reduced gut motility; abnormalities in the heart conduction system.
Hyperkalemia (plasma K^+ greater than 5.5 mEq/l)	Renal failure, hypertonic dehydration, massive tissue damage, Addison's disease	Partial depolarization of membranes leads to temporary increase in excitability of nerve and muscle cells; further reduction in polarization causes cells to lose ability to respond to stimuli; impairment of the heart conduction system and eventual cardiac arrest.

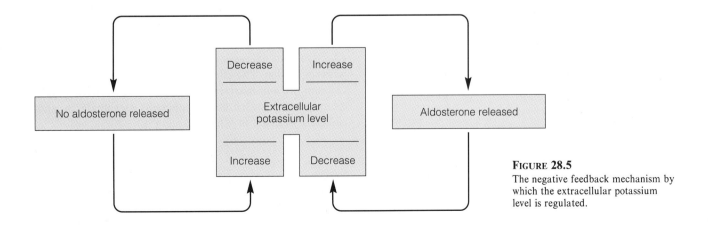

FIGURE 28.5
The negative feedback mechanism by which the extracellular potassium level is regulated.

TABLE 28.5   Some Causes and Effects of Calcium Imbalances

Imbalance	Causes	Effects
Hypocalcemia (total plasma Ca^{2+} less than 8.5 mg/dl or ionic Ca^{2+} less than 4.0 mg/dl)	Excess of phosphates can precipitate calcium; lack of plasma proteins fails to carry adequate bound calcium, and plasma protein deficiencies result from liver or kidney disease or starvation; decreased parathormone secretion, vitamin D deficiency, hyperventilation by increasing plasma pH.	Partial depolarization of nerve and muscle cell membranes leads to hyperexcitability. Hyperexcitability is displayed by muscle cramps, convulsions, mental confusion, and tetany (continuous muscle contraction). Tetany prevents normal breathing movements and can cause severe dyspnea and death. Increased cardiac excitability causes a delay between depolarization and repolarization of the ventricles; it also reduces the strength of contractions.
Hypercalcemia (total plasma Ca^{2+} greater than 12 mg/dl or ionic Ca^{2+} greater than 5.6 mg/dl)	Increased parathormone secretion, malignant tumors (that are thought to secrete parathormone), excessive intake of vitamin D, hyperthyroidism, Addison's disease, certain bone diseases, and immobilization (that leads to excessive bone reabsorption).	Excessive active transport of Ca^{2+} out of cells, which interferes with entry of Na^+ when cells are stimulated and leads to hyperpolarization of nerve and muscle cell membranes. Decreased excitability leads to fatigue, listlessness, sleepiness and even coma, muscle weakness, reduced gut motility, sometimes abdominal pain and anorexia. Dehydration may also occur as Ca^{2+} interferes with the reabsorption of Na^+ and the concentration of urine. $CaHPO_4$ precipitates in tissues and especially in the collecting ducts of the kidneys and the walls of arteries.

Ca^{2+} freely diffuses between plasma and interstitial fluid producing an interstitial calcium concentration of about 6 mg/dl—60 percent of that in the plasma.

In plasma, an equilibrium between protein-bound calcium and free calcium is influenced by the plasma pH.

$$Ca\text{-protein} + 2H^+ \leftrightarrow Ca^{2+} + 2H\text{-protein}$$

Decreasing plasma pH drives the reaction to the right as excess H^+ replaces Ca^{2+} on plasma proteins; increasing plasma pH drives the reaction to the left as H^+ is released from proteins and Ca^{2+} attaches to them. Plasma contains relatively more free Ca^{2+} when slightly acidic and relatively less Ca^{2+} when slightly alkaline.

Intracellular fluids normally contain only very low concentrations ($10^{-7}M$) of free Ca^{2+}; however, a variety of cellular processes are activated by the influx of Ca^{2+} into stimulated cells.

Phosphate concentration is relatively high (15 mEq/l) in intracellular fluids and much lower (2 mEq/l) in extracellular fluids. In cells, phosphates participate in phosphorylation of glucose and nucleotides such as ADP. In extracellular fluids, phosphates, which are about 80 percent dibasic phosphate (HPO_4^{2-}) and 20 percent monobasic phosphate ($H_2PO_4^-$), help to maintain a normal blood pH. As with calcium, the relative proportions of phosphates are affected by the pH of extracellular fluids.

$$HPO_4^{2-} + H^+ \leftrightarrow H_2PO_4^-$$

Decreasing plasma pH drives the reaction to the right as H^+ from plasma binds to HPO_4^{2-}, whereas increasing plasma pH drives the reaction to the left as H^+ enter plasma. Unlike many changes in electrolyte concentrations, changes in extracellular phosphate concentrations—from one-fourth to four times the normal—cause no problems. (Parathormone does decrease plasma phosphate concentration, as was explained in Chapter 16.)

When calcium ions increase, phosphate ions decrease, and *vice versa,* but the product of their concentrations remains constant (at about 35). Should the total exceed that which can dissolve in plasma, $CaHPO_4$ will precipitate and deplete body fluids of the ions. People who frequently drink carbonated beverages, which contain large amounts of phosphate, may fail to absorb Ca^{2+} because it binds to phosphate in the gut before being absorbed.

Calcium Imbalances

In contrast to phosphates, changes in extracellular calcium ion levels have profound effects on many physiological processes. Ca^{2+} regulate cellular processes (Chapter 3), muscle function (Chapter 9), neurotransmitter release (Chapter 11), blood clotting (Chapter 17), and cardiac function (Chapter 18). Very small Ca^{2+} concentrations are effective, and excesses or deficiencies seriously interfere with regulation. The effects of excesses and deficiencies of calcium are summarized in Table 28.5. Disorders of calcium metabolism include osteoporosis, osteomalacia, and rickets, discussed in other chapters.

TABLE 28.6   Some Causes and Effects of Imbalances in Magnesium and Chloride

Imbalance	Causes	Effects
Hypomagnesemia (plasma Mg^{2+} less than 1.5 mEq/l)	Excessive loss from excess aldosterone, thiazide diuretics, or polyuria; inadequate intake from malnutrition or alcoholism; inadequate absorption from excess calcium that competes for transport, diarrhea, vomiting, hypoparathyroidism, and malabsorption disorders.	Muscle cramps, tetany, muscle weakness; tachycardia and heart arrhythmias; psychological changes, ataxia, and convulsions; hypertension.
Hypermagnesemia (plasma Mg^{2+} greater than 2.5 mEq/l)	Renal failure, hyperparathyroidism, inadvertent overdose of magnesium in therapy.	Bradycardia and hypotension as Mg^{2+} blocks sympathetic stimulation of heart and blood vessels; nausea and vomiting; lethargy, depressed respiration, and decreased reflex activity; death from respiratory or cardiac arrest.
Hypochloremia (plasma Cl^- less than 95 mEq/l)	Prolonged vomiting, and in infants prolonged diarrhea; excessive use of diuretics; metabolic alkalosis	No specific effects known. Hypochloremia is usually associated with alkalosis.
Hyperchloremia (plasma Cl^- greater than 115 mEq/l)	Depletion of bicarbonate in extracellular fluids as a result of metabolic acidosis.	No specific effects known. Hyperchloremia is usually associated with acidosis. Some recent evidence suggests that high chloride intake may contribute to hypertension.

Other Electrolytes

Body fluids contain numerous other electrolytes—magnesium, chloride, sulfate, nitrate, lactate, urate, and trace minerals. We will consider only magnesium and chloride.

Magnesium

The magnesium ion concentration is high in intracellular fluids, quite low in extracellular fluids, and highly variable in urine depending on diet (Table 28.3). Fruits, whole grains, and other foods from plants contain more magnesium than foods from animals. Both magnesium and Ca^{2+} are actively transported in opposite directions by the same membrane pump, which maintains a high intracellular magnesium concentration and a high extracellular calcium concentration.

Magnesium has important functions in both intracellular and interstitial fluids. In cells, it is a cofactor for many enzymes, especially those that catalyze reactions involving ATP. In interstitial fluid, it blocks acetylcholine release at synapses and neuromuscular junctions, decreasing excitability of both nerve and muscle cells. This property is applied medically by administering magnesium sulfate to prevent seizures.

In addition to active transport at cell membranes, magnesium also is regulated by the kidneys, probably in a system involving aldosterone and negative feedback like that regulating potassium excretion. Magnesium usually is well regulated, and imbalances are rare. Possible causes and effects of such imbalances are summarized in Table 28.6.

Chloride

The chloride ion concentration is low in intracellular fluids and high in extracellular fluids. It varies in urine according to the amount ingested, but is usually higher than in plasma. Though most cell membranes are relatively impermeable to chloride ions, the ions can be actively transported or follow electrical gradients created by sodium ion transport. Also, in chloride shift, negatively charged bicarbonate ions move across a membrane and chloride ions move in the opposite direction, equalizing the charge (Chapter 22).

Chloride concentrations are maintained by membrane phenomena (impermeability, active transport, and chloride shift) and by kidney excretion. When aldosterone causes sodium reabsorption, chloride ions also are reabsorbed along the electrical gradient and, thereby, indirectly regulated. Excess chloride ions are excreted. Like magnesium, chloride usually is well regulated, and imbalances are rare. Some causes and effects of chloride imbalances are summarized in Table 28.6.

ACID-BASE REGULATION

The slightly acidic to neutral pH of intracellular fluids (6.7 to 7.0) is due to acids produced as cells carry out their metabolic activities. Cells also release carbon dioxide and other wastes that move from interstitial fluids to plasma. In spite of these wastes, the pH of such fluids stays within a narrow range of 7.35 to 7.45 because of quick-acting buffers and acid-base regulating mechanisms.

FIGURE 28.6
Buffer action.

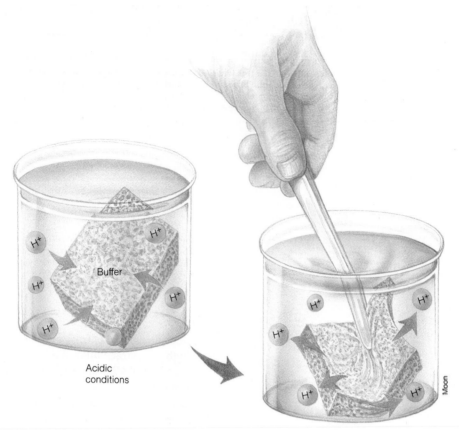

Acidic
conditions

Basic
conditions

Buffers

A **buffer** is a substance that resists pH change, usually by taking up or releasing H^+ (Figure 28.6). Recall that pH is a measure of H^+ concentration. The human body has three important buffer systems—bicarbonate-carbonic acid, phosphates, and proteins. The **bicarbonate-carbonic acid buffer system** consists of a weak acid called carbonic acid (H_2CO_3) and its salt called sodium bicarbonate ($NaHCO_3$). The salt of an acid is formed by replacing hydrogen with another cation, such as sodium. This buffer system acts mainly in plasma. The **phosphate buffer system** consists of two salts of phosphoric acid (NaH_2PO_4 and Na_2HPO_4). It acts mainly in intracellular fluids and in urine, not in plasma. The **protein buffer system** consists of carboxyl and amino groups of amino acids not involved in peptide bonds and, therefore, free to donate and accept H^+, respectively. Hemoglobin and plasma proteins are important blood buffers.

Buffers can resist pH change even in the presence of a strong acid or strong base. For example, hydrochloric acid (HCl) is a strong acid—it completely ionizes into H^+ and Cl^-—and sodium hydroxide (NaOH) is a strong base—it completely ionizes into Na^+ and OH^-. The OH^- greatly increases the pH of a solution because it acts as a

H^+ acceptor; that is, it removes H^+ from solution. Carbonic acid is a weak acid—only a small proportion of carbonic acid molecules ionize to form H^+ and HCO_3^-. Sodium bicarbonate acts as a weak base—it ionizes in solution to form Na^+ and HCO_3^-. The Na^+ does not affect pH, but bicarbonate ions are weak H^+ acceptors that remove some H^+ from solution.

Now let us see how the bicarbonate-carbonic acid system reduces the effects of strong acids or bases on the pH of a solution (Figure 28.7). When a strong acid and sodium bicarbonate react, a new salt and a weak acid are formed.

$$HCl + NaHCO_3 \rightarrow NaCl + H_2CO_3$$

The salt has no effect on pH, and carbonic acid releases far fewer H^+ into the solution than did hydrochloric acid. The number of free H^+ is reduced and the solution's pH is unchanged or held in a narrow range. Similarly, when a strong base and a weak acid react, water and a weak base are formed.

$$NaOH + H_2CO_3 \rightarrow H_2O + NaHCO_3$$

The water has no effect on pH and the sodium bicarbonate accepts far fewer H^+ than the strong base. Fewer H^+ are removed from the solution and the pH is held within a

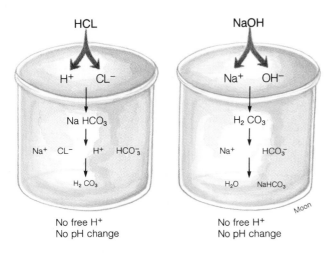

No free H⁺
No pH change

No free H⁺
No pH change

FIGURE 28.7
How the bicarbonate buffer system works.

narrow range. For the pH to remain stable, the solution must contain enough of the buffer to react with all the strong acid or base in the solution.

In the phosphate buffer system, sodium dihydrogen phosphate (NaH_2PO_4) behaves as a weak acid and disodium hydrogen phosphate (Na_2HPO_4) behaves as a weak base. The effects of this system on strong acids and bases are shown in the following reactions.

$$HCl + Na_2HPO_4 \leftrightarrow NaCl + NaH_2PO_4$$

(weak base)　　　　　(weak acid)

$$NaOH + NaH_2PO_4 \leftrightarrow H_2O + Na_2HPO_4$$

(weak acid)　　　　　(weak base)

Proteins buffer by taking up or releasing H^+, thereby stabilizing the H^+ concentration in a solution. The carboxyl group of an amino acid can gain or lose H^+.

The amino group of an amino acid can gain or lose H^+ or it can gain or lose OH^-.

All buffering reactions are easily reversible—they can go in either direction. Because the globin of hemoglobin and plasma proteins can hold or release either H^+ or OH^-, they act as a powerful buffer system in the blood.

As should be clear from the preceding examples, buffers are extremely important in regulating the pH of body fluids. They help to assure that excesses of acids or bases exert a minimal effect on body fluids until they can be removed by the lungs or kidneys. The actions of buffers are summarized in Table 28.7.

Mechanisms of Acid-Base Balance

Body fluids are said to be in **acid-base balance** when their pH is in a normal range and they contain neither an excess nor a deficiency of H^+. As we shall see, acid-base balance is maintained by both respiratory and nonrespiratory mechanisms.

The body's greatest single threat to acid-base balance is CO_2, which is continuously produced by cells. When it dissolves in plasma, it produces carbonic acid as the following reaction goes to the right:

$$CO_2 + H_2O \leftrightarrow H_2CO_3 \leftrightarrow H^+ + HCO_3^-$$

TABLE 28.7　　The Actions of Buffer Systems

System	Actions
Bicarbonate/ carbonic acid	Bicarbonate reacts with a strong acid to form a salt and a weak acid: $HCl + NaHCO_3 \rightarrow NaCl + H_2CO_3$
	Carbonic acid reacts with a strong base to form water and a weak base: $NaOH + H_2CO_3 \rightarrow H_2O + NaHCO_3$
Phosphate	Sodium monohydrogen phosphate reacts with a strong acid to form a salt and a weak acid: $HCl + Na_2HPO_4 \rightarrow NaCl + NaH_2PO_4$
	Sodium dihydrogen phosphate reacts with a strong base to form water and a weak base: $NaOH + NaH_2PO_4 \rightarrow H_2O + Na_2HPO_4$
Proteins	Amino acids in the globin of hemoglobin or in plasma proteins can gain or lose either H^+ or OH^-. By doing so they can have significant effects on the blood pH (see equations in text).

To prevent the plasma from becoming acidic, CO_2 must be removed continuously, causing the reaction to go to the left. Under normal physiological conditions, CO_2 is removed from the plasma as blood passes through lung capillaries, where CO_2 diffuses into alveoli and is exhaled. This causes HCO_3^- and H^+ to form more CO_2 (and water), and more CO_2 diffuses out of the plasma. After CO_2 leaves the blood and H^+ leaves blood buffers, the blood can again circulate through tissues, pick up new CO_2, and carry it to the lungs with no significant change in the blood pH. This constitutes respiratory regulation of acid-base balance.

A mechanism of nonrespiratory regulation of acid-base balance is H^+ secretion from kidney tubules (Figure 28.8). In this process, CO_2 passes from peritubular fluids into epithelial cells in the tubule wall. Within the epithelial cells, carbonic anhydrase causes CO_2 to combine with water to form carbonic acid. Some of the acid ionizes into HCO_3^- and H^+. H^+ is actively transported from the epithelial cell into the tubule lumen. HCO_3^- is returned to the interstitial fluid accompanied by Na^+. Some Na^+ and HCO_3^- diffuse into the blood.

Once in the lumen of the kidney tubule, H^+ can combine with ammonia from amino acid deamination or with phosphate ions. Kidney filtrate typically contains an abundance of phosphates that form a buffer system in urine. When H^+ combines with any substance, it no longer contributes to acidity. Secretion of H^+ into the kidney filtrate removes acid from the blood. Binding of H^+ to ammonia or a phosphate buffer prevents urine acidity. Only on a very acidic diet does enough H^+ enter the filtrate to overwhelm buffers and allow the urine pH to drop as low as 5.

The major problem in maintaining acid-base balance under normal physiological conditions is to prevent acidosis from CO_2 and H^+ accumulation. Both the lungs and kidneys can respond to either decreases or increases in the blood pH. When the blood pH begins to drop, the respiratory rate and the rate of H^+ excretion increase. Substances that acidify the blood are removed more rapidly. Conversely, when the blood pH begins to rise, the respiratory rate and the rate of H^+ excretion decrease. CO_2 and H^+ accumulate, counteracting the pH increase. By continuously adjusting the rate of removal of acidic substances, the body maintains the blood pH in the normal range.

As we have seen, buffers and lung and kidney functions operate together to maintain acid-base balance, but they have different capacities and different limitations. Buffer systems react rapidly but can become overloaded by the accumulation of acid or base. The lungs can respond relatively rapidly to either an increase or decrease in plasma CO_2 concentration, acting to return the concentration to normal. They, too, can be overloaded, especially when respiratory disorders slow gas diffusion or interfere with regulation of the rate and depth of breathing. The kidneys provide both the slowest and the most complete regulation of acid-base balance. The kidneys respond slowly because they must remove H^+ by active transport, but they are thorough because they can adjust the blood concentrations of many plasma solutes.

Carbonic acid always ionizes, or dissociates, to a certain degree and forms a predictable concentration of H^+ and HCO_3^-.

$$H_2CO_3 \leftrightarrow H^+ + HCO_3^-$$

This degree of dissociation is the **dissociation constant,** K, shown below, using brackets to represent concentrations.

$$K = \frac{[H^+]\,[HCO_3^-]}{[H_2CO_3]}$$

Carbonic acid, a weak acid, dissociates only slightly— about one per million acid molecules ionizes.

$$K = \frac{1 \times 1}{1,000,000} = \frac{1}{10^6} = 10^{-6}$$

Chemists define **pK** as the negative log of the dissociation constant. When $K = 10^{-6}$, the pK = 6. This is analogous to pH, the negative log of the H^+ concentration. When $[H^+] = 10^{-7}$, pH = 7.

The relationship between the pH and the pK of an acid is expressed by the **Henderson-Hasselbalch equation,** shown here for carbonic acid.

$$pH = pK + \log \frac{[HCO_3^-]}{[H_2CO_3]}$$

Substituting a pH of 7.4 for normal blood and a pK of 6.1 (the actual pK of carbonic acid), we find the log of the ratio of the concentrations of bicarbonate to carbonic acid is 1.3.

$$7.4 = 6.1 + \log \frac{[HCO_3^-]}{[H_2CO_3]}$$

$$1.3 = \log \frac{[HCO_3^-]}{[H_2CO_3]}$$

From a logarithm table, we find that the antilog of 1.3 = 20. This means that in blood at normal pH, the ratio of bicarbonate to carbonic acid is 20:1.

$$\frac{20}{1} = \frac{[HCO_3^-]}{[H_2CO_3]}$$

We can think of H_2CO_3 as CO_2 dissolved in water and substitute CO_2 for H_2CO_3 in the above equation.

$$\frac{20}{1} = \frac{[HCO_3^-]}{[CO_2]}$$

The combined actions of buffers, lungs, and kidneys maintain the arterial blood pH between 7.35 and 7.45 and the HCO_3^- to CO_2 ratio near 20:1. Venous blood has a lower pH because it carries more CO_2. Various clinical conditions cause the blood pH to fall outside the normal range and disturb acid-base balance.

CLINICAL APPLICATIONS

Disturbances in Fluid Volume

Fluid volume disturbances include **negative water balance** in which water output exceeds water intake and **positive water balance** in which water intake exceeds water output.

Dehydration

Negative water balance is caused by decreased fluid intake or increased fluid loss, such as excretion of much dilute urine, burns that leak plasma, vomiting, diarrhea, excessive sweating, and hemorrhage. These conditions lead to **dehydration,** a lack of body water, and usually to losses of electrolytes and other solutes. Dehydration first affects extracellular compartments. When relatively more water than solutes is lost, the remaining fluid becomes hyperosmotic. Intracellular fluid moves by osmosis to extracellular compartments. All compartments then have less than their normal volume but are only slightly hyperosmotic. Regardless of the cause, dehydration is made worse by **obligatory fluid loss** in normal kidney function. Though the kidneys can conserve fluid and excrete a concentrated urine, they cannot function without excreting some water along with waste solutes.

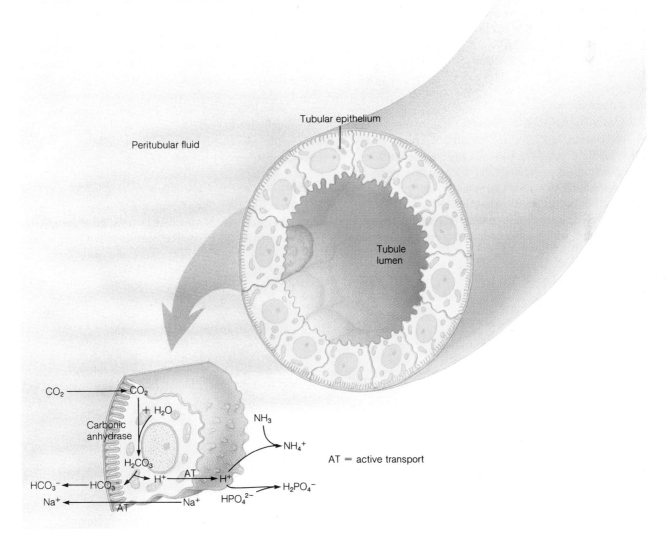

FIGURE 28.8
Tubular secretion of hydrogen ions.

Severe dehydration—in untreated diabetes insipidus, burns over large body areas, and persistent vomiting and diarrhea—can reduce blood volume to a life-threatening degree. This decreases blood pressure, cardiac output, and tissue perfusion, and can damage the brain and other vital organs. Conditions that cause dehydration also can cause electrolyte and acid-base imbalances. For example, excessive sweating can deplete sodium, and prolonged vomiting can deplete hydrochloric acid and raise body fluid pH.

Compared to adults the bodies of infants are more susceptible to dehydration because of the following:

1. They have a higher surface-to-volume ratio (Figure 28.9) and lose relatively more fluid from evaporation.
2. They have relatively more water than the bodies of adults.
3. They have immature kidneys with limited urine-concentrating ability.

In short, fluid-regulating mechanisms must keep more water in tissues in spite of the greater evaporative loss and immature kidney function. It is not surprising then, that an infant might take in 20 percent of its weight in fluid daily.

Water Intoxication

Positive water balance leads to **water intoxication,** or too much body water and a corresponding decrease in osmolarity of fluid compartments. Many people inadvertently suffer from water intoxication after profuse sweating by quenching their thirst with pure water, tea, coffee, soft drinks, or other beverages hyposmotic to sweat. Failure to replace lost electrolytes leads to diarrhea and abdominal and leg cramps, which can occur many hours after the sweating. Though hypotonic beverages increase extracellular fluid volume, they make the fluid even more hyposmotic and some water moves by osmosis into cells, decreasing intracellular osmolarity and causing cellular swelling.

Edema

Edema, another water balance disorder, is an increase in extracellular fluid volume without a change in osmolarity. Characterized by a puffy face and swollen abdomen, ankles, and fingers, edema usually is caused by excess sodium. Excess sodium in a body fluid draws water into a compartment until an osmotic balance is reached. We have already seen how edema results from hypertension and congestive heart failure in Chapter 20.

Acid-Base Imbalances

Disturbances in acid-base balance can cause **acidosis** (as″id-o′sis), an excess of H^+, or **alkalosis** (al″kah-lo′sis), a deficiency of H^+. Either condition can be respiratory or metabolic (nonrespiratory), depending on its cause. Disorders of the respiratory system, including control centers, can cause respiratory acidosis or alkalosis. Metabolic disorders such as accumulation of ketone bodies in untreated diabetes or ammonia in hyperammonemia can cause metabolic acidosis or alkalosis, respectively. Because such disorders can be caused by vomiting, diarrhea, kidney disorders, and other nonmetabolic problems, the term nonrespiratory is more descriptive than metabolic.

Compensation

Whenever an acid-base imbalance develops, the body's regulatory mechanisms operate to stabilize or *compensate* for the imbalance. **Complete compensation** occurs when the body completely counteracts the imbalance: pH,

FIGURE 28.9
Surface-to-volume ratio (S/V) is an object's surface area (S) in square units divided by its volume (V) in cubic units. (a) For a small cube, 1 unit on a side, representing an infant's body, S = 6, V = 1, and S/V = 6.0. (b) For a large cube, 4 units on a side, representing an adult's body, S = 6 x 16 (the area of each surface) = 96, V = 4 × 4 × 4 = 64, and S/V = 96/64 = 1.5. S/V is much greater in (a) than in (b); other factors being equal, water loss is much greater in (a) than in (b).

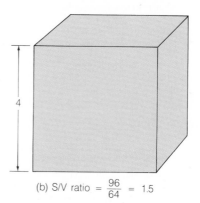
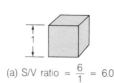

(a) S/V ratio $= \dfrac{6}{1} = 6.0$

(b) S/V ratio $= \dfrac{96}{64} = 1.5$

HCO_3^- and CO_2 concentrations, and the ratio between them are restored to normal. In most acid-base imbalances, the condition that caused the imbalance impairs normal regulation and **incomplete compensation** occurs: the HCO_3^- to CO_2 ratio and pH are restored to normal at the expense of maintaining both HCO_3^- and CO_2 concentrations above or below normal. The mechanism and the effect of incomplete compensation differ depending on the nature of the imbalance. Acidosis, alkalosis, and compensation are summarized in Table 28.8.

Respiratory Acidosis

Impaired ability of the lungs to remove CO_2 allows it to accumulate in the blood in **respiratory acidosis.** Part of the CO_2 combines with water to form H_2CO_3, some of which ionizes, releasing H^+ and lowering the blood pH. The acid-base imbalance in respiratory acidosis is excess CO_2 that leads to excess H^+.

Any factor that interferes with gas exchange between blood and alveoli can cause respiratory acidosis, including fluid accumulation in the alveoli in pneumonia, impaired exhalation in emphysema, obstruction of respiratory passageways in asthma and bronchitis, and reduction in respiratory surface area in any disorder that destroys alveoli. Narcotic suppression of or damage to respiratory centers can reduce the respiratory rate and lead to acidosis.

When lung disease causes respiratory acidosis, the kidneys compensate. They excrete H^+ and return enough HCO_3^- to the blood to balance the excess CO_2 that remains. The 20:1 HCO_3^- to CO_2 ratio is restored and the blood pH returns to 7.4 (according to the Henderson-Hasselbalch equation), but *both HCO_3^- and CO_2 are above normal.*

Respiratory Alkalosis

If the lungs remove CO_2 from the blood too rapidly, **respiratory alkalosis** occurs as more carbonic acid forms CO_2 and water and more H^+ and HCO_3^- are used to make carbonic acid. Blood H^+ decreases, and blood pH increases. The imbalance in respiratory alkalosis is deficient CO_2 that leads to deficient H^+.

TABLE 28.8 Acidosis, Alkalosis, and Compensation

	Acidosis	Alkalosis
General	Assume pH = 7.1 Given pK = 6.1, then $\log \dfrac{[HCO_3^-]}{[CO_2]} = 1.0$ antilog of 1.0 = 10 $\dfrac{[HCO_3^-]}{[CO_2]} = \dfrac{10}{1}$ (half the normal ratio)	Assume pH = 7.7 Given pK = 6.1, then $\log \dfrac{[HCO_3^-]}{[CO_2]} = 1.6$ antilog of 1.6 = 40 $\dfrac{[HCO_3^-]}{[CO_2]} = \dfrac{40}{1}$ (twice the normal ratio)
Respiratory Compensation	Caused by impaired ability of lungs to rid the body of CO_2. *Problem: excess CO_2 and, thus, excess H^+.* Accomplished by kidney because respiratory system malfunction caused the condition and is incapable of correcting it. Kidney excretes more H^+ than normal and returns more HCO_3^- to blood until a blood pH of 7.4 is restored. The HCO_3^- is retained in blood until its concentration is 20 times the CO_2 concentration. Both HCO_3^- and CO_2 are present in excess in incomplete compensation.	Caused by excessive removal of CO_2 as in hyperventilation. *Problem: deficiency of CO_2 and, thus, deficient H^+.* Kidney excretes less H^+ than normal and returns less HCO_3^- to blood until a blood pH of 7.4 is restored. Only enough HCO_3^- is present in the blood to restore the ratio of 20:1 with CO_2. Because the CO_2 concentration is below normal so is the concentration of HCO_3^- below normal in incomplete compensation.
Metabolic Compensation	Caused by any of several metabolic disorders such as excesses of metabolic acids, but results in a decrease in bicarbonate. *Problem: deficiency of HCO_3^- and, thus, excess H^+.* Accomplished by the respiratory system because it is not impaired in metabolic disorders and can compensate the condition more rapidly than the kidney. Lungs release more CO_2 into exhaled air until the blood pH is returned to 7.4 and the ratio of HCO_3^- to CO_2 is restored to 20:1. Because of the original deficiency of HCO_3^-, the concentrations of both HCO_3^- and CO_2 are below normal in incomplete compensation.	Caused by any of several metabolic disorders such as loss of acids but results in an increase in bicarbonate. *Problem: excess of HCO_3^- and, thus, deficient H^+.* Lungs release smaller amounts of CO_2 into exhaled air until the blood pH is returned to 7.4 and the ratio of HCO_3^- to CO_2 is restored to 20:1. Because of the excess of HCO_3^-, the concentrations of both HCO_3^- and CO_2 are above normal in incomplete compensation.

Respiratory alkalosis, less common than acidosis, can be caused by anxiety, aspirin overdose, or a change from low to high altitude. Anxiety and aspirin act directly on respiratory centers, whereas altitude lowers blood oxygen. All these conditions produce hyperventilation and the removal of greater than normal amounts of carbon dioxide.

Because the lungs cannot compensate, the kidneys do so by secreting less H^+ and reabsorbing less HCO_3^-. The 20:1 HCO_3^- to CO_2 ratio is restored so the blood pH returns to normal, but *both HCO_3^- and CO_2 are below normal.*

Metabolic Acidosis

Accumulation of acids and ketone bodies often causes **metabolic acidosis.** Plasma HCO_3^- is depleted by neutralizing H^+ and this taxes the blood's buffering capacity, allowing H^+ to accumulate and lowers the blood pH. The imbalance in metabolic acidosis is deficient HCO_3^- that leads to excess H^+.

Metabolic acidosis includes all types of acidosis *not* caused by excess CO_2, but excessive fat metabolism in untreated diabetes mellitus is a common cause. Low carbohydrate diets, starvation, renal tubular defects that impair H^+ secretion and/or HCO_3^- reabsorption, and loss of HCO_3^- in severe diarrhea can cause acidosis. The combination of dehydration and acidosis from severe diarrhea causes many childhood deaths, especially where emergency medical care is not available.

The lungs usually act quickly in metabolic acidosis to compensate by removing CO_2 from the blood. The 20:1 HCO_3^- to CO_2 ratio is restored, the blood pH returns to normal, but *both HCO_3^- and CO_2 are below normal.*

Metabolic Alkalosis

Depletion of acids can lead to **metabolic alkalosis** as HCO_3^- binds most free H^+ and the blood pH increases. The imbalance in metabolic alkalosis is excess HCO_3^- that leads to deficient H^+.

Relatively rare metabolic alkalosis can be caused by taking too much antacid, loss of Cl^-, and excess aldosterone secretion. Antacids used to treat ulcers or gastritis neutralize acids and increase blood HCO_3^-. Vomiting causes loss of stomach acid, which, in turn, depletes H^+ and Cl^- in extracellular fluids. H^+ remains depleted and Cl^- is replaced by HCO_3^-. Excess aldosterone causes Na^+ reabsorption and H^+ secretion. Metabolic alkalosis and magnesium deficiency may combine to cause the *delirium tremens* (de-ler'e-um tre'menz) of alcohol withdrawal. Diuretic use and abuse can lead to *contraction alkalosis,* in which much water is lost and H^+ secretion is excessive. This leaves an excess of HCO_3^- dissolved in a smaller water volume.

The lungs compensate for excess HCO_3^- by removing CO_2 from the blood more slowly. The 20:1 ratio of HCO_3^- to CO_2 is restored, the blood pH returns to normal, but *both HCO_3^- and CO_2 are above normal.*

Diabetic ketoacidosis (DKA), a severe life-threatening complication of insulin-dependent diabetes, involves fluid, acid-base, and electrolyte imbalances. Insulin deficiency allows blood glucose to rise and fat to be metabolized. Glucose spills into the urine, causing polyuria, loss of potassium and other electrolytes, and dehydration. Dehydration lowers blood volume and blood pressure. Fat metabolism causes ketoacidosis. Deep rapid breathing, called Kussmaul breathing, helps to compensate for acidosis. With lowered blood volume, a given amount of acid causes a greater than usual drop in blood pH.

Emergency treatment is needed to prevent coma. Intravenous fluids are given to raise the blood volume and blood pressure. Insulin added to IV fluids prevents further breakdown of lipids and helps cells take in glucose and lipids. If renal function is adequate, it will restore acid-base balance, but lactated Ringer's solution may be given. As rehydration is completed, potassium chloride is added to the fluids. When these measures succeed, the patient's condition stabilizes in about eight hours.

Summary of Acid-Base Imbalances

All acid-base imbalances involve an excess or deficiency of H^+, but the causes of such imbalances differ. In respiratory acidosis, H^+ excess is due to an *excess of CO_2.* In respiratory alkalosis, H^+ deficiency is due to a *deficiency of CO_2.* In metabolic acidosis, H^+ excess is due to a *deficiency of HCO_3^-.* In metabolic alkalosis, H^+ deficiency is due to an *excess of HCO_3^-.* The HCO_3^- to CO_2 ratio is high in acidosis and low in alkalosis.

Incomplete compensation fairly quickly restores the 20:1 ratio and returns the pH to 7.4. Complete compensation, which restores normal HCO_3^- and CO_2 concentrations, occurs slowly and may be impossible if lung or kidney disease severely impairs compensating mechanisms. Though healthy lungs act quickly to compensate for acid-base imbalances of metabolic origin, healthy kidneys assist at a slower pace by appropriately adjusting the rates of H^+ secretion and HCO_3^- reabsorption.

Acidosis or alkalosis can be life threatening. When the blood pH drops below 7.0, depressed central nervous system function leads to coma. If not treated, the patient will remain comatose and die from respiratory failure. When the blood pH rises above 7.8, the whole nervous system becomes hyperactive. Peripheral nerve hyperactivity spreads from forearm muscles, to facial muscles, and then to all muscles. Central nervous system hyperactivity can cause seizures and convulsions. Death usually is from respiratory muscle tetany.

CLINICAL TERMS
antacid (ant-as'id)   a substance that neutralizes acid
effusion (ef-u'zhun)   escape of fluid into an organ or tissue
hypervolemia (hi''per-vol-e'me-ah)   increased blood volume
hypovolemia (hi''po-vol-e'me-ah)   decreased blood volume

FLUID THERAPY

Fluid therapy is the administration of fluids orally or parenterally (by any route other than the digestive tract) to correct a fluid imbalance. It is used to regulate total body fluid and electrolyte, acid, and base concentrations.

Fluid imbalances caused by mild, acute illnesses often can be treated with oral fluids at home. Fever causes water loss from accelerated breathing and Na^+ loss from sweating. Diarrhea leads to loss of water, Na^+, Cl^-, HCO_3^-, and any nutrients or electrolytes that are not absorbed before they are excreted. A patient can replace these fluids by drinking water, but other beverages help to replace electrolytes. Beef broth replaces Na^+ and Cl^- and carbonated beverages replace HCO_3^-. Beverages containing sugar supply readily digestible carbohydrate for energy.

Fluid imbalances caused by severe disorders unmanageable with oral fluids are treated parenterally, usually in a hospital. Such disorders include shock; burns; consequences of surgery or trauma; or respiratory, renal, or metabolic diseases that alter fluid, electrolyte, or acid-base balance. The most often used technique is the intravenous (IV) drip—the slow infusion of sterile fluid through a needle or catheter inserted into a vein.

Before initiating fluid therapy, the exact nature of imbalances should be determined by laboratory tests and careful patient observation. Observable effects of fluid deficit include: (1) weight loss, (2) dry membranes and cracked lips, (3) lowered blood pressure, (4) decreased urinary output, or (5) decreased skin turgor (pinched skin remains folded). Observable effects of fluid overload are: (1) a bounding pulse, (2) rales and other abnormal breathing sounds, possibly from fluid in the lungs, (3) excessive urinary output, or (4) edema. Potassium deficit is suggested by certain cardiac arrhythmias, and sodium deficit by viscous saliva and cold, clammy skin.

To plan fluid therapy, one must know normal daily requirements for fluid, electrolytes, and nutrients. An adult with a normal body temperature needs about 30 ml water/kg body weight, 100 mEq of each Na^+ and Cl^-, and 8 to 20 mEq Mg^{2+}. Fluid volume must be increased by about 15 percent for every 1° C rise in temperature, and this fluid should contain sufficient NaCl (about 0.3 percent) to replace losses in sweat.

If a patient cannot eat, nutrients should be supplied intravenously. An adult receiving no nutrients loses 70 to 85 g of protein a day from body tissues. Giving 100 to 150 g/day of glucose (dextrose) in intravenous fluids supplies 340 to 510 kilocalories and reduces protein loss by about one-half. If eating is suspended for more than three days, more nutrients should be given intravenously. Glucose can be increased to supply 700 to 1300 kilocalories, with insulin if needed to increase cellular uptake. To prevent protein loss, 40 g of protein hydrolysate (individual amino acids) and water soluble vitamins can be added to the fluid. Administering such nutrient-laden fluids constitutes **peripheral hyperalimentation.**

When no food can be taken for eight days or more, **total parental nutrition** (TPN) can be used. In TPN, the patient receives a hyperosmolar solution of protein hydrolysate in a 20 to 25 percent glucose solution, with vitamins and electrolytes added as appropriate. Because such a solution irritates peripheral veins, it is infused into a subclavian or other large vein. TPN is used when tube feeding is impossible because of gastrointestinal lesions or when the patient is in a hypermetabolic state and large amounts of nutrients are needed to repair tissue damage from trauma, burns, or severe infections. TPN supplies about 900 kilocalories per liter. A 10 percent fat emulsion can be given at a rate of 500 ml every other day to supply linoleic acid and as a source of high-calorie nutrients. Fat emulsions are not given on a long-term basis, and their long-term effects are not yet known.

The treatment of acid-base disorders requires careful monitoring of blood pH and concentrations of bicarbonate and blood gases, especially CO_2. Acidosis can be treated with an intravenous lactated Ringer's solution, which contains lactate, calcium, and other electrolytes. Lactate metabolized in the liver counteracts acidity, but care must be used not to overtreat. Overtreatment can lead to alkalosis and tetany—a risk that is reduced by calcium lactate in solutions. Alkalosis can be treated with potassium chloride. Because of careful blood pH monitoring in modern medical treatment, acid-base imbalances and their overtreatment has been greatly reduced in recent years.

A special problem in managing body fluids is the so-called **third space phenomenon,** the entrapment of large quantities of fluid in traumatized tissues after severe injuries or extensive abdominal or pelvic surgery. Such sequestered fluid, which cannot easily exchange ions as interstitial fluids and plasma do, is said to occupy a third extracellular fluid compartment, or third space. Sequestered third-space fluid decreases blood volume and leads to hypotension or shock. This problem is treated with plasma proteins or colloids that will draw fluid back into the bloodstream.

Finally, when fluid therapy is used, one must monitor fluid volumes taken in and excreted, and check the blood regularly for electrolyte and acid-base balance. Careful attention to these problems contributes to a patient's comfort and recovery. It can make the difference between life and death.

Questions
1. What kinds of conditions require fluid therapy?
2. Which fluid therapy can be given at home, and which require hospitalization?
3. How can dehydration and fluid overload be distinguished?
4. How can acidosis and alkalosis be distinguished?
5. What is TPN, and under what conditions is it indicated?

Chapter Summary

Physiology of Fluids, Electrolytes, Acids, and Bases

- Homeostasis of body fluids is maintained by a great number of regulatory mechanisms that operate at all levels from the cellular to the systemic.

Regulation of Body Fluids

- The body's main fluid compartments are intracellular, interstitial, and plasma. Other fluids include lymph, cerebrospinal fluid, and fluid in joint and eye cavities.
- Water balance means fluid intake equals fluid output.
- Electrolyte balance is regulated by active transport and by hormones. Total fluid volume in the body is determined by intake and excretion, and is regulated by antidiuretic hormone, aldosterone, atrial peptides, and neural factors.

Regulation of Specific Electrolytes

- Electrolytes of particular importance are sodium, potassium, chloride, bicarbonate, and calcium. Sodium is regulated by antidiuretic hormone and aldosterone. Potassium is regulated by aldosterone and, indirectly, by the renin-angiotensin mechanism. Chloride is regulated mainly by movement of other ions and bicarbonate with acid-base balance. Calcium ions are regulated by parathormone, calcitonin, and vitamin D (Chapter 16).

Acid-Base Regulation

- Acid-base regulation involves buffers, or substances that resist pH change, and activities of the lungs and kidneys.
- Buffers, which include the bicarbonate-carbonic acid, phosphates, and proteins, help to maintain normal body fluid pH by taking up H^+ when pH drops, and releasing H^+ when pH rises.
- The lungs and the kidneys are the main organs in acid-base balance. The lungs remove more CO_2 from the blood when the pH becomes acid, and retain CO_2 when the pH becomes alkaline. The kidneys excrete H^+ and reabsorb HCO_3^- ions when the pH becomes acidic, and retain H^+ when the pH becomes alkaline.

Clinical Applications

- Fluid balance disturbances include dehydration, water intoxication, and edema.
- Acidosis and alkalosis can be caused by respiratory or metabolic (nonrespiratory) disorders. These conditions and the body's mechanisms of compensation are summarized in Table 28.8.

Questions and Problems

The questions at the end of each chapter are numbered to correspond with the objectives listed at the beginning of the chapter. Italics indicate that a question requires critical thinking skills beyond simple factual recall.

Questions

1. *What kinds of mechanisms help to maintain homeostasis under different environmental conditions?*
2. (a) What is the composition of fluid in each of the body's three main fluid compartments?
   (b) Where are these compartments located?
3. *What mechanisms contribute to regulating fluid volume, and how do they work?*
4. *(a) What mechanisms are involved in regulating sodium and potassium concentrations in body fluids, and what disorders result from improper regulation?*
   (b) What mechanisms are involved in regulating calcium and phosphate, and what disorders result from improper regulation?
   (c) What mechanisms are involved in regulating other electrolytes, and what disorders result from improper regulation?
5. (a) What is a buffer, and what kinds of buffers are found in the body?
   (b) What are the specific effects of each of the body's three main buffer systems?
6. *What mechanisms are involved in maintaining acid-base balance?*
7. *What are the causes and effects of dehydration, water intoxication, and edema, and how does the body compensate for these conditions?*
8. *How do acidosis and alkalosis differ, and how do respiratory and metabolic (nonrespiratory) causes of these disorders differ?*
9. *What is the cause, nature, and most likely means of compensation for each of the following: (a) respiratory acidosis, (b) respiratory alkalosis, (c) metabolic acidosis, and (d) metabolic alkalosis.*

Problems

1. How would your body cope if you were lost in a hot desert without water?
2. How would your body cope with the following:
   (a) drinking fairly large amounts of alcohol, (b) constant nibbling at salted nuts and potato chips, (c) drinking several gallons of water at one time, and (d) drinking excessive quantities of very sweet carbonated beverages?
3. Prepare a report on total parenteral nutrition, and explain how it maintains nutritional, fluid, and electrolyte balance.
4. Infants regularly drink 20 percent of their body weight in fluid daily. How much would you have to drink to do this? How does this compare with your normal fluid intake?

Genetic and reproductive engineering, life-support systems for the critically ill, birth control, abortion, and other choices about life and death raise difficult ethical issues. All of us will make decisions about such issues at certain times, and health professionals will deal with them on a daily basis.

Ethical decision-making techniques are available to help us deal with social and ethical implications of problems like those cited above. We can make value judgments rationally, even when we have strong feelings about an issue if we are consciously aware of the values we hold. We can avoid making judgments at an emotional level without carefully considering our alternatives. The ethical decision-making techniques presented here help people to clearly identify their values and to make decisions about social and ethical issues consistent with those values.

A CASE STUDY

A couple has a son with Duchenne muscular dystrophy (DMD). They want another child only if they can be assured the second child will not have DMD. In genetic counseling, they learned that DMD is inherited as a sex-linked recessive characteristic (Chapter 30). Since neither of them has the disease, it is likely that the mother carries a DMD gene but her genes could be normal and the son's DMD due to a spontaneous mutation.

The couple considered the following actions:

1. To refrain from having another child.
2. To have fetal blood taken during the next pregnancy and have enzyme studies done on the blood to determine whether the fetus is affected.
3. To have amniocentesis (Chapter 29) performed and the sex of the child determined (only a male child is likely to be affected).
4. To do nothing and accept the 25 percent risk of having another affected child. (The probability of having a male child is 0.5 and the probability of a male being affected is 0.5; the chance of both occurring simultaneously is the product of the probabilities—$0.5 \times 0.5 = .25$, or 25 percent.)

Can you think of other courses of action?

The couple then considered the consequences of various actions:

1. Refraining from having another child: (a) avoids the possibility of having an affected child, (b) deprives the parents of having another child of their own, (c) prevents transmitting the DMD gene to a female child who would be normal but might someday have an affected son.
2. Having fetal blood used for enzyme studies: (a) subjects the fetus and the mother to a potentially dangerous procedure, (b) leaves the parents with the decision of whether to abort an affected fetus, (c) leaves the parents with the worry that the test result was in error and the child is affected, and (d) incurs significant medical expense.
3. Having amniocentesis and sex determination: (a) if the fetus is female, assures the parents that the child will be normal (except for the risk of other defects that exists in any pregnancy), (b) if the fetus is male, creates the need to decide whether to abort a fetus that has a 50 percent chance of being affected, (c) subjects the mother and fetus to a small risk from amniocentesis, and (d) incurs moderate medical expense.
4. Doing nothing and accepting the 25 percent risk of having an affected child: (a) does not solve the stated problem, (b) avoids extra expense during the pregnancy but may incur significant expense if the child is affected, (c) avoids the risks of prenatal diagnostic procedures, and (d) may result in the birth of a second affected child.

Can you think of other consequences? What are the consequences of other courses of action you may have proposed?

Finally, the couple considered the following value judgments to clarify their own judgments: Parents have the right to try to avoid having a child with a genetic disease. Parents should accept any child they conceive. The possibility of having another child with a genetic disease ought—or ought not—to influence one's actions. The possibility of passing a defective gene to future

ETHICAL DECISION MAKING

generations should—or should not—be considered. Human rights include—or do not include—the right to have children. Prenatal diagnosis is the right of every pregnant woman. No fetus should be subjected to prenatal testing. Abortion is unacceptable under any conditions, acceptable if the fetus is known to be defective, acceptable if the fetus has a 50 percent chance of being defective, or is the right of any woman. A child deserves to be wanted, or must learn to accept being unwanted.

Which of your values are expressed here? What other values pertain to the problem? Among your values, some are more important than others. Like most people, you probably value money, but you would not decide this issue by how much it costs. List your values in their order of importance in solving this problem.

If you already had a child with a genetic disease, what would you do? Decide before you read on.

The couple in our example decided that in the next pregnancy they would request amniocentesis to determine the sex of the fetus, intending to have a male fetus aborted. They rejected fetal blood testing as too great a risk to the fetus. As it happened, the fetus was female and the couple has a healthy girl.

A Procedure for Ethical Decision Making

The following steps illustrate a procedure for ethical decision making.

State the problem. The first step in ethical decision making, like any other decision making, is to state the problem clearly.

Determine possible courses of action. Decision makers should think carefully and creatively about possible courses of action. It is important that all reasonable possibilities be included.

Determine the consequences of each course of action. After identifying courses of action, list all the consequences of each course of action— good, bad, and indifferent. At this stage, make no value judgments about any consequences.

State values involved in each course of action. You probably have your own positive or negative attitude about each of the courses of action proposed in this example. Make value statements (reasons for your attitudes) about each course of action.

Rank order values. List your values pertaining to a problem, then place them in rank order, with the one you feel most strongly about first, and proceeding through the list until you have ranked all your value statements. If you have fewer than 10 statements, you probably need to think more carefully.

Decide on a personal course of action. Use the values most important to you to evaluate each course of action, and choose the action most consistent with those values. Could you live comfortably with your decision? If not, find which of your values is in conflict with the decision and modify the decision accordingly.

Social Policy Decisions

Though some bioethical decisions involve individuals, many involve larger groups—families, communities, or whole countries. In addition to emotional comfort (whether one can live comfortably with a decision), two other criteria can be used to assess the validity of a decision—universality and proportionate good. The criterion of **universality** asks whether the result would be acceptable if everyone in a similar situation made the same decision. The criterion of **proportionate good** asks whether the decision results in the greatest good for the most people.

Suppose that you and other members of a committee at the hospital where you work must decide what services should be made available to all parents who already have a child with a genetic disease. Is amniocentesis to be available to all? What about studies of fetal blood? Under what circumstances will abortions be available? Should any of these services be mandatory? Who will pay for the services? Who will support the unwanted children of women who desire, but do not obtain, abortions?

As in any situation that requires a group consensus, you may be unable to persuade other members of the group to accept your solutions. Identify trade-offs you would be willing to make. For example, though you prefer a decision that is compatible with all the values on your rank-ordered list, you might offer a compromise that sacrifices some values low on your list to arrive at a solution consistent with values high on your list.

Application of these ethical decision-making techniques should be of help to you in coping with the increasing number of ethical issues facing health professionals—and all citizens—in today's complex world.

Problems

1. Apply the ethical decision-making procedure to a problem that personally concerns you.
2. Identify three social or ethical problems from among articles in your daily newspaper. Make a list of 10 values that pertain to each problem and put each list in your own rank order.
3. Use the procedures described here to decide what to do if (a) your grandfather is being kept alive on life-support systems and appears to have little chance of recovering from a severe cerebrovascular accident that has destroyed his speech and left one side of his body completely paralyzed, (b) your sister was in an automobile accident and has been in a coma and on life-support systems for six months, (c) your brother and his wife are childless and have been offered the opportunity to adopt an infant with AIDS, and (d) some friends have a child with Down's syndrome and they have asked your advice on whether they should have another child.
4. With your instructor's approval, divide the class into several groups of six to eight people. Have each group use ethical decision making to develop a public policy regarding (a) care for AIDS victims, (b) who gets the use of life-support systems and under what conditions, (c) allocation of dialysis equipment when there are more patients needing dialysis than available equipment can handle, (d) how to allocate $200,000 for public health when agencies have requested $100,000 for well-baby and immunization clinics, $50,000 for prenatal care, $100,000 for prevention of drug use and addict rehabilitation, and $75,000 for visiting nurse programs for homebound elderly people.
5. If you are employed, use ethical decision making to deal with a job-related problem.

CONTINUITY OF LIFE

29

REPRODUCTIVE SYSTEM

815

ORGANIZATION AND GENERAL FUNCTIONS

For all living things, some means of reproduction is essential for the survival of the species. In humans, reproduction can be more than a mere survival strategy. It can be part of how we express our most human feelings—loving, caring, and nurturing.

Reproduction is the creation of new individuals of the same species. In humans, it is sexual; that is, it requires both a male and a female parent. The primary reproductive organs are the **gonads** (go'nadz)—ovaries in the female and testes in the male. Gonads produce sex hormones and **gametes** (gam'ētz)—ova in females and sperm in males. Each reproductive system also has ducts that transport gametes and glands that release secretions into the ducts. Males have a penis through which sperm are deposited into the female reproductive tract during sexual intercourse. Females have a uterus in which offspring develop, and mammary glands that produce milk after a birth has occurred.

Sex hormones (Figure 29.1), which differ in males and females by only small chemical differences, regulate function in gonads and accessory reproductive organs. They also stimulate maturation of reproductive organs and development of **secondary sexual characteristics**—male musculature and hair pattern, and female breasts and body contours—during puberty. **Puberty** is a period of three to five years beginning between age 10 and 14, during which fertility is attained. Puberty probably results from maturation of the hypothalamus and the secretion of gonadotropin-releasing hormone (GnRH).

Sex hormones—testosterone, estradiol, and progesterone—are synthesized from cholesterol or acetyl-CoA and transported in the blood bound to specific transport globulins or in loose association with plasma albumins.

These steroid hormones diffuse through membranes and bind to nuclear receptors in target cells. Hormone-receptor complexes act on DNA to induce transcription and protein synthesis. Hormone molecules that do not enter target cells are removed from the blood mostly by the liver and, to a small extent, directly by the kidneys. The liver converts steroid hormones to inactive substances, excreting some with bile and releasing some into the blood for excretion by the kidneys.

Sexual reproduction contributes to species survival by greatly increasing genetic variability in three ways: (1) creating many possible combinations of chromosomes in eggs and sperm, (2) varying genetic information on a given chromosome, and (3) mixing information from two organisms in the union of an egg and sperm. Changes in DNA called mutations account for yet another means of increasing genetic variability. Considering only item (1) above, the possible combinations of 23 pairs of chromosomes in an egg or sperm is 2^{23}—more than 8 million!

Genetic variability is important in several ways. It assures gene combinations that allow some members of a species to adapt to almost any environmental change. It gives each human, except for identical twins, a unique set of genetic characteristics. Finally, it accounts for many small variations in body function and in individual responses to diseases and treatments.

DEVELOPMENT

Though an infant's genetic sex is determined at conception, undifferentiated reproductive structures appear in both male and female embryos (Figure 29.2). Gonads develop from mesoderm medial to the kidneys and look the same in males and females until the eighth week of development. In female embryos, the cortex of gonads becomes the ovaries, and in male embryos, the medulla of

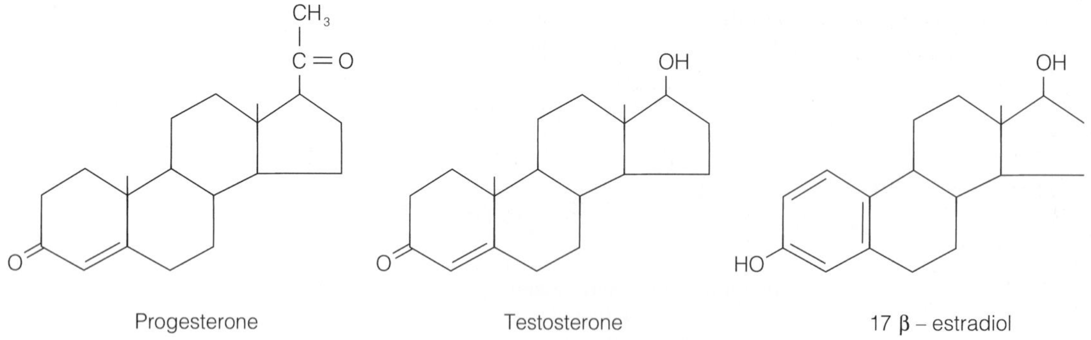

Progesterone Testosterone 17 β – estradiol

FIGURE 29.1
Steroid hormones.

gonads becomes the testes. As female development proceeds, **müllerian ducts** become uterine ducts, which fuse in the midline to form the uterus and vagina. As male development proceeds, **wolffian ducts** (mesonephric ducts remaining from degenerating kidneys) become male genital ducts. Male development is fostered by chorionic gonadotropin and a Y chromosome, which cause the developing testes to secrete testosterone. Without testosterone, even embryos with Y chromosomes develop female genital ducts. Conversely, female embryos, lacking testosterone, develop female ducts; thus, sex is determined genetically and influenced by testosterone.

Up to about the eighth week of development, undifferentiated external structures consist mainly of a **genital tubercle** (Figure 29.3). Testosterone, when converted to dihydrotestosterone, also fosters development of male external structures. The anterior genital tubercle develops into the **penis** and urethral folds close around the urethra at the raphe, leaving an opening only at the tip of the penis.

Labioscrotal swellings enlarge, fuse, and form pouches that become the **scrotum** (skro'tum), into which the testes descend a month or two before birth. In the female, lacking dihydrotestosterone, the anterior genital tubercle becomes the **clitoris** (kli'tor-is), and urethral folds around the urogenital sinus become the **labia minora** (la'be-ah mi-no'rah). Labioscrotal swellings become the **labia majora** (mah-jo'rah), which surround other genital organs and urethral and vaginal openings.

MALE REPRODUCTIVE ANATOMY

The male reproductive system (Figure 29.4) consists of the testes, which lie in the scrotum, a system of ducts that transport sperm, glands that contribute their secretions to semen (the fluid in which the sperm are suspended), and the penis.

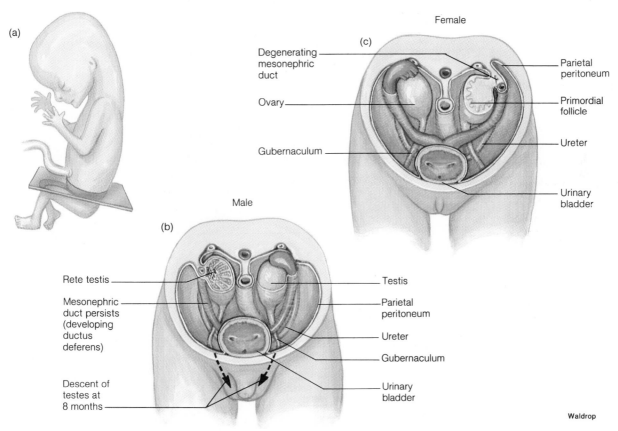

Waldrop

FIGURE 29.2
The development of the internal reproductive systems: (*a*) an embryo at 10 weeks showing level of sections of (*b*) a male embryo and (*c*) a female embryo.

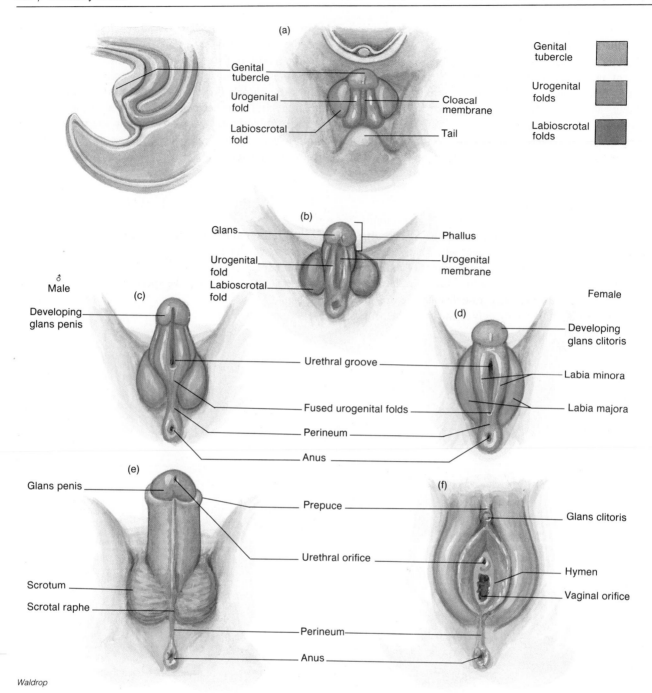

Waldrop

FIGURE 29.3
The development of the external reproductive structures: (*a*) an
embryo at 6 weeks and (*b*) an embryo at 8 weeks in sagittal view; and
(*c*) distinctly male structures and (*d*) distinctly female structures at
10 weeks; and (*e*) male and (*f*) female structures at 12 weeks.

(a)

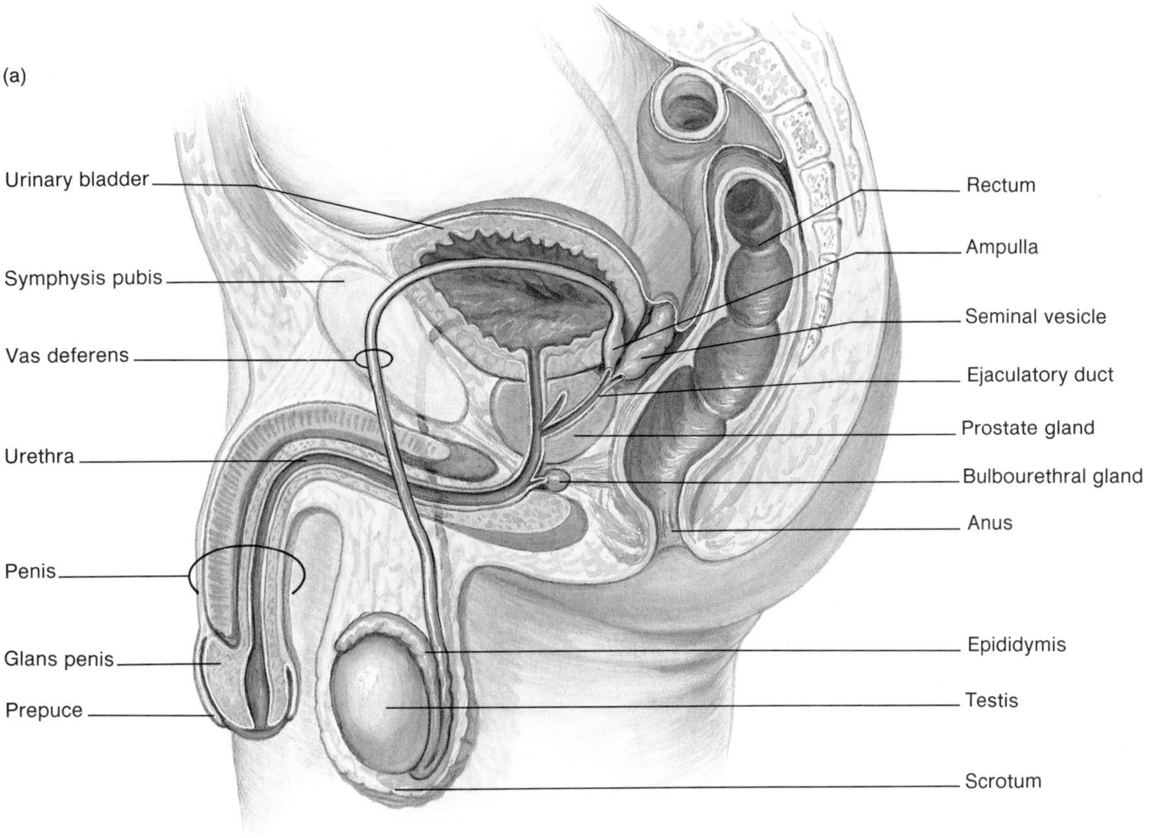

Urinary bladder

Symphysis pubis

Vas deferens

Urethra

Penis

Glans penis

Prepuce

Rectum

Ampulla

Seminal vesicle

Ejaculatory duct

Prostate gland

Bulbourethral gland

Anus

Epididymis

Testis

Scrotum

(b)

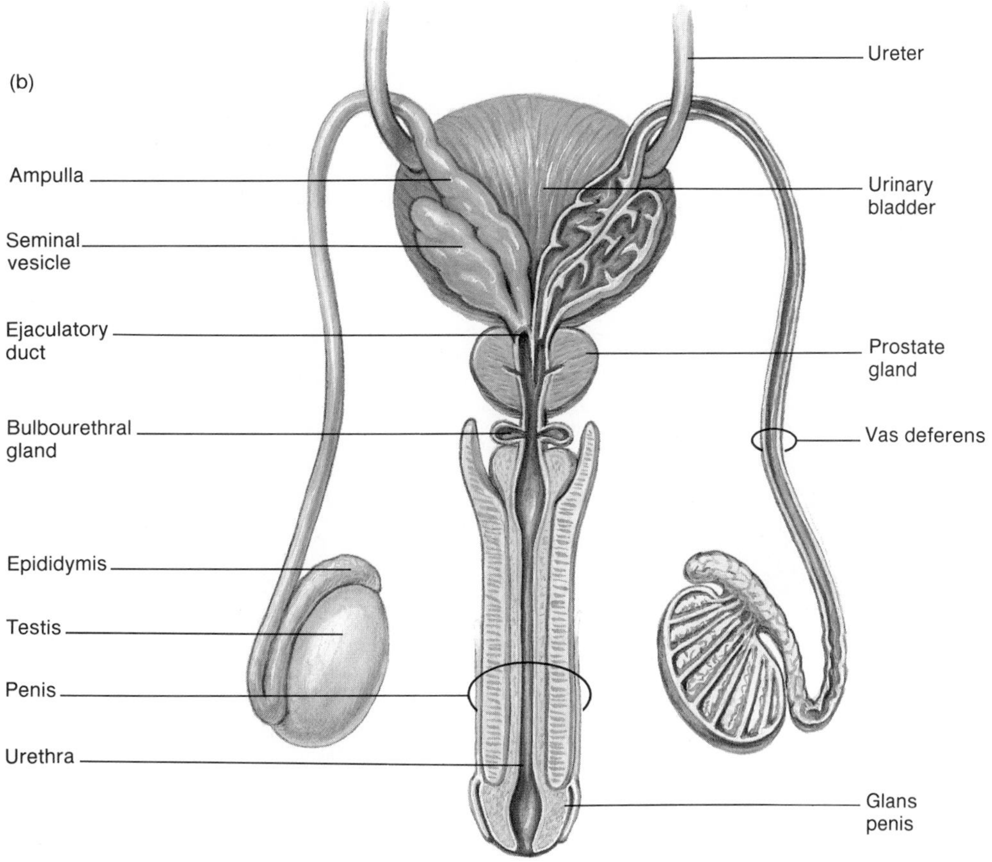

Ampulla

Seminal vesicle

Ejaculatory duct

Bulbourethral gland

Epididymis

Testis

Penis

Urethra

Ureter

Urinary bladder

Prostate gland

Vas deferens

Glans penis

FIGURE 29.4
The male reproductive system (*a*) in sagittal section, and (*b*) dissected in anterior view.

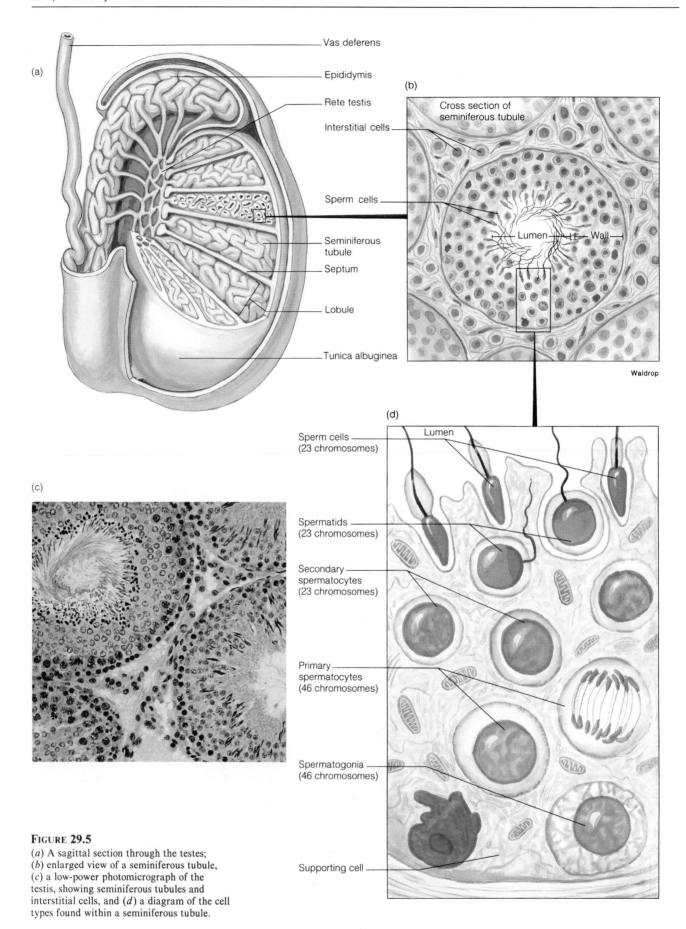

(a) Vas deferens

Epididymis

Rete testis

(b) Cross section of seminiferous tubule

Interstitial cells

Sperm cells

Lumen   Wall

Seminiferous tubule

Septum

Lobule

Tunica albuginea

Waldrop

(c)

(d)

Lumen

Sperm cells (23 chromosomes)

Spermatids (23 chromosomes)

Secondary spermatocytes (23 chromosomes)

Primary spermatocytes (46 chromosomes)

Spermatogonia (46 chromosomes)

Supporting cell

FIGURE 29.5

(a) A sagittal section through the testes;
(b) enlarged view of a seminiferous tubule,
(c) a low-power photomicrograph of the
testis, showing seminiferous tubules and
interstitial cells, and (d) a diagram of the cell
types found within a seminiferous tubule.

Testes and Scrotum

The adult **testes** (tes'tēz) are oval organs about 5 cm long and 2.5 cm in diameter, surrounded by a connective tissue *tunica albuginea* (too'nik-ah al''bu-jin'e-ah) (Figure 29.5a and b). Inside each testis, septa separate several hundred lobules, and each lobule contains one to three convoluted (folded and coiled) **seminiferous** (se''min-if'er-us) **tubules** (Figure 29.5c and d).

A seminiferous tubule contains **germinal epithelium** with many spermatogonia, cells that divide throughout reproductive years and give rise to sperm. It also contains

Sertoli's (ser-to'lez) **cells** and **interstitial cells.** Sertoli's cells extend from the basement membrane to the lumen and probably supply nutrients, enzymes, and maybe hormones to spermatids that attach to them while becoming mature sperm. Interstitial cells (cells of Leydig) secrete testosterone.

A mature spermatozoan, often referred to simply as a sperm (Figure 29.6) has an acrosome, a head, a middle piece, and a tail (flagellum). The **acrosome** (ak'ro-sōm), derived from the Golgi apparatus, contains hyaluronidase and proteolytic enzymes that help a sperm to digest its

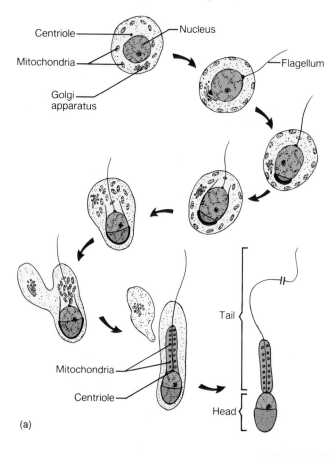

(a)

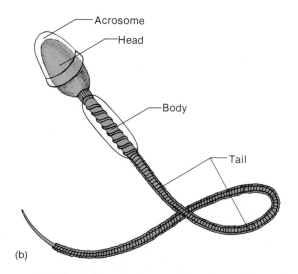

(b)

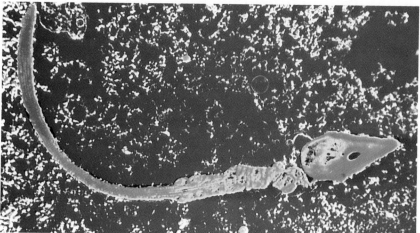

(c)

FIGURE 29.6

(*a*) Development of sperm, (*b*) mature sperm, and (*c*) false color scanning electron micrograph of a single human sperm.

way into an ovum. The head contains the nucleus and the middle piece contains mitochondria. The long flagellum, which develops from the centrioles, uses energy from its relatively large supply of ATP to move.

After mature sperm reach the lumen and pass through the coiled portion of a seminiferous tubule, they pass into a **straight tubule.** From the straight tubule, the sperm pass through a network called the *rete* (re′te) *testis,* and on through the **efferent ducts.** On leaving the testes, sperm enter the epididymis, a body of connective tissue in which the *ductus epididymis* is embedded, as described below.

The **scrotum** consists of superficial skin and underlying **dartos** (dar′toz), which consists of fascia and smooth muscle. A median septum divides the scrotum into two separate sacs, each sac containing one testis. The normal temperature of the testes in the scrotum is about 2° C lower than the internal body temperature—the ideal temperature for developing sperm. When the body is chilled, both the dartos and the *cremaster* (kre-mas′ter) *muscles,* one of which suspends each testis, bring the testes closer to the pelvic cavity. This allows the testes to absorb heat from the pelvic area and helps prevent damage to the sperm from chilling. Conversely, wearing tight-fitting clothing can increase the temperature of the testes and decrease sperm production.

A month or two before birth, the testes descend into the scrotum (Figure 29.7), pulling along blood vessels and nerves. They push ahead the **tunica vaginalis** (vaj′′in-a′lis), a membrane that covers the testes in the scrotum. Connective tissue and smooth muscle form a cord, the **gubernaculum** (gu′′ber-nak′u-lum), that connects each testis

and epididymis to the scrotum. Shortening of the gubernaculum as the fetus grows helps guide each testis into the scrotum. In about 3 percent of males, one or both testes fail to descend into the scrotum before birth. In this condition, called **cryptorchidism** (krip-tor′kid-izm), the testes descend spontaneously during the first year of life in about 80 percent of cases, and by puberty in another 10 percent of the cases. When descent does not occur spontaneously by puberty, hormones or surgery are used to cause it. Cryptorchidism leads to sterility because the high temperature of the abdominal cavity destroys developing sperm.

Male Reproductive Ducts

Other ducts of the male reproductive system carry sperm out of the testis and out of the body, and during intercourse, into the female reproductive tract. In addition to the ductus epididymis, these ducts include the paired ductus deferens and ejaculatory ducts and the unpaired urethra, which also is part of the urinary system (refer back to Figure 29.4).

The **ductus epididymis** (ep′′id-id′im-is) consists of a coiled tubule between the efferent ductules of the testis and the ductus deferens. It extends from a head on the superior surface of the testis along the posterior surface to a tail, where it joins the ductus deferens. The epididymis itself is about 4 cm long, but the coiled duct within it is about 6 m long. During ejaculation (sperm release), smooth muscle in the epididymis wall contracts and pushes sperm into the ductus deferens.

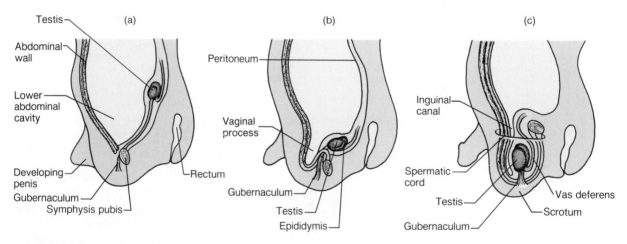

FIGURE 29.7
The descent of the testes into the scrotum: (*a*) testis before beginning descent, (*b*) tunica vaginalis being pushed ahead of testis as gubernaculum guides movement, and (*c*) testis arriving in scrotum.

The **ductus deferens** (def'er-enz), or **vas deferens,** is about 45 cm long and has a slightly wider diameter than the epididymis. Its walls contain three layers of smooth muscle. The ductus deferens carries sperm from the epididymis to the ejaculatory duct. The ductus deferens joins blood vessels, lymphatics, nerves, and the cremaster muscle to form the **spermatic** (sper-mat'ik) **cord,** which passes through the inguinal canal, over the pubic bone, and enters the pelvic cavity. In the pelvic cavity, the ductus deferens passes over and behind the urinary bladder, joins the duct of the seminal vesicle, and becomes the **ejaculatory** (e-jak'u-lah-to''re) **duct.** Peristaltic contractions of the ductus deferens move sperm through its lumen. The paired ejaculatory ducts, which lie inferior to the bladder, receive sperm from the testes and secretions from seminal vesicles and can eject them into the urethra.

The **urethra,** a common passageway for both sperm and urine, consists of three regions. The *prostatic urethra* passes from the bladder through the prostate gland, where the ejaculatory ducts join it. The *membranous urethra* passes from the prostatic region through the *urogenital diaphragm,* the muscular floor of the pelvic cavity. The terminal *penile* (pe'nīl) *urethra* lies outside the pelvic cavity and passes through the penis.

The **penis,** in addition to conducting urine, is the male copulatory organ. It deposits sperm into the female reproductive tract during sexual intercourse. The penis contains cylindrical masses of erectile tissue (Figure 29.8)— two dorsal *corpora* (kor'po-rah) *cavernosa* and a ventral *corpus spongiosum* (spon''je-o'sum)—surrounded by fibrous tissue. The corpus spongiosum, which contains the urethra, is enlarged at the distal end of the penis to form the **glans penis.** The penis is covered with skin, and the fold of skin over the glans is called the **prepuce** (pre'pūs), or foreskin.

Male Reproductive Glands

Glands that contribute their secretions to fluid ejaculated with the sperm include the seminal vesicles, the prostate gland, and the bulbourethral glands. The ductus deferens also contributes some fluid.

The **seminal** (sem'in-al) **vesicles** are saclike structures near the base of the bladder; their ducts join the ductus deferens to form the ejaculatory duct. Their viscous secretions contain fructose, which may nourish sperm, and prostaglandins, which may help to regulate reproductive function.

The **prostate gland** is a single large gland that surrounds the urethra. Its several small ducts secrete into the urethra a milky fluid containing a little citric acid, some lipids, a few enzymes, and enough bicarbonate ions to make the fluid alkaline. Prostate secretions aid in sperm motility and help to neutralize acid in urethral urine residue and vaginal secretions.

A pair of **bulbourethral** (bul''bo-u-re'thral) **glands,** or Cowper's glands, are located in the floor of the pelvic cavity. Their secretions, which contain mucus, empty through ducts into the membranous urethra.

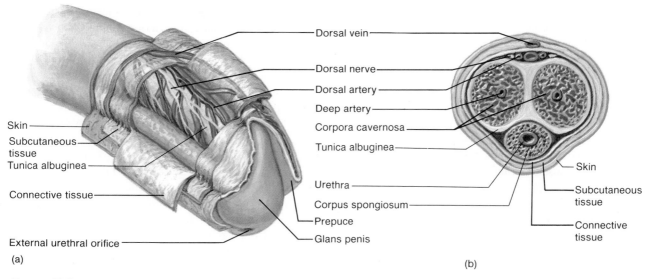

(a)

(b)

FIGURE 29.8
The penis: (*a*) dissection of layers, and (*b*) cross section.

MALE REPRODUCTIVE PHYSIOLOGY

The physiology of the male reproductive system concerns the action of testosterone and the production of sperm and semen. It also deals with erection of the penis and ejaculation of semen.

Testosterone and Its Regulation

Testosterone and **dihydrotestosterone** are responsible for the development and maintenance of masculine characteristics. During embryonic development, chorionic gonadotropin from the placenta stimulates the testes to secrete testosterone from about the seventh week of development until a few weeks after birth. At the onset of puberty, gonadotropin-releasing hormone (GnRH) causes the anterior pituitary gland to release luteinizing hormone (LH), which stimulates interstitial cells to make testosterone and dihydrotestosterone. Male hormone secretion rises sharply during puberty, peaks in the early twenties, declines gradually until the early forties, and more rapidly thereafter. Secretion stays at about one-fifth the peak rate even at age 80. High dihydrotestosterone acts by negative feedback to suppress LH and GnRH release.

During puberty, testosterone stimulates the growth and development of male sex organs and the development of male secondary sexual characteristics—male body hair pattern, muscular development, skeletal growth, and basal metabolic rate. Dihydrotestosterone controls hair growth, which occurs not only in the pubic area, but also on the face, chest, limbs, and axillae (armpits). Increased protein synthesis accounts for a higher metabolic rate, muscular development, thicker skin, and probably an increase in red blood cell numbers. Increased melanin synthesis deepens the skin hue. Bone matrix synthesis and mineral deposition account for bone growth. Enlargement of the larynx—especially the thyroid cartilage, or "Adam's apple"—and corresponding lengthening of the vocal cords deepens the voice.

Once the male reproductive system has completed development during puberty, male hormones are still needed to maintain mature organs and produce sperm. Testosterone increases muscle strength by increasing muscle-cell size, and by increasing the amount of glucose they take up and phosphorylate and the amount of glycogen they store. It stimulates the pituitary gland to release growth hormone and prepares the heart to grow in response to it. It also fosters sperm maturation by stimulating Sertoli's cells. Dihydrotestosterone acts on the prostate gland and sebaceous glands. The skin of acne patients contains 2 to 20 times the normal amount of this hormone.

Spermatogenesis and Its Regulation

Spermatogenesis (sper″mat-o-jen′es-is), the process by which functional sperm are produced (Figure 29.9a), takes place in the seminiferous tubules from puberty until testosterone secretion becomes too low to stimulate it. Lining the inner walls of the seminiferous tubules are great numbers of undifferentiated **spermatogonia** (sper″mat-o-go′ne-ah), which continuously divide by mitosis. Their progeny grow and migrate toward the tubule lumen, becoming **primary spermatocytes** (sper-mat′o-sītz). To form sperm, each primary spermatocyte replicates its DNA once and undergoes two meiotic divisions (Figure 29.9b). In the first division, after DNA replication, each homologous pair of replicated chromosomes synapse, or join together, temporarily forming a **tetrad** (tet′rad) of four chromatids. Each chromatid contains the DNA equivalent to one chromosome. When these cells divide, one of each pair of chromosomes goes to each new cell. The progeny cells called **secondary spermatocytes** have half the normal number of chromosomes, but each chromosome contains two chromatids. This constitutes **reduction division.**

Secondary spermatocytes undergo the second meiotic division to form **spermatids** (sper′mat-idz). During this division, the chromatids separate and each progeny cell receives a single copy of each chromosome. These cells are **haploid** (hap′loid), or 1n, cells because they have only one copy of 22 autosomal (nonsex) chromosomes and either an X or Y sex chromosome. Spermatids differentiate, acquire flagella, lose most of their cytoplasm, and become mature gametes called **spermatozoa** (sper″mat-o-zo′ah) in a process called **spermiogenesis** (sper″me-o-jen′es-is).

Sertoli's cells, which aid spermiogenesis, are connected together by tight junctions to form a continuous layer called the **blood-testis barrier** around the perimeter of seminiferous tubules. Substances from the blood reach spermatids only by passing through Sertoli's cell cytoplasm. Though testosterone acts directly on spermatocytes undergoing meiosis, FSH binds to Sertoli's cell receptors, causing those cells to mediate sperm maturation. FSH may be needed to initiate sperm maturation but may not be necessary to maintain spermatogenesis once it is started.

Carefully regulated release of GnRH, LH, and testosterone are needed for normal spermatogenesis (Figure 29.10). When the blood testosterone level rises, testosterone acts by negative feedback to inhibit GnRH release. As GnRH decreases, LH decreases and interstitial cells are not stimulated. When the testosterone level is too low to inhibit it, GnRH secretion again releases LH.

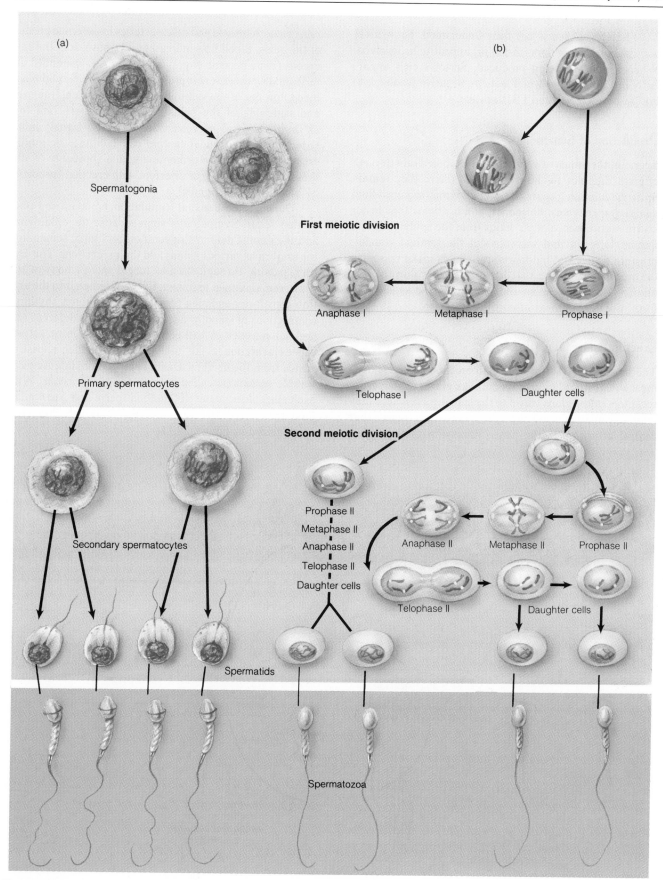

(a)

(b)

Spermatogonia

First meiotic division

Anaphase I   Metaphase I   Prophase I

Primary spermatocytes

Telophase I   Daughter cells

Second meiotic division

Secondary spermatocytes

Prophase II
Metaphase II
Anaphase II   Anaphase II   Metaphase II   Prophase II
Telophase II
Daughter cells

Telophase II   Daughter cells

Spermatids

Spermatozoa

FIGURE 29.9

(*a*) Spermatogenesis, and (*b*) meiosis.

FSH regulation is not fully understood, but GnRH and the hormone **inhibin** (in-hib'in) appear to be involved. When sperm are being produced, Sertoli's cells apparently release inhibin, which acts by negative feedback to inhibit both GnRH and FSH release.

Production of Semen

Sperm in the lumen of a seminiferous tubule drain through efferent ductules to the epididymis, where they remain until ejaculation occurs. The epididymis secretes fluid containing enzymes, hormones, and nutrients that foster sperm maturation. Sperm, fluids from the epididymis and ductus deferens, and secretions of the seminal vesicles, prostate gland, and bulbourethral glands make up **semen** (se'men), the mixture that leaves the penis during ejaculation. Secretions from the prostate gland apparently foster the development of motility in sperm.

Erection and Ejaculation

The most direct factor leading to erection of the penis is mechanical stimulation of its bulblike glans. Signals from receptors in the glans travel to the spinal cord, where they inhibit sympathetic and elicit parasympathetic signals that control arterioles in the penis. Parasympathetic signals relax smooth muscle and dilate arteries in cavernous bodies of the penis. Blood flow increases and sinusoids of the cavernous tissue become engorged with blood. Distended arteries and sinusoids exert pressure on the venules and block blood flow out of the sinusoids. The resulting enlargement and stiffening of the penis constitutes an **erection** (er-ek'shun). Central nervous system (CNS) signals influenced by psychological factors can facilitate or inhibit erection. Alcohol can promote erection (probably by obliterating inhibitions) or interfere with erection (probably by blocking CNS signals).

During intense sexual stimulation, emission and ejaculation occur. Sympathetic signals cause the epididymis and the ductus deferens to contract and force semen into the urethra as various glands add their secretions. The urethra, passing through the less turgid spongy body of the penis, remains open to semen. Entry of semen into the urethra constitutes **emission** (e-mish'un).It initiates rhythmic contractions of smooth muscle in the ductus deferens and skeletal muscles in the pelvic floor. These contractions cause **ejaculation** (e-jak"u-la'shun), the sudden release of semen from the urethra. During ejaculation, intense pleasurable sensations called **orgasm** (or'gazm) occur. With each ejaculation about 400 million sperm and 2 to 6 ml of fluid are ejected. Erection, emission, and ejaculation are summarized in Figure 29.11.

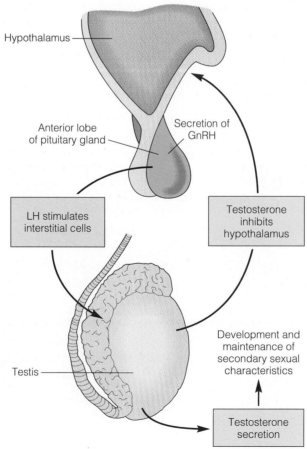

FIGURE 29.10

The regulation of testosterone secretion.

MALE REPRODUCTIVE DISORDERS

Disorders of the prostate gland account for a large proportion of male reproductive disorders. **Prostatitis** (pros"tah-ti'tis), inflammation of the prostate gland, causes swelling that blocks the urethra and make urination painful and difficult. Untreated prostatitis can lead to

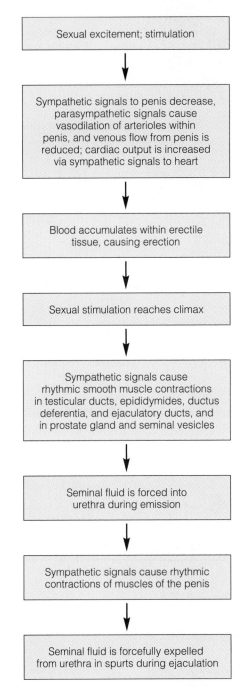

FIGURE 29.11
The regulation of erection, emission, and ejaculation.

severe back pressure that damages nephrons. Incomplete bladder emptying increases the risk of recurrent urinary tract infections. After age 55, benign hypertrophy (enlargement) of the prostate gland is common, though its cause is unknown. It usually is treated by transurethral resection, surgical removal of enough of the prostate to enlarge the urethra. **Prostate cancer,** a common malignancy accounting for 2 to 3 percent of male deaths, is stimulated by testosterone. It can be treated by removing the testes to prevent testosterone release or with estrogens to counteract the testosterone. Prostate cancer metastasizes to bone, but the above treatments sometimes arrest the disease and allow at least temporary healing of bones.

    Impotence (im'po-tens), the inability to achieve and/or sustain an erection long enough to complete sexual intercourse, can be caused by psychological factors, neurological or vascular disorders, and syphilis. **Sterility,** the inability to fertilize an ovum, occurs when the sperm count is less than 20 million per ml of semen, or when more than 20 percent of the sperm are abnormal or have reduced motility. Causes of sterility include gonorrhea and other infections, low-testosterone secretion, and poor nutrition that allows degeneration of reproductive ducts.

FEMALE REPRODUCTIVE ANATOMY

The female reproductive system (Figures 29.12 and 29.13) consists of ovaries, uterine tubes, uterus, vagina, and external genitalia. Because of their role in nourishing offspring, the mammary glands are included with the female reproductive system.

Ovaries

The ovaries are paired, almond-shaped glands about 3.5 cm long, 2 cm wide, and 1 cm thick. They are located in the upper pelvic cavity. Each ovary is covered by cuboidal epithelium called the **germinal epithelium** and underlying connective tissue called the *tunica albuginea.* The main body of an ovary, the **stroma** (stro'mah), consists of a dense outer cortex and a less dense inner medulla. During female development, primary oocytes migrate to the ovary and become surrounded by follicular cells, forming **primordial** (pri-mor'de-al) **follicles.** Such follicles persist unchanged until puberty, when FSH stimulates development of a few follicles in each ovary every month.

    The ovaries are anchored by ligaments to other female reproductive organs (Figure 29.13). The **ovarian ligament** attaches the ovary to the uterus and to the **suspensory ligament.** A membrane called the **mesovarium** (mes"o-va're-um) attaches each ovary to the **broad ligament.**

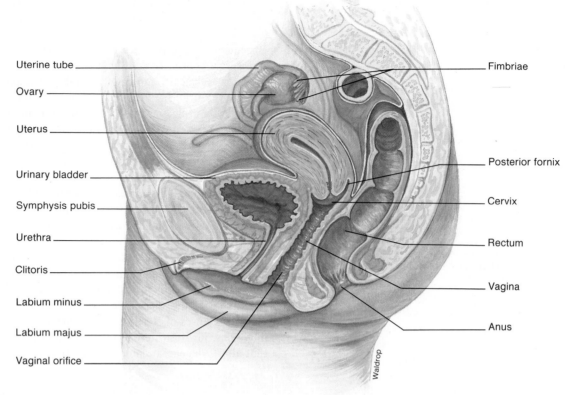

FIGURE 29.12
The female reproductive system, in sagittal view.

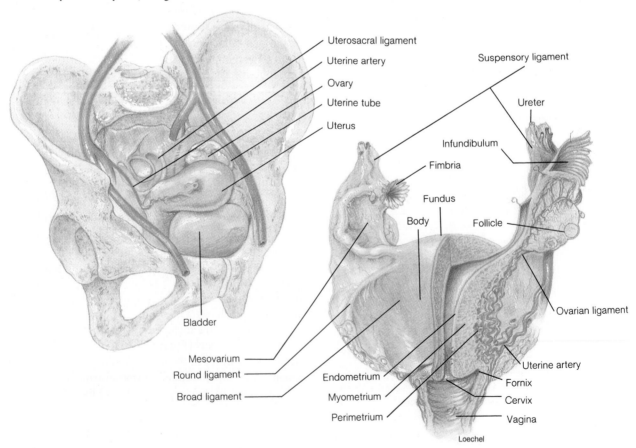

FIGURE 29.13
The female reproductive system and its ligaments in anterior view.

Uterine Tubes, Uterus, Vagina, and External Genitalia

In addition to the ovaries, the uterine tubes, uterus, and vagina form the internal female reproductive system (refer back to Figure 29.12). The paired uterine tubes extend from the ovaries to the uterus. The vagina is a muscular tube that leads from the uterus to the outside of the body.

Each **uterine** (u′ter-īn) **tube** (fallopian tube or oviduct) is attached to the uterus at one end and flares into a funnel-shaped **infundibulum** (in″fun-dib′u-lum) at the other end. **Fimbriae** (fim′bri-e), fingerlike projections of the infundibulum, help move an ovum from the ovary into the uterine tubes. The uterine tubes, supported by a part of the broad ligament, are about 10 cm long. Their walls consist of an outer serous membrane continuous with the peritoneum, a middle layer of smooth muscle, and an inner epithelium with some ciliated cells. The ovum is propelled through the uterine tube by peristaltic contractions of the smooth muscle and maybe by beating of cilia.

The **uterus** (u′ter-us) is a pear-shaped organ lying between the urinary bladder and the rectum. Prior to pregnancy, it is about the size of a clenched fist. It is held in place by three paired ligaments and two single ligaments (Figure 29.13). Except for the fibrous bands of connective tissue that form the **round ligaments,** all ligaments arise from folds in the peritoneum. The round ligaments contain nerves and blood vessels and attach the uterus to the external genitalia. The **broad ligaments** suspend the uterus and partition the pelvic cavity. The **uterosacral** (u″ter-o-sa′kral) **ligaments** attach the uterus to the sacrum on either side of the rectum. The **posterior ligament** attaches the uterus to the rectum, forming the deep **rectouterine** (rek″to-u′ter-in) **pouch** (pouch of Douglas) where pus sometimes collects during pelvic infections. The **anterior ligament** attaches the uterus to the bladder, forming the shallow **vesicouterine** (ves″ik-o-u′ter-īn) **pouch.**

The uterus (Figure 29.13) is divided into a rounded superior **fundus,** a central body, a constricted **isthmus** (is′mus), and an inferior **cervix** (ser′viks). The **internal ostium** divides the cavity inside the uterus into the superior **uterine cavity** and the inferior **cervical** (ser′vik-al) **canal.** The **external ostium** marks the transition from cervix to vagina.

The uterine wall consists of three layers (Figure 29.13). The outer serous **perimetrium** (per-i-me′tre-um) is parietal peritoneum continuous with the broad ligament. It covers the surface of the uterus except for an area near the cervix. The middle **myometrium** (mi-o-me′tre-um) consists of smooth muscle and accounts for most of the wall thickness. The inner **endometrium** (en-do-me′tre-um) is a mucous membrane. Its surface *stratum functionalis* (funk″she-o-na′lis) is sloughed during menstruation, and its *stratum basalis* (bas-a′lis) remains intact.

The **vagina** (vah-ji′nah) is a tube about 10 cm long that extends from the cervix to the outside of the body (refer back to Figure 29.12). It provides a passageway for

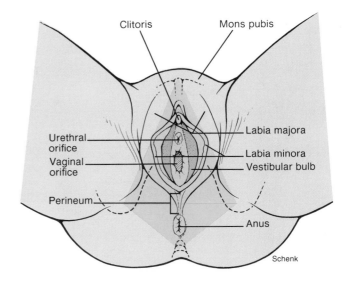

FIGURE 29.14
The female external genitalia.

the menstrual flow, serves as a receptacle for sperm during intercourse, and forms part of the birth canal. It is lined with mucous membrane and has thick, easily stretched transverse rugae (folds) in its muscular wall. A recess called the **fornix** (for′niks) surrounds the area where the cervix projects into the vagina. The posterior fornix is deeper than the anterior and lateral fornices. The **vaginal** (vaj′in-al) **orifice** is partially covered by a membrane called the **hymen** (hi′men) in most virgins. In a condition called imperforate hymen, the hymen completely closes off the vaginal orifice and must be removed to allow menstrual flow.

The female **external genitalia** (jen″it-a′le-ah) are also called the vulva or pudendum (Figure 29.14). The anteriormost **mons pubis** (monz pu′bis) is a fatty area covered with skin and pubic hair. Posterior to the mons pubis is the **clitoris,** which terminates in the **glans clitoris** surrounded by the **prepuce.** These structures are homologous with those of the penis. The clitoris is erectile tissue and the glans clitoris is especially sensitive to stimulation. Such sensitivity accounts for female **orgasm.**

Posterior to these structures in the midline are the urethral opening and the vaginal orifice surrounded by two pairs of skin folds. Separated from the vaginal orifice by the **vestibule** are the **labia minora,** skin folds covered with stratified squamous epithelium. A pair of **Bartholin's** (bar′to-linz) **glands** secrete mucus into the vestibule. The **labia majora** are thicker, more lateral folds that are partly covered with pubic hair and are homologous to the scrotum. The **perineum** (per″in-e′um) is the external area from the pubic symphysis to the coccyx; the clinical perineum is the area between the vagina and the anus, where tears often occur during childbirth.

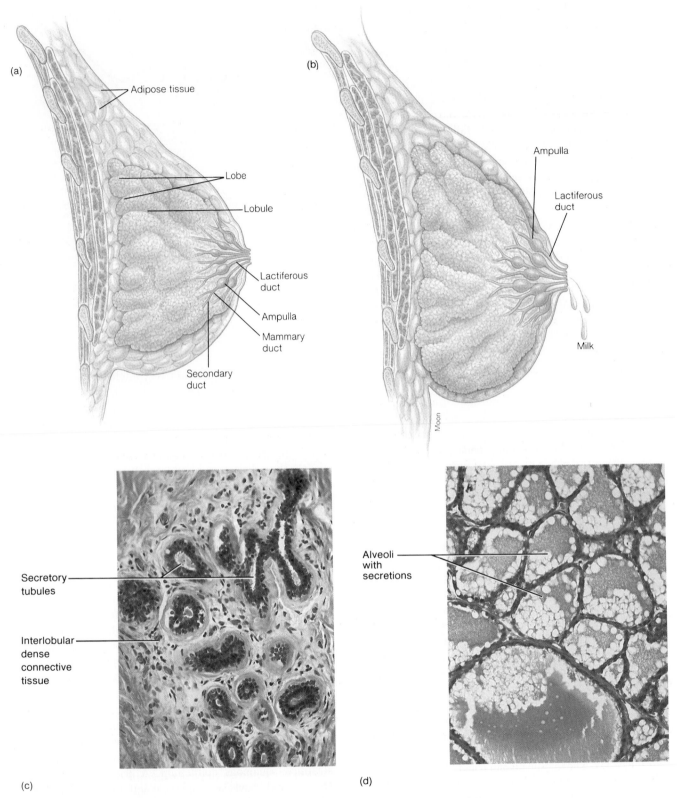

(a)

Adipose tissue

Lobe

Lobule

Lactiferous
duct

Ampulla

Mammary
duct

Secondary
duct

(b)

Ampulla

Lactiferous
duct

Milk

(c)

Secretory
tubules

Interlobular
dense
connective
tissue

(d)

Alveoli
with
secretions

FIGURE 29.15

The mammary gland. (*a*) Nonlactating, (*b*) lactating, and
(*c–d*) corresponding photomicrographs.

Breasts and Mammary Glands

The **mammary** (mam'er-e) **glands** (Figure 29.15) are modified sweat glands surrounded by adipose tissue and covered with skin. Together, these tissues form breasts, which lie over pectoral muscles. At puberty, the mammary glands begin to develop under the stimulation of estradiol and other hormones in females, but they remain undeveloped in males. Each breast has a nipple surrounded by a pigmented area called the **areola** (ar-e'o-lah). The areolar surface is rough because of modified sebaceous glands beneath the surface. Internally, the mammary gland contains 15 to 20 lobes, each of which contains many lobules of glandular tissues imbedded in **fatty stroma.** Variation in mammary gland size is due to the amount of fatty stroma and has no relation to milk production. Within each lobule are many small alveoli, each containing numerous secretory cells.

As milk is secreted, it enters **secondary tubules.** Several secondary tubules come together to form a **mammary duct** and just beneath the nipple, these ducts expand to form **ampullae. Lactiferous** (lak-tif'er-us) **ducts** carry the milk from the ampullae to the nipple. Most lactiferous ducts drain one lobe, but a few drain two lobes.

FEMALE REPRODUCTIVE PHYSIOLOGY

Female reproductive processes include oogenesis (production of ova), cyclic changes in both the ovaries and uterus, and the response to sexual stimulation. Special processes occur during pregnancy and lactation.

Oogenesis

Oogenesis (o''o-jen'es-is) is the production of ova, and like spermatogenesis, it involves meiosis (Figure 29.16). During embryonic development before the birth of a human female, several million **oogonia** (o''o-go'ne-ah), cells capable of producing ova, migrate from the yolk sac to gonads that become ovaries. Many oogonia undergo **atresia** (ah-tre'ze-ah), a kind of degeneration, and surviving cells start through meiosis and become **primary oocytes** (o'o-sītz). At birth, about 2 million remain, and by puberty, only 400,000—half in each ovary—remain. All the ova a woman can ever produce come from these primary oocytes, which have chromosomes in tetrads as a result of arrested meiosis.

At puberty, a few oocytes begin to develop each month. The primary oocyte completes division forming two cells with equal amounts of DNA—a **secondary oocyte** that contains most of the cytoplasm and a **first polar body** that contains mainly DNA. The main function of polar bodies is to get rid of excess DNA while a single, viable ovum forms from the oocyte.

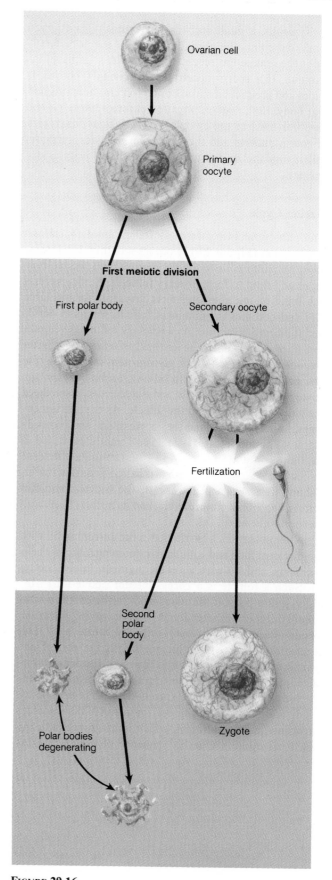

FIGURE 29.16
Oogenesis.

Typically, a single secondary oocyte is released each month and enters the uterine tube where it can be fertilized if the female has recently had, or will soon have, intercourse. If a sperm penetrates a secondary oocyte, it divides and produces a mature **ovum** (o'vum) and a **second polar body.** Both contain unpaired chromosomes, but the ovum has most of the cytoplasm. Its nucleus fuses with the sperm nucleus and the cytoplasm provides nutrients until it implants in the uterus. Unfertilized oocytes degenerate.

Ovarian Cycle

After puberty, an ovarian cycle of follicle development and degeneration (Figure 29.17) occurs monthly. FSH stimulates follicular cells to divide, mature, and secrete hormones and fluid to form **primary follicles.** Among several such follicles, one competes most favorably for available FSH and becomes the dominant follicle. Partially developed follicles degenerate. As a primary follicle develops a fluid-filled cavity called the **antrum** (an'trum), it becomes a **secondary follicle.** A thick glycoprotein membrane, the **zona pellucida** (zo'nah pel-lu'sid-ah), forms between the oocyte and the follicular cells, and a connective tissue **theca** (the'kah) develops around the follicle. As the antrum enlarges, follicle cells around the oocyte divide and protrude into it. The now-mature follicle, sometimes called a **graafian** (graf'e-an) **follicle,** will soon rupture and release the oocyte. Release of the oocyte is called **ovulation** (ov"u-la'shun). If a sperm penetrates it, the oocyte completes meiosis, becomes a mature ovum, and its nucleus fuses with the sperm nucleus.

After ovulation, LH stimulates the antrum to fill with a semiclotted fluid and stimulates remaining follicular cells to enlarge and fill with a yellow pigment, **lutein** (loo'te-in). Such a follicle is a **corpus luteum** (loo'te-um)—literally, yellow body. If the ovum is not fertilized, the corpus luteum degenerates after about 10 days, leaving a scarlike **corpus albicans** (al'bik-anz)—literally, white body. Degeneration of the corpus luteum marks the end of one ovarian cycle, but another soon begins. If the ovum is fertilized, the corpus luteum enlarges and persists as an endocrine gland for the first six to eight weeks of pregnancy.

In addition to releasing ova, maturing follicles secrete **estrogens,** mainly **estradiol,** which helps to regulate the menstrual cycle, a monthly sequence of events that includes sloughing part of the uterine lining. After ovulation, corpus luteum cells produce estrogens and **progesterone.** Progesterone is essential to maintaining pregnancy, but it also helps to regulate the menstrual cycle.

Only small amounts of estradiol are released before puberty. After puberty, estradiol secretion increases to about 20 times the prepubertal level. Estradiol stimulates reproductive organs to grow and become functional, and develops secondary sexual characteristics, such as the female pattern of fat distribution. The vaginal epithelium becomes more resistant to infections and trauma. The uterus doubles or triples in size in the first few years after puberty, and its endometrium proliferates. Ciliated epithelial cells of uterine tubes increase in number and activity of cilia and assist ova in reaching the uterus. Estradiol also stimulates duct development and fat deposition in mammary glands.

In addition to its effects on the reproductive system, estradiol affects the skeleton, skin, metabolism, and electrolyte balance. It stimulates bone growth during the pubertal growth spurt and then stops growth by causing epiphyses to fuse with the shafts of the long bones. Women usually are not as tall as men because estradiol has a faster more potent effect on epiphyseal fusion than testosterone. Estradiol causes skin to become soft and smooth, but thicker and more vascular than in childhood. Growth of pubic hair and the development of axillary sweat glands probably are controlled, not by estradiol, but by adrenal androgens, which also account for acne in females. Estradiol increases the metabolic rate—but by only about one-third as much as testosterone, and it causes a smaller increase in total body protein than testosterone. As a result of these differences, women generally have slower metabolism, smaller bodies, less muscle, and more body fat than men. Fat is deposited in the breasts, thighs, and buttocks, producing feminine body contours.

Progesterone prepares the uterus for implantation of a fertilized ovum and prepares the secretory portions of the mammary glands for milk production. It causes uterine tubes to release secretions that help to nourish the ovum, and it suppresses uterine contractions during pregnancy. Progesterone slightly increases protein catabolism, making amino acids from body tissues available for protein synthesis in a growing fetus if dietary nutrients do not satisfy the need.

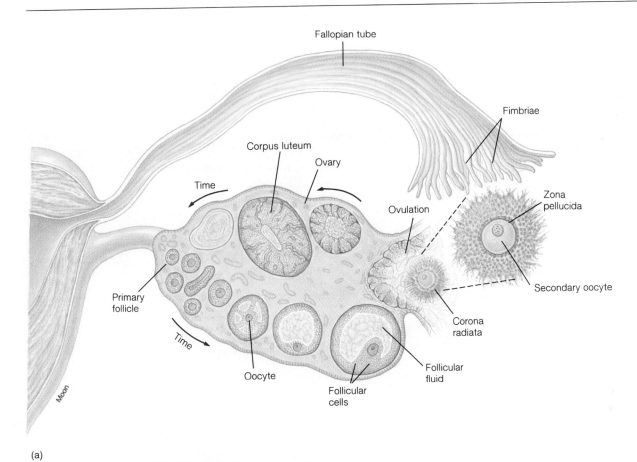

(a)

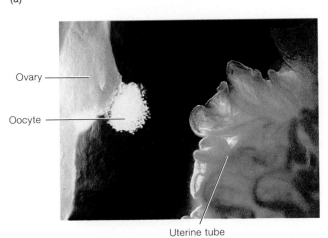

(b)

FIGURE 29.17
(*a*) Diagram of ovarian cycle with stages located in cross section of ovary, and (*b*) photomicrograph of ovulation.

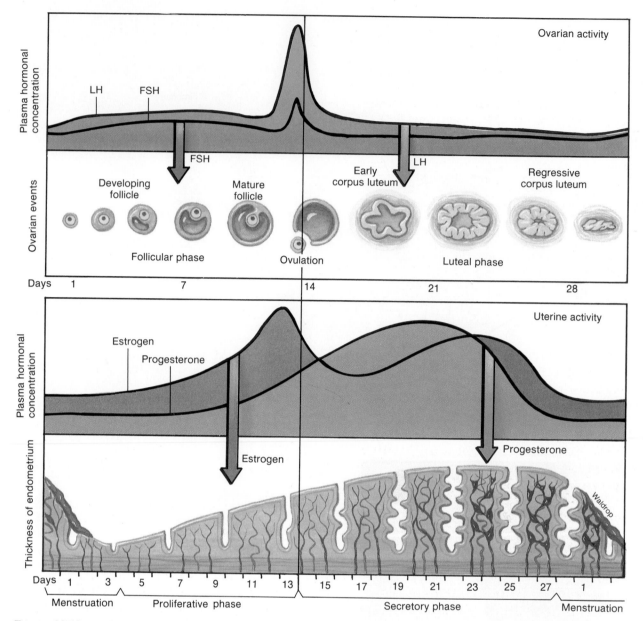

FIGURE 29.18
Main events in the menstrual cycle related to the ovarian cycle and hormone concentrations.

The Menstrual Cycle and Its Regulation

Uterine functions are closely coordinated with ovarian functions by several hormones in the monthly female **menstrual** (men'stroo-al) **cycle** (Figure 29.18). In this cycle, **menstruation** (men''stroo-a'shun) is the shedding of the **stratum functionalis** (surface portion) of the endometrium and the bleeding that accompanies it. Blood loss ranges from 25 to 65 ml, and clotting is prevented by fibrinolysin degrading fibrin as it forms. **Menarche** (men-ar'ke), the age at which menstruation begins, marks the beginning of ovarian and uterine functions. Hypothalamic maturation leads to episodic GnRH secretion, which causes the anterior pituitary to secrete pulses of FSH and LH.

A menstrual cycle lasts from the beginning of one menstrual flow to the beginning of the next flow. A typical cycle lasts about 28 days, but variations from 24 to 35 days are very common. The time from ovulation to menstruation is usually 14 days, and variation in cycle length more often occurs between menstruation and ovulation. Differences in how much GnRH, FSH, LH, estrogen, and progesterone are secreted account for some variations, but stress—leaving home to go to college or work, final exams, or a family crisis—also account for variations in cycle length. Stress probably causes signals to be relayed from other parts of the brain to the hypothalamus, where they alter GnRH release.

Body fat is an important factor in normal menstrual cycles. Studies of athletic and nonathletic girls and women show that the body must contain at least 17 percent fat for menarche to occur and about 22 percent fat to maintain reproductive capacity. The combination of hard physical work and low body fat may delay menarche, decrease fertility, and hasten menopause. It also may lower the risk of breast and cervical cancer.

> Synchronization of menstrual cycles in all-female environments—women's colleges, convents, and prisons—has been reported. One possible explanation for this phenomenon is that chemical signals pass from one woman to another. Exactly what chemicals may be involved is not known but they must have the capacity to influence one or more hormones that normally regulate the menstrual cycle. Whatever the signal, the presence of men appears to disrupt it and desynchronize the cycles.

Preovulatory Phase

The time during menstruation and about 10 days thereafter until ovulation is the **preovulatory** (pre-ov′u-la-to-re) **phase,** or follicular phase, of the menstrual cycle. Blood estrogens and progesterone are at their lowest levels at the onset of menstruation. FSH secreted into the blood initiates follicle development and a dominant follicle matures while the others degenerate. FSH activity increases faster than FSH plasma concentration, probably because FSH receptors increase in number as follicular cells divide. A positive feedback situation develops in which follicle cells secrete increasing quantities of estradiol, and estradiol stimulates follicle development.

Like FSH, LH remains nearly constant until estradiol secretion reaches a peak on the last day of the preovulatory phase. Estradiol then acts by positive feedback to cause surges of LH and GnRH. GnRH then stimulates secretion of both LH and FSH. The result is a large midcycle LH surge and a lesser FSH surge.

Within 24 hours of the midcycle LH surge, ovulation occurs, usually on about day 14 of a normal menstrual cycle. Though the sudden LH surge appears to trigger ovulation, how it does so is not known. LH may stimulate production of follicular enzymes, which digest the thin surface membrane of a bulging follicle.

Changes in the uterine lining during the menstrual cycle are under the direct control of estradiol and progesterone. During the preovulatory phase, the uterus is in its **proliferative phase.** Estradiol stimulates proliferation by cell division of the stratum functionalis and growth of its spiral arteries. It also may cause endometrial cells to develop progesterone receptors.

The preovulatory phase of the menstrual cycle and its regulation is summarized in Figure 29.19.

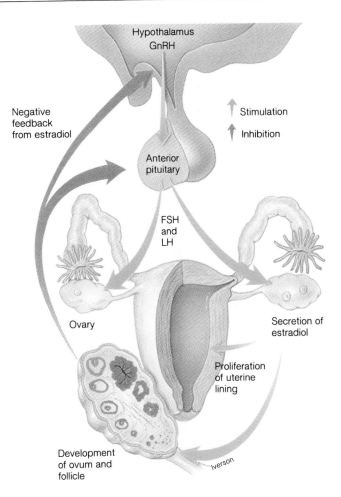

FIGURE 29.19
Summary of the preovulatory portion of the menstrual cycle.

Postovulatory Phase

Events after ovulation and before the next phase of bleeding make up the **postovulatory** (post-ov′u-la-to-re) **phase,** or luteal phase, of the menstrual cycle. The oocyte reaches the uterus, and follicular cells form a corpus luteum, which secretes large quantities of estradiol and progesterone. These hormones act by negative feedback to suppress release of GnRH, FSH, and LH. The uterus is in its **secretory phase.** Progesterone causes the endometrium and cervix to develop mucous glands and secrete mucus, and endometrial cells to accumulate glycogen in preparation for the implantation of a fertilized ovum. If implantation occurs, progesterone maintains the endometrium and inhibits uterine contractions during pregnancy.

If implantation does not occur, the corpus luteum regresses; estradiol and progesterone drop to their lowest levels; and GnRH, FSH, and LH are freed from negative feedback. Sloughing of the stratum functionalis during menstruation seems to be initiated by the precipitous drop

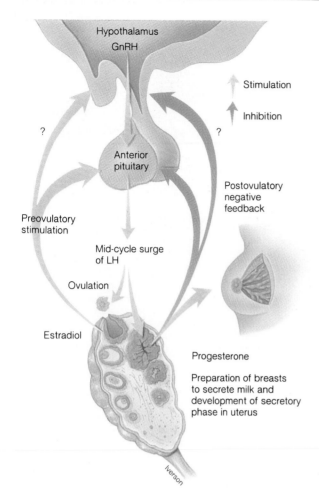

Stimulation

Inhibition

?

?

Hypothalamus
GnRH

Anterior
pituitary

Preovulatory
stimulation

Postovulatory
negative
feedback

Mid-cycle surge
of LH

Ovulation

Estradiol

Progesterone

Preparation of breasts
to secrete milk and
development of secretory
phase in uterus

Iverson

FIGURE 29.20
Summary of ovulation and the postovulatory portion of the menstrual cycle.

TABLE 29.1	Actions of Hormones Involved in the Female Reproductive Cycle
Hormone	**Actions**
Follicle-stimulating hormone (FSH)	Stimulation of follicle growth
	Stimulation of estradiol release
Estradiol (low level)	Increase in sensitivity of anterior pituitary to GnRH
	Stimulation of anterior pituitary to emphasize LH synthesis over FSH synthesis
	Inhibition of LH release
	Proliferation of cells of uterine lining
Estradiol (high level)	Inhibition of FSH release
	Stimulation of LH release
Luteinizing hormone (LH)	Induction of ovulation
	Stimulation of follicle growth
	Formation and maintenance of corpus luteum (source of more estradiol and progesterone)
	Stimulation of progesterone release
Progesterone (high level)	Further development of uterus in preparation for implantation
	Increase in size and sensitivity of breasts (with estradiol)
	Inhibition of LH release
	Inhibition of further follicle development
Progesterone (low level)	Degeneration of corpus luteum
	Increase in release of FSH (thus cycle begins again)

in progesterone, which causes blood vessel spasms, ischemia, and eventual necrosis in the stratum functionalis. Even as menstruation occurs, FSH stimulates follicle development and a new cycle begins.

Ovulation and the postovulatory portion of the menstrual cycle are summarized in Figure 29.20. Actions of hormones in the female reproductive cycle are summarized in Table 29.1.

Menopause

Menstrual cycles normally continue at more or less regular intervals from puberty to menopause, except during pregnancy. **Menopause** (men′o-pawz), the cessation of menstruation, occurs around age 50 with the gradual cessation of ovarian function. Follicles fail to develop, estradiol and progesterone secretion decreases, and menstrual periods become irregular and eventually cease. Estradiol production decreases markedly throughout most of the fifth decade of a woman's life in spite of the fact that more FSH, LH, and GnRH are secreted than before menopause. Decreasing estradiol provides less and less negative feedback inhibition of the other hormones. "Hot flashes," periods of intense warmth resulting from the dilation of skin arterioles, during menopause are associated with LH pulses. Such pulses elicit alpha adrenergic signals that stimulate thermoregulatory neurons and neurons that release GnRH.

A more serious consequence of postmenopausal estrogen deficiency is the increased likelihood of osteoporosis (Chapter 6), in which loss of vertebral bone mass can occur at a rate of 6 percent per year. Estrogen replacement therapy (administration of estrogens to compensate for the deficiency) is the best available means of preventing or minimizing bone loss. Bone cells are known to have estrogen receptors, so estrogen probably acts on them directly, and it may also stimulate calcitonin secretion. Calcitonin decreases both bone reabsorption and the plasma calcium level. Decreased plasma calcium releases parathormone, which increases vitamin D synthesis. Increasing vitamin D increases calcium absorption and raises

1. In the shower:
Examine your breasts during bath or shower; hands glide easier over wet skin. Fingers flat, move gently over every part of each breast. Use right hand to examine left breast, left hand for right breast. Check for any lump, hard knot, or thickening.

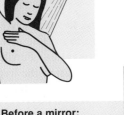

3. Lying down:
To examine your right breast, put a pillow or folded towel under your right shoulder. Place right hand behind your head—this distributes breast tissue more evenly on the chest. With left hand, fingers flat, press gently in small circular motions around an imaginary clock face. Begin at outermost top of your right breast for 12 o'clock, then move to 1 o'clock, and so on around the circle back to 12. A ridge of firm tissue in the lower curve of each breast is normal. Then move in an inch, toward the nipple, keep circling to examine every part of your breast, including nipple. This requires at least three more circles. Now slowly repeat procedure on your left breast with a pillow under your left shoulder and left hand behind your head. Notice how your breast structure feels.

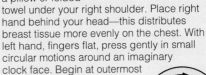

2. Before a mirror:
Inspect your breasts with arms at your sides. Next, raise your arms high overhead. Look for any changes in contour of each breast, a swelling, dimpling of skin, or changes in the nipple.
Then, rest palms on hips and press down firmly to flex your chest muscles. Left and right breast will not exactly match—few women's breasts do.
Regular inspection shows what is normal for you and will give you confidence in your examination.

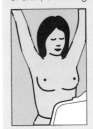

Finally, squeeze the nipple of each breast gently between thumb and index finger. Any discharge, clear or bloody, should be reported to your doctor immediately.

FIGURE 29.21
Monthly self-examination of breasts is an important way to find malignancies while they are still treatable.
Source: American Cancer Society.

the plasma calcium level. Dietary calcium supplements and exercise offer slight protection against osteoporosis compared to estrogen replacement therapy, but estrogens increase the risk of breast and endometrial cancer and thromboses.

Female Response to Sexual Stimulation

Psychological factors and mechanical stimulation of erotic tissues in the clitoris, other parts of the external genitalia, and breasts lead to sexual arousal. Signals from erotic tissue are transmitted to the spinal cord and then to the cerebrum. Parasympathetic signals cause engorgement of blood vessels beneath the clitoris and, during intercourse, cause the vaginal walls to tighten around the penis. These signals also cause Bartholin's glands to secrete a copious quantity of mucus, which lubricates the vagina and increases its massaging effect on the penis. Tightening of the vagina around the penis heightens sensations for both the male and the female.

FEMALE REPRODUCTIVE DISORDERS

A variety of mild to severe, or even life-threatening, disorders affect female reproductive organs. Among common disorders, the most serious are breast and cervical cancer, and the less serious include ovarian cysts, endometriosis, several disorders related to the menstrual cycle, and infertility.

Breast and cervical cancer are leading causes of death from cancer among women. **Breast cancer,** rarely seen before age 30, increases in incidence after menopause. Because breast cancer is difficult to detect, monthly self-examination of the breasts is essential (Figure 29.21) and regular mammograms also are now recommended. Though most lumps are not malignant, any lump should be examined by a physician immediately. Together, these practices maximize the chance that a malignancy will be detected before it has metastasized. **Cervical cancer,** a relatively slow-growing cancer often can be detected by an annual Papanicolaou's test (Pap smear), in which cells from the cervix are examined for malignant changes. Both breast and cervical cancer can be treated by surgery, radiation therapy, and chemotherapy.

Ovarian cysts are ovarian tumors filled with blood or follicular fluid. They can arise from follicular tissue, corpus luteum, or even endometrial cells that have migrated from the uterine tube to the ovarian surface. Some rupture or regress (get smaller) over a four- to eight-week period; many are removed surgically.

Endometriosis (en"do-me"tre-o'sis), the presence of endometrial tissue outside the uterus, is believed to arise from a genetic error that allows cells in other sites to behave like endometrium. Hormones cause such tissue to proliferate and slough along with the stages of the menstrual cycle. Endometriosis, which causes premenstrual and menstrual pain, disappears after menopause but often requires treatment in younger women. The drug danazol, a testosterone derivative, suppresses estrogen secretion, but has undesirable side effects such as increasing blood LDLs, promoting body hair growth, and deepening the voice. Nafarelin mimics GnRH action and decreases estrogen secretion by garbling regulatory information. It has fewer side effects, but can cause hot flashes. When drugs fail to control endometriosis, the ovaries are sometimes surgically removed.

Common menstrual disorders include amenorrhea, excessive or abnormal bleeding, and dysmenorrhea. **Amenorrhea** (am-en"or-e'ah) is the absence of menstruation. Primary amenorrhea is a condition in which menstruation has never occurred. It usually is due to a genetic or birth defect, but can be due to an endocrine disorder. Secondary amenorrhea is the cessation of menstruation in a woman who has previously menstruated. It is caused by stress, endocrine disorders, or anorexia nervosa, but pregnancy should be ruled out before other causes are considered. Excessive bleeding between menstrual periods or after menopause can be caused by endocrine disorders or cancer. **Dysmenorrhea** (dis"men-o-re'ah), painful menstruation or "cramps," affects from 30 to 50 percent of all women of childbearing age. Uterine tumors and endometriosis are responsible for a few cases, but the cause of many cases is unknown. Some evidence shows that prostaglandins, which appear to help initiate both menstruation and labor contractions, may be responsible for dysmenorrhea. They cause uterine contractions and uterine ischemia, which probably is the direct cause of the pain.

Premenstrual (pre-men'stroo-al) **syndrome** (PMS) is a name for an assortment of physical and emotional symptoms that appear in the luteal phase of the menstrual cycle and that disappear within a day of the onset of menstruation. Physical symptoms include fatigue, headache, bloating, breast tenderness and swelling, and acne. Emotional symptoms include anxiety, hostility, anger, and depression. Possible causes of PMS include too little progesterone or endogenous opiate peptides, disordered brain catecholamine metabolism, or too much prolactin or prostaglandins.

Infertility, the inability to become pregnant, can be due to failure to ovulate or to an anatomical factor that prevents fertilization. Fertility drugs induce ovulation, probably by releasing the hypothalamus and anterior pituitary gland from inhibition by estradiol. This allows GnRH to increase FSH secretion, and FSH stimulates follicle development and ovulation. Women treated with such drugs often have twins and other multiple births; the drugs must allow simultaneous development of several follicles. Anatomical disorders such as blockage of the uterine tubes sometimes can be corrected surgically.

FERTILIZATION

The union of egg and sperm, or fertilization, usually occurs in a uterine tube near the ovary. Sperm, having been introduced into the vagina during sexual intercourse, move by flagella, contractions of perineal muscles, and possibly by chemical attraction. Cilia and contractions of the uterine tube move the ovum toward the sperm. Fertilized ova normally reach the uterus before they implant, but a few remain in the uterine tube or enter the abdominal cavity. Implantation in any site other than the uterus constitutes an **ectopic** (ek-top'ik) **pregnancy,** which must be surgically terminated immediately to prevent complications such as uterine tube rupture and hemorrhage.

Fertilization usually occurs within 24 hours after ovulation and four to six hours after intercourse, events associated with fertilization itself take at least 30 minutes. When sperm reach an oocyte they attach loosely to its zona pellucida. Egg-binding proteins in the sperm cell membrane bind specifically and tenaciously with sperm receptors in the zona pellucida, activating lysosomal enzymes in the acrosome. Sperm with activated enzymes digest their way to the zona pellucida, leaving their tails behind, and stop in a narrow space between the zona pellucida and the oocyte plasma membrane. The first sperm to fuse with the oocyte plasma membrane causes a sudden membrane change (probably depolarization) that prevents any other sperm from fusing. It also releases oocyte enzymes that harden the zona pellucida and keep it from binding sperm.

This mechanism assures that an egg will be fertilized by only one sperm, thereby maintaining a normal chromosome number in a fertilized egg.

Entry of a sperm causes the oocyte to complete meiosis and become a mature ovum. The **male pronucleus** (pro-nu'kle-us) of the sperm fuses with the **female pronucleus** of the ovum to form a **zygote** (zi'gōt), the first cell of a developing embryo.

PREGNANCY

From fertilization (or conception) to birth, human pregnancy lasts about 280 days, or 40 weeks, but durations one week longer or shorter are common. Infants born before the 28th week of pregnancy rarely survive. Ovulation and menstruation cease for the duration of pregnancy.

Implantation

Within 36 hours of fertilization, the zygote begins to divide by mitosis and divides several times during the three days that it remains in the uterine tube and is nourished by tubular secretions. After reaching the uterus, the cell mass remains free in the uterine cavity for another three days, continuing to divide and being nourished by secretions from endometrial cells. **Implantation,** or attachment of the embryo to the uterine wall, occurs about a week after fertilization. By this time, the embryo is a **blastocyst** (blas'to-sist), a multicellular, hollow ball (Chapter 30).

Trophoblast (trof'o-blast) **cells** that surround the young embryo perform four functions: (1) They secrete proteolytic enzymes that digest surface endometrium. (2) They phagocytize endometrial cells and make nutrients from them available to the blastocyst. (3) They aid implantation by forming attachments to deeper layers of the endometrium. (4) After implantation, they proliferate rapidly over the uterine surface as the embryonic part of the placenta. At the same time, trophoblastic cells are proliferating, progesterone stimulates endometrium and other uterine tissue to differentiate and become more vascularized, and eventually form the maternal part of the placenta (Chapter 30).

Hormonal Regulation of Pregnancy

Hormones that help to maintain pregnancy include hCG, estradiol, estriol, progesterone, relaxin, and human chorionic somatomammotropin (hCS). Certain embryonic cells produce **human chorionic gonadotropin** (ko''re-on'ik go''nad-o-tro'pin), or hCG, an LH-like hormone that

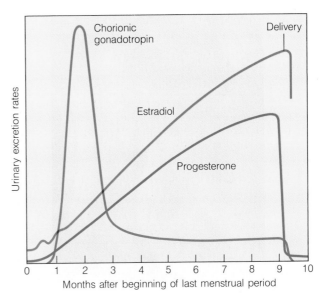

FIGURE 29.22
Urinary excretion of hormones that maintain pregnancy.

causes the corpus luteum to double in size and continue to secrete estradiol and progesterone. These hormones maintain the uterine lining and prevent menstruation during pregnancy. Secretion of hCG persists for five to six weeks until the placenta can produce estrogen and progesterone. The hCG excreted in the early weeks of pregnancy is the basis for most pregnancy tests.

During the first six weeks of pregnancy, the corpus luteum, under the influence of hCG, produces estradiol and progesterone. As hCG production drops, the placenta starts to secrete estrogens and progesterone. Placental estrogens are about 10 percent estradiol and 90 percent estriol, a less active kind of estrogen. Secretion of these hormones increases throughout pregnancy, reaching a peak concentration 50 times that of the menstrual cycle. A few weeks before the onset of labor, progesterone reaches a peak 10 times that in the menstrual cycle. Typical hormone concentrations during pregnancy are reflected in urine as shown in Figure 29.22.

Estrogens and progesterone have special functions during pregnancy. Estrogens cause the uterus, breasts, and external genitalia to enlarge. Progesterone stimulates development of the breasts and maternal part of the placenta and uterine tube secretions. It suppresses uterine muscle contractility.

Relaxin gradually relaxes pelvic ligaments so that by the time of birth the pelvic outlet is larger and slightly flexible. These changes ease passage of the fetus through the birth canal.

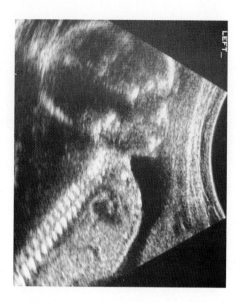

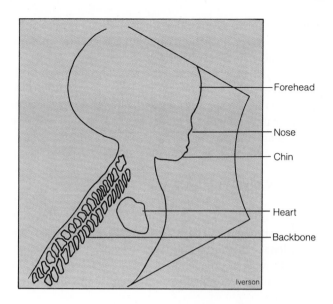

FIGURE 29.23

Sonogram of a near-term fetus in utero, with a diagram showing the orientation of the fetus.

Along with other hormones, hCS stimulates breast development and, like growth hormone, maintains positive nitrogen balance, mobilizes maternal fats for energy, and maintains maternal blood glucose concentration to meet needs of both mother and fetus.

Images produced by sound waves called sonograms can be used to study the fetus after about the third month of pregnancy. They show fetal position, head diameter relative to the pelvic outlet, or how many fetuses are developing (Figure 29.23). Such information is useful in deciding whether a cesarean delivery (delivery through an incision in the abdomen) is required and in planning for multiple births.

Normal Changes in Mother's Body

Along with the hormonal effects just described and cessation of menstruation, other changes occur in the mother's body during pregnancy. Her metabolic rate increases by 5 to 10 percent, providing proteins and energy for her own needs and those of the developing embryo. Blood volume increases 1 l and tissue fluids by 2 l, largely by sodium and water retention fostered by estrogen, progesterone, and maybe renin, aldosterone, and antidiuretic hormone. The additional blood compensates for the volume of placental vessels and maintains blood pressure.

Many women lose a few pounds early in pregnancy, possibly because of nausea, but they eventually gain weight. Over the duration of a pregnancy, most women gain about 20 lbs—fetus about 7 lbs (3.2 kg); amniotic fluid, placenta, and fetal membranes about 4 lbs (1.8 kg); increased weight of the uterus and breasts about 4 lbs (1.8 kg); and fluid accumulation about 5 lbs (2.0 kg). When weight gain exceeds this amount, it usually is due to fat deposition or excessive fluid accumulation. Good nutrition is important during pregnancy to assure that the needs of the fetus are met without depriving the mother. Pregnant women should avoid weight-loss diets.

Mild nausea called "morning sickness" is a common early sign of pregnancy. It poses no great risk, but can be of great concern to the mother. Severe persistent nausea called **hyperemesis gravidarum** (hi″per-em′es-is grav″id-a′rum) can sufficiently deplete nutrients to jeopardize the health of both mother and fetus. The cause of nausea in pregnancy is unknown, but possible causes include high hCG or high estrogen levels and breakdown products from trophoblastic activity.

A fetus is immunologically incompatible with its mother because it receives genes from its father that produce proteins foreign to the mother. Why this situation does not lead to the mother's body rejecting the fetus like it would a foreign skin graft has puzzled immunologists. It may be that certain, poorly understood properties of the trophoblast prevent immune reactions and that progesterone acts as an immunosuppressant. Another immunosuppressant called uromodulin has recently been isolated from the urine of pregnant women. Though rejection of the fetus is prevented in most pregnancies, some spontaneous abortions and stillbirths have been attributed to immunological reactions. These reactions sometimes involve AB blood antigens.

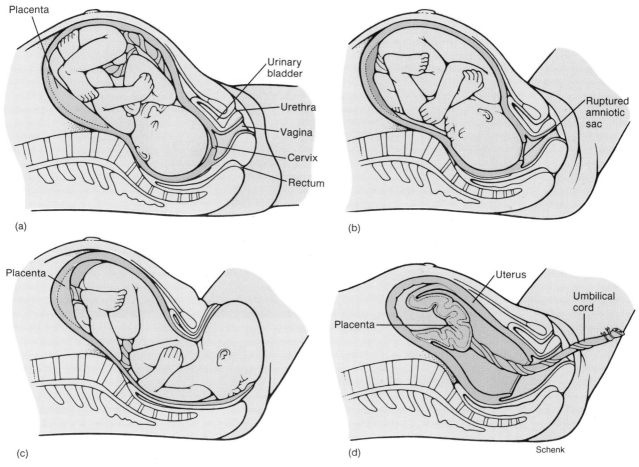

(a)

Placenta
Urinary bladder
Urethra
Vagina
Cervix
Rectum

(b)

Ruptured amniotic sac

(c)

Placenta

(d)

Uterus
Umbilical cord
Placenta

Schenk

FIGURE 29.24
Positions of the fetus during labor: (*a*) normal position of a full-term fetus just prior to the onset of labor, (*b*) position of fetus during first stage (cervix partially dilated), (*c*) position of fetus during the early part of the second stage, and (*d*) placenta separating from uterine wall in the third stage.

Good prenatal care is important to detect and treat abnormal conditions. **Preeclampsia** (pre″e-klamp′se-ah) begins with excessive fluid retention and proceeds to hypertension and protein excretion. It may be due to an immunological reaction in which fluid retention is associated with inflammation or altered kidney function. Limiting the mother's salt intake reduces fluid accumulation, but how salt relates to the immunologic reaction is not known. **Eclampsia** (ek-lamp′se-ah), which can develop from preeclampsia, strikes shortly before or after delivery with widespread spasmodic blood vessel contractions. Kidney and liver functions are impaired, hypertension is severe, and convulsions usually are followed by coma. Vasodilator drugs and diuretics reverse vascular spasms and lowers the blood pressure in 95 percent of cases. The other 5 percent die in spite of vigorous treatment.

PARTURITION

Parturition (par″tu-rish′un) is the process of giving birth (Figure 29.24). It is initiated and controlled by hormonal and mechanical factors, and occurs in three stages, called the stages of labor.

Initiation and Regulation of Labor

Progesterone secretion decreases in the last few weeks of pregnancy and eventually fails to inhibit the action of estrogens. Estrogens then stimulate uterine contractions. Prostaglandins, fetal oxytocin, and increased numbers of oxytocin receptors in uterine muscle make the uterus progressively more excitable until it finally begins to contract rhythmically. Although initiation of labor is not completely understood, positive feedback is an important factor. Periodic episodes of spontaneous uterine contractions become stronger toward the end of pregnancy, and those contractions push the fetus's head against the cervix and stretch it. They also usually rupture the amnion, a fluid-filled membrane that surrounds the fetus. Stretching the cervix initiates a reflex wave of contraction over the body of the uterus. Each such wave further stretches the cervix and elicits a stronger contraction, thereby accelerating labor by positive feedback. As uterine contractions accelerate, reflexes from the birth canal via the spinal cord cause strong abdominal muscle contractions. These contractions work with uterine contractions to push the fetus through the cervix. A woman can voluntarily increase abdominal muscle contractions and speed her own

FIGURE 29.25
Factors that initiate and accelerate labor.

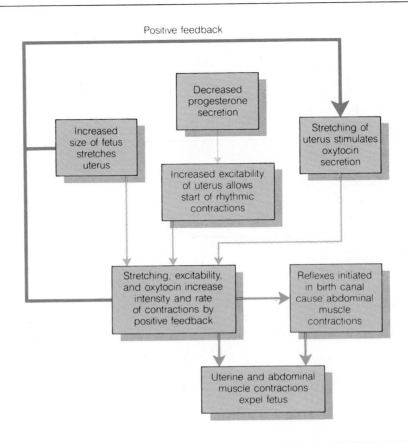

labor. Mechanical pressure from a vigorously moving fetus or twins also accelerates labor and sometimes bring it on two or three weeks early.

Uterine contractions begin in the fundus and spread through the body of the uterus to the cervix, forcing the fetus toward the cervix. Early mild contractions occur about once every 30 minutes, but as labor proceeds they increase in frequency and intensity until strong contractions occur every one to three minutes. Relaxation between contractions is essential to allow blood to circulate through the placenta to nourish the fetus and the uterine muscle itself. If the uterine muscle remained contracted, the fetus would become anoxic (lack oxygen) and the uterine muscle ischemic. Factors involved in the initiation and acceleration of labor are summarized in Figure 29.25.

Stages of Labor

During the first and longest stage of labor, the cervix dilates to the diameter of the fetus's head. This stage lasts 8 to 24 hours for a woman's first delivery, but is much shorter in subsequent deliveries. Previous dilation apparently prepares the cervix for later rapid dilation. The second stage of labor, delivery of the fetus, consists of moving the infant through the birth canal. It lasts about 30 minutes in a first delivery and may be only three to five minutes in subsequent deliveries. The third stage of labor is the separation and expulsion of the placenta. As the placenta separates, the uterine muscle contracts strongly,

closing off endometrial blood vessels that have supplied the placenta. In a normal delivery, blood loss does not exceed 350 ml.

The fetal heart rate is monitored during labor to detect abnormal conditions such as oxygen deprivation, which is indicated by a fetal heart rate below 120 or above 160 beats per minute. The fetus becomes oxygen deprived when the umbilical cord is compressed or labor prolonged. Prolonged labor (more than 24 hours without progress toward delivery) threatens the fetus because the placenta may separate from the uterine wall and cut off fetal oxygen before birth can occur.

When the birth canal is too small for the infant to pass through without tearing, an **episiotomy** (ep-iz″e-ot′om-e), an incision in the clinical perineum can prevent traumatic tearing. The need for episiotomies is decreasing because mothers learn through childbirth classes to control abdominal contractions and breathing, and fathers learn to support and coach them.

A **neonate** (ne′o-nāt), or newborn infant, receives several kinds of routine care (Figure 29.26). The umbilical cord is tied to prevent bleeding and cut. Mucus is aspirated (removed by suction) from the nose and throat. The baby is weighed, measured, footprints made for identification, and its physiological status rated by the APGAR scale (Chapter 30).

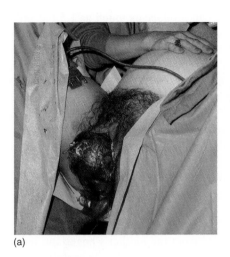
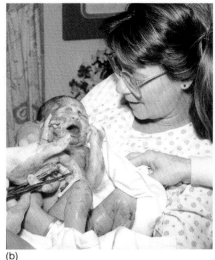
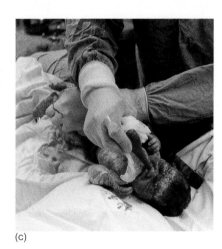

(a)   (b)   (c)

FIGURE 29.26
A new baby arrives: (*a*) the baby's head emerges, (*b*) the mother holds baby while the umbilical cord is clamped and cut, (*c*) the baby's airway is cleared.

Not every pregnancy terminates in the birth of a healthy infant. **Spontaneous abortion** (miscarriage) occurs in about 10 percent of pregnancies, usually within the first four months. The embryo usually has died from a severe developmental defect and the body naturally gets rid of it. Uterine or hormonal abnormalities also can cause spontaneous abortions.

Also, not every baby is born head first. The most common abnormal orientation, seen in about 5 percent of births, is a **breech birth** in which the buttocks are the presenting part. In a breech birth, passage through the birth canal is more difficult and labor may be prolonged or a cesarean delivery may be necessary.

LACTATION

Lactation (lak-ta′shun), secretion of milk from the **mammary glands,** normally occurs following the birth of an infant and continues as long as milk is regularly removed. As milk is secreted from secretory cells within the glands, it passes through mammary ducts, ampullae, and lactiferous ducts.

Mammary Gland Development

At puberty, estradiol, progesterone, prolactin, and growth hormone stimulate development of secretory portions and ducts of the mammary glands. During pregnancy, progesterone and estradiol further stimulate mammary tissues and prepare them for milk secretion. Prolactin-releasing hormone (PRH) and the absence of prolactin-inhibiting hormone (PIH) greatly increase prolactin secretion. Prolactin, which initiates milk secretion, is released in ever-increasing amounts throughout preg-

nancy, but its action is inhibited by placental estrogens and progesterone. Inhibition is removed with the placenta and milk secretion begins.

Initiation of Lactation

Suckling of the infant is the primary stimulus for prolactin secretion and it, thus, leads to lactation. Suckling stimulates mechanoreceptors in the nipples and signals from them go to the hypothalamus. These signals probably cause PRH secretion and inhibit PIH secretion. The anterior pituitary secretes prolactin, which stimulates the mammary glands to synthesize milk for the next feeding. As long as suckling occurs periodically, milk secretion also continues unless it is interrupted by pregnancy.

Sometimes infants (male or female) are born with mammary glands producing a little milk called "witches' milk." Fetal breast development and milk secretion occur as hormones that stimulate the mother's mammary glands also reach the fetus. Removed from the influences of maternal hormones at birth, milk secretion soon ceases and breast development regresses.

Ejection of Milk

In addition to the above hormones, oxytocin also is involved in milk release, as explained in Chapter 16. The infant's suckling sends neural signals that release oxytocin, which, in turn, causes specialized **myoepithelial** (mi″o-ep″ĭ-the′le-al) **cells** around the alveoli to contract and push milk into ampullae and lactiferous ducts. This process, called **milk let-down,** allows the infant to obtain milk. Oxytocin also causes uterine contractions felt by many lactating mothers that increase the rate at which the uterus returns to its normal size. Though milk ejection can occur reflexively, psychological factors can contribute to or interfere with milk secretion and let-down.

An infant's suckling inhibits FSH and LH release from the hypothalamus and suppresses ovulation in some women. Because of this phenomenon, breast-feeding may prevent conception, but it is not a reliable means of birth control.

Factors that regulate lactation are shown in Figure 29.27.

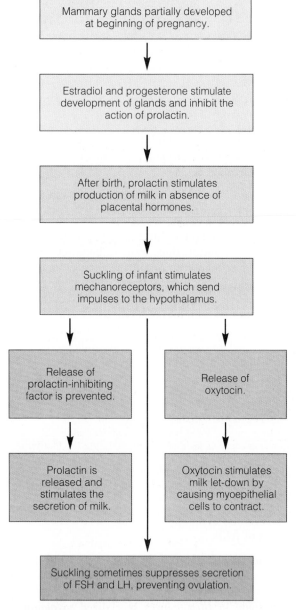

Figure 29.27
Summary of factors that regulate lactation. Psychological factors also can contribute to or interfere with the production of milk.

Effects of Lactation on the Mother's Body

A lactating mother can produce as much as 1.5 l of milk per day. Compared to cow's milk, human milk contains one and one-half times as much lactose, but only half as much protein and about one-third as much mineral such as calcium and phosphate. **Colostrum** (ko-los'trum), the clear yellow fluid secreted during the first few days after a birth, is especially high in protein, minerals, and vitamin A, but low in fat and lactose.

Lactation makes excessive demands on the mother's metabolic processes. To add nutrients to milk, her body uses energy to synthesize as much as 20 g of protein, 50 g of fat, and 100 g of lactose daily. The lactose is synthesized from the mother's glucose supply. Milk also contains 2 to 3 g each of calcium and phosphate per day, and these minerals are removed from the mother's bones and teeth if not available from her diet. Because milk secretion uses large quantities of nutrients, the mother needs a diet high in protein, minerals, and energy.

SEXUALLY TRANSMITTED DISEASES

Along with AIDS (Interlude 4), other sexually transmitted diseases (STDs) pose serious public health problems. No vaccines are available for most STDs and immunity usually does not develop from having such a disease. Some causative agents are resistant to certain antibiotics. So far, new antibiotics have been found as strains develop resistance to older ones, but this may not always be possible. The only way to prevent STDs is to avoid exposure to them.

Gonorrhea (gon''or-e'ah), a disease caused by the bacterium *Neisseria gonorrhoeae*, is characterized by a purulent discharge from the penis or vagina, along with pain around the genitals, and burning on urination. Gonorrhea, which means "flow of seed," was coined by the ancient Greek physician Galen, who mistook pus for semen. The bacterium makes a toxin that damages uterine tube mucosa, and ridding the body of such bacteria is difficult because they sometimes survive phagocytosis and multiply in phagocytic cells. Gonorrhea is transmitted sexually by people who harbor the organism with or without symptoms. As many as half of the women with gonorrhea develop *pelvic inflammatory disease* (PID) as the infection spreads throughout the pelvic cavity. Sterility can follow PID as uterine tubes are occluded by scarring. Infants get *ophthalmia neonatorum* (of-thal'me-ah ne''o-na-to'rum) from passing through the birth canal of an infected mother. Such infections can be prevented by instilling silver nitrate or an appropriate antibiotic into the eyes at delivery. Children can become infected from sexual molestation and from dirty bed linens—as occurred in at least one hospital.

The small bacterium *Chlamydia trachomatis*, which grows inside cells, now causes more sexually transmitted infections than any other bacterium. After one to three weeks incubation, a scanty watery urethral discharge characteristic of *nongonococcal urethritis* (non-gon''o-kok'al u''rĕ-thri'tis), or NGU, can appear. Many such infections produce no symptoms but some cause PID. Both men and women can have NGU, and infants can become infected as they are born. Erythromycin, but not silver nitrate, protects infants against such infection. NGU also is caused by *Mycoplasma hominis* and *Ureaplasma urealyticum*, which accounts for 60 percent of infections leading to infertility. NGU is responsible for fetal deaths, recurrent miscarriages, prematurity, and low birth weight—itself a leading cause of neonatal death. Tetracycline is effective against most of these organisms except for resistant strains that can be treated with erythromycin and spectinomycin.

Now less common in the United States than gonorrhea, **syphilis** (sif'il-is) is caused by the spirochete *Treponema pallidum*. Transmission usually is sexual, but can occur via saliva and other body fluids. After infection, the organisms multiply and spread through the body for two to six weeks. Inflammatory responses at entry sites and other sites produce **chancres** (shang'kerz), small, hard, painless lesions with no discharge. Chancres disappear in four to six weeks, ending the first stage of syphilis, and patients may think they have recovered.

In the second stage, symptoms—copper-colored rash on palms and soles and other skin eruptions—come and go for up to five years and the patient remains contagious. The disease can become latent for life or can give rise to the tertiary stage, in which damage to blood vessels and heart valves, thickening of meninges, an unsteady gait, and eventual paralysis and dementia occur. **Congenital syphilis** is transmitted across the placenta from mother to baby.

Immunologic tests and tests for tissue damage are available for diagnosis. Syphilis usually is treated with penicillin, but tetracycline and erythromycin also are effective. The longer the patient has had syphilis, the more important continued treatment and testing for eradication of organisms becomes.

Over 20 million Americans now have **genital herpes** (her'pēz) and half a million new cases appear each year, making this incurable disease the most common STD. The **herpes simplex virus type 2** (HSV-2), causes most cases, but **herpes simplex virus type 1** (HSV-1), which ordinarily causes fever blisters, can cause genital herpes, especially in patients who engage in oral sex. Both have a 4- to 10-day incubation period and cause identical lesions. Vesicles appear on the mucous membranes of the labia, vagina, and cervix in females, and on the penis and foreskin in males. Urethritis, a watery discharge, intense pain, and itching

also occur. Initial infections can be asymptomatic or can cause painful, localized lesions that heal completely in two to three weeks without scarring unless a secondary bacterial infection develops. Most adults have herpes viruses antibodies, but only 10 to 15 percent have had symptoms.

Within two weeks of an active infection, the viruses travel via sensory neurons to ganglia, where they can remain dormant for life or later reactivate. Reactivated viruses move along the axon to the epithelial cells where they replicate, causing recurrent lesions—smaller lesions that shed fewer viruses and heal more rapidly than primary lesions. Most patients have five to seven recurrences (reactivations)—some having as few as one or two recurrences and others periodic recurrences for life. Patients are contagious anytime lesions are present and sometimes continuously regardless of the presence of lesions. Promiscuous sexual practices and ignorance or lack of concern about disease transmission have greatly increased the number of herpes cases. Babies most often become infected during delivery, but they are highly susceptible to HSV infection and should not be cared for by infected people. Infected mothers who must care for their infants should scrupulously follow sanitary procedures.

In recent years, several drugs have been used with varying degrees of success in treating HSV infections. Acyclovir prevents the spread of lesions, decreases viral shedding, and shortens lesion healing time, but doesn't always prevent recurrences.

Condylomas (kon''dil-o'mahz), or *genital warts,* most often occur in sexually promiscuous, young adults. Two-thirds of the sexual partners of infected individuals also develop warts. In males, warts appear on the penis, anus, and perineum; and in females, warts appear on the vagina, cervix, perineum, and anus. Both herpes and wart viruses may be involved in the development of cervical cancer, but how they might foster a malignancy is not known.

CLINICAL TERMS

curettage (ku''ret-ahzh') using a scraping instrument to remove tissue from a cavity such as the uterus

laparoscopy (lap''ar-os'ko-pe) visual examination of the interior of the abdominal or pelvic cavity using a lighted instrument inserted through a small incision

leukorrhea (loo''ko-re'ah) a viscid whitish discharge from the uterine cavity

lochia (lo'ke-ah) the vaginal discharge that occurs for a week or two following childbirth

metrorrhagia (me''tror-a'je-ah) abnormal uterine hemorrhage, especially between menses

puerpera (pu-er'per-ah) a woman who has just given birth

salpingitis (sal''pin-ji'tis) inflammation of a uterine tube

uterine prolapse (pro-laps') sinking of the uterus toward, or into, the vagina

vaginitis (vaj''in-i'tis) inflammation of the vagina

Essay
Birth Control

Humans have tried to control their own fertility since at least the beginning of recorded history. Over much of that time, birth control, the intentional prevention of fertilization or implantation, has been controversial. Views range from opposition to distributing any birth control information to objections from women who resent having the full burden of hazardous treatments.

The modern birth control movement was launched in England in 1822 by Francis Place, who recommended that working people limit their numbers so they could command living wages. Resistance was strong in the United States,

where the 1873 Comstock Law made it a criminal offense for anyone—even a physician—to transport in interstate commerce medicine to prevent conception or cause abortion. This law stayed in force until the mid-twentieth century, when Margaret Sanger worked in New York City tenements and crusaded to make birth control information freely available. The battle to make contraceptives available was long and difficult, and controversy and violence still exists over contraceptives for teenagers and the legality of abortion.

Birth control methods differ in physiological effects (Figure 29.28). They also vary in effectiveness, side effects, cost, various health factors, and long-term effects (Table 29.2).

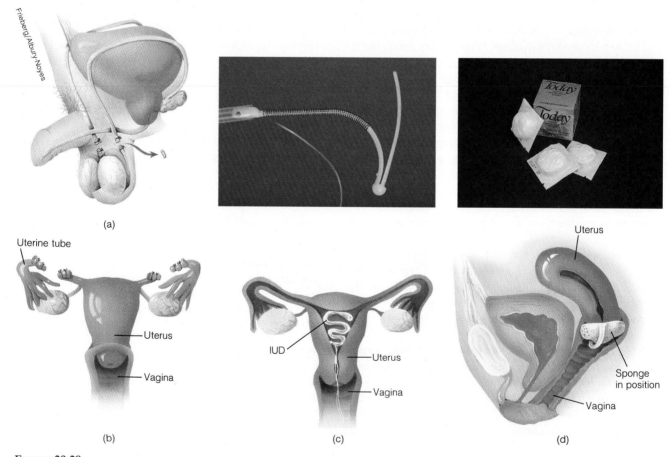

Figure 29.28
Contraceptives: (*a*) vasectomy, (*b*) tubal ligation, (*c*) IUD, (*d*) sponge, (*e*) diaphragm, (*f*) oral contraceptive, (*g*) spermicidal substance, and (*h*) condom.

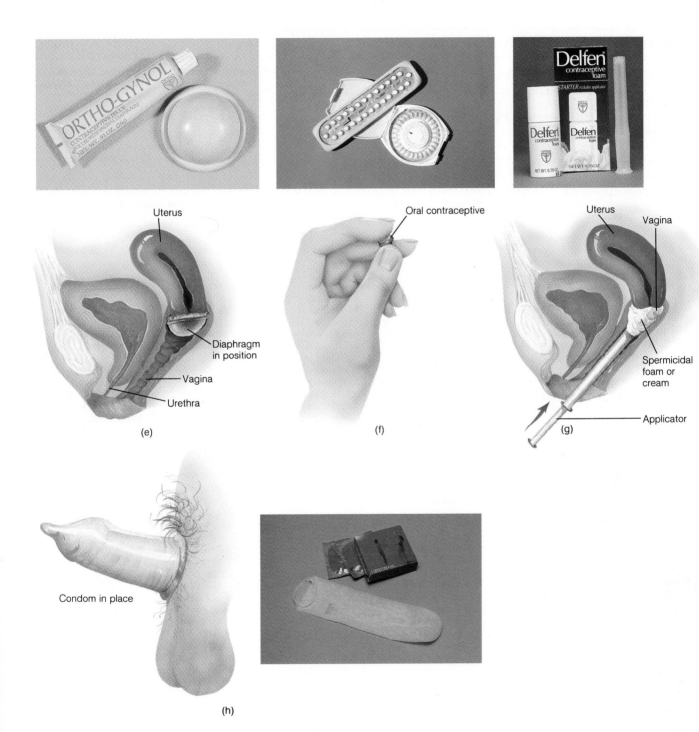

Uterus

Diaphragm in position

Vagina

Urethra

(e)

Oral contraceptive

(f)

Uterus

Vagina

Spermicidal foam or cream

Applicator

(g)

Condom in place

(h)

TABLE 29.2 Characteristics of Birth Control Methods

Method	Effectiveness	Advantages and Disadvantages	Side Effects	Health Factors to Consider	Long-Term Effect on Ability to Have Children
Vasectomy	Virtually 100%	One time procedure for male; does not require hospitalization; should be considered permanent; other contraceptive must be used for a few months until all sperm are out of reproductive tract.	Complications in less than 4% of cases can include infection and inflammatory reaction.	Autoimmune reactions, testicular cysts.	Procedure is not always reversible.
Tubal ligation	Virtually 100%	One time procedure for female; does require hospitalization, usually in outpatient facility; should be considered permanent.	Complications are rare but include infection, bleeding, and injury to other organs.	Surgery carries some risk and varies with the general health of the patient.	Procedure is not always reversible.
Oral contraceptives	Less than 1 pregnancy in 100 woman-years* with the estrogen-progesterone pill; 2–3 pregnancies in 100 woman-years with progesterone only pill.	No action needed except to take pill, which must be taken regularly; combination is taken for 21 days each month, and progesterone is taken continuously.	Tender breasts, nausea, loss or gain of weight; can cause blood clots; increases risk of heart attack, especially in smokers; high blood pressure and gallbladder disease in a few users.	Women who smoke should not use the pill, nor should those who already have heart disease, blood clots, or breast cancer. Migraine headaches, depression, fibroids of the uterus, kidney disease, asthma, high blood pressure, diabetes, and epilepsy may be made worse by the pill.	Using the pill does not seem to prevent later pregnancy, but user should wait a few months after stopping the pill before becoming pregnant; after childbirth nursing mothers should not use the pill because the drugs appear in the milk and long-range effects on the infant are not known.
Intrauterine device (IUD)	Effectiveness depends on whether IUD stays in place; 1–6 pregnancies per 100 woman-years.	Must be inserted by a physician and checked annually, though woman can check for strings extending into vagina; may cause discomfort when inserted; may cause cramps and heavier menstrual flow; can be expelled without woman knowing.	Complications are relatively infrequent but include anemia, pregnancy outside uterus, infection, and perforation of the uterus.	Women who have or have had the following conditions should discuss them with a physician before deciding to use the IUD: cancer of cervix or uterus, heavy or irregular bleeding, pelvic infection, abortion, miscarriage, VD, severe cramps, anemia, fainting, vaginal discharge, or suspicious or abnormal Pap smear.	Pelvic infection in some IUD users may result in future inability to have children.

*A woman-year refers to one woman for one year. Pregnancies per 100 woman-years usually means that a certain number of pregnancies occurred among 100 women over a period of one year.

Method	Effectiveness	Advantages and Disadvantages	Side Effects	Health Factors to Consider	Long-Term Effect on Ability to Have Children
Diaphragm (with spermicidal cream, foam, or jelly)	Effectiveness depends on how correctly the method is used; 2–20 pregnancies in 100 woman-years.	Diaphragm and spermicide must be inserted no more than 2 hours before intercourse; no effect on chemical or physical processes of the body; must be left in place for 6 hours after intercourse; must be fitted by a physician and refitted after childbirth or abortion; requires instruction in insertion technique; some women whose cervix protrudes abnormally into the vagina cannot use a diaphragm.	A few women are allergic to the rubber of the diaphragm or to the spermicidal jelly.	None	None
Sponge	Similar to diaphragm in all respects, except can be worn up to 48 hours.				
Condom	Effectiveness depends on how correctly the method is used; 3–36 pregnancies in 100 woman-years.	Easily available, requires no long-term planning; also protects against VD; some users feel the condom reduces the pleasure of intercourse; male must interrupt foreplay and put condom in place before sexual entry into the woman; condom can slip or tear and release sperm into the vagina.	A few men are allergic to rubber condoms, but can use those made of lamb cecum.	None	None
Spermicidal foam, cream, or jelly used alone	Effectiveness depends on how correctly the method is used; 2–36 pregnancies in 100 woman-years.	Easy to obtain and use; must be used 1 hour or less before intercourse; must be left in vagina 6–8 hours after intercourse.	Allergic reaction may cause irritation of penis or vagina; can be corrected by changing brands.	None	None
Rhythm method	Effectiveness depends on how correctly the method is used; 11–25 pregnancies in 100 woman-years using temperature or vaginal secretion with calendar.	No drugs or devices needed; careful record must be kept to predict approximate time of ovulation; temperature must be taken on arising to determine when ovulation occurs (temperature rises), or record of mucus in vaginal secretions must be kept (mucus increases after ovulation); method is difficult to use if cycles are irregular.	No physical effects, but because couple must abstain from intercourse for a long time prior to and shortly after ovulation, the method can create pressures on a couple's relationship.	None	None

Vasectomy and tubal ligation disrupt the continuity of ducts that carry sperm and ova, respectively, so fertilization cannot occur. They require minor surgery and may not be reversible.

Oral contraceptives suppress ovulation by interfering with FSH and LH secretion, possibly preventing the midcycle LH surge. Either estrogens or progestins (steroids having progesterone-like actions) inhibit ovulation, but may cause abnormal bleeding. Today's oral contraceptives are made of synthetic hormones that resist destruction by the liver and inhibit ovulation in smaller doses. They include the estrogens ethynyl estradiol and mestranol and the progestins norethindrone ethynodiol and norgestrel. Birth control pills with some estrogen-progestin combination are taken from the beginning of menstruation 20 or more days until after ovulation would have occurred even in unusually long cycles.

Oral contraceptives containing very low doses of estrogens and progestins may, instead of inhibiting ovulation, render cervical mucus lethal to sperm, cause contractions that expel the ovum, or prevent implantation by impairing endometrial development or altering the ovum's time in the uterine tube.

Birth control pills can cause serious side effects and about 4 per 100,000 users die each year from them—about half the proportion that die from complications of pregnancy. The most serious side effects are blood lipid elevation and blood clots lodging in the brain or lungs, especially in women who smoke. Birth control pills also increase the risk of cervical and skin cancer and decrease the risk of ovarian and uterine cancer.

The mechanism of action of the intrauterine device (IUD) is not known, but it probably causes persistent inflammation in which toxic breakdown products of intrauterine neutrophils kill sperm. The major risk is perforation of the uterus, which occurs in about 1 per 1000 users, usually at the time of insertion. Surgery is required to remove an IUD that has perforated the uterus.

Several devices mechanically or chemically prevent fertilization. The diaphragm, sponge, and condom mechanically prevent sperm from entering the uterus. Spermicide in the sponge, and spermicidal foams, creams, or jellies used with or without a diaphragm chemically damage sperm. Except for condoms, women must take responsibility for these methods. None cause serious health hazards, but some spermicides inflame vaginal membranes.

The rhythm method is based on detecting ovulation and abstaining from intercourse during the period that fertilization could occur. Progesterone secretion causes about a 0.5° C increase in a woman's basal body temperature about a day after ovulation (Figure 29.29). A woman determines her basal (resting) body temperature on awakening and before getting out of bed. Changes in the properties of vaginal mucus also signal that ovulation has occurred. A major disadvantage of this method is that it detects ovulation only *after* it has occurred. Because sperm can fertilize an ovum up to three days after intercourse and an ovum remains viable for about one day, a couple must abstain from intercourse from at least three days before to one day after ovulation. The rhythm method fails to provide the information needed to do this.

Abortion and the "morning-after pill" terminate development after fertilization has occurred. Abortion remains controversial and opinions range from support of it as every woman's right, to objection to it under any circumstances. The most common method for first trimester abortions is vacuum aspiration—dilating the cervix and using a tube attached to a suction pump to remove the uterine contents. After the first trimester, various methods are used to stimulate the uterus to contract and expel the fetus.

Chapter Summary

Organization and General Functions

- Male reproductive organs include testes, ducts, and glands; female reproductive organs include ovaries, ducts, uterus, vagina, and mammary glands. Both male and female organs produce hormones and gametes, and the uterus provides a place for the embryo to develop.

Development

- The gonads, ducts, and external genitalia begin as undifferentiated organs. A Y chromosome and testosterone cause male reproductive organs to develop. Without a Y chromosome, female organs develop.

Male Reproductive Anatomy

- The testes are small glands located in the scrotum that contain many seminiferous tubules and spermatogonia, which give rise to sperm. A mature sperm has an acrosome with enzymes and a flagellum. Sertoli's cells provide nutrients, enzymes, and possibly hormones for sperm maturation.
- Sperm pass from seminiferous tubules through the straight tubule, rete testis, and efferent ducts, to the epididymis.
- The scrotum consist of skin and the dartos. Its temperature is about 2° C lower than the internal body temperature—ideal for developing sperm. Though testes develop in the abdominal cavity, they descend into the scrotum before birth. Cryptorchidism leads to sterility as the high abdominal temperature destroys sperm.

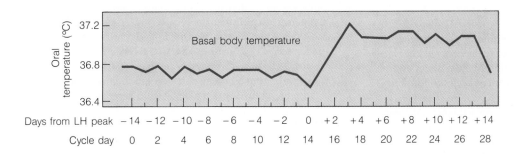

FIGURE 29.29
A basal temperature chart.

The morning-after pill is used when a woman is at risk of pregnancy from rape or incest, or when her physical or emotional condition might cause the pregnancy to have grave results. This pill, containing diethylstilbestrol (DES), causes uterine contractions in early pregnancy. In the 1940s and 1950s, DES—then thought to prevent uterine contractions—was given to prevent miscarriage in established pregnancies. Many daughters of women who received DES have since died of cancer of the reproductive organs.

Two methods of contraception used in other countries have not been approved for use in the United States. Injection of Depo-Provera, a progesterone-like drug, inhibits follicle development. Many South American women have received injections two to four times a year for several years without serious complications. This drug is not approved for use in the United States as a contraceptive, but it is used to treat inoperable endometrial cancer and to decrease sex drive in pedophiles (child molesters). This latter action appears to result from suppression of sexual arousal centers in the limbic system and hypothalamus. Pills containing gossypol, which impairs the ability of the testes to produce sperm, have been used by Chinese men for more than a decade with a 99

percent success rate. The only side effect is a decrease in serum potassium, apparently because of the drug's mildly toxic effects on the kidneys.

Immunocontraception, the use of vaccines to immunize against reproductive hormones, is under development. Vaccines that elicit antibodies against hCG prevent hCG from fostering establishment of a pregnancy. A vaccine against GnRH, which would be effective in both men and women, also is being developed. Some of these vaccines are under clinical trial in India.

Choosing to use contraceptives is, or course, a personal matter. Most couples who wish to use contraceptives should be able to find a suitable one among the variety available.

Questions

1. What are some landmarks in the history of birth control?
2. Which methods of birth control are most and least reliable?
3. Which contraceptives have undesirable side effects, and what are those effects?
4. Which birth control methods place responsibility on the female? On the male?

- Other male ducts include paired ductus deferens and ejaculatory ducts, and the unpaired urethra. Ejaculatory ducts eject sperm into the urethra. The penis conducts urine and is the male copulatory organ. It is covered with smooth skin and the skin over the glans is called the prepuce, or foreskin.
- Male glands include seminal vesicles, which secrete fructose and prostaglandins; the prostate gland, which secretes lipids, enzymes, and bicarbonate ions; and the bulbourethral glands, which secrete mucus.

Male Reproductive Physiology

- GnRH stimulates release of the pituitary hormones, FSH stimulates sperm production, LH stimulates interstitial cells to produce testosterone and also fosters development of sperm.
- Semen consists of sperm and the secretions of the seminal vesicles and prostate and bulbourethral glands.
- Erection, emission, and ejaculation are summarized in Figure 29.11.

Male Reproductive Disorders

- Prostatitis is an inflammation of the prostate gland. Prostate cancer, a common male malignancy, is stimulated by testosterone and treated with estrogens or surgery. Impotence is the inability to maintain erection during sexual intercourse. Sterility is the inability to fertilize an ovum.

Female Reproductive Anatomy

- The ovaries are small glands that produces female sex hormones and ova in follicles, and uterine tubes extend from the ovaries to the uterus. All these organs are held in place by ligaments. Fimbrae of each infundibulum direct ova into the uterine tubes. Peristalsis and cilia sweep ova through tubes.
- The uterus has a fundus, central body, isthmus, and cervix. Its walls consist of three layers—the outer serous perimetrium, the middle muscular myometrium, and the inner endometrium. The endometrium is a mucous membrane of which the functionalis is sloughed during menstruation and the basalis remains intact.
- The vagina, a muscular tube that leads from the uterus to the outside of the body, provides a passageway for the menstrual flow, serves as a receptacle for sperm during intercourse, and forms part of the birth canal. The vaginal orifice is partially covered by a membrane called the hymen.
- The female external genitalia (vulva or pudendum) include the fatty mons pubis, the sexually responsive clitoris, the vestibule, Bartholin's glands, and skin folds called the labia minora and labia majora.
- The mammary glands are modified sweat glands that develop at puberty in females under the stimulation of estradiol and other hormones. Each of the many lobes has numerous small alveoli, consisting of secretory cells capable of secreting milk.

Female Reproductive Physiology

- Ovarian follicles contain ova and produce hormones. GnRH stimulates release of FSH and LH. FSH stimulates follicle development and LH stimulates ovulation and progesterone secretion. After ovulation, the corpus luteum produces hormones if fertilization occurs, and degenerates if it does not.
- The menstrual cycle consists of ovarian and uterine events that delivers an ovum to the uterine cavity once every month and provides conditions for its development as shown in Figure 29.18.
- The preovulatory and postovulatory phases are summarized in Figures 29.19 and 29.20, respectively. Menstrual cycles occur from menarche to menopause, except when disrupted by pregnancy.
- In addition to reproductive functions, estradiol stimulates growth and development of female structures, bone growth at puberty and fusion of epiphyses, softening and vascularization of skin, and characteristic feminine fat deposition. Progesterone maintains pregnancy, stimulates development of secretory portion of the mammary glands, suppresses uterine contractions, mobilizes proteins, and causes sodium and water retention.

- Sexual arousal is produced by psychological factors and mechanical stimulation of erotic tissues in the clitoris, other parts of the external genitalia, and breasts. Female orgasm probably facilitates conception.

Female Reproductive Disorders

- Breast and cervical cancer are leading causes of death among women. Breast self-examination, mammograms, and "Pap" smears help to detect malignancies while they are treatable. Ovarian cysts are fluid filled tumors. Endometriosis is due to endometrial tissue outside the uterus. Menstrual disorders include amenorrhea, excessive or abnormal bleeding, dysmenorrhea, and premenstrual syndrome. Infertility is the inability to become pregnant.

Fertilization

- Fertilization is the union of the male and female pronuclei. Muscular contractions during intercourse, contractions of the uterine tubes, beating of flagella on sperm, and beating of cilia in the uterine tubes foster fertilization.

Pregnancy

- Pregnancy lasts about 280 days and is maintained by estradiol and progesterone from the corpus luteum and placental hormones. Chorionic gonadotropin regulates the production of estradiol and progesterone. Changes in the mother's body include an increased metabolic rate, retention of sodium and fluids in greater quantities than in the nonpregnant state, and adaptation to the fetus, usually without an immunological response.

Parturition

- Parturition is initiated by hormonal and mechanical factors. Estradiol and mechanical pressure of the fetus stimulate uterine contractions. Cervical stretching acts by positive feedback to cause uterine contractions, which cause more cervical stretching.
- In the first stage of labor, the cervix dilates; in the second stage, the fetus is expelled; and in the third stage, the placenta detaches and is expelled.

Lactation

- The mammary glands secrete milk for the nourishment of the infant after birth. During pregnancy, the mammary glands increase in size and are prepared to secrete milk. Regulation of lactation is summarized in Figure 29.27.

Sexually Transmitted Diseases

- Sexually transmitted diseases include AIDS, gonorrhea, nongonococcal urethritis, syphilis, and genital herpes. Most are difficult to treat and are prevented by avoiding sexual contact with infected people.

QUESTIONS AND PROBLEMS

The questions at the end of each chapter are numbered to correspond with the objectives listed at the beginning of the chapter. Italics indicate that a question requires critical thinking skills beyond simple factual recall.

Questions

1. (a) What are the functions of the male and female reproductive systems?
   (b) Name the major organs of each reproductive system.
2. (a) Describe the differentiation of the male and female reproductive systems, noting homologies between the two.
   (b) *What are the roles of the Y chromosome and testosterone in the development of the reproductive system?*
3. (a) What different types of cells are found in the testes, and what are their functions?
   (b) What are the characteristics of each of the ducts in the male reproductive system?
   (c) *Trace the pathway of sperm through the testes and ducts of the male reproductive system.*
   (d) Describe the location and secretions of the glands of the male reproductive system.
4. What are the functions of testosterone?
5. (a) *What are the major events in spermatogenesis, and what is the significance of each?*
   (b) *How is spermatogenesis regulated?*
6. (a) How is semen formed?
   (b) What does semen contain?
7. (a) *What processes are involved in erection and ejaculation?*
   (b) *How are erection and ejaculation controlled?*
8. How is function disturbed in the following disorders of the male reproductive system: (a) prostatitis, (b) prostate cancer, (c) impotence, and (d) sterility?
9. (a) Describe the structure of the ovaries.
   (b) How does the structure of the uterine tubes help to transport ova?
   (c) Describe the structure of the uterus, including the layers of its walls.
   (d) What are the main parts of the vagina and external genitalia?
   (e) Describe the structure of the mammary glands.
10. *How does oogenesis differ from spermatogenesis?*
11. (a) *What events occur in the ovarian cycle, and what is their significance?*
    (b) *How is the ovarian cycle controlled?*
12. (a) What is menarche?
    (b) *What events occur in the preovulatory phase of the menstrual cycle, and what is their significance?*
    (c) *What events occur during ovulation, and why is ovulation significant?*
    (d) *What events occur in the postovulatory phase of the menstrual cycle, and what is their significance?*

13. (a) *How is the menstrual cycle regulated?*
    (b) *How is the menstrual cycle disrupted by pregnancy?*
    (c) *What events occur in menopause, and how do they cause the menstrual cycle to cease?*
14. What are the main events in the female sexual response?
15. How do the following conditions disturb function: (a) breast and cervical cancer, (b) ovarian cysts, (c) endometriosis, (d) other menstrual disorders, and (e) infertility?
16. (a) What events occur in fertilization?
    (b) *What is the significance of fertilization?*
    (c) *Why is it important that one sperm fertilize an ovum?*
17. (a) *How do hormone secretions change in pregnancy?*
    (b) *How does pregnancy change a woman's body?*
18. (a) *What happens in each of the three stages of labor?*
    (b) *Explain how labor is initiated and regulated?*
19. (a) *What events occur in breast development, and how are they regulated?*
    (b) *How is lactation initiated?*
    (c) *How is milk ejection regulated?*
    (d) What are the effects of lactation on the mother's body?
20. (a) Summarize the main points about AIDS in Interlude 4.
    (b) What are causes and effects of gonorrhea, nongonococcal urethritis, syphilis, and genital herpes?
    (c) *Why is avoiding infection so important in controlling sexually transmitted diseases?*

Problems

1. Research the topic of prostaglandins in reproductive function, and if possible, make a report to the class.
2. Survey the literature for advances in one of the following areas: (a) treatment or immunization for genital herpes, (b) toxemia in pregnancy, or (c) new forms of birth control.

30

OBJECTIVES

1. Explain the basic principles of genetics and the nature of mutations.

2. Summarize the major events in prenatal development.
3. Explain the functions of embryonic membranes and the placenta.
4. Describe the stresses of being born.
5. Distinguish between monozygotic and dizygotic twins, and list factors that contribute to multiple births.
6. Explain how teratogens and chromosomal abnormalities affect development.

7. Briefly summarize the physiological aspects of genetic screening and genetic and reproductive engineering.

8. Describe how neonates differ physiologically from adults.
9. Describe how infants and children differ physiologically from adults.
10. Describe changes that occur during puberty, and explain how adolescent physiology differs from adult physiology.

11. Briefly describe factors that affect aging.
12. Explain how body systems change with age and how aging can be counteracted.
13. Evaluate theories of aging.

GENETICS, DEVELOPMENT, AND LIFE STAGES

GENETICS

In the last chapter we saw how eggs and sperm, with their unpaired chromosomes, carry genetic information to the first cell of a new individual. In this brief review of **genetics** (jen-et'ikz), the study of inheritance, we will consider basic principles and see how they apply to inherited diseases.

Basic Principles of Genetics

Each of a cell's proteins, including its enzymes, is synthesized according to information in a specific segment of DNA called a **gene** (jēn). Together, all genes on all chromosomes constitute the information a person inherits from his or her parents—information that sets limits on body structure, function, and behavior. Most genes occur in two or more forms called **alleles** (al-ēlz'). For example, a gene *S* directs synthesis of a normal polypeptide of hemoglobin. Another gene *s* directs synthesis of an abnormal polypeptide of hemoglobin in people with sickle cell anemia (Chapter 17). Cells have two copies of most genes, one on each member of a pair of chromosomes. If these two copies carry the same information (*SS* or *ss*), the alleles are said to be **homozygous** (ho''mo-zi'gus). If the copies are different (*Ss*), the alleles are said to be **heterozygous** (het''er-o-zi'gus).

The combination of alleles is the **genotype** (jen'o-tīp); the characteristic expressed is the **phenotype** (fe'no-tīp). For example, genotype *SS* produces normal erythrocytes, and genotype *ss* produces erythrocytes that can take on a sickle shape under certain conditions. Genotype *Ss* produces normal erythrocytes because the *S* allele is a **dominant** allele, and when dominant alleles are present they are expressed in the phenotype. The *s* allele is a **recessive** (re-ses'iv) allele, and recessive alleles are expressed only in the absence of dominant alleles. Blood tests are available to determine both genotype and phenotype for these particular alleles.

In human cells, the 22 pairs of homologous chromosomes, called the **autosomal** (aw''to-so'mal) **chromosomes**, control most inherited characteristics and the nonhomologous **sex chromosomes** control sex-linked inheritance. Most genes involved in sex-linked inheritance are on the X chromosome, but a Y chromosome is necessary for normal male development. Because males have only one X chromosome, any recessive alleles on that chromosome will be expressed. When males have an allele for color blindness (X^c) they are colorblind (genotype Y X^c). (Note that a superscript C for a normal gene and a *c* for a color blindness gene are added to X to indicate the information is carried on the X chromosome.) Because females have two X chromosomes, they need two copies of a recessive gene for the recessive characteristic to appear. Women who have a normal allele (X^C) and a color blindness allele (X^c) have normal color vision (genotype $X^C X^c$)

TABLE 30.1	Basic Terminology of Genetics
Term	**Definition**
Allele	One of two or more alternate forms of a gene that occupies a particular locus (site) on a particular chromosome
Homozygous	A condition in which the alleles of a gene on the two homologous chromosomes are alike
Heterozygous	A condition in which the alleles of a gene on the two homologous chromosomes are not alike
Genotype	The combination of the alleles at a given locus carried by an individual; the genetic makeup of the individual for all genes
Phenotype	The outward appearance of the individual with respect to a particular characteristic (or all characteristics) determined by the genotype
Dominant	An allele that is expressed in the phenotype when it is carried either homozygously or heterozygously
Recessive	An allele that is expressed in the phenotype only in the absence of a dominant gene.
Carrier	A heterozygous individual of normal phenotype who has in his or her genotype a hidden recessive allele and can, thus, transmit the recessive allele to offspring
Autosomal inheritance	Inheritance of characteristics that are carried on one of the 22 pairs (in humans) of autosomal, not sex, chromosomes
Sex-linked inheritance	Inheritance of characteristics carried on the X chromosome (or rarely on the Y chromosome).

but can transmit the color blindness allele to their offspring. The basic terminology of genetics is summarized in Table 30.1.

> The number of a person's ancestors doubles with every generation—two parents, four grandparents, eight great grandparents and so on—back through the history of the human race. It has been estimated that, assuming no overlap or relatedness to an ancestor in more than one way, one need go back only 200,000 years, or about 8000 generations, to find the number of one's ancestors in a particular generation equivalent to the entire human population at that time. This population provided the gene pool from which all of our genes are derived.

Many genetic diseases are due to alleles that lead to an absent or defective enzyme or other protein (Table 30.2). Prospective parents' genotypes for a particular characteristic often can be determined by laboratory tests or by studying family members. Such information is used to calculate the probability of their having a child with a certain genetic defect. For example, if both parents carry

TABLE 30.2   Selected Human Genetic Diseases

Disease and Inheritance Pattern*	Description
Achondroplasia (AD)	Dwarfism, with especially short limbs and adult height of about 4 ft. Defect has a high mutation rate and can appear in families never before affected.
Albinism (AR)	Lack of an enzyme to convert amino acid tyrosine to the pigment melanin. Individuals have white hair and very fair skin; the iris of the eye appears pink due to blood vessels unobscured by normal pigmentation.
Brachydactyly (AD)	Individuals have very short, stubby fingers.
Bruton agammaglobulinemia (XR)	Almost complete inability to form antibodies to infectious agents. Affected individuals require lifelong gamma globulin treatment and remain susceptible to infections.
Cystic fibrosis (AR)	Excess mucus blocks respiratory passages and impairs release of digestive enzymes. Affected individuals who receive intensive therapy sometimes live to adulthood.
Duchenne muscular dystrophy (XR)	Muscle wasting usually begins by age six and leads to paralysis by early adulthood. High rate of mutation leads to frequent appearance of disease in previously unaffected families.
Huntington's disease (AD)	Onset from 10 to 60 years of age, but often not until after the affected individual has had children. Progressive deterioration of nervous system leads to uncontrollable movements, personality changes, mental impairment, and, ultimately, death. Carriers of the gene can be identified.
Lesch-Nyhan syndrome (XR)	Enzyme defect causes overproduction of uric acid. The affected individual is mentally retarded and engages in self-mutilation. Uric acid accumulation leads to kidney damage, but this effect can now be minimized by drug treatment. Tests are available to detect carriers.
Marfan's syndrome (AD)	Produces long spidery appendages, eye defects, and aortic disease. It has been speculated that Abraham Lincoln was mildly affected by this disease.
Phenylketonuria (AR)	Enzyme defect causes inability to metabolize the amino acid phenylalanine. Mental retardation results unless the patient is placed on a low-phenylalanine diet in the first few weeks of life. Tests are available to detect this defect in neonates.
Sickle cell anemia (AR)	Enzyme defect causes production of abnormal hemoglobin. Effects are chronic anemia and crises with extreme pain in joints, blockage of blood vessels by sickle-shaped red blood cells, and need for transfusions. Heterozygous individuals are resistant to the disease malaria.
Tay-Sachs disease (AR)	Defect in lipid metabolism. Results in progressive deterioration of nerve tissue in an apparently normal infant until death at about age three or four. Tests are available to identify carriers and to detect affected fetuses prenatally.

*Inheritance patterns of genetic diseases:

AD = autosomal dominant

AR = autosomal recessive

XR = sex-linked recessive

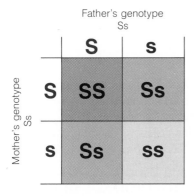

Probability ¼ SS: ½ Ss: ¼ ss

FIGURE 30.1
Probability of each genotype among offspring of parents who are carriers of the sickle cell anemia gene.

Recent research has made it possible to locate certain genes in the human **genome** (je'nŏm), the whole of genetic material in a human cell. One such technique depends on finding certain easily detected base sequences that act as markers. Even when the disease allele itself cannot be found, finding a nearby marker can allow a disease to be traced through a family. It is now technologically possible to identify base sequences in DNA molecules—even the 3 billion bases in the human genome. Just such a project, the Human Genome Project, has been proposed with cost estimates from $300 million (the cost of a B-1 bomber) to $3 billion (the cost of a space shuttle). The project has been divided into small components and some are underway. Eventually, it will identify every human gene and find out what each does. Genetic defects associated with diseases could be identified. People could know their risks of bearing children with defects, and ways to repair defects might be developed.

the sickle cell anemia gene (genotype *Ss*), they have a probability of 0.5 (a 50–50 chance) of transmitting allele *S* to each child. They also have a probability of 0.5 of transmitting allele *s*. Allelic combinations and probabilities are shown in Figure 30.1. *These probabilities remain the same for each child regardless of how many children the couple has.*

Mutations

Any change in the base sequences in DNA constitutes a **mutation** (mu-ta'shun), and mutations provide a source of genetic variability. To understand the effects of mutations,

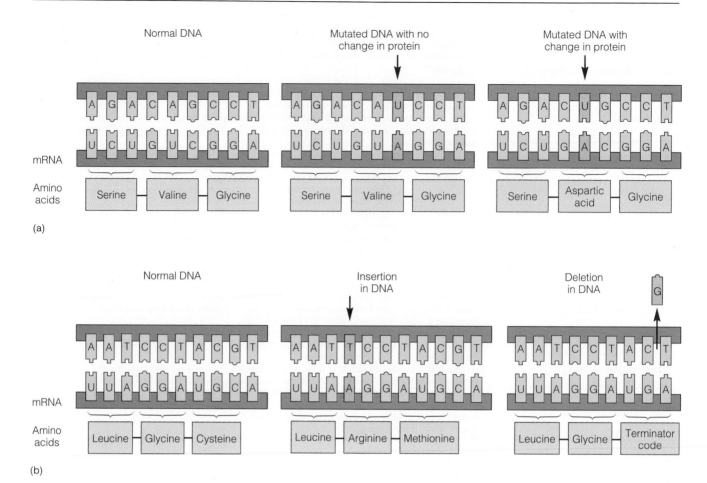

FIGURE 30.2

(a) The effects of a point mutation. Arrow indicates the substituted base in DNA. (b) The effects of frameshift mutations. Insertions and deletions are indicated by arrows. The deletion mutation happens to create a DNA sequence for which mRNA contains a terminator code at the third codon. Protein synthesis stops after the formation of a dipeptide.

recall that protein synthesis is directed by codons consisting of base triplets of mRNA transcribed from DNA (Chapter 3). We can represent a small segment of DNA with a sentence consisting of three-letter words:

OUR BIG BAD DOG BIT OUR OLD FAT CAT.

A **point mutation** is the substitution of a single base as a segment of DNA is replicated:

OUR BIG BAF DOG BIT OUR OLD FAT CAT.

Except for the word "baf," the sentence still has meaning. As shown in Figure 30.2a, such a change may or may not specify a different amino acid. If it does, the protein will differ from its normal makeup by one amino acid—exactly the kind of error that produces sickle cell anemia. The amino acid valine is substituted for glutamic acid in a polypeptide of hemoglobin. If, in spite of the change, the RNA codon still specifies the normal amino acid, the protein is made without error.

A **frameshift mutation** consists of adding or deleting one or more bases in DNA. Adding a letter (X) to our sentence produces:

OUR XBI GBA DDO GBI TOU ROL DFA TCA T.

Deleting a letter (B) yields:

OUR IGB ADD OGB ITO URO LDF ATC AT.

In both examples, the sentence has become nonsense. Such changes alter all base triplets beyond the mutation, making nonsense of the code. As shown in Figure 30.2b, the resulting protein will be vastly different from the normal protein. Many frameshift mutations insert a termination codon in mRNA so that only a short nonfunctional polypeptide, and not a whole protein, is made.

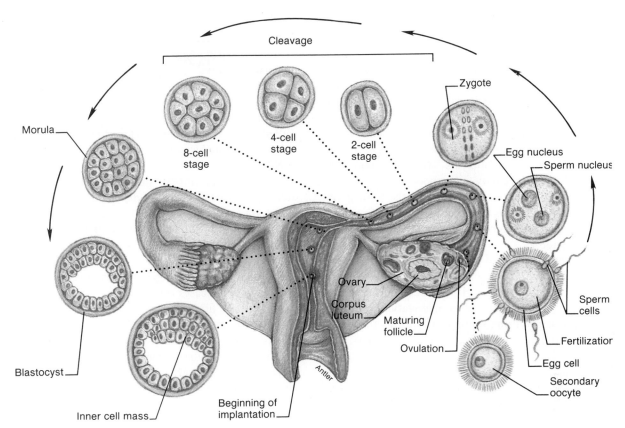

FIGURE 30.3
An overview of the developmental stages of an embryo from
fertilization to early differentiation following implantation.

Various environmental agents called **mutagens** (mu′ta-jenz) can cause mutations. Chemical mutagens include nitrous acid, mustard gases, and base analogs such as bromouracil. Other mutagens include X ray and ultraviolet radiation. The ways mutations alter body structure and function are varied, but most modify structural proteins or enzymes. A mutation in an egg or a sperm is passed to all the cells of the new individual. This process accounts for the initial appearance of genetic diseases in families where the disease has never been seen before. A mutation in body cells affects only the progeny of the mutated cell in a particular individual and is not transmitted from one generation to the next. Cancer can result from such mutations.

Combining DNA from two different organisms creates **recombinant** (re-kom′bin-ant) **DNA**. When a segment of human DNA is inserted into a strand of bacterial DNA, the bacterium makes all its own proteins and human protein too. If the human DNA directs insulin synthesis, the bacterium synthesizes insulin. This technique is now used to cause microorganisms to make insulin and several other biologically active molecules. It may someday be used to replace defective genes with normal ones.

PRENATAL DEVELOPMENT

In Chapter 29, we considered spermatogenesis, oogenesis, fertilization, and effects of pregnancy on the mother. Here we will consider embryonic development from fertilization to birth.

Major Developmental Events

The zygote (fertilized ovum) undergoes mitotic divisions known as **cleavage** (kle′vij) and the number of cells in the embryo doubles with each division. A single cell becomes two cells, two cells become four, four cells become eight, and so on until a solid ball of cells, the **morula** (mor′u-lah), is formed. These divisions occur as the embryo moves through the uterine tube (Figure 30.3). About three days after ovulation, the morula reaches the uterus, and its cells form a hollow ball called a **blastocyst** (blas′to-sist).

Within a week of fertilization, the blastocyst consists of an outer **trophoblast** (trof′o-blast), an **inner cell mass**, and a cavity called the **blastocoele** (blas′to-sēl). The **zona pellucida**, which has covered the blastocyst, is shed. The trophoblast touches the endometrium and its cells secrete a protease that digests into the stratum basalis. Trophoblast cells proliferate using nutrients from uterine cells until implantation is complete in four to six days, and uterine tissue covers the blastocyst.

In the second week, cells of the inner cell mass continue to divide by mitosis and two cavities surrounded by embryonic membranes—the **yolk sac** and the **amnion** (am′ne-on)—form. Between the cavities is the **embryonic** (em″bre-on′ik) **disk**, from which the embryo itself develops. It contains two germ layers—**ectoderm** from cells nearest the amniotic cavity and **endoderm** from cells nearest the gut cavity. In the third week, the third germ layer, **mesoderm**, develops between the layers of ectoderm and endoderm. Tissues derived from each germ layer were listed in Chapter 4.

By the end of the third week, **somites** (so′mītz) have begun to develop. In the fourth week, ectoderm folds inward on the embryo's posterior surface and forms the **neural tube** by a process called **neurulation** (nu″roola′shun). The neural tube subsequently forms the brain and spinal cord. Rudiments of eyes, ears, nose, lungs, and limbs also appear and the embryo is about 4 mm long.

Differentiation, the process by which genetically identical embryonic cells give rise to different tissues, occurs over the next several weeks. Gene activation plays an important role in differentiation. Though cells have a complete genome, only about 5 percent of the genes function in any particular cell. Genes needed to synthesize epinephrine function in cells of the adrenal medulla, but not in bone, muscle, or most other cells. Genes needed to synthesize hemoglobin are active only in bone marrow, where erythrocytes form. Such activation may involve proteins called histones that inhibit transcription until other proteins, called nonhistones, activate specific portions of the DNA.

The human embryo takes on clearly human form during the second month of development. The brain and heart are among the first of the body's organs to become recognizable. By the end of the second month, the basic structure of all the organs and systems is laid down and the embryo has become a **fetus** (fe′tus).

Organ structure is refined during the third month. As we shall see, diseases contracted and drugs taken by the mother in the first three months (while organs are being formed) are most likely to cause damage. After the third month, fetal development involves mainly growth and maturation. Body size from implantation to week 15 of development are shown in Figure 30.4. The events in early development are summarized in Table 30.3 and Figure 30.5.

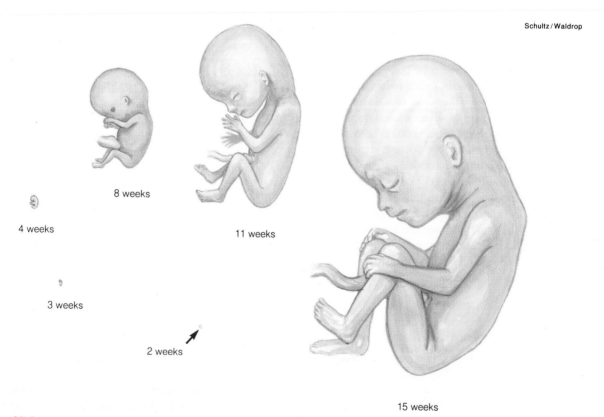

Schultz / Waldrop

8 weeks

4 weeks

11 weeks

3 weeks

2 weeks

15 weeks

Figure 30.4
Changes in body size from implantation to the fifteenth week of development.

TABLE 30.3   Events of Early Human Development

Age in Weeks	Length in Millimeters	Description
2.5	1.5	Neural groove indicated; blood islands formed; embryonic disk flat
4	5.0	All somites present; limb buds indicated; heart prominent; eye, jaws, thyroid present; gut tube differentiated; liver, pancreas, lungs, kidney, nerves forming
5	8.0	Tail prominent; umbilical cord organized; mouth and pharyngeal glands form; intestine elongates; genital ridge forms
6	12.0	Head becomes dominant in size; limbs recognizable; gonads form; heart has general definitive form; cartilage formation begins
8	23.0	Nose forms; digits recognizable; tongue muscles well formed; gut structure further differentiates; liver large; testis or ovary recognizable as such; main blood vessels have typical plan; first indication of bone formation; definitive muscles of head, trunk, limbs formed
12	56.0	Sex readily determined by external inspection; tooth primordia formed; fusion of palate complete; blood formation in bone marrow begins, blood vessels well formed; ossification spreading, with many bones outlined; brain and spinal cord attain general structural form
16	112.0	Face looks human; head hair appears; muscles become spontaneously active; gastric glands formed; kidney has typical form; skin glands form; eye, ear, nose approach typical appearance
20–40 (5–9 months)	160.0–350.0	Body lean but has "baby" proportions (month 6); eyelids open (7); testes descend into scrotum (8); fat collects (8–9); tooth formation continues; tonsils, appendix, spleen acquire typical structure; fingernails form (8); spinal cord myelinization begins (5); brain myelinization begins (9); retina of eye complete and light-sensitive (7); taste sense present (8)

From *Biology* by Jensen, W. A., Heinrich, B., Wake, D. B., and Wake, M. H., © 1979 by Wadsworth, Inc. Reprinted by permission of the publisher.

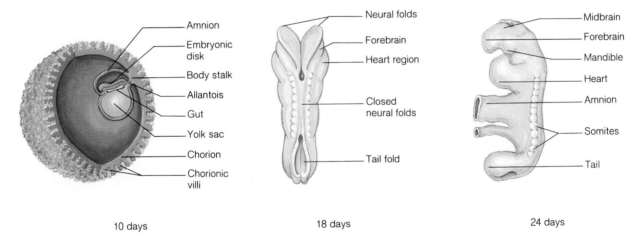

10 days   18 days   24 days

FIGURE 30.5
Events in the development of an embryo from 10 to 24 days.

Schultz/Waldrop

Embryonic Membranes and Placenta

The embryo has four membranes, each surrounding a specific cavity (Figure 30.6). The amnion forms the **amniotic** (am''ne-ot'ik) **cavity,** which contains the embryo and **amniotic fluid,** a kind of shock distributor for the embryo. It protects from mechanical injury like the white of an egg protects the fragile yolk, and it fosters symmetrical development by allowing the embryo to float. If the embryo rested on the uterine wall, pressure on one side could distort the body or stunt limb growth. The amnion, or "bag of waters," ruptures during labor. The **yolk sac,** which has little yolk in humans, supplies cells that later give rise to eggs or sperm before it becomes a nonfunctional part of the umbilical cord. (In birds and reptiles, it supplies nutrients for the entire embryonic period.) The **allantois** (al-an'to-is) forms the **allantoic** (al''an-to'ik) **sac,** which collects wastes in reptiles and birds and which gives rise to blood vessels that connect the fetus with the placenta in humans and other mammals. The **chorion** (ko're-on), derived from the trophoblast, surrounds the entire developing structure—embryo, amnion, and other membranes—and becomes the embryonic part of the placenta.

By the end of the third month, the placenta and umbilical cord are completely developed and fully functional (Figure 30.6). In the **placenta** (plah-sen'tah), maternal blood vessels form sinuses in intervillous spaces around **chorionic villi,** which contain fetal capillaries. This arrangement allows nutrients and oxygen to enter fetal blood and allows carbon dioxide and wastes to enter maternal blood. Fetal and maternal blood do not normally mix.

The **umbilical cord,** surrounded by amnion, contains two umbilical arteries and one umbilical vein. It also contains connective tissue and mucus, called **Wharton's jelly,** made by the allantois. The umbilical cord is cut at birth and the mark left after its stub degenerates is the **umbilicus,** or navel.

At delivery, the placenta detaches from the endometrium and leaves the uterus as the "afterbirth." Fetal blood sometimes enters maternal sinuses before detachment is complete. If this blood contains Rh antigen and the mother's blood does not, the mother's body can become sensitized to it (Chapter 21). Placental detachment cuts off fetal oxygen, so it is important that the newborn begin breathing before detachment occurs.

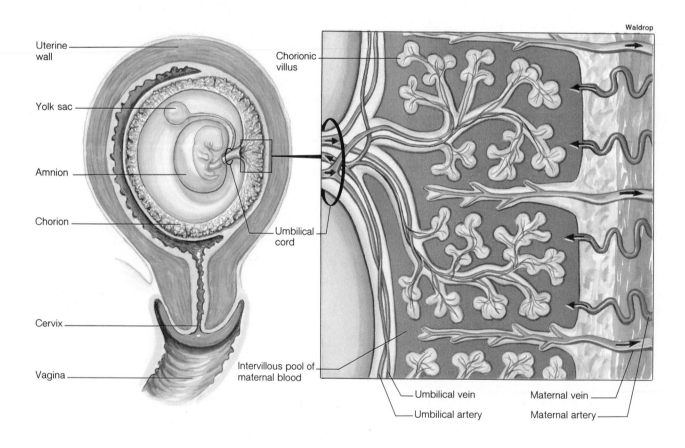

FIGURE 30.6
Embryonic membranes and placenta.

Stress of Being Born

During vaginal delivery, most infants develop a mild hypoxia like that in a sprinter after a run. Blood levels of epinephrine (E) and norepinephrine (NE) begin rising during labor, peaking at 10 times the normal concentration at birth. Though severe hypoxia can trigger E and NE secretion, pressure on the infant's head during uterine contractions, and not hypoxia, seems to cause such secretion in vaginal deliveries (because similar increases are not seen in cesarean deliveries).

Benefits of E and NE are numerous. They help to counteract hypoxia by increasing blood flow to vital organs and improving lung function. They hasten absorption of any fluid present in the lungs at birth, raise surfactant production, increase lung compliance, and dilate bronchioles. E and NE increase the metabolic rate and release nutrients into the blood by degrading glycogen to glucose and fats to fatty acids. Finally, E and NE increase the newborn's alertness and may play a role in its bonding to the mother.

Multiple Births

Of multiple births among humans, twins occur naturally about once in 80 births, triplets in less than one in 5000 births, and quadruplets in only one in 500,000 births. Fertility drugs that stimulate ovulation have recently increased these frequencies. Twins can be monozygotic (identical) or dizygotic (fraternal). **Monozygotic** (mon″o-zi-got′ik) **twins** arise from a single zygote, in which cells separate and give rise to two embryos before or at the blastocyst stage. Incomplete separation results in Siamese twins. Monozygotic twins are genetically identical and always of the same gender. **Dizygotic** (di″zi-got′ik) **twins** develop from two separate zygotes and are no more alike than other siblings except that they are born at the same time. Other multiple births can include both genetically identical and nonidentical individuals.

How multiple births occur is not completely understood. Monozygotic twinning occurs at random at about the same rate in all human populations. Having identical twins does not increase a couple's chance of other multiple births. Dizygotic twinning, on the other hand, does seem to "run in families." Such twins are often born to older women and women with several children.

When more than one embryo implants, each usually has its own chorion and amnion except that monozygotic twins have a single chorion and can have one or two amnions. Twins often differ in size at birth because one has a larger placenta and a more abundant nutrient and oxygen supply than the other.

Effects of Teratogens on Development

Teratogens (ter′at-o-jenz), agents that cause malformation in a developing embryo or fetus, include various drugs, hormones, and infectious agents. Such agents cause most damage before the fetus is three months old, the particular damage depending on what structures were most vulnerable when it was present. Many teratogens interfere with cell division or with differentiation of tissues. Often a teratogen causes abnormalities in several body systems that happen to be vulnerable to its effects.

Most antibiotics and many other drugs taken by the mother cross the placenta with ease and damage the fetus. Tetracycline concentrates in bone, mottling and weakening its structure. Chloramphenicol depresses blood cell formation in pregnant women and probably in fetuses too. Sulfonamides cause anemia by interfering with folic acid synthesis and jaundice by interfering with bilirubin transport. Salicylates (aspirin) and dicumarol (an anticoagulant) can cause fetal bleeding and sometimes anatomical malformations. Thiouracil, used to treat hyperthyroidism, can cause fetal goiter. Some years ago, the tranquilizer thalidomide caused many infants to be born with malformed or missing limbs. Alcohol, heroin, and methadone (used to treat heroin addiction) produce addiction in infants, who go through withdrawal symptoms after birth. Alcohol use during pregnancy leads to fetal alcohol syndrome with its assortment of physical and mental disturbances.

Certain hormones give rise to birth defects. Hydrocortisone, used to treat inflammatory disorders, can cause cleft palate. Diethylstilbestrol (DES) causes masculinization of a female fetus, male genital abnormalities, and sometimes vaginal cancer (Chapter 29). Though DES is no longer given to women, it may be added to animal feed to stimulate growth. How much passes in meat to pregnant women and fetuses remains to be learned.

Maternal infections during pregnancy can lead to birth defects. The rubella (German measles) virus causes congenital heart disease, blindness, deafness, and inflammatory reactions in many fetal organs. Syphilis and genital herpes can be transmitted to a fetus from an infected mother. All of these viruses damage the nervous system.

Smoking during pregnancy reduces the infant's birth weight. It can cause microencephaly (markedly small head and brain size).

Vitamin deficiencies during pregnancy are associated with certain birth defects: riboflavin with cleft palate, vitamin B_{12} with hydrocephalus and other central nervous system defects, vitamin D with skeletal abnormalities, and folic acid with limb abnormalities, hydrocephalus, harelip, and eye abnormalities. Vitamin A deficiency may cause a variety of heart and blood vessel defects and kidney, lung, and eye abnormalities.

Chromosomal Abnormalities

Alterations in the configuration or number of chromosomes constitute **chromosomal abnormalities,** which usually are more extensive than mutations. They include deletion, translocation, trisomy, and abnormal numbers of chromosomes. The last is common in spontaneously aborted fetuses.

Deletion is the loss of a chromosome segment. Using letters to represent genes, a chromosome might have genes ABCDEFGHIJK before a deletion and only ABCDEFGH after a deletion with IJK having been lost. The "cat cry" syndrome (*cri du chat*), in which the infant has a mewing, catlike cry, is due to deletion of a segment of chromosome 5. (Pairs of autosomal chromosomes are numbered 1 through 22.) Cat cry syndrome causes severe mental retardation, gastrointestinal and cardiovascular defects, and an abnormal glottis and larynx that accounts for the unusual cry.

Translocation is the shifting of a DNA segment from one chromosome to another. It interferes with chromosome pairing and can cause information loss at break points. Translocation is one cause of Down's syndrome (discussed below).

Trisomy (tri'so-me) is the presence of three copies of a chromosome—as in trisomy 21, which is also a cause of Down's syndrome (Figure 30.7). Trisomy can arise by **nondisjunction** (non''dis-junk'shun) when replicated chromosomes fail to separate during meiosis and one gamete receives two copies of a chromosome and another gamete receives none. In Down's syndrome, the error is in the sperm in about one-fourth of cases, and in the ovum, in the remainder. Incidence of Down's syndrome increases with maternal age from 1 per 1000 births up to age 30 to 1 per 100 births at age 40, suggesting that age of the ovum may contribute to nondisjunction. The risk also increases when paternal age is over 55. Parents who have one child with Down's syndrome often worry that they will have another. The risk is high when the cause is translocation, but only 1 percent greater than the risk of any defect associated with parental age when the cause is nondisjunction.

Victims of Down's syndrome are mentally retarded (usually severely) and are prone to respiratory disease, cataracts, and leukemia. With modern care, they have a life expectancy of about 40 years. The recent trend toward home rather than institutional care allows them maximal mental and physical development.

Several chromosomal abnormalities result from defective sex chromosome transmission. Individuals with **Klinefelter's** (klīn'fel-terz) **syndrome** (Figure 30.8) have an extra X chromosome (XXY) received from either parent through nondisjunction. They have male genitals, usually are sterile, and may develop breasts and elongated limbs. Their mental development may be slightly or significantly below normal. Individuals with **Turner's syndrome** (Figure 30.9) have a single X chromosome (XO) received from either parent. They are female, sterile, usually have poor spatial visualization, and may be below normal in intelligence.

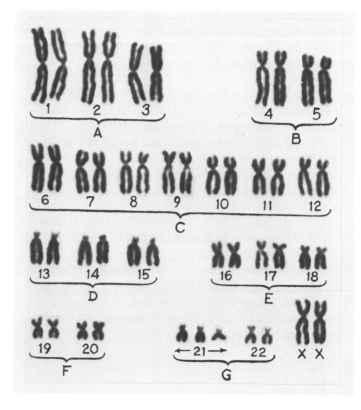

FIGURE 30.7
A photograph and a karyotype of a female child with Down's syndrome. A karyotype is a display of all the chromosomes in a cell.

GENETIC SCREENING AND ENGINEERING

Advances in understanding the causes of genetic and developmental disorders have led to important applications. Prospective parents and neonates can be screened for many genetic defects. Parents can obtain counseling about their risk of having a child with a birth defect and about how to cope with raising a handicapped child. Physicians have several ways to manipulate genetic material and reproductive processes. In the United States, 1 in 10 liveborn infants has an inherited disease and 40 to 50 percent of all spontaneous abortions are due to chromosomal abnormalities. About 25 percent of all hospital beds, and 30 percent in pediatric wards, are occupied by victims of genetic disorders. Preventing and ameliorating genetic disease is an important part of modern health care.

Genetic Screening

The identification of genetic defects and the genes that transmit them is called **genetic screening.** Specific tests are available for use prenatally or with prospective parents to identify: over 100 inborn metabolic errors, over 30 monogenic disorders, many chromosomal abnormalities, and a variety of structural defects in embryos. A common test for phenylketonuria (PKU) involves finding an abnormal product of phenylalanine metabolism in blood or urine. When PKU is detected in the first few weeks of life, the infant is placed on a low-phenylalanine diet and mental development proceeds almost normally.

Screening to detect carriers of Tay-Sachs disease has been particularly successful. This lipid storage disease is due to a defective enzyme that leads to neurological deterioration and death by about age four. It is rare except in descendants of Ashkenazi Jews (Jews of eastern European ancestry), where it occurs with a frequency of 1:3600 births. Factors that have made screening successful are as follows:

1. The gene occurs almost exclusively in a small, defined group.
2. A simple, reliable, relatively inexpensive test detects carriers of the autosomal recessive gene.
3. Both parents must transmit the gene to have an affected child.
4. Prenatal diagnosis can identify an affected fetus early enough for safe termination of the pregnancy if that option is chosen.
5. Most potential carriers are well educated and interested in avoiding the disease.
6. Educational efforts have increased awareness of screening, and reduced resistance to learning that one carries a harmful gene.

When couples at risk of having a child with a genetic defect visit a genetic counselor, they learn the nature of the disease and the risk of its occurring. Few realize that 1 to 2 percent of all pregnancies result in the birth of a child with some defect. Many have difficulty accepting the fact that, though healthy, they can transmit a genetic disease to a child.

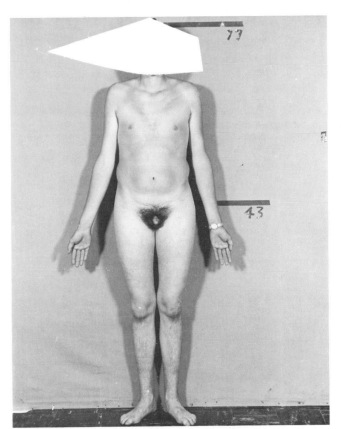

FIGURE 30.8
Person with Klinefelter's syndrome.

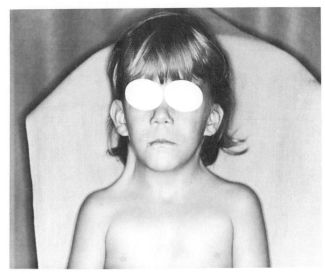

FIGURE 30.9
Child with Turner's syndrome.

Fetal cells can be obtained for prenatal screening by amniocentesis or chorionic villus biopsy (Figure 30.10). **Amniocentesis** (am″ne-o-sen-te′sis) involves withdrawing amniotic fluid, usually 14 to 16 weeks after conception when removing 25 ml of a 200 ml volume does not risk collapsing the amniotic cavity. The fetus and placenta are located by high-frequency sound waves (ultrasound), minimizing the likelihood of accidental damage. Amniotic fluid contains a few fetal cells, which are cultured for several weeks to provide large enough samples to test for genetic defects. A **chorionic villus biopsy** involves in-serting a catheter through the cervix and removing a villus from the chorion by suction or cutting. Biopsy can be done earlier than amniocentesis, but runs a small risk of initi-ating premature labor. Because biopsy provides many fetal cells, tests can be done immediately and results known as early as week 10 of a pregnancy.

Tests on fetal cells detect enzyme defects and other metabolic disorders and some abnormalities in chromo-some structure and number. They also identify the sex of the fetus so the likelihood of sex-linked disorders can be assessed. A female fetus is less likely to have a sex-linked disorder.

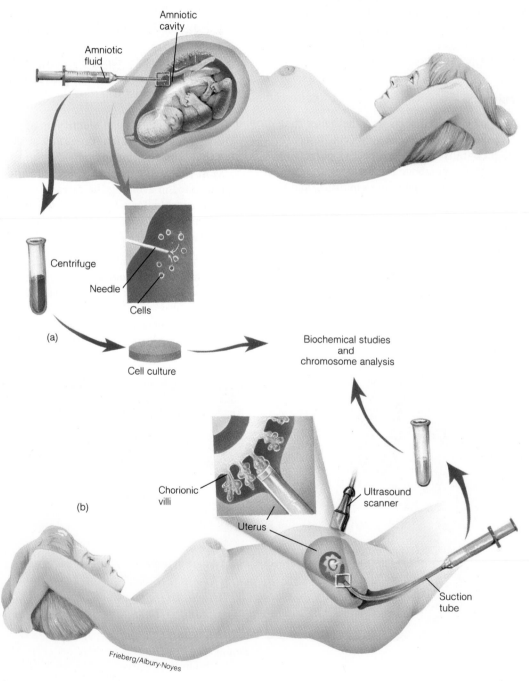

FIGURE 30.10
(*a*) Amniocentesis, and
(*b*) chorionic villi sampling.

Genetic and Reproductive Engineering

The manipulation of genes, **genetic engineering,** and the manipulation of reproduction, **reproductive engineering,** allow some formerly childless couples to have children. Reproductive engineering (artificial insemination and in vitro fertilization) has come into use, whereas genetic engineering (gene transfer and cloning) is still experimental. Genetic engineering involving recombinant DNA and microbes was described earlier.

Artificial insemination, depositing sperm in a woman's vagina by a means other than intercourse, was first recorded in the *Talmud,* a Jewish religious book, and has been in continuous use since the end of the eighteenth century. Sperm from an anonymous donor usually are used and may be supplemented by sperm from the male partner if available. This technique is used when a man does not ejaculate, has too few viable sperm, or carries a harmful allele that he does not wish to transmit to a child.

First performed in England in 1978, **in vitro fertilization** involves removing ova from a woman's ovary and transferring them to carefully prepared sterile media. There, the ova are fertilized by sperm and allowed to develop for a short time. The developing embryo is inserted into the uterus, where it develops normally and the infant is born naturally. This procedure is used when uterine tubes do not allow fertilization to occur normally.

Techniques for **gene transfer,** moving of genes from one cell to another, are limited to nonhuman organisms at present. In theory, they could be used to replace a defective gene with a normal one in embryonic cells early in development. A whole chromosome, or even a whole nucleus, might be replaced.

Cloning, the development of many identical cells from a single cell by mitosis, can be used to produce genetically identical copies of certain organisms (Figure 30.11). Unicellular organisms, reproducing asexually, normally produce clones. Cloning in multicellular organisms typically

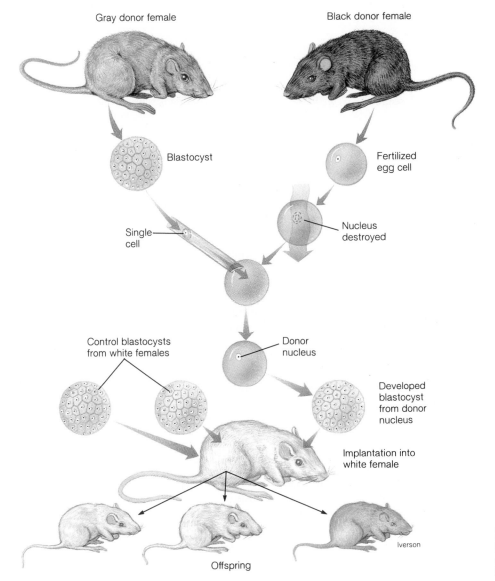

Gray donor female

Black donor female

Blastocyst

Fertilized egg cell

Nucleus destroyed

Single cell

Donor nucleus

Control blastocysts from white females

Developed blastocyst from donor nucleus

Implantation into white female

Iverson

Offspring

FIGURE 30.11
One method by which cloning can be used to produce a new individual.

begins with an unfertilized ovum, the nucleus of which is replaced by a nucleus with paired chromosomes. Many ova can be treated, and the resulting organisms are genetically identical to the one whose cells were used.

> Given that in vitro fertilization is now in widespread use, replacing the nucleus of the ovum with the nucleus of a body cell is not entirely beyond the realm of possibility. Many complex ethical issues are raised by this and other possible applications of genetic and reproductive engineering. The need to resolve these issues becomes increasingly urgent as procedures become more available. Should in vitro fertilization be available to all childless couples regardless of their ability to pay? Given our overpopulated world, should special procedures be used to increase the population? How would you decide these issues? What other important issues do you think are raised by genetic and reproductive engineering?

VARIATIONS AT DEVELOPMENTAL LIFE STAGES

From birth to adulthood, the human body moves through a new set of developmental stages. These stages include the neonatal period, infancy, childhood, and adolescence.

The Neonate

The **neonatal period** is the first four weeks after birth. Less than optimal gas exchange during delivery leaves most newborn infants with mild anoxia and acidosis from carbon dioxide accumulating in the blood. As the baby's breathing becomes vigorous, homeostatic mechanisms correct these problems.

Neonates differ from adults in several ways. Their bodies are less able to regulate temperature, their kidneys are less able to concentrate urine, and their immunity to disease is lower. Also, their breathing rates (30 to 50 breaths per minute) and heart rates (120 to 140 beats per minute) are faster and their blood pressure (75/40 torr) is lower than that of adults. Neonates are especially vulnerable to temperature fluctuations, dehydration, electrolyte imbalances, and infections. Daily urine volume accounts for as much as 6 percent of body weight compared with less than 2 percent for adults. Though the neonate receives passive immunity from placental transfer and breast milk, its limited antibody production makes it susceptible to infections. Finally, immature regulatory mechanisms make neonates more subject to respiratory failure than normal adults.

The neonate has just experienced birth—one of the most traumatic experiences of a lifetime. The APGAR rating (Table 30.4) is used to assess the physiological condition of the neonate at one minute and five minutes after birth. Each characteristic—**a**ppearance, **p**ulse, **g**rimace, **a**ctivity, and **r**espiratory effort—is assessed on a scale of zero to two. The sum of the ratings helps to determine whether the newborn infant needs special assistance to stay alive.

> Over the years, opinion has vacillated about the advisability of circumcision (removal of the foreskin). One view is that good hygiene renders circumcision unnecessary and that, because it is usually done in neonates without anesthesia, it inflicts unnecessary pain and risk of infection soon after birth, a stressful event in itself. The other view is that circumcision may have benefits such as decreasing the incidence of urinary tract infections in infancy and reducing the risk of penile cancer in men and cervical cancer in their mates. Unfortunately, better studies using more cases are needed to determine whether these findings are valid.

Infancy and Childhood

Infancy (from four weeks to two years of age) and **childhood** (two years of age to puberty) are periods of growth and maturation. Genetic endowment, nutritional status, hormones, and a number of other factors determine how fast a person grows and how body parts mature.

TABLE 30.4 The APGAR Rating Scale for Neonates

Characteristic	Rating		
	0	*1*	*2*
A Appearance (skin)	Pale or blue	Body pink, limbs blue	Whole body pink
P Pulse (heart rate)	None	Less than 100	More than 100
G Grimace (reflex response to stimulation of sole of foot)	No response	Facial grimace and some movement	Facial grimace and extensive movement
A Activity (muscle tone)	Flaccid	Extremities partially flexed	Good muscle tone and active movement
R Respiratory effort	None	Irregular and slow, no crying	Rapid and generally regular, strong cry

The physiology of infants and children differs from that of adults in several ways. The heart rate drops with age from 120–140 beats per minute in early infancy to 80–90 beats per minute in adolescence. Blood pressure increases from about 80/50 torr in infancy to about 110/70 torr in adolescence. Over the same period, respiratory rate decreases from 20–30 breaths per minute to 14–16 breaths per minute. Total urine volume increases; but it becomes a smaller proportion of body weight as kidneys better concentrate urine. Motor and mental developments of infancy and childhood are summarized in Table 30.5.

In a normal 24-hour day, infants sleep about 16 hours at birth, 14 hours at three weeks of age, and 13 hours at one year. The daily duration of sleep continues a gradual decline to about eight hours by the late teens and the distribution of sleep periods through the day is altered dramatically. In the early weeks, an infant is awake almost as much at night as in the daytime (as any new parent knows). By 12 weeks, daytime waking and nighttime sleeping have become reasonably well established. About half a neonate's sleep appears to be REM sleep, but this decreases to about 20 percent (the adult value) by about age two. Slow-wave sleep is not detectable in infants until about three months of age, probably because the parts of the thalamus and cerebral cortex responsible for generating this sleep stage are immature.

Adolescence

Childhood ends and **adolescence** begins with the onset of puberty. Puberty is characterized by a sudden growth spurt followed by maturation of reproductive organs and development of secondary sexual characteristics. What triggers the onset of puberty has puzzled researchers. It now appears that GnRH secretion, which initiates sexual maturation, is triggered by a not-yet-understood change in the brain that activates a specific gene. Whatever this change is, it can be inhibited by melatonin from the pineal gland. Because melatonin secretion occurs mainly at night, night blood samples are necessary to detect changes in its secretion rate. Such samples from hospitalized children show a progressive decline in melatonin from peak levels in children under five to one-fourth the peak level by the end of puberty.

Pubertal changes in females include a growth spurt between ages 10½ and 13 and development of sexual characteristics in the following order: Breast budding, the first stage of puberty in females, occurs between 9 and 13 and at an average age of 11.2 years. Breasts and nipples enlarge, and areolae become pigmented. The pelvis widens, pubic hair begins to grow, and apocrine glands develop in the axillae. Menstruation begins at an average age of 13—about two years after the growth spurt. Ovulation occurs

TABLE 30.5 Summary of Development During Preschool Years

Abilities Seen in 75 % of Infants and Children at Age:	Gross Motor	Fine Motor	Language and Social
2 months	Lifts head when placed on stomach	Follows objects with eyes	Makes noncrying sounds Smiles responsively
4 months	Holds head steady when placed in sitting position Rolls over	Grasps objects placed in hand	Laughs and squeals Smiles spontaneously
6 months	Bears some weight on legs	Reaches for objects	Makes speechlike sounds
9 months	Sits without support Stands holding on	Passes objects hand to hand Uses thumb and finger to pick up small objects	Imitates speech sounds Feeds self crackers Plays peek-a-boo Works to get to toy that is out of reach
1 year	Walks holding on to furniture Stands momentarily	Bangs two cubes together	Says dada or mama Plays pat-a-cake
1½ years	Walks well Stoops and regains standing position without falling	Stacks two cubes Scribbles spontaneously Imitates movements of others	Says three words other than mama and dada Drinks from cup Uses spoon fairly well
2 years	Walks up stairs Can walk backward Kicks ball and throws ball overhand	Stacks four cubes	Combines two different words Follows simple directions Removes garments
3 years	Jumps in place Pedals tricycle	Stacks eight cubes Copies "o" Makes vertical line	Uses plurals Washes and dries hands Puts on clothing
4 years	Does broad jump Balances on one foot for 5 seconds	Copies "+" Picks longer line of two presented	Plays simple games Gives name Recognizes colors Comprehends prepositions Dresses without supervision
5 years	Hops on one foot Catches bounced ball most times	Copies stacks of cubes Draws three part human figure	Defines words sometimes Comprehends simple opposite analogies

in only about half the menstrual cycles during the first year, and becomes a regular monthly event about seven years after menarche.

Males experience a pubertal growth spurt between ages 12½ and 15 and then typically develop sexual characteristics in the following order: The penis and testes enlarge. Hair begins to grow first in the pubic area, then the axillae, upper lip, legs, and chest, and finally on other parts of the face. The larynx increases in size and the voice deepens. The first ejaculation occurs spontaneously, usually a few months after the appearance of pubic hair—this and similar ejaculations typically occur at night and are called **nocturnal emissions.**

In addition to changes in reproductive organs, physiological changes occur in other systems in both sexes: both the growth rate and metabolic rate decrease, the heart rate and respiratory rate decline, and blood pressure increases to adult levels. Although much development of the brain and other neural structures occurs before birth and during infancy, the reticular formation and some cortical association areas do not become fully myelinated until the late twenties. Except for this myelination, all body systems reach physiological maturity by about age 25.

During the various stages in human development, growth is accompanied by changes in body proportion (Figure 30.12). Head length is nearly half the body length in a two-month embryo, but only one-eighth the body length by adulthood. Also, lymphoid, neural, and reproductive tissues show different rates of growth than the body in general (Figure 30.13). Lymphoid tissues reach a peak size at the onset of puberty and decline thereafter. The brain grows rapidly in infancy and early childhood, and the reproductive organs grow rapidly in adolescence.

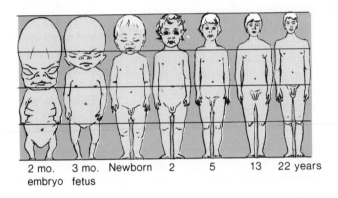

FIGURE 30.12
Changing proportions of the human body during prenatal and postnatal growth. All stages are drawn to the same total height.

From *Human Embryology* by L. Patten (redrawn from Scammon). Copyright © 1933 McGraw-Hill, Inc.

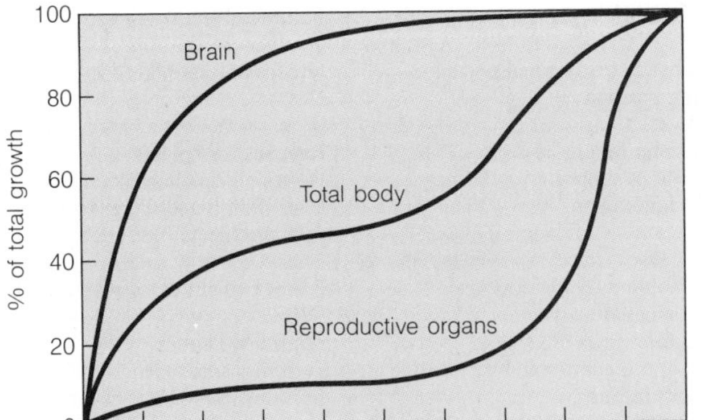

2 mo. embryo 3 mo. fetus Newborn 2 5 13 22 years

FIGURE 30.13
Relative growth rates of different parts of the body.

AGING

Aging has been variously defined as: (1) progressive decline in cellular efficiency that occurs after maturation, (2) all changes that occur in living organisms with the passage of time, or (3) the sum total of all changes with time that lead to functional impairment and death. A principal aspect of aging is the decreased ability to adapt to a changing environment—to maintain homeostasis. As changes create stress in the elderly, homeostatic balance becomes more precarious; eventually, stress surpasses the body's capacity to adjust and death ensues.

> Who among us wants to grow old? As children and adolescents, we often wish we were older; but after 30, we find ourselves wishing we could put on the brakes and slow down the aging process. In ourselves and others, we readily observe the various physical and mental changes and the increasing susceptibility to illnesses inherent in aging. We also see that the elderly are often ignored and sometimes mistreated; yet we will inevitably age and experience the biological and social insults of aging. As future providers of health care, many users of this book will deal professionally with elderly clients. To provide better care in a more humane environment, and to approach our own senior years with greater health and dignity, it seems appropriate—even essential—to conclude our study of human anatomy and physiology with a consideration of aging and death.

The Aging Process and Factors in It

Determining which bodily changes are due to aging and which to disease is difficult. In general, changes that occur to some degree over time in nearly all adults are consequences of aging—graying of hair, loss of sensory acuity, skin wrinkling, restriction of joint movement, and bone fragility. In contrast, diseases that increase in incidence with age—cirrhosis of the liver, heart disease, stroke, nephritis, emphysema, and some cancers—can be caused by sedentary life-style, infections, alcohol, smoking, and other agents—not to aging *per se*.

Physiological processes regulated by several control mechanisms are less susceptible to the effects of aging than those with a few control mechanisms. For example, acid-base balance and blood sugar, which have multiple controls, generally remain stable into old age. Locomotion, heart, kidney, lung and brain functions, and immunological competence, which probably have fewer control mechanisms, are more susceptible to degenerative changes.

TABLE 30.6 Biologic Function Changes between the Ages of 30 and 70

Biologic Function	Change
Work capacity (%)	↓25 to 30
Cardiac output (%)	↓30
Maximum heart rate (beats min)	↓24
Blood pressure (torr)	
Systolic	↑10 to 40
Diastolic	↑ 5 to 10
Respiration (%)	
Vital capacity	↓40 to 50
Residual volume	↑30 to 50
Basal metabolic rate (%)	↓ 8 to 12
Musculature (%)	
Muscle mass	↓25 to 30
Hand grip strength	↓25 to 30
Nerve conduction velocity (%)	↓10 to 15
Flexibility (%)	↓20 to 30
Bone (%)	
Women	↓25 to 30
Men	↓15 to 20
Renal function (%)	↓30 to 50

From Smith, et al., "Physical Activity Prescription for the Older Adult" in *The Physician and Sportsmedicine,* 11:92. Copyright © 1983 McGraw-Hill Healthcare Group.

Changes in Systems with Aging

Systems vary in the rate at which they show aging (Table 30.6). Sense organs are among the first to show signs of aging. In the eyes, accommodation gradually diminishes from early adulthood through the fifties; dark adaptation first declines at age 20 and continues to decline throughout life, but adequate vitamin A can slow this process. Auditory acuity declines relatively rapidly from adolescence to about age 50 and at a slower rate thereafter. Avoiding loud noises can minimize hearing loss.

Loss of muscular strength accelerates with age as biochemical changes in the muscle fibers themselves and in neuromuscular junctions occur; exercise can slow these processes. Lost muscle tissue is replaced by connective tissue and fat, and neural signal transmission becomes less efficient. Bones lose minerals and become less dense and osteoporosis develops, especially in white women. Joint cartilage erodes and contributes to osteoarthritis, a disorder so common some consider it a normal part of aging.

Cardiovascular function, including cardiac output and cardiac reserve, decline. Heart muscle is subject to the same changes that decrease the strength of contractions in other muscles. The heart becomes less responsive to neurotransmitters and activity of its pacemaker system is reduced. Though heart rate and blood pressure increase in stress, cardiac output may not meet the demands of extreme exertion. A gradual, steady rise in systolic and diastolic pressures is associated with lower baroreceptor sensitivity, though renin-angiotensin activity also declines. Atherosclerosis accelerates the rate of decline in cardiovascular function. Hypertension and diabetes increase the likelihood of heart disease in the elderly, but high blood pressure and high blood glucose may force more glucose and oxygen into brain tissue through atherosclerotic vessels. Low blood pressure increases resistance to heart disease, especially in women.

In the urinary system, decreases occur in renal blood flow, creatinine clearance, ability to concentrate urine, and ammonium excretion. Kidney atrophy in advanced age can be due to ischemia from blood vessel sclerosis. Basement membranes in glomeruli and in tubules thicken, especially in individuals with diabetes. These changes make maintaining water and electrolyte balance increasingly difficult. Urine retention from decreased bladder tone increases the incidence of urinary tract infections.

In females after menopause, the uterus and vagina get smaller and the ovaries become fibrous and nonfunctional. In the male reproductive system, changes are more gradual; but after about age 50, testosterone secretion and sperm production decrease at a fairly rapid rate. Prostate disorders increase in frequency.

Among digestive functions, motility and secretion decrease with age. Reduced esophageal peristalsis and sphincter relaxation can make swallowing difficult. Gastric gland atrophy and mucosal thinning can lead to gastric ulcers, pernicious anemia, and recurrent gastritis. In spite of small decreases in function, the small intestine has enormous reserve capacity for absorption. Even as the appetite decreases, nutrient absorption is adequate. The liver's decreased ability to detoxify foreign substances, combined with the numerous medications taken by some elderly people, often leads to drug reactions. Amyloidosis, degeneration of the islets of Langerhans, may be an immunological disorder and probably is not due to aging. As skin loses elastic fibers, it wrinkles, and as it loses sulfhydryl cross-linkages between collagen fibers, it becomes less pliable (Chapter 5).

In the respiratory system, decreased elasticity of both lung tissue and the chest wall and decreased alveolar surface area impair breathing. Though total lung volume remains nearly constant, movement of air in and out of the lungs decreases. As alveoli change from spheres to shallow sacs, gas exchange decreases. Oxygen enters blood at a rate of about 4 l/min at age 20, but only about 1.5 l/min at age 75. Declining blood oxygen is a primary factor in lower heart, skeletal muscle, and central nervous system performance. Preserving tissue oxygenation is a primary problem in maintaining homeostasis in the elderly.

Research on brain aging is limited to animal studies and observations on human brains after death. For example, rats show the greatest decline in brain cell numbers during the first 100 days of life. From then to 650 days (old age in a rat), only small numbers of cells are lost. A stimulating environment leads to thickening of cerebral cortex in young rats and slows brain cell death. If one can generalize from rats to humans, a stimulating environment for young humans would limit later decreases in mental functions. Autopsies of aged human brains show senile plaques (areas of degeneration) that resemble amyloidosis. Severe mental impairment in the elderly (senile dementia) includes loss of memory for recent events, gradual deterioration of cognitive functions (such as reading, writing, speaking, and calculating), and sometimes psychological disturbances (such as irritability, delusions, and emotional lability).

As the immune system becomes less reactive to foreign antigens and more reactive to body tissues, disease resistance declines and autoimmune diseases and cancer are augmented.

Ways to Counteract Aging

Some of us have the luxury of looking for ways to counteract aging because we have lived long enough to develop degenerative diseases. In industrialized countries, life expectancy has risen from about 50 for a person born in the early 1900s to about 75 for a person born in the 1980s. Nearly half of all deaths in the United States in the early 1900s were due to infectious diseases such as pneumonia, influenza, tuberculosis, and communicable diseases of childhood. Today, such diseases account for less than 3 percent of all deaths, while cardiovascular diseases account for nearly half. Cancer, reported to cause about 4 percent of deaths in the early 1900s, now accounts for 22 percent because of better diagnosis and greater longevity.

The most important factors in counteracting aging are good nutrition and exercise. Conversely, the main factors in degenerative diseases are obesity and lack of physical fitness.

Good nutrition requires adequate daily intake of protein, vitamins, minerals, and other nutrients (Chapter 26). It also requires limiting dietary saturated fats and cholesterol. Recent evidence suggests that glucose tolerance declines at a rate of about 5 mg/dl per decade. Reducing dietary sugars, which elevate blood glucose most rapidly, and eating more complex carbohydrates might minimize the effects of high blood glucose (Chapter 25). Finally, underfeeding (limiting intake to 1500 to 2000 kcal/day) may delay or prevent events associated with aging.

Exercise, which uses at least 2000 kcal per week, reduces the risk of cardiovascular disease by increasing performance of the heart and lungs. It also raises blood HDLs and lowers LDLs, probably by increasing blood lipoprotein lipase. Weight-bearing exercise, such as walking, dancing, or even standing for two to four hours, reduces the risk of osteoporosis.

Theories of Aging

There may be as many theories of aging as there are investigators studying aging, but better evidence is available for some theories than for others. A few of the better known theories are summarized here.

According to the **cytological aging theory,** the body is progressively damaged by cumulative effects of injuries to cells or macromolecules within them. Possible causes of aging include ionizing radiation, liberation of highly reactive free radicals, and accumulation of lipofuscin, a pigment found in cells that have lost their ability to divide.

The **theory of genetically programmed aging** asserts that genes are turned on and off to control aging, as they are to control growth and development. This theory is supported by the observation that cells in cultures, except for cancer cells, divide a finite number of times and die. Senescence genes may code for molecules that inhibit DNA synthesis and block cell division. Fibronectin, a growth inhibitor abundant in senescent cells and almost lacking in cancer cells, may be such a molecule.

The **fat metabolism theory** suggests that aging can be due to a decreasing ability to metabolize fats, which may be triggered by excess food intake early in life. In rats, food restriction slows the decline in fat metabolism and underfed rats far outlive rats permitted to eat freely. Underfeeding also may slow aging of the immune system or decrease free radical formation. Glucagon and adrenalin normally promote fat metabolism, but during aging, fat cells become less responsive to them and fats accumulate in blood and cells. If these findings apply to humans, they might account for increases in heart disease and obesity with aging.

The **glucose hypothesis of aging** notes that glucose can spontaneously bind (without enzymes) to cellular proteins and form abnormal glycoproteins. Such glycoproteins may account for collagen stiffening, cataract formation, and basement membrane thickening. If abnormal glycoproteins foster aging, removing them could slow it. The drug aminoguanidine blocks glycoprotein linkage in collagen in the laboratory and is being tried in humans to prevent complications of diabetes. Another treatment might be to stimulate activity of phagocytes that normally engulf abnormal glycoproteins, but that become less efficient with age.

According to the **growth hormone theory of aging,** lack of growth hormone accounts for the effects of aging in about half the elderly population. It accounts for 20 to 50 percent of shrinkage of muscle, liver, kidneys, and spleen between age 30 and age 70. Tissue shrinkage, bone mineral loss, and adipose tissue accumulation accelerate after age 50 at the same time growth hormone secretion usually slows down. If this theory is correct, growth hormone, which can now be synthesized by genetic technology, might be used to slow the aging process.

The **glucocorticoid hypothesis** is based on the idea that stress may accelerate aging and that ability to adapt to stress decreases with age. The adrenal cortex secretes excess ACTH in response to stress—making more energy available; increasing cardiovascular tone and alertness; and inhibiting growth, immune and inflammatory responses, and reproduction. Oversecretion in stress leads to excess corticosteroid secretion, loss of ACTH receptors, and loss of hippocampal neurons that are target cells for corticosteroids. Certain hippocampal neurons appear to be responsible for negative feedback signals that inhibit CRF secretion. Without such signals, ACTH secretion cannot be turned off and more severe disorders—immunosuppression; muscle atrophy; osteoporosis; atherosclerosis; and excess blood calcium, glucose, and lipids—develop. Though this hypothesis is based on work with rats, it probably applies to humans where hippocampal neurons are likewise the target cells of corticosteroids. It is supported by the observation that victims of Alzheimer's disease (Chapter 12) have increased glucocorticoid secretion and hippocampal damage.

None of these theories is proven. As we learn more about aging, treatments to slow or stop it may become possible. It is certain that aging and disease will be more clearly separated and that treatment of diseases also may increase life expectancy.

CLINICAL TERMS

abortion (ab-or'shun)   premature expulsion of a fetus from the uterus

abruptio placentae (ab-rup'she-o plas-en'ta-e)   premature detachment of the placenta from the uterine wall

hydatidiform (hi''dat-id'if-orm) **mole**   a benign tumor of the placenta

hydramnios (hi-dram'ne-os)   an excess of amniotic fluid

hypospadias (hi''po-spa'de-as)   congenital opening of the urethra on the underside of the penis or into the vagina

meconium (mek-o'ne-um)   a dark green substance in the intestine of a full-term fetus

placenta previa (plas-en'tah pre've-ah)   a condition in which the placenta is attached near the cervix and subject to rupture

ESSAY
DEATH AND DYING

The simplest definition of death is the loss of spontaneous heartbeat and respiration. Now that these functions can be maintained mechanically, death is defined as a 24-hour absence of any cerebral function, as shown by a "flat" electroencephalogram. Even when vital functions of heartbeat, respiration, and urine formation have stopped, many cells remain alive up to an hour until their oxygen and nutrients are depleted. Organs can be removed for transplant after vital functions have ceased if they are maintained by life-support systems.

Death often results from the failure of a particular organ such as the heart or kidneys. The effects of one organ's failure spread and other organs also fail. For example, in a patient dying of kidney failure (assuming an artificial kidney is not in use), toxic wastes accumulate in the blood, which damage brain cells and impair heart function. Eventually, consciousness is lost, and the heart stops beating.

Until recently, dying people were cared for at home by their families, but today more dying people are cared for in hospices (homes for dying patients), nursing homes, or hospitals. Common in Switzerland, Austria, and England, hospices are increasing in number in the United States. A hospice has special characteristics. First, the staff is mainly concerned with making a person's last days as comfortable and painfree as possible and with providing opportunities to discuss death and the fear of it that most people have. Second, family members are encouraged to be with a dying person and are helped to accept the inevitable death. Third, the atmosphere of a hospice is homelike, and staff members take time to talk at length with the patients. Patients know that they are dying but have the assurance that they will be made as comfortable as possible and will receive emotional support.

When a hospice is not available, many of the same services can be provided by a sensitive family that has the strength to care for a dying relative at home. In some communities, nurses and social workers are available to counsel dying patients and their families. Courses are available in some locations for people who want to learn to assist dying patients. Even the recent trend toward discussing death and dying in a more open way helps to make people more able to cope with their own or another's death.

Now that hospitals have respirators, heart-lung machines, artificial kidneys, and other life-support devices, people can be kept alive who otherwise would die. If a patient has a chance of recovering, such devices are of great value, but if not, they can add to the suffering of the patient and family.

Many ethical issues are created by the availability of life-support systems. Which patients will receive such services? If a patient who is dependent on life-support systems for survival appears to have no chance of recovery, can the machines be disconnected? Who will make such decisions, and what criteria will be used? What legal actions might be taken against persons or institutions responsible for such decisions? If more life-support systems are needed than are available, how will the people to use the systems be chosen? Because using such equipment is generally extremely expensive, who will pay for such services? And what concern should there be for the wishes of the patient?

In the minds of many people, death is deprived of dignity when it occurs in a hospital room surrounded with machines that maintain failing organs. Finding answers to ethical questions about dying and the use of life-support systems is a major challenge facing health care workers and all citizens today.

Questions

1. How has modern medicine changed the definition of death?
2. Why are life-support systems sometimes kept operating after death has occurred?
3. What are the relative advantages and disadvantages of allowing a person to die at home, at a hospice, or at a hospital?
4. How would you define death with dignity?
5. Hold a classroom debate regarding ethical questions related to the use of life-support systems for dying patients.

Chapter Summary

Genetics

- Basic genetic terms are summarized in Table 30.1.
- Genetic changes arise from mutations caused by agents called mutagens.

Prenatal Development

- Major events in prenatal development include formation of the zygote, blastocyst, embryonic disk, primary germ layers, and organs.
- A teratogen is any agent that can cause malformations in the developing embryo or fetus. Infectious agents, hormones, and certain kinds of medications frequently act as teratogens. Teratogens often cause heart defects, abnormalities of the reproductive system and limbs, deafness, and mental retardation.
- Chromosomal abnormalities in number or composition are responsible for Down's syndrome and many other defects.

Genetic Screening and Engineering

- Genetic screening is used to detect the presence of defective genes, genetic engineering is the manipulation of genetic material, and reproductive engineering is the manipulation of reproduction.

Variations at Developmental Life Stages

- Compared to adults, neonates are less able to regulate body temperature and to concentrate urine, have less immunological competence, and have higher heart and respiratory rates and lower blood pressure.
- Growth and maturation occur in infancy and childhood. Heart and respiratory rates decrease and the blood pressure increases in these stages. Motor and mental development occur.
- Changes during puberty include rapid growth, maturation of reproductive organs, and development of secondary sexual characteristics. The heart rate, respiratory rate, and blood pressure reach adult values during late adolescence.

Aging

- Aging may be defined as a decreased ability to adapt to a changing environment. It occurs at different rates in different organs and any organ may show changes due to aging or disease.
- Factors that affect the aging process include degenerative changes in regulatory mechanisms and biochemical changes within organs. Several theories of aging are described briefly.

Questions and Problems

The questions at the end of each chapter are numbered to correspond with the objectives listed at the beginning of the chapter. Italics indicate that a question requires critical thinking skills beyond simple factual recall.

Questions

1. (a) Use the terms in Table 30.1 to explain the basic principles of genetics.
    (b) *What are some common genetic disorders, and how are they inherited?*
    (c) What is a mutation, and how do frameshift and point mutations differ?
    (d) *What are some causes of mutations?*
2. (a) What are the major events in prenatal development?
    (b) In what order do the above events occur?
3. (a) What are the functions of each embryonic membrane?
    (b) How is the placenta formed, and what are its functions?
4. *What factors contribute to stress during birth?*
5. *How do monozygotic and dizygotic twins differ, and what factors are thought to contribute to twinning?*
6. (a) *What are the likely effects of each of three teratogens on fetuses?*
    (b) *What are chromosomal abnormalities, and how can they affect fetuses?*
7. (a) What is involved in genetic screening and in genetic and reproductive engineering?
    (b) *What kinds of physiological disorders might they help to prevent?*
8. *How do neonates differ physiologically from adults?*
9. *How do infants and children differ physiologically from adults?*
10. (a) *How do adolescents differ physiologically from adults?*
    (b) *How do events in puberty differ in males and females?*
11. *What is aging, and what factors affect it?*
12. (a) *How do body systems change with age?*
    (b) *How can nutrition and exercise help to delay aging?*
13. Briefly describe the major theories of aging.

Problems

1. Research the topic of embryological differentiation, and explain how it might limit the physiological processes that an organ can carry out.
2. Find out what is meant by the phrase "ontogeny recapitulates phylogeny."
3. What are the advantages of using bacteria to produce human proteins via recombinant DNA?
4. An antiaging drug is currently being evaluated by the Food and Drug Administration. The possibility of having such a drug poses ethical and social questions. Consider the following: (a) How would one find out whether it prolongs productive life or prolongs a life plagued with degenerative changes of aging? (b) Could the drug be made available to all, or only to those who could afford it, and who would decide? (c) Should there be research on the drug using human subjects? (d) What effects would such a drug have on life-styles, education, use of leisure, pension plans, quality of life for the elderly, and burdens of the elderly on the young?
5. How do the death of eggs and sperm that have not become part of a zygote compare with the death of a zygote?

The following set of illustrations includes medial sections, horizontal sections, and regional dissections of human cadavers. These photographs will help you visualize the spatial and proportional relationships between the major anatomic structures of actual specimens. The photographs can also serve as the basis for a review of the information you have gained from your study of the human organism.

HUMAN CADAVERS

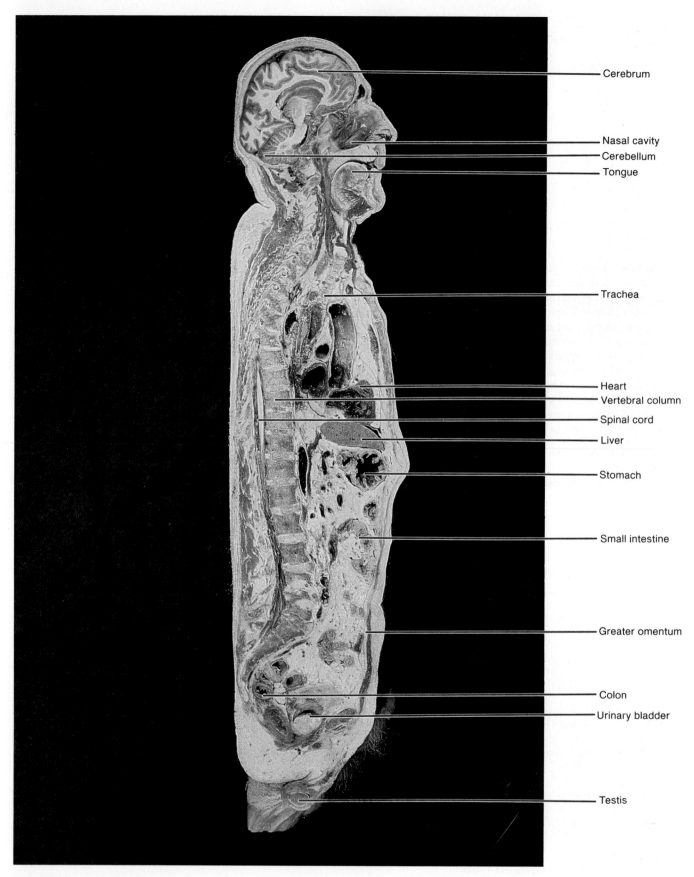

Cerebrum

Nasal cavity
Cerebellum
Tongue

Trachea

Heart
Vertebral column
Spinal cord
Liver

Stomach

Small intestine

Greater omentum

Colon
Urinary bladder

Testis

PLATE 11
Median section of the head and trunk.

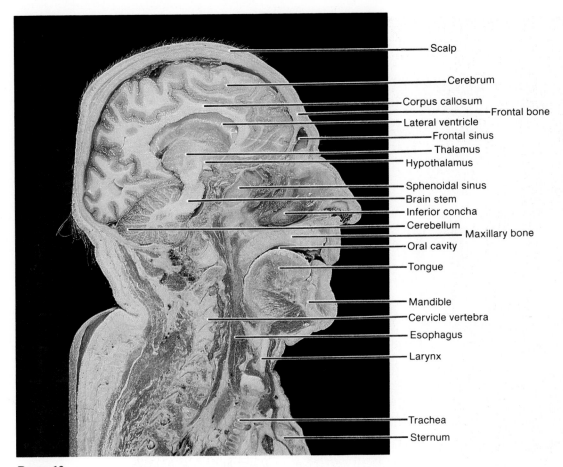

Scalp
Cerebrum
Corpus callosum
Frontal bone
Lateral ventricle
Frontal sinus
Thalamus
Hypothalamus
Sphenoidal sinus
Brain stem
Inferior concha
Cerebellum
Maxillary bone
Oral cavity
Tongue
Mandible
Cervicle vertebra
Esophagus
Larynx
Trachea
Sternum

PLATE 12
Median section of the head and neck.

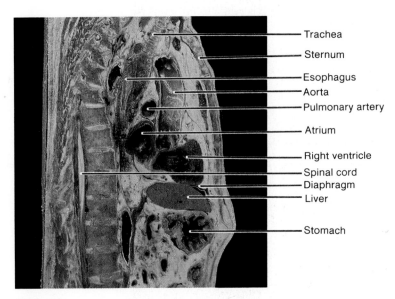

Trachea
Sternum
Esophagus
Aorta
Pulmonary artery
Atrium
Right ventricle
Spinal cord
Diaphragm
Liver
Stomach

PLATE 13
Viscera of the thoracic cavity, medial section.

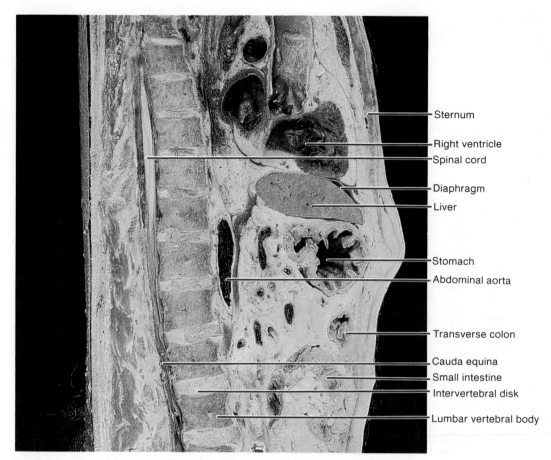

— Sternum
— Right ventricle
— Spinal cord
— Diaphragm
— Liver
— Stomach
— Abdominal aorta
— Transverse colon
— Cauda equina
— Small intestine
— Intervertebral disk
— Lumbar vertebral body

PLATE 14
Viscera of the abdominal cavity, medial section.

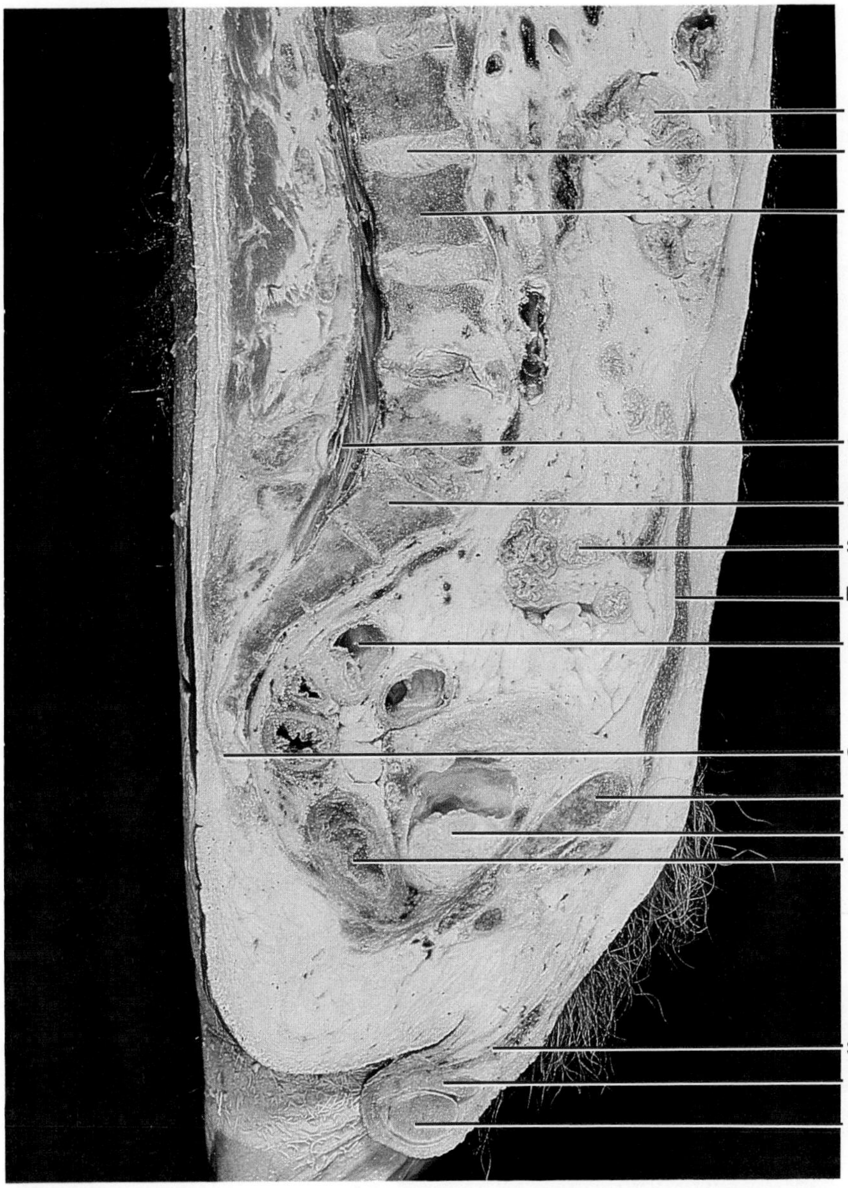

Small intestine
Intervertebral disk
Lumbar vertebral body

Cauda equina
Sacrum
Small intestine
Rectus abdominus
Sigmoid colon

Coccyx
Symphysis pubis
Urinary bladder
Rectum

Spermatic cord
Epididymis
Testis

PLATE 15
Viscera of the pelvic cavity, medial section.

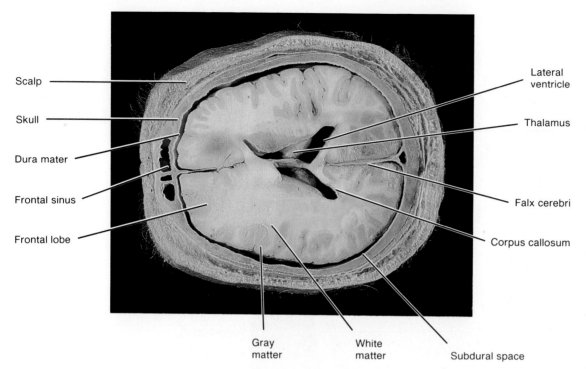

Scalp

Skull

Dura mater

Frontal sinus

Frontal lobe

Lateral ventricle

Thalamus

Falx cerebri

Corpus callosum

Gray matter

White matter

Subdural space

PLATE 16
Horizontal section of the head above the eyes, superior view.

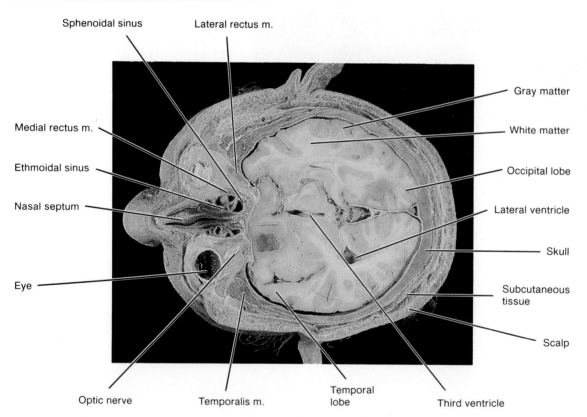

Sphenoidal sinus

Lateral rectus m.

Medial rectus m.

Ethmoidal sinus

Nasal septum

Eye

Gray matter

White matter

Occipital lobe

Lateral ventricle

Skull

Subcutaneous tissue

Scalp

Optic nerve

Temporalis m.

Temporal lobe

Third ventricle

PLATE 17
Horizontal section of the head at the level of the eyes, superior view.

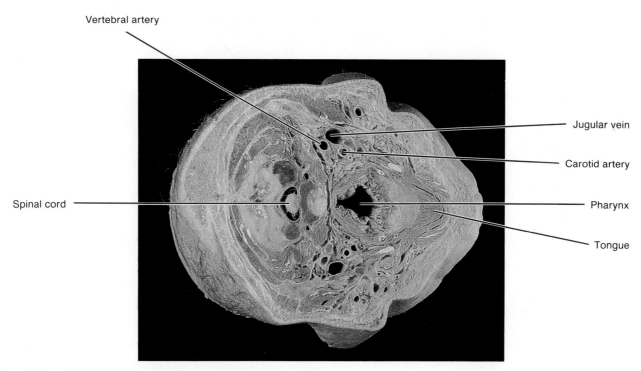

Vertebral artery

Jugular vein

Carotid artery

Spinal cord

Pharynx

Tongue

PLATE 18
Horizontal section of the neck, inferior view.

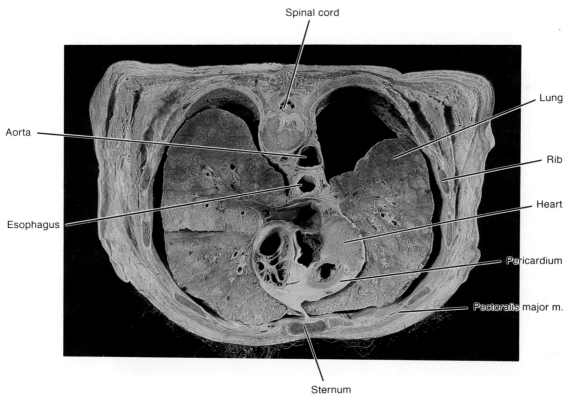

Spinal cord

Lung

Aorta

Rib

Heart

Esophagus

Pericardium

Pectoralis major m.

Sternum

PLATE 19
Horizontal section of the thorax through the base of the heart, inferior
view.

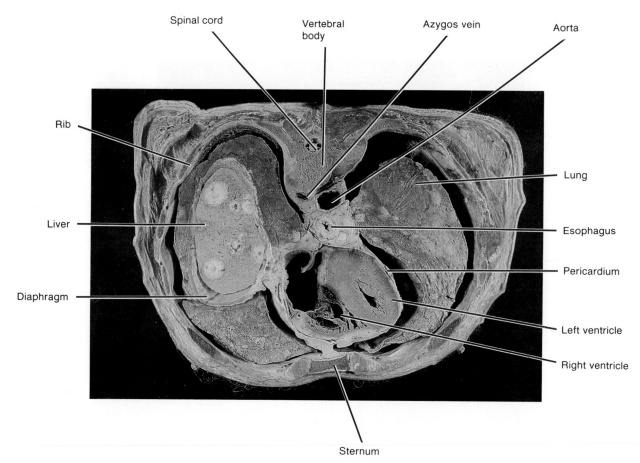

Spinal cord

Vertebral body

Azygos vein

Aorta

Rib

Lung

Liver

Esophagus

Pericardium

Diaphragm

Left ventricle

Right ventricle

Sternum

PLATE 20

Horizontal section of the thorax through the heart, inferior view.

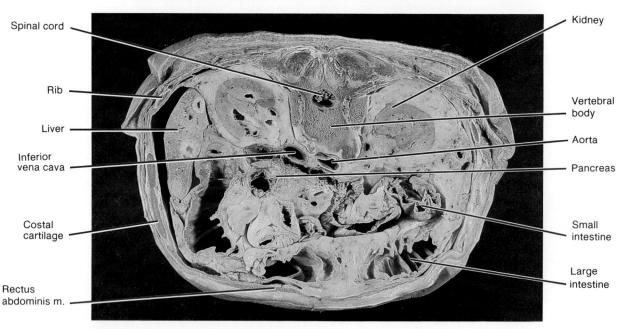

Spinal cord

Kidney

Rib

Vertebral body

Liver

Aorta

Inferior vena cava

Pancreas

Costal cartilage

Small intestine

Large intestine

Rectus abdominis m.

PLATE 21

Horizontal section of the abdomen through the kidneys, inferior view.

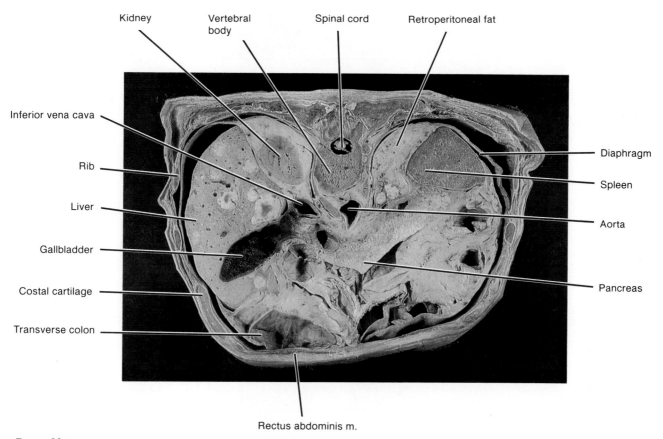

Kidney
Vertebral body
Spinal cord
Retroperitoneal fat

Inferior vena cava

Rib

Liver

Gallbladder

Costal cartilage

Transverse colon

Diaphragm

Spleen

Aorta

Pancreas

Rectus abdominis m.

PLATE 22
Horizontal section of the abdomen through the pancreas, inferior view.

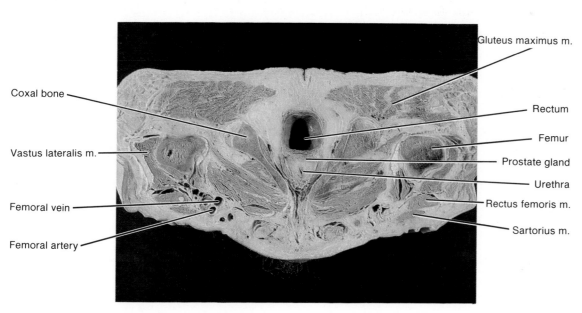

Gluteus maximus m.

Coxal bone

Vastus lateralis m.

Femoral vein

Femoral artery

Rectum

Femur

Prostate gland

Urethra

Rectus femoris m.

Sartorius m.

PLATE 23
Horizontal section of the male pelvic cavity, superior view.

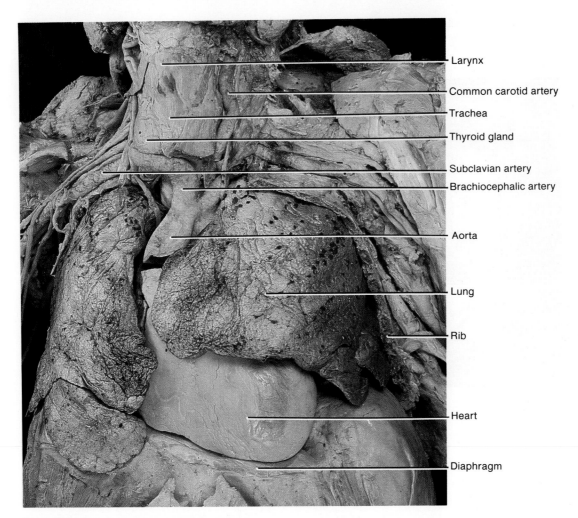

Larynx

Common carotid artery

Trachea

Thyroid gland

Subclavian artery

Brachiocephalic artery

Aorta

Lung

Rib

Heart

Diaphragm

PLATE 24
Thoracic viscera, ventral view. (Brachiocephalic vein has been
removed to expose the aorta.)

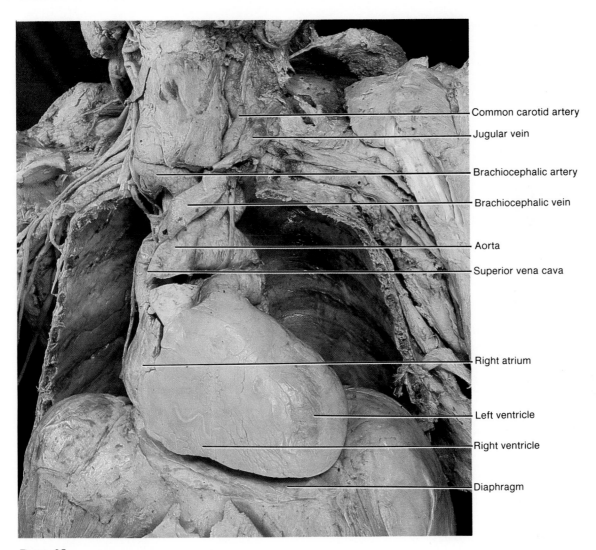

— Common carotid artery
— Jugular vein
— Brachiocephalic artery
— Brachiocephalic vein
— Aorta
— Superior vena cava
— Right atrium
— Left ventricle
— Right ventricle
— Diaphragm

PLATE 25
Thorax with the lungs removed, ventral view.

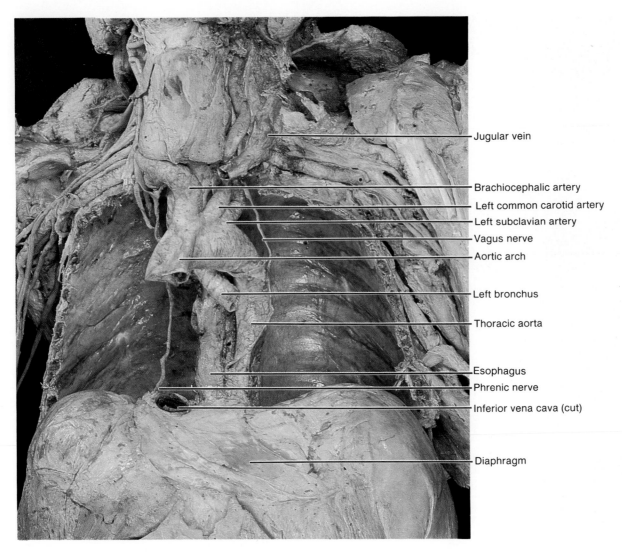

- Jugular vein
- Brachiocephalic artery
- Left common carotid artery
- Left subclavian artery
- Vagus nerve
- Aortic arch
- Left bronchus
- Thoracic aorta
- Esophagus
- Phrenic nerve
- Inferior vena cava (cut)
- Diaphragm

PLATE 26
Thorax with the heart and lungs removed, ventral view.

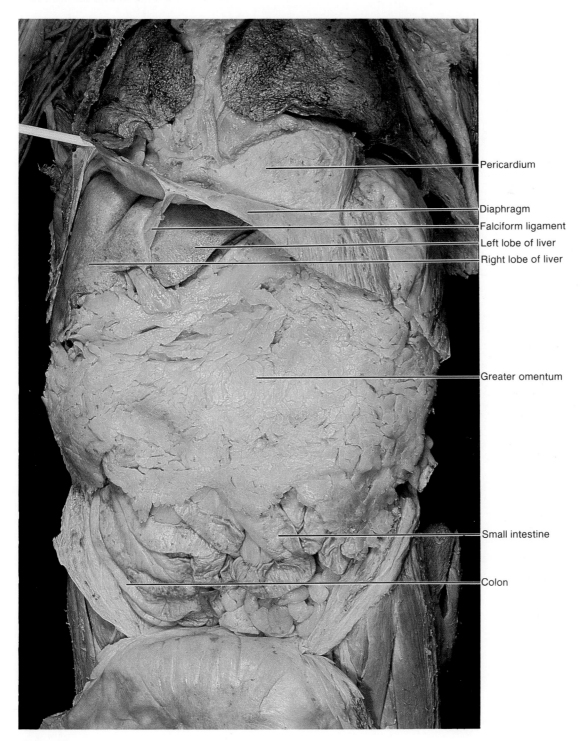

Pericardium

Diaphragm

Falciform ligament

Left lobe of liver

Right lobe of liver

Greater omentum

Small intestine

Colon

PLATE 27
Abdominal viscera, ventral view.

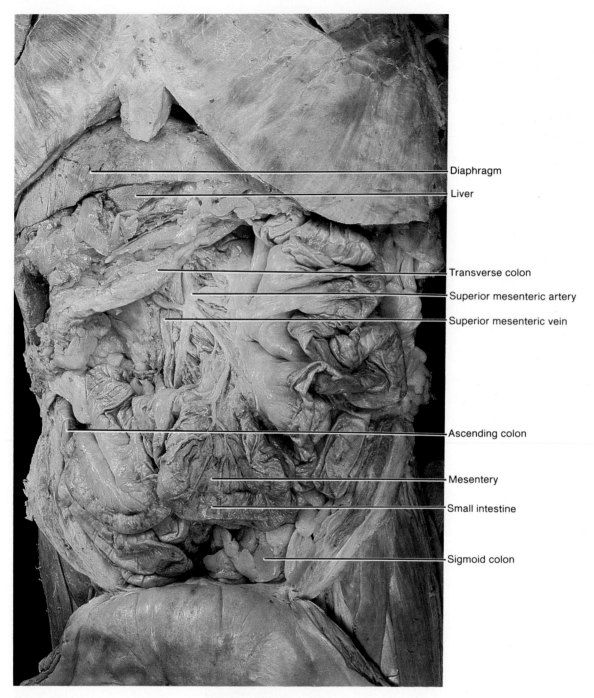

Diaphragm

Liver

Transverse colon

Superior mesenteric artery

Superior mesenteric vein

Ascending colon

Mesentery

Small intestine

Sigmoid colon

PLATE 28

Abdominal viscera with the greater omentum removed. (Small intestine has been displaced to the left.)

Readings

Chapter 1

Asimov, I. 1964. *A short history of biology.* Garden City, N.Y.: Natural History Press.
————. 1982. *Asimov's biographical encyclopedia of science and technology.* New York: Doubleday. (Interesting reading about physiologists and other scientists.)
Borror, D. J. 1960. *Dictionary of word roots and combining forms.* Palo Alto, Calif.: Mayfield.
Cannon, W. B. 1932. *The wisdom of the body.* New York: W. W. Norton. (Cannon's classic work on homeostasis.)
Dubos, R. 1974. *Beast or angel?* New York: Charles Scribner's Sons.
Harre, R. 1981. *Great scientific experiments.* Oxford, England: Oxford University Press. (Good discussions of some experiments that led to significant advances in science.)
Karlen, A. 1984. *Napoleon's glands and other ventures in biohistory.* New York: Warner Books. (An interesting book on how human physiology has affected the course of history.)
Katzir, A. 1989. Optical fibers in medicine. *Scientific American* 260, no. 5 (May):120.
Marx, J. L. 1988. The 1988 Nobel Prize for physiology or medicine. *Science* 242 (October 28):516.
Mayr, O. 1970. The origins of feedback control. *Scientific American* 223, no. 4:111.
Singer, S. and Hilgard, H. R. 1978. *The biology of people.* San Francisco: W. H. Freeman. (An evolutionary perspective on human physiology.)

Chapter 2

Caplan, A. I. 1984. Cartilage. *Scientific American* 251, no. 4 (October):84. (The properties of cartilage protein are described in detail.)
Clark, I. A. 1986. Tissue damage caused by free oxygen radicals. *Pathology* 18 (April):181 . (A discussion of how one way radiation damages cells.)
Cobb, C. E. 1989. Living with radiation. *National Geographic* 175, no. 4 (April):403. (An excellent review of the principles of radiation and its benefits and hazards.)
Darnell, J. E., Jr. 1985. RNA. *Scientific American* 253, no. 4 (October):68.
Doolittle, R. F. 1985. Proteins. *Scientific American* 253, no. 4 (October):88.
Felsenfeld, G. 1985. DNA. *Scientific American* 253, no. 4 (October):58 .
Hall, E. J. et al. 1988. Basic radiobiology. *American Journal of Clinical Oncology* 11, no. 3 (June):220.
James, M. L. and Schreck, J. O. 1982. *General, organic, and biological chemistry.* Lexington, Mass.: D. C. Heath. (A good basic chemistry text.)
Marx, J. L. 1987. Oxygen free radicals linked to many diseases. *Science* 235 (January 30):529.
Montgomery, R.; Dryer, R. L.; Conway, T. W.; and Spector, A. A. 1983. *Biochemistry: a case-oriented approach.* 4th ed. St. Louis: C. V. Mosby. (A good reference for further information about biochemical processes in the human body.)

Chapter 3

Allen, R. D. 1987. The microtubule as an intracellular engine. *Scientific American* 256, no. 2 (February):42.
Bretscher, M. S. 1985. The molecules of the cell membrane. *Scientific American* 253, no. 4 (October):100.
Brodsky, F. M. 1988. Living with clathrin: Its role in intracellular membrane traffic. *Science* 242 (December 9):1396.

Edwards, D. D. 1988. A microscopic movable feat. *Science News* 133 (January 23):53. (A readable article about cell movements.)
————. 1988. Stopping the deadly invasion of cancer. *Science News* 133 (January 16):37. (Some new approaches to cancer therapy.)
Feldman, M. and Eisenbach, L. 1988. What makes a tumor cell metastatic? *Scientific American* 259, no. 5 (November):60.
Kartner, N. and Ling, V. 1989. Multidrug resistance in cancer. *Scientific American* 260, no. 3 (March):44.
Mahendra, K. J. 1988. *Introduction to biological membranes.* New York: John Wiley and Sons. (A detailed account of what is known about membranes.)
Marx, J. 1988. Cell growth control takes balance. *Science* 239 (February 26):975. (Considers excess stimulation and deficient inhibition of cell division.)
————. 1989. How cancer cells spread in the body. *Science* 244 (April 14):147.
————. 1989. How DNA viruses may cause cancer. *Science* 243 (February 24):1012.
————. 1989. The cell cycle coming under control. *Science* 245 (July 21):252.
Pool, R. 1990. Closing the gap between proteins and DNA. *Science* 248 (June 29):1609).
Radman, M. and Wagner, R. 1988. The high fidelity of DNA duplication. *Scientific American* 259, no. 2 (August):40.
Rosenberg, S. A. 1990. Adoptive immunotherapy for cancer. *Scientific American* 262, no. 5 (May):62.
Ross, J. 1989. The turnover of messenger RNA. *Scientific American* 260, no. 4 (April):48.
Rothman, J. E. 1985. The compartmental organization of the Golgi apparatus. *Scientific American* 253, no. 3 (September):74.
Weintraub, H. M. 1990. Antisense RNA and DNA. *Scientific American* 262, no. 1 (January):40.

Chapter 4

Cormack, D. H. 1987. *Ham's histology.* 9th ed. Philadelphia: J. B. Lippincott. (A well written histology text.)
Craiemyle, M. B. L. 1986. *Color atlas of histology.* Chicago: Year Book Medical Publishers. (Attractive illustrations.)
Eyre, D. R. 1980. Collagen: Molecular diversity in the body's protein scaffold. *Science* 207 (March 21):1315.
Hertzberg, E. L.; Lawrence, T. S.; and Gilula, N. B. 1981. Gap junctional communication. *Annual Review of Physiology* 43:479.
Kuehl, F. A. and Egan, R. W. 1980. Prostaglandins, arachidonic acid, and inflammation. *Science* 210 (November 28):978.
Leeson, C. R.; Leeson, T. S.; and Paparo, A. A. 1985. *Textbook of histology.* 5th ed. Philadelphia: Saunders.
Meyer, D. B. and Ross, M. H. 1985. *Laboratory guide for human histology.* Detroit: Wayne State University Press.
Miller, J. A. 1985. Between the cells: control by glue. *Science News* 128 (July 20):36.
Ross, M. H. and Reith, E. 1985. *Histology: A text and atlas.* New York: Harper and Row.
Satir, B. 1975. The final steps in secretion. *Scientific American* 233, no. 4 (October):29.
Staehelin, L. A. and Hull, B. E. 1978. Junctions between living cells. *Scientific American* 238, no. 5 (May):141.
Vuorio, E. 1986. Connective tissue diseases: Mutations of collagen genes. *Annals of Clinical Research* 18:234.

INTERLUDE 1

Baker, J. J. W. 1968. *Hypothesis, prediction, and implication in biology.* Reading, Mass.: Addison-Wesley.

Harre, R. 1983. *Great scientific experiments.* Oxford, Eng.: Oxford University Press.

CHAPTER 5

Demling, R. H. 1985. Burns. *New England Journal of Medicine* 313, no. 22 (November 28):1389.

Joffe, S. N. and Schroder, T. 1987. Lasers in general surgery. *Advances in Surgery* 20:125.

Kligman, L. H. and Kligman, A. M. 1986. The nature of photoaging: Its prevention and repair. *Photodermatology* 3:215.

Maugh, T. H. 1978. Hair: A diagnostic tool to complement blood serum and urine. *Science* 202 (December 22):1271.

National Institutes of Health. 1978. Burn care research. *News and Features from NIH.* Bethesda, Md.: U.S. Department of Health, Education, and Welfare (November).

Pihl, R. O. and Parkes, M. 1977. Hair elements content in learning disabled children. *Science* 198 (October 14):204.

Vahlquist, A. and Berne, Berit. 1986. Sunlight, vitamin A, and the skin. *Photodermatology* 3:203.

Wise, R. 1988. Wrestling with wrinkles. *Science News* 134 (September 24):200.

Young, A. R. 1987. The sunburn cell. *Photodermatology* 4:127.

CHAPTER 6

Bora, F. W. 1987. Joint physiology, cartilage metabolism, and the etiology of osteoarthritis. *Hand Clinics* 3, no. 3 (August):325.

Christakos, S. and Norman, A. W. 1978. Vitamin D_3–induced calcium binding protein in bone tissue. *Science* 202 (October 6):70.

Hamerman, D. 1989. The biology of osteoarthritis. *New England Journal of Medicine* 320, no. 20 (May 18):1322.

Hayward, M. and Fiedler-Nagy, C. 1987. Mechanisms of bone loss: Rheumatoid arthritis, periodontal disease, and osteoporosis. *Agents and Actions* 22, no. 3/4:251.

Kolata, G. 1986. How important is dietary calcium in preventing osteoporosis? *Science* 233 (August 1):519.

Parfitt, A. M. 1987. Bone and plasma calcium homeostasis. *Bone* 8, Suppl. 1:S1.

Resnick, D. 1988. Paget disease of bone: Current status and a look back to 1943 and earlier. *American Journal of Roentgenology* 150 (February):249.

Vuorio, E. 1986. Connective tissue diseases: Mutations of collagen genes. *Annals of Clinical Research* 18:234.

CHAPTER 7

Evans, F. G. 1957. *Atlas of human anatomy.* Towata, N.J.: Rowman and Littlefield.

Gray, H. 1973. *Anatomy of the human body.* 29th American ed. ed. C. M. Goss. Philadelphia: Lea and Febiger.

Levin, S. M. 1981. The icosahedron as a biological support system. Presented at the 34th Annual Conference on Engineering in Medicine and Biology, Houston, Texas, September 21–23.

Moore, K. L. 1985. *Clinically oriented anatomy.* 2nd ed. Baltimore: Williams and Wilkins.

Schlossberg, L. 1977. *The Johns Hopkins atlas of human functional anatomy.* ed. G. D. Zuidema. Baltimore: Johns Hopkins Press.

Snell, R. S. 1978. *Atlas of clinical anatomy.* Boston: Little, Brown.

CHAPTER 8

Gray, H. 1973. *Anatomy of the human body.* 29th American ed. ed. C. M. Goss. Philadelphia: Lea and Febiger.

Hamerman, D. 1989. The biology of osteoarthritis. *New England Journal of Medicine* 320, no. 20 (May 18):1322.

Harris, E. D., Jr. 1990. Rheumatoid arthritis. *New England Journal of Medicine* 332, no. 18 (May 3):1277.

Ross, M. H. and Reith, E. 1985. *Histology: A text and atlas.* New York: Harper and Row.

Schlossbert, L. 1978. *The Johns Hopkins atlas of human functional anatomy.* ed. G. D. Zuidema. Baltimore: Johns Hopkins Press.

Sonstegard, D. A.; Matthews, L. S.; and Kaufer, H. 1978. The surgical replacement of the human knee joint. *Scientific American* 238, no. 1 (January):44.

Vaughan, J. A. 1984. Rheumatoid arthritis: Evidence of a defect in T-cell function. *Hospital Practice* 19:101.

Woodley, D. E.; Harris, E. J.; Jr.; Mainardi, C. L.; and Brickerhoff, C. E. 1978. Collagenase immunolocalization in cultures of rheumatoid synovial cells. *Science* 200 (May 19):773.

CHAPTER 9

Appell, H. J. 1986. Skeletal muscle atrophy during immobilization. *International Journal of Sports Medicine* 7:1. (A readable article.)

Arnason, B. G. W. 1987. Myasthenia gravis conference: A summation. *Annals of the New York Academy of Sciences* 505:683. (A summary of current research findings and questions.)

Bolton, T. B. 1986. Calcium metabolism in vascular smooth muscle. *British Medical Bulletin* 42, no. 4:421.

Brenner, B. 1987. Mechanical and structural approaches to correlation of cross-bridge action in muscle with actomyosin. *Annual Review of Physiology* 49:655.

Carafoli, E. and Penniston, J. T. 1985. The calcium signal. *Scientific American* 253, no. 5 (November):70.

Cormack, D. H. 1987. *Ham's histology.* 9th ed. Philadelphia: Lippincott.

Cureton, K. J.; Collins, M. A.; Will, D. W.; and McElhannon, F. M. 1988. Muscle hypertrophy in men and women. *Medicine and Science in Sports and Exercise* 20, no. 4:338.

Edwards, D. D. 1988. Muscular dystrophy protein identified. *Science News* 133 (January 2):4. (A short summary of the status of research on muscular dystrophy.)

Herbison, G. J. and Talbot, J. M. 1985. Muscle atrophy during space flight: Research needs and opportunities. *The Physiologist* 28, no. 6:520.

Katz, A. and Sahlin, K. 1988. Regulation of lactic acid production during exercise. *Journal of Applied Physiology* 65, no. 2:509.

Lullmann, H. and Ziegler, A. 1987. Calcium, cell membrane, and excitation-contraction coupling. *Journal of Cardiovascular Pharmacology* 10, Suppl. 1:52.

Marshall, E. 1988. The drug of champions. *Science* 242 (October 14):183.

Stull, J. T.; Kamm, K. E.; and Taylor, D. A. 1988. Calcium control of smooth muscle contractility. *American Journal of Medical Sciences* 296, no. 4:241.

Sturek, M. and Hermsmeyer, K. 1986. Calcium and sodium channels in spontaneously contracting vascular muscle cells. *Science* 233 (July 25):475.

CHAPTER 10

Astrand, P. O. and Rodahl, K. 1977. *Textbook of work physiology: physiological basis of exercise.* 2nd ed. New York: McGraw-Hill.

Fackelmann, K. A. 1990. Hormone may restore muscle in elderly. *Science News* 138 (July 14):23.

Gallistel, C. R. 1980. From muscles to motivation. *Scientific American* 68 (July/August):398.

Hinson, M. M. 1973. *Kinesiology.* 2nd ed. Dubuque, Iowa: Wm. C. Brown.

Holloszy, J. O. 1976. Adaptations of muscular tissue to training. *Progress in Cardiovascular Diseases* 18, no. 6:445.

Merriman, J. E. 1978. Exercise prescriptions for apparently healthy individuals and for cardiac patients. In Exercise and the heart, ed. N. K. Wenger. *Cardiovascular Clinician* 9, no. 3:81.

Moore, K. L. 1985. *Clinically oriented anatomy.* 2nd ed. Baltimore: Williams and Wilkins.

Schlossberg, L. 1977. *The Johns Hopkins atlas of human functional anatomy.* ed. G. D. Zuidema. Baltimore: Johns Hopkins Press.

INTERLUDE 2

Beil, L. 1988. Of joints and juveniles. *Science News* 134 (September 17):190.

Cantwell, J. D. 1985. Cardiovascular aspects of running. *Clinics in Sports Medicine* 4, no. 4 (October):627.

Chillag, S. A. 1986. Endurance athletes: Physiologic changes and nonorthopedic problems. *Southern Medical Journal* 79, no. 10 (October):1264.

Herring, S. A. and Hilson, K. L. 1987. Introduction to overuse injuries. *Clinics in Sports Medicine* 6, no. 2 (April):225.

Krucoff, C. 1988. You don't have to be a fanatic to be fit. *Washington Post/Health* (November 1).

Ozolin, P. Blood flow in the extremities of athletes. *International Journal of Sports Medicine* 7:117.

Rippe, J. M. et al. 1988. Walking for health and fitness. *Journal of the American Medical Association* 259, no. 18 (May 13):2720.

CHAPTER 11

Bioelectricity References

Bassett, C. 1982. Pulsing electromagnetic fields: A new method to modify cell behavior in calcified and non-calcified tissues. *Calcified Tissue International* 34:1.

Becker, R. O. and Marino, A. A. 1982. *Electromagnetism and Life.* Albany: State University of New York Press.

Borgens, R. B. 1984. Endogenous ionic currents traverse intact and damaged bone. *Science* 225:478–82.

Goldstein, N. N. and Free, M. J. 1979. *Foundations of physiological instrumentation.* Springfield: C. C. Thomas.

Illingworth, C. M. and Banker, C. M. 1980. Measurement of electrical currents during the regeneration of amputation finger tips in children. *Clinical Physical and Physiological Measurements* 1, no. 1:87–89.

Mannheimer, J. S. and Lampe, G. N. 1984. *Clinical Transcutaneous Electrical Nerve Stimulation.* Philadelphia: Davis Co.

Marino, A. A., ed. 1988. *Modern Bioelectricity.* New York: Marcel Dekker.

O'Connor, M. E. and Lovely, R. H., eds. 1988. *Electromagnetic Fields and Neurobehavioral Function.* New York: Alan R. Liss.

Robinson, K. R. and McCaig. 1980. Electrical fields, calcium gradients and cell growth. In *Growth Regulation by Ion Fluxes,* ed. H. L. Leffert, vol. 339:132–38. New York: New York Academy of Sciences.

Spanswick, R. M. and Bennett, A. B. 1988. Electrogenic ion transport in higher plants. In *Electrogenic transport: fundamental principles and physiological implications,* ed. M. P. Blaustein and M. L. Leiberman. New York: Raven.

United States Congress. 1987. House. Committee on Interior and Insular Affairs, Subcommittee on Water and Power Resources. Oversight hearing on health effects of transmission lines.

General References

Bloom, F. E. 1988. Neurotransmitters: Past, present, and future directions. *FASEB Journal* 2, no. 1 (January):32. (A good review of a complex topic.)

Dick, G. and Gay, D. 1988. Multiple sclerosis— Autoimmune or microbial? *Journal of Infection* 16:25.

DiPolo, R. and Beauge, L. 1987. The squid axon as a model for studying plasma membrane mechanisms for calcium regulation. *Hypertension* 10, Suppl. 1:1–15.

Dunant, Y. and Israel, M. 1985. The release of acetylcholine. *Scientific American* 252, no. 4 (April):58.

Fackelmann, K. A. 1990. Myelin on the mend. *Science News* 137 (April 7):218.

Fine, A. 1986. Transplantation in the central nervous system. *Scientific American* 255, no. 2 (August):52.

Gunning, R. 1987. Increased numbers of ion channels promoted by intracellular second messenger. *Science* 235 (January 2):80.

Kimelberg, H. K. and Norenberg, M. D. 1989. Astrocytes. *Scientific American* 260, no. 4 (April):66.

Kromer, L. F. 1987. Nerve growth factor treatment after brain injury prevents neuronal death. *Science* 235 (January 9):214.

Levi-Montalcini, R. 1987. The nerve-growth factor 35 years later. *Science* 237 (September 4):1154.

Marx, J. L. 1986. Nerve growth factor acts in brain. *Science* 232 (June 13):1341.

———. 1990. Marijuana receptor gene cloned. *Science* 249 (August 10):624.

Miller, R. J. 1987. Multiple calcium channels and neuronal function. *Science* 235 (January 2):46.

Nicoll, R. A. 1988. The coupling of neurotransmitter receptors to ion channels in the brain. *Science* 241 (July 29):545.

Nottebohm, F. 1989. From birdsong to neurogenesis. *Scientific American* 260, no. 2 (February):74.

Stanfield, P. R. 1986. Voltage-dependent calcium cannnels of excitable membranes. *British Medical Bulletin* 42, no. 4:359.

Thoenen, H. and Edgar, D. 1985. Neurotropic factors. *Science* 229 (July 19):238.

Wernig, A. and Herrera, A. A. 1986. Sprouting and remodeling at the nerve-muscle junction. *Progress in Neurobiology* 27:251.

Zucker, R. S. and Lando, L. 1986. Mechanism of transmitter release: Voltage hypothesis and calcium hypothesis. *Science* 231 (February 7):574.

CHAPTER 12

Alkon, D. L. 1989. Memory storage and neural systems. *Scientific American* 261, no. 1 (July):42.

Aoki, C. and Siekevitz, P. 1988. Plasticity in brain development. *Scientific American* 259, no. 6 (December):56.

Barinaga, M. 1990. The tide of memory, turning. *Science* 248 (June 29):1603.

Barnes, D. M. 1987. Defect in Alzheimer's is on chromosome 21. *Science* 235 (February 20):846.

Bower, B. 1988. Chaotic connections. *Science News* 133 (January 23):58. (Readable new ideas about learning and memory.)

———. 1988. Epileptic PET probes. *Science News* 133 (April 30):280. (Radical surgery for epilepsy.)

Castaño, E. M. and Frangione, B. 1988. Human amyloidosis, Alzheimer disease, and related disorders. *Laboratory Investigations* 58, no. 2:122.

Cohen, M. L. 1987. Neural tube defects: Epidemiology, detection, and prevention. *Journal of Obstetrical, Gynecological, and Neonatal Nursing* 16, no. 2:105. (A useful discussion of relatively common birth defects.)

Delabar, J. et al. 1987. Beta amyloid gene duplication in Alzheimer's disease and karyotypically normal Down syndrome. *Science* 235 (March 13):1390.

Gabel, S. 1987. Informational processing in rapid-eye movement sleep: Possible neurological, neuropsychological, and clinical correlates. *Journal of Nervous and Mental Disease* 175, no. 4 (April):193.

Gazzaniga, M. S. 1989. Organization of the human brain. *Science* 245 (September 1):957.

Gibb, W. R. G. and Lees, A. J. 1986. The restless legs syndrome. *Postgraduate Medical Journal* 62:329.

Gibbons, A. 1990. New maps of the human brain. *Science* 249 (July 13):122.

Goelet, P. et al. 1986. The long and short of long-term memory—a molecular framework. *Nature* 322 (July 21):419. (A short, readable article on memory.)

Kinsbourne, M. 1987. Brain mechanisms and memory. *Human Neurobiology* 6:81.

Kolata, G. 1987. Manic-depression gene tied to chromosome 11. *Science* 235 (March 6):1139. (An account of research progress in relating mental illness to biochemical defects.)

Lemire, R. J. 1988. Neural tube defects. *Journal of the American Medical Association* 259, no. 4 (January 22/29):588.

Marx, J. 1987. Role of Alzheimer's protein is tangled. *Science* 238 (December 4):1352. (One kind of Alzheimer's disease may be inherited.)

Miller, J. A. 1987. Crazy-quilt brain. *BioScience* 37, no. 10 (November):701. (A readable summary of relationships between cerebral cortex structure, perception, and thought.)

Mishkin, M. and Appenzeller, T. 1987. The anatomy of memory. *Scientific American* 256, no. 6 (June):80.

Pennisi, E. 1989. Neurobiology gets computational. *BioScience* 39, no. 5 (May):283.

Roberts, G. W. and Crow, T. J. 1987. The neuropathology of schizophrenia—a progress report. *British Medical Bulletin* 43, no. 3:599.

Selkoe, D. J. 1990. Deciphering Alzheimer's disease: The amyloid precursor protein yields new clues. *Science* 248 (June 1):1058.

Spector, R. and Johanson, C. E. 1989. The mammalian choroid plexus. *Scientific American* 261, no. 5 (November):68.

Tulving, E. 1989. Remembering and knowing the past. *American Scientist* 77 (July–August):361.

Wurtman, R. J. 1988. Effects of dietary amino acids, carbohydrates, and choline on neurotransmitter synthesis. *Mount Sinai Journal of Medicine* 55, no. 1 (January):75.

CHAPTER 13

Ditmars, D. M. and Houin, H. P. 1986. Carpal tunnel syndrome. *Hand Clinics* 2, no. 3 (August):525.

Duvoisin, R. C. 1986. Etiology of Parkinson's disease: Current concepts. *Clinical Neuropharmacology* 9, Suppl. 1:S3.

Edstrom, L. and Grimby, L. 1986. Effect of exercise on the motor unit. *Muscle and Nerve* 9 (February):104.

Lee, R. G. 1987. Physiology of the basal ganglia and pathophysiology of Parkinson's disease. *Canadian Journal of Neurological Science* 14:373.

Levin, R. 1986. Age factors loom in parkinsonian research. *Science* 234 (December 5):1200.

———. 1987. Brain grafts benefit Parkinson's patients. *Science* 236 (April 10):149.

———. 1987. Dramatic results with brain grafts. *Science* 237 (July 17):245.

Morell, P. and Norton, W. T. 1980. Myelin. *Scientific American* 242, no. 5 (May):88.

Poggio, T. and Koch, C. 1987. Synapses that compute motion. *Scientific American* 256, no. 5 (May):46.

Sieb, R. A. 1987. Skeletomotor subsystems. *Medical Hypotheses* 24:303.

Tetrud, J. W. and Langston, J. W. 1989. The effect of deprenyl (Selegiline) on the natural history of Parkinson's disease. *Science* 245 (August 4):519.

Weiss, R. 1990. Drug reduces paralysis after spinal injury. *Science News* 137 (April 7):212.

CHAPTER 14

Burnstock, G. 1986. Autonomic neuromuscular junctions: Current developments and future directions. *Journal of Anatomy* 146:1.

Diz, D. I.; Barnes, K. L.; and Ferrario, C. M. 1987. Functional characteristics of neuropeptides in the dorsal medulla oblongata and vagus nerve. *Federation Proceedings* 46, no. 1 (January):30.

Einhorn, D.; Young, J. B.; and Landsberg, L. 1982. Hypotensive effect of fasting: Possible involvement of the sympathetic nervous system and endogenous opiates. *Science* 217 (August 20):727.

Ekman, P.; Levenson, R. W.; and Priesen, W. V. 1983. Autonomic nervous system activity distinguishes among emotions. *Science* 221 (September 16):1208.

Ewing, D. J. and Clarke, B. F. 1986. Autonomic neuropathy: Its diagnosis and prognosis. *Clinics in Endocrinology and Metabolism* 15, no. 4 (November):855.

Frishman, W. H. 1982. Beta-adrenergic blockade in clinical practice. *Hospital Practice* 17, no. 9 (September):57.

Giesecke, M. E. 1987. Panic disorder in university students: A review. *Journal of American College Health* 36, no. 3 (November):149. (Episodic panic attacks may be due to sudden bursts of autonomic activity.)

Marley, P. and Livett, B. G. 1987. Neuropeptides in the autonomic nervous system. *CRC Critical Reviews in Clinical Neurobiology* 1, no. 3:201.

Sheppard, D. 1987. Physiology of the parasympathetic nervous system of the lung. *Postgraduate Medical Journal* 63, Suppl. 1:21.

Vingerhoets, A. J. J. M. 1985. The role of the parasympathetic division of the autonomic nervous system in stress and the emotions. *International Journal of Psychosomatics* 32, no. 3:28.

CHAPTER 15

Besson, J. and Chaouch, A. 1987. Peripheral and spinal mechanisms of nociception. *Physiological Reviews* 67, no. 1:67.

Boulant, J. A. and Dean, J. B. 1986. Temperature receptors in the central nervous system. *Annual Review of Physiology* 48:639.

Brou, P. et al. 1986. The colors of things. *Scientific American* 255, no. 3 (September):84. (Some new ideas about color vision.)

Edmeads, J. G. 1988. Migraine. *Canadian Medical Association Journal* 138, no. 2:107.

Finke, R. A. 1986. Mental imagery and the visual system. *Scientific American* 254, no. 3 (March):88.

Girvin, J. P. 1988. Current status of artificial vision by electrocortical stimulation. *Canadian Journal of Neurological Sciences* 15, no. 11:58.

Harrison, M. and Cotanch, P. H. 1987. Pain: Advances and issues in critical care. *Nursing Clinics of North America* 22, no. 3 (September):691.

Harrison, R. V. 1987. Cochlear implants: A review of the principles and important physiological factors. *Journal of Otolaryngology* 16, no. 5:268. (Progress is being made in the treatment of nerve deafness.)

He, L. 1987. Involvement of endogenous opioid peptides in acupuncture analgesia. *Pain* 31:99.

Kiester, E., Jr. 1987. Doctors close in on the mechanisms behind headache. *Smithsonian* 18, no. 9 (December):175.

Livingstone, M. and Hubel, D. 1988. Segregation of form, color, movement, and depth: Anatomy, physiology, and perception. *Science* 240 (May 6):740.

Pittaway, K. M. and Hill, R. G. 1987. Cholecystokinin and pain. *Pain and Headache* 9:213.

Schnapf, J. L. and Baylor, D. A. 1987. How photoreceptors respond to light. *Scientific American* 256, no. 4 (April):40.

Snyder, S. H. 1986. Neuronal receptors. *Annual Review of Physiology* 48:461.

Spray, D. C. 1986. Cutaneous temperature receptors. *Annual Review of Physiology* 48:625.

Suberg, S. N. and Watkins, L. R. 1987. Interaction of cholecystokinin and opioids in pain modulation. *Pain and Headache* 9:247.

Stryer, L. 1987. The molecules of visual excitation. *Scientific American* 257, no. 1 (July):42. (A readable account of how rods convert light to neural signals.)

Vaughan, C. 1988. A new view of vision. *Science News* 134 (July 23):58.

Weisburd, S. 1988. Computer scents. *Science News* 133 (January 9):27. (A new way to study the sense of smell.)

Chapter 16

Barnes, D. M. 1986. Steroids may influence changes in mood. *Science* 232:1344.

Bello, M. 1985. Testing the effects of growth hormone-releasing hormone. *Research Resources* 9, no. 10:1. (A publication of the Public Health Service.)

Evans, R. M. 1988. The steroid and thyroid hormone receptor superfamily. *Science* 240 (May 13):889. (Important new findings about mechanisms of hormone action.)

Flier, J. S.; Underhill, L. H.; and Laragh, J. H. 1985. Atrial natriuretic hormone, the renin-aldosterone axis, and blood pressure-electrolyte homeostasis. *New England Journal of Medicine* 313, no. 21 (November 21):1330.

Gibbs, D. M. 1986. Vasopressin and oxytocin: Hypothalamic modulators of the stress response: A review. *Psychoneuroendocrinology* 11, no. 2:131.

Hall, A. 1990. The cellular functions of small GTP-binding proteins. *Science* 249 (August 10):635.

Hanley, R. M. and Steiner, A. L. 1989. The second-messenger system for peptide hormones. *Hospital Practice* 24, no. 8 (August 15):59. (Well-written, up-to-date article.)

Kohrle, J.; Brabant, G.; and Hesch, R. D. 1987. Metabolism of thyroid hormones. *Hormone Research* 26:58.

Laragh, J. H. 1985. Atrial natriuretic hormone, the renin-aldosterone axis, and blood pressure-electrolyte homeostasis. *New England Journal of Medicine* 313, no. 21 (November 21):1330. (A good overview of some important regulatory mechanisms.)

Lechan, R. M. 1987. Neuroendocrinology of pituitary hormone regulation. *Endocrinology and Metabolism Clinics* 16, no. 3:475.

Majewska, M. D. 1987. Steroids and brain activity: Essential dialogue between body and mind. *Biochemical Pharmacology* 22:3781. (Steroids appear to act at cell membranes as well as in nuclei.)

McGiff, J. C. 1987. Arachidonic acid metabolism. *Preventive Medicine* 16:503.

Rasmussen, H. 1989. The cycling of calcium as an intracellular messenger. *Scientific American* 261, no. 4 (October):66. (Interesting explanation of how calcium works with other messengers.)

Seyle, H. 1973. The evolution of the stress concept. *American Scientist* 61 (November–December):692. (Seyle's own description of his work.)

Tracey, K. J.; Lowry, S. F.; and Cerami, A. 1988. Cachectin: A hormone that triggers acute shock and chronic cachexia. *Journal of Infectious Diseases* 157, no. 3:413.

Weiss, R. 1988. Bone density drops with thyroid therapy. *Science News* 133 (June 4):359. (An interesting news item.)

Interlude 3

Ayensu, E. S. and Whitfield, P., eds. 1981. *The rhythms of life.* New York: Crown Publishers.

Brady, J., ed. 1982. *Biological timekeeping.* Cambridge, Eng.: Cambridge University Press.

Brown, F. A., Jr. and Graeber, R. C., eds. 1982. *Rhythmic aspects of behavior.* Hillsdale, N. J.: Lawrence Erlbaum Associates.

Cziesler, C. A. et al. 1986. Bright light resets the human circadian pacemaker independent of the timing of the sleep-wake cycle. *Science* 233 (August 8):667.

Griffiths, R. A. 1986. Natural environmental clues and circadian rhythms of behavior—a perspective. *Chronobiology International* 3, no. 4:247.

Hrushesky, W. J. M. 1985. Circadian timing of cancer chemotherapy. *Science* 228 (April 5):73.

Levy, A. J. et al. 1987. Antidepressant and circadian phase-shifting effects of light. *Science* 235 (January 16):352.

Medon, P. J. and Holshouser, M. H. 1984. Pharmacological implications of biological rhythms. *U. S. Pharmacist* 9, no. 9 (September):43.

Minors, D. S. and Waterhouse, J. M. 1981. *Circadian rhythms and the human.* Littleton, Mass.: John Wright. (A good introduction to circadian rhythms.)

———. 1987. Chronobiochemistry: An overview of circadian rhythms and some applications to clinical medicine and biochemistry. *Annals of Clinical Biochemistry* 24, suppl.:S1.

Moore-Ede, M. C.; Sulzman, F. M.; and Fuller, C. A. 1982. *The clocks that time us: Physiology of the circadian timing system.* Cambridge, Mass.: Harvard University Press.

Reppert, S. M.; Weaver, D. R, Rivkees, S. A.; and Stopa, E. G. 1988. Putative melatonin receptors in a human biological clock. *Science* 242 (October 7):78.

Takahashi, J. S. and Zatz, M. 1982. Regulation of circadian rhythmicity. *Science* 217 (September 17):1104.

Chapter 17

Cluose, L. H. and Comp, P. C. 1986. The regulation of hemostasis: The protein C system. *New England Journal of Medicine* 314, no. 20:1298.

Collen, D. and Lijnen, H. R. 1986. Tissue-type plasminogen activator. *Haemostasis* 16, suppl. 3:25. (TPA is an important component of emergency treatment for heart attack victims.)

Esmon, C. T. 1987. The regulation of natural anticoagulant pathways. *Science* 235 (March 13):1348.

Faithfull, N. S. 1987. Fluorocarbons. *Anaesthesia* 42:234. (Describes the current status of work with artificial blood.)

Golde, D. W. and Gasson, J. C. 1988. Hormones that stimulate the growth of blood cells. *Scientific American* 259, no. 1 (July):62. (A major advance in understanding blood cells with great potential for therapy.)

Gonias, S. L. and Pizzo, S. V. 1986. The biochemistry of hemostasis. *Clinical Laboratory Haematology* 8:281.

Haber, E.; Quertemous, T.; Matsueda, G. R.; and Runge, M. S. 1989. Innovative approaches to plaminogen activator therapy. *Science* 243 (January 6):51.

Jacobson, E. J. 1988. Chronic mononucleosis—it almost never happens. *Postgraduate Medicine* 83, no. 1:56.

Kennedy, J. W. 1987. Streptokinase for the treatment of acute myocardial infarction: A brief review of randomized trials. *Journal of the American College of Cardiology* 10, no. 5, suppl.B (November):28B.

Koury, M. J. and Bondurant, M. C. 1990. Erythropoietin retards DNA breakdown and prevents programmed death in erythroid progenitor cells. *Science* 248 (April 20):378.

Loupe, D. E. 1989. Breaking the sickle cycle. *Science News* 136 (December 2):360.

Marder, V. J. and Francis, C. W. 1987. *Physiological balance of haemostasis and bleeding. Drugs* 33, suppl. 3:13.

Talbot, J. M. and Fisher, K. D. 1986. Influence of space flight on red blood cells. *Federation Proceedings* 45:2285.

Tobi, M. and Straus, S. E. 1988. Chronic mononucleosis—a legitimate diagnosis. *Postgraduate Medicine* 83, no. 1:69.

Vaughan, C. 1988. New name and identity for mysterious Epstein-Barr syndrome. *Science News* 133 (March 12):167. (Is Epstein-Barr syndrome real?)

Weiss, R. 1988. New therapy blocks newborn jaundice. *Science News* 133 (April 16):247. (Jaundice will be preventable if new treatment becomes accepted.)

Chapter 18

Acker, M. A. et al. 1987. Skeletal muscle as the potential power source for a cardiovascular pump: Assessment in vivo. *Science* 236 (April 17):324. (An exciting possibility for providing assistance to defective hearts.)

Hoffman, B. F. and Dangman, K. H. 1987. Mechanisms for cardiac arrhythmias. *Experientia* 43:1049.

Lakatta, E. G. 1987. Starling's law of the heart is explained by an intimate interaction of muscle length and myofilament calcium activation. *Journal of the American College of Cardiology* 10, no. 5 (November):1157.

Marx, J. 1990. Holding the line against heart disease. *Science* 248 (June 22):1491.

Noble, D. 1987. Experimental and theoretical work on excitation and excitation-contraction coupling in the heart. *Experientia* 43:1146.

Robinson, T. F.; Factor, S. M.; and Sonnenblick, E. H. 1986. The heart as a suction pump. *Scientific American* 254, no. 6 (June):84. (A new model to explain heart filling—one that contradicts the Frank-Starling law.)

Simmons, K. 1984. Implantable assist pump—final heart option? *Journal of the American Medical Association* 251, no. 6 (February 10):700.

Smith, T. W. 1988. Digitalis: Mechanisms of action and clinical use. *New England Journal of Medicine* 318, no. 6 (February 11):358.

Strauss, M. J. 1984. The political history of the artificial heart. *The New England Journal of Medicine* 310, no. 5 (February 2):332.

Surawicz, B. 1987. Contributions of cellular electrophysiology to the understanding of the electrocardiogram. *Experientia* 43:1061.

Vaughan-Jones, R. D. 1986. Excitation and contraction in the heart: The role of calcium. *British Medical Bulletin* 42, no. 4:413.

Weiss, P. L. 1990. Calcium channels dwindle in old hearts. *Science News* 138 (August 18):100.

CHAPTER 19

Clarkson, T. B. et al. 1987. Mechanisms of atherogenesis. *Circulation* 76, suppl. 1:1–20.

Gray, H. 1973. *Anatomy of the human body.* 29th American ed. ed. C. M. Goss. Philadelphia: Lea and Febiger.

Heymann, M. A.; Iwamoto, H. S.; and Rudolph, A. M. 1981. Factors affecting changes in the neonatal systemic circulation. *Annual Review of Physiology* 43:371.

Johansen, K. 1982. Aneurysms. *Scientific American* 247, no. 1 (July):110.

Kolata, G. 1984. Lowered cholesterol decreases heart disease. *Science* 223 (January 27):381.

Kontos, H. A. 1981. Regulation of cerebral circulation. *Annual Review of Physiology* 43:397.

Moore, K. L. 1985. *Clinically oriented anatomy.* 2nd ed. Baltimore: Williams and Wilkins.

Olsson, R. A. 1981. Local factors regulating cardiac and skeletal muscle blood flow. *Annual Review of Physiology* 43:385.

Rudolph, A. M. 1979. Fetal and neonatal pulmonary circulation. *Annual Review of Physiology* 41:383.

Sparks, H. V. and Belloni, F. L. 1978. Local vascular regulation. *Annual Review of Physiology* 40:67.

CHAPTER 20

Barnes, D. M. 1987. Drug may protect brains of heart attack victims. *Science* 235:632.

Byrne, D. G. 1987. Personality, life events, and cardiovascular disease. *Journal of Psychosomatic Research* 31, no. 6:661.

Cantwell, J. D. 1985. Cardiovascular aspects of running. *Clinics in Sports Medicine* 4, no. 4 (October):627.

Conway, J. 1986. Blood pressure and heart rate variability. *Journal of Hypertension* 4:261.

Eron, C. 1988. Young hearts. *Science News* 134 (October 8):234. (A readable article on preventing heart attacks.)

Folkman, J. and Klagsbrun, M. 1987. Angiogenic factors. *Science* 235 (January 23):442.

Goetz, K. L. 1988. Physiology and pathophysiology of atrial peptides. *American Journal of Physiology* 254, Endocrinology and Metabolism, no. 17:E1.

Heine, H. and Weiss, M. 1987. Life stress and hypertension. *European Heart Journal* 8, suppl. B:45.

Hopkins, P. N. and Williams, R. R. 1986. Identification and relative weight of cardiovascular risk factors. *Cardiology Clinics* 4, no. 1:3.

Kannel, W. B. and Sytkowski, P. A. 1987. Atherosclerosis risk factors. *Pharmacology and Therapeutics* 32:207.

Kruth, H. S. 1985. Platelet-mediated cholesterol accumulation in cultured aortic smooth muscle cells. *Science* 227 (March 8):1243.

Leitschuh, M. and Chobanian, A. 1987. Vascular changes in hypertension. *Medical Clinics of North America* 71, no. 5:827.

Moe, G. W. and Armstrong, P. W. 1988. Congestive heart failure. *Canadian Medical Association Journal* 138 (April 15):689.

Reichen, J. and Lebrec, D. 1987. The effect of drugs on the portal circulation. *Journal of Hepatology* 5:235.

Robertson, D.; Tseng, C.; and Appalsamy, M. 1988. Smoking and mechanisms of cardiovascular control. *American Heart Journal* 115:258.

CHAPTER 21

Ada, G. L. and Nossal, G. 1987. The clonal-selection theory. *Scientific American* 257, no. 2 (August):62.

Amato, I. 1989. Teaching antibodies new tricks. *Science News* 136 (September 2):152. (Some interesting applications for catalytic antibodies.)

American Association for the Advancement of Science. 1990. *Science* 248 (June 15), *Immunology Special Issue.*

American Medical Association. 1987. *Journal of the American Medical Association* 258, no. 20 (November 27), Immunology Special Issue.

Centers for Disease Control. 1983. General recommendations on immunization: Recommendations of the Immunization Practices' Advisory Committee. *Annals of Internal Medicine* 98:615.

Cohen, I. R. 1988. The self, the world, and autoimmunity. *Scientific American* 258, no. 4 (April):52.

Cowen, R. 1990. Protecting tissue from inflammatory attack. *Science News* 138 (July 14):23.

Edelson, R. L. and Fink, J. M. 1985. The immunologic function of skin. *Scientific American* 252, no. 6 (June):46.

Frank, M. M. 1987. Current concepts: Complement in the pathophysiology of human disease. *New England Journal of Medicine* 316, no. 24 (June 11):1525.

Gray, H. M.; Sette, A.; and Buus, S. 1989. How T cells see antigen. *Scientific American* 261, no. 5 (November):56. (A readable discussion of recent advances in the understanding of T cell function.)

Hall, S. S. 1989. Psychoneuroimmunology. *Smithsonian* 20, no. 3:62.

Hochberg, M. C. 1989. NSAIDs: Mechanisms and pathways of action. *Hospital Practice* 24, no. 3:185.

Koshland, D. E. 1987. Frontiers in immunology. *Science* 238 (November 20):1023.

Lerner, R. A. and Tramontano. 1988. Catalytic antibodies. *Scientific American* 258, no. 3 (March):58.

Maddaus, M. A.; Ahrenholz, D.; and Simmons, R. L. 1988. The biology of peritonitis and implications for treatment. *Surgical Clinics of North America* 68, no. 2 (April):431.

Marrack, P. and Kappler, J. 1986. The T cell and its receptor. *Scientific American* 254, no. 2 (February):36.

Marx, J. L. 1985. Antibodies made to order. *Science* 229 (August 2):455.

———. 1987. Histocompatibility restriction explained. *Science* 235:843.

———. 1988. What T cells see and how they see it. *Science* 242 (November 11):863.

———. 1990. Taming rogue immune reactions. *Science* 249 (July 20):246.

McClain, B. R. 1990. Meet the T cell antigen receptor. *American Biology Teacher* 52, no. 5 (May):276.

Stites, D. P.; Stobo, J. D.; and Wells, J. V. 1987. *Basic and clinical immunology.* Los Altos, Calif.: Appleton and Lange.

Thomas, L. 1983. *The youngest science: Notes of a medicine watcher.* New York: Viking.

Tipton, W. R. 1987. Immunotherapy for allergic diseases. *Primary Care* 14, no. 3 (September):623.

Tonegawa, S. 1985. The molecules of the immune system. *Scientific American* 253, no. 4 (October):122.

Weber, R. W. 1987. Allergens. *Primary Care* 14, no. 3 (September):435.

Weiss, S. J. 1989. Tissue destruction by neutrophils. *New England Journal of Medicine* 320, no. 6:365.

Young, J. D. and Cohn, Z. A. 1988. How killer cells kill. *Scientific American* 258, no. 1 (January):38.

INTERLUDE 4

AIDS Special Issue. 1988. *Science* 239 (February 5):573–621.

American Medical Association. 1987. *Journal of the American Medical Association* 258, no. 20 (November 27), Immunology Special Issue.

Barinaga, M. 1990. AIDS meeting: Unexpected progress. *Science* 248 (June 29):1596.

Barnes, D. M. 1988. Obstacles to an AIDS vaccine. *Science* 240 (May 6):719.

Cowen, R. 1990. The shell game. *Science News* 138 (July 28):56. (Explains how a cold virus might help conquer AIDS.)

Fackelmann, K. A. 1990. Lymphoma enigma. *Science News* 138 (August 18):104.

Gallo, R. C. 1987. The AIDS virus. *Scientific American* 256, no. 1 (January):46.

Loschen, D. J. 1988. Protecting against HIV exposure in family practice. *American Family Physician* 37, no. 1 (January):213.

Shilts, R. 1988. *And the band played on: Politics, people, and the AIDS epidemic.* New York: Viking.

CHAPTER 22

Bower, B. 1989. Talk of ages. *Science News* 136 (July 8):24. (An interesting discussion of how the hyoid bone relates to the larynx and speech.)

Bramble, D. M. and Carrier, D. R. 1983. Running and breathing in mammals. *Science* 219 (January 21):251.

Cohen, M. I. 1981. Central determinants of respiratory rhythm. *Annual Review of Physiology* 43:91.

Downing, S. E. and Lee, J. C. 1980. Nervous control of the pulmonary circulation. *Annual Review of Physiology* 42:199.

Frieherr, G. 1985. Respiratory distress syndrome: Searching for causes and improved treatments. *Research Resources Reporter* 9, no. 2:1. National Institutes of Health.

Jones, J. G. 1987. Mechanisms of some pulmonary effects of general anesthesia. *British Journal of Hospital Medicine* 38, no. 5:472.

Kresch, M. J. and Gross, I. 1987. The biochemistry of fetal lung development. *Clinics in Perinatology* 14, no. 3:481.

Lin, Y. 1988. Applied physiology of diving. *Sports Medicine* 5:41.

Naeye, R. L. 1980. Sudden infant death. *Scientific American* 242, no. 4 (April):56.

Orlowski, J. P. 1987. Drowning, near-drowning, and ice-water submersions. *Pediatric Clinics of North America* 34, no. 1 (February):75.

Reid, G. M. 1987. Sudden infant death syndrome: Congenital copper deficiency. *Medical Hypotheses* 24:167.

Sachs, D. P. L. 1986. Cigarette smoking: Health effects and cessation strategies. *Clinics in Geriatric Medicine* 2, no. 2 (May):337.

Wagner, W. W., Jr.; Latham, L. P.; Gillespie, M. N.; Guenther, J. P.; and Capen, R. L. 1982. Direct measurement of pulmonary capillary transit times. *Science* 218 (October 22):379.

Weiss, R. 1988. Surfactant therapy: New questions arise. *Science News* 134 (October 15):245.

Welton, A. F.; O'Donnell, M.; and Morgan, D. W. 1987. The physiology and biochemistry of normal and diseased lung. *Advances in Clinical Chemistry* 26:293.

West, J. B. 1984. Human physiology at extreme altitude on Mount Everest. *Science* 223 (February 24):784.

CHAPTER 23

Bouchier, I. A. D. 1988. Nonsurgical treatment of gallstones: Many contenders but who will win the crown? *Gut* 29:137.

Carlson, G. L.; Li, B. U. K.; Bass, P.; and Olsen, W. A. 1983. A bean alpha-amylase inhibitor formulation (starch blocker) is ineffective in man. *Science* 219 (January 28):393.

Cheromcha, D. P. and Hyman, P. E. 1988. Neonatal necrotizing enterocolitis. *Digestive Diseases and Sciences* 33, suppl. 3:78S.

Fackelmann, K. A. 1989. Hidden heart hazards. *Science News* 136 (September 16):184. (Do high blood insulin levels foretell heart disease?)

Fernald, G. W. and Boat, T. F. 1987. Cystic fibrosis: Overview. *Seminars in Roentgenology* 22, no. 2 (April):87.

Friedman, S. L. and Bisell, D. M. 1990. Hepatic fibrosis: New insights into pathogenesis. *Hospital Practice* 25, no. 5 (May 15):43.

Frizzell, R. A.; Rechkemmer, G.; and Shoemaker, R. L. 1986. Altered regulation of airway epithelial cell chloride channels in cystic fibrosis. *Science* 233 (August 1):558.

Levitan, I. B. 1989. The basic defect in cystic fibrosis. *Science* 244 (June 23):1423.

Makhlouf, G. M. 1990. Neural and hormonal regulation of function in the gut. *Hospital Practice* 25, no. 2 (February 15):79.

McNicholl, B. 1986. Coeliac disease: Ecology, life history, and management. *Human Nutrition: Applied Nutrition* 40A, suppl. 1:55.

McPherson, M. A. and Dormer, R. L. 1987. The molecular and biochemical basis of cystic fibrosis. *Bioscience Reports* 7, no. 3:167.

Milla, P. J. 1986. Intestinal motility and its disorders. *Clinics in Gastroenterology* 15, no. 1 (January):121 .

Moog, F. 1981. The lining of the small intestine. *Scientific American* 245, no. 5 (November):154.

Rehfeld, J. F. 1980. Cholecystokinin as satiety signal. *International Journal of Obesity* 5:465.

Stricker, E. M. 1984. Biological basis of hunger and satiety: Therapeutic implications. *Nutritional Reviews* 42, no. 10:333.

Strober, S. P. J. W.; Quinn, T. C.; and Danovitch, S. H. 1987. Crohn's disease: New concepts of pathogenesis and current approaches to treatment. *Digestive Diseases and Sciences* 32, no. 11 (November):1297.

Williams, R. C. 1990. Periodontal disease. *New England Journal of Medicine* 322, no. 6 (February 8):373.

CHAPTER 24

Brown, M. S. and Goldstein, J. L. 1986. A receptor-mediated pathway for cholesterol homeostasis. *Science* 232 (April 4):34.

Dietschy, J. M. 1990. LDL cholesterol: Its regulation and manipulation. *Hospital Practice* 25, no. 6 (June 15):67.

Edwards, D. D. 1988. Pattern B . . . another genetic heart risk? *Science News* 133 (January 30):69.

Mahley, R. W. 1988. Apolipoprotein E: Cholesterol transport protein with expanding role in cell biology. *Science* 240 (April 29):622.

Montgomery, R. et al. 1983. *Biochemistry: A case-oriented approach*. St. Louis: C.V. Mosby. (A good reference for more detailed information about human metabolism and metabolic disorders.)

Moore, P. B. 1988. The ribosome returns. *Nature* 331 (January 21):223.

Wilson, F. A. 1990. Modern approaches to bile acid transport proteins. *Hospital Practice* 25, no. 4 (April 15):95.

CHAPTER 25

Anderson, J. W. and Geil, P. B. 1988. New perspectives in nutrition management of diabetes mellitus. *American Journal of Medicine* 85, suppl. 5A (November 28):159.

Atkinson, M. A. and Maclaran, N. K. 1990. What causes diabetes? *Scientific American* 263, no. 1 (July):62.

Bower, B. 1988. Alcoholism's elusive genes. *Science News* 134 (July 30):74.

————. 1988. Intoxicating habits. *Science News* 134 (August 6):88.

Davidson, M. B. 1987. Effect of growth hormone on carbohydrate and lipid metabolism. *Endocrine Reviews* 8, no. 2:115.

Duggan, J. P. and Booth, D. A. 1986. Obesity, overeating, and rapid gastric emptying in rats with ventromedial hypothalamic lesions. *Science* 231 (February 7):609.

Glieberman, L. and Harburg, E. 1986. Alcohol usage and blood pressure: A review. *Human Biology* 58, no. 1 (February):1.

Hochachka, P. W. 1986. Defense strategies against hypoxia and hypothermia. *Science* 231 (January 17):234.

Hue, L. 1987. Gluconeogenesis and its regulation. *Diabetes/Metabolism Reviews* 3, no. 1:111.

Keesey, R. E. 1988. The body-weight set point. *Postgraduate Medicine* 83, no. 6:114. (The body-weight set point may explain why people who lose weight, often quickly regain it.)

Kirk, E. and Truex, L. 1987. Effects of alcohol on the central nervous system: Implications for the neuroscience nurse. *Journal of Neuroscience Nursing* 19, no. 6 (December):326.

Kolata, G. B. 1986. Obese children: A growing problem. *Science* 232 (April 4):20.

————. 1987. Diabetics should lose weight, avoid diet fads. *Science* 235 (January 9):163.

Krane, E. J. 1987. Diabetic ketoacidosis. *Pediatric and Adolescent Endocrinology* 34, no. 4 (August):935.

Laposata, E. A. and Lange, L. G. 1986. Presence of nonoxidative ethanol metabolism in human organs commonly damaged by ethanol abuse. *Science* 231 (January 31):497.

Manson, J. E. et al. 1990. A prospective study of obesity and risk of coronary heart disease in women. *New England Journal of Medicine* 322, no. 13 (March 29):882.

Mole, P. A. et al. 1989. Exercise reverses depressed metabolic rate produced by severe caloric restriction. *Medicine and Science in Sports and Exercise* 21, no. 1:29.

Taylor, N. A. S. Eccrine sweat glands: Adaptations to physical training and heat acclimation. *Sports Medicine* 3:387.

CHAPTER 26

Anderson, J. W. 1987. Dietary fiber, lipids, and atherosclerosis. *American Journal of Cardiology* 60:17G.

Cohen, L. A. 1987. Diet and cancer. *Scientific American* 257, no. 5 (November):42 .

Coleman, E. 1987. *Eating for endurance*. Palo Alto, Calif.: Bull Publishing Co.

Ernst, N. D. et al. 1988. The national cholesterol education program: Implications for dietetic practitioners from the adult treatment panel recommendations. *Journal of the American Dietetic Association* 88:1401. (Guidelines for controlling blood cholesterol.)

Fernstrom, J. D. 1987. Food-induced changes in brain serotonin synthesis: Is there a relationship to appetite for specific macronutrients? *Appetite* 8:163.

Food and Nutrition Board. 1989. *Recommended dietary allowances*. 10th ed. Washington: National Academy of Sciences.

Gums, J. G. 1987. Clinical significance of magnesium. *Drug Intelligence and Clinical Pharmacology* 21:240.

Jukes, T. H. 1988. Adventures with vitamins. *Journal of the American College of Nutrition* 7, no. 2:93. (Autobiographic account of a scientist's life work.)

Kral, J. G. and Heymsfield, S. 1987. Morbid obesity: Definitions, epidemiology, and methodological problems. *Gastroenterology Clinics of North America* 16, no. 2:197.

Kris-Etherton, P. M. et al. 1988. The national cholesterol education program: The effect of diet on plasma lipids, lipoproteins, and coronary heart disease. *Journal of the American Dietetic Association* 88:1373.

Lappe, F. M. 1971. *Diet for a small planet*. New York: Ballantine Books.

Lemon, P. W. R. 1987. Protein and exercise: Update 1987. *Medicine and Science in Sports and Exercise* 19, no. 5:S179.

Liu, C. and Mangham, K. B. 1987. The time-calorie displacement approach to obesity. *The Alabama Journal of Medical Sciences* 24, no. 4:451.

McKenzie, A. 1989. A tangle of fibers. *Science News* 136 (November 25):344. (A good summary of the effects of different kinds of dietary fiber.)

Miller, G. T., Jr. 1975. *Living in the environment*. Belmont, Calif.: Wadsworth.

Morgan, S. L. 1987. Eating by the book. *The Alabama Journal of Medical Sciences* 24, no. 4:446.

National Research Council. 1982. *Diet, nutrition, and cancer*. Washington: National Academy of Sciences.

Oldfield, J. E. 1987. The two faces of selenium. *Journal of Nutrition* 117:2002.

Raloff, J. 1989. Dioxin: Paper's trace. *Science News* 135 (February 18):104.

————. 1990. Beyond oat bran. *Science News* 137 (May 26):330.

Surgeon General's Report of Nutrition and Health. 1988. Washington: U.S. Government Printing Office.

Watson, R. R. and Leonard, T. K. 1986. Selenium and vitamins A, E, and C: Nutrients with cancer prevention properties. *Journal of the American Dietetic Association* 86:505.

Weinsier, R. L. 1987. Etiology, complications, and treatment of obesity. *The Alabama Journal of Medical Sciences* 24, no. 4:435.

Worthington-Roberts, B. S. 1981. *Contemporary developments in nutrition*. St. Louis: Mosby.

Wurtman, R. J. and Wurtman, J. J. 1989. Carbohydrates and depression. *Scientific American* 260, no. 1 (January):68.

CHAPTER 27

Cantin, M. and Genest, J. 1986. The heart as an endocrine gland. *Scientific American* 254, no. 2 (February):76.

de Bold, A. J. 1985. Atrial natriuretic factor: A hormone produced by the heart. *Science* 230 (November 15):767.

DeFronzo, R. A. and Thier, S. C. 1982. Inherited tubule disorders. *Hospital Practice* 17, no. 2 (February):111.

Genest, J. and Cantin, M. 1986. Regulation of body fluid volume: The atrial natriuretic factor. *News in Physiological Sciences* 1 (February):3.

Goetz, K. L. 1988. Physiology and pathophysiology of atrial peptides. *American Journal of Physiology* 254, Endocrinology and Metabolism 17:E1.

Khanna, R. and Oreopoulos, D. G. 1986. Dialysis: Continuous ambulatory peritoneal dialysis and hemodialysis. *Clinics in Endocrinology and Metabolism* 15, no. 4 (November):823.

Laragh, J. H. 1985. Atrial natriuretic hormone, the renin-aldosterone acix, and blood pressure-electrolyte homeostasis. *New England Journal of Medicine* 313, no. 21 (November 21):1330.

Mende, C. W. 1990. Current issues in diuretic therapy. *Hospital Practice* 25, Suppl. 1 (January):15.

Re, R. N. 1987. The renin-angiotensin systems. *Medical Clinics of North America* 71, no. 5 (September):877.

Ribeiro, A. B. et al. 1985. The renin-angiotensin system in the control of systemic arterial pressure. *Drugs* 30, suppl. 1:6.

Tilkian, S. M.; Conover, M. B.; and Tilkian, A. G. 1979. *Clinical implications of laboratory tests.* St. Louis: Mosby.

CHAPTER 28

Atkinson, D. E. and Bourke, E. 1987. Metabolic aspects of the regulation of systemic pH. *American Journal of Physiology* 252, Renal Fluid Electrolyte Physiology 21:F947.

Better, O. S. 1987. Impaired fluid and electrolyte balance in hot climates. *Kidney International* 32, suppl. 21:S97.

Groer, M. W. 1987. *Physiology and pathophysiology of the body fluids.* St. Louis: Mosby.

Hyneck, M. L. 1985. Simple acid-base disorders. *American Journal of Hospital Pharmacy* 42 (September):1992.

Javaheri, S. and Kazemi, H. 1987. Metabolic alkalosis and hypoventilation in humans. *American Review of Respiratory Diseases* 136:1011.

Rasmussen, H. 1989. The cycling of calcium as an intracellular messenger. *Scientific American* 261, no. 4 (October):66. (Provides detailed information on various functions of calcium ions.)

Steardo, L. and Nathanson, J. A. 1987. Brain barrier tissues: End organs for atriopeptins. *Science* 235 (January 23):470.

Weiss, M. H. 1985. Cerebral edema. *Acute Care* 11:187.

INTERLUDE 5

Forrow, L.; Wartman, S. A.; and Brock, D. W. 1988. Science, ethics, and the making of clinical decisions. *Journal of the American Medical Association* 259, no. 21 (June 3):3161.

Hastings Center Studies and Reports. Institute of Society, Ethics, and the Life Sciences. Hastings-on-Hudson, N.Y.

Hildreth, E. A. 1990. 'Workup' for a bioethical problem. *Hospital Practice* 25, no. 1 (January 15):86.

Kennedy Center for Bioethics. Washington, D. C.: Georgetown University. (Various publications.)

Kieffer, G. H. 1979. Can bioethics be taught? *American Biology Teacher* 41, no. 3 (March):176.

Raths, L. E.; Harmin, M.; and Simon, S. B. 1966. *Values and teaching.* Columbus, Ohio: Charles E. Merrill.

Reisman, E. C. 1988. Ethical issues confronting nurses. *Nursing Clinics of North America* 23, no. 4:789.

Weiss, R. 1988. Forbidding fruits of fetal-cell research. *Science News* 134 (November 5):296.

CHAPTER 29

Barbo, D. M. 1987. The physiology of the menopause. *Medical Clinics of North America* 7, no. 1:11.

Chao, S. 1987. The effects of lactation on ovulation and fertility. *Clinics in Perinatology* 14, no. 1:39.

Clarke, I. J. 1987. Control of GnRH secretion. *Journal of Reproduction and Fertility,* suppl. 34:1.

Edwards, D. D. 1988. Beating breast cancer. *Science News* 133 (May 14):314.

Frieherr, G. 1986. Tapping the medical potential of ultrasound. *Research Resources Reporter* 10, no. 1 (January):1. (A National Institutes of Health Publication).

Hardin, G., ed. 1964. *Population, evolution, and birth control.* San Francisco: W.H. Freeman.

Hayashi, R. H. 1985. Normal and abnormal labor: Mechanisms for its control. *Birth Defects: Original Article Series* 21, no. 5:175.

Hogan, R. M. 1985. Human sexuality: A nursing perspective. Norwalk, Conn.: Appleton-Century-Crofts.

Jones, K. L.; Shainberg, L. W.; and Byer, C. O. 1985. *Dimensions of human sexuality.* Dubuque, Iowa: Wm. C. Brown.

Keye, W. R. 1987. General evaluation of premenstrual symptoms. *Clinical Obstetrics and Gynecology* 30, no. 2:396.

Lindsay, R. 1987. The menopause: Sex steroids and osteoporosis. *Clinical Obstetrics and Gynecology* 30, no. 2:847.

Mishell, D. R. 1989. Contraception. *New England Journal of Medicine* 320, no. 12 (March 23):777. (An evaluation of the effectiveness and risks of various contraceptive methods.)

Mooradian, A. D.; Morley, J. E.; and Korenman, S. G. 1987. Biological actions of androgens. *Endocrine Reviews* 8, no. 1:1.

Muchmore, A. V. and Decker, J. M. 1985. A unique 85-kilodalton immunosuppressive glycoprotein isolated from urine of pregnant women. *Science* 229 (August 2):479.

Strickler, R. C. 1987. Endocrine hypotheses for the etiology of premenstrual syndrome. *Clinical Obstetrics and Gynecology* 30, no. 2:377.

Talwar, G. P. and Gaur, A. 1987. Recent developments in immunocontraception. *American Journal of Obstetrics and Gynecology* 157 (4, pt. 2):1075.

Warner, C. M.; Brownell, M. S.; and Ewoldsen, M. A. 1988. Why aren't embryos immunologically rejected by their mothers? *Biology of Reproduction* 38:17.

Wassarman, P. M. 1988. Fertilization in mammals. *Scientific American* 259, no. 6 (December):78.

Whitehead, M. I. 1987. The menopause. *The Practitioner* 231:37.

CHAPTER 30

Bower, B. 1989. Drinking while pregnant risks child's IQ. *Science News* 135 (February 4):68.

Cerami, A.; Vlassara, H.; and Brownlee, M. 1987. Glucose and aging. *Scientific American* 156, no. 5 (May):90.

Cohn, J. P. 1987. The molecular biology of aging. *BioScience* 37, no. 2 (February):99.

Conn, R. B. 1985. *Current diagnosis.* Philadelphia: W. B. Saunders.

Lagercrantz, H. and Slotkin, T. A. 1986. The "stress" of being born. *Scientific American* 254, no. 4 (April):100.

Mandler, J. M. 1990. A new perspective on cognitive development. *American Scientist* 78 (May–June):236.

Marx, J. L. 1988. Are aging and death programmed in our genes? *Science* 242 (October 7):33.

———. 1988. Sexual responses are—almost—all in the brain. *Science* 241 (August 19):903.

Rae, P. M. M. and McClelland, A. 1988. DNA sequence analysis of the human genome. *American Biotechnology Laboratory* 1, no. 6 (January):8. (A discussion of the issues surrounding the human genome project.)

Reiter, E. O. 1987. Neuroendocrine control processes: Pubertal onset and progression. *Journal of Adolescent Health Care* 8:479.

Rudman, D. 1985. Growth hormone, body composition, and aging. *Journal of the American Geriatrics Society* 33, no. 11:800.

Schroeder, S. A. et al., eds. 1988. *Current medical diagnosis and treatment.* Norwalk, Conn.: Appleton-Lange.

Slap, G. B. 1986. Normal physiological and psychological growth in the adolescent. *Journal of Adolescent Health Care* 7:13S.

Smith, E. L.; Smith, P. E.; and Gilligan, C. 1988. Diet, exercise, and chronic disease patterns in older adults. *Nutrition Reviews* 46, no. 2 (February):52.

Stern, N. and Tuck, M. L. 1986. Homeostatic fragility in the elderly. *Cardiology Clinics* 4, no. 2 (May):201.

Weiss, R. 1990. Uneven inheritance. *Science News* 138 (July 7):8.

White, R. and Lalouel, J. 1988. Chromosome mapping with DNA markers. *Scientific American* 258, no. 2 (February):40.

GLOSSARY

Many of the boldface terms in this glossary are followed by a phonetic spelling in parentheses. These pronunciation aids usually come from *Dorland's Illustrated Medical Dictionary*. The following rules are taken from this dictionary and will help in using its phonetic spelling system.

1. An unmarked vowel ending a syllable (an open syllable) is long; thus *ma* represents the pronunciation of *may, ne*, that of *knee; ri*, of *wry; so*, of *sew; too*, of *two;* and *vu*, of *view*.

2. An unmarked vowel in a syllable ending with a consonant (a closed syllable) is short; thus *kat* represents *cat; bed, bed; hit, hit; not, knot; foot, foot;* and *kusp, cusp*.

3. A long vowel in a closed syllable is indicated by a macron; thus *māt* stands for *mate; sēd*, for *seed; bīl*, for *bile; mōl*, for *mole; fūm*, for *fume;* and *fōōl*, for *fool*.

4. A short vowel that ends or itself constitutes a syllable is indicated by a breve; thus *ĕ-fekt'* for *effect, ĭ-mūn'* for *immune*, and *ŏ-klōōamod'* for *occlude*.

Primary (') and secondary (") accents are shown in polysyllabic words. Unstressed syllables are followed by hyphens.

A

abdomen (ab'do-men)   Portion of the body between the diaphragm and the pelvis.

abdominal (ab-dom'in-al)   Pertaining to the abdomen.

abdominopelvic (ab-dom''-in-o-pel'vik)   Pertaining to the abdominal and pelvic regions.

abduction (ab-dukt'shun)   Movement of a body part away from the midline.

absorption (ab-sorp'shun)   Taking in of substances.

absorptive (ab-sorp'tīv)   Pertaining to absorption.

acceleratory (ak-sel'er-a-tor''e) **reflex**   Process that maintains balance during starting, stopping, and turning motions.

accommodation (ah-kom-o-da'shun)   Adjustment of the focal distance of the eyes.

acetylcholine (as''e-til-ko'lēn)   A neurotransmitter released from many axons, including those that control skeletal muscles.

acetyl-coenzyme A (as''et'il ko-en'zīm a)   A coenzyme needed to prepare molecules to enter certain metabolic pathways.

Achilles (ah-kil'ez) **tendon**   A heel tendon.

acid   A substance that can donate hydrogen ions.

acid-base balance   The process by which the pH of body fluids is maintained in a normal range.

acidosis (as''id-o'sis)   The condition of having too low a blood pH.

acinar (as'ĭ-nar)   Having an acinus.

acinus (ah'sin-us)   A small grapelike cluster.

acne (ak'ne)   An inflammation of sebaceous glands.

acquired immune deficiency syndrome (AIDS)   A viral disease that severely impairs the immune system.

acquired immunity   Resistance to disease obtained from another's antibodies.

acromioclavicular (ak-ro''me-o-klav-ik'u-lar)   Pertaining to the acromion process of the scapula and the clavicle.

acrosome (ak'ro-sōm)   Dense anterior part of a sperm containing enzymes that help it to penetrate an ovum.

actin (ak'tin)   A contractile protein.

action   A movement produced by one or more muscles.

action potential   An altered electrical potential across the cell membrane of an excitable cell.

action potential with plateau   A special prolonged potential displayed by certain smooth muscle cells.

active immunity   Resistance to disease obtained by the immune system responding to a microorganism or a vaccine.

active transport   Transport of a substance against a gradient using an enzyme, carrier molecule, and cellular energy.

acute   Sudden in onset and severe.

acyl (as'il)   A fatty acid.

Adam's apple   A prominent thyroid cartilage.

adaptation   A decrease in the excitability of sensory receptors in response to continuous, constant intensity stimulation.

adduction   Movement toward the body's midline.

adenohypophysis (ad''en-o-hi-pof'is-is)   The anterior pituitary gland.

adenoid (ad'en-oid)   An enlarged pharyngeal tonsil.

adenosine (ad-en'o-sēn) **triphosphate** (ATP)   The body's main energy storage molecule.

adenylate cyclase (ad-en'il-āt si'klās)   A membrane-bound enzyme that converts ATP to cAMP.

adipose (ad'ĭ-pōs)   Related to fat.

adolescence (ad-o-les'ence)   The period between the onset of puberty and adulthood.

adrenal (ad-re'nal)   Above the kidney; a gland lying above the kidney.

adrenalin (ad-ren'ah-lin)   A hormone secreted by the adrenal glands.

adrenergic (ad''ren-er'jik)   Pertaining to an axon that releases norepinephrine (a neurotransmitter similar to adrenalin, or epinephrine).

adrenocorticotrophic (ad-re''no-kor''te-ko-trof'ik) **hormone**   A hormone that causes the adrenal cortex to secrete hormones; corticotropin.

adsorptive endocytosis (ad-sorp′tiv en″do-si-to′sis) Entry of a substance into a cell after attaching to the cell membrane.

adventitia (ad″ven-tish′e-ah) Outermost connective tissue covering an organ or blood vessel.

aerobic (a′ro-bik) With oxygen.

afferent (af′er-ent) Movement toward a structure.

agammaglobulinemia (a-gam″ah-glob″u-lin-e′me-ah) An immunodeficiency due to a lack of B lymphocytes.

agglutinin (ag-loo′tin-in) An antibody in an agglutination reaction.

agglutinogen (ag″loo-tin′o-jen) An antigen that elicits an agglutination reaction.

aging The process of growing old.

agonist (ag′o-nist) The prime mover in a group of muscles.

agranular leukocyte (a-gran′u-lar lu′ko-sīt) A white blood cell lacking cytoplasmic granules.

AIDS related complex Disease with mild AIDS symptoms.

alarm reaction Mechanism by which the body adjusts to stress.

albicans (al′bik-anz) White.

albumin (al-bu′min) A small protein found mainly in blood plasma.

alcoholism Disease associated with excessive intake of alcohol.

aldosterone (al-dos′ter-ōn) An adrenocortical hormone that increases sodium reabsorption.

alkaline (al′kah-lin) Basic, pertaining to a substance that can accept hydrogen ions.

alkalosis (al″kah-lo′sis) The condition of having too high a blood pH.

allantoic (al″an-to′ik) Pertaining to the allantois.

allantois (al-an′to-is) A fetal membrane that contributes to the formation of the placenta.

allele (al-ēl′) One of a set of genes with different information about a characteristic that can occupy a particular site on a chromosome.

allergen (al′er-jen) A substance capable of producing allergy.

allergy (al′er-je) Unusual sensitivity to a substance that is harmless to most people in the same concentration.

all-or-none law A rule stating that when a neuron responds to a stimulus, it conducts a signal (impulse) at maximum strength.

allosteric (al″o-ster′ik) Other site or shape.

alpha wave A brain wave seen in an awake, relaxed state.

alveolar (al-ve′o-lar) Pertaining to an alveolus.

alveolus (al-ve′o-lus) (1) A saclike structure in the lungs and in secretory portions of some glands; (2) a tooth socket.

Alzheimer's (altz′hi-merz) **disease** A degenerative neurological disorder that causes loss of memory and behavioral changes.

amenorrhea (am-en″or-e′ah) Absence of menstruation.

amino (ah-me′no) **acid** A molecule containing both acid and amino functional groups.

aminopeptidase (am″in-o-pep′tid-ās) An enzyme that breaks peptide bonds from the amino end of a protein.

amniocentesis (am″ne-o-sen-te′sis) A procedure for removing a sample of amniotic fluid from around a developing fetus.

amnion (am′ne-on) A membrane that forms a fluid-filled bag surrounding a developing fetus.

amniotic (am″ne-ot′ik) Pertaining to the amnion.

amphiarthrosis (am″fe-ar-thro′sis) An immovable or slightly movable joint at which articular surfaces are connected by cartilage.

amplitude (am′pleh-tood) The strength of a signal; the intensity of a sound.

ampulla (am-pul′lah) A dilation at the end of a passageway.

amygdaloid (am-ig′dal-oid) Almond-shaped.

amylase (am′il-ās) A starched-digesting enzyme.

anabolic (an-ah-bol′ik) Pertaining to anabolism.

anabolic steroid (an-ah-bol′ik ste′roid) A synthetic hormone that increases the size of muscles.

anabolism (ah-nab′o-lizm″) Synthetic processes that require energy.

anaerobic (an-er-o′bik) Without oxygen.

anaphase (an′ah-fāz) A stage in mitosis during which chromosomes move apart.

anaphylaxis (an″af-il-aks′is) A severe allergic reaction to a substance that one has been sensitized to previously.

anastomosis (an-as″to-mo′sis) A connection between blood vessels or nerves.

anatomical dead space Respiratory passageways in which gas exchange does not occur.

anatomical position Body in standing position with palms turned the same direction as the face.

anatomy The study of structure.

androgen (an′dro-jen) A molecule having male hormone activity.

anemia (an-ēm′e-ah) A hemoglobin deficiency.

anencephaly (an″en-sef′al-e) A condition in which the brain is lacking.

anesthetic (an″es-thet′ik) Agent that produces loss of sensation.

aneurysm (an′u-rizm) A saclike dilation in the wall of an artery.

angina pectoris (an′jin-ah pek′tor-is) Severe, constrictive, suffocating pain in the chest associated with ischemic heart disease.

angiogenesis (an″je-o-jen′es-is) The development of blood vessels.

angiogenic (an″je-o-jen′ik) Pertaining to angiogenesis.

angioplasty (an″je-o-plas′te) Surgery on blood vessels.

angiotensin (an″je-o-ten′sin) A substance that constricts blood vessels and stimulates aldosterone release.

angiotensinogen (an″je-o-ten′sin-o-jen) Inactive precursor of angiotensin.

anion (an′i-on) A negatively charged ion.

anorexia nervosa (an″o-rek′se-ah ner-vo′sah) A serious neurological disorder in which a person loses weight and becomes emaciated.

antagonist (an-tag′on-ist) A muscle that opposes an agonist.

anterior Toward the front, ventral.

antibody (an′ti-bod″e) A protein elicited by an antigen that can react with and inactivate the antigen.

anticoagulant (an″ti-ko-ag′u-lant) A substance that prevents blood clotting.

anticodon (an-te-ko′don) A set of three bases on transfer RNA that fit with a particular codon on messenger RNA.

antidiuretic (an″te-di-ur-et′ik) **hormone (ADH)** A hypothalamic hormone stored in the posterior pituitary gland that conserves fluid by causing the kidneys to return more water to the blood.

antigen (an′ti-jen) A substance that elicits a response from the immune system.

antigenic determinant The part of a molecule that elicits an immunologic response.

antigen presenting cell (APC) A macrophage or other cells that process and present an antigen to a B or T lymphocyte.

antiviral protein A protein produced by cells stimulated by interferon.

antrum (an′trum) A cavity or chamber.

anuria (ah-nu′re-ah) Without urine.

anus (a′nus) The opening through which wastes leave the digestive tract.

aorta (a-or′tah) The large artery carrying blood from the left ventricle to other arteries of the systemic circulation.

aortic (a-or′tik) Pertaining to the aorta.

apical (ap′ik-l) Pertaining to or located at the apex or tip of a structure.

aplastic (a-plas′tik) Having no tendency to develop new tissue.

apneusis (ap-nu′sis) The arrest of breathing during inspiration.

apneustic (ap-nu′stik) Pertaining to apneusis.

apocrine (ap′o-krīn) A gland that loses part of its substance with its secretions.

apoferritin (ap″o-fer′it-in) An iron-carrying protein.

aponeurosis (ap″o-nu-ro′sis) A fibrous sheet of connective tissue to which muscles attach.

appendicitis (ah-pen-deh-si′tis) Inflammation of the appendix.

appendicular (ap''en-dik'u-lar) Pertaining to the appendages.

aqueduct (ak'we-dukt) A canal that conducts liquid.

aqueous Watery.

arachnoid (ar-ak'noid) Spiderlike; pertaining to a delicate middle meninges covering the brain and spinal cord.

arbor vitae Treelike pattern of white matter in the cerebellum.

arcuate (ar'ku-āt) Arch-shaped.

areola (ar-e'o-lah) Pigmented region surrounding the nipple of a mammary gland.

argentaffin (ar-jen'taf-in) Cells in the stomach lining that secrete histamine and serotonin.

arrestin (ah-res'tin) Protein that prevents opsin from binding to transducin in the dark.

arrhythmia (ah-rith'me-ah) A loss of rhythm, especially in the conduction of signals through the heart.

arteriogram (ar-te're-o-gram'') An image of an artery.

arteriole (ar-te're-ōl) A blood vessel that carries blood from an artery to capillaries.

artery (ar'ter-e) A large blood vessel that carries blood away from the heart.

arthrosis (ar-thro'sis) A joint.

articular (ar-tik'u-lar) Pertaining to a joint.

articulation (ar-tik''u-la'shun) A joint.

artificial insemination A procedure for introducing sperm into a female reproductive tract without sexual intercourse.

artificially acquired active immunity Resistance to disease obtained by stimulating the immune system with a vaccine.

artificially acquired passive immunity Resistance to disease temporarily obtained by receiving another's antibodies.

artificial pacemaker A device that stimulates the heart so as to maintain a regular heart rate.

arytenoid (ar''et-e'noid) Pitcher-shaped.

ascites (ah-si'tēz) Accumulation of fluid in the abdominal cavity.

ascorbic (ah-skor'bik) **acid** Vitamin C.

association fiber A neural tract that carries signals between parts of the cerebrum.

association neuron A neuron that relays signals from sensory to motor neurons, especially in the spinal cord.

asthma (az'mah) A condition in which bronchioles constrict and cause difficulty in breathing.

astigmatism (as-tig'mat-izm) A condition of blurred vision because of irregular curvature of one or more refractive surfaces in the eye.

astrocyte (as'tro-sīt) A star-shaped neuroglial cell in the central nervous system.

atelectasis (at''el-ek'tas-is) Collapse of alveoli in the lungs or their failure to expand at birth.

atherosclerosis (ath''er-o-skler-o'sis) Hardening and obstruction of arteries by plaque deposits.

atlantoaxial (at-lan''to-ax'e-al) Pertaining to the atlas and axis, the two superiormost vertebrae.

atlanto-occipital (at-lan''to-ok-sip'it-al) Pertaining to the uppermost vertebra and the joint it forms with the occipital region of the skull.

atom (at'om) Smallest particle of an element.

atomic (ah-tom'ik) **number** The number of protons in an atom.

atomic weight A number approximately equal to the sum of an atom's protons and neutrons.

atopy (at'o-pe) A kind of hypersensitivity.

atresia (ah-tre'ze-ah) (1) Absence of an opening; (2) degeneration of oocytes.

atrial Pertaining to the atria of the heart.

atrial natriuretic hormone (ANH) A substance secreted by the atria that accelerates sodium excretion.

atrioventricular (a'tre-o-ven-trik'u-lar) Pertaining to an atrium and ventricle of the heart.

atrium (a'tre-um) A chamber or opening.

atrophy (at'ro-fe) A decrease in size.

attenuation (ah-ten''u-a'shun) Dampening of an effect.

auditory (aw-dit-o're) Pertaining to hearing.

auricle (aw'rik-l) An earlike appendage.

autoantibody (aw''to-an'tib-od''e) An antibody produced against some substance normally present in the body.

autoimmune (aw''to-im-ūn') **disorder** A condition in which autoantibodies are produced.

autonomic nervous system A part of the nervous system that regulates function of internal organs and involuntary processes.

autoregulation Self-regulation.

autosomal (aw''to-so'mal) Pertaining to paired (nonsex) chromosomes and the genetic information on them.

AV node An aggregation of conductive tissue in the atrioventricular septum of the heart.

axial (ak'se-al) Pertaining to the body's axis or bones of the head and trunk as parts of the axial skeleton.

axoaxonic (ak''so-ak-son'ik) **synapse** A site at which an axon of one neuron relays a signal to the axon of another neuron.

axodendritic (ak''so-den-drit'ik) **synapse** A very common site at which an axon of one neuron relays a signal to the dendrite of another neuron.

axon (ax'on) The part of a neuron that normally carries signals away from the cell body toward another neuron.

axon terminal The end of an axon where neurotransmitter is released.

axosomatic (ak''so-mat'ik) **synapse** A site at which an axon of one neuron relays a signal to the cell body of another neuron.

B

Babinski (bab-in'ske) **reflex** A reflex in which the toes are dorsiflexed when the sole of the foot is scratched.

ball-and-socket joint A joint in which a ball-shaped articular surface of one bone fits into a socket-shaped articular surface of another bone.

baroreceptor (bar''o-re-sep'tor) Pressure detector.

Bartholin's (bar'to-linz) **gland** Vulvovaginal gland.

basal metabolic rate (BMR) Amount of energy used to maintain life in an awake, resting individual.

basal metabolism The process of using energy from nutrients to maintain life in an awake, resting individual.

basal nucleus One of several aggregations of cell bodies in the base of the cerebrum.

base A chemical substance that can accept hydrogen ions or react with an acid to form a salt.

basilar membrane Membrane in the inner ear associated with sound receptors.

basophil (bas'o-fil) A leukocyte having granular cytoplasm and able to be stained with a basic dye.

benign (be-nīn') Not malignant, favorable for recovery.

beta blocker A drug that interferes with sympathetic signals that would normally stimulate beta receptors.

beta lipotropin (lip''o-tro'pin) A molecule from which endorphins are derived.

beta oxidation A metabolic pathway that oxidizes fatty acids.

beta reduction A metabolic pathway that synthesizes fatty acids.

beta wave A brain wave seen during mental alertness.

bicuspid (bi-kus'pid) Having two points or cusps.

bilateral Pertaining to both sides of the body; having left and right sides.

bile Fluid secreted by liver that aids in digestion by emulsifying fats.

bilirubin (bil''i-roo'bin) A red bile pigment from the breakdown of hemoglobin.

biliverdin (bil''i-ver'din) A green bile pigment from the breakdown of hemoglobin.

binding site A site at which a particular molecule binds to a membrane or other structure.

binocular vision Vision with two eyes that allows perception of three-dimensionality.

biocybernetic (bi-o-si''ber-net'ik) Pertaining to an automatic self-controlling system in a living organism.

bioenergetics (bi'o-en-er-jet'iks) The study of energy changes in living organisms.

biofeedback A process in which signals indicating levels of autonomic processes are made perceptible and usable to control the process.

biological clock The regulatory mechanism for biorhythms.

biotin (bi′o-tin) A vitamin needed in fat synthesis.

bipolar Having two poles or processes.

blastocoele (blas′to-sēl) The cavity in a blastula.

blastocyst (blas′to-sist) Hollow ball of cells comprising an early stage in embryonic development.

blind spot Region of the retina lacking light receptors where optic nerve fibers leave the eye.

blocking antibody Antibody that binds to allergen.

blood Fluid circulated by the heart through a closed system of vessels.

blood-brain barrier A specialized capillary structure that limits passage of substances from the blood into the tissues of the brain.

B lymphocyte A lymphocyte that gives rise to plasma cells, which, in turn, produce antibodies.

Bohr effect Tendency of the high concentration of oxygen in the lungs to cause hemoglobin to release carbon dioxide.

bolus (bo′lus) A mass.

Bowman's capsule Glomerular capsule.

Boyle's law A law stating that a gas exerts a pressure inversely proportional to its volume.

brachial (bra′ke-al) Pertaining to the arm.

brachiocephalic (brak″e-o-se-fal′ik) Pertaining to the arm and head.

bradycardia (brad″e-kar′de-ah) An abnormally slow heart rate.

bradykinin (brad′e-ki′nin) A polypeptide having potent vasodilating action.

brain stem Portions of the brain that relay signals to and from the cerebrum, cerebellum, and other brain structures.

brain wave An electrical signal detectable from the scalp that represents brain activity.

breech birth Delivery of an infant in which the buttocks are the presenting part.

Broca's (bro′kahz) **motor speech area** A functional area of the cerebrum in which thoughts are translated into speech.

bronchial (brong′ke-al) Pertaining to bronchi.

bronchiole (brong′ke-ōl) Small tubes in the lungs.

bronchitis (brong-ki′tis) Inflammation of bronchi.

bronchus (brong′kus) One of many tubes (bronchi) leading from the trachea to the lung.

brown fat Fat containing a very high quantity of energy found around organs in newborn infants.

Brown-Sequard (sa-kār′) **syndrome** Partial paralysis seen in patients with injury to one side of the spinal cord.

Brunner's gland A mucous gland of the duodenum.

buffer A substance that resists pH change in a solution.

bulbourethral (bul″bo-u-re′thral) **gland** A male gland that adds secretions to semen.

bulimia (bu-lim′e-ah) Binge eating, usually followed by self-induced vomiting.

bulk flow Streaming flow of molecules that allows them to move faster than by diffusion.

bursa (bur′sah) A sac filled with synovial fluid found at pressure points or near joints.

bursa of Fabricius (fab-ris′e-us) A structure found in birds in which differentiation of B lymphocytes was first identified.

bursitis (bur-si′tis) Inflammation of a bursa.

C

cachexia (kak-ek′se-ah) Weight loss, wasting, and weakness seen especially in cancer patients.

calcaneus (kal-ka′ne-us) Heelbone.

calciferol (kal-sif′er-ol) A steroid with vitamin D activity.

calcification (kal″si-fi-ka′shun) Deposition of calcium salts in an organic matrix.

calcitonin (kal″sit-o′nin) A hormone that increases blood calcium.

callus (kal′us) A thickened area.

calmodulin (kal-mod′u-lin) An intracellular calcium carrier molecule.

calorie (kal′or-e) The quantity of heat required to raise the temperature of 1 gram of water 1° C.

calyx (ka′liks) A cup-shaped organ or cavity.

canaliculus (kan″ah-lik′u-lus) A tiny canal.

canal of Schlemm Passageway in the anterior eye that drains aqueous humor.

cancellous (kan-sel′us) Spongy.

capacitance (kap-as′it-ans) **vessel** One of many blood vessels (veins) that together contain the largest volume of blood.

capillary (kap-il-a′re) A small blood vessel that connects an arteriole with a venule.

carbaminohemoglobin (kar-bam″in-o-hem″o-glo′bin) Hemoglobin carrying carbon dioxide.

carbohydrate (kar-bo-hi′drāt) An organic compound containing several alcohol groups and at least one aldehyde or ketone.

carbohydrate-craving obesity (CCO) An overweight condition in which the patient has a persistent desire for sugars and starches.

carboxypeptidase (kar-box-e-pep′tid-ās) A proteolytic enzyme that acts on peptide bonds at the carboxyl end of a peptide.

carcinogen (kar-sin′o-jen) An agent that can induce cancer.

cardiac (kar′de-ak) Pertaining to the heart.

cardiac tamponade (kar′de-ak tam″pon-ād′) Compression of the heart due to accumulation of fluid or blood in the pericardial sac.

cardioinhibitory (kar″de-o-in-hib′it-or-e) Slowing the heart rate.

cardiologist A physician who specializes in treating heart disease.

cardiopulmonary resuscitation (kar″de-o-pul′mon-ar-e re-sus-sit-a′shun) (CPR) A procedure for maintaining minimum blood flow and gas exchange in a patient whose heart and lungs have stopped functioning.

cardiovascular (kar″de-o-vas′ku-lar) Pertaining to the heart and blood vessels.

carotene (kar′o-tēn) A yellow substance, usually having Vitamin A activity.

carotid (car-ot′id) **body** Structure at the branching of carotid arteries that contains chemoreceptors.

carpal (kar′pal) A bone of the wrist.

carrier (1) A molecule that transfers or transports another substance; (2) a person who can transmit an unexpressed gene to offspring.

carrier saturation A condition in which all carrier molecules are occupied with the substance they carry.

cartilage (kar′ti-lāj) A firm, resilient connective tissue.

cartilaginous Pertaining to or consisting of cartilage.

catabolic (kah-tah-bol′ik) Pertaining to catabolism.

catabolism (kat-ab′o-lizm) The breakdown of molecules for obtaining energy.

catalyst (kat′ah-list) A substance that increases the rate of reaction.

cataract (kat′ah-rakt) An opacity of the lens of the eye.

catecholamine (kat″eh-kol′ah-mēn) A group of amines that act as chemical messengers; dopamine, epinephrine, and norepinephrine.

cation (kat′i-on) A positively charged ion.

cauda equina (kau-dah e-kwi′nah) The bundle of spinal nerve roots that extend below the end of the spinal cord; literally, horse's tail.

CCK-PZ (cholecystokinin-pancreozymin) An enteric hormone that stimulates release of bile from the gallbladder and enzymes from the pancreas.

cecum (se′kum) Blind pouch.

celiac (se′le-ak) Pertaining to the abdomen.

cell The basic functional unit of living organisms.

cell cycle A repetitive sequence of events involving DNA replication and cell division.

cell-mediated immunity Resistance to disease involving direct destruction of antigenic cells.

cell membrane The lipid and protein structure that forms the boundary of a cell.

cell theory A theory stating that all living things are composed of cells.

cementum (se-men′tum) The covering of the root of a tooth.

central nervous system (CNS) The brain and spinal cord.

central venous pressure Pressure in veins near the heart.

centriole (sen'tre-ōl) One of a pair of structures that participate in forming the mitotic spindle.

centrum (sen'trum) Any anatomical center.

cephalic (sef-al'ik) Pertaining to or toward the head; in humans, superior.

cerebellar (ser-e-bel'ar) Pertaining to the cerebellum.

cerebellum (ser''e-bel'um) A portion of the brain behind the cerebrum and above the pons concerned with the coordination of movements.

cerebral (ser'e-bral) Pertaining to the cerebrum.

cerebrospinal fluid (CSF) A clear fluid found in spaces within and around the central nervous system.

cerebrovascular (ser''ĕ-bro-vas'ku-lar) Pertaining to blood vessels of the brain.

cerebrum (ser'e-brum) The largest part of the human brain, the cortex of which receives and responds to sensory signals and carries out mental processes.

ceruloplasmin (se-roo''lo-plaz'min) A copper transport protein in the blood.

cerumen (sĕ-roo'men) Earwax.

ceruminous (sĕ-roo'mĭ-nus) Pertaining to earwax or glands that secrete it.

cervical (ser'vik-al) Pertaining to a neck.

cervix (ser'viks) Neck; narrow end of the uterus where it joins the vagina.

chancre (shang'ker) Skin lesion, such as one found where organisms that cause syphilis enter the body.

chemiosmotic (kem''e-os-mot'ik) **theory** An explanation of energy capture in mitochondria.

chemoreceptor (ke''mo-re-sep'tor) A receptor that responds to certain chemical substances.

chemotaxis (ke''mo-tak'sis) The process by which chemical stimuli attract or repel; a process that attracts some leukocytes to sites of injury.

chenodeoxycholic (ken''o-de-ok''se-kol'ik) **acid** A bile acid.

Cheyne-Stokes (chan'stōks) **breathing** Alternating episodes of abnormally rapid and abnormally slow breathing.

chiasma (ki-as'mah) A crossing over of fibers.

chief cell A pepsin-secreting cell of a gastric gland in the stomach.

chloride shift The movement of chloride ions down an electrical gradient toward a positively charged region.

chlorolabe (klor'o-lāb) A pigment in cones sensitive to green light.

cholecystectomy (ko'le-sis-tek'to-me) Surgical removal of the gallbladder.

cholesteryl esterase (ko-les'ter-il es'ter-ās) An enzyme that breaks ester bonds between cholesterol and a fatty acid.

cholic (ko'lik) **acid** A bile acid.

cholinergic (ko''lin-er'jik) Pertaining to a neuron whose terminals release the neurotransmitter acetylcholine.

cholinesterase (ko''lin-es'ter-ās) An enzyme that breaks down acetylcholine.

cholinesterase inhibitor A substance that blocks the action of cholinesterase.

chondroblast (kon'dro-blast) A cell that produces cartilage.

chondrocyte (kon'dro-sīt) A cell found in mature cartilage.

chordae tendineae (kor'de ten-din'e) Tough bands of connective tissue that attach cusps of A-V valves to papillary muscles within the heart's ventricles.

chorion (ko're-on) The outermost fetal membrane.

chorionic villi Tufts of fetal blood vessels that exchange substances with maternal blood.

chorionic villus biopsy A procedure for obtaining fetal tissue for analysis during development.

choroid (ko'roid) Skinlike; highly vascular middle layer of the eyeball.

choroid plexus (ko'roid plek'sus) One of several vascular projections that secrete cerebrospinal fluid into brain ventricles.

chromatin (kro'mah-tin) Nuclear material that becomes identifiable as distinct chromosomes during cell division.

chromatolysis (kro''mat-ol'is-is) Dissolution of chromatin.

chromosomal abnormality Any modification of the configuration of DNA in chromosomes.

chromosome (kro'mo-sōm) One of 46 structures consisting of DNA and protein found in the nucleus of a human cell.

chronic (kron'ik) Long continued, not acute.

chrononcology (kron-on-kol'o-je) The use of rhythms of cell division to schedule cancer therapy.

chronotherapy (kron-o-ther'a-pe) The use of any kind of rhythms to schedule therapy.

chronotropic (kron''o-trop'ik) Pertaining to the effects of time.

chylomicron (ki-lo-mi'kron) A particle consisting of lipids and protein made in the intestinal mucosa and released into lacteals.

chyme (kīm) A semiliquid food mass leaving the stomach.

chymotrypsin (ki''mo-trip'sin) A proteolytic enzyme from the pancreas.

chymotrypsinogen (ki''mo-trip-sin'o-jen) An inactive form of chymotrypsin.

ciliary (sil'e-er''e) **body** Anteriormost portion of the choroid layer of the eye; contains ciliary muscles that participate in accommodation.

cilium (sil'e-um) One of many tiny hairlike projections found on some epithelial cells.

cingulate (sin'gu-lāt) Pertaining to a bundle of fibers.

circadian (sir-ka'de-an) Pertaining to daily rhythms; literally, about a day.

circle of Willis A ring of blood vessels at the base of the brain that provide alternate circulatory pathways through the brain.

circumduction (sir-kum-duk'shun) Movement in a circle.

circus rhythm A movement traveling in a circular fashion.

cirrhosis (sir-ro'sis) A liver disorder in which liver cells are replaced with connective tissue; literally, orange.

cisterna (sis-ter'na) Cavity or reservoir.

cisterna chyli (sis-ter'nah ki'li) Large lymphatic vessel that drains lymph from the legs and abdominal organs.

clavicle (klav'ik-l) Bone extending from the sternum to the scapula.

clearance The rate at which a substance is removed from the blood by the kidneys.

cleavage (kle'vij) Division into two equal parts; the process by which a zygote becomes a multicellular ball of cells.

clitoris (kli'tor-is) Small erectile organ homologous to the penis located in the anterior vulva.

clonal selection theory A theory that explains how lymphocytes become sensitized to particular antigens and how the immune system acquires tolerance for self.

clone (klōn) A group of identical cells descending from a single parent cell.

cloning The process of forming a clone.

clonus Alternating rigidity and relaxation in a spasm.

closed reduction The realignment of bones in a fracture without surgery.

clot retraction The shrinking of a blood clot.

coarctation (ko''ark-ta'shun) Pressing together or constriction.

coccygeal (kok-sij'e-al) Pertaining to the coccyx.

coccyx (kok'siks) The caudal end of the spinal column.

cochlea (kok'le-ah) Snail-shaped bony portion of the inner ear.

cochlear (kok'le-ar) Pertaining to the cochlea.

codon (ko'don) A sequence of three bases in messenger RNA derived from DNA and specifying the placement of an amino acid in a protein.

colipase (ko-li'pās) An enzyme that aids a lipase.

colitis (ko-li'tis) Inflammation of the colon.

collagen (kol'ah-jen) A fibrous protein found in connective tissue.

collateral circulation Alternate pathways of blood vessels.

collecting duct One of many ducts in the kidneys that receives filtrate from kidney tubules.

colloid (kol'oid) A particle in a colloidal dispersion; literally, resembling glue.

colloidal (kol-oid'al) **dispersion** A state of matter in which small particles are suspended in a medium.

colon (ko'lon) The large intestine from the cecum to the rectum.

colostrum (kol-os'trum) The first fluid expressed from a mammary gland after childbirth.

columnar Cylindrical in shape.

comminuted (kom'i-nūt''ed) Broken into small pieces, as in a bone fracture.

commissural (kom-mis'u-ral) **fiber** One of a bundle of nerve fibers that connects the left and right sides of the brain and spinal cord.

complement (kom'ple-ment) A set of plasma enzymes that contributes to the body's nonspecific defenses.

complementarity (kom-ple-men-tar'it-e) **of structure and function** A concept that structures and their functions are closely related and mutually reinforcing.

complementary (kom-ple-men'ta-re) **base pairing** The binding together of certain bases in nucleic acids.

compliance (kom-pli'antz) The ability of a structure to conform to a certain shape.

compound A substance that consists of two or more elements combined in definite proportion.

compression (kom-pres'shun) Pressing together.

concha (kong'kah) A shell-shaped structure, such as is seen in the bones of the nasal cavity.

conduction deafness Loss of hearing due to impairment of the transmission of vibrations to sound receptors.

conduction system A set of fibers in the heart through which signals coordinate atrial and ventricular contractions.

condyle (kon'dīl) A large rounded surface at the end of a bone.

condyloid (kon'dil-oid) Resembling a condyle.

condyloma (kon-dil-o'mah) Genital wart.

cone A light receptor that responds to a certain color.

congenital (kon-jen'it-al) Present at birth.

congestive heart failure Loss of pumping capacity of the heart accompanied by accumulation of excess tissue fluids.

conjunctiva (kon''junk-ti'vah) A mucous membrane lining the eyelids and covering the anterior surface of the eyeball.

connective tissue A kind of tissue consisting of fibrocytes and certain other cells imbedded in ground substance deposited in an organic fibrous matrix.

consciousness Awareness of signals from the sense organs.

contractile protein A protein that participates in shortening a muscle or causing it to develop tension.

contractility The ability to shorten or develop tension.

contraction alkalosis An abnormal increase in the blood pH associated with a decrease in body fluid volume.

contraction cycle The repetitive actions of actin and myosin in a muscle filament as it develops tension.

contracture (kon-trak'tūr) Permanent shortening of a muscle.

contralateral (kon''tra-lat'er-al) On the opposite side.

controlled output The output of a body process regulated by an automatic control system.

controller A structure that exerts control.

control variable A condition or event that is prevented from changing for the duration of an experiment.

conus medullaris (ko'nus med'u-la''ris) The terminal end of the spinal cord.

convergence Coming together.

coracohumeral (kor''ah-ko-hu'mer-al) Pertaining to the coracoid process of the scapula and the humerus.

core body temperature The temperature deep within the body.

Cori cycle A metabolic pathway involving lactic acid.

cornea (kor'ne-ah) A transparent portion of the anterior surface of the eye.

coronal (ko-ro'nal) Crownlike; pertaining to the crown of the head.

coronary (kor-on-a're) Pertaining to the heart; circling like a crown.

coronoid (kor'o-noid) A sharp process on a bone; literally, like a crow's beak.

corpus albicans (al'bik-anz) A scar on an ovary that remains after the degeneration of a corpus luteum; literally, white body.

corpus callosum (kor'pus kah-lo'sum) A mass of myelinated neural fibers carrying signals between the two cerebral hemispheres.

corpus luteum (lu'te-um) Cells of an ovarian follicle that remain after ovulation and produce hormones; literally, yellow body.

cortex (kor'tex) Outer portion of an organ; literally, bark.

cortical nephron (nef'ron) A functional unit of a kidney located mainly in the cortex.

corticosterone (kor''tik-os'ter-ōn) A steroid hormone from the cortex of the adrenal gland.

cortisol (kor'tis-ol) An adrenocortical hormone that helps to regulate carbohydrate metabolism, also counteracts inflammation.

costal (kos'tal) Pertaining to a rib.

co-transport Movement of two substances across a membrane on the same carrier.

coumarin (koo'mar-in) An anticoagulant.

countercurrent mechanism A mechanism in which fluids flowing out of the system affect those flowing in.

covalent bond A chemical bond in which two atoms share electrons.

coxal (kok'sal) **bone** Bone of pelvic girdle.

cramp A painful spasmodic muscle contraction.

cranial (kra'ne-al) Pertaining to the cranium.

craniosacral (kra''ne-o-sa'kral) Pertaining to the cranial and sacral regions.

cranium (kra'ne-um) Skull bones that encase the brain.

creatine phosphate An energy storage molecule found in muscle.

creatinine (kre-at'in-in) A breakdown product of creatine excreted at a constant rate in the urine.

cribriform (krib'rif-orm) Perforated like a sieve.

cricoid (krik'oid) Ring-shaped.

crista (kris'tah) Ridge or crest.

cross-bridge The end of a myosin filament that binds to actin during muscle contraction.

cross-matching Comparison of donor and prospective recipient bloods.

cruciate (kroo'she-āt) Shaped like a cross.

cryostat (kri'o-stat) A device that maintains a very low temperature.

cryptorchidism (krip-tor'kid-izm) Failure of the testes to descend into the scrotum.

cuboidal (ku-boi'dal) Cube-shaped.

cuneate (ku-ne-āt) Wedge-shaped.

cuneiform (ku-ne'if-orm) Wedge-shaped.

cupula (ku'pu-lah) A domelike, cup-shaped structure.

cuspid (kus'pid) A tapering projection.

cuticle (ku'tik-l) The outer layer of the skin, especially around nails.

cyanocobalamin (si''an-o-ko-bal'am-in) A vitamin transported by intrinsic factor and needed for normal cell division.

cyanolabe (si-an'o-lāb'') A cone pigment that responds to blue light.

cybernetic (si-ber-net'ik) **system** An automatic, self-controlling system.

cystic (sis'tik) Pertaining to a cyst or to the urinary bladder or gallbladder.

cystic fibrosis An inherited disorder that causes thick mucous secretions, especially in respiratory system and pancreas.

cystitis (sis-ti'tis) Inflammation of the urinary bladder.

cytokinesis (si-to-kin-e'sis) Division of the cytoplasm of a cell.

cytoplasm The substance of a cell, excluding its nucleus.

cytoskeleton The organelles that form an internal framework within a cell.

cytosol (si'to-sol) The fluid part of cytoplasm in which organelles are suspended.

cytotoxic Harmful to cells.

D

Dalton's law A law stating that each gas in a mixture exerts a partial pressure independent of other gases.

deamination (de-am''in-a'shun) Removal of an amino group.

decibel (des′ib-el) A unit on the logarithmic scale of sound intensity.

deciduous (de-sid′u-us) **teeth** Teeth that are not permanent; "baby teeth."

decompression sickness A disorder caused by nitrogen bubbles in the tissues due to too rapid decrease in pressure; the bends.

decussate (de-kus′āt) Cross midline of the body.

defecation (def″e-ka′shun) The passage of contents of the rectum outside the body.

deferens (def′er-enz) A duct that carries sperm away from testis.

defibrillation (de-fib″ril-a′shun) The use of an electrical current to terminate an extremely fast heart rate.

deglutition (deg-loo-tish′un) The act of swallowing.

dehydration (de-hi-dra′shun) Removal of water.

deletion The loss of one or more bases from a strand of DNA.

delta wave A brain wave associated with deep sleep.

deltoid (del′toid) Like the Greek letter delta; triangular.

denaturation (de-nat-ur-a′shun) A change in the shape and properties of a protein molecule.

dendrite (den′drīt) A cytoplasmic process of a neuron that typically receives signals from other neurons.

dendrodendritic (den″dro-den-drit′ik) **synapse** A junction between two neurons in which a dendrite of one neuron relays signals to a dendrite of another neuron.

denervation atrophy (at′ro-fe) Wasting of a muscle because of a lack of nerve stimulation.

dentin (den′tin) Bonelike substance of a tooth beneath the surface.

deoxyribonuclease (de-ok″se-ri″bo-nu′kle-ās) An enzyme that digests DNA.

deoxyribonucleic (de-ok″se-ri″bo-nu-kla′ik) **acid** (DNA) A nucleic acid found in chromosomes that stores and transmits genetic information from one generation to the next.

depression (1) A lowering movement; (2) an emotional disorder.

depressor An agent that slows or inhibits a process.

dermal papilla (pah-pil′ah) An upward projection of the dermis into the epidermis.

dermis A thick layer of the skin underlying the epidermis.

desmosome (des′mo-sōm) A structure that holds cells together.

diabetes insipidus (di-ah-be′tez in-sip′id-us) A disorder caused by a deficiency of antidiuretic hormone in which large quantities of dilute urine are produced.

diabetes mellitus (di-ah-be′tez mel-i′tus) A disorder caused by a deficiency or inactivity of insulin in which glucose accumulates in the blood and urine.

dialysis (di-al′is-is) The separation of small molecules from larger ones as smaller ones pass through a selectively permeable membrane.

diapedesis (di″ah-ped-e′sis) A process by which leukocytes squeeze between the cells of capillary walls.

diaphragm (di′ah-fram) A thin wall or partition, as between the abdominal and thoracic cavities.

diaphysis (di-af′i-sis) The shaft of a long bone.

diarrhea Excessively frequent bowel movements.

diarthrosis (di″ar-thro′sis) A freely movable joint.

diastole (di-as′to-le) Dilation; a period between contractions in the heart.

diastolic (di″as-tol′ik) Pertaining to diastole.

differential white cell count The expression of relative numbers of different kinds of leukocytes in a blood sample.

differentiation (dif-er-ent′she-a′shun) The specialization of structures in the process of embryonic development.

diffuse endocrine system (DES) Different kinds of hormone-secreting cells in various locations throughout the body.

DiGeorge syndrome An immunodeficiency caused by the absence of T lymphocytes.

digestion The breakdown of large molecules into smaller ones.

digestive Pertaining to digestion.

diopter (di-op′ter) A measure of the light-bending strength of a lens.

dipeptidase (di-pep′tid-ās) An enzyme that breaks dipeptides into amino acids.

dipeptide (di-pep′tīd) A molecule consisting of two amino acids held together by a peptide bond.

diploe (dip′lo-e) The loose osseous tissue between the two outer plates of a cranial bone.

disaccharide (di-sak′ar-īd) A molecule consisting of two sugar (saccharide) units held together by a glycosidic bond.

dislocation The displacement of a body part, especially a bone.

dissociation (dis-so-she-a′shun) **constant** (K) A measure of the degree to which a chemical compound separates into components (usually ions).

distal (dis′tal) Farthest from the point of origin of a structure.

distress alarm reaction The body's response to a stressful situation.

disuse atrophy (at′ro-fe) The wasting of a muscle because of lack of use.

diuresis (di″u-re′sis) Increased urine production.

diuretic (di″u-ret′ik) An agent that causes diuresis.

divergence Movement spreading in different directions.

dizygotic (di″zi-got′ik) From two separate zygotes.

DNA polymerase (pol-im′er-ās) An enzyme involved in the synthesis of DNA.

DNA replication Synthesis of new DNA according to the information in existing DNA.

dominant In genetics, a characteristic that appears in the phenotype whenever the allele for it is present in the genotype.

dopamine (do′pah-mēn) A neurotransmitter and precursor of norepinephrine.

dorsal (dor′sal) Pertaining to the back; posterior in humans.

dorsiflexion (dor″si-flek′shun) Movement of the toes and foot toward the shin at the ankle joint.

dual innervation The presence of both sympathetic and parasympathetic innervation in an organ.

duct A tube.

ductus (duk′tus) A tube.

duodenal (du″o-de′nal) Pertaining to the duodenum.

duodenum (du″o-de′num) A short portion of the small intestine connected to the stomach and receiving secretions of the liver and pancreas.

dural (dur′al) Pertaining to the dura mater.

dura mater (du′rah ma′ter) The tough outermost meninges surrounding the brain and spinal cord and other meninges.

dynamic equilibrium (1) The maintenance of balance when the head and body are suddenly rotated; (2) the maintenance of nearly constant internal conditions within the body in spite of continuously occurring small changes.

dynorphin (dīn-or′fin) A polypeptide similar to an enkephalin.

dyslexia (dis-lek′se-ah) Inability to read with understanding.

dysmenorrhea (dis″men-o-re′ah) Painful menstruation.

E

eccrine (ek′rin) Excretory.

eclampsia (ek-lamp′se-ah) A severe toxemia of late pregnancy and delivery involving fluid and electrolyte imbalances and sometimes convulsions.

ectoderm (ek′to-derm) The outermost embryonic germ layer.

ectopic (ek-top′ik) Not in the normal place.

edema (ĕ-de′mah) Accumulation of excess fluid in the tissues.

efferent (ef′er-ent) Leading away from.

ejaculation (e-jak″u-la′shun) Ejection of semen.

ejaculatory (e-jak′u-lah-to″re) Pertaining to ejaculation.

elasticity The ability to be stretched and return to original shape.

elastic recoil Return to original shape after stretching.

electrocardiogram (e-lek″tro-kar′de-o-gram″) (ECG) A record of electrical changes detected on the surface of the body associated with heart contractions.

electroencephalogram (e-lek″tro-en-sef′ah-lo-gram″) (EEG) A record of electrical changes detected on the scalp associated with brain activity.

electrolyte (e-lek′tro-līt) Any substance that ionizes and conducts electricity.

electron A negatively charged particle that moves around the nucleus of an atom.

electron transport system A set of enzymes and coenzymes in the cristae of mitochondria that move electrons from substrates to oxygen.

eleidin (el-e′i-din) A precursor to keratin found in the stratum lucidum of skin.

element A fundamental unit of matter.

elevation A raising movement.

ellipsoid (e-lip′soid) **joint** A joint whose articular surfaces are oval in shape.

embolus (em′bo-lus) A blood clot moving in a blood vessel.

embryonic (em″bre-on′ik) **disk** The cells of a trophoblast that give rise to the embryo.

emission (e-mish′un) A discharge, especially of semen.

emmetropia (em-et-ro′pe-ah) Normal vision, neither nearsighted nor farsighted.

emotion A state of feeling; the affective aspect of consciousness.

emphysema (em″fis-e′mah) A disorder in which the alveoli are enlarged by destruction or dilation of their walls.

emulsification (e-mul″sĕ-fĭ-ka′shun) Process by which fat droplets in foods are broken into smaller particles by the action of bile salts.

enamel Hard covering of the exposed part of a tooth.

end-diastolic (end-di-as-tol′ik) **volume** (EDV) The volume of blood in a ventricle at the end of its relaxation period.

endergonic (end″er-gon′ik) Using energy.

endocardium (en″do-kar′de-um) The epithelial lining of the heart.

endochondral (en″do-kon′dral) Within cartilage.

endocrine (en′do-krin) Pertaining to a ductless gland.

endocytosis (en-do-si-to′sis) Movement of particles into cells.

endoderm (en′do-derm) The inner embryonic germ layer.

endogenous (en-doj′en-us) Originating within an organism.

endolymph (en′do-limf) Fluid within the membranous labyrinth of the inner ear.

endometriosis (en″do-me′tre-o′sis) Presence of endometrial tissue in abnormal places.

endometrium (en-do-me′tre-um) Epithelial and connective tissue lining the uterus.

endomysium (en-do-mi′se-um) Connective tissue covering individual muscle fibers.

endoneurium (en-do-nu′re-um) Connective tissue covering individual nerve fibers.

endoplasmic reticulum (en″do-plas′mik rĕ-tik′u-lum) A network of membranous vesicles within a cell.

endorphin (en-dor′fin) A brain peptide that binds to opiate receptors.

endosteum (en-dos′te-um) Membrane that lines the marrow cavity of bones.

endothelium (en″do-the′le-um) Epithelial lining of blood and lymph vessels.

end plate potential The potential difference at the interface of a neuron and muscle.

enkephalin (en-kef′al-in) A brain peptide derived from endorphin that binds to opiate receptors.

enteric (en-ter′ik) Pertaining to the intestine.

enterogastric (en″ter-o-gas′trik) **reflex** A neural signal activated by the presence of excessively acidic or fatty chyme in the small intestine that slows stomach peristalsis.

enterohepatic (en″ter-o-hep-at′ik) **circulation** Movement by which bile salts are returned to the liver and resecreted in bile.

enterokinase (en″ter-o-ki′nās) A proteolytic enzyme from the intestinal mucosa.

entrainment (en-trān′ment) Synchronization of the biological clock by environmental factors.

entropy (en′tro-pe) Tendency toward disorder.

enzyme (en′zīm) A protein that catalyzes a chemical reaction.

eosinophil (e″o-sin′o-fil) A granular leukocyte capable of being stained with the dye eosin.

ependymal (ep-en′dim-al) **cell** Neuroglia cells that line the brain ventricles and central canal of the spinal cord.

epicardium (ep″ĭ-kar′de-um) A connective tissue that forms the outermost layer of the heart and lines the pericardial sac.

epicondyle (ep″ik-on′dīl) A protrusion on a bone to which muscles attach.

epidermis (ep″ĭ-der′mis) Outer epithelial layer of the skin.

epididymis (ep″id-id′im-is) A coiled duct between the testis and the ductus deferens.

epidural (ep″ĭ-du′ral) Outside the dura mater.

epigastric (ep″ĭ-gas′trik) Pertaining to the upper middle portion of the abdomen.

epiglottis (ep″ĭ-glot′is) Elastic cartilage closing the glottis.

epimysium (ep-e-mi′se-um) Connective tissue covering a muscle.

epinephrine (ep-e-nef′rin) Main hormone secreted by the adrenal medulla.

epineurium (ep-e-nu′re-um) Connective tissue covering an entire nerve.

epiphyseal (ep″ĭ-fiz′e-al) Pertaining to an epiphysis of a bone or to the adjacent growth region.

epiphysis (ĕ-pif′ĭ-sis) The end region of a long bone.

epiploic (ep″ĭ-plo′ik) Pertaining to an omentum.

episiotomy (ep-iz″e-ot′om-e) A surgical incision of the vulvar orifice to prevent tearing during childbirth.

epithelial (ep-e-the′le-al) Pertaining to epithelium.

epithelium (ep-e-the′le-um) A tissue that lines hollow organs or covers surfaces.

eponychium (ep″o-nik′e-um) Skin fold over the root of a nail.

erection (ĕ-rek′shun) Rigid state of the penis.

erythrocyte (er-ith′ro-sīt) Red blood cell.

erythrolabe (er-ith′ro-lāb) A cone pigment sensitive to red light.

erythropoiesis (er-ith″ro-poi-e′sis) Production of red blood cells.

erythropoietin (er-ith″ro-poi′et-in) A substance secreted by the kidney that stimulates production of red blood cells.

esophageal (ĕ-sof″ah-ge′al) Pertaining to the esophagus.

esophagus (e-sof′ag-us) Muscular tube in the digestive tract between the pharynx and stomach.

essential amino acid An amino acid that must be present in the diet because the body cannot make it.

essential fatty acid A fatty acid that must be present in the diet because the body cannot make it.

essential hypertension (hi-per-ten′shun) Elevated arterial blood pressure of unknown cause.

estradiol (es-tra-di′ol) The main human estrogen.

estrogen (es′tro-jen) A kind of female hormone that stimulates development of sex organs and secondary sexual characteristics.

ethmoid (eth′moid) Sievelike.

Eustachian (u-sta′ke-an) **tube** A passageway between the middle ear and the pharynx.

eustress (u′stres) A productive form of stress.

evaporation The process of changing from liquid to gaseous form.

eversion Turning outward.

exchange vessel Blood vessel (capillary) through which substances are exchanged between blood and tissues.

excitability Ability to respond to a stimulus.

excitation-contraction coupling The connection between neural signals and muscle contraction.

excitatory postsynaptic potential (EPSP) Partial depolarization of a postsynaptic membrane.

excretion Elimination of a waste product.

exergonic (ek″ser-gon′ik) Giving off energy.

exhaustion stage Stage at which the body has failed to cope with stress.

exocrine (ex'o-krin)   Pertaining to a gland with ducts.

exocytosis (eks''o-si-to'sis)   The release of particles from cells.

exogenous (eks-oj'en-us)   Originating outside the body.

experimental variable   A condition or event under study in a scientific experiment.

expiration   Exhaling; breathing out.

expiratory reserve volume   The volume of gas that can be forcibly exhaled after normal exhalation.

extensibility   Capability of being extended.

extension   Movement that increases the angle between two bones.

external   On the outside.

exteroceptor (eks''ter-o-sep'tor)   A sensory receptor that detects changes in the environment.

extracellular (ex''tra-sel'u-lar)   Outside a cell.

extrafusal (ex''tra-fu'sal)   Outside a muscle spindle.

extrapyramidal (ex-tra-pi-ram'id-al) **tract**   Bundles of motor fibers in the spinal cord that do not pass through the pyramids.

extrinsic   Originating outside the body.

F

facilitated diffusion   Diffusion down a gradient aided by a carrier molecule, but not requiring cellular energy.

falciform (fal'sif-orm)   Sickle-shaped.

fallopian (fal-o'pe-an) **tube**   Uterine tube.

fascia (fash'e-ah)   Fibrous connective tissue sheath around muscles and beneath skin.

fascicle (fas'ik-l)   A bundle of fibers.

fasciculation (fas-ik''u-la'shun)   Involuntary twitching of a bundle of muscle fibers.

fasciculus (fah-sik'u-lus)   A bundle of fibers.

fatigue   Temporary loss of power.

fatty acid   A long hydrocarbon chain with an organic acid group at one end.

fauces (faw'sēz)   Passage from the mouth to the pharynx.

feces (fe'sēz)   Material expelled from the rectum through the anus.

feedback   An aspect of a system whereby the output of a process affects the process.

female pronucleus (pro-nu'kle-us)   The nucleus of an ovum that fuses with the nucleus of a sperm.

femoral (fem'or-al)   Pertaining to a femur.

femur (fe'mur)   Thigh bone.

fenestrated (fen'es-tra''ted)   Pierced with one or more openings; having windows.

ferritin (fer'it-in)   A molecule consisting of the protein apoferritin and iron.

fertilized ovum   An ovum that has been penetrated by a sperm.

fetus (fe'tus)   An unborn child from two months development to birth.

fever   An elevated body temperature.

fibrillation (fib-ril-a'shun)   Uncoordinated contraction of some of the fibers of a muscle.

fibrin (fi'brin)   A protein that forms a network in a blood clot.

fibrinogen (fi-brin'o-jen)   Inactive fibrin.

fibroblast (fi'bro-blast)   A connective tissue cell that synthesizes fibers and ground substance.

fibrous tunic   Outer layer of the eye.

fibula (fib'u-lah)   Lateral bone of leg.

filtration   Passage of a fluid through a membrane because of a mechanical pressure.

filum terminale (fi'lum ter-min-a'le)   Fibrous attachment between the spinal cord and the coccyx.

fimbria (fim'bri-ah)   Fringelike structure.

fissure   A slit or furrow.

flagellum (flah-jel'um)   A movable hairlike process on a cell.

flavin adenine dinucleotide (fla'vin ad'en-ēn di-nu'kle-o-tīd) (FAD)   A hydrogen-carrying coenzyme.

flexion   A movement that decreases the angle between two bones.

flexure (flek'sher)   Turn or bend.

fluid-mosaic model   A description of the molecular structure of a cell membrane.

fluid pinocytosis (pi''no-si-to'sis)   Cell drinking.

fluid regulation   The maintenance of fluid volumes in the body within normal ranges.

flutter   Extremely rapid, ineffective heart contractions.

focal length   Distance from lens to point at which it focuses light rays.

focal point   Point at which light rays passing through a lens focus, or converge.

folacin (fol'ah-sin)   A vitamin that helps transfer single carbon groups.

follicle-stimulating hormone (FSH)   A hormone that stimulates maturation of ova and sperm.

fontanel (fon''tah-nel')   A membranous region between cranial bones in an infant.

foramen (fo-ra'men)   Opening or hole.

forced expiratory volume   The volume of gas that can be exhaled after a normal exhalation.

fornix (for'niks)   A vaultlike space.

fossa (fos'ah)   A hollow depression.

fovea centralis (fo've-ah sen-tra'lis)   A pit in the retina having only cone receptors; region of greatest visual acuity.

fracture   Breaking of a part, particularly a bone.

frameshift mutation   A change in DNA sequence caused by adding or deleting bases.

Frank-Starling law   A law stating that stroke volume increases in proportion to the stretching of heart muscle fibers.

free-running rhythm   A biological rhythm that occurs under constant conditions and that is not affected by ordinary environmental changes.

frenulum (fren'u-lum)   A small fold in a membrane that limits the movement of an organ or part, such as the membrane under the tongue.

frequency   Number of times an event occurs in a given period, such as vibrations per second of sound waves.

frontal (frun'tal)   Pertaining to the forehead.

frostbite   Freezing of tissue.

fulcrum (ful'krum)   Fixed point about which a lever moves.

functional group   The part of a molecule that participates in a chemical reaction.

fundus (fun'dus)   The base of an organ farthest from its outlet.

G

gallbladder   A sac on the underside of the liver where bile is stored.

gallstone   Deposit of particles in gallbladder.

gamete (gam'ēt)   An ovum or sperm.

gamma-aminobutyric (gam'ah ah-me'no-bu-tir'ik) **acid**   (GABA) An inhibitory neurotransmitter of the central nervous system.

ganglion (gang'le-on)   An aggregation of cell bodies in the peripheral nervous system.

gastric (gas'trik)   Pertaining to the stomach.

gastrin (gas'trin)   A hormone from the stomach that circulates in the blood to the stomach and stimulates HCl secretion.

gastritis (gas-tri'tis)   Inflammation of the stomach.

gastroesophageal (gas''tro-e-sof''ah-je'al)   Pertaining to the stomach and esophagus.

gastroferrin (gas''tro-fer'in)   An iron-binding protein in the stomach.

gastroileal (gas''tro-il'e-al)   Pertaining to stomach and ileum.

gel   The liquid state of a colloidal dispersion.

gene (jēn)   Functional unit of heredity; the part of a chromosome that transmits a particular hereditary characteristic.

general adaptation syndrome   A group of changes that appear in animals under stress.

generator potential   The initial depolarization of a sensory receptor.

genetic code   The three-base sequences in messenger RNA derived from DNA that determine the order of amino acids in proteins.

genetic engineering   The use of procedures designed by humans that alter genetic information.

genetics (jen-et'ikz)   The study of heredity.

genetic screening   The use of procedures to determine whether certain genetic characteristics are present in prospective parents, fetuses, or newborns.

geniculate (jen-ik′u-lāt)   Bent, like a knee.

genital herpes (her′pēs)   A sexually transmitted viral infection.

genitalia (jen″it-a′le-ah)   Genital, or sex, organs.

genital tubercle   A sexually undifferentiated protrusion from which certain parts of male or female external genitalia develop.

genome (je′nōm)   An organism's entire complement of DNA.

genotype (jen′o-tīp)   The combination of alleles of a gene carried by a particular individual; all the genes carried by a particular individual.

germinal center   Aggregation of cells in a lymph node that give rise to lymphocytes.

germinal epithelium   Epithelial cells that give rise to gamete-producing cells.

gingiva (jin-ji′vah)   Gums.

gladiolus (glah-di′o-lus)   A part of the sternum; sword-shaped.

glans (glanz)   A conical structure forming the tip of the penis or clitoris.

glaucoma (glaw-ko′mah)   A disorder in which aqueous humor accumulates and produces excessive intraocular pressure.

glenohumeral (glen″o-hu′mer-al)   Pertaining to the glenoid fossa and humerus.

glenoid (gle′noid)   Socketlike.

gliadin (gli′ad-in)   A protein in wheat and some other grains.

gliding joint   A joint in which articular surfaces glide over one another.

glioma (gli-o′ma)   Tumor of neuroglia.

globin (glo′bin)   A kind of protein found in hemoglobin and certain other biological molecules.

globulin (glob′u-lin)   A globular shaped protein, including many in the plasma.

glomerular (glo-mer′u-lar)   Pertaining to a glomerulus.

glomerular filtration rate (GFR)   The rate at which fluid leaves the blood at the glomerulus and enters the kidney filtrate.

glomerulonephritis (glo-mer″u-lo-nef-ri′tis)   An inflammation of the glomeruli of the kidneys.

glomerulus (glo-mer′u-lus)   A tuft of capillaries surrounded by a glomerular capsule.

glottis (glot′is)   A slitlike opening into the larynx.

glucagon (gloo′kah-gon)   A hormone that raises blood glucose.

glucocorticoid (gloo-ko-kor′tĭ-koid)   A hormone that helps to regulate carbohydrate metabolism.

gluconeogenesis (glu-ko′ne-o-jen′e-sis)   A metabolic pathway that makes glucose from noncarbohydrate substances.

glucose sparing   The metabolism of fats by many cells, thereby conserving blood glucose for cells that cannot metabolize fats.

gluten (gloo′ten)   A protein found in wheat and some other grains.

glycine (gli′sēn)   The simplist amino acid.

glycogenesis (gli″ko-jen′is-is)   The synthesis of glycogen.

glycogenolysis (gli″ko-jen-ol′is-is)   The breakdown of glycogen.

glycolipid (gli″ko-lip′id)   A molecule containing both carbohydrate and lipid.

glycolysis (gli-kol′is-is)   The breakdown of glucose to pyruvic acid.

glycoprotein (gli″ko-pro′te-in)   A molecule containing both carbohydrate and protein.

Golgi (gol′je) **apparatus**   A cluster of membranous vesicles found in cells that completes synthesis of secretions.

Golgi tendon organ   A proprioceptor in a tendon.

gomphosis (gom-fo′sis)   A joint in which a spike of bone fits into a socket, such as a tooth and its socket.

gonad (go′nad)   An ovum or sperm.

gonorrhea (gon″or-e′ah)   A sexually transmitted disease caused by *Neisseria gonorrhoeae.*

gout (gowt)   A joint inflammation caused by accumulation of uric acid.

graafian (graf′e-an) **follicle**   An ovarian follicle.

gracile (gras′il)   Slender or delicate.

gradient (gra′di-ent)   The rate of change in the magnitude of a variable, such as pressure or concentration.

graft   A tissue transplanted from one site to another.

graft-versus-host disease   A disorder in which graft cells react immunologically and destroy host cells.

gram molecular weight   A quantity of a substance equal to its molecular weight in grams.

granular leukocyte   A white blood cell with granular cytoplasm.

gray matter   Unmyelinated tissue, mainly cell bodies, in the central nervous system.

ground substance   A protein-polysaccharide substance deposited among fibers of connective tissue.

growth hormone-hypothalamic mechanism   A mechanism that helps to regulate metabolism.

guanidine (gwan′id-ēn) **triphosphate** (GTP)   An energy storage molecule.

gubernaculum (gu″ber-nak′u-lum)   A band of connective tissue that guides a movement, such as the descent of a testis.

gustatory (gus′tat-o″re)   Pertaining to the sense of taste.

gyrus (ji′rus)   A fold, such as a convolution of the cerebrum.

H

habit-forming   The quality that causes addicts to spend their time seeking a drug.

habituation (hab-it-u-a′shun)   Gradual adaptation to a continuing stimulus.

hair cell   A kind of sensory receptor that is stimulated by movement of a thin projection.

haploid (hap′loid)   Having one of each pair of chromosomes.

hapten (hap′ten)   A small molecule that can act as an antigenic determinant when combined with a larger molecule.

haustra (haws′trah)   A saclike protrusion of the wall of the large intestine.

haustral (haws′tral) **churning**   Motility of the large intestine that moves its contents from one haustra to the next.

Haversian (ha-ver′shan) **system**   A set of concentric lamellae surrounding a canal in compact bone.

heart block   An arrhythmia in which conduction is disrupted.

heart failure   The inability of the heart to pump sufficient blood to supply the tissues of the body with nutrients and remove their wastes.

heart murmur   An abnormal sound associated with turbulence around a leaky valve.

heliocotrema   A helix; the passage at the tip of the cochlea that connects the scala tympani and the scala vestibuli.

hematocrit (hem-at′o-krit)   The proportion of a volume of blood occupied by red blood cells.

hematoma (hem″at-o′mah)   An accumulation of effused blood.

hematopoiesis (he-mat″o-poi-e′sis)   The formation of blood cells.

hematopoietic (hem″ah-to-poi-et′ik)   Pertaining to hematopoiesis.

heme (hēm)   An iron-containing pigment that binds to oxygen and that forms part of hemoglobin.

hemisection   Severing of the left or right side of the cord.

hemodialysis (he″mo-di-al′is-is)   The removal of substances from the blood by dialysis.

hemoglobin (he′mo-glo″bin)   The oxygen-carrying protein in erythrocytes.

hemolysis (he-mol′is-is)   Breakdown of erythrocytes with the release of hemoglobin.

hemolytic (he″mo-lit′ik)   Pertaining to hemolysis.

hemolytic disease of the newborn   A disorder in which a previously sensitized Rh negative mother's anti-Rh antibodies destroy erythrocytes in an Rh-positive fetus.

hemophilia (he″mo-fil′e-ah)   An inherited absence of a blood clotting factor.

hemopoiesis (he″mo-poi-e′sis) The formation of blood cells.

hemopoietic (he″mo-poi-et′ik) Pertaining to hemopoiesis, or hematopoiesis.

hemorrhage (hem′or-aj) Loss of a significant quantity of blood.

hemorrhagic (hem″or-aj′ik) Pertaining to a hemorrhage.

hemorrhoids (hem′or-oidz) Enlarged blood vessels in the rectal mucosa.

hemosiderin (he″mo-sid′er-in) An iron storage molecule.

hemostasis (he″mo-sta′sis) The arrest of bleeding.

Henderson-Hasselbalch equation An equation that relates pH to properties of an acid and its salt.

Henry's law Gases dissolve in liquids in proportion to their partial pressures and solubilities.

heparin (hep′ar-in) An anticoagulant produced in several body tissues.

hepatic (he-pat′ik) Pertaining to the liver.

hepatitis (hep″ah-ti′tis) An inflammation of the liver.

Hering-Breuer reflex A reflex that protects the lungs against excessive stretching.

hernia (her′ne-ah) Rupture of a muscle layer.

herpes (her′pēz) A kind of viral infection.

hertz A unit of vibration frequency.

heterogeneity (het″er-o-jen-e′it-e) Diversity.

heterozygous (het″er-o-zi′gus) Having unlike alleles at corresponding loci of a pair of chromosomes.

hiatal (hi-a′tal) Pertaining to a gap or fissure.

hilum (hi′lum) Region of an organ where blood vessels and nerves enter and leave.

hinge joint A joint in which the articulating surfaces move relative to each other like the parts of a hinge.

hippocampal (hip″o-kam′pal) Pertaining to the hippocampus.

hippocampus (hip″o-kam′pus) A part of the limbic system found in the temporal region and concerned with emotion and memory.

hirudin (hi-ru′din) An anticoagulant secreted by leeches.

histamine (his′tah-mēn) A substance from injured cells that causes vasodilation, bronchial constriction, and other symptoms of an allergic reaction.

histiocyte (his′te-o-sīt) A tissue macrophage.

histologist (his-tol′o-jist) A scientist who studies tissues.

histology (his-tol′o-je) The study of tissues.

holocrine (ho′lo-krin) A gland that releases whole cells and their products.

homeostasis (ho″me-o-sta′-sis) The maintenance of conditions within a narrow, tolerable range.

homeostatic (ho″me-o-stat′ik) **system** A control system that helps to maintain homeostasis.

homozygous (ho″mo-zi′gus) Having like alleles at corresponding loci of a pair of chromosomes.

hormone (hor′mōn) A regulatory substance produced by an endocrine cell and transported in the blood to its target cells.

host-versus-graft disease A disorder in which host cells react immunologically and destroy graft cells.

human chorionic gonadotropin (ko″re-on′ik go″nad-o-tro′pin) A hormone from the placenta that stimulates hormone secretion from the corpus luteum.

human leukocyte antigen (HLA) One of several antigens that gives a person's cells their unique identity; such antigens are used in tissue typing to match organ donor and recipient.

humerus (hu′mer-us) Bone of the upper arm.

humor Fluid.

humoral (hu′mor-al) **immunity** Disease resistance produced by antibodies.

Huntington's chorea A dominant hereditary disorder that leads to degeneration of the nervous system.

hyaline (hi′ah-lin) **membrane disease** (HMD) A disorder in newborns in which alveoli collapse because of a lack of surfactant.

hybridoma (hi″brid-o′mah) A cell made by fusion of parts of two cells.

hydrocephalus (hi-dro-sef′al-us) Excessive fluid in brain ventricles; literally, water on the brain.

hydrogen bond A relatively weak covalent bond that binds hydrogen and other elements, especially oxygen or nitrogen, and is found in proteins and nucleic acids.

hydrolysis (hi-drol′i-sis) The splitting of large molecules with the addition of water.

hydrophilic (hi-dro-fil′ik) Water loving.

hydrophobic (hi-dro-fo′bik) Water fearing.

hydrostatic pressure Force exerted by a fluid.

hydroxyapatite (hi-drok″se-ap′ah-tīt) A mineral that comprises much of the substance of bone.

hymen (hi′men) An easily ruptured membrane over the opening to the vagina.

hyperemesis gravidarum (hi″per-em′ĕ-sis grav″id-a′rum) Excessive vomiting during pregnancy.

hyperextension Extension of a joint beyond its normal range of movement.

hyperglycemia (hi-per-gli-sem′e-ah) Abnormally high blood glucose.

hypermetropia (hi″per-me-tro′pe-ah) Farsightedness.

hyperosmotic (hi″per-os-mot′ik) Having higher osmotic pressure (relative to a cell).

hypersensitivity (hi″per-sen′sit-iv″it-e) Abnormal overreactivity, as occurs in allergy and certain other immune reactions.

hypertension (hi″per-ten′shun) Excessively high blood pressure.

hyperthermia (hi″per-ther′me-ah) An abnormally high body temperature.

hypertonic (hi″per-ton′ik) Producing movement of water out of cells.

hypertrophy (hi-per′trof-e) Increase in the size of an organ, usually by increase in the size of its cells.

hypochondriac (hi″po-kon′dre-ak) Pertaining to the abdominal region beneath the ribs; having a morbid anxiety about health.

hypodermis (hi-po-der′mis) Tissue beneath the skin.

hypogastric (hi″po-gas′trik) Pertaining to the abdominal region below the stomach.

hypoglossal (hi-po-glos′al) Under the tongue.

hypoglycemia (hi″po-gli-se′me-ah) Abnormally low blood glucose.

hypoglycemic (hi″po-gli-se′mik) An agent that lowers blood glucose.

hyponychium (hi-po-nik′e-um) Cornified epithelium under the free border of a nail.

hypophysis (hi-pof′is-is) The pituitary gland.

hyposensitization (hi″po-sen″sit-i-za′shun) Reduction of sensitivity, as through allergy shots.

hyposmotic (hi″pos-mot′ik) Having lower osmotic pressure (relative to a cell).

hypotension (hi″po-ten′shun) Excessively low blood pressure.

hypothalamus (hi′po-thal′am-us) A portion of the brain that provides the main connection between the nervous and endocrine systems.

hypothermia (hi″po-ther′me-ah) An abnormally low body temperature.

hypothesis A proposed explanation.

hypotonic (hi-po-ton′ik) Producing movement of water into cells.

hypovolemic (hi″po-vo-le′mik) Having below normal volume.

hypoxemia (hi-pox-em′e-ah) A deficiency of oxygen in the blood.

hypoxia (hi-pok′se-ah) A deficiency of oxygen in cells.

I

IgA An immunoglobulin found in secretions.

IgD An immunoglobulin of unknown function.

IgE An immunoglobulin associated with allergy.

IgG An immunoglobulin found mainly in the blood and primarily responsible for resisting infection.

IgM A multiunit immunoglobulin most abundant early in an immune response.

ileocecal (il″e-o-se′kal) Pertaining to the ileum and cecum.

ileostomy (il″e-os′to-me)   The making of an artificial opening into the ileum.

ileum (il′e-um)   The lower part of the small intestine.

iliac (il′e-ak)   Pertaining to the ilium.

ilium (il-e-um)   A bone of the pelvis.

immune (im-mūn′)   Resistant to disease.

immune complex disorder   A kind of hypersensitivity in which antigen-antibody complexes damage tissues.

immunity (im-mūn′it-e)   Resistance to disease.

immunization (im-mun-iz-a′shun)   A procedure to create immunity.

immunodeficiency (im″u-no-def-ish′en-se)   A lack of normal immune function.

immunoglobulin (im″u-no-glob′u-lin)   A protein that binds with a foreign substance; an antibody that binds with an antigen.

immunology (im-un-ol′o-je)   The study of immunity.

immunosuppression (im″u-no-sup-resh′un)   The lessening of an immune response.

immunotoxin   An antibody with a toxic drug attached.

imperforate anus   Failure of the tissue closing the embryonic anus to degenerate.

implantation   The attachment of an embryo to the endometrium of the uterus.

impotence (im′po-tens)   Inability to copulate, usually failure of a male to maintain an erection.

incisor (in-si′zer)   A tooth shaped for cutting.

incontinence (in-kon′ten-ens)   Inability to control the release of urine from the urinary bladder.

incus (ing′kus)   Anvil; the middle one of three ear bones.

infancy   Period from 1 month to 2 years of age.

inferior   Below or beneath.

infertility   Inability to produce offspring.

inflammation   Localized defensive response to tissue injury.

infundibulum (in″fun-dib′u-lum)   Funnel-shaped passage or structure.

ingestion   The taking of food or fluid into the stomach.

inguinal (ing′win-al)   Pertaining to the groin.

inhibitory postsynaptic potential (IPSP)   A potential that makes generating a signal at a synapse more difficult.

innate   Present at birth.

innervation   Nerve supply.

inotropic (in″o-trop′ik)   Pertaining to muscle fibers.

insensible   Not perceptible.

insertion   Attachment of a muscle to a bone at its most movable end.

inspiration   Breathing in.

inspiratory capacity   The total volume of gas that can be inhaled.

inspiratory reserve volume   The volume of gas that can be forcibly inhaled after normal inhalation.

insulin (in′su-lin)   A hormone from the pancreas that lowers blood glucose.

insulin-glucagon mechanism   A mechanism that helps to regulate metabolism.

insulin shock   A circulatory insufficiency due to an insulin overdose that suddenly lowers blood glucose.

integral   Inseparable part of.

integumentary (in-teg-u-men′tar-e)   Pertaining to the skin.

intensity   Degree of tension or activity.

interatrial (in″ter-a′tre-al)   Between the atria of the heart.

intercalated (in-ter′kah-lat-ed)   Inserted between.

interferon (in″ter-fe′ron)   A protein released by cells infected with a virus, which causes adjacent cells to make an antiviral protein.

interleukin (in-ter-lu′kin)   A substance that enhances an immune reaction.

interlobar   Between lobes.

interlobular   Between lobules.

intermediate   A molecule produced within a metabolic pathway.

internal   On the inside.

internal environment   The environment of cells within the body.

interneuron   A neuron that relays signals between a sensory and a motor neuron, typically found in the spinal cord.

internodal   Between nodes.

interosseous   Situated between bones.

interphase   Stage of the cell cycle during which the cell is not dividing.

interstitial (in-ter-stish′al)   Pertaining to spaces between cells.

interventricular (in-ter-ven-trik′u-lar)   Between the ventricles of the heart.

intestinal   Pertaining to the intestine.

intra-alveolar (in-trah-al-ve′o-lar)   Within the alveoli of the lungs.

intracellular   Within a cell.

intrafusal   Modified muscle fibers that detect stretching within a muscle spindle.

intramembranous (in″trah-mem′brah-nus)   Within a membrane.

intrapleural   Within the pleural cavity.

intrathoracic   Within the thorax, or chest.

intrinsic   Entirely within.

intrinsic factor   Substance produced by the gastric mucosa that is necessary for the transport and absorption of vitamin B_{12}.

inversion   Turning inward; a rearrangement in the sequence of nucleotides in DNA of a chromosome.

in vitro (in ve′tro)   Outside the body; literally, in glass.

ion   A charged atom or group of atoms.

ionic bond   A chemical bond through which atoms are held together by the attraction of unlike charges.

ipsilateral (ip″si-lat′er-al)   On the same side.

iris   Muscular diaphragm anterior to the lens of the eye that controls the amount of light entering the eye.

ischemia (is-ke′me-ah)   Reduction in blood flow to an area.

ischium (is′ke um)   A posterior bone of the pelvic girdle.

islet of Langerhans (lahn′ger-hanz)   Cluster of hormone-secreting cells in the pancreas.

isomer (i′so-mer)   A molecule having the same number and variety of atoms as another molecule, but arranged differently.

isometric (i-so-met′rik)   Same length.

isosmotic (i″sos-mot′ik)   Having the same osmotic pressure.

isotonic (i″so-ton′ik)   Having the same tonicity.

isotope (i′so-tōp)   An atom of an element having a different number of neutrons than certain other atoms of the same element.

isovolumetric (i″so-vol-u-met′rik)   Having the same volume.

isozyme (i-so-zīm)   Isomer of an enzyme.

isthmus (isth′mus)   Neck or constriction.

J

jaundice (jawn′dis)   Yellowishness of skin and membranes due to excess bile pigments in the blood.

jejunum (jĕ-joo′num)   The middle section of the small intestine.

joint   A region of union between two or more bones.

juxtaglomerular (juks-tah-glo-mer′u-lar)   Pertaining to or near a glomerulus.

juxtamedullary (juks″tah-med′ul-ar-e)   Pertaining to or near the medulla, usually referring to the kidney.

K

keratinocyte (kĕr-at′i-no-sīt)   A skin cell that contains keratin.

keratohyalin (ker″ah-to-hi′ah-lin)   A substance that gives skin a translucent appearance.

kernicterus (ker-nik′ter-us)   Deposition of bile pigments in brain tissue.

ketone (ke′tōn) **body**   Acidic molecule produced during excessive metabolism of fat.

ketosis (ke-to′sis)   Accumulation of ketone bodies in blood and urine.

kilocalorie (kcal)   Amount of heat required to raise the temperature of 1 kilogram of water 1° C.

kinesthetic (ken″es-thet′ik)   Pertaining to the sensing of movement.

kinetic   Energy of motion.

kinin (ki′nin)   A substance that stimulates events in the inflammatory process.

Klinefelter's (klīn′fel-terz) **syndrome**   A condition in which XXY sex chromosomes are present.

Korotkoff's (ko-rot′kofs) **sound** A sound heard through a stethoscope during the measurement of blood pressure.

Krebs cycle A sequence of reactions that oxidize acetyl-CoA; citric acid cycle; tricarboxylic acid cycle.

Kupffer's (koop-fers) **cell** A phagocytic cell in the walls of liver sinusoids.

kwashiorkor (kwash-e-or′kor) Protein deficiency, usually in young children.

L

labia (la′be-ah) Lip-shaped structures.

labyrinth (lab′eh-rinth) A maze.

lacrimal (lak′rim-al) Pertaining to tears.

lactase (lak′tās) An enzyme that digests lactose (milk sugar).

lactation (lak-ta′shun) The production and secretion of milk.

lacteal (lak′te-al) A lymph vessel within a villus of the small intestine.

lactiferous (lak-tif′er-us) Producing or conveying milk.

lacuna (lah-ku′nah) A cavity.

lambdoidal (lam-doid′al) Pertaining to a ridge and suture of the skull; literally like the Greek letter lambda.

lamella (lah-mel′ah) Thin layer.

lamina (lam′in-ah) Thin, flat plate.

lamina propria (lam′in-ah pro′pre-ah) A connective tissue layer underlying the epithelium of the intestinal mucosa.

laminar (lam′in-ar) Made up of thin plates.

lanugo (lan-u′go) Fine hair on the body of a fetus.

laryngopharynx (lar-ing″o-far′inks) Pertaining to the larynx and pharynx.

laryngotracheal (lar-ing″o-tra′ke-al) Pertaining to the larynx and trachea.

larynx (lar′inks) Voice box.

latent period In muscle physiology, the time between the application of a stimulus and the contraction of a muscle.

lateral On the side.

lateral inhibition Sharpening pitch detection by suppressing hair cells adjacent to those stimulated.

lateralization Variation in function in the two sides of a bilateral structure, especially the cerebrum.

law of adequate stimulus A law stating that a receptor responds only if it receives a sufficiently strong stimulus.

law of LaPlace A law stating that the distending pressure in a hollow object equals the tension in the wall divided by the radius of curvature of the object.

learning A change in behavior in response to environmental stimuli.

lecithin (les′ith-in) A kind of phospholipid found in animal tissues.

left-to-right shunt The flow of blood from the left to the right side of the heart.

lens Transparent, biconcave structure behind the iris of the eye that changes shape to focus the eye on far or near objects.

leukemia (lu-ke′me-ah) A malignant increase in the numbers of leukocytes.

leukocyte (loo′ko-sīt) A white blood cell.

leukocytosis (loo-ko-si-to′sis) **promoting** (LP) **factor** A substance that increases the attraction of leukocytes to a site of injury.

lever A mechanical device consisting of a rod, fulcrum, weight, and source of energy applied to the rod.

ligament (lig′ah-ment) Fibrous connective tissue that attaches bones together.

ligand (li′gand) A substance that binds to a receptor.

limbic (lim′bik) **system** Portion of the brain mainly associated with emotions.

linea alba (lin′e-ah al′bah) Connective tissue on the midline of the anterior abdominal wall; literally, white line.

lingual (ling′gwal) Pertaining to the tongue.

lipase (li′pās) An enzyme that digests lipids.

lipid Fats and fatlike substances.

lipoprotein A molecule containing both lipid and protein.

lobule A small lobe.

loop of Henle A U-shaped segment of a nephron that helps to concentrate sodium chloride in peritubular fluid.

lumbar (lum′bar) Pertaining to the loin region.

lumbosacral (lum-bo-sa′kral) Pertaining to the loin and sacral regions.

lunula (loo′nu-lah) Whitish crescent at the base of a nail; literally, half moon.

lutein (loo′te-in) A yellow pigment.

luteinizing (loo′te-in-īz″ing) **hormone** (LH) A hormone that stimulates ovulation and other reproductive processes.

luteum (loo′te-um) Yellow.

lymph (limf) Interstitial fluid that has entered a lymphatic vessel.

lymphatic (lim-fat′ik) Pertaining to lymph or a vessel that carries it.

lymph node An aggregation of lymphatic tissue interposed along the path of a lymphatic vessel.

lymphocyte (lim′fo-sīt) A leukocyte capable of participating in an immune response.

lymphoid (lim′foid) Resembling lymph or lymphatic tissue.

lymphokine (lim′fo-kīn) A substance that stimulates lymphocytes.

lysosome (li′so-sōm) Membrane-bound organelle that releases digestive enzymes.

lysozyme (li′so-zīm) An enzyme in tears that helps to destroy microbes.

M

macrophage (mak′ro-fāj) A large phagocytic cell found in connective tissue.

macula (mak′u-lah) A structure in the inner ear that contains sensory receptors for static equilibrium.

macula densa (mak′u-lah den′sah) Modified cells of the kidney's distal convoluted tubules associated with the juxtaglomerular apparatus.

major histocompatibility (his″to-kom-pat″ib-il′it-e) **complex** (MHC) **proteins** Proteins that account for antigenic individuality of a person's cells.

male pronucleus (pro-nu′kle-us) The nucleus from a sperm that has penetrated an ovum.

malignancy (mah-lig′nan-se) A tendency to progress in virulence; a cancerous growth.

malignant (mal-ig′nant) Having qualities of malignancy.

malleus (mal′e-us) Hammer; the outermost bone of the middle ear.

malnutrition Ill health due to an inadequate diet.

maltase (mawl′tās) Enzyme that digests maltose, a disaccharide derived from starch.

mammary (mam′er-e) **gland** A gland whose cells synthesize and secrete milk.

mandibular (man-dib′u-lar) Pertaining to a mandible.

manic-depressive psychosis (man′ik-de-pres′iv si-ko′sis) A severe mental disorder with wide mood swings.

manubrium (man-u′bre-um) A handle; the upper part of the sternum.

marasmus (mah-raz′mus) Near-starvation malnutrition.

marrow Fatty substance found in cavities inside bones.

mast cell A connective tissue cell that releases histamine in allergic reactions.

mastoid (mas′toid) Nipple-shaped; part of the temporal bone behind the ear.

matrix The fibrous structure in which ground substance of connective tissue is deposited.

matter That which has substance.

maxilla (mak-sil′ah) A bone containing sockets for upper teeth.

maxillary (mak-sil-ar′e) Pertaining to a maxilla.

mean arterial pressure (MAP) The average of systolic and diastolic pressures in an artery.

mechanoreceptor (mek″an″o-re-sep′tor) A receptor that responds to mechanical pressure.

median In the middle.

mediastinum (me″de-as-ti′num) Region of the thoracic cavity between the lungs, containing the heart, trachea, and part of the esophagus.

medulla (meh-dul′ah) The core or inner part of an organ.

medulla oblongata (med-ul′ah ob″long-gah′tah) The portion of the brain continuous with the spinal cord.

medullary (med′u-la″re) Pertaining to the core or inner part of an organ.

melanin (mel′ah-nin) A dark brown pigment found in hair and skin.

melanocyte (mel′ah-no-sīt) A cell that synthesizes melanin.

melanoma (mel′ah-no′mah) A kind of cancer common in skin.

melatonin (mel-ah-to′nin) A secretion of the pineal gland.

membrane potential An electrical potential difference between the inside and outside of a membrane.

memory The process of storing and recalling previous experiences.

menarche (men-ar′ke) The onset of menstruation.

Meniere's (men″e-ārz) **disease** An inflammation of the semicircular canals that leads to sight, hearing, and balance disorders.

meninges (men-in′jēz) Membranes that enclose the brain and spinal cord.

menopause (men′o-pawz) The cessation of menstruation.

menstrual (men′stroo-al) **cycle** A repetitive sequence of events in women of reproductive age, involving ovulation and preparation of the uterus for implantation.

menstruation (men-stroo-a′shun) The periodic discharge of blood, tissue debris, and fluid from the uterus.

meridional (mĕ-rid′e-o-nal) **fibers** Muscle fibers that attach suspensory ligament to choroid.

merocrine (mer′o-krīn) A gland that discharges secretions by exocytosis.

mesencephalon (mes-en-sef′a-lon) A middle region of the embryonic brain.

mesenchyme (mes′eng-kīm) Embryonic mesodern.

mesenteric (mes′en-ter″ik) Pertaining to a mesentery.

mesentery (mes′en-ter″e) A membrane consisting of two layers of peritoneum that suspends an abdominal organ.

mesoderm (mes′o-derm) The middle of three embryonic germ layers.

mesonephros (mes″o-nef′ros) A temporary embryonic kidney.

mesovarium (mes″o-va′re-um) A fold in the peritoneum that hold an ovary in place.

messenger RNA (mRNA) A nucleic acid that carries information in the form of codons for the synthesis of a protein.

metabolic acidosis A lowering of blood pH because of a metabolic disorder.

metabolic alkalosis A raising of blood pH because of a metabolic disorder.

metabolic rate The rate at which nutrients are oxidized.

metabolic water Water derived from the oxidation of foodstuffs.

metabolism (met-ab′o-lizm) The sum total of all chemical reactions in the body.

metacarpal (met″ah-kar′pal) A bone of the palm of the hand.

metanephros (met″ah-nef′ros) The embryonic kidney from which the adult functional kidney is derived.

metaphase A stage in mitosis during which chromosomes align along the middle of a cell.

metastasis (me-tas′tah-sis) The transfer of disease from one organ to another.

metatarsal (met″ah-tar′sal) A bone of the foot.

micelle (mi-sel′) A small fat droplet found in chyme.

microfilament A small, hollow protein fiber found within the cytoplasm of a cell.

microglia (mi-krog′le-ah) Small supporting cells of the central nervous system.

microtubule A cylindrical organelle that contributes to a cell's mitotic spindle.

microvillus (mi″kro-vil′us) A tiny cytoplasmic projection from the surface of intestinal epithelial cells.

micturition (mik″tu-rish′un) Urination.

migraine (mi′grān) Affecting half of the head.

mineral An inorganic substance.

mineralocorticoid A hormone that regulates mineral metabolism.

minute respiratory volume The volume of gas moved into and out of the lungs per minute.

miscarriage Expulsion of a fetus before it is capable of independent life.

mitochondrion (mi″to-kon′dre-on) An organelle in which oxidative and energy-capturing processes occur.

mitosis The division of a nucleus into two identical nuclei.

mitral (mi′tral) **valve** A bicuspid valve between the left atrium and ventricle of the heart.

mixed nerve A nerve that contains both sensory and motor fibers.

mixture A combination of two or more substances in any proportions in which the substances do not lose their properties.

modiolus (mo-di′o-lus) A supporting structure of the cochlea.

molar (mo′lar) (1) A large grinding tooth; (2) pertaining to the concentration of a solution.

mole A gram molecular weight.

molecule The smallest quantity of a substance that retains its chemical properties.

monoclonal (mon″o-klo′nal) **antibody** An antibody to a selected antigen made by the cells of a clone.

monocyte (mon′o-sīt) A large, phagocytic, agranular leukocyte.

monosaccharide (mon-o-sak′ar-īd) A simple sugar.

monozygotic (mon″o-zi-got′ik) From the same zygote.

mons pubis (monz pu′bis) Fatty area covered with pubic hair.

morula (mor′u-lah) A solid ball of cells formed in early embryological development.

motility Ability to move.

motion sickness Nausea and dizziness due to a temporary disturbance in semicircular canal function because of changes in movement.

motor end plate The portion of sarcolemma lying beneath nerve endings.

motor neuron A neuron that carries signals toward a muscle or gland.

motor unit A motor neuron and all the muscle fibers it innervates.

mucin (mu′sin) A glycoprotein found in ground substance and mucous secretions.

mucociliary (mu″ko-sil′e-ar″e) **escalator** The mechanism by which cilia and mucus of the respiratory tract move debris toward the pharynx.

mucosa (mu-ko′sah) Mucous membrane that lines cavities and tubes.

mucous (mu′kus) Pertaining to mucus.

multiple sclerosis (skle-ro′sis) A disease of the nervous system in which patches of tissue in the brain and spinal cord become hardened (sclerotic).

muscle spindle A spindle-shaped proprioceptor in a skeletal muscle.

muscular dystrophy (dis′tro-fe) A disease in which muscles undergo progressive degeneration.

muscularis (mus″ku-la′ris) (1) A muscular layer in the wall of an organ; (2) the muscular portion of the intestinal mucosa.

mutagen (mu′ta-jen) An agent that can cause changes in DNA.

mutation (mu-ta′shun) A change in the genetic information (DNA) in a cell.

myasthenia gravis (mi″as-the′ne-ah gra′vis) Progressive weakening of muscles because of an autoimmune reaction at motor end plates.

myelin (mi′el-in) An insulating substance deposited around axons by Schwann cells in the peripheral nervous system and oligodendrocytes in the central nervous system.

myelinated Having myelin.

myeloid (mi′el-oid) Pertaining to or derived from bone marrow.

myenteric (mi″en-ter′ik) Pertaining to the muscle layer of the intestine.

myocardial infarction (MI) Damage to heart muscle cells from interrupted blood supply.

myocardium (mi″o-kar′de-um) The middle muscular layer of the heart.

myoepithelial (mi″o-ep″i-the′le-al) Pertaining to epithelial cells that can contract.

myofibril (mi″o-fi′bril) A contractile fiber in a muscle cell.

myofilament (mi″o-fil′a-ment) A protein molecule that makes up part of a myofibril.

myogenic (mi″o-jen′ik) Originating in muscle tissue.

myoglobin (mi″o-glo′bin) A pigmented protein in muscle tissue that binds oxygen.

myokinase (mi″o-kin′ās) An enzyme that makes ATP and AMP from two molecules of ADP.

myometrium (mi″o-me′tre-um) The middle muscular layer of the uterus.

myoneural (mi″o-nu′ral) **junction** The joining of nerve and muscle tissue.

myopia (mi-o′pe-ah) Nearsightedness.

myosin (mi′o-sin) A protein that makes up the thick filaments of a myofibril.

myotome (mi′o-tōm) A block of mesoderm that gives rise to muscle.

N

naloxone (nal-oks′ōn) A drug that counteracts opium overdoses.

narcosis (nar-ko′sis) Profound unconsciousness.

nares (na′rēs) Nostrils.

nasal Pertaining to the nose.

nasopharynx (na″zo-far′inks) Pertaining to the nose and throat.

natriuresis (nat-re-u-re′sis) Stimulation of sodium excretion.

naturally acquired active immunity Immunity from having a disease.

naturally acquired passive immunity Immunity from receiving antibodies across the placenta or in breast milk.

negative feedback A mechanism by which the output of a system suppresses or inhibits activity of the system.

neocortex (ne″o-kor′teks) The most recently evolved portion of the cerebrum.

neonatal (ne-o-na′tal) **period** The period from birth to one month of age.

neonate (ne′o-nāt) A newborn infant.

nephrogenic (nef″ro-jen′ik) Arising from a kidney.

nephron (nef′ron) The functional unit of a kidney.

nerve A cordlike structure consisting primarily of axons of neurons covered with connective sheaths; conveys signals in the nervous system.

nerve deafness Impairment of hearing due to damage to sound receptors or nerve fibers leading from them to the brain.

net filtration pressure Pressure pushing substances out of a blood vessel.

net protein utilization The proportion of protein ingested that can actually be used by cells.

neural crest Cells derived from the neural tube that give rise to sensory neurons, the sympathetic nervous system, and the adrenal medulla.

neuralgia (nu-ral′je-ah) Pain associated with a nerve.

neural oscillator A process that controls a kind of movement or other behavior.

neural tube A tube formed from invagination of ectoderm that gives rise to the nervous system.

neurilemma (nu″ril-em′mah) The cell membrane of a Schwann cell.

neuritis (nu-ri′tis) Inflammation of a nerve.

neurofibrillary (nu-ro-fi′bril-a-re) **tangles** Masses of disorderly neural fibers found in the brains of Alzheimer patients.

neuroglia (nu-rog′le-ah) Supporting cells of the nervous system.

neuroglial (nu-rog′le-al) Pertaining to supporting cells of the nervous system.

neurohypophysis (nu″ro-hi-pof′is-is) The posterior, neural portion of the pituitary gland.

neuromuscular (nu″ro-mus′ku-lar) Pertaining to the association between the nervous and muscular systems.

neuronal (nu-ro′nal) **pool** A complex set of synapses in the central nervous system.

neuropeptide (nu-ro-pep′tīd) A chain of amino acids having some influence on neural function.

neurotransmitter (nu″ro-trans-mit′er) A chemical substance released from one neuron that transmits a signal to another neuron at a synapse.

neurulation (nu″roo-la′shun) Formation of the neural tube during embryonic development

neutron (nu′tron) An uncharged particle in the nucleus of an atom.

neutrophil (nu′tro-fil) A granular leukocyte that fails to stain with either acidic or basic stains.

niacin (ni′as-in) A B vitamin used to make the coenzyme NAD.

nicotinamide adenine dinucleotide (NAD) A coenzyme that carries hydrogen atoms or electrons in oxidation-reduction reactions.

Nissl (nis′l) **granule** RNA-containing granules and endoplasmic reticulum in nerve cell bodies and dendrites.

nitrogen balance A condition in which nitrogen entering the body equals nitrogen leaving it.

nociceptor (no″se-sep′tor) A receptor that responds to painful stimuli.

node of Ranvier (rahn-ve-a′) A gap in the myelin sheath of an axon.

nondisjunction (non″dis-junk′shun) Failure of replicated chromosomes to separate.

nonpolar Uncharged.

noradrenalin (nor-ad-ren′a-lin) Norepinephrine.

norepinephrine (nor″ep-ĕ-nef′rin) A neurotransmitter of the sympathetic division of the autonomic nervous system and of some brain neurons.

nuclear (nu′kle-ar) Pertaining to the nucleus.

nucleic (nu-kle′ik) **acid** A polymer of nitrogenous bases, five carbon sugars, and phosphates; DNA or RNA.

nucleolus (nu-kle′o-lus) An RNA-containing body within a nucleus.

nucleoplasm (nu′kle-o-plazm″) The substance of a nucleus.

nucleotide (nu′kle-o-tīd) A molecular unit containing a nitrogenous base, a five-carbon sugar, and one or more phosphates.

nucleus (nu′kle-us) (1) A cell's control center; (2) the central part of an atom; (3) an aggregation of cell bodies in the central nervous system.

nucleus pulposus (pul-po′sis) A jellylike material in the center of an intervertebral disk.

nutrition The act of providing all substances needed for good health through food ingestion.

O

obesity An excessive amount of fat.

obstruction The act of blocking or clogging.

occipital (ok-sip′it-al) Pertaining to the back of the head.

olfactory (ol-fak′to-re) Pertaining to the sense of smell.

oligodendrocyte (ol″ig-o-den′dro-sīt) Myelin-producing cell of the central nervous system.

oliguria (ol″ig-u′re-ah) Production of only a small quantity of urine.

omentum (o-men′tum) A double layer of mesentery between certain abdominal organs.

oncotic (ong-kot′ik) **pressure** Osmotic pressure created by the presence of protein molecules.

oocyte (o′o-sīt) A cell from which an ovum develops.

oogenesis (o-o-jen′is-is) The process by which an ovum is produced.

oogonia (o-o-go′ne-ah) Female germ cells.

open reduction Surgical repair of a fracture.

operating set point Hypothetical average value about which a regulated condition varies.

ophthalmic (of-thal′mik) Pertaining to the eye.

opsin (op′sin) A protein associated with the response to light.

optic (op′tik) Pertaining to the eye or properties of light.

optic chiasma (ki-az′mah) A site anterior to the pituitary where optic nerve fibers cross from one side of the body to the other.

optic disk Region of retina where optic nerve fibers leave the eyeball; blind spot.

oral Pertaining to the mouth.

organ A structure composed of several kinds of tissues capable of carrying out particular functions; a component of a system.

organelle (or″gah-nel′)   A tiny functional unit within a cell.

organic   Pertaining to carbon-containing substances.

organizational complexity   A concept pertaining to the levels of structure and function in an organism.

organ of Corti (kor′te)   The inner ear structure that contains sound receptors.

orgasm (or′gazm)   An intense pleasurable culmination of sexual intercourse.

origin   The least movable attachment of a muscle.

oropharynx (o″ro-far′inks)   The portion of the pharynx that communicates with the mouth.

oscillation   Repetitive variation over the same range.

osmolarity   The osmotic concentration of a solution determined by the number of osmotically active particles it contains.

osmoreceptor (oz″mo-re-cep′tor)   A receptor that detects changes in osmolarity.

osmosis   Movement of water across a membrane from a region of lower to higher solvent concentration; diffusion of water from its own higher to lower concentration.

osmotic diuresis   Increased urine volume because of the large number of osmotically active particles in the kidney filtrate.

osmotic pressure   The pressure produced by osmosis.

ossification (os″i-fi-ka′shun)   The formation of bone.

osteoarthritis   A chronic degenerative disease, usually affecting several joints.

osteoblast (os′te-o-blast)   A cell that forms bone matrix.

osteoclast (os′te-o-klast)   A cell that digests bone matrix.

osteocyte (os′te-o-sīt)   A cell occupying a lacuna in bone.

osteogenesis (os″te-o-jen′e-sis)   The process of bone formation.

osteoid (os′te-oid)   Resembling bone.

osteomalacia (os″te-o-mah-la′she-ah)   Softening of the bones in adults as a result of a vitamin D deficiency.

osteon (os′te-on)   A system of cells, matrix, and passages making a unit of compact bone.

osteoporosis (os″te-o-po-ro′sis)   Abnormal porousness of bone.

otolith (o′to-lith)   A small particle of calcium carbonate associated with receptors for static equilibrium; an ear stone.

oval window   A membrane-covered opening between the middle and inner ear to which the stapes transmits vibrations.

ovarian (o-var′e-an)   Pertaining to an ovary.

ovary (o′var-e)   A female gonad.

oviduct (o′veh-dukt)   Uterine tube or fallopian tube.

ovulation (ov″u-la′shun)   Release of an ovum from a follicle.

ovum (o′vum)   A female gamete.

oxidation   Addition of oxygen or loss of electrons or hydrogen.

oxidative phosphorylation   Capture of energy in ATP as molecules are oxidized.

oxygen debt   The amount of oxygen required to oxidize metabolites produced anaerobically during strenuous activity.

oxyhemoglobin (ok″se-he″mo-glo′bin)   Hemoglobin-carrying oxygen.

oxytocin (ok″se-to′sin)   A hormone that stimulates uterine contraction.

P

pacemaker   A group of spontaneously discharging cells that excite other cells, especially those of the sinoatrial node of the heart.

Pacinian (pah-sin′e-an) **corpuscle**   A receptor that responds to pressure.

palate (pal′at)   A flat plate forming the roof of the mouth.

palatine (pal′ah-tīn)   Pertaining to the palate.

palpebra (pal′pe-brah)   Eyelid.

pancreas (pan′kre-as)   A digestive gland that produces enzymes and hormones.

pancreatic (pan″kre-at′ik)   Pertaining to the pancreas.

pancreatitis (pan″kre-at-i′tis)   Inflammation of the pancreas.

pantothenic (pan″to-then′ik) **acid**   A B vitamin used to make coenzyme A.

papilla (pah-pil′ah)   A nipplelike projection.

papillary (pap′il-er-e) **muscles**   Projections of cardiac muscle attached to chordae tendinae.

Pap smear   A sample of cells from the cervix for microscopic examination to detect cervical cancer.

paraplegia (par-ah-ple′je-ah)   Paralysis of the legs or lower part of the body.

parasympathetic (par″a-sim-path-et′ik) **division**   A component of the autonomic nervous system that regulates visceral function and returns functions to normal levels after stressful situations.

parasympatholytic (par″ah-sim″pah-tho-lit′ik)   Pertaining to substances that block or counteract sympathetic signals.

parasympathomimetic (par″ah-sim″pah-tho-mi-met′ik)   Pertaining to substances that mimic the action of the parasympathetic nervous system.

parathormone (par″ah-thor′mōn) (PTH)   A hormone from the parathyroid gland that decreases blood calcium.

parathyroid (par″a-thi′roid) **glands**   Glands imbedded in the thyroid gland.

parietal (par-i′et-al)   Pertaining to the wall of a cavity.

Parkinson's disease   Muscle rigidity and tremors due to a dopamine deficiency in the brain.

parotid (par-ot′id) **gland**   A large salivary gland located inferior to the ear.

partial pressure   The pressure exerted by one gas in a mixture of gases.

parturition (par-tu-rish′un)   Childbirth.

passive immunity   Temporary immunity acquired from another's antibodies.

passive transport   Any process that causes movement of substances without the expenditure of energy by the organism.

patella (pat-el′ah)   Bone of the kneecap.

patent (pa′tent)   Open.

pattern generator   A control mechanism for repetitive movements.

pedicle (ped′ik-el)   A bony process that connects the lamina and centrum of a vertebra.

peduncle (pe-dung′k′l)   Certain tracts within the brain.

pelvic (pel′vik)   Pertaining to the pelvis.

penis (pe′nis)   The male copulatory organ.

pentose phosphate pathway   A metabolic pathway that metabolizes glucose, producing five-carbon sugars and reduced NADP.

pepsin (pep′sin)   An enzyme that digests protein in the stomach.

peptide (pep′tīd) **bond**   A chemical bond that holds amino acids together.

perception   Conscious interpretation of signals from sensory receptors.

perfluorocarbon (per-flu-ro-kar′bon)   An oxygen-carrying substance used in synthetic blood.

perforin (per′fo-rin)   A cytotoxic protein.

perfusion (per-fu′shun)   The passage of blood through vessels.

pericardial (per″i-kar′de-al)   Around the heart.

pericardium (per″i-kar′de-um)   A sac around the heart.

perichondrium (per″e-kon′dre-um)   A connective tissue membrane that covers cartilage.

perikaryon (per-e-ka′re-on)   Substance surrounding a nucleus; the cell body of a neuron.

perilymph (per′e-limf)   A clear fluid in the osseous labyrinth that surrounds the membranous labyrinth in the inner ear.

perimetrium (per-i-me′tre-um)   The outer layer of the uterus.

perimysium (per-i-mi′se-um)   A connective tissue sheath that covers bundles of skeletal muscles.

perineum (per″in-e′um)   A region around the pelvic outlet.

perineurium (per-e-nu′re-um)   A connective tissue sheath that covers bundles of nerve fibers.

periodontal (per″e-o-don′tal)   Situated around a tooth.

periosteum (per-e-os′te-um)   A membrane covering the surface of bone.

peripheral   Outer.

peristalsis (per″is-tal′sis)   Wavelike contractions that propel substances along tubular structures.

peritoneal (per″e-to-ne′al) Pertaining to the peritoneum.

peritoneum (per″e-to-ne′um) A membrane that covers organs and lines the abdominal cavity.

peritubular (per″e-tu′bu-lar) Around a tubule, as in the kidney.

pernicious (per-nish′us) **anemia** Anemia caused by a deficiency of intrinsic factor and, therefore, vitamin B$_{12}$.

peroxisome (pĕ-roks′ĭ-sōm) An organelle containing oxidative enzymes.

petrous (pet′rus) Resembling a rock.

Peyer's patch An elevated area of lymphoid tissue of the mucosa in the small intestine.

pH The negative logarithm of the hydrogen ion concentration.

phagocytosis (fag″o-si-to′sis) The engulfment and digestion of particles by a scavenger cell.

phalanges (fa-lan′jēz) Bones of fingers and toes.

phantom pain Pain perceived as originating in a body part that is no longer present, such as an amputated limb.

pharyngeal (far-in′je-al) Pertaining to the pharynx.

pharyngoesophageal (far-ing″go-e-sof′ah je″al) Pertaining to the pharynx and esophagus.

pharynx (far′inks) Throat.

phase shift Advance or delay in a circadian rhythm.

phenotype (fen′o-tīp) The appearance of an individual with respect to one or all inherited characteristics.

phenylketonuria (fen″il-ke″to-nu′re-ah) A genetic defect in phenylalanine metabolism that leads to mental retardation if untreated.

phonation (fo-na′shun) The making of vocal sounds.

phosphocreatine (fos″fo-kre′at-in) An energy storage molecule found in muscle.

phospholipid (fos″fo-lip′id) A lipid containing glycerol, fatty acids, and phosphoric acid.

phosphorylation (fos″fōr-il-a′shun) Addition of a phosphate group to a molecule.

photon The smallest unit of light energy.

photoreceptor (fo″to-re-sep′tor) A receptor that responds to light.

physiological dead space volume The volume of gas in respiratory passageways not exposed to gas exchange membranes.

physiological dependence The need for a drug to prevent withdrawal symptoms.

physiology The study of function of a living organism.

pia (pi′ah) **mater** A delicate membrane covering the surface of the brain and spinal cord.

pica (pi′kah) A craving for nonfood substances.

piloerection (pi″lo-e-rek′shun) The standing on end of hairs.

pilus (pi′lus) A hair.

pineal (pin′e-al) **gland** A gland between the cerebral hemispheres concerned with regulating circadian rhythms.

pinna (pin′nah) The part of the ear projecting from the head.

pitch A quality of sound determined by the frequency of vibrations.

pK The negative log of the dissociation constant of an acid.

placenta (plah-sen′tah) Structure attached to the uterine wall that nourishes a developing fetus.

plantar (plan′tar) Pertaining to the sole of the foot.

plantarflexion (plan-tar-flek′shun) Movement of the foot and toes downward away from the shin.

plaque (plak) A sheetlike deposit.

plasma (plaz′mah) The fluid portion of blood including inactive clotting factors.

plasma cell A cell derived from a B lymphocyte that produces antibodies.

plasma membrane Cell membrane.

plasmin (plaz′min) An enzyme that dissolves blood clots.

plasminogen (plaz-min′o-jen) An inactive form of plasmin.

plasticity The quality of being formable or changeable.

platelet A fragment of a megakaryocyte that circulates in blood and participates in blood clotting reactions.

pleural (ploor′al) Pertaining to the lungs.

plexus (plek′sus) A network, usually of nerves or blood vessels.

plicae circularis (pli′ke sir-ku-la′ris) Transverse folds in the mucosa of the small intestine.

pneumonia (nu-mo′ne-ah) An inflammation of the lungs.

pneumotaxic (nu″mo-taks′ik) **area** A region of the brain involved in regulating breathing.

pneumothorax (nu-mo-tho′raks) The presence of air in the thorax, which usually causes a lung to collapse.

podocyte (pod′o-sīt) Epithelial cell with footlike processes found in the glomerular capsule where its processes wrap around capillaries.

point mutation A modification of a single base in a molecule of DNA.

polar compound A molecule that has a charged area.

poliomyelitis (po″le-o-mi″el-i′tis) Inflammation of motor cells in the anterior horn of the spinal cord.

polycythemia (pol-e-si-the′me-ah) An abnormal excess of erythrocytes.

polydipsia (pol″e-dip′se-ah) Excessive fluid intake.

polymer (pol′im-er) A high molecular weight compound consisting of repeating units.

polymorphonuclear (pol″e-mor″fo-nu′kle-ar) **leukocyte (PMNL)** A white blood cell with an irregularly shaped nucleus.

polyphagia (pol″e-fa′je-ah) Excessive eating.

polysaccharide (pol″e-sak′ar-īd) Molecule consisting of many saccharide units.

polyuria (pol″e-u′re-ah) Excessive urine production.

pons A part of the brain stem associated with the cerebellum.

pontine (pon′tēn) Pertaining to the pons.

portal (por′tal) Pertaining to blood circulation from one set of capillaries to another.

portal triad A set of three vessels (branches of the hepatic artery, hepatic portal vein, and a bile duct) located around the periphery of liver lobules.

positive feedback A signal that accelerates change.

postabsorptive Pertaining to metabolism after absorption of a meal is completed.

posterior Toward the rear, dorsal in humans.

postganglionic (post-gang-le-on′ik) Referring to a neuron encountered after a signal passes through a ganglion; the second neuron in an autonomic pathway.

postovulatory (post-ov′u-la-to-re) After ovulation.

postsynaptic (post″sin-ap′tik) Referring to a neuron that receives a signal at a synapse.

potential energy Energy due to position that can be released, as in a rock at the top of a hill.

precapillary sphincter Structure in an arteriole that regulates the flow of blood through a capillary bed.

preeclampsia (pre″e-klamp′se-ah) A toxemia that develops during pregnancy.

preganglionic (pre-gang-le-on′ik) Referring to a neuron encountered before a signal passes through a ganglion; the first neuron in an autonomic pathway.

premenstrual (pre-men′stroo-al) **syndrome (PMS)** An assortment of symptoms that occur together a few days before the onset of menstruation.

preovulatory (pre-ov′u-la-to-re) Before ovulation.

prepuce (pre′pūs) Foreskin of the penis.

presbyopia (pres″be-o′pe-ah) Loss of accommodation of the lens; literally, elder vision.

pressor Effect tending to increase pressure.

pressure Stress due to compression.

presynaptic (pre″sin-ap′tik) Referring to a neuron that conveys a signal to a synapse.

primary First in order of importance.

primary follicle An early stage in the development of an ovarian follicle.

prime mover A muscle that directly causes a particular movement.

primordial (pri-mor′de-al) Original or primitive.

principle of forward conduction A rule stating that signals travel along axons toward the next neuron in a pathway.

procarboxypeptidase (pro″kar-bok-se-pep′tid-ās) An inactive form of a proteolytic enzyme.

process (1) An extension or outgrowth; (2) a mode of action.

product A substance formed in a chemical reaction or process.

progesterone (pro-jes′ter-ōn) A hormone that helps to maintain pregnancy.

projection fiber A neural fiber that connects the cerebral cortex with other parts of the central nervous system.

prolactin (pro-lak′tin) A hormone that stimulates milk secretion.

prolapse A falling down or sinking.

proliferative phase A phase of uterine change in which cells of the endometrium divide and increase in number.

pronation (pro-na′shun) (1) Rotation of the forearm, causing the palm of the hand to face backward; (2) placing the abdomen downward, as lying in a prone position.

pronephros (pro-nef′ros) The most primitive embryonic kidney.

pronucleus (pro-nu′kle-us) A male or female cell nucleus that contributes to a zygote.

properdin (pro′per-din) **pathway** A sequence of reactions by which complement is activated.

prophase The first stage of mitosis during which the chromosomes become distinct.

proprioceptor (pro″pre-o-sep′tor) A sensory receptor in a muscle, joint, or tendon that detects position or movement.

prosencephalon An anterior part of the embryonic brain.

prostaglandin (pros″tah-glan′din) A substance derived from the fatty acid arachidonic acid that acts as a chemical messenger.

prostate (pros′tāt) A gland surrounding the male urethra.

prostatic (pros-tat′ik) Pertaining to the prostate gland.

prostatitis (pros″tah-ti′tis) Inflammation of the prostate gland.

protein A macromolecule consisting of amino acids.

prothrombin (pro-throm′bin) An inactive form of thrombin.

proton A positively charged particle in the nucleus of an atom.

protoplasm The substance of a cell; literally, first formed substance.

protraction Movement of the mandible forward.

proximal Nearest the body or nearest a point of attachment.

psychoneuroimmunology A new scientific field that considers the combined effects of psychological, neurological, and immunological factors.

puberty A period during which sexual maturity is achieved.

pubis (pu′bis) An anterior bone of the pelvic girdle.

pubofemoral Pertaining to the pubic and femoral regions.

pulmonary (pul-mon-a′re) Pertaining to the lungs or blood vessels that carry blood to and from gas exchange membranes.

pulp cavity A chamber inside a tooth that contains blood vessels and nerves.

pulse Rhythmic expansion and contraction of an artery caused by the pumping action of the heart.

pulse pressure Difference between highest and lowest pressure in an artery caused by the pumping action of the heart.

Purkinje (pur-kin′je) **cell** A kind of neuron in the cerebellum that has a very large number of synapses with other cells.

Purkinje fiber The terminal ends of fibers of the heart's conduction system.

pus A product of inflammation consisting mainly of debris from dead leukocytes and microorganisms.

putative (pu′ta-tīv) Presumed or thought likely.

pyelonephritis (pi″el-o-nef-ri′tis) An inflammation of the kidney and its pelvis.

pyloric (pi-lor′ik) Pertaining to the pylorus.

pylorus (pi-lo′rus) The region of the stomach that attaches to the small intestine.

pyramid A cone-shaped eminence on an organ.

pyramidal (pi-ram′id-al) **tract** A bundle of motor nerve fibers passing through the pyramids of the medulla and controlling voluntary movements.

pyrogen (pi′ro-jen) A substance that causes an increase in body temperature.

Q

QRS complex A portion of an electrocardiogram associated with ventricular contraction.

quadrate (kwod′rāt) Four sided.

quadriplegia (quad-rip-le′je-ah) Paralysis of all four limbs.

R

radial (ra′de-al) Pertaining to the radius.

radiation Divergence from a center; giving off electromagnetic waves and particles.

radioimmunoassay (ra″de-o-im″u-no-as′a) Measurement of the concentration of a substance using antibodies that bind to the substance and emit radiation.

radius (ra′de-us) Smaller bone of the forearm.

raphe (ra′fe) A seamlike ridge.

rapid-eye-movement (REM) sleep A sleep interval during which the eyeballs move and the EEG resembles wakefulness; paradoxical sleep.

rarefaction A decrease in density.

Rathke's (rahth′kez) **pouch** An embryological structure that gives rise to the anterior pituitary.

reactant A substance that enters into a chemical reaction.

receptor (1) A sense organ, the peripheral endings of sensory nerves; (2) a specific site or chemical configuration on a cell membrane or inside a cell with which a specific substance (hormone or neurotransmitter) can bind to alter cell function.

recessive (re-ses′iv) In genetics, a characteristic that appears in the phenotype only when the recessive allele is the only one present in the genotype.

reciprocal innervation The neural connections whereby one process is slowed as another is accelerated, such as in the contraction of one muscle and the simultaneous relaxation of its antagonist.

recombinant (re-kom′bin-ant) **DNA** The union (in the laboratory) of DNA from two different species.

recruitment A gradual increase in the intensity of a contraction by activating additional motor units.

rectouterine (rek-to-u′ter-in) Behind the uterus.

rectum (rek′tum) The terminal portion of the digestive tract that leads into the anal canal.

rectus (rek′tus) Straight.

red muscle Skeletal muscle containing relatively large amounts of myoglobin.

red nucleus A midbrain nucleus that gives rise to the rubrospinal tract.

red pulp The portion of the spleen that contains numerous blood sinuses.

reduction (1) The gain of an electron or hydrogen, or the loss of oxygen; (2) the realignment of a fractured bone.

referred pain Pain perceived to have come from a site other than that of its actual origin.

reflex (re′fleks) An involuntary response to the stimulation of a receptor.

refraction The bending of light rays as they pass from a medium having one density to a medium having a different density.

relaxin (re-laks′in) A hormone from the corpus luteum of pregnancy.

remission The abatement of symptoms of a disease or the period during which it occurs.

renal (re′nal) Pertaining to the kidney.

renal clearance rate The rate at which a substance is removed, or cleared, from the blood.

renal failure Inability of the kidneys to adequately remove wastes from the blood.

renal glycosuria The presence of glucose in the urine because of the kidney's inability to return it to the blood.

renal hypertension Elevated blood pressure because of constriction of a renal artery.

renal threshold The maximum concentration of a substance the kidneys can return to the blood.

renin (ren′in) A secretion from the kidney that activates angiotensinogen to angiotensin I.

renin-angiotensin (an-je-o-ten′sin) **mechanism** A control mechanism that increases blood pressure and blood volume when either falls below normal.

replication (rep-li-ka′shun) Duplication.

reproduction The process by which offspring are formed.

reproductive engineering A class of procedures that modifies the reproductive process.

residual volume A volume of gas remaining in the lungs after normal exhalation.

resistance Opposition to flow, as in blood vessels.

resistance stage A period during which the body successfully overcomes stress.

resistance vessel A blood vessel (artery or arteriole) that resists stretching when under pressure.

respiration (res″per-a′shun) A process that includes ventilation (breathing) and the exchange of gases between the environment and the body.

respiratory (res″per-a-to′re) Pertaining to respiration.

respiratory acidosis A lowering of blood pH because of a respiratory disorder.

respiratory alkalosis A raising of blood pH because of a respiratory disorder.

respiratory bronchiole (brong′ke-ōl) A small, thin-walled passageway capable of exchanging gases.

respiratory center A group of neurons in the brain stem that regulate respiration.

respiratory distress syndrome of the newborn A condition of labored breathing and impaired gas exchange because of surfactant deficiency.

respiratory quotient (RQ) The ratio of carbon dioxide released to oxygen consumed.

reticular (re-tik-u-lar) Pertaining to a net or meshwork.

reticular activating system (RAS) A structure in the brain stem involved in maintaining consciousness.

retina (ret′in-ah) The innermost layer of the eye, which contains light receptors; nervous tunic.

retinal (ret′in-al) A carotenoid pigment involved in vision.

retraction Backward movment, as exemplified by the mandible.

retroperitoneal (re″tro-per″it-o-ne′al) Behind the peritoneum.

Reye's (rīz) **syndrome** A severe neurological disorder sometimes following a viral infection and made more likely by using aspirin to treat fever.

rheumatoid (roo′mat-oid) **arthritis** An autoimmune inflammatory disease affecting many tissues, but especially joints.

rhodopsin (ro-dop′sin) A light-sensitive protein found in rods of the retina.

rhombencephalon A posterior part of the embryonic brain.

rhythm The periodicity of a recurring event.

rhythmicity (rith-mis′it-e) **area** Region of medulla that maintains resting level breathing.

riboflavin (ri″bo-fla′vin) A heat-labile B vitamin used to make the coenzyme FAD.

ribonuclease (ri″bo-nu′kle-ās) An enzyme that digests RNA.

ribonucleic (ri″bo-nu-kle′ik) **acid** (RNA) A nucleic acid involved in protein synthesis.

ribosomal (ri-bo-so′mal) **RNA** (rRNA) A nucleic acid that forms part of ribosomes.

ribosome (ri-bo-sōm′) An organelle consisting of ribonucleic acid and protein that serves as a site for protein synthesis.

rickets A childhood disorder in which the bones fail to harden because of a calcium deficiency.

right-to-left shunt The flow of blood from the right to the left side of the heart without going through the pulmonary circuit.

rigor mortis (ri′gor mor′tis) Rigidity or stiffening of the muscles following death.

rod A receptor in the retina that responds to dim light but not to color.

root A base or foundation.

rotation Movement of a part about its own axis.

round window A membrane-covered opening between the middle and inner ear.

rubella (ru-bel′ah) A viral infection (German measles).

ruga (roo′gah) A ridge or fold.

S

saccadic (sah′kād′ik) Eye movements that follow a moving object.

saccule (sak′ūl) A little sac, especially as in the vestibule of the inner ear.

sacral (sa′kral) Pertaining to the sacrum.

sacroiliac (sak-ro-il′e-ak) Pertaining to the sacrum and ilium.

sacrum (sa′krum) Bone consisting of fine fused vertebrae that form the posterior attachment to the pelvic girdle.

saddle joint A joint with saddle-shaped articulating surfaces.

sagittal (saj′it-al) Dividing right and left sides.

salivary (sal′eh-ver-e) Pertaining to saliva or the glands that produce it.

saltatory (sal′tat-or″e) Leaping.

SA node Portion of the heart's conduction system that normally initiates contractions.

sarcolemma (sar″ko-lem′ah) The cell membrane of a muscle cell.

sarcomere (sar′ko-mēr) A contractile unit of skeletal muscle.

sarcoplasm (sar′ko-plazm) The nonfibrillar substance of a muscle cell.

sarcoplasmic reticulum (sar″ko-plaz′mik re-tik′u-lum) A network of tubules and vesicles associated with the myofibrils of a striated muscle cell.

satiety center A hypothalamic center concerned with the regulation of food intake.

saturated fatty acid A fatty acid having no double bonds in the carbon chain and being filled or saturated with hydrogen.

saturation State of having all chemical affinities satisfied.

scab A crust on a superficial wound.

scala tympani (ska′lah tim-pan′e) The inferior canal in the cochlea.

scala vestibuli (ska′lah ves-tib′u-le) The superior canal in the cochlea.

scapula (skap′u-lah) Large bone in the posterior shoulder area.

scar Connective tissue that has replaced tissue unable to replace itself after an injury.

schizophrenia (skiz″o-fre′ne-ah) A mental disorder having a complex set of disturbances in thinking and feeling.

Schwann cell A supporting cell that produces myelin in the peripheral nervous system.

sciatic (si-at′ik) Pertaining to the hip.

sclerotome (skle′ro-tōm) An embryonic tissue from which vertebrae and ribs develop.

scrotum (skro′tum) A pouch containing the testes.

scurvy A disease caused by a vitamin C deficiency.

seasonal affective disorder (SAD) Depression attributed to reduced intensity and duration of light in fall and winter.

sebaceous (se-ba′shus) Pertaining to sebum.

sebum (se′bum) A substance composed of oils and epithelial cell debris exuded from sebaceous glands.

secondary Being second in time, place, or importance.

secretagogue (se-kre′tā-gog) A substance that stimulates digestive secretions.

secretin (se-kre′tin) A hormone from the intestinal mucosa that stimulates secretion of pancreatic fluid and bile.

secretion (1) A cell product; the process of releasing a substance from a cell. (2) The active transport of substances from the blood to the kidney filtrate.

secretory phase The part of the menstrual cycle during which glands develop in the uterine mucosa.

segmentation (seg-men-ta′shun) (1) Splitting into segments; (2) Alternating contraction of segments of the intestine.

selectively permeable A property of membranes allowing some substances to pass through while preventing the passage of others.

selenosis (se″len-o′sis) A disorder caused by too much selenium in the body.

sella turcica (sel′ah tur′sik-ah) A saddle-shaped depression in the sphenoid bone that contains the pituitary gland; literally, Turkish saddle.

semen (se′men) A mixture of sperm and secretions emitted from the male reproductive system.

semicircular canal One of three pairs of fluid-filled tubes in the inner ear that detect head movements.

semilunar (sem″e-lun′ar) Half-moon shaped.

seminal (sem′in-al) **vesicle** A convoluted saclike structure near the ductus deferens that contributes secretions to semen.

seminiferous (se″min-if′er-us) **tubule** A coiled tubule within a testis in which sperm are produced.

senile (se′nīl) Pertaining to old age.

sensation An impression conveyed.

sensitization The process of rendering a cell of the immune system sensitive to a foreign substance.

sensor A device that responds to a stimulus; a component of a control system that detects changes in the system.

sensory Pertaining to sensation.

seroreversion (se′ro-re-ver′shun) The loss of a once present positive result when testing serum.

serosa (se-ro′sah) A membrane that lines body cavities and secretes a watery substance.

serotonin (ser″o-to′nin) A substance secreted as a neurotransmitter in the brain and as a hormone in the gut.

serrated Saw-toothed.

Sertoli's (ser-to′lez) **cell** A kind of cell in seminiferous tubules that nourishes developing sperm.

serum (se′rum) The fluid part of blood after formed elements and clotting factors have been removed.

sesamoid (ses′ah-moid) Resembling a sesame seed; pertaining to a bone formed in a tendon.

severe combined immunodeficiency disease (SCID) An absence of immunity caused by a deficiency of both B and T lymphocytes.

sickle cell anemia An inherited anemia in which erythrocytes assume a sickle shape under low-oxygen conditions.

sigmoid (sig′moid) S-shaped.

signal Something that conveys information.

sinus (si′nus) A recess or cavity.

sinus bradycardia (brad-e-kar′de-ah) A slow heart rate originating in the SA node.

sinusitis (si″nus-i′tis) An inflammation of the sinuses.

sinusoid (si′nus-oid) A blood vessel similar to a capillary but having a larger diameter.

sinus tachycardia (tak-e-kar′de-ah) A rapid heart rate originating in the SA node.

sliding filament theory An explanation of how myofilaments move during muscle contraction.

smooth muscle A kind of muscle found in the walls of hollow organs and blood vessels.

sodium-potassium pump A mechanism that actively transports sodium ions out of cells and potassium ions back into them against natural gradients that cause the ions to move in the opposite direction.

sol The liquid state of a colloidal dispersion.

solute A dissolved substance.

solution A liquid containing dissolved substances.

solvent A substance in which other substances dissolve.

soma (so′mah) Cell body.

somatic (so-mat′ik) Pertaining to the body.

somatostatin (so″mat-o-stat′in) Growth hormone inhibiting hormone.

somatotropin (so″mat-o-tro′pin) Growth hormone.

somite (so′mīt) An embryonic mesodermal segment.

spasm A sudden, violent, involuntary muscle contraction.

specific heat The amount of heat needed to raise the temperature of a specific volume of substance 1° C.

specificity The quality of being specific.

spectrin (spek′trin) A protein that contributes to the flexibility of erythrocyte membranes.

spermatic (sper-mat′ik) **cord** A cord extending from the scrotum to the inguinal ligament containing the ductus deferens, blood vessels, lymphatics, and nerves.

spermatid (sper′mah-tid) An immature spermatozoan.

spermatocyte (sper′mah-to-sīt) A cell that gives rise to a sperm.

spermatogenesis (sper″mah-to-jen′es-is) The process by which sperm are formed.

spermatogonium (sper″mah-to-go′ne-um) Undifferentiated male germ cells.

spermatozoan (sper″mah-to-zo′an) Male gamete, sperm.

spermiogenesis (sper″me-o-jen′es-is) The maturation of spermatids to sperm.

sphenoid (sfe′noid) Winglike, as in the skull bone of that name.

sphincter (sfingk′ter) A ringlike muscle that closes a natural orifice.

sphygmomanometer (sfig″mo-man-om′et-er) An instrument used to measure blood pressure.

spicule (spik′ūl) A needle-shaped structure.

spinal (spi′nal) Pertaining to the vertebral column.

spinal shock A temporary condition following spinal injury during which spinal reflexes below the injury are lost.

spirometer (spi-rom′et-er) A device to measure gas volumes.

spirometry (spi-rom′et-re) The measurement of volumes of gas entering or leaving the lungs.

splanchnic (splank′nik) Pertaining to the viscera.

spontaneous abortion The expelling of a fetus from the uterus without external influence.

sprain A joint injury in which surrounding tissue may be damaged but the joint is not dislocated.

sprue (sproo) A disease in which the lining of the gastrointestinal tract is inflamed and partially destroyed.

squamous (skwa′mus) Scalelike.

stapes (sta′pēz) The bone of the middle ear that transmits vibrations to the oval window.

static equilibrium The maintenance of balance when the head is stationary.

stenosis (sten-o′sis) A narrowing.

stereoisomer (sta-re-o-is′om-er) A compound containing the same number and variety of atoms as another compound, but in a different spatial arrangement.

sterility The inability to produce offspring.

sternoclavicular Pertaining to the sternum and clavicle.

steroid (ste′roid) A lipid having a complex four-ring structure.

stethoscope (steth′o-skōp) An instrument used for listening to sounds inside the body.

stimulus (stim′-u-lus) A change in the environment that in some way modifies the activity of a receptor.

stomodeum (sto″mo-de′um) An evagination of the ectoderm from which the mouth and upper pharynx form.

strain A stretching of tissues around a joint.

strangulation Choking; occlusion.

stratum (stra′tum) A layer, usually of tissue.

streptokinase (strep″to-ki′nās) A substance used to digest blood clots in coronary arteries.

stress A condition of the body produced by a variety of injurious agents.

stressor (stres′or) That which produces stress.

stretch reflex Contraction of a muscle following stimulation of stretch receptors in the muscle or its tendon.

striation Stripe.

stroke volume Blood volume ejected by one ventricle during a single contraction.

stroma (stro′mah) A framework usually consisting of connective tissue and supporting an organ.

styloid (sti′loid) A slender process like a stylus.

subarachnoid (sub-ar-ak′noid) Beneath the arachnoid membrane.

subclavian (sub-kla′ve-an) Beneath the clavicle.

subconscious Imperfectly or partially conscious.

subdural (sub-dur′al) Beneath the dura mater.

sublingual (sub-ling′gwal) Beneath the tongue.

submandibular (sub″man-dib′u-lar) Beneath the mandible.

submucosa (sub″mu-ko′sa) Beneath the intestinal mucosa.

subneural (sub-nu′ral) Beneath a nerve.

substantia nigra (sub-stan′she-ah ni′grah) A pigmented nucleus; one of the basal nuclei.

subthalamic (sub-thal-am′ik) Beneath the thalamus.

sucrase (soo′krās) An enzyme that digests sucrose.

sudden infant death syndrome (SIDS) Sudden death of an apparently healthy infant without known cause.

sudoriferous (su″dor-if′er-us) Sweat-secreting.

sulcus (sul′kus) A groove or furrow.

summation An adding of effects.

supination (soo″pin-a′shun) Rotation of the forearm so that the palm of the hand faces forward in the anatomical position.

suprachiasmatic (soo-prah-ki-as-mat′ik) **nuclei** (SCN) Cell bodies in the brain stem that help to regulate circadian rhythms.

supraorbital (soo″prah-or′bit-al) Above the orbit of the eye.

surface tension Resistance to rupture possessed by the surface film of a liquid.

surface-to-volume ratio The surface area of a structure divided by its volume.

surfactant (sur-fak′tant) A substance that reduces surface tension.

suspensory Serving to hold up.

sutural (su′tu-ral) Pertaining to a suture.

sutural bone A small accessory bone sometimes found between larger skull bones.

suture (su′tūr) An immovable fibrous joint.

sympathetic chain ganglion An aggregation of cell bodies of postsynaptic neurons of the sympathetic division.

sympathetic division The part of the autonomic nervous system that responds to stressful situations.

sympatholytic (sim″pah-tho-lit′ik) Blocking or counteracting sympathetic signals.

sympathomimetic (sim″pah″tho-mi-met′ik) Mimicking the action of the sympathetic division, or a drug that does so.

symphysis (sim′fis-is) A slightly movable cartilaginous joint.

synapse (sin′aps) A junction between neurons across which a neurotransmitter diffuses.

synaptic (sin-ap′tik) Pertaining to a synapse.

synarthrosis (sin″ar-thro′sis) An immovable joint with no intervening tissue between the bones.

synchondrosis (sin-kon-dro′sis) A slightly movable or immovable joint in which bones are connected by cartilage.

syncytium (sin-sit′e-um) A group of cells that have lost the membranes that formerly separated them.

syndesmosis (sin″des-mo′sis) A joint in which two bones are bound together by fibrous connective tissue.

synergist (sin′er-jist) A muscle that assists a prime mover.

synostosis (sin″os-to′sis) A joint in which bones are joined by bony material, such as in the ossification of cartilage.

synovial (sin-o′ve-al) Pertaining to a freely movable joint.

synovial (sin-o′ve-al) **joint** A freely movable joint.

syphilis (sif′il-is) A sexually transmitted disease caused by the spirochete *Treponema pallidum.*

systemic Pertaining to the whole body.

systemic lupus erythematosus (lu′pus er-ithe″mah-to′sus) An autoimmune disease often characterized by a butterfly-shaped rash over the nose and cheeks.

systole (sis′to-le) Contraction.

T

tachycardia (tak″e-kar′de-ah) A rapid heart rate.

talus (tal′us) An ankle bone.

target cell A cell that responds to a particular hormone.

tarsal (tar′sal) A bone of the ankle.

taste bud A structure on the tongue containing receptor cells for taste.

taurine (taw′rēn) An amino acid derivative that conjugates with bile acids.

T cell T lymphocytes.

tear A fluid release from tear glands.

tectorial (tek-to′re-al) **membrane** A membrane overlying the hair cells of the organ of Corti.

telophase The last stage of mitosis during which nuclei reform.

template Pattern.

temporal (tem′por-al) Pertaining to time; a lobe of the brain that contains olfactory and auditory areas.

temporomandibular (tem″po-ro-man-dib′u-lar) Pertaining to the temporal bone and mandibles.

tendon A cord of fibrous connective tissue that holds a muscle to a bone.

tendon organ A sensory receptor in a tendon that responds to stretching.

tendon sheath A synovial membrane found around certain tendons.

tension A pulling force.

teratogen (ter′at-o-jen) An agent that can cause defective embryonic development.

testis (tes′tis) A male gonad.

testosterone (tes-tos′ter-ōn) A male hormone.

tetanus (tet′an-us) (1) A sustained contraction produced by repeated stimulation of a muscle; (2) lockjaw, a disease caused by a bacterial toxin.

tetrad (tet′rad) A set of four copies of the same chromosome.

tetralogy of Fallot (tet-ral′o-je of fal-o′) A set of four congenital heart defects that often occur together.

thalamus (thal′am-us) Gray matter near the anterior end of the brain stem.

thalassemia A kind of anemia caused by deficient alpha or beta chains of hemoglobin.

theca (the′kah) Sheath; the covering of an ovarian follicle.

theory A hypothesis, usually supported by a significant amount of evidence.

theory of immune surveillance A possible explanation of how malignant cells are found and destroyed.

thermogenesis (ther″mo-jen′es-is) The generation of heat.

thermoreceptor (ther″mo-re-sep′tor) A receptor that detects changes in temperature.

theta wave A brain wave seen in children and in some adult brain disorders.

thiamine (thi′am-in) A water-soluble B vitamin.

thirst The desire for water or other fluid.

thirst center A nucleus in the hypothalamus that responds to osmotic changes in the blood, causing drinking behavior when the osmotic pressure increases.

thoracic (tho-ras′ik) Pertaining to the chest.

thoracolumbar (tho″rak-o-lum′bar) Pertaining to the chest and lumbar regions.

thorax (tho′raks) The chest.

thoroughfare channel A capillary through which blood passes at all times.

thrombin An enzyme that converts fibrinogen to fibrin in the blood-clotting mechanism.

thrombocyte (throm′bo-sīt) A platelet.

thrombocytopenia (throm″bo-si″to-pe′ne-ah) A platelet deficiency.

thrombus (throm′bus) A blood clot attached to a blood vessel wall.

thymosin (thi′mo-sin) A hormone secreted by the thymus gland.

thymus (thi′mus) **gland** A gland that processes T lymphocytes before it regresses during puberty.

thyroid gland A gland in the throat that produces hormones that regulate metabolism.

thyroid-stimulating hormone (TSH) A hormone that causes the thyroid gland to secrete hormones.

tibia (tib′e-ah) Weight-bearing bone of the leg.

tidal volume The volume of gas moving into and out of the lungs in normal resting breathing.

timbre (tam'br)   The musical quality of a sound.

tinnitus (tin-i'tus)   Ringing in the ears.

tissue   An aggregation of similar cells and their intercellular substances.

tissue plasminogen activator (tPA)   A substance secreted by many tissues that converts plasminogen to plasmin.

tissue thromboplastin (throm''bo-plas'tin)   A substance released by injured tissue that initiates the extrinsic blood clotting mechanism.

T lymphocyte   A thymus-processed lymphocyte that differentiates into one of several kinds of T cells.

tocopherol (to-kof'er-ol)   A substance with vitamin E activity.

tolerance   The requirement for larger doses to produce the same result.

tonicity (to-nis'i-te)   The degree to which conditions cause fluids to move into or out of cells.

tonsil   An aggregate of lymphatic tissue in the pharynx.

tonus (ton'us)   A slight, continuous contraction of muscle.

torr   A unit of pressure equal to that which will support a column of mercury 1 mm tall.

torticollis (tor''tik-ol'is)   Wryneck.

total parental nutrition (TPN)   The supplying of all required nutrients by a route other than the digestive tract (usually intravenously).

trabecula (trah-bek'u-lah)   A septum; a spicule of spongy bone.

trace element   A chemical element normally found in the body in very small amounts.

trachea (tra'ke-ah)   Passage from the larynx to the bronchi.

tracheoesophageal fistula (tra''ke-o-e-sof'ah-je-al fis'tu-lah)   Developmental defect that allows food to enter the lungs.

tract   A bundle of myelinated neurons in the brain or spinal cord.

transamination (trans-am-in-a'shun)   Transfer of an amino group from one molecule to another.

transcription   The process of transferring coded genetic information from DNA to mRNA.

transducin (trans-du'sin)   Enzyme involved in visual process.

transduction   The conversion of a signal from one type to another, as in converting a sensory signal from a receptor to an electrical signal in a neuron.

transferrin (trans-fer'rin)   A plasma protein that transports iron.

transfer RNA (tRNA)   RNA that carries amino acids and places them in specific sites in a growing peptide chain.

translation   The process by which information in mRNA codons is used to determine the sequence of amino acids in a protein.

translocation   The transfer of a portion of a chromosome from its normal location to a location on another chromosome.

transplantation   Moving graft tissue to a new host.

transplant rejection   An immunologic reaction that destroys transplanted tissue.

transverse   Placed crosswise.

transverse (T) tubule   A tubule running crosswise of myofibrils in skeletal muscle that conveys a signal from the sarcolemma to the myofibrils.

treppe (trep'eh)   A gradual increase in the extent of muscular contraction following rapidly repeated stimulation; literally, staircase.

triacylglycerol (tri-as''il-gli'ser-ol)   A lipid consisting of glycerol and three fatty acids.

triad   Any group of three.

tricuspid (tri-kus'pid)   Having three points or cusps.

triglyceride   A triacylglycerol.

trigone (tri'gōn)   A triangle; the triangle formed in the urinary bladder by the openings of the ureters and urethra.

trismus (triz'mus)   A tetanic spasm of jaw muscles; lockjaw.

trisomy (tri'so-me)   Containing three copies of a chromosome.

trochanter (tro-kan'ter)   Large, round bony process.

trochlea (trok'le-ah)   Pulley.

trophoblast (trof'o-blast)   Outer layer of a blastocyst, which establishes connection with maternal tissue and gives rise to chorionic villi.

tropic (tro'pik)   Influencing another organ or process.

tropomyosin (tro''po-mi'o-sin)   A muscle protein involved in regulating the contraction process.

troponin (tro-po'nin)   A muscle protein involved in regulating the contraction process.

trypsin (trip'sin)   A proteolytic enzyme released from the pancreas.

trypsinogen (trip-sin'o-jen)   An inactive form of trypsin.

tubercle (too'ber-k'l)   Small, rounded protrusion.

tuberosity (too''ber-os'it-e)   Protrusion on a bone to which muscles attach.

tubulin (tu'bu-lin)   A protein found in intracellular microtubules.

tumor necrosis (ne-kro'sis) **factor**   A substance that causes tumor cells to degenerate and die.

tunic   An external coat.

tunica (too'nik-ah)   A coat or layer.

Turner's syndrome   A condition in which only a single X chromosome (without another X or Y chromosome) is present.

turnover   Reuse of a substance after breakdown.

T wave   A component of an electrocardiogram associated with ventricular repolarization.

twitch   Response of a muscle to a single stimulus.

tympanic (tim-pan'ik) **membrane**   Eardrum; membrane dividing the external and middle ear.

U

ulcerative (ul''ser-a'tiv)   Pertaining to an ulcer.

ulna (ul'nah)   Larger bone of the forearm.

ultrafiltrate   Filtrate formed under high pressure.

ultrasound   Sound waves of high frequency.

umbilical (um-bil'i-kal)   Pertaining to the umbilicus.

umbilicus (um-bil'i-kus)   Site of attachment of the umbilical cord to fetus; navel.

unsaturated fatty acid   Fatty acid with hydrogen atoms replaced by one or more double bonds in the carbon chain.

unstirred water layer   A region near the intestinal mucosa where water molecules remain relatively stationary.

urea cycle   A metabolic pathway for the synthesis of urea.

ureter (u-re'ter)   Tube that carries urine from the kidney to the urinary bladder.

ureteric (u''rĕ-ter'ik)   Pertaining to the ureter.

urethra (u-re'thrah)   Tube that carries urine from the urinary bladder outside the body.

uridine (ur'id-ēn) **triphosphate** (UTP)   A high energy molecule.

urinary bladder   Stretchable sac where urine is stored.

uterine (u'ter-in)   Pertaining to the uterus.

uterine tube   A tube that carries ova to the uterus.

uterosacral (u''ter-o-sa'kral)   Pertaining to the region of the uterus and sacrum.

uterus (u'ter-us)   A hollow, pearshaped organ in which a fetus develops.

utricle (u'tre-k'l)   Large chamber in the vestibule of the ear that contains receptors for equilibrium.

uvea (u've-ah)   The middle layer of the eyeball; vascular tunic.

V

vagina (vah-ji'nah)   Passageway from the uterus.

vaginal (vaj'in-al)   Pertaining to the vagina.

valence   The positive or negative charge on an ion.

valve   A structure in a passageway which prevents reflux (backflow).

variable   An event or condition that can change.

varicose (var'ik-ōs)   Unnaturally swollen.

varicosity (var″ik-os′it-ē)   A varicose condition.

vasa recta (va′sah rek′tah)   Straight vessels, especially capillaries found in the renal pyramids.

vasa vasorum (va′sah vas-o′rum)   Blood vessels within the walls of larger blood vessels.

vascular (vas′ku-lar)   Pertaining to or full of blood vessels.

vas deferens (def′er-enz)   A duct that carries sperm from the testes to the ejaculatory duct.

vasoconstriction (va″so-kon-strik′shun)   A narrowing of the lumen of a blood vessel.

vasodilation (vas″o-di-la′shun)   A widening of the lumen of a blood vessel.

vasomotor   Pertaining to regulation of blood vessel diameter.

vasomotor center   An area in the brain stem that regulates the diameter of blood vessels, especially arterioles.

vein (vān)   A blood vessel carrying blood to the heart.

vena cava (ve′nah ka′vah)   A large vein near the heart.

venous (ve′nus)   Pertaining to veins.

ventilation   Exchange of air between the lungs and the environment.

ventral (ven′tral)   Pertaining to or toward the belly.

ventricle (ven′trik-l)   A small cavity.

venule (ven′ūl)   A small vein connecting capillaries with a large vein.

vermiform (ver′mi-form)   Worm-shaped.

vermis (ver′mis)   Worm.

vesicouterine (ves″ik-o-u′ter-īn)   Pertaining to the bladder and uterus.

vestibular (ves-tib′u-lar)   Pertaining to a vestibule.

vestibule (ves′tib-ul)   A space or cavity at the entrance of a canal.

villikinin (vil″ī-ki′nin)   A mucosal hormone that stimulates movement of villi.

villus (vil′us)   A vascular tuft.

viscera (vis′er-ah)   Internal organs.

visceral (vis′er-al)   Pertaining to the viscera.

visceroceptor (vis″er-o-sep′tor)   A receptor in or near an internal organ.

viscosity   Tendency of a fluid to resist flowing.

visual area   A region of the occipital lobe of the cerebrum where signals from the retina are received and interpreted.

vital capacity   The largest volume of gas that can be expired after a maximal inspiration.

vitamin A   A vitamin needed to make visual pigments and maintain epithelial cells.

vitamin D   A vitamin that fosters calcium absorption.

vitamin E   A vitamin that acts as an antioxidant.

vitamin K   A vitmain needed for normal blood clotting.

vitreous (vit′re-us)   Glassy.

vocal   Pertaining to the voice.

Volkmann's canal   Channel that carries blood vessels through the matrix of bone to Haversian systems.

vomer (vo′mer)   A shovel-shaped bone in the nasal septum.

W

Wallerian degeneration   Disintegration of the axon of an injured neuron distal to the point of injury.

water balance   A state in which water intake equals water output.

Wharton's jelly   A soft, pulpy connective tissue forming the matrix of the umbilical cord.

white matter   Myelinated nerve fibers in the central nervous system.

white muscle   Muscle containing relatively little myoglobin.

white pulp   Aggregations of lymphocytes in the spleen.

withdrawal reflex   A reflex causing flexion and removal of a limb from the source of a painful stimulus.

X

xanthine (zan′thēn)   A purine that inhibits breakdown of cAMP.

xiphoid (zi′foid)   Sword-shaped; a process of the sternum.

Y

yolk sac   Bag containing nutrients in many embryos.

Z

zeitgeber (tsīt′ge-ber)   An external factor that regulates a body rhythm; literally, time giver.

zona fasciculata (zo′nah fas-sik-u-lat′ah)   Middle layer of the adrenal cortex.

zona glomerulosa (zo′nah glom-er-u-lo′sa)   Outer layer of the adrenal cortex.

zona pellucida (zo′nah pel-lu′sid-ah)   Translucent layer surrounding an oocyte in an ovarian follicle.

zona reticularis (zo′nah rē-tik″u-lar′is)   Inner layer of the adrenal cortex.

zygomatic (zi-go-mat′ik)   Pertaining to the cheekbone.

zygote (zi′gōt)   Single cell formed by the union of an ovum and a sperm; the first cell of a new individual.

FIGURES

Chapter 1

Fig. 1.4: From Kent M. Van De Graaff and Stuart Ira Fox, *Concepts of Human Anatomy and Physiology,* 2d ed. Copyright © 1989 Wm. C. Brown Publishers, Dubuque, Iowa. All Rights Reserved. Reprinted by permission; **Fig. 1.10:** From John W. Hole, Jr., *Human Anatomy and Physiology,* 5th ed. Copyright © 1990 Wm. C. Brown Publishers, Dubuque, Iowa. All Rights Reserved. Reprinted by permission.

Chapter 2

Fig. 2.31: From Stuart Ira Fox, *Human Physiology,* 3d ed. Copyright © 1990 Wm. C. Brown Publishers, Dubuque, Iowa. All Rights Reserved. Reprinted by permission.

Chapter 3

Fig. 3.5: From Stuart Ira Fox, *Human Physiology,* 3d ed. Copyright © 1990 Wm. C. Brown Publishers, Dubuque, Iowa. All Rights Reserved. Reprinted by permission; **Fig. 3.17b:** From Leland G. Johnson, *Biology,* 2d ed. Copyright © 1987 Wm. C. Brown Publishers, Dubuque, Iowa. All Rights Reserved. Reprinted by permission; **Fig. 3.17d:** From John W. Hole, Jr., *Human Anatomy and Physiology,* 5th ed. Copyright © 1990 Wm. C. Brown Publishers, Dubuque, Iowa. All Rights Reserved. Reprinted by permission; **Fig. 3.18a-b:** From Ross M. Durham, *Human Physiology: Functions of the Human Body.* Copyright © 1989 Wm. C. Brown Publishers, Dubuque, Iowa. All Rights Reserved. Reprinted by permission; **Fig. 3.20:** From Leland G. Johnson, *Biology,* 2d ed. Copyright © 1987 Wm. C. Brown Publishers, Dubuque, Iowa. All Rights Reserved. Reprinted by permission; **Fig. 3.22:** From Kent M. Van De Graaff and Stuart Ira Fox, *Concepts of Human Anatomy and Physiology,* 2d ed. Copyright © 1989 Wm. C. Brown Publishers, Dubuque, Iowa. All Rights Reserved. Reprinted by permission; **Fig. 3.25:** From Stuart Ira Fox, *Human Physiology,* 3d ed. Copyright © 1990 Wm. C. Brown Publishers, Dubuque, Iowa. All Rights Reserved. Reprinted by permission.

Chapter 4

Fig. 4.4: From Kent M. Van De Graaff and Stuart Ira Fox, *Concepts of Human Anatomy and Physiology,* 2d ed. Copyright © 1989 Wm. C. Brown Publishers, Dubuque, Iowa. All Rights Reserved. Reprinted by permission; **Fig. 4.7:** From Kent M. Van De Graaff and Stuart Ira Fox, *Concepts of Human Anatomy and Physiology,* 2d ed. Copyright © 1989 Wm. C. Brown Publishers, Dubuque, Iowa. All Rights Reserved. Reprinted by permission.

Chapter 5

Fig. 5.2: From John W. Hole, Jr., *Human Anatomy and Physiology,* 5th ed. Copyright © 1990 Wm. C. Brown Publishers, Dubuque, Iowa. All Rights Reserved. Reprinted by permission; **Fig. 5.3b:** From John W. Hole, Jr., *Human Anatomy and Physiology,* 5th ed. Copyright © 1990 Wm. C. Brown Publishers, Dubuque, Iowa. All Rights Reserved. Reprinted by permission; **Fig. 5.4a:** From John W. Hole, Jr., *Human Anatomy and Physiology,* 4th ed. Copyright © 1987 Wm. C. Brown Publishers, Dubuque, Iowa. All Rights Reserved. Reprinted by permission; **Fig. 5.6a:** From John W. Hole, Jr., *Human Anatomy and Physiology,* 5th ed. Copyright © 1990 Wm. C. Brown Publishers, Dubuque, Iowa. All Rights Reserved. Reprinted by permission; **Fig. 5.7:** From Kent M. Van De Graaff and Stuart Ira Fox, *Concepts of Human Anatomy and Physiology,* 2d ed. Copyright © 1989 Wm. C. Brown Publishers, Dubuque, Iowa. All Rights Reserved. Reprinted by permission; **Fig. 5.11:** From Kent M. Van De Graaff and Stuart Ira Fox, *Concepts of Human Anatomy and Physiology,* 2d ed. Copyright © 1989 Wm. C. Brown Publishers, Dubuque, Iowa. All Rights Reserved. Reprinted by permission.

Chapter 7

Fig. 7.1: From Kent M. Van De Graaff, *Human Anatomy,* 2d ed. Copyright © 1988 Wm. C. Brown Publishers, Dubuque, Iowa. All Rights Reserved. Reprinted by permission; **Fig. 7.2:** From Kent M. Van De Graaff, *Human Anatomy,* 2d ed. Copyright © 1988 Wm. C. Brown Publishers, Dubuque, Iowa. All Rights Reserved. Reprinted by permission; **Fig. 7.3:** From Kent M. Van De Graaff, *Human Anatomy,* 2d ed. Copyright © 1988 Wm. C. Brown Publishers, Dubuque, Iowa. All Rights Reserved. Reprinted by permission; **Fig. 7.4:** From Kent M. Van De Graaff, *Human Anatomy,* 2d ed. Copyright © 1988 Wm. C. Brown Publishers, Dubuque, Iowa. All Rights Reserved. Reprinted by permission; **Fig. 7.8:** From John W. Hole, Jr., *Human Anatomy and Physiology,* 5th ed. Copyright © 1990 Wm. C. Brown Publishers, Dubuque, Iowa. All Rights Reserved. Reprinted by permission; **Fig. 7.10:** From John W. Hole, Jr., *Human Anatomy and Physiology,* 5th ed. Copyright © 1990 Wm. C. Brown Publishers, Dubuque, Iowa. All Rights Reserved. Reprinted by permission; **Fig. 7.13:** From Kent M. Van De Graaff, *Human Anatomy,* 2d ed. Copyright © 1988 Wm. C. Brown Publishers, Dubuque, Iowa. All Rights Reserved. Reprinted by permission; **Fig. 7.18:** From John W. Hole, Jr., *Human Anatomy and Physiology,* 5th ed. Copyright © 1990 Wm. C. Brown Publishers, Dubuque, Iowa. All Rights Reserved. Reprinted by permission; **Fig. 7.19:** From John W. Hole, Jr., *Human Anatomy and Physiology,* 5th ed. Copyright © 1990 Wm. C. Brown Publishers, Dubuque, Iowa. All Rights Reserved. Reprinted by permission.

Chapter 8

Fig. 8.10b-c: From John W. Hole, Jr., *Human Anatomy and Physiology,* 5th ed. Copyright © 1990 Wm. C. Brown Publishers, Dubuque, Iowa. All Rights Reserved. Reprinted by permission; **Fig. 8.11:** From John W. Hole, Jr., *Human Anatomy and Physiology,* 5th ed. Copyright © 1990 Wm. C. Brown Publishers, Dubuque, Iowa. All Rights Reserved. Reprinted by permission; **Fig. 8.13:** From John W. Hole, Jr., *Human Anatomy and Physiology,* 5th ed. Copyright © 1990 Wm. C. Brown Publishers, Dubuque, Iowa. All Rights Reserved. Reprinted by permission; **Fig. 8.14:** From John W. Hole, Jr., *Human Anatomy and Physiology,* 5th ed. Copyright © 1990 Wm. C. Brown Publishers, Dubuque, Iowa. All Rights Reserved. Reprinted by permission.

Chapter 9

Fig. 9.1: From John W. Hole, Jr., *Human Anatomy and Physiology,* 5th ed. Copyright © 1990 Wm. C. Brown Publishers, Dubuque, Iowa. All Rights Reserved. Reprinted by permission; **Fig. 9.2a:** From John W. Hole, Jr., *Human Anatomy and Physiology,* 4th ed. Copyright © 1987 Wm. C. Brown Publishers, Dubuque, Iowa. All Rights Reserved. Reprinted by permission; **Fig. 9.2b:** From John W. Hole, Jr., *Human Anatomy and Physiology,* 5th ed. Copyright © 1990 Wm. C. Brown Publishers, Dubuque, Iowa. All Rights Reserved. Reprinted by permission; **Fig. 9.7a:** From John W. Hole, Jr., *Human Anatomy and Physiology,* 5th ed. Copyright © 1990 Wm. C. Brown Publishers, Dubuque, Iowa. All Rights Reserved. Reprinted by permission; **Fig. 9.9 (bottom):** From Kent M. Van De Graaff and Stuart Ira Fox, *Concepts of Human Anatomy and Physiology,* 2d ed. Copyright © 1989 Wm. C. Brown Publishers, Dubuque, Iowa. All Rights Reserved. Reprinted by permission; **Fig. 9.11:** From John W. Hole, Jr., *Human Anatomy and Physiology,* 4th ed. Copyright © 1987 Wm. C. Brown Publishers, Dubuque, Iowa. All Rights Reserved. Reprinted by permission; **Fig. 9.12 (top):** From Stuart Ira Fox, *Human Physiology,* 2d ed. Copyright © 1987 Wm. C. Brown Publishers, Dubuque, Iowa. All Rights Reserved. Reprinted by permission

CREDITS

ILLUSTRATORS

Laurel Antler
3.17d, 17.3, 30.3.

Ayres Associates
3.22.

Sam Collins
11.21.

Chris Creek
4.7, 27.1.

John Frieberg/Mary Albury-Noyes
7.12, 15.21, 20.5, 21.18, 23.12, 29.28a-h, 30.10.

FineLine Illustrations, Inc.
1.2, 1.10, 1.11, 1.12b, 1.13, 1.14, 1.15, 1.16, 2.21b, 2.24, 2.25, 2.26, 2.31, 3.5, 3.20, 4.2, 4.3; Interlude 1.2, 1.3a-c; 9.5, 9.8, 9.10, 9.11, 9.13, 9.16, 9.17; Interlude 2.3; 11.12, 11.14, 11.26, 12.11, 13.8, 13.11, 15.2b, 15.10, 15.11, 15.20, 15.27, 16.10, 16.12; Interlude 3.1, 3.3, 3.4a; 18.8, 18.10b,c, 18.11, 20.12, 21.7b, 21.10, 21.13, 22.1, 22.7, 22.11, 22.14, 22.16, 22.19, 27.3c, 27.12, 28.3, 28.9, 29.22, 29.25, 29.29, 30.1, 30.13.

Floyd Hosmer
1.3, 1.5, 1.7, 3.6, 11.7, 11.8, 11.10, 12.18, 12.34, 12.35, 13.4, 14.8, 14.9, 14.10, 15.15, 15.16, 15.32, 16.1, 16.20, 19.2, 19.25, 22.5, 22.6, 22.8, 22.9, 23.3, 23.4, 23.25, 27.5a.

Illustrious, Inc.
1.6, 2.2–2.6, 2.8–2.19, 2.21a, 2.22, 2.23, 2.27–2.30, 3.2, 3.12b,c, 3.13b, 3.19, 3.23–3.26, 4.1, 4.5, 5.5, 6.7, 7.13, 7.27, 7.28, 9.9, 9.12, 9.14, 9.15, 10.4, 10.21, 11.1, 11.5, 11.6, 11.11, 11.13, 11.16, 11.18, 11.20, 11.24, 12.13, 12.23, 12.26, 12.28, 13.10, 13.13, 13.14, 13.15, 14.4, 15.9, 15.19, 15.22–15.24, 15.28, 15.37, 16.5, 16.6, 16.9, 16.11, 16.15, 16.19, 16.23, 16.24, 16.27, 16.28; Interlude 3.2, 3.4b, 3.5; 17.4, 17.6, 17.13, 17.14, 17.16, 18.14, 19.18b, 19.24, 20.4, 20.8, 20.15, 20.16, 20.18, 21.3, 22.17, 22.21, 23.1, 23.8, 23.11, 23.17, 23.21, 23.23, 23.24, 24.1–24.17, 25.1–25.7, 25.11, 25.12, 26.1–26.4, 27.4b, 27.8, 27.14, 27.15, 27.19; text art, page 801; 28.2, 28.5, 29.1, 29.10, 29.11, 29.21, 29.27, 30.2

Carlyn Iverson

Provided coordination services for the new illustration program and rendered the following figures:

1.9, 3.1, 3.3, 3.4*a,c*, 3.8, 3.14*b*, 3.15*a*, 3.21, 12.1, 12.3, 12.10, 12.20, 12.21, 13.2, 15.4, 15.18, 15.26, 15.35, 15.36, 16.3, 16.4, 17.7, 18.12, 20.1–20.3, 20.6, 20.7, 20.13, 20.20*b*, 21.1, 21.4, 21.5*b*, 21.6, 21.8, 21.9, 21.11, 21.12, 21.14–21.16, 21.19, 21.20, 22.12, 22.13, 23.16, 23.18–23.20, 27.6, 27.7, 29.19, 29.20, 29.23, 30.11.

Keith Kasnot

9.3, 9.4, 9.6, 12.4, 12.9, 12.15, 15.14*a*, 15.31, 20.10, 20.11, 22.15, 27.20.

Michael King

13.18, 15.13*b*, 16.25, 16.26, 18.3, 18.17.

Ruth Krabach

6.1–6.6, 6.10, 6.12, 7.5–7.7, 7.9, 7.11, 7.14–7.18, 7.19, 7.20*a,b*, 7.21, 7.22*a,b*–7.26, 8.1–8.3, 8.5–8.15, 13.6, 13.7, 13.9, 15.25, 17.12, 21.2, 22.10.

Bill Loechel

4.11, 15.3*a*, 16.18, 17.8, 18.4, 19.1, 19.8, 19.9*a,b*, 19.15, 23.5, 23.9, 23.10*a,b*, 23.13*a,b*, 23.14*a,b*, 23.15, 23.22, 27.3*a,b*, 27.4*a*, 27.18, 29.13.

Patrick Lynch

19.10, 19.17.

Robert Margulies

19.04.

Robert Margulies/Tom Waldrop

10.1, 10.2, 10.23, 19.5, 19.6*b*, 19.7, 19.12, 19.14, 19.18*a*, 23.6.

Nancy Marshburn

11.4, 17.2, 17.5*a*.

Ron McClean

29.7.

Medical Media

4.15, 5.1, 22.2, 27.9, 27.10, 29.9, 29.16.

Steve Moon

1.4*a.b*, 2.34, 3.7, 3.9, 4.4, 4.6, 4.8–4.10, 4.12, 5.2, 5.3*b*, 5.4*a*, 5.7, 11.2, 11.9, 11.15, 11.17*a,b*, 11.19, 11.22, 11.23, 12.33, 13.1, 14.1, 14.3, 14.5–14.7, 15.2*a*, 15.8*a*, 15.12*a,b*, 15.13*a*, 15.17*a*, 15.34, 16.17, 18.1, 18.5, 18.6, 18.13, 18.16, 19.3, 19.20, 20.9, 20.14, 22.3, 22.18, 23.7, 28.1, 28.4, 28.6–28.8, 29.15*a,b*, 29.17*a*.

Diane Nelson & Associates

7.2, 7.3*a,b*, 7.4*a,b*, 7.8, 7.10, 17.1, 18.7, 18.9, 19.6*a*, 19.13, 19.16, 19.23*a*, 20.19*b*, 29.6*b*.

Felecia Paras

4.13, 5.10, 10.5, 12.22, 12.31, 12.32*a*, 13.3, 15.7, 19.19, 21.5*a*, 22.4, 23.2.

Precision Graphics

3.18.

Mildred Rinehart

5.6*a*, 22.6*a*.

Mike Schenk

7.01, 12.6, 13.05, 15.06, 29.14, 29.24.

Tom Sims

27.13.

Tom Sims/Mike Schenk

5.11.

Lois Schultz/Tom Waldrop

10.9, 12.8, 12.29*a*, 16.7, 19.22, 30.4, 30.5.

Tom Waldrop

1.8, 9.1, 9.2*a,b*, 9.7*a*, 10.3, 10.6, 10.7, 10.8, 10.10–10.20*a,b*, 10.21, 10.22, 11.3*a*, 12.5, 12.7, 12.24, 12.25, 12.27, 12.30, 14.2, 15.5, 15.29, 15.30, 16.2, 16.8, 16.22, 18.2, 19.21, 27.2, 27.17*a,b*, 29.2, 29.3, 29.4, 29.5*a,b,d*, 29.8, 29.12, 29.18, 30.6.

John Walters and Associates

3.17*b*, 12.16*b*, 27.11.

Photo Credits

Part Openers

One: Richard Anderson; **Two:** Sand/King/The Image Bank; **Four:** Vladimir Lange, M.D./The Image Bank; **Five:** Alexander Tsiaras/Photo Researchers, Inc.

Chapter 1

Figure 1.1: Will & Deni McIntyre/Photo Researchers, Inc.; **1.12*a*:** Alfred Wolf/Photo Researchers, Inc.

Chapter 2

Figure 2.1*a,b*: Russ Lappa/Photo Researchers, Inc.; **2.7:** John Bova/Photo Researchers, Inc.; **2.20:** Richard Gross; **2.32*a*:** Richard Anderson; **2.32*b*:** U.S. Navy/Photo Researchers, Inc.; **2.33:** Richard Anderson; **2.34*a*:** Journalism Services, Inc.

Chapter 3

Figure 3.4*b*: Biophoto Associates/Photo Researchers, Inc.; **3.7*b*:** S. J. Singer; **3.10*a*:** Richard Chao; **3.10*b*:** E. G. Pollack; **3.11:** Courtesy Kleberg Cytogenetics Lab Inst. for Molecular Genetics, Baylor College of Medicine; **3.12*a*:** Dr. Keith Porter; **3.13*a*:** BioPhoto Associates; **3.14*a*:** Keith Porter; **3.15*b*:** Elias Lazarides; **3.16:** Carolyn Chambers; **3.16*b*:** W. H. Freeman; **3.17*a*:** D. Phillips/Visuals Unlimited; **3.17*c*:** Gordon Leedale/BioPhoto Associates; **3.18*b*:** Dr. Keith Porter; **3.21*b-e*:** Ed Reschke; **3.27*a*:** BioPhoto Associates/Photo Researchers, Inc.; **3.28:** Ulrike Welsch/Photo Researchers, Inc.

Chapter 4

Figure 4.6*a,b*: Edwin Reschke; **4.6*c*:** Manfred Kage/Peter Arnold, Inc.; **4.6*d*:** Taurus Photos; **4.6*e*:** Edwin Reschke; **4.6*f*:** Visuals Unlimited; **4.6*g*:** Edwin Reschke; **4.8*a*:** Harold Benson; **4.9*a*:** Harold Benson; **4.10*a*:** Harold Benson; **4.11*b*:** Ed Reschke; **4.12*a*:** Ed Reschke; **4.13*b*:** © Manfred Kage/Peter Arnold, Inc.; **4.14:** John Watney/Science Source/Photo Researchers, Inc.; **I1.1:** James Prince/Photo Researchers, Inc.

Chapter 5

Figure 5.3*a*: Edwin Reschke/Peter Arnold, Inc.; **5.4*b*:** Michael Abbey/Science Source/Photo Researchers, Inc.; **5.5*a* (all):** Richard Anderson: **5.6*b*:** Dr. Kerry L. Openshaw; **5.8*a,b*:** BioPhoto Associates/Photo Researchers, Inc.; **5.9:** "The Contributions of UVA & UVB to Connective Tissue Damage in Hairless Mice" by Lorraine H. Kligman, Frank J. Akin and Albert M. Kligman. *Journal of Investigative Dermatology* 84: 272–276, 1985.

Chapter 6

Figure 6.8*a,b*: BioPhoto Associates/Photo Researchers, Inc.; **6.9:** Richard Anderson; **6.11:** Hudgens & Hudgens/Richard Anderson; **6.13:** Richard Anderson

Chapter 7

Figure 7.20: Eastman Kodak; **7.22*c*:** Martin Rotker/Taurus Photos

Chapter 8

Figure 8.4: Richard Hutchings/Photo Researchers, Inc.; **8.16:** Science Photo Library/Photo Researchers, Inc.; **8.17:** Richard Anderson; **8.18*a*:** James Stevenson/Science Photo Library/Photo Researchers, Inc.; **8.18*b*:** Richard Anderson

Chapter 9

Figure 9.2*c*: H. E. Huxley; **9.7*b*:** John Cunningham/Visuals Unlimited; **9.18:** © Art Stein/Photo Researchers, Inc.; **9.19:** Richard Anderson; **9.20:** BioPhoto Associates/Photo Researchers, Inc.; **10.6*a*:** Hudgens & Hudgens/Richard Anderson; **10.6*b,c*:** Richard Gross; **10.6*d*:** Susan Johns/Photo Researchers, Inc.

ColorPlates

Plate 1–28: William C. Brown Company Publishers. **I2.1*a*:** Tim Davis/Photo Researchers, Inc.; **I2.1*b*:** Tim Davis/Photo Researchers, Inc.; **I2.2:** Benton/Photo Researchers, Inc.; **I2.4:** Michael P. Gadomski/Photo Researchers, Inc.

Chapter 11

Figure 11.3*b*: John Hubbard; **11.17*c*:** Don Fawcett/Photo Researchers, Inc.; **11.25*a*:** Richard Anderson

Chapter 12

Figure 12.2: Alfred I. duPont Institute; **12.12*a-d*:** Richard Anderson; **12.14:** Bob Coyle; **12.16*a*:** Larry Mulvehill/Photo Researchers, Inc.; **12.17*a,b*:** Richard Anderson; **12.19 (1–4):** Richard Anderson; **12.19 (5):** SIU/Photo Researchers, Inc.; **12.25*c*:** Utah Valley Hospital, Dept. of Radiology; **12.29*b*:** Richard Anderson; **12.32*b*:** Per H. Kjeldson; **12.36:** Finkle/Custom Medical Stock Photography

Chapter 13

Figure 13.1*b*: W. H. Freeman/Tissues & Organs SEM by Kessel and Kardon; **13.12:** NASA/Science Source/Photo Researchers, Inc.; **13.16:** Tom Hollyman/Photo Researchers, Inc.; **13.17 (all):** Richard Anderson; **13.19:** Will & Deni McIntyre/Photo Researchers, Inc.; **13.20 & 13.21:** Richard Anderson

Chapter 15

Figure 15.1: © James L. Shaffer; **15.3*b*:** Ed Reschke; **15.8*b&c*:** Victor Eichler; **15.14*b*:** Carroll Weiss/Camera M. D. Studios; **15.17*b*:** Per H. Kjeldsen; **15.30*b*:** Fred Hossler/Visuals Unlimited; **15.33:** Stephen Clark

Chapter 16

Figure 16.13*a*: Richard Hutchings/Photo Researchers, Inc.; **16.13*b*:** Bettina Cirone/Photo Researchers, Inc.; **16.14:** Martin Rotker/Taurus Photos; **16.16*a*:** Martin Rotker/Taurus Photos; **16.16*b*:** Lester Bergman & Associates; **16.16*c*:** Lester Bergman & Associates; **16.21*a,b*:** Custom Medical Stock; **16.22*b*:** Ed Reschke

Chapter 17

Figure 17.5*b*: Bill Longcore/Science Source/Photo Researchers, Inc.; **17.9*a*:** Eric Grave/Photo Researchers, Inc.; **17.9*b*:** Bill Longcore/Photo Researchers, Inc.; **17.10*a*:** W. Rosenberg/Biological Photo Service; **17.10*b*:** Ed Reschke/Peter Arnold, Inc.; **17.10*c*:** A. Owczorak/Taurus Photos; **17.10*d,e*:** Ed Reschke; **17.11:** Victor Eichler; **17.15:** Manfred Kage/Peter Arnold, Inc.; **17.17:** Richard Anderson

Chapter 18

Figure 18.10*a*: Larry Mulvehill/Photo Researchers, Inc.; **18.10*b*:** Richard Anderson; **18.15*a*:** Jim Shaffer; **18.15*b*:** Eastman Kodak; **18.15*c*:** Richard Anderson; **18.15*d,e*:** James L. Shaffer

Chapter 19

Figure 19.2*a*: Edwin Reschke; **19.2*b*:** Victor Eichler; **19.2*d*:** W. H. Freeman & Company/Tissues & Organs SEM by Kessel and Kardon; **19.7*a*:** J. Derkmeyer, J. G. Lambers, and J. M. F. Landsmerr *Practische Ontleedkunde* BOHN, SCHELTEMA, HOLKEMA; **19.8*b*:** From Ballinger, P. W.: Merrill's Atlas of Radiographic Positions and Radiologic Procedures, 7/e, 1991, Mosby-Year Book, Inc.; **19.11*a,b*:** Richard Anderson; **19.21*b*:** Igaku Shoin, LTD; **19.21*c*:** Ed Reschke; **19.23*b*:** John D. Cunningham/Visuals Unlimited; **19.26*a,b*:** Carroll Weiss/Camera M.D. Studios

Chapter 20

Figure 20.17: Richard Anderson; **20.19a:** Donald S. Baim, Hurst et al: THE HEART 5/e McGraw Hill Book Company; **20.20a:** Richard Anderson

Chapter 21

Figure 21.7a: A. Olson/Scripps Clinic; **21.8a:** Rosenthal, Alan S., *New England Journal of Medicine;* **21.12b,c:** Andrejs Liepins; **21.13a1:** Guy Gillette/Photo Researchers, Inc.; **21.13b:** Bill Bachman/Photo Researchers, Inc.; **21.15a:** Custom Medical Stock Photography.

I4.1a: Science Photo Library/Photo Researchers, Inc.; **I4.1b:** CNRI/Science Photo Library/Photo Researchers, Inc.; **I4.2:** Dept. of Medical Photography, St. Stephen's Hospital, London/ Science Photo Library/Photo Researchers, Inc.; **I4.3a:** Hank Morgan/Science Source/Photo Researchers, Inc.; **I4.3b:** Custom Medical Stock

Chapter 22

Figure 22.6b: Igaku-Shoin Ltd.; **22.9b:** John Watney Photo Library; **22.9d:** Victor Eickler; **22.14a:** Ed Lettau/Photo Researchers, Inc.; **22.20:** K. Bayer/ Visuals Unlimited; **22.22a,b:** Victor Eickler; **22.23:** Prince/Science Source/Photo Researchers, Inc.

Chapter 23

Figure 23.8c: Dr. Karl M. Francis; **23.10a2:** Utah Valley Hospital/Dept. of Radiology; **23.10b2:** Edwin Reschke; **23.25a:** Lester V. Bergman & Associates

Chapter 24

Figure 24.18: Carroll Weiss/Camera M. D. Studios

Chapter 25

Figures 25.8, 25.9: Richard Anderson; **25.10:** Dr. Randy A. Sansone/Wright State University

Chapter 27

Figure 27.5b: Don Fawcett/Photo Researchers, Inc.; **27.16:** Per H. Kjeldsen; **27.21:** Custom Medical Stock

Chapter 29

Figure 29.5c: Biophoto Associates/Photo Researchers, Inc.; **29.6c:** CNRI/Science Photo Library/Photo Researchers, Inc.; **29.15a2:** Biophoto Associates/Photo Researchers, Inc.; **29.15b2:** Biophoto Associates/Photo Researchers, Inc.; **29.17b:** Landrum Shettles; **29.26a:** Susan Leavines/ Photo Researchers, Inc.; **29.26b:** Fred McConnaughey/Photo Researchers, Inc.; **29.26c:** Charles Belinky/Photo Researchers, Inc.; **29.28c:** Ray Ellis/Photo Researchers, **29.28d:** James L. Shaffer; **29.28e:** Bob Coyle; **29.28f:** James L. Shaffer; **29.28g,h:** Bob Coyle

Chapter 30

Figure 30.7a: M. Coleman/Visuals Unlimited; **30.7b:** © Biophoto Associates/Photo Researchers, Inc.; **30.8, 30.9:** Armed Forces Institute of Pathology

INDEX

A

AAT. *See* Alpha-1 antitrypsin (AAT)
Abdomen, 553, 884, 885, 889, 890
Abdominal cavity, 14, 880
Abdominal wall, muscles of, 248–49
Abdominopelvic cavity, 14
Abducens nerve, 366
Abduction, 191, 193
Abortion, 873
Abrasion, 135
Abruptio placentae, 873
Absolute refractory period, 299
Absorption, 133, 664, 670, 681–83
  of iron, 498
  of vitamins, 683
Absorptive metabolism, 718–22
Abstraction, 340
Acceleratory forces, 584
Acceleratory reflexes, 444
Accessory nerve, 366
Accommodation, 423–25
A cells, 472
Acetabular labrum, 200, 202
Acetabulum, 178
Acetylcholine, 308, 392
Acetyl-coenzyme A, 699, 700
Achalasia, 691
Achilles, 258
Achilles tendon, 258
Aching pain, 412
Achlorhydria, 691
Achondroplasia, 151, 857
Acid phosphatase, 146
Acid-base balance, 791–808
Acidosis, 705, 804, 805, 806
Acids
  amino. *See* Amino acids
  ascorbic, 146, 750
  aspartic, 45, 88
  bases and pH, 36–37
  bile, 672
  chenodeoxycholic, 672
  cholic, 672
  citric, 699–701
  DNA. *See* Deoxyribonucleic acid (DNA)
  fatty, 42, 564, 741–42
  folic, 55
  GABA. *See* Gamma-aminobutyric acid (GABA)
  glutamic, 88
  HCl. *See* Hydrochloric acid (HCl)
  hyaluronic, 104, 190
  hydrochloric, 21
  LDH. *See* Lactic acid dehydrogenase (LDH)
  lipoic, 750
  LSD. *See* Lysergic acid diethylamide (LSD)
  nucleic. *See* Nucleic acids
  organic, 38
  PABA. *See* Para-aminobenzoic acid (PABA)
  palmitoleic, 42
  pantothenic, 747
  phosphoric, 50
  physiology of, 792
  RNA. *See* Ribonucleic acid (RNA)
  stearic, 42
  tricarboxylic, 699–701
Acinar glands, 102, 103
Acne, 134
Acquired immune deficiency syndrome (AIDS), 624–26
  and AIDS related complex (ARC), 625
  and antibodies, 607, 608, 610
  and AZT, 625–26
  diagnosis and treatment, 625–26
  and ELISA, 625
  and HIV, 624
  and risk, 626
  and vaccines, 626

Acquired immunity, 611
Acrobats, 190, 191
Acromegaly, 151, 463
Acromioclavicular joint, 198
Acromion process, 174
ACTH. *See* Adrenocorticotropic hormone (ACTH)
Actin, 79, 213
Action
  chronotropic and inotropic, 528
  involuntary and voluntary, 108
  potential. *See* Action potential
  program of, 375
  pumping. *See* Heart
  of skeletal muscle, 238, 261–65
Action potential, 216, 296, 298–301, 408, 521–22. *See also* Action
  and heart, 521–22
  propagation of, 300
  properties of, 307
  with a plateau, 228
Activation energy, 53
Activation, mass. *See* Mass activation
Active immunity, 611
Active site, 54, 55
Active transport, 68, 70–73, 75, 770
Acupuncture, 445
Acute, 502
Acyl carrier protein, 705
Adam's apple, 632
Adaptation, 408–9
  and general adaptation syndrome, 477
  light and dark, 428
Addiction, 22, 358, 446
Addison's disease, 470, 487
Additives, food, 754, 755
Adduction, 191, 193
Adductor brevis, 252
Adductor longus, 252
Adductor magnus, 252
Adenine, 49, 50
Adenohypophysis, 451, 456, 457, 458–61
Adenoids, 632
Adenosine diphosphate (ADP), 53, 224–26
Adenosine monophosphate (AMP), 224
Adenosine triphosphate (ATP), 49, 50, 53, 64, 68, 70, 79, 81, 88, 90
  and ATPase, 72, 214, 221, 225, 228
  energy yields, 704
  and fatigue, 222
  and muscle metabolism, 224–26
  and rigor mortis, 217
  and sliding filament theory, 213–16, 219
Adenylate cyclase, 455
Adequate stimulus, 408
Adhering junctions, 98
Adipose tissue, 104, 106
Adolescence, 869–70
ADP. *See* Adenosine diphosphate (ADP)
Adrenal cortex, 451, 466, 468–69
Adrenal glands, 466–70
Adrenal medulla, 451, 466, 467
Adrenal sex hormones, 468
Adrenaline, 467
Adrenergic effector organs, 402
Adrenergic effects, 392
Adrenergic neurons, 392
Adrenocorticotropic hormone (ACTH), 458, 459, 460, 469, 470, 477, 478, 484, 487, 610
Adsorptive endocytosis, 73, 75
Aerobic exercise, 277
Aerobic metabolism, 224
Aerobics Research, 279
Afferent lymphatic vessels, 559
Afferent neurons, 287, 290
After-hyperpolarization, 298
Afterload, 528
Agammaglobulinemia, 619

Age
  and aging, 871–83
  and disease risk factors, 590
  photo-. *See* Photoaging
  and skeletal differences, 182–83
Agglutinins, 615
Agglutinogens, 615
Aggression, 344
Aging. *See* Age
Agonist, 240
Agranular leukocytes, 501, 502
Agranulocytosis, 509
Agraphia, 357
AIDS. *See* Acquired immune deficiency syndrome (AIDS)
AIDS related complex (ARC). *See* Acquired immune deficiency syndrome (AIDS)
Air volumes, 642–43
Airway resistance, 640–41
Alanine, 87, 88
Alarm reaction, 477
Albinism, 132, 857
Albumins, 49, 494
Albuminuria, 785
Alcohol, 38
Alcoholism, 22, 358, 734–35
Aldehyde, 38
Aldosterone, 468, 469, 796
Alexia, 357
Algae, 315
Alkaline, 36
Alkaline phosphatase, 146
Alkalosis, 804, 805–6
Alleles, 856
Allergen, 612
Allergies, 414, 612
All-or-none principle, 223, 296, 300
Allosteric enzymes, 711
Allosteric site, 55, 711
Alopecia, 135
Alpha-endorphin, 445, 446
Alpha-fetoprotein, 321
Alpha globulins, 494
Alpha interferon, 597
Alpha particles, 58
Alpha waves, 336
Alpha-1 antitrypsin (AAT), 653
Altitude, and respiration, 650–51
Altruism, 23
Alveolar connective tissue, 104
Alveolar glands, 102, 103
Alveolar macrophage, 596
Alveolar volume, 642–43
Alveoli, 164, 636, 667
Alzheimer's disease, 356, 357
Amenorrhea, 838
Ametropia, 444
Amino acid pool, 708
Amino acids, 44, 46, 740–41, 759–60
  and absorption, 682
  and endorphins and enkephalins, 445–46
  nonpolar and polar, 45
  structure, 45
Amino group, 38, 45
Ammonemia, 712
Amnesia, 357
Amniocentesis, 866
Amniotic cavity, 98
AMP. *See* Adenosine monophosphate (AMP)
Amphetamines, 358
Amphiarthrosis, 188
Amplification, enzyme, 456
Amplitude, 307, 434
Ampulla, 444
Ampulla of Vater, 674
Amputation, 185
Amygdala, 343
Amygdaloid nucleus, 328, 329
Amylase, 674
Amyotrophic, 314
Anabolic steroids, 231

Medical Abbreviations*

ad lib as desired
a.c. before meals
b.i.d. twice a day
B.M.R. basal metabolic rate
B.P. blood pressure
BUN blood urea nitrogen
CAT computerized axial tomography
c.b.c. complete blood count
CCU coronary care unit
CPR cardiopulmonary resuscitation
CSF cerebrospinal fluid
CVP central venous pressure
d a day
D and C dilation and curettage
ECG, EKG electrocardiogram
Feb. dur while the fever lasts
G.I. gastrointestinal
gtt. drops
h hour
HCT hematocrit
Hb, Hgb hemoglobin
h.s. at bedtime
ICU intensive care unit
I.M. intramuscularly
IPPB intermittent positive pressure breathing
I.V. intravenous
MRI magnetic resonance imaging
Noct. at night
NPO nothing by mouth
O.D. in the right eye
O.S. in the left eye
O.U. in each eye
p.c. after meals
P.O. by mouth
p.r.n. as needed
q.d. every day
q.h. every hour
q.i.d. four times a day
R.B.C. red blood cell
Sig. label
sp. gr. specific gravity
SQ subcutaneous injection
s.s. one-half
stat. immediately
sum. take
t.i.d. three times a day
ung. ointment
W.B.C. white blood cell

*Abbreviations as shown in *Dorland's Medical Dictionary*, 27 ed., which notes that periods do not necessarily indicate proper usage.

Metric-Apothecary Equivalents

1 gram (g) =
0.001 kilogram (kg)
1000 milligrams (mg)
1,000,000 micrograms (μg)
15.4 grains (gr)
0.032 ounce (oz)

1 liter (l) =
1000 milliliters (ml)
1000 cubic centimeters (cc)
2.1 pints (pt)
270 fluid drams (fl dr)
34 fluid ounces (fl oz)

1 meter (m) =
0.001 kilometers (km)
100 centimeters (cm)
1000 millimeters (mm)

Apothecary-Metric Equivalents

1 grain (gr) =
0.017 dram (dr)
0.002 ounce (oz)
0.0002 pound (lb)
0.065 gram (g)
65 milligrams (mg)

1 minim (min) =
0.017 fluid dram (fl dr)
0.002 fluid ounce (fl oz)
0.0001 pint (pt)
0.06 milliliter (ml)
0.06 cubic centimeter (cc)

Household Equivalents

1 teaspoon (tsp) =
5 milliliters (ml)
5 cubic centimeters (cc)
1 fluid dram (fl dr)

1 tablespoon (tbsp) =
15 milliliters
15 cubic centimeters (cc)
0.5 fluid ounce (fl oz)
3 teaspoons

Fahrenheit-Celsius Temperature Conversion

°F	°C	
95.0	35.0	
95.5	35.2	Hypothermia
96.0	35.5	
96.5	35.8	
97.0	36.1	
97.5	36.4	
98.0	36.7	
98.5	37.0	
98.6	**37.0**	**Normal**
99.0	37.2	
99.5	37.5	
100.0	37.8	
100.5	38.0	
101.0	38.3	
101.5	38.6	
102.0	38.9	
102.5	39.2	Fever
103.0	39.5	
103.5	39.8	
104.0	40.0	
104.5	40.3	
105.0	40.6	

$°F = 9/5 \ (°C + 32)$
$°C = 5/9 \ (°F - 32)$